# RESIDENTIAL/ LIGHT COMMERCIAL COST DATA 1985

**Publisher**
Robert Sturgis Godfrey

**Editor-in-chief**
Albert J. Sauerbier

**Senior Editor**
Dean E. Entwistle

**Contributing Editors**
Allan B. Cleveland
Nancy B. Delano
Paul A. Fellini
Jane M. Hart
F. William Horsley
John T. Kenny
Dwayne R. Lehigh
William D. Mahoney
Melville J. Mossman
John J. Moylan
Jeannene D. Murphy
Kornelis Smit
Arthur Thornley
John T. Trainor
James P. Treichler

**Graphics**
Carl W. Linde
Harold L. Jordan

Means

# FOREWORD

## GENERAL INTRODUCTION

Use this book to develop fast, accurate figures for Residential and Light Commercial building cost estimates. On the following pages are three groups of data needed to get that job done: Systems Costs, Square Foot Costs and Unit Prices.

As an estimating guide, this book can help overcome some of the biggest estimating problems facing estimators in the 1980s. If used carefully, "Residential/Light Commercial Cost Data" can help you avoid out-of-date prices. No longer will it be necessary to guess total "man-hours," labor charges or material prices. More importantly, this book can help an estimator avoid overlooking any components that could make cost projections inaccurate.

## SYSTEMS COSTS

Included inside are 100 illustrated building systems. From foundation to roof, windows to interior finish — all the materials needed to develop each system cost are listed. In all cases, totals will add up to whatever the individual building's design calls for. All of these systems are defined completely, including the necessary components needed to create custom-made construction systems. Using the "Systems" approach, it is easy to see where all the components are in a project, and where all costs will come from. Convenient tables help factor total costs. Labor expense and contractor's overhead are readily available. Selective price sheets are conveniently located to accommodate whatever degree of sophistication a project requires.

## SQUARE FOOT COSTS

Square Foot Costs are the cornerstone of building cost analysis. Knowing the total cost per square foot of a proposed structure can help with conceptual estimating, feasibility studies, component depreciation estimates, and replacement or reproduction cost estimates. Use these figures to verify bids, and to update known construction costs. Square Foot Costs for typical building types enable the estimator to factor prices for different models. Each page lists one particular building type by size and then shows a breakdown of the most common building components for that type of construction. This enables the estimator to develop Square Foot Costs from the foundation to the roof. Square Foot estimates can help to separate building component costs and aid in cost comparison.

## UNIT PRICE COSTS

The convenient Unit Price Cost section lists more than 100 pages of data. This is the type of detailed cost information that has been used to develop the Systems and Square Foot sections of the book. Use Unit Prices to analyze specific costs, or to develop custom systems. This in-depth listing is also designed to aid in the development of unit price estimates.

## PRICE VARIATIONS

The prices listed in this book are developed specifically for the housing and light commercial construction industry. All cost information is constantly updated by the staff of R. S. Means Co., Inc. Users will find that differences in wage rates, labor efficiency, labor restrictions and material prices will occur due to varying local conditions. The total construction cost index is listed by zip code for 635 U.S and Canadian cities, on pages 413 to 417. This zip code factor index is based on cost in effect on Jan. 1, 1985.

Local, regional or national shortages of construction materials can influence material costs and cause delays with corresponding increases in indirect job costs.

No guaranty or warranty is made by Robert Snow Means Co., Inc. or the editors as to the correctness or sufficiency of the information in this book. Robert Snow Means Co., Inc. or the editors assume no responsibility or liability in connection with the use of this book.

## FACTORS AFFECTING COST

**QUALITY:** The prices given in this book are in accordance with National Building Codes and represent good, sound construction.

**OVERTIME:** No allowance has been made for overtime, although this cost can be factored into any estimate by adjusting the labor costs.

**PRODUCTIVITY:** The man-hour figures for installation are based on an eight-hour day in daylight hours.

**SIZE OF JOB:** This book is intended for use by those involved primarily in Residential and Light Commercial construction costing less than $400,000. This includes the construction of homes, row houses, townhouses, condominiums, retail stores, offices, warehouses, apartments and motels. With reasonable exercise of judgment, this same information can be used for repair and remodeling projects. For extensive remodeling projects, consult R. S. Means' "Repair and Remodeling Cost Data". The cost information contained here does not cover high-rise construction, nor the large-scale concrete and steel work associated with those structures. Consult the R. S. Means publication "Building Construction Cost Data" for cost information pertaining to that type of construction.

**LOCATION:** Material prices are for metropolitan areas. Beyond a 20-mile radius of large cities, extra transportation charges increase material costs slightly. This may be offset by lower wage rates, but these two factors should be given consideration in preparing an estimate, especially if the job site is remote. Highly specialized subcontract items may require high travel and per diem expenses for craftsmen.

**OTHER FACTORS:** Other factors which affect costs are weather, season of year, contractor management, local labor restrictions, building code requirements, natural disasters, availability of adequate energy, as well as availability of skilled labor and building materials. The general building conditions which influence the "in place" costs of all building items are another cost factor. Substitute materials and construction methods may increase installed cost above those anticipated. These factors are difficult to evaluate, and leave a margin of error that is unavoidable and increasingly important.

## LABOR

Labor rates used in this book are for non-union labor and have been rounded to the nearest five cents. It is possible to devise figures for different parts of the U.S. and Canada by using the location factors found on page 413. A detailed breakdown of the labor rate (or "Billing Rate") for each trade is shown inside the back cover.

It may be noticed that in some cases labor costs over the past several years have held steady or even decreased in spite of the increase in hourly wages. This is a result of new mechanical equipment being used in the field or new methods of handling materials. Factory prefabrication of components and many other innovations have reduced both helper's and skilled mechanic's time. Severe competition between subcontractors may tend to keep bid prices from rising as rapidly as the labor rates.

Precise Labor costs for the various trades that make up the crews used in the Unit Price section of this book are shown in the table of Standard Crews on pages 405 through 412. The costs for each element in a particular crew are listed in each crew listing. All figures are totaled at the bottom. All costs are shown three ways. One column lists the bare labor and equipment rate (the Base Rate plus fringe benefits). The next column lists the "Billing Rate" (the rate with the installing contractor's overhead and profit added). The third column lists the cost per man-hour. Local labor rates may be substituted to arrive at precise local daily crew costs.

All Trades shown in the Unit Price section are listed inside the back cover. Bare cost, cost including installing contractor's overhead and profit, and cost per man-hour are shown.

The "Standard Crews" and "Trades" listings are developed using the same labor rates shown inside the back cover of this book. That table explains the exact percentage markups used to develop the Billing Rate. A different percentage mark-up is applied to each trade to arrive at the total hourly and daily rates which include the installing contractor's overhead and profit. That final figure is called the Billing Rate.

The Base Rate Including Fringe Benefits (Bare Cost) does not include insurance, taxes or installing contractor's overhead and profit. The Billing rate includes all of these additional costs (usually an additional 50% to 77% more than the Base Rate).

Labor costs listed in the Unit Price section of this book, under the heading "LABOR," are Bare Costs. Bare Costs reflect the Base Labor Rate Plus Fringe Benefits. Total costs for each unit price listed under the heading "TOTAL INCL. O&P" reflect the Billing Rate.

The Systems and Square Foot sections do not list Bare Costs. All figures listed under the heading "INST." (Installation) in the Systems section, and under "LABOR" in the Square Foot section, reflect the full Billing Rate cost for one particular area of Residential and Light Commercial construction. Therefore, it is NOT necessary to add a percentage to allow for the installing contractor's overhead and profit when using the Systems and Square Foot sections.

## CONSTRUCTION EQUIPMENT COSTS

The power equipment required for each crew listed in the Unit Price section is included in the particular crew cost. Each piece of power equipment required is listed with the crew in the Crew Listing. Daily equipment costs are usually the sum of the weekly bare rental rate divided by 5 days plus the hourly operating cost times 8 hours.

[(WEEKLY RENTAL RATE/5 DAYS) + (HOURLY OPERATING COST x 8 HOURS) = BARE DAILY EQUIPMENT COST FOR CREWS]

Division 1.5 in the Unit Price section contains a listing of power equipment and accessories normally encountered in Residential and Light Commercial work.

In order to establish the daily equipment cost including the installing contractor's overhead and profit shown in the right-hand column in the crew listing, 10% was added to the Bare Cost figure. This is generally enough to allow for the contractor's mark-up on equipment costs. In addition, the total equipment costs for each crew are divided by the total daily man-hours of the crew to determine the equipment cost per man-hour for the crew. This hourly cost is listed both as a Bare Cost and with Overhead and Profit in order to simplify estimating based solely on man-hours.

## MATERIAL COSTS

Material costs used are those compiled specifically for the Residential and Light Commercial construction industry by the staff of R.S. Means. These are Jan. 1, 1985 national average material costs including delivery to the job site. To adjust material costs in this book for any specific location, use the zip code index found on page 413.

## HOW THIS BOOK IS ARRANGED

**SYSTEMS COSTS:** The first part of this book is the Systems section. The format used breaks cost information into ten sections that represent the major functional elements that go into a building. The editors have analyzed the variables for each section and developed costs for custom-made designs. Each part of the Systems section has instructions for using the tables plus sketches illustrating numerous systems. These sketches are for illustration purposes only and are not necessarily drawn to scale. An outline of a typical Systems page layout and an explanation of System prices is located on page 2.

**SQUARE FOOT COSTS:** The second part of the book lists Square Foot Costs. The format used divides Residential and Light Commercial construction into basic Building Classes. Each page in this section divides the building into the ten most common components of construction. This section begins with an introduction that defines all building classes and types. The introduction begins on page 208.

**UNIT PRICE:** Part Three of the book lists Unit Prices which follow the 16 Divisions of the Uniform Construction Index (UCI) format. The introduction to this section, including an outline of a typical page in the listing, begins on page 256.

## CITY ADJUSTMENTS

To modify the prices in this book, use the zip code factors found on page 413. All factors are based on January 1, 1985.

## OVERHEAD, PROFIT AND CONTINGENCIES

### GENERAL CONDITIONS

The "Systems" and "Square Foot" sections of this book use costs that include the installing contractor's overhead and profit (O&P). An allowance covering the general contractor's mark-up must be added to these figures. The general contractor can include this price in the bid with a normal mark-up ranging from 5% to 15%. The mark-up depends on economic conditions plus the supervision and troubleshooting expected by the general contractor. For purposes of this book, it is best for a general contractor to add an allowance of 10% to the figures in the Systems and Square Foot Cost sections.

In the Unit Price section, the extreme right-hand column of each chart gives the "Total Including O&P." These figures contain the original installing subcontractor's O&P (in other words, the contractor doing the work), therefore it is necessary for a general contractor to add a percentage for all subcontracted items.

### OVERHEAD AND PROFIT

Breakdowns in Division 10 of the Systems section, starting on page 201 give details on overhead. For Systems costs and Square Foot costs, simply add 10% to the estimate for general contractor's profit.

After reviewing O&P with the tables in Division 10 of the Systems section, the general contractor should then be able to tell from a close look at records what the overhead cost will be. Whatever profit and contingency percentages that circumstances dictate can then be added.

### SUBCONTRACTORS

Increasingly, considerable portions of construction projects are being subcontracted. In fact, the percentage done by "subs" may run over 50% on some projects. Since the workmen employed by these companies do nothing but install their particular product, they soon become experts in that line. The result is, installation by these firms is accomplished so efficiently that the total in-place cost, even with the subcontractor's O&P, is no more (and often less) than if the principal contractor had handled the installation directly.

### CONTINGENCIES

Estimates that include an allowance for contingencies generally require a margin to allow for unforeseen construction difficulties. If drawings are final and only field contingencies are being considered, 2% to 3% is probably sufficient and often nothing need be added. As far as the contract is concerned, future changes in plans will be covered by extras. The contractor should consider inflationary price trends and possible material shortages during the course of the job. The escalation factor is dependent upon both economic conditions and the anticipated time between the estimate and actual center-line construction. If drawings are not complete or approved, or if a budget cost is required before proceeding with a project, it is wise to add 5% to 10%. Contingencies are therefore a matter of judgment. Additional allowances are shown in Division 10 of the Systems section.

### ESTIMATING PRECISION

In making an estimate, ignore the cents column. Just list totals to the nearest dollar. The cents will average up in a column of figures. An estimate of $80,323.37 is unrealistic. A figure of $80,325 is certainly more sensible. The rounded figure of $80,300 is better, and just as likely to be right.

If an estimator follows that simple instruction, the time saved is tremendous with an added important advantage. Using round figures leaves the mind free to exercise judgment and common sense, rather than being overcome and befuddled by a mass of computations.

When finished, make a rough check of the big items for correct location of the decimal point. This is important. A large error can creep in if the person doing the estimate writes down $300 instead of $3,000. Also check the lists to be sure that any large items have not been omitted. A common error is to overlook, for example, heating or finished flooring, or to forget painting — or otherwise commit a gross omission. No amount of accuracy in prices can compensate for such an oversight.

It is important to keep bare costs and costs that already include the subcontractor's overhead and profit separate since different mark-ups will have to be applied to each category. Organize estimating procedures to minimize confusion and simplify checking, and to insure against omissions and/or duplications. By using printed forms listing the usual building items, the chance of omissions and duplications is lessened, but never completely eliminated.

## THE COMPANY AND THE EDITORS

Robert S. Means Co., Inc., is a 43-year old firm actively engaged in construction cost publishing and consulting throughout North America. The company helps you cut down on the time and energy you use when organizing and utilizing cost data. A thoroughly experienced and highly qualified staff of over ninety professionals at R. S. Means works daily at collecting, analyzing and disseminating reliable cost information for your needs.

Each editor contributing to this publication is a full-time Means employee. All these staff members have years of practical contractor experience and thorough engineering training prior to joining the firm. Each Means construction expert contributes to the maintenance of a complete, continually-updated construction cost data system.

Constant flow of new products means that even the most sophisticated construction professional finds little time to examine and evaluate all the construction cost possibilities when they are so many and diverse. R. S. Means fills the gap between theoretical and practical knowledge in the field of construction engineering. The Means organization is always prepared to assist you and help in the solution of construction problems through the services of its four major divisions: Construction and Cost Data Publishing, Computer Software Services, Consulting, and Educational Seminars.

## CONSTRUCTION AND COST DATA PUBLISHING

**BUILDING CONSTRUCTION COST DATA,** published annually for 43 years. Contains over 20,000 unit prices for building construction items broken down into material, labor, and total costs as well as total costs including subcontractors' overhead and profit.

**MEANS MECHANICAL COST DATA,** published annually. Offers the benefits of expansive mechanical unit and systems cost coverage as well as comprehensive labor and productivity information. Contains all plumbing and HVAC systems plus a wide range of energy-related data. Many illustrations.

**MEANS ELECTRICAL COST DATA,** published annually. Highly-detailed treatment of electrical unit and systems costs, including design reference tables and demolition figures, illustrated estimating procedures adaptable to customized projects and an efficient man-hour, materials and labor cost format for fast scheduling and materials procurement.

**REPAIR & REMODELING COST DATA,** published annually. An illustrated book detailing construction costs for conversion, reuse, fire-damage repair and renovation for residential, commercial and industrial construction.

**MEANS SQUARE FOOT COSTS,** published annually. A budgeting cost resource and replacement cost guide with over 6,000 costs for typical buildings and attachments applicable to residential, commercial, industrial and institutional structures. Examples, illustrations, photographs.

**MEANS SYSTEMS COSTS,** published annually. Offers thousands of detailed systems costs with hundreds of drawings, explanations, component breakdowns and tables. Uses the expanded and improved "UNIFORMAT" numbering system, assisting in the conceptual phase of estimating any job.

**MEANS SITE WORK COST DATA,** published annually. Provides answers to the most demanding site work estimating questions. Complete unit and systems costs enable you to confidently handle every aspect of site work estimating, from excavation to planting the grass. Illustrated.

**LABOR RATES FOR THE CONSTRUCTION INDUSTRY,** published annually. Contains detailed union labor rates and factors for 46 construction trades in more than 300 U.S. and Canadian cities. Ideally suited for making comparisons between cities, combinations of cities and for rapidly determining wage scales above or below national averages. Includes historical wage rates and a five-year national average chart.

**INTERIOR COST DATA,** published annually. Provides up-to-date cost facts for solving the complex estimating problems of modern interior construction. Included are detailed material, equipment and labor costs for hardware, custom interior work and furnishings for both new and remodeling commercial and industrial construction.

**BOATING COST GUIDE,** published annually. Provides average installed cost for tasks involved in the repair, maintenance, outfitting and improvement of power and sail boats up to 65 feet in length. Specifications and cost data are derived from the actual experience of marine service firms and suppliers, coast-to-coast.

**MEANS HISTORICAL COST INDEXES,** published annually. Indexes historical construction costs for each of 209 United States and Canadian cities from 1940 through 1985. Derived from years of research, this manual gives functional and reliable assistance in all historical cost estimating activities.

**MEANS CONSTRUCTION COST INDEXES,** published quarterly, every January, April, July and October. This handy report provides cost adjustment factors for the preparation of more precise estimates no matter how late in the year. It is also the ideal method for making continuous cost revisions on on-going projects as the year progresses.

**MEANS MAN-HOUR STANDARDS,** a working encyclopedia of labor productivity information. Offers detailed description of crews, man-hours with units, daily output, convenient U.C.I. format, many alternate man-hour selections, introductory guidance and information.

**SQUARE FOOT ESTIMATING,** provides user with the expertise to perform conceptual and design stage estimates. Demonstrates how to minimize time and effort to achieve accurate estimates. Offers methodology for reliability in square foot costing. Illustrated.

**MEANS SCHEDULING MANUAL,** discusses today's most advanced methods for scheduling and managing building construction projects in the mid-sized range. PERT, CPM, precedence scheduling, computer-based control, preparing and using the network are all included. Illustrated with charts, tables and graphs.

**ESTIMATING FOR THE GENERAL CONTRACTOR,** a comprehensive book designed to give estimating a refreshing breath of new ideas, insights and better techniques. Presents logical explanations, detailed examples, tables and graphs conveying all components of estimating activities - from analyzing a project's desirability through the final bid for the job.

**MEANSCO FORMS** are preprinted forms designed for architects, contractors and owners. The forms facilitate construction estimating, cost control and appraisals.

## COMPUTER SOFTWARE SERVICES

**GALAXY (an automated quantity takeoff and pricing system)** —simplifies the process of doing a takeoff using the technology of electronic drafting systems. You can reduce the time it takes to do an estimate as much as 50%, and more for many projects, thanks to GALAXY'S electronic digitizer table. Designed for use with the STAR General Estimating Program, GALAXY provides you with access to your own files of known costs as well as any of the R.S. Means standard cost information files of up-to-date unit price information.

**STAR (Means General Estimating Program)** — helps develop fast and accurate estimates and manage vital cost data. Easy to use, STAR reduces errors, provides more accurate estimates and enables you to increase productivity and improve profit margins. STAR allows you to customize estimates to specific requirements, and is available for both daily output and man-hours analysis. STAR is designed to access Means Unit Price Cost files listed below.

**UNIT PRICE COST FILES** — complete cost files on floppy diskettes. Contains item description, crew makeup, daily productivity, unit of measure and unit prices for material, labor, equipment, and total for Divisions 1-16, and separate in-depth files for Building Construction, Repair/Remodeling, Site Work, Mechanical, Electrical, Residential/Light Commercial and Interiors.

**COMET** — designed for the construction and facilities professional with a growing need for customized reports, COMET Report Writer Sorting program will help you create winning proposals in a convenient form to meet the needs of any required format. This program enables you to calculate, accumulate and print individual items for estimating data in the STAR and GALAXY programs.

**VEGA** — The VEGA Scheduling Program is designed to create a realistic schedule early in the conceptual stage, or later on during the bidding and negotiating phase. VEGA can provide you with a checklist of the different items involved in the project, and can even estimate resource excesses or predict the target date. You can determine resource allocation schedules based upon availability while estimating duration times and cost data for each selected activity. VEGA will also provide a printout of the schedule, which includes early start and early finish, late start and late finish, critical path and resource leveling.

The VEGA Scheduling Program communicates with both STAR and GALAXY programs - it's an absolute must for the professional who has to stay on top of his scheduling situation.

## CONSULTING SERVICES

Through direct consultation with clients associated with the construction industry, R. S. Means performs a wide range of assignments. As a client, you may retain segments of our staff for detailed estimates, appraisals, and assistance for large or unique projects in related areas. Typical assignments include: Feasibility Studies, Estimating Services, Appraising Services, Cost Engineering Research, Independent Cost Evaluations, Energy Conservation Programs, Insurance Claim Settlement Programs, Quantity Surveys, and Value Engineering Studies.

## EDUCATIONAL SEMINARS

The Means Continuing Education Department conducts professional seminars throughout the year in selected North American cities, and also provides on-site programs for corporations and others. All Means seminars focus on the latest estimating and management approaches, including computerized estimating and scheduling systems as applied specifically to the construction industry. The program covers the following topics:

**UNIT PRICE ESTIMATING** — shows how today's advanced estimating techniques and cost information sources can be used to efficiently develop comprehensive, detailed estimates for projects of any size. Course material includes *Building Construction Cost Data* and a workbook.

**REPAIR & REMODELING ESTIMATING** — demonstrates how to produce detailed and accurate cost estimates for building reuse, conversion, repair, and remodeling. Course participants use *Repair & Remodeling Cost Data* and a seminar workbook for definitive estimating.

**SQUARE FOOT COST ESTIMATING** — provides expert guidance for those needing extremely rapid estimates based on logical cost relationships of major systems. Course participants can perfect their skills for conceptual stage estimating, budget planning, building planning, building valuation, and replacement cost studies. Seminar presentation is based on *Means Square Foot Costs, Means Systems Costs,* and a workbook.

**COMPUTERS IN CONSTRUCTION** — prepared for those interested in using computers for construction activities, this seminar describes and demonstrates the latest available hardware and software products for construction cost estimating, project scheduling, and accounting applications. Actual demonstrations of the Means line of construction-oriented software — as well as other construction and business software — highlight this Seminar. Optimum use is made of *Building Construction Cost Data*

**SCHEDULING & PROJECT MANAGEMENT** — explains how to make efficient use of manpower, equipment, and money to ensure timely completion of projects. *Means Scheduling Manual, Building Construction Cost Data,* and a comprehensive workbook assist attendees in applying the most successful management techniques.

# RESIDENTIAL/ LIGHT COMMERCIAL COST DATA

**SYSTEMS**

**SQUARE FOOT**

**UNIT PRICE**

# HOW TO USE SYSTEMS PAGES

**Illustration**
Each building system is accompanied by a detailed, illustrated description. Each individual component is labeled. Every element involved in the total system function is shown.

**Quantities**
Each material in a system is shown with the quantity required for the system unit. For example, the rafters in this system have 1.170 L.F. per S.F. of ceiling area.

**Unit of Measure**
In the three right-hand columns, each cost figure is adjusted to agree with the unit of measure for the entire system. In this case, COST PER SQUARE FOOT (S.F.) is the common unit of measure. NOTE: In addition, under the UNIT heading, all the elements of each system are defined in relation to the product as a selling commodity. For example, "fascia board" is defined in linear feet, instead of in board feet.

**Description**
Each page includes a brief outline of any special conditions to be used when pricing a system. All units of measure are defined here.

**System Definition**
Not only are all components broken down for each system, but alternative components can be found on the page opposite. Simply insert any chosen new element into the chart to develop a custom system.

**Man-hours**
Total man-hours for a system can be found by simply multiplying QUANTITY times MAN-HOURS. The resulting figure is the total man-hours needed to complete the system.
(QUANTITY x MAN-HOURS = TOTAL SYSTEM MAN-HOURS)

**Materials**
This column contains the MATERIAL COST of each element. These cost figures include 10% for handling.

**Installation**
Labor rates include both the INSTALLATION COST of the contractor and the standard contractor's O&P. On the average, the LABOR COST will be 59.6% over and above BARE LABOR COST.

**Shaded Totals**
These convenient shaded columns provide the necessary system cost totals. TOTAL SYSTEM COST can be derived by multiplying the SHADED TOTAL times each system's SQUARE FOOT ESTIMATE. (SHADED TOTAL x SQUARE FEET = TOTAL SYSTEM COST)

**Work Sheet**
Using the SELECTIVE PRICE SHEET on the page opposite each system, it is possible to create estimates with alternative items for any number of systems.

**Total**
MATERIAL COST + INSTALLATION COST = TOTAL. Work on table from left to right across cost columns to derive totals.

## 3 | FRAMING — 12 | Gable End Roof Framing Systems

| SYSTEM DESCRIPTION | QUAN. | UNIT | MAN HOURS | MAT. | INST. | TOTAL |
|---|---|---|---|---|---|---|
| **2" X 8" RAFTERS, 16" O.C., 4/12 PITCH** | | | | | | |
| Rafters, 2" x 8", 16" O.C., 4/12 pitch | 1.170 | L.F. | .025 | 64 | 54 | 1.18 |
| Ceiling joists, 2" x 6", 16" O.C. | 1.000 | L.F. | .015 | 40 | 31 | 71 |
| Ridge board, 2" x 8" | .050 | L.F. | .001 | 03 | 02 | 05 |
| Fascia board, 2" x 8" | .100 | L.F. | .007 | 05 | 15 | 20 |
| Rafter tie, 1" x 4", 4' O.C. | .060 | L.F. | | 01 | 01 | 02 |
| Soffit nailer (outrigger), 2" x 4", 24" O.C. | .170 | L.F. | .004 | 07 | 08 | 15 |
| Sheathing, exterior, plywood, CDX, 1/2" thick | 1.170 | S.F. | .012 | 48 | 27 | 75 |
| Furring strips, 1" x 3", 16" O.C. | 1.000 | L.F. | .023 | 10 | 45 | 55 |
| TOTAL | | | .087 | 1.78 | 1.83 | 3.61 |
| **2" X 6" RAFTERS, 16" O.C., 4/12 PITCH** | | | | | | |
| Rafters, 2" x 6", 16" O.C., 4/12 pitch | 1.170 | L.F. | .022 | 47 | 44 | 91 |
| Ceiling joists, 2" x 4", 16" O.C. | 1.000 | L.F. | .013 | 28 | 26 | 54 |
| Ridge board, 2" x 6" | .050 | L.F. | | 02 | 02 | 04 |
| Fascia board, 2" x 6" | .100 | L.F. | .005 | 04 | 11 | 15 |
| Rafter tie, 1" x 4", 4' O.C. | .060 | L.F. | | 01 | 01 | 02 |
| Soffit nailer (outrigger), 2" x 4", 24" O.C. | .170 | L.F. | .004 | 07 | 08 | 15 |
| Sheathing, exterior, plywood, CDX, 1/2" thick | 1.170 | S.F. | .012 | 48 | 27 | 75 |
| Furring strips, 1" x 3", 16" O.C. | 1.000 | L.F. | .023 | 10 | 45 | 55 |
| TOTAL | | | .079 | 1.47 | 1.64 | 3.11 |

The cost of this system is based on the square foot of plan area. All quantities have been adjusted accordingly.

| DESCRIPTION | QUAN. | UNIT | MAN HOURS | MAT. | INST. | TOTAL |
|---|---|---|---|---|---|---|
| | | | | | | |

# SYSTEMS COSTS

Use this section of the book to develop total cost estimates for the individual building systems found in residential and light commercial construction.

Totals derived here can be compared with alternative systems, and the results can be used to analyze how individual components will affect total project cost.

With every system described in this section, a selective price sheet is included. The user can add new components to any system by using this sheet and the list of components on the page facing each system.

The systems described here are complete, working models. All the different versions of each system are taken into account and grouped together for easy cost comparison.

The information used to derive labor and material costs reflects the work of R. S. Means' staff of engineers who have assembled the book with the residential housing contractor/home builder and light commercial contractor in mind.

For a project of any scale, simply find the system and then develop size estimates. The cost tables (see example to the left of this page) do the rest. Find material and labor costs with no guesswork. Estimate man-hours with confidence.

All aspects of a construction project — excavation through finish work — are presented in the first nine sections. In Section 10, the user can develop accurate cost figures to estimate overhead and profit.

The cost tables are easy to use and they are versatile. The "Man-Hour Totals" listed here enable the user to estimate the time it will take to complete each system. Take, for example, the system described to the left of this page. Assume an estimate is needed to build a 4/12 pitch roof, using 2" x 6" rafters. Go to the tables to derive the MAN-HOURS PER SQUARE FOOT needed to complete the system. In this case that figure is .079 man-hours per square foot. Multiply that figure times the square foot measurement of building (use 1000 square feet as an example) to derive the TOTAL MAN-HOURS needed to build the system. (.079 x 1000 SQ. FT. = 79 TOTAL MAN-HOURS)

Anyone with unique labor rates can develop their own cost totals by using their own labor rates applied against the man-hours listed in these tables. All labor rates that have been used to derive costs are listed inside the back cover. Installation costs are based on the "Billing Rate" which includes fringe benefits and the installing contractor's overhead and profit for non-union labor.

## TABLE OF CONTENTS

| Section | | Page |
|---|---|---|
| Section 1 | Site Work | 5 |
| Section 2 | Foundations | 21 |
| Section 3 | Framing | 33 |
| Section 4 | Exterior Walls | 61 |
| Section 5 | Roofing | 103 |
| Section 6 | Interiors | 125 |
| Section 7 | Specialties | 147 |
| Section 8 | Mechanical | 161 |
| Section 9 | Electrical | 195 |
| Section 10 | Overhead & Profit | 201 |

# SECTION 1
# SITE WORK

| System Number | Title | Page |
|---|---|---|
| **1-04** | Footing Excavation | 6 |
| **1-08** | Foundation Excavation | 8 |
| **1-12** | Utility Trenching | 10 |
| **1-16** | Sidewalk | 12 |
| **1-20** | Driveway | 14 |
| **1-24** | Septic | 16 |
| **1-60** | Chain Link Fence | 18 |
| **1-64** | Wood Fence | 19 |

# 1 SITE WORK — 04 Footing Excavation Systems

*Backfill*
*Excavate*

| SYSTEM DESCRIPTION | QUAN. | UNIT | MAN HOURS | COST EACH MAT. | COST EACH INST. | COST EACH TOTAL |
|---|---|---|---|---|---|---|
| **BUILDING, 24' X 38', 4' DEEP** | | | | | | |
| Clear and strip, dozer, light trees, 30' from building | .190 | Acre | 11.400 | | 384.75 | 384.75 |
| Excavate, backhoe | 174.000 | C.Y. | 7.656 | | 360.18 | 360.18 |
| Backfill, dozer, 4' lifts, no compaction | 87.000 | C.Y. | .870 | | 65.25 | 65.25 |
| Rough grade, dozer, 30' from building | 87.000 | C.Y. | 6.525 | | 240.99 | 240.99 |
| TOTAL | | | 26.451 | | 1051.17 | 1051.17 |
| **BUILDING, 26' X 46', 4' DEEP** | | | | | | |
| Clear and strip, dozer, light trees, 30' from building | .210 | Acre | 12.600 | | 425.25 | 425.25 |
| Excavate, backhoe | 201.000 | C.Y. | 8.844 | | 416.07 | 416.07 |
| Backfill, dozer, 4' lifts, no compaction | 100.000 | C.Y. | 1.000 | | 75.00 | 75.00 |
| Rough grade, dozer, 30' from building | 100.000 | C.Y. | 7.500 | | 277.00 | 277.00 |
| TOTAL | | | 29.944 | | 1193.32 | 1193.32 |
| **BUILDING, 26' X 60', 4' DEEP** | | | | | | |
| Clear and strip, dozer, light trees, 30' from building | .240 | Acre | 14.400 | | 486.00 | 486.00 |
| Excavate, backhoe | 240.000 | C.Y. | 10.560 | | 496.80 | 496.80 |
| Backfill, dozer, 4' lifts, no compaction | 120.000 | C.Y. | 1.200 | | 90.00 | 90.00 |
| Rough grade, dozer, 30' from building | 120.000 | C.Y. | 9.000 | | 332.40 | 332.40 |
| TOTAL | | | 35.160 | | 1405.20 | 1405.20 |
| **BUILDING, 30' X 66', 4' DEEP** | | | | | | |
| Clear and strip, dozer, light trees, 30' from building | .260 | Acre | 15.600 | | 526.50 | 526.50 |
| Excavate, backhoe | 268.000 | C.Y. | 11.792 | | 554.76 | 554.76 |
| Backfill, dozer, 4' lifts, no compaction | 134.000 | C.Y. | 1.340 | | 100.50 | 100.50 |
| Rough grade, dozer, 30' from building | 134.000 | C.Y. | 10.050 | | 371.18 | 371.18 |
| TOTAL | | | 38.782 | | 1552.94 | 1552.94 |

The costs in this system are on a cost each basis.
Quantities are based on 1'-0" clearance on each side of footing.

| DESCRIPTION | QUAN. | UNIT | MAN HOURS | COST EACH MAT. | COST EACH INST. | COST EACH TOTAL |
|---|---|---|---|---|---|---|
| | | | | | | |
| | | | | | | |
| | | | | | | |

# Footing Excavation Price Sheet

| | QUAN. | UNIT | MAN. HOURS | COST EACH MAT | COST EACH INST. | TOTAL |
|---|---|---|---|---|---|---|
| Clear and grub, medium brush, 30' from building, 24' x 38' | .190 | Acre | 9.000 | | 303.75 | 303.75 |
| 26' x 46' | .210 | Acre | 10.200 | | 344.25 | 344.25 |
| 26' x 60' | .240 | Acre | 11.400 | | 384.75 | 384.75 |
| 30' x 66' | .260 | Acre | 12.600 | | 425.25 | 425.25 |
| Light trees, to 6" dia. cut & chip, 24' x 38' | .190 | Acre | 11.400 | | 384.75 | 384.75 |
| 26' x 46' | .210 | Acre | 12.600 | | 425.25 | 425.25 |
| 26' x 60' | .240 | Acre | 14.400 | | 486.00 | 486.00 |
| 30' x 66' | .260 | Acre | 15.600 | | 526.50 | 526.50 |
| Medium trees, to 10" dia. cut & chip, 24' x 38' | .190 | Acre | 13.028 | | 437.00 | 437.00 |
| 26' x 46' | .210 | Acre | 14.399 | | 483.00 | 483.00 |
| 26' x 60' | .240 | Acre | 16.456 | | 552.00 | 552.00 |
| 30' x 66' | .260 | Acre | 17.828 | | 598.00 | 598.00 |
| Excavation, footing, 24' x 38', 2' deep | 68.000 | C.Y. | 2.992 | | 140.76 | 140.76 |
| 4' deep | 174.000 | C.Y. | 7.656 | | 360.18 | 360.18 |
| 8' deep | 384.000 | C.Y. | 16.896 | | 794.88 | 794.88 |
| 26' x 46', 2' deep | 79.000 | C.Y. | 3.476 | | 163.53 | 163.53 |
| 4' deep | 201.000 | C.Y. | 8.844 | | 416.07 | 416.07 |
| 8' deep | 404.000 | C.Y. | 17.776 | | 836.28 | 836.28 |
| 26' x 60', 2' deep | 94.000 | C.Y. | 4.136 | | 194.58 | 194.58 |
| 4' deep | 240.000 | C.Y. | 10.560 | | 496.80 | 496.80 |
| 8' deep | 483.000 | C.Y. | 21.252 | | 999.81 | 999.81 |
| 30' x 66', 2' deep | 105.000 | C.Y. | 4.620 | | 217.35 | 217.35 |
| 4' deep | 268.000 | C.Y. | 11.792 | | 554.76 | 554.76 |
| 8' deep | 539.000 | C.Y. | 23.716 | | 1115.73 | 1115.73 |
| Backfill, 24' x 38', 2' lifts, no compaction | 34.000 | C.Y. | .340 | | 25.50 | 25.50 |
| Compaction, air tamped | 34.000 | C.Y. | 2.278 | | 163.54 | 163.54 |
| 4' lifts, no compaction | 87.000 | C.Y. | .870 | | 65.25 | 65.25 |
| Compaction, air tamped | 87.000 | C.Y. | 5.829 | | 418.47 | 418.47 |
| 8' lifts, no compaction | 192.000 | C.Y. | 1.920 | | 144.00 | 144.00 |
| Compaction, air tamped | 192.000 | C.Y. | 12.864 | | 923.52 | 923.52 |
| 26' x 46', 2' lifts, no compaction | 40.000 | C.Y. | .400 | | 30.00 | 30.00 |
| Compaction, air tamped | 40.000 | C.Y. | 2.680 | | 192.40 | 192.40 |
| 4' lifts, no compaction | 100.000 | C.Y. | 1.000 | | 75.00 | 75.00 |
| Compaction, air tamped | 100.000 | C.Y. | 6.700 | | 481.00 | 481.00 |
| 8' lifts, no compaction | 202.000 | C.Y. | 2.020 | | 151.50 | 151.50 |
| Compaction, air tamped | 202.000 | C.Y. | 13.534 | | 971.62 | 971.62 |
| 26' x 60', 2' lifts, no compaction | 47.000 | C.Y. | .470 | | 35.25 | 35.25 |
| Compaction, air tamped | 47.000 | C.Y. | 3.149 | | 226.07 | 226.07 |
| 4' lifts, no compaction | 120.000 | C.Y. | 1.200 | | 90.00 | 90.00 |
| Compaction, air tamped | 120.000 | C.Y. | 8.040 | | 577.20 | 577.20 |
| 8' lifts, no compaction | 242.000 | C.Y. | 2.420 | | 181.50 | 181.50 |
| Compaction, air tamped | 242.000 | C.Y. | 16.214 | | 1164.02 | 1164.02 |
| 30' x 66', 2' lifts, no compaction | 53.000 | C.Y. | .530 | | 39.75 | 39.75 |
| Compaction, air tamped | 53.000 | C.Y. | 3.551 | | 254.93 | 254.93 |
| 4' lifts, no compaction | 134.000 | C.Y. | 1.340 | | 100.50 | 100.50 |
| Compaction, air tamped | 134.000 | C.Y. | 8.978 | | 644.54 | 644.54 |
| 8' lifts, no compaction | 269.000 | C.Y. | 2.690 | | 201.75 | 201.75 |
| Compaction, air tamped | 269.000 | C.Y. | 18.023 | | 1293.89 | 1293.89 |
| Rough grade, 30' from building, 24' x 38' | 87.000 | C.Y. | 6.525 | | 240.99 | 240.99 |
| 26' x 46' | 100.000 | C.Y. | 7.500 | | 277.00 | 277.00 |
| 26' x 60' | 120.000 | C.Y. | 9.000 | | 332.40 | 332.40 |
| 30' x 66' | 134.000 | C.Y. | 10.050 | | 371.18 | 371.18 |

# 1 | SITE WORK — 08 | Foundation Excavation Systems

*Backfill*

*Excavate*

| SYSTEM DESCRIPTION | QUAN. | UNIT | MAN HOURS | COST EACH MAT. | COST EACH INST. | COST EACH TOTAL |
|---|---|---|---|---|---|---|
| **BUILDING, 24' X 38', 8' DEEP** | | | | | | |
| Clear & grub, dozer, medium brush, 30' from building | .190 | Acre | 4.750 | | 285.00 | 285.00 |
| Excavate, track loader, 1-1/2 C.Y. bucket | 550.000 | C.Y. | 11.550 | | 511.50 | 511.50 |
| Backfill, dozer, 8' lifts, no compaction | 180.000 | C.Y. | 1.800 | | 135.00 | 135.00 |
| Rough grade, dozer, 30' from building | 280.000 | C.Y. | 2.800 | | 210.00 | 210.00 |
| TOTAL | | | 20.900 | | 1141.50 | 1141.50 |
| **BUILDING, 26' X 46', 8' DEEP** | | | | | | |
| Clear & grub, dozer, medium brush, 30' from building | .210 | Acre | 5.250 | | 315.00 | 315.00 |
| Excavate, track loader, 1-1/2 C.Y. bucket | 672.000 | C.Y. | 14.112 | | 624.96 | 624.96 |
| Backfill, dozer, 8' lifts, no compaction | 220.000 | C.Y. | 2.200 | | 165.00 | 165.00 |
| Rough grade, dozer, 30' from building | 340.000 | C.Y. | 3.400 | | 255.00 | 255.00 |
| TOTAL | | | 24.962 | | 1359.96 | 1359.96 |
| **BUILDING, 26' X 60', 8' DEEP** | | | | | | |
| Clear & grub, dozer, medium brush, 30' from building | .240 | Acre | 6.000 | | 360.00 | 360.00 |
| Excavate, track loader, 1-1/2 C.Y. bucket | 829.000 | C.Y. | 17.409 | | 770.97 | 770.97 |
| Backfill, dozer, 8' lifts, no compaction | 270.000 | C.Y. | 2.700 | | 202.50 | 202.50 |
| Rough grade, dozer, 30' from building | 420.000 | C.Y. | 4.200 | | 315.00 | 315.00 |
| TOTAL | | | 30.309 | | 1648.47 | 1648.47 |
| **BUILDING, 30' X 66', 8' DEEP** | | | | | | |
| Clear & grub, dozer, medium brush, 30' from building | .260 | Acre | 6.500 | | 390.00 | 390.00 |
| Excavate, track loader, 1-1/2 C.Y. bucket | 990.000 | C.Y. | 20.790 | | 920.70 | 920.70 |
| Backfill dozer, 8' lifts, no compaction | 320.000 | C.Y. | 3.200 | | 240.00 | 240.00 |
| Rough grade, dozer, 30' from building | 500.000 | C.Y. | 5.000 | | 375.00 | 375.00 |
| TOTAL | | | 35.490 | | 1925.70 | 1925.70 |

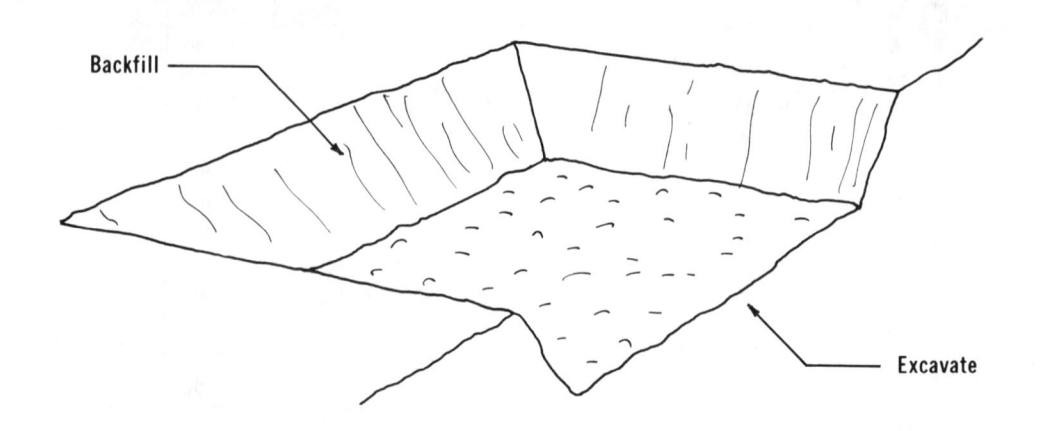

The costs in this system are on a cost each basis.
Quantities are based on 1'-0" clearance beyond footing projection.

| DESCRIPTION | QUAN. | UNIT | MAN HOURS | COST EACH MAT. | COST EACH INST. | COST EACH TOTAL |
|---|---|---|---|---|---|---|
| | | | | | | |
| | | | | | | |
| | | | | | | |

# Foundation Excavation Price Sheet

| | QUAN. | UNIT | MAN HOURS | COST EACH MAT. | COST EACH INST. | TOTAL |
|---|---|---|---|---|---|---|
| Clear & grub, medium brush, 30' from building, 24' x 38' | .190 | Acre | 4.750 | | 285.00 | 285.00 |
| 26' x 46' | .210 | Acre | 5.250 | | 315.00 | 315.00 |
| 26' x 60' | .240 | Acre | 6.000 | | 360.00 | 360.00 |
| 30' x 66' | .260 | Acre | 6.500 | | 390.00 | 390.00 |
| Light trees, to 6" dia. cut & chip, 24' x 38' | .190 | Acre | 11.400 | | 384.75 | 384.75 |
| 26' x 46' | .210 | Acre | 12.600 | | 425.25 | 425.25 |
| 26' x 60' | .240 | Acre | 14.400 | | 486.00 | 486.00 |
| 30' x 66' | .260 | Acre | 15.600 | | 526.50 | 526.50 |
| Medium trees, to 10" dia. cut & chip, 24' x 38' | .190 | Acre | 13.028 | | 437.00 | 437.00 |
| 26' x 46' | .210 | Acre | 14.399 | | 483.00 | 483.00 |
| 26' x 60' | .240 | Acre | 9.800 | | 355.65 | 355.65 |
| 30' x 66' | .260 | Acre | 9.680 | | 453.69 | 453.69 |
| Excavation, basement, 24' x 38', 2' deep | 98.000 | C.Y. | 2.058 | | 91.14 | 91.14 |
| 4' deep | 220.000 | C.Y. | 4.620 | | 204.60 | 204.60 |
| 8' deep | 550.000 | C.Y. | 11.550 | | 511.50 | 511.50 |
| 26' x 46', 2' deep | 123.000 | C.Y. | 12.300 | | 444.93 | 444.93 |
| 4' deep | 274.000 | C.Y. | 12.056 | | 533.91 | 533.91 |
| 8' deep | 672.000 | C.Y. | 14.112 | | 624.96 | 624.96 |
| 26' x 60', 2' deep | 157.000 | C.Y. | 15.700 | | 580.54 | 580.54 |
| 4' deep | 345.000 | C.Y. | 15.180 | | 724.91 | 724.91 |
| 8' deep | 829.000 | C.Y. | 17.409 | | 770.97 | 770.97 |
| 30' x 66', 2' deep | 192.000 | C.Y. | 8.448 | | 405.39 | 405.39 |
| 4' deep | 419.000 | C.Y. | 8.799 | | 389.67 | 389.67 |
| 8' deep | 990.000 | C.Y. | 20.790 | | 920.70 | 920.70 |
| Backfill, 24' x 38', 2' lifts, no compaction | 32.000 | C.Y. | .320 | | 24.00 | 24.00 |
| Compaction, air tamped | 32.000 | C.Y. | 2.144 | | 153.92 | 153.92 |
| 4' lifts, no compaction | 72.000 | C.Y. | .720 | | 54.00 | 54.00 |
| Compaction, air tamped | 72.000 | C.Y. | 4.824 | | 346.32 | 346.32 |
| 8' lifts, no compaction | 180.000 | C.Y. | 1.800 | | 135.00 | 135.00 |
| Compaction, air tamped | 180.000 | C.Y. | 12.060 | | 865.80 | 865.80 |
| 26' x 46', 2' lifts, no compaction | 40.000 | C.Y. | .400 | | 30.00 | 30.00 |
| Compaction, air tamped | 40.000 | C.Y. | 2.680 | | 192.40 | 192.40 |
| 4' lifts, no compaction | 90.000 | C.Y. | .900 | | 67.50 | 67.50 |
| Compaction, air tamped | 90.000 | C.Y. | 6.030 | | 432.90 | 432.90 |
| 8' lifts, no compaction | 220.000 | C.Y. | 2.200 | | 165.00 | 165.00 |
| Compacton, air tamped | 220.000 | C.Y. | 14.740 | | 1058.20 | 1058.20 |
| 26' x 60', 2' lifts, no compaction | 50.000 | C.Y. | .500 | | 37.50 | 37.50 |
| Compaction, air tamped | 50.000 | C.Y. | 3.350 | | 240.50 | 240.50 |
| 4' lifts, no compaction | 110.000 | C.Y. | 1.100 | | 82.50 | 82.50 |
| Compaction, air tamped | 110.000 | C.Y. | 7.370 | | 529.10 | 529.10 |
| 8' lifts, no compaction | 270.000 | C.Y. | 2.700 | | 202.50 | 202.50 |
| Compaction, air tamped | 270.000 | C.Y. | 18.090 | | 1298.70 | 1298.70 |
| 30' x 66', 2' lifts, no compaction | 60.000 | C.Y. | .600 | | 45.00 | 45.00 |
| Compaction, air tamped | 60.000 | C.Y. | 4.020 | | 288.60 | 288.60 |
| 4' lifts, no compaction | 130.000 | C.Y. | 1.300 | | 97.50 | 97.50 |
| Compaction, air tamped | 130.000 | C.Y. | 8.710 | | 625.30 | 625.30 |
| 8' lifts, no compaction | 320.000 | C.Y. | 3.200 | | 240.00 | 240.00 |
| Compaction, air tamped | 320.000 | C.Y. | 21.440 | | 1539.20 | 1539.20 |
| Rough grade, 30' from building, 24' x 38' | 87.000 | C.Y. | 6.525 | | 240.99 | 240.99 |
| 26' x 46' | 100.000 | C.Y. | 7.500 | | 277.00 | 277.00 |
| 26' x 60' | 120.000 | C.Y. | 9.000 | | 332.40 | 332.40 |
| 30' x 66' | 134.000 | C.Y. | 10.050 | | 371.18 | 371.18 |

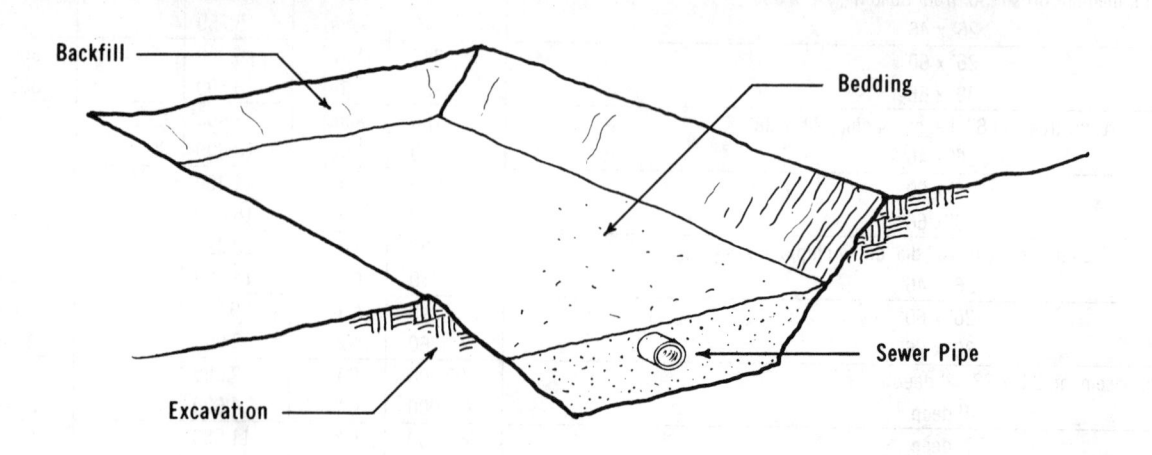

# 1 | SITE WORK    12 | Utility Trenching Systems

| SYSTEM DESCRIPTION | QUAN. | UNIT | MAN HOURS | MAT. | INST. | TOTAL |
|---|---|---|---|---|---|---|
| **2' DEEP** | | | | | | |
| Excavation, backhoe | .296 | C.Y. | .023 | | .80 | .80 |
| Bedding, sand | .111 | C.Y. | .059 | .79 | 1.21 | 2.00 |
| Utility, sewer, 6" cast iron | 1.000 | L.F. | .203 | 5.61 | 4.14 | 9.75 |
| Backfill, incl. compaction | .185 | C.Y. | .043 | | .69 | .69 |
| TOTAL | | | .328 | 6.40 | 6.84 | 13.24 |
| **4' DEEP** | | | | | | |
| Excavation, backhoe | .889 | C.Y. | .071 | | 2.39 | 2.39 |
| Bedding, sand | .111 | C.Y. | .059 | .79 | 1.21 | 2.00 |
| Utility, sewer, 6" cast iron | 1.000 | L.F. | .203 | 5.61 | 4.14 | 9.75 |
| Backfill, incl. compaction | .778 | C.Y. | .182 | | 2.90 | 2.90 |
| TOTAL | | | .515 | 6.40 | 10.64 | 17.04 |
| **6' DEEP** | | | | | | |
| Excavation, backhoe | 1.770 | C.Y. | .141 | | 4.76 | 4.76 |
| Bedding, sand | .111 | C.Y. | .059 | .79 | 1.21 | 2.00 |
| Utility, sewer, 6" cast iron | 1.000 | L.F. | .203 | 5.61 | 4.14 | 9.75 |
| Backfill, incl. compaction | 1.660 | C.Y. | .390 | | 6.19 | 6.19 |
| TOTAL | | | .793 | 6.40 | 16.30 | 22.70 |
| **8' DEEP** | | | | | | |
| Excavation, backhoe | 2.960 | C.Y. | .236 | | 7.96 | 7.96 |
| Bedding, sand | .111 | C.Y. | .059 | .79 | 1.21 | 2.00 |
| Utility, sewer, 6" cast iron | 1.000 | L.F. | .203 | 5.61 | 4.14 | 9.75 |
| Backfill, incl. compaction | 2.850 | C.Y. | .669 | | 10.63 | 10.63 |
| TOTAL | | | 1.167 | 6.40 | 23.94 | 30.34 |

The costs in this system are based on a cost per linear foot of trench, and based on 2' wide at bottom of trench up to 6' deep.

| DESCRIPTION | QUAN. | UNIT | MAN HOURS | MAT. | INST. | TOTAL |
|---|---|---|---|---|---|---|
| | | | | | | |
| | | | | | | |
| | | | | | | |

# Utility Trenching Price Sheet

| Item | QUAN. | UNIT | MAN HOURS | MAT. | INST. | TOTAL |
|---|---|---|---|---|---|---|
| Excavation, bottom of trench 2' wide, 2' deep | .296 | C.Y. | .023 | | .80 | .80 |
| 4' deep | .889 | C.Y. | .095 | | 2.69 | 2.69 |
| 6' deep | 1.770 | C.Y. | .141 | | 4.76 | 4.76 |
| 8' deep | 2.960 | C.Y. | .236 | | 7.96 | 7.96 |
| Bedding, sand, bottom of trench, 2' wide, no compaction, pipe, 2" diameter | .070 | C.Y. | .037 | .50 | .76 | 1.26 |
| 4" diameter | .084 | C.Y. | .044 | .60 | .92 | 1.52 |
| 6" diameter | .105 | C.Y. | .055 | .75 | 1.15 | 1.90 |
| 8" diameter | .122 | C.Y. | .065 | .87 | 1.33 | 2.20 |
| Compacted, pipe, 2" diameter | .074 | C.Y. | .039 | .53 | .81 | 1.34 |
| 4" diameter | .092 | C.Y. | .049 | .66 | 1.00 | 1.66 |
| 6" diameter | .111 | C.Y. | .059 | .79 | 1.21 | 2.00 |
| 8" diameter | .129 | C.Y. | .068 | .92 | 1.41 | 2.33 |
| 3/4" stone, bottom of trench 2' wide, pipe, 4" diameter | .082 | C.Y. | .043 | .59 | .89 | 1.48 |
| 6" diameter | .099 | C.Y. | .052 | .71 | 1.08 | 1.79 |
| 3/8" stone, bottom of trench 2' wide pipe, 4" diameter | .084 | C.Y. | .044 | .60 | .92 | 1.52 |
| 6" diameter | .102 | C.Y. | .054 | .73 | 1.11 | 1.84 |
| Utilities, drainage & sewerage, asbestos cement, 6" diameter | 1.000 | L.F. | .073 | 3.30 | 1.52 | 4.82 |
| 8" diameter | 1.000 | L.F. | .074 | 4.29 | 1.81 | 6.10 |
| Bituminous fiber, 4" diameter | 1.000 | L.F. | .042 | 1.38 | .92 | 2.30 |
| 6" diameter | 1.000 | L.F. | .047 | 2.75 | 1.03 | 3.78 |
| 8" diameter | 1.000 | L.F. | .053 | 6.55 | 1.15 | 7.70 |
| Concrete, nonreinforced, 6" diameter | 1.000 | L.F. | .091 | 2.48 | 1.89 | 4.37 |
| 8" diameter | 1.000 | L.F. | .093 | 2.70 | 2.26 | 4.96 |
| PVC, SDR 33.5, 4" diameter | 1.000 | L.F. | .064 | .72 | 1.33 | 2.05 |
| SDR 35, 6" diameter | 1.000 | L.F. | .069 | 1.38 | 1.43 | 2.81 |
| SDR 33.5, 8" diameter | 1.000 | L.F. | .072 | 2.42 | 1.50 | 3.92 |
| Vitrified clay, 4" diameter | 1.000 | L.F. | .091 | 1.76 | 1.89 | 3.65 |
| 6" diameter | 1.000 | L.F. | .120 | 2.75 | 2.50 | 5.25 |
| 8" diameter | 1.000 | L.F. | .140 | 3.85 | 3.40 | 7.25 |
| Gas & service, polyethylene, 1-1/4" diameter | 1.000 | L.F. | .060 | .56 | 1.26 | 1.82 |
| Steel sched. 40, 1" diameter | 1.000 | L.F. | .107 | 2.42 | 2.30 | 4.72 |
| 2" diameter | 1.000 | L.F. | .114 | 3.85 | 2.45 | 6.30 |
| Subdrainage, asbestos cement pipe class 4000 perf., 4" diameter | 1.000 | L.F. | .062 | 1.98 | 1.29 | 3.27 |
| 6" diameter | 1.000 | L.F. | .063 | 3.14 | 1.32 | 4.46 |
| Bituminous fiber, perforated, 3" diameter | 1.000 | L.F. | .040 | 1.16 | .87 | 2.03 |
| 4" diameter | 1.000 | L.F. | .042 | 1.32 | .92 | 2.24 |
| 5" diameter | 1.000 | L.F. | .044 | 2.31 | .97 | 3.28 |
| 6" diameter | 1.000 | L.F. | .047 | 2.86 | 1.03 | 3.89 |
| Porous wall concrete, 4" diameter | 1.000 | L.F. | .072 | 1.43 | 1.50 | 2.93 |
| Vitrified clay, perforated, 4" diameter | 1.000 | L.F. | .060 | 1.98 | 1.26 | 3.24 |
| 6" diameter | 1.000 | L.F. | .076 | 3.36 | 1.59 | 4.95 |
| Water service, copper, type K, 3/4" | 1.000 | L.F. | .114 | .97 | 2.51 | 3.48 |
| 1" diameter | 1.000 | L.F. | .141 | 1.31 | 3.07 | 4.38 |
| PVC, 3/4" diameter | 1.000 | L.F. | .133 | .64 | 2.93 | 3.57 |
| 1" diameter | 1.000 | L.F. | .141 | .80 | 3.10 | 3.90 |
| Backfill, bottom of trench 2' wide no compact, 2' deep, pipe, 2" diameter | .226 | C.Y. | .053 | | .84 | .84 |
| 4" diameter | .212 | C.Y. | .049 | | .79 | .79 |
| 6" diameter | .185 | C.Y. | .043 | | .69 | .69 |
| 4' deep, pipe, 2" diameter | .819 | C.Y. | .192 | | 3.05 | 3.05 |
| 4" diameter | .805 | C.Y. | .189 | | 3.00 | 3.00 |
| 6" diameter | .778 | C.Y. | .182 | | 2.90 | 2.90 |
| 6' deep, pipe, 2" diameter | 1.700 | C.Y. | .399 | | 6.34 | 6.34 |
| 4" diameter | 1.690 | C.Y. | .397 | | 6.30 | 6.30 |
| 6" diameter | 1.660 | C.Y. | .390 | | 6.19 | 6.19 |
| 8' deep, pipe, 2" diameter | 2.890 | C.Y. | .679 | | 10.78 | 10.78 |
| 4" diameter | 2.870 | C.Y. | .674 | | 10.71 | 10.71 |
| 6" diameter | 2.850 | C.Y. | .669 | | 10.63 | 10.63 |

# 1 | SITE WORK — 16 | Sidewalk Systems

| SYSTEMS DESCRIPTION | QUAN. | UNIT | MAN HOURS | COST PER S.F. MAT. | COST PER S.F. INST. | COST PER S.F. TOTAL |
|---|---|---|---|---|---|---|
| **ASPHALT SIDEWALK SYSTEM, 3' WIDE WALK** | | | | | | |
| Gravel fill, 4" deep | 1.000 | S.F. | | .11 | .02 | .13 |
| Compact fill | .012 | C.Y. | .001 | | .02 | .02 |
| Handgrade | 1.000 | S.F. | .012 | | .19 | .19 |
| Walking surface, bituminous paving, 2" thick | 1.000 | S.F. | .008 | .37 | .16 | .53 |
| Edging, brick, laid on edge | .670 | L.F. | .079 | .94 | 1.40 | 2.34 |
| TOTAL | | | .100 | 1.42 | 1.79 | 3.21 |
| **CONCRETE SIDEWALK SYSTEM, 3' WIDE WALK** | | | | | | |
| Gravel fill, 4" deep | 1.000 | S.F. | | .11 | .02 | .13 |
| Compact fill | .012 | C.Y. | .001 | | .02 | .02 |
| Handgrade | 1.000 | S.F. | .012 | | .19 | .19 |
| Walking surface, concrete, 4" thick | 1.000 | S.F. | .040 | .87 | .72 | 1.59 |
| Edging, brick, laid on edge | .670 | L.F. | .079 | .94 | 1.40 | 2.34 |
| TOTAL | | | .132 | 1.92 | 2.35 | 4.27 |
| **PAVERS, BRICK SIDEWALK SYSTEM, 3' WIDE WALK** | | | | | | |
| Sand base fill, 4" deep | 1.000 | S.F. | | .09 | .03 | .12 |
| Compact fill | .012 | C.Y. | .001 | | .02 | .02 |
| Handgrade | 1.000 | S.F. | .012 | | .19 | .19 |
| Walking surface, brick pavers | 1.000 | S.F. | .145 | 1.93 | 2.57 | 4.50 |
| Edging, redwood, untreated, 1" x 4" | .670 | L.F. | .031 | .46 | .67 | 1.13 |
| TOTAL | | | .189 | 2.48 | 3.48 | 5.96 |

The costs in this system are based on a cost per square foot of sidewalk area. Concrete used is 3000 p.s.i.

| DESCRIPTION | QUAN. | UNIT | MAN HOURS | COST PER S.F. MAT. | COST PER S.F. INST. | COST PER S.F. TOTAL |
|---|---|---|---|---|---|---|
| | | | | | | |
| | | | | | | |
| | | | | | | |
| | | | | | | |
| | | | | | | |

# Sidewalk Price Sheet

| | QUAN. | UNIT | MAN HOURS | COST PER S.F. MAT. | COST PER S.F. INST. | TOTAL |
|---|---|---|---|---|---|---|
| Base, crushed stone, 3" deep | 1.000 | S.F. | .001 | .20 | .03 | .23 |
| 6" deep | 1.000 | S.F. | .001 | .41 | .05 | .46 |
| 9" deep | 1.000 | S.F. | .002 | .62 | .06 | .68 |
| 12" deep | 1.000 | S.F. | .002 | .81 | .08 | .89 |
| Bank run gravel, 6" deep | 1.000 | S.F. | | .17 | .03 | .20 |
| 9" deep | 1.000 | S.F. | .001 | .25 | .05 | .30 |
| 12" deep | 1.000 | S.F. | .001 | .33 | .06 | .39 |
| Compact base, 3" deep | .009 | C.Y. | | | .02 | .02 |
| 6" deep | .012 | C.Y. | .001 | | .03 | .03 |
| 9" deep | .018 | C.Y. | .002 | | .05 | .05 |
| Handgrade | 1.000 | S.F. | .012 | | .19 | .19 |
| Surface, brick, pavers dry joints, laid flat, running bond | 1.000 | S.F. | .145 | 1.93 | 2.57 | 4.50 |
| Basket weave | 1.000 | S.F. | .168 | 2.79 | 2.96 | 5.75 |
| Herringbone | 1.000 | S.F. | .174 | 2.79 | 3.06 | 5.85 |
| Laid on edge, running bond | 1.000 | S.F. | .229 | 3.03 | 4.02 | 7.05 |
| Mortar jts. laid flat, running bond | 1.000 | S.F. | .174 | 2.32 | 3.08 | 5.40 |
| Basket weave | 1.000 | S.F. | .201 | 3.35 | 3.55 | 6.90 |
| Herringbone | 1.000 | S.F. | .208 | 3.35 | 3.67 | 7.02 |
| Laid on edge, running bond | 1.000 | S.F. | .274 | 3.64 | 4.82 | 8.46 |
| Bituminous paving, 1-1/2" thick | .111 | S.F. | .005 | .26 | .11 | .37 |
| 2" thick | .111 | S.F. | .008 | .37 | .16 | .53 |
| 2-1/2" thick | .111 | S.F. | .010 | .46 | .20 | .66 |
| Sand finish, 3/4" thick | .111 | S.F. | .002 | .13 | .04 | .17 |
| 1" thick | .111 | S.F. | .002 | .19 | .06 | .25 |
| Concrete, reinforced, broom finish, 4" thick | 1.000 | S.F. | .040 | .87 | .72 | 1.59 |
| 5" thick | 1.000 | S.F. | .044 | 1.05 | .78 | 1.83 |
| 6" thick | 1.000 | S.F. | .047 | 1.21 | .84 | 2.05 |
| Crushed stone, white marble, 3" thick | 1.000 | S.F. | .009 | .47 | .15 | .62 |
| Bluestone, 3" thick | 1.000 | S.F. | .009 | .10 | .15 | .25 |
| Flagging, bluestone, 1" | 1.000 | S.F. | .198 | 1.98 | 3.47 | 5.45 |
| 1-1/2" | 1.000 | S.F. | .188 | 3.58 | 3.32 | 6.90 |
| Slate, natural cleft, 3/4" | 1.000 | S.F. | .174 | 1.43 | 3.07 | 4.50 |
| Random rect., 1/2" | 1.000 | S.F. | .152 | 3.36 | 2.69 | 6.05 |
| Granite blocks | 1.000 | S.F. | .174 | 4.68 | 3.07 | 7.75 |
| Edging, corrugated aluminum, 4", 3' wide walk | .666 | L.F. | .004 | .24 | .10 | .34 |
| 4' wide walk | .500 | L.F. | .003 | .20 | .08 | .28 |
| 6", 3' wide walk | .666 | L.F. | .004 | .24 | .10 | .34 |
| 4' wide walk | .500 | L.F. | .003 | .20 | .08 | .28 |
| Redwood-cedar-cypress, 1" x 4", 3' wide walk | .666 | L.F. | .021 | .47 | .44 | .91 |
| 4' wide walk | .500 | L.F. | .016 | .35 | .33 | .68 |
| 1" x 6", 3' wide walk | .666 | L.F. | .026 | .59 | .55 | 1.14 |
| 4' wide walk | .500 | L.F. | .020 | .44 | .41 | .85 |
| Brick, dry joints, 3' wide walk | .666 | L.F. | .079 | .94 | 1.40 | 2.34 |
| 4' wide walk | .500 | L.F. | .059 | .70 | 1.05 | 1.75 |
| Mortar joints, 3' wide walk | .666 | L.F. | .095 | 1.13 | 1.68 | 2.81 |
| 4' wide walk | .500 | L.F. | .071 | .84 | 1.25 | 2.09 |

# 1 | SITE WORK  20 | Driveway Systems

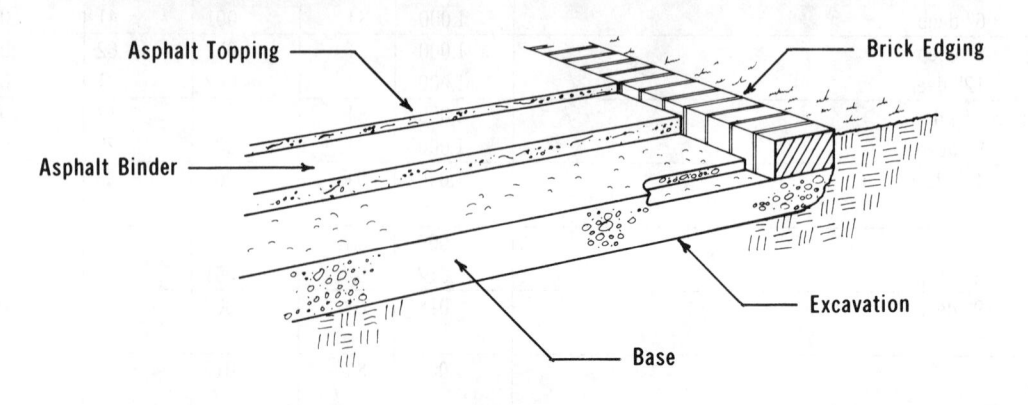

| SYSTEMS DESCRIPTION | QUAN. | UNIT | MAN HOURS | MAT. | INST. | TOTAL |
|---|---|---|---|---|---|---|
| **ASPHALT DRIVEWAY TO 10' WIDE** | | | | | | |
| Excavation, driveway to 10' wide, 6" deep | .019 | C.Y. | | | .02 | .02 |
| Base, 6" crushed stone | 1.000 | S.F. | .001 | .41 | .05 | .46 |
| Handgrade base | 1.000 | S.F. | .012 | | .19 | .19 |
| Surface, asphalt, 2" thick base, 1" topping | 1.000 | S.F. | .002 | .19 | .06 | .25 |
| Edging, brick pavers | .200 | L.F. | .023 | .28 | .42 | .70 |
| TOTAL | | | .038 | .88 | .74 | 1.62 |
| **CONCRETE DRIVEWAY TO 10' WIDE** | | | | | | |
| Excavation, driveway to 10' wide, 6" deep | .019 | C.Y. | | | .02 | .02 |
| Base, 6" crushed stone | 1.000 | S.F. | .001 | .41 | .05 | .46 |
| Handgrade base | 1.000 | S.F. | .012 | | .19 | .19 |
| Surface, concrete, 4" thick | 1.000 | S.F. | .040 | .87 | .72 | 1.59 |
| Edging, brick pavers | .200 | L.F. | .023 | .28 | .42 | .70 |
| TOTAL | | | .076 | 1.56 | 1.40 | 2.96 |
| **PAVERS, BRICK DRIVEWAY TO 10' WIDE** | | | | | | |
| Excavation, driveway to 10' wide, 6" deep | .019 | C.Y. | | | .02 | .02 |
| Base, 6" sand | 1.000 | S.F. | | .14 | .05 | .19 |
| Handgrade base | 1.000 | S.F. | .012 | | .19 | .19 |
| Surface, pavers, brick laid flat, running bond | 1.000 | S.F. | .145 | 1.93 | 2.57 | 4.50 |
| Edging, redwood, untreated, 2" x 4" | .200 | L.F. | .009 | .14 | .20 | .34 |
| TOTAL | | | .166 | 2.21 | 3.03 | 5.24 |

| DESCRIPTION | QUAN. | UNIT | MAN HOURS | MAT. | INST. | TOTAL |
|---|---|---|---|---|---|---|
| | | | | | | |
| | | | | | | |
| | | | | | | |
| | | | | | | |
| | | | | | | |

# Driveway Price Sheet

| | QUAN. | UNIT | MAN. HOURS | COST PER S.F. MAT. | COST PER S.F. INST. | TOTAL |
|---|---|---|---|---|---|---|
| Excavation, by machine, 10' wide, 6" deep | .019 | C.Y. | | | .02 | .02 |
| 12" deep | .037 | C.Y. | .001 | | .05 | .05 |
| 18" deep | .055 | C.Y. | .001 | | .07 | .07 |
| 20' wide, 6" deep | .019 | C.Y. | | | .02 | .02 |
| 12" deep | .037 | C.Y. | .001 | | .05 | .05 |
| 18" deep | .055 | C.Y. | .001 | | .07 | .07 |
| Base, crushed stone, 10' wide, 3" deep | 1.000 | S.F. | .001 | .20 | .03 | .23 |
| 6" deep | 1.000 | S.F. | .001 | .41 | .05 | .46 |
| 9" deep | 1.000 | S.F. | .002 | .62 | .06 | .68 |
| 20' wide, 3" deep | 1.000 | S.F. | .001 | .20 | .03 | .23 |
| 6" deep | 1.000 | S.F. | .001 | .41 | .05 | .46 |
| 9" deep | 1.000 | S.F. | .002 | .62 | .06 | .68 |
| Bank run gravel, 10' wide, 3" deep | 1.000 | S.F. | | .09 | .01 | .10 |
| 6" deep | 1.000 | S.F. | | .17 | .03 | .20 |
| 9" deep | 1.000 | S.F. | .001 | .25 | .05 | .30 |
| 20' wide, 3" deep | 1.000 | S.F. | | .09 | .01 | .10 |
| 6" deep | 1.000 | S.F. | | .17 | .03 | .20 |
| 9" deep | 1.000 | S.F. | .001 | .25 | .05 | .30 |
| Handgrade, 10' wide | 1.000 | S.F. | .012 | | .19 | .19 |
| 20' wide | 1.000 | S.F. | .012 | | .19 | .19 |
| Surface, asphalt, 10' wide, 3/4" topping, 1" base | 1.000 | S.F. | .004 | .37 | .09 | .46 |
| 2" base | 1.000 | S.F. | .006 | .46 | .14 | .60 |
| 1" topping, 1" base | 1.000 | S.F. | .004 | .43 | .11 | .54 |
| 2" base | 1.000 | S.F. | .006 | .52 | .16 | .68 |
| 20' wide, 3/4" topping, 1" base | 1.000 | S.F. | .004 | .37 | .09 | .46 |
| 2" base | 1.000 | S.F. | .006 | .46 | .14 | .60 |
| 1" topping, 1" base | 1.000 | S.F. | .004 | .43 | .11 | .54 |
| 2" base | 1.000 | S.F. | .006 | .52 | .16 | .68 |
| Concrete, 10' wide, 4" thick | 1.000 | S.F. | .040 | .87 | .72 | 1.59 |
| 6" thick | 1.000 | S.F. | .047 | 1.21 | .84 | 2.05 |
| 20' wide, 4" thick | 1.000 | S.F. | .040 | .87 | .72 | 1.59 |
| 6" thick | 1.000 | S.F. | .047 | 1.21 | .84 | 2.05 |
| Paver, brick 10' wide dry joints, running bond, laid flat | 1.000 | S.F. | .145 | 1.93 | 2.57 | 4.50 |
| Laid on edge | 1.000 | S.F. | .229 | 3.03 | 4.02 | 7.05 |
| Mortar joints, laid flat | 1.000 | S.F. | .174 | 2.32 | 3.08 | 5.40 |
| Laid on edge | 1.000 | S.F. | .274 | 3.64 | 4.82 | 8.46 |
| 20' wide, running bond, dry jts., laid flat | 1.000 | S.F. | .145 | 1.93 | 2.57 | 4.50 |
| Laid on edge | 1.000 | S.F. | .229 | 3.03 | 4.02 | 7.05 |
| Mortar joints, laid flat | 1.000 | S.F. | .174 | 2.32 | 3.08 | 5.40 |
| Laid on edge | 1.000 | S.F. | .274 | 3.64 | 4.82 | 8.46 |
| Crushed stone, 10' wide, white marble, 3" | 1.000 | S.F. | .009 | .47 | .15 | .62 |
| Bluestone, 3" | 1.000 | S.F. | .009 | .10 | .15 | .25 |
| 20' wide, white marble, 3" | 1.000 | S.F. | .009 | .47 | .15 | .62 |
| Bluestone, 3" | 1.000 | S.F. | .009 | .10 | .15 | .25 |
| Soil cement, 10' wide | 1.000 | S.F. | .004 | .17 | .30 | .47 |
| 20' wide | 1.000 | S.F. | .004 | .17 | .30 | .47 |
| Granite blocks, 10' wide | 1.000 | S.F. | .174 | 4.68 | 3.07 | 7.75 |
| 20' wide | 1.000 | S.F. | .174 | 4.68 | 3.07 | 7.75 |
| Asphalt block, solid 1-1/4" thick | 1.000 | S.F. | .119 | 2.54 | 2.10 | 4.64 |
| Solid 3" thick | 1.000 | S.F. | .123 | 3.30 | 2.20 | 5.50 |
| Edging, brick, 10' wide | .200 | L.F. | .023 | .28 | .42 | .70 |
| 20' wide | .100 | L.F. | .011 | .14 | .21 | .35 |
| Redwood, untreated 2" x 4", 10' wide | .200 | L.F. | .009 | .14 | .20 | .34 |
| 20' wide | .100 | L.F. | .004 | .07 | .10 | .17 |
| Granite, 4 1/2" x 12" straight, 10' wide | .200 | L.F. | .037 | .92 | .96 | 1.88 |
| 20' wide | .100 | L.F. | .018 | .46 | .48 | .94 |
| Finishes, asphalt sealer, 10' wide | 1.000 | S.F. | .023 | .34 | .39 | .73 |
| 20' wide | 1.000 | S.F. | .023 | .34 | .39 | .73 |
| Concrete, exposed aggregate 10' wide | 1.000 | S.F. | .013 | .07 | .23 | .30 |
| 20' wide | 1.000 | S.F. | .013 | .07 | .23 | .30 |

# 1 SITE WORK — 24 Septic Systems

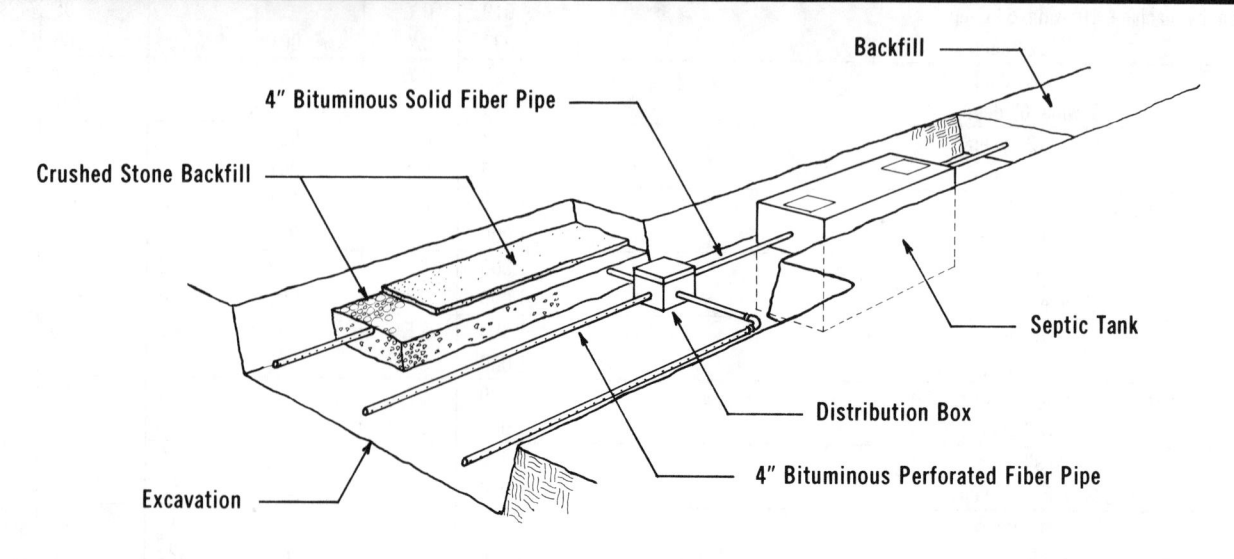

*Diagram labels: Backfill, 4" Bituminous Solid Fiber Pipe, Crushed Stone Backfill, Septic Tank, Distribution Box, 4" Bituminous Perforated Fiber Pipe, Excavation*

| SYSTEM DESCRIPTION | QUAN. | UNIT | MAN HOURS | MAT | INST. | TOTAL |
|---|---|---|---|---|---|---|
| **SEPTIC SYSTEM WITH 1000 S.F. LEACHING FIELD, 1000 GALLON TANK** | | | | | | |
| Tank, 1000 gallon, concrete | 1.000 | Ea. | 3.500 | 302.50 | 87.50 | 390.00 |
| Distribution box, concrete | 1.000 | Ea. | 1.000 | 27.50 | 21.50 | 49.00 |
| 4" bituminous fiber pipe | 25.000 | L.F. | 1.050 | 34.50 | 23.00 | 57.50 |
| Tank and field excavation | 119.000 | C.Y. | 13.090 | | 539.07 | 539.07 |
| Crushed stone backfill | 76.000 | C.Y. | 12.160 | 585.20 | 292.60 | 877.80 |
| Backfill with excavated material | 36.000 | C.Y. | .360 | | 27.00 | 27.00 |
| Building paper | 125.000 | S.Y. | 1.100 | 16.50 | 27.50 | 44.00 |
| 4" bituminous fiber perforated pipe | 145.000 | L.F. | 6.090 | 191.40 | 133.40 | 324.80 |
| 4" pipe fittings | 2.000 | Ea. | 2.280 | 14.96 | 45.04 | 60.00 |
| TOTAL | | | 40.630 | 1172.56 | 1196.61 | 2369.17 |
| **SEPTIC SYSTEM WITH 2 LEACHING PITS, 1000 GALLON TANK** | | | | | | |
| Tank, 1000 gallon, concrete | 1.000 | Ea. | 3.500 | 302.50 | 87.50 | 390.00 |
| Distribution box, concrete | 1.000 | Ea. | 1.000 | 27.50 | 21.50 | 49.00 |
| 4" bituminous fiber pipe | 75.000 | L.F. | 3.150 | 103.50 | 69.00 | 172.50 |
| Excavation for tank only | 20.000 | C.Y. | 2.200 | | 90.60 | 90.60 |
| Crushed stone backfill | 10.000 | C.Y. | 1.600 | 77.00 | 38.50 | 115.50 |
| Backfill with excavated material | 55.000 | C.Y. | .550 | | 41.25 | 41.25 |
| Pits, 6' diameter, including excavation and stone backfill | 2.000 | Ea. | | 1100.00 | | 1100.00 |
| TOTAL | | | 12.000 | 1610.50 | 348.35 | 1958.85 |

The costs in this system include all necessary piping and excavation.

| DESCRIPTION | QUAN. | UNIT | MAN HOURS | MAT. | INST. | TOTAL |
|---|---|---|---|---|---|---|
| | | | | | | |
| | | | | | | |
| | | | | | | |
| | | | | | | |
| | | | | | | |

# Septic Systems Price Sheet

| | QUAN. | UNIT | MAN. HOURS | COST EACH MAT. | COST EACH INST. | TOTAL |
|---|---|---|---|---|---|---|
| Tank, precast concrete, 1000 gallon | 1.000 | Ea. | 3.500 | 302.50 | 87.50 | 390.00 |
| 2000 gallon | 1.000 | Ea. | 5.600 | 583.00 | 137.00 | 720.00 |
| Distribution box, concrete, 5 outlets | 1.000 | Ea. | 1.000 | 27.50 | 21.50 | 49.00 |
| 12 outlets | 1.000 | Ea. | 2.000 | 192.50 | 42.50 | 235.00 |
| 4" pipe, bituminous fiber, solid | 25.000 | L.F. | 1.050 | 34.50 | 23.00 | 57.50 |
| Tank and field excavation, 1000 S.F. field | 119.000 | C.Y. | 13.090 | | 539.07 | 539.07 |
| 2000 S.F. field | 190.000 | C.Y. | 20.900 | | 860.70 | 860.70 |
| Tank excavation only, 1000 gallon tank | 20.000 | C.Y. | 2.200 | | 90.60 | 90.60 |
| 2000 gallon tank | 32.000 | C.Y. | 3.520 | | 144.96 | 144.96 |
| Backfill, crushed stone 1000 S.F. field | 76.000 | C.Y. | 12.160 | 585.20 | 292.60 | 877.80 |
| 2000 S.F. field | 140.000 | C.Y. | 22.400 | 1078.00 | 539.00 | 1617.00 |
| Backfill with excavated material, 1000 S.F. field | 36.000 | C.Y. | .360 | | 27.00 | 27.00 |
| 2000 S.F. field | 60.000 | C.Y. | .600 | | 45.00 | 45.00 |
| 6' diameter pits | 55.000 | C.Y. | .550 | | 41.25 | 41.25 |
| 3' diameter pits | 42.000 | C.Y. | .420 | | 31.50 | 31.50 |
| Building paper, 1000 S.F. field | 125.000 | S.Y. | 2.200 | 33.00 | 55.00 | 88.00 |
| 2000 S.F. field | 250.000 | S.Y. | 4.400 | 66.00 | 110.00 | 176.00 |
| 4" pipe, bituminous fiber, perforated, 1000 S.F. field | 145.000 | L.F. | 6.090 | 191.40 | 133.40 | 324.80 |
| 2000 S.F. field | 265.000 | L.F. | 11.130 | 349.80 | 243.80 | 593.60 |
| Pipe fittings, bituminous fiber, 1000 S.F. field | 2.000 | Ea. | 2.280 | 14.96 | 45.04 | 60.00 |
| 2000 S.F. field | 4.000 | Ea. | 4.560 | 29.92 | 90.08 | 120.00 |
| Leaching pit, including excavation and stone backfill, 3' diameter | 1.000 | Ea. | | 412.50 | | 412.50 |
| 6' diameter | 1.000 | Ea. | | 550.00 | | 550.00 |
| Tank, precast concrete, 1000 gallon | | | | | | |

# 1 | SITE WORK — 60 | Chain Link Fence Price Sheet

| SYSTEM DESCRIPTION | QUAN. | UNIT | MAN. HOURS | MAT. | INST. | TOTAL |
|---|---|---|---|---|---|---|
| Chain link fence | | | | | | |
|   Galv.9ga. wire, 1-5/8"post 10'O.C., 1-3/8"top rail, 2"corner post, 3'high | 1.000 | L.F. | .130 | 2.75 | 2.19 | 4.94 |
|     4' high | 1.000 | L.F. | .141 | 3.19 | 2.41 | 5.60 |
|     6' high | 1.000 | L.F. | .209 | 4.13 | 3.52 | 7.65 |
|     Add for gate 3' wide 1-3/8" frame 3' high | 1.000 | Ea. | 1.040 | 27.50 | 17.50 | 45.00 |
|       4' high | 1.000 | Ea. | 1.200 | 36.30 | 20.70 | 57.00 |
|       6' high | 1.000 | Ea. | 1.600 | 55.00 | 27.00 | 82.00 |
|     Add for gate 4' wide 1-3/8" frame 3' high | 1.000 | Ea. | 1.090 | 36.30 | 18.70 | 55.00 |
|       4' high | 1.000 | Ea. | 1.330 | 48.40 | 22.60 | 71.00 |
|       6' high | 1.000 | Ea. | 2.000 | 72.60 | 32.40 | 105.00 |
|   Alum.9ga. wire, 1-5/8"post, 10'O.C., 1-3/8"top rail, 2"corner post,3'high | 1.000 | L.F. | .130 | 3.30 | 2.20 | 5.50 |
|     4' high | 1.000 | L.F. | .141 | 3.85 | 2.40 | 6.25 |
|     6' high | 1.000 | L.F. | .209 | 4.95 | 3.55 | 8.50 |
|     Add for gate 3' wide 1-3/8" frame 3' high | 1.000 | Ea. | 1.040 | 33.00 | 18.00 | 51.00 |
|       4' high | 1.000 | Ea. | 1.200 | 44.00 | 20.00 | 64.00 |
|       6' high | 1.000 | Ea. | 1.600 | 66.00 | 27.00 | 93.00 |
|     Add for gate 4' wide 1-3/8" frame 3' high | 1.000 | Ea. | 1.090 | 44.00 | 18.00 | 62.00 |
|       4' high | 1.000 | Ea. | 1.330 | 58.30 | 22.70 | 81.00 |
|       6' high | 1.000 | Ea. | 2.000 | 88.00 | 32.00 | 120.00 |
|   Vinyl 9ga. wire, 1-5/8"post 10'O.C., 1-3/8"top rail, 2"corner post,3'high | 1.000 | L.F. | .130 | 3.08 | 2.17 | 5.25 |
|     4' high | 1.000 | L.F. | .141 | 3.52 | 2.38 | 5.90 |
|     6' high | 1.000 | L.F. | .209 | 4.40 | 3.55 | 7.95 |
|     Add for gate 3' wide 1-3/8" frame 3' high | 1.000 | Ea. | 1.040 | 36.30 | 17.70 | 54.00 |
|       4' high | 1.000 | Ea. | 1.200 | 48.40 | 20.60 | 69.00 |
|       6' high | 1.000 | Ea. | 1.600 | 73.70 | 26.30 | 100.00 |
|     Add for gate 4' wide 1-3/8" frame 3' high | 1.000 | Ea. | 1.090 | 48.40 | 18.60 | 67.00 |
|       4' high | 1.000 | Ea. | 1.330 | 64.90 | 22.10 | 87.00 |
|       6' high | 1.000 | Ea. | 2.000 | 96.80 | 33.20 | 130.00 |
| Tennis court chain link fence, 10' high | | | | | | |
|   Galv.11ga.wire, 2"post 10'O.C., 1-3/8"top rail, 2-1/2"corner post | 1.000 | L.F. | .253 | 7.59 | 4.26 | 11.85 |
|     Add for gate 3' wide 1-3/8" frame | 1.000 | Ea. | 2.400 | 99.00 | 41.00 | 140.00 |
|   Alum.11ga.wire, 2"post 10'O.C., 1-3/8"top rail, 2-1/2"corner post | 1.000 | L.F. | .253 | 8.80 | 4.25 | 13.05 |
|     Add for gate 3' wide 1-3/8" frame | 1.000 | Ea. | 2.400 | 121.00 | 39.00 | 160.00 |
|   Vinyl 11ga.wire, 2"post 10' O.C., 1-3/8"top rail, 2-1/2"corner post | 1.000 | L.F. | .253 | 9.79 | 4.26 | 14.05 |
|     Add for gate 3' wide 1-3/8" frame | 1.000 | Ea. | 2.400 | 132.00 | 43.00 | 175.00 |
| Railings, commercial | | | | | | |
|   Aluminum balcony rail, 1-1/2" posts with pickets | 1.000 | L.F. | .164 | 27.50 | 4.50 | 32.00 |
|     With expanded metal panels | 1.000 | L.F. | .164 | 44.00 | 4.00 | 48.00 |
|     With porcelain enamel panel inserts | 1.000 | L.F. | .164 | 55.00 | 4.00 | 59.00 |
|   Mild steel, ornamental rounded top rail | 1.000 | L.F. | .164 | 27.50 | 4.50 | 32.00 |
|     As above, but pitch down stairs | 1.000 | L.F. | .183 | 27.50 | 4.50 | 32.00 |
|   Steel pipe, welded, 1-1/2" round, painted | 1.000 | L.F. | .160 | 22.00 | 4.00 | 26.00 |
|     Galvanized | 1.000 | L.F. | .160 | 33.00 | 4.00 | 37.00 |
|   Residential, stock units, mild steel, deluxe | 1.000 | L.F. | .102 | 11.00 | 2.60 | 13.60 |
|     Economy | 1.000 | L.F. | .102 | 8.80 | 2.60 | 11.40 |

# 1 | SITE WORK — 64 | Wood Fence Price Sheet

| SYSTEM DESCRIPTION | QUAN. | UNIT | MAN HOURS | MAT. | INST. | TOTAL |
|---|---|---|---|---|---|---|
| Basketweave, 3/8"x4" boards, 2"x4" stringers on spreaders, 4"x4" posts | | | | | | |
| No. 1 cedar, 6' high | 1.000 | L.F. | .150 | 14.03 | 2.52 | 16.55 |
| Treated pine, 6' high | 1.000 | L.F. | .160 | 7.15 | 2.70 | 9.85 |
| Board fence, 1"x4" boards, 2"x4" rails, 4"x4" posts | | | | | | |
| Preservative treated, 2 rail, 3' high | 1.000 | L.F. | .166 | 4.57 | 2.78 | 7.35 |
| 4' high | 1.000 | L.F. | .178 | 4.95 | 3.00 | 7.95 |
| 3 rail, 5' high | 1.000 | L.F. | .185 | 6.16 | 3.14 | 9.30 |
| 6' high | 1.000 | L.F. | .192 | 6.71 | 3.24 | 9.95 |
| Western cedar, No. 1, 2 rail, 3' high | 1.000 | L.F. | .166 | 4.84 | 2.81 | 7.65 |
| 3 rail, 4' high | 1.000 | L.F. | .178 | 9.19 | 3.01 | 12.20 |
| 5' high | 1.000 | L.F. | .185 | 11.83 | 3.12 | 14.95 |
| 6' high | 1.000 | L.F. | .192 | 14.30 | 3.25 | 17.55 |
| No. 1 cedar, 2 rail, 3' high | 1.000 | L.F. | .166 | 11.55 | 2.80 | 14.35 |
| 4' high | 1.000 | L.F. | .178 | 13.20 | 3.00 | 16.20 |
| 3 rail, 5' high | 1.000 | L.F. | .185 | 14.58 | 3.12 | 17.70 |
| 6' high | 1.000 | L.F. | .192 | 15.95 | 3.25 | 19.20 |
| Shadow box, 1"x6" boards, 2"x4" rails, 4"x4" posts | | | | | | |
| Fir, pine or spruce, treated, 3 rail, 6' high | 1.000 | L.F. | .160 | 6.60 | 2.70 | 9.30 |
| No. 1 cedar, 3 rail, 4' high | 1.000 | L.F. | .178 | 12.54 | 3.01 | 15.55 |
| 6' high | 1.000 | L.F. | .185 | 14.03 | 3.12 | 17.15 |
| Open rail, split rails, No. 1 cedar, 2 rail, 3' high | 1.000 | L.F. | .150 | 3.14 | 2.51 | 5.65 |
| 3 rail, 4' high | 1.000 | L.F. | .160 | 4.13 | 2.72 | 6.85 |
| No. 2 cedar, 2 rail, 3' high | 1.000 | L.F. | .150 | 1.65 | 2.54 | 4.19 |
| 3 rail, 4' high | 1.000 | L.F. | .160 | 1.93 | 2.70 | 4.63 |
| Open rail, rustic rails, No. 1 cedar, 2 rail, 3' high | 1.000 | L.F. | .150 | 2.75 | 2.55 | 5.30 |
| 3 rail, 4' high | 1.000 | L.F. | .160 | 3.41 | 2.69 | 6.10 |
| No. 2 cedar, 2 rail, 3' high | 1.000 | L.F. | .150 | 1.76 | 2.54 | 4.30 |
| 3 rail, 4' high | 1.000 | L.F. | .160 | 2.31 | 2.69 | 5.00 |
| Rustic picket, molded pine pickets, 2 rail, 3' high | 1.000 | L.F. | .171 | 2.59 | 2.91 | 5.50 |
| 3 rail, 4' high | 1.000 | L.F. | .196 | 2.98 | 3.35 | 6.33 |
| No. 1 cedar, 2 rail, 3' high | 1.000 | L.F. | .171 | 5.28 | 2.92 | 8.20 |
| 3 rail, 4' high | 1.000 | L.F. | .196 | 6.07 | 3.36 | 9.43 |
| Picket fence, fir, pine or spruce, preserved, treated | | | | | | |
| 2 rail, 3' high | 1.000 | L.F. | .171 | 3.03 | 2.87 | 5.90 |
| 3 rail, 4' high | 1.000 | L.F. | .185 | 3.52 | 3.13 | 6.65 |
| Western cedar, 2 rail, 3' high | 1.000 | L.F. | .171 | 3.52 | 2.88 | 6.40 |
| 3 rail, 4' high | 1.000 | L.F. | .185 | 3.85 | 3.10 | 6.95 |
| No. 1 cedar, 2 rail, 3' high | 1.000 | L.F. | .171 | 7.98 | 2.87 | 10.85 |
| 3 rail, 4' high | 1.000 | L.F. | .185 | 9.08 | 3.12 | 12.20 |
| Stockade, No. 1 cedar, 3-1/4" rails, 6' high | 1.000 | L.F. | .150 | 10.73 | 2.52 | 13.25 |
| 8' high | 1.000 | L.F. | .155 | 16.78 | 2.62 | 19.40 |
| No. 2 cedar, treated rails, 6' high | 1.000 | L.F. | .150 | 4.40 | 2.55 | 6.95 |
| Treated pine, treated rails, 6' high | 1.000 | L.F. | .960 | 6.88 | 16.12 | 23.00 |
| Gates, No. 2 cedar, picket, 3'-6" wide 4' high | 1.000 | Ea. | .511 | 37.40 | 8.60 | 46.00 |
| No. 2 cedar, rustic round, 3' wide, 3' high | 1.000 | Ea. | .511 | 47.00 | 9.00 | 56.00 |
| No. 2 cedar, stockade screen, 3'-6" wide, 6' high | 1.000 | Ea. | .686 | 36.30 | 11.70 | 48.00 |
| General, wood, 3'-6" wide, 4' high | 1.000 | Ea. | .565 | 35.20 | 7.80 | 43.00 |
| 6' high | 1.000 | Ea. | .706 | 44.00 | 12.00 | 56.00 |

# SECTION 2
# FOUNDATIONS

| System Number | Title | Page |
|---|---|---|
| **2-04** | Footing | 22 |
| **2-08** | Block Wall | 24 |
| **2-12** | Concrete Wall | 26 |
| **2-16** | Wood Wall Foundation | 28 |
| **2-20** | Floor Slab | 30 |

# 2 | FOUNDATIONS    04 | Footing Systems

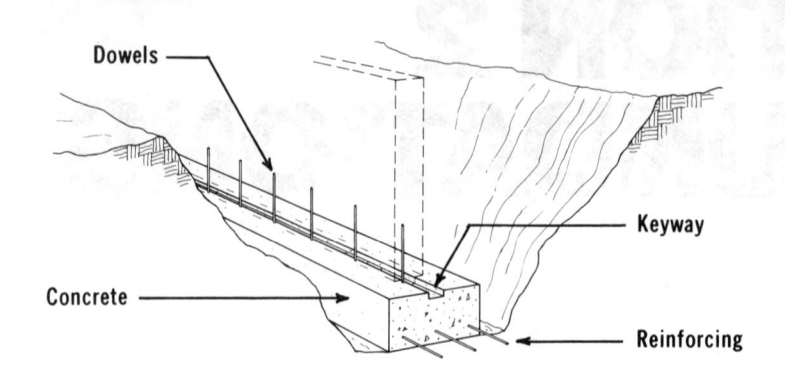

| SYSTEM DESCRIPTION | QUAN. | UNIT | MAN HOURS | MAT | INST. | TOTAL |
|---|---|---|---|---|---|---|
| **8" THICK BY 18" WIDE FOOTING** | | | | | | |
| Concrete, 3000 psi | .040 | C.Y. | | 2.04 | | 2.04 |
| Place concrete, direct chute | .040 | C.Y. | .016 | | .29 | .29 |
| Forms, footing, 4 uses | 1.330 | SFCA | .102 | .48 | 1.98 | 2.46 |
| Reinforcing, 1/2" diameter bars, 2 each | 1.380 | Lb. | .011 | .36 | .23 | .59 |
| Keyway, 2" x 4", beveled, 4 uses | 1.000 | L.F. | .015 | .08 | .29 | .37 |
| Dowels, 1/2" diameter bars, 2' long, 6' O.C. | .166 | Ea. | .021 | .16 | .47 | .63 |
| TOTAL | | | .165 | 3.12 | 3.26 | 6.38 |
| **12" THICK BY 24" WIDE FOOTING** | | | | | | |
| Concrete, 3000 psi | .070 | C.Y. | | 3.57 | | 3.57 |
| Place concrete, direct chute | .070 | C.Y. | .028 | | .50 | .50 |
| Forms, footing, 4 uses | 2.000 | SFCA | .154 | .72 | 2.98 | 3.70 |
| Reinforcing, 1/2" diameter bars, 2 each | 1.380 | Lb. | .011 | .36 | .23 | .59 |
| Keyway, 2" x 4", beveled, 4 uses | 1.000 | L.F. | .015 | .08 | .29 | .37 |
| Dowels, 1/2" diameter bars, 2' long, 6' O.C. | .166 | Ea. | .021 | .16 | .47 | .63 |
| TOTAL | | | .229 | 4.89 | 4.47 | 9.36 |
| **12" THICK BY 36" WIDE FOOTING** | | | | | | |
| Concrete, 3000 psi | .110 | C.Y. | | 5.61 | | 5.61 |
| Place concrete, direct chute | .110 | C.Y. | .044 | | .79 | .79 |
| Forms, footing, 4 uses | 2.000 | SFCA | .154 | .72 | 2.98 | 3.70 |
| Reinforcing, 1/2" diameter bars, 2 each | 1.380 | Lb. | .011 | .36 | .23 | .59 |
| Keyway, 2" x 4", beveled, 4 uses | 1.000 | L.F. | .015 | .08 | .29 | .37 |
| Dowels, 1/2" diameter bars, 2' long, 6' O.C. | .166 | Ea. | .021 | .16 | .47 | .63 |
| TOTAL | | | .245 | 6.93 | 4.76 | 11.69 |

Cost per L.F.

The footing costs in this system are on a cost per linear foot basis

| DESCRIPTION | QUAN. | UNIT | MAN HOURS | MAT. | INST. | TOTAL |
|---|---|---|---|---|---|---|
| | | | | | | |
| | | | | | | |
| | | | | | | |
| | | | | | | |
| | | | | | | |

Cost per S.F.

# Footing Price Sheet

| | QUAN. | UNIT | MAN HOURS | COST PER L.F. MAT. | COST PER L.F. INST. | TOTAL |
|---|---|---|---|---|---|---|
| Concrete, 8" thick by 18" wide footing | | | | | | |
|     2000 psi concrete | .040 | C.Y. | | 1.88 | | 1.88 |
|     2500 psi concrete | .040 | C.Y. | | 1.92 | | 1.92 |
|     3000 psi concrete | .040 | C.Y. | | 2.04 | | 2.04 |
|     3500 psi concrete | .040 | C.Y. | | 2.08 | | 2.08 |
|     4000 psi concrete | .040 | C.Y. | | 2.16 | | 2.16 |
|   12" thick by 24" wide footing | | | | | | |
|     2000 psi concrete | .070 | C.Y. | | 3.29 | | 3.29 |
|     2500 psi concrete | .070 | C.Y. | | 3.36 | | 3.36 |
|     3000 psi concrete | .070 | C.Y. | | 3.57 | | 3.57 |
|     3500 psi concrete | .070 | C.Y. | | 3.64 | | 3.64 |
|     4000 psi concrete | .070 | C.Y. | | 3.78 | | 3.78 |
|   12" thick by 36" wide footing | | | | | | |
|     2000 psi concrete | .110 | C.Y. | | 5.17 | | 5.17 |
|     2500 psi concrete | .110 | C.Y. | | 5.28 | | 5.28 |
|     3000 psi concrete | .110 | C.Y. | | 5.61 | | 5.61 |
|     3500 psi concrete | .110 | C.Y. | | 5.72 | | 5.72 |
|     4000 psi concrete | .110 | C.Y. | | 5.94 | | 5.94 |
| Place concrete, 8" thick by 18" wide footing, direct chute | .040 | C.Y. | .016 | | .29 | .29 |
|   Pumped concrete | .040 | C.Y. | .025 | | .68 | .68 |
|   Crane & bucket | .040 | C.Y. | .028 | | .78 | .78 |
|   12" thick by 24" wide footing, direct chute | .070 | C.Y. | .028 | | .50 | .50 |
|   Pumped concrete | .070 | C.Y. | .044 | | 1.19 | 1.19 |
|   Crane & bucket | .070 | C.Y. | .049 | | 1.36 | 1.36 |
|   12" thick by 36" wide footing, direct chute | .110 | C.Y. | .044 | | .79 | .79 |
|   Pumped concrete | .110 | C.Y. | .070 | | 1.87 | 1.87 |
|   Crane & bucket | .110 | C.Y. | .078 | | 2.13 | 2.13 |
| Forms, 8" thick footing, 1 use | 1.330 | SFCA | .139 | 1.17 | 2.70 | 3.87 |
|   4 uses | 1.330 | SFCA | .102 | .48 | 1.98 | 2.46 |
|   12" thick footing, 1 use | 2.000 | SFCA | .210 | 1.76 | 4.06 | 5.82 |
|   4 uses | 2.000 | SFCA | .154 | .72 | 2.98 | 3.70 |
| Reinforcing, 3/8" diameter bar, 1 each | .400 | Lb. | .003 | .10 | .07 | .17 |
|   2 each | .800 | Lb. | .006 | .21 | .13 | .34 |
|   3 each | 1.200 | Lb. | .009 | .31 | .21 | .52 |
|   1/2" diameter bar, 1 each | .700 | Lb. | .005 | .18 | .12 | .30 |
|   2 each | 1.380 | Lb. | .011 | .36 | .23 | .59 |
|   3 each | 2.100 | Lb. | .016 | .55 | .35 | .90 |
|   5/8" diameter bar, 1 each | 1.040 | Lb. | .008 | .27 | .18 | .45 |
|   2 each | 2.080 | Lb. | .016 | .54 | .35 | .89 |
| Keyway, beveled, 2" x 4", 1 use | 1.000 | L.F. | .030 | .16 | .58 | .74 |
|   2 uses | 1.000 | L.F. | .022 | .12 | .44 | .56 |
|   2" x 6", 1 use | 1.000 | L.F. | .032 | .20 | .62 | .82 |
|   2 uses | 1.000 | L.F. | .024 | .15 | .47 | .62 |
| Dowels, 2 feet long, 6' O.C., 3/8" bar | .166 | Ea. | .018 | .13 | .42 | .55 |
|   1/2" bar | .166 | Ea. | .021 | .16 | .47 | .63 |
|   5/8" bar | .166 | Ea. | .024 | .19 | .54 | .73 |
|   3/4" bar | .166 | Ea. | .025 | .24 | .56 | .80 |

# 2 | FOUNDATIONS  08 | Block Wall Systems

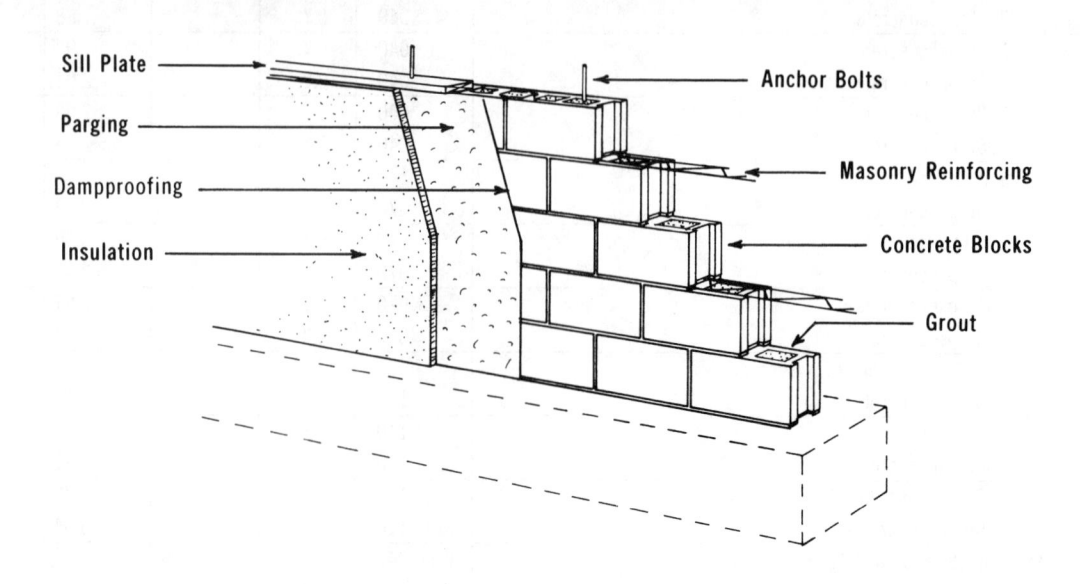

| SYSTEM DESCRIPTION | QUAN. | UNIT | MAN HOURS | MAT. | INST. | TOTAL |
|---|---|---|---|---|---|---|
| **8" WALL, GROUTED, FULL HEIGHT** | | | | | | |
| Concrete block, 8" x 16" x 8" | 1.000 | S.F. | | 1.49 | 1.41 | 2.90 |
| Masonry reinforcing, every second course | .750 | L.F. | .002 | .18 | .04 | .22 |
| Parging, plastering with portland cement plaster, 1 coat | 1.000 | S.F. | .012 | .18 | .23 | .41 |
| Dampproofing, bituminous coating, 1 coat | 1.000 | S.F. | .012 | .09 | .23 | .32 |
| Insulation, 1" rigid polystyrene | 1.000 | S.F. | .010 | .40 | .20 | .60 |
| Grout, solid, pumped | 1.000 | S.F. | .038 | .62 | .85 | 1.47 |
| Anchor bolts, 1/2" diameter, 8" long, 4' O.C. | .060 | Ea. | .002 | .03 | .05 | .08 |
| Sill plate, 2" x 4", treated | .250 | L.F. | .006 | .10 | .13 | .23 |
| TOTAL | | | .083 | 3.09 | 3.14 | 6.23 |
| **12" WALL, GROUTED, FULL HEIGHT** | | | | | | |
| Concrete block, 8" x 16" x 12" | 1.000 | S.F. | | 2.18 | 2.24 | 4.42 |
| Masonry reinforcing, every second course | .750 | L.F. | .003 | .19 | .06 | .25 |
| Parging, plastering with portland cement plaster, 1 coat | 1.000 | S.F. | .012 | .18 | .23 | .41 |
| Dampproofing, bituminous coating, 1 coat | 1.000 | S.F. | .012 | .09 | .23 | .32 |
| Insulation, 1" rigid polystyrene | 1.000 | S.F. | .010 | .40 | .20 | .60 |
| Grout, solid, pumped | 1.000 | S.F. | .040 | 1.01 | .91 | 1.92 |
| Anchor bolts, 1/2" diameter, 8" long, 4' O.C. | .060 | Ea. | .002 | .03 | .05 | .08 |
| Sill plate, 2" x 4", treated | .250 | L.F. | .006 | .10 | .13 | .23 |
| TOTAL | | | .085 | 4.18 | 4.05 | 8.23 |

The costs in this system are based on a square foot of wall. Do not subtract for window or door openings.

| DESCRIPTION | QUAN. | UNIT | MAN HOURS | MAT. | INST. | TOTAL |
|---|---|---|---|---|---|---|
| | | | | | | |
| | | | | | | |
| | | | | | | |
| | | | | | | |
| | | | | | | |

# Block Wall Systems

| | QUAN. | UNIT | MAN HOURS | MAT. | INST. | TOTAL |
|---|---|---|---|---|---|---|
| Concrete, block, 8" x 16" x, 6" thick | 1.000 | S.F. | .075 | 1.21 | 1.37 | 2.58 |
| 8" thick | 1.000 | S.F. | .077 | 1.49 | 1.41 | 2.90 |
| 10" thick | 1.000 | S.F. | .080 | 2.20 | 1.45 | 3.65 |
| 12" thick | 1.000 | S.F. | .126 | 2.18 | 2.24 | 4.42 |
| Solid block, 8" x 16" x, 6" thick | 1.000 | S.F. | .076 | 1.65 | 1.39 | 3.04 |
| 8" thick | 1.000 | S.F. | .080 | 1.98 | 1.47 | 3.45 |
| 10" thick | 1.000 | S.F. | .100 | 2.48 | 1.83 | 4.31 |
| 12" thick | 1.000 | S.F. | .131 | 2.97 | 2.33 | 5.30 |
| Masonry reinforcing, wire strips, to 8" wide, every course | 1.500 | L.F. | .004 | .36 | .08 | .44 |
| Every 2nd course | .750 | L.F. | .002 | .18 | .04 | .22 |
| Every 3rd course | .500 | L.F. | .001 | .12 | .03 | .15 |
| Every 4th course | .400 | L.F. | .001 | .10 | .02 | .12 |
| Wire strips to 12" wide, every course | 1.500 | L.F. | .006 | .38 | .12 | .50 |
| Every 2nd course | .750 | L.F. | .003 | .19 | .06 | .25 |
| Every 3rd course | .500 | L.F. | .002 | .13 | .04 | .17 |
| Every 4th course | .400 | L.F. | .001 | .10 | .03 | .13 |
| Parging, plastering with portland cement plaster, 1 coat | 1.000 | S.F. | .012 | .18 | .23 | .41 |
| 2 coats | 1.000 | S.F. | .019 | .28 | .35 | .63 |
| Dampproofing, bituminous, brushed on, 1 coat | 1.000 | S.F. | .012 | .09 | .23 | .32 |
| 2 coats | 1.000 | S.F. | .016 | .17 | .30 | .47 |
| Sprayed on, 1 coat | 1.000 | S.F. | .010 | .13 | .19 | .32 |
| 2 coats | 1.000 | S.F. | .016 | .17 | .30 | .47 |
| Troweled on, 1/16" thick | 1.000 | S.F. | .016 | .55 | .31 | .86 |
| 1/8" thick | 1.000 | S.F. | .020 | 1.10 | .38 | 1.48 |
| 1/2" thick | 1.000 | S.F. | .023 | 4.73 | .42 | 5.15 |
| Insulation, rigid, fiberglass, 1.5#/C.F., unfaced | | | | | | |
| 1-1/2" thick R 6.2 | 1.000 | S.F. | .008 | .28 | .16 | .44 |
| 2" thick R 8.5 | 1.000 | S.F. | .009 | .41 | .17 | .58 |
| 3" thick R 13 | 1.000 | S.F. | .010 | .62 | .19 | .81 |
| Foamglass, 1-1/2" thick R 2.64 | 1.000 | S.F. | .011 | 1.56 | .21 | 1.77 |
| 2" thick R 5.26 | 1.000 | S.F. | .011 | 2.11 | .23 | 2.34 |
| Perlite, 1" thick R 2.77 | 1.000 | S.F. | .011 | .31 | .21 | .52 |
| 2" thick R 5.55 | 1.000 | S.F. | .011 | .55 | .22 | .77 |
| Polystyrene, extruded, 1" thick R 5.4 | 1.000 | S.F. | .001 | .04 | .02 | .06 |
| 2" thick R 10.8 | 1.000 | S.F. | .011 | .86 | .22 | 1.08 |
| Molded 1" thick R 3.85 | 1.000 | S.F. | .011 | .17 | .20 | .37 |
| 2" thick R 7.7 | 1.000 | S.F. | .011 | .34 | .22 | .56 |
| Urethane, 1" thick, R 5.8 | 1.000 | S.F. | .011 | .45 | .21 | .66 |
| 2" thick R 11.7 | 1.000 | S.F. | .011 | .79 | .23 | 1.02 |
| Grout, concrete block cores, 6" thick | 1.000 | S.F. | .028 | .47 | .63 | 1.10 |
| 8" thick | 1.000 | S.F. | .038 | .62 | .85 | 1.47 |
| 10" thick | 1.000 | S.F. | .039 | .81 | .88 | 1.69 |
| 12" thick | 1.000 | S.F. | .040 | 1.01 | .91 | 1.92 |
| Anchor bolts, 2' on center, 1/2" diameter, 8" long | .120 | Ea. | .004 | .06 | .09 | .15 |
| 12" long | .120 | Ea. | .005 | .07 | .10 | .17 |
| 3/4" diameter, 8" long | .120 | Ea. | .006 | .16 | .12 | .28 |
| 12" long | .120 | Ea. | .006 | .19 | .13 | .32 |
| 4' on center, 1/2" diameter, 8" long | .060 | Ea. | .002 | .03 | .05 | .08 |
| 12" long | .060 | Ea. | .002 | .04 | .05 | .09 |
| 3/4" diameter, 8" long | .060 | Ea. | .003 | .08 | .06 | .14 |
| 12" long | .060 | Ea. | .003 | .10 | .06 | .16 |
| Sill plates, treated, 2" x 4" | .250 | L.F. | .006 | .10 | .13 | .23 |
| 4" x 4" | .250 | L.F. | .006 | .21 | .14 | .35 |

# 2 | FOUNDATIONS — 12 | Concrete Wall Systems

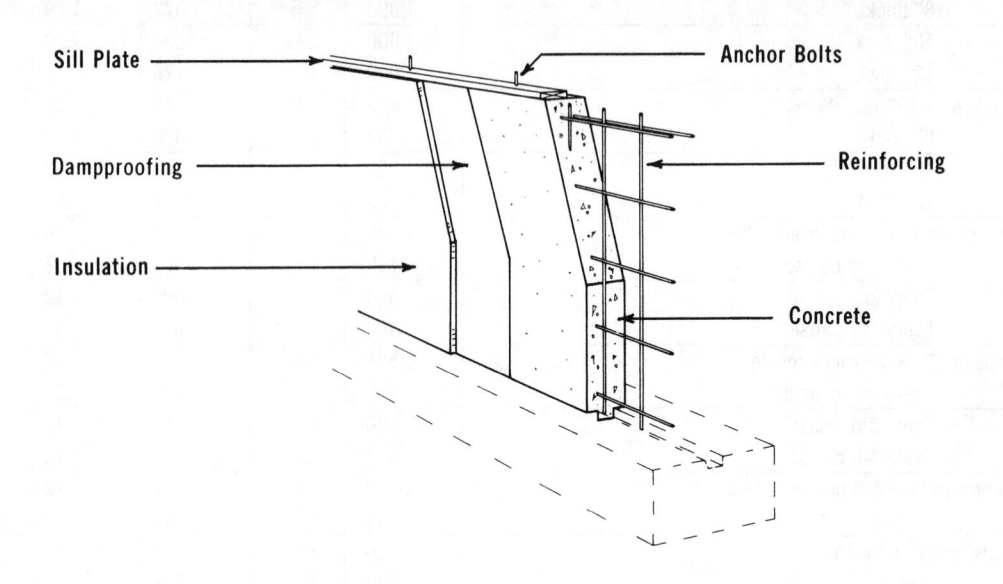

*Diagram labels: Sill Plate, Anchor Bolts, Dampproofing, Reinforcing, Insulation, Concrete*

| SYSTEM DESCRIPTION | QUAN. | UNIT | MAN HOURS | MAT. | INST. | TOTAL |
|---|---|---|---|---|---|---|
| **8" THICK, POURED CONCRETE WALL** | | | | | | |
| Concrete, 8" thick, 3000 psi | .025 | C.Y. | | 1.28 | | 1.28 |
| Forms, prefabricated plywood, 4 uses per month | 2.000 | SFCA | .106 | .56 | 2.12 | 2.68 |
| Reinforcing, light | .670 | Lb. | .003 | .19 | .07 | .26 |
| Placing concrete, direct chute | .025 | C.Y. | .013 | | .24 | .24 |
| Dampproofing, brushed on, 2 coats | 1.000 | S.F. | .016 | .17 | .30 | .47 |
| Rigid insulation, 1" polystyrene | 1.000 | S.F. | .010 | .40 | .20 | .60 |
| Anchor bolts, 1/2" diameter, 12" long, 4' O.C. | .060 | Ea. | .002 | .04 | .05 | .09 |
| Sill plates, 2" x 4", treated | .250 | L.F. | .006 | .10 | .13 | .23 |
| TOTAL | | | .157 | 2.74 | 3.11 | 5.85 |
| **12" THICK, POURED CONCRETE WALL** | | | | | | |
| Concrete, 12" thick, 3000 psi | .040 | C.Y. | | 2.04 | | 2.04 |
| Forms, prefabricated plywood, 4 uses per month | 2.000 | SFCA | .106 | .56 | 2.12 | 2.68 |
| Reinforcing, light | 1.000 | Lb. | .005 | .28 | .11 | .39 |
| Placing concrete, direct chute | .040 | C.Y. | .019 | | .35 | .35 |
| Dampproofing, brushed on, 2 coats. | 1.000 | S.F. | .016 | .17 | .30 | .47 |
| Rigid insulation, 1" polystyrene | 1.000 | S.F. | .010 | .40 | .20 | .60 |
| Anchor bolts, 1/2" diameter, 12" long, 4' O.C. | .060 | Ea. | .002 | .04 | .05 | .09 |
| Sill plates, 2" x 4" treated | .250 | L.F. | .006 | .10 | .13 | .23 |
| TOTAL | | | .164 | 3.59 | 3.26 | 6.85 |

The costs in this system are based on sq. ft. of wall. Do not subtract for window and door openings. The costs assume a 4' high wall.

| DESCRIPTION | QUAN. | UNIT | MAN HOURS | MAT. | INST. | TOTAL |
|---|---|---|---|---|---|---|
| | | | | | | |
| | | | | | | |
| | | | | | | |
| | | | | | | |
| | | | | | | |

# Concrete Wall Price Sheet

| | QUAN. | UNIT | MAN HOURS | COST PER S.F. MAT. | COST PER S.F. INST. | TOTAL |
|---|---|---|---|---|---|---|
| Concrete, 8" wall, concrete, 2500 psi | .025 | C.Y. | | 1.20 | | 1.20 |
| 3000 psi | .025 | C.Y. | | 1.28 | | 1.28 |
| 3500 psi | .025 | C.Y. | | 1.30 | | 1.30 |
| 4500 psi | .025 | C.Y. | | 1.40 | | 1.40 |
| 10" wall, concrete, 2500 psi | .030 | C.Y. | | 1.44 | | 1.44 |
| 3000 psi | .030 | C.Y. | | 1.53 | | 1.53 |
| 3500 psi | .030 | C.Y. | | 1.56 | | 1.56 |
| 4500 psi | .030 | C.Y. | | 1.68 | | 1.68 |
| 12" wall, concrete, 2500 psi | .040 | C.Y. | | 1.92 | | 1.92 |
| 3000 psi | .040 | C.Y. | | 2.04 | | 2.04 |
| 3500 psi | .040 | C.Y. | | 2.08 | | 2.08 |
| 4500 psi | .040 | C.Y. | | 2.24 | | 2.24 |
| Formwork, prefabricated plywood, 1 use per month | 2.000 | SFCA | .106 | 1.62 | 2.12 | 3.74 |
| 4 uses per month | 2.000 | SFCA | .106 | .56 | 2.12 | 2.68 |
| Job built forms, 1 use per month | 2.000 | SFCA | .260 | 2.56 | 5.20 | 7.76 |
| 4 uses per month | 2.000 | SFCA | .190 | 1.12 | 3.82 | 4.94 |
| Reinforcing, 8" wall, light reinforcing | .670 | Lb. | .003 | .19 | .07 | .26 |
| Heavy reinforcing | 1.500 | Lb. | .007 | .42 | .17 | .59 |
| 10" wall, light reinforcing | .850 | Lb. | .004 | .24 | .09 | .33 |
| Heavy reinforcing | 2.000 | Lb. | .010 | .56 | .22 | .78 |
| 12" wall light reinforcing | 1.000 | Lb. | .005 | .28 | .11 | .39 |
| Heavy reinforcing | 2.250 | Lb. | .011 | .63 | .25 | .88 |
| Placing concrete, 8" wall, direct chute | .025 | C.Y. | .013 | | .24 | .24 |
| Pumped concrete | .025 | C.Y. | .018 | | .50 | .50 |
| Crane & bucket | .025 | C.Y. | .020 | | .55 | .55 |
| 10" wall, direct chute | .030 | C.Y. | .015 | | .29 | .29 |
| Pumped concrete | .030 | C.Y. | .022 | | .60 | .60 |
| Crane & bucket | .030 | C.Y. | .024 | | .66 | .66 |
| 12" wall, direct chute | .040 | C.Y. | .019 | | .35 | .35 |
| Pumped concrete | .040 | C.Y. | .026 | | .72 | .72 |
| Crane & bucket | .040 | C.Y. | .028 | | .78 | .78 |
| Dampproofing, bituminous, brushed on, 1 coat | 1.000 | S.F. | .012 | .09 | .23 | .32 |
| 2 coats | 1.000 | S.F. | .016 | .17 | .30 | .47 |
| Sprayed on, 1 coat | 1.000 | S.F. | .010 | .13 | .19 | .32 |
| 2 coats | 1.000 | S.F. | .016 | .17 | .30 | .47 |
| Troweled on, 1/16" thick | 1.000 | S.F. | .016 | .55 | .31 | .86 |
| 1/8" thick | 1.000 | S.F. | .020 | 1.10 | .38 | 1.48 |
| 1/2" thick | 1.000 | S.F. | .023 | 4.73 | .42 | 5.15 |
| Insulation rigid, fiberglass, 1.5#/C.F., unfaced | | | | | | |
| 1-1/2" thick, R 6.2 | 1.000 | S.F. | .008 | .28 | .16 | .44 |
| 2" thick, R 8.3 | 1.000 | S.F. | .009 | .41 | .17 | .58 |
| 3" thick, R 12.4 | 1.000 | S.F. | .010 | .62 | .19 | .81 |
| Foamglass, 1-1/2" thick R 2.64 | 1.000 | S.F. | .011 | 1.56 | .21 | 1.77 |
| 2" thick R 5.26 | 1.000 | S.F. | .011 | 2.11 | .23 | 2.34 |
| Perlite, 1" thick R 2.77 | 1.000 | S.F. | .011 | .31 | .21 | .52 |
| 2" thick R 5.55 | 1.000 | S.F. | .011 | .55 | .22 | .77 |
| Polystyrene, extruded, 1" thick R 5.40 | 1.000 | S.F. | .010 | .40 | .20 | .60 |
| 2" thick R 10.8 | 1.000 | S.F. | .011 | .86 | .22 | 1.08 |
| Molded, 1" thick R 3.85 | 1.000 | S.F. | .011 | .17 | .20 | .37 |
| 2" thick R 7.70 | 1.000 | S.F. | .011 | .34 | .22 | .56 |
| Anchor bolts, 2' on center, 1/2" diameter, 8" long | .120 | Ea. | .004 | .06 | .09 | .15 |
| 12" long | .120 | Ea. | .005 | .07 | .10 | .17 |
| 3/4" diameter, 8" long | .120 | Ea. | .006 | .16 | .12 | .28 |
| 12" long | .120 | Ea. | .006 | .19 | .13 | .32 |
| Sill plates, treated lumber, 2" x 4" | .250 | L.F. | .006 | .10 | .13 | .23 |
| 4" x 4" | .250 | L.F. | .006 | .21 | .14 | .35 |

# 2 | FOUNDATIONS     16 | Wood Wall Foundation Systems

Diagram labels: Top Plates, Sheathing, Studs, Asphalt Paper, Insulation, Vapor Barrier, Bottom Plate

| SYSTEM DESCRIPTION | QUAN. | UNIT | MAN HOURS | MAT. | INST. | TOTAL |
|---|---|---|---|---|---|---|
| **2" X 4" STUDS, 16" O.C., WALL** | | | | | | |
| Studs, 2" x 4", 16" O.C., treated | 1.000 | L.F. | .015 | .40 | .31 | .71 |
| Plates, double top plate, single bottom plate, treated, 2" x 4" | .750 | L.F. | .011 | .30 | .23 | .53 |
| Sheathing, 1/2", exterior grade, CDX, treated | 1.000 | S.F. | .014 | .45 | .29 | .74 |
| Asphalt paper, 15# roll | 1.100 | S.F. | .002 | .03 | .06 | .09 |
| Vapor barrier, 4 mil polyethylene | 1.000 | S.F. | .002 | .06 | .04 | .10 |
| Insulation, batts, fiberglass, 3-1/2" thick, R 11 | 1.000 | S.F. | .005 | .20 | .10 | .30 |
| TOTAL | | | .049 | 1.44 | 1.03 | 2.47 |
| **2" X 6" STUDS, 16" O.C., WALL** | | | | | | |
| Studs, 2" x 6", 16" O.C., treated | 1.000 | L.F. | .017 | .59 | .35 | .94 |
| Plates, double top plate, single bottom plate, treated, 2" x 6" | .750 | L.F. | .012 | .44 | .27 | .71 |
| Sheathing, 5/8" exterior grade, CDX, treated | 1.000 | S.F. | .015 | .67 | .31 | .98 |
| Asphalt paper, 15# roll | 1.100 | S.F. | .002 | .03 | .06 | .09 |
| Vapor barrier, 4 mil polyethylene | 1.000 | S.F. | .002 | .06 | .04 | .10 |
| Insulation, batts, fiberglass, 6" thick. R 19 | 1.000 | S.F. | .006 | .33 | .12 | .45 |
| TOTAL | | | .054 | 2.12 | 1.15 | 3.27 |
| **2" X 8" STUDS, 16" O.C., WALL** | | | | | | |
| Studs, 2" x 8", 16" O.C. treated | 1.000 | L.F. | .020 | .88 | .41 | 1.29 |
| Plates, double top plate, single bottom plate, treated, 2" x 8" | .750 | L.F. | .015 | .66 | .31 | .97 |
| Sheathing, 3/4" exterior grade, CDX, treated | 1.000 | S.F. | .016 | .70 | .33 | 1.03 |
| Asphalt paper, 15# roll | 1.100 | S.F. | .002 | .03 | .06 | .09 |
| Vapor barrier, 4 mil polyethylene | 1.000 | S.F. | .002 | .06 | .04 | .10 |
| Insulation, batts, fiberglass, 9" thick, R 30 | 1.000 | S.F. | .006 | .53 | .11 | .64 |
| TOTAL | | | .061 | 2.86 | 1.26 | 4.12 |

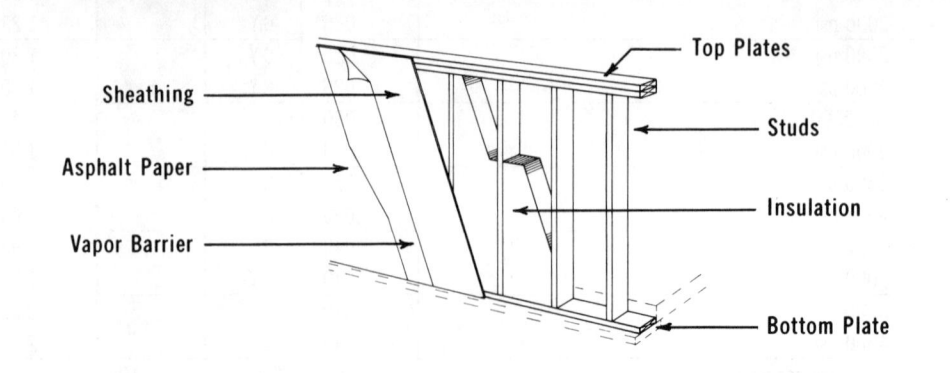

The costs in this system are based on a sq. ft. of wall area. Do not subtract for window or door openings. The costs assume a 4' high wall.

| DESCRIPTION | QUAN. | UNIT | MAN HOURS | MAT. | INST. | TOTAL |
|---|---|---|---|---|---|---|
| | | | | | | |
| | | | | | | |
| | | | | | | |
| | | | | | | |
| | | | | | | |

# Wood Wall Foundation Price Sheet

| | QUAN. | UNIT | MAN HOURS | COST PER S.F. MAT. | COST PER S.F. INST. | TOTAL |
|---|---|---|---|---|---|---|
| Studs, treated, 2" x 4", 12" O.C. | 1.000 | L.F. | .018 | .50 | .39 | .89 |
| 16" O.C. | 1.000 | L.F. | .015 | .40 | .31 | .71 |
| 2" x 6", 12" O.C. | 1.000 | L.F. | .021 | .74 | .44 | 1.18 |
| 16" O.C. | 1.000 | L.F. | .017 | .59 | .35 | .94 |
| 2" x 8", 12" O.C. | 1.000 | L.F. | .025 | 1.10 | .51 | 1.61 |
| 16" O.C. | 1.000 | L.F. | .020 | .88 | .41 | 1.29 |
| Plates, treated, double top single bottom, 2" x 4" | .750 | L.F. | .011 | .30 | .23 | .53 |
| 2" x 6" | .750 | L.F. | .012 | .44 | .27 | .71 |
| 2" x 8" | .750 | L.F. | .015 | .66 | .31 | .97 |
| Sheathing, treated exterior grade CDX, 1/2" thick | 1.000 | S.F. | .014 | .45 | .29 | .74 |
| 5/8" thick | 1.000 | S.F. | .015 | .67 | .31 | .98 |
| 3/4" thick | 1.000 | S.F. | .016 | .70 | .33 | 1.03 |
| Asphalt paper, 15# roll | 1.100 | S.F. | .002 | .03 | .06 | .09 |
| Vapor barrier, polyethylene, 4 mil | 1.000 | S.F. | .002 | .06 | .04 | .10 |
| 10 mil | 1.000 | S.F. | .002 | .07 | .04 | .11 |
| Insulation, rigid, fiberglass, 1.5#/C.F., unfaced | | | | | | |
| 1-1/2" thick, R 6.2 | 1.000 | S.F. | .008 | .28 | .16 | .44 |
| 2" thick, R 8.3 | 1.000 | S.F. | .009 | .41 | .17 | .58 |
| 3" thick, R 12.4 | 1.000 | S.F. | .010 | .63 | .20 | .83 |
| Foamglass 1 1/2" thick, R 2.64 | 1.000 | S.F. | .011 | 1.56 | .21 | 1.77 |
| 2" thick, R 5.26 | 1.000 | S.F. | .011 | 2.11 | .23 | 2.34 |
| Perlite 1" thick, R 2.77 | 1.000 | S.F. | .011 | .31 | .21 | .52 |
| 2" thick, R 5.55 | 1.000 | S.F. | .011 | .55 | .22 | .77 |
| Polystyrene, extruded, 1" thick, R 5.40 | 1.000 | S.F. | .010 | .40 | .20 | .60 |
| 2" thick, R 10.8 | 1.000 | S.F. | .011 | .86 | .22 | 1.08 |
| Molded 1" thick, R 3.85 | 1.000 | S.F. | .011 | .17 | .20 | .37 |
| 2" thick, R 7.7 | 1.000 | S.F. | .011 | .34 | .22 | .56 |
| Urethane 1" thick, R 5.8 | 1.000 | S.F. | .011 | .45 | .21 | .66 |
| 2" thick, R 11.7 | 1.000 | S.F. | .011 | .79 | .23 | 1.02 |
| Nonrigid, batts, fiberglass, paper backed, 3-1/2" thick roll, R 11 | 1.000 | S.F. | .005 | .20 | .10 | .30 |
| 6", R 19 | 1.000 | S.F. | .006 | .33 | .12 | .45 |
| 9", R 30 | 1.000 | S.F. | .006 | .53 | .11 | .64 |
| 12", R 38 | 1.000 | S.F. | .006 | .72 | .11 | .83 |
| Mineral fiber, paper backed, 3-1/2", R 13 | 1.000 | S.F. | .005 | .53 | .10 | .63 |
| 6", R 19 | 1.000 | S.F. | .005 | .64 | .10 | .74 |
| 10", R 30 | 1.000 | S.F. | .006 | 1.01 | .12 | 1.13 |

2

# 2 | FOUNDATIONS    20 | Floor Slab Systems

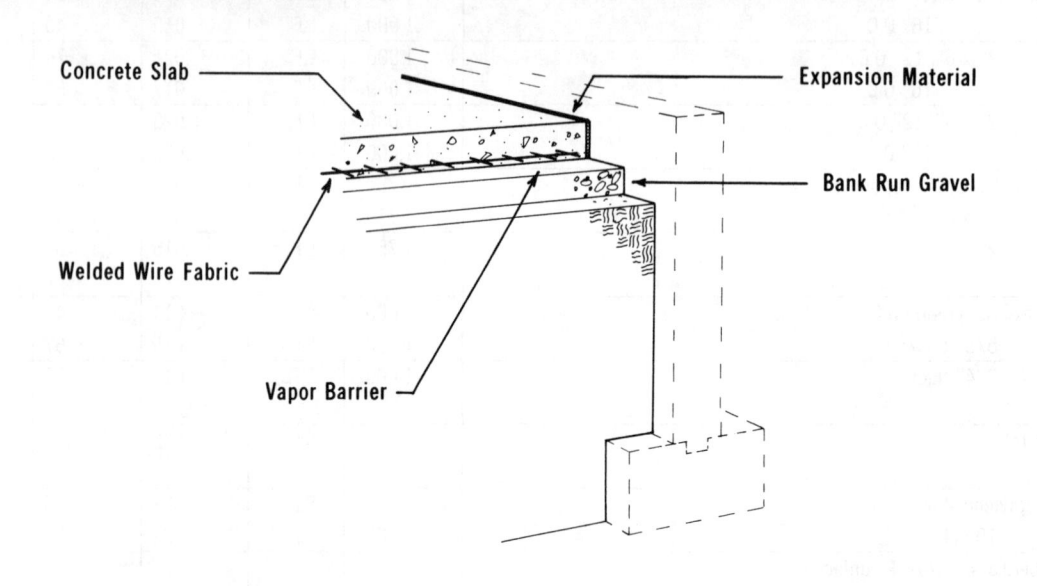

*Diagram labels: Concrete Slab, Expansion Material, Bank Run Gravel, Welded Wire Fabric, Vapor Barrier*

| SYSTEM DESCRIPTION | QUAN. | UNIT | MAN HOURS | MAT. | INST. | TOTAL |
|---|---|---|---|---|---|---|
| **4" THICK SLAB** | | | | | | |
| Concrete, 4" thick, 3000 psi concrete | .012 | C.Y. | | .61 | | .61 |
| Place concrete, direct chute | .012 | C.Y. | .005 | | .09 | .09 |
| Bank run gravel, 4" deep | 1.000 | S.F. | | .13 | .02 | .15 |
| Polyethylene vapor barrier, .006" thick | 1.000 | S.F. | .002 | .06 | .04 | .10 |
| Edge forms, expansion material | .100 | L.F. | .005 | .02 | .10 | .12 |
| Welded wire fabric, 6 x 6, 10/10 (W1.4/W1.4) | 1.100 | S.F. | .005 | .10 | .11 | .21 |
| Steel trowel finish | 1.000 | S.F. | .015 | | .32 | .32 |
| TOTAL | | | .033 | .92 | .68 | 1.60 |
| **6" THICK SLAB** | | | | | | |
| Concrete, 6" thick, 3000 psi concrete | .019 | C.Y. | | .97 | | .97 |
| Place concrete, direct chute | .019 | C.Y. | .008 | | .15 | .15 |
| Bank run gravel, 4" deep | 1.000 | S.F. | | .13 | .02 | .15 |
| Polyethylene vapor barrier, .006" thick | 1.000 | S.F. | .002 | .06 | .04 | .10 |
| Edge forms, expansion material | .100 | L.F. | .005 | .02 | .10 | .12 |
| Welded wire fabric, 6 x 6, 10/10 (W1.4/W1.4) | 1.100 | S.F. | .005 | .10 | .11 | .21 |
| Steel trowel finish | 1.000 | S.F. | .015 | | .32 | .32 |
| TOTAL | | | .036 | 1.28 | .74 | 2.02 |

The slab costs in this section are based on a cost per square foot of floor area.

| DESCRIPTION | QUAN. | UNIT | MAN HOURS | MAT. | INST. | TOTAL |
|---|---|---|---|---|---|---|
| | | | | | | |
| | | | | | | |
| | | | | | | |
| | | | | | | |
| | | | | | | |

# Floor Slab Price Sheet

| | QUAN. | UNIT | MAN HOURS | MAT. | INST. | TOTAL |
|---|---|---|---|---|---|---|
| Concrete, 4" thick slab, 2000 psi concrete | .012 | C.Y. | | .56 | | .56 |
| 2500 psi concrete | .012 | C.Y. | | .58 | | .58 |
| 3000 psi concrete | .012 | C.Y. | | .61 | | .61 |
| 3500 psi concrete | .012 | C.Y. | | .62 | | .62 |
| 4000 psi concrete | .012 | C.Y. | | .65 | | .65 |
| 4500 psi concrete | .012 | C.Y. | | .67 | | .67 |
| 5" thick slab, 2000 psi concrete | .015 | C.Y. | | .71 | | .71 |
| 2500 psi concrete | .015 | C.Y. | | .72 | | .72 |
| 3000 psi concrete | .015 | C.Y. | | .77 | | .77 |
| 3500 psi concrete | .015 | C.Y. | | .78 | | .78 |
| 4000 psi concrete | .015 | C.Y. | | .81 | | .81 |
| 4500 psi concrete | .015 | C.Y. | | .84 | | .84 |
| 6" thick slab, 2000 psi concrete | .019 | C.Y. | | .89 | | .89 |
| 2500 psi concrete | .019 | C.Y. | | .91 | | .91 |
| 3000 psi concrete | .019 | C.Y. | | .97 | | .97 |
| 3500 psi concrete | .019 | C.Y. | | .99 | | .99 |
| 4000 psi concrete | .019 | C.Y. | | 1.03 | | 1.03 |
| 4500 psi concrete | .019 | C.Y. | | 1.06 | | 1.06 |
| Place concrete, 4" slab, direct chute | .012 | C.Y. | .005 | | .09 | .09 |
| Pumped concrete | .012 | C.Y. | .006 | | .17 | .17 |
| Crane & bucket | .012 | C.Y. | .006 | | .19 | .19 |
| 5" slab, direct chute | .015 | C.Y. | .006 | | .12 | .12 |
| Pumped concrete | .015 | C.Y. | .007 | | .21 | .21 |
| Crane & bucket | .015 | C.Y. | .008 | | .24 | .24 |
| 6" slab, direct chute | .019 | C.Y. | .008 | | .15 | .15 |
| Pumped concrete | .019 | C.Y. | .010 | | .27 | .27 |
| Crane & bucket | .019 | C.Y. | .011 | | .30 | .30 |
| Gravel, bank run, 4" deep | 1.000 | S.F. | | .13 | .02 | .15 |
| 6" deep | 1.000 | S.F. | | .17 | .03 | .20 |
| 9" deep | 1.000 | S.F. | .001 | .25 | .05 | .30 |
| 12" deep | 1.000 | S.F. | .001 | .33 | .06 | .39 |
| 3/4" crushed stone, 3" deep | 1.000 | S.F. | .001 | .20 | .03 | .23 |
| 6" deep | 1.000 | S.F. | .001 | .41 | .05 | .46 |
| 9" deep | 1.000 | S.F. | .002 | .62 | .06 | .68 |
| 12" deep | 1.000 | S.F. | .002 | .81 | .08 | .89 |
| Vapor barrier polyethylene, .004" thick | 1.000 | S.F. | .002 | .06 | .04 | .10 |
| .006" thick | 1.000 | S.F. | .002 | .06 | .04 | .10 |
| Edge forms, expansion material, 4" thick slab | .100 | L.F. | .003 | .01 | .07 | .08 |
| 6" thick slab | .100 | L.F. | .005 | .02 | .10 | .12 |
| Welded wire fabric 6 x 6, 10/10 (W1.4/W1.4) | 1.100 | S.F. | .005 | .10 | .11 | .21 |
| 6 x 6, 6/6 (W2.9/W2.9) | 1.100 | S.F. | .006 | .19 | .13 | .32 |
| 4 x 4, 10/10 (W1.4/W1.4) | 1.100 | S.F. | .005 | .12 | .13 | .25 |
| Finish concrete, screed finish | 1.000 | S.F. | .009 | | .16 | .16 |
| Float finish | 1.000 | S.F. | .011 | | .24 | .24 |
| Steel trowel, for resilient floor | 1.000 | S.F. | .013 | | .28 | .28 |
| For finished floor | 1.000 | S.F. | .015 | | .32 | .32 |

# SECTION 3 FRAMING

| System Number | Title | Page |
|---|---|---|
| **3-04** | Floor (Wood) | 34 |
| **3-06** | Floor (Steel) | 36 |
| **3-08** | Exterior Wall | 38 |
| **3-12** | Gable End Roof | 40 |
| **3-16** | Truss Roof | 42 |
| **3-20** | Hip Roof | 44 |
| **3-24** | Gambrel Roof | 46 |
| **3-28** | Mansard Roof | 48 |
| **3-32** | Shed/Flat Roof | 50 |
| **3-36** | Flat Roof (Steel) | 52 |
| **3-40** | Gable Dormer | 54 |
| **3-44** | Shed Dormer | 56 |
| **3-48** | Partition | 58 |

# 3 | FRAMING  04 | Floor Framing Systems (Wood)

*Diagram labels: Box Sill, Sheathing, Bridging, Furring, Girder, Wood Joists, Box Sill*

| SYSTEM DESCRIPTION | QUAN. | UNIT | MAN HOURS | MAT. | INST. | TOTAL |
|---|---|---|---|---|---|---|
| **2" X 8", 16" O.C.** | | | | | | |
| Wood joists, 2" x 8", 16" O.C. | 1.000 | L.F. | .017 | .55 | .35 | .90 |
| Bridging, 1" x 3", 6' O.C. | .080 | Pr. | .004 | .03 | .10 | .13 |
| Box sills, 2" x 8" | .150 | L.F. | .002 | .08 | .06 | .14 |
| Girder, including lally columns, 3- 2" x 8" | .125 | L.F. | .015 | .26 | .32 | .58 |
| Sheathing, plywood, subfloor, 5/8" CDX | 1.000 | S.F. | .012 | .45 | .24 | .69 |
| Furring, 1" x 3", 16" O.C. | 1.000 | L.F. | .023 | .10 | .45 | .55 |
| TOTAL | | | .073 | 1.47 | 1.52 | 2.99 |
| **2" X 10", 16" O.C.** | | | | | | |
| Wood joists, 2" x 10", 16" O.C. | 1.000 | L.F. | .018 | .77 | .36 | 1.13 |
| Bridging, 1" x 3", 6' O.C. | .080 | Pr. | .004 | .03 | .10 | .13 |
| Box sills, 2" x 10" | .150 | L.F. | .002 | .12 | .05 | .17 |
| Girder, including lally columns, 3-2" x 10" | .125 | L.F. | .016 | .33 | .35 | .68 |
| Sheathing, plywood, subfloor, 5/8" CDX | 1.000 | S.F. | .012 | .45 | .24 | .69 |
| Furring, 1" x 3", 16" O.C. | 1.000 | L.F. | .023 | .10 | .45 | .55 |
| TOTAL | | | .075 | 1.80 | 1.55 | 3.35 |
| **2" X 12", 16" O.C.** | | | | | | |
| Wood joists, 2" x 12", 16" O.C. | 1.000 | L.F. | .018 | .94 | .37 | 1.31 |
| Bridging, 1" x 3", 6' O.C. | .080 | Pr. | .004 | .03 | .10 | .13 |
| Box sills, 2" x 12" | .150 | L.F. | .002 | .14 | .06 | .20 |
| Girder, including lally columns, 3-2" x 12" | .125 | L.F. | .017 | .40 | .36 | .76 |
| Sheathing, plywood, subfloor, 5/8" CDX | 1.000 | S.F. | .012 | .45 | .24 | .69 |
| Furring, 1" x 3", 16" O.C. | 1.000 | L.F. | .023 | .10 | .45 | .55 |
| TOTAL | | | .076 | 2.06 | 1.58 | 3.64 |

Floor costs on this page are given on a cost per square foot basis.

| DESCRIPTION | QUAN. | UNIT | MAN HOURS | MAT. | INST. | TOTAL |
|---|---|---|---|---|---|---|
| | | | | | | |
| | | | | | | |
| | | | | | | |
| | | | | | | |

# Floor Framing Price Sheet (Wood)

| Item | QUAN. | UNIT | MAN HOURS | MAT. | INST. | TOTAL |
|---|---|---|---|---|---|---|
| Joists, #2 or better, pine, 2" x 4", 12" O.C. | 1.250 | L.F. | .016 | .35 | .33 | .68 |
| 16" O.C. | 1.000 | L.F. | .013 | .28 | .26 | .54 |
| 2" x 6", 12" O.C. | 1.250 | L.F. | .018 | .50 | .39 | .89 |
| 16" O.C. | 1.000 | L.F. | .015 | .40 | .31 | .71 |
| 2" x 8", 12" O.C. | 1.250 | L.F. | .021 | .69 | .44 | 1.13 |
| 16" O.C. | 1.000 | L.F. | .017 | .55 | .35 | .90 |
| 2" x 10", 12" O.C. | 1.250 | L.F. | .022 | .96 | .45 | 1.41 |
| 16" O.C. | 1.000 | L.F. | .018 | .77 | .36 | 1.13 |
| 2" x 12", 12" O.C. | 1.250 | L.F. | .022 | 1.18 | .46 | 1.64 |
| 16" O.C. | 1.000 | L.F. | .018 | .94 | .37 | 1.31 |
| Bridging, wood 1" x 3", joists 12" O.C. | .100 | Pr. | .006 | .04 | .13 | .17 |
| 16" O.C. | .080 | Pr. | .004 | .03 | .10 | .13 |
| Metal, galvanized, joists 12" O.C. | .100 | Pr. | .006 | .10 | .12 | .22 |
| 16" O.C. | .080 | Pr. | .004 | .08 | .09 | .17 |
| Compression type, joists 12" O.C. | .100 | Pr. | .004 | .09 | .08 | .17 |
| 16" O.C. | .080 | Pr. | .003 | .07 | .07 | .14 |
| Box sills, #2 or better pine, 2" x 4" | .150 | L.F. | .001 | .04 | .04 | .08 |
| 2" x 6" | .150 | L.F. | .002 | .06 | .05 | .11 |
| 2" x 8" | .150 | L.F. | .002 | .08 | .06 | .14 |
| 2" x 10" | .150 | L.F. | .002 | .12 | .05 | .17 |
| 2" x 12" | .150 | L.F. | .002 | .14 | .06 | .20 |
| Girders including lally columns, 3 pieces spiked together, 2" x 8" | .125 | L.F. | .015 | .26 | .32 | .58 |
| 2" x 10" | .125 | L.F. | .016 | .33 | .35 | .68 |
| 2" x 12" | .125 | L.F. | .017 | .40 | .36 | .76 |
| Solid girders, 3" x 8" | .040 | L.F. | .003 | .11 | .09 | .20 |
| 3" x 10" | .040 | L.F. | .004 | .12 | .10 | .22 |
| 3" x 12" | .040 | L.F. | .004 | .13 | .11 | .24 |
| 4" x 8" | .040 | L.F. | .004 | .12 | .10 | .22 |
| 4" x 10" | .040 | L.F. | .004 | .14 | .11 | .25 |
| 4" x 12" | .040 | L.F. | .005 | .15 | .12 | .27 |
| Steel girders, bolted & including fabrication, wide flange shapes | | | | | | |
| 12" deep, 14#/L.F. | 1.000 | L.F. | .078 | 7.55 | 2.95 | 10.50 |
| 10" deep, 15#/L.F. | 1.000 | L.F. | .078 | 8.25 | 2.95 | 11.20 |
| 8" deep, 10#/L.F. | 1.000 | L.F. | .078 | 5.50 | 2.95 | 8.45 |
| 6" deep, 9#/L.F. | 1.000 | L.F. | .078 | 4.95 | 2.95 | 7.90 |
| 5" deep, 16#/L.F. | 1.000 | L.F. | .078 | 9.05 | 2.95 | 12.00 |
| Sheathing, plywood exterior grade CDX, 1/2" thick | 1.000 | S.F. | .011 | .41 | .22 | .63 |
| 5/8" thick | 1.000 | S.F. | .012 | .45 | .24 | .69 |
| 3/4" thick | 1.000 | S.F. | .013 | .51 | .26 | .77 |
| Boards, 1" x 8" laid regular | 1.000 | S.F. | .016 | .83 | .32 | 1.15 |
| Laid diagonal | 1.000 | S.F. | .019 | .83 | .38 | 1.21 |
| 1" x 10" laid regular | 1.000 | S.F. | .015 | .81 | .30 | 1.11 |
| Laid diagonal | 1.000 | S.F. | .018 | .81 | .37 | 1.18 |
| Furring, 1" x 3", 12" O.C. | 1.250 | L.F. | .028 | .13 | .56 | .69 |
| 16" O.C. | 1.000 | L.F. | .023 | .10 | .45 | .55 |
| 24" O.C. | .750 | L.F. | .017 | .08 | .33 | .41 |

# 3 | FRAMING  06 | Floor Framing Systems (Steel)

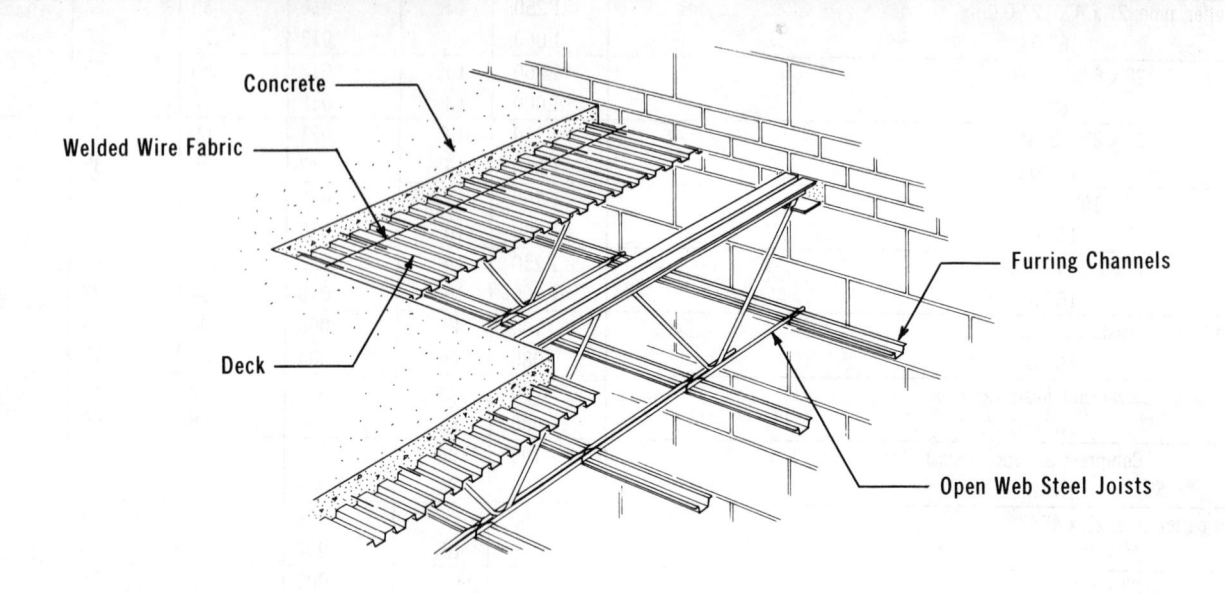

*Diagram labels: Concrete, Welded Wire Fabric, Deck, Furring Channels, Open Web Steel Joists*

| SYSTEM DESCRIPTION | QUAN. | UNIT | MAN HOURS | COST PER S.F. MAT. | COST PER S.F. INST. | TOTAL |
|---|---|---|---|---|---|---|
| **20' SPAN, 100 PSF LIVE LOAD** | | | | | | |
| Open web steel joists, 16" deep, 7.9#/L.F., 24" O.C. | .550 | L.F. | .023 | 1.44 | .89 | 2.33 |
| Decking, slab form, 28 gauge, 9/16" deep, galvanized | 1.100 | S.F. | .006 | .46 | .17 | .63 |
| Concrete, 3000 psi, 2-1/2" thick | .008 | C.Y. | | .41 | | .41 |
| Place concrete, pumped | .008 | C.Y. | .004 | | .12 | .12 |
| Welded wire fabric, 6 x 6, 10/10 (W1.4/W1.4) | 1.100 | S.F. | .005 | .10 | .11 | .21 |
| Steel trowel finish | 1.000 | S.F. | .015 | | .32 | .32 |
| Edge form, 18 gauge | .100 | L.F. | .003 | .01 | .07 | .08 |
| Ceiling furring, 3/4" channels, galvanized, 24" O.C. | 1.000 | S.F. | .019 | .19 | .36 | .55 |
| TOTAL | | | .075 | 2.61 | 2.04 | 4.65 |
| **30' SPAN, 100 PSF LIVE LOAD** | | | | | | |
| Open web steel joists, 22" deep, 12.3#/L.F., 24" O.C. | .530 | L.F. | .018 | 2.15 | .68 | 2.83 |
| Decking, slab form, 28 gauge, 9/16" deep, galvanized | 1.100 | S.F. | .006 | .46 | .17 | .63 |
| Concrete, 3000 psi, 2-1/2" thick | .008 | C.Y. | | .41 | | .41 |
| Place concrete, pumped | .008 | C.Y. | .004 | | .12 | .12 |
| Welded wire fabric, 6 x 6 10/10 (W1.4/W1.4) | 1.100 | S.F. | .005 | .10 | .11 | .21 |
| Steel trowel finish | 1.000 | S.F. | .015 | | .32 | .32 |
| Edge form, 18 gauge | .067 | L.F. | .005 | .02 | .10 | .12 |
| Ceiling furring, 3/4" channels, galvanized, 24" O.C. | 1.000 | S.F. | .019 | .19 | .36 | .55 |
| TOTAL | | | .072 | 3.33 | 1.86 | 5.19 |

| DESCRIPTION | QUAN. | UNIT | MAN HOURS | COST PER S.F. MAT. | COST PER S.F. INST. | TOTAL |
|---|---|---|---|---|---|---|
| | | | | | | |
| | | | | | | |
| | | | | | | |
| | | | | | | |
| | | | | | | |

# Floor Framing Price Sheet (Steel)

| | QUAN. | UNIT | MAN HOURS | COST PER S.F. MAT. | COST PER S.F. INST. | TOTAL |
|---|---|---|---|---|---|---|
| Joists 24" O.C., 15' span, 40 psf live load, 12" deep, 6.5#/L.F. | .530 | L.F. | .028 | 1.14 | 1.07 | 2.21 |
| 75 psf live load, 14" deep, 7.4#/L.F. | .530 | L.F. | .028 | 1.29 | 1.07 | 2.36 |
| 100 psf live load, 16" deep, 7.9#/L.F. | .530 | L.F. | .028 | 1.38 | 1.07 | 2.45 |
| 125 psf live load, 16" deep, 7.9#/L.F. | .530 | L.F. | .028 | 1.38 | 1.07 | 2.45 |
| 20' span, 40 psf live load, 14" deep, 5.5#/L.F. | .550 | L.F. | .023 | 1.00 | .89 | 1.89 |
| 75 psf live load, 14" deep, 6.5#/L.F. | .550 | L.F. | .023 | 1.18 | .89 | 2.07 |
| 100 psf live load, 16" deep, 6.6#/L.F. | .550 | L.F. | .023 | 1.20 | .89 | 2.09 |
| 125 psf live load, 16" deep, 7.9#/L.F. | .550 | L.F. | .023 | 1.43 | .89 | 2.32 |
| 25' span, 40 psf live load, 18" deep, 6.6#/L.F. | .520 | L.F. | .019 | 1.13 | .73 | 1.86 |
| 75 psf live load, 20" deep, 8.4#/L.F. | .520 | L.F. | .019 | 1.44 | .73 | 2.17 |
| 100 psf live load, 18" deep, 10.4#/L.F. | .520 | L.F. | .019 | 1.78 | .73 | 2.51 |
| 125 psf live load, 20" deep, 10.7#/L.F. | .520 | L.F. | .019 | 1.84 | .73 | 2.57 |
| 30' span, 40 psf live load, 20' deep, 8.4#/L.F. | .530 | L.F. | .018 | 1.47 | .68 | 2.15 |
| 75 psf live load, 24" deep, 10.3#/L.F. | .530 | L.F. | .018 | 1.80 | .68 | 2.48 |
| 100 psf live load, 24" deep, 11.5#/L.F. | .530 | L.F. | .018 | 2.01 | .68 | 2.69 |
| 125 psf live load, 22" deep, 12.3#/L.F. | .530 | L.F. | .018 | 2.15 | .68 | 2.83 |
| 35' span, 40 psf live load, 24" deep, 10.3#/L.F. | .510 | L.F. | .016 | 1.73 | .62 | 2.35 |
| 75 psf live load, 26" deep, 11.9#/L.F. | .510 | L.F. | .016 | 2.00 | .62 | 2.62 |
| 100 psf live load, 26" deep, 12.8#/L.F. | .510 | L.F. | .016 | 2.15 | .62 | 2.77 |
| 125 psf live load, 26" deep, 12.8#/L.F. | .510 | L.F. | .016 | 2.15 | .62 | 2.77 |
| 40' span, 40 psf live load, 24" deep, 12.7#/L.F. | .530 | L.F. | .018 | 2.22 | .70 | 2.92 |
| 75 psf live load, 26" deep, 14.8#/L.F. | .530 | L.F. | .018 | 2.59 | .70 | 3.29 |
| 100 psf live load, 28" deep, 17.1#/L.F. | .530 | L.F. | .018 | 2.99 | .70 | 3.69 |
| 125 psf live load, 24" deep, 19.0#/L.F. | .530 | L.F. | .018 | 3.32 | .70 | 4.02 |
| Decking slab form, 28 gauge 9/16" deep, uncoated | 1.100 | S.F. | .006 | .43 | .16 | .59 |
| Galvanized | 1.100 | S.F. | .006 | .46 | .17 | .63 |
| 24 gauge, 1-5/16" deep, uncoated | 1.100 | S.F. | .006 | .61 | .17 | .78 |
| Galvanized | 1.100 | S.F. | .006 | .67 | .17 | .84 |
| Concrete, 2-1/2" thick, 2500 psi | .008 | C.Y. | | .38 | | .38 |
| 3000 psi | .008 | C.Y. | | .41 | | .41 |
| 3500 psi | .008 | C.Y. | | .42 | | .42 |
| 4000 psi | .008 | C.Y. | | .43 | | .43 |
| Place concrete, truck unloading | .008 | C.Y. | .003 | | .06 | .06 |
| Crane and bucket | .008 | C.Y. | .005 | | .15 | .15 |
| Pumped | .008 | C.Y. | .004 | | .12 | .12 |
| Welded wire fabric, 6 x 6, 10/10 (W1.4/W1.4) | 1.100 | S.F. | .005 | .10 | .11 | .21 |
| 6/6 (W2.9/W2.9) | 1.100 | S.F. | .006 | .19 | .13 | .32 |
| 4 x 4, 10/10 (W1.4/W1.4) | 1.100 | S.F. | .005 | .12 | .13 | .25 |
| Finish concrete, screed finish | 1.000 | S.F. | .009 | | .16 | .16 |
| Darby finish | 1.000 | S.F. | .011 | | .19 | .19 |
| Float finish | 1.000 | S.F. | .011 | | .24 | .24 |
| Broom finish | 1.000 | S.F. | .012 | | .26 | .26 |
| Steel trowel finish, for tile | 1.000 | S.F. | .013 | | .28 | .28 |
| For finish floor | 1.000 | S.F. | .015 | | .32 | .32 |
| Edge form, 18 gauge, 15' span | .133 | L.F. | .007 | .02 | .14 | .16 |
| 20' span | .100 | L.F. | .005 | .02 | .10 | .12 |
| 25' span | .080 | L.F. | .004 | .01 | .09 | .10 |
| 30' span | .067 | L.F. | .003 | .01 | .07 | .08 |
| 35' span | .057 | L.F. | .003 | .01 | .06 | .07 |
| 40' span | .050 | L.F. | .002 | .01 | .05 | .06 |
| Furring ceiling, 3/4" galvanized chanels, 12" O.C. | 1.000 | S.F. | .038 | .30 | .73 | 1.03 |
| 16" O.C. | 1.000 | S.F. | .028 | .25 | .53 | .78 |
| 24" O.C. | 1.000 | S.F. | .019 | .19 | .36 | .55 |
| 1-1/2" galvanized channels, 12" O.C. | 1.000 | S.F. | .042 | .37 | .81 | 1.18 |
| 16" O.C. | 1.000 | S.F. | .031 | .32 | .59 | .91 |
| 24" O.C. | 1.000 | S.F. | .021 | .24 | .40 | .64 |

# 3 | FRAMING    08 | Exterior Wall Framing Systems

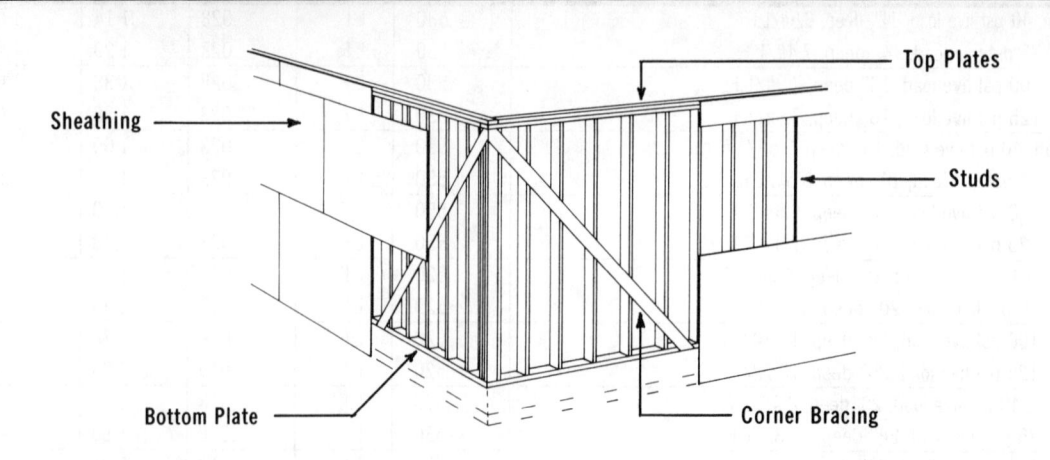

Diagram labels: Sheathing, Top Plates, Studs, Bottom Plate, Corner Bracing

| SYSTEM DESCRIPTION | QUAN. | UNIT | MAN HOURS | MAT. | INST. | TOTAL |
|---|---|---|---|---|---|---|
| **2" X 4", 16" O.C.** | | | | | | |
| 2" x 4" studs, 16" O.C. | 1.000 | L.F. | .015 | .25 | .32 | .57 |
| Plates, 2" x 4", double top, single bottom | .375 | L.F. | .005 | .09 | .12 | .21 |
| Corner bracing, let-in, 1" x 6" | .063 | L.F. | .003 | .01 | .07 | .08 |
| Sheathing, 1/2" plywood, CDX | 1.000 | S.F. | .014 | .41 | .29 | .70 |
| TOTAL | | | .037 | .76 | .80 | 1.56 |
| **2" X 4", 24" O.C.** | | | | | | |
| 2" x 4" studs, 24" O.C. | .750 | L.F. | .011 | .19 | .24 | .43 |
| Plates, 2" x 4", double top, single bottom | .375 | L.F. | .005 | .09 | .12 | .21 |
| Corner bracing, let-in, 1" x 6" | .063 | L.F. | .002 | .01 | .05 | .06 |
| Sheathing, 1/2" plywood, CDX | 1.000 | S.F. | .014 | .41 | .29 | .70 |
| TOTAL | | | .033 | .70 | .70 | 1.40 |
| **2" X 6", 16" O.C.** | | | | | | |
| 2" x 6" studs, 16" O.C. | 1.000 | L.F. | .018 | .40 | .36 | .76 |
| Plates, 2" x 6", double top, single bottom | .375 | L.F. | .006 | .15 | .14 | .29 |
| Corner bracing, let-in, 1" x 6" | .063 | L.F. | .003 | .01 | .07 | .08 |
| Sheathing, 1/2" plywood, CDX | 1.000 | S.F. | .014 | .41 | .29 | .70 |
| TOTAL | | | .042 | .97 | .86 | 1.83 |
| **2" X 6", 24" O.C.** | | | | | | |
| 2" x 6" studs, 24" O.C. | .750 | L.F. | .013 | .30 | .27 | .57 |
| Plates, 2" x 6", double top, single bottom | .375 | L.F. | .006 | .15 | .14 | .29 |
| Corner bracing, let-in, 1" x 6" | .063 | L.F. | .002 | .01 | .05 | .06 |
| Sheathing, 1/2" plywood, CDX | 1.000 | S.F. | .014 | .41 | .29 | .70 |
| TOTAL | | | .036 | .87 | .75 | 1.62 |

The wall costs on this page are given in cost per square foot of wall.
For window and door openings see next page.

| DESCRIPTION | QUAN. | UNIT | MAN HOURS | MAT. | INST. | TOTAL |
|---|---|---|---|---|---|---|
| | | | | | | |
| | | | | | | |
| | | | | | | |

# Exterior Wall Framing Price Sheet

| | QUAN. | UNIT | MAN HOURS | COST PER S.F. MAT. | COST PER S.F. INST. | TOTAL |
|---|---|---|---|---|---|---|
| Studs, #2 or better, 2" x 4", 12" O.C. | 1.250 | L.F. | .018 | .31 | .40 | .71 |
| 16" O.C. | 1.000 | L.F. | .015 | .25 | .32 | .57 |
| 24" O.C. | .750 | L.F. | .011 | .19 | .24 | .43 |
| 32" O.C. | .600 | L.F. | .009 | .15 | .19 | .34 |
| 2" x 6", 12" O.C. | 1.250 | L.F. | .022 | .50 | .45 | .95 |
| 16" O.C. | 1.000 | L.F. | .018 | .40 | .36 | .76 |
| 24" O.C. | .750 | L.F. | .013 | .30 | .27 | .57 |
| 32" O.C. | .600 | L.F. | .010 | .24 | .22 | .46 |
| 2" x 8", 12" O.C. | 1.250 | L.F. | .025 | .79 | .51 | 1.30 |
| 16" O.C. | 1.000 | L.F. | .020 | .63 | .41 | 1.04 |
| 24" O.C. | .750 | L.F. | .015 | .47 | .31 | .78 |
| 32" O.C. | .600 | L.F. | .012 | .38 | .24 | .62 |
| Plates, #2 or better, double top, single bottom, 2" x 4" | .375 | L.F. | .005 | .09 | .12 | .21 |
| 2" x 6" | .375 | L.F. | .006 | .15 | .14 | .29 |
| 2" x 8" | .375 | L.F. | .007 | .24 | .15 | .39 |
| Corner bracing, let-in 1" x 6" boards, studs, 12" O.C. | .070 | L.F. | .003 | .01 | .08 | .09 |
| 16" O.C. | .063 | L.F. | .003 | .01 | .07 | .08 |
| 24" O.C. | .063 | L.F. | .002 | .01 | .05 | .06 |
| 32" O.C. | .057 | L.F. | .001 | .01 | .04 | .05 |
| Let-in steel ("T" shape), studs, 12" O.C. | .070 | L.F. | | .04 | .02 | .06 |
| 16" O.C. | .063 | L.F. | | .04 | .01 | .05 |
| 24" O.C. | .063 | L.F. | | .04 | .01 | .05 |
| 32" O.C. | .057 | L.F. | | .03 | .02 | .05 |
| Sheathing, plywood CDX, 3/8" thick | 1.000 | S.F. | .013 | .32 | .27 | .59 |
| 1/2" thick | 1.000 | S.F. | .014 | .41 | .29 | .70 |
| 5/8" thick | 1.000 | S.F. | .015 | .45 | .31 | .76 |
| 3/4" thick | 1.000 | S.F. | .016 | .51 | .33 | .84 |
| Boards, 1" x 6", laid regular | 1.000 | S.F. | .025 | .80 | .51 | 1.31 |
| Laid diagonal | 1.000 | S.F. | .027 | .80 | .56 | 1.36 |
| 1" x 8", laid regular | 1.000 | S.F. | .021 | .80 | .43 | 1.23 |
| Laid diagonal | 1.000 | S.F. | .021 | .80 | .43 | 1.23 |
| Wood fiber, regular, no vapor barrier, 1/2" thick | 1.000 | S.F. | .013 | .44 | .27 | .71 |
| 5/8" thick | 1.000 | S.F. | .013 | .55 | .27 | .82 |
| Asphalt impregnated 25/32" thick | 1.000 | S.F. | .013 | .29 | .27 | .56 |
| 1/2" thick | 1.000 | S.F. | .013 | .24 | .28 | .52 |
| Polystyrene, regular, 3/4" thick | 1.000 | S.F. | .010 | .40 | .20 | .60 |
| 2" thick | 1.000 | S.F. | .011 | .86 | .22 | 1.08 |
| Fiberglass, foil faced, 1" thick | 1.000 | S.F. | .008 | .81 | .17 | .98 |
| 2" thick | 1.000 | S.F. | .009 | 1.25 | .19 | 1.44 |

# Window & Door Openings

| | QUAN. | UNIT | MAN HOURS | COST EACH MAT. | COST EACH INST. | TOTAL |
|---|---|---|---|---|---|---|
| The following costs are to be added to the total costs of the wall for each opening. Do not subtract the area of the openings. | | | | | | |
| Headers, 2" x 6" double, 2' long | 4.000 | L.F. | .124 | 1.60 | 2.52 | 4.12 |
| 3' long | 6.000 | L.F. | .186 | 2.40 | 3.78 | 6.18 |
| 4' long | 8.000 | L.F. | .248 | 3.20 | 5.04 | 8.24 |
| 5' long | 10.000 | L.F. | .310 | 4.00 | 6.30 | 10.30 |
| 2" x 8" double, 4' long | 8.000 | L.F. | .376 | 4.40 | 8.24 | 12.64 |
| 5' long | 10.000 | L.F. | .470 | 5.50 | 10.30 | 15.80 |
| 6' long | 12.000 | L.F. | .564 | 6.60 | 12.36 | 18.96 |
| 8' long | 16.000 | L.F. | .752 | 8.80 | 16.48 | 25.28 |
| 2" x 10" double, 4' long | 8.000 | L.F. | .400 | 6.16 | 8.64 | 14.80 |
| 6' long | 12.000 | L.F. | .600 | 9.24 | 12.96 | 22.20 |
| 8' long | 16.000 | L.F. | .800 | 12.32 | 17.28 | 29.60 |
| 10' long | 20.000 | L.F. | 1.000 | 15.40 | 21.60 | 37.00 |
| 2" x 12" double, 8' long | 16.000 | L.F. | .736 | 15.04 | 14.88 | 29.92 |
| 12' long | 24.000 | L.F. | 1.104 | 22.56 | 22.32 | 44.88 |

# 3 | FRAMING
# 12 | Gable End Roof Framing Systems

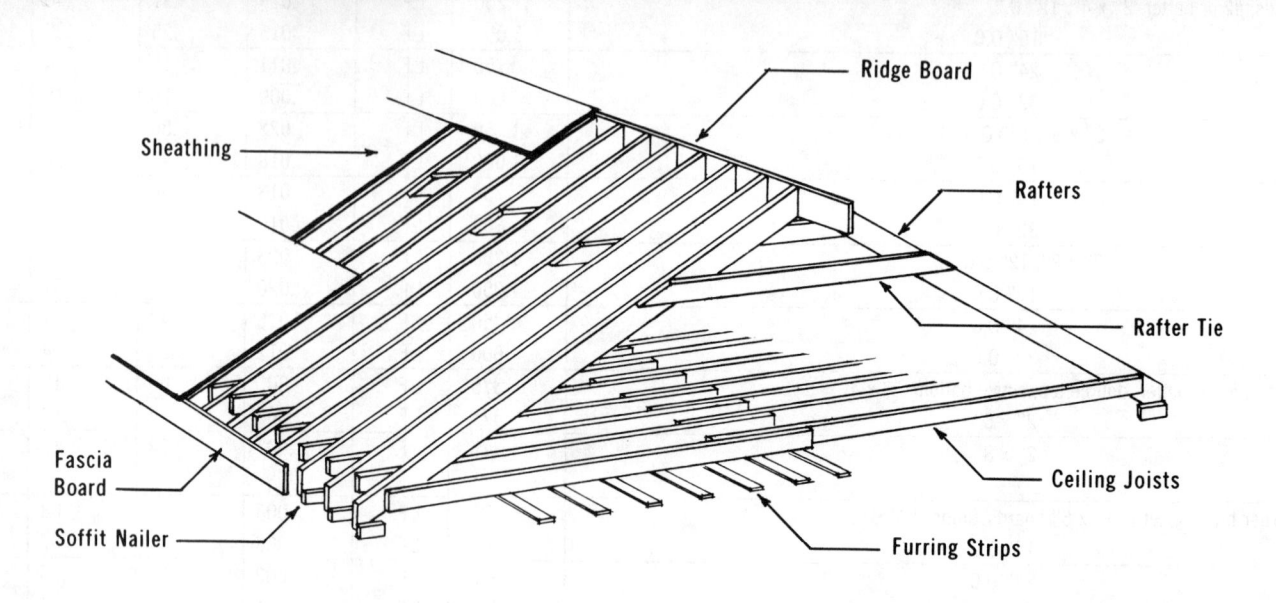

| SYSTEM DESCRIPTION | QUAN. | UNIT | MAN HOURS | MAT. | INST. | TOTAL |
|---|---|---|---|---|---|---|
| **2" X 8" RAFTERS, 16" O.C., 4/12 PITCH** | | | | | | |
| Rafters, 2" x 8", 16" O.C., 4/12 pitch | 1.170 | L.F. | .025 | .64 | .54 | 1.18 |
| Ceiling joists, 2" x 6", 16" O.C. | 1.000 | L.F. | .015 | .40 | .31 | .71 |
| Ridge board, 2" x 8" | .050 | L.F. | .001 | .03 | .02 | .05 |
| Fascia board, 2" x 8" | .100 | L.F. | .007 | .05 | .15 | .20 |
| Rafter tie, 1" x 4", 4' O.C. | .060 | L.F. | | .01 | .01 | .02 |
| Soffit nailer (outrigger), 2" x 4", 24" O.C. | .170 | L.F. | .004 | .07 | .08 | .15 |
| Sheathing, exterior, plywood, CDX, 1/2" thick | 1.170 | S.F. | .012 | .48 | .27 | .75 |
| Furring strips, 1" x 3", 16" O.C. | 1.000 | L.F. | .023 | .10 | .45 | .55 |
| TOTAL | | | .087 | 1.78 | 1.83 | 3.61 |
| **2" X 6" RAFTERS, 16" O.C., 4/12 PITCH** | | | | | | |
| Rafters, 2" x 6", 16" O.C., 4/12 pitch | 1.170 | L.F. | .022 | .47 | .44 | .91 |
| Ceiling joists, 2" x 4", 16" O.C. | 1.000 | L.F. | .013 | .28 | .26 | .54 |
| Ridge board, 2" x 6" | .050 | L.F. | | .02 | .02 | .04 |
| Fascia board, 2" x 6" | .100 | L.F. | .005 | .04 | .11 | .15 |
| Rafter tie, 1" x 4", 4' O.C. | .060 | L.F. | | .01 | .01 | .02 |
| Soffit nailer (outrigger), 2" x 4", 24" O.C. | .170 | L.F. | .004 | .07 | .08 | .15 |
| Sheathing, exterior, plywood, CDX, 1/2" thick | 1.170 | S.F. | .012 | .48 | .27 | .75 |
| Furring strips, 1" x 3", 16" O.C. | 1.000 | L.F. | .023 | .10 | .45 | .55 |
| TOTAL | | | .079 | 1.47 | 1.64 | 3.11 |

The cost of this system is based on the square foot of plan area. All quantities have been adjusted accordingly.

| DESCRIPTION | QUAN. | UNIT | MAN HOURS | MAT. | INST. | TOTAL |
|---|---|---|---|---|---|---|
| | | | | | | |
| | | | | | | |
| | | | | | | |
| | | | | | | |
| | | | | | | |

# Gable End Roof Framing Price Sheet

| | QUAN. | UNIT | MAN HOURS | COST PER S.F. MAT. | COST PER S.F. INST. | TOTAL |
|---|---|---|---|---|---|---|
| Rafters, #2 or better, 16" O.C., 2" x 6", 4/12 pitch | 1.170 | L.F. | .022 | .47 | .44 | .91 |
| 8/12 pitch | 1.330 | L.F. | .031 | .53 | .64 | 1.17 |
| 2" x 8", 4/12 pitch | 1.170 | L.F. | .025 | .64 | .54 | 1.18 |
| 8/12 pitch | 1.330 | L.F. | .038 | .73 | .79 | 1.52 |
| 2" x 10", 4/12 pitch | 1.170 | L.F. | .029 | .88 | .61 | 1.49 |
| 8/12 pitch | 1.330 | L.F. | .042 | 1.00 | .88 | 1.88 |
| 24" O.C., 2" x 6", 4/12 pitch | .940 | L.F. | .017 | .38 | .35 | .73 |
| 8/12 pitch | 1.060 | L.F. | .025 | .42 | .51 | .93 |
| 2" x 8", 4/12 pitch | .940 | L.F. | .020 | .52 | .43 | .95 |
| 8/12 pitch | 1.060 | L.F. | .030 | .58 | .63 | 1.21 |
| 2" x 10", 4/12 pitch | .940 | L.F. | .023 | .71 | .48 | 1.19 |
| 8/12 pitch | 1.060 | L.F. | .033 | .80 | .69 | 1.49 |
| Ceiling joist, #2 or better, 2" x 4", 16" O.C. | 1.000 | L.F. | .013 | .28 | .26 | .54 |
| 24" O.C. | .750 | L.F. | .009 | .21 | .20 | .41 |
| 2" x 6", 16" O.C. | 1.000 | L.F. | .015 | .40 | .31 | .71 |
| 24" O.C. | .750 | L.F. | .011 | .30 | .23 | .53 |
| 2" x 8", 16" O.C. | 1.000 | L.F. | .017 | .55 | .35 | .90 |
| 24" O.C. | .750 | L.F. | .012 | .41 | .27 | .68 |
| 2" x 10", 16" O.C. | 1.000 | L.F. | .018 | .77 | .36 | 1.13 |
| 24" O.C. | .750 | L.F. | .013 | .58 | .27 | .85 |
| Ridge board, #2 or better, 1" x 6" | .050 | L.F. | .001 | .01 | .01 | .02 |
| 1" x 8" | .050 | L.F. | .001 | .01 | .02 | .03 |
| 1" x 10" | .050 | L.F. | .001 | .02 | .02 | .04 |
| 2" x 6" | .050 | L.F. | .001 | .02 | .02 | .04 |
| 2" x 8" | .050 | L.F. | .001 | .03 | .02 | .05 |
| 2" x 10" | .050 | L.F. | .001 | .04 | .02 | .06 |
| Fascia board, #2 or better, 1" x 6" | .100 | L.F. | .003 | .03 | .08 | .11 |
| 1" x 8" | .100 | L.F. | .004 | .03 | .10 | .13 |
| 1" x 10" | .100 | L.F. | .005 | .04 | .10 | .14 |
| 2" x 6" | .100 | L.F. | .005 | .04 | .12 | .16 |
| 2" x 8" | .100 | L.F. | .007 | .05 | .15 | .20 |
| 2" x 10" | .100 | L.F. | .008 | .08 | .18 | .26 |
| Rafter tie, #2 or better, 4' O.C., 1" x 4" | .060 | L.F. | .001 | .01 | .01 | .02 |
| 1" x 6" | .060 | L.F. | .001 | .01 | .01 | .02 |
| 2" x 4" | .060 | L.F. | .001 | .01 | .01 | .02 |
| 2" x 6" | .060 | L.F. | .001 | .01 | .02 | .03 |
| Soffit nailer (outrigger), 2" x 4", 16" O.C. | .220 | L.F. | .005 | .09 | .10 | .19 |
| 24" O.C. | .170 | L.F. | .004 | .07 | .08 | .15 |
| 2" x 6", 16" O.C. | .220 | L.F. | .006 | .10 | .12 | .22 |
| 24" O.C. | .170 | L.F. | .004 | .08 | .09 | .17 |
| Sheathing, plywood CDX, 4/12 pitch, 3/8" thick | 1.170 | S.F. | .011 | .37 | .25 | .62 |
| 1/2" thick | 1.170 | S.F. | .012 | .48 | .27 | .75 |
| 5/8" thick | 1.170 | S.F. | .014 | .53 | .29 | .82 |
| 8/12 pitch, 3/8" | 1.330 | S.F. | .013 | .43 | .27 | .70 |
| 1/2" thick | 1.330 | S.F. | .014 | .55 | .30 | .85 |
| 5/8" thick | 1.330 | S.F. | .015 | .60 | .33 | .93 |
| Boards, 4/12 pitch roof, 1" x 6" | 1.170 | S.F. | .025 | .94 | .53 | 1.47 |
| 1" x 8" | 1.170 | S.F. | .021 | .94 | .44 | 1.38 |
| 8/12 pitch roof, 1" x 6" | 1.330 | S.F. | .029 | 1.06 | .62 | 1.68 |
| 1" x 8" | 1.330 | S.F. | .023 | 1.06 | .51 | 1.57 |
| Furring, 1" x 3", 12" O.C. | 1.200 | L.F. | .027 | .12 | .54 | .66 |
| 16" O.C. | 1.000 | L.F. | .023 | .10 | .45 | .55 |
| 24" O.C. | .800 | L.F. | .018 | .08 | .36 | .44 |

3

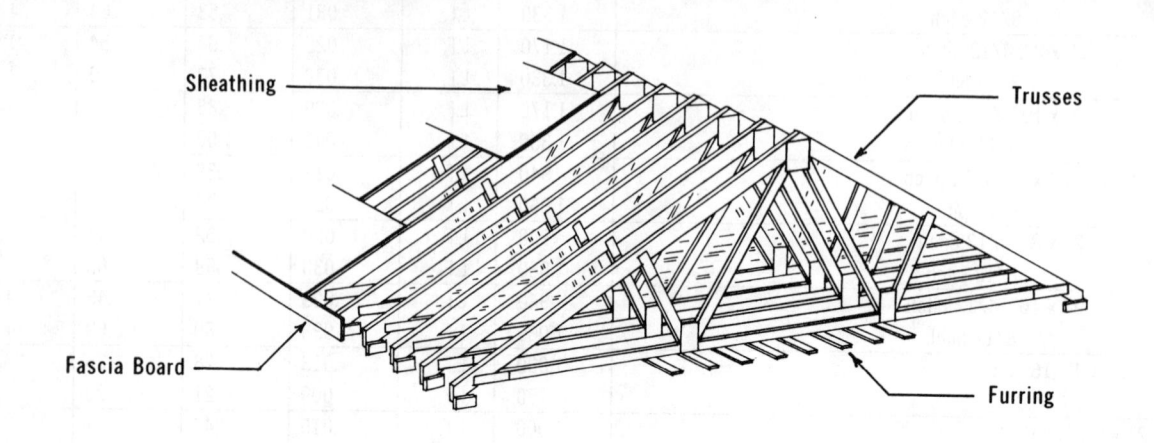

# 3 | FRAMING

## 16 | Truss Roof Framing Systems

*Sheathing, Trusses, Fascia Board, Furring*

| SYSTEM DESCRIPTION | QUAN. | UNIT | MAN HOURS | COST PER S.F. MAT. | COST PER S.F. INST. | COST PER S.F. TOTAL |
|---|---|---|---|---|---|---|
| **TRUSS, 16" O.C., 4/12 PITCH, INCLUDING 1' OVERHANG, 26' SPAN** | | | | | | |
| Truss, 40# loading, 16" O.C., 4/12 pitch, 26' span | .030 | Ea. | .001 | 1.25 | .52 | 1.77 |
| Fascia board, 2" x 6" | .100 | L.F. | .005 | .04 | .11 | .15 |
| Sheathing, exterior, plywood, CDX, 1/2" thick | 1.170 | S.F. | .012 | .48 | .27 | .75 |
| Furring, 1" x 3", 16" O.C. | 1.000 | L.F. | .023 | .10 | .45 | .55 |
| TOTAL | | | .041 | 1.87 | 1.35 | 3.22 |
| **TRUSS, 16" O.C., 8/12 PITCH, INCLUDING 1' OVERHANG, 26' SPAN** | | | | | | |
| Truss, 40# loading, 16" O.C., 8/12 pitch, 26' span | .030 | Ea. | .001 | 1.39 | .59 | 1.98 |
| Fascia board, 2" x 6" | .100 | L.F. | .005 | .04 | .11 | .15 |
| Sheathing, exterior, plywood, CDX, 1/2" thick | 1.330 | S.F. | .014 | .55 | .30 | .85 |
| Furring, 1" x 3", 16" O.C. | 1.000 | L.F. | .023 | .10 | .45 | .55 |
| TOTAL | | | .043 | 2.08 | 1.45 | 3.53 |
| **TRUSS, 24" O.C., 4/12 PITCH, INCLUDING 1' OVERHANG, 26' SPAN** | | | | | | |
| Truss, 40# loading, 24" O.C., 4/12 pitch, 26' span | .020 | Ea. | .001 | .84 | .34 | 1.18 |
| Fascia board, 2" x 6" | .100 | L.F. | .005 | .04 | .11 | .15 |
| Sheathing, exterior, plywood, CDX, 1/2" thick | 1.170 | S.F. | .012 | .48 | .27 | .75 |
| Furring, 1" x 3", 16" O.C. | 1.000 | L.F. | .023 | .10 | .45 | .55 |
| TOTAL | | | .041 | 1.46 | 1.17 | 2.63 |
| **TRUSS, 24" O.C., 8/12 PITCH, INCLUDING 1' OVERHANG, 26' SPAN** | | | | | | |
| Truss, 40# loading, 24" O.C., 8/12 pitch, 26' span | .020 | Ea. | .001 | .92 | .40 | 1.32 |
| Fascia board, 2" x 6" | .100 | L.F. | .005 | .04 | .11 | .15 |
| Sheathing, exterior, plywood, CDX, 1/2" thick | 1.330 | S.F. | .014 | .55 | .30 | .85 |
| Furring, 1" x 3", 16" O.C. | 1.000 | L.F. | .023 | .10 | .45 | .55 |
| TOTAL | | | .043 | 1.61 | 1.26 | 2.87 |

The cost of this system is based on the square foot of plan area. A one Foot overhang is included.

| DESCRIPTION | QUAN. | UNIT | MAN HOURS | COST PER S.F. MAT. | COST PER S.F. INST. | COST PER S.F. TOTAL |
|---|---|---|---|---|---|---|
| | | | | | | |
| | | | | | | |
| | | | | | | |
| | | | | | | |

# Truss Roof Framing Price Sheet

| | QUAN. | UNIT | MAN HOURS | MAT. | INST. | TOTAL |
|---|---|---|---|---|---|---|
| Truss, 40# load including 1' overhang, 4/12 pitch, 24' span, 16" O.C. | .033 | Ea. | .021 | 1.16 | .56 | 1.72 |
| 24" O.C. | .022 | Ea. | .014 | .77 | .37 | 1.14 |
| 26' span, 16" O.C. | .030 | Ea. | .021 | 1.25 | .52 | 1.77 |
| 24" O.C. | .020 | Ea. | .014 | .84 | .34 | 1.18 |
| 28' span, 16" O.C. | .027 | Ea. | .020 | 1.34 | .50 | 1.84 |
| 24" O.C. | .019 | Ea. | .014 | .94 | .35 | 1.29 |
| 32' span, 16" O.C. | .024 | Ea. | .019 | 1.43 | .47 | 1.90 |
| 24" O.C. | .016 | Ea. | .012 | .95 | .31 | 1.26 |
| 36' span, 16" O.C. | .022 | Ea. | .019 | 1.74 | .46 | 2.20 |
| 24" O.C. | .015 | Ea. | .013 | 1.19 | .31 | 1.50 |
| 8/12 pitch, 24' span, 16" O.C. | .033 | Ea. | .023 | 1.38 | .60 | 1.98 |
| 24" O.C. | .022 | Ea. | .015 | .92 | .40 | 1.32 |
| 26' span, 16" O.C. | .030 | Ea. | .023 | 1.39 | .59 | 1.98 |
| 24" O.C. | .020 | Ea. | .015 | .92 | .40 | 1.32 |
| 28' span, 16" O.C. | .027 | Ea. | .022 | 1.54 | .57 | 2.11 |
| 24" O.C. | .019 | Ea. | .015 | 1.09 | .39 | 1.48 |
| 32' span, 16" O.C. | .024 | Ea. | .021 | 1.50 | .54 | 2.04 |
| 24" O.C. | .016 | Ea. | .014 | 1.00 | .36 | 1.36 |
| 36' span, 16" O.C. | .022 | Ea. | .021 | 1.60 | .53 | 2.13 |
| 24" O.C. | .015 | Ea. | .014 | 1.09 | .37 | 1.46 |
| Fascia board, #2 or better, 1" x 6" | .100 | L.F. | .003 | .03 | .08 | .11 |
| 1" x 8" | .100 | L.F. | .004 | .03 | .10 | .13 |
| 1" x 10" | .100 | L.F. | .005 | .04 | .10 | .14 |
| 2" x 6" | .100 | L.F. | .005 | .04 | .12 | .16 |
| 2" x 8" | .100 | L.F. | .007 | .05 | .15 | .20 |
| 2" x 10" | .100 | L.F. | .008 | .08 | .18 | .26 |
| Sheathing, plywood CDX, 4/12 pitch, 3/8" thick | 1.170 | S.F. | .011 | .37 | .25 | .62 |
| 1/2" thick | 1.170 | S.F. | .012 | .48 | .27 | .75 |
| 5/8" thick | 1.170 | S.F. | .014 | .53 | .29 | .82 |
| 8/12 pitch, 3/8" thick | 1.330 | S.F. | .013 | .43 | .27 | .70 |
| 1/2" thick | 1.330 | S.F. | .014 | .55 | .30 | .85 |
| 5/8" thick | 1.330 | S.F. | .015 | .60 | .33 | .93 |
| Boards, 4/12 pitch, 1" x 6" | 1.170 | S.F. | .025 | .94 | .53 | 1.47 |
| 1" x 8" | 1.170 | S.F. | .021 | .94 | .44 | 1.38 |
| 8/12 pitch, 1" x 6" | 1.330 | S.F. | .029 | 1.06 | .62 | 1.68 |
| 1" x 8" | 1.330 | S.F. | .023 | 1.06 | .51 | 1.57 |
| Furring, 1" x 3", 12" O.C. | 1.200 | L.F. | .027 | .12 | .54 | .66 |
| 16" O.C. | 1.000 | L.F. | .023 | .10 | .45 | .55 |
| 24" O.C. | .800 | L.F. | .018 | .08 | .36 | .44 |

43

# 3 | FRAMING — 20 | Hip Roof Framing System

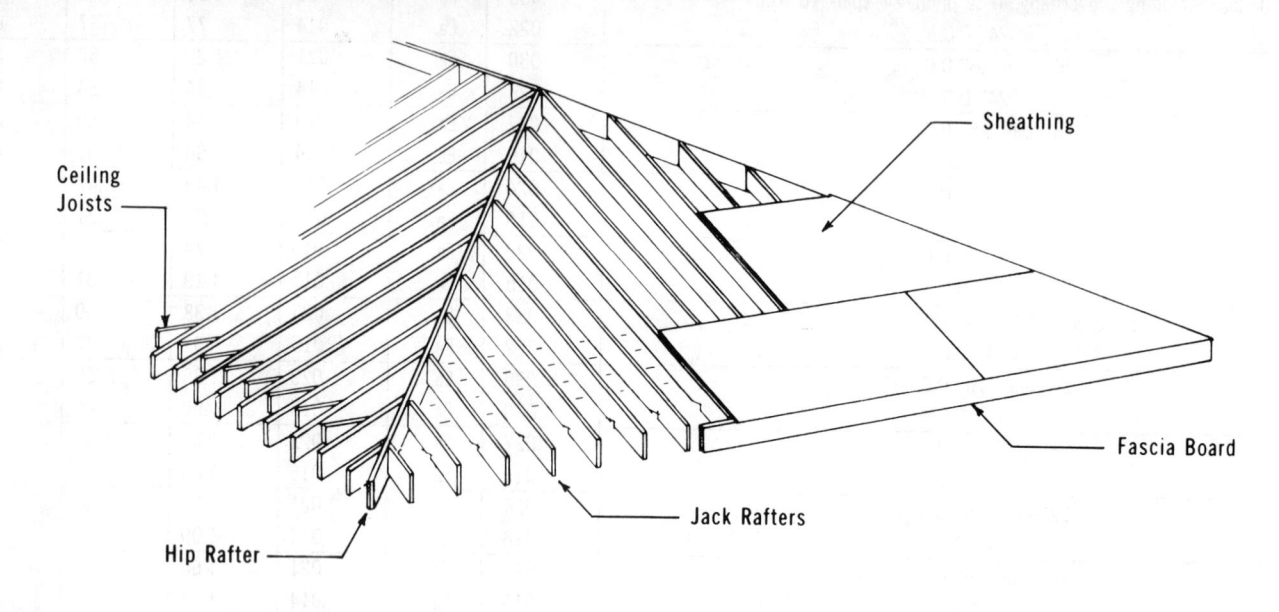

| SYSTEM DESCRIPTION | QUAN. | UNIT | MAN HOURS | MAT. | INST. | TOTAL |
|---|---|---|---|---|---|---|
| **2" X 8", 16" O.C., 4/12 PITCH** | | | | | | |
| Hip rafters, 2" x 8", 4/12 pitch | .160 | L.F. | .004 | .09 | .08 | .17 |
| Jack rafters, 2" x 8", 16" O.C., 4/12 pitch | 1.430 | L.F. | .047 | .79 | .95 | 1.74 |
| Ceiling joists, 2" x 6", 16" O.C. | 1.000 | L.F. | .015 | .40 | .31 | .71 |
| Fascia board, 2" x 8" | .220 | L.F. | .015 | .12 | .32 | .44 |
| Soffit nailer (outrigger), 2" x 4", 24" O.C. | .220 | L.F. | .005 | .09 | .10 | .19 |
| Sheathing, 1/2" exterior plywood, CDX | 1.570 | S.F. | .017 | .64 | .36 | 1.00 |
| Furring strips, 1" x 3", 16" O.C. | 1.000 | L.F. | .023 | .10 | .45 | .55 |
| TOTAL | | | .126 | 2.23 | 2.57 | 4.80 |
| **2" X 6", 16" O.C., 4/12 PITCH** | | | | | | |
| Hip rafters, 2" x 6", 4/12 pitch | .160 | L.F. | .003 | .06 | .07 | .13 |
| Jack rafters, 2" x 6", 16" O.C., 4/12 pitch | 1.430 | L.F. | .038 | .57 | .77 | 1.34 |
| Ceiling joists, 2" x 4", 16" O.C. | 1.000 | L.F. | .013 | .28 | .26 | .54 |
| Fascia board, 2" x 6" | .220 | L.F. | .012 | .09 | .25 | .34 |
| Soffit nailer (outrigger), 2" x 4", 24" O.C. | .220 | L.F. | .005 | .09 | .10 | .19 |
| Sheathing, 1/2" exterior plywood, CDX | 1.570 | S.F. | .017 | .64 | .36 | 1.00 |
| Furring strips, 1" x 3", 16" O.C. | 1.000 | L.F. | .023 | .10 | .45 | .55 |
| TOTAL | | | .111 | 1.83 | 2.26 | 4.09 |

The cost of this system is based on S.F. of plan area. Measurement is the area under the hip roof only. See gable roof system for added costs.

| DESCRIPTION | QUAN. | UNIT | MAN HOURS | MAT. | INST. | TOTAL |
|---|---|---|---|---|---|---|
| | | | | | | |
| | | | | | | |
| | | | | | | |
| | | | | | | |
| | | | | | | |
| | | | | | | |

# Hip Roof Framing Price Sheet

| | QUAN. | UNIT | MAN HOURS | MAT. | INST. | TOTAL |
|---|---|---|---|---|---|---|
| Hip rafters, #2 or better, 2" x 6", 4/12 pitch | .160 | L.F. | .003 | .06 | .07 | .13 |
| 8/12 pitch | .210 | L.F. | .005 | .08 | .12 | .20 |
| 2" x 8", 4/12 pitch | .160 | L.F. | .004 | .09 | .08 | .17 |
| 8/12 pitch | .210 | L.F. | .006 | .12 | .13 | .25 |
| 2" x 10", 4/12 pitch | .160 | L.F. | .004 | .12 | .09 | .21 |
| 8/12 pitch roof | .210 | L.F. | .007 | .16 | .15 | .31 |
| Jack rafters, #2 or better, 16" O.C., 2" x 6", 4/12 pitch | 1.430 | L.F. | .038 | .57 | .77 | 1.34 |
| 8/12 pitch | 1.800 | L.F. | .061 | .72 | 1.24 | 1.96 |
| 2" x 8", 4/12 pitch | 1.430 | L.F. | .047 | .79 | .95 | 1.74 |
| 8/12 pitch | 1.800 | L.F. | .075 | .99 | 1.53 | 2.52 |
| 2" x 10", 4/12 pitch | 1.430 | L.F. | .051 | 1.07 | 1.05 | 2.12 |
| 8/12 pitch | 1.800 | L.F. | .082 | 1.35 | 1.69 | 3.04 |
| 24" O.C., 2" x 6", 4/12 pitch | 1.150 | L.F. | .031 | .46 | .62 | 1.08 |
| 8/12 pitch | 1.440 | L.F. | .048 | .58 | .99 | 1.57 |
| 2" x 8", 4/12 pitch | 1.150 | L.F. | .037 | .63 | .77 | 1.40 |
| 8/12 pitch | 1.440 | L.F. | .060 | .79 | 1.23 | 2.02 |
| 2" x 10", 4/12 pitch | 1.150 | L.F. | .041 | .86 | .84 | 1.70 |
| 8/12 pitch | 1.440 | L.F. | .066 | 1.08 | 1.35 | 2.43 |
| Ceiling joists, #2 or better, 2" x 4", 16" O.C. | 1.000 | L.F. | .013 | .28 | .26 | .54 |
| 24" O.C. | .750 | L.F. | .009 | .21 | .20 | .41 |
| 2" x 6", 16" O.C. | 1.000 | L.F. | .015 | .40 | .31 | .71 |
| 24" O.C. | .750 | L.F. | .011 | .30 | .23 | .53 |
| 2" x 8", 16" O.C. | 1.000 | L.F. | .017 | .55 | .35 | .90 |
| 24" O.C. | .750 | L.F. | .012 | .41 | .27 | .68 |
| 2" x 10", 16" O.C. | 1.000 | L.F. | .018 | .77 | .36 | 1.13 |
| 24" O.C. | .750 | L.F. | .013 | .58 | .27 | .85 |
| Fascia board, #2 or better, 1" x 6" | .220 | L.F. | .008 | .07 | .17 | .24 |
| 1" x 8" | .220 | L.F. | .010 | .08 | .20 | .28 |
| 1" x 10" | .220 | L.F. | .011 | .09 | .23 | .32 |
| 2" x 6" | .220 | L.F. | .012 | .10 | .25 | .35 |
| 2" x 8" | .220 | L.F. | .015 | .12 | .32 | .44 |
| 2" x 10" | .220 | L.F. | .019 | .17 | .40 | .57 |
| Soffit nailer (outrigger), 2" x 4", 16" O.C. | .280 | L.F. | .006 | .11 | .13 | .24 |
| 24" O.C. | .220 | L.F. | .005 | .09 | .10 | .19 |
| 2" x 8", 16" O.C. | .280 | L.F. | .008 | .11 | .18 | .29 |
| 24" O.C. | .220 | L.F. | .006 | .09 | .14 | .23 |
| Sheathing, plywood CDX, 4/12 pitch, 3/8" thick | 1.570 | S.F. | .015 | .50 | .33 | .83 |
| 1/2" thick | 1.570 | S.F. | .017 | .64 | .36 | 1.00 |
| 5/8" thick | 1.570 | S.F. | .018 | .71 | .39 | 1.10 |
| 8/12 pitch, 3/8" thick | 1.900 | S.F. | .019 | .61 | .40 | 1.01 |
| 1/2" thick | 1.900 | S.F. | .020 | .78 | .44 | 1.22 |
| 5/8" thick | 1.900 | S.F. | .022 | .86 | .47 | 1.33 |
| Boards, 4/12 pitch, 1" x 6" boards | 1.450 | S.F. | .031 | 1.16 | .67 | 1.83 |
| 1" x 8" boards | 1.450 | S.F. | .026 | 1.16 | .55 | 1.71 |
| 8/12 pitch, 1" x 6" boards | 1.750 | S.F. | .038 | 1.40 | .81 | 2.21 |
| 1" x 8" boards | 1.750 | S.F. | .031 | 1.40 | .67 | 2.07 |
| Furring, 1" x 3", 12" O.C. | 1.200 | L.F. | .027 | .12 | .54 | .66 |
| 16" O.C. | 1.000 | L.F. | .023 | .10 | .45 | .55 |
| 24" O.C. | .800 | L.F. | .018 | .08 | .36 | .44 |

# 3 | FRAMING
## 24 | Gambrel Roof Framing System

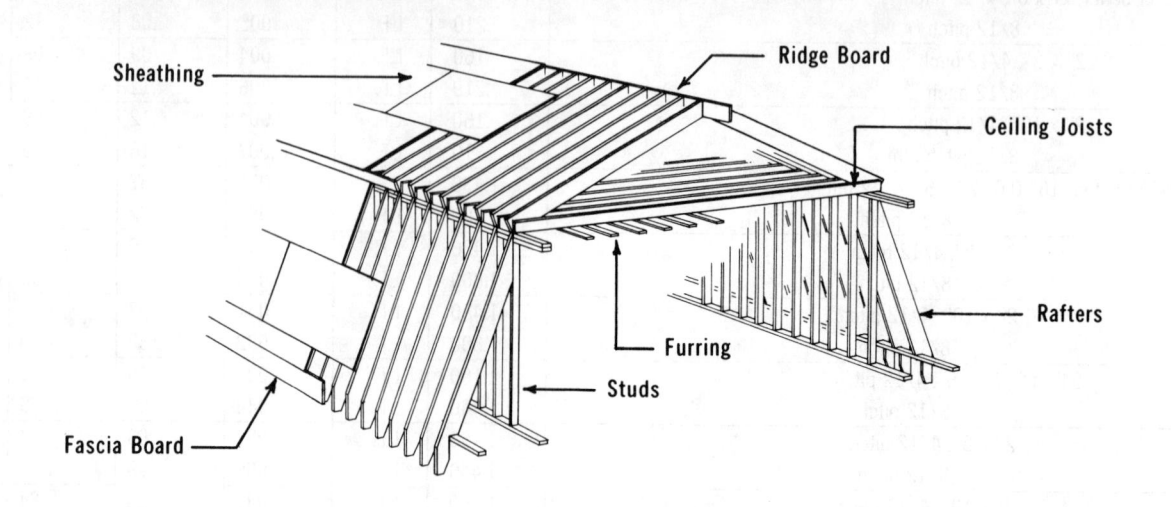

*Diagram labels: Sheathing, Ridge Board, Ceiling Joists, Rafters, Furring, Studs, Fascia Board*

| SYSTEM DESCRIPTION | QUAN. | UNIT | MAN HOURS | MAT. | INST. | TOTAL |
|---|---|---|---|---|---|---|
| **2" X 6" RAFTERS, 16" O.C.** | | | | | | |
| Roof rafters, 2" x 6", 16" O.C. | 1.430 | L.F. | .034 | .57 | .69 | 1.26 |
| Ceiling joists, 2" x 6", 16" O.C. | .710 | L.F. | .010 | .28 | .22 | .50 |
| Stud wall, 2" x 4", 16" O.C., including plates | .790 | L.F. | .012 | .20 | .26 | .46 |
| Furring strips, 1" x 3", 16" O.C. | .710 | L.F. | .016 | .07 | .32 | .39 |
| Ridge board, 2" x 8" | .050 | L.F. | .001 | .03 | .02 | .05 |
| Fascia board, 2" x 6" | .100 | L.F. | .005 | .04 | .12 | .16 |
| Sheathing, exterior grade plywood, 1/2" thick | 1.450 | S.F. | .015 | .59 | .34 | .93 |
| TOTAL | | | .093 | 1.78 | 1.97 | 3.75 |
| **2" X 8" RAFTERS, 16" O.C.** | | | | | | |
| Roof rafters, 2" x 8", 16" O.C. | 1.430 | L.F. | .041 | .79 | .84 | 1.63 |
| Ceiling joists, 2" x 6", 16" O.C. | .710 | L.F. | .010 | .28 | .22 | .50 |
| Stud wall, 2" x 4", 16" O.C., including plates | .790 | L.F. | .012 | .20 | .26 | .46 |
| Furring strips, 1" x 3", 16" O.C. | .710 | L.F. | .016 | .07 | .32 | .39 |
| Ridge board, 2" x 8" | .050 | L.F. | .001 | .03 | .02 | .05 |
| Fascia board, 2" x 8" | .100 | L.F. | .007 | .05 | .15 | .20 |
| Sheathing, exterior grade plywood, 1/2" thick | 1.450 | S.F. | .015 | .59 | .34 | .93 |
| TOTAL | | | .102 | 2.01 | 2.15 | 4.16 |

The cost of this system is based on the square foot of plan area on the first floor.

| DESCRIPTION | QUAN. | UNIT | MAN HOURS | MAT. | INST. | TOTAL |
|---|---|---|---|---|---|---|
| | | | | | | |
| | | | | | | |
| | | | | | | |
| | | | | | | |
| | | | | | | |
| | | | | | | |

# Gambrel Roof Framing Price Sheet

| | QUAN. | UNIT | MAN HOURS | COST PER S.F. MAT. | COST PER S.F. INST. | TOTAL |
|---|---|---|---|---|---|---|
| Roof rafters, #2 or better, 2" x 6", 16" O.C. | 1.430 | L.F. | .034 | .57 | .69 | 1.26 |
| 24" O.C. | 1.140 | L.F. | .027 | .46 | .54 | 1.00 |
| 2" x 8", 16" O.C. | 1.430 | L.F. | .041 | .79 | .84 | 1.63 |
| 24" O.C. | 1.140 | L.F. | .033 | .63 | .67 | 1.30 |
| 2" x 10", 16" O.C. | 1.430 | L.F. | .045 | 1.07 | .95 | 2.02 |
| 24" O.C. | 1.140 | L.F. | .036 | .86 | .75 | 1.61 |
| Ceiling joist, #2 or better, 2" x 4", 16" O.C. | .710 | L.F. | .009 | .20 | .18 | .38 |
| 24" O.C. | .570 | L.F. | .007 | .16 | .15 | .31 |
| 2" x 6", 16" O.C. | .710 | L.F. | .010 | .28 | .22 | .50 |
| 24" O.C. | .570 | L.F. | .008 | .23 | .17 | .40 |
| 2" x 8", 16" O.C. | .710 | L.F. | .012 | .39 | .25 | .64 |
| 24" O.C. | .570 | L.F. | .009 | .31 | .20 | .51 |
| Stud wall, #2 or better, 2" x 4", 16" O.C. | .790 | L.F. | .012 | .20 | .26 | .46 |
| 24" O.C. | .630 | L.F. | .010 | .16 | .21 | .37 |
| 2" x 6", 16" O.C. | .790 | L.F. | .014 | .32 | .28 | .60 |
| 24" O.C. | .630 | L.F. | .011 | .25 | .23 | .48 |
| Furring, 1" x 3", 16" O.C. | .710 | L.F. | .016 | .07 | .32 | .39 |
| 24" O.C. | .590 | L.F. | .013 | .06 | .26 | .32 |
| Ridge board, #2 or better, 1" x 6" | .050 | L.F. | | .01 | .01 | .02 |
| 1" x 8" | .050 | L.F. | | .01 | .02 | .03 |
| 1" x 10" | .050 | L.F. | | .02 | .02 | .04 |
| 2" x 6" | .050 | L.F. | | .02 | .02 | .04 |
| 2" x 8" | .050 | L.F. | .001 | .03 | .02 | .05 |
| 2" x 10" | .050 | L.F. | .001 | .04 | .02 | .06 |
| Fascia board, #2 or better, 1" x 6" | .100 | L.F. | .003 | .03 | .08 | .11 |
| 1" x 8" | .100 | L.F. | .004 | .03 | .10 | .13 |
| 1" x 10" | .100 | L.F. | .005 | .04 | .10 | .14 |
| 2" x 6" | .100 | L.F. | .005 | .04 | .12 | .16 |
| 2" x 8" | .100 | L.F. | .007 | .05 | .15 | .20 |
| 2" x 10" | .100 | L.F. | .008 | .08 | .18 | .26 |
| Sheathing, plywood, exterior grade CDX, 3/8" thick | 1.450 | S.F. | .014 | .46 | .31 | .77 |
| 1/2" thick | 1.450 | S.F. | .015 | .59 | .34 | .93 |
| 5/8" thick | 1.450 | S.F. | .017 | .65 | .37 | 1.02 |
| 3/4" thick | 1.450 | S.F. | .018 | .74 | .39 | 1.13 |
| Boards, 1" x 6", laid regular | 1.450 | S.F. | .031 | 1.16 | .67 | 1.83 |
| Laid diagonal | 1.450 | S.F. | .036 | 1.16 | .74 | 1.90 |
| 1" x 8", laid regular | 1.450 | S.F. | .026 | 1.16 | .55 | 1.71 |
| Laid diagonal | 1.450 | S.F. | .031 | 1.16 | .67 | 1.83 |

# 3 | FRAMING — 28 | Mansard Roof Framing System

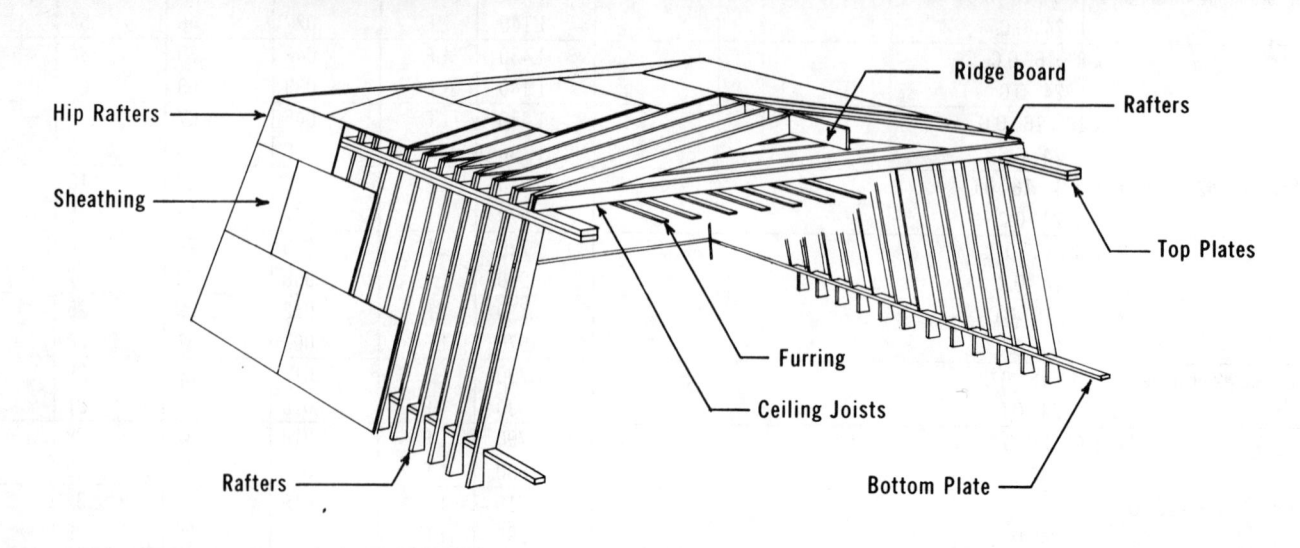

*Diagram labels: Hip Rafters, Sheathing, Rafters, Ridge Board, Rafters, Top Plates, Furring, Ceiling Joists, Bottom Plate*

| SYSTEM DESCRIPTION | QUAN. | UNIT | MAN HOURS | MAT. | INST. | TOTAL |
|---|---|---|---|---|---|---|
| **2" X 6" RAFTERS, 16" O.C.** | | | | | | |
| Roof rafters, 2" x 6", 16" O.C. | 1.210 | L.F. | .032 | .48 | .67 | 1.15 |
| Rafter plates, 2" x 6", double top, single bottom | .364 | L.F. | .009 | .15 | .20 | .35 |
| Ceiling joists, 2" x 4", 16" O.C. | .920 | L.F. | .011 | .26 | .24 | .50 |
| Hip rafter, 2" x 6" | .070 | L.F. | .002 | .03 | .04 | .07 |
| Jack rafter, 2" x 6", 16" O.C. | 1.000 | L.F. | .039 | .40 | .80 | 1.20 |
| Ridge board, 2" x 6" | .018 | L.F. | | .01 | | .01 |
| Sheathing, exterior grade plywood, 1/2" thick | 2.210 | S.F. | .024 | .91 | .50 | 1.41 |
| Furring strips, 1" x 3", 16" O.C. | .920 | L.F. | .021 | .09 | .42 | .51 |
| TOTAL | | | .138 | 2.33 | 2.87 | 5.20 |
| **2" X 8" RAFTERS, 16" O.C.** | | | | | | |
| Roof rafters, 2" x 8", 16" O.C. | 1.210 | L.F. | .039 | .67 | .82 | 1.49 |
| Rafter plates, 2" x 8", double top, single bottom | .364 | L.F. | .012 | .20 | .25 | .45 |
| Ceiling joists, 2" x 6", 16" O.C. | .920 | L.F. | .013 | .37 | .28 | .65 |
| Hip rafter, 2" x 8" | .070 | L.F. | .002 | .04 | .05 | .09 |
| Jack rafter, 2" x 8", 16" O.C. | 1.000 | L.F. | .048 | .55 | .98 | 1.53 |
| Ridge board, 2" x 8" | .018 | L.F. | | .01 | .01 | .02 |
| Sheathing, exterior grade plywood, 1/2" thick | 2.210 | S.F. | .024 | .91 | .50 | 1.41 |
| Furring strips, 1" x 3", 16" O.C. | .920 | L.F. | .021 | .09 | .42 | .51 |
| TOTAL | | | .159 | 2.84 | 3.31 | 6.15 |

The cost of this system is based on the square foot of plan area.

| DESCRIPTION | QUAN. | UNIT | MAN HOURS | MAT. | INST. | TOTAL |
|---|---|---|---|---|---|---|
| | | | | | | |
| | | | | | | |
| | | | | | | |
| | | | | | | |
| | | | | | | |
| | | | | | | |

# Mansard Roof Framing Price Sheet

| | QUAN. | UNIT | MAN HOURS | MAT. | INST. | TOTAL |
|---|---|---|---|---|---|---|
| Roof rafters, #2 or better, 2" x 6", 16" O.C. | 1.210 | L.F. | .032 | .48 | .67 | 1.15 |
| 24" O.C. | .970 | L.F. | .026 | .39 | .53 | .92 |
| 2" x 8", 16" O.C. | 1.210 | L.F. | .039 | .67 | .82 | 1.49 |
| 24" O.C. | .970 | L.F. | .032 | .53 | .66 | 1.19 |
| 2" x 10", 16" O.C. | 1.210 | L.F. | .045 | .91 | .93 | 1.84 |
| 24" O.C. | .970 | L.F. | .036 | .73 | .74 | 1.47 |
| Rafter plates, #2 or better double top, single bottom, 2" x 6" | .364 | L.F. | .009 | .15 | .20 | .35 |
| 2" x 8" | .364 | L.F. | .012 | .20 | .25 | .45 |
| 2" x 10" | .364 | L.F. | .013 | .27 | .28 | .55 |
| Ceiling joist, #2 or better, 2" x 4", 16" O.C. | .092 | L.F. | .011 | .26 | .24 | .50 |
| 24" O.C. | .740 | L.F. | .009 | .21 | .19 | .40 |
| 2" x 6", 16" O.C. | .920 | L.F. | .013 | .37 | .28 | .65 |
| 24" O.C. | .740 | L.F. | .011 | .30 | .23 | .53 |
| 2" x 8", 16" O.C. | .920 | L.F. | .015 | .51 | .32 | .83 |
| 24" O.C. | .740 | L.F. | .012 | .41 | .26 | .67 |
| Hip rafter, #2 or better, 2" x 6" | .070 | L.F. | .002 | .03 | .04 | .07 |
| 2" x 8" | .070 | L.F. | .002 | .04 | .05 | .09 |
| 2" x 10" | .070 | L.F. | .002 | .05 | .06 | .11 |
| Jack rafter, #2 or better, 2" x 6", 16" O.C. | 1.000 | L.F. | .039 | .40 | .80 | 1.20 |
| 24" O.C. | .800 | L.F. | .031 | .32 | .64 | .96 |
| 2" x 8", 16" O.C. | 1.000 | L.F. | .048 | .55 | .98 | 1.53 |
| 24" O.C. | .800 | L.F. | .038 | .44 | .78 | 1.22 |
| Ridge board, #2 or better, 1" x 6" | .018 | L.F. | | .01 | | .01 |
| 1" x 8" | .018 | L.F. | | .01 | | .01 |
| 1" x 10" | .018 | L.F. | | .01 | | .01 |
| 2" x 6" | .018 | L.F. | | .01 | | .01 |
| 2" x 8" | .018 | L.F. | | .01 | .01 | .02 |
| 2" x 10" | .018 | L.F. | | .01 | .01 | .02 |
| Sheathing, plywood exterior grade CDX, 3/8" thick | 2.210 | S.F. | .022 | .71 | .46 | 1.17 |
| 1/2" thick | 2.210 | S.F. | .024 | .91 | .50 | 1.41 |
| 5/8" thick | 2.210 | S.F. | .026 | .99 | .56 | 1.55 |
| 3/4" thick | 2.210 | S.F. | .028 | 1.13 | .59 | 1.72 |
| Boards, 1" x 6", laid regular | 2.210 | S.F. | .048 | 1.77 | 1.01 | 2.78 |
| Laid diagonal | 2.210 | S.F. | .055 | 1.77 | 1.13 | 2.90 |
| 1" x 8", laid regular | 2.210 | S.F. | .039 | 1.77 | .84 | 2.61 |
| Laid diagonal | 2.210 | S.F. | .048 | 1.77 | 1.01 | 2.78 |
| Furring, 1" x 3", 12" O.C. | 1.150 | L.F. | .026 | .12 | .51 | .63 |
| 24" O.C. | .740 | L.F. | .017 | .07 | .34 | .41 |

# Mansard Roof Framing Price Sheet

| | QUAN. | UNIT | MAN HOURS | MAT. | INST. | TOTAL |
|---|---|---|---|---|---|---|
| Roof rafters, #2 or better, 2" x 6", 16" O.C. | | | | | | |

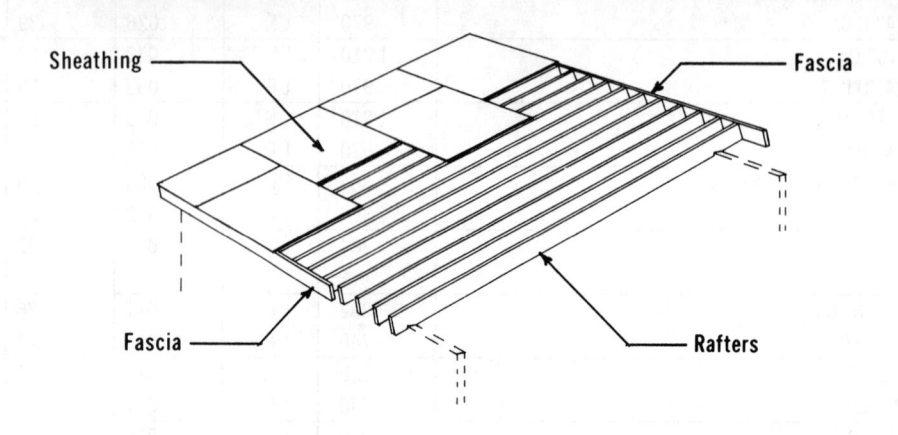

# 3 | FRAMING  32 | Shed/Flat Roof Framing System

| SYSTEM DESCRIPTION | QUAN. | UNIT | MAN HOURS | MAT. | INST. | TOTAL |
|---|---|---|---|---|---|---|
| **2" X 6", 16" O.C., 4/12 PITCH** | | | | | | |
| Rafters, 2" x 6", 16" O.C., 4/12 pitch | 1.170 | L.F. | .022 | .47 | .44 | .91 |
| Fascia, 2" x 6" | .100 | L.F. | .005 | .04 | .12 | .16 |
| Bridging, 1" x 3", 6' O.C. | .080 | Pr. | .004 | .03 | .10 | .13 |
| Sheathing, exterior grade plywood, 1/2" thick | 1.230 | S.F. | .013 | .50 | .29 | .79 |
| TOTAL | | | .044 | 1.04 | .95 | 1.99 |
| **2" X 6", 24" O.C., 4/12 PITCH** | | | | | | |
| Rafters, 2" x 6", 24" O.C., 4/12 pitch | .940 | L.F. | .017 | .38 | .35 | .73 |
| Fascia, 2" x 6" | .100 | L.F. | .005 | .04 | .12 | .16 |
| Bridging, 1" x 3", 6' O.C. | .060 | Pr. | .003 | .03 | .07 | .10 |
| Sheathing, exterior grade plywood, 1/2" thick | 1.230 | S.F. | .013 | .50 | .29 | .79 |
| TOTAL | | | .038 | .95 | .83 | 1.78 |
| **2" X 8", 16" O.C., 4/12 PITCH** | | | | | | |
| Rafters, 2" x 8", 16" O.C., 4/12 pitch | 1.170 | L.F. | .025 | .64 | .54 | 1.18 |
| Fascia, 2" x 8" | .100 | L.F. | .007 | .05 | .15 | .20 |
| Bridging, 1" x 3", 6' O.C. | .080 | Pr. | .004 | .03 | .10 | .13 |
| Sheathing, exterior grade plywood, 1/2" thick | 1.230 | S.F. | .013 | .50 | .29 | .79 |
| TOTAL | | | .049 | 1.22 | 1.08 | 2.30 |
| **2" X 8", 24" O.C., 4/12 PITCH** | | | | | | |
| Rafters, 2" x 8", 24" O.C., 4/12 pitch | .940 | L.F. | .020 | .52 | .43 | .95 |
| Fascia, 2" x 8" | .100 | L.F. | .007 | .05 | .15 | .20 |
| Bridging, 1" x 3", 6' O.C. | .060 | Pr. | .003 | .03 | .07 | .10 |
| Sheathing, exterior grade plywood, 1/2" thick | 1.230 | S.F. | .013 | .50 | .29 | .79 |
| TOTAL | | | .044 | 1.10 | .94 | 2.04 |

The cost of this system is based on the square foot of plan area. A 1' overhang is assumed. No ceiling joists or furring are included.

| DESCRIPTION | QUAN. | UNIT | MAN HOURS | MAT. | INST. | TOTAL |
|---|---|---|---|---|---|---|
| | | | | | | |
| | | | | | | |
| | | | | | | |

# Shed/Flat Roof Framing Price Sheet

| Description | QUAN. | UNIT | MAN HOURS | MAT. | INST. | TOTAL |
|---|---|---|---|---|---|---|
| Rafters, #2 or better, 16" O.C., 2" x 4", 0 - 4/12 pitch | 1.170 | L.F. | .016 | .35 | .33 | .68 |
| 5/12 - 8/12 pitch | 1.330 | L.F. | .024 | .40 | .48 | .88 |
| 2" x 6", 0 - 4/12 pitch | 1.170 | L.F. | .022 | .47 | .44 | .91 |
| 5/12 - 8/12 pitch | 1.330 | L.F. | .031 | .53 | .64 | 1.17 |
| 2" x 8", 0 - 4/12 pitch | 1.170 | L.F. | .025 | .64 | .54 | 1.18 |
| 5/12 - 8/12 pitch | 1.330 | L.F. | .038 | .73 | .79 | 1.52 |
| 2" x 10", 0 - 4/12 pitch | 1.170 | L.F. | .029 | .88 | .61 | 1.49 |
| 5/12 - 8/12 pitch | 1.330 | L.F. | .042 | 1.00 | .88 | 1.88 |
| 24" O.C., 2" x 4", 0 - 4/12 pitch | .940 | L.F. | .013 | .28 | .27 | .55 |
| 5/12 - 8/12 pitch | 1.060 | L.F. | .019 | .32 | .38 | .70 |
| 2" x 6", 0 - 4/12 pitch | .940 | L.F. | .017 | .38 | .35 | .73 |
| 5/12 - 8/12 pitch | 1.060 | L.F. | .025 | .42 | .51 | .93 |
| 2" x 8", 0 - 4/12 pitch | .940 | L.F. | .020 | .52 | .43 | .95 |
| 5/12 - 8/12 pitch | 1.060 | L.F. | .030 | .58 | .63 | 1.21 |
| 2" x 10", 0 - 4/12 pitch | .940 | L.F. | .023 | .71 | .48 | 1.19 |
| 5/12 - 8/12 pitch | 1.060 | L.F. | .033 | .80 | .69 | 1.49 |
| Fascia, #2 or better, 1" x 4" | .100 | L.F. | .002 | .02 | .06 | .08 |
| 1" x 6" | .100 | L.F. | .003 | .03 | .08 | .11 |
| 1" x 8" | .100 | L.F. | .004 | .03 | .10 | .13 |
| 1" x 10" | .100 | L.F. | .005 | .04 | .10 | .14 |
| 2" x 4" | .100 | L.F. | .004 | .04 | .10 | .14 |
| 2" x 6" | .100 | L.F. | .005 | .04 | .12 | .16 |
| 2" x 8" | .100 | L.F. | .007 | .05 | .15 | .20 |
| 2" x 10" | .100 | L.F. | .008 | .08 | .18 | .26 |
| Bridging, wood 6' O.C., 1" x 3", rafters, 16" O.C. | .080 | Pr. | .004 | .03 | .10 | .13 |
| 24" O.C. | .060 | Pr. | .003 | .03 | .07 | .10 |
| Metal, galvanized, rafters, 16" O.C. | .080 | Pr. | .004 | .08 | .09 | .17 |
| 24" O.C. | .060 | Pr. | .003 | .08 | .07 | .15 |
| Compression type, rafters, 16" O.C. | .080 | Pr. | .003 | .07 | .07 | .14 |
| 24" O.C. | .060 | Pr. | .002 | .06 | .04 | .10 |
| Sheathing, plywood, exterior grade, 3/8" thick, flat 0 - 4/12 pitch | 1.230 | S.F. | .012 | .39 | .26 | .65 |
| 5/12 - 8/12 pitch | 1.330 | S.F. | .013 | .43 | .27 | .70 |
| 1/2" thick, flat 0 - 4/12 pitch | 1.230 | S.F. | .013 | .50 | .29 | .79 |
| 5/12 - 8/12 pitch | 1.330 | S.F. | .014 | .55 | .30 | .85 |
| 5/8" thick, flat 0 - 4/12 pitch | 1.230 | S.F. | .014 | .55 | .31 | .86 |
| 5/12 - 8/12 pitch | 1.330 | S.F. | .015 | .60 | .33 | .93 |
| 3/4" thick, flat 0 - 4/12 pitch | 1.230 | S.F. | .015 | .63 | .33 | .96 |
| 5/12 - 8/12 pitch | 1.330 | S.F. | .017 | .68 | .36 | 1.04 |
| Boards, 1" x 6", laid regular, flat 0 - 4/12 pitch | 1.230 | S.F. | .027 | .98 | .57 | 1.55 |
| 5/12 - 8/12 pitch | 1.330 | S.F. | .041 | 1.06 | .84 | 1.90 |
| Laid diagonal, flat 0 - 4/12 pitch | 1.230 | S.F. | .030 | .98 | .63 | 1.61 |
| 5/12 - 8/12 pitch | 1.330 | S.F. | .043 | 1.06 | .92 | 1.98 |
| 1" x 8", laid regular, flat 0 - 4/12 pitch | 1.230 | S.F. | .022 | .98 | .47 | 1.45 |
| 5/12 - 8/12 pitch | 1.330 | S.F. | .033 | 1.06 | .70 | 1.76 |
| Laid diagonal, flat 0 - 4/12 pitch | 1.230 | S.F. | .027 | .98 | .57 | 1.55 |
| 5/12 - 8/12 pitch | 1.330 | S.F. | .035 | 1.06 | .75 | 1.81 |

# 3 | FRAMING   36 | Flat Roof Systems (Steel)

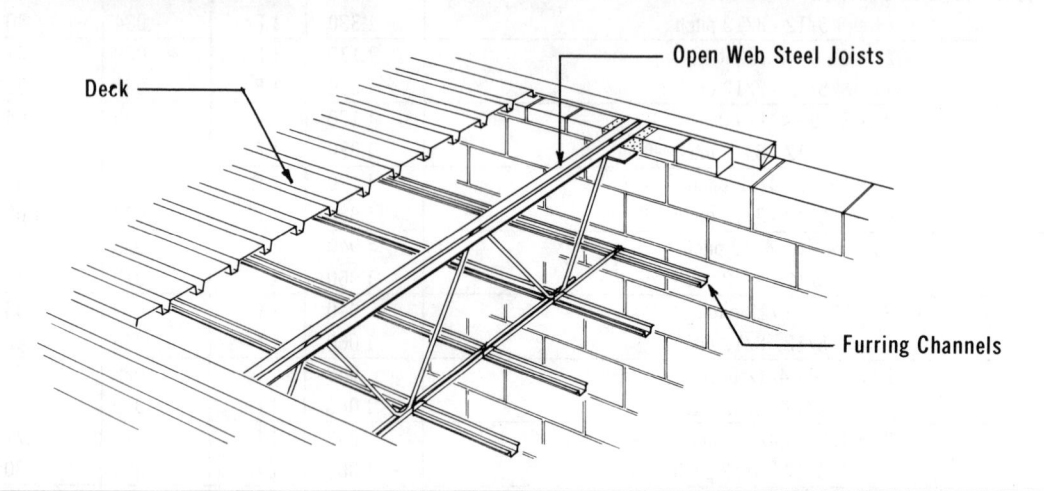

*Deck — Open Web Steel Joists — Furring Channels*

| SYSTEM DESCRIPTION | QUAN. | UNIT | MAN HOURS | MAT. | INST. | TOTAL |
|---|---|---|---|---|---|---|
| **20' SPAN, 40 PSF SUPERIMPOSED LOAD** | | | | | | |
| Open web steel joists, 14" deep, 6.5#/L.F., 5' O.C. | .200 | L.F. | .008 | .43 | .32 | .75 |
| Decking, open type, 1-1/2" deep, 22 gauge | 1.100 | S.F. | .007 | .85 | .20 | 1.05 |
| Ceiling furring, 3/4" channels, galvanized, 24" O.C. | 1.000 | S.F. | .019 | .19 | .36 | .55 |
| TOTAL | | | .034 | 1.47 | .88 | 2.35 |
| **25' SPAN, 40 PSF SUPERIMPOSED LOAD** | | | | | | |
| Open web steel joists, 18" deep, 8.0#/L.F., 5' O.C. | .200 | L.F. | .007 | .53 | .28 | .81 |
| Decking, open type, 1-1/2" deep, 22 gauge | 1.100 | S.F. | .007 | .85 | .20 | 1.05 |
| Ceiling furring, 3/4" channels, galvanized, 24" O.C. | 1.000 | S.F. | .019 | .19 | .36 | .55 |
| TOTAL | | | .033 | 1.57 | .84 | 2.41 |
| **30' SPAN, 40 PSF SUPERIMPOSED LOAD** | | | | | | |
| Open web steel joists, 22" deep, 9.7#/L.F., 5' O.C. | .200 | L.F. | .006 | .64 | .26 | .90 |
| Decking, open type, 1-1/2" deep, 22 gauge | 1.100 | S.F. | .007 | .85 | .20 | 1.05 |
| Ceiling furring, 3/4" channels, galvanized, 24" O.C. | 1.000 | S.F. | .019 | .19 | .36 | .55 |
| TOTAL | | | .032 | 1.68 | .82 | 2.50 |
| **35' SPAN, 40 PSF SUPERIMPOSED LOAD** | | | | | | |
| Open web steel joists, 24" deep, 11.5#/L.F., 5' O.C. | .200 | L.F. | .006 | .76 | .24 | 1.00 |
| Decking, open type, 1-1/2" deep, 22 gauge | 1.100 | S.F. | .007 | .85 | .20 | 1.05 |
| Ceiling furring, 3/4" channels, galvanized, 24" O.C. | 1.000 | S.F. | .019 | .19 | .36 | .55 |
| TOTAL | | | .032 | 1.80 | .80 | 2.60 |

| DESCRIPTION | QUAN. | UNIT | MAN HOURS | MAT. | INST. | TOTAL |
|---|---|---|---|---|---|---|
| | | | | | | |
| | | | | | | |
| | | | | | | |

# Flat Roof Price Sheet (Steel)

| | QUAN. | UNIT | MAN HOURS | COST PER S.F. MAT. | COST PER S.F. INST. | TOTAL |
|---|---|---|---|---|---|---|
| Joists 5' O.C., 15' span, 20 psf live load, 10" deep, 5.0#/L.F. | .200 | L.F. | .010 | .33 | .40 | .73 |
| 30 psf superimposed load, 10" deep, 5.0#/L.F. | .200 | L.F. | .010 | .33 | .40 | .73 |
| 40 psf superimposed load, 10" deep, 5.0#L.F. | .200 | L.F. | .010 | .33 | .40 | .73 |
| 50 psf superimposed load, 12" deep, 5.2#/L.F. | .200 | L.F. | .010 | .34 | .40 | .74 |
| 20' span, 20 psf superimposed load, 12" deep, 5.2#/L.F. | .200 | L.F. | .008 | .34 | .32 | .66 |
| 30 psf superimposed load, 14" deep, 5.5#/L.F. | .200 | L.F. | .008 | .36 | .32 | .68 |
| 40 psf superimposed load, 14" deep, 6.5#/L.F. | .200 | L.F. | .008 | .43 | .32 | .75 |
| 50 psf superimposed load, 14" deep, 6.5#/L.F. | .200 | L.F. | .008 | .43 | .32 | .75 |
| 25' span, 20 psf superimposed load, 16" deep, 6.6#/L.F. | .200 | L.F. | .007 | .44 | .28 | .72 |
| 30 psf superimposed load, 16" deep, 7.8#/L.F. | .200 | L.F. | .007 | .51 | .28 | .79 |
| 40 psf superimposed load, 18" deep, 8.0#/L.F. | .200 | L.F. | .007 | .53 | .28 | .81 |
| 50 psf superimposed load, 18" deep, 8.0#/L.F. | .200 | L.F. | .007 | .53 | .28 | .81 |
| 30' span, 20 psf superimposed load, 18" deep, 8.0#/L.F. | .200 | L.F. | .006 | .53 | .26 | .79 |
| 30 psf superimposed load, 20" deep, 8.4#/L.F. | .200 | L.F. | .006 | .55 | .26 | .81 |
| 40 psf superimposed load, 22" deep, 9.7#/L.F. | .200 | L.F. | .006 | .64 | .26 | .90 |
| 50 psf superimposed load, 24" deep, 10.3#/L.F. | .200 | L.F. | .006 | .68 | .26 | .94 |
| 35' span, 20 psf superimposed load, 22" deep, 9.7#/L.F. | .200 | L.F. | .006 | .64 | .24 | .88 |
| 30 psf superimposed load, 24" deep, 10.3#/L.F. | .200 | L.F. | .006 | .68 | .24 | .92 |
| 40 psf superimposed load, 24" deep, 11.5#/L.F. | .200 | L.F. | .006 | .76 | .24 | 1.00 |
| 50 psf superimposed load, 24" deep, 12.7#/L.F. | .200 | L.F. | .006 | .84 | .24 | 1.08 |
| 40' span, 20 psf superimposed load, 24" deep, 11.5#/L.F. | .200 | L.F. | .007 | .76 | .26 | 1.02 |
| 30 psf superimposed load, 24" deep, 12.7#/L.F. | .200 | L.F. | .007 | .84 | .26 | 1.10 |
| 40 psf superimposed load, 28" deep, 13.5#/L.F. | .200 | L.F. | .007 | .89 | .26 | 1.15 |
| 50 psf superimposed load, 28" deep, 13.5#/L.F. | .200 | L.F. | .007 | .89 | .26 | 1.15 |
| Decking, open type, 1-1/2" deep, 22 gauge | 1.100 | S.F. | .007 | .85 | .20 | 1.05 |
| 20 gauge | 1.100 | S.F. | .008 | .97 | .23 | 1.20 |
| 18 gauge | 1.100 | S.F. | .008 | 1.33 | .24 | 1.57 |
| Furring, ceiling, 3/4" galvanized channels, 12" O.C. | 1.000 | S.F. | .038 | .30 | .73 | 1.03 |
| 16" O.C. | 1.000 | S.F. | .028 | .25 | .53 | .78 |
| 24" O.C. | 1.000 | S.F. | .019 | .19 | .36 | .55 |
| 1-1/2" galvanized channels, 12" O.C. | 1.000 | S.F. | .042 | .37 | .81 | 1.18 |
| 16" O.C. | 1.000 | S.F. | .031 | .32 | .59 | .91 |
| 24" O.C. | 1.000 | S.F. | .021 | .24 | .40 | .64 |

# Beams, Girders & Columns

| | QUAN. | UNIT | MAN HOURS | COST PER L.F. MAT. | COST PER L.F. INST. | TOTAL |
|---|---|---|---|---|---|---|
| Beams & girders, bolted, including fabrication | | | | | | |
| Wide flange shape, 12" deep, 14#/L.F. | 1.000 | L.F. | .078 | 7.55 | 2.95 | 10.50 |
| 10" deep, 15#/L.F. | 1.000 | L.F. | .078 | 8.25 | 2.95 | 11.20 |
| 8" deep, 10#/L.F. | 1.000 | L.F. | .078 | 5.50 | 2.95 | 8.45 |
| 6" deep, 9#/L.F. | 1.000 | L.F. | .078 | 4.95 | 2.95 | 7.90 |
| 5" deep, 16#/L.F. | 1.000 | L.F. | .078 | 9.05 | 2.95 | 12.00 |
| Columns, to 15', bolted or welded, including fabrication | | | | | | |
| Pipe, 3" diameter, 7.58#/L.F. | 1.000 | L.F. | .030 | 5.00 | 1.22 | 6.22 |
| 4" diameter, 10.79#/L.F. | 1.000 | L.F. | .043 | 7.12 | 1.73 | 8.85 |
| 6" diameter, 18.97#/L.F. | 1.000 | L.F. | .018 | 12.52 | .76 | 13.28 |
| Tube, rectangular, 8" x 6" x 1/4", 22.42#/L.F. | 1.000 | L.F. | .112 | 20.18 | 4.26 | 24.44 |
| 6" x 4" x 1/4", 15.62#/L.F. | 1.000 | L.F. | .078 | 14.06 | 2.97 | 17.03 |
| 5" x 3" x 1/4", 12.21#/L.F. | 1.000 | L.F. | .061 | 10.99 | 2.32 | 13.31 |

# 3 | FRAMING    40 | Gable Dormer Framing Systems

| SYSTEM DESCRIPTION | QUAN. | UNIT | MAN HOURS | MAT. | INST. | TOTAL |
|---|---|---|---|---|---|---|
| **2" X 6", 16" O.C.** | | | | | | |
| Dormer rafter, 2" x 6", 16" O.C. | 1.330 | L.F. | .035 | .53 | .73 | 1.26 |
| Ridge board, 2" x 6" | .280 | L.F. | .005 | .11 | .10 | .21 |
| Trimmer rafters, 2" x 6" | .880 | L.F. | .016 | .35 | .34 | .69 |
| Wall studs & plates, 2" x 4", 16" O.C. | 3.160 | L.F. | .056 | .79 | 1.17 | 1.96 |
| Fascia, 2" x 6" | .220 | L.F. | .012 | .09 | .25 | .34 |
| Valley rafter, 2" x 6", 16" O.C. | .280 | L.F. | .008 | .11 | .18 | .29 |
| Cripple rafter, 2" x 6", 16" O.C. | .560 | L.F. | .021 | .22 | .45 | .67 |
| Headers, 2" x 6", doubled | .670 | L.F. | .020 | .27 | .42 | .69 |
| Ceiling joist, 2" x 4", 16" O.C. | 1.000 | L.F. | .013 | .28 | .26 | .54 |
| Sheathing, exterior grade plywood, 1/2" thick | 3.610 | S.F. | .050 | 1.48 | 1.05 | 2.53 |
| TOTAL | | | .236 | 4.23 | 4.95 | 9.18 |
| **2" X 8", 16" O.C.** | | | | | | |
| Dormer rafter, 2" x 8", 16" O.C. | 1.330 | L.F. | .043 | .73 | .91 | 1.64 |
| Ridge board, 2" x 8" | .280 | L.F. | .005 | .15 | .12 | .27 |
| Trimmer rafter, 2" x 8" | .880 | L.F. | .019 | .48 | .41 | .89 |
| Wall studs & plates, 2" x 4", 16" O.C. | 3.160 | L.F. | .056 | .79 | 1.17 | 1.96 |
| Fascia, 2" x 8" | .220 | L.F. | .015 | .12 | .32 | .44 |
| Valley rafter, 2" x 8", 16" O.C. | .280 | L.F. | .010 | .15 | .22 | .37 |
| Cripple rafter, 2" x 8", 16" O.C. | .560 | L.F. | .026 | .31 | .55 | .86 |
| Headers, 2" x 8", doubled | .670 | L.F. | .031 | .37 | .69 | 1.06 |
| Ceiling joist, 2" x 4", 16" O.C. | 1.000 | L.F. | .013 | .28 | .26 | .54 |
| Sheathing,, exterior grade plywood, 1/2" thick | 3.610 | S.F. | .050 | 1.48 | 1.05 | 2.53 |
| TOTAL | | | .268 | 4.86 | 5.70 | 10.56 |

The cost in this system is based on the square foot of plan area. The measurement being the plan area of the dormer only.

| DESCRIPTION | QUAN. | UNIT | MAN HOURS | MAT. | INST. | TOTAL |
|---|---|---|---|---|---|---|
| | | | | | | |
| | | | | | | |
| | | | | | | |
| | | | | | | |
| | | | | | | |
| | | | | | | |

# Gable Dormer Framing Price Sheet

| | QUAN. | UNIT | MAN HOURS | COST PER S.F. MAT. | COST PER S.F. INST. | TOTAL |
|---|---|---|---|---|---|---|
| Dormer rafters, #2 or better, 2" x 4", 16" O.C. | 1.330 | L.F. | .028 | .43 | .58 | 1.01 |
| 24" O.C. | 1.060 | L.F. | .022 | .34 | .47 | .81 |
| 2" x 6", 16" O.C. | 1.330 | L.F. | .035 | .53 | .73 | 1.26 |
| 24" O.C. | 1.060 | L.F. | .028 | .42 | .59 | 1.01 |
| 2" x 8", 16" O.C. | 1.330 | L.F. | .043 | .73 | .91 | 1.64 |
| 24" O.C. | 1.060 | L.F. | .034 | .58 | .72 | 1.30 |
| Ridge board, #2 or better, 1" x 4" | .280 | L.F. | .002 | .04 | .06 | .10 |
| 1" x 6" | .280 | L.F. | .003 | .06 | .06 | .12 |
| 1" x 8" | .280 | L.F. | .004 | .07 | .09 | .16 |
| 2" x 4" | .280 | L.F. | .004 | .09 | .08 | .17 |
| 2" x 6" | .280 | L.F. | .005 | .11 | .10 | .21 |
| 2" x 8" | .280 | L.F. | .005 | .15 | .12 | .27 |
| Trimmer rafters, #2 or better, 2" x 4" | .880 | L.F. | .013 | .28 | .27 | .55 |
| 2" x 6" | .880 | L.F. | .016 | .35 | .34 | .69 |
| 2" x 8" | .880 | L.F. | .019 | .48 | .41 | .89 |
| 2" x 10" | .880 | L.F. | .022 | .66 | .46 | 1.12 |
| Wall studs & plates, #2 or better, 2" x 4" studs, 16" O.C. | 3.160 | L.F. | .056 | .79 | 1.17 | 1.96 |
| 24" O.C. | 2.800 | L.F. | .050 | .70 | 1.04 | 1.74 |
| 2" x 6" studs, 16" O.C. | 3.160 | L.F. | .063 | 1.26 | 1.30 | 2.56 |
| 24" O.C. | 2.800 | L.F. | .056 | 1.12 | 1.15 | 2.27 |
| Fascia, #2 or better, 1" x 4" | .220 | L.F. | .006 | .05 | .13 | .18 |
| 1" x 6" | .220 | L.F. | .007 | .06 | .16 | .22 |
| 1" x 8" | .220 | L.F. | .009 | .07 | .19 | .26 |
| 2" x 4" | .220 | L.F. | .010 | .08 | .22 | .30 |
| 2" x 6" | .220 | L.F. | .013 | .10 | .28 | .38 |
| 2" x 8" | .220 | L.F. | .015 | .12 | .32 | .44 |
| Valley rafter, #2 or better, 2" x 4" | .280 | L.F. | .006 | .09 | .14 | .23 |
| 2" x 6" | .280 | L.F. | .008 | .11 | .18 | .29 |
| 2" x 8" | .280 | L.F. | .010 | .15 | .22 | .37 |
| 2" x 10" | .280 | L.F. | .011 | .21 | .24 | .45 |
| Cripple rafter, #2 or better, 2" x 4", 16" O.C. | .560 | L.F. | .017 | .18 | .36 | .54 |
| 24" O.C. | .450 | L.F. | .014 | .14 | .29 | .43 |
| 2" x 6", 16" O.C. | .560 | L.F. | .021 | .22 | .45 | .67 |
| 24" O.C. | .450 | L.F. | .017 | .18 | .36 | .54 |
| 2" x 8", 16" O.C. | .560 | L.F. | .026 | .31 | .55 | .86 |
| 24" O.C. | .450 | L.F. | .021 | .25 | .44 | .69 |
| Headers, #2 or better double header, 2" x 4" | .670 | L.F. | .016 | .22 | .34 | .56 |
| 2" x 6" | .670 | L.F. | .020 | .27 | .42 | .69 |
| 2" x 8" | .670 | L.F. | .031 | .37 | .69 | 1.06 |
| 2" x 10" | .670 | L.F. | .033 | .52 | .72 | 1.24 |
| Ceiling joist, #2 or better, 2" x 4", 16" O.C. | 1.000 | L.F. | .013 | .28 | .26 | .54 |
| 24" O.C. | .800 | L.F. | .010 | .22 | .21 | .43 |
| 2" x 6", 16" O.C. | 1.000 | L.F. | .015 | .40 | .31 | .71 |
| 24" O.C. | .800 | L.F. | .012 | .32 | .25 | .57 |
| Sheathing, plywood exterior grade, 3/8" thick | 3.610 | S.F. | .046 | 1.16 | .97 | 2.13 |
| 1/2" thick | 3.610 | S.F. | .050 | 1.48 | 1.05 | 2.53 |
| 5/8" thick | 3.610 | S.F. | .054 | 1.62 | 1.12 | 2.74 |
| 3/4" thick | 3.610 | S.F. | .057 | 1.84 | 1.19 | 3.03 |
| Boards, 1" x 6", laid regular | 3.610 | S.F. | .090 | 2.89 | 1.84 | 4.73 |
| Laid diagonal | 3.610 | S.F. | .097 | 2.89 | 2.02 | 4.91 |
| 1" x 8", laid regular | 3.610 | S.F. | .075 | 3.01 | 1.55 | 4.56 |
| Laid diagonal | 3.610 | S.F. | .085 | 3.01 | 1.78 | 4.79 |

3

# 3 | FRAMING — 44 | Shed Dormer Framing Systems

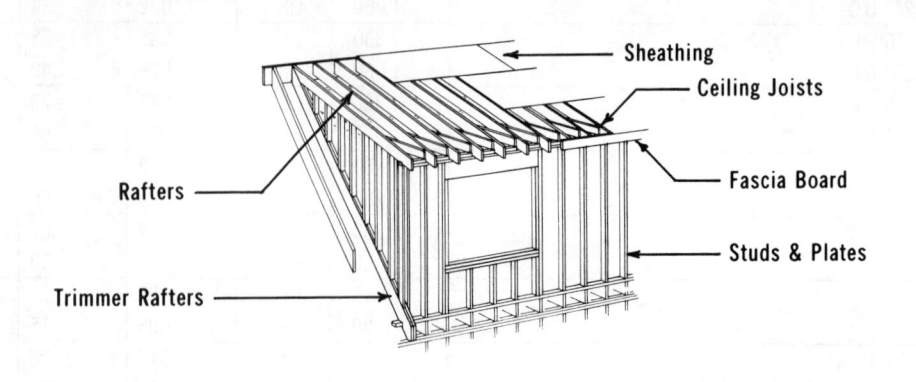

*Diagram labels: Sheathing, Ceiling Joists, Fascia Board, Studs & Plates, Rafters, Trimmer Rafters*

| SYSTEM DESCRIPTION | QUAN. | UNIT | MAN HOURS | MAT. | INST. | TOTAL |
|---|---|---|---|---|---|---|
| **2" X 6" RAFTERS, 16" O.C.** | | | | | | |
| Dormer rafter, 2" x 6", 16" O.C. | 1.080 | L.F. | .029 | .43 | .60 | 1.03 |
| Trimmer rafter, 2" x 6" | .400 | L.F. | .007 | .16 | .15 | .31 |
| Studs & plates, 2" x 4", 16" O.C. | 2.750 | L.F. | .049 | .69 | 1.02 | 1.71 |
| Fascia, 2" x 6" | .250 | L.F. | .013 | .10 | .28 | .38 |
| Ceiling joist, 2" x 4", 16" O.C. | 1.000 | L.F. | .013 | .28 | .26 | .54 |
| Sheathing, exterior grade plywood, CDX, 1/2" thick | 2.940 | S.F. | .041 | 1.21 | .85 | 2.06 |
| TOTAL | | | .153 | 2.87 | 3.16 | 6.03 |
| **2" X 8" RAFTERS, 16" O.C.** | | | | | | |
| Dormer rafter, 2" x 8", 16" O.C. | 1.080 | L.F. | .035 | .59 | .74 | 1.33 |
| Trimmer rafter, 2" x 8" | .400 | L.F. | .008 | .22 | .18 | .40 |
| Studs & plates, 2" x 4", 16" O.C. | 2.750 | L.F. | .049 | .69 | 1.02 | 1.71 |
| Fascia, 2" x 8" | .250 | L.F. | .017 | .14 | .36 | .50 |
| Ceiling joist, 2" x 6", 16" O.C. | 1.000 | L.F. | .015 | .40 | .31 | .71 |
| Sheathing, exterior grade plywood, CDX, 1/2" thick | 2.940 | S.F. | .041 | 1.21 | .85 | 2.06 |
| TOTAL | | | .167 | 3.25 | 3.46 | 6.71 |
| **2" X 10" RAFTERS, 16" O.C.** | | | | | | |
| Dormer rafter, 2" x 10", 16" O.C. | 1.080 | L.F. | .041 | .81 | .83 | 1.64 |
| Trimmer rafter, 2" x 10" | .400 | L.F. | .010 | .30 | .21 | .51 |
| Studs & plates, 2" x 4", 16" O.C. | 2.750 | L.F. | .049 | .69 | 1.02 | 1.71 |
| Fascia, 2" x 10" | .250 | L.F. | .022 | .19 | .45 | .64 |
| Ceiling joist, 2" x 6", 16" O.C. | 1.000 | L.F. | .015 | .40 | .31 | .71 |
| Sheathing, exterior grade plywood, CDX, 1/2" thick | 2.940 | S.F. | .041 | 1.21 | .85 | 2.06 |
| TOTAL | | | .178 | 3.60 | 3.67 | 7.27 |

The cost in this system is based on the square foot of plan area. The measurement is the plan area of the dormer only.

| DESCRIPTION | QUAN. | UNIT | MAN HOURS | MAT. | INST. | TOTAL |
|---|---|---|---|---|---|---|
| | | | | | | |
| | | | | | | |
| | | | | | | |
| | | | | | | |
| | | | | | | |

# Shed Dormer Framing Price Sheet

| | QUAN. | UNIT | MAN HOURS | COST PER S.F. MAT. | COST PER S.F. INST. | TOTAL |
|---|---|---|---|---|---|---|
| Dormer rafters, #2 or better, 2" x 4", 16" O.C. | 1.080 | L.F. | .023 | .35 | .47 | .82 |
| 24" O.C. | .860 | L.F. | .018 | .28 | .37 | .65 |
| 2" x 6", 16" O.C. | 1.080 | L.F. | .029 | .43 | .60 | 1.03 |
| 24" O.C. | .860 | L.F. | .023 | .34 | .48 | .82 |
| 2" x 8", 16" O.C. | 1.080 | L.F. | .035 | .59 | .74 | 1.33 |
| 24" O.C. | .860 | L.F. | .028 | .47 | .59 | 1.06 |
| 2" x 10", 16" O.C. | 1.080 | L.F. | .041 | .81 | .83 | 1.64 |
| 24" O.C. | .860 | L.F. | .032 | .65 | .66 | 1.31 |
| Trimmer rafter, #2 or better, 2" x 4" | .400 | L.F. | .006 | .13 | .12 | .25 |
| 2" x 6" | .400 | L.F. | .007 | .16 | .15 | .31 |
| 2" x 8" | .400 | L.F. | .008 | .22 | .18 | .40 |
| 2" x 10" | .400 | L.F. | .010 | .30 | .21 | .51 |
| Studs & plates, #2 or better, 2" x 4", 16" O.C. | 2.750 | L.F. | .049 | .69 | 1.02 | 1.71 |
| 24" O.C. | 2.200 | L.F. | .039 | .55 | .81 | 1.36 |
| 2" x 6", 16" O.C. | 2.750 | L.F. | .055 | 1.10 | 1.13 | 2.23 |
| 24" O.C. | 2.200 | L.F. | .044 | .88 | .90 | 1.78 |
| Fascia, #2 or better, 1" x 4" | .250 | L.F. | .006 | .05 | .13 | .18 |
| 1" x 6" | .250 | L.F. | .007 | .06 | .16 | .22 |
| 1" x 8" | .250 | L.F. | .009 | .07 | .19 | .26 |
| 2" x 4" | .250 | L.F. | .010 | .08 | .22 | .30 |
| 2" x 6" | .250 | L.F. | .013 | .10 | .28 | .38 |
| 2" x 8" | .250 | L.F. | .017 | .14 | .36 | .50 |
| Ceiling joist, #2 or better, 2" x 4", 16" O.C. | 1.000 | L.F. | .013 | .28 | .26 | .54 |
| 24" O.C. | .800 | L.F. | .010 | .22 | .21 | .43 |
| 2" x 6", 16" O.C. | 1.000 | L.F. | .015 | .40 | .31 | .71 |
| 24" O.C. | .800 | L.F. | .012 | .32 | .25 | .57 |
| 2" x 8", 16" O.C. | 1.000 | L.F. | .017 | .55 | .35 | .90 |
| 24" O.C. | .800 | L.F. | .013 | .44 | .28 | .72 |
| Sheathing, plywood exterior grade, 3/8" thick | 2.940 | S.F. | .038 | .94 | .79 | 1.73 |
| 1/2" thick | 2.940 | S.F. | .041 | 1.21 | .85 | 2.06 |
| 5/8" thick | 2.940 | S.F. | .044 | 1.32 | .91 | 2.23 |
| 3/4" thick | 2.940 | S.F. | .047 | 1.50 | .97 | 2.47 |
| Boards, 1" x 6", laid regular | 2.940 | S.F. | .073 | 2.35 | 1.50 | 3.85 |
| Laid diagonal | 2.940 | S.F. | .079 | 2.35 | 1.65 | 4.00 |
| 1" x 8", laid regular | 2.940 | S.F. | .061 | 2.35 | 1.27 | 3.62 |
| Laid diagonal | 2.940 | S.F. | .070 | 2.45 | 1.47 | 3.92 |

# Window Openings

| | QUAN. | UNIT | MAN HOURS | COST EACH MAT. | COST EACH INST. | TOTAL |
|---|---|---|---|---|---|---|
| The following are to be added to the total cost of the dormers for window openings. Do not subtract window area from the stud wall quantities. | | | | | | |
| Headers, 2" x 6" doubled, 2' long | 4.000 | L.F. | .124 | 1.60 | 2.52 | 4.12 |
| 3' long | 6.000 | L.F. | .186 | 2.40 | 3.78 | 6.18 |
| 4' long | 8.000 | L.F. | .248 | 3.20 | 5.04 | 8.24 |
| 5' long | 10.000 | L.F. | .310 | 4.00 | 6.30 | 10.30 |
| 2" x 8" doubled, 4' long | 8.000 | L.F. | .376 | 4.40 | 8.24 | 12.64 |
| 5' long | 10.000 | L.F. | .470 | 5.50 | 10.30 | 15.80 |
| 6' long | 12.000 | L.F. | .564 | 6.60 | 12.36 | 18.96 |
| 8' long | 16.000 | L.F. | .752 | 8.80 | 16.48 | 25.28 |
| 2" x 10" doubled, 4' long | 8.000 | L.F. | .400 | 6.16 | 8.64 | 14.80 |
| 6' long | 12.000 | L.F. | .600 | 9.24 | 12.96 | 22.20 |
| 8' long | 16.000 | L.F. | .800 | 12.32 | 17.28 | 29.60 |
| 10' long | 20.000 | L.F. | 1.000 | 15.40 | 21.60 | 37.00 |

# 3 | FRAMING  48 | Partition Framing System

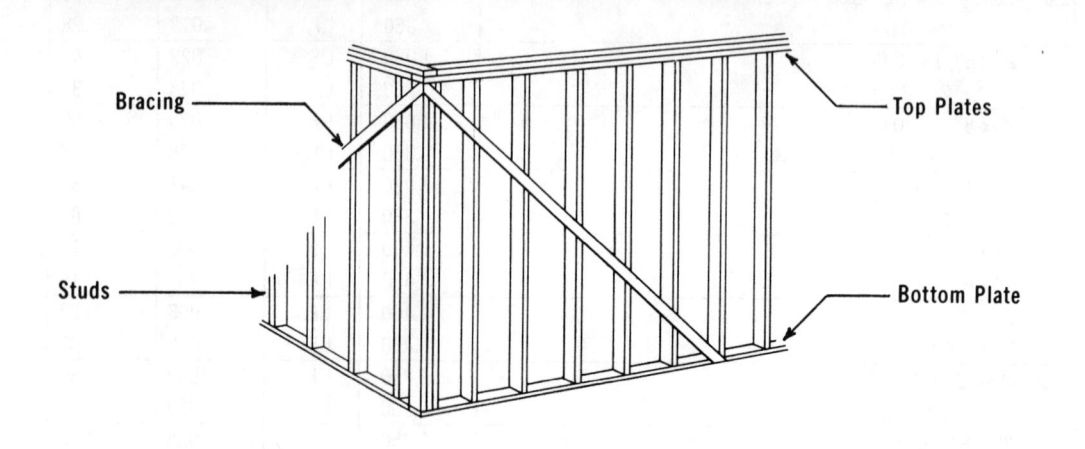

*Bracing*  
*Studs*  
*Top Plates*  
*Bottom Plate*

| SYSTEM DESCRIPTION | QUAN. | UNIT | MAN HOURS | MAT. | INST. | TOTAL |
|---|---|---|---|---|---|---|
| **2" X 4", 16" O.C.** | | | | | | |
| 2" x 4" studs, #2 or better, 16" O.C. | 1.000 | L.F. | .015 | .25 | .32 | .57 |
| Plates, double top, single bottom | .375 | L.F. | .005 | .09 | .12 | .21 |
| Cross bracing, let-in, 1" x 6" | .080 | L.F. | .004 | .02 | .08 | .10 |
| TOTAL | | | .024 | .36 | .52 | .88 |
| **2" X 4", 24" O.C.** | | | | | | |
| 2" x 4" studs, #2 or better, 24" O.C. | .800 | L.F. | .012 | .20 | .26 | .46 |
| Plates, double top, single bottom | .375 | L.F. | .005 | .09 | .12 | .21 |
| Cross bracing, let-in, 1" x 6" | .080 | L.F. | .002 | .02 | .05 | .07 |
| TOTAL | | | .020 | .31 | .43 | .74 |
| **2" X 6", 16" O.C.** | | | | | | |
| 2" x 6" studs, #2 or better, 16" O.C. | 1.000 | L.F. | .018 | .40 | .36 | .76 |
| Plates, double top, single bottom | .375 | L.F. | .006 | .15 | .14 | .29 |
| Cross bracing, let-in, 1" x 6" | .080 | L.F. | .004 | .02 | .08 | .10 |
| TOTAL | | | .028 | .57 | .58 | 1.15 |
| **2" X 6", 24" O.C.** | | | | | | |
| 2" x 6" studs, #2 or better, 24" O.C. | .800 | L.F. | .014 | .32 | .29 | .61 |
| Plates, double top, single bottom | .375 | L.F. | .006 | .15 | .14 | .29 |
| Cross bracing, let-in, 1" x 6" | .080 | L.F. | .002 | .02 | .05 | .07 |
| TOTAL | | | .023 | .49 | .48 | .97 |

The costs in this system are based on a square foot of wall area. Do not subtract for door or window openings. For openings see next page.

| DESCRIPTION | QUAN. | UNIT | MAN HOURS | MAT. | INST. | TOTAL |
|---|---|---|---|---|---|---|
| | | | | | | |
| | | | | | | |
| | | | | | | |

# Partition Framing Price Sheet

| | QUAN. | UNIT | MAN HOURS | COST PER S.F. MAT. | COST PER S.F. INST. | TOTAL |
|---|---|---|---|---|---|---|
| Wood studs, #2 or better, 2" x 4", 12" O.C. | 1.250 | L.F. | .018 | .31 | .40 | .71 |
| 16" O.C. | 1.000 | L.F. | .015 | .25 | .32 | .57 |
| 24" O.C. | .800 | L.F. | .012 | .20 | .26 | .46 |
| 32" O.C. | .650 | L.F. | .009 | .16 | .21 | .37 |
| 2" x 6", 12" O.C. | 1.250 | L.F. | .022 | .50 | .45 | .95 |
| 16" O.C. | 1.000 | L.F. | .018 | .40 | .36 | .76 |
| 24" O.C. | .800 | L.F. | .014 | .32 | .29 | .61 |
| 32" O.C. | .650 | L.F. | .011 | .26 | .23 | .49 |
| Plates, #2 or better double top single bottom, 2" x 4" | .375 | L.F. | .005 | .09 | .12 | .21 |
| 2" x 6" | .375 | L.F. | .006 | .15 | .14 | .29 |
| 2" x 8" | .375 | L.F. | .006 | .21 | .13 | .34 |
| Cross bracing, let-in, 1" x 6" boards studs, 12" O.C. | .080 | L.F. | .005 | .02 | .11 | .13 |
| 16" O.C. | .080 | L.F. | .004 | .02 | .08 | .10 |
| 24" O.C. | .080 | L.F. | .002 | .02 | .05 | .07 |
| 32" O.C. | .080 | L.F. | .002 | .01 | .05 | .06 |
| Let-in steel (T shaped) studs, 12" O.C. | .080 | L.F. | .001 | .06 | .03 | .09 |
| 16" O.C. | .080 | L.F. | .001 | .05 | .02 | .07 |
| 24" O.C. | .080 | L.F. | .001 | .05 | .02 | .07 |
| 32" O.C. | .080 | L.F. | | .04 | .02 | .06 |
| Steel straps studs, 12" O.C. | .080 | L.F. | .001 | .05 | .02 | .07 |
| 16" O.C. | .080 | L.F. | .001 | .05 | .02 | .07 |
| 24" O.C. | .080 | L.F. | .001 | .05 | .02 | .07 |
| 32" O.C. | .080 | L.F. | | .04 | .02 | .06 |
| Metal studs, load bearing 24" O.C., 20 ga. galv., 2-1/2" wide | 1.000 | S.F. | .016 | .40 | .32 | .72 |
| 3-5/8" wide | 1.000 | S.F. | .017 | .42 | .34 | .76 |
| 4" wide | 1.000 | S.F. | .018 | .46 | .37 | .83 |
| 6" wide | 1.000 | S.F. | .019 | .52 | .38 | .90 |
| 16 ga., 2-1/2" wide | 1.000 | S.F. | .017 | .96 | .34 | 1.30 |
| 3-5/8" wide | 1.000 | S.F. | .018 | .96 | .36 | 1.32 |
| 4" wide | 1.000 | S.F. | .019 | 1.07 | .38 | 1.45 |
| 6" wide | 1.000 | S.F. | .019 | 1.27 | .39 | 1.66 |
| 8" wide | 1.000 | S.F. | .020 | 1.61 | .41 | 2.02 |
| Nonload bearing 24" O.C., 25 ga. galv., 1-5/8" wide | 1.000 | S.F. | .016 | .18 | .32 | .50 |
| 2-1/2" wide | 1.000 | S.F. | .016 | .20 | .33 | .53 |
| 3-5/8" wide | 1.000 | S.F. | .017 | .23 | .34 | .57 |
| 4" wide | 1.000 | S.F. | .018 | .28 | .36 | .64 |
| 6" wide | 1.000 | S.F. | .018 | .34 | .37 | .71 |
| 20 ga., 2-1/2" wide | 1.000 | S.F. | .016 | .24 | .33 | .57 |
| 3-5/8" wide | 1.000 | S.F. | .017 | .28 | .34 | .62 |
| 4" wide | 1.000 | S.F. | .018 | .31 | .36 | .67 |
| 6" wide | 1.000 | S.F. | .018 | .36 | .38 | .74 |

# Window & Door Openings

| | QUAN. | UNIT | MAN HOURS | COST EACH MAT. | COST EACH INST. | TOTAL |
|---|---|---|---|---|---|---|
| The following costs are to be added to the total costs of the walls. Do not subtract openings from total wall area. | | | | | | |
| Headers, 2" x 6" double, 2' long | 4.000 | L.F. | .124 | 1.60 | 2.52 | 4.12 |
| 3' long | 6.000 | L.F. | .186 | 2.40 | 3.78 | 6.18 |
| 4' long | 8.000 | L.F. | .248 | 3.20 | 5.04 | 8.24 |
| 5' long | 10.000 | L.F. | .310 | 4.00 | 6.30 | 10.30 |
| 2" x 8" double, 4' long | 8.000 | L.F. | .376 | 4.40 | 8.24 | 12.64 |
| 5' long | 10.000 | L.F. | .470 | 5.50 | 10.30 | 15.80 |
| 6' long | 12.000 | L.F. | .564 | 6.60 | 12.36 | 18.96 |
| 8' long | 16.000 | L.F. | .752 | 8.80 | 16.48 | 25.28 |
| 2" x 10" double, 4' long | 8.000 | L.F. | .400 | 6.16 | 8.64 | 14.80 |
| 6' long | 12.000 | L.F. | .600 | 9.24 | 12.96 | 22.20 |
| 8' long | 16.000 | L.F. | .800 | 12.32 | 17.28 | 29.60 |
| 10' long | 20.000 | L.F. | 1.000 | 15.40 | 21.60 | 37.00 |
| 2" x 12" double, 8' long | 16.000 | L.F. | .736 | 15.04 | 14.88 | 29.92 |
| 12' long | 24.000 | L.F. | 1.104 | 22.56 | 22.32 | 44.88 |

# SECTION 4
# EXTERIOR WALLS

| System Number | Title | Page |
|---|---|---|
| 4-02 | Masonry Block Wall | 62 |
| 4-04 | Brick/Stone Veneer | 64 |
| 4-08 | Wood Siding | 66 |
| 4-12 | Shingle Siding | 68 |
| 4-16 | Metal & Plastic Siding | 70 |
| 4-20 | Insulation | 72 |
| 4-28 | Double Hung Window | 74 |
| 4-32 | Casement Window | 76 |
| 4-36 | Awning Window | 78 |
| 4-40 | Sliding Window | 80 |
| 4-44 | Bow/Bay Window | 82 |
| 4-48 | Fixed Window | 84 |
| 4-52 | Entrance Door | 86 |
| 4-53 | Sliding Door | 88 |
| 4-54 | Commercial Entrance | 90 |
| 4-55 | Commercial Overhead Door | 92 |
| 4-56 | Residential Overhead Door | 94 |
| 4-58 | Aluminum Window | 96 |
| 4-60 | Storm Door & Window | 98 |
| 4-64 | Shutters/Blinds | 99 |
| 4-80 | Storefront | 100 |

# 4 | EXTERIOR WALLS  02 | Masonry Block Wall Systems

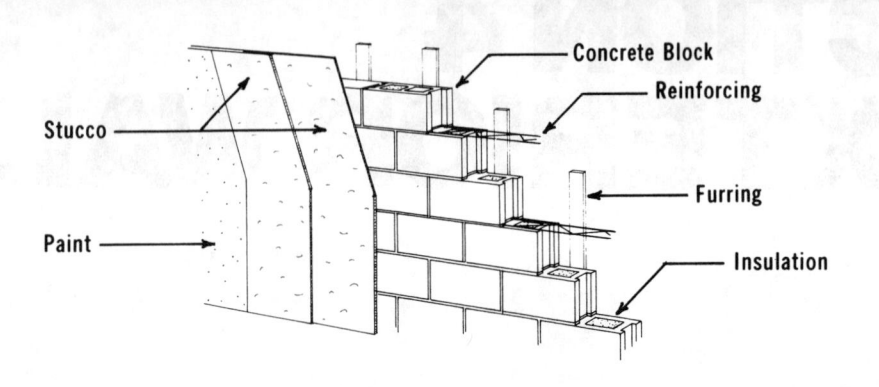

| SYSTEM DESCRIPTION | QUAN. | UNIT | MAN HOURS | MAT. | INST. | TOTAL |
|---|---|---|---|---|---|---|
| **6" THICK CONCRETE BLOCK WALL** | | | | | | |
| 6" thick concrete block, 6" x 8" x 16" | 1.000 | S.F. | .084 | 1.09 | 1.90 | 2.99 |
| Masonry reinforcing, truss strips every other course | .625 | L.F. | .001 | .15 | .03 | .18 |
| Furring, 1" x 3", 16" O.C. | 1.000 | L.F. | .016 | .10 | .32 | .42 |
| Masonry insulation, poured vermiculite | 1.000 | S.F. | .013 | .51 | .26 | .77 |
| Stucco, 2 coats | 1.000 | S.F. | .068 | .19 | 1.30 | 1.49 |
| Masonry paint, 2 coats | 1.000 | S.F. | .022 | .18 | .41 | .59 |
| TOTAL | | | .204 | 2.22 | 4.22 | 6.44 |
| **8" THICK CONCRETE BLOCK WALL** | | | | | | |
| 8" thick concrete block, 8" x 8" x 16" | 1.000 | S.F. | .089 | 1.32 | 2.03 | 3.35 |
| Masonry reinforcing, truss strips every other course | .625 | L.F. | .001 | .15 | .03 | .18 |
| Furring, 1" x 3", 16" O.C. | 1.000 | L.F. | .016 | .10 | .32 | .42 |
| Masonry insulation, poured vermiculite | 1.000 | S.F. | .017 | .67 | .35 | 1.02 |
| Stucco, 2 coats | 1.000 | S.F. | .068 | .19 | 1.30 | 1.49 |
| Masonry paint, 2 coats | 1.000 | S.F. | .022 | .18 | .41 | .59 |
| TOTAL | | | .213 | 2.61 | 4.44 | 7.05 |
| **12" THICK CONCRETE BLOCK WALL** | | | | | | |
| 12" thick concrete block, 12" x 8" x 16" | 1.000 | S.F. | .147 | 1.99 | 2.61 | 4.60 |
| Masonry reinforcing, truss strips every other course | .625 | L.F. | .002 | .16 | .05 | .21 |
| Furring, 1" x 3", 16" O.C. | 1.000 | L.F. | .016 | .10 | .32 | .42 |
| Masonry insulation, poured vermiculite | 1.000 | S.F. | .026 | .99 | .51 | 1.50 |
| Stucco, 2 coats | 1.000 | S.F. | .068 | .19 | 1.30 | 1.49 |
| Masonry paint, 2 coats | 1.000 | S.F. | .022 | .18 | .41 | .59 |
| TOTAL | | | .281 | 3.61 | 5.20 | 8.81 |

Costs for this system are based on a square foot of wall area. Do not subtract for window openings.

| DESCRIPTION | QUAN. | UNIT | MAN HOURS | MAT. | INST. | TOTAL |
|---|---|---|---|---|---|---|
| | | | | | | |
| | | | | | | |
| | | | | | | |
| | | | | | | |
| | | | | | | |

# Masonry Block Wall Price Sheet

| | QUAN. | UNIT | MAN HOURS | MAT. | INST. | TOTAL |
|---|---|---|---|---|---|---|
| Block concrete, 8" x 16" regular, 4" thick | 1.000 | S.F. | .077 | .88 | 1.77 | 2.65 |
| 6" thick | 1.000 | S.F. | .084 | 1.09 | 1.90 | 2.99 |
| 8" thick | 1.000 | S.F. | .089 | 1.32 | 2.03 | 3.35 |
| 10" thick | 1.000 | S.F. | .093 | 1.94 | 2.11 | 4.05 |
| 12" thick | 1.000 | S.F. | .147 | 1.99 | 2.61 | 4.60 |
| Solid block, 4" thick | 1.000 | S.F. | .080 | 1.23 | 1.45 | 2.68 |
| 6" thick | 1.000 | S.F. | .085 | 1.54 | 1.55 | 3.09 |
| 8" thick | 1.000 | S.F. | .092 | 2.04 | 1.67 | 3.71 |
| 10" thick | 1.000 | S.F. | .135 | 2.67 | 2.42 | 5.09 |
| 12" thick | 1.000 | S.F. | .151 | 2.97 | 2.68 | 5.65 |
| Lightweight, 4" thick | 1.000 | S.F. | .076 | 1.02 | 1.73 | 2.75 |
| 6" thick | 1.000 | S.F. | .081 | 1.31 | 1.86 | 3.17 |
| 8" thick | 1.000 | S.F. | .087 | 1.54 | 1.98 | 3.52 |
| 10" thick | 1.000 | S.F. | .090 | 2.31 | 2.06 | 4.37 |
| 12" thick | 1.000 | S.F. | .142 | 2.37 | 2.17 | 4.54 |
| Split rib profile, 4" thick | 1.000 | S.F. | .094 | 1.64 | 2.21 | 3.85 |
| 6" thick | 1.000 | S.F. | .101 | 2.08 | 2.34 | 4.42 |
| 8" thick | 1.000 | S.F. | .107 | 2.67 | 2.48 | 5.15 |
| 10" thick | 1.000 | S.F. | .162 | 2.66 | 2.92 | 5.58 |
| 12" thick | 1.000 | S.F. | .181 | 2.96 | 3.24 | 6.20 |
| Masonry reinforcing wire truss strips, every course, 8" block | 1.375 | L.F. | .004 | .33 | .07 | .40 |
| 12" block | 1.375 | L.F. | .005 | .34 | .11 | .45 |
| Every other course, 8" block | .625 | L.F. | .001 | .15 | .03 | .18 |
| 12" block | .625 | L.F. | .002 | .16 | .05 | .21 |
| Furring, wood, 1" x 3", 12" O.C. | 1.250 | L.F. | .020 | .13 | .40 | .53 |
| 16" O.C. | 1.000 | L.F. | .016 | .10 | .32 | .42 |
| 24" O.C. | .800 | L.F. | .012 | .08 | .26 | .34 |
| 32" O.C. | .640 | L.F. | .010 | .06 | .21 | .27 |
| Steel, 3/4" channels, 12" O.C. | 1.250 | L.F. | .034 | .30 | .65 | .95 |
| 16" O.C. | 1.000 | L.F. | .030 | .25 | .58 | .83 |
| 24" O.C. | .800 | L.F. | .023 | .19 | .44 | .63 |
| 32" O.C. | .640 | L.F. | .018 | .15 | .35 | .50 |
| Masonry insulation, vermiculite or perlite poured 4" thick | 1.000 | S.F. | .008 | .33 | .17 | .50 |
| 6" thick | 1.000 | S.F. | .013 | .50 | .26 | .76 |
| 8" thick | 1.000 | S.F. | .017 | .67 | .35 | 1.02 |
| 10" thick | 1.000 | S.F. | .021 | .82 | .41 | 1.23 |
| 12" thick | 1.000 | S.F. | .026 | .99 | .51 | 1.50 |
| Block inserts polystyrene, 6" thick | 1.000 | S.F. | | .77 | | .77 |
| 8" thick | 1.000 | S.F. | | .77 | | .77 |
| 10" thick | 1.000 | S.F. | | .93 | | .93 |
| 12" thick | 1.000 | S.F. | | .99 | | .99 |
| Stucco, 1 coat | 1.000 | S.F. | .056 | .15 | 1.08 | 1.23 |
| 2 coats | 1.000 | S.F. | .068 | .19 | 1.30 | 1.49 |
| 3 coats | 1.000 | S.F. | .081 | .22 | 1.53 | 1.75 |
| Painting, 1 coat | 1.000 | S.F. | .016 | .13 | .30 | .43 |
| 2 coats | 1.000 | S.F. | .022 | .18 | .41 | .59 |
| Primer & 1 coat | 1.000 | S.F. | .017 | .23 | .32 | .55 |
| 2 coats | 1.000 | S.F. | .023 | .29 | .42 | .71 |
| Lath, metal lath expanded 2.5 lb/S.Y., painted | 1.000 | S.F. | .010 | .20 | .20 | .40 |
| Galvanized | 1.000 | S.F. | .011 | .22 | .22 | .44 |

4

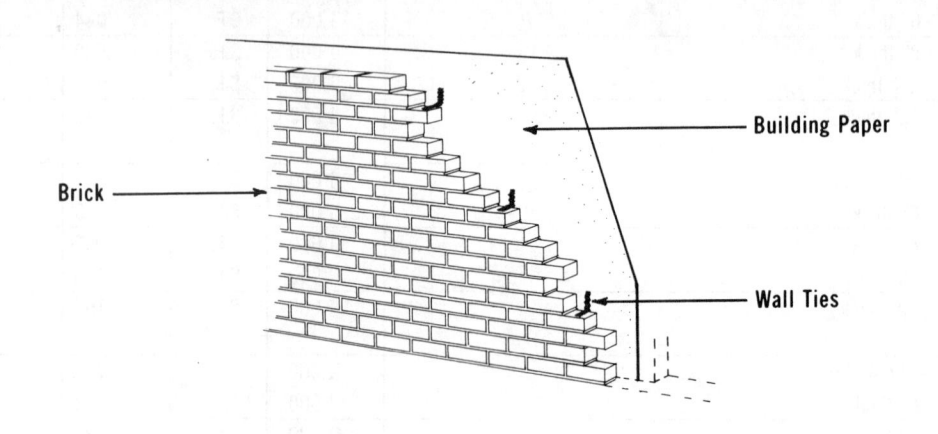

# 4 | EXTERIOR WALLS  04 | Brick/Stone Veneer Systems

*Brick* → *Building Paper* → *Wall Ties*

| SYSTEM DESCRIPTION | QUAN. | UNIT | MAN HOURS | COST PER S.F. MAT. | COST PER S.F. INST. | COST PER S.F. TOTAL |
|---|---|---|---|---|---|---|
| **COMMON BRICK $140 PER THOUSAND** | | | | | | |
| Brick, common, running bond, brick $140 per thousand | 1.000 | S.F. | .183 | 1.29 | 3.33 | 4.62 |
| Wall ties, 7/8" x 7", 24 gauge | 1.000 | Ea. | .008 | .03 | .15 | .18 |
| Building paper, #15 asphalt | 1.100 | S.F. | .002 | .03 | .06 | .09 |
| Trim, pine, painted | .125 | L.F. | .005 | .04 | .10 | .14 |
| TOTAL | | | .198 | 1.39 | 3.64 | 5.03 |
| **COMMON BRICK $215 PER THOUSAND** | | | | | | |
| Brick, common, running bond, brick $215 per thousand | 1.000 | S.F. | .200 | 1.87 | 3.63 | 5.50 |
| Wall ties, 7/8" x 7", 24 gauge | 1.000 | Ea. | .008 | .03 | .15 | .18 |
| Building paper, #15 asphalt | 1.100 | S.F. | .002 | .03 | .06 | .09 |
| Trim, pine, painted | .125 | L.F. | .005 | .04 | .10 | .14 |
| TOTAL | | | .215 | 1.97 | 3.94 | 5.91 |
| **BUFF OR GREY FACE BRICK $250 PER THOUSAND** | | | | | | |
| Brick, buff or grey $250 per thousand | 1.000 | S.F. | .200 | 2.20 | 3.65 | 5.85 |
| Wall ties, 7/8" x 7", 24 gauge | 1.000 | Ea. | .008 | .03 | .15 | .18 |
| Building paper, #15 asphalt | 1.100 | S.F. | .002 | .03 | .06 | .09 |
| Trim, pine, painted | .125 | L.F. | .005 | .04 | .10 | .14 |
| TOTAL | | | .215 | 2.30 | 3.96 | 6.26 |
| **STONE WORK, ROUGH STONE, AVERAGE** | | | | | | |
| Stone work, rough stone, average | 1.000 | S.F. | .194 | 6.45 | 3.43 | 9.88 |
| Wall ties, 7/8" x 7", 24 gauge | 1.000 | Ea. | .008 | .03 | .15 | .18 |
| Building paper, #15 asphalt | 1.100 | S.F. | .002 | .03 | .06 | .09 |
| Trim, pine, painted | .125 | L.F. | .005 | .04 | .10 | .14 |
| TOTAL | | | .210 | 6.55 | 3.74 | 10.29 |

The costs in this system are based on a square foot of wall area. Do not subtract area for window & door openings.

| DESCRIPTION | QUAN. | UNIT | MAN HOURS | COST PER S.F. MAT. | COST PER S.F. INST. | COST PER S.F. TOTAL |
|---|---|---|---|---|---|---|
| | | | | | | |
| | | | | | | |
| | | | | | | |

# Brick/Stone Veneer Price Sheet

| | QUAN. | UNIT | MAN HOURS | MAT. | INST. | TOTAL |
|---|---|---|---|---|---|---|
| Brick, common standard running bond brick | 1.000 | S.F. | .183 | 1.29 | 3.33 | 4.62 |
| $200 per thousand | 1.000 | S.F. | .191 | 1.76 | 3.49 | 5.25 |
| $215 per thousand | 1.000 | S.F. | .200 | 1.87 | 3.63 | 5.50 |
| Buff or grey faced, brick $250 per thousand, running bond | 1.000 | S.F. | .200 | 2.20 | 3.65 | 5.85 |
| Header every 6th course | 1.000 | S.F. | .238 | 2.70 | 4.30 | 7.00 |
| English bond | 1.000 | S.F. | .314 | 3.52 | 5.73 | 9.25 |
| Flemish bond | 1.000 | S.F. | .215 | 2.42 | 3.93 | 6.35 |
| Common bond | 1.000 | S.F. | .293 | 3.08 | 5.32 | 8.40 |
| Stack bond | 1.000 | S.F. | .200 | 2.20 | 3.65 | 5.85 |
| Jumbo, $900 per thousand, running bond | 1.000 | S.F. | .101 | 3.47 | 1.83 | 5.30 |
| Norman, $385 per thousand, running bond | 1.000 | S.F. | .138 | 2.33 | 2.51 | 4.84 |
| Norwegian, $450 per thousand, running bond | 1.000 | S.F. | .117 | 2.26 | 2.13 | 4.39 |
| Economy, $370 per thousand, running bond | 1.000 | S.F. | .142 | 2.20 | 2.58 | 4.78 |
| Engineer, $255 per thousand, running bond | 1.000 | S.F. | .169 | 1.98 | 3.07 | 5.05 |
| Roman, $460 per thousand, running bond | 1.000 | S.F. | .176 | 3.53 | 3.22 | 6.75 |
| Utility, $650 per thousand, running bond | 1.000 | S.F. | .098 | 2.53 | 1.78 | 4.31 |
| Glazed, $750 per thousand, running bond | 1.000 | S.F. | .210 | 6.11 | 3.79 | 9.90 |
| Stone work, rough stone, average | 1.000 | S.F. | .194 | 6.45 | 3.43 | 9.88 |
| Maximum | 1.000 | S.F. | .291 | 9.63 | 5.12 | 14.75 |
| Wall ties, galvanized, corrugated 7/8" x 7", 24 gauge | 1.000 | Ea. | .008 | .03 | .15 | .18 |
| 16 gauge | 1.000 | Ea. | .008 | .11 | .15 | .26 |
| Cavity wall, every 3rd course 6" long Z type, 1/8" diameter | 1.330 | L.F. | .010 | .23 | .18 | .41 |
| 3/16" diameter | 1.330 | L.F. | .010 | .09 | .20 | .29 |
| 8" long, Z type, 3/16" diameter | 1.330 | L.F. | .010 | .11 | .20 | .31 |
| Copper weld | 1.330 | L.F. | .010 | .16 | .20 | .36 |
| Building paper, aluminum and craft laminated foil, 1 side | 1.000 | S.F. | .004 | .03 | .09 | .12 |
| 2 sides | 1.000 | S.F. | .004 | .07 | .08 | .15 |
| #15 asphalt paper | 1.100 | S.F. | .002 | .03 | .06 | .09 |
| Polyethylene, .002" thick | 1.000 | S.F. | .002 | .03 | .05 | .08 |
| .004" thick | 1.000 | S.F. | .002 | .06 | .04 | .10 |
| .006" thick | 1.000 | S.F. | .002 | .06 | .04 | .10 |
| .008" thick | 1.000 | S.F. | .002 | .06 | .04 | .10 |
| .010" thick | 1.000 | S.F. | .002 | .07 | .04 | .11 |
| Trim, 1" x 4", cedar | .125 | L.F. | .005 | .09 | .10 | .19 |
| Fir | .125 | L.F. | .005 | .04 | .10 | .14 |
| Redwood | .125 | L.F. | .005 | .09 | .10 | .19 |
| White pine | .125 | L.F. | .005 | .04 | .10 | .14 |

# 4 | EXTERIOR WALLS  08 | Wood Siding Systems

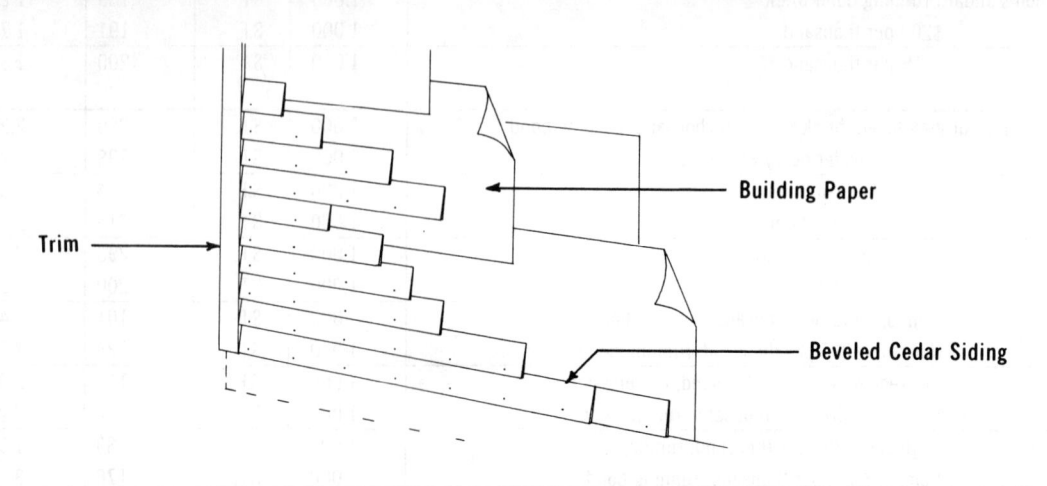

*Trim*
*Building Paper*
*Beveled Cedar Siding*

| SYSTEM DESCRIPTION | QUAN. | UNIT | MAN HOURS | COST PER S.F. MAT. | COST PER S.F. INST. | COST PER S.F. TOTAL |
|---|---|---|---|---|---|---|
| **1/2" X 6" BEVELED CEDAR SIDING, "A" GRADE** | | | | | | |
| 1/2" x 6" beveled cedar siding | 1.000 | S.F. | .028 | 1.39 | .54 | 1.93 |
| #15 asphalt felt paper | 1.100 | S.F. | .002 | .03 | .06 | .09 |
| Trim, cedar | .125 | L.F. | .005 | .09 | .10 | .19 |
| Paint, primer & 2 coats | 1.000 | S.F. | .017 | .19 | .31 | .50 |
| TOTAL | | | .052 | 1.70 | 1.01 | 2.71 |
| **1/2" X 8" BEVELED CEDAR SIDING, "A" GRADE** | | | | | | |
| 1/2" x 8" beveled cedar siding | 1.000 | S.F. | .023 | 1.30 | .45 | 1.75 |
| #15 asphalt felt paper | 1.100 | S.F. | .002 | .03 | .06 | .09 |
| Trim, cedar | .125 | L.F. | .005 | .09 | .10 | .19 |
| Paint, primer & 2 coats | 1.000 | S.F. | .017 | .19 | .31 | .50 |
| TOTAL | | | .047 | 1.61 | .92 | 2.53 |
| **1" X 4" TONGUE & GROOVE, VERTICAL, REDWOOD, VERTICAL GRAIN** | | | | | | |
| 1" x 4" tongue & groove, vertical, redwood | 1.000 | S.F. | .034 | 2.75 | .65 | 3.40 |
| #15 asphalt felt paper | 1.100 | S.F. | .002 | .03 | .06 | .09 |
| Trim, redwood | .125 | L.F. | .005 | .09 | .10 | .19 |
| Sealer, 1 coat, stain, 1 coat | 1.000 | S.F. | .012 | .13 | .22 | .35 |
| TOTAL | | | .053 | 3.00 | 1.03 | 4.03 |
| **1" X 6" TONGUE & GROOVE, VERTICAL, REDWOOD, VERTICAL GRAIN** | | | | | | |
| 1" x 6" tongue & groove, vertical, redwood | 1.000 | S.F. | .024 | 2.59 | .46 | 3.05 |
| #15 asphalt felt paper | 1.100 | S.F. | .002 | .03 | .06 | .09 |
| Trim, redwood | .125 | L.F. | .005 | .09 | .10 | .19 |
| Sealer, 1 coat, stain, 1 coat | 1.000 | S.F. | .012 | .13 | .22 | .35 |
| TOTAL | | | .043 | 2.84 | .84 | 3.68 |

The costs in this system are based on a square foot of wall area. Do not subtract area for door or window openings.

| DESCRIPTION | QUAN. | UNIT | MAN HOURS | COST PER S.F. MAT. | COST PER S.F. INST. | COST PER S.F. TOTAL |
|---|---|---|---|---|---|---|
| | | | | | | |
| | | | | | | |
| | | | | | | |

# Wood Siding Price Sheet

| | QUAN. | UNIT | MAN HOURS | COST PER S.F. MAT. | COST PER S.F. INST. | TOTAL |
|---|---|---|---|---|---|---|
| Siding, beveled cedar, "B" grade, 1/2" x 6" | 1.000 | S.F. | .028 | 1.30 | .54 | 1.84 |
| 1/2" x 8" | 1.000 | S.F. | .023 | 1.16 | .44 | 1.60 |
| "A" grade, 1/2" x 6" | 1.000 | S.F. | .028 | 1.39 | .54 | 1.93 |
| 1/2" x 8" | 1.000 | S.F. | .023 | 1.30 | .45 | 1.75 |
| Clear grade, 1/2" x 6" | 1.000 | S.F. | .028 | 1.51 | .54 | 2.05 |
| 1/2" x 8" | 1.000 | S.F. | .023 | 1.44 | .45 | 1.89 |
| Redwood, clear vertical grain, 1/2" x 6" | 1.000 | S.F. | .028 | 1.62 | .54 | 2.16 |
| 1/2" x 8" | 1.000 | S.F. | .023 | 1.38 | .44 | 1.82 |
| Clear all heart vertical grain, 1/2" x 6" | 1.000 | S.F. | .028 | 1.74 | .54 | 2.28 |
| 1/2" x 8" | 1.000 | S.F. | .023 | 1.51 | .44 | 1.95 |
| Siding board & batten, cedar, "B" grade, 1" x 10" | 1.000 | S.F. | .020 | 1.10 | .39 | 1.49 |
| 1" x 12" | 1.000 | S.F. | .017 | 1.16 | .34 | 1.50 |
| Redwood, clear vertical grain, 1" x 10" | 1.000 | S.F. | .020 | 2.31 | .39 | 2.70 |
| 1" x 12" | 1.000 | S.F. | .017 | 2.50 | .34 | 2.84 |
| White pine, #2 & better, 1" x 10" | 1.000 | S.F. | .020 | .51 | .39 | .90 |
| 1" x 12" | 1.000 | S.F. | .017 | .53 | .35 | .88 |
| Siding vertical, tongue & groove, cedar "B" grade, 1" x 4" | 1.000 | S.F. | .034 | 1.45 | .65 | 2.10 |
| 1" x 6" | 1.000 | S.F. | .024 | 1.39 | .46 | 1.85 |
| 1" x 8" | 1.000 | S.F. | .024 | 1.32 | .46 | 1.78 |
| 1" x 10" | 1.000 | S.F. | .021 | 1.28 | .41 | 1.69 |
| "A" grade, 1" x 4" | 1.000 | S.F. | .034 | 1.56 | .65 | 2.21 |
| 1" x 6" | 1.000 | S.F. | .024 | 1.53 | .46 | 1.99 |
| 1" x 8" | 1.000 | S.F. | .024 | 1.45 | .46 | 1.91 |
| 1" x 10" | 1.000 | S.F. | .021 | 1.39 | .41 | 1.80 |
| Clear vertical grain, 1" x 4" | 1.000 | S.F. | .034 | 2.26 | .65 | 2.91 |
| 1" x 6" | 1.000 | S.F. | .024 | 2.11 | .46 | 2.57 |
| 1" x 8" | 1.000 | S.F. | .024 | 2.02 | .46 | 2.48 |
| 1" x 10" | 1.000 | S.F. | .021 | 1.99 | .41 | 2.40 |
| Redwood, clear vertical grain, 1" x 4" | 1.000 | S.F. | .034 | 2.75 | .65 | 3.40 |
| 1" x 6" | 1.000 | S.F. | .024 | 2.59 | .46 | 3.05 |
| 1" x 8" | 1.000 | S.F. | .024 | 2.37 | .46 | 2.83 |
| 1" x 10" | 1.000 | S.F. | .021 | 2.37 | .41 | 2.78 |
| Clear all heart vertical grain, 1" x 4" | 1.000 | S.F. | .034 | 2.86 | .65 | 3.51 |
| 1" x 6" | 1.000 | S.F. | .024 | 2.70 | .46 | 3.16 |
| 1" x 8" | 1.000 | S.F. | .024 | 2.53 | .46 | 2.99 |
| 1" x 10" | 1.000 | S.F. | .021 | 2.48 | .41 | 2.89 |
| White pine, 1" x 8" | 1.000 | S.F. | .024 | .51 | .46 | .97 |
| Siding plywood, texture 1-11 cedar, 3/8" thick | 1.000 | S.F. | .024 | 1.09 | .49 | 1.58 |
| 5/8" thick | 1.000 | S.F. | .024 | 1.44 | .49 | 1.93 |
| Redwood, 3/8" thick | 1.000 | S.F. | .024 | 1.09 | .49 | 1.58 |
| 5/8" thick | 1.000 | S.F. | .024 | 1.27 | .48 | 1.75 |
| Fir, 3/8" thick | 1.000 | S.F. | .024 | .42 | .48 | .90 |
| 5/8" thick | 1.000 | S.F. | .024 | .59 | .49 | 1.08 |
| Southern yellow pine, 3/8" thick | 1.000 | S.F. | .024 | .42 | .48 | .90 |
| 5/8" thick | 1.000 | S.F. | .024 | .59 | .49 | 1.08 |
| Hard board, 7/16" thick primed, plain finish | 1.000 | S.F. | .021 | .37 | .44 | .81 |
| Board finish | 1.000 | S.F. | .021 | .64 | .44 | 1.08 |
| Polyvinyl coated, 3/8" thick | 1.000 | S.F. | .021 | .91 | .44 | 1.35 |
| 5/8" thick | 1.000 | S.F. | .024 | .59 | .49 | 1.08 |
| Paper, #15 asphalt felt | 1.100 | S.F. | .002 | .03 | .06 | .09 |
| Trim, cedar | .125 | L.F. | .005 | .09 | .10 | .19 |
| Fir | .125 | L.F. | .005 | .04 | .10 | .14 |
| Redwood | .125 | L.F. | .005 | .09 | .10 | .19 |
| White pine | .125 | L.F. | .005 | .04 | .10 | .14 |
| Painting, primer, & 1 coat | 1.000 | S.F. | .012 | .13 | .22 | .35 |
| 2 coats | 1.000 | S.F. | .017 | .19 | .31 | .50 |
| Stain, sealer, & 1 coat | 1.000 | S.F. | .013 | .12 | .24 | .36 |
| 2 coats | 1.000 | S.F. | .019 | .17 | .34 | .51 |

# 4 | EXTERIOR WALLS  12 | Shingle Siding Systems

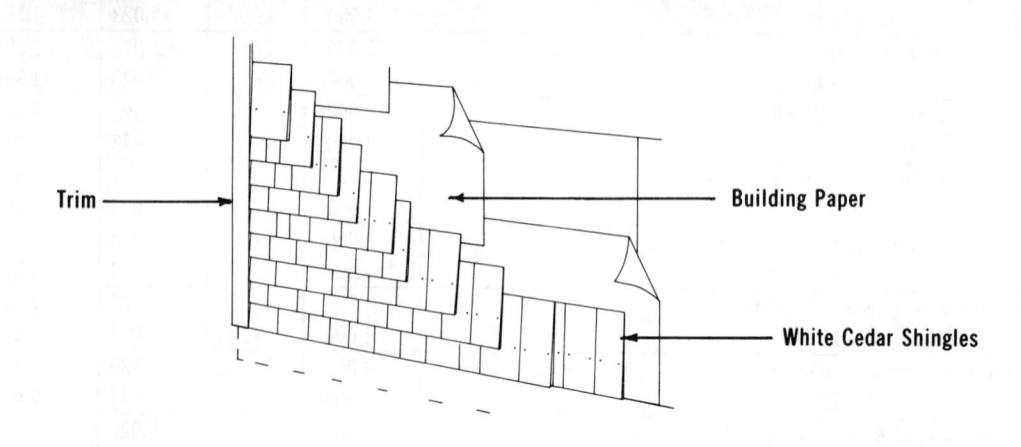

Trim → | ← Building Paper
← White Cedar Shingles

| SYSTEM DESCRIPTION | QUAN. | UNIT | MAN HOURS | MAT. | INST. | TOTAL |
|---|---|---|---|---|---|---|
| **WHITE CEDAR SHINGLES, 5" EXPOSURE** | | | | | | |
| White cedar shingles, 16" long, grade "A", 5" exposure | 1.000 | S.F. | .033 | .88 | .67 | 1.55 |
| #15 asphalt felt paper | 1.100 | S.F. | .002 | .03 | .05 | .08 |
| Trim, cedar | .125 | S.F. | .005 | .09 | .10 | .19 |
| Paint, primer & 1 coat | 1.000 | S.F. | .013 | .12 | .24 | .36 |
| TOTAL | | | .053 | 1.12 | 1.06 | 2.18 |
| **NO. 1 PERFECTIONS, 5-1/2" EXPOSURE** | | | | | | |
| No. 1 perfections, red cedar, 5-1/2" exposure | 1.000 | S.F. | .029 | 1.21 | .59 | 1.80 |
| #15 asphalt felt paper | 1.100 | S.F. | .002 | .03 | .05 | .08 |
| Trim, cedar | .125 | S.F. | .005 | .09 | .10 | .19 |
| Stain, sealer & 1 coat | 1.000 | S.F. | .013 | .12 | .24 | .36 |
| TOTAL | | | .049 | 1.45 | .98 | 2.43 |
| **RESQUARED & REBUTTED PERFECTIONS, 5-1/2" EXPOSURE** | | | | | | |
| Resquared & rebutted perfections, 5-1/2" exposure | 1.000 | S.F. | .026 | 1.22 | .53 | 1.75 |
| #15 asphalt felt paper | 1.100 | S.F. | .002 | .03 | .05 | .08 |
| Trim, cedar | .125 | S.F. | .005 | .09 | .10 | .19 |
| Stain, sealer & 1 coat | 1.000 | S.F. | .013 | .12 | .24 | .36 |
| TOTAL | | | .046 | 1.46 | .92 | 2.38 |
| **HAND-SPLIT SHAKES, 8-1/2" EXPOSURE** | | | | | | |
| Hand-split red cedar shakes, 18" long, 8-1/2" | 1.000 | S.F. | .040 | .92 | .78 | 1.70 |
| #15 asphalt felt paper | 1.100 | S.F. | .002 | .03 | .05 | .08 |
| Trim, cedar | .125 | S.F. | .005 | .09 | .10 | .19 |
| Stain, sealer & 1 coat | 1.000 | S.F. | .013 | .12 | .24 | .36 |
| TOTAL | | | .060 | 1.16 | 1.17 | 2.33 |

The costs in this system are based on a square foot of wall area. Do not subtract area for door or window openings.

| DESCRIPTION | QUAN. | UNIT | MAN HOURS | MAT. | INST. | TOTAL |
|---|---|---|---|---|---|---|
| | | | | | | |
| | | | | | | |
| | | | | | | |

# Shingle Siding Price Sheet

| | QUAN. | UNIT | MAN HOURS | COST PER S.F. MAT. | COST PER S.F. INST. | TOTAL |
|---|---|---|---|---|---|---|
| Shingles wood, white cedar 16" long, "A" grade, 5" exposure | 1.000 | S.F. | .033 | .88 | .67 | 1.55 |
| 7" exposure | 1.000 | S.F. | .029 | .79 | .61 | 1.40 |
| 8-1/2" exposure | 1.000 | S.F. | .032 | .62 | .62 | 1.24 |
| 10" exposure | 1.000 | S.F. | .028 | .54 | .55 | 1.09 |
| "B" grade, 5" exposure | 1.000 | S.F. | .033 | .63 | .67 | 1.30 |
| 7" exposure | 1.000 | S.F. | .023 | .44 | .47 | .91 |
| 8-1/2" exposure | 1.000 | S.F. | .019 | .38 | .40 | .78 |
| 10" exposure | 1.000 | S.F. | .016 | .32 | .33 | .65 |
| Fire retardant, "A" grade, 5" exposure | 1.000 | S.F. | .068 | 3.07 | 1.38 | 4.45 |
| 7" exposure | 1.000 | S.F. | .061 | 2.76 | 1.25 | 4.01 |
| 8-1/2" exposure | 1.000 | S.F. | .055 | 2.45 | 1.11 | 3.56 |
| 10" exposure | 1.000 | S.F. | .048 | 2.15 | .97 | 3.12 |
| "B" grade, 5" exposure | 1.000 | S.F. | .068 | 2.82 | 1.38 | 4.20 |
| 7" exposure | 1.000 | S.F. | .048 | 1.97 | .97 | 2.94 |
| 8-1/2" exposure | 1.000 | S.F. | .041 | 1.69 | .83 | 2.52 |
| 10" exposure | 1.000 | S.F. | .034 | 1.41 | .69 | 2.10 |
| No. 1 perfections red cedar, 18" long, 5-1/2" exposure | 1.000 | S.F. | .029 | 1.21 | .59 | 1.80 |
| 7" exposure | 1.000 | S.F. | .035 | 1.21 | .69 | 1.90 |
| 8-1/2" exposure | 1.000 | S.F. | .032 | 1.09 | .62 | 1.71 |
| 10" exposure | 1.000 | S.F. | .024 | .85 | .48 | 1.33 |
| Fire retardant, 5" exposure | 1.000 | S.F. | .058 | 3.59 | 1.16 | 4.75 |
| 7" exposure | 1.000 | S.F. | .071 | 3.40 | 1.40 | 4.80 |
| 8-1/2" exposure | 1.000 | S.F. | .064 | 3.06 | 1.26 | 4.32 |
| 10" exposure | 1.000 | S.F. | .049 | 2.38 | .98 | 3.36 |
| Resquared & rebutted, 5-1/2" exposure | 1.000 | S.F. | .026 | 1.22 | .53 | 1.75 |
| 7" exposure | 1.000 | S.F. | .024 | 1.10 | .48 | 1.58 |
| 8-1/2" exposure | 1.000 | S.F. | .021 | .98 | .42 | 1.40 |
| 10" exposure | 1.000 | S.F. | .018 | .85 | .38 | 1.23 |
| Fire retardant, 5" exposure | 1.000 | S.F. | .053 | 3.43 | 1.07 | 4.50 |
| 7" exposure | 1.000 | S.F. | .048 | 3.09 | .97 | 4.06 |
| 8-1/2" exposure | 1.000 | S.F. | .042 | 2.75 | .85 | 3.60 |
| 10" exposure | 1.000 | S.F. | .041 | 2.34 | .82 | 3.16 |
| Hand-split, red cedar, 24" long, 7" exposure | 1.000 | S.F. | .044 | 1.68 | .84 | 2.52 |
| 8-1/2" exposure | 1.000 | S.F. | .038 | 1.44 | .72 | 2.16 |
| 10" exposure | 1.000 | S.F. | .032 | 1.20 | .60 | 1.80 |
| 12" exposure | 1.000 | S.F. | .025 | .96 | .48 | 1.44 |
| Fire retardant, 7" exposure | 1.000 | S.F. | .089 | 4.93 | 1.72 | 6.65 |
| 8-1/2" exposure | 1.000 | S.F. | .076 | 4.23 | 1.47 | 5.70 |
| 10" exposure | 1.000 | S.F. | .064 | 3.52 | 1.23 | 4.75 |
| 12" exposure | 1.000 | S.F. | .051 | 2.82 | .98 | 3.80 |
| 18" long, 5" exposure | 1.000 | S.F. | .068 | 1.57 | 1.32 | 2.89 |
| 7" exposure | 1.000 | S.F. | .048 | 1.11 | .93 | 2.04 |
| 8-1/2" exposure | 1.000 | S.F. | .040 | .92 | .78 | 1.70 |
| 10" exposure | 1.000 | S.F. | .036 | .83 | .70 | 1.53 |
| Fire retardant, 5" exposure | 1.000 | S.F. | .136 | 6.34 | 2.67 | 9.01 |
| 7" exposure | 1.000 | S.F. | .096 | 4.48 | 1.88 | 6.36 |
| 8-1/2" exposure | 1.000 | S.F. | .080 | 3.73 | 1.57 | 5.30 |
| 10" exposure | 1.000 | S.F. | .072 | 3.35 | 1.42 | 4.77 |
| Paper, #15 asphalt felt | 1.100 | S.F. | .002 | .03 | .05 | .08 |
| Trim, cedar | .125 | S.F. | .005 | .09 | .10 | .19 |
| Fir | .125 | S.F. | .005 | .04 | .10 | .14 |
| Redwood | .125 | S.F. | .005 | .09 | .10 | .19 |
| White pine | .125 | S.F. | .005 | .04 | .10 | .14 |
| Painting, primer, & 1 coat | 1.000 | S.F. | .012 | .13 | .22 | .35 |
| 2 coats | 1.000 | S.F. | .017 | .19 | .31 | .50 |
| Staining, sealer, & 1 coat | 1.000 | S.F. | .013 | .12 | .24 | .36 |
| 2 coats | 1.000 | S.F. | .019 | .17 | .34 | .51 |

# 4 | EXTERIOR WALLS  16 | Metal & Plastic Siding

*Diagram labels: Alum. Trim, Building Paper, Alum. Horizontal Siding, Backer Insulation Board*

| SYSTEM DESCRIPTIION | QUAN. | UNIT | MAN HOURS | COST PER S.F. MAT. | COST PER S.F. INST. | COST PER S.F. TOTAL |
|---|---|---|---|---|---|---|
| **ALUMINUM CLAPBOARD SIDING, 8" WIDE, WHITE** | | | | | | |
| Aluminum horizontal siding, 8" clapboard | 1.000 | S.F. | .036 | .99 | .74 | 1.73 |
| Backer, insulation board | 1.000 | S.F. | .008 | .28 | .16 | .44 |
| Trim, aluminum | .600 | L.F. | .015 | .47 | .32 | .79 |
| Paper, #15 asphalt felt | 1.100 | S.F. | .002 | .03 | .06 | .09 |
| TOTAL | | | .061 | 1.77 | 1.28 | 3.05 |
| **ALUMINUM VERTICAL BOARD & BATTEN, WHITE** | | | | | | |
| Aluminum vertical board & batten | 1.000 | S.F. | .031 | 1.05 | .63 | 1.68 |
| Backer insulation board | 1.000 | S.F. | .008 | .28 | .16 | .44 |
| Trim, aluminum | .600 | L.F. | .015 | .47 | .32 | .79 |
| Paper, #15 asphalt felt | 1.100 | S.F. | .002 | .03 | .06 | .09 |
| TOTAL | | | .056 | 1.83 | 1.17 | 3.00 |
| **VINYL CLAPBOARD SIDING, 8" WIDE, WHITE** | | | | | | |
| PVC vinyl horizontal siding, 8" clapboard | 1.000 | S.F. | .032 | .61 | .66 | 1.27 |
| Backer, insulation board | 1.000 | S.F. | .008 | .28 | .16 | .44 |
| Trim, vinyl | .600 | L.F. | .013 | .49 | .28 | .77 |
| Paper, #15 asphalt felt | 1.100 | S.F. | .002 | .03 | .06 | .09 |
| TOTAL | | | .056 | 1.41 | 1.16 | 2.57 |
| **VINYL VERTICAL BOARD & BATTEN, WHITE** | | | | | | |
| PVC vinyl vertical board & batten | 1.000 | S.F. | .029 | .64 | .59 | 1.23 |
| Backer, insulation board | 1.000 | S.F. | .008 | .28 | .16 | .44 |
| Trim, vinyl | .600 | L.F. | .013 | .49 | .28 | .77 |
| Paper, #15 asphalt felt | 1.100 | S.F. | .002 | .03 | .06 | .09 |
| TOTAL | | | .053 | 1.44 | 1.09 | 2.53 |

The costs in this system are on a square foot of wall basis. Do not subtract openings from wall area.

| DESCRIPTION | QUAN. | UNIT | MAN HOURS | COST PER S.F. MAT. | COST PER S.F. INST. | COST PER S.F. TOTAL |
|---|---|---|---|---|---|---|
| | | | | | | |
| | | | | | | |
| | | | | | | |

# Metal & Plastic Siding Price Sheet

| | QUAN. | UNIT | MAN HOURS | COST PER S.F. MAT. | COST PER S.F. INST. | TOTAL |
|---|---|---|---|---|---|---|
| Siding, aluminum .024" thick, smooth, 8" wide, white | 1.000 | S.F. | .036 | .99 | .74 | 1.73 |
| Color | 1.000 | S.F. | .036 | 1.04 | .74 | 1.78 |
| Double 4" pattern, 8" wide, white | 1.000 | S.F. | .036 | 1.05 | .73 | 1.78 |
| Color | 1.000 | S.F. | .036 | 1.10 | .73 | 1.83 |
| Double 5" pattern, 10" wide, white | 1.000 | S.F. | .032 | 1.10 | .66 | 1.76 |
| Color | 1.000 | S.F. | .032 | 1.15 | .66 | 1.81 |
| Embossed, single, 8" wide, white | 1.000 | S.F. | .036 | 1.05 | .73 | 1.78 |
| Color | 1.000 | S.F. | .036 | 1.10 | .73 | 1.83 |
| Double 4" pattern, 8" wide, white | 1.000 | S.F. | .036 | 1.08 | .74 | 1.82 |
| Color | 1.000 | S.F. | .036 | 1.13 | .74 | 1.87 |
| Double 5" pattern, 10" wide, white | 1.000 | S.F. | .032 | 1.16 | .66 | 1.82 |
| Color | 1.000 | S.F. | .032 | 1.21 | .66 | 1.87 |
| Alum siding with insulation board, smooth, 8" wide, white | 1.000 | S.F. | .048 | 1.03 | .98 | 2.01 |
| Color | 1.000 | S.F. | .048 | 1.08 | .98 | 2.06 |
| Double 4" pattern, 8" wide, white | 1.000 | S.F. | .048 | 1.07 | .98 | 2.05 |
| Color | 1.000 | S.F. | .048 | 1.12 | .98 | 2.10 |
| Double 5" pattern, 10" wide, white | 1.000 | S.F. | .043 | 1.10 | .89 | 1.99 |
| Color | 1.000 | S.F. | .043 | 1.15 | .89 | 2.04 |
| Embossed, single, 8" wide, white | 1.000 | S.F. | .048 | 1.05 | .97 | 2.02 |
| Color | 1.000 | S.F. | .048 | 1.10 | .97 | 2.07 |
| Double 4" pattern, 8" wide, white | 1.000 | S.F. | .048 | 1.08 | .98 | 2.06 |
| Color | 1.000 | S.F. | .048 | 1.13 | .98 | 2.11 |
| Double 5" pattern, 10" wide, white | 1.000 | S.F. | .043 | 1.10 | .89 | 1.99 |
| Color | 1.000 | S.F. | .043 | 1.15 | .89 | 2.04 |
| Aluminum, shake finish, 10" wide, white | 1.000 | S.F. | .032 | 1.32 | .66 | 1.98 |
| Color | 1.000 | S.F. | .032 | 1.37 | .66 | 2.03 |
| Aluminum, vertical, 12" wide, white | 1.000 | S.F. | .031 | 1.05 | .63 | 1.68 |
| Color | 1.000 | S.F. | .031 | 1.10 | .63 | 1.73 |
| Vinyl siding, 8" wide, smooth, white | 1.000 | S.F. | .032 | .61 | .66 | 1.27 |
| Color | 1.000 | S.F. | .032 | .66 | .66 | 1.32 |
| 10" wide, smooth, white | 1.000 | S.F. | .029 | .63 | .59 | 1.22 |
| Color | 1.000 | S.F. | .029 | .68 | .59 | 1.27 |
| Double 4" pattern, 8" wide, white | 1.000 | S.F. | .032 | .66 | .66 | 1.32 |
| Color | 1.000 | S.F. | .032 | .71 | .66 | 1.37 |
| Double 5" pattern, 10" wide, white | 1.000 | S.F. | .029 | .68 | .60 | 1.28 |
| Color | 1.000 | S.F. | .029 | .73 | .60 | 1.33 |
| Embossed, single, 8" wide, white | 1.000 | S.F. | .032 | .66 | .66 | 1.32 |
| Color | 1.000 | S.F. | .032 | .71 | .66 | 1.37 |
| 10" wide, white | 1.000 | S.F. | .029 | .68 | .60 | 1.28 |
| Color | 1.000 | S.F. | .029 | .73 | .60 | 1.33 |
| Double 4" pattern, 8" wide, white | 1.000 | S.F. | .032 | .70 | .67 | 1.37 |
| Color | 1.000 | S.F. | .032 | .75 | .67 | 1.42 |
| Double 5" pattern, 10" wide, white | 1.000 | S.F. | .029 | .73 | .59 | 1.32 |
| Color | 1.000 | S.F. | .029 | .78 | .59 | 1.37 |
| Vinyl, shake finish, 10" wide, white | 1.000 | S.F. | .029 | .72 | .59 | 1.31 |
| Color | 1.000 | S.F. | .029 | .77 | .59 | 1.36 |
| Vinyl, vertical, double 5" pattern, 10" wide, white | 1.000 | S.F. | .029 | .64 | .59 | 1.23 |
| Color | 1.000 | S.F. | .029 | .69 | .59 | 1.28 |
| Backer board, installed in siding panels 8" or 10" wide | 1.000 | S.F. | .008 | .28 | .16 | .44 |
| 4' x 8' sheets, polystyrene, 3/4" thick | 1.000 | S.F. | .010 | .40 | .20 | .60 |
| 4' x 8' fiberboard, plain | 1.000 | S.F. | .008 | .28 | .16 | .44 |
| Foil faced | 1.000 | S.F. | .013 | .20 | .25 | .45 |
| Trim, aluminum, white | .600 | L.F. | .015 | .47 | .32 | .79 |
| Color | .600 | L.F. | .015 | .49 | .32 | .81 |
| Vinyl, white | .600 | L.F. | .013 | .49 | .28 | .77 |
| Color | .600 | L.F. | .014 | .50 | .29 | .79 |
| Paper, #15 asphalt felt | 1.100 | S.F. | .002 | .03 | .06 | .09 |
| Kraft paper, plain | 1.100 | S.F. | .004 | .03 | .10 | .13 |
| Foil backed | 1.100 | S.F. | .004 | .08 | .09 | .17 |

# 4 | EXTERIOR WALLS　20 | Insulation Systems

| DESCRIPTION | QUAN. | UNIT | MAN HOURS | COST PER S.F. MAT. | COST PER S.F. INST. | TOTAL |
|---|---|---|---|---|---|---|
| Poured insulation, cellulose fiber, R3.8 per inch | 1.000 | S.F. | .003 | .04 | .06 | .10 |
| Fiberglass, R4.0 per inch | 1.000 | S.F. | .003 | .02 | .07 | .09 |
| Mineral wool, R3.0 per inch | 1.000 | S.F. | .003 | .07 | .06 | .13 |
| Polystyrene, R4.0 per inch | 1.000 | S.F. | .003 | .12 | .07 | .19 |
| Vermiculite, R2.7 per inch | 1.000 | S.F. | .003 | .13 | .06 | .19 |
| Perlite, R2.7 per inch | 1.000 | S.F. | .003 | .13 | .06 | .19 |
| Reflective, aluminum foil on kraft paper, foil one side R9 | 1.000 | S.F. | .004 | .03 | .09 | .12 |
| Multilayered with air spaces, 2 ply, R14 | 1.000 | S.F. | .004 | .13 | .08 | .21 |
| 3 ply, R17 | 1.000 | S.F. | .005 | .18 | .10 | .28 |
| 5 ply, R22 | 1.000 | S.F. | .005 | .26 | .11 | .37 |
| Rigid insulation, fiberglass, unfaced, | | | | | | |
| 1-1/2" thick, R6.2 | 1.000 | S.F. | .008 | .28 | .16 | .44 |
| 2" thick, R8.3 | 1.000 | S.F. | .009 | .41 | .17 | .58 |
| 2-1/2" thick, R10.3 | 1.000 | S.F. | .010 | .62 | .19 | .81 |
| 3" thick, R12.4 | 1.000 | S.F. | .010 | .62 | .19 | .81 |
| Foil faced, 1" thick, R4.3 | 1.000 | S.F. | .008 | .81 | .17 | .98 |
| 1-1/2" thick, R6.2 | 1.000 | S.F. | .009 | .85 | .17 | 1.02 |
| 2" thick, R8.7 | 1.000 | S.F. | .009 | 1.25 | .19 | 1.44 |
| 2-1/2" thick, R10.9 | 1.000 | S.F. | .010 | 1.41 | .19 | 1.60 |
| 3" thick, R13.0 | 1.000 | S.F. | .011 | 1.68 | .21 | 1.89 |
| Foam glass, 1-1/2" thick R2.64 | 1.000 | S.F. | .011 | 1.56 | .21 | 1.77 |
| 2" thick R5.26 | 1.000 | S.F. | .011 | 2.11 | .23 | 2.34 |
| Perlite, 1" thick R2.77 | 1.000 | S.F. | .011 | .31 | .21 | .52 |
| 2" thick R5.55 | 1.000 | S.F. | .011 | .55 | .22 | .77 |
| Polystyrene, extruded, blue, 2.2#/C.F., 3/4" thick R4 | 1.000 | S.F. | .010 | .40 | .20 | .60 |
| 1-1/2" thick R8.1 | 1.000 | S.F. | .011 | .77 | .22 | .99 |
| 2" thick R10.8 | 1.000 | S.F. | .011 | .86 | .22 | 1.08 |
| Molded bead board, white, 1" thick R3.85 | 1.000 | S.F. | .011 | .17 | .20 | .37 |
| 1-1/2" thick, R5.6 | 1.000 | S.F. | .011 | .29 | .21 | .50 |
| 2" thick, R7.7 | 1.000 | S.F. | .011 | .34 | .22 | .56 |
| Urethane, no backing, 1/2" thick, R2.9 | 1.000 | S.F. | .010 | .23 | .20 | .43 |
| 1" thick, R5.8 | 1.000 | S.F. | .011 | .45 | .21 | .66 |
| 1-1/2" thick, R8.7 | 1.000 | S.F. | .011 | .66 | .22 | .88 |
| 2" thick, R11.7 | 1.000 | S.F. | .011 | .79 | .23 | 1.02 |
| Fire resistant, 1/2" thick, R2.9 | 1.000 | S.F. | .010 | .29 | .20 | .49 |
| 1" thick, R5.8 | 1.000 | S.F. | .011 | .55 | .21 | .76 |
| 1-1/2" thick, R8.7 | 1.000 | S.F. | .011 | .84 | .21 | 1.05 |
| 2" thick, R11.7 | 1.000 | S.F. | .011 | .99 | .22 | 1.21 |
| Non-rigid insulation, batts | | | | | | |
| Fiberglass, kraft faced, 3-1/2" thick, R11, 11" wide | 1.000 | S.F. | .005 | .20 | .10 | .30 |
| 15" wide | 1.000 | S.F. | .005 | .20 | .10 | .30 |
| 23" wide | 1.000 | S.F. | .005 | .20 | .10 | .30 |
| 6" thick, R19, 11" wide | 1.000 | S.F. | .006 | .33 | .12 | .45 |
| 15" wide | 1.000 | S.F. | .006 | .33 | .12 | .45 |
| 23" wide | 1.000 | S.F. | .006 | .33 | .12 | .45 |
| 9" thick, R30, 15" wide | 1.000 | S.F. | .006 | .53 | .11 | .64 |
| 23" wide | 1.000 | S.F. | .006 | .53 | .11 | .64 |
| 12" thick, R38, 15" wide | 1.000 | S.F. | .006 | .72 | .11 | .83 |
| 23" wide | 1.000 | S.F. | .006 | .72 | .11 | .83 |
| Fiberglass, foil faced, 3-1/2" thick, R11, 15" wide | 1.000 | S.F. | .005 | .22 | .10 | .32 |
| 23" wide | 1.000 | S.F. | .005 | .22 | .10 | .32 |
| 6" thick, R19, 15" thick | 1.000 | S.F. | .005 | .35 | .10 | .45 |
| 23" wide | 1.000 | S.F. | .005 | .35 | .10 | .45 |
| 9" thick, R30, 15" wide | 1.000 | S.F. | .006 | .57 | .12 | .69 |
| 23" wide | 1.000 | S.F. | .006 | .57 | .12 | .69 |

# Insulation Systems

| Insulation Systems | QUAN. | UNIT | MAN HOURS | COST PER S.F. MAT. | COST PER S.F. INST. | TOTAL |
|---|---|---|---|---|---|---|
| Non-rigid insulation batts | | | | | | |
|   Fiberglass unfaced, 3-1/2" thick, R11, 15" wide | 1.000 | S.F. | .005 | .18 | .09 | .27 |
|     23" wide | 1.000 | S.F. | .005 | .18 | .09 | .27 |
|   6" thick, R19, 15" wide | 1.000 | S.F. | .006 | .32 | .11 | .43 |
|     23" wide | 1.000 | S.F. | .006 | .32 | .11 | .43 |
|   9" thick, R19, 15" wide | 1.000 | S.F. | .007 | .51 | .13 | .64 |
|     23" wide | 1.000 | S.F. | .007 | .51 | .13 | .64 |
|   12" thick, R38, 15" wide | 1.000 | S.F. | .007 | .68 | .14 | .82 |
|     23" wide | 1.000 | S.F. | .007 | .68 | .14 | .82 |
| | | | | | | |
|   Mineral fiber batts, 3" thick, R11 | 1.000 | S.F. | .005 | .53 | .10 | .63 |
|     3-1/2" thick, R13 | 1.000 | S.F. | .005 | .53 | .10 | .63 |
|     6" thick, R19 | 1.000 | S.F. | .005 | .64 | .10 | .74 |
|     6-1/2" thick, R22 | 1.000 | S.F. | .005 | .64 | .10 | .74 |
|     10" thick, R30 | 1.000 | S.F. | .006 | 1.01 | .12 | 1.13 |

**4**

| Insulation Systems | QUAN. | UNIT | MAN HOURS | COST PER S.F. MAT. | COST PER S.F. INST. | TOTAL |
|---|---|---|---|---|---|---|
| Non-rigid insulation batts | | | | | | |
|   Fiberglass unfaced, 3-1/2" thick, R11, 15" wide | | | | | | |

# 4 | EXTERIOR WALLS  28 | Double Hung Window Systems

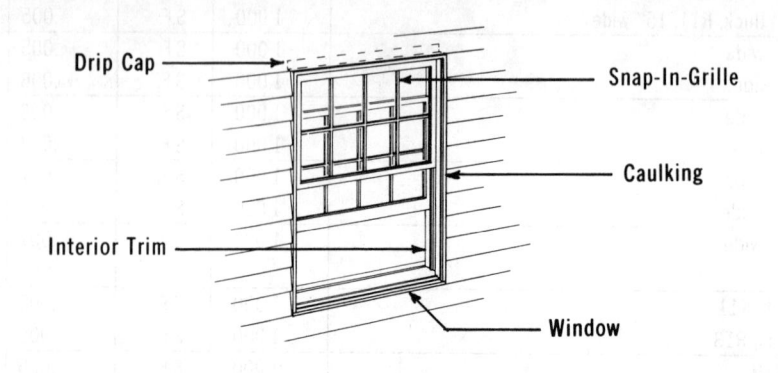

*Drip Cap, Snap-In-Grille, Caulking, Interior Trim, Window*

| SYSTEM DESCRIPTION | QUAN. | UNIT | MAN HOURS | MAT. | INST. | TOTAL |
|---|---|---|---|---|---|---|
| **BUILDER'S QUALITY WOOD WINDOW 2' X 3', DOUBLE HUNG** | | | | | | |
| Window, wood primed, builder's quality, 2' x 3', insulating glass | 1.000 | Ea. | .800 | 100.10 | 14.90 | 115.00 |
| Trim, interior casing | 11.000 | L.F. | .363 | 5.94 | 7.15 | 13.09 |
| Paint, interior & exterior, primer & 2 coats | 2.000 | FACE | .888 | 2.46 | 16.44 | 18.90 |
| Caulking | 10.000 | L.F. | .310 | .80 | 6.00 | 6.80 |
| Snap-in grille | 1.000 | Set | .333 | 23.10 | 6.90 | 30.00 |
| Drip cap, metal | 2.000 | L.F. | .040 | .28 | .78 | 1.06 |
| TOTAL | | | 2.734 | 132.68 | 52.17 | 184.85 |
| **PLASTIC CLAD WOOD WINDOW 3' X 4', DOUBLE HUNG** | | | | | | |
| Window, wood, plastic clad, premium, 3' x 4', insulating glass | 1.000 | Ea. | .889 | 176.00 | 19.00 | 195.00 |
| Trim, interior casing | 15.000 | L.F. | .495 | 8.10 | 9.75 | 17.85 |
| Paint, interior, primer & 2 coats | 1.000 | FACE | .800 | 1.89 | 14.81 | 16.70 |
| Caulking | 14.000 | L.F. | .434 | 1.12 | 8.40 | 9.52 |
| Snap-in grille | 1.000 | Set | .333 | 23.10 | 6.90 | 30.00 |
| TOTAL | | | 2.951 | 210.21 | 58.86 | 269.07 |
| **METAL CLAD WOOD WINDOW, 3' X 5', DOUBLE HUNG** | | | | | | |
| Window, wood, metal clad, deluxe, 3' x 5', insulating glass | 1.000 | Ea. | 1.000 | 275.00 | 20.00 | 295.00 |
| Trim, interior casing | 17.000 | L.F. | .561 | 9.18 | 11.05 | 20.23 |
| Paint, interior, primer & 2 coats | 1.000 | FACE | .800 | 1.89 | 14.81 | 16.70 |
| Caulking | 16.000 | L.F. | .496 | 1.28 | 9.60 | 10.88 |
| Snap-in grille | 1.000 | Set | .364 | 42.90 | 7.10 | 50.00 |
| Drip cap, metal | 3.000 | L.F. | .060 | .42 | 1.17 | 1.59 |
| TOTAL | | | 3.281 | 330.67 | 63.73 | 394.40 |

The cost of this system is on a cost per each window basis.

| DESCRIPTION | QUAN. | UNIT | MAN HOURS | MAT. | INST. | TOTAL |
|---|---|---|---|---|---|---|
| | | | | | | |
| | | | | | | |
| | | | | | | |
| | | | | | | |
| | | | | | | |

# Double Hung Window Price Sheet

| | QUAN. | UNIT | MAN HOURS | COST EACH MAT. | COST EACH INST. | TOTAL |
|---|---|---|---|---|---|---|
| Windows, double-hung, builder's quality, 2' x 3', single glass | 1.000 | Ea. | .800 | 67.10 | 15.90 | 83.00 |
| Insulating glass | 1.000 | Ea. | .800 | 100.10 | 14.90 | 115.00 |
| 3' x 4', single glass | 1.000 | Ea. | .889 | 90.20 | 19.80 | 110.00 |
| Insulating glass | 1.000 | Ea. | .889 | 143.00 | 17.00 | 160.00 |
| 4' x 4'-6", single glass | 1.000 | Ea. | 1.000 | 108.90 | 21.10 | 130.00 |
| Insulating glass | 1.000 | Ea. | 1.000 | 176.00 | 19.00 | 195.00 |
| Plastic clad premium insulating glass, 2'-6" x 3' | 1.000 | Ea. | .800 | 148.50 | 16.50 | 165.00 |
| 3' x 3'-6" | 1.000 | Ea. | .800 | 165.00 | 15.00 | 180.00 |
| 3' x 4' | 1.000 | Ea. | .889 | 176.00 | 19.00 | 195.00 |
| 3' x 4'-6" | 1.000 | Ea. | .889 | 192.50 | 17.50 | 210.00 |
| 3' x 5' | 1.000 | Ea. | 1.000 | 214.50 | 20.50 | 235.00 |
| 3'-6" x 6' | 1.000 | Ea. | 1.000 | 242.00 | 18.00 | 260.00 |
| Metal clad deluxe insulating glass, 2'-6" x 3' | 1.000 | Ea. | .800 | 187.00 | 18.00 | 205.00 |
| 3' x 3'-6" | 1.000 | Ea. | .800 | 214.50 | 15.50 | 230.00 |
| 3' x 4' | 1.000 | Ea. | .889 | 231.00 | 19.00 | 250.00 |
| 3' x 4'-6" | 1.000 | Ea. | .889 | 247.50 | 17.50 | 265.00 |
| 3' x 5' | 1.000 | Ea. | 1.000 | 275.00 | 20.00 | 295.00 |
| 3'-6" x 6' | 1.000 | Ea. | 1.000 | 308.00 | 22.00 | 330.00 |
| Trim, interior casing window, 2' x 3' | 11.000 | L.F. | .363 | 5.94 | 7.15 | 13.09 |
| 2'-6" x 3' | 12.000 | L.F. | .396 | 6.48 | 7.80 | 14.28 |
| 3' x 3'-6" | 14.000 | L.F. | .462 | 7.56 | 9.10 | 16.66 |
| 3' x 4' | 15.000 | L.F. | .495 | 8.10 | 9.75 | 17.85 |
| 3' x 4'-6" | 16.000 | L.F. | .528 | 8.64 | 10.40 | 19.04 |
| 3' x 5' | 17.000 | L.F. | .561 | 9.18 | 11.05 | 20.23 |
| 3'-6" x 6' | 20.000 | L.F. | .660 | 10.80 | 13.00 | 23.80 |
| 4' x 4'-6" | 18.000 | L.F. | .594 | 9.72 | 11.70 | 21.42 |
| Paint or stain, interior or exterior, 2' x 3' window, 1 coat | 1.000 | FACE | .333 | 1.01 | 6.14 | 7.15 |
| 2 coats | 1.000 | FACE | .400 | 1.09 | 7.41 | 8.50 |
| Primer & 1 coat | 1.000 | FACE | .381 | 1.17 | 7.03 | 8.20 |
| Primer & 2 coats | 1.000 | FACE | .444 | 1.23 | 8.22 | 9.45 |
| 3' x 4' window, 1 coat | 1.000 | FACE | .533 | 1.67 | 9.88 | 11.55 |
| 2 coats | 1.000 | FACE | .615 | 1.74 | 11.36 | 13.10 |
| Primer & 1 coat | 1.000 | FACE | .571 | 1.82 | 10.58 | 12.40 |
| Primer & 2 coats | 1.000 | FACE | .800 | 1.89 | 14.81 | 16.70 |
| 4' x 4'-6" window, 1 coat | 1.000 | FACE | .533 | 1.67 | 9.88 | 11.55 |
| 2 coats | 1.000 | FACE | .615 | 1.74 | 11.36 | 13.10 |
| Primer & 1 coat | 1.000 | FACE | .571 | 1.82 | 10.58 | 12.40 |
| Primer & 2 coats | 1.000 | FACE | .800 | 1.89 | 14.81 | 16.70 |
| Caulking, window, 2' x 3' | 10.000 | L.F. | .310 | .80 | 6.00 | 6.80 |
| 2'-6" x 3' | 11.000 | L.F. | .341 | .88 | 6.60 | 7.48 |
| 3' x 3'-6" | 13.000 | L.F. | .403 | 1.04 | 7.80 | 8.84 |
| 3' x 4' | 14.000 | L.F. | .434 | 1.12 | 8.40 | 9.52 |
| 3' x 4'-6" | 15.000 | L.F. | .465 | 1.20 | 9.00 | 10.20 |
| 3' x 5' | 16.000 | L.F. | .496 | 1.28 | 9.60 | 10.88 |
| 3'-6" x 6' | 19.000 | L.F. | .589 | 1.52 | 11.40 | 12.92 |
| 4' x 4'-6" | 17.000 | L.F. | .527 | 1.36 | 10.20 | 11.56 |
| Grilles, glass size to, 16" x 24" per sash | 1.000 | Set | .333 | 23.10 | 6.90 | 30.00 |
| 32" x 32" per sash | 1.000 | Set | .364 | 42.90 | 7.10 | 50.00 |
| Drip cap, aluminum, 2' long | 2.000 | L.F. | .040 | .28 | .78 | 1.06 |
| 3' long | 3.000 | L.F. | .060 | .42 | 1.17 | 1.59 |
| 4' long | 4.000 | L.F. | .080 | .56 | 1.56 | 2.12 |
| Wood, 2' long | 2.000 | L.F. | .066 | 1.08 | 1.30 | 2.38 |
| 3' long | 3.000 | L.F. | .099 | 1.62 | 1.95 | 3.57 |
| 4' long | 4.000 | L.F. | .132 | 2.16 | 2.60 | 4.76 |

# 4 | EXTERIOR WALLS — 32 | Casement Window Systems

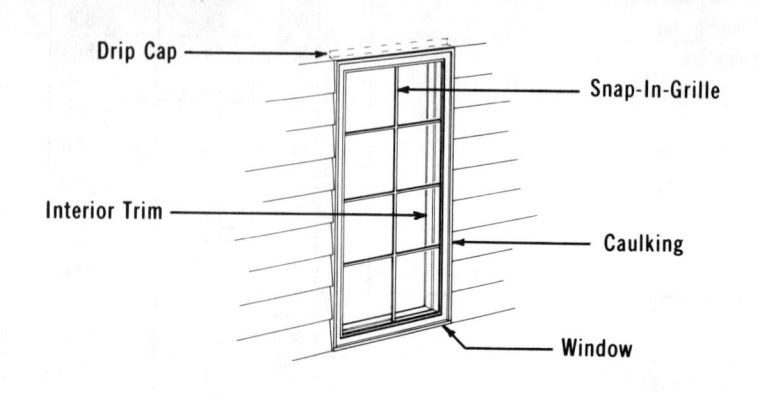

*Diagram labels: Drip Cap, Snap-In-Grille, Interior Trim, Caulking, Window*

| SYSTEM DESCRIPTION | QUAN. | UNIT | MAN HOURS | MAT. | INST. | TOTAL |
|---|---|---|---|---|---|---|
| **BUILDER'S QUALITY WOOD WINDOW, 2' BY 3', CASEMENT** | | | | | | |
| Window, wood primed, builder's quality, 2' x 3', insulating glass | 1.000 | Ea. | .800 | 110.00 | 15.00 | 125.00 |
| Trim, interior casing | 11.000 | L.F. | .363 | 5.94 | 7.15 | 13.09 |
| Paint, interior & exterior, primer & 2 coats | 2.000 | FACE | .888 | 2.46 | 16.44 | 18.90 |
| Caulking | 10.000 | L.F. | .310 | .80 | 6.00 | 6.80 |
| Snap-in grille | 1.000 | Ea. | .267 | 21.78 | 5.22 | 27.00 |
| Drip cap, metal | 2.000 | L.F. | .040 | .28 | .78 | 1.06 |
| TOTAL | | | 2.668 | 141.26 | 50.59 | 191.85 |
| **PLASTIC CLAD WOOD WINDOW, 2' X 4', CASEMENT** | | | | | | |
| Window, wood, plastic clad, premium, 2' x 4', insulating glass | 1.000 | Ea. | .889 | 154.00 | 16.00 | 170.00 |
| Trim, interior casing | 13.000 | L.F. | .429 | 7.02 | 8.45 | 15.47 |
| Paint, interior, primer & 2 coats | 1.000 | FACE | .444 | 1.23 | 8.22 | 9.45 |
| Caulking | 12.000 | L.F. | .372 | .96 | 7.20 | 8.16 |
| Snap-in grille | 1.000 | Ea. | .267 | 21.78 | 5.22 | 27.00 |
| TOTAL | | | 2.401 | 184.99 | 45.09 | 230.08 |
| **METAL CLAD WOOD WINDOW, 2' X 5', CASEMENT** | | | | | | |
| Window, wood, metal clad, deluxe, 2' x 5', insulating glass | 1.000 | Ea. | 1.000 | 220.00 | 20.00 | 240.00 |
| Trim, interior casing | 15.000 | L.F. | .495 | 8.10 | 9.75 | 17.85 |
| Paint, interior, primer & 2 coats | 1.000 | FACE | .800 | 1.89 | 14.81 | 16.70 |
| Caulking | 14.000 | L.F. | .434 | 1.12 | 8.40 | 9.52 |
| Snap-in grille | 1.000 | Ea. | .286 | 48.40 | 5.60 | 54.00 |
| Drip cap, metal | 12.000 | L.F. | .040 | .28 | .78 | 1.06 |
| TOTAL | | | 3.055 | 279.79 | 59.34 | 339.13 |

The cost of this system is on a cost per each window basis.

| DESCRIPTION | QUAN. | UNIT | MAN. HOURS | MAT. | INST. | TOTAL |
|---|---|---|---|---|---|---|
| | | | | | | |
| | | | | | | |
| | | | | | | |
| | | | | | | |
| | | | | | | |

# Casement Window Price Sheet

| | QUAN. | UNIT | MAN HOURS | COST EACH MAT. | COST EACH INST. | TOTAL |
|---|---|---|---|---|---|---|
| Window, casement, builders quality, 2' x 3', single glass | 1.000 | Ea. | .800 | 88.00 | 17.00 | 105.00 |
| Insulating glass | 1.000 | Ea. | .800 | 110.00 | 15.00 | 125.00 |
| 2' x 4'-6", single glass | 1.000 | Ea. | .889 | 115.50 | 19.50 | 135.00 |
| Insulating glass | 1.000 | Ea. | .889 | 148.50 | 16.50 | 165.00 |
| 2' x 6', single glass | 1.000 | Ea. | 1.000 | 132.00 | 18.00 | 150.00 |
| Insulating glass | 1.000 | Ea. | 1.000 | 176.00 | 19.00 | 195.00 |
| Plastic clad premium insulating glass, 2' x 3' | 1.000 | Ea. | .800 | 242.00 | 82.00 | 160.00 |
| 2' x 4' | 1.000 | Ea. | .889 | 154.00 | 16.00 | 170.00 |
| 2' x 5' | 1.000 | Ea. | 1.000 | 176.00 | 19.00 | 195.00 |
| 2' x 6' | 1.000 | Ea. | 1.000 | 214.50 | 20.50 | 235.00 |
| Metal clad deluxe insulating glass, 2' x 3' | 1.000 | Ea. | .800 | 374.00 | 184.00 | 190.00 |
| 2' x 4' | 1.000 | Ea. | .888 | 192.50 | 17.50 | 210.00 |
| 2' x 5' | 1.000 | Ea. | 1.000 | 220.00 | 20.00 | 240.00 |
| 2' x 6' | 1.000 | Ea. | 1.000 | 264.00 | 21.00 | 285.00 |
| Trim, interior casing window, 2' x 3' | 11.000 | L.F. | .363 | 5.94 | 7.15 | 13.09 |
| 2' x 4' | 13.000 | L.F. | .429 | 7.02 | 8.45 | 15.47 |
| 2' x 4'-6" | 14.000 | L.F. | .462 | 7.56 | 9.10 | 16.66 |
| 2' x 5' | 15.000 | L.F. | .495 | 8.10 | 9.75 | 17.85 |
| 2' x 6' | 17.000 | L.F. | .561 | 9.18 | 11.05 | 20.23 |
| Paint or stain, interior or exterior, 2' x 3' window, 1 coat | 1.000 | FACE | .333 | 1.01 | 6.14 | 7.15 |
| 2 coats | 1.000 | FACE | .400 | 1.09 | 7.41 | 8.50 |
| Primer & 1 coat | 1.000 | FACE | .381 | 1.17 | 7.03 | 8.20 |
| Primer & 2 coats | 1.000 | FACE | .444 | 1.23 | 8.22 | 9.45 |
| 2' x 4' window, 1 coat | 1.000 | FACE | .333 | 1.01 | 6.14 | 7.15 |
| 2 coats | 1.000 | FACE | .400 | 1.09 | 7.41 | 8.50 |
| Primer & 1 coat | 1.000 | FACE | .381 | 1.17 | 7.03 | 8.20 |
| Primer & 2 coats | 1.000 | FACE | .444 | 1.23 | 8.22 | 9.45 |
| 2' x 6' window, 1 coat | 1.000 | FACE | .533 | 1.67 | 9.88 | 11.55 |
| 2 coats | 1.000 | FACE | .615 | 1.74 | 11.36 | 13.10 |
| Primer & 1 coat | 1.000 | FACE | .571 | 1.82 | 10.58 | 12.40 |
| Primer & 2 coats | 1.000 | FACE | .800 | 1.89 | 14.81 | 16.70 |
| Caulking, window, 2' x 3' | 10.000 | L.F. | .310 | .80 | 6.00 | 6.80 |
| 2' x 4' | 12.000 | L.F. | .372 | .96 | 7.20 | 8.16 |
| 2' x 4'-6" | 13.000 | L.F. | .403 | 1.04 | 7.80 | 8.84 |
| 2' x 5' | 14.000 | L.F. | .434 | 1.12 | 8.40 | 9.52 |
| 2' x 6' | 16.000 | L.F. | .496 | 1.28 | 9.60 | 10.88 |
| Grilles, glass size, to 20" x 36" | 1.000 | Ea. | .267 | 21.78 | 5.22 | 27.00 |
| To 20" x 56" | 1.000 | Ea. | .286 | 48.40 | 5.60 | 54.00 |
| Drip cap, metal, 2' long | 2.000 | L.F. | .040 | .28 | .78 | 1.06 |
| Wood, 2' long | 2.000 | L.F. | .066 | 1.08 | 1.30 | 2.38 |

4

# 4 | EXTERIOR WALLS    36 | Awning Window System

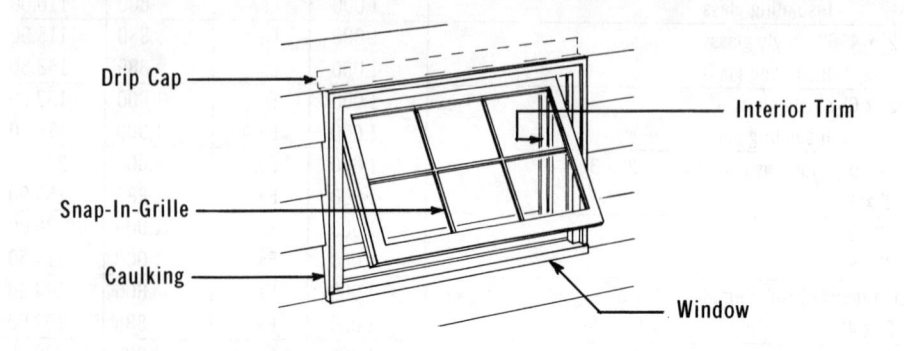

| SYSTEM DESCRIPTION | QUAN. | UNIT | MAN HOURS | MAT. | INST. | TOTAL |
|---|---|---|---|---|---|---|
| **BUILDER'S QUALITY WOOD, WINDOW, 34" X 22", AWNING** | | | | | | |
| Window, wood primed, builder's quality, 34" x 22", insulating glass | 1.000 | Ea. | .800 | 83.60 | 15.40 | 99.00 |
| Trim, interior casing | 10.500 | L.F. | .346 | 5.67 | 6.83 | 12.50 |
| Paint, interior & exterior, primer & 2 coats | 2.000 | FACE | .888 | 2.46 | 16.44 | 18.90 |
| Caulking | 9.500 | L.F. | .294 | .76 | 5.70 | 6.46 |
| Snap-in grille | 1.000 | Ea. | .267 | 10.67 | 5.23 | 15.90 |
| Drip cap, metal | 3.000 | L.F. | .060 | .42 | 1.17 | 1.59 |
| TOTAL | | | 2.656 | 103.58 | 50.77 | 154.35 |
| **PLASTIC CLAD WOOD WINDOW, 40" X 28", AWNING** | | | | | | |
| Window, wood, plastic clad, premium, 40" x 28", insulating glass | 1.000 | Ea. | .889 | 159.50 | 15.50 | 175.00 |
| Trim interior casing | 13.500 | L.F. | .445 | 7.29 | 8.78 | 16.07 |
| Paint, interior, primer & 2 coats | 1.000 | FACE | .444 | 1.23 | 8.22 | 9.45 |
| Caulking | 12.500 | L.F. | .387 | 1.00 | 7.50 | 8.50 |
| Snap-in grille | 1.000 | Ea. | .267 | 10.67 | 5.23 | 15.90 |
| TOTAL | | | 2.433 | 179.69 | 45.23 | 224.92 |
| **METAL CLAD WOOD WINDOW, 48" X 36", AWNING** | | | | | | |
| Window, wood, metal clad, deluxe, 48" x 36", insulating glass | 1.000 | Ea. | 1.000 | 247.50 | 17.50 | 265.00 |
| Trim, interior casing | 15.000 | L.F. | .495 | 8.10 | 9.75 | 17.85 |
| Paint, interior, primer & 2 coats | 1.000 | FACE | .800 | 1.89 | 14.81 | 16.70 |
| Caulking | 14.000 | L.F. | .434 | 1.12 | 8.40 | 9.52 |
| Snap-in grille | 1.000 | Ea. | .286 | 14.52 | 5.48 | 20.00 |
| Drip cap, metal | 4.000 | L.F. | .080 | .56 | 1.56 | 2.12 |
| TOTAL | | | 3.095 | 273.69 | 57.50 | 331.19 |

The cost of this system is on a cost per each window basis.

| DESCRIPTION | QUAN. | UNIT | MAN HOURS | MAT. | INST. | TOTAL |
|---|---|---|---|---|---|---|
| | | | | | | |
| | | | | | | |
| | | | | | | |
| | | | | | | |
| | | | | | | |
| | | | | | | |

# Awning Window Price Sheet

| | QUAN. | UNIT | MAN HOURS | COST EACH MAT. | COST EACH INST. | TOTAL |
|---|---|---|---|---|---|---|
| Windows, awning, builder's quality, 34" x 22", single glass | 1.000 | Ea. | .800 | 63.80 | 15.20 | 79.00 |
| Insulating glass | 1.000 | Ea. | .800 | 83.60 | 15.40 | 99.00 |
| 40" x 28", single glass | 1.000 | Ea. | .889 | 78.10 | 16.90 | 95.00 |
| Insulating glass | 1.000 | Ea. | .889 | 108.90 | 16.10 | 125.00 |
| 48" x 36", single glass | 1.000 | Ea. | 1.000 | 96.80 | 18.20 | 115.00 |
| Insulating glass | 1.000 | Ea. | 1.000 | 132.00 | 18.00 | 150.00 |
| Plastic clad premium insulating glass, 34" x 22" | 1.000 | Ea. | .800 | 126.50 | 13.50 | 140.00 |
| 40" x 22" | 1.000 | Ea. | .800 | 137.50 | 17.50 | 155.00 |
| 36" x 28" | 1.000 | Ea. | .889 | 154.00 | 16.00 | 170.00 |
| 40" x 28" | 1.000 | Ea. | .889 | 159.50 | 15.50 | 175.00 |
| 48" x 28" | 1.000 | Ea. | 1.000 | 165.00 | 20.00 | 185.00 |
| 48" x 36" | 1.000 | Ea. | 1.000 | 198.00 | 22.00 | 220.00 |
| Metal clad deluxe insulating glass, 34" x 22" | 1.000 | Ea. | .800 | 165.00 | 15.00 | 180.00 |
| 40" x 22" | 1.000 | Ea. | .800 | 170.50 | 14.50 | 185.00 |
| 36" x 28" | 1.000 | Ea. | .889 | 181.50 | 18.50 | 200.00 |
| 40" x 28" | 1.000 | Ea. | .889 | 192.50 | 17.50 | 210.00 |
| 48" x 28" | 1.000 | Ea. | 1.000 | 209.00 | 21.00 | 230.00 |
| 48" x 36" | 1.000 | Ea. | 1.000 | 247.50 | 17.50 | 265.00 |
| Trim, interior casing window, 34" x 22" | 10.500 | L.F. | .346 | 5.67 | 6.83 | 12.50 |
| 40" x 22" | 11.500 | L.F. | .379 | 6.21 | 7.48 | 13.69 |
| 36" x 28" | 12.500 | L.F. | .412 | 6.75 | 8.13 | 14.88 |
| 40" x 28" | 13.500 | L.F. | .445 | 7.29 | 8.78 | 16.07 |
| 48" x 28" | 14.500 | L.F. | .478 | 7.83 | 9.43 | 17.26 |
| 48" x 36" | 15.000 | L.F. | .495 | 8.10 | 9.75 | 17.85 |
| Paint or stain, interior or exterior, 34" x 22", 1 coat | 1.000 | FACE | .333 | 1.01 | 6.14 | 7.15 |
| 2 coats | 1.000 | FACE | .400 | 1.09 | 7.41 | 8.50 |
| Primer & 1 coat | 1.000 | FACE | .381 | 1.17 | 7.03 | 8.20 |
| Primer & 2 coats | 1.000 | FACE | .444 | 1.23 | 8.22 | 9.45 |
| 36" x 28", 1 coat | 1.000 | FACE | .333 | 1.01 | 6.14 | 7.15 |
| 2 coats | 1.000 | FACE | .400 | 1.09 | 7.41 | 8.50 |
| Primer & 1 coat | 1.000 | FACE | .381 | 1.17 | 7.03 | 8.20 |
| Primer & 2 coats | 1.000 | FACE | .444 | 1.23 | 8.22 | 9.45 |
| 48" x 36", 1 coat | 1.000 | FACE | .533 | 1.67 | 9.88 | 11.55 |
| 2 coats | 1.000 | FACE | .615 | 1.74 | 11.36 | 13.10 |
| Primer & 1 coat | 1.000 | FACE | .571 | 1.82 | 10.58 | 12.40 |
| Primer & 2 coats | 1.000 | FACE | .800 | 1.89 | 14.81 | 16.70 |
| Caulking, window, 34" x 22" | 9.500 | L.F. | .294 | .76 | 5.70 | 6.46 |
| 40" x 22" | 10.500 | L.F. | .325 | .84 | 6.30 | 7.14 |
| 36" x 28" | 11.500 | L.F. | .356 | .92 | 6.90 | 7.82 |
| 40" x 28" | 12.500 | L.F. | .387 | 1.00 | 7.50 | 8.50 |
| 48" x 28" | 13.500 | L.F. | .418 | 1.08 | 8.10 | 9.18 |
| 48" x 36" | 14.000 | L.F. | .434 | 1.12 | 8.40 | 9.52 |
| Grilles, glass size, to 28" by 16" | 1.000 | Ea. | .267 | 10.67 | 5.23 | 15.90 |
| To 44" by 24" | 1.000 | Ea. | .286 | 14.52 | 5.48 | 20.00 |
| Drip cap, aluminum, 3' long | 3.000 | L.F. | .060 | .42 | 1.17 | 1.59 |
| 3'-6" long | 3.500 | L.F. | .070 | .49 | 1.37 | 1.86 |
| 4' long | 4.000 | L.F. | .080 | .56 | 1.56 | 2.12 |
| Wood, 3' long | 3.000 | L.F. | .099 | 1.62 | 1.95 | 3.57 |
| 3'-6" long | 3.500 | L.F. | .115 | 1.89 | 2.28 | 4.17 |
| 4' long | 4.000 | L.F. | .132 | 2.16 | 2.60 | 4.76 |

# 4 | EXTERIOR WALLS   40 | Sliding Window Systems

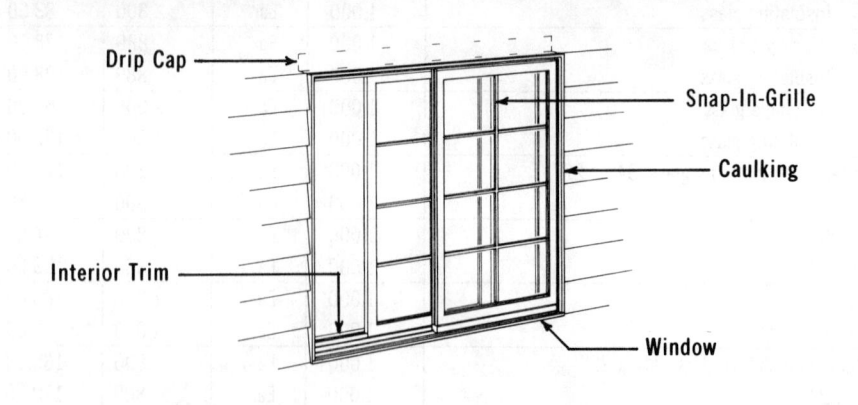

Diagram labels: Drip Cap, Snap-In-Grille, Caulking, Interior Trim, Window

| SYSTEM DESCRIPTION | QUAN. | UNIT | MAN HOURS | MAT. | INST. | TOTAL |
|---|---|---|---|---|---|---|
| **BUILDER'S QUALITY WOOD WINDOW, 3' X 2', SLIDING** | | | | | | |
| Window, wood primed, builder's quality, 3' x 2', insulating glass | 1.000 | Ea. | .800 | 93.50 | 16.50 | 110.00 |
| Trim, interior casing | 11.000 | L.F. | .363 | 5.94 | 7.15 | 13.09 |
| Paint, interior & exterior, primer & 2 coats | 2.000 | FACE | .888 | 2.46 | 16.44 | 18.90 |
| Caulking | 10.000 | L.F. | .310 | .80 | 6.00 | 6.80 |
| Snap-in grille | 1.000 | Set | .333 | 26.40 | 6.60 | 33.00 |
| Drip cap, metal | 3.000 | L.F. | .060 | .42 | 1.17 | 1.59 |
| TOTAL | | | 2.754 | 129.52 | 53.86 | 183.38 |
| **PLASTIC CLAD WOOD WINDOW, 4' X 3'-6", SLIDING** | | | | | | |
| Window, wood, plastic clad, premium, 4' x 3'-6", insulating glass | 1.000 | Ea. | .888 | 231.00 | 19.00 | 250.00 |
| Trim, interior casing | 16.000 | L.F. | .528 | 8.64 | 10.40 | 19.04 |
| Paint, interior, primer & 2 coats | 1.000 | FACE | .800 | 1.89 | 14.81 | 16.70 |
| Caulking | 17.000 | L.F. | .527 | 1.36 | 10.20 | 11.56 |
| Snap-in grille | 1.000 | Set | .333 | 26.40 | 6.60 | 33.00 |
| TOTAL | | | 3.076 | 269.29 | 61.01 | 330.30 |
| **METAL CLAD WOOD WINDOW, 6' X 5', SLIDING** | | | | | | |
| Window, wood, metal clad, deluxe, 6' x 5', insulating glass | 1.000 | Ea. | 1.000 | 500.50 | 19.50 | 520.00 |
| Trim, interior casing | 23.000 | L.F. | .759 | 12.42 | 14.95 | 27.37 |
| Paint, interior, primer & 2 coats | 1.000 | FACE | 1.000 | 3.92 | 18.08 | 22.00 |
| Caulking | 22.000 | L.F. | .682 | 1.76 | 13.20 | 14.96 |
| Snap-in grille | 1.000 | Set | .364 | 44.00 | 7.00 | 51.00 |
| Drip cap, metal | 6.000 | L.F. | .120 | .84 | 2.34 | 3.18 |
| TOTAL | | | 3.925 | 563.44 | 75.07 | 638.51 |

The cost of this system is on a cost per each window basis.

| DESCRIPTION | QUAN. | UNIT | MAN HOURS | MAT. | INST. | TOTAL |
|---|---|---|---|---|---|---|
| | | | | | | |
| | | | | | | |
| | | | | | | |
| | | | | | | |
| | | | | | | |

# Sliding Window Price Sheet

| | QUAN. | UNIT | MAN HOURS | COST EACH MAT. | COST EACH INST. | TOTAL |
|---|---|---|---|---|---|---|
| Windows, sliding, builder's quality, 3' x 2', single glass | 1.000 | EA | .800 | 275.00 | 191.00 | 84.00 |
| Insulating glass | 1.000 | EA | .800 | 93.50 | 16.50 | 110.00 |
| 4' x 3'-6", single glass | 1.000 | Ea. | .889 | 148.50 | 16.50 | 165.00 |
| Insulating glass | 1.000 | Ea. | .889 | 132.00 | 18.00 | 150.00 |
| 6' x 5', single glass | 1.000 | Ea. | 1.000 | 154.00 | 21.00 | 175.00 |
| Insulating glass | 1.000 | Ea. | 1.000 | 214.50 | 20.50 | 235.00 |
| Plastic clad premium insulating glass, 3' x 3' | 1.000 | Ea. | .800 | 181.50 | 13.50 | 195.00 |
| 4' x 3'-6" | 1.000 | Ea. | .888 | 231.00 | 19.00 | 250.00 |
| 5' x 4' | 1.000 | Ea. | .888 | 275.00 | 15.00 | 290.00 |
| 6' x 5' | 1.000 | Ea. | 1.000 | 390.50 | 19.50 | 410.00 |
| Metal clad deluxe insulating glass, 3' x 3' | 1.000 | Ea. | .800 | 236.50 | 13.50 | 250.00 |
| 4' x 3'-6" | 1.000 | Ea. | .889 | 291.50 | 18.50 | 310.00 |
| 5' x 4' | 1.000 | Ea. | .888 | 352.00 | 18.00 | 370.00 |
| 6' x 5' | 1.000 | Ea. | 1.000 | 489.50 | 19.50 | 520.00 |
| Trim, interior casing, window, 3' x 2' | 11.000 | L.F. | .363 | 5.94 | 7.15 | 13.09 |
| 3' x 3' | 13.000 | L.F. | .429 | 7.02 | 8.45 | 15.47 |
| 4' x 3'-6" | 16.000 | L.F. | .528 | 8.64 | 10.40 | 19.04 |
| 5' x 4' | 19.000 | L.F. | .627 | 10.26 | 12.35 | 22.61 |
| 6' x 5' | 23.000 | L.F. | .759 | 12.42 | 14.95 | 27.37 |
| Paint or stain, interior or exterior, 3' x 2' window, 1 coat | 1.000 | FACE | .333 | 1.01 | 6.14 | 7.15 |
| 2 coats | 1.000 | FACE | .400 | 1.09 | 7.41 | 8.50 |
| Primer & 1 coat | 1.000 | FACE | .381 | 1.17 | 7.03 | 8.20 |
| Primer & 2 coats | 1.000 | FACE | .444 | 1.23 | 8.22 | 9.45 |
| 4' x 3'-6" window, 1 coat | 1.000 | FACE | .533 | 1.67 | 9.88 | 11.55 |
| 2 coats | 1.000 | FACE | .615 | 1.74 | 11.36 | 13.10 |
| Primer & 1 coat | 1.000 | FACE | .571 | 1.82 | 10.58 | 12.40 |
| Primer & 2 coats | 1.000 | FACE | .800 | 1.89 | 14.81 | 16.70 |
| 6' x 5' window, 1 coat | 1.000 | FACE | .667 | 3.49 | 12.31 | 15.80 |
| 2 coats | 1.000 | FACE | .889 | 3.63 | 16.37 | 20.00 |
| Primer & 1 coat | 1.000 | FACE | .727 | 3.77 | 13.43 | 17.20 |
| Primer & 2 coats | 1.000 | FACE | 1.000 | 3.92 | 18.08 | 22.00 |
| Caulking, window, 3' x 2' | 10.000 | L.F. | .310 | .80 | 6.00 | 6.80 |
| 3' x 3' | 12.000 | L.F. | .372 | .96 | 7.20 | 8.16 |
| 4' x 3'-6" | 15.000 | L.F. | .465 | 1.20 | 9.00 | 10.20 |
| 5' x 4' | 18.000 | L.F. | .558 | 1.44 | 10.80 | 12.24 |
| 6' x 5' | 22.000 | L.F. | .682 | 1.76 | 13.20 | 14.96 |
| Grilles, glass size, to 14" x 36" | 1.000 | Set | .333 | 26.40 | 6.60 | 33.00 |
| To 36" x 36" | 1.000 | Set | .364 | 44.00 | 7.00 | 51.00 |
| Drip cap, aluminum, 3' long | 3.000 | L.F. | .060 | .42 | 1.17 | 1.59 |
| 4' long | 4.000 | L.F. | .080 | .56 | 1.56 | 2.12 |
| 5' long | 5.000 | L.F. | .100 | .70 | 1.95 | 2.65 |
| 6' long | 6.000 | L.F. | .120 | .84 | 2.34 | 3.18 |
| Wood, 3' long | 3.000 | L.F. | .099 | 1.62 | 1.95 | 3.57 |
| 4' long | 4.000 | L.F. | .132 | 2.16 | 2.60 | 4.76 |
| 5' long | 5.000 | L.F. | .165 | 2.70 | 3.25 | 5.95 |
| 6' long | 6.000 | L.F. | .198 | 3.24 | 3.90 | 7.14 |

4

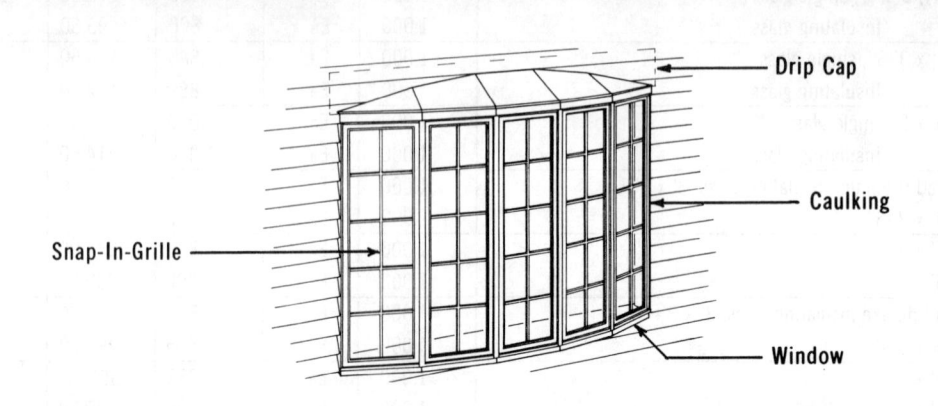

# 4 | EXTERIOR WALLS  44 | Bow/Bay Window Systems

*Labeled diagram: Drip Cap, Caulking, Window, Snap-In-Grille*

| SYSTEM DESCRIPTION | QUAN. | UNIT | MAN HOURS | MAT. | INST. | TOTAL |
|---|---|---|---|---|---|---|
| **AWNING TYPE BOW WINDOW, BUILDER'S QUALITY, 8' X 5'** | | | | | | |
| Window, wood primed, builder's quality, 8' x 5', insulating glass | 1.000 | Ea. | 1.600 | 539.00 | 31.00 | 570.00 |
| Trim, interior casing | 27.000 | L.F. | .891 | 14.58 | 17.55 | 32.13 |
| Paint, interior & exterior, primer & 1 coat | 2.000 | FACE | 2.000 | 7.84 | 36.16 | 44.00 |
| Drip cap, vinyl | 1.000 | Ea. | .533 | 55.00 | 10.00 | 65.00 |
| Caulking | 26.000 | L.F. | .806 | 2.08 | 15.60 | 17.68 |
| Snap-in grilles | 1.000 | Set | 1.068 | 87.12 | 20.88 | 108.00 |
| TOTAL | | | 6.898 | 705.62 | 131.19 | 836.81 |
| **CASEMENT TYPE BOW WINDOW, PLASTIC CLAD, 10' X 6'** | | | | | | |
| Window, wood, plastic clad, premium, 10' x 6', insulating glass | 1.000 | Ea. | 2.290 | 1210.00 | 40.00 | 1250.00 |
| Trim, interior casing | 33.000 | L.F. | 1.089 | 17.82 | 21.45 | 39.27 |
| Paint, interior, primer & 1 coat | 1.000 | FACE | 1.600 | 3.78 | 29.62 | 33.40 |
| Drip cap, vinyl | 1.000 | Ea. | .615 | 61.60 | 12.40 | 74.00 |
| Caulking | 32.000 | L.F. | .992 | 2.56 | 19.20 | 21.76 |
| Snap-in grilles | 1.000 | Set | 1.335 | 108.90 | 26.10 | 135.00 |
| TOTAL | | | 7.921 | 1404.66 | 148.77 | 1553.43 |
| **DOUBLE HUNG TYPE, METAL CLAD, 9' X 5'** | | | | | | |
| Window, wood, metal clad, deluxe, 9' x 5', insulating glass | 1.000 | Ea. | 2.670 | 1347.50 | 52.50 | 1400.00 |
| Trim, interior casing | 29.000 | L.F. | .957 | 15.66 | 18.85 | 34.51 |
| Paint, interior, primer & 1 coat | 1.000 | FACE | 1.600 | 3.78 | 29.62 | 33.40 |
| Drip cap, vinyl | 1.000 | Set | .615 | 61.60 | 12.40 | 74.00 |
| Caulking | 28.000 | L.F. | .868 | 2.24 | 16.80 | 19.04 |
| Snap-in grilles | 1.000 | Set | 1.068 | 87.12 | 20.88 | 108.00 |
| TOTAL | | | 7.778 | 1517.90 | 151.05 | 1668.95 |

The cost of this system is on a cost per each window basis.

| DESCRIPTION | QUAN. | UNIT | MAN HOURS | MAT. | INST. | TOTAL |
|---|---|---|---|---|---|---|
| | | | | | | |
| | | | | | | |
| | | | | | | |
| | | | | | | |
| | | | | | | |

# Bow/Bay Window Price Sheet

| | QUAN. | UNIT | MAN HOURS | COST EACH MAT. | COST EACH INST. | TOTAL |
|---|---|---|---|---|---|---|
| Windows, bow awning type, builder's quality, 8' x 5', single glass | 1.000 | Ea. | 1.600 | 467.50 | 32.50 | 500.00 |
| Insulating glass | 1.000 | Ea. | 1.600 | 539.00 | 31.00 | 570.00 |
| 12' x 6', single glass | 1.000 | Ea. | 2.670 | 808.50 | 51.50 | 860.00 |
| Insulating glass | 1.000 | Ea. | 2.670 | 885.50 | 54.50 | 940.00 |
| Plastic clad premium insulating glass, 6' x 4' | 1.000 | Ea. | 1.600 | 704.00 | 31.00 | 735.00 |
| 9' x 4' | 1.000 | Ea. | 2.000 | 935.00 | 40.00 | 975.00 |
| 10' x 5' | 1.000 | Ea. | 2.290 | 1210.00 | 40.00 | 1250.00 |
| 12' x 6' | 1.000 | Ea. | 2.670 | 1485.00 | 40.00 | 1525.00 |
| Metal clad deluxe insulating glass, 6' x 4' | 1.000 | Ea. | 1.600 | 885.50 | 29.50 | 915.00 |
| 9' x 4' | 1.000 | Ea. | 2.000 | 1188.00 | 37.00 | 1225.00 |
| 10' x 5' | 1.000 | Ea. | 2.290 | 1567.50 | 32.50 | 1600.00 |
| 12' x 6' | 1.000 | Ea. | 2.670 | 1925.00 | 50.00 | 1975.00 |
| Bow casement type, builder's quality, 8' x 5', single glass | 1.000 | Ea. | 1.600 | 429.00 | 31.00 | 460.00 |
| Insulating glass | 1.000 | Ea. | 1.600 | 544.50 | 30.50 | 575.00 |
| 12' x 6', single glass | 1.000 | Ea. | 2.670 | 638.00 | 52.00 | 690.00 |
| Insulating glass | 1.000 | Ea. | 2.670 | 852.50 | 52.50 | 905.00 |
| Plastic clad premium insulating glass, 8' x 5' | 1.000 | Ea. | 1.600 | 819.50 | 30.50 | 850.00 |
| 10' x 5' | 1.000 | Ea. | 2.000 | 1083.50 | 41.50 | 1125.00 |
| 10' x 6' | 1.000 | Ea. | 2.290 | 1210.00 | 40.00 | 1250.00 |
| 12' x 6' | 1.000 | Ea. | 2.670 | 1375.00 | 50.00 | 1425.00 |
| Metal clad deluxe insulating glass, 8' x 5' | 1.000 | Ea. | 1.600 | 1050.50 | 24.50 | 1075.00 |
| 10' x 5' | 1.000 | Ea. | 2.000 | 1347.50 | 27.50 | 1375.00 |
| 10' x 6' | 1.000 | Ea. | 2.290 | 1485.00 | 40.00 | 1525.00 |
| 12' x 6' | 1.000 | Ea. | 2.670 | 1815.00 | 60.00 | 1875.00 |
| Bow double hung type, builder's quality, 8' x 4', single glass | 1.000 | Ea. | 1.600 | 632.50 | 32.50 | 665.00 |
| Insulating glass | 1.000 | Ea. | 1.600 | 704.00 | 31.00 | 735.00 |
| 9' x 5', single glass | 1.000 | Ea. | 2.670 | 687.50 | 52.50 | 740.00 |
| Insulating glass | 1.000 | Ea. | 2.670 | 797.50 | 52.50 | 850.00 |
| Plastic clad premium insulating glass, 7' x 4' | 1.000 | Ea. | 1.600 | 825.00 | 30.00 | 855.00 |
| 8' x 4' | 1.000 | Ea. | 2.000 | 935.00 | 40.00 | 975.00 |
| 8' x 5' | 1.000 | Ea. | 2.290 | 880.00 | 45.00 | 925.00 |
| 9' x 5' | 1.000 | Ea. | 2.670 | 1034.00 | 41.00 | 1075.00 |
| Metal clad deluxe insulating glass, 7' x 4' | 1.000 | Ea. | 1.600 | 1094.50 | 30.50 | 1125.00 |
| 8' x 4' | 1.000 | Ea. | 2.000 | 1127.50 | 47.50 | 1175.00 |
| 8' x 5' | 1.000 | Ea. | 2.290 | 1237.50 | 37.50 | 1275.00 |
| 9' x 5' | 1.000 | Ea. | 2.670 | 1347.50 | 52.50 | 1400.00 |
| Trim, interior casing, window, 7' x 4' | 23.000 | L.F. | .759 | 12.42 | 14.95 | 27.37 |
| 8' x 5' | 27.000 | L.F. | .891 | 14.58 | 17.55 | 32.13 |
| 10' x 6' | 33.000 | L.F. | 1.089 | 17.82 | 21.45 | 39.27 |
| 12' x 6' | 37.000 | L.F. | 1.221 | 19.98 | 24.05 | 44.03 |
| Paint or stain, interior, or exterior, 7' x 4' window, 1 coat | 1.000 | FACE | .667 | 3.49 | 12.31 | 15.80 |
| Primer & 1 coat | 1.000 | FACE | .727 | 3.77 | 13.43 | 17.20 |
| 8' x 5' window, 1 coat | 1.000 | FACE | .667 | 3.49 | 12.31 | 15.80 |
| Primer & 1 coat | 1.000 | FACE | .727 | 3.77 | 13.43 | 17.20 |
| 10' x 6' window, 1 coat | 1.000 | FACE | 1.066 | 3.34 | 19.76 | 23.10 |
| Primer & 1 coat | 1.000 | FACE | 1.600 | 3.78 | 29.62 | 33.40 |
| 12' x 6' window, 1 coat | 1.000 | FACE | 1.334 | 6.98 | 24.62 | 31.60 |
| Primer & 1 coat | 1.000 | FACE | 1.454 | 7.54 | 26.86 | 34.40 |
| Drip cap, vinyl moulded window, 7' long | 1.000 | Ea. | .533 | 55.00 | 10.00 | 65.00 |
| 8' long | 1.000 | Ea. | .533 | 55.00 | 10.00 | 65.00 |
| 10' long | 1.000 | Ea. | .615 | 61.60 | 12.40 | 74.00 |
| 12' long | 1.000 | Ea. | .615 | 61.60 | 12.40 | 74.00 |
| Caulking, window, 7' x 4' | 22.000 | L.F. | .682 | 1.76 | 13.20 | 14.96 |
| 8' x 5' | 26.000 | L.F. | .806 | 2.08 | 15.60 | 17.68 |
| 10' x 6' | 32.000 | L.F. | .992 | 2.56 | 19.20 | 21.76 |
| 12' x 6' | 36.000 | L.F. | 1.116 | 2.88 | 21.60 | 24.48 |
| Grilles, window, 7' x 4' | 1.000 | Set | .801 | 65.34 | 15.66 | 81.00 |
| 8' x 5' | 1.000 | Set | 1.068 | 87.12 | 20.88 | 108.00 |
| 10' x 6' | 1.000 | Set | 1.335 | 108.90 | 26.10 | 135.00 |
| 12' x 6' | 1.000 | Set | 1.602 | 130.68 | 31.32 | 162.00 |

# 4 | EXTERIOR WALLS  48 | Fixed Window System

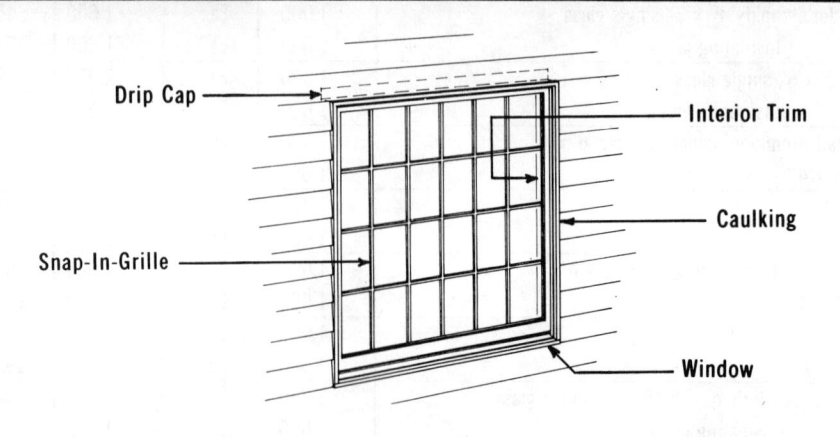

Diagram labels: Drip Cap, Snap-In-Grille, Interior Trim, Caulking, Window

| SYSTEM DESCRIPTION | QUAN. | UNIT | MAN HOURS | MAT. | INST. | TOTAL |
|---|---|---|---|---|---|---|
| **BUILDER'S QUALITY PICTURE WINDOW, 4' X 4'** | | | | | | |
| Window, wood, primed, builder's quality, 4' x 4', insulating glass | 1.000 | Ea. | 1.330 | 198.00 | 27.00 | 225.00 |
| Trim, interior casing | 17.000 | L.F. | .561 | 9.18 | 11.05 | 20.23 |
| Paint, interior & exterior, primer & 2 coats | 2.000 | FACE | 1.600 | 3.78 | 29.62 | 33.40 |
| Caulking | 16.000 | L.F. | .496 | 1.28 | 9.60 | 10.88 |
| Snap-in grille | 1.000 | Ea. | .267 | 44.00 | 5.00 | 49.00 |
| Drip cap, metal | 4.000 | L.F. | .080 | .56 | 1.56 | 2.12 |
| TOTAL | | | 4.334 | 256.80 | 83.83 | 340.63 |
| **PLASTIC CLAD WOOD WINDOW, 4'-6" X 6'-6"** | | | | | | |
| Window, wood, plastic clad, premium, 4'-6" x 6'-6", insulating glass | 1.000 | Ea. | 1.450 | 297.00 | 28.00 | 325.00 |
| Trim, interior casing | 23.000 | L.F. | .759 | 12.42 | 14.95 | 27.37 |
| Paint, interior, primer & 2 coats | 1.000 | FACE | .800 | 1.89 | 14.81 | 16.70 |
| Caulking | 22.000 | L.F. | .682 | 1.76 | 13.20 | 14.96 |
| Snap-in grille | 1.000 | Ea. | .267 | 44.00 | 5.00 | 49.00 |
| TOTAL | | | 3.958 | 357.07 | 75.96 | 433.03 |
| **METAL CLAD WOOD WINDOW, 6'-6" X 6'-6"** | | | | | | |
| Window, wood, metal clad, deluxe, 6'-6" x 6'-6", insulating glass | 1.000 | Ea. | 1.600 | 676.50 | 33.50 | 710.00 |
| Trim interior casing | 27.000 | L.F. | .891 | 14.58 | 17.55 | 32.13 |
| Paint, interior, primer & 2 coats | 1.000 | FACE | 1.000 | 3.92 | 18.08 | 22.00 |
| Caulking | 26.000 | L.F. | .806 | 2.08 | 15.60 | 17.68 |
| Snap-in grille | 1.000 | Ea. | .267 | 44.00 | 5.00 | 49.00 |
| Drip cap, metal | 6.500 | L.F. | .130 | .91 | 2.54 | 3.45 |
| TOTAL | | | 4.694 | 741.99 | 92.27 | 834.26 |

The cost of this system is on a cost per each window basis.

| DESCRIPTION | QUAN. | UNIT | MAN HOURS | MAT. | INST. | TOTAL |
|---|---|---|---|---|---|---|
| | | | | | | |
| | | | | | | |
| | | | | | | |
| | | | | | | |
| | | | | | | |

# Fixed Window Price Sheet

| | QUAN. | UNIT | MAN HOURS | MAT. | INST. | TOTAL |
|---|---|---|---|---|---|---|
| Window, picture, builder's quality, 4' x 4', single glass | 1.000 | Ea. | 1.330 | 170.50 | 24.50 | 195.00 |
| Insulating glass | 1.000 | Ea. | 1.330 | 198.00 | 27.00 | 225.00 |
| 4' x 4'-6", single glass | 1.000 | Ea. | 1.450 | 187.00 | 28.00 | 215.00 |
| Insulating glass | 1.000 | Ea. | 1.450 | 214.50 | 30.50 | 245.00 |
| 5' x 4', single glass | 1.000 | Ea. | 1.450 | 203.50 | 26.50 | 230.00 |
| Insulating glass | 1.000 | EA | 1.450 | 242.00 | 28.00 | 270.00 |
| 6' x 4'-6", single glass | 1.000 | Ea. | 1.600 | 225.50 | 29.50 | 255.00 |
| Insulating glass | 1.000 | Ea. | 1.600 | 269.50 | 30.50 | 300.00 |
| Plastic clad premium insulating glass, 4' x 4' | 1.000 | Ea. | 1.330 | 209.00 | 26.00 | 235.00 |
| 4'-6" x 6'-6" | 1.000 | Ea. | 1.450 | 297.00 | 28.00 | 325.00 |
| 5'-6" x 6'-6" | 1.000 | Ea. | 1.600 | 451.00 | 29.00 | 480.00 |
| 6'-6" x 6'-6" | 1.000 | Ea. | 1.600 | 533.50 | 31.50 | 565.00 |
| Metal clad deluxe insulating glass, 4' x 4' | 1.000 | Ea. | 1.330 | 264.00 | 26.00 | 290.00 |
| 4'-6" x 6'-6" | 1.000 | Ea. | 1.450 | 379.50 | 30.50 | 410.00 |
| 5'-6" x 6'-6" | 1.000 | Ea. | 1.600 | 572.00 | 33.00 | 605.00 |
| 6'-6" x 6'-6" | 1.000 | Ea. | 1.600 | 676.50 | 33.50 | 710.00 |
| Trim interior casing, window, 4' x 4' | 17.000 | L.F. | .561 | 9.18 | 11.05 | 20.23 |
| 4'-6" x 4'-6" | 19.000 | L.F. | .627 | 10.26 | 12.35 | 22.61 |
| 5'-0" x 4'-0" | 19.000 | L.F. | .627 | 10.26 | 12.35 | 22.61 |
| 4'-6" x 6'-6" | 23.000 | L.F. | .759 | 12.42 | 14.95 | 27.37 |
| 5'-6" x 6'-6" | 25.000 | L.F. | .825 | 13.50 | 16.25 | 29.75 |
| 6'-6" x 6'-6" | 27.000 | L.F. | .891 | 14.58 | 17.55 | 32.13 |
| Paint or stain, interior or exterior, 4' x 4' window, 1 coat | 1.000 | FACE | .533 | 1.67 | 9.88 | 11.55 |
| 2 coats | 1.000 | FACE | .615 | 1.74 | 11.36 | 13.10 |
| Primer & 1 coat | 1.000 | FACE | .571 | 1.82 | 10.58 | 12.40 |
| Primer & 2 coats | 1.000 | FACE | .800 | 1.89 | 14.81 | 16.70 |
| 4'-6" x 6'-6" window, 1 coat | 1.000 | FACE | .533 | 1.67 | 9.88 | 11.55 |
| 2 coats | 1.000 | FACE | .615 | 1.74 | 11.36 | 13.10 |
| Primer & 1 coat | 1.000 | FACE | .571 | 1.82 | 10.58 | 12.40 |
| Primer & 2 coats | 1.000 | FACE | .800 | 1.89 | 14.81 | 16.70 |
| 6'-6" x 6'-6" window, 1 coat | 1.000 | FACE | .667 | 3.49 | 12.31 | 15.80 |
| 2 coats | 1.000 | FACE | .889 | 3.63 | 16.37 | 20.00 |
| Primer & 1 coat | 1.000 | FACE | .727 | 3.77 | 13.43 | 17.20 |
| Primer & 2 coats | 1.000 | FACE | 1.000 | 3.92 | 18.08 | 22.00 |
| Caulking, window, 4' x 4' | 16.000 | L.F. | .496 | 1.28 | 9.60 | 10.88 |
| 4'-6" x 4'-6" | 18.000 | L.F. | .558 | 1.44 | 10.80 | 12.24 |
| 5'-0" x 4'-0" | 18.000 | L.F. | .558 | 1.44 | 10.80 | 12.24 |
| 4'-6" x 6'-6" | 22.000 | L.F. | .682 | 1.76 | 13.20 | 14.96 |
| 5'-6" x 6'-6" | 24.000 | L.F. | .744 | 1.92 | 14.40 | 16.32 |
| 6'-6" x 6'-6" | 26.000 | L.F. | .806 | 2.08 | 15.60 | 17.68 |
| Grilles, glass size, to 48" x 48" | 1.000 | Ea. | .267 | 44.00 | 5.00 | 49.00 |
| To 60" x 68" | 1.000 | Ea. | .286 | 71.50 | 5.50 | 77.00 |
| Drip cap, aluminum, 4' long | 4.000 | L.F. | .080 | .56 | 1.56 | 2.12 |
| 4'-6" long | 4.500 | L.F. | .090 | .63 | 1.76 | 2.39 |
| 5' long | 5.000 | L.F. | .100 | .70 | 1.95 | 2.65 |
| 6' long | 6.000 | L.F. | .120 | .84 | 2.34 | 3.18 |
| Wood, 4' long | 4.000 | L.F. | .132 | 2.16 | 2.60 | 4.76 |
| 4'-6" long | 4.500 | L.F. | .148 | 2.43 | 2.93 | 5.36 |
| 5' long | 5.000 | L.F. | .165 | 2.70 | 3.25 | 5.95 |
| 6' long | 6.000 | L.F. | .198 | 3.24 | 3.90 | 7.14 |

**4**

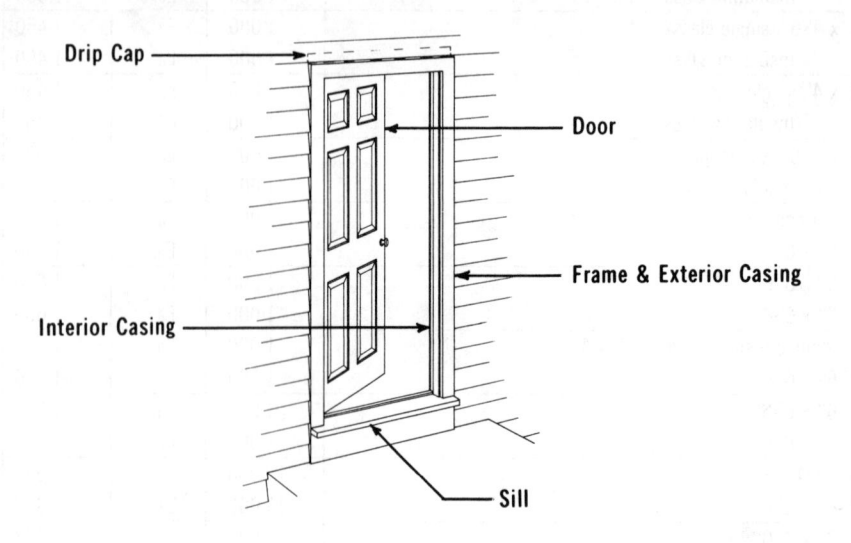

# 4 | EXTERIOR WALLS  52 | Entrance Door Systems

*Diagram of entrance door showing: Drip Cap, Door, Frame & Exterior Casing, Interior Casing, Sill*

| SYSTEM DESCRIPTION | QUAN. | UNIT | MAN HOURS | MAT. | INST. | TOTAL |
|---|---|---|---|---|---|---|
| **COLONIAL, 6 PANEL, 3' X 6'-8", WOOD** | | | | | | |
| Door, 3' x 6'-8" x 1-3/4" thick, pine, 6 panel colonial | 1.000 | Ea. | 1.070 | 176.00 | 24.00 | 200.00 |
| Frame, pine, 5-13/16" deep, including exterior casing & drip cap | 17.000 | L.F. | .731 | 51.68 | 14.79 | 66.47 |
| Interior casing, 2-1/2" wide | 18.000 | L.F. | .594 | 9.72 | 11.70 | 21.42 |
| Sill, 8/4 x 8" deep | 3.000 | L.F. | .480 | 17.67 | 9.78 | 27.45 |
| Butt hinges, brass, 4-1/2" x 4-1/2" | 1.500 | Pr. | | 12.90 | | 12.90 |
| Lockset | 1.000 | Ea. | .800 | 16.50 | 15.50 | 32.00 |
| Weatherstripping, metal, spring type, bronze | 1.000 | Set | 1.050 | 9.74 | 20.26 | 30.00 |
| Paint, interior & exterior, primer & 2 coats | 2.000 | FACE | 4.580 | 9.36 | 84.64 | 94.00 |
| TOTAL | | | 9.205 | 303.57 | 180.67 | 484.24 |
| **SOLID CORE BIRCH, FLUSH, 3' X 6'-8"** | | | | | | |
| Door, 3' x 6'-8", 1-3/4" thick, birch, flush solid core | 1.000 | Ea. | 1.070 | 110.00 | 20.00 | 130.00 |
| Frame, pine, 5-13/16" deep, including exterior casing & drip cap | 17.000 | L.F. | .731 | 51.68 | 14.79 | 66.47 |
| Interior casing, 2-1/2" wide | 18.000 | L.F. | .594 | 9.72 | 11.70 | 21.42 |
| Sill, 8/4 x 8" deep | 3.000 | L.F. | .480 | 17.67 | 9.78 | 27.45 |
| Butt hinges, brass, 4-1/2" x 4-1/2" | 1.500 | Pr. | | 12.90 | | 12.90 |
| Lockset | 1.000 | Ea. | .800 | 16.50 | 15.50 | 32.00 |
| Weatherstripping, metal, spring type, bronze | 1.000 | Set | 1.050 | 9.74 | 20.26 | 30.00 |
| Paint, interior & exterior, primer & 2 coats | 2.000 | FACE | 3.200 | 8.80 | 59.20 | 68.00 |
| TOTAL | | | 7.925 | 237.01 | 151.23 | 388.24 |

These systems are on a cost per each door basis.

| DESCRIPTION | QUAN. | UNIT | MAN HOURS | MAT. | INST. | TOTAL |
|---|---|---|---|---|---|---|
| | | | | | | |
| | | | | | | |
| | | | | | | |
| | | | | | | |
| | | | | | | |

# Entrance Door Price Sheet

| | QUAN. | UNIT | MAN HOURS | COST EACH MAT. | COST EACH INST. | TOTAL |
|---|---|---|---|---|---|---|
| Door exterior wood 1-3/4" thick, pine, dutch door, 2'-8" x 6'-8" minimum | 1.000 | Ea. | 1.330 | 192.50 | 27.50 | 220.00 |
| Maximum | 1.000 | Ea. | 1.600 | 236.50 | 33.50 | 270.00 |
| 3'-0" x 6'-8", minimum | 1.000 | Ea. | 1.330 | 203.50 | 26.50 | 230.00 |
| Maximum | 1.000 | Ea. | 1.600 | 253.00 | 32.00 | 285.00 |
| Colonial, 6 panel, 2'-8" x 6'-8" | 1.000 | Ea. | 1.000 | 159.50 | 20.50 | 180.00 |
| 3'-0" x 6'-8" | 1.000 | Ea. | 1.070 | 176.00 | 24.00 | 200.00 |
| 8 panel, 2'-6" x 6'-8" | 1.000 | Ea. | 1.000 | 203.50 | 21.50 | 225.00 |
| 3'-0" x 6'-8" | 1.000 | Ea. | 1.070 | 220.00 | 20.00 | 240.00 |
| Flush birch solid core, 2'-8" x 6'-8" | 1.000 | Ea. | 1.000 | 108.90 | 21.10 | 130.00 |
| 3'-0" x 6'-8" | 1.000 | Ea. | 1.070 | 110.00 | 20.00 | 130.00 |
| Porch door, 2'-8" x 6'-8" | 1.000 | Ea. | 1.000 | 148.50 | 21.50 | 170.00 |
| 3'-0" x 6'-8" | 1.000 | Ea. | 1.066 | 154.00 | 21.00 | 175.00 |
| Hand carved mahogany, 2'-8" x 6'-8" | 1.000 | Ea. | 1.070 | 154.00 | 21.00 | 175.00 |
| 3'-0" x 6'-8" | 1.000 | Ea. | 1.070 | 176.00 | 24.00 | 200.00 |
| Rosewood, 2'-8" x 6'-8" | 1.000 | Ea. | 1.070 | 374.00 | 21.00 | 395.00 |
| 3'-0" x 6-8" | 1.000 | Ea. | 1.070 | 407.00 | 23.00 | 430.00 |
| Door, metal clad wood 1-3/8" thick raised panel, 2'-8" x 6'-8" | 1.000 | Ea. | 1.066 | 148.50 | 21.50 | 170.00 |
| 3'-0" x 6'-8" | 1.000 | Ea. | 1.066 | 154.00 | 21.00 | 175.00 |
| Deluxe metal door, 2'-8" x 6'-8" | 1.000 | Ea. | 1.230 | 610.50 | 24.50 | 635.00 |
| 3'-0" x 6'-8" | 1.000 | Ea. | 1.230 | 643.50 | 26.50 | 670.00 |
| Frame, pine, including exterior trim & drip cap, 5/4, x 4-9/16" deep | 17.000 | L.F. | .731 | 46.07 | 14.79 | 60.86 |
| 5-13/16" deep | 17.000 | L.F. | .731 | 51.68 | 14.79 | 66.47 |
| 6-9/16" deep | 17.000 | L.F. | .731 | 58.65 | 14.96 | 73.61 |
| Safety glass lites | 1.000 | Ea. | | 17.60 | | 17.60 |
| Interior casing, 2'-8" x 6'-8" door | 18.000 | L.F. | .594 | 9.72 | 11.70 | 21.42 |
| 3'-0" x 6'-8" door | 19.000 | L.F. | .627 | 10.26 | 12.35 | 22.61 |
| Sill, oak, 8/4 x 8" deep | 3.000 | L.F. | .480 | 17.67 | 9.78 | 27.45 |
| 8/4 x 10" deep | 3.000 | L.F. | .534 | 22.77 | 10.98 | 33.75 |
| Butt hinges, steel plated, 4-1/2" x 4-1/2", plain | 1.500 | Pr. | | 12.90 | | 12.90 |
| Ball bearing | 1.500 | Pr. | | 28.43 | | 28.43 |
| Bronze, 4-1/2" x 4-1/2", plain | 1.500 | Pr. | | 51.00 | | 51.00 |
| Ball bearing | 1.500 | Pr. | | 78.00 | | 78.00 |
| Lockset, minimum | 1.000 | Ea. | .800 | 16.50 | 15.50 | 32.00 |
| Maximum | 1.000 | Ea. | .800 | 88.00 | 17.00 | 105.00 |
| Weatherstripping, metal, interlocking, zinc | 1.000 | Set | 2.670 | 20.02 | 51.98 | 72.00 |
| Bronze | 1.000 | Set | 2.670 | 33.00 | 52.00 | 85.00 |
| Spring type, bronze | 1.000 | Set | 1.050 | 9.74 | 20.26 | 30.00 |
| Rubber, minimum | 1.000 | Set | 1.050 | 4.02 | 20.98 | 25.00 |
| Maximum | 1.000 | Set | 1.140 | 4.57 | 22.43 | 27.00 |
| Felt minimum | 1.000 | Set | .571 | 1.65 | 11.20 | 12.85 |
| Maximum | 1.000 | Set | .615 | 1.93 | 12.02 | 13.95 |
| Paint or stain, flush door, interior or exterior, 1 coat | 2.000 | FACE | 1.230 | 5.36 | 22.74 | 28.10 |
| 2 coats | 2.000 | FACE | 2.280 | 7.70 | 42.30 | 50.00 |
| Primer & 1 coat | 2.000 | FACE | 2.280 | 7.70 | 42.30 | 50.00 |
| Primer & 2 coats | 2.000 | FACE | 3.200 | 8.80 | 59.20 | 68.00 |
| Paneled door, interior & exterior, 1 coat | 2.000 | FACE | 1.778 | 5.80 | 32.90 | 38.70 |
| 2 coats | 2.000 | FACE | 3.200 | 8.26 | 59.74 | 68.00 |
| Primer & 1 coat | 2.000 | FACE | 3.200 | 8.26 | 59.74 | 68.00 |
| Primer & 2 coats | 2.000 | FACE | 4.580 | 9.36 | 84.64 | 94.00 |

# 4 | EXTERIOR WALLS  53 | Sliding Door Systems

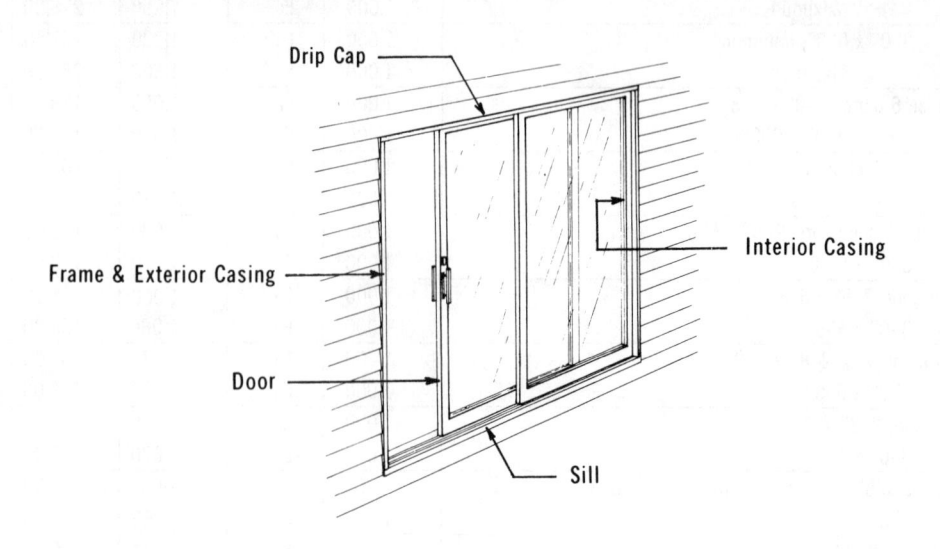

*Drip Cap*
*Frame & Exterior Casing*
*Interior Casing*
*Door*
*Sill*

| SYSTEM DESCRIPTION | QUAN. | UNIT | MAN HOURS | MAT. | INST. | TOTAL |
|---|---|---|---|---|---|---|
| **WOOD SLIDING DOOR, 8' WIDE, PREMIUM** | | | | | | |
| Wood 5/8" thick, tempered insul. glass, 8' wide, premium | 1.000 | Ea. | 5.330 | 880.00 | 105.00 | 985.00 |
| Interior casing | 22.000 | L.F. | .726 | 11.88 | 14.30 | 26.18 |
| Exterior casing | 22.000 | L.F. | .880 | 23.10 | 17.16 | 40.26 |
| Sill, oak, 8/4 x 8" deep | 8.000 | L.F. | 1.280 | 47.12 | 26.08 | 73.20 |
| Drip cap | 8.000 | L.F. | .160 | 1.12 | 3.12 | 4.24 |
| Paint, interior & exterior, primer & 2 coats | 2.000 | FACE | 2.992 | 9.68 | 55.44 | 65.12 |
| TOTAL | | | 11.368 | 972.90 | 221.10 | 1194.00 |
| **ALUMINUM SLIDING DOOR, 8' WIDE, PREMIUM** | | | | | | |
| Aluminum 5/8" thick, tempered insul. glass, 8' wide, premium | 1.000 | Ea. | 5.330 | 401.50 | 103.50 | 505.00 |
| Interior casing | 22.000 | L.F. | .726 | 11.88 | 14.30 | 26.18 |
| Exterior casing | 22.000 | L.F. | .880 | 23.10 | 17.16 | 40.26 |
| Sill, oak, 8/4 x 8" deep | 8.000 | L.F. | 1.280 | 47.12 | 26.08 | 73.20 |
| Drip cap | 8.000 | L.F. | .160 | 1.12 | 3.12 | 4.24 |
| Paint, interior & exterior, primer & 2 coats | 2.000 | FACE | 1.496 | 4.84 | 27.72 | 32.56 |
| TOTAL | | | 9.872 | 489.56 | 191.88 | 681.44 |

The cost of this system is on a cost per each door basis.

| DESCRIPTION | QUAN. | UNIT | MAN HOURS | MAT. | INST. | TOTAL |
|---|---|---|---|---|---|---|
| | | | | | | |
| | | | | | | |
| | | | | | | |
| | | | | | | |
| | | | | | | |
| | | | | | | |
| | | | | | | |

# Sliding Door Price Sheet

| | QUAN. | UNIT | MAN HOURS | COST EACH MAT. | COST EACH INST. | TOTAL |
|---|---|---|---|---|---|---|
| Sliding door wood 5/8" thick, tempered insul. glass, 6' wide, premium | 1.000 | Ea. | 4.000 | 748.00 | 77.00 | 825.00 |
| Economy | 1.000 | Ea. | 4.000 | 385.00 | 80.00 | 465.00 |
| 8' wide, wood premium | 1.000 | Ea. | 5.330 | 880.00 | 105.00 | 985.00 |
| Economy | 1.000 | Ea. | 5.330 | 660.00 | 105.00 | 765.00 |
| 12' wide, wood premium | 1.000 | Ea. | 6.400 | 1375.00 | 125.00 | 1500.00 |
| Economy | 1.000 | Ea. | 6.400 | 913.00 | 137.00 | 1050.00 |
| Aluminum 5/8" thick, tempered insul. glass, 6' wide, premium | 1.000 | Ea. | 4.000 | 341.00 | 79.00 | 420.00 |
| Economy | 1.000 | Ea. | 4.000 | 286.00 | 79.00 | 365.00 |
| 8' wide, premium | 1.000 | Ea. | 5.330 | 401.50 | 103.50 | 505.00 |
| Economy | 1.000 | Ea. | 5.330 | 341.00 | 104.00 | 445.00 |
| 12' wide, premium | 1.000 | Ea. | 6.400 | 456.50 | 123.50 | 580.00 |
| Economy | 1.000 | Ea. | 6.400 | 401.50 | 123.50 | 525.00 |
| Interior casing, 6' wide door | 20.000 | L.F. | .660 | 10.80 | 13.00 | 23.80 |
| 8' wide door | 22.000 | L.F. | .726 | 11.88 | 14.30 | 26.18 |
| 12' wide door | 26.000 | L.F. | .858 | 14.04 | 16.90 | 30.94 |
| Exterior casing, 6' wide door | 20.000 | L.F. | .800 | 21.00 | 15.60 | 36.60 |
| 8' wide door | 22.000 | L.F. | .880 | 23.10 | 17.16 | 40.26 |
| 12' wide door | 26.000 | L.F. | 1.040 | 27.30 | 20.28 | 47.58 |
| Sill, oak, 8/4 x 8" deep, 6' wide door | 6.000 | L.F. | .960 | 35.34 | 19.56 | 54.90 |
| 8' wide door | 8.000 | L.F. | 1.280 | 47.12 | 26.08 | 73.20 |
| 12' wide door | 12.000 | L.F. | 1.920 | 70.68 | 39.12 | 109.80 |
| 8/4 x 10" deep, 6' wide door | 6.000 | L.F. | 1.068 | 45.54 | 21.96 | 67.50 |
| 8' wide door | 8.000 | L.F. | 1.424 | 60.72 | 29.28 | 90.00 |
| 12' wide door | 12.000 | L.F. | 2.136 | 91.08 | 43.92 | 135.00 |
| Drip cap, 6' wide door | 6.000 | L.F. | .120 | .84 | 2.34 | 3.18 |
| 8' wide door | 8.000 | L.F. | .160 | 1.12 | 3.12 | 4.24 |
| 12' wide door | 12.000 | L.F. | .240 | 1.68 | 4.68 | 6.36 |
| Paint or stain, interior & exterior, 6' wide door, 1 coat | 2.000 | FACE | 1.760 | 4.65 | 32.00 | 36.65 |
| 2 coats | 2.000 | FACE | 2.160 | 6.40 | 39.20 | 45.60 |
| Primer & 1 coat | 2.000 | FACE | 1.920 | 6.40 | 35.20 | 41.60 |
| Primer & 2 coats | 2.000 | FACE | 2.720 | 8.80 | 50.40 | 59.20 |
| 8' wide door, 1 coat | 2.000 | FACE | 1.936 | 5.12 | 35.20 | 40.32 |
| 2 coats | 2.000 | FACE | 2.376 | 7.04 | 43.12 | 50.16 |
| Primer & 1 coat | 2.000 | FACE | 2.112 | 7.04 | 38.72 | 45.76 |
| Primer & 2 coats | 2.000 | FACE | 2.992 | 9.68 | 55.44 | 65.12 |
| 12' wide door, 1 coat | 2.000 | FACE | 2.288 | 6.05 | 41.60 | 47.65 |
| 2 coats | 2.000 | FACE | 2.808 | 8.32 | 50.96 | 59.28 |
| Primer & 1 coat | 2.000 | FACE | 2.496 | 8.32 | 45.76 | 54.08 |
| Primer & 2 coats | 2.000 | FACE | 3.536 | 11.44 | 65.52 | 76.96 |
| Aluminum door, trim only, interior & exterior, 6' door, 1 coat | 2.000 | FACE | .880 | 2.33 | 16.00 | 18.33 |
| 2 coats | 2.000 | FACE | 1.080 | 3.20 | 19.60 | 22.80 |
| Primer & 1 coat | 2.000 | FACE | .960 | 3.20 | 17.60 | 20.80 |
| Primer & 2 coats | 2.000 | FACE | 1.360 | 4.40 | 25.20 | 29.60 |
| 8' wide door, 1 coat | 2.000 | FACE | .968 | 2.56 | 17.60 | 20.16 |
| 2 coats | 2.000 | FACE | 1.188 | 3.52 | 21.56 | 25.08 |
| Primer & 1 coat | 2.000 | FACE | 1.056 | 3.52 | 19.36 | 22.88 |
| Primer & 2 coats | 2.000 | FACE | 1.496 | 4.84 | 27.72 | 32.56 |
| 12' wide door, 1 coat | 2.000 | FACE | 1.144 | 3.03 | 20.80 | 23.83 |
| 2 coats | 2.000 | FACE | 1.404 | 4.16 | 25.48 | 29.64 |
| Primer & 1 coat | 2.000 | FACE | 1.248 | 4.16 | 22.88 | 27.04 |
| Primer & 2 coats | 2.000 | FACE | 1.768 | 5.72 | 32.76 | 38.48 |

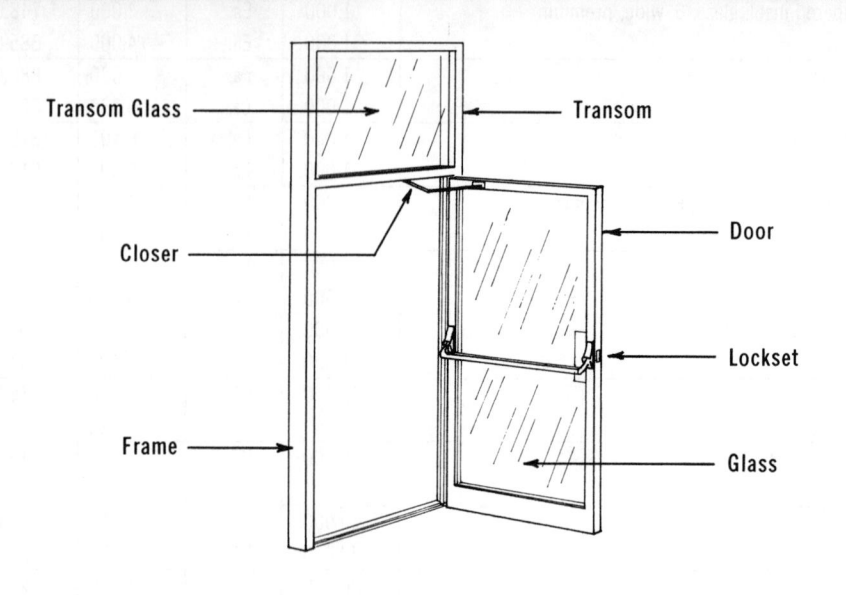

# 4 | EXTERIOR WALLS  54 | Commercial Entrance Systems

| SYSTEM DESCRIPTION | QUAN. | UNIT | MAN HOURS | MAT. | INST. | TOTAL |
|---|---|---|---|---|---|---|
| **MILL FINISH, ALUMINUM ENTRANCE DOOR WITH TRANSOM, 3'-0" X 10'-0"** | | | | | | |
| Door, mill finish, aluminum, narrow stile, including hinges | 1.000 | Ea. | 5.330 | 269.50 | 110.50 | 380.00 |
| Frame, aluminum, mill finish, 3'-0" x 7'-0" | 1.000 | Ea. | 2.290 | 137.50 | 52.50 | 190.00 |
| Glass, 1/4" thick, clear, tempered | 20.000 | S.F. | 2.660 | 64.00 | 52.00 | 116.00 |
| Transom, aluminum, mill finish, 3'-0" x 3'-0" | 1.000 | Ea. | .200 | 66.00 | 5.00 | 71.00 |
| Transom glass, 1/4" thick, clear, plain | 8.000 | S.F. | 1.064 | 25.60 | 20.80 | 46.40 |
| Closer, medium service | 1.000 | Ea. | 1.230 | 61.55 | 24.00 | 85.55 |
| Lockset, heavy duty, minimum | 1.000 | Ea. | 1.000 | 99.00 | 21.00 | 120.00 |
| TOTAL | | | 13.774 | 723.15 | 285.80 | 1008.95 |
| **BRONZE ALUMINUM, ENTRANCE DOOR WITH TRANSOM, 3'-0" X 10'-0"** | | | | | | |
| Door, bronze, aluminum, narrow stile, including hinges | 1.000 | Ea. | 5.330 | 319.00 | 111.00 | 430.00 |
| Frame, aluminum, anodized, 3'-0" x 7'-0" | 1.000 | Ea. | 2.285 | 181.50 | 53.50 | 235.00 |
| Glass, 1/4" thick, clear, tempered | 20.000 | S.F. | 2.660 | 64.00 | 52.00 | 116.00 |
| Transom, aluminum, anodized, 3'-0" x 3'-0" | 1.000 | Ea. | .200 | 77.00 | 5.00 | 82.00 |
| Transom glass, 1/4" thick, clear, plain | 8.000 | S.F. | 1.064 | 25.60 | 20.80 | 46.40 |
| Closer, medium service | 1.000 | Ea. | 1.230 | 61.55 | 24.00 | 85.55 |
| Lockset, heavy duty, minimum | 1.000 | Ea. | 1.000 | 99.00 | 21.00 | 120.00 |
| TOTAL | | | 13.769 | 827.65 | 287.30 | 1114.95 |

| DESCRIPTION | QUAN. | UNIT | MAN HOURS | MAT. | INST. | TOTAL |
|---|---|---|---|---|---|---|
| | | | | | | |
| | | | | | | |
| | | | | | | |
| | | | | | | |
| | | | | | | |

# Commercial Entrance Price Sheet

| | QUAN. | UNIT | MAN HOURS | COST EACH MAT. | COST EACH INST. | TOTAL |
|---|---|---|---|---|---|---|
| Doors, aluminum narrow stile, 3'-0" x 7'-0", mill finish | 1.000 | Ea. | 5.330 | 269.50 | 110.50 | 380.00 |
| Bronze finish | 1.000 | Ea. | 5.330 | 319.00 | 111.00 | 430.00 |
| Black finish | 1.000 | Ea. | 5.330 | 341.00 | 109.00 | 450.00 |
| 3'-6" x 7'-0", mill finish | 1.000 | Ea. | 5.330 | 302.50 | 107.50 | 410.00 |
| Bronze finish | 1.000 | Ea. | 5.330 | 357.50 | 107.50 | 465.00 |
| Black finish | 1.000 | Ea. | 5.330 | 385.00 | 110.00 | 495.00 |
| Steel, hollow metal flush, 20 ga., 1-3/4" x 7'-0" x, 2'-6" | 1.000 | Ea. | 1.066 | 143.00 | 22.00 | 165.00 |
| 2'-8" | 1.000 | Ea. | 1.066 | 154.00 | 21.00 | 175.00 |
| 3'-0" | 1.000 | Ea. | 1.142 | 165.00 | 25.00 | 190.00 |
| 3'-6" | 1.000 | Ea. | 1.142 | 181.50 | 23.50 | 205.00 |
| 16 ga., 1-3/4" x 7'-0" x 2'-6" | 1.000 | Ea. | 1.066 | 176.00 | 24.00 | 200.00 |
| 2'-8" | 1.000 | Ea. | 1.066 | 181.50 | 23.50 | 205.00 |
| 3'-0" | 1.000 | Ea. | 1.142 | 192.50 | 22.50 | 215.00 |
| 3'-6" | 1.000 | Ea. | 1.142 | 214.50 | 25.50 | 240.00 |
| Fire door, "A" label, 1-3/4" x 7'-0" x 2'-6" | 1.000 | Ea. | 1.066 | 203.50 | 21.50 | 225.00 |
| 2'-8" | 1.000 | Ea. | 1.066 | 214.50 | 20.50 | 235.00 |
| 3'-0" | 1.000 | Ea. | 1.142 | 231.00 | 24.00 | 255.00 |
| 3'-6" | 1.000 | Ea. | 1.142 | 253.00 | 22.00 | 275.00 |
| Frame, aluminum, mill finish, 7'-0" x 3'-0" wide | 1.000 | Ea. | 2.290 | 126.50 | 52.20 | 180.00 |
| 3'-6" | 1.000 | Ea. | 2.290 | 154.00 | 51.00 | 205.00 |
| 6'-0" | 1.000 | Ea. | 2.670 | 137.50 | 59.50 | 197.00 |
| 7'-0" | 1.000 | Ea. | 2.670 | 170.50 | 59.50 | 230.00 |
| Bronze or black, 7'-0" x 3'-0" wide | 1.000 | Ea. | 2.290 | 181.50 | 53.50 | 235.00 |
| 3'-6" | 1.000 | Ea. | 2.290 | 187.00 | 53.00 | 240.00 |
| 6'-0" | 1.000 | Ea. | 2.670 | 170.50 | 59.50 | 230.00 |
| 7'-0" | 1.000 | Ea. | 2.670 | 203.50 | 61.50 | 265.00 |
| Steel, 16 ga., 4-3/4" x 7'-0" x 2'-6" | 1.000 | Ea. | 1.000 | 63.80 | 20.20 | 84.00 |
| 2'-8" | 1.000 | Ea. | 1.000 | 66.00 | 21.00 | 87.00 |
| 3'-0" | 1.000 | Ea. | 1.066 | 66.00 | 22.00 | 88.00 |
| 3'-6" | 1.000 | Ea. | 1.066 | 66.00 | 22.00 | 88.00 |
| 5'-0" | 1.000 | Ea. | 1.142 | 71.50 | 23.50 | 95.00 |
| 5'-4" | 1.000 | Ea. | 1.142 | 73.70 | 23.30 | 97.00 |
| 6'-0" | 1.000 | Ea. | 1.142 | 77.00 | 23.00 | 100.00 |
| 7'-0" | 1.000 | Ea. | 1.230 | 81.40 | 23.60 | 105.00 |
| Glass, for door, 7'-0" x 3'-0", tempered, clear | 20.000 | S.F. | 2.660 | 64.00 | 52.00 | 116.00 |
| Tinted | 20.000 | S.F. | 2.660 | 83.60 | 52.40 | 136.00 |
| Insulated, tempered, clear | 20.000 | S.F. | 4.260 | 138.60 | 83.40 | 222.00 |
| Tinted | 20.000 | S.F. | 4.260 | 164.00 | 83.00 | 247.00 |
| 7'-0" x 3'-6", tempered, clear | 23.500 | Ea. | 3.125 | 75.20 | 61.10 | 136.30 |
| Tinted | 23.500 | S.F. | 3.125 | 98.23 | 61.57 | 159.80 |
| Insulated, tempered, clear | 23.500 | S.F. | 5.005 | 162.86 | 97.99 | 260.85 |
| Tinted | 23.500 | S.F. | 5.005 | 192.70 | 97.53 | 290.23 |
| Transom, aluminum, mill finish, 3'-0" x 3'-0" | 1.000 | Ea. | .200 | 66.00 | 5.00 | 71.00 |
| 3'-6" | 1.000 | Ea. | .200 | 68.20 | 4.80 | 73.00 |
| Bronze or black, 3'-0" x 3"-0" | 1.000 | Ea. | .200 | 77.00 | 5.00 | 82.00 |
| 3'-6" | 1.000 | Ea. | .200 | 82.50 | 4.50 | 87.00 |
| Transom glass, 3"-0" x 3'-0", plain, clear | 8.000 | S.F. | 1.064 | 25.60 | 20.80 | 46.40 |
| Tinted | 8.000 | S.F. | 1.064 | 33.44 | 20.96 | 54.40 |
| 3'-0" x 3'-6", plain, clear | 9.500 | S.F. | 1.263 | 30.40 | 24.70 | 55.10 |
| Tinted | 9.500 | S.F. | 1.263 | 39.71 | 24.89 | 64.60 |
| Closer, surface mounted, average duty | 1.000 | Ea. | 1.230 | 61.55 | 24.00 | 85.55 |
| Heavy duty | 1.000 | Ea. | 1.230 | 69.25 | 24.30 | 93.55 |
| Concealed, heavy duty, head | 1.000 | Ea. | 1.454 | 93.50 | 26.50 | 120.00 |
| Floor | 1.000 | Ea. | 3.636 | 110.00 | 70.00 | 180.00 |
| Lockset, heavy duty, knob type for steel doors, minimum | 1.000 | Ea. | .800 | 49.50 | 15.50 | 65.00 |
| Maximum | 1.000 | Ea. | .800 | 121.00 | 14.00 | 135.00 |
| Concealed, for aluminum store doors | 1.000 | Ea. | 1.000 | 99.00 | 21.00 | 120.00 |

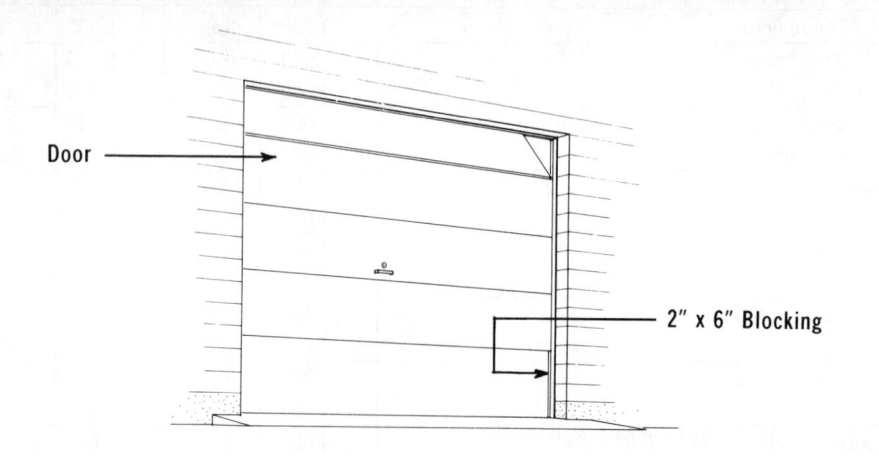

# 4 | EXTERIOR WALLS  55 | Comm'l Overhead Door Systems

*Door* → (illustration)
*2" x 6" Blocking* →

| SYSTEM DESCRIPTION | QUAN. | UNIT | MAN HOURS | MAT. | INST. | TOTAL |
|---|---|---|---|---|---|---|
| **WOOD, OVERHEAD DOOR, 10'-0" X 10'-0"** | | | | | | |
| Door, stock wood, heavy duty, 10'-0" x 10'-0" | 1.000 | Ea. | 8.890 | 583.00 | 172.00 | 755.00 |
| Blocking, 2" x 6" | 30.000 | L.F. | 1.080 | 12.00 | 22.20 | 34.20 |
| Opener, electric trolley operator | 1.000 | Ea. | 4.000 | 550.00 | 80.00 | 630.00 |
| Paint, primer & 1 coat | 1.000 | Ea. | 4.800 | 12.39 | 89.61 | 102.00 |
| TOTAL | | | 18.770 | 1157.39 | 363.81 | 1521.20 |
| **WOOD, OVERHEAD DOOR, 14'-0" X 14'-0"** | | | | | | |
| Door, stock wood, heavy duty, 14'-0" x 14'-0" | 1.000 | Ea. | 12.310 | 1320.00 | 230.00 | 1550.00 |
| Blocking, 2" x 6" | 42.000 | L.F. | 1.512 | 16.80 | 31.08 | 47.88 |
| Opener, electric trolley operator | 1.000 | Ea. | 4.000 | 550.00 | 80.00 | 630.00 |
| Paint, primer & 1 coat | 1.000 | Ea. | 6.400 | 16.52 | 119.48 | 136.00 |
| TOTAL | | | 24.222 | 1903.32 | 460.56 | 2363.88 |
| **FIBERGLASS AND ALUMINUM, 12'-0" X 12'-0"** | | | | | | |
| Door, fiberglass and aluminum, heavy duty, 12'-0" x 12'-0" | 1.000 | Ea. | 10.670 | 946.00 | 204.00 | 1150.00 |
| Blocking, 2" x 6" | 36.000 | L.F. | 1.296 | 14.40 | 26.64 | 41.04 |
| Opener, electric trolley operator | 1.000 | Ea. | 4.000 | 550.00 | 80.00 | 630.00 |
| TOTAL | | | 15.966 | 1510.40 | 310.64 | 1821.04 |
| **STEEL, 24 GUAGE, 12'-0" X 12'-0"** | | | | | | |
| Door, steel, 24 guage, 12'-0" x 12'-0" | 1.000 | Ea. | 10.670 | 753.50 | 206.50 | 960.00 |
| Blocking, 2" x 6" | 36.000 | L.F. | 1.296 | 14.40 | 26.64 | 41.04 |
| Opener, electric trolley operator | 1.000 | Ea. | 4.000 | 550.00 | 80.00 | 630.00 |
| Paint, primer & 1 coat | 1.000 | Ea. | 5.600 | 14.46 | 104.54 | 119.00 |
| TOTAL | | | 21.566 | 1332.36 | 417.68 | 1750.04 |

| DESCRIPTION | QUAN. | UNIT | MAN HOURS | MAT. | INST. | TOTAL |
|---|---|---|---|---|---|---|
| | | | | | | |
| | | | | | | |
| | | | | | | |

# Comm'l Overhead Door Price Sheet

| Item | QUAN. | UNIT | MAN HOURS | COST EACH MAT. | COST EACH INST. | TOTAL |
|---|---|---|---|---|---|---|
| Doors, fiberglass and aluminum, heavy duty, 12'-0" x 12'-0" | 1.000 | Ea. | 10.670 | 946.00 | 204.00 | 1150.00 |
| 20'-0" x 20'-0" chain hoist | 1.000 | Ea. | 32.000 | 2310.00 | 615.00 | 2925.00 |
| Steel, 24 guage, 8'-0" x 8'-0", plain | 1.000 | Ea. | 8.000 | 346.50 | 158.50 | 505.00 |
| Insulated | 1.000 | Ea. | 8.000 | 487.30 | 158.50 | 645.80 |
| 10'-0" x 10'-0", plain | 1.000 | Ea. | 8.890 | 522.50 | 172.50 | 695.00 |
| Insulated | 1.000 | Ea. | 8.890 | 742.50 | 172.50 | 915.00 |
| 12'-0" x 12'-0", plain | 1.000 | Ea. | 10.670 | 753.50 | 206.50 | 960.00 |
| Insulated | 1.000 | Ea. | 10.670 | 1070.30 | 206.50 | 1276.80 |
| 20'-0" x 14'-0", chain hoist, plain | 1.000 | Ea. | 22.860 | 1732.50 | 442.50 | 2175.00 |
| Insulated | 1.000 | Ea. | 22.860 | 2348.50 | 442.50 | 2791.00 |
| Wood, heavy duty, 1-3/4" thick, 8'-0" x 8'-0" | 1.000 | Ea. | 8.000 | 418.00 | 157.00 | 575.00 |
| 10'-0" x 10'-0" | 1.000 | Ea. | 8.890 | 583.00 | 172.00 | 755.00 |
| 12'-0" x 12'-0" | 1.000 | Ea. | 10.670 | 836.00 | 214.00 | 1050.00 |
| Chain hoist, 14'-0" x 14'-0" | 1.000 | Ea. | 12.310 | 1320.00 | 230.00 | 1550.00 |
| 12'-0" x 16'-0" | 1.000 | Ea. | 16.000 | 1347.50 | 302.50 | 1650.00 |
| 20'-0" x 8'-0" | 1.000 | Ea. | 20.000 | 1127.50 | 397.50 | 1525.00 |
| 20'-0" x 16'-0" | 1.000 | Ea. | 26.670 | 2585.00 | 515.00 | 3100.00 |
| Blocking, 2" x 6", 8'-0" x 8'-0" door | 24.000 | L.F. | .864 | 9.60 | 17.76 | 27.36 |
| 10'-0" x 10'-0" door | 30.000 | L.F. | 1.080 | 12.00 | 22.20 | 34.20 |
| 12'-0" x 12'-0" door | 36.000 | L.F. | 1.296 | 14.40 | 26.64 | 41.04 |
| 12'-0" x 16'-0" door | 44.000 | L.F. | 1.584 | 17.60 | 32.56 | 50.16 |
| 20'-0" x 8'-0" door | 48.000 | L.F. | 1.728 | 19.20 | 35.52 | 54.72 |
| 20'-0" x 20'-0" door | 60.000 | L.F. | 2.160 | 24.00 | 44.40 | 68.40 |
| Opener, electric trolley operator, to 14'-0" x 14'-0" door | 1.000 | Ea. | 4.000 | 550.00 | 80.00 | 630.00 |
| Over 14'-0" x 14'-0" door | 1.000 | Ea. | 8.000 | 660.00 | 155.00 | 815.00 |
| Paint, 1 coat, 8'-0" x 8'-0" door | 1.000 | Ea. | 2.222 | 7.25 | 41.13 | 48.38 |
| 10'-0" x 10'-0" door | 1.000 | Ea. | 2.667 | 8.70 | 49.35 | 58.05 |
| 12'-0" x 12'-0" door | 1.000 | Ea. | 3.111 | 10.15 | 57.58 | 67.73 |
| 12'-0" x 16'-0" door | 1.000 | Ea. | 3.556 | 11.60 | 65.80 | 77.40 |
| 20'-0" x 8'-0" door | 1.000 | Ea. | 3.333 | 10.88 | 61.68 | 72.56 |
| 20'-0" x 20'-0" door | 1.000 | Ea. | 5.334 | 17.40 | 98.70 | 116.10 |
| Primer & 1 coat, 8'-0" x 8'-0" door | 1.000 | Ea. | 4.000 | 10.33 | 74.67 | 85.00 |
| 10'-0" x 10'-0" door | 1.000 | Ea. | 4.800 | 12.39 | 89.61 | 102.00 |
| 12'-0" x 12'-0" door | 1.000 | Ea. | 5.600 | 14.46 | 104.54 | 119.00 |
| 12'-0" x 16'-0" door | 1.000 | Ea. | 6.400 | 16.52 | 119.48 | 136.00 |
| 20'-0" x 8'-0" door | 1.000 | Ea. | 6.000 | 15.49 | 112.01 | 127.50 |
| 20'-0" x 20'-0" door | 1.000 | Ea. | 9.600 | 24.78 | 179.22 | 204.00 |

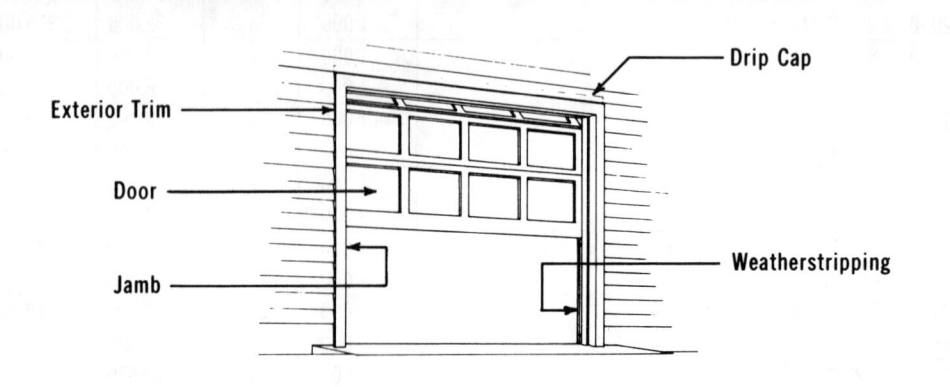

# 4 | EXTERIOR WALLS  56 | Residential Garage Door Systems

| SYSTEM DESCRIPTION | QUAN. | UNIT | MAN HOURS | MAT. | INST. | TOTAL |
|---|---|---|---|---|---|---|
| **OVERHEAD, SECTIONAL GARAGE DOOR, 9' X 7'** | | | | | | |
| Wood, overhead sectional door, standard, including hardware, 9' x 7' | 1.000 | Ea. | 2.000 | 220.00 | 40.00 | 260.00 |
| Jamb & header blocking, 2" x 6" | 25.000 | L.F. | .900 | 10.00 | 18.50 | 28.50 |
| Exterior trim | 25.000 | L.F. | 1.000 | 26.25 | 19.50 | 45.75 |
| Paint, interior & exterior, primer & 2 coats | 2.000 | FACE | 9.160 | 18.72 | 169.28 | 188.00 |
| Weatherstripping, molding type | 1.000 | Set | .759 | 12.42 | 14.95 | 27.37 |
| Drip cap | 9.000 | L.F. | .180 | 1.26 | 3.51 | 4.77 |
| TOTAL | | | 13.999 | 288.65 | 265.74 | 554.39 |
| **OVERHEAD, SECTIONAL GARAGE DOOR, 16' X 7'** | | | | | | |
| Wood, overhead sectional door, standard, including hardware 16' x 7' | 1.000 | Ea. | 2.670 | 440.00 | 50.00 | 490.00 |
| Jamb & header blocking, 2" x 6" | 30.000 | L.F | 1.080 | 12.00 | 22.20 | 34.20 |
| Exterior trim | 30.000 | L.F. | 1.200 | 31.50 | 23.40 | 54.90 |
| Paint, interior & exterior, primer & 2 coats | 2.000 | FACE | 13.740 | 28.08 | 253.92 | 282.00 |
| Weatherstripping, molding type | 1.000 | Set | .990 | 16.20 | 19.50 | 35.70 |
| Drip cap | 16.000 | L.F. | .320 | 2.24 | 6.24 | 8.48 |
| TOTAL | | | 20.000 | 530.02 | 375.26 | 905.28 |
| **OVERHEAD, SWING-UP TYPE, GARAGE DOOR, 16' X 7'** | | | | | | |
| Wood, overhead, swing-up type, standard, including hardware 16' x 7' | 1.000 | Ea. | 2.670 | 330.00 | 50.00 | 380.00 |
| Jamb & header blocking, 2" x 6" | 30.000 | L.F. | 1.080 | 12.00 | 22.20 | 34.20 |
| Exterior trim | 30.000 | L.F. | 1.200 | 31.50 | 23.40 | 54.90 |
| Paint, interior & exterior, primer & 2 coats | 2.000 | FACE | 13.740 | 28.08 | 253.92 | 282.00 |
| Weatherstripping, molding type | 1.000 | Set | .990 | 16.20 | 19.50 | 35.70 |
| Drip cap | 16.000 | L.F. | .320 | 2.24 | 6.24 | 8.48 |
| TOTAL | | | 20.000 | 420.02 | 375.26 | 795.28 |

This system is on a cost per each door basis.

| DESCRIPTION | QUAN. | UNIT | MAN HOURS | MAT. | INST. | TOTAL |
|---|---|---|---|---|---|---|
| | | | | | | |
| | | | | | | |
| | | | | | | |
| | | | | | | |
| | | | | | | |

# Residen'l Garage Door Price Sheet

| Description | QUAN. | UNIT | MAN HOURS | MAT. | INST. | TOTAL |
|---|---|---|---|---|---|---|
| Overhead sectional including hardware, fiberglass, 9' x 7', standard | 1.000 | Ea. | 2.000 | 313.50 | 41.50 | 355.00 |
| Deluxe | 1.000 | Ea. | 2.000 | 346.50 | 38.50 | 385.00 |
| 16' x 7', standard | 1.000 | Ea. | 2.670 | 522.50 | 52.50 | 575.00 |
| Deluxe | 1.000 | Ea. | 2.670 | 577.50 | 52.50 | 630.00 |
| Hardboard, 9' x 7', standard | 1.000 | Ea. | 2.000 | 203.50 | 41.50 | 245.00 |
| Deluxe | 1.000 | Ea. | 2.000 | 280.50 | 39.50 | 320.00 |
| 16' x 7', standard | 1.000 | Ea. | 2.670 | 302.50 | 52.50 | 355.00 |
| Deluxe | 1.000 | Ea. | 2.670 | 572.00 | 53.00 | 625.00 |
| Metal, 9' x 7', standard | 1.000 | Ea. | 2.000 | 253.00 | 37.00 | 290.00 |
| Deluxe | 1.000 | Ea. | 2.000 | 401.50 | 38.50 | 440.00 |
| 16' x 7', standard | 1.000 | Ea. | 2.670 | 368.50 | 51.50 | 420.00 |
| Deluxe | 1.000 | Ea. | 2.670 | 764.50 | 50.50 | 815.00 |
| Wood, 9' x 7', standard | 1.000 | Ea. | 2.000 | 220.00 | 40.00 | 260.00 |
| Deluxe | 1.000 | Ea. | 2.000 | 660.00 | 40.00 | 700.00 |
| 16' x 7', standard | 1.000 | Ea. | 2.670 | 440.00 | 50.00 | 490.00 |
| Deluxe | 1.000 | Ea. | 2.670 | 852.50 | 52.50 | 905.00 |
| Overhead swing-up type including hardware, fiberglass, 9' x 7', standard | 1.000 | Ea. | 2.000 | 203.50 | 41.50 | 245.00 |
| Deluxe | 1.000 | Ea. | 2.000 | 313.50 | 41.50 | 355.00 |
| 16' x 7', standard | 1.000 | Ea. | 2.670 | 352.00 | 53.00 | 405.00 |
| Deluxe | 1.000 | Ea. | 2.670 | 423.50 | 51.50 | 475.00 |
| Hardboard, 9' x 7', standard | 1.000 | Ea. | 2.000 | 231.00 | 39.00 | 270.00 |
| Deluxe | 1.000 | Ea. | 2.000 | 302.50 | 37.50 | 340.00 |
| 16' x 7', standard | 1.000 | Ea. | 2.670 | 357.50 | 52.50 | 410.00 |
| Deluxe | 1.000 | Ea. | 2.670 | 467.50 | 52.50 | 520.00 |
| Metal, 9' x 7', standard | 1.000 | Ea. | 2.000 | 225.50 | 39.50 | 265.00 |
| Deluxe | 1.000 | Ea. | 2.000 | 368.50 | 41.50 | 410.00 |
| 16' x 7', standard | 1.000 | Ea. | 2.670 | 396.00 | 54.00 | 450.00 |
| Deluxe | 1.000 | Ea. | 2.670 | 660.00 | 50.00 | 710.00 |
| Wood, 9' x 7', standard | 1.000 | Ea. | 2.000 | 170.50 | 39.50 | 210.00 |
| Deluxe | 1.000 | Ea. | 2.000 | 302.50 | 37.50 | 340.00 |
| 16' x 7', standard | 1.000 | Ea. | 2.670 | 330.00 | 50.00 | 380.00 |
| Deluxe | 1.000 | Ea. | 2.670 | 478.50 | 51.50 | 530.00 |
| Jamb & header blocking, 2" x 6", 9' x 7' door | 25.000 | L.F. | .900 | 10.00 | 18.50 | 28.50 |
| 16' x 7' door | 30.000 | L.F. | 1.080 | 12.00 | 22.20 | 34.20 |
| 2" x 8", 9' x 7' door | 25.000 | L.F. | 1.000 | 13.75 | 20.50 | 34.25 |
| 16' x 7' door | 30.000 | L.F. | 1.200 | 16.50 | 24.60 | 41.10 |
| Exterior trim, 9' x 7' door | 25.000 | L.F. | 1.000 | 26.25 | 19.50 | 45.75 |
| 16' x 7' door | 30.000 | L.F. | 1.200 | 31.50 | 23.40 | 54.90 |
| Paint or stain, interior & exterior, 9' x 7' door, 1 coat | 1.000 | FACE | 3.556 | 11.60 | 65.80 | 77.40 |
| 2 coats | 1.000 | FACE | 6.400 | 16.52 | 119.48 | 136.00 |
| Primer & 1 coat | 1.000 | FACE | 6.400 | 16.52 | 119.48 | 136.00 |
| Primer & 2 coats | 1.000 | FACE | 9.160 | 18.72 | 169.28 | 188.00 |
| 16' x 7' door, 1 coat | 1.000 | FACE | 5.334 | 17.40 | 98.70 | 116.10 |
| 2 coats | 1.000 | FACE | 9.600 | 24.78 | 179.22 | 204.00 |
| Primer & 1 coat | 1.000 | FACE | 9.600 | 24.78 | 179.22 | 204.00 |
| Primer & 2 coats | 1.000 | FACE | 13.740 | 28.08 | 253.92 | 282.00 |
| Weatherstripping, molding type, 9' x 7' door | 1.000 | Set | .759 | 12.42 | 14.95 | 27.37 |
| 16' x 7' door | 1.000 | Set | .990 | 16.20 | 19.50 | 35.70 |
| Drip cap, 9' door | 9.000 | L.F. | .180 | 1.26 | 3.51 | 4.77 |
| 16' door | 16.000 | L.F. | .320 | 2.24 | 6.24 | 8.48 |
| Garage door opener, economy | 1.000 | Ea. | 1.000 | 176.00 | 19.00 | 195.00 |
| Deluxe, including remote control | 1.000 | Ea. | 1.000 | 264.00 | 21.00 | 285.00 |

4

# 4 | EXTERIOR WALLS  58 | Aluminum Window Systems

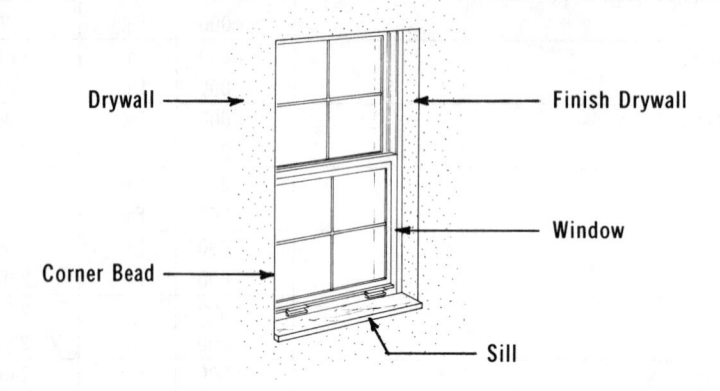

Drywall → | ← Finish Drywall
Corner Bead → | ← Window
↓ Sill

| SYSTEM DESCRIPTION | QUAN. | UNIT | MAN HOURS | MAT. | INST. | TOTAL |
|---|---|---|---|---|---|---|
| **SINGLE HUNG, 2' X 3' OPENING** | | | | | | |
| Window, single hung, 2' x 3' opening, enameled, insulating glass | 1.000 | Ea. | 1.600 | 126.50 | 38.50 | 165.00 |
| Blocking, 1" x 3" furring strip nailers | 10.000 | L.F. | .150 | 1.00 | 2.80 | 3.80 |
| Drywall, 1/2" thick, standard | 5.000 | S.F. | .045 | 1.15 | .85 | 2.00 |
| Corner bead, 1" x 1", galvanized steel | 8.000 | L.F. | .224 | .72 | 4.32 | 5.04 |
| Finish drywall, tape and finish corners inside and outside | 16.000 | L.F. | .240 | .64 | 4.64 | 5.28 |
| Sill, slate | 2.000 | L.F. | .376 | 9.52 | 6.68 | 16.20 |
| TOTAL | | | 2.635 | 139.53 | 57.79 | 197.32 |
| **SLIDING, 3' X 2' OPENING** | | | | | | |
| Window, sliding, 3' x 2' opening, enameled, insulating glass | 1.000 | Ea. | 1.600 | 84.70 | 35.30 | 120.00 |
| Blocking, 1" x 3" furring strip nailers | 10.000 | L.F. | .150 | 1.00 | 2.80 | 3.80 |
| Drywall, 1/2" thick, standard | 5.000 | S.F. | .045 | 1.15 | .85 | 2.00 |
| Corner bead, 1" x 1", galvanized steel | 7.000 | L.F. | .196 | .63 | 3.78 | 4.41 |
| Finish drywall, tape and finish corners inside and outside | 14.000 | L.F. | .210 | .56 | 4.06 | 4.62 |
| Sill, slate | 3.000 | L.F. | .564 | 14.28 | 10.02 | 24.30 |
| TOTAL | | | 2.765 | 102.32 | 56.81 | 159.13 |
| **AWNING, 3'-1" X 3'-2"** | | | | | | |
| Window awning, 3'-1" x 3'-2" opening, enameled, insulating glass | 1.000 | Ea. | 1.600 | 115.50 | 34.50 | 150.00 |
| Blocking, 1" x 3" furring strip, nailers | 12.500 | L.F. | .187 | 1.25 | 3.50 | 4.75 |
| Drywall, 1/2" thick, standard | 4.500 | S.F. | .040 | 1.04 | .76 | 1.80 |
| Corner bead, 1" x 1", galvanized steel | 9.250 | L.F. | .259 | .83 | 5.00 | 5.83 |
| Finish drywall, tape and finish corners, inside and outside | 18.500 | L.F. | .277 | .74 | 5.37 | 6.11 |
| Sill, slate | 3.250 | L.F. | .611 | 15.47 | 10.86 | 26.33 |
| TOTAL | | | 2.974 | 134.83 | 59.99 | 194.82 |

| DESCRIPTION | QUAN. | UNIT | MAN HOURS | MAT. | INST. | TOTAL |
|---|---|---|---|---|---|---|
| | | | | | | |
| | | | | | | |
| | | | | | | |
| | | | | | | |
| | | | | | | |

# Aluminum Window Price Sheet

| | QUAN. | UNIT | MAN HOURS | COST EACH MAT. | COST EACH INST. | TOTAL |
|---|---|---|---|---|---|---|
| Window, aluminum, awning, 3'-1" x 3'-2", standard glass | 1.000 | Ea. | 1.600 | 115.50 | 34.50 | 150.00 |
| Insulating glass | 1.000 | Ea. | 1.600 | 115.50 | 34.50 | 150.00 |
| 4'-5" x 5'-3", standard glass | 1.000 | Ea. | 2.000 | 159.50 | 45.50 | 205.00 |
| Insulating glass | 1.000 | Ea. | 2.000 | 181.50 | 43.50 | 225.00 |
| Casement, 3'-1" x 3'-2", standard glass | 1.000 | Ea. | 1.600 | 170.50 | 34.50 | 205.00 |
| Insulating glass | 1.000 | Ea. | 1.600 | 159.50 | 35.50 | 195.00 |
| Single hung, 2' x 3', standard glass | 1.000 | Ea. | 1.600 | 77.00 | 38.00 | 115.00 |
| Insulating glass | 1.000 | Ea. | 1.600 | 126.50 | 38.50 | 165.00 |
| 2'-8" x 6'-8", standard glass | 1.000 | Ea. | 2.000 | 165.00 | 45.00 | 210.00 |
| Insulating glass | 1.000 | Ea. | 2.000 | 209.00 | 46.00 | 255.00 |
| 3'-4" x 5'-0", standard glass | 1.000 | Ea. | 1.780 | 106.70 | 38.30 | 145.00 |
| Insulating glass | 1.000 | Ea. | 1.780 | 143.00 | 42.00 | 185.00 |
| Sliding, 3' x 2', standard glass | 1.000 | Ea. | 1.600 | 50.60 | 36.40 | 87.00 |
| Insulating glass | 1.000 | Ea. | 1.600 | 84.70 | 35.30 | 120.00 |
| 5' x 3', standard glass | 1.000 | Ea. | 1.780 | 88.00 | 42.00 | 130.00 |
| Insulating glass | 1.000 | Ea. | 1.780 | 143.00 | 42.00 | 185.00 |
| 8' x 4', standard glass | 1.000 | Ea. | 2.670 | 126.50 | 58.50 | 185.00 |
| Insulating glass | 1.000 | Ea. | 2.670 | 253.00 | 62.00 | 315.00 |
| Blocking, 1" x 3" furring, opening 3' x 2' | 10.000 | L.F. | .150 | 1.00 | 2.80 | 3.80 |
| 3' x 3' | 12.500 | L.F. | .187 | 1.25 | 3.50 | 4.75 |
| 3' x 5' | 16.000 | L.F. | .240 | 1.60 | 4.48 | 6.08 |
| 4' x 4' | 16.000 | L.F. | .240 | 1.60 | 4.48 | 6.08 |
| 4' x 5' | 18.000 | L.F. | .270 | 1.80 | 5.04 | 6.84 |
| 4' x 6' | 20.000 | L.F. | .300 | 2.00 | 5.60 | 7.60 |
| 4' x 8' | 24.000 | L.F. | .360 | 2.40 | 6.72 | 9.12 |
| 6'-8" x 2'-8" | 19.000 | L.F. | .285 | 1.90 | 5.32 | 7.22 |
| Drywall, 1/2" thick, standard, opening 3' x 2' | 5.000 | S.F. | .045 | 1.15 | .85 | 2.00 |
| 3' x 3' | 6.000 | S.F. | .054 | 1.38 | 1.02 | 2.40 |
| 3' x 5' | 8.000 | S.F. | .072 | 1.84 | 1.36 | 3.20 |
| 4' x 4' | 8.000 | S.F. | .072 | 1.84 | 1.36 | 3.20 |
| 4' x 5' | 9.000 | S.F. | .081 | 2.07 | 1.53 | 3.60 |
| 4' x 6' | 10.000 | S.F. | .090 | 2.30 | 1.70 | 4.00 |
| 4' x 8' | 12.000 | S.F. | .108 | 2.76 | 2.04 | 4.80 |
| 6'-8" x 2' | 9.500 | S.F. | .085 | 2.19 | 1.61 | 3.80 |
| Corner bead, 1" x 1", galvanized steel, opening 3' x 2' | 7.000 | L.F. | .196 | .63 | 3.78 | 4.41 |
| 3' x 3' | 9.000 | L.F. | .252 | .81 | 4.86 | 5.67 |
| 3' x 5' | 11.000 | L.F. | .308 | .99 | 5.94 | 6.93 |
| 4' x 4' | 12.000 | L.F. | .336 | 1.08 | 6.48 | 7.56 |
| 4' x 5' | 13.000 | L.F. | .364 | 1.17 | 7.02 | 8.19 |
| 4' x 6' | 14.000 | L.F. | .392 | 1.26 | 7.56 | 8.82 |
| 4' x 8' | 16.000 | L.F. | .448 | 1.44 | 8.64 | 10.08 |
| 6'-8" x 2' | 15.000 | L.F. | .420 | 1.35 | 8.10 | 9.45 |
| Tape and finish corners, inside and outside, opening 3' x 2' | 14.000 | L.F. | .210 | .56 | 4.06 | 4.62 |
| 3' x 3' | 18.000 | L.F. | .270 | .72 | 5.22 | 5.94 |
| 3' x 5' | 22.000 | L.F. | .330 | .88 | 6.38 | 7.26 |
| 4' x 4' | 24.000 | L.F. | .360 | .96 | 6.96 | 7.92 |
| 4' x 5' | 26.000 | L.F. | .390 | 1.04 | 7.54 | 8.58 |
| 4' x 6' | 28.000 | L.F. | .420 | 1.12 | 8.12 | 9.24 |
| 4' x 8' | 32.000 | L.F. | .480 | 1.28 | 9.28 | 10.56 |
| 6'-8" x 2' | 30.000 | L.F. | .450 | 1.20 | 8.70 | 9.90 |
| Sill, slate, 2' long | 2.000 | L.F. | .376 | 9.52 | 6.68 | 16.20 |
| 3' long | 3.000 | L.F. | .564 | 14.28 | 10.02 | 24.30 |
| 4' long | 4.000 | L.F. | .752 | 19.04 | 13.36 | 32.40 |
| Wood, 1-5/8" x 5-1/8", 2' long | 2.000 | L.F. | .128 | 3.30 | 2.50 | 5.80 |
| 3' long | 3.000 | L.F. | .192 | 4.95 | 3.75 | 8.70 |
| 4' long | 4.000 | L.F. | .256 | 6.60 | 5.00 | 11.60 |

4

# 4 | EXTERIOR WALLS  60 | Storm Door & Window Systems

Aluminum Window — Aluminum Door

| SYSTEM DESCRIPTION | QUAN. | UNIT | MAN HOURS | MAT. | INST. | TOTAL |
|---|---|---|---|---|---|---|
| Storm door, aluminum, combination, storm & screen, anodized, 2'-6" x 6'-8" | 1.000 | Ea. | 1.070 | 88.00 | 22.00 | 110.00 |
| 2'-8" x 6'-8" | 1.000 | Ea. | 1.140 | 93.50 | 21.50 | 115.00 |
| 3'-0" x 6'-8" | 1.000 | Ea. | 1.140 | 99.00 | 21.00 | 120.00 |
| Mill finish, 2'-6" x 6'-8" | 1.000 | Ea. | 1.070 | 77.00 | 22.00 | 99.00 |
| 2'-8" x 6'-8" | 1.000 | Ea. | 1.140 | 83.60 | 21.40 | 105.00 |
| 3'-0" x 6'-8" | 1.000 | Ea. | 1.140 | 85.80 | 24.20 | 110.00 |
| Painted, 2'-6" x 6'-8" | 1.000 | Ea. | 1.070 | 88.00 | 22.00 | 110.00 |
| 2'-8" x 6'-8" | 1.000 | Ea. | 1.140 | 93.50 | 21.50 | 115.00 |
| 3'-0" x 6'-8" | 1.000 | Ea. | 1.140 | 96.80 | 23.20 | 120.00 |
| Wood, combination, storm & screen, crossbuck, 2'-6" x 6'-9" | 1.000 | Ea. | 1.450 | 137.50 | 27.50 | 165.00 |
| 2'-8" x 6'-9" | 1.000 | Ea. | 1.600 | 137.50 | 32.50 | 170.00 |
| 3'-0" x 6'-9" | 1.000 | Ea. | 1.780 | 143.00 | 37.00 | 180.00 |
| Full lite, 2'-6" x 6'-9" | 1.000 | Ea. | 1.450 | 126.50 | 28.50 | 155.00 |
| 2'-8" x 6'-9" | 1.000 | Ea. | 1.600 | 126.50 | 33.50 | 160.00 |
| 3'-0" x | 1.000 | Ea. | 1.780 | 132.00 | 38.00 | 170.00 |
| Windows, aluminum, combination storm & screen, basement, 1'-10" x 1'-0" | 1.000 | Ea. | .533 | 11.33 | 10.67 | 22.00 |
| 2'-9" x 1'-6" | 1.000 | Ea. | .533 | 17.00 | 11.00 | 28.00 |
| 3'-4" x 2'-0" | 1.000 | Ea. | .533 | 20.41 | 10.59 | 31.00 |
| Double hung, anodized, 2'-0" x 3'-5" | 1.000 | Ea. | .533 | 42.90 | 11.10 | 54.00 |
| 2'-6" x 5'-0" | 1.000 | Ea. | .571 | 46.20 | 11.80 | 58.00 |
| 4'-0" x 6'-0" | 1.000 | Ea. | .640 | 58.30 | 12.70 | 71.00 |
| Painted, 2'-0" x 3'-5" | 1.000 | Ea. | .533 | 40.70 | 11.30 | 52.00 |
| 2'-6" x 5'-0" | 1.000 | Ea. | .571 | 42.90 | 12.10 | 55.00 |
| 4'-0" x 6'-0" | 1.000 | Ea. | .640 | 55.00 | 13.00 | 68.00 |
| Fixed window, anodized, 4'-6" x 4'-6" | 1.000 | Ea. | .640 | 77.00 | 13.00 | 90.00 |
| 5'-8" x 4'-6" | 1.000 | Ea. | .800 | 90.20 | 14.80 | 105.00 |
| Painted, 4'-6" x 4'-6" | 1.000 | Ea. | .640 | 70.40 | 13.60 | 84.00 |
| 5'-8" x 4'-6" | 1.000 | Ea. | .800 | 80.30 | 16.70 | 97.00 |

# 4 | EXTERIOR WALLS   64 | Shutters/Blinds Systems

**Aluminum Louvered**

**Wood Louvered**

**Polystyrene Raised Panel**

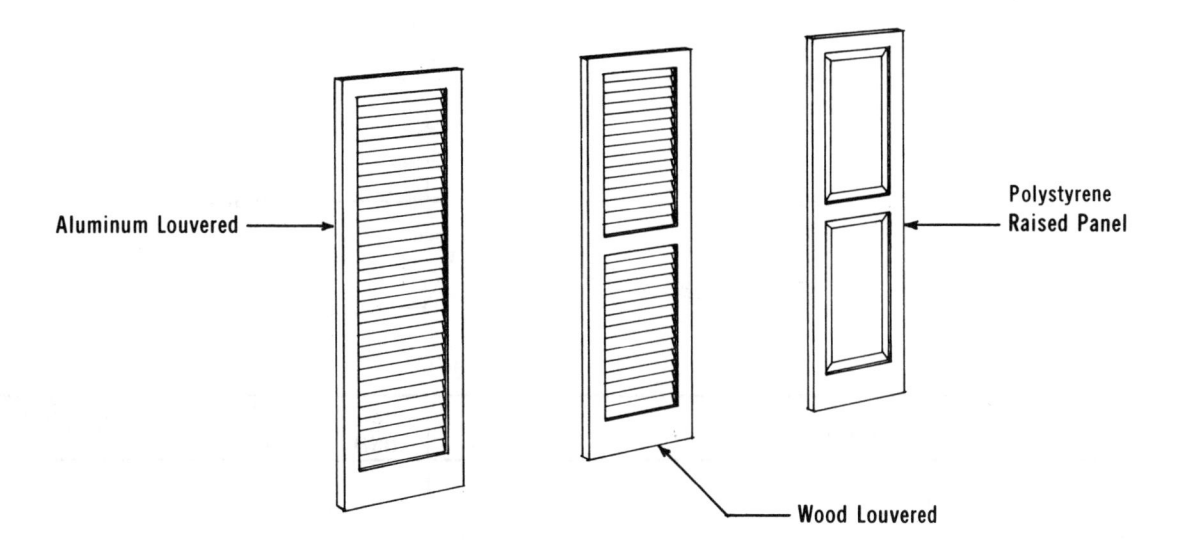

| SYSTEM DESCRIPTION | QUAN. | UNIT | MAN HOURS | COST PER PAIR MAT. | INST. | TOTAL |
|---|---|---|---|---|---|---|
| Shutters, exterior blinds, aluminum, louvered, 1'-4" wide, 3"-0" long | 1.000 | Set | .800 | 26.40 | 15.60 | 42.00 |
| 4'-0" long | 1.000 | Set | .800 | 30.80 | 15.20 | 46.00 |
| 5'-4" long | 1.000 | Set | .800 | 34.10 | 15.90 | 50.00 |
| 6'-8" long | 1.000 | Set | .889 | 45.10 | 16.90 | 62.00 |
| Wood, louvered, 1'-2" wide, 3'-3" long | 1.000 | Set | .800 | 39.60 | 15.40 | 55.00 |
| 4'-7" long | 1.000 | Set | .800 | 53.90 | 16.10 | 70.00 |
| 5'-3" long | 1.000 | Set | .800 | 48.40 | 15.60 | 64.00 |
| 1'-6" wide, 3'-3" long | 1.000 | Set | .800 | 45.10 | 15.90 | 61.00 |
| 4'-7" long | 1.000 | Set | .800 | 60.50 | 15.50 | 76.00 |
| Polystyrene, solid raised panel, 3'-3" wide, 3'-0" long | 1.000 | Set | .800 | 57.20 | 15.80 | 73.00 |
| 3'-11" long | 1.000 | Set | .800 | 67.10 | 15.90 | 83.00 |
| 5'-3" long | 1.000 | Set | .800 | 80.30 | 15.70 | 96.00 |
| 6'-8" long | 1.000 | Set | .889 | 106.70 | 18.30 | 125.00 |
| Polystyrene, louvered, 1'-2" wide, 3'-3" long | 1.000 | Set | .800 | 23.10 | 15.90 | 39.00 |
| 4'-7" long | 1.000 | Set | .800 | 30.80 | 15.20 | 46.00 |
| 5'-3" long | 1.000 | Set | .800 | 35.20 | 15.80 | 51.00 |
| 6'-8" long | 1.000 | Set | .889 | 46.20 | 17.80 | 64.00 |
| Vinyl, louvered, 1'-2" wide, 4'-7" long | 1.000 | Set | .720 | 26.73 | 13.77 | 40.50 |
| 1'-4" x 6'-8" long | 1.000 | Set | .889 | 45.10 | 16.90 | 62.00 |

# 4 | EXTERIOR WALLS | 80 | Storefront Systems

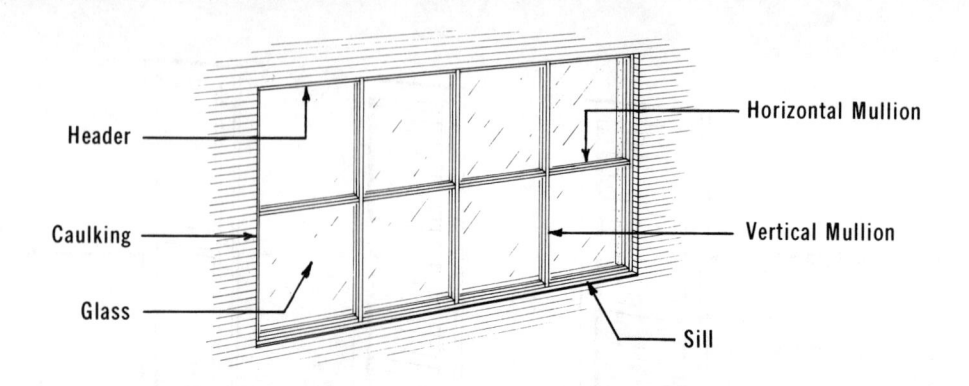

| SYSTEM DESCRIPTION | QUAN. | UNIT | MAN HOURS | COST PER S.F. MAT. | COST PER S.F. INST. | COST PER S.F. TOTAL |
|---|---|---|---|---|---|---|
| **MILL FINISH, 2" X 4-1/2" STOCK, 4' X 4' GLASS SIZE** | | | | | | |
| Header, mill finish, 2" x 4-1/2" deep | .200 | L.F. | .042 | 1.17 | .83 | 2.00 |
| Sill, mill finish, 2" x 4-1/2" deep | .125 | L.F. | .026 | .75 | .52 | 1.27 |
| Mullion, 2" x 4-1/2" deep, 4'-0" x 4'-0" grid | .350 | L.F. | .080 | 2.62 | 1.56 | 4.18 |
| Screw spline joints | .100 | Ea. | | .29 | | .29 |
| Caulking | .300 | L.F. | .009 | .04 | .19 | .23 |
| Glass, 1" thick insulating | 1.000 | S.F. | .213 | 6.93 | 4.17 | 11.10 |
| TOTAL | | | .371 | 11.80 | 7.27 | 19.07 |
| **BRONZE FINISH, 2" X 4-1/2" STOCK, 4' X 4' GLASS SIZE** | | | | | | |
| Header, bronze finish, 2" x 4-1/2" deep | .200 | L.F. | .042 | 1.17 | .83 | 2.00 |
| Sill, bronze finish, 2" x 4-1/2" deep | .125 | L.F. | .026 | .75 | .52 | 1.27 |
| Mullion, 2" x 4-1/2" deep, 4'-0" x 4'-0" grid | .350 | L.F. | .074 | 2.66 | 1.45 | 4.11 |
| Screw spline joints | .100 | Ea. | | .29 | | .29 |
| Caulking | .300 | L.F. | .009 | .04 | .19 | .23 |
| Glass, 1" thick insulating | 1.000 | S.F. | .213 | 6.93 | 4.17 | 11.10 |
| TOTAL | | | .366 | 11.84 | 7.16 | 19.00 |
| **BLACK FINISH, 2" X 4-1/2" STOCK, 4' X 4' GLASS SIZE** | | | | | | |
| Header, black finish, 2" x 4-1/2" deep | .200 | L.F. | .042 | 1.27 | .83 | 2.10 |
| Sill, black finish, 2" x 4-1/2" deep | .125 | L.F. | .026 | .81 | .52 | 1.33 |
| Mullion, 2" x 4-1/2" deep, 4'-0" x 4'-0" grid | .350 | L.F. | .074 | 2.85 | 1.46 | 4.31 |
| Screw spline joints | .100 | Ea. | | .29 | | .29 |
| Caulking | .300 | L.F. | .009 | .04 | .19 | .23 |
| Glass, 1" thick insulating | 1.000 | S.F. | .213 | 6.93 | 4.17 | 11.10 |
| TOTAL | | | .366 | 12.19 | 7.17 | 19.36 |

| DESCRIPTION | QUAN. | UNIT | MAN HOURS | COST PER S.F. MAT. | COST PER S.F. INST. | COST PER S.F. TOTAL |
|---|---|---|---|---|---|---|
| | | | | | | |
| | | | | | | |
| | | | | | | |
| | | | | | | |
| | | | | | | |

# Storefront Price Sheet

| | QUAN. | UNIT | MAN HOURS | COST PER S.F. MAT. | COST PER S.F. INST. | TOTAL |
|---|---|---|---|---|---|---|
| Headers, 1-3/4" x 4", mill finish | .200 | L.F. | .040 | .83 | .78 | 1.61 |
| Bronze finish | .200 | L.F. | .040 | .83 | .78 | 1.61 |
| Black finish | .200 | L.F. | .040 | .90 | .78 | 1.68 |
| 1-3/4" x 4-1/2" mill finish | .200 | L.F. | .040 | .97 | .78 | 1.75 |
| Bronze finish | .200 | L.F. | .040 | .97 | .78 | 1.75 |
| Black finish | .200 | L.F. | .040 | 1.04 | .78 | 1.82 |
| 2" x 4-1/2", mill finish | .200 | L.F. | .042 | 1.17 | .83 | 2.00 |
| Bronze finish | .200 | L.F. | .042 | 1.17 | .83 | 2.00 |
| Black finish | .200 | L.F. | .042 | 1.27 | .83 | 2.10 |
| 2-1/4" x 4-1/2", mill finish | .200 | L.F. | .043 | 1.73 | .84 | 2.57 |
| Bronze finish | .200 | L.F. | .042 | 1.73 | .83 | 2.56 |
| Black finish | .200 | L.F. | .042 | 1.87 | .83 | 2.70 |
| Sills, 1-3/4" x 4", mill finish | .125 | L.F. | .025 | .53 | .48 | 1.01 |
| Bronze finish | .125 | L.F. | .025 | .53 | .49 | 1.02 |
| Black finish | .125 | L.F. | .025 | .57 | .49 | 1.06 |
| 1-3/4" x 4-1/2" mill finish | .125 | L.F. | .025 | .62 | .49 | 1.11 |
| Bronze finish | .125 | L.F. | .025 | .62 | .49 | 1.11 |
| Black finish | .125 | L.F. | .025 | .67 | .49 | 1.16 |
| 2" x 4-1/2", mill finish | .125 | L.F. | .026 | .75 | .52 | 1.27 |
| Bronze finish | .125 | L.F. | .026 | .75 | .52 | 1.27 |
| Black finish | .125 | L.F. | .026 | .81 | .52 | 1.33 |
| 2-1/4" x 4-1/2", mill finish | .125 | L.F. | .027 | 1.08 | .53 | 1.61 |
| Bronze finish | .125 | L.F. | .026 | 1.08 | .52 | 1.60 |
| Black finish | .125 | L.F. | .026 | 1.17 | .52 | 1.69 |
| Mullion, 4'-0" x 4'-0" grid, 1-3/4" x 4", mill finish | .350 | L.F. | .070 | 2.00 | 1.36 | 3.36 |
| Bronze finish | .350 | L.F. | .070 | 2.00 | 1.36 | 3.36 |
| Black finish | .350 | L.F. | .070 | 2.16 | 1.36 | 3.52 |
| 1-3/4" x 4-1/2" mill finish | .350 | L.F. | .070 | 2.08 | 1.37 | 3.45 |
| Bronze finish | .350 | L.F. | .070 | 2.10 | 1.37 | 3.47 |
| Black finish | .350 | L.F. | .070 | 2.25 | 1.37 | 3.62 |
| 2" x 4-1/2", mill finish | .350 | L.F. | .080 | 2.62 | 1.56 | 4.18 |
| Bronze finish | .350 | L.F. | .074 | 2.66 | 1.45 | 4.11 |
| Black finish | .350 | L.F. | .074 | 2.85 | 1.46 | 4.31 |
| 2-1/4" x 4-1/2", mill finish | .350 | L.F. | .081 | 3.45 | 1.57 | 5.02 |
| Bronze finish | .350 | L.F. | .074 | 2.39 | 1.44 | 3.83 |
| Black finish | .350 | L.F. | .074 | 2.56 | 1.45 | 4.01 |
| 8' x 8' grid, 1-3/4" x 4", mill finish | .093 | L.F. | .018 | .53 | .36 | .89 |
| Bronze finish | .093 | L.F. | .018 | .53 | .36 | .89 |
| Black finish | .093 | L.F. | .018 | .57 | .36 | .93 |
| 1-3/4" x 4-1/2" mill finish | .093 | L.F. | .018 | .55 | .37 | .92 |
| Bronze finish | .093 | L.F. | .018 | .56 | .36 | .92 |
| Black finish | .093 | L.F. | .018 | .60 | .36 | .96 |
| 2" x 4-1/2", mill finish | .093 | L.F. | .021 | .70 | .41 | 1.11 |
| Bronze finish | .093 | L.F. | .019 | .71 | .38 | 1.09 |
| Black finish | .093 | L.F. | .019 | .76 | .38 | 1.14 |
| 2-1/4" x 4-1/2", mill finish | .093 | L.F. | .021 | .92 | .41 | 1.33 |
| Bronze finish | .093 | L.F. | .019 | .63 | .39 | 1.02 |
| Black finish | .093 | L.F. | .019 | .68 | .38 | 1.06 |
| Screw spline joints, 4'-0" x 4'-0" grid | .100 | Ea. | | .29 | | .29 |
| 8'-0" x 8'-0" grid | .040 | Ea. | | .12 | | .12 |
| Caulking | .300 | L.F. | .009 | .04 | .19 | .23 |
| Glass, plate glass, 1/4" thick, clear, plain | 1.000 | S.F. | .133 | 1.38 | 2.59 | 3.97 |
| Tempered | 1.000 | S.F. | .133 | 3.20 | 2.60 | 5.80 |
| Tinted, plain | 1.000 | S.F. | .133 | 1.82 | 2.59 | 4.41 |
| Tempered | 1.000 | S.F. | .133 | 4.18 | 2.62 | 6.80 |
| Insulating, 1/2" thick, clear | 1.000 | S.F. | .168 | 2.64 | 3.23 | 5.87 |
| Tinted | 1.000 | S.F. | .168 | 5.83 | 3.27 | 9.10 |
| 1" thick, clear | 1.000 | S.F. | .213 | 6.93 | 4.17 | 11.10 |
| Tinted | 1.000 | S.F. | .213 | 8.20 | 4.15 | 12.35 |

# SECTION 5 ROOFING

| System Number | Title | Page |
|---|---|---|
| 5-04 | Gable End Roofing | 104 |
| 5-08 | Hip Roof Roofing | 106 |
| 5-12 | Gambrel Roofing | 108 |
| 5-16 | Mansard Roofing | 110 |
| 5-20 | Shed Roofing | 112 |
| 5-24 | Gable Dormer Roofing | 114 |
| 5-28 | Shed Dormer Roofing | 116 |
| 5-32 | Skylight/Sky Window | 118 |
| 5-34 | Built-up Roofing | 120 |
| 5-36 | Single-Ply Roofing | 122 |

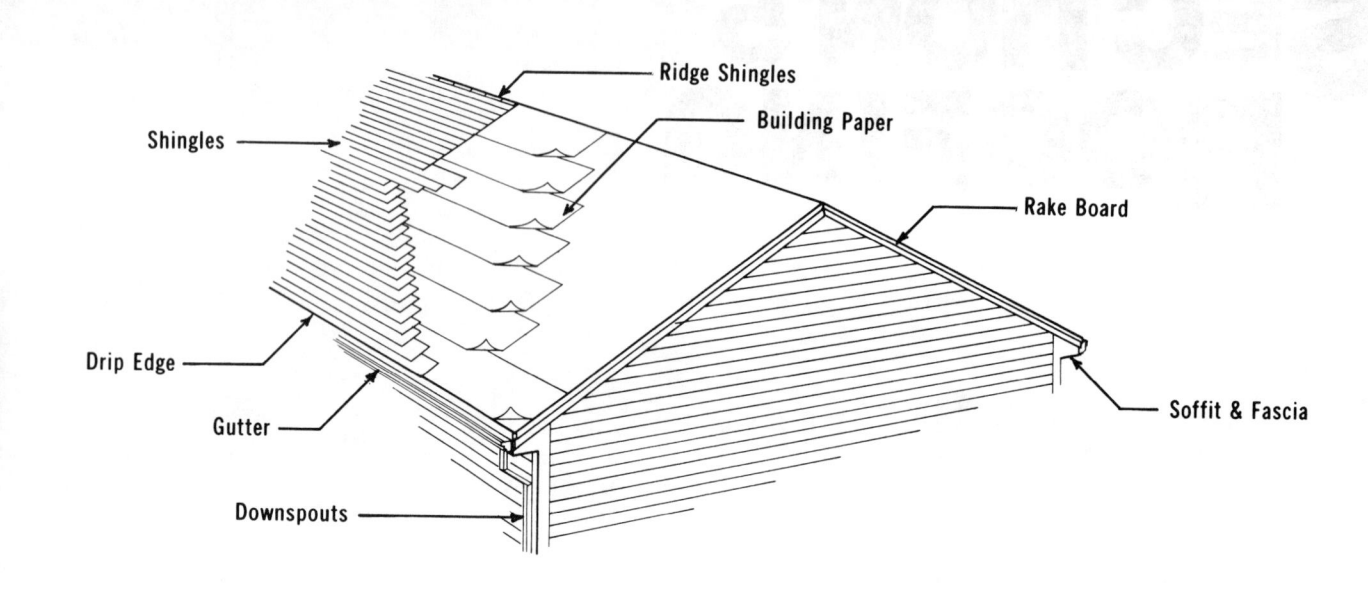

# 5 | ROOFING  04 | Gable End Roofing Systems

| SYSTEM DESCRIPTION | QUAN. | UNIT | MAN HOURS | MAT. | INST. | TOTAL |
|---|---|---|---|---|---|---|
| **ASPHALT, ROOF SHINGLES, CLASS A** | | | | | | |
| Shingles, asphalt standard, inorganic class A, 210-235 lb./sq. | 1.160 | S.F. | .017 | .37 | .34 | .71 |
| Drip edge, metal, 5" girth | .150 | L.F. | .003 | .03 | .05 | .08 |
| Building paper, #15 felt | 1.300 | S.F. | .001 | .04 | .03 | .07 |
| Ridge shingles, asphalt | .042 | L.F. | .001 | .03 | .02 | .05 |
| Soffit & fascia, white painted aluminum, 1' overhang | .083 | L.F. | .012 | .22 | .24 | .46 |
| Rake trim, painted, 1" x 6" | .040 | L.F. | .003 | .03 | .06 | .09 |
| Gutter, seamless, aluminum painted | .083 | L.F. | .005 | .08 | .12 | .20 |
| Downspouts, aluminum painted | .035 | L.F. | .001 | .03 | .03 | .06 |
| TOTAL | | | .043 | .83 | .89 | 1.72 |
| **WOOD, CEDAR SHINGLES NO. 1 PERFECTIONS** | | | | | | |
| Shingles, wood, cedar, No. 1 perfections, | 1.160 | S.F. | .034 | 1.45 | .71 | 2.16 |
| Drip edge, metal, 5" girth | .150 | L.F. | .003 | .03 | .05 | .08 |
| Building paper, #15 felt | 1.300 | S.F. | .001 | .04 | .03 | .07 |
| Ridge shingles, cedar | .042 | L.F. | .001 | .05 | .02 | .07 |
| Soffit & fascia, white painted aluminum, 1' overhang | .083 | L.F. | .012 | .22 | .24 | .46 |
| Rake trim, painted, 1" x 6" | .040 | L.F. | .003 | .03 | .06 | .09 |
| Gutter, seamless, aluminum, painted | .083 | L.F. | .005 | .08 | .12 | .20 |
| Downspouts, aluminum, painted | .035 | L.F. | .001 | .03 | .03 | .06 |
| TOTAL | | | .060 | 1.93 | 1.26 | 3.19 |

The prices in these systems are based on a square foot of plan area. All quantities have been adjusted accordingly.

| DESCRIPTION | QUAN. | UNIT | MAN HOURS | MAT. | INST. | TOTAL |
|---|---|---|---|---|---|---|
| | | | | | | |
| | | | | | | |
| | | | | | | |
| | | | | | | |
| | | | | | | |

# Gable End Roofing Price Sheet

| | QUAN. | UNIT | MAN HOURS | COST PER S.F. MAT. | COST PER S.F. INST. | TOTAL |
|---|---|---|---|---|---|---|
| Shingles, asphalt standard inorganic class A 210-235 lb./sq., 4/12 pitch | 1.160 | S.F. | .017 | .37 | .34 | .71 |
| 8/12 pitch | 1.330 | S.F. | .018 | .40 | .37 | .77 |
| Laminated, multilayered, 240-260 lb./sq., 4/12 pitch | 1.160 | S.F. | .021 | .67 | .41 | 1.08 |
| 8/12 pitch | 1.330 | S.F. | .023 | .73 | .44 | 1.17 |
| Premium laminated, multilayered, 260-300 lb./sq., 4/12 pitch | 1.160 | S.F. | .027 | 1.11 | .51 | 1.62 |
| 8/12 pitch | 1.330 | S.F. | .029 | 1.20 | .56 | 1.76 |
| Clay tile, Spanish tile, red, 4/12 pitch | 1.160 | S.F. | .053 | 3.12 | 1.02 | 4.14 |
| 8/12 pitch | 1.330 | S.F. | .057 | 3.37 | 1.12 | 4.49 |
| Mission tile, red, 4/12 pitch | 1.160 | S.F. | .083 | 6.15 | 1.65 | 7.80 |
| 8/12 pitch | 1.330 | S.F. | .090 | 6.66 | 1.79 | 8.45 |
| French tile, red, 4/12 pitch | 1.160 | S.F. | .071 | 12.01 | 1.49 | 13.50 |
| 8/12 pitch | 1.330 | S.F. | .077 | 13.01 | 1.62 | 14.63 |
| Slate, Buckingham, Virginia, black, 4/12 pitch | 1.160 | S.F. | .054 | 7.52 | 1.06 | 8.58 |
| 8/12 pitch | 1.330 | S.F. | .059 | 8.15 | 1.15 | 9.30 |
| Vermont, black or grey, 4/12 pitch | 1.160 | S.F. | .054 | 8.32 | 1.04 | 9.36 |
| 8/12 pitch | 1.330 | S.F. | .059 | 9.01 | 1.13 | 10.14 |
| Wood, No. 1 perfections, 16", 5" exposure, 4/12 pitch | 1.160 | S.F. | .038 | 1.76 | .76 | 2.52 |
| 8/12 pitch | 1.330 | S.F. | .041 | 1.90 | .83 | 2.73 |
| Fire retardant, 4/12 pitch | 1.160 | S.F. | .037 | 2.50 | .73 | 3.23 |
| 8/12 pitch | 1.330 | S.F. | .042 | 2.87 | .82 | 3.69 |
| 18", 5" exposure, 4/12 pitch | 1.160 | S.F. | .034 | 1.45 | .71 | 2.16 |
| 8/12 pitch | 1.330 | S.F. | .037 | 1.57 | .77 | 2.34 |
| Fire retardant, 4/12 pitch | 1.160 | S.F. | .034 | 2.85 | .69 | 3.54 |
| 8/12 pitch | 1.330 | S.F. | .037 | 3.09 | .75 | 3.84 |
| Resquared & rebutted, 18", 6" exposure, 4/12 pitch | 1.160 | S.F. | .032 | 1.47 | .63 | 2.10 |
| 8/12 pitch | 1.330 | S.F. | .034 | 1.59 | .69 | 2.28 |
| Fire retardant, 4/12 pitch | 1.160 | S.F. | .032 | 2.65 | .65 | 3.30 |
| 8/12 pitch | 1.330 | S.F. | .034 | 2.87 | .71 | 3.58 |
| Wood shakes hand split, 24" long, 10" exposure, 4/12 pitch | 1.160 | S.F. | .038 | 1.44 | .72 | 2.16 |
| 8/12 pitch | 1.330 | S.F. | .041 | 1.56 | .78 | 2.34 |
| Fire retardant, 4/12 pitch | 1.160 | S.F. | .038 | 2.79 | .75 | 3.54 |
| 8/12 pitch | 1.330 | S.F. | .041 | 3.02 | .82 | 3.84 |
| 18" long 8" exposure, 4/12 pitch | 1.160 | S.F. | .048 | 1.11 | .93 | 2.04 |
| 8/12 pitch | 1.330 | S.F. | .052 | 1.20 | 1.01 | 2.21 |
| Fire retardant, 4/12 pitch | 1.160 | S.F. | .048 | 3.37 | .95 | 4.32 |
| 8/12 pitch | 1.330 | S.F. | .052 | 3.65 | 1.03 | 4.68 |
| Drip edge, metal, 5" girth | .150 | L.F. | .003 | .03 | .05 | .08 |
| 8" girth | .150 | L.F. | .003 | .04 | .05 | .09 |
| Building paper, #15 asphalt felt | 1.300 | S.F. | .001 | .04 | .03 | .07 |
| Ridge shingles, asphalt | .042 | L.F. | .001 | .03 | .02 | .05 |
| Clay | .042 | L.F. | .001 | .16 | .03 | .19 |
| Slate | .042 | L.F. | .001 | .17 | .03 | .20 |
| Wood, shingles | .042 | L.F. | .001 | .05 | .02 | .07 |
| Shakes | .042 | L.F. | .001 | .05 | .02 | .07 |
| Soffit & fascia, aluminum, vented, 1' overhang | .083 | L.F. | .012 | .22 | .24 | .46 |
| 2' overhang | .083 | L.F. | .013 | .44 | .27 | .71 |
| Vinyl, vented, 1' overhang | .083 | L.F. | .011 | .12 | .23 | .35 |
| 2' overhang | .083 | L.F. | .012 | .25 | .24 | .49 |
| Wood, board fascia, plywood soffit, 1' overhang | .083 | L.F. | .003 | .01 | .08 | .09 |
| 2' overhang | .083 | L.F. | .005 | .02 | .11 | .13 |
| Rake trim, painted, 1" x 6" | .040 | L.F. | .003 | .03 | .06 | .09 |
| 1" x 8" | .040 | L.F. | .003 | .05 | .07 | .12 |
| Gutter, 5" box, aluminum, seamless, painted | .083 | L.F. | .005 | .08 | .12 | .20 |
| Vinyl | .083 | L.F. | .006 | .12 | .12 | .24 |
| Downspout, 2" x 3", aluminum, one story house | .035 | L.F. | .001 | .02 | .03 | .05 |
| Two story house | .060 | L.F. | .002 | .04 | .05 | .09 |
| Vinyl, one story house | .035 | L.F. | .001 | .03 | .03 | .06 |
| Two story house | .060 | L.F. | .002 | .04 | .05 | .09 |

# 5 | ROOFING  08 | Hip Roof Roofing Systems

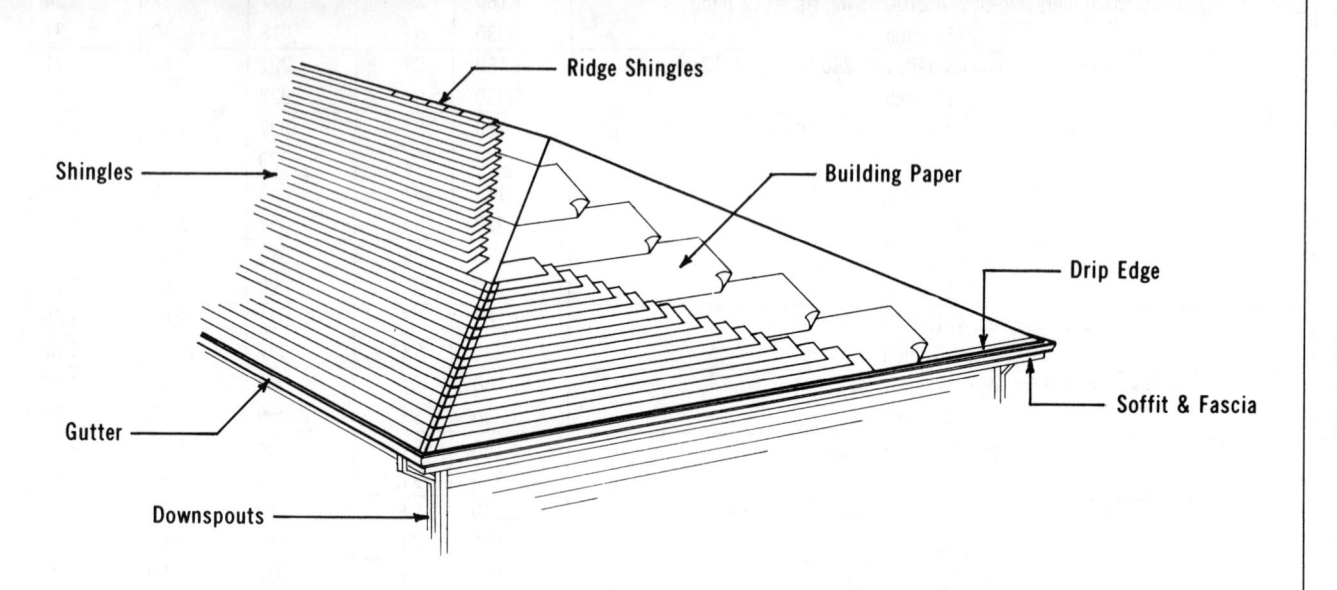

*Diagram of hip roof labeled with: Ridge Shingles, Shingles, Building Paper, Drip Edge, Soffit & Fascia, Gutter, Downspouts.*

| SYSTEM DESCRIPTION | QUAN. | UNIT | MAN HOURS | MAT. | INST. | TOTAL |
|---|---|---|---|---|---|---|
| **ASPHALT, ROOF SHINGLES, CLASS A** | | | | | | |
| Shingles, asphalt std. inorganic class A 210-235 lb./sq. 4/12 pitch | 1.570 | S.F. | .023 | .49 | .45 | .94 |
| Drip edge, metal, 5" girth | .122 | L.F. | .002 | .02 | .05 | .07 |
| Building paper, #15 asphalt felt | 1.800 | S.F. | .002 | .05 | .05 | .10 |
| Ridge shingles, asphalt | .075 | L.F. | .001 | .05 | .04 | .09 |
| Soffit & fascia, white painted aluminum, 1' overhang | .120 | L.F. | .017 | .32 | .35 | .67 |
| Gutter, seamless, aluminum, painted | .120 | L.F. | .008 | .12 | .18 | .30 |
| Downspouts, aluminum, painted | .035 | L.F. | .001 | .03 | .03 | .06 |
| TOTAL | | | .056 | 1.08 | 1.15 | 2.23 |
| **WOOD, CEDAR SHINGLES, NO. 1 PERFECTIONS** | | | | | | |
| Shingles, wood, cedar, No. 1 perfections, 5" exposure, 4/12 pitch | 1.570 | S.F. | .046 | 1.94 | .94 | 2.88 |
| Drip edge, metal, 5" girth | .122 | L.F. | .002 | .02 | .05 | .07 |
| Building paper, #15 asphalt felt | 1.800 | S.F. | .002 | .05 | .05 | .10 |
| Ridge shingles, wood, cedar | .075 | L.F. | .002 | .09 | .04 | .13 |
| Soffit & fascia, white painted aluminum, 1' overhang | .120 | L.F. | .017 | .32 | .35 | .67 |
| Gutter, seamless, aluminum, painted | .120 | L.F. | .008 | .12 | .18 | .30 |
| Downspouts, aluminum, painted | .035 | L.F. | .001 | .03 | .03 | .06 |
| TOTAL | | | .080 | 2.57 | 1.64 | 4.21 |

The prices in these systems are based on a square foot of plan area. All quantities have been adjusted accordingly.

| DESCRIPTION | QUAN. | UNIT | MAN HOURS | MAT. | INST. | TOTAL |
|---|---|---|---|---|---|---|
| | | | | | | |
| | | | | | | |
| | | | | | | |
| | | | | | | |
| | | | | | | |

# Hip Roof Roofing Price Sheet

| Description | QUAN. | UNIT | MAN HOURS | MAT. | INST. | TOTAL |
|---|---|---|---|---|---|---|
| Shingles, asphalt standard inorganic class A 210-235 lb./sq., 4/12 pitch | 1.570 | S.F. | .023 | .49 | .45 | .94 |
| 8/12 pitch | 1.850 | S.F. | .027 | .59 | .53 | 1.12 |
| Laminated, multilayered, 240-260 lb./sq., 4/12 pitch | 1.570 | S.F. | .028 | .90 | .54 | 1.44 |
| 8/12 pitch | 1.850 | S.F. | .033 | 1.07 | .64 | 1.71 |
| Premium laminated, multilayered, 260-300 lb./sq., 4/12 pitch | 1.570 | S.F. | .036 | 1.48 | .68 | 2.16 |
| 8/12 pitch | 1.850 | S.F. | .043 | 1.76 | .81 | 2.57 |
| Clay tile, Spanish tile, red, 4/12 pitch | 1.570 | S.F. | .071 | 4.15 | 1.37 | 5.52 |
| 8/12 pitch | 1.850 | S.F. | .084 | 4.93 | 1.63 | 6.56 |
| Mission tile, red, 4/12 pitch | 1.570 | S.F. | .111 | 8.20 | 2.20 | 10.40 |
| 8/12 pitch | 1.850 | S.F. | .132 | 9.74 | 2.61 | 12.35 |
| French tile, red, 4/12 pitch | 1.570 | S.F. | .094 | 16.02 | 1.98 | 18.00 |
| 8/12 pitch | 1.850 | S.F. | .112 | 19.02 | 2.36 | 21.38 |
| Slate, Buckingham, Virginia, black, 4/12 pitch | 1.570 | S.F. | .073 | 10.03 | 1.41 | 11.44 |
| 8/12 pitch | 1.850 | S.F. | .086 | 11.91 | 1.68 | 13.59 |
| Vermont, black or grey, 4/12 pitch | 1.570 | S.F. | .073 | 11.09 | 1.39 | 12.48 |
| 8/12 pitch | 1.850 | S.F. | .086 | 13.17 | 1.65 | 14.82 |
| Wood, No. 1 perfections, 16", 5" exposure, 4/12 pitch | 1.570 | S.F. | .051 | 2.34 | 1.02 | 3.36 |
| 8/12 pitch | 1.850 | S.F. | .060 | 2.78 | 1.21 | 3.99 |
| Fire retardant, 4/12 pitch | 1.570 | S.F. | .051 | 3.38 | .98 | 4.36 |
| 8/12 pitch | 1.850 | S.F. | .060 | 3.99 | 1.16 | 5.15 |
| 18", 5" exposure, 4/12 pitch | 1.570 | S.F. | .046 | 1.94 | .94 | 2.88 |
| 8/12 pitch | 1.850 | S.F. | .055 | 2.30 | 1.12 | 3.42 |
| Fire retardant, 4/12 pitch | 1.570 | S.F. | .046 | 3.80 | .92 | 4.72 |
| 8/12 pitch | 1.850 | S.F. | .055 | 4.51 | 1.10 | 5.61 |
| Resquared & rebutted, 18" 6" exposure, 4/12 pitch | 1.570 | S.F. | .042 | 1.95 | .85 | 2.80 |
| 8/12 pitch | 1.850 | S.F. | .050 | 2.32 | 1.01 | 3.33 |
| Fire retardant, 4/12 pitch | 1.570 | S.F. | .042 | 3.54 | .86 | 4.40 |
| 8/12 pitch | 1.850 | S.F. | .050 | 4.20 | 1.03 | 5.23 |
| Wood shakes hand split, 24" long, 10" exposure, 4/12 pitch | 1.570 | S.F. | .051 | 1.92 | .96 | 2.88 |
| 8/12 pitch | 1.850 | S.F. | .060 | 2.28 | 1.14 | 3.42 |
| Fire retardant, 4/12 pitch | 1.570 | S.F. | .051 | 3.71 | 1.01 | 4.72 |
| 8/12 pitch | 1.850 | S.F. | .060 | 4.41 | 1.20 | 5.61 |
| 18" long 8" exposure, 4/12 pitch | 1.570 | S.F. | .064 | 1.48 | 1.24 | 2.72 |
| 8/12 pitch | 1.850 | S.F. | .076 | 1.76 | 1.47 | 3.23 |
| Fire retardant, 4/12 pitch | 1.570 | S.F. | .064 | 4.49 | 1.27 | 5.76 |
| 8/12 pitch | 1.850 | S.F. | .076 | 5.33 | 1.51 | 6.84 |
| Drip edge, metal, 5" girth | .122 | L.F. | .002 | .02 | .05 | .07 |
| 8" girth | .122 | L.F. | .002 | .03 | .05 | .08 |
| Building paper, #15 asphalt felt | 1.800 | S.F. | .002 | .05 | .05 | .10 |
| Ridge shingles, asphalt | .075 | L.F. | .001 | .05 | .04 | .09 |
| Clay | .075 | L.F. | .003 | .29 | .06 | .35 |
| Slate | .075 | L.F. | .003 | .30 | .06 | .36 |
| Wood, shingles | .075 | L.F. | .002 | .09 | .04 | .13 |
| Shakes | .075 | L.F. | .002 | .09 | .04 | .13 |
| Soffit & fascia, aluminum, vented, 1' overhang | .120 | L.F. | .017 | .32 | .35 | .67 |
| 2' overhang | .120 | L.F. | .019 | .64 | .39 | 1.03 |
| Vinyl, vented, 1' overhang | .120 | L.F. | .015 | .18 | .33 | .51 |
| 2' overhang | .120 | L.F. | .017 | .36 | .35 | .71 |
| Wood, board fascia, plywood soffit, 1' overhang | .120 | L.F. | .003 | .01 | .08 | .09 |
| 2' overhang | .120 | L.F. | .005 | .02 | .11 | .13 |
| Gutter, 5" box, aluminum, seamless, painted | .120 | L.F. | .008 | .12 | .18 | .30 |
| Vinyl | .120 | L.F. | .008 | .18 | .17 | .35 |
| Downspout, 2" x 3", aluminum, one story house | .035 | L.F. | .001 | .03 | .03 | .06 |
| Two story house | .060 | L.F. | .002 | .04 | .05 | .09 |
| Vinyl, one story house | .035 | L.F. | .001 | .02 | .03 | .05 |
| Two story house | .060 | L.F. | .002 | .04 | .05 | .09 |

# 5 | ROOFING   12 | Gambrel Roofing Systems

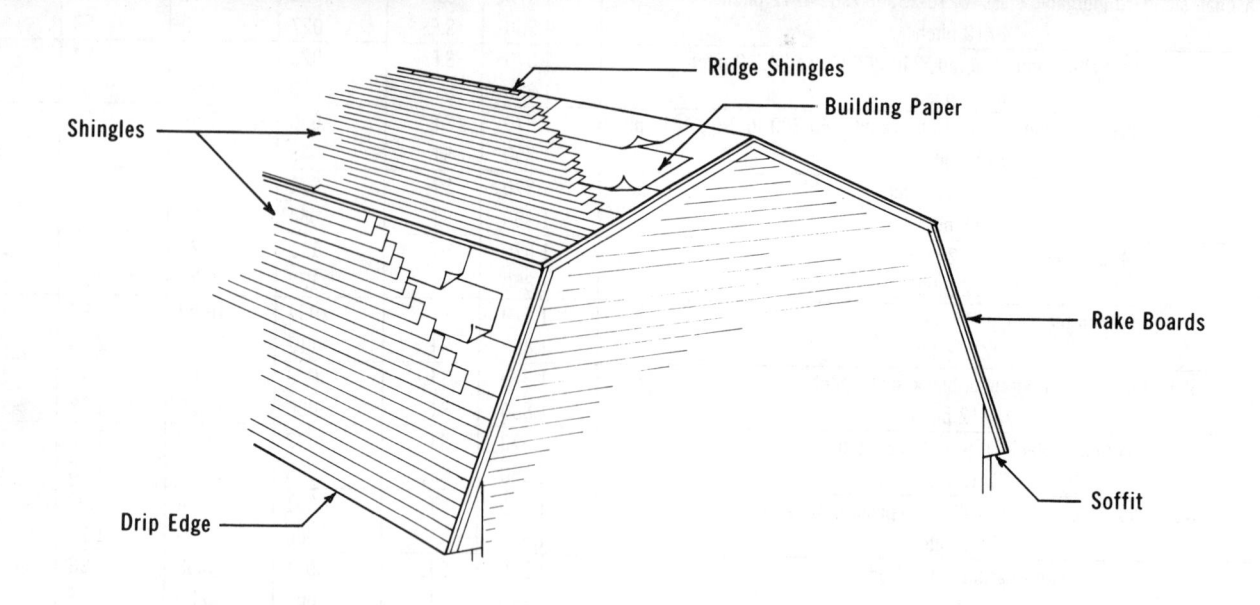

*Diagram labels: Shingles, Ridge Shingles, Building Paper, Rake Boards, Soffit, Drip Edge*

| SYSTEM DESCRIPTION | QUAN. | UNIT | MAN HOURS | COST PER S.F. MAT. | COST PER S.F. INST. | COST PER S.F. TOTAL |
|---|---|---|---|---|---|---|
| **ASPHALT, ROOF SHINGLES, CLASS A** | | | | | | |
| Shingles, asphalt standard inorganic class A 210-235 lb./sq. | 1.450 | S.F. | .021 | .46 | .43 | .89 |
| Drip edge, metal, 5" girth | .146 | L.F. | .002 | .02 | .06 | .08 |
| Building paper, #15 asphalt felt | 1.500 | S.F. | .002 | .04 | .04 | .08 |
| Ridge shingles, asphalt | .042 | L.F. | .001 | .03 | .02 | .05 |
| Soffit & fascia, painted aluminum, 1' overhang | .083 | L.F. | .012 | .22 | .24 | .46 |
| Rake, trim, painted, 1" x 6" | .063 | L.F. | .005 | .04 | .11 | .15 |
| Gutter, seamless, aluminum, painted | .083 | L.F. | .005 | .08 | .12 | .20 |
| Downspouts, aluminum, painted | .042 | L.F. | .001 | .03 | .04 | .07 |
| TOTAL | | | .049 | .92 | 1.06 | 1.98 |
| **WOOD, CEDAR SHINGLES, NO. 1 PERFECTIONS** | | | | | | |
| Shingles, wood, cedar, No. 1 perfections, 5" exposure | 1.450 | S.F. | .043 | 1.82 | .88 | 2.70 |
| Drip edge, metal, 5" girth | .146 | L.F. | .002 | .02 | .06 | .08 |
| Building paper, #15 asphalt felt | 1.500 | S.F. | .002 | .04 | .04 | .08 |
| Ridge shingles, wood | .042 | L.F. | .001 | .05 | .02 | .07 |
| Soffit & fascia, white painted aluminum, 1' overhang | .083 | L.F. | .012 | .22 | .24 | .46 |
| Rake, trim, painted, 1" x 6" | .063 | L.F. | .005 | .04 | .10 | .14 |
| Gutter, seamless, aluminum, painted | .083 | L.F. | .005 | .08 | .12 | .20 |
| Downspouts, aluminum, painted | .042 | L.F. | .001 | .03 | .04 | .07 |
| TOTAL | | | .071 | 2.30 | 1.50 | 3.80 |

The prices in this system are based on a square foot of plan area. All quantities have been adjusted accordingly.

| DESCRIPTION | QUAN. | UNIT | MAN HOURS | COST PER S.F. MAT. | COST PER S.F. INST. | COST PER S.F. TOTAL |
|---|---|---|---|---|---|---|
| | | | | | | |
| | | | | | | |
| | | | | | | |
| | | | | | | |
| | | | | | | |

# Gambrel Roofing Price Sheet

| | QUAN. | UNIT | MAN HOURS | COST PER S.F. MAT. | COST PER S.F. INST. | TOTAL |
|---|---|---|---|---|---|---|
| Shingles, asphalt standard inorganic class A 210-235 lb./sq. | 1.450 | S.F. | .021 | .46 | .43 | .89 |
| Laminated, multilayered, 240-260 lb./sq. | 1.450 | S.F. | .026 | .84 | .51 | 1.35 |
| Premium laminated, multilayered, 260-300 lb./sq. | 1.450 | S.F. | .034 | 1.39 | .64 | 2.03 |
| Slate, Buckingham, Virginia, black | 1.450 | S.F. | .068 | 9.41 | 1.32 | 10.73 |
| Vermont, black or grey | 1.450 | S.F. | .068 | 10.40 | 1.30 | 11.70 |
| Wood, No. 1 perfections, 16", 5" exposure, plain | 1.450 | S.F. | .048 | 2.19 | .96 | 3.15 |
| Fire retardant | 1.450 | S.F. | .048 | 2.59 | 1.14 | 3.73 |
| 18", 6" exposure, plain | 1.450 | S.F. | .043 | 1.82 | .88 | 2.70 |
| Fire retardant | 1.450 | S.F. | .043 | 3.56 | .87 | 4.43 |
| Resquared & rebutted, 18", 6" exposure, plain | 1.450 | S.F. | .040 | 1.83 | .80 | 2.63 |
| Fire retardant | 1.450 | S.F. | .040 | 3.32 | .81 | 4.13 |
| Shakes, hand split, 24" long, 10" exposure, plain | 1.450 | S.F. | .048 | 1.80 | .90 | 2.70 |
| Fire retardant | 1.450 | S.F. | .048 | 3.48 | .95 | 4.43 |
| 18" long, 8" exposure, plain | 1.450 | S.F. | .060 | 1.39 | 1.16 | 2.55 |
| Fire retardant | 1.450 | S.F. | .060 | 4.21 | 1.19 | 5.40 |
| Drip edge, metal, 5" girth | .146 | L.F. | .002 | .02 | .06 | .08 |
| 8" girth | .146 | L.F. | .002 | .04 | .05 | .09 |
| Building paper, #15 asphalt felt | 1.500 | S.F. | .002 | .04 | .04 | .08 |
| Ridge shingles, asphalt | .042 | L.F. | .001 | .03 | .02 | .05 |
| Slate | .042 | L.F. | .001 | .17 | .03 | .20 |
| Wood, shingles | .042 | L.F. | .001 | .05 | .02 | .07 |
| Shakes | .042 | L.F. | .001 | .05 | .02 | .07 |
| Soffit & fascia, aluminum, vented, 1' overhang | .083 | L.F. | .012 | .22 | .24 | .46 |
| 2' overhang | .083 | L.F. | .013 | .44 | .27 | .71 |
| Vinyl vented, 1' overhang | .083 | L.F. | .011 | .12 | .23 | .35 |
| 2' overhang | .083 | L.F. | .012 | .25 | .24 | .49 |
| Wood board fascia, plywood soffit, 1' overhang | .083 | L.F. | .003 | .01 | .08 | .09 |
| 2' overhang | .083 | L.F. | .005 | .02 | .11 | .13 |
| Rake trim, painted, 1" x 6" | .063 | L.F. | .005 | .04 | .11 | .15 |
| 1" x 8" | .063 | L.F. | .007 | .05 | .15 | .20 |
| Gutter, 5" box, aluminum, seamless, painted | .083 | L.F. | .005 | .08 | .12 | .20 |
| Vinyl | .083 | L.F. | .006 | .12 | .12 | .24 |
| Downspout 2" x 3", aluminum, one story house | .042 | L.F. | .001 | .03 | .03 | .06 |
| Two story house | .070 | L.F. | .002 | .04 | .07 | .11 |
| Vinyl, one story house | .042 | L.F. | .001 | .03 | .03 | .06 |
| Two story house | .070 | L.F. | .002 | .04 | .07 | .11 |

# 5 | ROOFING  16 | Mansard Roofing Systems

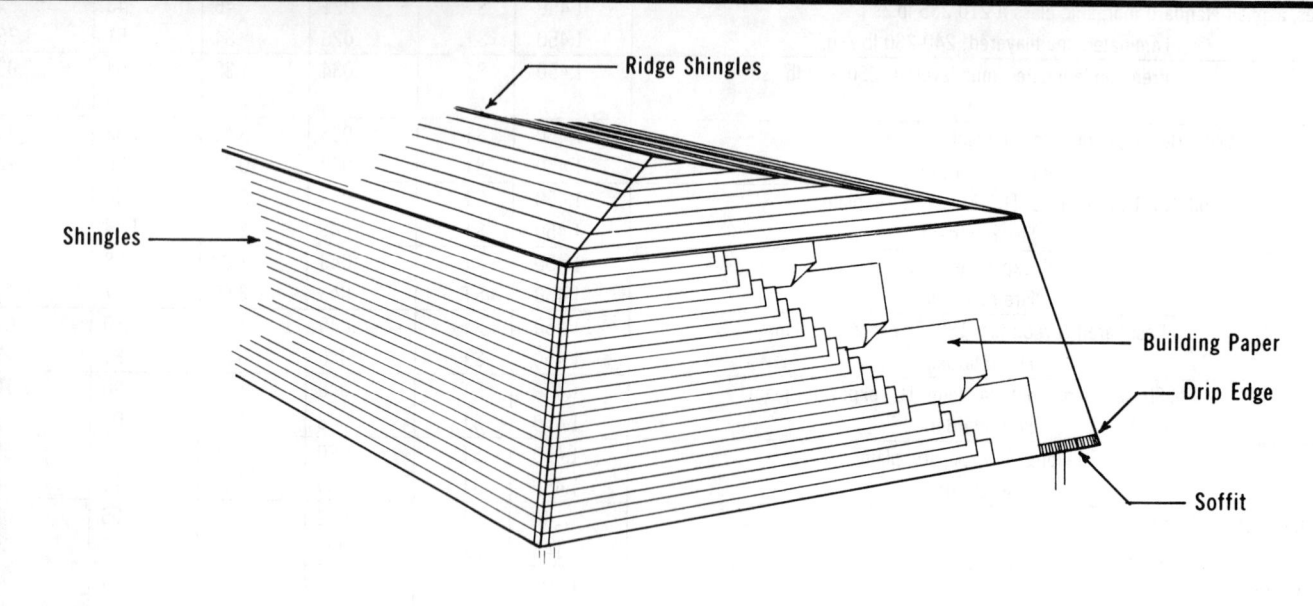

*Labels on diagram: Ridge Shingles, Shingles, Building Paper, Drip Edge, Soffit*

| SYSTEM DESCRIPTION | QUAN. | UNIT | MAN HOURS | COST PER S.F. MAT. | COST PER S.F. INST. | TOTAL |
|---|---|---|---|---|---|---|
| **ASPHALT, ROOF SHINGLES, CLASS A** | | | | | | |
| Shingles, asphalt standard inorganic class A 210-235 lb./sq. | 2.210 | S.F. | .031 | .68 | .62 | 1.30 |
| Drip edge, metal, 5" girth | .122 | L.F. | .002 | .02 | .05 | .07 |
| Building paper, #15 asphalt felt | 2.300 | S.F. | .003 | .07 | .06 | .13 |
| Ridge shingles, asphalt | .090 | L.F. | .002 | .06 | .05 | .11 |
| Soffit & fascia, white painted aluminum, 1' overhang | .122 | L.F. | .017 | .32 | .36 | .68 |
| Gutter, seamless, aluminum, painted | .122 | L.F. | .008 | .12 | .18 | .30 |
| Downspouts, aluminum, painted | .042 | L.F. | .001 | .03 | .04 | .07 |
| TOTAL | | | .064 | 1.30 | 1.36 | 2.66 |
| **WOOD, CEDAR SHINGLES, NO. 1 PERFECTIONS** | | | | | | |
| Shingles, wood, cedar, No. 1 perfections, 5" exposure | 2.210 | S.F. | .064 | 2.66 | 1.30 | 3.96 |
| Drip edge, metal, 5" girth | .122 | L.F. | .002 | .02 | .05 | .07 |
| Building paper, #15 asphalt felt | 2.300 | S.F. | .003 | .07 | .06 | .13 |
| Ridge shingles, wood | .090 | L.F. | .002 | .11 | .04 | .15 |
| Soffit & fascia, white painted aluminum, 1' overhang | .122 | L.F. | .017 | .32 | .36 | .68 |
| Gutter, seamless, aluminum, painted | .122 | L.F. | .008 | .12 | .18 | .30 |
| Downspouts, aluminum, painted | .042 | L.F. | .001 | .03 | .04 | .07 |
| TOTAL | | | .097 | 3.33 | 2.03 | 5.36 |

The prices in these systems are based on a square foot of plan area. All quantities have been adjusted accordingly.

| DESCRIPTION | QUAN. | UNIT | MAN HOURS | COST PER S.F. MAT. | COST PER S.F. INST. | TOTAL |
|---|---|---|---|---|---|---|
| | | | | | | |
| | | | | | | |
| | | | | | | |
| | | | | | | |
| | | | | | | |

# Mansard Roofing Price Sheet

| | QUAN. | UNIT | MAN HOURS | COST PER S.F. MAT. | COST PER S.F. INST. | TOTAL |
|---|---|---|---|---|---|---|
| Shingles, asphalt standard inorganic class A 210-235 lb./sq. | 2.210 | S.F. | .031 | .68 | .62 | 1.30 |
| Laminated, multilayered, 240-260 lb./sq. | 2.210 | S.F. | .039 | 1.23 | .75 | 1.98 |
| Premium laminated, multilayered, 260-300 lb./sq. | 2.210 | S.F. | .050 | 2.03 | .94 | 2.97 |
| Slate, Buckingham, Virginia, black | 2.210 | S.F. | .100 | 13.79 | 1.94 | 15.73 |
| Vermont, black or grey | 2.210 | S.F. | .100 | 15.25 | 1.91 | 17.16 |
| Wood, No. 1 perfections, 16" 5" exposure, plain | 2.210 | S.F. | .070 | 3.22 | 1.40 | 4.62 |
| Fire retardant | 2.210 | S.F. | .070 | 4.75 | 1.40 | 6.15 |
| 18" 6" exposure, plain | 2.210 | S.F. | .064 | 2.66 | 1.30 | 3.96 |
| Fire retardant | 2.210 | S.F. | .064 | 5.23 | 1.26 | 6.49 |
| Resquared & rebutted, 18" 6" exposure, plain | 2.210 | S.F. | .058 | 2.69 | 1.16 | 3.85 |
| Fire retardant | 2.210 | S.F. | .058 | 4.86 | 1.19 | 6.05 |
| Shakes, hand split, 24" long 10" exposure, plain | 2.210 | S.F. | .070 | 2.64 | 1.32 | 3.96 |
| Fire retardant | 2.210 | S.F. | .070 | 5.11 | 1.38 | 6.49 |
| 18" long 8" exposure, plain | 2.210 | S.F. | .088 | 2.03 | 1.71 | 3.74 |
| Fire retardant | 2.210 | S.F. | .088 | 6.17 | 1.75 | 7.92 |
| Drip edge, metal, 5" girth | .122 | S.F. | .002 | .02 | .05 | .07 |
| 8" girth | .122 | S.F. | .002 | .03 | .05 | .08 |
| Building paper, #15 asphalt felt | 2.300 | S.F. | .003 | .07 | .06 | .13 |
| Ridge shingles, asphalt | .090 | L.F. | .002 | .06 | .05 | .11 |
| Slate | .090 | L.F. | .003 | .36 | .07 | .43 |
| Wood, shingles | .090 | L.F. | .002 | .11 | .04 | .15 |
| Shakes | .090 | L.F. | .002 | .11 | .04 | .15 |
| Soffit & fascia, aluminum vented, 1' overhang | .122 | L.F. | .017 | .32 | .36 | .68 |
| 2' overhang | .122 | L.F. | .019 | .65 | .40 | 1.05 |
| Vinyl vented, 1' overhang | .122 | L.F. | .016 | .18 | .33 | .51 |
| 2' overhang | .122 | L.F. | .017 | .36 | .37 | .73 |
| Wood board fascia, plywood soffit, 1' overhang | .122 | L.F. | .012 | .14 | .26 | .40 |
| 2' overhang | .122 | L.F. | .019 | .20 | .38 | .58 |
| Gutter, 5" box, aluminum, seamless, painted | .122 | L.F. | .008 | .12 | .18 | .30 |
| Vinyl | .122 | L.F. | .008 | .18 | .17 | .35 |
| Downspout 2" x 3", aluminum, one story house | .042 | L.F. | .001 | .03 | .03 | .06 |
| Two story house | .070 | L.F. | .002 | .04 | .07 | .11 |
| Vinyl, one story house | .042 | L.F. | .001 | .03 | .03 | .06 |
| Two story house | .070 | L.F. | .002 | .04 | .07 | .11 |

# 5 | ROOFING — 20 | Shed Roofing Systems

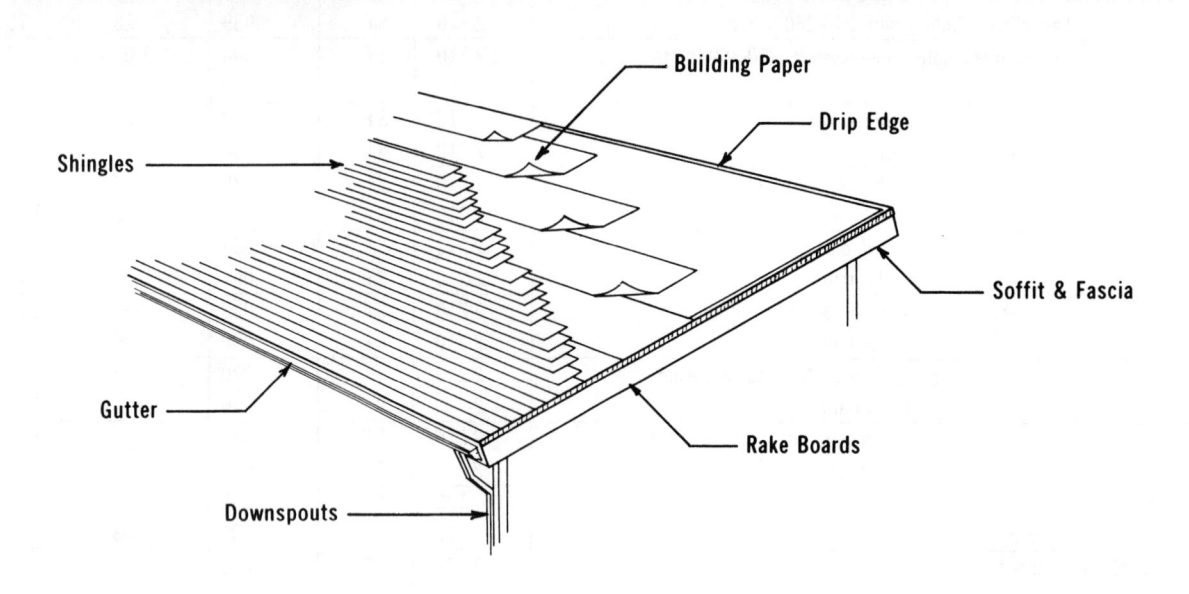

*Diagram labels: Shingles, Building Paper, Drip Edge, Soffit & Fascia, Rake Boards, Gutter, Downspouts*

| SYSTEM DESCRIPTION | QUAN. | UNIT | MAN HOURS | COST PER S.F. MAT. | COST PER S.F. INST. | COST PER S.F. TOTAL |
|---|---|---|---|---|---|---|
| **ASPHALT, ROOF SHINGLES, CLASS A** | | | | | | |
| Shingles, asphalt std. inorganic class A 210-235 lb./sq. 4/12 pitch | 1.230 | S.F. | .018 | .40 | .37 | .77 |
| Drip edge, metal, 5" girth | .100 | L.F. | .002 | .02 | .04 | .06 |
| Building paper, #15 asphalt felt | 1.300 | S.F. | .001 | .04 | .03 | .07 |
| Soffit & fascia, white painted aluminum, 1' overhang | .080 | L.F. | .011 | .21 | .24 | .45 |
| Rake trim, painted, 1" x 6" | .043 | L.F. | .003 | .03 | .06 | .09 |
| Gutter, seamless, aluminum, painted | .040 | L.F. | .002 | .04 | .06 | .10 |
| Downspouts, painted aluminum | .020 | L.F. | .001 | .02 | .01 | .03 |
| TOTAL | | | .038 | .76 | .81 | 1.57 |
| **WOOD, CEDAR SHINGLES, NO. 1 PERFECTIONS** | | | | | | |
| Shingles, wood, cedar, No. 1 perfections, 5" exposure, 4/12 pitch | 1.230 | S.F. | .034 | 1.45 | .71 | 2.16 |
| Drip edge, metal, 5" girth | .100 | L.F. | .002 | .02 | .04 | .06 |
| Building paper, #15 asphalt felt | 1.300 | S.F. | .001 | .04 | .03 | .07 |
| Soffit & fascia, white painted aluminum, 1' overhang | .080 | L.F. | .011 | .21 | .24 | .45 |
| Rake trim, painted, 1" x 6" | .043 | L.F. | .003 | .03 | .06 | .09 |
| Gutter, seamless, aluminum, painted | .040 | L.F. | .002 | .04 | .06 | .10 |
| Downspouts, painted aluminum | .020 | L.F. | .001 | .02 | .01 | .03 |
| TOTAL | | | .054 | 1.81 | 1.15 | 2.96 |

The prices in these systems are based on a square foot of plan area. All quantities have been adjusted accordingly.

| DESCRIPTION | QUAN. | UNIT | MAN HOURS | COST PER S.F. MAT. | COST PER S.F. INST. | COST PER S.F. TOTAL |
|---|---|---|---|---|---|---|
| | | | | | | |
| | | | | | | |
| | | | | | | |
| | | | | | | |
| | | | | | | |

# Shed Roofing Price Sheet

| | QUAN. | UNIT | MAN HOURS | COST PER S.F. MAT. | COST PER S.F. INST. | TOTAL |
|---|---|---|---|---|---|---|
| Shingles, asphalt standard inorganic class A 210-235 lb./sq., 4/12 pitch | 1.230 | S.F. | .017 | .37 | .34 | .71 |
| 8/12 pitch | 1.330 | S.F. | .018 | .40 | .37 | .77 |
| Laminated, multilayered, 240-260 lb./sq. 4/12 pitch | 1.230 | S.F. | .021 | .67 | .41 | 1.08 |
| 8/12 pitch | 1.330 | S.F. | .023 | .73 | .44 | 1.17 |
| Premium laminated, multilayered, 260-300 lb./sq. 4/12 pitch | 1.230 | S.F. | .027 | 1.11 | .51 | 1.62 |
| 8/12 pitch | 1.330 | S.F. | .029 | 1.20 | .56 | 1.76 |
| Clay tile, Spanish tile, red, 4/12 pitch | 1.230 | S.F. | .053 | 3.12 | 1.02 | 4.14 |
| 8/12 pitch | 1.330 | S.F. | .057 | 3.37 | 1.12 | 4.49 |
| Mission tile, red, 4/12 pitch | 1.230 | S.F. | .083 | 6.15 | 1.65 | 7.80 |
| 8/12 pitch | 1.330 | S.F. | .090 | 6.66 | 1.79 | 8.45 |
| French tile, red, 4/12 pitch | 1.230 | S.F. | .071 | 12.01 | 1.49 | 13.50 |
| 8/12 pitch | 1.330 | S.F. | .077 | 13.01 | 1.62 | 14.63 |
| Slate, Buckingham, Virginia, black, 4/12 pitch | 1.230 | S.F. | .054 | 7.52 | 1.06 | 8.58 |
| 8/12 pitch | 1.330 | S.F. | .059 | 8.15 | 1.15 | 9.30 |
| Vermont, black or grey, 4/12 pitch | 1.230 | S.F. | .054 | 8.32 | 1.04 | 9.36 |
| 8/12 pitch | 1.330 | S.F. | .059 | 9.01 | 1.13 | 10.14 |
| Wood, No. 1 perfections, 16", 5" exposure, 4/12 pitch | 1.230 | S.F. | .038 | 1.76 | .76 | 2.52 |
| 8/12 pitch | 1.330 | S.F. | .041 | 1.90 | .83 | 2.73 |
| Fire retardant, 4/12 pitch | 1.230 | S.F. | .038 | 3.33 | .78 | 4.11 |
| 8/12 pitch | 1.330 | S.F. | .041 | 3.58 | .87 | 4.45 |
| 18" 6" exposure, 4/12 pitch | 1.230 | S.F. | .034 | 1.45 | .71 | 2.16 |
| 8/12 pitch | 1.330 | S.F. | .037 | 1.57 | .77 | 2.34 |
| Fire retardant, 4/12 pitch | 1.230 | S.F. | .034 | 2.85 | .69 | 3.54 |
| 8/12 pitch | 1.330 | S.F. | .037 | 3.09 | .75 | 3.84 |
| Resquared & rebutted, 18" 6" exposure, 4/12 pitch | 1.230 | S.F. | .032 | 1.47 | .63 | 2.10 |
| 8/12 pitch | 1.330 | S.F. | .034 | 1.59 | .69 | 2.28 |
| Fire retardant, 4/12 pitch | 1.230 | S.F. | .032 | 2.65 | .65 | 3.30 |
| 8/12 pitch | 1.330 | S.F. | .034 | 2.87 | .71 | 3.58 |
| Wood shakes, hand split, 24" long 10" exposure, 4/12 pitch | 1.230 | S.F. | .038 | 1.44 | .72 | 2.16 |
| 8/12 pitch | 1.330 | S.F. | .041 | 1.56 | .78 | 2.34 |
| Fire retardant, 4/12 pitch | 1.230 | S.F. | .038 | 2.79 | .75 | 3.54 |
| 8/12 pitch | 1.330 | S.F. | .041 | 3.02 | .82 | 3.84 |
| 18" long 8" exposure, 4/12 pitch | 1.230 | S.F. | .048 | 1.11 | .93 | 2.04 |
| 8/12 pitch | 1.330 | S.F. | .052 | 1.20 | 1.01 | 2.21 |
| Fire retardant, 4/12 pitch | 1.230 | S.F. | .048 | 3.37 | .95 | 4.32 |
| 8/12 pitch | 1.330 | S.F. | .052 | 3.65 | 1.03 | 4.68 |
| Drip edge, metal, 5" girth | .100 | L.F. | .002 | .02 | .04 | .06 |
| 8" girth | .100 | L.F. | .002 | .02 | .04 | .06 |
| Building paper, #15 asphalt felt | 1.300 | S.F. | .001 | .04 | .03 | .07 |
| Soffit & fascia, aluminum vented, 1' overhang | .080 | L.F. | .011 | .21 | .24 | .45 |
| 2' overhang | .080 | L.F. | .012 | .43 | .26 | .69 |
| Vinyl vented, 1' overhang | .080 | L.F. | .010 | .12 | .22 | .34 |
| 2' overhang | .080 | L.F. | .011 | .24 | .24 | .48 |
| Wood board fascia, plywood soffit, 1' overhang | .080 | L.F. | .009 | .09 | .20 | .29 |
| 2' overhang | .080 | L.F. | .014 | .14 | .28 | .42 |
| Rake, trim, painted, 1" x 6" | .043 | L.F. | .003 | .03 | .06 | .09 |
| 1" x 8" | .043 | L.F. | .003 | .03 | .06 | .09 |
| Gutter, 5" box, aluminum, seamless, painted | .040 | L.F. | .002 | .04 | .06 | .10 |
| Vinyl | .040 | L.F. | .002 | .06 | .06 | .12 |
| Downspout 2" x 3", aluminum, one story house | .020 | L.F. | | .01 | .02 | .03 |
| Two story house | .020 | L.F. | .001 | .02 | .03 | .05 |
| Vinyl, one story house | .020 | L.F. | | .01 | .02 | .03 |
| Two story house | .020 | L.F. | .001 | .02 | .03 | .05 |

# 5 | ROOFING

## 24 | Gable Dormer Roofing Systems

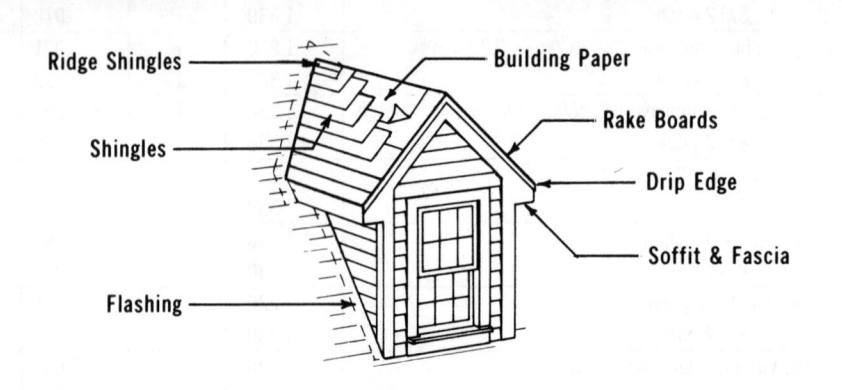

| SYSTEM DESCRIPTION | QUAN. | UNIT | MAN HOURS | MAT. | INST. | TOTAL |
|---|---|---|---|---|---|---|
| **ASPHALT, ROOF SHINGLES, CLASS A** | | | | | | |
| Shingles, asphalt standard inorganic class A 210-235 lb./sq | 1.400 | S.F. | .020 | .43 | .40 | .83 |
| Drip edge, metal, 5" girth | .220 | L.F. | .004 | .04 | .08 | .12 |
| Building paper, #15 asphalt felt | 1.500 | S.F. | .002 | .04 | .04 | .08 |
| Ridge shingles, asphalt | .280 | L.F. | .006 | .20 | .13 | .33 |
| Soffit & fascia, aluminum, vented | .220 | L.F. | .031 | .58 | .65 | 1.23 |
| Flashing, aluminum, mill finish, .013" thick | 1.500 | S.F. | .082 | .45 | 1.82 | 2.27 |
| TOTAL | | | .145 | 1.74 | 3.12 | 4.86 |
| **WOOD, CEDAR, NO. 1 PERFECTIONS** | | | | | | |
| Shingles, wood, No. 1 perfections, 16" long, 5" exposure | 1.400 | S.F. | .040 | 1.69 | .83 | 2.52 |
| Drip edge, metal, 5" girth | .220 | L.F. | .004 | .04 | .08 | .12 |
| Building paper, #15 asphalt felt | 1.500 | S.F. | .002 | .04 | .04 | .08 |
| Ridge shingles, wood | .280 | L.F. | .008 | .33 | .15 | .48 |
| Soffit & fascia, aluminum, vented | .220 | L.F. | .031 | .58 | .65 | 1.23 |
| Flashing, aluminum, mill finish, .013" thick | 1.500 | S.F. | .082 | .45 | 1.82 | 2.27 |
| TOTAL | | | .167 | 3.13 | 3.57 | 6.70 |
| **SLATE, BUCKINGHAM, BLACK** | | | | | | |
| Shingles, Buckingham, Virginia, black | 1.400 | S.F. | .063 | 8.78 | 1.23 | 10.01 |
| Drip edge, metal, 5" girth | .220 | L.F. | .004 | .04 | .08 | .12 |
| Building paper, #15 asphalt felt | 1.500 | S.F. | .002 | .04 | .04 | .08 |
| Ridge shingles, slate | .280 | L.F. | .011 | 1.11 | .22 | 1.33 |
| Soffit & fascia, aluminum, vented | .220 | L.F. | .031 | .58 | .65 | 1.23 |
| Flashing, copper, 16 oz. | 1.500 | S.F. | .105 | 1.83 | 2.30 | 4.13 |
| TOTAL | | | .216 | 12.38 | 4.52 | 16.90 |

The prices in these systems are based on a square foot of plan area under the dormer roof.

| DESCRIPTION | QUAN. | UNIT | MAN HOURS | MAT. | INST. | TOTAL |
|---|---|---|---|---|---|---|
| | | | | | | |
| | | | | | | |
| | | | | | | |
| | | | | | | |
| | | | | | | |

# Gable Dormer Roofing Price Sheet

| | QUAN. | UNIT | MAN HOURS | COST PER S.F. MAT. | COST PER S.F. INST. | TOTAL |
|---|---|---|---|---|---|---|
| Shingles, asphalt standard inorganic class A 210-235 lb./sq. | 1.400 | S.F. | .020 | .43 | .40 | .83 |
| Laminated, multilayered, 240-260 lb./sq. | 1.400 | S.F. | .024 | .79 | .47 | 1.26 |
| Premium laminated, multilayered, 260-300 lb./sq. | 1.400 | S.F. | .032 | 1.29 | .60 | 1.89 |
| Clay tile, Spanish tile, red | 1.400 | S.F. | .062 | 3.63 | 1.20 | 4.83 |
| Mission tile, red | 1.400 | S.F. | .097 | 7.18 | 1.92 | 9.10 |
| French tile, red | 1.400 | S.F. | .083 | 14.01 | 1.74 | 15.75 |
| Slate, Buckingham, Virginia, black | 1.400 | S.F. | .063 | 8.78 | 1.23 | 10.01 |
| Vermont, black or grey | 1.400 | S.F. | .063 | 9.70 | 1.22 | 10.92 |
| Wood, No. 1 perfections, 16" long, 5" exposure | 1.400 | S.F. | .044 | 2.05 | .89 | 2.94 |
| Fire retardant | 1.400 | S.F. | .044 | 3.87 | .91 | 4.78 |
| 18" long, 5" exposure | 1.400 | S.F. | .040 | 1.69 | .83 | 2.52 |
| Fire retardant | 1.400 | S.F | .040 | 3.33 | .80 | 4.13 |
| Resquared & rebutted, 18" long, 5" exposure | 1.400 | S.F. | .037 | 1.71 | .74 | 2.45 |
| Fire retardant | 1.400 | S.F. | .037 | 3.10 | .75 | 3.85 |
| Shakes hand split, 24" long, 10" exposure | 1.400 | S.F. | .044 | 1.68 | .84 | 2.52 |
| Fire retardant | 1.400 | S.F. | .044 | 3.25 | .88 | 4.13 |
| 18" long, 8" exposure | 1.400 | S.F. | .056 | 1.29 | 1.09 | 2.38 |
| Fire retardant | 1.400 | S.F. | .056 | 3.93 | 1.11 | 5.04 |
| Drip edge, metal, 5" girth | .220 | L.F. | .004 | .04 | .08 | .12 |
| 8" girth | .220 | L.F. | .004 | .05 | .09 | .14 |
| Building paper, #15 asphalt felt | 1.500 | S.F. | .002 | .04 | .04 | .08 |
| Ridge shingles, asphalt | .280 | L.F. | .006 | .20 | .13 | .33 |
| Clay | .280 | L.F. | .011 | 1.08 | .22 | 1.30 |
| Slate | .280 | L.F. | .011 | 1.11 | .22 | 1.33 |
| Wood | .280 | L.F. | .008 | .33 | .15 | .48 |
| Soffit & fascia, aluminum, vented | .220 | L.F. | .031 | .58 | .65 | 1.23 |
| Vinyl, vented | .220 | L.F. | .029 | .33 | .60 | .93 |
| Wood, board fascia, plywood soffit | .220 | L.F. | .026 | .27 | .51 | .78 |
| Flashing, aluminum, .013" thick | 1.500 | S.F. | .082 | .45 | 1.82 | 2.27 |
| .032" thick | 1.500 | S.F. | .082 | 1.28 | 1.81 | 3.09 |
| .040" thick | 1.500 | S.F. | .082 | 2.15 | 1.83 | 3.98 |
| .050" thick | 1.500 | S.F. | .082 | 2.64 | 1.83 | 4.47 |
| Copper, 16 oz. | 1.500 | S.F. | .105 | 1.83 | 2.30 | 4.13 |
| 20 oz. | 1.500 | S.F. | .109 | 2.31 | 2.40 | 4.71 |
| 24 oz. | 1.500 | S.F. | .114 | 2.76 | 2.52 | 5.28 |
| 32 oz. | 1.500 | S.F. | .120 | 3.66 | 2.66 | 6.32 |

# 5 | ROOFING
## 28 | Shed Dormer Roofing Systems

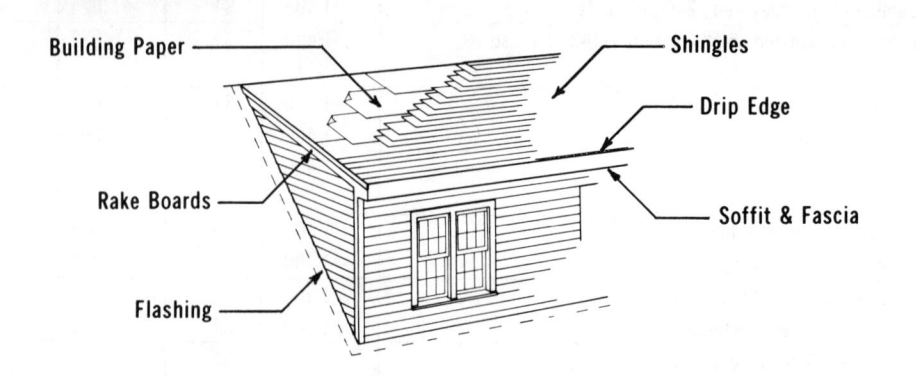

| SYSTEM DESCRIPTION | QUAN. | UNIT | MAN HOURS | MAT. | INST. | TOTAL |
|---|---|---|---|---|---|---|
| **ASPHALT, ROOF SHINGLES, CLASS A** | | | | | | |
| Shingles, asphalt standard inorganic class A 210-235 lb./sq. | 1.100 | S.F. | .015 | .34 | .31 | .65 |
| Drip edge, aluminum, 5" girth | .250 | L.F. | .005 | .04 | .09 | .13 |
| Building paper, #15 asphalt felt | 1.200 | S.F. | .001 | .03 | .04 | .07 |
| Soffit & fascia, aluminum, vented, 1' overhang | .250 | L.F. | .036 | .66 | .74 | 1.40 |
| Flashing, aluminum, mill finish, 0.013" thick | .800 | L.F. | .044 | .24 | .97 | 1.21 |
| TOTAL | | | .102 | 1.31 | 2.15 | 3.46 |
| **WOOD, CEDAR, NO. 1 PERFECTIONS** | | | | | | |
| Shingles, wood, cedar, #1 perfections, 5" exposure | 1.100 | S.F. | .032 | 1.33 | .65 | 1.98 |
| Drip edge, aluminum, 5" girth | .250 | L.F. | .005 | .04 | .09 | .13 |
| Building paper, #15 asphalt felt | 1.200 | S.F. | .001 | .03 | .04 | .07 |
| Soffit & fascia, aluminum, vented, 1' overhang | .250 | L.F. | .036 | .66 | .74 | 1.40 |
| Flashing, aluminum, mill finish, 0.013" thick | .800 | L.F. | .044 | .24 | .97 | 1.21 |
| TOTAL | | | .118 | 2.30 | 2.49 | 4.79 |
| **SLATE, BUCKINGHAM, BLACK** | | | | | | |
| Shingles, slate, Buckingham, black | 1.100 | S.F. | .050 | 6.90 | .97 | 7.87 |
| Drip edge, aluminum, 5" girth | .250 | L.F. | .005 | .04 | .09 | .13 |
| Building paper, #15 asphalt felt | 1.200 | S.F. | .001 | .03 | .04 | .07 |
| Soffit & fascia, aluminum, vented, 1' overhang | .250 | L.F. | .036 | .66 | .74 | 1.40 |
| Flashing, copper, 16 oz. | .800 | L.F. | .056 | .98 | 1.22 | 2.20 |
| TOTAL | | | .149 | 8.61 | 3.06 | 11.67 |

The prices in this system are based on a square foot of plan area under the dormer roof.

| DESCRIPTION | QUAN. | UNIT | MAN HOURS | MAT. | INST. | TOTAL |
|---|---|---|---|---|---|---|
| | | | | | | |
| | | | | | | |
| | | | | | | |
| | | | | | | |
| | | | | | | |

# Shed Dormer Roofing Price Sheet

| | QUAN. | UNIT | MAN HOURS | COST PER S.F. MAT. | COST PER S.F. INST. | TOTAL |
|---|---|---|---|---|---|---|
| Shingles, asphalt standard inorganic class A 210-235 lb./sq. | 1.100 | S.F. | .015 | .34 | .31 | .65 |
| Laminated, multilayered, 240-260 lb./sq. | 1.100 | S.F. | .019 | .62 | .37 | .99 |
| Premium laminated, multilayered, 260-300 lb./sq. | 1.100 | S.F. | .025 | 1.02 | .47 | 1.49 |
| Clay tile, Spanish tile, red | 1.100 | S.F. | .048 | 2.86 | .94 | 3.80 |
| Mission tile, red | 1.100 | S.F. | .076 | 5.64 | 1.51 | 7.15 |
| French tile, red | 1.100 | S.F. | .065 | 11.01 | 1.37 | 12.38 |
| Slate, Buckingham, Virginia, black | 1.100 | S.F. | .050 | 6.90 | .97 | 7.87 |
| Vermont, black or grey | 1.100 | S.F. | .050 | 7.62 | .96 | 8.58 |
| Wood, No. 1 perfections, 16" long, 5" exposure | 1.100 | S.F. | .035 | 1.61 | .70 | 2.31 |
| Fire retardant | 1.100 | S.F. | .035 | 3.04 | .71 | 3.75 |
| 18", 5" exposure | 1.100 | S.F. | .032 | 1.33 | .65 | 1.98 |
| Fire retardant | 1.100 | S.F. | .032 | 2.61 | .64 | 3.25 |
| Resquared & rebutted, 18" long, 5" exposure | 1.100 | S.F. | .029 | 1.34 | .59 | 1.93 |
| Fire retardant | 1.100 | S.F. | .029 | 2.43 | .60 | 3.03 |
| Shakes hand split, 24" long, 10" exposure | 1.100 | S.F. | .035 | 1.32 | .66 | 1.98 |
| Fire retardant | 1.100 | S.F. | .035 | 2.55 | .70 | 3.25 |
| 18" long, 8" exposure | 1.100 | S.F. | .044 | 1.02 | .85 | 1.87 |
| Fire retardant | 1.100 | S.F. | .044 | 3.09 | .87 | 3.96 |
| Drip edge, metal, 5" girth | .250 | L.F. | .005 | .04 | .09 | .13 |
| 8" girth | .250 | L.F. | .005 | .06 | .10 | .16 |
| Building paper, #15 asphalt felt | 1.200 | S.F. | .001 | .03 | .04 | .07 |
| Soffit & fascia, aluminum, vented | .250 | L.F. | .036 | .66 | .74 | 1.40 |
| Vinyl, vented | .250 | L.F. | .033 | .37 | .69 | 1.06 |
| Wood, board fascia, plywood soffit | .250 | L.F. | .030 | .31 | .59 | .90 |
| Flashing, aluminum, .013" thick | .800 | L.F. | .044 | .24 | .97 | 1.21 |
| .032" thick | .800 | L.F. | .044 | .68 | .97 | 1.65 |
| .040" thick | .800 | L.F. | .044 | 1.14 | .98 | 2.12 |
| .050" thick | .800 | L.F. | .044 | 1.41 | .97 | 2.38 |
| Copper, 16 oz. | .800 | L.F. | .056 | .98 | 1.22 | 2.20 |
| 20 oz. | .800 | L.F. | .058 | 1.23 | 1.28 | 2.51 |
| 24 oz. | .800 | L.F. | .060 | 1.47 | 1.35 | 2.82 |
| 32 oz. | .800 | L.F. | .064 | 1.95 | 1.42 | 3.37 |
| Shingles, asphalt standard inorganic class A 210-235 lb./sq. | | | | | | |

# 5 | ROOFING  32 | Skylight/Skywindow Systems

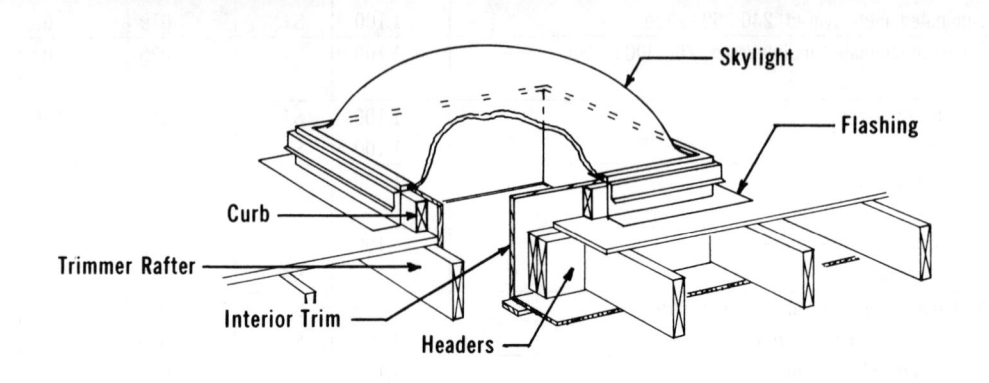

| SYSTEM DESCRIPTION | QUAN. | UNIT | MAN HOURS | MAT. | INST. | TOTAL |
|---|---|---|---|---|---|---|
| **SKYLIGHT, FIXED, 32" X 32"** | | | | | | |
| Skylight, fixed bubble, insulating, 32" x 32" | 1.000 | Ea. | 1.422 | 98.56 | 28.02 | 126.58 |
| Trimmer rafters, 2" x 6" | 28.000 | L.F. | .532 | 11.20 | 10.64 | 21.84 |
| Headers, 2" x 6" | 6.000 | L.F. | .186 | 2.40 | 3.78 | 6.18 |
| Curb, 2" x 4" | 12.000 | L.F. | .156 | 3.36 | 3.12 | 6.48 |
| Flashing, aluminum, .013" thick | 13.500 | S.F. | .742 | 4.05 | 16.34 | 20.39 |
| Trim, interior casing, painted | 12.000 | L.F. | .684 | 7.32 | 13.08 | 20.40 |
| TOTAL | | | 3.722 | 126.89 | 74.98 | 201.87 |
| **SKYLIGHT, FIXED, 48" X 48"** | | | | | | |
| Skylight, fixed bubble, insulating, 48" x 48" | 1.000 | Ea. | 1.296 | 184.80 | 25.60 | 210.40 |
| Trimmer rafters, 2" x 6" | 28.000 | L.F. | .532 | 11.20 | 10.64 | 21.84 |
| Headers, 2" x 6" | 8.000 | L.F. | .248 | 3.20 | 5.04 | 8.24 |
| Curb, 2" x 4" | 16.000 | L.F. | .208 | 4.48 | 4.16 | 8.64 |
| Flashing, aluminum, .013" thick | 16.000 | S.F. | .880 | 4.80 | 19.36 | 24.16 |
| Trim, interior casing, painted | 16.000 | L.F. | .912 | 9.76 | 17.44 | 27.20 |
| TOTAL | | | 4.076 | 218.24 | 82.24 | 300.48 |
| **SKYWINDOW, OPERATING, 24" X 48"** | | | | | | |
| Skywindow, operating, thermopane glass, 24" x 48" | 1.000 | Ea. | 3.200 | 424.60 | 65.40 | 490.00 |
| Trimmer rafters, 2" x 6" | 28.000 | L.F. | .532 | 11.20 | 10.64 | 21.84 |
| Headers, 2" x 6" | 8.000 | L.F. | .186 | 2.40 | 3.78 | 6.18 |
| Curb, 2" x 4" | 14.000 | L.F. | .182 | 3.92 | 3.64 | 7.56 |
| Flashing, aluminum, .013" thick | 14.000 | S.F. | .770 | 4.20 | 16.94 | 21.14 |
| Trim, interior casing, painted | 14.000 | L.F. | .798 | 8.54 | 15.26 | 23.80 |
| TOTAL | | | 5.668 | 454.86 | 115.66 | 570.52 |

The prices in these systems are on a cost each basis.

| DESCRIPTION | QUAN. | UNIT | MAN HOURS | MAT. | INST. | TOTAL |
|---|---|---|---|---|---|---|
| | | | | | | |
| | | | | | | |
| | | | | | | |
| | | | | | | |
| | | | | | | |

# Skylight/Skywindow Price Sheet

| | QUAN. | UNIT | MAN HOURS | COST EACH MAT. | COST EACH INST. | TOTAL |
|---|---|---|---|---|---|---|
| Skylight, fixed bubble insulating, 24" x 24" | 1.000 | Ea. | .800 | 55.44 | 15.76 | 71.20 |
| 32" x 32" | 1.000 | Ea. | 1.422 | 98.56 | 28.02 | 126.58 |
| 32" x 48" | 1.000 | Ea. | .863 | 123.19 | 17.07 | 140.26 |
| 48" x 48" | 1.000 | Ea. | 1.296 | 184.80 | 25.60 | 210.40 |
| Ventilating bubble insulating, 36" x 36" | 1.000 | Ea. | 2.670 | 383.90 | 51.10 | 435.00 |
| 52" x 52" | 1.000 | Ea. | 2.670 | 513.70 | 51.30 | 565.00 |
| 28" x 52" | 1.000 | Ea. | 3.200 | 424.60 | 65.40 | 490.00 |
| 36" x 52" | 1.000 | Ea. | 3.200 | 451.00 | 64.00 | 515.00 |
| Skywindow operating, thermopane glass, 24" x 48" | 1.000 | Ea. | 3.200 | 466.40 | 63.60 | 530.00 |
| 32" x 48" | 1.000 | Ea. | 3.560 | 456.50 | 68.50 | 525.00 |
| Trimmer rafters, 2" x 6" | 28.000 | L.F. | .532 | 11.20 | 10.64 | 21.84 |
| 2" x 8" | 28.000 | L.F. | .616 | 15.40 | 12.88 | 28.28 |
| 2" x 10" | 28.000 | L.F. | .700 | 21.00 | 14.56 | 35.56 |
| Headers, 24" window, 2" x 6" | 4.000 | L.F. | .124 | 1.60 | 2.52 | 4.12 |
| 2" x 8" | 4.000 | L.F. | .188 | 2.20 | 4.12 | 6.32 |
| 2" x 10" | 4.000 | L.F. | .200 | 3.08 | 4.32 | 7.40 |
| 32" window, 2" x 6" | 6.000 | L.F. | .186 | 2.40 | 3.78 | 6.18 |
| 2" x 8" | 6.000 | L.F. | .282 | 3.30 | 6.18 | 9.48 |
| 2" x 10" | 6.000 | L.F. | .300 | 4.62 | 6.48 | 11.10 |
| 48" window, 2" x 6" | 8.000 | L.F. | .248 | 3.20 | 5.04 | 8.24 |
| 2" x 8" | 8.000 | L.F. | .376 | 4.40 | 8.24 | 12.64 |
| 2" x 10" | 8.000 | L.F. | .400 | 6.16 | 8.64 | 14.80 |
| Curb 2" x 4", skylight, 24" x 24" | 8.000 | L.F. | .104 | 2.24 | 2.08 | 4.32 |
| 32" x 32" | 12.000 | L.F. | .156 | 3.36 | 3.12 | 6.48 |
| 32" x 48" | 14.000 | L.F. | .182 | 3.92 | 3.64 | 7.56 |
| 48" x 48" | 16.000 | L.F. | .208 | 4.48 | 4.16 | 8.64 |
| Flashing, aluminum .013" thick, skylight, 24" x 24" | 9.000 | S.F. | .495 | 2.70 | 10.89 | 13.59 |
| 32" x 32" | 13.500 | S.F. | .742 | 4.05 | 16.34 | 20.39 |
| 32" x 48" | 14.000 | S.F. | .770 | 4.20 | 16.94 | 21.14 |
| 48" x 48" | 16.000 | S.F. | .880 | 4.80 | 19.36 | 24.16 |
| Copper 16 oz., skylight, 24" x 24" | 9.000 | S.F. | .630 | 10.98 | 13.77 | 24.75 |
| 32" x 32" | 13.500 | S.F. | .945 | 16.47 | 20.66 | 37.13 |
| 32" x 48" | 14.000 | S.F. | .980 | 17.08 | 21.42 | 38.50 |
| 48" x 48" | 16.000 | S.F. | 1.120 | 19.52 | 24.48 | 44.00 |
| Trim, interior casing painted, 24" x 24" | 8.000 | L.F. | .456 | 4.88 | 8.72 | 13.60 |
| 32" x 32" | 12.000 | L.F. | .684 | 7.32 | 13.08 | 20.40 |
| 32" x 48" | 14.000 | L.F. | .798 | 8.54 | 15.26 | 23.80 |
| 48" x 48" | 16.000 | L.F. | .912 | 9.76 | 17.44 | 27.20 |

# 5 | ROOFING     34 | Built-up Roofing Systems

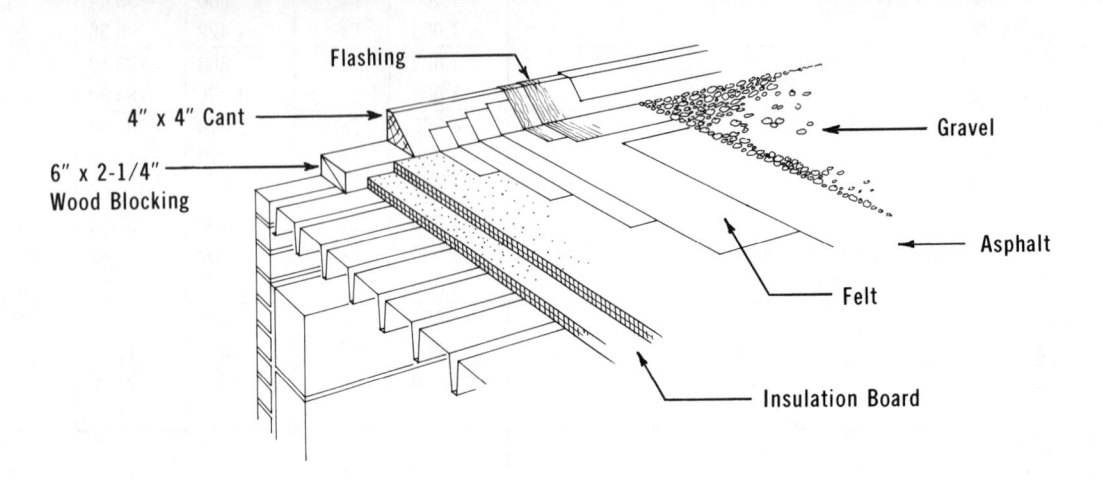

*Diagram labels: Flashing; 4" x 4" Cant; 6" x 2-1/4" Wood Blocking; Gravel; Asphalt; Felt; Insulation Board*

| SYSTEM DESCRIPTION | QUAN. | UNIT | MAN HOURS | MAT. | INST. | TOTAL |
|---|---|---|---|---|---|---|
| **ASPHALT, ORGANIC, 4-PLY, INSULATED DECK** | | | | | | |
| Membrane, asphalt, 4-plies #15 felt, gravel surfacing | 1.000 | S.F. | .025 | .35 | .52 | .87 |
| Insulation board, 2-layers of 1-1/16" glass fiber | 2.000 | S.F. | .016 | .86 | .30 | 1.16 |
| Wood blocking, treated, 6" x 2-1/4" & 4" x 4" cant | .040 | L.F. | .005 | .08 | .11 | .19 |
| Flashing, aluminum, 0.040" thick | .050 | S.F. | .002 | .07 | .06 | .13 |
| TOTAL | | | .049 | 1.36 | .99 | 2.35 |
| **ASPHALT, INORGANIC, 3-PLY, INSULATED DECK** | | | | | | |
| Membrane, asphalt, 3-plies type IV glass felt, gravel surfacing | 1.000 | S.F. | .025 | .42 | .52 | .94 |
| Insulation board, 2-layers of 1-1/16" glass fiber | 2.000 | S.F. | .016 | .86 | .30 | 1.16 |
| Wood blocking, treated, 6" x 2-1/4" & 4" x 4" cant | .040 | L.F. | .005 | .08 | .11 | .19 |
| Flashing, aluminum, 0.040" thick | .050 | S.F. | .002 | .07 | .06 | .13 |
| TOTAL | | | .049 | 1.43 | .99 | 2.42 |
| **COAL TAR, ORGANIC, 4-PLY, INSULATED DECK** | | | | | | |
| Membrane, coal tar, 4-plies #15 felt, gravel surfacing | 1.000 | S.F. | .025 | .87 | .53 | 1.40 |
| Insulation board, 2-layers of 1-1/16" glass fiber | 2.000 | S.F. | .016 | .86 | .30 | 1.16 |
| Wood blocking, treated, 6" x 2-1/4" & 4" x 4" cant | .040 | L.F. | .005 | .08 | .11 | .19 |
| Flashing, aluminum, 0.040" thick | .050 | S.F. | .002 | .07 | .06 | .13 |
| TOTAL | | | .049 | 1.88 | 1.00 | 2.88 |
| **COAL TAR, INORGANIC, 3-PLY, INSULATED DECK** | | | | | | |
| Membrane, coal tar, 3-plies type IV glass felt, gravel surfacing | 1.000 | S.F. | .028 | .75 | .55 | 1.30 |
| Insulation board, 2-layers of 1-1/16" glass fiber | 2.000 | S.F. | .016 | .86 | .30 | 1.16 |
| Wood blocking, treated, 6" x 2-1/4" & 4" x 4" cant | .040 | L.F. | .005 | .08 | .11 | .19 |
| Flashing, aluminum, 0.040" thick | .050 | S.F. | .002 | .07 | .06 | .13 |
| TOTAL | | | .052 | 1.76 | 1.02 | 2.78 |

| DESCRIPTION | QUAN. | UNIT | MAN HOURS | MAT. | INST. | TOTAL |
|---|---|---|---|---|---|---|
| | | | | | | |
| | | | | | | |
| | | | | | | |

# Built-Up Roofing Price Sheet

| | QUAN. | UNIT | MAN HOURS | COST PER S.F. MAT. | COST PER S.F. INST. | TOTAL |
|---|---|---|---|---|---|---|
| Membrane, asphalt, 4-plies #15 organic felt, gravel surfacing | 1.000 | S.F. | .025 | .35 | .52 | .87 |
| Asbestos base sheet & 3-plies #15 asbestos felt | 1.000 | S.F. | .025 | .58 | .52 | 1.10 |
| 3-plies type IV glass fiber felt | 1.000 | S.F. | .025 | .42 | .52 | .94 |
| 4-plies type IV glass fiber felt | 1.000 | S.F. | .028 | .52 | .58 | 1.10 |
| Coal tar, 4-plies #15 organic felt, gravel surfacing | 1.000 | S.F. | .025 | .87 | .53 | 1.40 |
| 4-plies asbestos felt | 1.000 | S.F. | .025 | 1.05 | .50 | 1.55 |
| 3-plies type IV glass fiber felt | 1.000 | S.F. | .028 | .75 | .55 | 1.30 |
| 4-plies type IV glass fiber felt | 1.000 | S.F. | .025 | .92 | .53 | 1.45 |
| Roll, asphalt, 1-ply #15 organic felt, 2-plies mineral surfaced | 1.000 | S.F. | .020 | .42 | .42 | .84 |
| 3-plies type IV glass fiber, 1-ply mineral surfaced | 1.000 | S.F. | .022 | .64 | .46 | 1.10 |
| Insulation boards, glass fiber, 1-1/16" thick | 1.000 | S.F. | .008 | .43 | .15 | .58 |
| 1-5/8" thick | 1.000 | S.F. | .008 | .66 | .15 | .81 |
| 1-7/8" thick | 1.000 | S.F. | .008 | .70 | .16 | .86 |
| 2-1/4" thick | 1.000 | S.F. | .010 | .76 | .19 | .95 |
| Expanded perlite, 1" thick | 1.000 | S.F. | .010 | .30 | .19 | .49 |
| 1-1/2" thick | 1.000 | S.F. | .010 | .44 | .19 | .63 |
| 2" thick | 1.000 | S.F. | .011 | .59 | .22 | .81 |
| Fiberboard, 1" thick | 1.000 | S.F. | .010 | .29 | .19 | .48 |
| 1-1/2" thick | 1.000 | S.F. | .010 | .40 | .19 | .59 |
| 2" thick | 1.000 | S.F. | .010 | .55 | .19 | .74 |
| Phenolic foam, 1-3/16" thick | 1.000 | S.F. | .010 | .29 | .19 | .48 |
| 1-3/4" thick | 1.000 | S.F. | .019 | .56 | .37 | .93 |
| 2" thick | 1.000 | S.F. | .010 | 1.21 | .19 | 1.40 |
| 3" thick | 1.000 | S.F. | .010 | 1.82 | .19 | 2.01 |
| Urethane, 1" thick | 1.000 | S.F. | .008 | .45 | .15 | .60 |
| 1-1/2" thick | 1.000 | S.F. | .008 | .55 | .15 | .70 |
| 2" thick | 1.000 | S.F. | .010 | .67 | .19 | .86 |
| 2-1/2" thick | 1.000 | S.F. | .010 | .78 | .19 | .97 |
| Glass fiber/urethane composite, 1-11/16" thick | 1.000 | S.F. | .008 | .61 | .15 | .76 |
| 2" thick | 1.000 | S.F. | .010 | .72 | .19 | .91 |
| 2-5/8" thick | 1.000 | S.F. | .010 | .89 | .19 | 1.08 |
| Perlite/urethane composite, 1-1/4" thick | 1.000 | S.F. | .008 | .66 | .15 | .81 |
| 1-1/2" thick | 1.000 | S.F. | .008 | .69 | .16 | .85 |
| 2-1/2" thick | 1.000 | S.F. | .011 | .83 | .20 | 1.03 |
| 3" thick | 1.000 | S.F. | .011 | 1.00 | .22 | 1.22 |
| Extruded polystyrene, 1" thick | 1.000 | S.F. | .005 | .31 | .10 | .41 |
| 2" thick | 1.000 | S.F. | .006 | .61 | .12 | .73 |
| 3" thick | 1.000 | S.F. | .008 | .92 | .16 | 1.08 |
| Wood blocking, treated, 6" x 2" & 4" x 4" cant | .040 | L.F. | .002 | .05 | .05 | .10 |
| 6" x 4-1/2" & 4" x 4" cant | .040 | L.F. | .005 | .08 | .11 | .19 |
| 6" x 5" & 4" x 4" cant | .040 | L.F. | .006 | .10 | .14 | .24 |
| Flashing, aluminum, 0.019" thick | .050 | S.F. | .002 | .03 | .07 | .10 |
| 0.032" thick | .050 | S.F. | .002 | .04 | .06 | .10 |
| 0.040" thick | .050 | S.F. | .002 | .07 | .06 | .13 |
| Copper sheets, 16 oz., under 500 lbs. | .050 | S.F. | .003 | .06 | .08 | .14 |
| Over 500 lbs. | .050 | S.F. | .002 | .06 | .06 | .12 |
| 20 oz., under 500 lbs. | .050 | S.F. | .003 | .08 | .08 | .16 |
| Over 500 lbs. | .050 | S.F. | .002 | .08 | .06 | .14 |
| Lead-coated copper, 16 oz. | .050 | S.F. | .009 | .40 | .22 | .62 |
| 20 oz. | .050 | S.F. | .004 | .30 | .10 | .40 |
| Galvanized steel, 25 gauge | .050 | S.F. | .002 | .06 | .05 | .11 |
| 24 gauge | .050 | S.F. | .002 | .07 | .06 | .13 |
| 20 gauge | .050 | S.F. | .002 | .09 | .06 | .15 |
| 16 gauge | .050 | S.F. | .002 | .11 | .06 | .17 |
| Stainless steel, 32 gauge | .050 | S.F. | .002 | .09 | .06 | .15 |
| 28 gauge | .050 | S.F. | .002 | .12 | .05 | .17 |
| 26 gauge | .050 | S.F. | .002 | .14 | .06 | .20 |
| 24 gauge | .050 | S.F. | .002 | .18 | .05 | .23 |

# 5 | ROOFING  36 | Single-Ply Roofing System

*Wood Blocking - 6" x 2-1/2"*
*Flashing*
*Membrane*
*Insulation Board*

| SYSTEM DESCRIPTION | QUAN. | UNIT | MAN HOURS | COST PER S.F. MAT. | COST PER S.F. INST. | COST PER S.F. TOTAL |
|---|---|---|---|---|---|---|
| **CHLOROSULFONATED POLYETHYLENE - HYPALON (CSPE)** | | | | | | |
| Membrane, 35 mil CSPE, fully adhered | 1.000 | S.F. | .011 | 1.41 | .23 | 1.64 |
| Insulation board, 2-1/2" perlite/urethane composite | 1.000 | S.F. | .011 | .83 | .20 | 1.03 |
| Wood blocking, treated, 6" x 2-1/2" | .040 | L.F. | .002 | .04 | .06 | .10 |
| Flashing, aluminum, 0.040" thick | .080 | S.F. | .002 | .07 | .06 | .13 |
| TOTAL | | | .027 | 2.35 | .55 | 2.90 |
| **ETHYLENE PROPYLENE DIENE MONUMER (EPDM)** | | | | | | |
| Membrane, 45 mil EPDM, loose-laid & ballasted w/stone (10 psf) | 1.000 | S.F. | .006 | 1.16 | .11 | 1.27 |
| Insulation board, 2-1/2" perlite/urethane composite | 1.000 | S.F. | .011 | .83 | .20 | 1.03 |
| Wood blocking, treated, 6" x 2-1/2" | .040 | L.F. | .002 | .04 | .06 | .10 |
| Flashing, aluminum, 0.040" thick | .080 | S.F. | .002 | .07 | .06 | .13 |
| TOTAL | | | .022 | 2.10 | .43 | 2.53 |
| **MODIFIED BITUMEN** | | | | | | |
| Membrane, 160 mil mod. bitumen w/asphalt emulsion coating | 1.000 | S.F. | .020 | 3.00 | .42 | 3.42 |
| Insulation board, 2 layers of 2-1/4" glass fiber | 2.000 | S.F. | .020 | 1.52 | .38 | 1.90 |
| Wood blocking, treated, 6" x 4-1/2" & 4" x 4" cant | .040 | L.F. | .005 | .08 | .11 | .19 |
| Flashing, aluminum, 0.040" thick | .080 | S.F. | .002 | .07 | .06 | .13 |
| TOTAL | | | .048 | 4.67 | .97 | 5.64 |
| **REINFORCED POLYVINYL CHLORIDE (PVC)** | 1.000 | S.F. | | | | |
| Membrane, 48 mil reinforced PVC, partially attached | 1.000 | S.F. | .008 | 2.62 | .16 | 2.78 |
| Insulation board, 2-1/2" perlite/urethane composite | 1.000 | S.F. | .011 | .83 | .20 | 1.03 |
| Wood blocking, treated, 6" x 2-1/2" | .040 | L.F. | .002 | .04 | .06 | .10 |
| Flashing, aluminum, 0.040" thick | .080 | S.F. | .002 | .07 | .06 | .13 |
| TOTAL | | | .024 | 3.56 | .48 | 4.04 |

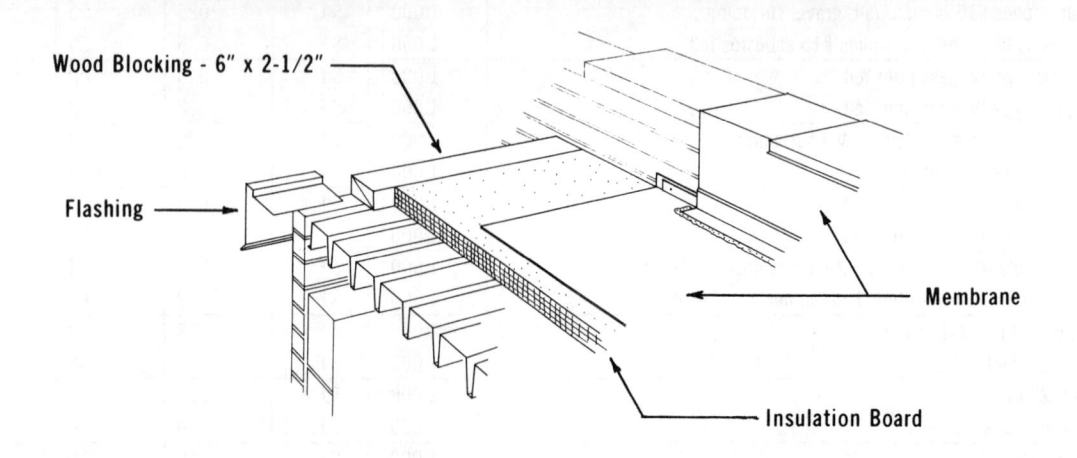

| DESCRIPTION | QUAN. | UNIT | MAN. HOURS | COST PER S.F. MAT. | COST PER S.F. INST. | COST PER S.F. TOTAL |
|---|---|---|---|---|---|---|
| | | | | | | |
| | | | | | | |
| | | | | | | |

# Single-Ply Roofing Price Sheet

| | QUAN. | UNIT | MAN HOURS | COST PER S.F. MAT. | COST PER S.F. INST. | TOTAL |
|---|---|---|---|---|---|---|
| Membrane-chlorinated polyethylene (CPE), 40 mil, partially adhered | 1.000 | S.F. | .008 | 1.62 | .16 | 1.78 |
| Chlorosulfonated polyethylene-hypalon (CSPE), 35 mil fully adhered | 1.000 | S.F. | .011 | 1.41 | .23 | 1.64 |
| 40 mil, partially adhered with batten strips | 1.000 | S.F. | .008 | 1.77 | .16 | 1.93 |
| Elastomer modified asphalt, 150 mil, partially adhered-torch | 1.000 | S.F. | .016 | 2.81 | .33 | 3.14 |
| Hot mopped with asphalt | 1.000 | S.F. | .016 | 2.86 | .33 | 3.19 |
| Fully adhered-torch | 1.000 | S.F. | .020 | 3.19 | .41 | 3.60 |
| Hot mopped with asphalt | 1.000 | S.F. | .020 | 3.25 | .41 | 3.66 |
| EPDM, 45 mil, loose-laid & ballasted with stone (10 psf) | 1.000 | S.F. | .006 | 1.16 | .11 | 1.27 |
| Partially adhered with batten strips | 1.000 | S.F. | .008 | 1.32 | .16 | 1.48 |
| Fully adhered with adhesive | 1.000 | S.F. | .011 | 1.60 | .23 | 1.83 |
| 55 mil, loose-laid & ballasted with stone (10 psf) | 1.000 | S.F. | .006 | .66 | .12 | .78 |
| Partially adhered to plates | 1.000 | S.F. | .008 | 2.15 | .16 | 2.31 |
| Fully adhered with adhesive | 1.000 | S.F. | .011 | .88 | .24 | 1.12 |
| 60 mil, loose-laid & ballasted with stone (10 psf) | 1.000 | S.F. | .006 | 1.42 | .12 | 1.54 |
| Partially adhered with bar anchors | 1.000 | S.F. | .008 | 1.60 | .16 | 1.76 |
| Fully adhered with adhesive | 1.000 | S.F. | .011 | 1.84 | .23 | 2.07 |
| Modified bitumen, 120 mils, fully adhered with solvent | 1.000 | S.F. | .014 | 1.97 | .30 | 2.27 |
| 160 mils, emulsion coating, loose-laid & ballasted, stone (10 psf) | 1.000 | S.F. | .013 | 2.48 | .25 | 2.73 |
| Partially adhered, torch welding | 1.000 | S.F. | .016 | 2.73 | .33 | 3.06 |
| Fully adhered, torch welding | 1.000 | S.F. | .020 | 3.00 | .42 | 3.42 |
| Neoprene, 60 mils, partially adhered w/fasteners | 1.000 | S.F. | .008 | 1.82 | .16 | 1.98 |
| Fully adhered w/adhesive | 1.000 | S.F. | .011 | 1.54 | .24 | 1.78 |
| Polyvinyl chloride(PVC),45 mil,loose-laid & ballasted,stone (10 psf) | 1.000 | S.F. | .006 | 1.60 | .11 | 1.71 |
| Partially adhered, with fasteners | 1.000 | S.F. | .008 | 2.16 | .16 | 2.32 |
| 48 mils, loose-laid & ballasted, stone/pavers (10 psf) | 1.000 | S.F. | .006 | .77 | .12 | .89 |
| Reinforced PVC, 48 mil, loose-laid & ballasted, stone (10 psf) | 1.000 | S.F. | .006 | 2.08 | .12 | 2.20 |
| Partially adhered with fasteners | 1.000 | S.F. | .008 | 2.62 | .16 | 2.78 |
| Fully adhered with adhesive | 1.000 | S.F. | .011 | 2.86 | .24 | 3.10 |
| Insulation board-glass fiber, 1-5/8" thick | 1.000 | S.F. | .008 | .66 | .15 | .81 |
| 2-1/4" thick | 1.000 | S.F. | .010 | .76 | .19 | .95 |
| Expanded perlite, 1-1/2" thick | 1.000 | S.F. | .010 | .44 | .19 | .63 |
| 2" thick | 1.000 | S.F. | .011 | .59 | .22 | .81 |
| Phenolic foam, 1-3/8" thick | 1.000 | S.F. | .042 | 2.24 | .81 | 3.05 |
| 2" thick | 1.000 | S.F. | .010 | 1.82 | .19 | 2.01 |
| Urethane, 1-1/2" thick | 1.000 | S.F. | .008 | .55 | .15 | .70 |
| 2" thick | 1.000 | S.F. | .010 | .67 | .19 | .86 |
| Glass fiber/urethane composite, 2" thick | 1.000 | S.F. | .010 | .72 | .19 | .91 |
| 2-5/8" thick | 1.000 | S.F. | .010 | .89 | .19 | 1.08 |
| Perlite/urethane composite, 2-1/2" thick | 1.000 | S.F. | .011 | .83 | .20 | 1.03 |
| 3" thick | 1.000 | S.F. | .011 | 1.00 | .22 | 1.22 |
| Extruded polystyrene, 2" thick | 1.000 | S.F. | .006 | .61 | .12 | .73 |
| 3" thick | 1.000 | S.F. | .008 | .92 | .16 | 1.08 |
| Wood blocking, treated - 6" x 4-1/2" w/4" x 4" cant | .040 | L.F. | .005 | .08 | .11 | .19 |
| 6" x 2" | .040 | L.F. | .001 | .02 | .03 | .05 |
| 6" x 4" | .040 | L.F. | .002 | .04 | .06 | .10 |
| 6" x 6" | .040 | L.F. | .005 | .08 | .12 | .20 |
| Flashing, aluminum, 0.032" thick | .080 | S.F. | .002 | .04 | .06 | .10 |
| 0.040" thick | .080 | S.F. | .002 | .07 | .06 | .13 |
| CPE clad metal | .080 | S.F. | .001 | .10 | .03 | .13 |
| Bonded CSPE | .080 | S.F. | .001 | .09 | .03 | .12 |
| Hypalon clad metal | .080 | S.F. | .001 | .14 | .03 | .17 |
| Cured EPDM | .080 | S.F. | .001 | .06 | .03 | .09 |
| Uncured EPDM | .080 | S.F. | .001 | .05 | .03 | .08 |
| Neoprene, 60 mil | .080 | S.F. | .001 | .11 | .03 | .14 |
| Uncured neoprene | .080 | S.F. | .001 | .10 | .03 | .13 |
| Self-curing neoprene | .080 | S.F. | .001 | .11 | .03 | .14 |
| PVC coated metal | .080 | S.F. | .001 | .07 | .03 | .10 |

# SECTION 6
# INTERIORS

| System Number | Title | Page |
|---|---|---|
| **6-04** | Drywall & Thincoat Wall | 126 |
| **6-08** | Drywall & Thincoat Ceiling | 128 |
| **6-12** | Plaster & Stucco Wall | 130 |
| **6-16** | Plaster & Stucco Ceiling | 132 |
| **6-18** | Suspended Ceiling | 134 |
| **6-20** | Interior Door | 136 |
| **6-24** | Closet Door | 138 |
| **6-30** | Commercial Door | 140 |
| **6-60** | Carpet | 142 |
| **6-64** | Flooring | 143 |
| **6-90** | Stairways | 144 |

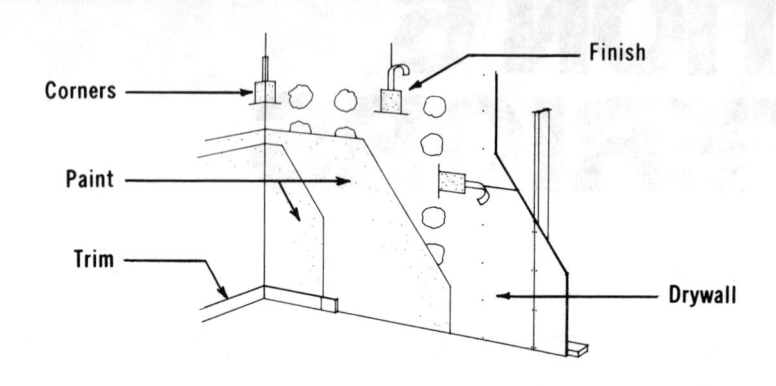

# 6 | INTERIORS — 04 | Drywall & Thincoat Wall Systems

| SYSTEM DESCRIPTION | QUAN. | UNIT | MAN HOURS | MAT. | INST. | TOTAL |
|---|---|---|---|---|---|---|
| **1/2" SHEETROCK, TAPED & FINISHED** | | | | | | |
| Drywall, 1/2" thick, standard | 1.000 | S.F. | .009 | .23 | .17 | .40 |
| Finish, taped & finished joints | 1.000 | S.F. | .008 | .01 | .16 | .17 |
| Corners, taped & finished, 32 L.F. per 12' x 12' room | .083 | L.F. | .001 | .01 | .02 | .03 |
| Painting, primer & 2 coats | 1.000 | S.F. | .009 | .10 | .16 | .26 |
| Trim, baseboard, painted | .125 | L.F. | .008 | .12 | .14 | .26 |
| TOTAL | | | .035 | .47 | .65 | 1.12 |
| **THINCOAT, SKIM-COAT, ON 1/2" BACKER DRYWALL** | | | | | | |
| Drywall, 1/2" thick, thincoat backer | 1.000 | S.F. | .009 | .23 | .17 | .40 |
| Thincoat plaster | 1.000 | S.F. | .013 | .08 | .25 | .33 |
| Corners, taped & finished, 32 L.F. per 12' x 12' room | .083 | L.F. | .001 | .01 | .02 | .03 |
| Painting, primer & 2 coats | 1.000 | S.F. | .009 | .10 | .16 | .26 |
| Trim, baseboard, painted | .125 | L.F. | .008 | .12 | .14 | .26 |
| TOTAL | | | .040 | .54 | .74 | 1.28 |
| **5/8" SHEETROCK, TAPED & FINISHED** | | | | | | |
| Drywall, 5/8" thick, standard | 1.000 | S.F. | .009 | .25 | .19 | .44 |
| Finish, taped & finished joints | 1.000 | S.F. | .008 | .01 | .16 | .17 |
| Corners, taped & finished, 32 L.F. per 12' x 12' room | .083 | L.F. | .001 | .01 | .02 | .03 |
| Painting, primer & 2 coats | 1.000 | S.F. | .009 | .10 | .16 | .26 |
| Trim, baseboard, painted | .125 | L.F. | .008 | .12 | .14 | .26 |
| TOTAL | | | .035 | .49 | .67 | 1.16 |

The costs in this system are based on a square foot of wall.
Do not deduct for openings.

| DESCRIPTION | QUAN. | UNIT | MAN HOURS | MAT. | INST. | TOTAL |
|---|---|---|---|---|---|---|
| | | | | | | |
| | | | | | | |
| | | | | | | |
| | | | | | | |
| | | | | | | |

# Drywall & Thincoat Wall Price Sheet

| | QUAN. | UNIT | MAN HOURS | COST PER S.F. MAT. | COST PER S.F. INST. | TOTAL |
|---|---|---|---|---|---|---|
| Drywall-sheetrock, 1/2" thick, standard | 1.000 | S.F. | .009 | .23 | .17 | .40 |
|     Fire resistant | 1.000 | S.F. | .009 | .26 | .18 | .44 |
|         Water resistant | 1.000 | S.F. | .009 | .30 | .17 | .47 |
|     5/8" thick, standard | 1.000 | S.F. | .009 | .25 | .19 | .44 |
|         Fire resistant | 1.000 | S.F. | .009 | .29 | .18 | .47 |
|         Water resistant | 1.000 | S.F. | .009 | .33 | .18 | .51 |
|     Drywall backer for thincoat system, 1/2" thick | 1.000 | S.F. | .009 | .23 | .17 | .40 |
|         5/8" thick | 1.000 | S.F. | .009 | .25 | .19 | .44 |
| Finish drywall, taped & finished | 1.000 | S.F. | | .01 | .16 | .17 |
|     Texture spray | 1.000 | S.F. | .011 | .10 | .21 | .31 |
|     Thincoat plaster, including tape | 1.000 | S.F. | .013 | .08 | .25 | .33 |
| Corners drywall, taped & finished, 32 L.F. per 4' x 4' room | .250 | L.F. | .003 | .01 | .07 | .08 |
|     6' x 6' room | .110 | L.F. | .001 | .01 | .03 | .04 |
|     10' x 10' room | .100 | L.F. | .001 | .01 | .02 | .03 |
|     12' x 12' room | .083 | L.F. | .001 | .01 | .02 | .03 |
|     16' x 16' room | .063 | L.F. | | .01 | .01 | .02 |
|     Thincoat system, 32 L.F. per 4' x 4' room | .250 | L.F. | .003 | .02 | .06 | .08 |
|     6' x 6' room | .110 | L.F. | .001 | .01 | .03 | .04 |
|     10' x 10' room | .100 | L.F. | .001 | .01 | .02 | .03 |
|     12' x 12' room | .083 | L.F. | .001 | .01 | .02 | .03 |
|     16' x 16' room | .063 | L.F. | | .01 | .01 | .02 |
| Painting, primer, & 1 coat | 1.000 | S.F. | .006 | .07 | .11 | .18 |
|     & 2 coats | 1.000 | S.F. | .009 | .10 | .16 | .26 |
|     Wallpaper, $9/double roll | 1.000 | S.F. | .013 | .18 | .23 | .41 |
|     $14/double roll | 1.000 | S.F. | .015 | .36 | .29 | .65 |
|     $20/double roll | 1.000 | S.F. | .018 | .78 | .35 | 1.13 |
|     Tile, ceramic adhesive thin set, 4-1/4" x 4-1/4" tiles | 1.000 | S.F. | .089 | 1.68 | 1.51 | 3.19 |
|     6" x 6" tiles | 1.000 | S.F. | .080 | 1.90 | 1.36 | 3.26 |
|     Pregrouted sheets | 1.000 | S.F. | .067 | 2.21 | 1.13 | 3.34 |
| Trim, painted or stained, baseboard | .125 | L.F. | .008 | .12 | .14 | .26 |
|     Base shoe | .125 | L.F. | .003 | .14 | .13 | .27 |
|     Chair rail | .125 | L.F. | .006 | .08 | .12 | .20 |
|     Cornice molding | .125 | L.F. | .006 | .05 | .11 | .16 |
|     Cove base, vinyl | .125 | L.F. | .003 | .05 | .06 | .11 |
| Paneling not including furring or trim | | | | | | |
|     Plywood, prefinished, 1/4" thick, 4' x 8' sheets, vert. grooves | | | | | | |
|         Birch faced, minimum | 1.000 | S.F. | .032 | .56 | .66 | 1.22 |
|         Average | 1.000 | S.F. | .038 | .87 | .78 | 1.65 |
|         Maximum | 1.000 | S.F. | .046 | 1.07 | .93 | 2.00 |
|         Mahogany, African | 1.000 | S.F. | .040 | 1.29 | .82 | 2.11 |
|         Philippine (lauan) | 1.000 | S.F. | .032 | .39 | .65 | 1.04 |
|         Oak or cherry, minimum | 1.000 | S.F. | .032 | 1.24 | .66 | 1.90 |
|         Maximum | 1.000 | S.F. | .040 | 2.12 | .82 | 2.94 |
|         Rosewood | 1.000 | S.F. | .050 | 8.97 | 1.03 | 10.00 |
|         Teak | 1.000 | S.F. | .040 | 2.35 | .82 | 3.17 |
|         Chestnut | 1.000 | S.F. | .043 | 3.37 | .87 | 4.24 |
|         Pecan | 1.000 | S.F. | .040 | 1.34 | .82 | 2.16 |
|         Walnut, minimum | 1.000 | S.F. | .032 | 2.01 | .66 | 2.67 |
|         Maximum | 1.000 | S.F. | .040 | 2.97 | .82 | 3.79 |

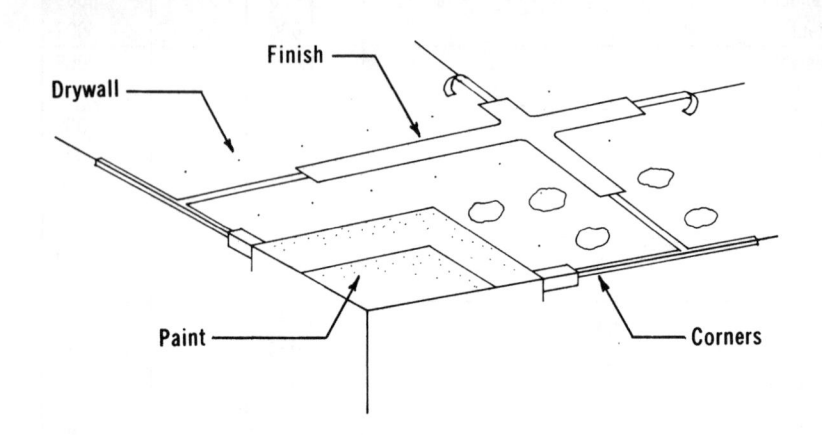

# 6 | INTERIORS  08 | Drywall & Thincoat Ceiling Systems

| SYSTEM DESCRIPTION | QUAN. | UNIT | MAN HOURS | MAT. | INST. | TOTAL |
|---|---|---|---|---|---|---|
| **1/2" SHEETROCK, TAPED & FINISHED** | | | | | | |
| Drywall, 1/2" thick, standard | 1.000 | S.F. | .009 | .23 | .25 | .48 |
| Finish, taped & finished | 1.000 | S.F. | .008 | .01 | .16 | .17 |
| Corners, taped & finished, 12' x 12' room | .330 | L.F. | .004 | .01 | .10 | .11 |
| Paint, primer & 2 coats | 1.000 | S.F. | .009 | .10 | .16 | .26 |
| TOTAL | | | .030 | .35 | .67 | 1.02 |
| **THINCOAT, SKIM COAT ON 1/2" BACKER DRYWALL** | | | | | | |
| Drywall, 1/2" thick, thincoat backer | 1.000 | S.F. | .009 | .23 | .25 | .48 |
| Thincoat plaster | 1.000 | S.F. | .013 | .08 | .25 | .33 |
| Corners, taped & finished, 12' x 12' room | .330 | L.F. | .004 | .01 | .10 | .11 |
| Paint, primer & 2 coats | 1.000 | S.F. | .009 | .10 | .16 | .26 |
| TOTAL | | | .035 | .42 | .76 | 1.18 |
| **WATER-RESISTANT SHEETROCK, 1/2" THICK, TAPED & FINISHED** | | | | | | |
| Drywall, 1/2" thick, water-resistant | 1.000 | S.F. | .009 | .30 | .25 | .55 |
| Finish, taped & finished | 1.000 | S.F. | .008 | .01 | .16 | .17 |
| Corners, taped & finished, 12' x 12' room | .330 | L.F. | .004 | .01 | .10 | .11 |
| Paint, primer & 2 coats | 1.000 | S.F. | .009 | .10 | .16 | .26 |
| TOTAL | | | .030 | .42 | .67 | 1.09 |
| **5/8" SHEETROCK, TAPED & FINISHED** | | | | | | |
| Drywall, 5/8" thick, standard | 1.000 | S.F. | .009 | .25 | .27 | .52 |
| Finish, taped & finished | 1.000 | S.F. | .008 | .01 | .16 | .17 |
| Corners, taped & finished, 12' x 12' room | .330 | L.F. | .004 | .01 | .10 | .11 |
| Paint, primer & 2 coats | 1.000 | S.F. | .009 | .10 | .16 | .26 |
| TOTAL | | | .030 | .37 | .69 | 1.06 |

The costs in this system are based on a square foot of ceiling basis.

| DESCRIPTION | QUAN. | UNIT | MAN HOURS | MAT. | INST. | TOTAL |
|---|---|---|---|---|---|---|
| | | | | | | |
| | | | | | | |
| | | | | | | |

# Drywall & Thincoat Ceiling Price Sheet

| | QUAN. | UNIT | MAN HOURS | COST PER S.F. MAT. | COST PER S.F. INST. | TOTAL |
|---|---|---|---|---|---|---|
| Drywall-sheetrock, 1/2" thick, standard | 1.000 | S.F. | .009 | .23 | .25 | .48 |
| Fire resistant | 1.000 | S.F. | .009 | .26 | .26 | .52 |
| Water resistant | 1.000 | S.F. | .009 | .30 | .25 | .55 |
| 5/8" thick, standard | 1.000 | S.F. | .009 | .25 | .27 | .52 |
| Fire resistant | 1.000 | S.F. | .009 | .29 | .26 | .55 |
| Water resistant | 1.000 | S.F. | .009 | .33 | .26 | .59 |
| Drywall backer for thincoat system, 1/2" thick | 1.000 | S.F. | .018 | .49 | .35 | .84 |
| 5/8" thick | 1.000 | S.F. | .018 | .51 | .37 | .88 |
| Finish drywall, taped & finished | 1.000 | S.F. | .008 | .01 | .16 | .17 |
| Texture spray | 1.000 | S.F. | .011 | .10 | .21 | .31 |
| Thincoat plaster | 1.000 | S.F. | .013 | .08 | .25 | .33 |
| Corners, taped & finished, 4' x 4' room | 1.000 | L.F. | .015 | .04 | .29 | .33 |
| 6' x 6' room | .670 | L.F. | .010 | .03 | .19 | .22 |
| 10' x 10' room | .400 | L.F. | .006 | .02 | .11 | .13 |
| 12' x 12' room | .330 | L.F. | .004 | .01 | .10 | .11 |
| 16' x 16' room | .187 | L.F. | .002 | .01 | .05 | .06 |
| Thincoat system, 4' x 4' room | 1.000 | L.F. | .013 | .08 | .25 | .33 |
| 6' x 6' room | .670 | L.F. | .008 | .05 | .17 | .22 |
| 10' x 10' room | .400 | L.F. | .005 | .03 | .10 | .13 |
| 12' x 12' room | .330 | L.F. | .004 | .03 | .08 | .11 |
| 16' x 16' room | .187 | L.F. | .002 | .01 | .05 | .06 |
| Painting, primer & 1 coat | 1.000 | S.F. | .006 | .07 | .11 | .18 |
| & 2 coats | 1.000 | S.F. | .009 | .10 | .16 | .26 |
| Wallpaper, $9/double roll | 1.000 | S.F. | .013 | .18 | .23 | .41 |
| $14/double roll | 1.000 | S.F. | .015 | .36 | .29 | .65 |
| $20/double roll | 1.000 | S.F. | .018 | .78 | .35 | 1.13 |
| Tile, ceramic adhesive thin set, 4-1/4" x 4-1/4" tiles | 1.000 | S.F. | .089 | 1.68 | 1.51 | 3.19 |
| 6" x 6" tiles | 1.000 | S.F. | .080 | 1.90 | 1.36 | 3.26 |
| Pregrouted sheets | 1.000 | S.F. | .067 | 2.21 | 1.13 | 3.34 |

# 6 | INTERIORS    12 | Plaster & Stucco Wall Systems

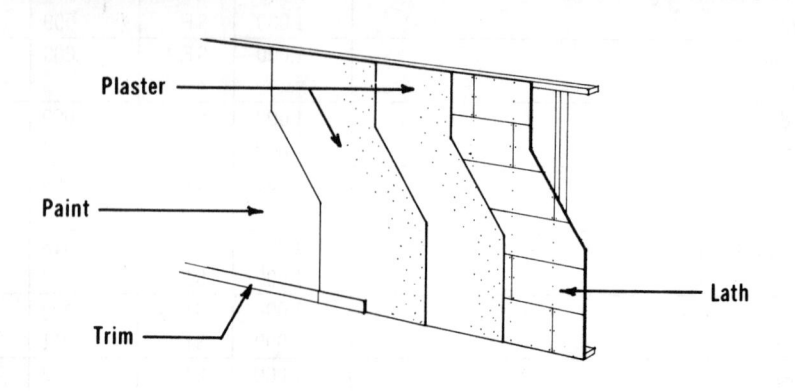

| SYSTEM DESCRIPTION | QUAN. | UNIT | MAN HOURS | MAT. | INST. | TOTAL |
|---|---|---|---|---|---|---|
| **PLASTER ON GYPSUM LATH** | | | | | | |
| Plaster, gypsum or perlite, 2 coats | 1.000 | S.F. | .042 | .32 | .80 | 1.12 |
| Lath, 3/8" gypsum | 1.000 | S.F. | .010 | .24 | .20 | .44 |
| Corners, expanded metal, 32 L.F. per 12' x 12' room | .083 | L.F. | .002 | .01 | .04 | .05 |
| Painting, primer & 2 coats | 1.000 | S.F. | .009 | .10 | .16 | .26 |
| Trim, baseboard, painted | .125 | L.F. | .008 | .12 | .14 | .26 |
| TOTAL | | | .071 | .79 | 1.34 | 2.13 |
| **PLASTER ON METAL LATH** | | | | | | |
| Plaster, gypsum or perlite, 2 coats | 1.000 | S.F. | .042 | .32 | .80 | 1.12 |
| Lath, 2.5 Lb. diamond, metal | 1.000 | S.F. | .010 | .20 | .20 | .40 |
| Corners, expanded metal, 32 L.F. per 12' x 12' room | .083 | L.F. | .002 | .01 | .04 | .05 |
| Painting, primer & 2 coats | 1.000 | S.F. | .009 | .10 | .16 | .26 |
| Trim, baseboard, painted | .125 | L.F. | .008 | .12 | .14 | .26 |
| TOTAL | | | .071 | .75 | 1.34 | 2.09 |
| **STUCCO ON METAL LATH** | | | | | | |
| Stucco, 2 coats | 1.000 | S.F. | .041 | .24 | .78 | 1.02 |
| Lath, 2.5 Lb. diamond, metal | 1.000 | S.F. | .010 | .20 | .20 | .40 |
| Corners, expanded metal, 32 L.F. per 12' x 12' room | .083 | L.F. | .002 | .01 | .04 | .05 |
| Painting, primer & 2 coats | 1.000 | S.F. | .009 | .10 | .16 | .26 |
| Trim, baseboard, painted | .125 | L.F. | .008 | .12 | .14 | .26 |
| TOTAL | | | .070 | .67 | 1.32 | 1.99 |

The costs in these systems are based on a per square foot of wall area.
Do not deduct for openings.

| DESCRIPTION | QUAN. | UNIT | MAN HOURS | MAT. | INST. | TOTAL |
|---|---|---|---|---|---|---|
| | | | | | | |
| | | | | | | |
| | | | | | | |
| | | | | | | |
| | | | | | | |

# Plaster & Stucco Wall Price Sheet

| | QUAN. | UNIT | MAN HOURS | COST PER S.F. MAT. | COST PER S.F. INST. | TOTAL |
|---|---|---|---|---|---|---|
| Plaster, gypsum or perlite, 2 coats | 1.000 | S.F. | .042 | .32 | .80 | 1.12 |
| 3 coats | 1.000 | S.F. | .051 | .44 | .97 | 1.41 |
| Lath, gypsum, standard, 3/8" thick | 1.000 | S.F. | .010 | .24 | .20 | .44 |
| 1/2" thick | 1.000 | S.F. | .013 | .29 | .24 | .53 |
| Fire resistant, 1/2" thick | 1.000 | S.F. | .013 | .34 | .24 | .58 |
| 5/8" thick | 1.000 | S.F. | .014 | .36 | .27 | .63 |
| Metal, diamond, 2.5 Lb. | 1.000 | S.F. | .010 | .20 | .20 | .40 |
| 3.4 Lb. | 1.000 | S.F. | .012 | .24 | .23 | .47 |
| Rib, 2.75 Lb. | 1.000 | S.F. | .012 | .28 | .22 | .50 |
| 3.4 Lb. | 1.000 | S.F. | .013 | .29 | .24 | .53 |
| Corners, expanded metal, 32 L.F. per 4' x 4' room | .250 | L.F. | .007 | .02 | .14 | .16 |
| 6' x 6' room | .110 | L.F. | .003 | .01 | .06 | .07 |
| 10' x 10' room | .100 | L.F. | .002 | .01 | .05 | .06 |
| 12' x 12' room | .083 | L.F. | .002 | .01 | .04 | .05 |
| 16' x 16' room | .063 | L.F. | .001 | .01 | .03 | .04 |
| Painting, primer & 1 coats | 1.000 | S.F. | .006 | .07 | .11 | .18 |
| Primer & 2 coats | 1.000 | S.F. | .009 | .10 | .16 | .26 |
| Wallpaper, $9/double roll | 1.000 | S.F. | .013 | .18 | .23 | .41 |
| $14/double roll | 1.000 | S.F. | .015 | .36 | .29 | .65 |
| $20/double roll | 1.000 | S.F. | .018 | .78 | .35 | 1.13 |
| Tile, ceramic thin set, 4-1/4" x 4-1/4" tiles | 1.000 | S.F. | .089 | 1.68 | 1.51 | 3.19 |
| 6" x 6" tiles | 1.000 | S.F. | .080 | 1.90 | 1.36 | 3.26 |
| Pregrouted sheets | 1.000 | S.F. | .067 | 2.21 | 1.13 | 3.34 |
| Trim, painted or stained, baseboard | .125 | L.F. | .008 | .12 | .14 | .26 |
| Base shoe | .125 | L.F. | .003 | .14 | .13 | .27 |
| Chair rail | .125 | L.F. | .006 | .08 | .12 | .20 |
| Cornice molding | .125 | L.F. | .006 | .05 | .11 | .16 |
| Cove base, vinyl | .125 | L.F. | .003 | .05 | .06 | .11 |
| Paneling not including furring or trim | | | | | | |
| Plywood, prefinished, 1/4" thick, 4' x 8' sheets, vert. grooves | | | | | | |
| Birch faced, minimum | 1.000 | S.F. | .032 | .56 | .66 | 1.22 |
| Average | 1.000 | S.F. | .038 | .87 | .78 | 1.65 |
| Maximum | 1.000 | S.F. | .046 | 1.07 | .93 | 2.00 |
| Mahogany, African | 1.000 | S.F. | .040 | 1.29 | .82 | 2.11 |
| Philippine (lauan) | 1.000 | S.F. | .032 | .39 | .65 | 1.04 |
| Oak or cherry, minimum | 1.000 | S.F. | .032 | 1.24 | .66 | 1.90 |
| Maximum | 1.000 | S.F. | .040 | 2.12 | .82 | 2.94 |
| Rosewood | 1.000 | S.F. | .050 | 8.97 | 1.03 | 10.00 |
| Teak | 1.000 | S.F. | .040 | 2.35 | .82 | 3.17 |
| Chestnut | 1.000 | S.F. | .043 | 3.37 | .87 | 4.24 |
| Pecan | 1.000 | S.F. | .040 | 1.34 | .82 | 2.16 |
| Walnut, minimum | 1.000 | S.F. | .032 | 2.01 | .66 | 2.67 |
| Maximum | 1.000 | S.F. | .040 | 2.97 | .82 | 3.79 |

# 6 | INTERIORS  16 | Plaster & Stucco Ceiling Systems

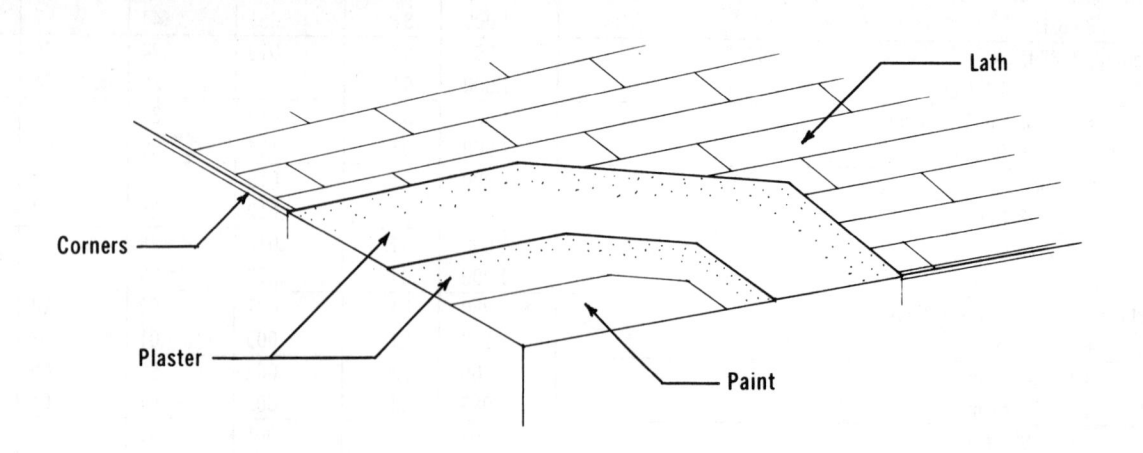

| SYSTEM DESCRIPTION | QUAN. | UNIT | MAN HOURS | COST PER S.F. MAT. | COST PER S.F. INST. | COST PER S.F. TOTAL |
|---|---|---|---|---|---|---|
| **PLASTER ON GYPSUM LATH** | | | | | | |
| Plaster, gypsum or perlite, | 1.000 | S.F. | .048 | .32 | .92 | 1.24 |
| Lath, 3/8" gypsum | 1.000 | S.F. | .050 | .24 | .97 | 1.21 |
| Corners, expanded metal, 12' x 12' room | .330 | L.F. | .009 | .03 | .18 | .21 |
| Painting, primer & 2 coats | 1.000 | S.F. | .009 | .10 | .16 | .26 |
| TOTAL | | | .116 | .69 | 2.23 | 2.92 |
| **PLASTER ON METAL LATH** | | | | | | |
| Plaster, gypsum or perlite, 2 coats | 1.000 | S.F. | .048 | .32 | .92 | 1.24 |
| Lath, 2.5 Lb. diamond, metal | 1.000 | S.F. | .012 | .20 | .23 | .43 |
| Corners, expanded metal, 12' x 12' room | .330 | L.F. | .009 | .03 | .18 | .21 |
| Painting, primer & 2 coats | 1.000 | S.F. | .009 | .10 | .16 | .26 |
| TOTAL | | | .078 | .65 | 1.49 | 2.14 |
| **STUCCO ON GYPSUM LATH** | | | | | | |
| Stucco, 2 coats | 1.000 | S.F. | .052 | .24 | .98 | 1.22 |
| Lath, 3/8" gypsum | 1.000 | S.F. | .050 | .24 | .97 | 1.21 |
| Corners, expanded metal, 12' x 12' room | .330 | L.F. | .009 | .03 | .18 | .21 |
| Painting, primer & 2 coats | 1.000 | S.F. | .009 | .10 | .16 | .26 |
| TOTAL | | | .120 | .61 | 2.29 | 2.90 |
| **STUCCO ON METAL LATH** | | | | | | |
| Stucco, 2 coats | 1.000 | S.F. | .052 | .24 | .98 | 1.22 |
| Lath, 2.5 Lb. diamond, metal | 1.000 | S.F. | .012 | .20 | .23 | .43 |
| Corners, expanded metal, 12' x 12' room | .330 | L.F. | .009 | .03 | .18 | .21 |
| Painting, primer & 2 coats | 1.000 | S.F. | .009 | .10 | .16 | .26 |
| TOTAL | | | .082 | .57 | 1.55 | 2.12 |

The costs in these systems are based on a square foot of ceiling area.

| DESCRIPTION | QUAN. | UNIT | MAN HOURS | COST PER S.F. MAT. | COST PER S.F. INST. | COST PER S.F. TOTAL |
|---|---|---|---|---|---|---|
| | | | | | | |
| | | | | | | |
| | | | | | | |

# Plaster & Stucco Ceiling Price Sheet

| | QUAN. | UNIT | MAN HOURS | COST PER S.F. MAT. | COST PER S.F. INST. | TOTAL |
|---|---|---|---|---|---|---|
| Plaster, gypsum or perlite, 2 coats | 1.000 | S.F. | .048 | .32 | .92 | 1.24 |
| 3 coats | 1.000 | S.F. | .051 | .44 | .97 | 1.41 |
| Lath, gypsum, standard, 3/8" thick | 1.000 | S.F. | .050 | .24 | .97 | 1.21 |
| 1/2" thick | 1.000 | S.F. | .051 | .26 | .99 | 1.25 |
| Fire resistant, 1/2" thick | 1.000 | S.F. | .053 | .34 | 1.01 | 1.35 |
| 5/8" thick | 1.000 | S.F. | .054 | .36 | 1.04 | 1.40 |
| Metal, diamond, 2.5 Lb. | 1.000 | S.F. | .012 | .20 | .23 | .43 |
| 3.4 Lb. | 1.000 | S.F. | .015 | .24 | .29 | .53 |
| Rib, 2.75 Lb. | 1.000 | S.F. | .012 | .28 | .22 | .50 |
| 3.4 Lb. | 1.000 | S.F. | .013 | .29 | .24 | .53 |
| Corners, expanded metal, 32 L.F. per 4' x 4' room | 1.000 | L.F. | .028 | .09 | .54 | .63 |
| 6' x 6' room | .670 | L.F. | .018 | .06 | .36 | .42 |
| 10' x 10' room | .400 | L.F. | .011 | .04 | .21 | .25 |
| 12' x 12' room | .330 | L.F. | .009 | .03 | .18 | .21 |
| 16' x 16' room | .187 | L.F. | .005 | .02 | .10 | .12 |
| Painting, primer & 1 coat | 1.000 | S.F. | .006 | .07 | .11 | .18 |
| Primer & 2 coats | 1.000 | S.F. | .009 | .10 | .16 | .26 |

6

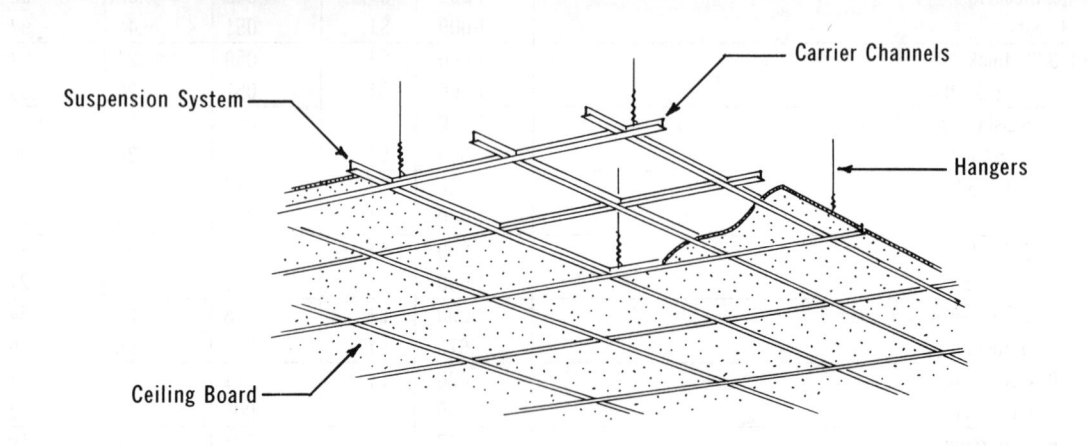

# 6 | INTERIORS   18 | Suspended Ceiling Systems

| SYSTEM DESCRIPTION | QUAN. | UNIT | MAN HOURS | MAT. | INST. | TOTAL |
|---|---|---|---|---|---|---|
| **2' X 2' GRID, FILM FACED FIBERGLASS, 5/8" THICK** | | | | | | |
| Suspension system, 2' x 2' grid, T bar | 1.000 | S.F. | .012 | .42 | .24 | .66 |
| Ceiling board, film faced fiberglass, 5/8" thick | 1.000 | S.F. | .012 | .34 | .23 | .57 |
| Carrier channels, 1-1/2" x 3/4" | 1.000 | S.F. | .017 | .12 | .33 | .45 |
| Hangers, #12 wire | 1.000 | S.F. | .001 | .02 | .03 | .05 |
| TOTAL | | | .042 | .90 | .83 | 1.73 |
| **2' X 4' GRID, FILM FACED FIBERGLASS, 5/8" THICK** | | | | | | |
| Suspension system, 2' x 4' grid, T bar | 1.000 | S.F. | .010 | .40 | .19 | .59 |
| Ceiling board, film faced fiberglass, 5/8" thick | 1.000 | S.F. | .012 | .34 | .23 | .57 |
| Carrier channels, 1-1/2" x 3/4" | 1.000 | S.F. | .017 | .12 | .33 | .45 |
| Hangers, #12 wire | 1.000 | S.F. | .001 | .02 | .03 | .05 |
| TOTAL | | | .040 | .88 | .78 | 1.66 |
| **2' X 2' GRID, MINERAL FIBER, REVEAL EDGE, 1" THICK** | | | | | | |
| Suspension system, 2' x 2' grid, T bar | 1.000 | S.F. | .012 | .42 | .24 | .66 |
| Ceiling board, mineral fiber, reveal edge, 1" thick | 1.000 | S.F. | .013 | 1.32 | .26 | 1.58 |
| Carrier channels, 1-1/2" x 3/4" | 1.000 | S.F. | .017 | .12 | .33 | .45 |
| Hangers, #12 wire | 1.000 | S.F. | .001 | .02 | .03 | .05 |
| TOTAL | | | .043 | 1.88 | .86 | 2.74 |
| **2' X 4' GRID, MINERAL FIBER, REVEAL EDGE, 1" THICK** | | | | | | |
| Suspension system, 2' x 4' grid, T bar | 1.000 | S.F. | .010 | .40 | .19 | .59 |
| Ceiling board, mineral fiber, reveal edge, 1" thick | 1.000 | S.F. | .013 | 1.32 | .26 | 1.58 |
| Carrier channels, 1-1/2" x 3/4" | 1.000 | S.F. | .017 | .12 | .33 | .45 |
| Hangers, #12 wire | 1.000 | S.F. | .001 | .02 | .03 | .05 |
| TOTAL | | | .041 | 1.86 | .81 | 2.67 |

# Suspended Ceiling Price Sheet

| | QUAN. | UNIT | MAN HOURS | MAT. | INST. | TOTAL |
|---|---|---|---|---|---|---|
| Suspension systems, T bar, 2' x 2' grid | 1.000 | S.F. | .012 | .42 | .24 | .66 |
| 2' x 4' grid | 1.000 | S.F. | .010 | .40 | .19 | .59 |
| Concealed Z bar, 12" module | 1.000 | S.F. | .015 | .51 | .30 | .81 |
| Ceiling boards, fiberglass, film faced, 2' x 2' or 2' x 4', 5/8" thick | 1.000 | S.F. | .012 | .34 | .23 | .57 |
| 3/4" thick | 1.000 | S.F. | .016 | .45 | .31 | .76 |
| 3" thick thermal R11 | 1.000 | S.F. | .016 | .81 | .32 | 1.13 |
| Glass cloth faced, 3/4" thick | 1.000 | S.F. | .016 | .83 | .31 | 1.14 |
| 1" thick | 1.000 | S.F. | .016 | .94 | .31 | 1.25 |
| 1-1/2" thick, nubby face | 1.000 | S.F. | .016 | 1.29 | .31 | 1.60 |
| Mineral fiber boards, 5/8" thick, aluminum face 2' x 2' | 1.000 | S.F. | .013 | .75 | .26 | 1.01 |
| 2' x 4' | 1.000 | S.F. | .012 | .75 | .24 | .99 |
| Standard faced, 2' x 2' or 2' x 4' | 1.000 | S.F. | .012 | .36 | .23 | .59 |
| Plastic coated face, 2' x 2' or 2' x 4' | 1.000 | S.F. | .020 | .40 | .39 | .79 |
| Fire rated, 2 hour rating, 5/8" thick | 1.000 | S.F. | .012 | .44 | .23 | .67 |
| Reveal edge, 2' x 2' or 2' x 4', painted, 1" thick | 1.000 | S.F. | .013 | 1.32 | .26 | 1.58 |
| 2" thick | 1.000 | S.F. | .015 | 2.27 | .28 | 2.55 |
| 2-1/2" thick | 1.000 | S.F. | .016 | 2.60 | .31 | 2.91 |
| 3" thick | 1.000 | S.F. | .018 | 2.88 | .35 | 3.23 |
| Luminous panels, prismatic, acrylic | 1.000 | S.F. | .020 | .80 | .39 | 1.19 |
| Polystyrene | 1.000 | S.F. | .020 | .46 | .39 | .85 |
| Flat or ribbed, acrylic | 1.000 | S.F. | .020 | 1.29 | .39 | 1.68 |
| Polystyrene | 1.000 | S.F. | .020 | .75 | .39 | 1.14 |
| Drop pan, white, acrylic | 1.000 | S.F. | .020 | 3.06 | .39 | 3.45 |
| Polystyrene | 1.000 | S.F. | .020 | 2.48 | .39 | 2.87 |
| Carrier channels, 4'-0" on center, 3/4" x 1-1/2" | 1.000 | S.F. | .017 | .12 | .33 | .45 |
| 1-1/2" x 3-1/2" | 1.000 | S.F. | .017 | .36 | .34 | .70 |
| Hangers, #12 wire | 1.000 | S.F. | .001 | .02 | .03 | .05 |

135

# 6 | INTERIORS  20 | Interior Door Systems

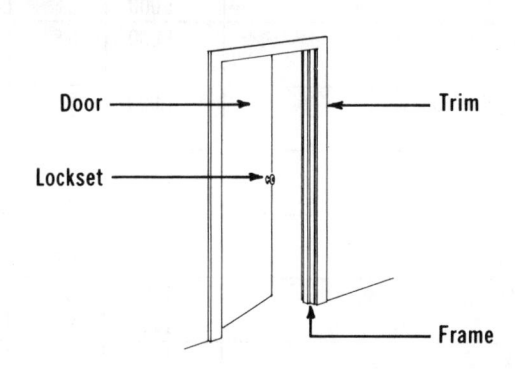

| SYSTEM DESCRIPTION | QUAN. | UNIT | MAN HOURS | MAT. | INST. | TOTAL |
|---|---|---|---|---|---|---|
| **LAUAN, FLUSH DOOR, HOLLOW CORE** | | | | | | |
| Door, flush, lauan, hollow core, 2'-8" wide x 6'-8" high | 1.000 | Ea. | .889 | 20.90 | 18.10 | 39.00 |
| Frame, pine, 4-5/8" jamb | 17.000 | L.F. | .731 | 23.97 | 14.79 | 38.76 |
| Trim, casing, painted | 34.000 | L.F. | 1.938 | 20.74 | 37.06 | 57.80 |
| Butt hinges, chrome, 3-1/2" x 3-1/2" | 1.500 | Pr. | | 25.05 | | 25.05 |
| Lockset, passage | 1.000 | Ea. | .500 | 8.03 | 13.97 | 22.00 |
| Paint, door & frame, primer & 2 coats | 2.000 | FACE | 3.200 | 8.80 | 59.20 | 68.00 |
| TOTAL | | | 7.258 | 107.49 | 143.12 | 250.61 |
| **BIRCH, FLUSH DOOR, HOLLOW CORE** | | | | | | |
| Door, flush, birch, hollow core, 2'-8" wide x 6'-8" high | 1.000 | Ea. | .889 | 29.70 | 18.30 | 48.00 |
| Frame, pine, 4-5/8" jamb | 17.000 | L.F. | .731 | 23.97 | 14.79 | 38.76 |
| Trim, casing, painted | 34.000 | L.F. | 1.938 | 20.74 | 37.06 | 57.80 |
| Butt hinges, chrome, 3-1/2" x 3-1/2" | 1.500 | Pr. | | 25.05 | | 25.05 |
| Lockset, passage | 1.000 | Ea. | .500 | 8.03 | 13.97 | 22.00 |
| Paint, door & frame, primer & 2 coats | 2.000 | FACE | 3.200 | 8.80 | 59.20 | 68.00 |
| TOTAL | | | 7.258 | 116.29 | 143.32 | 259.61 |
| **RAISED PANEL, SOLID, PINE DOOR** | | | | | | |
| Door, pine, raised panel, 2'-8" wide x 6'-8" high | 1.000 | Ea. | .889 | 103.40 | 16.60 | 120.00 |
| Frame, pine, 4-5/8" jamb | 17.000 | L.F. | .731 | 23.97 | 14.79 | 38.76 |
| Trim, casing, painted | 34.000 | L.F. | 1.938 | 20.74 | 37.06 | 57.80 |
| Butt hinges, bronze, 3-1/2" x 3-1/2" | 1.500 | Pr. | | 55.50 | | 55.50 |
| Lockset, passage | 1.000 | Ea. | .500 | 8.03 | 13.97 | 22.00 |
| Paint, door & frame, primer & 2 coats | 2.000 | FACE | 4.580 | 9.36 | 84.64 | 94.00 |
| TOTAL | | | 8.638 | 221.00 | 167.06 | 388.06 |

The costs in these systems are on a cost each door basis.

| DESCRIPTION | QUAN. | UNIT | MAN HOURS | MAT. | INST. | TOTAL |
|---|---|---|---|---|---|---|
| | | | | | | |
| | | | | | | |
| | | | | | | |
| | | | | | | |
| | | | | | | |

# Interior Door Price Sheet

| | QUAN. | UNIT | MAN HOURS | MAT. | INST. | TOTAL |
|---|---|---|---|---|---|---|
| Door, hollow core, lauan 1-3/8" thick, 6'-8" high x 1'-6" wide | 1.000 | Ea. | .842 | 17.60 | 17.40 | 35.00 |
| 2'-0" wide | 1.000 | Ea. | .889 | 17.60 | 18.40 | 36.00 |
| 2'-6" wide | 1.000 | Ea. | .889 | 19.80 | 18.20 | 38.00 |
| 2'-8" wide | 1.000 | Ea. | .889 | 20.90 | 18.10 | 39.00 |
| 3'-0" wide | 1.000 | Ea. | .941 | 22.00 | 19.00 | 41.00 |
| Birch 1-3/8" thick, 6'-8" high x 1'-6" wide | 1.000 | Ea. | .842 | 22.00 | 17.00 | 39.00 |
| 2'-0" wide | 1.000 | Ea. | .889 | 25.30 | 18.70 | 44.00 |
| 2'-6" wide | 1.000 | Ea. | .889 | 28.60 | 18.40 | 47.00 |
| 2'-8" wide | 1.000 | Ea. | .889 | 29.70 | 18.30 | 48.00 |
| 3'-0" wide | 1.000 | Ea. | .941 | 31.90 | 19.10 | 51.00 |
| Louvered pine 1-3/8" thick, 6'-8" high x 1'-6" wide | 1.000 | Ea. | .842 | 49.50 | 17.50 | 67.00 |
| 2'-0" wide | 1.000 | Ea. | .889 | 60.50 | 18.50 | 79.00 |
| 2'-6" wide | 1.000 | Ea. | .889 | 79.20 | 17.80 | 97.00 |
| 2'-8" wide | 1.000 | Ea. | .889 | 81.40 | 18.60 | 100.00 |
| 3'-0" wide | 1.000 | Ea. | .941 | 85.80 | 19.20 | 105.00 |
| Paneled pine 1-3/8" thick, 6'-8" high x 1'-6" wide | 1.000 | Ea. | .842 | 71.50 | 17.50 | 89.00 |
| 2'-0" wide | 1.000 | Ea. | .889 | 93.50 | 16.50 | 110.00 |
| 2'-6" wide | 1.000 | Ea. | .889 | 99.00 | 16.00 | 115.00 |
| 2'-8" wide | 1.000 | Ea. | .889 | 103.40 | 16.60 | 120.00 |
| 3'-0" wide | 1.000 | Ea. | .941 | 115.50 | 19.50 | 135.00 |
| Frame pine painted, 1'-6" thru 2'-0" wide door, 3-5/8" deep | 16.000 | L.F. | .688 | 17.92 | 14.08 | 32.00 |
| 4-5/8" deep | 16.000 | L.F. | .688 | 22.56 | 13.92 | 36.48 |
| 5-5/8" deep | 16.000 | L.F. | .688 | 26.88 | 14.08 | 40.96 |
| 2'-6" thru 3'0" wide door, 3-5/8" deep | 17.000 | L.F. | .731 | 19.04 | 14.96 | 34.00 |
| 4-5/8" deep | 17.000 | L.F. | .731 | 23.97 | 14.79 | 38.76 |
| 5-5/8" deep | 17.000 | L.F. | .731 | 28.56 | 14.96 | 43.52 |
| Trim, casing, painted, both sides, 1'-6" thru 2'-6" wide door | 32.000 | L.F. | 1.824 | 19.52 | 34.88 | 54.40 |
| 2'-6" thru 3'-0" wide door | 34.000 | L.F. | 1.938 | 20.74 | 37.06 | 57.80 |
| Butt hinges 3-1/2" x 3-1/2", steel plated, chrome | 1.500 | Pr. | | 25.05 | | 25.05 |
| Bronze | 1.500 | Pr. | | 55.50 | | 55.50 |
| Locksets, minimum | 1.000 | Ea. | .500 | 8.03 | 13.97 | 22.00 |
| Maximum | 1.000 | Ea. | 1.000 | 25.30 | 28.70 | 54.00 |
| Paint or stain both sides, flush door, 1'-6" thru 2'-0" wide | 2.000 | FACE | 2.560 | 7.04 | 47.36 | 54.40 |
| 2'-6" thru 3'-0" wide | 2.000 | FACE | 3.200 | 8.80 | 59.20 | 68.00 |
| Louvered door, 1'-6" thru 2'-0" wide | 2.000 | FACE | 3.664 | 7.49 | 67.71 | 75.20 |
| 2'-6" thru 3'-0" wide | 2.000 | FACE | 4.580 | 9.36 | 84.64 | 94.00 |
| Paneled door, 1'-6" thru 2'-0" wide | 2.000 | FACE | 3.664 | 7.49 | 67.71 | 75.20 |
| 2'-6" thru 3'-0" wide | 2.000 | FACE | 4.580 | 9.36 | 84.64 | 94.00 |

6

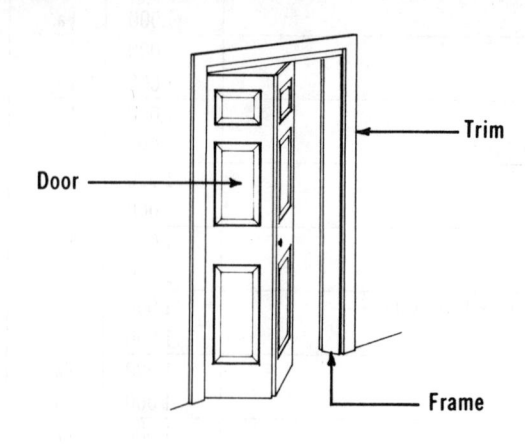

# 6 | INTERIORS  24 | Closet Door Systems

| SYSTEM DESCRIPTION | QUAN. | UNIT | MAN HOURS | MAT. | INST. | TOTAL |
|---|---|---|---|---|---|---|
| **BI-PASSING, FLUSH, LAUAN, HOLLOW CORE, 4'-0" X 6'-8"** | | | | | | |
| Door, flush, lauan, hollow core, 4'-0" x 6'-8" opening | 1.000 | Ea. | 1.330 | 68.20 | 27.80 | 96.00 |
| Frame, pine, 4-5/8" jamb | 18.000 | L.F. | .774 | 25.38 | 15.66 | 41.04 |
| Trim, both sides, casing, painted | 36.000 | L.F. | 2.052 | 21.96 | 39.24 | 61.20 |
| Paint, door & frame, primer & 2 coats | 2.000 | FACE | 3.200 | 8.80 | 59.20 | 68.00 |
| TOTAL | | | 7.356 | 124.34 | 141.90 | 266.24 |
| **BI-PASSING, FLUSH, BIRCH, HOLLOW CORE, 6'-0" X 6'-8"** | | | | | | |
| Door, flush, birch, hollow core, 6'-0" x 6'-8" opening | 1.000 | Ea. | 1.600 | 115.50 | 34.50 | 150.00 |
| Frame, pine, 4-5/8" jamb | 19.000 | L.F. | .817 | 26.79 | 16.53 | 43.32 |
| Trim, both sides, casing, painted | 38.000 | L.F. | 2.166 | 23.18 | 41.42 | 64.60 |
| Paint, door & frame, primer & 2 coats | 2.000 | FACE | 4.000 | 11.00 | 74.00 | 85.00 |
| TOTAL | | | 8.583 | 176.47 | 166.45 | 342.92 |
| **BI-FOLD, PINE, PANELED, 3'-0" X 6'-8"** | | | | | | |
| Door, pine, paneled, 3'-0" x 6'-8" opening | 1.000 | Ea. | 1.230 | 96.80 | 23.20 | 120.00 |
| Frame, pine, 4-5/8" jamb | 17.000 | L.F. | .731 | 23.97 | 14.79 | 38.76 |
| Trim, both sides, casing, painted | 34.000 | L.F. | 1.938 | 20.74 | 37.06 | 57.80 |
| Paint, door & frame, primer & 2 coats | 2.000 | FACE | 4.580 | 9.36 | 84.64 | 94.00 |
| TOTAL | | | 8.479 | 150.87 | 159.69 | 310.56 |
| **BI-FOLD, PINE, LOUVERED, 6'-0" X 6'-8"** | | | | | | |
| Door, pine, louvered, 6'-0" x 6'-8" opening | 1.000 | Ea. | 1.600 | 187.00 | 33.00 | 220.00 |
| Frame, pine, 4-5/8" jamb | 19.000 | L.F. | .817 | 26.79 | 16.53 | 43.32 |
| Trim, both sides, casing, painted | 38.000 | L.F. | 2.166 | 23.18 | 41.42 | 64.60 |
| Paint, door & frame, primer & 2 coats | 2.000 | FACE | 5.725 | 11.70 | 105.80 | 117.50 |
| TOTAL | | | 10.308 | 248.67 | 196.75 | 445.42 |

The costs in this system are on a cost each door basis.

| DESCRIPTION | QUAN. | UNIT | MAN HOURS | MAT. | INST. | TOTAL |
|---|---|---|---|---|---|---|
| | | | | | | |
| | | | | | | |
| | | | | | | |

# Closet Door Price Sheet

| | QUAN. | UNIT | MAN HOURS | MAT. | INST. | TOTAL |
|---|---|---|---|---|---|---|
| Doors, bi-passing, pine, louvered, 4'-0" x 6'-8" opening | 1.000 | Ea. | 1.330 | 121.00 | 29.00 | 150.00 |
| 6'-0" x 6'-8" opening | 1.000 | Ea. | 1.600 | 154.00 | 31.00 | 185.00 |
| Paneled, 4'-0" x 6'-8" opening | 1.000 | Ea. | 1.330 | 236.50 | 28.50 | 265.00 |
| 6'-0" x 6'-8" opening | 1.000 | Ea. | 1.600 | 286.00 | 34.00 | 320.00 |
| Flush, birch, hollow core, 4'-0" x 6'-8" opening | 1.000 | Ea. | 1.330 | 86.90 | 28.10 | 115.00 |
| 6'-0" x 6'-8" opening | 1.000 | Ea. | 1.600 | 115.50 | 34.50 | 150.00 |
| Flush, lauan, hollow core, 4'-0" x 6'-8" opening | 1.000 | Ea. | 1.330 | 68.20 | 27.80 | 96.00 |
| 6'-0" x 6'-8" opening | 1.000 | Ea. | 1.600 | 89.10 | 30.90 | 120.00 |
| Bi-fold, pine, louvered, 3'-0" x 6'-8" opening | 1.000 | Ea. | 1.230 | 96.80 | 23.20 | 120.00 |
| 6'-0" x 6'-8" opening | 1.000 | Ea. | 1.600 | 187.00 | 33.00 | 220.00 |
| Paneled, 3'-0" x 6'-8" opening | 1.000 | Ea. | 1.230 | 96.80 | 23.20 | 120.00 |
| 6'-0" x 6'-8" opening | 1.000 | Ea. | 1.600 | 187.00 | 33.00 | 220.00 |
| Flush, birch, hollow core, 3'-0" x 6'-8" opening | 1.000 | Ea. | 1.230 | 53.90 | 25.10 | 79.00 |
| 6'-0" x 6'-8" opening | 1.000 | Ea. | 1.600 | 97.90 | 32.10 | 130.00 |
| Flush, lauan, hollow core, 3'-0" x 6'8" opening | 1.000 | Ea. | 1.230 | 143.00 | 27.00 | 170.00 |
| 6'-0" x 6'-8" opening | 1.000 | Ea. | 1.600 | 286.00 | 34.00 | 320.00 |
| Frame pine, 3'-0" door, 3-5/8" deep | 17.000 | L.F. | .731 | 19.04 | 14.96 | 34.00 |
| 4-5/8" deep | 17.000 | L.F. | .731 | 23.97 | 14.79 | 38.76 |
| 5-5/8" deep | 17.000 | L.F. | .731 | 28.56 | 14.96 | 43.52 |
| 4'-0" door, 3-5/8" deep | 18.000 | L.F. | .774 | 20.16 | 15.84 | 36.00 |
| 4-5/8" deep | 18.000 | L.F. | .774 | 25.38 | 15.66 | 41.04 |
| 5-5/8" deep | 18.000 | L.F. | .774 | 30.24 | 15.84 | 46.08 |
| 6'-0" door, 3-5/8" deep | 19.000 | L.F. | .817 | 21.28 | 16.72 | 38.00 |
| 4-5/8" deep | 19.000 | L.F. | .817 | 26.79 | 16.53 | 43.32 |
| 5-5/8" deep | 19.000 | L.F. | .817 | 31.92 | 16.72 | 48.64 |
| Trim both sides, painted 3'-0" x 6'-8" door | 34.000 | L.F. | 1.938 | 20.74 | 37.06 | 57.80 |
| 4'-0" x 6'-8" door | 36.000 | L.F. | 2.052 | 21.96 | 39.24 | 61.20 |
| 6'-0" x 6'-8" door | 38.000 | L.F. | 2.166 | 23.18 | 41.42 | 64.60 |
| Paint or stain both sides, flush door & frame, 3'-0" x 6'-8" opening | 2.000 | FACE | 2.400 | 6.60 | 44.40 | 51.00 |
| 4'-0" x 6'-8" opening | 2.000 | FACE | 3.200 | 8.80 | 59.20 | 68.00 |
| 6'-0" x 6'-8" opening | 2.000 | FACE | 4.000 | 11.00 | 74.00 | 85.00 |
| Paneled door & frame, 3'-0" x 6'-8" opening | 2.000 | FACE | 3.435 | 7.02 | 63.48 | 70.50 |
| 4'-0" x 6'-8" opening | 2.000 | FACE | 4.580 | 9.36 | 84.64 | 94.00 |
| 6'-0" x 6'-8" opening | 2.000 | FACE | 5.725 | 11.70 | 105.80 | 117.50 |
| Louvered door & frame, 3'-0" x 6'-8" opening | 2.000 | FACE | 3.435 | 7.02 | 63.48 | 70.50 |
| 4'-0" x 6'-8" opening | 2.000 | FACE | 4.580 | 9.36 | 84.64 | 94.00 |
| 6'-0" x 6'-8" opening | 2.000 | FACE | 5.725 | 11.70 | 105.80 | 117.50 |

6

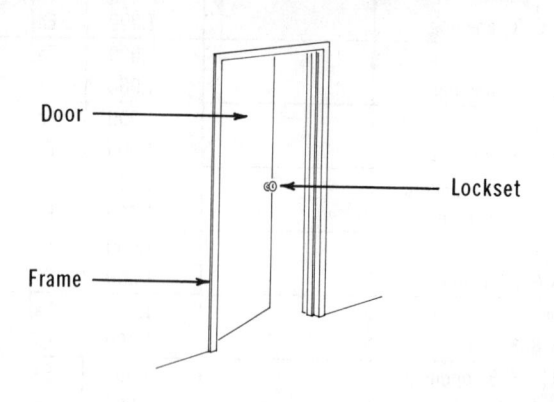

# 6 | INTERIORS  30 | Commercial Door Systems

| SYSTEM DESCRIPTION | QUAN. | UNIT | MAN HOURS | MAT. | INST. | TOTAL |
|---|---|---|---|---|---|---|
| **HOLLOW METAL DOOR, 3'-0" X 7'-0" X 1-3/4" THICK** | | | | | | |
| Door, interior, flush, 18 ga., 3'-0" x 7'-0" | 1.000 | Ea. | 1.142 | 176.00 | 24.00 | 200.00 |
| Frame, wrap-around, 18 ga., 4-3/4" x 3'-0" x 7'-0" | 1.000 | Ea. | 1.066 | 55.00 | 22.00 | 77.00 |
| Butt hinges, average frequency, steel, ball bearing | 1.500 | Pr. | | 28.43 | | 28.43 |
| Lockset, passage | 1.000 | Ea. | .800 | 74.80 | 15.20 | 90.00 |
| Painting, primer & 2 coats, per side | 2.000 | FACE | 3.200 | 8.80 | 59.20 | 68.00 |
| TOTAL | | | 6.208 | 343.03 | 120.40 | 463.43 |
| **FIRE RATED, METAL DOOR, 3"-0" X 7'-0" X 1-3/4" THICK** | | | | | | |
| Door, interior, flush, fire rated, "A" label, 3'-0" x 7'-0" | 1.000 | Ea. | 1.142 | 231.00 | 24.00 | 255.00 |
| Frame, fire rated, wrap around, 18 ga., 4-3/4" x 3'-0" x 7'-0" | 1.000 | Ea. | 1.066 | 71.50 | 21.50 | 93.00 |
| Butt hinges, average frequency, steel, ball bearing | 1.500 | Pr. | | 28.43 | | 28.43 |
| Lockset, passage | 1.000 | Ea. | .800 | 74.80 | 15.20 | 90.00 |
| Painting, primer & 2 coats, per side | 2.000 | FACE | 3.200 | 8.80 | 59.20 | 68.00 |
| TOTAL | | | 6.208 | 414.53 | 119.90 | 534.43 |
| **WOOD DOOR, METAL FRAME, 3'-0" X 7'-0" X 1-3/4" THICK** | | | | | | |
| Door, interior, flush wood, birch, 3'-0" x 7'-0" | 1.000 | Ea. | 1.142 | 105.60 | 24.40 | 130.00 |
| Frame, metal, wrap around, 18 ga., 4-3/4" x 3'-0" x 7'-0" | 1.000 | Ea. | 1.066 | 55.00 | 22.00 | 77.00 |
| Butt hinges, average frequency, steel, ball bearing | 1.500 | Pr. | | 28.43 | | 28.43 |
| Lockset passage | 1.000 | Ea. | .800 | 74.80 | 15.20 | 90.00 |
| Stain, 2 coats & sealer, per side | 2.000 | FACE | .798 | 7.14 | 14.28 | 21.42 |
| TOTAL | | | 3.005 | 270.97 | 75.88 | 346.85 |

| DECSRIPTION | QUAN. | UNIT | MAN HOURS | MAT. | INST. | TOTAL |
|---|---|---|---|---|---|---|
| | | | | | | |
| | | | | | | |
| | | | | | | |
| | | | | | | |
| | | | | | | |

# Commercial Door Price Sheet

| | QUAN. | UNIT | MAN HOURS | MAT. | INST. | TOTAL |
|---|---|---|---|---|---|---|
| Door, metal, flush, 20 ga., 1-3/4" x 7'-0" x 2'-6" | 1.000 | Ea. | 1.066 | 143.00 | 22.00 | 165.00 |
| 2'-8" | 1.000 | Ea. | 1.066 | 154.00 | 21.00 | 175.00 |
| 3'-0" | 1.000 | Ea. | 1.142 | 165.00 | 25.00 | 190.00 |
| 18 ga., 1-3/4" x 7'-0" x 2'-6" | 1.000 | Ea. | 1.066 | 170.50 | 19.50 | 190.00 |
| 2'-8" | 1.000 | Ea. | 1.066 | 176.00 | 24.00 | 200.00 |
| 3'-0" | 1.000 | Ea. | 1.142 | 176.00 | 24.00 | 200.00 |
| Fire rated, "A" label, 1-3/4" x 7'-0" x 2'-6" | 1.000 | Ea. | 1.066 | 203.50 | 21.50 | 225.00 |
| 2'-8" | 1.000 | Ea. | 1.066 | 214.50 | 20.50 | 235.00 |
| 3'-0" | 1.000 | Ea. | 1.042 | 231.00 | 24.00 | 255.00 |
| "B" label, 1-3/4" x 7'-0" x 2'-6" | 1.000 | Ea. | 1.066 | 181.50 | 23.50 | 205.00 |
| 2'-8" | 1.000 | Ea. | 1.066 | 187.00 | 23.00 | 210.00 |
| 3'-0" | 1.000 | Ea. | 1.042 | 192.50 | 22.50 | 215.00 |
| Wood, birch faced, 1-3/4" x 7'-0" x 2'-6" | 1.000 | Ea. | 1.066 | 97.90 | 22.10 | 120.00 |
| 2'-8" | 1.000 | Ea. | 1.066 | 101.20 | 23.80 | 125.00 |
| 3'-0" | 1.000 | Ea. | 1.042 | 105.60 | 24.40 | 130.00 |
| Oak faced, 1-3/4" x 7'-0" x 2'-6" | 1.000 | Ea. | 1.066 | 115.50 | 19.50 | 135.00 |
| 2'-8" | 1.000 | Ea. | 1.066 | 121.00 | 24.00 | 145.00 |
| 3'-0" | 1.000 | Ea. | 1.042 | 126.50 | 23.50 | 150.00 |
| Walnut faced, 1-3/4" x 7'-0" x 2'-6" | 1.000 | Ea. | 1.066 | 198.00 | 22.00 | 220.00 |
| 2'-8" | 1.000 | Ea. | 1.066 | 209.00 | 21.00 | 230.00 |
| 3'-0" | 1.000 | Ea. | 1.042 | 231.00 | 24.00 | 255.00 |
| Frame, metal, wrap-around, 18 ga., 4-3/4" x 7'-0" x 2'-6" | 1.000 | Ea. | 1.000 | 53.90 | 20.10 | 74.00 |
| 2'-8" | 1.000 | Ea. | 1.000 | 55.00 | 21.00 | 76.00 |
| 3'-0" | 1.000 | Ea. | 1.066 | 55.00 | 22.00 | 77.00 |
| 5'-0" | 1.000 | Ea. | 1.142 | 59.40 | 23.60 | 83.00 |
| 5'-4" | 1.000 | Ea. | 1.142 | 59.40 | 23.60 | 83.00 |
| 6'-0" | 1.000 | Ea. | 1.142 | 64.90 | 23.10 | 88.00 |
| 16 ga., 4-3/4" x 7'-0" x 2'-6" | 1.000 | Ea. | 1.000 | 63.80 | 20.20 | 84.00 |
| 2'-8" | 1.000 | Ea. | 1.000 | 66.00 | 21.00 | 87.00 |
| 3'-0" | 1.000 | Ea. | 1.066 | 66.00 | 22.00 | 88.00 |
| 5'-0" | 1.000 | Ea. | 1.142 | 71.50 | 23.50 | 95.00 |
| 5'-4" | 1.000 | Ea. | 1.142 | 73.70 | 23.30 | 97.00 |
| 6'-0" | 1.000 | Ea. | 1.142 | 77.00 | 23.00 | 100.00 |
| Fire rated, 4-3/4" x 7'-0" x 2'-6" | 1.000 | Ea. | 1.000 | 70.40 | 20.60 | 91.00 |
| 2'-8" | 1.000 | Ea. | 1.000 | 70.40 | 20.60 | 91.00 |
| 3'-0" | 1.000 | Ea. | 1.066 | 71.50 | 21.50 | 93.00 |
| 5'-0" | 1.000 | Ea. | 1.142 | 77.00 | 23.00 | 100.00 |
| 5'-4" | 1.000 | Ea. | 1.142 | 79.20 | 25.80 | 105.00 |
| 6'-0" | 1.000 | Ea. | 1.142 | 81.40 | 23.60 | 105.00 |
| Butt hinges, 4-1/2" x 4-1/2" high frequency, steel, ball bearing | 1.500 | Pr. | | 75.00 | | 75.00 |
| Bronze, ball bearing | 1.500 | Pr. | | 133.50 | | 133.50 |
| Average frequency, steel, ball bearing | 1.500 | Pr. | | 28.43 | | 28.43 |
| Bronze, ball bearing | 1.500 | Pr. | | 78.00 | | 78.00 |
| Low frequency, steel plain bearing | 1.500 | Pr. | | 12.90 | | 12.90 |
| Bronze plain bearing | 1.500 | Pr. | | 51.00 | | 51.00 |
| Lockset, heavy duty, passage | 1.000 | Ea. | .800 | 74.80 | 15.20 | 90.00 |
| Inner office | 1.000 | Ea. | .800 | 95.70 | 14.30 | 110.00 |
| Apartment | 1.000 | Ea. | .800 | 121.00 | 14.00 | 135.00 |
| Standard duty, passage | 1.000 | Ea. | .800 | 23.10 | 15.90 | 39.00 |
| Inner office | 1.000 | Ea. | .800 | 47.35 | 15.65 | 63.00 |
| Apartment | 1.000 | Ea. | .800 | 50.60 | 15.40 | 66.00 |
| Painting, door and frame, 1 coat, per side | 2.000 | FACE | 1.230 | 5.36 | 22.74 | 28.10 |
| 2 coats, per side | 2.000 | FACE | 2.280 | 7.70 | 42.30 | 50.00 |
| Primer & 1 coat, per side | 2.000 | FACE | 2.280 | 7.70 | 42.30 | 50.00 |
| 2 coats, per side | 2.000 | FACE | 3.200 | 8.80 | 59.20 | 68.00 |
| Staining, 1 coat, per side | 2.000 | FACE | .378 | 5.46 | 6.72 | 12.18 |
| 2 coats, per side | 2.000 | FACE | .546 | 5.88 | 10.08 | 15.96 |
| Sealer & 1 coat, per side | 2.000 | FACE | .588 | 5.88 | 10.92 | 16.80 |
| 2 coats, per side | 2.000 | FACE | .798 | 7.14 | 14.28 | 21.42 |

# 6 | INTERIORS  60 | Carpet Systems

| SYSTEM DESCRIPTION | QUAN. | UNIT | MAN HOURS | COST PER S.F. MAT. | COST PER S.F. INST. | TOTAL |
|---|---|---|---|---|---|---|
| Carpet, direct glue-down, acrylic 26 oz. | 1.000 | S.F. | .028 | 1.28 | .51 | 1.79 |
| 28 oz. | 1.000 | S.F. | .028 | 1.57 | .52 | 2.09 |
| 35 oz. | 1.000 | S.F. | .028 | 1.74 | .51 | 2.25 |
| Nylon, non anti-static, 15 oz. | 1.000 | S.F. | .028 | 1.02 | .52 | 1.54 |
| Anti-static, 17 oz. | 1.000 | S.F. | .028 | 1.21 | .52 | 1.73 |
| 20 oz. | 1.000 | S.F. | .028 | 1.29 | .51 | 1.80 |
| 22 oz. | 1.000 | S.F. | .028 | 1.34 | .52 | 1.86 |
| 24 oz. | 1.000 | S.F. | .028 | 1.45 | .52 | 1.97 |
| 26 oz. | 1.000 | S.F. | .028 | 1.55 | .52 | 2.07 |
| 28 oz. | 1.000 | S.F. | .028 | 1.66 | .52 | 2.18 |
| Needle bonded, 20 oz. | 1.000 | S.F. | .017 | 1.03 | .32 | 1.35 |
| Polypropylene, 15 oz. | 1.000 | S.F. | .017 | .66 | .32 | .98 |
| 22 oz. | 1.000 | S.F. | .022 | .92 | .42 | 1.34 |
| Scrim installed nylon spongeback carpet, 20 oz. | 1.000 | S.F. | .028 | 2.21 | .52 | 2.73 |
| 60 oz. | 1.000 | S.F. | .032 | 3.78 | .60 | 4.38 |
| Tile, foam backed, needle punch | 1.000 | S.F. | .017 | .80 | .33 | 1.13 |
| Tufted loop or shag | 1.000 | S.F. | .017 | 1.43 | .33 | 1.76 |
| Wool, 36 oz. | 1.000 | S.F. | .028 | 2.99 | .52 | 3.51 |
| Sponge backed, 36 oz. | 1.000 | S.F. | .017 | 2.95 | .32 | 3.27 |
| 42 oz. | 1.000 | S.F. | .028 | 3.30 | .52 | 3.82 |
| Padding, sponge rubber cushion, minimum | 1.000 | S.F. | .011 | .17 | .20 | .37 |
| Maximum | 1.000 | S.F. | .011 | .41 | .20 | .61 |
| Felt, 32 oz. to 56 oz., minimum | 1.000 | S.F. | .011 | .20 | .21 | .41 |
| Maximum | 1.000 | S.F. | .012 | .46 | .22 | .68 |
| Bonded urethane, 3/8" thick, minimum | 1.000 | S.F. | .010 | .17 | .19 | .36 |
| Maximum | 1.000 | S.F. | .012 | .36 | .22 | .58 |
| Prime urethane, 1/4" thick, minimum | 1.000 | S.F. | .010 | .17 | .19 | .36 |
| Maximum | 1.000 | S.F. | .012 | .57 | .22 | .79 |
| Stairs, for stairs, add to above carpet prices | 1.000 | Riser | .267 |  | 4.98 | 4.98 |
| Underlayment plywood, 3/8" thick | 1.000 | S.F. | .010 | .37 | .21 | .58 |
| 1/2" thick | 1.000 | S.F. | .011 | .43 | .23 | .66 |
| 5/8" thick | 1.000 | S.F. | .011 | .54 | .23 | .77 |
| 3/4" thick | 1.000 | S.F. | .012 | .59 | .25 | .84 |
| Particle board, 3/8" thick | 1.000 | S.F. | .010 | .23 | .21 | .44 |
| 1/2" thick | 1.000 | S.F. | .011 | .24 | .23 | .47 |
| 5/8" thick | 1.000 | S.F. | .011 | .27 | .23 | .50 |
| 3/4" thick | 1.000 | S.F. | .012 | .34 | .25 | .59 |
| Hardboard, 4' x 4', 0.215" thick | 1.000 | S.F. | .010 | .25 | .21 | .46 |

# 6 | INTERIORS — 64 | Flooring Systems

| SYSTEM DESCRIPTION | QUAN. | UNIT | MAN HOURS | MAT. | INST. | TOTAL |
|---|---|---|---|---|---|---|
| Resilient flooring, asphalt tile on concrete, 1/8" thick | | | | | | |
|   Color group B | 1.000 | S.F. | .015 | .68 | .28 | .96 |
|   Color group C & D | 1.000 | S.F. | .015 | .74 | .27 | 1.01 |
| Asphalt tile on wood subfloor, 1/8" thick | | | | | | |
|   Color group B | 1.000 | S.F. | .015 | .75 | .28 | 1.03 |
|   Color group C & D | 1.000 | S.F. | .015 | .81 | .27 | 1.08 |
| Vinyl composition tile, 12" x 12", 1/16" thick | 1.000 | S.F. | .015 | .58 | .28 | .86 |
|   Embossed | 1.000 | S.F. | .015 | .74 | .27 | 1.01 |
|   Marbleized | 1.000 | S.F. | .015 | .74 | .27 | 1.01 |
|   Plain | 1.000 | S.F. | .015 | .80 | .28 | 1.08 |
| .080" thick, embossed | 1.000 | S.F. | .015 | .86 | .27 | 1.13 |
|   Marbleized | 1.000 | S.F. | .015 | .92 | .28 | 1.20 |
|   Plain | 1.000 | S.F. | .015 | 1.09 | .28 | 1.37 |
| 1/8" thick, marbleized | 1.000 | S.F. | .015 | .94 | .27 | 1.21 |
|   Plain | 1.000 | S.F. | .015 | 1.16 | .27 | 1.43 |
| Vinyl tile, 12" x 12", .050" thick, minimum | 1.000 | S.F. | .019 | 1.35 | .35 | 1.70 |
|   Maximum | 1.000 | S.F. | .023 | 4.60 | .45 | 5.05 |
| 1/8" thick, minimum | 1.000 | S.F. | .019 | 1.99 | .35 | 2.34 |
|   Maximum | 1.000 | S.F. | .023 | 2.42 | .43 | 2.85 |
| 1/8" thick, solid colors | 1.000 | S.F. | .021 | 3.28 | .38 | 3.66 |
|   Florentine pattern | 1.000 | S.F. | .021 | 4.42 | .39 | 4.81 |
|   Marbleized or travertine pattern | 1.000 | S.F. | .021 | 8.25 | .40 | 8.65 |
| Vinyl sheet goods, backed, .070" thick, minimum | 1.000 | S.F. | .012 | .92 | .23 | 1.15 |
|   Maximum | 1.000 | S.F. | .025 | 2.17 | .46 | 2.63 |
| .093" thick, minimum | 1.000 | S.F. | .012 | 1.21 | .23 | 1.44 |
|   Maximum | 1.000 | S.F. | .025 | 2.71 | .46 | 3.17 |
| .125" thick, minimum | 1.000 | S.F. | .012 | 1.44 | .23 | 1.67 |
|   Maximum | 1.000 | S.F. | .025 | 3.47 | .46 | 3.93 |
| .250" thick, minimum | 1.000 | S.F. | .012 | 1.96 | .23 | 2.19 |
|   Maximum | 1.000 | S.F. | .025 | 5.02 | .48 | 5.50 |
| Wood, oak, finished in place, 25/32" x 2-1/2" clear | 1.000 | S.F. | .074 | 2.19 | 1.45 | 3.64 |
|   Select | 1.000 | S.F. | .074 | 2.00 | 1.45 | 3.45 |
|   No. 1 common | 1.000 | S.F. | .070 | 1.67 | 1.37 | 3.04 |
| Prefinished, oak, 2-1/2" wide | 1.000 | S.F. | .047 | 1.85 | .92 | 2.77 |
|   3-1/4" wide | 1.000 | S.F. | .043 | 1.85 | .84 | 2.69 |
|   Ranch plank, oak, random width | 1.000 | S.F. | .055 | 3.16 | 1.08 | 4.24 |
| Parquet, 5/16" thick, finished in place, oak, minimum | 1.000 | S.F. | .077 | 1.50 | 1.51 | 3.01 |
|   Maximum | 1.000 | S.F. | .107 | 6.12 | 2.11 | 8.23 |
|   Teak, minimum | 1.000 | S.F. | .077 | 3.08 | 1.51 | 4.59 |
|   Maximum | 1.000 | S.F. | .107 | 6.07 | 2.11 | 8.18 |
| Sleepers, treated, 16" O.C., 1" x 2" | 1.000 | S.F. | .007 | .10 | .14 | .24 |
|   1" x 3" | 1.000 | S.F. | .008 | .14 | .17 | .31 |
|   2" x 4" | 1.000 | S.F. | .011 | .40 | .21 | .61 |
|   2" x 6" | 1.000 | S.F. | .012 | .59 | .26 | .85 |
| Subfloor, plywood, 1/2" thick | 1.000 | S.F. | .011 | .41 | .22 | .63 |
|   5/8" thick | 1.000 | S.F. | .012 | .45 | .24 | .69 |
|   3/4" thick | 1.000 | S.F. | .013 | .51 | .26 | .77 |
| Ceramic tile, color group 2, 1" x 1" | 1.000 | S.F. | .087 | 2.07 | 1.48 | 3.55 |
|   2" x 2" or 2" x 1" | 1.000 | S.F. | .087 | 2.30 | 1.48 | 3.78 |

# 6 | INTERIORS  90 | Stairways

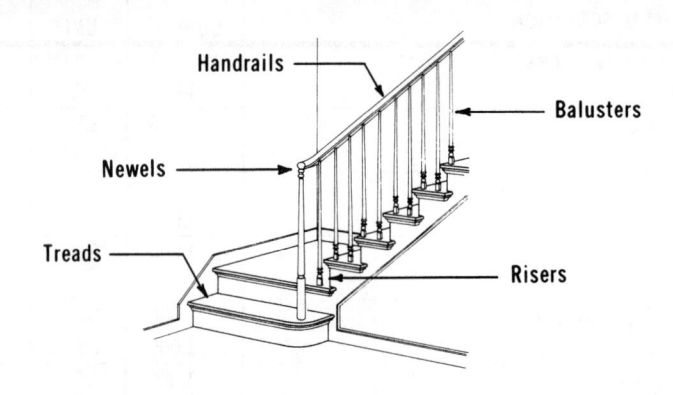

| SYSTEM DESCRIPTION | QUAN. | UNIT | MAN HOURS | MAT. | INST. | TOTAL |
|---|---|---|---|---|---|---|
| **7 RISERS, OAK TREADS, BOX STAIRS** | | | | | | |
| Treads, oak, 9-1/2" x 1-1/16" thick | 6.000 | Ea. | 2.664 | 69.66 | 50.34 | 120.00 |
| Risers, 3/4" thick, beech | 6.000 | Ea. | 2.625 | 68.67 | 51.03 | 119.70 |
| Balusters, birch, 30" high | 12.000 | Ea. | 3.432 | 45.00 | 67.20 | 112.20 |
| Newels, 3-1/4" wide | 2.000 | Ea. | 2.280 | 46.20 | 43.80 | 90.00 |
| Handrails, oak laminated | 7.000 | L.F. | .931 | 28.28 | 18.27 | 46.55 |
| Stringers, 2" x 10", 3 each | 21.000 | L.F. | 2.583 | 16.59 | 53.13 | 69.72 |
| TOTAL | | | 14.515 | 274.40 | 283.77 | 558.17 |
| **14 RISERS, OAK TREADS, BOX STAIRS** | | | | | | |
| Treads, oak, 9-1/2" x 1-1/16" thick | 13.000 | Ea. | 5.772 | 150.93 | 109.07 | 260.00 |
| Risers, 3/4" thick, beech | 13.000 | Ea. | 5.250 | 137.34 | 102.06 | 239.40 |
| Balusters, birch, 30" high | 26.000 | Ea. | 7.436 | 97.50 | 145.60 | 243.10 |
| Newels, 3-1/4" wide | 2.000 | Ea. | 2.280 | 46.20 | 43.80 | 90.00 |
| Handrails, oak, laminated | 14.000 | L.F. | 1.862 | 56.56 | 36.54 | 93.10 |
| Stringers, 2" x 10", 3 each | 42.000 | L.F. | 5.166 | 33.18 | 106.26 | 139.44 |
| TOTAL | | | 27.766 | 521.71 | 543.33 | 1065.04 |
| **14 RISERS, PINE TREADS, BOX STAIRS** | | | | | | |
| Treads, pine, 9-1/2" x 3/4" thick | 13.000 | Ea. | 5.772 | 51.48 | 112.97 | 164.45 |
| Risers, 3/4" thick, pine | 13.000 | Ea. | 5.082 | 42.00 | 99.54 | 141.54 |
| Balusters, pine, 30" high | 26.000 | Ea. | 7.436 | 82.68 | 144.82 | 227.50 |
| Newels, 3-1/4" wide | 2.000 | Ea. | 2.280 | 46.20 | 43.80 | 90.00 |
| Handrails, oak, laminated | 14.000 | L.F. | 1.862 | 56.56 | 36.54 | 93.10 |
| Stringers, 2" x 10", 3 each | 42.000 | L.F. | 5.166 | 33.18 | 106.26 | 139.44 |
| TOTAL | | | 27.598 | 312.10 | 543.93 | 856.03 |

| DESCRIPTION | QUAN. | UNIT | MAN HOURS | MAT. | INST. | TOTAL |
|---|---|---|---|---|---|---|
| | | | | | | |
| | | | | | | |
| | | | | | | |
| | | | | | | |
| | | | | | | |

# Stairway Price Sheet

| | QUAN. | UNIT | MAN HOURS | COST EACH MAT. | COST EACH INST. | TOTAL |
|---|---|---|---|---|---|---|
| Treads, oak, 1-1/16" x 9-1/2", 3' long, 7 riser stair | 6.000 | Ea. | 2.664 | 69.66 | 50.34 | 120.00 |
| 14 riser stair | 13.000 | Ea. | 5.772 | 150.93 | 109.07 | 260.00 |
| 1-1/16" x 11-1/2", 3' long, 7 riser stair | 6.000 | Ea. | 2.664 | 83.52 | 54.48 | 138.00 |
| 14 riser stair | 13.000 | Ea. | 5.772 | 180.96 | 118.04 | 299.00 |
| Pine, 3/4" x 9-1/2", 3' long, 7 riser stair | 6.000 | Ea. | 2.664 | 23.76 | 52.14 | 75.90 |
| 14 riser stair | 13.000 | Ea. | 5.772 | 51.48 | 112.97 | 164.45 |
| 3/4" x 11-1/4", 3' long, 7 riser stair | 6.000 | Ea. | 2.664 | 43.26 | 52.14 | 95.40 |
| 14 riser stair | 13.000 | Ea. | 5.772 | 93.73 | 112.97 | 206.70 |
| Risers, oak, 3/4" x 7-1/2" high, 7 riser stair | 6.000 | Ea. | 2.625 | 48.93 | 51.45 | 100.38 |
| 14 riser stair | 13.000 | Ea. | 5.250 | 97.86 | 102.90 | 200.76 |
| Beech, 3/4" x 7-1/2" high, 7 riser stair | 6.000 | Ea. | 2.625 | 68.67 | 51.03 | 119.70 |
| 14 riser stair | 13.000 | Ea. | 5.250 | 137.34 | 102.06 | 239.40 |
| Baluster, turned, 30" high, pine, 7 riser stair | 12.000 | Ea. | 3.432 | 38.16 | 66.84 | 105.00 |
| 14 riser stair | 26.000 | Ea. | 7.436 | 82.68 | 144.82 | 227.50 |
| 30" birch, 7 riser stair | 12.000 | Ea. | 3.432 | 45.00 | 67.20 | 112.20 |
| 14 riser stair | 26.000 | Ea. | 7.436 | 97.50 | 145.60 | 243.10 |
| 42" pine, 7 riser stair | 12.000 | Ea. | 3.552 | 48.60 | 69.60 | 118.20 |
| 14 riser stair | 26.000 | Ea. | 7.696 | 105.30 | 150.80 | 256.10 |
| 42" birch, 7 riser stair | 12.000 | Ea. | 3.552 | 72.00 | 69.60 | 141.60 |
| 14 riser stair | 26.000 | Ea. | 7.696 | 156.00 | 150.80 | 306.80 |
| Newels, 3-1/4" wide, starting, 7 riser stair | 2.000 | Ea. | 2.280 | 46.20 | 43.80 | 90.00 |
| 14 riser stair | 2.000 | Ea. | 2.280 | 46.20 | 43.80 | 90.00 |
| Landing, 7 riser stair | 2.000 | Ea. | 3.200 | 81.40 | 62.60 | 144.00 |
| 14 riser stair | 2.000 | Ea. | 3.200 | 81.40 | 62.60 | 144.00 |
| Handrails, oak, laminated, 7 riser stair | 7.000 | L.F. | .931 | 28.28 | 18.27 | 46.55 |
| 14 riser stair | 14.000 | L.F. | 1.862 | 56.56 | 36.54 | 93.10 |
| Stringers, fir, 2" x 10" 7 riser stair | 21.000 | L.F. | 2.583 | 16.59 | 53.13 | 69.72 |
| 14 riser stair | 42.000 | L.F. | 5.166 | 33.18 | 106.26 | 139.44 |
| 2" x 12", 7 riser stair | 21.000 | L.F. | 2.583 | 19.74 | 52.92 | 72.66 |
| 14 riser stair | 42.000 | L.F. | 5.166 | 39.48 | 105.84 | 145.32 |

# Special Stairways

| | QUAN. | UNIT | MAN HOURS | COST EACH MAT. | COST EACH INST. | TOTAL |
|---|---|---|---|---|---|---|
| Basement stairs, prefabricated, open risers | 1.000 | Flight | 4.000 | 110.00 | 80.00 | 190.00 |
| Curved stairways, 3'-3" wide, prefabricated oak, 9' high | 1.000 | Flight | 22.860 | 4372.50 | 452.50 | 4825.00 |
| 10' high | 1.000 | Flight | 22.860 | 4812.50 | 437.50 | 5250.00 |
| Open two sides, 9' high | 1.000 | Flight | 32.000 | 6957.50 | 617.50 | 7575.00 |
| 10' high | 1.000 | Flight | 32.000 | 7755.00 | 620.00 | 8375.00 |
| Spiral stairs, oak, 4'-6" diameter, prefabricated, 9' high | 1.000 | Flight | 10.670 | 2695.00 | 205.00 | 2900.00 |
| Aluminum, 5'-0" diameter stock unit | 1.000 | Flight | 9.954 | 2079.00 | 231.00 | 2310.00 |
| Custom unit | 1.000 | Flight | 9.954 | 3080.00 | 280.00 | 3360.00 |
| Cast iron, 4'-0" diameter, minimum | 1.000 | Flight | 9.954 | 1925.00 | 245.00 | 2170.00 |
| Maximum | 1.000 | Flight | 17.920 | 2849.00 | 441.00 | 3290.00 |
| Steel, industrial, pre-erected, 3'-6" wide, bar rail | 1.000 | Flight | 5.264 | 1540.00 | 140.00 | 1680.00 |
| Picket rail | 1.000 | Flight | 26.320 | 2079.00 | 651.00 | 2730.00 |

# SECTION 7
# SPECIALTIES

| System Number | Title | Page |
|---|---|---|
| **7-08** | Kitchen | 148 |
| **7-12** | Appliances | 150 |
| **7-16** | Bath Accessories | 151 |
| **7-24** | Masonry Fireplace | 152 |
| **7-30** | Prefabricated Fireplace | 154 |
| **7-32** | Greenhouse | 156 |
| **7-36** | Swimming Pool | 157 |
| **7-40** | Wood Deck | 158 |

# 7 | SPECIALTIES  08 | Kitchen Systems

| SYSTEM DESCRIPTION | QUAN. | UNIT | MAN HOURS | MAT. | INST. | TOTAL |
|---|---|---|---|---|---|---|
| **KITCHEN, ECONOMY GRADE** | | | | | | |
| Top cabinets, economy grade | 1.000 | L.F. | .170 | 30.62 | 2.98 | 33.60 |
| Bottom cabinets, economy grade | 1.000 | L.F. | .255 | 45.94 | 4.46 | 50.40 |
| Counter top, laminated plastic, post formed | 1.000 | L.F. | .267 | 8.80 | 5.20 | 14.00 |
| Blocking, wood, 2" x 4" | 1.000 | L.F. | .032 | .26 | .66 | .92 |
| Soffit, framing, wood, 2" x 4" | 4.000 | L.F. | .072 | 1.00 | 1.48 | 2.48 |
| Soffit drywall, painted | 2.000 | S.F. | .064 | .62 | 1.22 | 1.84 |
| TOTAL | | | .860 | 87.24 | 16.00 | 103.24 |
| **AVERAGE GRADE** | | | | | | |
| Top cabinets, average grade | 1.000 | L.F. | .213 | 38.28 | 3.72 | 42.00 |
| Bottom cabinets, average grade | 1.000 | L.F. | .319 | 57.42 | 5.58 | 63.00 |
| Counter top, laminated plastic, square edge, incl. backsplash | 1.000 | L.F. | .267 | 23.10 | 4.90 | 28.00 |
| Blocking, wood, 2" x 4" | 1.000 | L.F. | .032 | .26 | .66 | .92 |
| Soffit framing, wood, 2" x 4" | 4.000 | L.F. | .072 | 1.00 | 1.48 | 2.48 |
| Soffit drywall, painted | 2.000 | S.F. | .064 | .62 | 1.22 | 1.84 |
| TOTAL | | | .967 | 120.68 | 17.56 | 138.24 |
| **CUSTOM GRADE** | | | | | | |
| Top cabinets, custom grade | 1.000 | L.F. | .256 | 52.80 | 5.20 | 58.00 |
| Bottom cabinets, custom grade | 1.000 | L.F. | .384 | 79.20 | 7.80 | 87.00 |
| Counter top, laminated plastic, square edge, incl. backsplash | 1.000 | L.F. | .267 | 23.10 | 4.90 | 28.00 |
| Blocking, wood, 2" x 4" | 1.000 | L.F. | .032 | .26 | .66 | .92 |
| Soffit framing, wood, 2" x 4" | 4.000 | L.F. | .072 | 1.00 | 1.48 | 2.48 |
| Soffit drywall, painted | 2.000 | S.F. | .064 | .62 | 1.22 | 1.84 |
| TOTAL | | | 1.075 | 156.98 | 21.26 | 178.24 |

| DESCRIPTION | QUAN. | UNIT | MAN HOURS | MAT. | INST. | TOTAL |
|---|---|---|---|---|---|---|
| | | | | | | |
| | | | | | | |
| | | | | | | |
| | | | | | | |
| | | | | | | |

# Kitchen Price Sheet

| | QUAN. | UNIT | MAN HOURS | MAT. | INST. | TOTAL |
|---|---|---|---|---|---|---|
| Top cabinets, economy grade | 1.000 | L.F. | .170 | 30.62 | 2.98 | 33.60 |
| Average grade | 1.000 | L.F. | .213 | 38.28 | 3.72 | 42.00 |
| Custom grade | 1.000 | L.F. | .256 | 52.80 | 5.20 | 58.00 |
| Bottom cabinets, economy grade | 1.000 | L.F. | .255 | 45.94 | 4.46 | 50.40 |
| Average grade | 1.000 | L.F. | .319 | 57.42 | 5.58 | 63.00 |
| Custom grade | 1.000 | L.F. | .384 | 79.20 | 7.80 | 87.00 |
| Counter top, laminated plastic, 7/8" thick, no splash | 1.000 | L.F. | .267 | 14.03 | 5.22 | 19.25 |
| With backsplash | 1.000 | L.F. | .267 | 18.59 | 5.41 | 24.00 |
| 1-1/4" thick, no splash | 1.000 | L.F. | .286 | 16.28 | 5.72 | 22.00 |
| With backsplash | 1.000 | L.F. | .286 | 20.85 | 5.15 | 26.00 |
| Post formed, laminated plastic | 1.000 | L.F. | .267 | 8.80 | 5.20 | 14.00 |
| Marble, with backsplash, minimum | 1.000 | L.F. | .471 | 23.10 | 8.90 | 32.00 |
| Maximum | 1.000 | L.F. | .615 | 72.60 | 12.40 | 85.00 |
| Maple, solid laminated, no backsplash | 1.000 | L.F. | .286 | 28.60 | 5.40 | 34.00 |
| With backsplash | 1.000 | L.F. | .286 | 33.00 | 6.00 | 39.00 |
| Blocking, wood, 2" x 4" | 1.000 | L.F. | .032 | .26 | .66 | .92 |
| 2" x 6" | 1.000 | L.F. | .036 | .40 | .74 | 1.14 |
| 2" x 8" | 1.000 | L.F. | | .55 | .82 | 1.37 |
| Soffit framing, wood, 2" x 3" | 4.000 | L.F. | .064 | .80 | 1.32 | 2.12 |
| 2" x 4" | 4.000 | L.F. | .072 | 1.00 | 1.48 | 2.48 |
| Soffit, drywall, painted | 2.000 | S.F. | .064 | .62 | 1.22 | 1.84 |
| Paneling, standard | 2.000 | S.F. | .064 | 1.12 | 1.32 | 2.44 |
| Deluxe | 2.000 | S.F. | .092 | 2.14 | 1.86 | 4.00 |
| Sinks, porcelain on cast iron single bowl, 21" x 24" | 1.000 | Ea. | 13.420 | 215.33 | 264.67 | 480.00 |
| 21" x 30" | 1.000 | Ea. | 13.420 | 237.33 | 262.67 | 500.00 |
| Double bowl, 20" x 32" | 1.000 | Ea. | 14.570 | 253.83 | 286.17 | 540.00 |
| Stainless steel, single bowl, 16" x 20" | 1.000 | Ea. | 13.420 | 314.33 | 265.67 | 580.00 |
| 22" x 25" | 1.000 | Ea. | 13.420 | 336.33 | 263.67 | 600.00 |
| Double bowl, 20" x 32" | 1.000 | Ea. | 14.570 | 179.03 | 285.97 | 465.00 |

**7**

149

# 7 | SPECIALTIES — 12 | Appliance Systems

| SYSTEM DESCRIPTION | QUAN. | UNIT | MAN. HOURS | MAT. | INST. | TOTAL |
|---|---|---|---|---|---|---|
| All appliances include plumbing and electrical rough-in & hook-ups | | | | | | |
| Range, free standing, minimum | 1.000 | Ea. | 3.600 | 385.00 | 66.00 | 451.00 |
| Maximum | 1.000 | Ea. | 6.000 | 1353.00 | 98.00 | 1451.00 |
| Built-in, minimum | 1.000 | Ea. | 6.000 | 561.00 | 120.00 | 681.00 |
| Maximum | 1.000 | Ea. | 10.000 | 913.00 | 188.00 | 1101.00 |
| Counter top range, 4-burner, minimum | 1.000 | Ea. | 3.330 | 253.00 | 73.00 | 326.00 |
| Maximum | 1.000 | Ea. | 4.670 | 506.00 | 100.00 | 606.00 |
| Compactor, built-in, minimum | 1.000 | Ea. | 2.215 | 380.05 | 44.35 | 424.40 |
| Maximum | 1.000 | Ea. | 3.285 | 545.05 | 64.35 | 609.40 |
| Dishwasher, built-in, minimum | 1.000 | Ea. | 7.815 | 346.97 | 172.43 | 519.40 |
| Maximum | 1.000 | Ea. | 11.815 | 445.97 | 258.43 | 704.40 |
| Garbage disposer, minimum | 1.000 | Ea. | 4.800 | 76.96 | 105.54 | 182.50 |
| Maximum | 1.000 | Ea. | 6.930 | 219.96 | 152.54 | 372.50 |
| Microwave oven, minimum | 1.000 | Ea. | 2.615 | 248.05 | 56.35 | 304.40 |
| Maximum | 1.000 | Ea. | 4.615 | 1106.05 | 88.35 | 1194.40 |
| Range hood, ducted, minimum | 1.000 | Ea. | 5.855 | 98.25 | 119.35 | 217.60 |
| Maximum | 1.000 | Ea. | 7.985 | 285.25 | 167.35 | 452.60 |
| Ductless, minimum | 1.000 | Ea. | 3.815 | 85.25 | 77.35 | 162.60 |
| Maximum | 1.000 | Ea. | 5.945 | 272.25 | 125.35 | 397.60 |
| Refrigerator, 16 cu.ft., minimum | 1.000 | Ea. | 2.000 | 481.25 | 31.25 | 512.50 |
| Maximum | 1.000 | Ea. | 3.200 | 770.00 | 50.00 | 820.00 |
| 16 cu.ft. with icemaker, minimum | 1.000 | Ea. | 4.000 | 599.46 | 73.04 | 672.50 |
| Maximum | 1.000 | Ea. | 5.200 | 888.21 | 91.79 | 980.00 |
| 19 cu.ft., minimum | 1.000 | Ea. | 2.670 | 638.00 | 42.00 | 680.00 |
| Maximum | 1.000 | Ea. | 4.672 | 1116.50 | 73.50 | 1190.00 |
| 19 cu.ft. with icemaker, minimum | 1.000 | Ea. | 4.937 | 819.46 | 88.04 | 907.50 |
| Maximum | 1.000 | Ea. | 6.940 | 1297.96 | 119.54 | 1417.50 |
| Sinks, porcelain on cast iron single bowl, 21" x 24" | 1.000 | Ea. | 13.420 | 215.33 | 264.67 | 480.00 |
| 21" x 30" | 1.000 | Ea. | 13.420 | 237.33 | 262.67 | 500.00 |
| Double bowl, 20" x 32" | 1.000 | Ea. | 14.570 | 253.83 | 286.17 | 540.00 |
| Stainless steel, single bowl 16" x 20" | 1.000 | Ea. | 13.420 | 314.33 | 265.67 | 580.00 |
| 22" x 25" | 1.000 | Ea. | 13.420 | 336.33 | 263.67 | 600.00 |
| Double bowl, 20" x 32" | 1.000 | Ea. | 14.570 | 179.03 | 285.97 | 465.00 |
| Water heater, electric, 30 gallon | 1.000 | Ea. | 1.820 | 148.50 | 41.50 | 190.00 |
| 40 gallon | 1.000 | Ea. | 2.000 | 154.00 | 46.00 | 200.00 |
| Gas, 30 gallon | 1.000 | Ea. | 2.000 | 165.00 | 45.00 | 210.00 |
| 75 gallon | 1.000 | Ea. | 2.670 | 440.00 | 60.00 | 500.00 |
| Wall, packaged terminal heater/air conditioner cabinet, wall sleeve, Louver, electric heat, thermostat, manual changeover, 208V | | | | | | |
| 7000 BTUH cooling, 8800 BTU heating | 1.000 | Ea. | 2.670 | 693.00 | 52.00 | 745.00 |
| 9000 BTUH cooling, 13,900 BTU heating | 1.000 | Ea. | 3.200 | 720.50 | 64.50 | 785.00 |
| 11,300 BTUH cooling, 13,900 BTU heating | 1.000 | Ea. | 4.000 | 770.00 | 80.00 | 850.00 |
| 14,800 BTUH cooling, 13,900 BTU heating | 1.000 | Ea. | 5.330 | 803.00 | 107.00 | 910.00 |

# 7 SPECIALTIES — 16 Bath Accessories

| SYSTEM DESCRIPTION | QUAN. | UNIT | MAN. HOURS | MAT. | INST. | TOTAL |
|---|---|---|---|---|---|---|
| Curtain rods, stainless, 1" diameter, 3' long | 1.000 | Ea. | .369 | 42.42 | 7.23 | 49.65 |
| 5' long | 1.000 | Ea. | .615 | 70.70 | 12.05 | 82.75 |
| Grab bar, 1" diameter, 12" long | 1.000 | Ea. | .333 | 17.66 | 6.34 | 24.00 |
| 36" long | 1.000 | Ea. | .400 | 23.10 | 7.90 | 31.00 |
| 1-1/4" diameter, 12" long | 1.000 | Ea. | .382 | 20.31 | 7.29 | 27.60 |
| 36" long | 1.000 | Ea. | .460 | 26.57 | 9.08 | 35.65 |
| 1-1/2" diameter, 12" long | 1.000 | Ea. | .416 | 22.08 | 7.92 | 30.00 |
| 36" long | 1.000 | Ea. | .500 | 28.88 | 9.87 | 38.75 |
| Mirror, 18" x 24" | 1.000 | Ea. | .400 | 61.60 | 7.40 | 69.00 |
| 72" x 24" | 1.000 | Ea. | 1.330 | 198.00 | 27.00 | 225.00 |
| Medicine chest with mirror, 18" x 24" | 1.000 | Ea. | .400 | 80.30 | 7.70 | 88.00 |
| 36" x 24" | 1.000 | Ea. | .600 | 120.45 | 11.55 | 132.00 |
| Toilet tissue dispenser, surface mounted, minimum | 1.000 | Ea. | .267 | 18.81 | 5.19 | 24.00 |
| Maximum | 1.000 | Ea. | .400 | 28.22 | 7.78 | 36.00 |
| Flush mounted, minimum | 1.000 | Ea. | .293 | 20.69 | 5.71 | 26.40 |
| Maximum | 1.000 | Ea. | .427 | 30.10 | 8.30 | 38.40 |
| Towel bar, 18" long, minimum | 1.000 | Ea. | .278 | 15.05 | 5.75 | 20.80 |
| Maximum | 1.000 | Ea. | .348 | 18.81 | 7.19 | 26.00 |
| 24" long, minimum | 1.000 | Ea. | .313 | 16.93 | 6.47 | 23.40 |
| Maximum | 1.000 | Ea. | .382 | 20.69 | 7.91 | 28.60 |
| 36" long, minimum | 1.000 | Ea. | .381 | 19.80 | 7.20 | 27.00 |
| Maximum | 1.000 | Ea. | .419 | 21.78 | 7.92 | 29.70 |

# 7 | SPECIALTIES  24 | Masonry Fireplace Systems

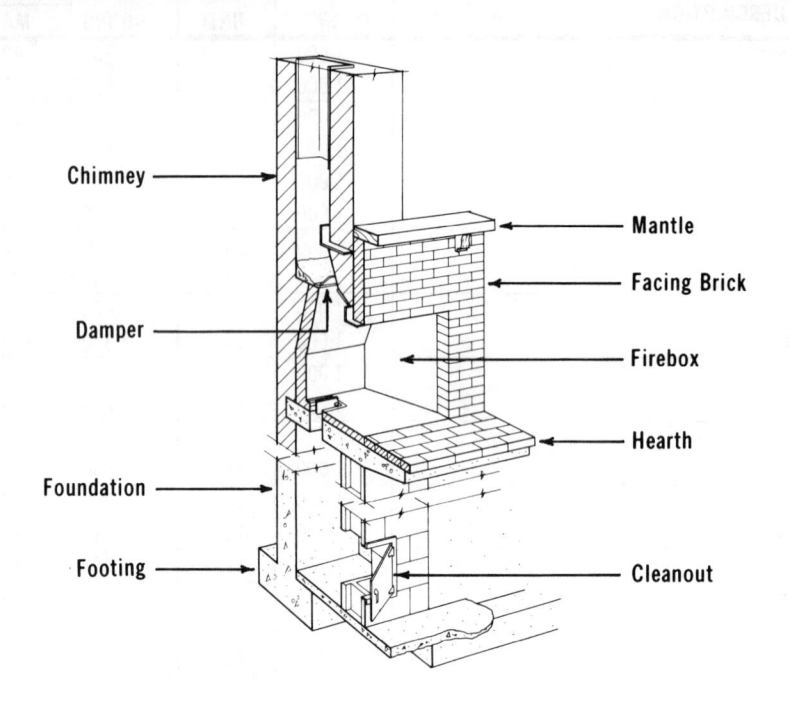

Labeled diagram: Chimney, Damper, Foundation, Footing, Mantle, Facing Brick, Firebox, Hearth, Cleanout.

| SYSTEM DESCRIPTION | QUAN. | UNIT | MAN HOURS | MAT. | INST. | TOTAL |
|---|---|---|---|---|---|---|
| **MASONRY FIREPLACE** | | | | | | |
| Footing, 8" thick, concrete, 4' x 7' | .700 | C.Y. | 1.561 | 43.89 | 36.61 | 80.50 |
| Foundation, concrete block, 32" x 60" x 4' deep | 1.000 | Ea. | 5.30 | 64.20 | 97.19 | 161.39 |
| Fireplace, brick firebox, 30" x 24" opening | 1.000 | Ea. | 40.000 | 247.50 | 707.50 | 955.00 |
| Damper, cast iron, 30" opening | 1.000 | Ea. | 1.330 | 49.50 | 26.50 | 76.00 |
| Facing brick, standard size brick, 6' x 5' | 30.000 | S.F. | 5.730 | 52.80 | 104.70 | 157.50 |
| Hearth, standard size brick, 3' x 6' | 1.000 | Ea. | 8.000 | 66.00 | 139.00 | 205.00 |
| Chimney, standard size brick, 8" x 12" flue, one story house | 12.000 | V.L.F. | 12.000 | 149.16 | 210.84 | 360.00 |
| Mantle, 4" x 8", wood | 6.000 | L.F. | 1.332 | 21.48 | 25.92 | 47.40 |
| Cleanout, cast iron, 8" x 8" | 1.000 | Ea. | .667 | 13.20 | 12.80 | 26.00 |
| TOTAL | | | 75.960 | 707.73 | 1361.06 | 2068.79 |

The costs in this system are on a cost each basis.

| DESCRIPTION | QUAN. | UNIT | MAN HOURS | MAT. | INST. | TOTAL |
|---|---|---|---|---|---|---|
| | | | | | | |
| | | | | | | |
| | | | | | | |
| | | | | | | |
| | | | | | | |
| | | | | | | |
| | | | | | | |
| | | | | | | |

# Masonry Fireplace Price Sheet

| | QUAN. | UNIT | MAN HOURS | COST EACH MAT. | COST EACH INST. | TOTAL |
|---|---|---|---|---|---|---|
| Footing 8" thick, 3' x 6' | .440 | C.Y. | .981 | 27.59 | 23.01 | 50.60 |
| 4' x 7' | .700 | C.Y. | 1.561 | 43.89 | 36.61 | 80.50 |
| 5' x 8' | 1.000 | C.Y. | 2.230 | 62.70 | 52.30 | 115.00 |
| 1' thick, 3' x 6' | .670 | C.Y. | 1.494 | 42.01 | 35.04 | 77.05 |
| 4' x 7' | 1.030 | C.Y. | 2.296 | 64.58 | 53.87 | 118.45 |
| 5' x 8' | 1.480 | C.Y. | 3.300 | 92.80 | 77.40 | 170.20 |
| Foundation-concrete block, 24" x 48", 4' deep | 1.000 | Ea. | 3.600 | 58.08 | 84.96 | 143.04 |
| 8' deep | 1.000 | Ea. | 7.200 | 116.16 | 169.92 | 286.08 |
| 24" x 60", 4' deep | 1.000 | Ea. | 4.200 | 67.76 | 99.12 | 166.88 |
| 8' deep | 1.000 | Ea. | 8.400 | 135.52 | 198.24 | 333.76 |
| 32" x 48", 4' deep | 1.000 | Ea. | 3.975 | 64.13 | 93.81 | 157.94 |
| 8' deep | 1.000 | Ea. | 7.950 | 128.26 | 187.62 | 315.88 |
| 32" x 60", 4' deep | 1.000 | Ea. | 4.500 | 72.60 | 106.20 | 178.80 |
| 8' deep | 1.000 | Ea. | 9.150 | 147.62 | 215.94 | 363.56 |
| 32" x 72", 4' deep | 1.000 | Ea. | 5.175 | 83.49 | 122.13 | 205.62 |
| 8' deep | 1.000 | Ea. | 10.350 | 166.98 | 244.26 | 411.24 |
| Fireplace, brick firebox 30" x 24" opening | 1.000 | Ea. | 40.000 | 247.50 | 707.50 | 955.00 |
| 48" x 30" opening | 1.000 | Ea. | 60.000 | 371.25 | 1061.25 | 1432.50 |
| Steel fire box with registers, 25" opening | 1.000 | Ea. | 26.670 | 473.00 | 475.00 | 948.00 |
| 48" opening | 1.000 | Ea. | 44.000 | 962.50 | 787.50 | 1750.00 |
| Damper, cast iron, 30" opening | 1.000 | Ea. | 1.330 | 49.50 | 26.50 | 76.00 |
| 36" opening | 1.000 | Ea. | 1.103 | 41.09 | 21.99 | 63.08 |
| Steel, 30" opening | 1.000 | Ea. | 1.330 | 44.00 | 26.00 | 70.00 |
| 36" opening | 1.000 | Ea. | 1.103 | 36.52 | 21.58 | 58.10 |
| Facing for fireplace, standard size brick, 6' x 5' | 30.000 | S.F. | 5.730 | 52.80 | 104.70 | 157.50 |
| 7' x 5' | 35.000 | S.F. | 6.685 | 61.60 | 122.15 | 183.75 |
| 8' x 6' | 48.000 | S.F. | 9.168 | 84.48 | 167.52 | 252.00 |
| Fieldstone, 6' x 5' | 30.000 | S.F. | 5.730 | 322.80 | 104.70 | 427.50 |
| 7' x 5' | 35.000 | S.F. | 6.685 | 376.60 | 122.15 | 498.75 |
| 8' x 6' | 48.000 | S.F. | 9.168 | 516.48 | 167.52 | 684.00 |
| Sheetrock on metal, studs, 6' x 5' | 30.000 | S.F. | .780 | 15.00 | 20.40 | 35.40 |
| 7' x 5' | 35.000 | S.F. | .910 | 17.50 | 23.80 | 41.30 |
| 8' x 6' | 48.000 | S.F. | 1.248 | 24.00 | 32.64 | 56.64 |
| Hearth, standard size brick, 3' x 6' | 1.000 | Ea. | 8.000 | 66.00 | 139.00 | 205.00 |
| 3' x 7' | 1.000 | Ea. | 9.280 | 76.56 | 161.24 | 237.80 |
| 3' x 8' | 1.000 | Ea. | 10.640 | 87.78 | 184.87 | 272.65 |
| Stone, 3' x 6' | 1.000 | Ea. | 8.000 | 74.75 | 139.00 | 213.75 |
| 3' x 7' | 1.000 | Ea. | 9.280 | 86.71 | 161.24 | 247.95 |
| 3' x 8' | 1.000 | Ea. | 10.640 | 99.42 | 184.87 | 284.29 |
| Chimney, standard size brick, 8" x 12" flue, one story house | 12.000 | V.L.F. | 12.000 | 149.16 | 210.84 | 360.00 |
| Two story house | 20.000 | V.L.F. | 20.000 | 248.60 | 351.40 | 600.00 |
| Mantle wood, beams, 4" x 8" | 6.000 | L.F. | 1.332 | 21.48 | 25.92 | 47.40 |
| 4" x 10" | 6.000 | L.F. | 1.374 | 26.76 | 26.94 | 53.70 |
| Ornate, prefabricated, 6' x 3'-6" opening, minimum | 1.000 | Ea. | 1.600 | 77.00 | 33.00 | 110.00 |
| Maximum | 1.000 | Ea. | 1.600 | 115.50 | 29.50 | 145.00 |
| Cleanout, door and frame, cast iron, 8" x 8" | 1.000 | Ea. | .667 | 13.20 | 12.80 | 26.00 |
| 12" x 12" | 1.000 | Ea. | .800 | 26.40 | 15.60 | 42.00 |

7

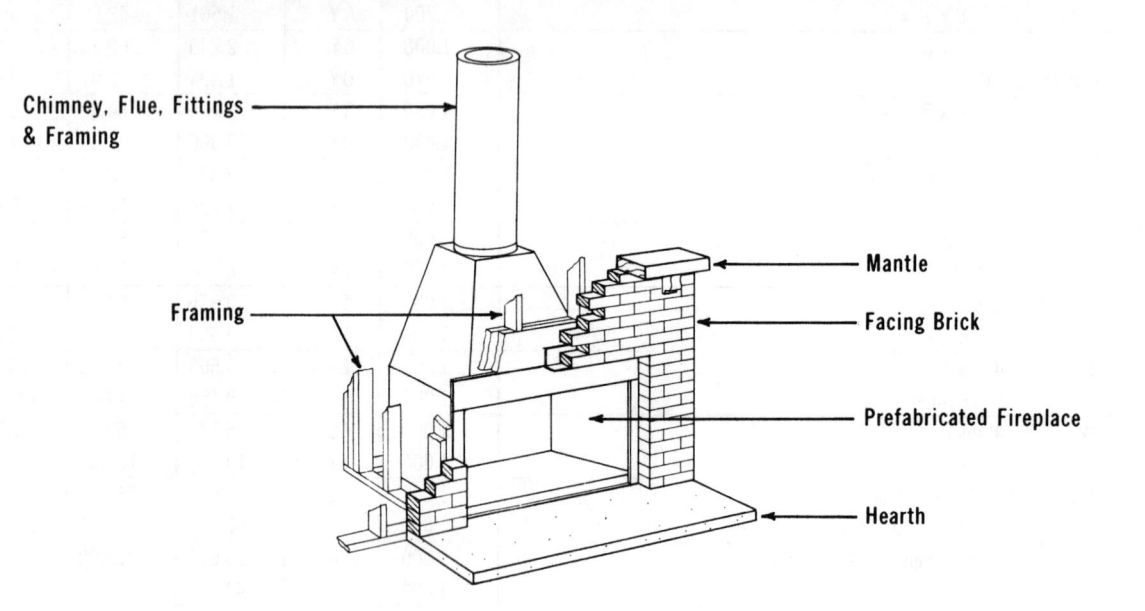

# 7 | SPECIALTIES — 30 | Prefabricated Fireplace Systems

Labels on diagram:
- Chimney, Flue, Fittings & Framing
- Framing
- Mantle
- Facing Brick
- Prefabricated Fireplace
- Hearth

| SYSTEM DESCRIPTION | QUAN. | UNIT | MAN HOURS | MAT. | INST. | TOTAL |
|---|---|---|---|---|---|---|
| **PREFABRICATED FIREPLACE** | | | | | | |
| Prefabricated fireplace, metal, minimum | 1.000 | Ea. | 6.150 | 533.50 | 126.50 | 660.00 |
| Framing, 2" x 4" studs, 6' x 5' | 35.000 | L.F. | .525 | 8.75 | 11.20 | 19.95 |
| Sheetrock, 1/2" fire resistant, 6' x 5' | 40.000 | S.F. | .360 | 10.80 | 13.60 | 24.40 |
| Facing, brick, standard size brick, 6' x 5' | 30.000 | S.F. | 5.730 | 52.80 | 104.70 | 157.50 |
| Hearth, standard size brick, 3' x 6' | 1.000 | Ea. | 8.000 | 66.00 | 139.00 | 205.00 |
| Chimney, one story house, framing, 2" x 4" studs | 80.000 | L.F. | 1.200 | 20.00 | 25.60 | 45.60 |
| Sheathing, plywood, 5/8" thick | 32.000 | S.F. | .768 | 46.08 | 15.68 | 61.76 |
| Flue, 10" metal, insulated pipe | 12.000 | V.L.F. | 3.996 | 142.56 | 79.44 | 222.00 |
| Fittings, ceiling support | 1.000 | Ea. | .667 | 59.40 | 13.60 | 73.00 |
| Fittings, joist shield | 1.000 | Ea. | .667 | 24.20 | 12.80 | 37.00 |
| Fittings, roof flashing | 1.000 | Ea. | .667 | 68.20 | 12.80 | 81.00 |
| Mantle beam, wood, 4" x 8" | 6.000 | L.F. | 1.332 | 21.48 | 25.92 | 47.40 |
| TOTAL | | | 30.062 | 1053.77 | 580.84 | 1634.61 |

The costs in this system are on a cost each basis.

| DESCRIPTION | QUAN. | UNIT | MAN HOURS | MAT. | INST. | TOTAL |
|---|---|---|---|---|---|---|
| | | | | | | |
| | | | | | | |
| | | | | | | |
| | | | | | | |
| | | | | | | |
| | | | | | | |
| | | | | | | |
| | | | | | | |
| | | | | | | |

# Prefabricated Fireplace Price Sheet

| | QUAN. | UNIT | MAN HOURS | MAT. | INST. | TOTAL |
|---|---|---|---|---|---|---|
| Prefabricated fireplace, minimum | 1.000 | Ea. | 6.150 | 533.50 | 126.50 | 660.00 |
| Average | 1.000 | Ea. | 8.000 | 1006.50 | 168.50 | 1175.00 |
| Maximum | 1.000 | Ea. | 8.890 | 3850.00 | 175.00 | 4025.00 |
| Framing, 2" x 4" studs, fireplace, 6' x 5' | 35.000 | L.F. | .525 | 8.75 | 11.20 | 19.95 |
| 7' x 5' | 40.000 | L.F. | .600 | 10.00 | 12.80 | 22.80 |
| 8' x 6' | 45.000 | L.F. | .675 | 11.25 | 14.40 | 25.65 |
| Sheetrock, 1/2" thick, fireplace, 6' x 5' | 40.000 | S.F. | .360 | 10.80 | 13.60 | 24.40 |
| 7' x 5' | 45.000 | S.F. | .405 | 12.15 | 15.30 | 27.45 |
| 8' x 6' | 50.000 | S.F. | .450 | 13.50 | 17.00 | 30.50 |
| Facing for fireplace, brick, 6' x 5' | 30.000 | S.F. | 5.730 | 52.80 | 104.70 | 157.50 |
| 7' x 5' | 35.000 | S.F. | 6.685 | 61.60 | 122.15 | 183.75 |
| 8' x 6' | 48.000 | S.F. | 9.168 | 84.48 | 167.52 | 252.00 |
| Fieldstone, 6' x 5' | 30.000 | S.F. | 5.730 | 315.30 | 104.70 | 420.00 |
| 7' x 5' | 35.000 | S.F. | 6.685 | 367.85 | 122.15 | 490.00 |
| 8' x 6' | 48.000 | S.F. | 9.168 | 504.48 | 167.52 | 672.00 |
| Hearth, standard size brick, 3' x 6' | 1.000 | Ea. | 8.000 | 66.00 | 139.00 | 205.00 |
| 3' x 7' | 1.000 | Ea. | 9.280 | 76.56 | 161.24 | 237.80 |
| 3' x 8' | 1.000 | Ea. | 10.640 | 87.78 | 184.87 | 272.65 |
| Stone, 3' x 6' | 1.000 | Ea. | 8.000 | 74.75 | 139.00 | 213.75 |
| 3' x 7' | 1.000 | Ea. | 9.280 | 86.71 | 161.24 | 247.95 |
| 3' x 8' | 1.000 | Ea. | 10.640 | 99.42 | 184.87 | 284.29 |
| Chimney, framing, 2" x 4", one story house | 80.000 | L.F. | 1.200 | 20.00 | 25.60 | 45.60 |
| Two story house | 120.000 | L.F. | 1.800 | 30.00 | 38.40 | 68.40 |
| Sheathing, plywood, 5/8" thick | 32.000 | S.F. | .768 | 46.08 | 15.68 | 61.76 |
| Stucco on plywood | 32.000 | S.F. | 1.128 | 28.49 | 22.51 | 51.00 |
| Flue, 10" metal pipe, insulated, one story house | 12.000 | V.L.F. | 3.996 | 142.56 | 79.44 | 222.00 |
| Two story house | 20.000 | V.L.F. | 6.660 | 237.60 | 132.40 | 370.00 |
| Fittings, ceiling support | 1.000 | Ea. | .667 | 59.40 | 13.60 | 73.00 |
| Fittings joist sheild, one story house | 1.000 | Ea. | .667 | 24.20 | 12.80 | 37.00 |
| Two story house | 2.000 | Ea. | 1.334 | 48.40 | 25.60 | 74.00 |
| Fittings roof flashing | 1.000 | Ea. | .667 | 68.20 | 12.80 | 81.00 |
| Mantle, wood beam, 4" x 8" | 6.000 | L.F. | 1.332 | 21.48 | 25.92 | 47.40 |
| 4" x 10" | 6.000 | L.F. | 1.374 | 26.76 | 26.94 | 53.70 |
| Ornate prefabricated, 6' x 3'-6" opening, minimum | 1.000 | Ea. | 1.600 | 77.00 | 33.00 | 110.00 |
| Maximum | 1.000 | Ea. | 1.600 | 115.50 | 29.50 | 145.00 |

# 7 | SPECIALTIES    32 | Greenhouse Systems

| SYSTEM DESCRIPTION | QUAN. | UNIT | MAN HOURS | MAT. | INST. | TOTAL |
|---|---|---|---|---|---|---|
| Economy, lean to, shell only, not incl. 2' stub wall, fndtn, flrs, heat 4'x10' | 1.000 | Ea. | 18.840 | 1320.00 | 360.00 | 1680.00 |
| 4' x 16' | 1.000 | Ea. | 26.234 | 1838.10 | 501.30 | 2339.40 |
| 4' x 24' | 1.000 | Ea. | 30.285 | 2121.90 | 578.70 | 2700.60 |
| 6' x 10' | 1.000 | Ea. | 16.560 | 1518.00 | 342.00 | 1860.00 |
| 6' x 16' | 1.000 | Ea. | 23.046 | 2112.55 | 475.95 | 2588.50 |
| 6' x 24' | 1.000 | Ea. | 29.808 | 2732.40 | 615.60 | 3348.00 |
| 8' x 10' | 1.000 | Ea. | 22.080 | 2024.00 | 456.00 | 2480.00 |
| 8' x 16' | 1.000 | Ea. | 38.419 | 3521.76 | 793.44 | 4315.20 |
| 8' x 24' | 1.000 | Ea. | 49.680 | 4554.00 | 1026.00 | 5580.00 |
| Free standing, 8' x 8' | 1.000 | Ea. | 17.344 | 2323.20 | 364.80 | 2688.00 |
| 8' x 16' | 1.000 | Ea. | 30.189 | 4043.82 | 634.98 | 4678.80 |
| 8' x 24' | 1.000 | Ea. | 39.024 | 5227.20 | 820.80 | 6048.00 |
| 10' x 10' | 1.000 | Ea. | 18.800 | 2860.00 | 340.00 | 3200.00 |
| 10' x 16' | 1.000 | Ea. | 24.064 | 3660.80 | 435.20 | 4096.00 |
| 10' x 24' | 1.000 | Ea. | 31.584 | 4804.80 | 571.20 | 5376.00 |
| 14' x 10' | 1.000 | Ea. | 20.720 | 3542.00 | 378.00 | 3920.00 |
| 14' x 16' | 1.000 | Ea. | 24.864 | 4250.40 | 453.60 | 4704.00 |
| 14' x 24' | 1.000 | Ea. | 33.314 | 5695.03 | 607.77 | 6302.80 |
| Standard, lean to, shell only, not incl. 2' stub wall, fndtn, flrs, heat 4'x10' | 1.000 | Ea. | 28.260 | 1980.00 | 540.00 | 2520.00 |
| 4' x 16' | 1.000 | Ea. | 39.375 | 2758.80 | 752.40 | 3511.20 |
| 4' x 24' | 1.000 | Ea. | 45.451 | 3184.50 | 868.50 | 4053.00 |
| 6' x 10' | 1.000 | Ea. | 24.840 | 2277.00 | 513.00 | 2790.00 |
| 6' x 16' | 1.000 | Ea. | 34.555 | 3167.56 | 713.64 | 3881.20 |
| 6' x 24' | 1.000 | Ea. | 44.712 | 4098.60 | 923.40 | 5022.00 |
| 8' x 10' | 1.000 | Ea. | 33.120 | 3036.00 | 684.00 | 3720.00 |
| 8' x 16' | 1.000 | Ea. | 57.628 | 5282.64 | 1190.16 | 6472.80 |
| 8' x 24' | 1.000 | Ea. | 74.520 | 6831.00 | 1539.00 | 8370.00 |
| Free standing, 8' x 8' | 1.000 | Ea. | 26.016 | 3484.80 | 547.20 | 4032.00 |
| 8' x 16' | 1.000 | Ea. | 45.284 | 6065.73 | 952.47 | 7018.20 |
| 8' x 24' | 1.000 | Ea. | 58.536 | 7840.80 | 1231.20 | 9072.00 |
| 10' x 10' | 1.000 | Ea. | 28.200 | 4290.00 | 510.00 | 4800.00 |
| 10' x 16' | 1.000 | Ea. | 36.096 | 5491.20 | 652.80 | 6144.00 |
| 10' x 24' | 1.000 | Ea. | 47.376 | 7207.20 | 856.80 | 8064.00 |
| 14' x 10' | 1.000 | Ea. | 31.080 | 5313.00 | 567.00 | 5880.00 |
| 14' x 16' | 1.000 | Ea. | 37.296 | 6375.60 | 680.40 | 7056.00 |
| 14' x 24' | 1.000 | Ea. | 49.979 | 8543.81 | 911.79 | 9455.60 |
| Deluxe, lean to, shell only, not incl. 2' stub wall, fndtn, flrs or heat, 4'x10' | 1.000 | Ea. | 20.640 | 3344.00 | 416.00 | 3760.00 |
| 4' x 16' | 1.000 | Ea. | 33.024 | 5350.40 | 665.60 | 6016.00 |
| 4' x 24' | 1.000 | Ea. | 49.536 | 8025.60 | 998.40 | 9024.00 |
| 6' x 10' | 1.000 | Ea. | 30.960 | 5016.00 | 624.00 | 5640.00 |
| 6' x 16' | 1.000 | Ea. | 49.536 | 8025.60 | 998.40 | 9024.00 |
| 6' x 24' | 1.000 | Ea. | 74.304 | 12038.40 | 1497.60 | 13536.00 |
| 8' x 10' | 1.000 | Ea. | 41.280 | 6688.00 | 832.00 | 7520.00 |
| 8' x 16' | 1.000 | Ea. | 66.048 | 10700.80 | 1331.20 | 12032.00 |
| 8' x 24' | 1.000 | Ea. | 99.072 | 16051.20 | 1996.80 | 18048.00 |
| Freestanding, 8' x 8' | 1.000 | Ea. | 18.624 | 4646.40 | 345.60 | 4992.00 |
| 8' x 16' | 1.000 | Ea. | 37.248 | 9292.80 | 691.20 | 9984.00 |
| 8' x 24' | 1.000 | Ea. | 55.872 | 13939.20 | 1036.80 | 14976.00 |
| 10' x 10' | 1.000 | Ea. | 29.100 | 7260.00 | 540.00 | 7800.00 |
| 10' x 16' | 1.000 | Ea. | 46.560 | 11616.00 | 864.00 | 12480.00 |
| 10' x 24' | 1.000 | Ea. | 69.840 | 17424.00 | 1296.00 | 18720.00 |
| 14' x 10' | 1.000 | Ea. | 40.740 | 10164.00 | 756.00 | 10920.00 |
| 14' x 16' | 1.000 | Ea. | 65.184 | 16262.40 | 1209.60 | 17472.00 |
| 14' x 24' | 1.000 | Ea. | 97.776 | 24393.60 | 1814.40 | 26208.00 |

# 7 | SPECIALTIES  36 | Swimming Pool Systems

| SYSTEM DESCRIPTION | QUAN. | UNIT | MAN HOURS | MAT. | INST. | TOTAL |
|---|---|---|---|---|---|---|
| Swimming pools, vinyl lined, metal sides, sand bottom, 12' x 28' | 1.000 | Ea. | | 4300.80 | | 4300.80 |
| 12' x 32' | 1.000 | Ea. | | 4745.60 | | 4745.60 |
| 12' x 36' | 1.000 | Ea. | | 5148.00 | | 5148.00 |
| 16' x 32' | 1.000 | Ea. | | 5725.44 | | 5725.44 |
| 16' x 36' | 1.000 | Ea. | | 6101.92 | | 6101.92 |
| 16' x 40' | 1.000 | Ea. | | 6403.04 | | 6403.04 |
| 20' x 36' | 1.000 | Ea. | | 6673.60 | | 6673.60 |
| 20' x 40' | 1.000 | Ea. | | 7040.00 | | 7040.00 |
| 20' x 44' | 1.000 | Ea. | | 7744.00 | | 7744.00 |
| 24' x 40' | 1.000 | Ea. | | 8448.00 | | 8448.00 |
| 24' x 44' | 1.000 | Ea. | | 9292.80 | | 9292.80 |
| 24' x 48' | 1.000 | Ea. | | 10137.60 | | 10137.60 |
| Vinyl lined, concrete sides, 12' x 28' | 1.000 | Ea. | | 6809.60 | | 6809.60 |
| 12' x 32' | 1.000 | Ea. | | 7612.80 | | 7612.80 |
| 12' x 36' | 1.000 | Ea. | | 8370.40 | | 8370.40 |
| 16' x 32' | 1.000 | Ea. | | 9547.84 | | 9547.84 |
| 16' x 36' | 1.000 | Ea. | | 10401.60 | | 10401.60 |
| 16' x 40' | 1.000 | Ea. | | 11180.80 | | 11180.80 |
| 28' x 36' | 1.000 | Ea. | | 12049.92 | | 12049.92 |
| 20' x 40' | 1.000 | Ea. | | 12800.00 | | 12800.00 |
| 20' x 44' | 1.000 | Ea. | | 14080.00 | | 14080.00 |
| 24' x 40' | 1.000 | Ea. | | 15360.00 | | 15360.00 |
| 24' x 44' | 1.000 | Ea. | | 16896.00 | | 16896.00 |
| 24' x 48' | 1.000 | Ea. | | 18432.00 | | 18432.00 |
| Gunite, bottom and sides, 12' x 28' | 1.000 | Ea. | | 8400.00 | | 8400.00 |
| 12' x 32' | 1.000 | Ea. | | 9202.50 | | 9202.50 |
| 12' x 36' | 1.000 | Ea. | | 9905.75 | | 9905.75 |
| 16' x 32' | 1.000 | Ea. | | 10858.25 | | 10858.25 |
| 16' x 36' | 1.000 | Ea. | | 11420.00 | | 11420.00 |
| 16' x 40' | 1.000 | Ea. | | 11805.00 | | 11805.00 |
| 20' x 36' | 1.000 | Ea. | | 12166.25 | | 12166.25 |
| 20' x 40' | 1.000 | Ea. | | 12800.00 | | 12800.00 |
| 20' x 44' | 1.000 | Ea. | | 14080.00 | | 14080.00 |
| 24' x 40' | 1.000 | Ea. | | 15360.00 | | 15360.00 |
| 24' x 44' | 1.000 | Ea. | | 17687.50 | | 17687.50 |
| 24' x 48' | 1.000 | Ea. | | 19295.00 | | 19295.00 |

# 7 | SPECIALTIES    40 | Wood Deck Systems

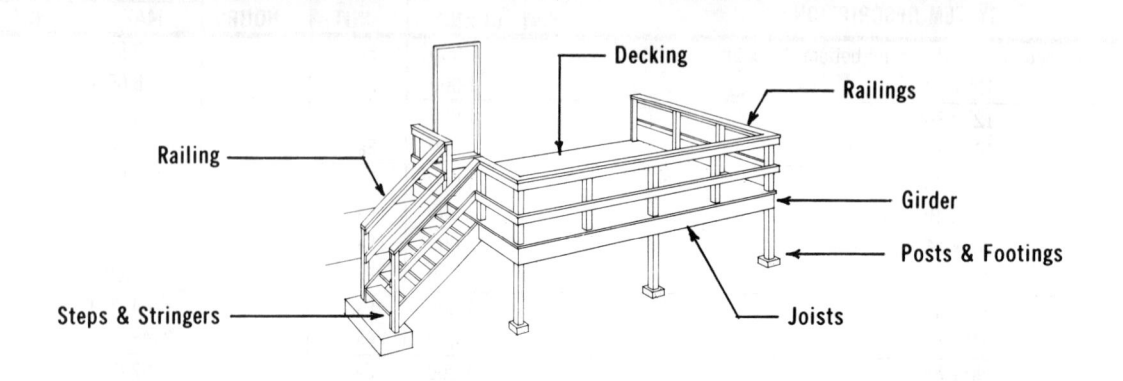

| SYSTEM DESCRIPTION | QUAN. | UNIT | MAN HOURS | MAT. | INST. | TOTAL |
|---|---|---|---|---|---|---|
| **8' X 12' DECK, PRESSURE TREATED LUMBER, JOISTS 16" O.C.** | | | | | | |
| Decking, 2" x 6" lumber | 2.080 | L.F. | .031 | 1.08 | .65 | 1.73 |
| Joists, 2" x 8", 16" O.C. | 1.000 | L.F. | .017 | .71 | .35 | 1.06 |
| Girder, 2" x 10" | .125 | L.F. | .002 | .13 | .04 | .17 |
| Posts, 4" x 4", including concrete footing | .250 | L.F. | .020 | .38 | .41 | .79 |
| Stairs, 2" x 10" stringers, 2" x 10" steps | 1.000 | Set | .020 | .55 | .40 | .95 |
| Railings, 2" x 4" | 1.000 | L.F. | .024 | .47 | .48 | .95 |
| TOTAL | | | .115 | 3.32 | 2.33 | 5.65 |
| **12' X 16' DECK, PRESSURE TREATED LUMBER, JOISTS 24" O.C.** | | | | | | |
| Decking, 2" x 6" | 2.080 | L.F. | .031 | 1.08 | .65 | 1.73 |
| Joists, 2" x 10", 24" O.C. | .800 | L.F. | .014 | .78 | .28 | 1.06 |
| Girder, 2" x 10" | .083 | L.F. | .001 | .08 | .03 | .11 |
| Posts, 4" x 4", including concrete footing | .122 | L.F. | .015 | .25 | .30 | .55 |
| Stairs, 2" x 10" stringers, 2" x 10" steps | 1.000 | Set | .012 | .33 | .24 | .57 |
| Railings, 2" x 4" | .670 | L.F. | .016 | .31 | .32 | .63 |
| TOTAL | | | .090 | 2.83 | 1.82 | 4.65 |
| **12' X 24' DECK, REDWOOD OR CEDAR, JOISTS 16" O.C.** | | | | | | |
| Decking, 2" x 6" redwood | 2.080 | L.F. | .027 | 2.14 | .69 | 2.83 |
| Joists, 2" x 10", 16" O.C. | 1.000 | L.F. | .018 | 1.82 | .36 | 2.18 |
| Girder, 2" x 10" | .083 | L.F. | .001 | .15 | .03 | .18 |
| Post, 4" x 4", including concrete footing | .111 | L.F. | .010 | .60 | .38 | .98 |
| Stairs, 2" x 10" stringers, 2" x 10" steps | 1.000 | Set | .012 | 1.34 | .24 | 1.58 |
| Railings, 2" x 4" | .540 | L.F. | .004 | .37 | .12 | .49 |
| TOTAL | | | .073 | 6.42 | 1.82 | 8.24 |

The costs in this system are on a square foot basis.

| DESCRIPTION | QUAN. | UNIT | MAN HOURS | MAT. | INST. | TOTAL |
|---|---|---|---|---|---|---|
| | | | | | | |
| | | | | | | |
| | | | | | | |
| | | | | | | |
| | | | | | | |

# Wood Deck Price Sheet

| | QUAN. | UNIT | MAN HOURS | COST PER S.F. MAT. | COST PER S.F. INST. | TOTAL |
|---|---|---|---|---|---|---|
| Decking, treated lumber, 1" x 4" | 3.430 | L.F. | .031 | 1.73 | .61 | 2.34 |
| 1" x 6" | 2.180 | L.F. | .032 | 1.82 | .64 | 2.46 |
| 2" x 4" | 3.200 | L.F. | .041 | 1.16 | .83 | 1.99 |
| 2" x 6" | 2.080 | L.F. | .031 | 1.08 | .65 | 1.73 |
| Redwood or cedar,, 1" x 4" | 3.430 | L.F. | .034 | 2.07 | .67 | 2.74 |
| 1" x 6" | 2.180 | L.F. | .035 | 2.16 | .70 | 2.86 |
| 2" x 4" | 3.200 | L.F. | .028 | 2.27 | .72 | 2.99 |
| 2" x 6" | 2.080 | L.F. | .027 | 2.14 | .69 | 2.83 |
| Joists for deck, treated lumber, 2" x 8", 16" O.C. | 1.000 | L.F. | .017 | .71 | .35 | 1.06 |
| 24" O.C. | .800 | L.F. | .013 | .57 | .28 | .85 |
| 2" x 10", 16" O.C. | 1.000 | L.F. | .018 | .97 | .36 | 1.33 |
| 24" O.C. | .800 | L.F. | .014 | .78 | .28 | 1.06 |
| Redwood or cedar, 2" x 8", 16" O.C. | 1.000 | L.F. | .015 | 1.41 | .30 | 1.71 |
| 24" O.C. | .800 | L.F. | .012 | 1.13 | .24 | 1.37 |
| 2" x 10", 16" O.C. | 1.000 | L.F. | .018 | 1.82 | .36 | 2.18 |
| 24" O.C. | .800 | L.F. | .014 | 1.46 | .28 | 1.74 |
| Girder for joists, treated lumber, 2" x 10", 8' x 12' deck | .125 | L.F. | .002 | .13 | .04 | .17 |
| 12' x 16' deck | .083 | L.F. | .001 | .08 | .03 | .11 |
| 12' x 24' deck | .083 | L.F. | .001 | .08 | .03 | .11 |
| Redwood or cedar, 2" x 10", 8' x 12' deck | .125 | L.F. | .002 | .23 | .04 | .27 |
| 12' x 16' deck | .083 | L.F. | .001 | .15 | .03 | .18 |
| 12' x 24' deck | .083 | L.F. | .001 | .15 | .03 | .18 |
| Posts, 4" x 4", including concrete footing, 8' x 12' deck | .250 | L.F. | .020 | .38 | .41 | .79 |
| 12' x 16' deck | .122 | L.F. | .015 | .25 | .30 | .55 |
| 12' x 24' deck | .111 | L.F. | .014 | .24 | .30 | .54 |
| Stairs 2" x 10" stringers, treated lumber, 8' x 12' deck | 1.000 | Set | .020 | .55 | .40 | .95 |
| 12' x 16' deck | 1.000 | Set | .012 | .33 | .24 | .57 |
| 12' x 24' deck | 1.000 | Set | .008 | .22 | .16 | .38 |
| Redwood or cedar, 8' x 12' deck | 1.000 | Set | .040 | 4.46 | .79 | 5.25 |
| 12' x 16' deck | 1.000 | Set | .020 | 2.23 | .40 | 2.63 |
| 12' x 24' deck | 1.000 | Set | .012 | 1.34 | .24 | 1.58 |
| Railings 2" x 4", treated lumber, 8' x 12' deck | 1.000 | L.F. | .024 | .47 | .48 | .95 |
| 12' x 16' deck | .670 | L.F. | .016 | .31 | .32 | .63 |
| 12' x 24' deck | .540 | L.F. | .012 | .25 | .26 | .51 |
| Redwood or cedar, 8' x 12' deck | 1.000 | L.F. | .008 | .69 | .22 | .91 |
| 12' x 16' deck | .670 | L.F. | .005 | .45 | .15 | .60 |
| 12' x 24' deck | .540 | L.F. | .004 | .37 | .12 | .49 |

**7**

# SECTION 8 MECHANICAL

| System Number | Title | Page |
|---|---|---|
| 8-04 | Two Fixture Lavatory | 162 |
| 8-08 | Two Fixture Lavatory | 164 |
| 8-12 | Three Fixture Bathroom | 166 |
| 8-16 | Three Fixture Bathroom | 168 |
| 8-20 | Three Fixture Bathroom | 170 |
| 8-24 | Three Fixture Bathroom | 172 |
| 8-28 | Three Fixture Bathroom | 174 |
| 8-32 | Three Fixture Bathroom | 176 |
| 8-36 | Four Fixture Bathroom | 178 |
| 8-40 | Four Fixture Bathroom | 180 |
| 8-44 | Five Fixture Bathroom | 182 |
| 8-50 | Lavatory | 184 |
| 8-60 | Gas Fired Heating/Cooling | 186 |
| 8-64 | Oil Fired Heating/Cooling | 188 |
| 8-68 | Hot Water Heating | 190 |
| 8-80 | Rooftop Heating/Cooling | 192 |

# 8 | MECHANICAL  04 | Two Fixture Lavatory Systems

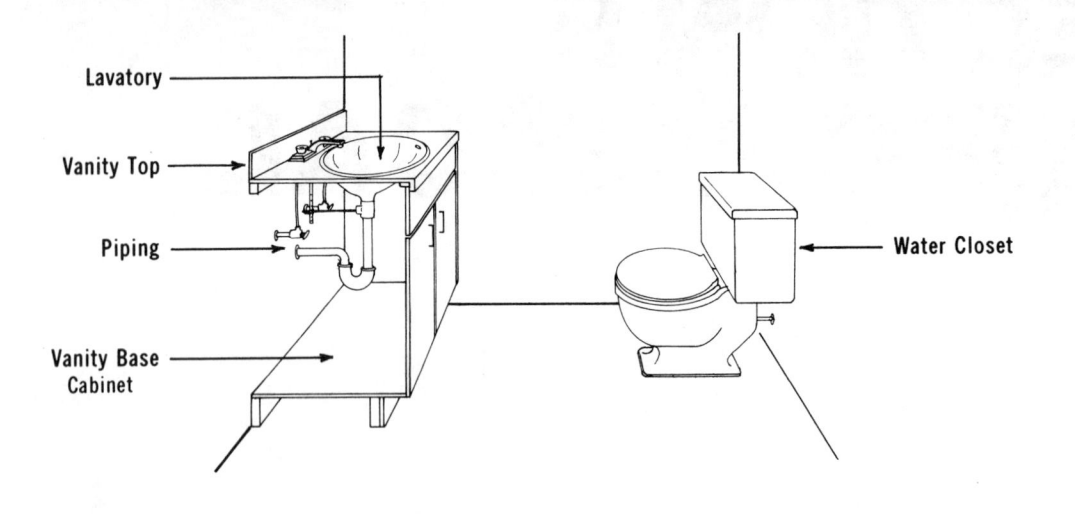

| SYSTEM DESCRIPTION | QUAN. | UNIT | MAN HOURS | MAT. | INST. | TOTAL |
|---|---|---|---|---|---|---|
| **LAVATORY INSTALLED WITH VANITY, PLUMBING IN 2 WALLS** | | | | | | |
| Water closet, floor mounted, 2 piece, close coupled, white | 1.000 | Ea. | 3.020 | 115.50 | 59.50 | 175.00 |
| Rough-in supply, waste and vent for water closet | 1.000 | Ea. | 2.862 | 35.35 | 58.19 | 93.54 |
| Lavatory, 20" x 18", P.E. cast iron white | 1.000 | Ea. | 2.500 | 110.00 | 50.00 | 160.00 |
| Rough-in supply, waste and vent for lavatory | 1.000 | Ea. | 3.578 | 34.52 | 73.68 | 108.20 |
| Piping, supply, copper 1/2" diameter, type "L" | 10.000 | L.F. | 1.510 | 8.40 | 33.00 | 41.40 |
| Waste, cast iron, 4" diameter, no-hub | 7.000 | L.F. | 2.072 | 39.83 | 40.67 | 80.50 |
| Vent, cast iron, 2" diameter, no-hub | 12.000 | L.F. | 3.204 | 39.72 | 62.88 | 102.60 |
| Vanity base cabinet, 2 door, 30" wide | 1.000 | Ea. | 1.630 | 110.00 | 30.00 | 140.00 |
| Vanity top, plastic & laminated, square edge | 2.670 | L.F. | .712 | 61.68 | 13.08 | 74.76 |
| TOTAL | | | 21.088 | 555.00 | 421.00 | 976.00 |
| **LAVATORY WITH WALL-HUNG LAVATORY, PLUMBING IN 2 WALLS** | | | | | | |
| Water closet, floor mounted, 2 piece close coupled, white | 1.000 | Ea. | 3.020 | 115.50 | 59.50 | 175.00 |
| Rough-in supply, waste and vent for water closet | 1.000 | Ea. | 2.862 | 35.35 | 58.19 | 93.54 |
| Lavatory, 20" x 18", P.E. cast iron, wall hung, white | 1.000 | Ea. | 2.000 | 126.50 | 38.50 | 165.00 |
| Rough-in supply, waste and vent for lavatory | 1.000 | Ea. | 3.578 | 34.52 | 73.68 | 108.20 |
| Piping, supply, copper 1/2" diameter, type "L" | 10.000 | L.F. | 1.510 | 8.40 | 33.00 | 41.40 |
| Waste, cast iron, 4" diameter, no-hub | 7.000 | L.F. | 2.072 | 39.83 | 40.67 | 80.50 |
| Vent, cast iron, 2" diameter, no hub | 12.000 | L.F. | 3.204 | 39.72 | 62.88 | 102.60 |
| Carrier, steel for studs, no arms | 1.000 | Ea. | 1.140 | 11.61 | 25.39 | 37.00 |
| TOTAL | | | 19.386 | 411.43 | 391.81 | 803.24 |

| DESCRIPTION | QUAN. | UNIT | MAN HOURS | MAT. | INST. | TOTAL |
|---|---|---|---|---|---|---|
| | | | | | | |
| | | | | | | |
| | | | | | | |
| | | | | | | |
| | | | | | | |

# Two Fixture Lavatory Price Sheet

| | QUAN. | UNIT | MAN HOURS | COST EACH MAT. | COST EACH INST. | TOTAL |
|---|---|---|---|---|---|---|
| Water closet, close coupled standard 2 piece, white | 1.000 | Ea. | 3.020 | 115.50 | 59.50 | 175.00 |
| Color | 1.000 | Ea. | 3.020 | 137.50 | 57.50 | 195.00 |
| One piece elongated bowl, white | 1.000 | Ea. | 3.020 | 407.00 | 58.00 | 465.00 |
| Color | 1.000 | Ea. | 3.020 | 528.00 | 57.00 | 585.00 |
| Low profile, one piece elongated bowl, white | 1.000 | Ea. | 3.020 | 627.00 | 58.00 | 685.00 |
| Color | 1.000 | Ea. | 3.020 | 825.00 | 60.00 | 885.00 |
| Rough-in for water closet | | | | | | |
| 1/2" copper supply, 4" cast iron waste, 2" cast iron vent | 1.000 | Ea. | 2.862 | 35.35 | 58.19 | 93.54 |
| 4" PVC waste, 2" PVC vent | 1.000 | Ea. | 3.711 | 22.52 | 74.87 | 97.39 |
| 4" copper waste, 2" copper vent | 1.000 | Ea. | 4.239 | 41.96 | 86.88 | 128.84 |
| 3" cast iron waste, 1-1/2" cast iron vent | 1.000 | Ea. | 2.734 | 31.12 | 55.87 | 86.99 |
| 3" PVC waste, 1-1/2" PVC vent | 1.000 | Ea. | 3.174 | 18.39 | 66.65 | 85.04 |
| 3" copper waste, 1-1/2" copper vent | 1.000 | Ea. | 4.406 | 38.99 | 88.75 | 127.74 |
| 1/2" PVC supply, 4" PVC waste, 2" PVC vent | 1.000 | Ea. | 3.825 | 26.78 | 77.27 | 104.05 |
| 3" PVC waste, 1-1/2" PVC vent | 1.000 | Ea. | 3.288 | 22.65 | 69.05 | 91.70 |
| 1/2" steel supply, 4" cast iron waste, 2" cast iron vent | 1.000 | Ea. | 2.718 | 38.65 | 55.01 | 93.66 |
| 4" cast iron waste, 2" steel vent | 1.000 | Ea. | 2.650 | 43.57 | 53.69 | 97.26 |
| 4" PVC waste, 2" PVC vent | 1.000 | Ea. | 3.567 | 25.82 | 71.69 | 97.51 |
| Lavatory, vanity top mounted, P.E. on cast iron 20" x 18" white | 1.000 | Ea. | 2.500 | 110.00 | 50.00 | 160.00 |
| Color | 1.000 | Ea. | 2.500 | 137.50 | 47.50 | 185.00 |
| Steel, enameled 10" x 17" white | 1.000 | Ea. | 2.500 | 66.00 | 49.00 | 115.00 |
| Color | 1.000 | Ea. | 2.500 | 69.30 | 50.70 | 120.00 |
| Vitreous china 20" x 16", white | 1.000 | Ea. | 2.500 | 165.00 | 50.00 | 215.00 |
| Color | 1.000 | Ea. | 2.500 | 176.00 | 49.00 | 225.00 |
| Wall hung, P.E. on cast iron, 20" x 18", white | 1.000 | Ea. | 2.000 | 126.50 | 38.50 | 165.00 |
| Color | 1.000 | Ea. | 2.000 | 159.50 | 40.50 | 200.00 |
| Vitreous china 19" x 17", white | 1.000 | Ea. | 2.000 | 79.20 | 40.80 | 120.00 |
| Color | 1.000 | Ea. | 2.000 | 91.30 | 38.70 | 130.00 |
| Rough-in supply waste and vent for lavatory | | | | | | |
| 1/2" copper supply, 2" cast iron waste, 1-1/2" cast iron vent | 1.000 | Ea. | 3.578 | 34.52 | 73.68 | 108.20 |
| 2" PVC waste, 1-1/2" PVC vent | 1.000 | Ea. | 3.970 | 19.64 | 83.56 | 103.20 |
| 2" copper waste, 1-1/2" copper vent | 1.000 | Ea. | 3.986 | 29.40 | 87.20 | 116.60 |
| 1-1/2" PVC waste, 1-1/4" PVC vent | 1.000 | Ea. | 3.646 | 18.56 | 79.64 | 98.20 |
| 1-1/2" copper waste, 1-1/4" copper vent | 1.000 | Ea. | 3.722 | 24.96 | 81.44 | 106.40 |
| 1/2" PVC supply, 2" PVC waste, 1-1/2" PVC vent | 1.000 | Ea. | 4.160 | 26.74 | 87.56 | 114.30 |
| 1-1/2" PVC waste, 1-1/4" PVC vent | 1.000 | Ea. | 3.836 | 25.66 | 83.64 | 109.30 |
| 1/2" steel supply, 2" cast iron waste, 1-1/2" cast iron vent | 1.000 | Ea. | 3.338 | 40.02 | 68.38 | 108.40 |
| 2" cast iron waste, 2" steel vent | 1.000 | Ea. | 3.338 | 45.30 | 68.30 | 113.60 |
| 2" PVC waste, 1-1/2" PVC vent | 1.000 | Ea. | 3.730 | 25.14 | 78.26 | 103.40 |
| 1-1/2" PVC waste, 1-1/4" PVC vent | 1.000 | Ea. | 3.406 | 24.06 | 74.34 | 98.40 |
| Piping, supply, 1/2" copper, type "L" | 10.000 | L.F. | 1.510 | 8.40 | 33.00 | 41.40 |
| 1/2" steel | 10.000 | L.F. | 1.270 | 13.90 | 27.70 | 41.60 |
| 1/2" PVC | 10.000 | L.F. | 1.700 | 15.50 | 37.00 | 52.50 |
| Waste, 4" cast iron | 7.000 | L.F. | 2.072 | 39.83 | 40.67 | 80.50 |
| 4" copper | 7.000 | L.F. | 4.669 | 58.52 | 88.48 | 147.00 |
| 4" PVC | 7.000 | L.F. | 3.297 | 26.60 | 64.75 | 91.35 |
| Vent, 2" cast iron | 12.000 | L.F. | 3.204 | 39.72 | 62.88 | 102.60 |
| 2" copper | 12.000 | L.F. | 3.996 | 35.52 | 87.48 | 123.00 |
| 2" PVC | 12.000 | L.F. | 4.176 | 18.24 | 81.96 | 100.20 |
| 2" steel | 12.000 | Ea. | 3.000 | 54.48 | 58.92 | 113.40 |
| Vanity base cabinet, 2 door, 24" x 30" | 1.000 | Ea. | 1.630 | 110.00 | 30.00 | 140.00 |
| 24" x 36" | 1.000 | Ea. | 2.000 | 154.00 | 41.00 | 195.00 |
| Vanity top, laminated plastic, square edge 25" x 32" | 2.670 | L.F. | .712 | 61.68 | 13.08 | 74.76 |
| 25" x 38" | 3.170 | L.F. | .846 | 73.23 | 15.53 | 88.76 |
| Post formed, laminated plastic, 25" x 32" | 2.670 | L.F. | .712 | 23.50 | 13.88 | 37.38 |
| 25" x 38" | 3.170 | L.F. | .846 | 27.90 | 16.48 | 44.38 |
| Cultured marble, 25" x 32" with bowl | 1.000 | Ea. | 2.500 | 85.80 | 49.20 | 135.00 |
| 25" x 38" with bowl | 1.000 | Ea. | 2.500 | 104.50 | 50.50 | 155.00 |
| Carrier for lavatory, steel for studs | 1.000 | Ea. | 1.140 | 11.61 | 25.39 | 37.00 |
| Wood 2" x 8" blocking | 1.330 | L.F. | .052 | .73 | 1.09 | 1.82 |

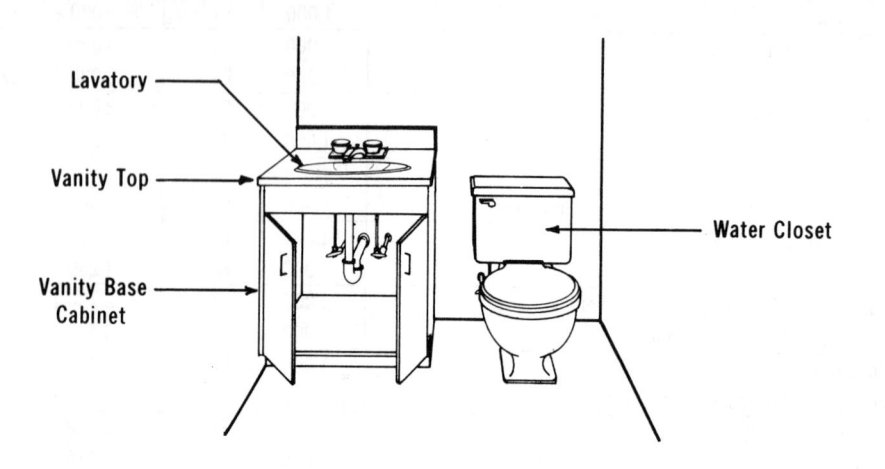

# 8 | MECHANICAL  08 | Two Fixture Lavatory Systems

| SYSTEM DESCRIPTION | QUAN. | UNIT | MAN HOURS | MAT. | INST. | TOTAL |
|---|---|---|---|---|---|---|
| **LAVATORY INSTALLED WITH VANITY, PLUMBING IN 1 WALL** | | | | | | |
| Water closet, floor mounted, 2 piece, close coupled, white | 1.000 | Ea. | 3.020 | 115.50 | 59.50 | 175.00 |
| Rough-in supply, waste and vent for water closet | 1.000 | Ea. | 2.862 | 35.35 | 58.19 | 93.54 |
| Lavatory, 20" x 18" P.E. cast iron, white | 1.000 | Ea. | 2.500 | 110.00 | 50.00 | 160.00 |
| Rough-in supply, waste and vent for lavatory | 1.000 | Ea. | 3.578 | 34.52 | 73.68 | 108.20 |
| Piping, supply, copper 1/2" diameter, "L" type | 6.000 | L.F. | .906 | 5.04 | 19.80 | 24.84 |
| Waste, cast iron, 4" diameter, no-hub | 4.000 | L.F. | 1.184 | 22.76 | 23.24 | 46.00 |
| Vent, cast iron, 2" diameter, no-hub | 6.000 | L.F. | 1.602 | 19.86 | 31.44 | 51.30 |
| Vanity base cabinet, 2 door 30" wide | 1.000 | Ea. | 1.630 | 110.00 | 30.00 | 140.00 |
| Vanity top, plastic & laminated, square edges | 2.670 | L.F. | .712 | 61.68 | 13.08 | 74.76 |
| TOTAL | | | 17.994 | 514.71 | 358.93 | 873.64 |
| **LAVATORY WITH WALL-HUNG LAVATORY, PLUMBING IN 1 WALL** | | | | | | |
| Water closet, floor mounted, 2 piece close coupled, white | 1.000 | Ea. | 3.020 | 115.50 | 59.50 | 175.00 |
| Rough-in supply, waste and vent for water closet | 1.000 | Ea. | 2.862 | 35.35 | 58.19 | 93.54 |
| Lavatory, 20" x 18", P.E. cast iron, wall hung, white | 1.000 | Ea. | 2.000 | 126.50 | 38.50 | 165.00 |
| Rough-in supply, waste and vent for lavatory | 1.000 | Ea. | 3.578 | 34.52 | 73.68 | 108.20 |
| Piping, supply, copper 1/2" diameter, type "L" | 6.000 | L.F. | .906 | 5.04 | 19.80 | 24.84 |
| Waste, cast iron, 4" diameter, no-hub | 4.000 | L.F. | 1.184 | 22.76 | 23.24 | 46.00 |
| Vent, cast iron, 2" diameter, no-hub | 6.000 | L.F. | 1.602 | 19.86 | 31.44 | 51.30 |
| Carrier, steel for studs, no arms | 1.000 | Ea. | 1.330 | 82.50 | 27.50 | 110.00 |
| TOTAL | | | 16.292 | 371.14 | 329.74 | 700.88 |

The costs in this system are on a cost each basis. All necessary piping is included.

| DESCRIPTION | QUAN. | UNIT | MAN HOURS | MAT. | INST. | TOTAL |
|---|---|---|---|---|---|---|
| | | | | | | |
| | | | | | | |
| | | | | | | |
| | | | | | | |
| | | | | | | |

# Two Fixture Lavatory Price Sheet

| | QUAN. | UNIT | MAN HOURS | COST EACH MAT. | COST EACH INST. | TOTAL |
|---|---|---|---|---|---|---|
| Water closet, close coupled, standard 2 piece, white | 1.000 | Ea. | 3.020 | 115.50 | 59.50 | 175.00 |
| Color | 1.000 | Ea. | 3.020 | 137.50 | 57.50 | 195.00 |
| One piece, elongated bowl, white | 1.000 | Ea. | 3.020 | 407.00 | 58.00 | 465.00 |
| Color | 1.000 | Ea. | 3.020 | 528.00 | 57.00 | 585.00 |
| Low profile, one piece elongated bowl, white | 1.000 | Ea. | 3.020 | 627.00 | 58.00 | 685.00 |
| Color | 1.000 | Ea. | 3.020 | 869.00 | 61.00 | 930.00 |
| Rough-in for water closet | | | | | | |
| 1/2" copper supply, 4" cast iron waste, 2" cast iron vent | 1.000 | Ea. | 2.862 | 35.35 | 58.19 | 93.54 |
| 4" PVC waste, 2" PVC vent | 1.000 | Ea. | 3.711 | 22.52 | 74.87 | 97.39 |
| 4" copper waste, 2" copper vent | 1.000 | Ea. | 4.239 | 41.96 | 86.88 | 128.84 |
| 3" cast iron waste, 1-1/2" cast iron vent | 1.000 | Ea. | 2.734 | 31.12 | 55.87 | 86.99 |
| 3" PVC waste, 1-1/2" PVC vent | 1.000 | Ea. | 3.174 | 18.39 | 66.65 | 85.04 |
| 3" copper waste, 1-1/2" copper vent | 1.000 | Ea. | 3.550 | 28.75 | 74.39 | 103.14 |
| 1/2" PVC supply, 4" PVC waste, 2" PVC vent | 1.000 | Ea. | 3.825 | 26.78 | 77.27 | 104.05 |
| 3" PVC waste, 1-1/2" PVC vent | 1.000 | Ea. | 3.288 | 22.65 | 69.05 | 91.70 |
| 1/2" steel supply, 4" cast iron waste, 2" cast iron vent | 1.000 | Ea. | 2.718 | 38.65 | 55.01 | 93.66 |
| 4" cast iron waste, 2" steel vent | 1.000 | Ea. | 2.650 | 43.57 | 53.69 | 97.26 |
| 4" PVC waste, 2" PVC vent | 1.000 | Ea. | 3.567 | 25.82 | 71.69 | 97.51 |
| Lavatory, vanity top mounted, P.E. on cast iron 20" x 18" white | 1.000 | Ea. | 2.500 | 110.00 | 50.00 | 160.00 |
| Color | 1.000 | Ea. | 2.500 | 137.50 | 47.50 | 185.00 |
| Steel enameled 20" x 17" white | 1.000 | Ea. | 2.500 | 66.00 | 49.00 | 115.00 |
| Color | 1.000 | Ea. | 2.500 | 69.30 | 50.70 | 120.00 |
| Vitreous china 20" x 16", white | 1.000 | Ea. | 2.500 | 165.00 | 50.00 | 215.00 |
| Color | 1.000 | Ea. | 2.500 | 176.00 | 49.00 | 225.00 |
| Wall hung, P.E., on cast iron 20" x 8", white | 1.000 | Ea. | 2.000 | 126.50 | 38.50 | 165.00 |
| Color | 1.000 | Ea. | 2.000 | 159.50 | 40.50 | 200.00 |
| Vitreous china 19" x 17", white | 1.000 | Ea. | 2.000 | 79.20 | 40.80 | 120.00 |
| Color | 1.000 | Ea. | 2.000 | 91.30 | 38.70 | 130.00 |
| Rough-in for lavatory | | | | | | |
| 1/2" copper supply, 2" cast iron waste, 1-1/2" cast iron vent | 1.000 | Ea. | 3.578 | 34.52 | 73.68 | 108.20 |
| 2" PVC waste, 1-1/2" PVC vent | 1.000 | Ea. | 3.970 | 19.64 | 83.56 | 103.20 |
| 2" copper waste, 1-1/2" copper vent | 1.000 | Ea. | 3.986 | 29.40 | 87.20 | 116.60 |
| 1-1/2" PVC waste, 1-1/4" PVC vent | 1.000 | Ea. | 3.646 | 18.56 | 79.64 | 98.20 |
| 1-1/2 copper waste, 1-1/4" copper vent | 1.000 | Ea. | 3.722 | 24.96 | 81.44 | 106.40 |
| 1/2" PVC supply, 2" PVC waste, 1-1/2" PVC vent | 1.000 | Ea. | 4.160 | 26.74 | 87.56 | 114.30 |
| 1-1/2" PVC waste, 1-1/4" PVC vent | 1.000 | Ea. | 3.836 | 25.66 | 83.64 | 109.30 |
| 1/2" steel supply, 2" cast iron waste, 1-1/2" cast iron vent | 1.000 | Ea. | 3.338 | 40.02 | 68.38 | 108.40 |
| 2" cast iron waste, 2" steel vent | 1.000 | Ea. | 3.338 | 45.30 | 68.30 | 113.60 |
| 2" PVC waste 1-1/2" PVC vent | 1.000 | Ea. | 3.730 | 25.14 | 78.26 | 103.40 |
| 1-1//2" PVC waste, 1-1/4" PVC vent | 1.000 | Ea. | 3.406 | 24.06 | 74.34 | 98.40 |
| Piping, supply, 1/2" copper, type "L" | 6.000 | L.F. | .906 | 5.04 | 19.80 | 24.84 |
| 1/2" steel | 6.000 | L.F. | .762 | 8.34 | 16.62 | 24.96 |
| 1/2" PVC | 6.000 | L.F. | 1.020 | 9.30 | 22.20 | 31.50 |
| Piping, waste, 4" cast iron | 4.000 | L.F. | 1.184 | 22.76 | 23.24 | 46.00 |
| 4" copper | 4.000 | L.F. | 2.668 | 33.44 | 50.56 | 84.00 |
| 4" PVC | 4.000 | L.F. | 1.884 | 15.20 | 37.00 | 52.20 |
| Piping, vent, 2" cast iron | 6.000 | L.F. | 1.602 | 19.86 | 31.44 | 51.30 |
| 2" copper | 6.000 | L.F. | 1.998 | 17.76 | 43.74 | 61.50 |
| 2" PVC | 6.000 | L.F. | 2.088 | 9.12 | 40.98 | 50.10 |
| 2" steel | 6.000 | L.F. | 1.500 | 27.24 | 29.46 | 56.70 |
| Vanity base cabinet 2 door, 24" x 30" | 1.000 | Ea. | 1.630 | 110.00 | 30.00 | 140.00 |
| 24" x 36" | 1.000 | Ea. | 2.000 | 154.00 | 41.00 | 195.00 |
| Vanity top laminated plastic, square edge 25" x 32" | 2.670 | L.F. | .712 | 61.68 | 13.08 | 74.76 |
| 25" x 38" | 3.170 | L.F. | .846 | 73.23 | 15.53 | 88.76 |
| Post formed, 25" x 32" | 2.670 | L.F. | .712 | 23.50 | 13.88 | 37.38 |
| 25" x 38" | 3.170 | L.F. | .846 | 27.90 | 16.48 | 44.38 |
| Cultured marble, 25" x 38" with bowl | 3.170 | Ea. | 2.500 | 85.80 | 49.20 | 135.00 |
| 25" x 38" with bowl | 3.170 | Ea. | 2.500 | 104.50 | 50.50 | 155.00 |
| Carrier for lavatory, steel for studs | 1.000 | Ea. | 1.140 | 11.61 | 25.39 | 37.00 |
| Wood 2" x 8" blocking | 1.330 | L.F. | .052 | .72 | 1.06 | 1.78 |

8

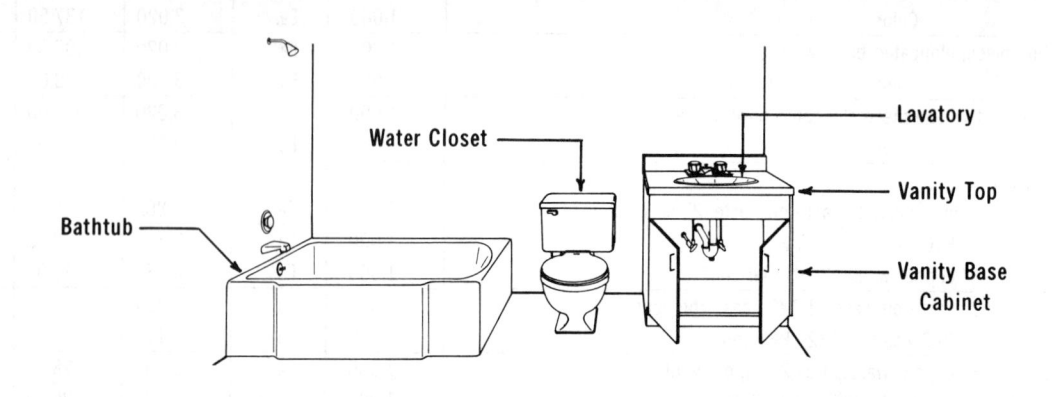

# 8 | MECHANICAL   12 | Three Fixture Bathroom Systems

| SYSTEM DESCRIPTION | QUAN. | UNIT | MAN HOURS | MAT. | INST. | TOTAL |
|---|---|---|---|---|---|---|
| **BATHROOM INSTALLED WITH VANITY** | | | | | | |
| Water closet, floor mounted, 2 piece, close coupled, white | 1.000 | Ea. | 3.020 | 115.50 | 59.50 | 175.00 |
| Rough-in supply, waste and vent for water closet | 1.000 | Ea. | 2.862 | 35.35 | 58.19 | 93.54 |
| Lavatory, 20" x 18", P.E. cast iron with accessories, white | 1.000 | Ea. | 2.500 | 110.00 | 50.00 | 160.00 |
| Rough-in supply, waste and vent for lavatory | 1.000 | Ea. | 3.510 | 34.16 | 72.44 | 106.60 |
| Bathtub, P.E. cast iron, 5' long with accessories, white | 1.000 | Ea. | 3.640 | 269.50 | 70.50 | 340.00 |
| Rough-in supply, waste and vent for bathtub | 1.000 | Ea. | 3.542 | 34.63 | 75.47 | 110.10 |
| Piping, supply, 1/2" copper | 20.000 | L.F. | 3.020 | 16.80 | 66.00 | 82.80 |
| Waste, 4" cast iron, no hub | 9.000 | L.F. | 2.664 | 51.21 | 52.29 | 103.50 |
| Vent, 2" galvanized steel | 6.000 | L.F. | 1.500 | 27.24 | 29.46 | 56.70 |
| Vanity base cabinet, 2 door, 30" wide | 1.000 | Ea. | 1.630 | 110.00 | 30.00 | 140.00 |
| Vanity top, plastic laminated square edge | 2.670 | L.F. | .712 | 48.03 | 13.38 | 61.41 |
| TOTAL | | | 28.600 | 852.42 | 577.23 | 1429.65 |
| **BATHROOM WITH WALL HUNG LAVATORY** | | | | | | |
| Water closet, floor mounted, 2 piece, close coupled, white | 1.000 | Ea. | 3.020 | 115.50 | 59.50 | 175.00 |
| Rough-in supply, waste and vent for water closet | 1.000 | Ea. | 2.862 | 35.35 | 58.19 | 93.54 |
| Lavatory, 20" x 18" P.E. cast iron, wall hung, white | 1.000 | Ea. | 2.000 | 126.50 | 38.50 | 165.00 |
| Rough-in supply, waste and vent for lavatory | 1.000 | Ea. | 3.510 | 34.16 | 72.44 | 106.60 |
| Bathtub, P.E. cast iron, 5' long with accessories, white | 1.000 | Ea. | 3.640 | 269.50 | 70.50 | 340.00 |
| Rough-in supply, waste and vent for bathtub | 1.000 | Ea. | 8.118 | 71.27 | 175.63 | 246.90 |
| Piping, supply, 1/2" copper | 20.000 | L.F. | 3.020 | 16.80 | 66.00 | 82.80 |
| Waste, 4" cast iron, no hub | 9.000 | L.F. | 2.664 | 51.21 | 52.29 | 103.50 |
| Vent, 2" galvanized steel | 6.000 | L.F. | 1.500 | 27.24 | 29.46 | 56.70 |
| Carrier, steel, for studs, no arms | 1.000 | Ea. | 1.140 | 11.61 | 25.39 | 37.00 |
| TOTAL | | | 28.614 | 736.24 | 585.30 | 1321.54 |

The costs in this system are a cost each basis, all necessary piping is included

| DESCRIPTION | QUAN. | UNIT | MAN HOURS | MAT. | INST. | TOTAL |
|---|---|---|---|---|---|---|
| | | | | | | |
| | | | | | | |
| | | | | | | |
| | | | | | | |
| | | | | | | |

# Three Fixture Bathroom Price Sheet

| | QUAN. | UNIT | MAN HOURS | COST EACH MAT. | COST EACH INST. | TOTAL |
|---|---|---|---|---|---|---|
| Water closet, close coupled standard 2 piece, white | 1.000 | Ea. | 3.020 | 115.50 | 59.50 | 175.00 |
| Color | 1.000 | Ea. | 3.020 | 137.50 | 57.50 | 195.00 |
| One piece, elongated bowl, white | 1.000 | Ea. | 3.020 | 407.00 | 58.00 | 465.00 |
| Color | 1.000 | Ea. | 3.020 | 528.00 | 57.00 | 585.00 |
| Low profile, one piece elongated bowl, white | 1.000 | Ea. | 3.020 | 627.00 | 58.00 | 685.00 |
| Color | 1.000 | Ea. | 3.020 | 825.00 | 60.00 | 885.00 |
| Rough-in for water closet | | | | | | |
| 1/2" copper supply, 4" cast iron waste, 2" cast iron vent | 1.000 | Ea. | 2.862 | 35.35 | 58.19 | 93.54 |
| 4" PVC/DWV waste, 2" PVC | 1.000 | Ea. | 3.711 | 22.52 | 74.87 | 97.39 |
| 4" copper waste, 2" copper vent | 1.000 | Ea. | 4.239 | 41.96 | 86.88 | 128.84 |
| 3" cast iron waste, 1-1/2" cast iron vent | 1.000 | Ea. | 2.734 | 31.12 | 55.87 | 86.99 |
| 3" PVC waste, 1-1/2" PVC vent | 1.000 | Ea. | 3.174 | 18.39 | 66.65 | 85.04 |
| 3" copper waste, 1-1/2" copper vent | 1.000 | Ea. | 3.550 | 28.75 | 74.39 | 103.14 |
| 1/2" PVC supply, 4" PVC waste, 2" PVC vent | 1.000 | Ea. | 3.825 | 26.78 | 77.27 | 104.05 |
| 3" PVC waste, 1-1/2" PVC supply | 1.000 | Ea. | 3.288 | 22.65 | 69.05 | 91.70 |
| 1/2" steel supply, 4" cast iron waste, 2" cast iron vent | 1.000 | Ea. | 2.718 | 38.65 | 55.01 | 93.66 |
| 4" cast iron waste, 2" steel vent | 1.000 | Ea. | 2.650 | 43.57 | 53.69 | 97.26 |
| 4" PVC waste, 2" PVC vent | 1.000 | Ea. | 3.567 | 25.82 | 71.69 | 97.51 |
| Lavatory, wall hung, P.E. cast iron 20" x 18", white | 1.000 | Ea. | 2.000 | 126.50 | 38.50 | 165.00 |
| Color | 1.000 | Ea. | 2.000 | 159.50 | 40.50 | 200.00 |
| Vitreous china 19" x 17", white | 1.000 | Ea. | 2.000 | 79.20 | 40.80 | 120.00 |
| Color | 1.000 | Ea. | 2.000 | 91.30 | 38.70 | 130.00 |
| Lavatory, for vanity top, P.E. cast iron 20" x 18"", white | 1.000 | Ea. | 2.500 | 110.00 | 50.00 | 160.00 |
| Color | 1.000 | Ea. | 2.500 | 137.50 | 47.50 | 185.00 |
| Steel, enameled 20" x 17", white | 1.000 | Ea. | 2.500 | 66.00 | 49.00 | 115.00 |
| Color | 1.000 | Ea. | 2.500 | 69.30 | 50.70 | 120.00 |
| Vitreous china 20" x 16", white | 1.000 | Ea. | 2.500 | 165.00 | 50.00 | 215.00 |
| Color | 1.000 | Ea. | 2.500 | 176.00 | 49.00 | 225.00 |
| Rough-in for lavatory | | | | | | |
| 1/2" copper supply, 1-1/2" C.I. waste, 1-1/2" C.I. vent | 1.000 | Ea. | 3.510 | 34.16 | 72.44 | 106.60 |
| 1-1/2" PVC waste, 1-1/4" PVC vent | 1.000 | Ea. | 3.646 | 18.56 | 79.64 | 98.20 |
| 1/2" steel supply, 1-1/4" cast iron waste, 1-1/4" steel vent | 1.000 | Ea. | 2.990 | 38.10 | 61.50 | 99.60 |
| 1-1/4" PVC waste, 1-1/4" PVC vent | 1.000 | Ea. | 3.406 | 23.90 | 74.50 | 98.40 |
| 1/2" PVC supply, 1-1/2" PVC waste, 1-1/2" PVC vent | 1.000 | Ea. | 3.836 | 25.82 | 83.48 | 109.30 |
| Bathtub, P.E. cast iron, 5' long corner with fittings, white | 1.000 | Ea. | 3.640 | 269.50 | 70.50 | 340.00 |
| Color | 1.000 | Ea. | 3.640 | 346.50 | 73.50 | 420.00 |
| Rough-in for bathtub | | | | | | |
| 1/2" copper supply, 4" cast iron waste, 1-1/2" copper vent | 1.000 | Ea. | 3.542 | 34.63 | 75.47 | 110.10 |
| 4" PVC waste, 1-1/2" PVC vent | 1.000 | Ea. | 3.991 | 24.96 | 83.99 | 108.95 |
| 1/2" steel supply, 4" cast iron waste, 1-1/2" steel vent | 1.000 | Ea. | 2.958 | 44.21 | 60.89 | 105.10 |
| 4" PVC waste, 1-1/2" PVC vent | 1.000 | Ea. | 3.751 | 30.46 | 78.69 | 109.15 |
| 1/2" PVC supply, 4" PVC waste, 1-1/2" PVC vent | 1.000 | Ea. | 4.181 | 32.06 | 87.99 | 120.05 |
| Piping, supply 1/2" copper | 20.000 | L.F. | 3.020 | 16.80 | 66.00 | 82.80 |
| 1/2" steel | 20.000 | L.F. | 2.540 | 27.80 | 55.40 | 83.20 |
| 1/2" PVC | 20.000 | L.F. | 3.400 | 31.00 | 74.00 | 105.00 |
| Piping, waste, 4" cast iron no hub | 9.000 | L.F. | 2.664 | 51.21 | 52.29 | 103.50 |
| 4" PVC/DWV | 9.000 | L.F. | 4.239 | 34.20 | 83.25 | 117.45 |
| 4" copper/DWV | 9.000 | L.F. | 6.003 | 75.24 | 113.76 | 189.00 |
| Piping, vent 2" cast iron no hub | 6.000 | L.F. | 1.602 | 19.86 | 31.44 | 51.30 |
| 2" copper/DWV | 6.000 | L.F. | 1.998 | 17.76 | 43.74 | 61.50 |
| 2" PVC/DWV | 6.000 | L.F. | 2.088 | 9.12 | 40.98 | 50.10 |
| 2" steel, galvanized | 6.000 | L.F. | 1.500 | 27.24 | 29.46 | 56.70 |
| Vanity base cabinet, 2 door, 24" x 30" | 1.000 | Ea. | 1.630 | 110.00 | 30.00 | 140.00 |
| 24" x 36" | 1.000 | Ea. | 2.000 | 154.00 | 41.00 | 195.00 |
| Vanity top, laminated plastic square edge 25" x 32" | 2.670 | L.F. | .712 | 48.03 | 13.38 | 61.41 |
| 25" x 38" | 3.160 | L.F. | .843 | 56.85 | 15.83 | 72.68 |
| Cultured marble, 25" x 32", with bowl | 1.000 | Ea. | 2.500 | 85.80 | 49.20 | 135.00 |
| 25" x 38", with bowl | 1.000 | Ea. | 2.500 | 104.50 | 50.50 | 155.00 |
| Carrier, for lavatory, steel for studs, no arms | 1.000 | Ea. | 1.140 | 11.61 | 25.39 | 37.00 |
| Wood, 2" x 8" blocking | 1.300 | L.F. | .052 | .72 | 1.06 | 1.78 |

# 8 | MECHANICAL  16 | Three Fixture Bathroom Systems

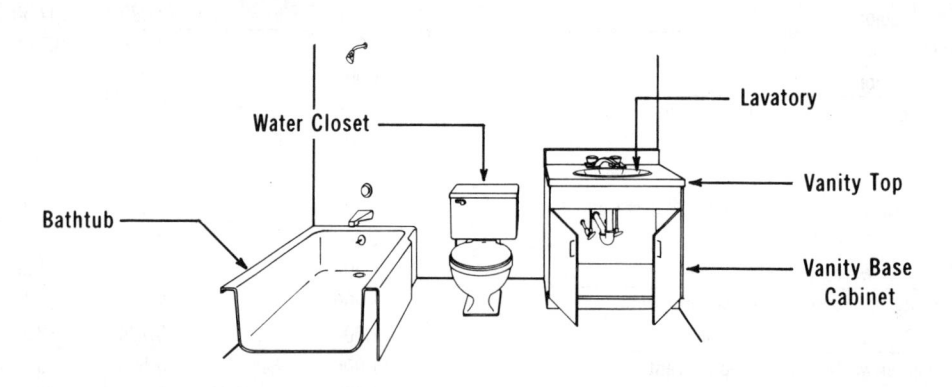

| SYSTEM DESCRIPTION | QUAN. | UNIT | MAN HOURS | MAT. | INST. | TOTAL |
|---|---|---|---|---|---|---|
| **BATHROOM WITH LAVATORY INSTALLED IN VANITY** | | | | | | |
| Water closet, floor mounted, 2 piece, close coupled, white | 1.000 | Ea. | 3.020 | 115.50 | 59.50 | 175.00 |
| Rough-in supply, waste and vent for water closet | 1.000 | Ea. | 2.862 | 35.35 | 58.19 | 93.54 |
| Lavatory, 20" x 18", P.E. cast iron with accessories, white | 1.000 | Ea. | 2.500 | 110.00 | 50.00 | 160.00 |
| Rough-in supply, waste and vent for lavatory | 1.000 | Ea. | 3.510 | 34.16 | 72.44 | 106.60 |
| Bathtub, P.E. cast iron 5' long with accessories, white | 1.000 | Ea. | 3.640 | 269.50 | 70.50 | 340.00 |
| Rough-in supply, waste and vent for bathtub | 1.000 | Ea. | 3.542 | 34.63 | 75.47 | 110.10 |
| Piping, supply, 1/2" copper | 10.000 | L.F. | 1.510 | 8.40 | 33.00 | 41.40 |
| Waste, 4" cast iron, no hub | 6.000 | L.F. | 1.776 | 34.14 | 34.86 | 69.00 |
| Vent, 2" galvanized steel | 6.000 | L.F. | 1.500 | 27.24 | 29.46 | 56.70 |
| Vanity base cabinet, 2 door, 30" wide | 1.000 | Ea. | 1.630 | 110.00 | 30.00 | 140.00 |
| Vanity top, plastic laminated square edge | 2.670 | L.F. | .712 | 48.03 | 13.38 | 61.41 |
| TOTAL | | | 26.202 | 826.95 | 526.80 | 1353.75 |
| **BATHROOM WITH WALL HUNG LAVATORY** | | | | | | |
| Water closet, floor mounted, 2 piece, close coupled, white | 1.000 | Ea. | 3.020 | 115.50 | 59.50 | 175.00 |
| Rough-in supply, waste and vent for water closet | 1.000 | Ea. | 2.862 | 35.35 | 58.19 | 93.54 |
| Lavatory, 20" x 18" P.E. cast iron, wall hung, white | 1.000 | Ea. | 2.000 | 126.50 | 38.50 | 165.00 |
| Rough-in supply, waste and vent for lavatory | 1.000 | Ea. | 3.510 | 34.16 | 72.44 | 106.60 |
| Bathtub, P.E. cast iron, 5' long with accessories, white | 1.000 | Ea. | 3.640 | 269.50 | 70.50 | 340.00 |
| Rough-in supply, waste and vent for bathtub | 1.000 | Ea. | 3.542 | 34.63 | 75.47 | 110.10 |
| Piping, supply, 1/2" copper | 10.000 | L.F. | 1.510 | 8.40 | 33.00 | 41.40 |
| Waste, 4" cast iron, no hub | 6.000 | L.F. | 1.776 | 34.14 | 34.86 | 69.00 |
| Vent, 2" galvanized steel | 6.000 | L.F. | 1.500 | 27.24 | 29.46 | 56.70 |
| Carrier, steel, for studs, no arms | 1.000 | Ea. | 1.140 | 11.61 | 25.39 | 37.00 |
| TOTAL | | | 24.500 | 697.03 | 497.31 | 1194.34 |

The costs in this system are on a cost each basis. All necessary piping is included.

| DESCRIPTION | QUAN. | UNIT | MAN HOURS | MAT. | INST. | TOTAL |
|---|---|---|---|---|---|---|
| | | | | | | |
| | | | | | | |
| | | | | | | |
| | | | | | | |
| | | | | | | |

# Three Fixture Bathroom Price Sheet

| Description | QUAN. | UNIT | MAN HOURS | MAT. | INST. | TOTAL |
|---|---|---|---|---|---|---|
| Water closet, close coupled standard 2 piece, white | 1.000 | Ea. | 3.020 | 115.50 | 59.50 | 175.00 |
| Color | 1.000 | Ea. | 3.020 | 137.50 | 57.50 | 195.00 |
| One piece elongated bowl, white | 1.000 | Ea. | 3.020 | 407.00 | 58.00 | 465.00 |
| Color | 1.000 | Ea. | 3.020 | 528.00 | 57.00 | 585.00 |
| Low profile, one piece elongated bowl, white | 1.000 | Ea. | 3.020 | 627.00 | 58.00 | 685.00 |
| Color | 1.000 | Ea. | 3.020 | 825.00 | 60.00 | 885.00 |
| Rough-in for water closet | | | | | | |
| 1/2" copper supply, 4" cast iron waste, 2" cast iron vent | 1.000 | Ea. | 2.862 | 35.35 | 58.19 | 93.54 |
| 4" PVC/DWV waste, 2" PVC vent | 1.000 | Ea. | 3.711 | 22.52 | 74.87 | 97.39 |
| 4" carrier waste, 2" copper vent | 1.000 | Ea. | 4.239 | 41.96 | 86.88 | 128.84 |
| 3" cast iron waste, 1-1/2" cast iron vent | 1.000 | Ea. | 2.734 | 31.12 | 55.87 | 86.99 |
| 3" PVC waste, 1-1/2" PVC vent | 1.000 | Ea. | 3.174 | 18.39 | 66.65 | 85.04 |
| 3" copper waste, 1-1/2" copper vent | 1.000 | Ea. | 3.550 | 28.75 | 74.39 | 103.14 |
| 1/2" PVC supply, 4" PVC waste, 2" PVC vent | 1.000 | Ea. | 3.825 | 26.78 | 77.27 | 104.05 |
| 3" PVC waste, 1-1/2" PVC supply | 1.000 | Ea. | 3.288 | 22.65 | 69.05 | 91.70 |
| 1/2" steel supply, 4" cast iron waste, 2" cast iron vent | 1.000 | Ea. | 2.718 | 38.65 | 55.01 | 93.66 |
| 4" cast iron waste, 2" steel vent | 1.000 | Ea. | 2.650 | 43.57 | 53.69 | 97.26 |
| 4" PVC waste, 2" PVC vent | 1.000 | Ea. | 3.567 | 25.82 | 71.69 | 97.51 |
| Lavatory, wall hung, PE cast iron 20" x 18", white | 1.000 | Ea. | 2.000 | 126.50 | 38.50 | 165.00 |
| Color | 1.000 | Ea. | 2.000 | 159.50 | 40.50 | 200.00 |
| Vitreous china 19" x 17", white | 1.000 | Ea. | 2.000 | 79.20 | 40.80 | 120.00 |
| Color | 1.000 | Ea. | 2.000 | 91.30 | 38.70 | 130.00 |
| Lavatory, for vanity top, PE cast iron 20" x 18", white | 1.000 | Ea. | 2.500 | 110.00 | 50.00 | 160.00 |
| Color | 1.000 | Ea. | 2.500 | 137.50 | 47.50 | 185.00 |
| Steel enameled 20" x 17", white | 1.000 | Ea. | 2.500 | 66.00 | 49.00 | 115.00 |
| Color | 1.000 | Ea. | 2.500 | 69.30 | 50.70 | 120.00 |
| Vitreous china 20" x 16", white | 1.000 | Ea. | 2.500 | 165.00 | 50.00 | 215.00 |
| Color | 1.000 | Ea. | 2.500 | 176.00 | 49.00 | 225.00 |
| Rough-in for lavatory | | | | | | |
| 1/2" copper supply, 1-1/2" cast iron waste, 1-1/2" cast iron vent | 1.000 | Ea. | 3.510 | 34.16 | 72.44 | 106.60 |
| 1-1/2" PVC waste, 1-1/4" PVC vent | 1.000 | Ea. | 3.646 | 18.56 | 79.64 | 98.20 |
| 1/2" steel supply, 1-1/4" cast iron waste, 1-1/4" steel vent | 1.000 | Ea. | 2.990 | 38.10 | 61.50 | 99.60 |
| 1-1/4" PVC waste, 1-1/4" PVC vent | 1.000 | Ea. | 3.406 | 23.90 | 74.50 | 98.40 |
| 1/2" PVC supply, 1-1/2" PVC waste, 1-1/2" PVC vent | 1.000 | Ea. | 3.836 | 25.82 | 83.48 | 109.30 |
| Bathtub, PE cast iron, 5' long corner with fittings, white | 1.000 | Ea. | 3.640 | 269.50 | 70.50 | 340.00 |
| Color | 1.000 | Ea. | 3.640 | 346.50 | 73.50 | 420.00 |
| Rough-in for bathtub | | | | | | |
| 1/2" copper supply, 4" cast iron waste, 1-1/2" copper vent | 1.000 | Ea. | 3.542 | 34.63 | 75.47 | 110.10 |
| 4" PVC waste, 1/2" PVC vent | 1.000 | Ea. | 3.991 | 24.96 | 83.99 | 108.95 |
| 1/2" steel supply, 4" cast iron waste, 1-1/2" steel vent | 1.000 | Ea. | 2.958 | 44.21 | 60.89 | 105.10 |
| 4" PVC waste, 1-1/2" PVC vent | 1.000 | Ea. | 3.751 | 30.46 | 78.69 | 109.15 |
| 1/2" PVC supply, 4" PVC waste, 1-1/2" PVC vent | 1.000 | Ea. | 4.181 | 32.06 | 87.99 | 120.05 |
| Piping supply, 1/2" copper | 10.000 | L.F. | 1.510 | 8.40 | 33.00 | 41.40 |
| 1/2" steel | 10.000 | L.F. | 1.270 | 13.90 | 27.70 | 41.60 |
| 1/2" PVC | 10.000 | L.F. | 1.700 | 15.50 | 37.00 | 52.50 |
| Piping waste, 4" cast iron no hub | 6.000 | L.F. | 1.776 | 34.14 | 34.86 | 69.00 |
| 4" PVC/DWV | 6.000 | L.F. | 2.826 | 22.80 | 55.50 | 78.30 |
| 4" copper/DWV | 6.000 | L.F. | 4.002 | 50.16 | 75.84 | 126.00 |
| Piping vent 2" cast iron no hub | 6.000 | L.F. | 1.602 | 19.86 | 31.44 | 51.30 |
| 2" copper/DWV | 6.000 | L.F. | 1.998 | 17.76 | 43.74 | 61.50 |
| 2" PVC/DWV | 6.000 | L.F. | 2.088 | 9.12 | 40.98 | 50.10 |
| 2" | 6.000 | L.F. | 1.500 | 27.24 | 29.46 | 56.70 |
| Vanity base cabinet, 2 door, 24" x 30" | 1.000 | Ea. | 1.630 | 110.00 | 30.00 | 140.00 |
| 24" x 36" | 1.000 | Ea. | 2.000 | 154.00 | 41.00 | 195.00 |
| Vanity top, laminated plastic square edge 25" x 32" | 2.670 | L.F. | .712 | 48.03 | 13.38 | 61.41 |
| 25" x 38" | 3.160 | L.F. | .843 | 56.85 | 15.83 | 72.68 |
| Cultured marble, 25" x 32", with bowl | 1.000 | Ea. | 2.500 | 85.80 | 49.20 | 135.00 |
| 25" x 38", with bowl | 1.000 | Ea. | 2.500 | 104.50 | 50.50 | 155.00 |
| Carrier, for lavatory, steel for studs, no arms | 1.000 | Ea. | 1.140 | 11.61 | 25.39 | 37.00 |
| Wood, 2" x 8" blocking | 1.300 | L.F. | .052 | .72 | 1.06 | 1.78 |

8

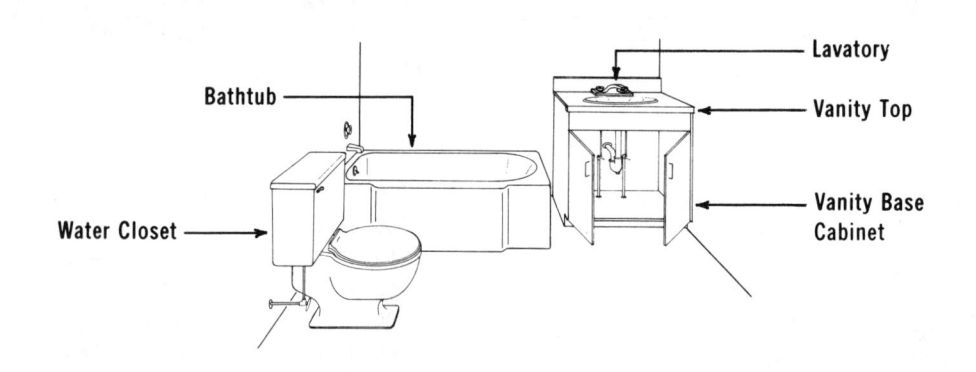

# 8 | MECHANICAL  20 | Three Fixture Bathroom Systems

| SYSTEM DESCRIPTION | QUAN. | UNIT | MAN HOURS | MAT. | INST. | TOTAL |
|---|---|---|---|---|---|---|
| **BATHROOM WITH LAVATORY INSTALLED IN VANITY** | | | | | | |
| Water closet, floor mounted, 2 piece, close coupled, white | 1.000 | Ea. | 3.020 | 115.50 | 59.50 | 175.00 |
| Rough-in supply, waste and vent for water closet | 1.000 | Ea. | 2.862 | 35.35 | 58.19 | 93.54 |
| Lavatory, 20" x 18", PE cast iron with accessories, white | 1.000 | Ea. | 2.500 | 110.00 | 50.00 | 160.00 |
| Rough-in supply, waste and vent for lavatory | 1.000 | Ea. | 3.510 | 34.16 | 72.44 | 106.60 |
| Bathtub, P.E. cast iron, 5' long with accessories, white | 1.000 | Ea. | 3.640 | 269.50 | 70.50 | 340.00 |
| Rough-in supply, waste and vent for bathtub | 1.000 | Ea. | 3.542 | 34.63 | 75.47 | 110.10 |
| Piping, supply, 1/2" copper | 32.000 | L.F. | 4.832 | 26.88 | 105.60 | 132.48 |
| Waste, 4" cast iron, no hub | 12.000 | L.F. | 3.552 | 68.28 | 69.72 | 138.00 |
| Vent, 2" galvanized steel | 6.000 | L.F. | 1.500 | 27.24 | 29.46 | 56.70 |
| Vanity base cabinet, 2 door, 30" wide | 1.000 | Ea. | 1.630 | 110.00 | 30.00 | 140.00 |
| Vanity top, plastic laminated square edge | 2.670 | L.F. | .712 | 48.03 | 13.38 | 61.41 |
| TOTAL | | | 31.300 | 879.57 | 634.26 | 1513.83 |
| **BATHROOM WITH WALL HUNG LAVATORY** | | | | | | |
| Water closet, floor mounted, 2 piece, close coupled, white | 1.000 | Ea. | 3.020 | 115.50 | 59.50 | 175.00 |
| Rough-in supply, waste and vent for water closet | 1.000 | Ea. | 2.862 | 35.35 | 58.19 | 93.54 |
| Lavatory, 20" x 18" P.E. cast iron, wall hung, white | 1.000 | Ea. | 2.000 | 126.50 | 38.50 | 165.00 |
| Rough-in supply, waste and vent for lavatory | 1.000 | Ea. | 3.510 | 34.16 | 72.44 | 106.60 |
| Bathtub, P.E. cast iron, 5' long with accessories, white | 1.000 | Ea. | 3.640 | 269.50 | 70.50 | 340.00 |
| Rough-in supply, waste and vent for bathtub | 1.000 | Ea. | 3.542 | 34.63 | 75.47 | 110.10 |
| Piping, supply, 1/2" copper | 32.000 | L.F. | 4.832 | 26.88 | 105.60 | 132.48 |
| Waste, 4" cast iron, no hub | 12.000 | L.F. | 3.552 | 68.28 | 69.72 | 138.00 |
| Vent, 2" galvanized steel | 6.000 | L.F. | 1.500 | 27.24 | 29.46 | 56.70 |
| Carrier steel, for studs, no arms | 1.000 | Ea. | 1.140 | 11.61 | 25.39 | 37.00 |
| TOTAL | | | 29.598 | 749.65 | 604.77 | 1354.42 |

The costs in this system are on a cost each basis. All necessary piping is included.

| DESCRIPTION | QUAN. | UNIT | MAN HOURS | MAT. | INST. | TOTAL |
|---|---|---|---|---|---|---|
| | | | | | | |
| | | | | | | |
| | | | | | | |
| | | | | | | |
| | | | | | | |

## Three Fixture Bathroom Price Sheet

| | QUAN. | UNIT | MAN HOURS | COST EACH MAT. | COST EACH INST. | TOTAL |
|---|---|---|---|---|---|---|
| Water closet, close coupled, standard 2 piece, white | 1.000 | Ea. | 3.020 | 115.50 | 59.50 | 175.00 |
| Color | 1.000 | Ea. | 3.020 | 137.50 | 57.50 | 195.00 |
| One piece, elongated bowl, white | 1.000 | Ea. | 3.020 | 407.00 | 58.00 | 465.00 |
| Color | 1.000 | Ea. | 3.020 | 528.00 | 57.00 | 585.00 |
| Low profile, one piece, elongated bowl, white | 1.000 | Ea. | 3.020 | 627.00 | 58.00 | 685.00 |
| Color | 1.000 | Ea. | 3.020 | 825.00 | 60.00 | 885.00 |
| Rough-in for water closet | | | | | | |
| 1/2" copper supply, 4" cast iron waste, 2" cast iron vent | 1.000 | Ea. | 2.862 | 35.35 | 58.19 | 93.54 |
| 4" PVC/DWV waste, 2" PVC vent | 1.000 | Ea. | 3.711 | 22.52 | 74.87 | 97.39 |
| 4" copper waste, 2" copper vent | 1.000 | Ea. | 4.239 | 41.96 | 86.88 | 128.84 |
| 3" cast iron waste, 1-1/2" cast iron vent | 1.000 | Ea. | 2.734 | 31.12 | 55.87 | 86.99 |
| 3" PVC waste, 1-1/2" PVC vent | 1.000 | Ea. | 3.174 | 18.39 | 66.65 | 85.04 |
| 3" copper waste, 1-1/2" copper vent | 1.000 | Ea. | 3.550 | 28.75 | 74.39 | 103.14 |
| 1/2" PVC supply, 4" PVC waste, 2" PVC vent | 1.000 | Ea. | 3.825 | 26.78 | 77.27 | 104.05 |
| 3" PVC waste, 1-1/2" PVC supply | 1.000 | Ea. | 3.288 | 22.65 | 69.05 | 91.70 |
| 1/2" steel supply, 4" cast iron waste, 2" cast iron vent | 1.000 | Ea. | 2.718 | 38.65 | 55.01 | 93.66 |
| 4" cast iron waste, 2" steel vent | 1.000 | Ea. | 2.650 | 43.57 | 53.69 | 97.26 |
| 4" PVC waste, 2" PVC vent | 1.000 | Ea. | 3.567 | 25.82 | 71.69 | 97.51 |
| Lavatory wall hung, P.E. cast iron, 20" x 18", white | 1.000 | Ea. | 2.000 | 126.50 | 38.50 | 165.00 |
| Color | 1.000 | Ea. | 2.000 | 159.50 | 40.50 | 200.00 |
| Vitreous china, 19" x 17", white | 1.000 | Ea. | 2.000 | 79.20 | 40.80 | 120.00 |
| Color | 1.000 | Ea. | 2.000 | 91.30 | 38.70 | 130.00 |
| Lavatory, for vanity top, P.E., cast iron, 20" x 18", white | 1.000 | Ea. | 2.500 | 110.00 | 50.00 | 160.00 |
| Color | 1.000 | Ea. | 2.500 | 137.50 | 47.50 | 185.00 |
| Steel, enameled, 20" x 17", white | 1.000 | Ea. | 2.500 | 66.00 | 49.00 | 115.00 |
| Color | 1.000 | Ea. | 2.500 | 69.30 | 50.70 | 120.00 |
| Vitreous china, 20" x 16", white | 1.000 | Ea. | 2.500 | 165.00 | 50.00 | 215.00 |
| Color | 1.000 | Ea. | 2.500 | 176.00 | 49.00 | 225.00 |
| Rough-in for lavatory | | | | | | |
| 1/2" copper supply, 1-1/2" C.I. waste, 1-1/2" C.I. vent | 1.000 | Ea. | 3.510 | 34.16 | 72.44 | 106.60 |
| 1-1/2" PVC waste, 1-1/4" PVC vent | 1.000 | Ea. | 3.646 | 18.56 | 79.64 | 98.20 |
| 1/2" steel supply, 1-1/4" cast iron waste, 1-1/4" steel vent | 1.000 | Ea. | 2.990 | 38.10 | 61.50 | 99.60 |
| 1-1/4" PVC waste, 1-1/4" PVC vent | 1.000 | Ea. | 3.406 | 23.90 | 74.50 | 98.40 |
| 1/2" PVC supply, 1-1/2" PVC waste, 1-1/2" PVC vent | 1.000 | Ea. | 3.836 | 25.82 | 83.48 | 109.30 |
| Bathtub, P.E. cast iron, 5' long corner with fittings, white | 1.000 | Ea. | 3.640 | 269.50 | 70.50 | 340.00 |
| Color | 1.000 | Ea. | 3.640 | 346.50 | 73.50 | 420.00 |
| Rough-in for bathtub | | | | | | |
| 1/2" copper supply, 4" cast iron waste, 1-1/2" copper vent | 1.000 | Ea. | 3.542 | 34.63 | 75.47 | 110.10 |
| 4" PVC waste, 1/2" PVC vent | 1.000 | Ea. | 3.991 | 24.96 | 83.99 | 108.95 |
| 1/2" steel supply, 4" cast iron waste, 1-1/2" steel vent | 1.000 | Ea. | 2.958 | 44.21 | 60.89 | 105.10 |
| 4" PVC waste, 1-1/2" PVC vent | 1.000 | Ea. | 3.751 | 30.46 | 78.69 | 109.15 |
| 1/2" PVC supply, 4" PVC waste, 1-1/2" PVC vent | 1.000 | Ea. | 4.181 | 32.06 | 87.99 | 120.05 |
| Piping, supply, 1/2" copper | 32.000 | L.F. | 4.832 | 26.88 | 105.60 | 132.48 |
| 1/2" steel | 32.000 | L.F. | 4.064 | 44.48 | 88.64 | 133.12 |
| 1/2" PVC | 32.000 | L.F. | 5.440 | 49.60 | 118.40 | 168.00 |
| Piping, waste, 4" cast iron no hub | 12.000 | L.F. | 3.552 | 68.28 | 69.72 | 138.00 |
| 4" PVC/DWV | 12.000 | L.F. | 5.652 | 45.60 | 111.00 | 156.60 |
| 4" copper/DWV | 12.000 | L.F. | 8.004 | 100.32 | 151.68 | 252.00 |
| Piping, vent, 2" cast iron no hub | 6.000 | L.F. | 1.602 | 19.86 | 31.44 | 51.30 |
| 2" copper/DWV | 6.000 | L.F. | 1.998 | 17.76 | 43.74 | 61.50 |
| 2" PVC/DWV | 6.000 | L.F. | 2.088 | 9.12 | 40.98 | 50.10 |
| 2" steel, galvanized | 6.000 | L.F. | 1.500 | 27.24 | 29.46 | 56.70 |
| Vanity base cabinet, 2 door, 24" x 30" | 1.000 | Ea. | 1.630 | 110.00 | 30.00 | 140.00 |
| 24" x 36" | 1.000 | Ea. | 2.000 | 154.00 | 41.00 | 195.00 |
| Vanity top, laminated plastic square edge, 25" x 32" | 2.670 | L.F. | .712 | 48.03 | 13.38 | 61.41 |
| 25" x 38" | 3.160 | L.F. | .843 | 56.85 | 15.83 | 72.68 |
| Cultured marble, 25" x 32", with bowl | 1.000 | Ea. | 2.500 | 85.80 | 49.20 | 135.00 |
| 25" x 38", with bowl | 1.000 | Ea. | 2.500 | 104.50 | 50.50 | 155.00 |
| Carrier, for lavatory, steel for studs, no arms | 1.000 | Ea. | 1.140 | 11.61 | 25.39 | 37.00 |
| Wood, 2" x 8" blocking | 1.300 | L.F. | .052 | .72 | 1.06 | 1.78 |

8

# 8 | MECHANICAL  24 | Three Fixture Bathroom Systems

| SYSTEM DESCRIPTION | QUAN. | UNIT | MAN HOURS | MAT. | INST. | TOTAL |
|---|---|---|---|---|---|---|
| **BATHROOM WITH LAVATORY INSTALLED IN VANITY** | | | | | | |
| Water closet, floor mounted, 2 piece, close coupled, white | 1.000 | Ea. | 3.020 | 115.50 | 59.50 | 175.00 |
| Rough-in supply waste and vent for water closet | 1.000 | Ea. | 2.862 | 35.35 | 58.19 | 93.54 |
| Lavatory, 20" x 18", P.E. cast iron with fittings, white | 1.000 | Ea. | 2.500 | 110.00 | 50.00 | 160.00 |
| Rough-in supply waste and vent for lavatory | 1.000 | Ea. | 3.510 | 34.16 | 72.44 | 106.60 |
| Bathtub, P.E. cast iron, 5' long corner with fittings, white | 1.000 | Ea. | 3.640 | 269.50 | 70.50 | 340.00 |
| Rough-in supply waste and vent for bathtub | 1.000 | Ea. | 3.542 | 34.63 | 75.47 | 110.10 |
| Piping supply, 1/2" copper | 3.200 | L.F. | 4.832 | 26.88 | 105.60 | 132.48 |
| Waste, 4" cast iron, no hub | 12.000 | L.F. | 3.552 | 68.28 | 69.72 | 138.00 |
| Vent, 2" steel, galvanized | 6.000 | L.F. | 1.500 | 27.24 | 29.46 | 56.70 |
| Vanity base cabinet, 2 door, 30" wide | 1.000 | Ea. | 1.630 | 110.00 | 30.00 | 140.00 |
| Vanity top, plastic laminated, square edge | 2.670 | L.F. | .712 | 61.68 | 13.08 | 74.76 |
| TOTAL | | | 31.300 | 893.22 | 633.96 | 1527.18 |
| **BATHROOM WITH WALL HUNG LAVATORY** | | | | | | |
| Water closet, floor mounted, 2 piece, close coupled, white | 1.000 | Ea. | 3.020 | 115.50 | 59.50 | 175.00 |
| Rough-in supply waste and vent for water closet | 1.000 | Ea. | 2.862 | 35.35 | 58.19 | 93.54 |
| Lavatory, 20" x 18", P.E. cast iron, with fittings, white | 1.000 | Ea. | 2.000 | 126.50 | 38.50 | 165.00 |
| Rough-in supply, waste and vent, lavatory | 1.000 | Ea. | 3.510 | 34.16 | 72.44 | 106.60 |
| Bathtub, P.E. cast iron, 5' long corner, with fittings, white | 1.000 | Ea. | 3.640 | 269.50 | 70.50 | 340.00 |
| Rough-in supply, waste and vent, bathtub | 1.000 | Ea. | 3.542 | 34.63 | 75.47 | 110.10 |
| Piping, supply, 1/2" copper | 32.000 | L.F. | 4.832 | 26.88 | 105.60 | 132.48 |
| Waste, 4" cast iron, no hub | 12.000 | L.F. | 3.552 | 68.28 | 69.72 | 138.00 |
| Vent, 2" steel, galvanized | 6.000 | L.F. | 1.500 | 27.24 | 29.46 | 56.70 |
| Carrier, steel, for studs, no arms | 1.000 | Ea. | 1.140 | 11.61 | 25.39 | 37.00 |
| TOTAL | | | 29.598 | 749.65 | 604.77 | 1354.42 |

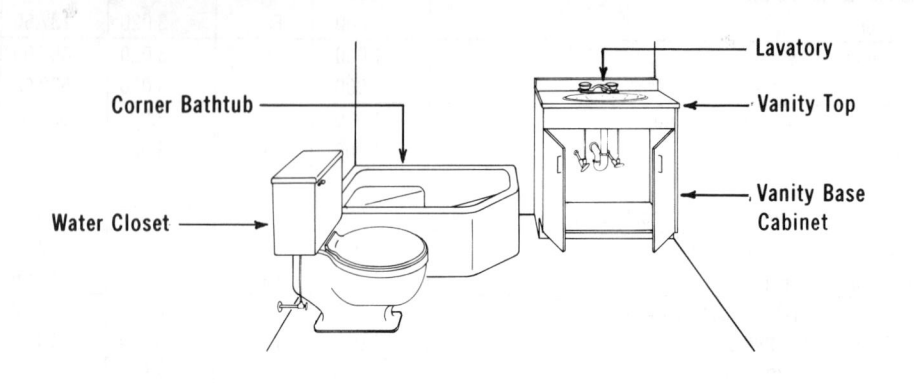

The costs in this system are on a cost each basis. All necessary piping is included.

| DESCRIPTION | QUAN. | UNIT | MAN HOURS | MAT. | INST. | TOTAL |
|---|---|---|---|---|---|---|
| | | | | | | |
| | | | | | | |
| | | | | | | |
| | | | | | | |
| | | | | | | |

# Three Fixture Bathroom Price Sheet

| | QUAN. | UNIT | MAN HOURS | MAT. COST EACH | INST. COST EACH | TOTAL |
|---|---|---|---|---|---|---|
| Water closet, close coupled, standard 2 piece, white | 1.000 | Ea. | 3.020 | 115.50 | 59.50 | 175.00 |
| Color | 1.000 | Ea. | 3.020 | 137.50 | 57.50 | 195.00 |
| One piece elongated bowl, white | 1.000 | Ea. | 3.020 | 407.00 | 58.00 | 465.00 |
| Color | 1.000 | Ea. | 3.020 | 528.00 | 57.00 | 585.00 |
| Low profile, one piece elongated bowl, white | 1.000 | Ea. | 3.020 | 627.00 | 58.00 | 685.00 |
| Color | 1.000 | Ea. | 3.020 | 869.00 | 61.00 | 930.00 |
| Rough-in for water closet | | | | | | |
| 1/2" copper supply, 4" cast iron waste, 2" cast iron vent | 1.000 | Ea. | 2.862 | 35.35 | 58.19 | 93.54 |
| 4" PVC/DWV waste, 2" PVC vent | 1.000 | Ea. | 3.711 | 22.52 | 74.87 | 97.39 |
| 4" copper waste, 2" copper vent | 1.000 | Ea. | 4.239 | 41.96 | 86.88 | 128.84 |
| 3" cast iron waste, 1-1/2" cast iron vent | 1.000 | Ea. | 2.734 | 31.12 | 55.87 | 86.99 |
| 3" PVC waste, 1-1/2" PVC vent | 1.000 | Ea. | 3.174 | 18.39 | 66.65 | 85.04 |
| 3" copper waste, 1-1/2" copper vent | 1.000 | Ea. | 3.550 | 28.75 | 74.39 | 103.14 |
| 1/2" PVC supply, 4" PVC waste, 2" PVC vent | 1.000 | Ea. | 3.825 | 26.78 | 77.27 | 104.05 |
| 3" PVC waste, 1-1/2" PVC supply | 1.000 | Ea. | 3.288 | 22.65 | 69.05 | 91.70 |
| 1/2" steel supply, 4" cast iron waste, 2" cast iron vent | 1.000 | Ea. | 2.718 | 38.65 | 55.01 | 93.66 |
| 4" cast iron waste, 2" steel vent | 1.000 | Ea. | 2.650 | 43.57 | 53.69 | 97.26 |
| 4" PVC waste, 2" PVC vent | 1.000 | Ea. | 3.567 | 25.82 | 71.69 | 97.51 |
| Lavatory, wall hung P.E. cast iron 20" x 18", white | 1.000 | Ea. | 2.000 | 126.50 | 38.50 | 165.00 |
| Color | 1.000 | Ea. | 2.000 | 159.50 | 40.50 | 200.00 |
| Vitreous china 19" x 17", white | 1.000 | Ea. | 2.000 | 79.20 | 40.80 | 120.00 |
| Color | 1.000 | Ea. | 2.000 | 91.30 | 38.70 | 130.00 |
| Lavatory, for vanity top, P.E., cast iron, 20" x 18", white | 1.000 | Ea. | 2.500 | 110.00 | 50.00 | 160.00 |
| Color | 1.000 | Ea. | 2.500 | 137.50 | 47.50 | 185.00 |
| Steel enameled 20" x 17", white | 1.000 | Ea. | 2.500 | 66.00 | 49.00 | 115.00 |
| Color | 1.000 | Ea. | 2.500 | 69.30 | 50.70 | 120.00 |
| Vitreous china 20" x 16", white | 1.000 | Ea. | 2.500 | 165.00 | 50.00 | 215.00 |
| Color | 1.000 | Ea. | 2.500 | 176.00 | 49.00 | 225.00 |
| Rough-in for lavatory | | | | | | |
| 1/2" copper supply, 1-1/2" cast iron waste, 1-1/2" cast iron vent | 1.000 | Ea. | 3.510 | 34.16 | 72.44 | 106.60 |
| 1-1/2" PVC waste, 1-1/4" PVC vent | 1.000 | Ea. | 3.646 | 18.56 | 79.64 | 98.20 |
| 1/2" steel supply, 1-1/4" cast iron waste, 1-1/4" steel vent | 1.000 | Ea. | 2.990 | 38.10 | 61.50 | 99.60 |
| 1-1/4" PVC waste, 1-1/4" PVC vent | 1.000 | Ea. | 3.406 | 23.90 | 74.50 | 98.40 |
| 1/2" PVC supply, 1-1/2" PVC waste, 1-1/2" PVC vent | 1.000 | Ea. | 3.836 | 25.82 | 83.48 | 109.30 |
| Bathtub, P.E. cast iron, 5' long corner with fittings, white | 1.000 | Ea. | 3.640 | 269.50 | 70.50 | 340.00 |
| Color | 1.000 | Ea. | 3.640 | 346.50 | 73.50 | 420.00 |
| Rough-in for bathtub | | | | | | |
| 1/2" copper supply, 4" cast iron waste, 1-1/2" copper vent | 1.000 | Ea. | 3.542 | 34.63 | 75.47 | 110.10 |
| 4" PVC waste, 1-1/2" PVC vent | 1.000 | Ea. | 3.991 | 24.96 | 83.99 | 108.95 |
| 1/2" steel supply, 4" cast iron waste, 1-1/2" steel vent | 1.000 | Ea. | 2.958 | 44.21 | 60.89 | 105.10 |
| 4" PVC waste, 1-1/2" PVC vent | 1.000 | Ea. | 3.751 | 30.46 | 78.69 | 109.15 |
| 1/2" PVC supply, 4" PVC waste, 1-1/2" PVC vent | 1.000 | Ea. | 4.181 | 32.06 | 87.99 | 120.05 |
| Piping, supply, 1/2" copper | 32.000 | L.F. | 4.832 | 26.88 | 105.60 | 132.48 |
| 1/2" steel | 32.000 | L.F. | 4.064 | 44.48 | 88.64 | 133.12 |
| 1/2" PVC | 32.000 | L.F. | 5.440 | 49.60 | 118.40 | 168.00 |
| Piping, waste, 4" cast iron, no hub | 12.000 | L.F. | 3.552 | 68.28 | 69.72 | 138.00 |
| 4" PVC/DWV | 12.000 | L.F. | 5.652 | 45.60 | 111.00 | 156.60 |
| 4" copper/DWV | 12.000 | L.F. | 8.004 | 100.32 | 151.68 | 252.00 |
| Piping, vent 2" cast iron, no hub | 6.000 | L.F. | 1.602 | 19.86 | 31.44 | 51.30 |
| 2" copper/DWV | 6.000 | L.F. | 1.998 | 17.76 | 43.74 | 61.50 |
| 2" PVC/DWV | 6.000 | L.F. | 2.088 | 9.12 | 40.98 | 50.10 |
| 2" steel, galvanized | 6.000 | L.F. | 1.500 | 27.24 | 29.46 | 56.70 |
| Vanity base cabinet, 2 door, 24" x 30" | 1.000 | Ea. | 1.630 | 110.00 | 30.00 | 140.00 |
| 24" x 36" | 1.000 | Ea. | 2.000 | 154.00 | 41.00 | 195.00 |
| Vanity top, laminated plastic square edge 25" x 32" | 2.670 | L.F. | .712 | 61.68 | 13.08 | 74.76 |
| 25" x 38" | 3.160 | L.F. | .843 | 73.00 | 15.48 | 88.48 |
| Cultured marble, 25" x 32", with bowl | 1.000 | Ea. | 2.500 | 85.80 | 49.20 | 135.00 |
| 25" x 38", with bowl | 1.000 | Ea. | 2.500 | 104.50 | 50.50 | 155.00 |
| Carrier, for lavatory, steel for studs, no arms | 1.000 | Ea. | 1.140 | 11.61 | 25.39 | 37.00 |
| Wood, 2" x 8" blocking | 1.300 | L.F. | .052 | .73 | 1.09 | 1.82 |

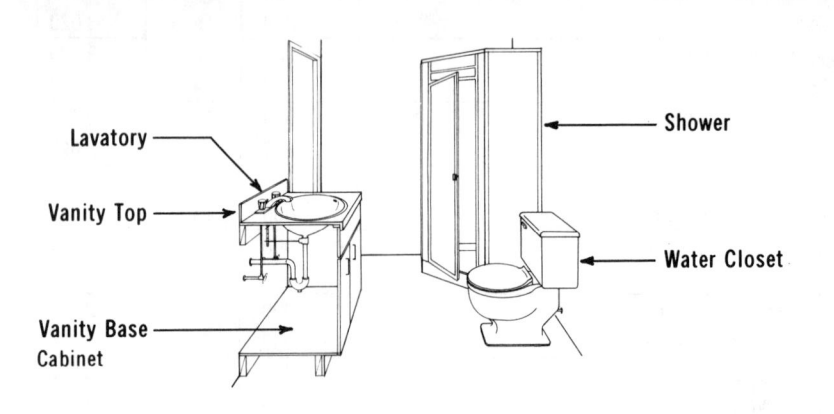

# 8 | MECHANICAL  28 | Three Fixture Bathroom Systems

| SYSTEM DESCRIPTION | QUAN. | UNIT | MAN HOURS | MAT. | INST. | TOTAL |
|---|---|---|---|---|---|---|
| **BATHROOM WITH SHOWER, LAVATORY INSTALLED IN VANITY** | | | | | | |
| Water closet, floor mounted, 2 piece, close coupled, white | 1.000 | Ea. | 3.020 | 115.50 | 59.50 | 175.00 |
| Rough-in supply, waste and vent for water closet | 1.000 | Ea. | 2.862 | 35.35 | 58.19 | 93.54 |
| Lavatory, 20" x 18" P.E. cast iron with fittings, white | 1.000 | Ea. | 2.500 | 110.00 | 50.00 | 160.00 |
| Rough-in supply, waste and vent | 1.000 | Ea. | 3.510 | 34.16 | 72.44 | 106.60 |
| Shower, steel enameled, stone base | 1.000 | Ea. | 8.000 | 286.00 | 159.00 | 445.00 |
| Rough-in supply, waste and vent | 1.000 | Ea. | 4.268 | 36.52 | 89.17 | 125.69 |
| Piping supply, 1/2" copper | 36.000 | L.F. | 6.342 | 35.28 | 138.60 | 173.88 |
| Waste 4" cast iron, no hub | 7.000 | L.F. | 2.960 | 56.90 | 58.10 | 115.00 |
| Vent 2" steel galvanized | 6.000 | L.F. | 2.250 | 40.86 | 44.19 | 85.05 |
| Vanity base 2 door, 30" wide | 1.000 | Ea. | 1.630 | 110.00 | 30.00 | 140.00 |
| Vanity top, plastic laminated, square edge | 2.170 | L.F. | .712 | 49.64 | 14.44 | 64.08 |
| TOTAL | | | 38.054 | 910.21 | 773.63 | 1683.84 |
| **BATHROOM WIH SHOWER, WALL HUNG LAVATORY** | | | | | | |
| Water closet, floor mounted, close coupled | 1.000 | Ea. | 3.020 | 115.50 | 59.50 | 175.00 |
| Rough-in supply, waste and vent for water closet | 1.000 | Ea. | 2.862 | 35.35 | 58.19 | 93.54 |
| Lavatory, 20" x 18" P.E. cast iron with fittings, white | 1.000 | Ea. | 2.000 | 126.50 | 38.50 | 165.00 |
| Rough-in supply, waste and vent for lavatory | 1.000 | Ea. | 3.510 | 34.16 | 72.44 | 106.60 |
| Shower, steel enameled, stone base, white | 1.000 | Ea. | 8.000 | 286.00 | 159.00 | 445.00 |
| Rough-in supply, waste and vent for shower | 1.000 | Ea. | 4.268 | 36.52 | 89.17 | 125.69 |
| Piping supply, 1/2" copper | 36.000 | L.F. | 6.342 | 35.28 | 138.60 | 173.88 |
| Waste, 4" cast iron, no hub | 7.000 | L.F. | 2.960 | 56.90 | 58.10 | 115.00 |
| Vent, 2" steel, galvanized | 6.000 | L.F. | 2.250 | 40.86 | 44.19 | 85.05 |
| Carrier, steel, for studs, no arms | 1.000 | Ea. | 1.140 | 11.61 | 25.39 | 37.00 |
| TOTAL | | | 36.352 | 778.68 | 743.08 | 1521.76 |

The costs in this system are on a cost each basis. All necessary piping is included.

| DESCRIPTION | QUAN. | UNIT | MAN HOURS | MAT. | INST. | TOTAL |
|---|---|---|---|---|---|---|
| | | | | | | |
| | | | | | | |
| | | | | | | |
| | | | | | | |
| | | | | | | |

# Three Fixture Bathroom Price Sheet

| | QUAN. | UNIT | MAN HOURS | COST EACH MAT. | COST EACH INST. | TOTAL |
|---|---|---|---|---|---|---|
| Water closet, close coupled, standard 2 piece, white | 1.000 | Ea. | 3.020 | 115.50 | 59.50 | 175.00 |
| Color | 1.000 | Ea. | 3.020 | 137.50 | 57.50 | 195.00 |
| One piece elongated bowl, white | 1.000 | Ea. | 3.020 | 407.00 | 58.00 | 465.00 |
| Color | 1.000 | Ea. | 3.020 | 528.00 | 57.00 | 585.00 |
| Low profile, one piece elongated bowl, white | 1.000 | Ea. | | | | |
| Color | 1.000 | Ea. | 3.020 | 825.00 | 60.00 | 885.00 |
| Rough-in for water closet | | | | | | |
| 1/2" copper supply, 4" cast iron waste, 2" cast iron vent | 1.000 | Ea. | 2.862 | 35.35 | 58.19 | 93.54 |
| 4" PVC/DWV waste, 2" PVC vent | 1.000 | Ea. | 3.711 | 22.52 | 74.87 | 97.39 |
| 4" copper waste, 2" copper vent | 1.000 | Ea. | 4.239 | 41.96 | 86.88 | 128.84 |
| 3" cast iron waste, 1-1/2" cast iron vent | 1.000 | Ea. | 2.734 | 31.12 | 55.87 | 86.99 |
| 3" PVC waste, 1-1/2" PVC vent | 1.000 | Ea. | 3.174 | 18.39 | 66.65 | 85.04 |
| 3" copper waste, 1-1/2" copper vent | 1.000 | Ea. | 3.550 | 28.75 | 74.39 | 103.14 |
| 1/2" PVC supply, 4" PVC waste, 2" PVC vent | 1.000 | Ea. | 3.825 | 26.78 | 77.27 | 104.05 |
| 3" PVC waste, 1-1/2" PVC supply | 1.000 | Ea. | 3.288 | 22.65 | 69.05 | 91.70 |
| 1/2" steel supply, 4" cast iron waste, 2" cast iron vent | 1.000 | Ea. | 2.718 | 38.65 | 55.01 | 93.66 |
| 4" cast iron waste, 2" steel vent | 1.000 | Ea. | 2.650 | 43.57 | 53.69 | 97.26 |
| 4" PVC waste, 2" PVC vent | 1.000 | Ea. | 3.567 | 25.82 | 71.69 | 97.51 |
| Lavatory, wall hung, P.E. cast iron 20" x 18", white | 1.000 | Ea. | 2.000 | 126.50 | 38.50 | 165.00 |
| Color | 1.000 | Ea. | 2.000 | 159.50 | 40.50 | 200.00 |
| Vitreous china 19" x 17", white | 1.000 | Ea. | 2.000 | 79.20 | 40.80 | 120.00 |
| Color | 1.000 | Ea. | 2.000 | 91.30 | 38.70 | 130.00 |
| Lavatory, for vanity top, P.E. cast iron 20" x 18", white | 1.000 | Ea. | 2.500 | 110.00 | 50.00 | 160.00 |
| Color | 1.000 | Ea. | 2.500 | 137.50 | 47.50 | 185.00 |
| Steel enameled 20" x 17", white | 1.000 | Ea. | 2.500 | 66.00 | 49.00 | 115.00 |
| Color | 1.000 | Ea. | 2.500 | 69.30 | 50.70 | 120.00 |
| Vitreous china 20" x 16", white | 1.000 | Ea. | 2.500 | 165.00 | 50.00 | 215.00 |
| Color | 1.000 | Ea. | 2.500 | 176.00 | 49.00 | 225.00 |
| Rough-in for lavatory | | | | | | |
| 1/2" copper supply, 1-1/2" cast iron waste, 1-1/2" cast iron vent | 1.000 | Ea. | 3.510 | 34.16 | 72.44 | 106.60 |
| 1-1/2" PVC waste, 1-1/2" PVC vent | 1.000 | Ea. | 3.646 | 18.56 | 79.64 | 98.20 |
| 1/2" steel supply, 1-1/4" cast iron waste, 1-1/4" steel vent | 1.000 | Ea. | 2.990 | 38.10 | 61.50 | 99.60 |
| 1-1/4" PVC waste, 1-1/4" PVC vent | 1.000 | Ea. | 3.406 | 24.06 | 74.34 | 98.40 |
| 1/2" PVC supply, 1-1/2" PVC waste, 1-1/2" PVC vent | 1.000 | Ea. | 3.836 | 25.82 | 83.48 | 109.30 |
| Shower, steel enameled stone base, 32" x 32", white | 1.000 | Ea. | 8.000 | 286.00 | 159.00 | 445.00 |
| Color | 1.000 | Ea. | 7.823 | 532.40 | 154.00 | 686.40 |
| 36" x 36" white | 1.000 | Ea. | 8.890 | 539.00 | 176.00 | 715.00 |
| Color | 1.000 | Ea. | 8.890 | 605.00 | 175.00 | 780.00 |
| Rough-in for shower | | | | | | |
| 1/2" copper supply, 4" cast iron waste, 1-1/2" copper vent | 1.000 | Ea. | 4.268 | 36.52 | 89.17 | 125.69 |
| 4" PVC waste, 1-1/2" PVC vent | 1.000 | Ea. | 4.771 | 23.85 | 99.59 | 123.44 |
| 1/2" steel supply, 4" cast iron waste, 1-1/2" steel vent | 1.000 | Ea. | 3.834 | 45.41 | 79.70 | 125.11 |
| 4" PVC waste, 1-1/2" PVC vent | 1.000 | Ea. | 4.387 | 32.65 | 91.11 | 123.76 |
| 1/2" PVC supply, 4" PVC waste, 1-1/2" PVC vent | 1.000 | Ea. | 5.075 | 35.21 | 105.99 | 141.20 |
| Piping, supply, 1/2" copper | 36.000 | L.F. | 6.342 | 35.28 | 138.60 | 173.88 |
| 1/2" steel | 36.000 | L.F. | 5.334 | 58.38 | 116.34 | 174.72 |
| 1/2" PVC | 36.000 | L.F. | 7.140 | 65.10 | 155.40 | 220.50 |
| Piping, waste, 4" cast iron no hub | 7.000 | L.F. | 2.960 | 56.90 | 58.10 | 115.00 |
| 4" PVC/DWV | 7.000 | L.F. | 4.710 | 38.00 | 92.50 | 130.50 |
| 4" copper/DWV | 7.000 | L.F. | 6.670 | 83.60 | 126.40 | 210.00 |
| Piping, vent, 2" cast iron no hub | 6.000 | L.F. | 2.403 | 29.79 | 47.16 | 76.95 |
| 2" copper/DWV | 6.000 | L.F. | 2.997 | 26.64 | 65.61 | 92.25 |
| 2" PVC/DWV | 6.000 | L.F. | 3.132 | 13.68 | 61.47 | 75.15 |
| 2" steel, galvanized | 6.000 | L.F. | 2.250 | 40.86 | 44.19 | 85.05 |
| Vanity base cabinet, 2 door, 24" x 30" | 1.000 | Ea. | 1.630 | 110.00 | 30.00 | 140.00 |
| 24" x 36" | 1.000 | Ea. | 2.000 | 154.00 | 41.00 | 195.00 |
| Vanity top, laminated plastic square edge, 25" x 32" | 2.170 | L.F. | .712 | 49.64 | 14.44 | 64.08 |
| 25" x 38" | 2.670 | L.F. | .846 | 58.93 | 17.15 | 76.08 |
| Carrier, for lavatory, steel for studs, no arms | 1.000 | Ea. | 1.140 | 11.61 | 25.39 | 37.00 |
| Wood, 2" x 8" blocking | 1.300 | L.F. | | .72 | 1.06 | 1.78 |

175

# 8 | MECHANICAL  32 | Three Fixture Bathroom Systems

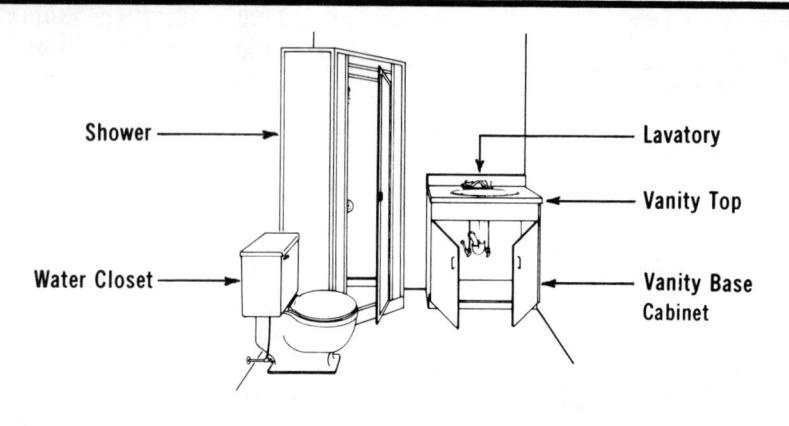

| SYSTEM DESCRIPTION | QUAN. | UNIT | MAN HOURS | MAT. | INST. | TOTAL |
|---|---|---|---|---|---|---|
| **BATHROOM WITH LAVATORY INSTALLED IN VANITY** | | | | | | |
| Water closet, floor mounted, 2 piece, close coupled, white | 1.000 | Ea. | 3.020 | 115.50 | 59.50 | 175.00 |
| Rough-in supply, waste and vent for water closet | 1.000 | Ea. | 2.862 | 35.35 | 58.19 | 93.54 |
| Lavatory, 20" x 18", P.E. cast iron with fittings, white | 1.000 | Ea. | 2.500 | 110.00 | 50.00 | 160.00 |
| Rough-in supply, waste and vent for lavatory | 1.000 | Ea. | 3.510 | 34.16 | 72.44 | 106.60 |
| Shower, steel enameled, stone base, corner, white | 1.000 | Ea. | 8.000 | 286.00 | 159.00 | 445.00 |
| Rough-in supply, waste and vent for shower | 1.000 | Ea. | 4.268 | 36.52 | 89.17 | 125.69 |
| Piping, supply, 1/2" copper | 36.000 | L.F. | 5.436 | 30.24 | 118.80 | 149.04 |
| Waste, 4" cast iron, no hub | 7.000 | L.F. | 2.072 | 39.83 | 40.67 | 80.50 |
| Vent, 2" steel, galvanized | 6.000 | L.F. | 1.500 | 27.24 | 29.46 | 56.70 |
| Vanity base, 2 door, 30" wide | 1.000 | Ea. | 1.630 | 110.00 | 30.00 | 140.00 |
| Vanity top, plastic laminated, square edge | 2.670 | L.F. | .712 | 48.03 | 13.38 | 61.41 |
| TOTAL | | | 35.510 | 872.87 | 720.61 | 1593.48 |
| **BATHROOM, WITH WALL HUNG LAVATORY** | | | | | | |
| Water closet, floor mounted, 2 piece, close coupled, white | 1.000 | Ea. | 3.020 | 115.50 | 59.50 | 175.00 |
| Rough-in supply, waste and vent for water closet | 1.000 | Ea. | 2.862 | 35.35 | 58.19 | 93.54 |
| Lavatory, wall hung, 20" x 18" P.E. cast iron with fittings, white | 1.000 | Ea. | 2.000 | 126.50 | 38.50 | 165.00 |
| Rough-in supply, waste and vent for lavatory | 1.000 | Ea. | 3.510 | 34.16 | 72.44 | 106.60 |
| Shower, steel enameled, stone base, corner, white | 1.000 | Ea. | 8.000 | 286.00 | 159.00 | 445.00 |
| Rough-in supply, waste and vent for shower | 1.000 | Ea. | 4.268 | 36.52 | 89.17 | 125.69 |
| Piping, supply, 1/2" copper | 36.000 | L.F. | 5.436 | 30.24 | 118.80 | 149.04 |
| Waste, 4" cast iron, no hub | 7.000 | L.F. | 2.072 | 39.83 | 40.67 | 80.50 |
| Vent, 2" steel, galvanized | 6.000 | L.F. | 1.500 | 27.24 | 29.46 | 56.70 |
| Carrier, steel, for studs, no arms | 1.000 | Ea. | 1.140 | 11.61 | 25.39 | 37.00 |
| TOTAL | | | 33.808 | 742.95 | 691.12 | 1434.07 |

The costs in this system are on a cost each basis. All necessary piping is included.

| DESCRIPTION | QUAN. | UNIT | MAN HOURS | MAT. | INST. | TOTAL |
|---|---|---|---|---|---|---|
| | | | | | | |
| | | | | | | |
| | | | | | | |
| | | | | | | |
| | | | | | | |

# Three Fixture Bathroom Price Sheet

| Description | QUAN. | UNIT | MAN HOURS | MAT. | INST. | TOTAL |
|---|---|---|---|---|---|---|
| Water closet, close coupled, standard 2 piece, white | 1.000 | Ea. | 3.020 | 115.50 | 59.50 | 175.00 |
| Color | 1.000 | Ea. | 3.020 | 137.50 | 57.50 | 195.00 |
| One piece elongated bowl, white | 1.000 | Ea. | 3.020 | 407.00 | 58.00 | 465.00 |
| Color | 1.000 | Ea. | 3.020 | 528.00 | 57.00 | 585.00 |
| Low profile one piece elongated bowl, white | 1.000 | Ea. | 3.020 | 627.00 | 58.00 | 685.00 |
| Color | 1.000 | Ea. | 3.624 | 1042.80 | 73.20 | 1116.00 |
| Rough-in for water closet | | | | | | |
| 1/2" copper supply, 4" cast iron waste, 2" cast iron vent | 1.000 | Ea. | 2.862 | 35.35 | 58.19 | 93.54 |
| 4" P.V.C./DWV waste, 2" PVC vent | 1.000 | Ea. | 3.711 | 22.52 | 74.87 | 97.39 |
| 4" copper waste, 2" copper vent | 1.000 | Ea. | 4.239 | 41.96 | 86.88 | 128.84 |
| 3" cast iron waste, 1-1/2" cast iron vent | 1.000 | Ea. | 2.734 | 31.12 | 55.87 | 86.99 |
| 3" PVC waste, 1-1/2" PVC vent | 1.000 | Ea. | 3.174 | 18.39 | 66.65 | 85.04 |
| 3" copper waste, 1-1/2" copper vent | 1.000 | Ea. | 3.550 | 28.75 | 74.39 | 103.14 |
| 1/2" P.V.C. supply, 4" P.V.C. waste, 2" P.V.C. vent | 1.000 | Ea. | 3.825 | 26.78 | 77.27 | 104.05 |
| 3" P.V.C. waste, 1-1/2" P.V.C. vent | 1.000 | Ea. | 3.288 | 22.65 | 69.05 | 91.70 |
| 1/2" steel supply, 4" cast iron waste, 2" cast iron vent | 1.000 | Ea. | 2.718 | 38.65 | 55.01 | 93.66 |
| 4" cast iron waste, 2" steel vent | 1.000 | Ea. | 2.650 | 43.57 | 53.69 | 97.26 |
| 4" P.V.C. waste, 2" P.V.C. vent | 1.000 | Ea. | 3.567 | 25.82 | 71.69 | 97.51 |
| Lavatory, wall hung P.E. cast iron 20" x 18", white | 1.000 | Ea. | 2.000 | 126.50 | 38.50 | 165.00 |
| Color | 1.000 | Ea. | 2.000 | 159.50 | 40.50 | 200.00 |
| Vitreous china 19" x 17", white | 1.000 | Ea. | 2.000 | 79.20 | 40.80 | 120.00 |
| Color | 1.000 | Ea. | 2.000 | 91.30 | 38.70 | 130.00 |
| Lavatory, for vanity top P.E. cast iron 20" x 18", white | 1.000 | Ea. | 2.500 | 110.00 | 50.00 | 160.00 |
| Color | 1.000 | Ea. | 2.500 | 137.50 | 47.50 | 185.00 |
| Steel enameled 20" x 17", white | 1.000 | Ea. | 2.500 | 66.00 | 49.00 | 115.00 |
| Color | 1.000 | Ea. | 2.500 | 69.30 | 50.70 | 120.00 |
| Vitreous china 20" x 16", white | 1.000 | Ea. | 2.500 | 165.00 | 50.00 | 215.00 |
| Color | 1.000 | Ea. | 2.500 | 176.00 | 49.00 | 225.00 |
| Rough-in for lavatory | | | | | | |
| 1/2" copper supply, 1-1/2" cast iron waste, 1-1/2" cast iron vent | 1.000 | Ea. | 3.510 | 34.16 | 72.44 | 106.60 |
| 1-1/2" P.V.C. waste, 1-1/2" P.V.C. vent | 1.000 | Ea. | 3.646 | 18.56 | 79.64 | 98.20 |
| 1/2" steel supply, 1-1/2" cast iron waste, 1-1/4" steel vent | 1.000 | Ea. | 2.990 | 38.10 | 61.50 | 99.60 |
| 1-1/2" P.V.C. waste, 1-1/4" P.V.C. vent | 1.000 | Ea. | 3.406 | 24.06 | 74.34 | 98.40 |
| 1/2" P.V.C. supply, 1-1/2" P.V.C. waste, 1-1/2" P.V.C. vent | 1.000 | Ea. | 3.836 | 25.82 | 83.48 | 109.30 |
| Shower, steel enameled stone base, 32" x 32", white | 1.000 | Ea. | 8.000 | 286.00 | 159.00 | 445.00 |
| Color | 1.000 | Ea. | 7.823 | 532.40 | 154.00 | 686.40 |
| 36" x 36", white | 1.000 | Ea. | 8.890 | 539.00 | 176.00 | 715.00 |
| Color | 1.000 | Ea. | 8.890 | 605.00 | 175.00 | 780.00 |
| Rough-in for shower | | | | | | |
| 1/2" copper supply, 2" cast iron waste, 1-1/2" copper vent | 1.000 | Ea. | 4.304 | 35.59 | 90.50 | 126.09 |
| 2" P.V.C. waste, 1-1/2" P.V.C. vent | 1.000 | Ea. | 4.771 | 23.85 | 99.59 | 123.44 |
| 1/2" steel supply, 2" cast iron waste, 1-1/2" steel vent | 1.000 | Ea. | 4.008 | 59.69 | 83.12 | 142.81 |
| 2" P.V.C. waste, 1-1/2" P.V.C. vent | 1.000 | Ea. | 4.387 | 32.65 | 91.11 | 123.76 |
| 1/2" P.V.C. supply, 2" P.V.C. waste, 1-1/2" P.V.C. vent | 1.000 | Ea. | 5.075 | 35.21 | 105.99 | 141.20 |
| Piping, supply, 1/2" copper | 36.000 | L.F. | 5.436 | 30.24 | 118.80 | 149.04 |
| 1/2" steel | 36.000 | L.F. | 4.572 | 50.04 | 99.72 | 149.76 |
| 1/2" P.V.C. | 36.000 | L.F. | 6.120 | 55.80 | 133.20 | 189.00 |
| Waste, 4" cast iron, no hub | 7.000 | L.F. | 2.072 | 39.83 | 40.67 | 80.50 |
| 4" P.V.C./DWV | 7.000 | L.F. | 3.297 | 26.60 | 64.75 | 91.35 |
| 4" copper/DWV | 7.000 | L.F. | 4.669 | 58.52 | 88.48 | 147.00 |
| Vent, 2" cast iron, no hub | 6.000 | L.F. | 1.998 | 17.76 | 43.74 | 61.50 |
| 2" copper/DWV | 6.000 | L.F. | 1.998 | 17.76 | 43.74 | 61.50 |
| 2" P.V.C./DWV | 6.000 | L.F. | 2.088 | 9.12 | 40.98 | 50.10 |
| 2" steel, galvanized | 6.000 | L.F. | 1.500 | 27.24 | 29.46 | 56.70 |
| Vanity base cabinet, 2 door, 24" x 30" | 1.000 | Ea. | 1.630 | 110.00 | 30.00 | 140.00 |
| 24" x 36" | 1.000 | Ea. | 2.000 | 154.00 | 41.00 | 195.00 |
| Vanity top, laminated plastic square edge, 25" x 32" | 2.670 | L.F. | .712 | 48.03 | 13.38 | 61.41 |
| 25" x 38" | 3.170 | L.F. | .846 | 57.03 | 15.88 | 72.91 |
| Carrier, for lavatory, steel, for studs, no arms | 1.000 | Ea. | 1.140 | 11.61 | 25.39 | 37.00 |
| Wood, 2" x 8" blocking | 1.300 | L.F. | .040 | .72 | 1.06 | 1.78 |

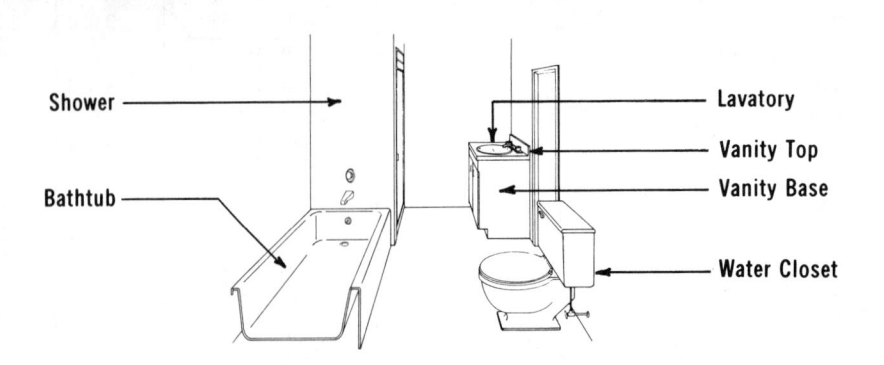

# 8 | MECHANICAL  36 | Four Fixture Bathroom Systems

| SYSTEM DESCRIPTION | QUAN. | UNIT | MAN HOURS | MAT. | INST. | TOTAL |
|---|---|---|---|---|---|---|
| **BATHROOM WITH LAVATORY INSTALLED IN VANITY** | | | | | | |
| Water closet, floor mounted, 2 piece, close coupled, white | 1.000 | Ea. | 3.020 | 115.50 | 59.50 | 175.00 |
| Rough-in supply, waste and vent for water closet | 1.000 | Ea. | 2.862 | 35.35 | 58.19 | 93.54 |
| Lavatory, 20" x 18" P.E. cast iron with fittings, white | 1.000 | Ea. | 2.500 | 110.00 | 50.00 | 160.00 |
| Shower, steel, enameled, stone base, corner, white | 1.000 | Ea. | 8.890 | 539.00 | 176.00 | 715.00 |
| Rough-in supply, waste and vent for lavatory and shower | 2.000 | Ea. | 9.832 | 91.28 | 204.20 | 295.48 |
| Bathtub, P.E. cast iron, 5' long with fittings, white | 1.000 | Ea. | 3.640 | 269.50 | 70.50 | 340.00 |
| Rough-in supply, waste and vent for bathtub | 1.000 | Ea. | 3.542 | 34.63 | 75.47 | 110.10 |
| Piping, supply, 1/2" copper | 42.000 | L.F. | 6.342 | 35.28 | 138.60 | 173.88 |
| Waste, 4" cast iron, no hub | 10.000 | L.F. | 2.960 | 56.90 | 58.10 | 115.00 |
| Vent, 2" steel galvanized | 13.000 | L.F. | 3.250 | 59.02 | 63.83 | 122.85 |
| Vanity base, 2 doors, 30" wide | 1.000 | Ea. | 1.630 | 110.00 | 30.00 | 140.00 |
| Vanity top, plastic laminated, square edge | 2.670 | L.F. | .712 | 48.03 | 13.38 | 61.41 |
| TOTAL | | | 49.180 | 1504.49 | 997.77 | 2502.26 |
| **BATHROOM WITH WALL HUNG LAVATORY** | | | | | | |
| Water closet, floor mounted, 2 piece, close coupled, white | 1.000 | Ea. | 3.020 | 115.50 | 59.50 | 175.00 |
| Rough-in supply, waste and vent for water closet | 1.000 | Ea. | 2.862 | 35.35 | 58.19 | 93.54 |
| Lavatory, 20" x 18" P.E. cast iron with fittings, white | 1.000 | Ea. | 2.000 | 126.50 | 38.50 | 165.00 |
| Shower, steel enameled, stone base, corner, white | 1.000 | Ea. | 8.890 | 539.00 | 176.00 | 715.00 |
| Rough-in supply, waste and vent for lavatory and shower | 2.000 | Ea. | 9.832 | 91.28 | 204.20 | 295.48 |
| Bathtub, P.E. cast iron, 5' long with fittings, white | 1.000 | Ea. | 3.640 | 269.50 | 70.50 | 340.00 |
| Rough-in supply, waste and vent for bathtub | 1.000 | Ea. | 3.542 | 34.63 | 75.47 | 110.10 |
| Piping, supply, 1/2" copper | 42.000 | L.F. | 6.342 | 35.28 | 138.60 | 173.88 |
| Waste, 4" cast iron, no hub | 10.000 | L.F. | 2.960 | 56.90 | 58.10 | 115.00 |
| Vent, 2" steel galvanized | 13.000 | L.F. | 3.250 | 59.02 | 63.83 | 122.85 |
| Carrier, steel, for studs, no arms | 1.000 | Ea. | 1.140 | 11.61 | 25.39 | 37.00 |
| TOTAL | | | 47.478 | 1374.57 | 968.28 | 2342.85 |

The costs in this system are on a cost each basis. All necessary piping is included.

| DESCRIPTION | QUAN. | UNIT | MAN HOURS | MAT. | INST. | TOTAL |
|---|---|---|---|---|---|---|
| | | | | | | |
| | | | | | | |
| | | | | | | |
| | | | | | | |

# Four Fixture Bathroom Price Sheet

| Description | QUAN. | UNIT | MAN HOURS | MAT. (COST EACH) | INST. (COST EACH) | TOTAL |
|---|---|---|---|---|---|---|
| Water closet, close coupled, standard 2 piece, white | 1.000 | Ea. | 3.020 | 115.50 | 59.50 | 175.00 |
| Color | 1.000 | Ea. | 3.020 | 137.50 | 57.50 | 195.00 |
| One piece elongated bowl, white | 1.000 | Ea. | 3.020 | 407.00 | 58.00 | 465.00 |
| Color | 1.000 | Ea. | 3.020 | 528.00 | 57.00 | 585.00 |
| Low profile, one piece elongated bowl, white | 1.000 | Ea. | 3.020 | 627.00 | 58.00 | 685.00 |
| Color | 1.000 | Ea. | 3.020 | 825.00 | 60.00 | 885.00 |
| Rough-in for water closet | | | | | | |
| 1/2" copper supply, 4" cast iron waste, 2" cast iron vent | 1.000 | Ea. | 2.862 | 35.35 | 58.19 | 93.54 |
| 4" PVC/DWV waste, 2" PVC vent | 1.000 | Ea. | 3.711 | 22.52 | 74.87 | 97.39 |
| 4" copper waste, 2" copper vent | 1.000 | Ea. | 4.239 | 41.96 | 86.88 | 128.84 |
| 3" cast iron waste, 1-1/2" cast iron vent | 1.000 | Ea. | 2.734 | 31.12 | 55.87 | 86.99 |
| 3" P.V.C. waste, 1-1/2" P.V.C. vent | 1.000 | Ea. | 3.174 | 18.39 | 66.65 | 85.04 |
| 3" copper waste, 1-1/2" copper vent | 1.000 | Ea. | 3.550 | 28.75 | 74.39 | 103.14 |
| 1/2" P.V.C. supply, 4" P.V.C. waste, 2" P.V.C. vent | 1.000 | Ea. | 3.825 | 26.78 | 77.27 | 104.05 |
| 3" P.V.C. waste, 1-1/2" P.V.C. vent | 1.000 | Ea. | 3.288 | 22.65 | 69.05 | 91.70 |
| 1/2" steel supply, 4" cast iron waste, 2" cast iron vent | 1.000 | Ea. | 2.718 | 38.65 | 55.01 | 93.66 |
| 4" cast iron waste, 2" steel vent | 1.000 | Ea. | 2.650 | 43.57 | 53.69 | 97.26 |
| 4" P.V.C. waste, 2" P.V.C. vent | 1.000 | Ea. | 3.567 | 25.82 | 71.69 | 97.51 |
| Lavatory, wall hung P.E. cast iron 20" x 18", white | 1.000 | Ea. | 2.000 | 126.50 | 38.50 | 165.00 |
| Color | 1.000 | Ea. | 2.000 | 159.50 | 40.50 | 200.00 |
| Vitreous china 19" x 17", white | 1.000 | Ea. | 2.000 | 79.20 | 40.80 | 120.00 |
| Color | 1.000 | Ea. | 2.000 | 91.30 | 38.70 | 130.00 |
| Lavatory for vanity top, P.E. cast iron 20" x 18", white | 1.000 | Ea. | 2.500 | 110.00 | 50.00 | 160.00 |
| Color | 1.000 | Ea. | 2.500 | 137.50 | 47.50 | 185.00 |
| Steel enameled, 20" x 17", white | 1.000 | Ea. | 2.500 | 66.00 | 49.00 | 115.00 |
| Color | 1.000 | Ea. | 2.500 | 69.30 | 50.70 | 120.00 |
| Vitreous china 20" x 16", white | 1.000 | Ea. | 2.500 | 165.00 | 50.00 | 215.00 |
| Color | 1.000 | Ea. | 2.500 | 176.00 | 49.00 | 225.00 |
| Shower, steel enameled stone base, 36" square, white | 1.000 | Ea. | 8.890 | 539.00 | 176.00 | 715.00 |
| Color | 1.000 | Ea. | 8.890 | 605.00 | 175.00 | 780.00 |
| Rough-in for lavatory or shower | | | | | | |
| 1/2" copper supply, 1-1/2" cast iron waste, 1-1/2" cast iron vent | 1.000 | Ea. | 4.916 | 45.64 | 102.10 | 147.74 |
| 1-1/2" P.V.C. waste, 1-1/4" P.V.C. vent | 1.000 | Ea. | 5.086 | 26.18 | 111.06 | 137.24 |
| 1/2" steel supply, 1-1/4" cast iron waste, 1-1/4" steel vent | 1.000 | Ea. | 4.252 | 52.88 | 87.98 | 140.86 |
| 1-1/4" P.V.C. waste, 1-1/4" P.V.C. vent | 1.000 | Ea. | 4.702 | 34.74 | 102.82 | 137.56 |
| 1/2" P.V.C. supply, 1-1/2" P.V.C. waste, 1-1/2" P.V.C. vent | 1.000 | Ea. | 5.390 | 37.70 | 117.30 | 155.00 |
| Bathtub, P.E. cast iron, 5' long with fittings, white | 1.000 | Ea. | 3.640 | 269.50 | 70.50 | 340.00 |
| Color | 1.000 | Ea. | 3.640 | 346.50 | 73.50 | 420.00 |
| Steel, enameled 5' long with fittings, white | 1.000 | Ea. | 2.910 | 181.50 | 58.50 | 240.00 |
| Color | 1.000 | Ea. | 2.910 | 192.50 | 57.50 | 250.00 |
| Rough-in for bathtub | | | | | | |
| 1/2" copper supply, 4" cast iron waste, 1-1/2" copper vent | 1.000 | Ea. | 3.542 | 34.63 | 75.47 | 110.10 |
| 4" P.V.C. waste, 1-1/2" P.V.C. vent | 1.000 | Ea. | 3.991 | 24.96 | 83.99 | 108.95 |
| 1/2" steel supply, 4" cast iron waste, 1-1/2" steel vent | 1.000 | Ea. | 2.958 | 44.21 | 60.89 | 105.10 |
| 4" P.V.C. waste, 1-1/2" P.V.C. vent | 1.000 | Ea. | 3.751 | 30.46 | 78.69 | 109.15 |
| 1/2" P.V.C. supply, 4" P.V.C. waste, 1-1/2" P.V.C. vent | 1.000 | Ea. | 4.181 | 32.06 | 87.99 | 120.05 |
| Piping, supply, 1/2" copper | 42.000 | L.F. | 6.342 | 35.28 | 138.60 | 173.88 |
| 1/2" steel | 42.000 | L.F. | 5.334 | 58.38 | 116.34 | 174.72 |
| 1/2" P.V.C. | 42.000 | L.F. | 7.140 | 65.10 | 155.40 | 220.50 |
| Waste, 4" cast iron, no hub | 10.000 | L.F. | 2.960 | 56.90 | 58.10 | 115.00 |
| 4" P.V.C./DWV | 10.000 | L.F. | 4.710 | 38.00 | 92.50 | 130.50 |
| 4" copper/DWV | 10.000 | Ea. | 6.670 | 83.60 | 126.40 | 210.00 |
| Vent 2" cast iron, no hub | 13.000 | L.F. | 3.471 | 43.03 | 68.12 | 111.15 |
| 2" copper/DWV | 13.000 | L.F. | 4.329 | 38.48 | 94.77 | 133.25 |
| 2" P.V.C./DWV | 13.000 | L.F. | 4.524 | 19.76 | 88.79 | 108.55 |
| 2" steel, galvanized | 13.000 | L.F. | 3.250 | 59.02 | 63.83 | 122.85 |
| Vanity base cabinet, 2 doors, 30" wide | 1.000 | Ea. | 1.630 | 110.00 | 30.00 | 140.00 |
| Vanity top, plastic laminated, square edge | 2.670 | L.F. | .712 | 48.03 | 13.38 | 61.41 |
| Carrier, steel for studs, no arms | 1.000 | Ea. | 1.140 | 11.61 | 25.39 | 37.00 |
| Wood, 2" x 8" blocking | 1.300 | L.F. | .040 | .72 | 1.06 | 1.78 |

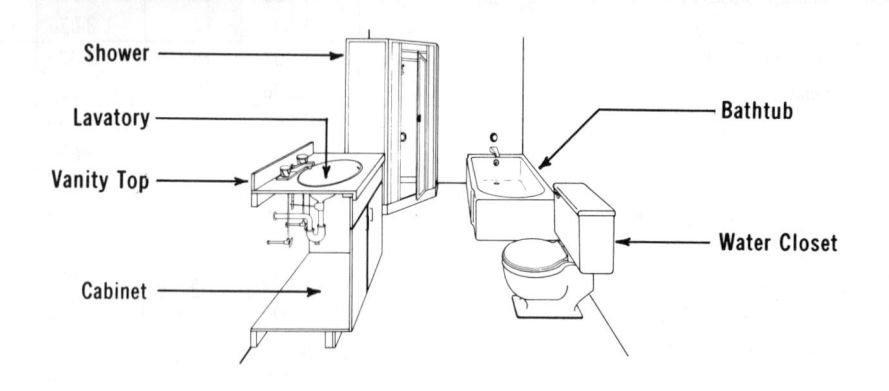

# 8 | MECHANICAL | 40 | Four Fixture Bathroom Systems

| SYSTEM DESCRIPTION | QUAN. | UNIT | MAN HOURS | MAT. | INST. | TOTAL |
|---|---|---|---|---|---|---|
| **BATHROOM WITH LAVATORY INSTALLED IN VANITY** | | | | | | |
| Water closet, floor mounted, 2 piece, close coupled, white | 1.000 | Ea. | 3.020 | 115.50 | 59.50 | 175.00 |
| Rough-in supply, waste and vent for water closet | 1.000 | Ea. | 2.862 | 35.35 | 58.19 | 93.54 |
| Lavatory, 20" x 18" P.E. cast iron with fittings, white | 1.000 | Ea. | 2.500 | 110.00 | 50.00 | 160.00 |
| Shower, steel, enameled, stone base, corner, white | 1.000 | Ea. | 8.890 | 539.00 | 176.00 | 715.00 |
| Rough-in supply waste and vent for lavatory and shower | 2.000 | Ea. | 9.832 | 91.28 | 204.20 | 295.48 |
| Bathtub, P.E. cast iron, 5' long with fittings, white | 1.000 | Ea. | 3.640 | 269.50 | 70.50 | 340.00 |
| Rough-in supply waste and vent for bathtub | 1.000 | Ea. | 3.542 | 34.63 | 75.47 | 110.10 |
| Piping supply, 1/2" copper | 42.000 | L.F. | 7.550 | 42.00 | 165.00 | 207.00 |
| Waste, 4" cast iron, no hub | 10.000 | L.F. | 4.440 | 85.35 | 87.15 | 172.50 |
| Vent, 2" steel galvanized | 13.000 | L.F. | 4.500 | 81.72 | 88.38 | 170.10 |
| Vanity base, 2 doors, 30" wide | 1.000 | Ea. | 1.630 | 110.00 | 30.00 | 140.00 |
| Vanity top, plastic laminated, square edge | 2.670 | L.F. | .712 | 49.64 | 14.44 | 64.08 |
| TOTAL | | | 53.118 | 1563.97 | 1078.83 | 2642.80 |
| **BATHROOM WITH WALL HUNG LAVATORY** | | | | | | |
| Water closet, floor mounted, 2 piece, close coupled, white | 1.000 | Ea. | 3.020 | 115.50 | 59.50 | 175.00 |
| Rough-in supply, waste and vent for water closet | 1.000 | Ea. | 2.862 | 35.35 | 58.19 | 93.54 |
| Lavatory, 20" x 18" P.E. cast iron with fittings, white | 1.000 | Ea. | 2.000 | 126.50 | 38.50 | 165.00 |
| Shower, steel enameled, stone base, corner, white | 1.000 | Ea. | 8.890 | 539.00 | 176.00 | 715.00 |
| Rough-in supply, waste and vent for lavatory and shower | 2.000 | Ea. | 9.832 | 91.28 | 204.20 | 295.48 |
| Bathtub, P.E. cast iron, 5" long with fittings, white | 1.000 | Ea. | 3.640 | 269.50 | 70.50 | 340.00 |
| Rough-in supply, waste and vent for bathtub | 1.000 | Ea. | 3.542 | 34.63 | 75.47 | 110.10 |
| Piping, supply, 1/2" copper | 42.000 | L.F. | 7.550 | 42.00 | 165.00 | 207.00 |
| Waste, 4" cast iron, no hub | 10.000 | L.F. | 4.440 | 85.35 | 87.15 | 172.50 |
| Vent, 2" steel galvanized | 13.000 | L.F. | 4.500 | 81.72 | 88.38 | 170.10 |
| Carrier, steel for studs, no arms | 1.000 | Ea. | 1.140 | 11.61 | 25.39 | 37.00 |
| TOTAL | | | 51.416 | 1432.44 | 1048.28 | 2480.72 |

The costs in this system are on a cost each basis. All necessary piping is included

| DESCRIPTION | QUAN. | UNIT | MAN. HOURS | MAT. | INST. | TOTAL |
|---|---|---|---|---|---|---|
| | | | | | | |
| | | | | | | |
| | | | | | | |
| | | | | | | |

# Four Fixture Bathroom Price Sheet

| | QUAN. | UNIT | MAN HOURS | COST EACH MAT. | COST EACH INST. | TOTAL |
|---|---|---|---|---|---|---|
| Water closet, close coupled, standard 2 piece, white | 1.000 | Ea. | 3.020 | 115.50 | 59.50 | 175.00 |
| Color | 1.000 | Ea. | 3.020 | 137.50 | 57.50 | 195.00 |
| One piece, elongated bowl, white | 1.000 | Ea. | 3.020 | 407.00 | 58.00 | 465.00 |
| Color | 1.000 | Ea. | 3.020 | 528.00 | 57.00 | 585.00 |
| Low profile, one piece elongated bowl, white | 1.000 | Ea. | 3.020 | 627.00 | 58.00 | 685.00 |
| Color | 1.000 | Ea. | 3.020 | 825.00 | 60.00 | 885.00 |
| Rough-in for water closet | | | | | | |
| 1/2" copper supply, 4" cast iron waste, 2" cast iron vent | 1.000 | Ea. | 2.862 | 35.35 | 58.19 | 93.54 |
| 4" PVC/DWV waste, 2" PVC vent | 1.000 | Ea. | 3.711 | 22.52 | 74.87 | 97.39 |
| 4" copper waste, 2" copper vent | 1.000 | Ea. | 4.239 | 41.96 | 86.88 | 128.84 |
| 3" cast iron waste, 1-1/2" cast iron vent | 1.000 | Ea. | 2.734 | 31.12 | 55.87 | 86.99 |
| 3" PVC waste, 1-1/2" PVC vent | 1.000 | Ea. | 3.174 | 18.39 | 66.65 | 85.04 |
| 3" PVC waste, 1-1/2" PVC vent | 1.000 | Ea. | 3.550 | 28.75 | 74.39 | 103.14 |
| 1/2" PVC supply, 4" PVC waste, 2" PVC vent | 1.000 | Ea. | 3.825 | 26.78 | 77.27 | 104.05 |
| 3" PVC waste, 1-1/2" PVC vent | 1.000 | Ea. | 3.288 | 22.65 | 69.05 | 91.70 |
| 1/2" steel supply, 4" cast iron waste, 2" cast iron vent | 1.000 | Ea. | 2.718 | 38.65 | 55.01 | 93.66 |
| 4" cast iron waste, 2" steel vent | 1.000 | Ea. | 2.650 | 43.57 | 53.69 | 97.26 |
| 4" PVC waste, 2" PVC vent | 1.000 | Ea. | 3.567 | 25.82 | 71.69 | 97.51 |
| Lavatory wall hung, P.E. cast iron 20" x 18", white | 1.000 | Ea. | 2.000 | 126.50 | 38.50 | 165.00 |
| Color | 1.000 | Ea. | 2.000 | 159.50 | 40.50 | 200.00 |
| Vitreous china 19" x 17", white | 1.000 | Ea. | 2.000 | 79.20 | 40.80 | 120.00 |
| Color | 1.000 | Ea. | 2.000 | 91.30 | 38.70 | 130.00 |
| Lavatory for vanity top, P.E. cast iron, 20" x 18", white | 1.000 | Ea. | 2.500 | 110.00 | 50.00 | 160.00 |
| Color | 1.000 | Ea. | 2.500 | 137.50 | 47.50 | 185.00 |
| Steel, enameled 20" x 17", white | 1.000 | Ea. | 2.500 | 66.00 | 49.00 | 115.00 |
| Color | 1.000 | Ea. | 2.500 | 69.30 | 50.70 | 120.00 |
| Vitreous china 20" x 16", white | 1.000 | Ea. | 2.500 | 165.00 | 50.00 | 215.00 |
| Color | 1.000 | Ea. | 2.500 | 176.00 | 49.00 | 225.00 |
| Shower, steel enameled, stone base 36" square, white | 1.000 | Ea. | 8.890 | 539.00 | 176.00 | 715.00 |
| Color | 1.000 | Ea. | 8.890 | 605.00 | 175.00 | 780.00 |
| Rough-in for lavatory and shower | | | | | | |
| 1/2" copper supply, 1-1/2" cast iron waste, 1-1/2" cast iron vent | 1.000 | Ea. | 9.832 | 91.28 | 204.20 | 295.48 |
| 1-1/2" PVC waste, 1-1/4" PVC vent | 1.000 | Ea. | 10.172 | 52.36 | 222.12 | 274.48 |
| 1/2" steel supply, 1-1/4" cast iron waste, 1-1/4" steel vent | 1.000 | Ea. | 8.504 | 105.76 | 175.96 | 281.72 |
| 1-1/4" PVC waste, 1-1/4" PVC vent | 1.000 | Ea. | 9.404 | 69.48 | 205.64 | 275.12 |
| 1/2" PVC supply, 1-1/2" PVC waste, 1-1/2" PVC vent | 1.000 | Ea. | 10.780 | 75.40 | 234.60 | 310.00 |
| Bathtub, P.E. cast iron, 5' long with fittings, white | 1.000 | Ea. | 3.640 | 269.50 | 70.50 | 340.00 |
| Color | 1.000 | Ea. | 3.640 | 346.50 | 73.50 | 420.00 |
| Steel enameled, 5' long with fittings, white | 1.000 | Ea. | 2.910 | 181.50 | 58.50 | 240.00 |
| Color | 1.000 | Ea. | 2.910 | 192.50 | 57.50 | 250.00 |
| Rough-in for bathtub | | | | | | |
| 1/2" copper supply, 4" cast iron waste, 1-1/2" copper vent | 1.000 | Ea. | 3.542 | 34.63 | 75.47 | 110.10 |
| 4" PVC waste, 1-1/2" PVC vent | 1.000 | Ea. | 3.991 | 24.96 | 83.99 | 108.95 |
| 1/2" steel supply, 4" cast iron waste, 1-1/2" steel vent | 1.000 | Ea. | 2.958 | 44.21 | 60.89 | 105.10 |
| 4" PVC waste, 1-1/2" PVC vent | 1.000 | Ea. | 3.751 | 30.46 | 78.69 | 109.15 |
| 1/2" PVC supply, 4" PVC waste, 1-1/2" PVC vent | 1.000 | Ea. | 4.181 | 32.06 | 87.99 | 120.05 |
| Piping supply, 1/2" copper | 42.000 | L.F. | 6.342 | 35.28 | 138.60 | 173.88 |
| 1/2" steel | 42.000 | L.F. | 5.334 | 58.38 | 116.34 | 174.72 |
| 1/2" PVC | 42.000 | L.F. | 7.140 | 65.10 | 155.40 | 220.50 |
| Piping, waste, 4" cast iron, no hub | 10.000 | L.F. | 3.848 | 73.97 | 75.53 | 149.50 |
| 4" PVC/DWV | 10.000 | L.F. | 6.123 | 49.40 | 120.25 | 169.65 |
| 4" copper/DWV | 10.000 | L.F. | 8.671 | 108.68 | 164.32 | 273.00 |
| Piping, vent, 2" cast iron, no hub | 13.000 | L.F. | 3.471 | 43.03 | 68.12 | 111.15 |
| 2" copper/DWV | 13.000 | L.F. | 4.329 | 38.48 | 94.77 | 133.25 |
| 2" PVC/DWV | 13.000 | L.F. | 4.524 | 19.76 | 88.79 | 108.55 |
| 2" steel, galvanized | 13.000 | L.F. | 3.250 | 59.02 | 63.83 | 122.85 |
| Vanity base cabinet, 2 doors, 30" wide | 1.000 | Ea. | 1.630 | 110.00 | 30.00 | 140.00 |
| Vanity top, plastic laminated, square edge | 3.160 | L.F. | .843 | 56.85 | 15.83 | 72.68 |
| Carrier, steel, for studs, no arms | 1.000 | Ea. | 1.140 | 11.61 | 25.39 | 37.00 |
| Wood, 2" x 8" blocking | 1.300 | L.F. | .040 | .72 | 1.06 | 1.78 |

# 8 | MECHANICAL  44 | Five Fixture Bathroom Systems

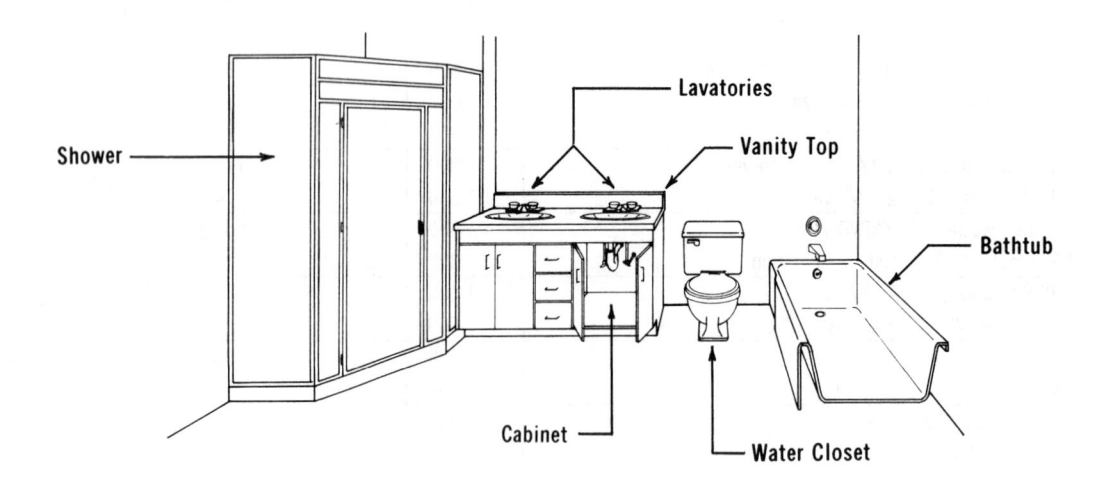

| SYSTEM DESCRIPTION | QUAN. | UNIT | MAN HOURS | MAT. | INST. | TOTAL |
|---|---|---|---|---|---|---|
| **BATHROOM WITH SHOWER, BATHTUB, LAVATORIES IN VANITY** | | | | | | |
| Water closet, floor mounted, 1 piece combination, white | 1.000 | Ea. | | | | |
| Rough-in supply, waste and vent for water closet | 1.000 | Ea. | 2.862 | 35.35 | 58.19 | 93.54 |
| Lavatory, 20" x 16", vitreous china oval, with fittings, white | 2.000 | Ea. | 5.000 | 330.00 | 100.00 | 430.00 |
| Shower, steel enameled, stone base, corner, white | 1.000 | Ea. | 8.890 | 539.00 | 176.00 | 715.00 |
| Rough-in supply waste and vent for lavatory and shower | 3.000 | Ea. | 10.530 | 102.48 | 217.32 | 319.80 |
| Bathtub, P.E. cast iron, 5' long with fittings, white | 1.000 | Ea. | 3.640 | 269.50 | 70.50 | 340.00 |
| Rough-in supply, waste and vent for bathtub | 1.000 | Ea. | 3.838 | 40.32 | 81.28 | 121.60 |
| Piping, supply, 1/2" copper | 42.000 | L.F. | 6.342 | 35.28 | 138.60 | 173.88 |
| Waste, 4" cast iron, no hub | 10.000 | L.F. | 2.960 | 56.90 | 58.10 | 115.00 |
| Vent, 2" steel galvanized | 13.000 | L.F. | 3.250 | 59.02 | 63.83 | 122.85 |
| Vanity base, 2 door, 24" x 48" | 1.000 | Ea. | 2.420 | 192.50 | 47.50 | 240.00 |
| Vanity top, plastic laminated, square edge | 4.170 | L.F. | 1.113 | 75.02 | 20.89 | 95.91 |
| TOTAL | | | 50.845 | 1735.37 | 1032.21 | 2767.58 |

The costs in this system are on a cost each basis. All necessary piping is included

| DESCRIPTION | QUAN. | UNIT | MAN HOURS | MAT. | INST. | TOTAL |
|---|---|---|---|---|---|---|
| | | | | | | |
| | | | | | | |
| | | | | | | |
| | | | | | | |
| | | | | | | |
| | | | | | | |
| | | | | | | |
| | | | | | | |

# Five Fixture Bathroom Price Sheet

| Description | QUAN. | UNIT | MAN HOURS | MAT. | INST. | TOTAL |
|---|---|---|---|---|---|---|
| Water closet, close coupled, standard 2 piece, white | 1.000 | Ea. | 3.020 | 115.50 | 59.50 | 175.00 |
| Color | 1.000 | Ea. | 3.020 | 137.50 | 57.50 | 195.00 |
| One piece elongated bowl, white | 1.000 | Ea. | 3.020 | 407.00 | 58.00 | 465.00 |
| Color | 1.000 | Ea. | 3.020 | 528.00 | 57.00 | 585.00 |
| Low profile, one piece elongated bowl, white | 1.000 | Ea. | 3.020 | 627.00 | 58.00 | 685.00 |
| Color | 1.000 | Ea. | 3.020 | 825.00 | 60.00 | 885.00 |
| Rough-in, supply, waste and vent for water closet | | | | | | |
| 1/2" copper supply, 4" cast iron waste, 2" cast iron vent | 1.000 | Ea. | 2.862 | 35.35 | 58.19 | 93.54 |
| 4" P.V.C./DWV waste, 2" P.V.C. vent | 1.000 | Ea. | 3.711 | 22.52 | 74.87 | 97.39 |
| 4" copper waste, 2" copper vent | 1.000 | Ea. | 4.239 | 41.96 | 86.88 | 128.84 |
| 3" cast iron waste, 1-1/2" cast iron vent | 1.000 | Ea. | 2.734 | 31.12 | 55.87 | 86.99 |
| 3" P.V.C. waste, 1-1/2" P.V.C. vent | 1.000 | Ea. | 3.174 | 18.39 | 66.65 | 85.04 |
| 3" copper waste, 1-1/2" copper vent | 1.000 | Ea. | 3.550 | 28.75 | 74.39 | 103.14 |
| 1/2" P.V.C. supply, 4" P.V.C. waste, 2" P.V.C. vent | 1.000 | Ea. | 3.825 | 26.78 | 77.27 | 104.05 |
| 3" P.V.C. waste, 1-1/2" P.V.C. supply | 1.000 | Ea. | 3.288 | 22.65 | 69.05 | 91.70 |
| 1/2" steel supply, 4" cast iron waste, 2" cast iron vent | 1.000 | Ea. | 2.718 | 38.65 | 55.01 | 93.66 |
| 4" cast iron waste, 2" steel vent | 1.000 | Ea. | 2.650 | 43.57 | 53.69 | 97.26 |
| 4" P.V.C. waste, 2" P.V.C. vent | 1.000 | Ea. | 3.567 | 25.82 | 71.69 | 97.51 |
| Lavatory, wall hung, P.E. cast iron 20" x 18", white | 2.000 | Ea. | 4.000 | 253.00 | 77.00 | 330.00 |
| Color | 2.000 | Ea. | 4.000 | 319.00 | 81.00 | 400.00 |
| Vitreous china, 19" x 17", white | 2.000 | Ea. | 4.000 | 158.40 | 81.60 | 240.00 |
| Color | 2.000 | Ea. | 4.000 | 182.60 | 77.40 | 260.00 |
| Lavatory, for vanity top, P.E. cast iron, 20" x 18", white | 2.000 | Ea. | 5.000 | 220.00 | 100.00 | 320.00 |
| Color | 2.000 | Ea. | 5.000 | 275.00 | 95.00 | 370.00 |
| Steel enameled 20" x 17", white | 2.000 | Ea. | 5.000 | 132.00 | 98.00 | 230.00 |
| Color | 2.000 | Ea. | 5.000 | 138.60 | 101.40 | 240.00 |
| Vitreous china 20" x 16", white | 2.000 | Ea. | 5.000 | 330.00 | 100.00 | 430.00 |
| Color | 2.000 | Ea. | 5.000 | 352.00 | 98.00 | 450.00 |
| Shower, steel enameled, stone base 36" square, white | 2.000 | Ea. | 8.890 | 539.00 | 176.00 | 715.00 |
| Color | 2.000 | Ea. | 8.890 | 605.00 | 175.00 | 780.00 |
| Rough-in for lavatory or shower | | | | | | |
| 1/2" copper supply, 1-1/2" cast iron waste, 1-1/2" cast iron vent | 3.000 | Ea. | 10.530 | 102.48 | 217.32 | 319.80 |
| 1-1/2" P.V.C. waste, 1-1/4" P.V.C. vent | 3.000 | Ea. | 10.938 | 55.68 | 238.92 | 294.60 |
| 1/2" steel supply, 1-1/4" cast iron waste, 1-1/4" steel vent | 3.000 | Ea. | 8.970 | 114.30 | 184.50 | 298.80 |
| 1-1/4" P.V.C. waste, 1-1/4" P.V.C. vent | 3.000 | Ea. | 10.218 | 71.70 | 223.50 | 295.20 |
| 1/2" P.V.C. supply, 1-1/2" P.V.C. waste, 1-1/2" P.V.C. vent | 3.000 | Ea. | 11.508 | 77.46 | 250.44 | 327.90 |
| Bathtub, P.E. cast iron 5' long with fittings, white | 1.000 | Ea. | 3.640 | 269.50 | 70.50 | 340.00 |
| Color | 1.000 | Ea. | 3.640 | 346.50 | 73.50 | 420.00 |
| Steel, enameled 5' long with fittings, white | 1.000 | Ea. | 2.910 | 181.50 | 58.50 | 240.00 |
| Color | 1.000 | Ea. | 2.910 | 192.50 | 57.50 | 250.00 |
| Rough-in for bathtub | | | | | | |
| 1/2" copper supply, 4" cast iron waste, 1-1/2" copper vent | 1.000 | Ea. | 3.838 | 40.32 | 81.28 | 121.60 |
| 4" P.V.C. waste, 1-1/2" P.V.C. vent | 1.000 | Ea. | 4.462 | 28.76 | 93.24 | 122.00 |
| 1/2" steel supply, 4" cast iron waste, 1-1/2" steel vent | 1.000 | Ea. | 3.254 | 49.90 | 66.70 | 116.60 |
| 4" P.V.C. waste, 1-1/2" P.V.C. vent | 1.000 | Ea. | 4.222 | 34.26 | 87.94 | 122.20 |
| 1/2" P.V.C. supply, 4" P.V.C. waste, 1-1/2" P.V.C. vent | 1.000 | Ea. | 4.652 | 35.86 | 97.24 | 133.10 |
| Piping, supply, 1/2" copper | 42.000 | L.F. | 6.342 | 35.28 | 138.60 | 173.88 |
| 1/2" steel | 42.000 | L.F. | 5.334 | 58.38 | 116.34 | 174.72 |
| 1/2" P.V.C. | 42.000 | L.F. | 7.140 | 65.10 | 155.40 | 220.50 |
| Piping, waste, 4" cast iron, no hub | 10.000 | L.F. | 2.960 | 56.90 | 58.10 | 115.00 |
| 4" P.V.C./DWV | 10.000 | L.F. | 4.710 | 38.00 | 92.50 | 130.50 |
| 4" copper/DWV | 10.000 | L.F. | 6.670 | 83.60 | 126.40 | 210.00 |
| Piping, vent, 2" cast iron, no hub | 13.000 | L.F. | 3.471 | 43.03 | 68.12 | 111.15 |
| 2" copper/DWV | 13.000 | L.F. | 4.329 | 38.48 | 94.77 | 133.25 |
| 2" P.V.C./DWV | 13.000 | L.F. | 4.524 | 19.76 | 88.79 | 108.55 |
| 2" steel, galvanized | 13.000 | L.F. | 3.250 | 59.02 | 63.83 | 122.85 |
| Vanity base cabinet, 2 doors, 24" x 48" | 1.000 | Ea. | 2.420 | 192.50 | 47.50 | 240.00 |
| Vanity top, plastic laminated, square edge | 4.170 | L.F. | 1.113 | 75.02 | 20.89 | 95.91 |
| Carrier, steel, for studs, no arms | 1.000 | Ea. | 1.140 | 11.61 | 25.39 | 37.00 |
| Wood, 2" x 8" blocking | 1.300 | L.F. | .040 | .72 | 1.06 | 1.78 |

8

# 8 | MECHANICAL    50 | Lavatory Systems

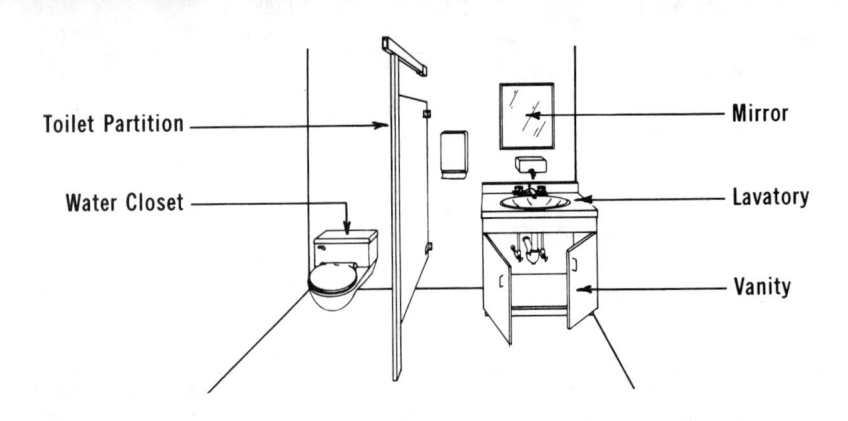

| SYSTEM DESCRIPTION | QUAN. | UNIT | MAN HOURS | MAT. | INST. | TOTAL |
|---|---|---|---|---|---|---|
| **TOILET ROOM SYSTEM WITH VANITY** | | | | | | |
| Water closet, tank, type, one piece, wall hung | 1.000 | Ea. | 3.020 | 467.50 | 57.50 | 525.00 |
| Rough-in, supply waste and vent for water closet | 1.000 | Ea. | 5.710 | 100.31 | 114.69 | 215.00 |
| Lavatory, vanity top, porcelain enamel on cast iron, 20" x 18" | 1.000 | Ea. | 2.500 | 110.00 | 50.00 | 160.00 |
| Rough-in, supply waste and vent for lavatory | 1.000 | Ea. | 5.520 | 74.62 | 110.38 | 185.00 |
| Partition, floor and ceiling mounted, painted metal | 1.000 | Ea. | 3.200 | 220.00 | 65.00 | 285.00 |
| Tissue dispenser, stainless steel, single roll | 1.000 | Ea. | .267 | 18.81 | 5.19 | 24.00 |
| Towel dispenser, stainless steel, surface mounted | 1.000 | Ea. | .500 | 45.10 | 9.90 | 55.00 |
| Soap dispenser, stainless steel, surface mounted | 1.000 | Ea. | .400 | 51.70 | 8.30 | 60.00 |
| Mirror, stainless steel, 3/4" frame, 18" x 24" | 1.000 | Ea. | .400 | 61.60 | 7.40 | 69.00 |
| Vanity, 21" deep, 36" wide | 1.000 | Ea. | 2.000 | 154.00 | 41.00 | 195.00 |
| TOTAL | | | 23.517 | 1303.64 | 469.36 | 1773.00 |
| **TOILET ROOM SYSTEM WITH WALL HUNG LAVATORY** | | | | | | |
| Water closet, tank type, one piece, wall hung | 1.000 | Ea. | 3.020 | 467.50 | 57.50 | 525.00 |
| Rough-in, supply waste and vent for water closet | 1.000 | Ea. | 5.710 | 100.31 | 114.69 | 215.00 |
| Lavatory, wall hung, porcelain enamel on cast iron, 20" x 18" | 1.000 | Ea. | 2.000 | 126.50 | 38.50 | 165.00 |
| Rough-in, supply waste and vent for lavatory | 1.000 | Ea. | 5.520 | 110.74 | 119.26 | 230.00 |
| Partition, floor and ceiling mounted, painted metal | 1.000 | Ea. | 3.200 | 220.00 | 65.00 | 285.00 |
| Tissue dispenser, stainless steel, single roll | 1.000 | Ea. | .267 | 18.81 | 5.19 | 24.00 |
| Towel dispenser, stainless steel, surface mounted | 1.000 | Ea. | .500 | 45.10 | 9.90 | 55.00 |
| Soap dispenser, stainless steel, surface mounted | 1.000 | Ea. | .400 | 51.70 | 8.30 | 60.00 |
| Mirror, stainless steel, 3/4" frame, 18" x 24" | 1.000 | Ea. | .400 | 61.60 | 7.40 | 69.00 |
| TOTAL | | | 21.017 | 1202.26 | 425.74 | 1628.00 |

| DESCRIPTION | QUAN. | UNIT | MAN HOURS | MAT. | INST. | TOTAL |
|---|---|---|---|---|---|---|
| | | | | | | |
| | | | | | | |
| | | | | | | |
| | | | | | | |
| | | | | | | |
| | | | | | | |

# Lavatory Price Sheet

| | QUAN. | UNIT | MAN HOURS | MAT. | INST. | TOTAL |
|---|---|---|---|---|---|---|
| Water closet, bowl only, vitreous china, bowl only | 1.000 | Ea. | 2.760 | 242.00 | 53.00 | 295.00 |
| Floor mounted | 1.000 | Ea. | 2.760 | 214.50 | 55.50 | 270.00 |
| Tank type, one piece, wall hung | 1.000 | Ea. | 3.020 | 467.50 | 57.50 | 525.00 |
| Floor mounted | 1.000 | Ea. | 3.020 | 407.00 | 58.00 | 465.00 |
| Two piece, wall hung | 1.000 | Ea. | 3.020 | 286.00 | 59.00 | 345.00 |
| Floor mounted | 1.000 | Ea. | 3.020 | 115.50 | 59.50 | 175.00 |
| Rough-in, supply waste and vent, bowl only | 1.000 | Ea. | 6.400 | 174.50 | 125.50 | 300.00 |
| Tank type | 1.000 | Ea. | 5.710 | 100.31 | 114.69 | 215.00 |
| Lavatory, vanity top, porcelain enamel on cast iron, 20" x 18" | 1.000 | Ea. | 2.500 | 110.00 | 50.00 | 160.00 |
| 26" x 18" | 1.000 | Ea. | 2.500 | 143.00 | 47.00 | 190.00 |
| 18" round | 1.000 | Ea. | 2.500 | 101.20 | 48.80 | 150.00 |
| Cultured marble, 19" x 17" | 1.000 | Ea. | 2.500 | 71.50 | 48.50 | 120.00 |
| 37" x 22" | 1.000 | Ea. | 2.500 | 104.50 | 50.50 | 155.00 |
| Vitreous china, 20" x 16" | 1.000 | Ea. | 2.500 | 165.00 | 50.00 | 215.00 |
| 20" x 17" | 1.000 | Ea. | 2.500 | 115.00 | 50.00 | 165.00 |
| 22" x 13" | 1.000 | Ea. | 2.500 | 110.00 | 50.00 | 160.00 |
| Wall hung, porcelain enamel on cast iron, 16" x 14" | 1.000 | Ea. | 2.000 | 181.50 | 38.50 | 220.00 |
| 20" x 18" | 1.000 | Ea. | 2.000 | 126.50 | 38.50 | 165.00 |
| Vitreous china, 18" x 14" | 1.000 | Ea. | 2.000 | 132.00 | 38.00 | 170.00 |
| 19" x 17" | 1.000 | Ea. | 2.000 | 79.20 | 40.80 | 120.00 |
| Rough-in, lavatory, waste and vent, vanity top | 1.000 | Ea. | 5.520 | 74.62 | 110.38 | 185.00 |
| Wall hung | 1.000 | Ea. | 5.520 | 110.74 | 119.26 | 230.00 |
| Partitions, ceiling hung, painted metal | 1.000 | Ea. | 4.000 | 220.00 | 80.00 | 300.00 |
| Plastic laminate on particle board | 1.000 | Ea. | 4.000 | 324.50 | 80.50 | 405.00 |
| Porcelain enamel | 1.000 | Ea. | 4.000 | 665.50 | 79.50 | 745.00 |
| Stainless steel | 1.000 | Ea. | 4.000 | 610.50 | 79.50 | 690.00 |
| Floor and ceiling anchored, painted metal | 1.000 | Ea. | 3.200 | 220.00 | 65.00 | 285.00 |
| Plastic laminate on particle board | 1.000 | Ea. | 3.200 | 385.00 | 65.00 | 450.00 |
| Porcelain enamel | 1.000 | Ea. | 3.200 | 676.50 | 63.50 | 740.00 |
| Stainless steel | 1.000 | Ea. | 3.200 | 610.50 | 64.50 | 675.00 |
| Tissue dispenser, stainless steel, single roll | 1.000 | Ea. | .267 | 18.81 | 5.19 | 24.00 |
| Double roll | 1.000 | Ea. | .333 | 30.80 | 6.20 | 37.00 |
| Towel dispenser, stainless steel, surface mounted | 1.000 | Ea. | .500 | 45.10 | 9.90 | 55.00 |
| Flush mounted, recessed | 1.000 | Ea. | .800 | 82.50 | 15.50 | 98.00 |
| Soap dispenser, chrome, surface mounted, liquid | 1.000 | Ea. | .400 | 51.70 | 8.30 | 60.00 |
| Flush mounted, recessed, stainless steel | 1.000 | Ea. | .800 | 59.40 | 15.60 | 75.00 |
| Mirror, stainless steel, 3/4" frame, 18" x 24" | 1.000 | Ea. | .400 | 61.60 | 7.40 | 69.00 |
| 36" x 24" | 1.000 | Ea. | .533 | 107.80 | 12.20 | 120.00 |
| Including 5" shelf, 18" x 24" | 1.000 | Ea. | .400 | 80.30 | 7.70 | 88.00 |
| 36" x 24" | 1.000 | Ea. | .533 | 148.50 | 11.50 | 160.00 |
| Vanity, 21" deep, 24" wide | 1.000 | Ea. | 1.450 | 103.40 | 26.60 | 130.00 |
| 36" wide | 1.000 | Ea. | 2.000 | 154.00 | 41.00 | 195.00 |
| 48" wide | 1.000 | Ea. | 2.420 | 192.50 | 47.50 | 240.00 |

8

# 8 | MECHANICAL  60 | Gas Heating/Cooling Systems

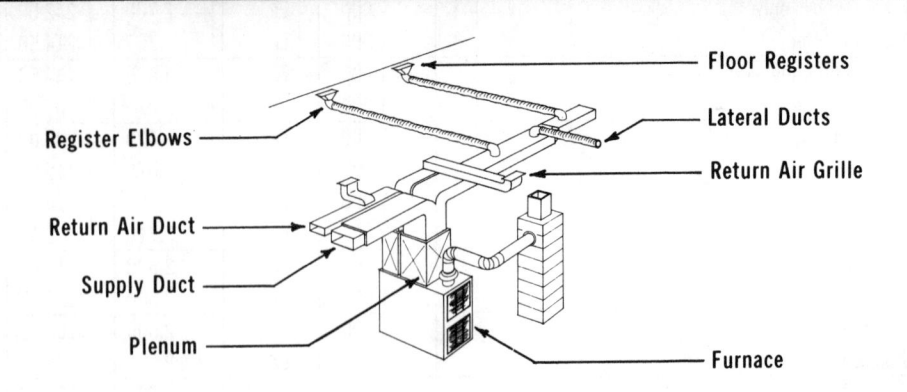

| SYSTEM DESCRIPTION | QUAN. | UNIT | MAN HOURS | MAT. | INST. | TOTAL |
|---|---|---|---|---|---|---|
| **HEATING ONLY, GAS FIRED HOT AIR, ONE ZONE, 1200 S.F. BUILDING** | | | | | | |
| Furnace, gas, up flow | 1.000 | Ea. | 4.710 | 418.00 | 92.00 | 510.00 |
| Intermittent pilot | 1.000 | Ea. | | 75.00 | | 75.00 |
| Supply duct, rigid fiberglass | 176.000 | L.F. | 12.144 | 82.72 | 248.16 | 330.88 |
| Return duct, sheet metal, galvanized | 158.000 | Lb. | 16.116 | 206.98 | 331.80 | 538.78 |
| Lateral ducts, 6" flexible fiberglass | 144.000 | L.F. | 8.928 | 220.32 | 175.68 | 396.00 |
| Register, elbows | 12.000 | Ea. | 3.204 | 102.96 | 63.24 | 166.20 |
| Floor registers, enameled steel | 12.000 | Ea. | 3.000 | 219.84 | 68.16 | 288.00 |
| Floor grille, return air | 2.000 | Ea. | .728 | 18.48 | 16.02 | 34.50 |
| Thermostat | 1.000 | Ea. | 1.000 | 44.00 | 22.00 | 66.00 |
| Plenum | 1.000 | Ea. | 1.000 | 49.50 | 19.50 | 69.00 |
| TOTAL | | | 50.830 | 1437.80 | 1036.56 | 2474.36 |
| **HEATING/COOLING, GAS FIRED FORCED AIR, ONE ZONE, 1200 S.F. BUILDING** | | | | | | |
| Furnace, including plenum, compressor, coil | 1.000 | Ea. | 14.720 | 2631.20 | 289.80 | 2921.00 |
| Intermittent pilot | 1.000 | Ea. | | 75.00 | | 75.00 |
| Supply duct, rigid fiberglass | 176.000 | L.F. | 12.144 | 82.72 | 248.16 | 330.88 |
| Return duct, sheet metal, galvanized | 158.000 | Lb. | 16.116 | 206.98 | 331.80 | 538.78 |
| Lateral duct, 6" flexible fiberglass | 144.000 | L.F. | 8.928 | 220.32 | 175.68 | 396.00 |
| Register elbows | 12.000 | Ea. | 3.204 | 102.96 | 63.24 | 166.20 |
| Floor registers, enameled steel | 12.000 | Ea. | 3.000 | 219.84 | 68.16 | 288.00 |
| Floor grille return air | 2.000 | Ea. | .728 | 18.48 | 16.02 | 34.50 |
| Thermostat | 1.000 | Ea. | 1.000 | 44.00 | 22.00 | 66.00 |
| Refrigeration piping | 25.000 | L.F. | | 120.00 | | 120.00 |
| TOTAL | | | 59.840 | 3721.50 | 1214.86 | 4936.36 |

The costs in these systems are based on complete system basis. For Larger buildings use the pricesheet on the opposite page.

| DESCRIPTION | QUAN. | UNIT | MAN HOURS | MAT. | INST. | TOTAL |
|---|---|---|---|---|---|---|
| | | | | | | |
| | | | | | | |
| | | | | | | |
| | | | | | | |
| | | | | | | |
| | | | | | | |

# Gas Heating/Cooling Price Sheet

| | QUAN. | UNIT | MAN HOURS | COST PER SYSTEM MAT. | INST. | TOTAL |
|---|---|---|---|---|---|---|
| Furnace, heating only, 100 MBH, area to 1200 S.F. | 1.000 | Ea. | 4.710 | 418.00 | 92.00 | 510.00 |
| 120 MBH, area to 1500 S.F. | 1.000 | Ea. | 5.000 | 467.50 | 97.50 | 565.00 |
| 160 MBH, area to 2000 S.F. | 1.000 | Ea. | 5.710 | 1001.00 | 124.00 | 1125.00 |
| 200 MBH, area to 2400 S.F. | 1.000 | Ea. | 6.150 | 1402.50 | 122.50 | 1525.00 |
| Heating/cooling, 100 MBH heat, 36 MBH cool, to 1200 S.F. | 1.000 | Ea. | 16.000 | 2860.00 | 315.00 | 3175.00 |
| 120 MBH heat, 42 MBH cool, to 1500 S.F. | 1.000 | Ea. | 18.460 | 3245.00 | 380.00 | 3625.00 |
| 144 MBH heat, 47 MBH cool, to 2000 S.F. | 1.000 | Ea. | 20.000 | 3327.50 | 422.50 | 3750.00 |
| 200 MBH heat, 60 MBH cool, to 2400 S.F. | 1.000 | S.F. | 34.290 | 3547.50 | 702.50 | 4250.00 |
| Intermittent pilot, 100 MBH furnace | 1.000 | Ea. | | 75.00 | | 75.00 |
| 200 MBH furnace | 1.000 | Ea. | | 75.00 | | 75.00 |
| Supply duct, rectangular, area to 1200 S.F., rigid fiberglass | 176.000 | S.F. | 12.144 | 82.72 | 248.16 | 330.88 |
| Sheet metal insulated | 228.000 | Lb. | 31.352 | 358.52 | 640.72 | 999.24 |
| Area to 1500 S.F., rigid fiberglass | 176.000 | S.F. | 12.144 | 82.72 | 248.16 | 330.88 |
| Sheet metal insulated | 228.000 | Lb. | 31.352 | 358.52 | 640.72 | 999.24 |
| Area to 2400 S.F., rigid fiberglass | 205.000 | S.F. | 14.145 | 96.35 | 289.05 | 385.40 |
| Sheet metal insulated | 271.000 | Lb. | 37.072 | 424.71 | 757.70 | 1182.41 |
| Round flexible, insulated 6" diameter, to 1200 S.F. | 156.000 | L.F. | 9.672 | 238.68 | 190.32 | 429.00 |
| To 1500 S.F. | 184.000 | L.F. | 11.408 | 281.52 | 224.48 | 506.00 |
| 8" diameter, to 2000 S.F. | 269.000 | L.F. | 23.941 | 497.65 | 473.44 | 971.09 |
| To 2400 S.F. | 248.000 | L.F. | 22.072 | 458.80 | 436.48 | 895.28 |
| Return duct, sheet metal galvanized, to 1500 S.F. | 158.000 | Lb. | 16.116 | 206.98 | 331.80 | 538.78 |
| To 2400 S.F. | 191.000 | Lb. | 19.482 | 250.21 | 401.10 | 651.31 |
| Lateral ducts, flexible round 6" insulated, to 1200 S.F. | 144.000 | L.F. | 8.928 | 220.32 | 175.68 | 396.00 |
| To 1500 S.F. | 172.000 | L.F. | 10.664 | 263.16 | 209.84 | 473.00 |
| To 2000 S.F. | 261.000 | L.F. | 16.182 | 399.33 | 318.42 | 717.75 |
| To 2400 S.F. | 300.000 | L.F. | 18.600 | 459.00 | 366.00 | 825.00 |
| Spiral steel insulated, to 1200 S.F. | 144.000 | L.F. | 20.122 | 423.58 | 402.44 | 826.02 |
| To 1500 S.F. | 172.000 | L.F. | 24.018 | 505.82 | 480.36 | 986.18 |
| To 2000 S.F. | 261.000 | L.F. | 36.451 | 767.59 | 729.02 | 1496.61 |
| To 2400 S.F. | 300.000 | L.F. | 41.940 | 882.60 | 838.80 | 1721.40 |
| Rectangular sheet metal galvanized insulated, to 1200 S.F. | 228.000 | Lb. | 39.126 | 415.98 | 796.20 | 1212.18 |
| To 1500 S.F. | 344.000 | Lb. | 54.040 | 590.72 | 1101.44 | 1692.16 |
| To 2000 S.F. | 522.000 | Lb. | 82.040 | 896.66 | 1672.12 | 2568.78 |
| To 2400 S.F. | 600.000 | Lb. | 94.320 | 1030.80 | 1922.40 | 2953.20 |
| Register elbows, to 1500 S.F. | 12.000 | Ea. | 3.204 | 102.96 | 63.24 | 166.20 |
| To 2400 S.F. | 14.000 | Ea. | 3.738 | 120.12 | 73.78 | 193.90 |
| Floor registers, enameled steel w/damper, to 1500 S.F. | 12.000 | Ea. | 3.000 | 219.84 | 68.16 | 288.00 |
| To 2400 S.F. | 14.000 | Ea. | 4.312 | 152.46 | 94.64 | 247.10 |
| Return air grille, area to 1500 S.F. 12" x 12" | 2.000 | Ea. | .728 | 18.48 | 16.02 | 34.50 |
| Area to 2400 S.F. 8" x 16" | 2.000 | Ea. | .444 | 16.83 | 10.17 | 27.00 |
| Area to 2400 S.F. 8" x 16" | 2.000 | Ea. | .728 | 18.48 | 16.02 | 34.50 |
| 16" x 16" | 1.000 | Ea. | .364 | 13.92 | 8.08 | 22.00 |
| Thermostat, manual, 1 set back | 1.000 | Ea. | 1.000 | 44.00 | 22.00 | 66.00 |
| Electric, timed, 1 set back | 1.000 | Ea. | 1.000 | 69.30 | 21.70 | 91.00 |
| 2 set back | 1.000 | Ea. | 1.000 | 70.40 | 21.60 | 92.00 |
| Plenum, heating only, 100 M.B.H. | 1.000 | Ea. | 1.000 | 49.50 | 19.50 | 69.00 |
| 120 MBH | 1.000 | Ea. | 1.000 | 49.50 | 19.50 | 69.00 |
| 160 MBH | 1.000 | Ea. | 1.000 | 49.50 | 19.50 | 69.00 |
| 200 MBH | 1.000 | Ea. | 1.000 | 49.50 | 19.50 | 69.00 |
| Refrigeration piping, 3/8" | 25.000 | L.F. | | 9.33 | | 9.33 |
| 3/4" | 25.000 | L.F. | | 18.50 | | 18.50 |
| 7/8" | 25.000 | L.F. | | 21.57 | | 21.57 |
| Diffusers, ceiling, 6" diameter, to 1500 S.F. | 10.000 | Ea. | 4.440 | 117.70 | 102.30 | 220.00 |
| To 2400 S.F. | 12.000 | Ea. | 6.000 | 155.76 | 132.24 | 288.00 |
| Floor, aluminum, adjustable, 2-1/4" x 12" to 1500 S.F. | 12.000 | Ea. | 3.000 | 86.52 | 65.88 | 152.40 |
| To 2400 S.F. | 14.000 | Ea. | 3.500 | 100.94 | 76.86 | 177.80 |
| Side wall, aluminum, adjustable, 8" x 4", to 1500 S.F. | 12.000 | Ea. | 3.000 | 190.08 | 61.92 | 252.00 |
| 5" x 10" to 2400 S.F. | 12.000 | Ea. | 3.696 | 244.92 | 79.08 | 324.00 |

# 8 | MECHANICAL     64 | Oil Fired Heating/Cooling Systems

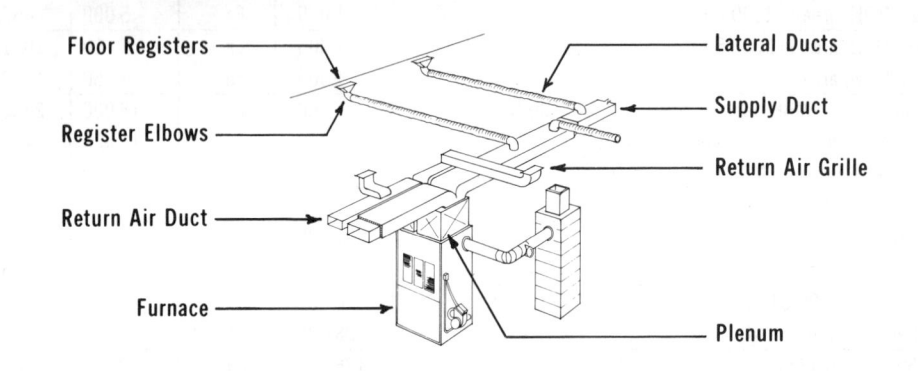

| SYSTEM DESCRIPTION | QUAN. | UNIT | MAN HOURS | MAT. | INST. | TOTAL |
|---|---|---|---|---|---|---|
| **HEATING ONLY, OIL FIRED HOT AIR, ONE ZONE, 1200 S.F. BUILDING** | | | | | | |
| Furnace, oil fired, atomizing gun type burner | 1.000 | Ea. | 4.570 | 660.00 | 90.00 | 750.00 |
| Oil piping to furnace | 1.000 | Ea. | 4.501 | 28.99 | 98.21 | 127.20 |
| Oil tank, 275 gallon, on legs | 1.000 | Ea. | 3.200 | 170.50 | 64.50 | 235.00 |
| Supply duct, rigid fiberglass | 176.000 | S.F. | 12.144 | 82.72 | 248.16 | 330.88 |
| Return duct, sheet metal, galvanized | 158.000 | Lb. | 16.116 | 206.98 | 331.80 | 538.78 |
| Lateral ducts, 6" flexible fiberglass | 144.000 | L.F. | 8.928 | 220.32 | 175.68 | 396.00 |
| Register elbows | 12.000 | Ea. | 3.204 | 102.96 | 63.24 | 166.20 |
| Floor register, enameled steel | 12.000 | Ea. | 3.000 | 219.84 | 68.16 | 288.00 |
| Floor grill, return air | 2.000 | Ea. | .728 | 18.48 | 16.02 | 34.50 |
| Thermostat | 1.000 | Ea. | 1.000 | 44.00 | 22.00 | 66.00 |
| TOTAL | | | 57.391 | 1754.79 | 1177.77 | 2932.56 |
| **HEATING/COOLING, OIL FIRED, FORCED AIR, ONE ZONE, 1200 S.F. BUILDING** | | | | | | |
| Furnace, including plenum, compressor, coil | 1.000 | Ea. | 16.000 | 2997.50 | 327.50 | 3325.00 |
| Oil piping to furnace | 1.000 | Ea. | 4.733 | 81.24 | 102.96 | 184.20 |
| Oil tank, 275 gallon on legs | 1.000 | Ea. | 3.200 | 170.50 | 64.50 | 235.00 |
| Supply duct, rigid fiberglass | 176.000 | S.F. | 12.144 | 82.72 | 248.16 | 330.88 |
| Return duct, sheet metal, galvanized | 158.000 | Lb. | 16.116 | 206.98 | 331.80 | 538.78 |
| Lateral ducts, 6" flexible fiberglass | 144.000 | L.F. | 8.928 | 220.32 | 175.68 | 396.00 |
| Register elbows | 12.000 | Ea. | 3.204 | 102.96 | 63.24 | 166.20 |
| Floor registers, enameled steel | 12.000 | Ea. | 3.000 | 219.84 | 68.16 | 288.00 |
| Floor grill, return air | 2.000 | Ea. | .728 | 18.48 | 16.02 | 34.50 |
| Refrigeration piping | 25.000 | L.F. | | 120.00 | | 120.00 |
| TOTAL | | | 68.053 | 4220.54 | 1398.02 | 5618.56 |

| DESCRIPTION | QUAN. | UNIT | MAN HOURS | MAT. | INST. | TOTAL |
|---|---|---|---|---|---|---|
| | | | | | | |
| | | | | | | |
| | | | | | | |
| | | | | | | |
| | | | | | | |
| | | | | | | |

# Oil Fired Heating/Cooling Price Sheet

| Description | QUAN. | UNIT | MAN HOURS | MAT. | INST. | TOTAL |
|---|---|---|---|---|---|---|
| Furnace, heating, 95.2 MBH, area to 1200 S.F. | 1.000 | Ea. | 4.570 | 660.00 | 90.00 | 750.00 |
| 123.2 MBH, area to 1500 S.F. | 1.000 | Ea. | 5.000 | 814.00 | 101.00 | 915.00 |
| 151.2 MBH, area to 2000 S.F. | 1.000 | Ea. | 5.330 | 1028.50 | 96.50 | 1125.00 |
| 200 MBH, area to 2400 S.F. | 1.000 | Ea. | 6.150 | 1292.50 | 132.50 | 1425.00 |
| Heating/cooling, 95.2 MBH heat, 36 MBH cool, to 1200 S.F. | 1.000 | Ea. | 16.000 | 2997.50 | 327.50 | 3325.00 |
| 112 MBH heat, 42 MBH cool, to 1500 S.F. | 1.000 | Ea. | 24.000 | 4496.25 | 491.25 | 4987.50 |
| 151 MBH heat, 47 MBH cool, to 2000 S.F. | 1.000 | Ea. | 20.800 | 3896.75 | 425.75 | 4322.50 |
| 184.8 MBH heat, 60 MBH cool, to 2400 S.F. | 1.000 | Ea. | 24.000 | 3712.50 | 487.50 | 4200.00 |
| Oil piping to furnace, 3/8" dia., copper | 1.000 | Ea. | 4.733 | 81.24 | 102.96 | 184.20 |
| Oil tank, on legs above ground, 275 gallons | 1.000 | Ea. | 3.200 | 170.50 | 64.50 | 235.00 |
| 550 gallons | 1.000 | Ea. | 4.000 | 643.50 | 81.50 | 725.00 |
| Below ground, 275 gallons | 1.000 | Ea. | 3.200 | 170.50 | 64.50 | 235.00 |
| 550 gallons | 1.000 | Ea. | 4.000 | 643.50 | 81.50 | 725.00 |
| 1000 gallons | 1.000 | Ea. | 8.000 | 896.50 | 178.50 | 1075.00 |
| Supply duct, rectangular, area to 1200 S.F., rigid fiberglass | 176.000 | S.F. | 12.144 | 82.72 | 248.16 | 330.88 |
| Sheet metal, insulated | 228.000 | Lb. | 31.352 | 358.52 | 640.72 | 999.24 |
| Area to 1500 S.F., rigid fiberglass | 176.000 | S.F. | 12.144 | 82.72 | 248.16 | 330.88 |
| Sheet metal, insulated | 228.000 | Lb. | 31.352 | 358.52 | 640.72 | 999.24 |
| Area to 2400 S.F., rigid fiberglass | 205.000 | S.F. | 14.145 | 96.35 | 289.05 | 385.40 |
| Sheet metal, insulated | 271.000 | Lb. | 37.072 | 424.71 | 757.70 | 1182.41 |
| Round flexible, insulated, 6" diameter to 1200 S.F. | 156.000 | L.F. | 9.672 | 238.68 | 190.32 | 429.00 |
| To 1500 S.F. | 184.000 | L.F. | 11.408 | 281.52 | 224.48 | 506.00 |
| 8" diameter to 2000 S.F. | 269.000 | L.F. | 23.941 | 497.65 | 473.44 | 971.09 |
| To 2400 S.F. | 269.000 | L.F. | 22.072 | 458.80 | 436.48 | 895.28 |
| Return duct, sheet metal galvanized, to 1500 S.F. | 158.000 | Lb. | 16.116 | 206.98 | 331.80 | 538.78 |
| To 2400 S.F. | 191.000 | Lb. | 19.482 | 250.21 | 401.10 | 651.31 |
| Lateral ducts, flexible round, 6", insulated to 1200 S.F. | 144.000 | L.F. | 8.928 | 220.32 | 175.68 | 396.00 |
| To 1500 S.F. | 172.000 | L.F. | 10.664 | 263.16 | 209.84 | 473.00 |
| To 2000 S.F. | 261.000 | L.F. | 16.182 | 399.33 | 318.42 | 717.75 |
| To 2400 S.F. | 300.000 | L.F. | 18.600 | 459.00 | 366.00 | 825.00 |
| Spiral steel, insulated to 1200 S.F. | 144.000 | L.F. | 20.122 | 423.58 | 402.44 | 826.02 |
| To 1500 S.F. | 172.000 | L.F. | 24.018 | 505.82 | 480.36 | 986.18 |
| To 2000 S.F. | 261.000 | L.F. | 36.451 | 767.59 | 729.02 | 1496.61 |
| To 2400 S.F. | 300.000 | L.F. | 41.940 | 882.60 | 838.80 | 1721.40 |
| Rectangular sheet metal galvanized insulated, to 1200 S.F. | 288.000 | Lb. | 45.246 | 494.58 | 922.20 | 1416.78 |
| To 1500 S.F. | 344.000 | Lb. | 54.040 | 590.72 | 1101.44 | 1692.16 |
| To 2000 S.F. | 522.000 | Lb. | 82.040 | 896.66 | 1672.12 | 2568.78 |
| To 2400 S.F. | 600.000 | Lb. | 94.320 | 1030.80 | 1922.40 | 2953.20 |
| Register elbows, to 1500 S.F. | 12.000 | Ea. | 3.204 | 102.96 | 63.24 | 166.20 |
| To 2400 S.F. | 14.000 | Ea. | 3.738 | 120.12 | 73.78 | 193.90 |
| Floor registers, enameled steel w/damper, to 1500 S.F. | 12.000 | Ea. | 3.000 | 219.84 | 68.16 | 288.00 |
| To 2400 S.F. | 14.000 | Ea. | 4.312 | 152.46 | 94.64 | 247.10 |
| Return air grille, area to 1500 S.F., 12" x 12" | 2.000 | Ea. | .728 | 18.48 | 16.02 | 34.50 |
| 12" x 24" | 1.000 | Ea. | .444 | 16.83 | 10.17 | 27.00 |
| Area to 2400 S.F., 8" x 16" | 2.000 | Ea. | .728 | 18.48 | 16.02 | 34.50 |
| 16" x 16" | 1.000 | Ea. | .364 | 13.92 | 8.08 | 22.00 |
| Thermostat, manual, 1 set back | 1.000 | Ea. | 1.000 | 44.00 | 22.00 | 66.00 |
| Electric, timed, 1 set back | 1.000 | Ea. | 1.000 | 69.30 | 21.70 | 91.00 |
| 2 set back | 1.000 | Ea. | 1.000 | 70.40 | 21.60 | 92.00 |
| Refrigeration piping, 3/8" | 25.000 | L.F. | | 9.33 | | 9.33 |
| 3/4" | 25.000 | L.F. | | 18.50 | | 18.50 |
| Diffusers, ceiling, 6" diameter, to 1500 S.F. | 10.000 | Ea. | 4.440 | 117.70 | 102.30 | 220.00 |
| To 2400 S.F. | 12.000 | Ea. | 6.000 | 155.76 | 132.24 | 288.00 |
| Floor, aluminum, adjustable, 2-1/4" x 12" to 1500 S.F. | 12.000 | Ea. | 3.000 | 86.52 | 65.88 | 152.40 |
| To 2400 S.F. | 14.000 | Ea. | 3.500 | 100.94 | 76.86 | 177.80 |
| Side wall, aluminum, adjustable, 8" x 4", to 1500 S.F. | 12.000 | Ea. | 3.000 | 190.08 | 61.92 | 252.00 |
| 5" x 10" to 2400 S.F. | 12.000 | Ea. | 3.696 | 244.92 | 79.08 | 324.00 |

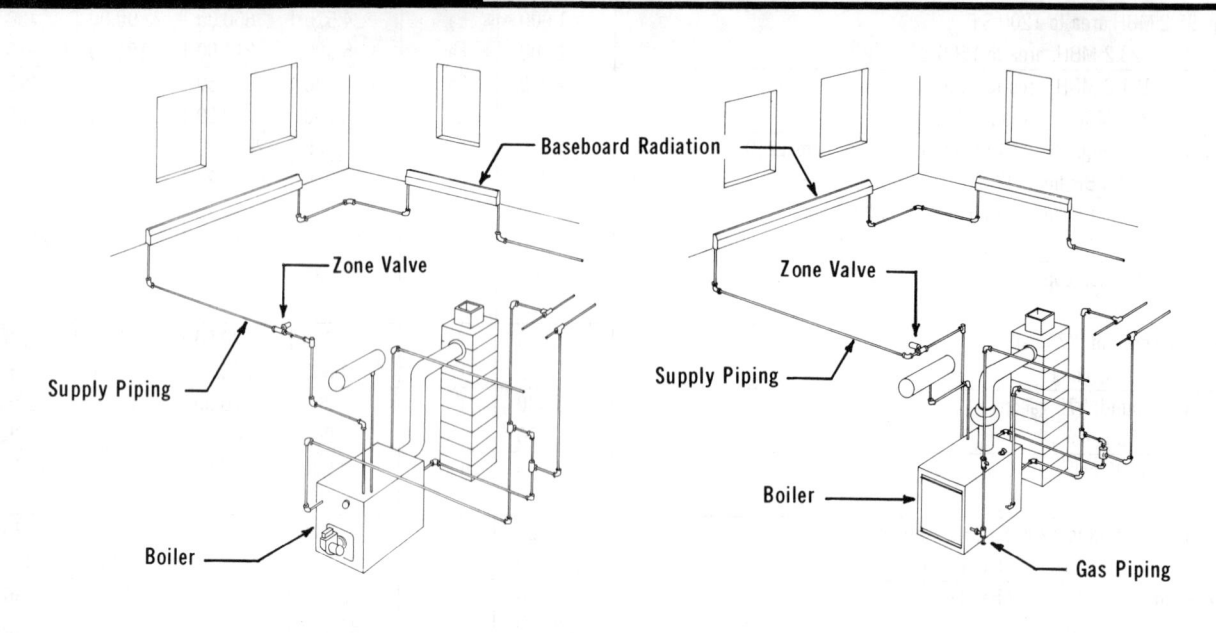

# 8 | MECHANICAL  68 | Hot Water Heating Systems

| SYSTEM DESCRIPTION | QUAN. | UNIT | MAN HOURS | MAT. | INST. | TOTAL |
|---|---|---|---|---|---|---|
| **OIL FIRED HOT WATER HEATING SYSTEM, AREA TO 1200 S.F.** | | | | | | |
| Boiler package, oil fired, 97 MBH, area to 1200 S.F. building | 1.000 | Ea. | 12.630 | 1430.00 | 270.00 | 1700.00 |
| Oil piping, 1/4" flexible copper tubing | 1.000 | Ea. | 4.539 | 35.59 | 99.11 | 134.70 |
| Oil tank, 275 gallon, with black iron filler pipe | 1.000 | Ea. | 3.200 | 170.50 | 64.50 | 235.00 |
| Supply piping, 3/4" copper tubing | 176.000 | L.F. | 32.032 | 214.72 | 700.48 | 915.20 |
| Supply fittings, copper 3/4" | 36.000 | Ea. | 15.156 | 11.88 | 331.92 | 343.80 |
| Supply valves, 3/4" | 2.000 | Ea. | .800 | 40.48 | 17.52 | 58.00 |
| Baseboard radiation, 3/4" | 106.000 | L.F. | 70.702 | 682.64 | 1394.96 | 2077.60 |
| Zone valve | 1.000 | Ea. | .400 | 34.10 | 8.90 | 43.00 |
| TOTAL | | | 139.459 | 2619.91 | 2887.39 | 5507.30 |
| **OIL FIRED HOT WATER HEATING SYSTEM, AREA TO 2400 S.F.** | | | | | | |
| Boiler package, oil fired, 225 MBH, area to 2400 S.F. building | 1.000 | Ea. | 17.140 | 2172.50 | 352.50 | 2525.00 |
| Oil piping, 3/8" flexible copper tubing | 1.000 | Ea. | 4.584 | 43.02 | 99.93 | 142.95 |
| Oil tank, 550 gallon, with black iron pipe filler pipe | 1.000 | EA.. | 4.000 | 643.50 | 81.50 | 725.00 |
| Supply piping, 3/4" copper tubing | 228.000 | L.F. | 41.496 | 278.16 | 907.44 | 1185.60 |
| Supply fittings, copper | 46.000 | Ea. | 19.366 | 15.18 | 424.12 | 439.30 |
| Supply valves | 2.000 | Ea. | .800 | 40.48 | 17.52 | 58.00 |
| Baseboard radiation | 212.000 | L.F. | 141.404 | 1365.28 | 2789.92 | 4155.20 |
| Zone valve | 1.000 | Ea. | .400 | 34.10 | 8.90 | 43.00 |
| TOTAL | | | 229.190 | 4592.22 | 4681.83 | 9274.05 |

The costs in this system are on a cost each basis. The costs represent a total cost for the system based on a gross square foot of plan area.

| DESCRIPTION | QUAN. | UNIT | MAN HOURS | MAT. | INST. | TOTAL |
|---|---|---|---|---|---|---|
| | | | | | | |
| | | | | | | |
| | | | | | | |
| | | | | | | |
| | | | | | | |

# Hot Water Heating Price Sheet

| | QUAN. | UNIT | MAN. HOURS | COST EACH MAT. | COST EACH INST. | TOTAL |
|---|---|---|---|---|---|---|
| Boiler, oil fired, 97 MBH, area to 1200 S.F. | 1.000 | Ea. | 12.630 | 1430.00 | 270.00 | 1700.00 |
| 118 MBH, area to 1500 S.F. | 1.000 | Ea. | 13.330 | 1457.50 | 267.50 | 1725.00 |
| 161 MBH, area to 2000 S.F. | 1.000 | Ea. | 16.000 | 1842.50 | 332.50 | 2175.00 |
| 215 MBH, area to 2400 S.F. | 1.000 | Ea. | 17.140 | 2172.50 | 352.50 | 2525.00 |
| Oil piping, (valve & filter), 3/8" copper | 1.000 | Ea. | 4.584 | 43.02 | 99.93 | 142.95 |
| 1/4" copper | 1.000 | Ea. | 4.539 | 35.59 | 99.11 | 134.70 |
| Oil tank, filler pipe and cap on legs, 275 gallon | 1.000 | Ea. | 3.200 | 170.50 | 64.50 | 235.00 |
| 550 gallon | 1.000 | Ea. | 4.000 | 643.50 | 81.50 | 725.00 |
| Buried underground, 275 gallon | 1.000 | Ea. | 3.200 | 170.50 | 64.50 | 235.00 |
| 550 gallon | 1.000 | Ea. | 4.000 | 643.50 | 81.50 | 725.00 |
| 1000 gallon | 1.000 | Ea. | 8.000 | 896.50 | 178.50 | 1075.00 |
| Supply piping copper, area to 1200 S.F., 1/2" tubing | 176.000 | L.F. | 26.576 | 147.84 | 580.80 | 728.64 |
| 3/4" tubing | 176.000 | L.F. | 32.032 | 214.72 | 700.48 | 915.20 |
| Area to 1500 S.F., 1/2" tubing | 186.000 | L.F. | 28.086 | 156.24 | 613.80 | 770.04 |
| 3/4" tubing | 186.000 | L.F. | 33.852 | 226.92 | 740.28 | 967.20 |
| Area to 2000 S.F., 1/2" tubing | 204.000 | L.F. | 30.804 | 171.36 | 673.20 | 844.56 |
| 3/4" tubing | 204.000 | L.F. | 37.128 | 248.88 | 811.92 | 1060.80 |
| Area to 2400 S.F., 1/2" tubing | 228.000 | L.F. | 34.428 | 191.52 | 752.40 | 943.92 |
| 3/4" tubing | 228.000 | L.F. | 41.496 | 278.16 | 907.44 | 1185.60 |
| Supply pipe fittings copper, area to 1200 S.F., 1/2" | 36.000 | Ea. | 14.400 | 5.04 | 315.36 | 320.40 |
| 3/4" | 36.000 | Ea. | 15.156 | 11.88 | 331.92 | 343.80 |
| Area to 1500 S.F., 1/2" | 40.000 | Ea. | 16.000 | 5.60 | 350.40 | 356.00 |
| 3/4" | 40.000 | Ea. | 16.840 | 13.20 | 368.80 | 382.00 |
| Area to 2000 S.F., 1/2" | 44.000 | Ea. | 17.600 | 6.16 | 385.44 | 391.60 |
| 3/4" | 44.000 | Ea. | 18.524 | 14.52 | 405.68 | 420.20 |
| Area to 2400, S.F., 1/2" | 46.000 | Ea. | 18.400 | 6.44 | 402.96 | 409.40 |
| 3/4" | 46.000 | Ea. | 19.366 | 15.18 | 424.12 | 439.30 |
| Supply valves, 1/2" pipe size | 2.000 | Ea. | .666 | 32.24 | 13.76 | 46.00 |
| 3/4" | 2.000 | Ea. | .800 | 40.48 | 17.52 | 58.00 |
| Baseboard radiation, area to 1200 S.F., 1/2" tubing | 106.000 | L.F. | 65.190 | 652.96 | 1286.84 | 1939.80 |
| 3/4" tubing | 106.000 | L.F. | 70.702 | 682.64 | 1394.96 | 2077.60 |
| Area to 1500 S.F., 1/2" tubing | 134.000 | L.F. | 82.410 | 825.44 | 1626.76 | 2452.20 |
| 3/4" tubing | 134.000 | L.F. | 89.378 | 862.96 | 1763.44 | 2626.40 |
| Area to 2000 S.F., 1/2" tubing | 178.000 | L.F. | 109.470 | 1096.48 | 2160.92 | 3257.40 |
| 3/4" tubing | 178.000 | L.F. | 118.726 | 1146.32 | 2342.48 | 3488.80 |
| Area to 2400 S.F., 1/2" tubing | 212.000 | L.F. | 130.380 | 1305.92 | 2573.68 | 3879.60 |
| 3/4" tubing | 212.000 | L.F. | 141.404 | 1365.28 | 2789.92 | 4155.20 |
| Zone valves, 1/2" tubing | 1.000 | Ea. | .400 | 34.10 | 8.90 | 43.00 |
| 3/4" tubing | 1.000 | Ea. | .400 | 34.10 | 8.90 | 43.00 |

# 8 | MECHANICAL  80 | Rooftop Systems

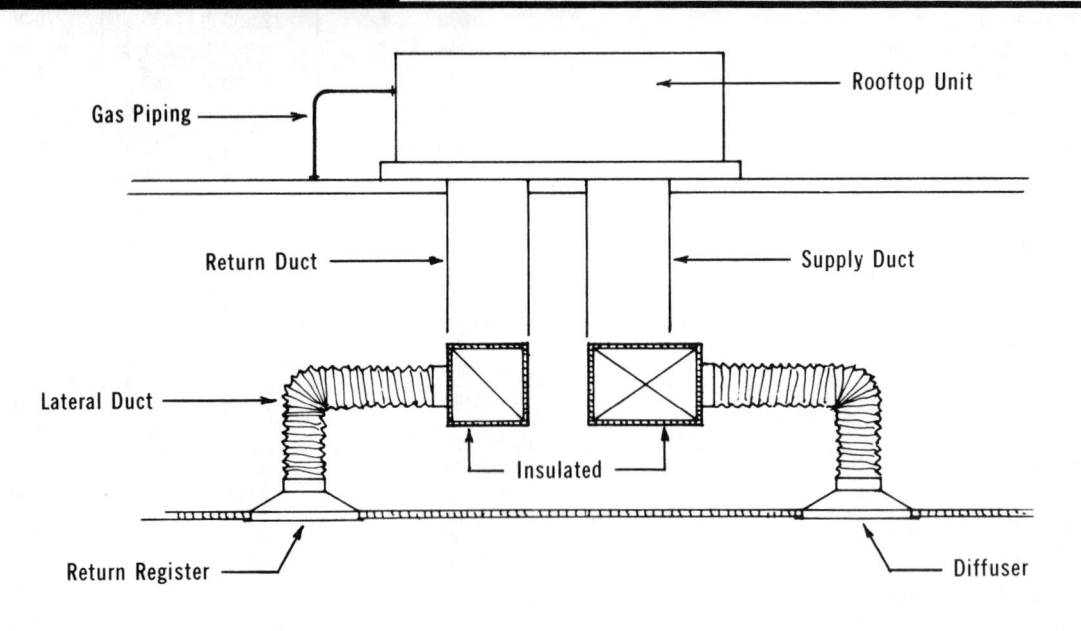

| SYSTEM DESCRIPTION | QUAN. | UNIT | MAN HOURS | MAT. | INST. | TOTAL |
|---|---|---|---|---|---|---|
| **ROOFTOP HEATING/COOLING UNIT, AREA TO 2000 S.F.** | | | | | | |
| Rooftop unit, single zone, electric cool, gas heat, to 2000 s.f. | 1.000 | Ea. | 28.570 | 3685.00 | 565.00 | 4250.00 |
| Gas piping | 34.500 | L.F. | 5.209 | 67.62 | 113.51 | 181.13 |
| Duct, supply and return, galvanized steel | 38.000 | Lb. | 3.876 | 49.78 | 79.80 | 129.58 |
| Insulation, ductwork | 33.000 | S.F. | 1.518 | 11.22 | 30.36 | 41.58 |
| Lateral duct, flexible duct 12" diameter, insulated | 72.000 | L.F. | 11.520 | 190.08 | 227.52 | 417.60 |
| Diffusers | 4.000 | Ea. | 4.560 | 572.00 | 108.00 | 680.00 |
| Return registers | 1.000 | Ea. | 1.000 | 48.40 | 21.60 | 70.00 |
| TOTAL | | | 56.253 | 4624.10 | 1145.79 | 5769.89 |
| **ROOFTOP HEATING/COOLING UNIT, AREA TO 5000 S.F.** | | | | | | |
| Rooftop unit, single zone, electric cool, gas heat, to 5000 s.f. | 1.000 | Ea. | 77.420 | 9487.50 | 1612.50 | 11100.00 |
| Gas piping | 86.250 | L.F. | 13.023 | 169.05 | 283.76 | 452.81 |
| Duct supply and return, galvanized steel | 95.000 | Lb. | 9.690 | 124.45 | 199.50 | 323.95 |
| Insulation, ductwork | 82.000 | S.F. | 3.772 | 27.88 | 75.44 | 103.32 |
| Lateral duct, flexible duct, 12" diameter, insulated | 180.000 | L.F. | 28.800 | 475.20 | 568.80 | 1044.00 |
| Diffusers | 10.000 | Ea. | 11.400 | 1430.00 | 270.00 | 1700.00 |
| Return registers | 3.000 | Ea. | 3.000 | 145.20 | 64.80 | 210.00 |
| TOTAL | | | 147.105 | 11859.28 | 3074.80 | 14934.08 |

| DESCRIPTION | QUAN. | UNIT | MAN HOURS | MAT. | INST. | TOTAL |
|---|---|---|---|---|---|---|
| | | | | | | |
| | | | | | | |
| | | | | | | |
| | | | | | | |
| | | | | | | |

# Rooftop Price Sheet

| | QUAN. | UNIT | MAN HOURS | COST EACH MAT. | COST EACH INST. | TOTAL |
|---|---|---|---|---|---|---|
| Rooftop unit, single zone, electric cool, gas heat to 2000 S.F. | 1.000 | Ea. | 28.570 | 3685.00 | 565.00 | 4250.00 |
| Area to 3000 S.F. | 1.000 | Ea. | 52.170 | 7122.50 | 1077.50 | 8200.00 |
| Area to 5000 S.F. | 1.000 | Ea. | 77.420 | 9487.50 | 1612.50 | 11100.00 |
| Area to 10000 S.F. | 1.000 | Ea. | 145.000 | 19580.00 | 3020.00 | 22600.00 |
| Gas piping, area 2000 through 4000 S.F. | 34.500 | L.F. | 5.209 | 67.62 | 113.51 | 181.13 |
| Area 5000 to 10000 S.F. | 86.250 | L.F. | 13.023 | 169.05 | 283.76 | 452.81 |
| Duct, supply and return, galvanized steel, to 2000 S.F. | 38.000 | Lb. | 3.876 | 49.78 | 79.80 | 129.58 |
| Area to 3000 S.F. | 57.000 | Lb. | 5.814 | 74.67 | 119.70 | 194.37 |
| Area to 5000 S.F. | 95.000 | Lb. | 9.690 | 124.45 | 199.50 | 323.95 |
| Area to 10000 S.F. | 190.000 | Lb. | 19.380 | 248.90 | 399.00 | 647.90 |
| Rigid fiberglass, area to 2000 S.F. | 33.000 | S.F. | 2.277 | 15.51 | 46.53 | 62.04 |
| Area to 3000 S.F. | 49.000 | S.F. | 3.381 | 23.03 | 69.09 | 92.12 |
| Area to 5000 S.F. | 82.000 | S.F. | 5.658 | 38.54 | 115.62 | 154.16 |
| Area to 10000 S.F. | 164.000 | S.F. | 11.316 | 77.08 | 231.24 | 308.32 |
| Insulation, supply and return, blanket type, area to 2000 S.F. | 33.000 | S.F. | 1.518 | 11.22 | 30.36 | 41.58 |
| Area to 3000 S.F. | 49.000 | S.F. | 2.254 | 16.66 | 45.08 | 61.74 |
| Area to 5000 S.F. | 82.000 | S.F. | 3.772 | 27.88 | 75.44 | 103.32 |
| Area to 10000 S.F. | 164.000 | S.F. | 7.544 | 55.76 | 150.88 | 206.64 |
| Lateral ducts, flexible round, 12" insulated, to 2000 S.F. | 72.000 | L.F. | 11.520 | 190.08 | 227.52 | 417.60 |
| Area to 3000 S.F. | 108.000 | L.F. | 17.280 | 285.12 | 341.28 | 626.40 |
| Area to 5000 S.F. | 180.000 | L.F. | 28.800 | 475.20 | 568.80 | 1044.00 |
| Area to 10000 S.F. | 360.000 | L.F. | 57.600 | 950.40 | 1137.60 | 2088.00 |
| Rectangular, galvanized steel, to 2000 S.F. | 239.000 | Lb. | 24.378 | 313.09 | 501.90 | 814.99 |
| Area to 3000 S.F. | 360.000 | Lb. | 36.720 | 471.60 | 756.00 | 1227.60 |
| Area to 5000 S.F. | 599.000 | Lb. | 61.098 | 784.69 | 1257.90 | 2042.59 |
| Area to 10000 S.F. | 998.000 | Lb. | 101.796 | 1307.38 | 2095.80 | 3403.18 |
| Diffusers, ceiling, 1 to 4 way blow, 24" x 24", to 2000 S.F. | 4.000 | Ea. | 4.560 | 572.00 | 108.00 | 680.00 |
| Area to 3000 S.F. | 6.000 | Ea. | 6.840 | 858.00 | 162.00 | 1020.00 |
| Area to 5000 S.F. | 10.000 | Ea. | 11.400 | 1430.00 | 270.00 | 1700.00 |
| Area to 10000 S.F. | 20.000 | Ea. | 22.800 | 2860.00 | 540.00 | 3400.00 |
| Return grilles, 24" x 24", to 2000 S.F. | 1.000 | Ea. | 1.000 | 48.40 | 21.60 | 70.00 |
| Area to 3000 S.F. | 2.000 | Ea. | 2.000 | 96.80 | 43.20 | 140.00 |
| Area to 5000 S.F. | 3.000 | Ea. | 3.000 | 145.20 | 64.80 | 210.00 |
| Area to 10000 S.F. | 5.000 | Ea. | 5.000 | 242.00 | 108.00 | 350.00 |

8

# SECTION 9
# ELECTRICAL

| System Number | Title | Page |
|---|---|---|
| 9-10 | Electric Service | 196 |
| 9-20 | Electric Heating | 197 |
| 9-30 | Wiring Devices | 198 |
| 9-40 | Light Fixtures | 199 |

# 9 | ELECTRICAL  10 | Electric Service Systems

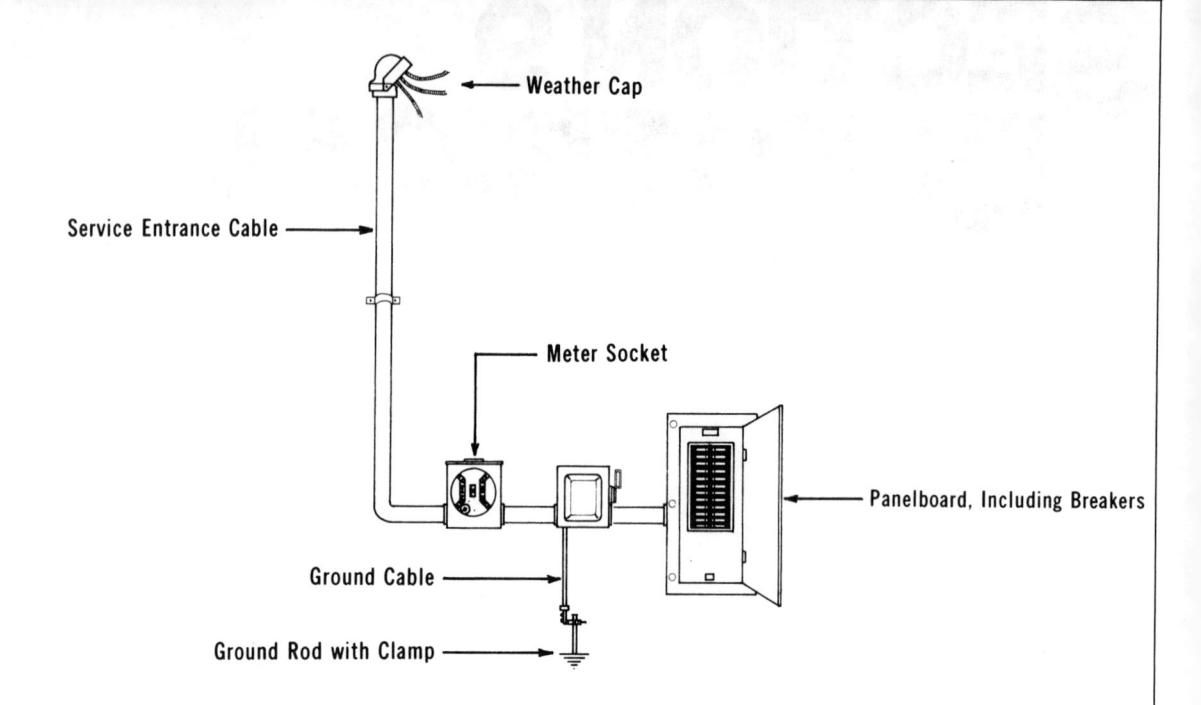

*Diagram labels: Weather Cap, Service Entrance Cable, Meter Socket, Ground Cable, Ground Rod with Clamp, Panelboard, Including Breakers*

| SYSTEM DESCRIPTION | QUAN. | UNIT | MAN HOURS | MAT. | INST. | TOTAL |
|---|---|---|---|---|---|---|
| **100 AMP SERVICE** | | | | | | |
| Weather cap | 1.000 | Ea. | .667 | 3.52 | 14.43 | 17.95 |
| Service entrance cable | 10.000 | L.F. | .760 | 12.70 | 16.50 | 29.20 |
| Meter socket | 1.000 | Ea. | 2.500 | 19.80 | 54.20 | 74.00 |
| Ground rod with clamp | 1.000 | Ea. | 1.510 | 16.50 | 32.50 | 49.00 |
| Ground cable | 5.000 | L.F. | .250 | 4.30 | 5.40 | 9.70 |
| Panel board, 12 circuit | 1.000 | Ea. | 6.670 | 94.60 | 145.40 | 240.00 |
| TOTAL | | | 12.357 | 151.42 | 268.43 | 419.85 |
| **200 AMP SERVICE** | | | | | | |
| Weather cap | 1.000 | Ea. | 1.000 | 11.00 | 22.00 | 33.00 |
| Service entrance cable | 10.000 | L.F. | 1.140 | 20.90 | 24.80 | 45.70 |
| Meter socket | 1.000 | Ea. | 4.210 | 29.70 | 90.30 | 120.00 |
| Ground rod with clamp | 1.000 | Ea. | 1.820 | 21.29 | 39.71 | 61.00 |
| Ground cable | 10.000 | L.F. | .500 | 8.60 | 10.80 | 19.40 |
| 3/4" EMT | 5.000 | L.F. | .310 | 1.75 | 6.70 | 8.45 |
| Panel board, 24 circuit | 1.000 | Ea. | 12.310 | 236.50 | 268.50 | 505.00 |
| TOTAL | | | 21.290 | 329.74 | 462.81 | 792.55 |
| **400 AMP SERVICE** | | | | | | |
| Weather cap | 2.000 | Ea. | 12.500 | 124.30 | 273.70 | 398.00 |
| Service entrance cable | 180.000 | L.F. | 5.760 | 156.60 | 124.20 | 280.80 |
| Meter socket | 1.000 | Ea. | 4.210 | 29.70 | 90.30 | 120.00 |
| Ground rod with clamp | 1.000 | Ea. | 2.070 | 24.48 | 45.12 | 69.60 |
| Ground cable | 20.000 | L.F. | .480 | 11.80 | 10.60 | 22.40 |
| 3/4" greenfield | 20.000 | L.F. | 1.000 | 7.00 | 21.80 | 28.80 |
| Current transformer cabinet | 1.000 | Ea. | 6.150 | 83.60 | 131.40 | 215.00 |
| Panel board, 42 circuit | 1.000 | Ea. | 33.330 | 1815.00 | 735.00 | 2550.00 |
| TOTAL | | | 65.500 | 2252.48 | 1432.12 | 3684.60 |

# 9 | ELECTRICAL   20 | Electric Perimeter Heating Systems

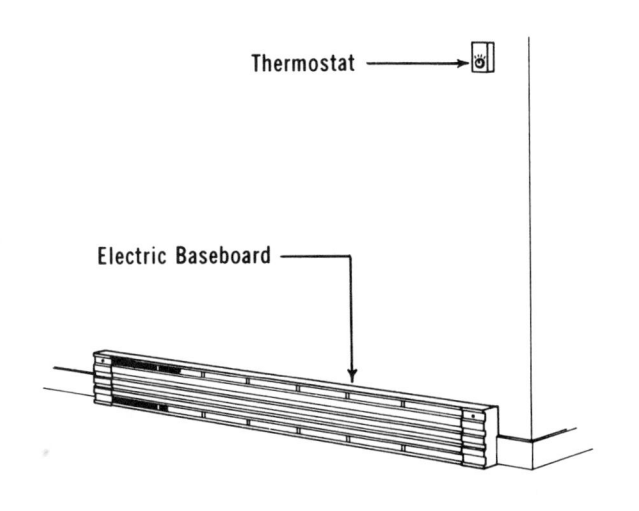

| SYSTEM DESCRIPTION | QUAN. | UNIT | MAN HOURS | MAT. | INST. | TOTAL |
|---|---|---|---|---|---|---|
| **4' BASEBOARD HEATER** | | | | | | |
| Electric baseboard heater, 4' long | 1.000 | Ea. | 1.190 | 47.30 | 25.70 | 73.00 |
| Thermostat, integral | 1.000 | Ea. | .500 | 16.50 | 10.50 | 27.00 |
| Romex, 12-3 with ground | 40.000 | L.F. | 1.600 | 8.40 | 34.80 | 43.20 |
| Panel board breaker, 20 Amp | 1.000 | Ea. | .300 | 5.08 | 6.62 | 11.70 |
| TOTAL | | | 3.590 | 77.28 | 77.62 | 154.90 |
| **6' BASEBOARD HEATER** | | | | | | |
| Electric baseboard heater, 6' long | 1.000 | Ea. | 1.600 | 70.40 | 34.60 | 105.00 |
| Thermostat, integral | 1.000 | Ea. | .500 | 16.50 | 10.50 | 27.00 |
| Romex, 12-3 with ground | 40.000 | L.F. | 1.600 | 8.40 | 34.80 | 43.20 |
| Panel board breaker, 20 Amp | 1.000 | Ea. | .400 | 6.78 | 8.82 | 15.60 |
| TOTAL | | | 4.100 | 102.08 | 88.72 | 190.80 |
| **8' BASEBOARD HEATER** | | | | | | |
| Electric baseboard heater, 8' long | 1.000 | Ea. | 2.000 | 94.60 | 45.40 | 140.00 |
| Thermostat, integral | 1.000 | Ea. | .500 | 16.50 | 10.50 | 27.00 |
| Romex, 12-3 with ground | 40.000 | L.F. | 1.600 | 8.40 | 34.80 | 43.20 |
| Panel board breaker, 20 Amp | 1.000 | Ea. | .500 | 8.47 | 11.03 | 19.50 |
| TOTAL | | | 4.600 | 127.97 | 101.73 | 229.70 |
| **10' BASEBOARD HEATER** | | | | | | |
| Electric baseboard heater, 10' long | 1.000 | Ea. | 2.420 | 107.80 | 52.20 | 160.00 |
| Thermostat, integral | 1.000 | Ea. | .500 | 16.50 | 10.50 | 27.00 |
| Romex, 12-3 with ground | 40.000 | L.F. | 1.600 | 8.40 | 34.80 | 43.20 |
| Panel board breaker, 20 Amp | 1.000 | Ea. | .750 | 12.71 | 16.54 | 29.25 |
| TOTAL | | | 5.270 | 145.41 | 114.04 | 259.45 |

The costs in this system are on a cost each basis and include all necessary conduit fittings.

| DESCRIPTION | QUAN. | UNIT | MAN HOURS | MAT. | INST. | TOTAL |
|---|---|---|---|---|---|---|
| | | | | | | |
| | | | | | | |

# 9 | ELECTRICAL    30 | Wiring Device Systems

The prices in this system are on a cost each basis and include 20 feet of wire and conduit (as necessary) for each device.

| SYSTEM DESCRIPTION | QUAN. | UNIT | MAN. HOURS | MAT. | INST. | TOTAL |
|---|---|---|---|---|---|---|
| Air conditioning receptacles | | | | | | |
|     Using non-metallic sheathed cable | 1.000 | Ea. | .800 | 6.60 | 17.40 | 24.00 |
|     Using BX cable | 1.000 | Ea. | .964 | 11.00 | 21.00 | 32.00 |
|     Using EMT conduit | 1.000 | Ea. | 1.190 | 11.00 | 26.00 | 37.00 |
| Disposal wiring | | | | | | |
|     Using non-metallic sheathed cable | 1.000 | Ea. | .889 | 5.50 | 19.50 | 25.00 |
|     Using BX cable | 1.000 | Ea. | 1.070 | 8.80 | 23.20 | 32.00 |
|     Using EMT conduit | 1.000 | Ea. | 1.330 | 8.80 | 29.20 | 38.00 |
| Dryer circuit | | | | | | |
|     Using non-metallic sheathed cable | 1.000 | Ea. | 1.450 | 14.30 | 31.70 | 46.00 |
|     Using BX cable | 1.000 | Ea. | 1.740 | 22.00 | 38.00 | 60.00 |
|     Using EMT conduit | 1.000 | Ea. | 2.160 | 16.50 | 46.50 | 63.00 |
| Duplex receptacles | | | | | | |
|     Using non-metallic sheathed cable | 1.000 | Ea. | .615 | 6.05 | 13.35 | 19.40 |
|     Using BX cable | 1.000 | Ea. | .741 | 9.90 | 16.10 | 26.00 |
|     Using EMT conduit | 1.000 | Ea. | .920 | 9.90 | 20.10 | 30.00 |
| Exhaust fan wiring | | | | | | |
|     Using non-metallic sheathed cable | 1.000 | Ea. | .800 | 5.50 | 17.50 | 23.00 |
|     Using BX cable | 1.000 | Ea. | .964 | 8.80 | 21.20 | 30.00 |
|     Using EMT conduit | 1.000 | Ea. | 1.190 | 8.80 | 26.20 | 35.00 |
| Furnace circuit & switch | | | | | | |
|     Using non-metallic sheathed cable | 1.000 | Ea. | 1.330 | 7.81 | 29.19 | 37.00 |
|     Using BX cable | 1.000 | Ea. | 1.600 | 12.10 | 34.90 | 47.00 |
|     Using EMT conduit | 1.000 | Ea. | 2.000 | 12.10 | 42.90 | 55.00 |
| Ground fault | | | | | | |
|     Using non-metallic sheathed cable | 1.000 | Ea. | 1.000 | 39.60 | 21.40 | 61.00 |
|     Using BX cable | 1.000 | Ea. | 1.210 | 45.10 | 25.90 | 71.00 |
|     Using EMT conduit | 1.000 | Ea. | 1.480 | 45.10 | 31.90 | 77.00 |
| Heater circuits | | | | | | |
|     Using non-metallic sheathed cable | 1.000 | Ea. | 1.000 | 5.50 | 21.50 | 27.00 |
|     Using BX cable | 1.000 | Ea. | 1.210 | 8.80 | 26.20 | 35.00 |
|     Using EMT conduit | 1.000 | Ea. | 1.480 | 8.80 | 32.20 | 41.00 |
| Lighting wiring | | | | | | |
|     Using non-metallic sheathed cable | 1.000 | Ea. | .500 | 5.50 | 10.85 | 16.35 |
|     Using BX cable | 1.000 | Ea. | .602 | 9.90 | 13.10 | 23.00 |
|     Using EMT conduit | 1.000 | Ea. | .748 | 9.90 | 16.10 | 26.00 |
| Range circuits | | | | | | |
|     Using non-metallic sheathed cable | 1.000 | Ea. | 2.000 | 33.00 | 43.00 | 76.00 |
|     Using BX cable | 1.000 | Ea. | 2.420 | 46.20 | 52.80 | 99.00 |
|     Using EMT conduit | 1.000 | Ea. | 2.960 | 28.60 | 64.40 | 93.00 |
| Switches, single pole | | | | | | |
|     Using non-metallic sheathed cable | 1.000 | Ea. | .500 | 5.94 | 10.86 | 16.80 |
|     Using BX cable | 1.000 | Ea. | .602 | 9.90 | 13.10 | 23.00 |
|     Using EMT conduit | 1.000 | Ea. | .748 | 9.90 | 16.10 | 26.00 |
| Switches, 3-way | | | | | | |
|     Using non-metallic sheathed cable | 1.000 | Ea. | .667 | 7.70 | 14.30 | 22.00 |
|     Using BX cable | 1.000 | Ea. | .800 | 13.20 | 17.80 | 31.00 |
|     Using EMT conduit | 1.000 | Ea. | 1.330 | 19.80 | 29.20 | 49.00 |
| Water heater | | | | | | |
|     Using non-metallic sheathed cable | 1.000 | Ea. | 1.600 | 7.70 | 34.30 | 42.00 |
|     Using BX cable | 1.000 | Ea. | 1.900 | 13.20 | 41.80 | 55.00 |
|     Using EMT conduit | 1.000 | Ea. | 2.350 | 13.20 | 50.80 | 64.00 |
| Weatherproof receptacle | | | | | | |
|     Using non-metallic sheathed cable | 1.000 | Ea. | 1.330 | 56.10 | 28.90 | 85.00 |
|     Using BX cable | 1.000 | Ea. | 1.600 | 60.50 | 34.50 | 95.00 |
|     Using EMT conduit | 1.000 | Ea. | 2.000 | 60.50 | 44.50 | 105.00 |

# 9 | ELECTRICAL  40 | Light Fixture Systems

The costs for these fixtures are on an each basis and include installation of the fixture only. For wiring and switches see systems sheet 9-30.

| DESCRIPTION | QUAN. | UNIT | MAN. HOURS | MAT. | INST. | TOTAL |
|---|---|---|---|---|---|---|
| Fluorescent strip, 4' long, 1 light, average | 1.000 | Ea. | .941 | 18.70 | 20.30 | 39.00 |
| Deluxe | 1.000 | Ea. | 1.129 | 22.44 | 24.36 | 46.80 |
| 2 lights, average | 1.000 | Ea. | 1.000 | 19.80 | 21.20 | 41.00 |
| Deluxe | 1.000 | Ea. | 1.200 | 23.76 | 25.44 | 49.20 |
| 8' long, 1 light, average | 1.000 | Ea. | 1.190 | 33.00 | 26.00 | 59.00 |
| Deluxe | 1.000 | Ea. | 1.428 | 39.60 | 31.20 | 70.80 |
| 2 lights, average | 1.000 | Ea. | 1.290 | 38.50 | 27.50 | 66.00 |
| Deluxe | 1.000 | Ea. | 1.548 | 46.20 | 33.00 | 79.20 |
| Surface mounted, 4' x 1', economy | 1.000 | Ea. | .912 | 31.68 | 19.52 | 51.20 |
| Average | 1.000 | Ea. | 1.140 | 39.60 | 24.40 | 64.00 |
| Deluxe | 1.000 | Ea. | 1.368 | 47.52 | 29.28 | 76.80 |
| 4' x 2', economy | 1.000 | Ea. | 1.208 | 52.80 | 26.40 | 79.20 |
| Average | 1.000 | Ea. | 1.510 | 66.00 | 33.00 | 99.00 |
| Deluxe | 1.000 | Ea. | 1.812 | 79.20 | 39.60 | 118.80 |
| Recessed, 4'x 1', 2 lamps, economy | 1.000 | Ea. | 1.120 | 30.80 | 24.40 | 55.20 |
| Average | 1.000 | Ea. | 1.400 | 38.50 | 30.50 | 69.00 |
| Deluxe | 1.000 | Ea. | 1.680 | 46.20 | 36.60 | 82.80 |
| 4' x 2', 4' lamps, economy | 1.000 | Ea. | 1.360 | 40.48 | 29.92 | 70.40 |
| Average | 1.000 | Ea. | 1.700 | 50.60 | 37.40 | 88.00 |
| Deluxe | 1.000 | Ea. | 2.040 | 60.72 | 44.88 | 105.60 |
| Incandescent, exterior, 150W, single spot | 1.000 | Ea. | .500 | 14.30 | 10.70 | 25.00 |
| Double spot | 1.000 | Ea. | 1.167 | 38.50 | 25.50 | 64.00 |
| Recessed, 100W, economy | 1.000 | Ea. | .800 | 31.68 | 17.12 | 48.80 |
| Average | 1.000 | Ea. | 1.000 | 39.60 | 21.40 | 61.00 |
| Deluxe | 1.000 | Ea. | 1.200 | 47.52 | 25.68 | 73.20 |
| 150W, economy | 1.000 | Ea. | .800 | 32.56 | 17.04 | 49.60 |
| Average | 1.000 | Ea. | 1.000 | 40.70 | 21.30 | 62.00 |
| Deluxe | 1.000 | Ea. | 1.200 | 48.84 | 25.56 | 74.40 |
| Surface mounted, 60W, economy | 1.000 | Ea. | .800 | 23.10 | 16.90 | 40.00 |
| Average | 1.000 | Ea. | 1.000 | 28.60 | 21.40 | 50.00 |
| Deluxe | 1.000 | Ea. | 1.190 | 57.20 | 25.80 | 83.00 |
| Mercury vapor, recessed, 2' x 2' with 250W DX lamp | 1.000 | Ea. | 2.500 | 231.00 | 54.00 | 285.00 |
| 2' x 2' with 400W DX lamp | 1.000 | Ea. | 2.760 | 242.00 | 58.00 | 300.00 |
| Surface mounted, 2' x 2' with 250W DX lamp | 1.000 | Ea. | 2.960 | 220.00 | 65.00 | 285.00 |
| 2' x 2' with 400W DX lamp | 1.000 | Ea. | 3.330 | 231.00 | 74.00 | 305.00 |
| High bay, single unit, 400W DX lamp | 1.000 | Ea. | 3.480 | 209.00 | 76.00 | 285.00 |
| Twin unit, 400W DX lamp | 1.000 | Ea. | 5.000 | 418.00 | 107.00 | 525.00 |
| Low bay, 250W DX lamp | 1.000 | Ea. | 2.500 | 258.50 | 56.50 | 315.00 |
| Metal halide, recessed 2' x 2' 250W | 1.000 | Ea. | 2.500 | 242.00 | 53.00 | 295.00 |
| 2' x 2', 400W | 1.000 | Ea. | 2.760 | 286.00 | 59.00 | 345.00 |
| Surface mounted, 2' x 2', 250W | 1.000 | Ea. | 2.960 | 231.00 | 64.00 | 295.00 |
| 2' x 2', 400W | 1.000 | Ea. | 3.330 | 275.00 | 70.00 | 345.00 |
| High bay, single, unit, 400W | 1.000 | Ea. | 3.480 | 242.00 | 73.00 | 315.00 |
| Twin unit, 400W | 1.000 | Ea. | 5.000 | 484.00 | 106.00 | 590.00 |
| Low bay, 250W | 1.000 | Ea. | 2.500 | 302.50 | 52.50 | 355.00 |

# SECTION 10 OVERHEAD & PROFIT

| System Number | Title | Page |
|---|---|---|
| | Introduction | 202 |
| | Examples | 203 |
| | Installer's Overhead and Profit | 204 |
| 10-A | Sales Tax | 204 |
| 10-B | Overtime | 205 |
| 10-C | Architectural Fees | 205 |
| 10-D | Unemployment & S.S. Taxes | 205 |
| 10-E | Builder's Risk Insurance | 206 |
| 10-F | Insurance Rates by Trade | 206 |
| 10-G | Insurance Rates by States | 207 |

# INTRODUCTION

The general conditions of cost estimates vary from contractor to contractor. Each job has its own unique overhead cost. Due to inflation and labor cost increases, these expenses are constantly changing from year to year. The factors involved in such overhead costs — insurance, taxes, or overtime — can be expressed in three ways; as a total cost, as a function of the total cost, or as a function of labor cost. Even when a reliable total cost has been developed, a contractor must estimate the company volume of work to be able to allocate this additional overhead cost.

The tables in this section can be used to develop these important cost figures. The first section explains the method for making these determinations. Tables 10-A through 10-G present data from which a contractor can determine the costs of maintaining his business during a construction project. Use these figures to develop an accurate estimate of the numerous overhead expenses involved with any type of residential or light commercial construction project. These tables can also be used to estimate a contractor's profit figures for a project of any size.

# 10 | OVERHEAD & PROFIT

# Step by Step Process For Overhead and Profit Estimation

After completion of an estimate from the other divisions of this book, the estimator is ready to determine other necessary COST ADJUSTMENTS. The adjustments which are not included in all other cost estimating systems MUST BE ADDED TO ANY COST ESTIMATE. These adjustments and additions include:

10-A Sales Tax
10-B Overtime
10-C Architectural Fees

Tables 10-A through 10-C provide the necessary cost adjustment information to make these adjustments and additions.

The COSTS developed in the SYSTEMS section of this book include MARK-UPS ON MATERIAL AND LABOR FOR HANDLING, OVERHEAD AND PROFIT. The mark-up on materials is 10%. The mark-up on labor is 59.6%.

That 59.6% figure, which is an over-all average for all the trades, includes workmen's compensation insurance, average fixed overhead related to the trades and the job, installer's operating overhead, and an allowance for profit. The percentages below were used in preparing that figure.

| | |
|---|---|
| Workers' Compensation Insurance | 9.0% |
| Average Fixed Overhead for all trades (including U.S. and State Unemployment, Social Security (FICA), builder's risk and public liabilities. | 13.8% |
| Installer's Operating Overhead | 26.8% |
| Installer's Profit | 10.0% |
| Total | 59.6% |

# Handling and Material Costs

There may be situations which require adjustments on the built-in allowances factored into HANDLING COST data found in this book. Other adjustments sometimes are needed to estimate OVERHEAD AND PROFIT. In these situations, the factor of 10% on MATERIAL and 59.6% on LABOR should be deducted and pertinent adjustments made using these tables:

10-D Unemployment and Social Security Taxes
10-E Builder's Risk Insurance
10-F Insurance Rates by Craft
10-G Insurance Rates by States

# 10 | OVERHEAD & PROFIT

## Estimating Overhead Costs

The cost of maintaining a business - the contractor's overhead - is often the one expense overlooked and improperly allocated to specific projects. Since the OPERATING OVERHEAD must be recovered during the year, some planning and budgeting is needed. The contractor must anticipate this overhead and also should have an estimate of the expected volume of business that will be billed during the year. OPERATING OVERHEAD can be calculated as a percentage of the JOB COST or as a percentage of LABOR COST. Other methods of allocating OPERATING OVERHEAD can be designed to meet the needs of individual projects.

OVERHEAD AND PROFIT FOR SUBCONTRACTORS will vary greatly. OVERHEAD in this case accounts for those expenses that a business must incur to operate. This OVERHEAD COST includes selling expenses, advertising, rental cost, vehicle and equipment cost, plus other costs that cannot be directly allocated to a specific project. These OVERHEAD EXPENSES are generally based on budgeted annual volume and anticipated economic conditions. The following example is typical of a project with numerous OVERHEAD COSTS. In this example, assume that the figures represent the overhead costs of a contractor building 10 homes per year at a cost of $70,000 each. This would result in an annual LABOR COST of approximately $280,000. His gross income sales each year would be $700,000.

### Example of Overhead Expenses

| | |
|---|---|
| Owner's salary (salesman/supervisor) | $50,000 |
| Secretary $300/week | 15,600 |
| Office rental $500/month | 6,000 |
| Telephone answering service | 900 |
| Office equipment $100/month | 1,200 |
| Accountant | 3,000 |
| Legal | 3,000 |
| Medical, Worker's Compensation (office personnel) | 9,200 |
| Advertising (phone listing, etc.) | 3,600 |
| Auto and truck expense | 9,000 |
| Association dues | 500 |
| Seminars, travel | 3,000 |
| Entertainment | 5,200 |
| Bad debts (1% of gross) | 7,000 |
| Total | $117,200 |

The contractor in this example now has two options for estimating overhead. One possibility is to develop the ratio of Total Overhead Expenses to Annual Labor Cost, and apply that ratio against labor costs. This can be prorated, or used to develop an annual overhead cost figure. In this example, the following ratio: $117,200 (Total Overhead)/$280,000 (Annual Labor Cost) = 42%

The second option is to develop the ratio of Total Overhead Expenses to Annual Gross Sales Income. This can be prorated, or used to develop another type of annual overhead cost figure. In this example, create the following ratio: $117,200 (Total Overhead)/$700,000 (Annual Gross Sales Income) = 17%.

## 10-A Sales Tax Table

State Sales Taxes on materials are tabulated on Table 10-A. Five states in the U.S. have no sales tax. Many states allow local jurisdictions to levy additional sales taxes. Of course, some projects may be exempt from sales tax.

| State | Tax | State | Tax | State | Tax | State | Tax |
|---|---|---|---|---|---|---|---|
| Alabama | 4% | Illinois | 5% | Montana | 0% | Rhode Island | 6% |
| Alaska | 0 | Indiana | 5 | Nebraska | 3.5 | South Carolina | 5 |
| Arizona | 5 | Iowa | 4 | Nevada | 5.75 | South Dakota | 4 |
| Arkansas | 4 | Kansas | 3 | New Hampshire | 0 | Tennessee | 5.5 |
| California | 4.75 | Kentucky | 5 | New Jersey | 6 | Texas | 4 |
| Colorado | 3 | Louisiana | 4 | New Mexico | 3.75 | Utah | 4.625 |
| Connecticut | 6 | Maine | 5 | New York | 4 | Vermont | 4 |
| Delaware | 0 | Maryland | 5 | North Carolina | 3 | Virginia | 3 |
| District of Columbia | 6 | Massachusetts | 5 | North Dakota | 4 | Washington | 6.5 |
| Florida | 5 | Michigan | 4 | Ohio | 5 | West Virginia | 5 |
| Georgia | 3 | Minnesota | 6 | Oklahoma | 3 | Wisconsin | 5 |
| Hawaii | 4 | Mississippi | 6 | Oregon | 0 | Wyoming | 3 |
| Idaho | 4 | Missouri | 4.125 | Pennsylvania | 6 | Average | 4.11% |

# 10 | OVERHEAD & PROFIT

## 10-B Overtime

Table 10-B shows the effective hourly cost of overtime over a prolonged period of time compared to the hourly payroll cost of a construction project. Short-term overtime does not involve as great a reduction of efficiency, and cost premiums would approach the PAYROLL COSTS rather than the EFFECTIVE HOURLY COSTS listed here. As the total hours per week are increased, more time is lost by absenteeism and accident rate increases.

| Days per Week | Hours per Day | Total Hours Worked | Actual Productive Hours | Production Efficiency | Payroll Cost per Hour Overtime after 40 hrs. @ 1-1/2 times | Payroll Cost per Hour Overtime after 40 hrs. @ 2 times | Effective Cost per Hour Overtime after 40 hrs. @ 1-1/2 times | Effective Cost per Hour Overtime after 40 hrs. @ 2 times |
|---|---|---|---|---|---|---|---|---|
|   | 8  | 40 | 40.0 | 100.0% | 100.0% | 100.0% | 100.0% | 100.0% |
|   | 9  | 45 | 43.4 | 96.5   | 105.6  | 111.1  | 109.4  | 115.2  |
| 5 | 10 | 50 | 46.5 | 93.0   | 110.0  | 120.0  | 118.3  | 129.0  |
|   | 11 | 55 | 49.2 | 89.5   | 113.6  | 127.3  | 127.0  | 142.3  |
|   | 12 | 60 | 51.6 | 86.0   | 116.7  | 133.3  | 135.7  | 155.0  |
|   | 8  | 48 | 46.1 | 96.0   | 108.3  | 116.7  | 112.8  | 121.5  |
|   | 9  | 54 | 48.9 | 90.6   | 113.0  | 125.9  | 124.7  | 139.1  |
| 6 | 10 | 60 | 51.1 | 85.2   | 116.7  | 133.3  | 137.0  | 156.6  |
|   | 11 | 66 | 52.7 | 79.8   | 119.7  | 139.4  | 149.9  | 174.6  |
|   | 12 | 72 | 53.6 | 74.4   | 122.2  | 144.4  | 164.2  | 194.0  |
|   | 8  | 56 | 48.8 | 87.1   | 114.3  | 128.6  | 131.1  | 147.5  |
|   | 9  | 63 | 52.2 | 82.8   | 118.3  | 136.5  | 142.7  | 164.8  |
| 7 | 10 | 70 | 55.0 | 78.5   | 121.4  | 142.9  | 154.5  | 181.8  |
|   | 11 | 77 | 57.1 | 74.2   | 124.0  | 148.1  | 167.3  | 199.6  |
|   | 12 | 84 | 58.7 | 69.9   | 126.2  | 152.4  | 180.6  | 218.1  |

## 10-C Architectural

Table 10-C shows typical percentage fees for architectural design. Adequate design work cannot be expected below these fee levels. The fees themselves may vary from those listed due to economic conditions. Rates can be interpolated horizontally and vertically. Various portions of the same project that require different rates should be adjusted proportionally.

| Building Type | 25 | 50 | 100 | 250 | 500 | 1,000 |
|---|---|---|---|---|---|---|
| Repetitive Housing, Warehouses | 13.0 | 11.0 | 9.0 | 8.0 | 7.0 | 6.2 |
| Apartments, Motels, Offices | 14.0 | 12.6 | 11.7 | 10.8 | 8.5 | 7.3 |
| Single Homes, Retail Stores | 18.0 | 16.0 | 14.0 | 12.8 | 11.9 | 10.9 |

Total Project Size in Thousands of Dollars

## 10-D Unemployment And Social Security Taxes

Mass. State Unemployment tax ranges from 1.5% to 5.7% plus a 1% solvency tax on the first $7000 of wages. Federal Unemployment tax is 3.5% of the first $7000 of wages. This is reduced by a credit for payment to the state. The minimum Federal Unemployment tax is .8% after all credits.

Combined rates in Mass. thus vary from 2.3% to 6.5% of the first $7000 of wages. Combined average U.S. rate is about 5.5% of the first $7000. Contractors with permanent workers will pay less since the average annual wages for skilled labor is $20.10 x 2000 hours or about $40,200 per year. The average combined rate for U.S. would thus be 5.5% x 7,000 ÷ 40,200 = .96% of total wages for permanent employees.

Rates not only vary from state to state but also with the experience rating of the contractor.

Social Security (FICA) for 1985 is 7.05% of wages up to $40,000.

# 10 | OVERHEAD & PROFIT

## 10-E Builder's Risk Insurance

Builder's risk insurance premiums (Table 10-E) are paid by the owner or the contractor. Blasting insurance, collapse insurance and underground insurance would raise the total insurance cost above those listed. A floater insurance policy to cover materials delivered to the job runs $1 to $1.50 per $100 value. Contractor equipment insurance runs $.50 to $2.50 per $100 value.

This table shows New England Builder's risk insurance rates in dollars per $100 value for $1,000 deductible. For $25,000 deductible, the rates can be reduced 13% to 34%. On contracts over $500,000, rates are lower than what are shown here. High liability limits may be difficult to obtain. Policies are written annually for the total value in place. For "All Risk" insurance (excluding flood, earthquake and other dangers) add $.022 to total rates shown in the table.

| Coverage | Frame Construction (Class 1) Range | Average | Brick Construction (Class 4) Range | Average | Fire Resistive (Class 6) Range | Average |
|---|---|---|---|---|---|---|
| Fire Insurance | $.165 to .378 | $.287 | $.100 to .177 | $.170 | $.059 to .168 | $.134 |
| Extended Coverage | .077 to .111 | .098 | .042 to .112 | .095 | .029 to .075 | .063 |
| Vandalism | .008 to .011 | .010 | .008 to .011 | .010 | .008 to .011 | .010 |
| Total Annual Rate | $.250 to .500 | $.395 | $.150 to .300 | $.275 | $.096 to .254 | $.207 |

## 10-F Insurance Rates by Trade

The table below lists the national averages for Worker's Compensation Insurance rates by trade and type of building. The average "Insurance Rate" is multiplied by the "% Building Cost" for each trade. This produces the "Workers' Compensation Cost" in terms of percentage of total labor cost. Add this to figures for each trade to determine the weighted average Workers' Compensation rate for the building types analyzed.

| Trade | Insurance Rate (% of Labor Cost) Range | Average | % of Building Cost Office Bldgs. | Schools & Apts. | Mfg. | Workers' Compensation Cost Office Bldgs. | Schools & Apts. | Mfg. |
|---|---|---|---|---|---|---|---|---|
| Excavation, Grading, etc. | 1.6% to 27.8% | 6.9% | 4.8% | 4.9% | 4.5% | .33% | .34% | .31% |
| Piles & Foundations | 3.5 to 44.3 | 16.6 | 7.1 | 5.2 | 8.7 | 1.18 | .86 | 1.44 |
| Concrete | 1.8 to 27.8 | 8.8 | 5.0 | 14.8 | 3.7 | .44 | 1.30 | .33 |
| Masonry | 1.4 to 18.7 | 7.2 | 6.9 | 7.5 | 1.9 | .50 | .54 | .14 |
| Structural Steel | 3.5 to 40.5 | 19.5 | 10.7 | 3.9 | 17.6 | 2.09 | .76 | 3.43 |
| Miscellaneous & Ornamental Metals | 1.2 to 15.7 | 6.6 | 2.8 | 4.0 | 3.6 | .18 | .26 | .24 |
| Carpentry & Millwork | 2.4 to 54.1 | 9.6 | 3.7 | 4.0 | 0.5 | .36 | .38 | .05 |
| Metal or Composition Siding | 1.7 to 16.4 | 7.4 | 2.3 | 0.3 | 4.3 | .17 | .02 | .32 |
| Roofing | 4.1 to 44.4 | 17.1 | 2.3 | 2.6 | 3.1 | .39 | .44 | .53 |
| Doors & Hardware | 1.4 to 15.9 | 5.5 | 0.9 | 1.4 | 0.4 | .05 | .08 | .02 |
| Sash & Glazing | 2.5 to 33.3 | 7.6 | 3.5 | 4.0 | 1.0 | .27 | .30 | .08 |
| Lath & Plaster | 1.6 to 22.9 | 7.2 | 3.3 | 6.9 | 0.8 | .24 | .50 | .06 |
| Tile, Marble & Floors | 1.0 to 18.7 | 5.2 | 2.6 | 3.0 | 0.5 | .14 | .16 | .03 |
| Acoustical Ceilings | 1.2 to 15.4 | 5.5 | 2.4 | 0.2 | 0.3 | .13 | .01 | .02 |
| Painting | 2.0 to 16.6 | 7.2 | 1.5 | 1.6 | 1.6 | .11 | .12 | .12 |
| Interior Partitions | 2.4 to 54.1 | 9.6 | 3.9 | 4.3 | 4.4 | .37 | .41 | .42 |
| Miscellaneous Items | 1.1 to 54.1 | 8.3 | 5.2 | 3.7 | 9.7 | .43 | .31 | .81 |
| Elevators | 1.3 to 15.6 | 5.3 | 2.1 | 1.1 | 2.2 | .11 | .06 | .12 |
| Sprinklers | 1.4 to 16.8 | 5.6 | 0.5 | — | 2.0 | .03 | — | .11 |
| Plumbing | 1.2 to 12.5 | 4.7 | 4.9 | 7.2 | 5.2 | .23 | .34 | .24 |
| Heat., Vent., Air Conditioning | 1.7 to 11.4 | 6.0 | 13.5 | 11.0 | 12.9 | .81 | .66 | .77 |
| Electrical | 1.1 to 13.1 | 3.9 | 10.1 | 8.4 | 11.1 | .39 | .33 | .43 |
| Total | 1.0% to 54.1% | — | 100.0% | 100.0% | 100.0% | 8.95% | 8.18% | 10.02% |
| Overall Weighted Average | | | | | | | 9.05% | |

# 10 | OVERHEAD & PROFIT

# 10-G Insurance Rates by State

The table below lists the weighted average Workers' Compensation base rate for each state. There is a "factor" column that relates each state to the national Worker's Compensation average of 9.0%.

| State | Weighted Average | Factor | State | Weighted Average | Factor | State | Weighted Average | Factor |
|---|---|---|---|---|---|---|---|---|
| Alabama | 6.5% | 72 | Kentucky | 5.8% | 64 | North Dakota | 7.1% | 79 |
| Alaska | 12.0 | 133 | Louisiana | 6.5 | 72 | Ohio | 7.1 | 79 |
| Arizona | 8.3 | 92 | Maine | 14.5 | 161 | Oklahoma | 8.2 | 91 |
| Arkansas | 7.5 | 83 | Maryland | 14.2 | 157 | Oregon | 18.2 | 202 |
| California | 10.6 | 118 | Massachusetts | 12.7 | 141 | Pennsylvania | 10.7 | 118 |
| Colorado | 8.4 | 93 | Michigan | 11.2 | 124 | Rhode Island | 12.5 | 138 |
| Connecticut | 16.8 | 187 | Minnesota | 11.2 | 124 | South Carolina | 6.4 | 71 |
| Delaware | 9.9 | 110 | Mississippi | 5.5 | 61 | South Dakota | 5.4 | 60 |
| District of Columbia | 16.7 | 186 | Missouri | 5.0 | 55 | Tennessee | 4.9 | 54 |
| Florida | 6.6 | 73 | Montana | 12.2 | 135 | Texas | 5.9 | 65 |
| Georgia | 5.5 | 61 | Nebraska | 6.0 | 66 | Utah | 5.2 | 57 |
| Hawaii | 25.5 | 283 | Nevada | 10.8 | 12 | Vermont | 6.2 | 69 |
| Idaho | 7.7 | 86 | New Hampshire | 13.5 | 150 | Virginia | 7.2 | 80 |
| Illinois | 10.9 | 121 | New Jersey | 7.7 | 85 | Washington | 6.9 | 76 |
| Indiana | 2.2 | 24 | New Mexico | 13.2 | 146 | West Virginia | 5.0 | 55 |
| Iowa | 6.5 | 72 | New York | 9.1 | 101 | Wisconsin | 6.8 | 75 |
| Kansas | 6.4 | 71 | North Carolina | 4.8 | 53 | Wyoming | 5.5 | 61 |

Weighted Average for U.S. is 9.0% of payroll = 100

# HOW TO USE SQUARE FOOT COST PAGES

### Components
Each page contains the ten components needed to develop the complete square foot cost of the dwelling specified. All components are defined with a description of the materials and/or task involved. Use cost figures from each component to estimate the cost per square foot of that section of the project.

### Specifications
Each page contains a definition of the parameters for a specific size and type of dwelling. Included are the square foot dimensions of the proposed building. LIVING AREA takes into account the number of floors and other factors needed to define a building's TOTAL SQUARE FOOTAGE. Perimeter and partition dimensions are defined in terms of linear feet.

### Line Totals
The extreme right-hand column lists the sum of two figures. Use this total to determine the sum of MATERIAL COST plus INSTALLATION COST. The result is a convenient total cost for each of the ten components.

### Man-hours
Use this column to determine the unit of measure in MAN-HOURS needed to perform a task. This figure will give the builder his MAN-HOURS PER SQUARE FOOT of building. The TOTAL MAN-HOURS PER COMPONENT are determined by multiplying the LIVING AREA times the MAN-HOURS listed on that line.
(TOTAL MAN-HOURS PER COMPONENT = LIVING AREA x MAN-HOURS)

**Average 1 Story**

| | Living Area | 1500 S.F. |
|---|---|---|
| | Ground Area | 1500 S.F. |
| | Perimeter | 160 L.F. |
| | Partitions | 190 L.F. |

| # | Component | Description | MAN-HOURS | MAT. | LABOR | TOTAL |
|---|---|---|---|---|---|---|
| 1 | Site Work | Site preparation for slab; trench 4' deep for foundation wall. | .048 | .51 | 1.74 | 2.25 |
| 2 | Foundations | Continuous concrete footing 8" deep x 18" wide; cast-in-place concrete wall, 8" thick, 4' deep, 2500 psi; waterproofing; draintile; 4" concrete slab on 4" crushed stone base, trowel finish. | .113 | 2.35 | 2.42 | 4.77 |
| 3 | Framing | 2" x 4" wood studs, 16" O.C.; 1/2" plywood sheathing; 2" x 6" rafters 16" O.C. with 1/2" plywood sheathing 4 in 12 pitch; 2" x 8" ceiling joists 16" O.C.; 1/2" wafer board subfloor on 1" x 2" wood sleepers 16" O.C. | .136 | 2.99 | 3.36 | 6.35 |
| 4 | Exterior Walls | Horizontal beveled wood siding; #15 felt building paper; 3-1/2" batt insulation; 14 wood double hung windows; 3 flush solid core wood exterior doors, storms and screens. | .098 | 4.07 | 2.00 | 6.07 |
| 5 | Roofing | 240# asphalt shingles; #15 felt building paper; aluminum flashing; 6" attic insulation. Aluminum gutters and downspouts. | .047 | 1.09 | 1.08 | 2.17 |
| 6 | Interiors | 1/2" drywall taped and finished, painted with primer and 1 coat; softwood baseboard and trim, painted with primer and 1 coat; finished hardwood floor 40%, carpet with underlayment 40%, vinyl tile with underlayment 15%, ceramic tile with underlayment 5%; 23 hollow core doors. | .251 | 4.80 | 5.33 | 10.13 |
| 7 | Specialties | Kitchen cabinets - 14 L.F. wall and base cabinets with laminated plastic counter top; medicine cabinet. | .009 | 1.17 | .16 | 1.33 |
| 8 | Mechanical | 1 lavatory, white, wall hung; 1 water closet, white; 1 bathtub with shower, porcelain enamel steel, white; 1 kitchen sink, stainless steel, single; 1 water heater, gas fired, 30 gal.; gas fired forced air heat. | .098 | 1.67 | 1.30 | 2.97 |
| 9 | Electrical | 200 Amp. service; romex wiring; incandescent lighting fixtures, switches, receptacles. | .041 | .48 | .87 | 1.35 |
| 10 | Overhead | Permit and plans. | | | .27 | .27 |
| | **Total** | | | **19.40** | **18.26** | **37.66** |

### Materials
This column gives the unit needed to develop the COST OF MATERIALS. Note: The figures given here are not BARE COSTS. Ten percent has been added to BARE MATERIAL COST to cover handling.

### Installation
The labor rates included here incorporate the overhead and profit costs for the installing contractor. The average mark-up used to create these figures is 59.6% over and above BARE LABOR COST including fringe benefits.

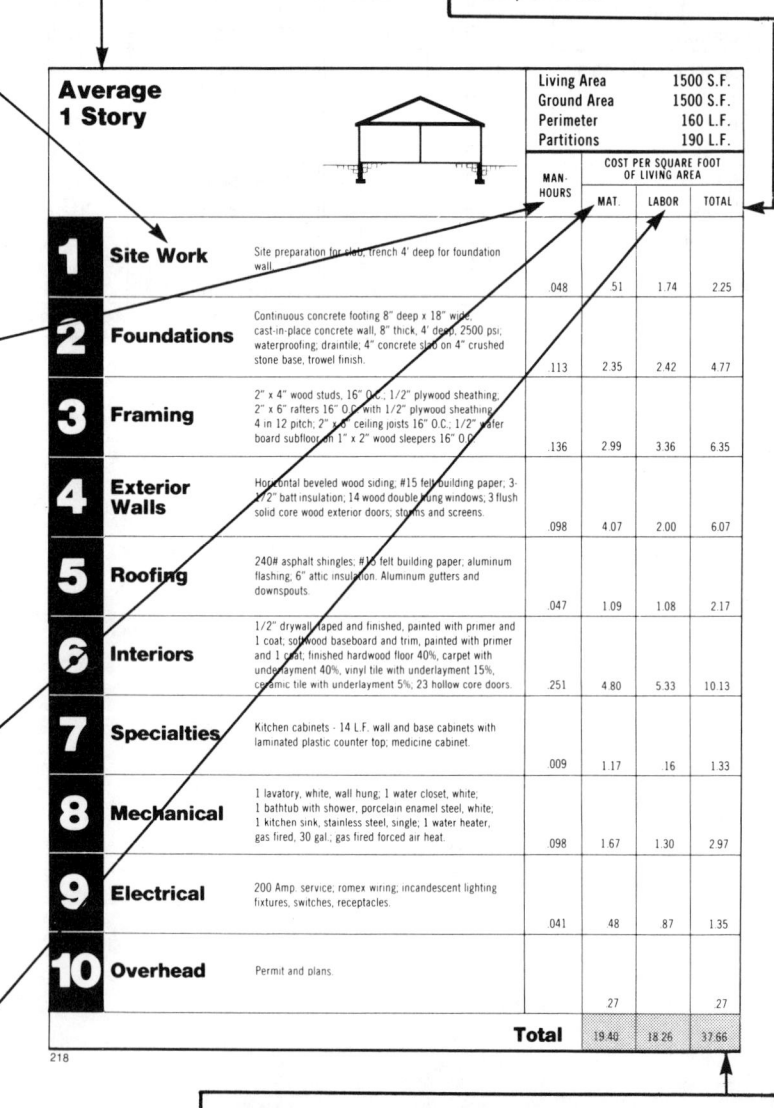

### Bottom Line Total
This figure is the complete square foot cost for the construction project. To determine TOTAL PROJECT COST, multiply the BOTTOM LINE TOTAL times the LIVING AREA. (TOTAL PROJECT COST = BOTTOM LINE TOTAL x LIVING AREA)

# SQUARE FOOT COSTS

Use this section of the book to develop quick and easy square foot costs for Residential and Light Industrial construction. With these tables, a builder can take a quick look at all the components that go into any individual project. Included in the description of each construction "type" described here are building size, and the length of partitions in linear feet. Man-hour costs for each project's completion are also listed. This is important for the modern builder who often relies on sub-contractors to accomplish many separate construction tasks.

For example: a 1500-square-foot house (a medium-size Ranch) will take .140 man-hours per square foot to frame, according to the tables in this section. The total man-hours needed are 210 hours, or two workers taking slightly more than 13 days to frame a medium Ranch.

All rates are developed using the assumption that the tradesmen used on the project are proficient in the craft they are involved in. This section of the book includes 20 different building types for residential housing — from small, economy homes to large, luxury residential construction. The tables also cover typical Light Industrial types such as row houses, townhouses, retail stores, office buildings, warehouses, apartment buildings and motels.

Each listing of square foot costs is broken down into the ten most common components of construction. Start the estimate with site work and continue through the listing to the mechanical and electrical costs. Each listing also shows typical overhead and profit costs for each building type. All costs are derived from the systems section of this book.

The bottom line figures derived here can be used for conceptual estimates, or as a quick check on more specific totals. It is then possible to go back to the systems section to develop more specific cost totals.

## TABLE OF CONTENTS

| Building Type | Page |
|---|---|
| **Residential** | |
| Economy | 213 |
| Average | 218 |
| Custom | 223 |
| Luxury | 228 |
| **Light Commercial** | |
| Apartment Building | 234 |
| Motel | 236 |
| Office Building | 240 |
| Retail Store | 244 |
| Row House/Townhouse | 248 |
| Warehouse | 252 |

# RESIDENTIAL BUILDING TYPES

### One Story
This is an example of a one-story dwelling. The living area of this type of residence is confined to the ground floor. The headroom in the attic is usually too low for use as a living area.

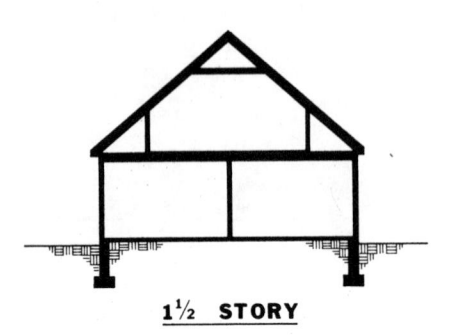

### One-and-a-half Story
The living area in the upper level of this type of residence is 50% to 90% of the ground floor. This is made possible by a combination of this design's high-peaked roof and/or dormers. Only the upper level area with a ceiling height of six feet or more is considered living area. The living area of this residence is the sum of the ground floor area plus the area on the second level with a ceiling height of six feet or more.

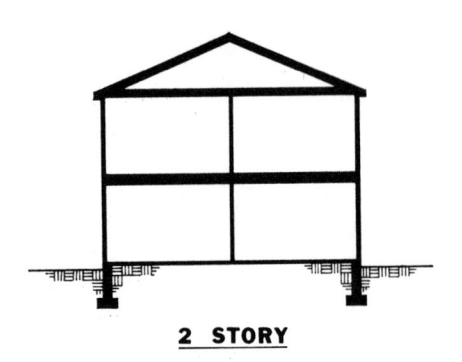

### Two Story
This type of residence has a second floor or upper level area which is equal or nearly equal to the ground floor area. The upper level of this type of residence can range from 90% to 110% of the ground floor area, depending on setbacks or overhangs. The living area is the sum of the ground floor area and the upper level floor area.

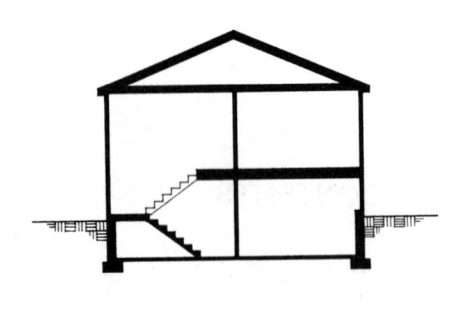

### Bi-level
This type of residence has two living areas. One is above the other. One area is about four feet below grade and the second is about four feet above grade. Both areas are equal in size. The lower level in this type of residence is originally designed and built to serve as a living area and not as a basement. Both levels have full ceiling heights. The living area is the sum of the lower level area plus the upper level area.

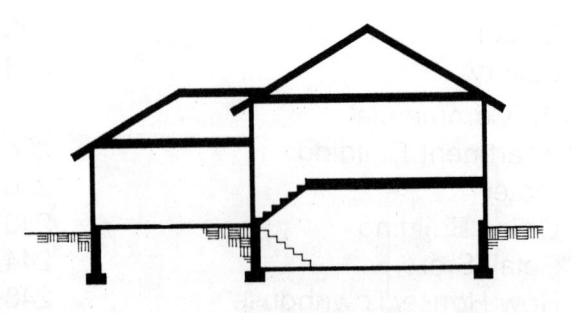

### Tri-level
This type of residence has three levels of living area. One is at grade level, the second is about four feet below grade, and the third is about four feet above grade. All levels are originally designed to serve as living areas. All levels have full ceiling heights. The living area is a sum of the areas of each of the three levels.

# BUILDING CLASSES

Given below are the four general definitions of building classes. Each building type — Economy, Average, Custom and Luxury — is common in residential construction. All four are used in this book to determine costs per square foot.

## Economy Class
An economy class residence is usually mass-produced from stock plans. The materials and workmanship are sufficient only to satisfy minimum building codes. Low construction cost is more important than distinctive features. Design is seldom other than square or rectangular.

## Average Class
An average class residence is simple in design and is built from standard designer plans. Materials and workmanship are average, but often exceed the minimum building codes. There are frequently special features that give the residence some distinctive characteristics.

## Custom Class
A custom class residence is usually built from a designer's plans which have been modified to give the building a distinction of design. Material and workmanship are generally above average with obvious attention given to construction details. Construction normally exceeds building code requirements.

## Luxury Class
A luxury class residence is built from an architect's plan for a specific owner. It is unique in design and workmanship. There are many special features, and construction usually exceeds all building codes. It is obvious that primary attention is placed on the owner's comfort and pleasure. Construction is supervised by an architect.

# Economy 1 Story

| | | |
|---|---|---|
| Living Area | 1200 S.F. |
| Ground Area | 1200 S.F. |
| Perimeter | 148 L.F. |
| Partitions | 160 L.F. |

| | | | MAN-HOURS | COST PER SQUARE FOOT OF LIVING AREA ||| 
|---|---|---|---|---|---|---|
| | | | | MAT. | LABOR | TOTAL |
| **1** | **Site Work** | Site preparation for slab; trench 4' deep for foundation wall. | .060 | .51 | 1.97 | 2.48 |
| **2** | **Foundations** | Continuous concrete footing 8" deep x 18" wide; 8" thick concrete block foundation wall, 4' deep; 4" concrete slab on 4" crushed stone base, trowel finish. | .131 | 2.46 | 2.27 | 4.73 |
| **3** | **Framing** | 2" x 4" wood studs, 16" O.C.; 1/2" insulation board sheathing; 2" x 4" wood truss roof 24" O.C. with 3/8" plywood sheathing, 4 in 12 pitch. | .098 | 2.64 | 2.55 | 5.19 |
| **4** | **Exterior Walls** | Horizontal aluminum siding; #15 felt building paper; wood double hung windows; 2 flush solid core wood exterior doors. | .110 | 3.65 | 2.24 | 5.89 |
| **5** | **Roofing** | 240# asphalt shingles; #15 felt building paper; aluminum flashing; 6" attic insulation. | .047 | 1.09 | 1.08 | 2.17 |
| **6** | **Interiors** | 1/2" drywall, taped and finished, painted with primer and 1 coat; softwood baseboard and trim, painted with primer and 1 coat; rubber backed carpeting 80%, asphalt tile 20%; 19 hollow core wood interior doors. | .243 | 4.25 | 5.32 | 9.57 |
| **7** | **Specialties** | Kitchen cabinets - 6 L.F. wall and base cabinets with laminated plastic counter top. | .004 | .43 | .07 | .50 |
| **8** | **Mechanical** | 1 lavatory, white, wall hung; 1 water closet, white; 1 bathtub, porcelain enamel steel, white; 1 kitchen sink, stainless steel, single; 1 water heater, gas fired, 30 gal.; gas fired forced air heat. | .086 | 2.09 | 1.62 | 3.71 |
| **9** | **Electrical** | 100 Amp. service; romex wiring; incandescent lighting fixtures, switches, receptacles. | .036 | .37 | .76 | 1.13 |
| **10** | **Overhead** | Permit and plans. | | .33 | | .33 |
| | | **Total** | | 17.82 | 17.88 | 35.70 |

# Economy 1½ Story

| | | Living Area | 1600 S.F. |
|---|---|---|---|
| | | Ground Area | 900 S.F. |
| | | Perimeter | 109 L.F. |
| | | Partitions | 185 L.F. |

| # | Category | Description | MAN-HOURS | MAT. | LABOR | TOTAL |
|---|---|---|---|---|---|---|
| 1 | Site Work | Site preparation for slab; trench 4' deep for foundation wall. | .041 | .48 | 1.41 | 1.89 |
| 2 | Foundations | Continuous concrete footing 8" deep x 18" wide, 8" thick concrete block foundation wall, 4' deep; 4" concrete slab on 4" crushed stone base, trowel finish. | .073 | 1.37 | 1.26 | 2.63 |
| 3 | Framing | 2" x 4" wood studs, 16" O.C.; 1/2" insulation board sheathing; 2" x 6" rafters 16" O.C. with 3/8" plywood sheathing, 8 in 12 pitch; 2" x 8" floor joists 16" O.C. with bridging and 1/2" plywood subfloor. | .090 | 2.04 | 2.22 | 4.26 |
| 4 | Exterior Walls | Horizontal aluminum siding; #15 felt building paper; wood double hung windows; 3 flush solid core wood exterior doors. | .077 | 2.61 | 1.55 | 4.16 |
| 5 | Roofing | 240# asphalt shingles; #15 felt building paper; aluminum flashing; 6" attic insulation. | .029 | .63 | .62 | 1.25 |
| 6 | Interiors | 1/2" drywall, taped and finished, painted with primer and 1 coat; softwood baseboard and trim, painted with primer and 1 coat; rubber backed carpeting 80%, asphalt tile 20%; 21 hollow core wood interior doors. | .204 | 3.63 | 4.44 | 8.07 |
| 7 | Specialties | Kitchen cabinets - 6 L.F. wall and base cabinets with laminated plastic counter top; stairs. | .020 | .61 | .12 | .73 |
| 8 | Mechanical | 1 lavatory, white, wall hung; 1 water closet, white; 1 bathtub, porcelain enamel steel, white; 1 kitchen sink, stainless steel, single; 1 water heater, gas fired, 30 gal.; gas fire forced air heat. | .079 | 1.60 | 1.22 | 2.82 |
| 9 | Electrical | 100 Amp. service; romex wiring; incandescent lighting fixtures, switches, receptacles. | .033 | .34 | .71 | 1.05 |
| 10 | Overhead | Permit and plans. | | .25 | | .25 |
| | **Total** | | | 13.56 | 13.55 | 27.11 |

# Economy 2 Story

| | Living Area | 2000 S.F. |
|---|---|---|
| | Ground Area | 1000 S.F. |
| | Perimeter | 130 L.F. |
| | Partitions | 265 L.F. |

| | | | MAN-HOURS | COST PER SQUARE FOOT OF LIVING AREA |||
|---|---|---|---|---|---|---|
| | | | | MAT. | LABOR | TOTAL |
| 1 | Site Work | Site preparation slab; trench 4' deep for foundation wall. | .034 | .38 | 1.20 | 1.58 |
| 2 | Foundations | Continuous concrete footing 8" deep x 18" wide; 8" thick concrete block foundation wall, 4' deep; 4" concrete slab on 4" crushed stone base, trowel finish. | .069 | 1.27 | 1.17 | 2.44 |
| 3 | Framing | 2" x 4" wood studs, 16" O.C.; 1/2" insulation board sheathing; 2" x 6" rafters 16" O.C. with 3/8" plywood sheathing, 4 in 12 pitch; 2" x 6" ceiling joists 16" O.C.; 2" x 8" floor joists 16" O.C. with bridging and 5/8" plywood subfloor. | .112 | 2.65 | 2.72 | 5.37 |
| 4 | Exterior Walls | Horizontal aluminum siding; #15 felt building paper; wood double hung windows; 3 flush solid core wood exterior doors. | .107 | 3.52 | 2.14 | 5.66 |
| 5 | Roofing | 240# asphalt shingles; #15 felt building paper; aluminum flashing; 6" attic insulation. | .024 | .55 | .54 | 1.09 |
| 6 | Interiors | 1/2" drywall, taped and finished, painted with primer and 1 coat; softwood baseboard and trim, painted with primer and 1 coat; rubber backed carpeting 80%, asphalt tile 20%; 29 hollow core wood doors. | .219 | 3.94 | 4.87 | 8.81 |
| 7 | Specialties | Kitchen cabinets - 6 L.F. wall and base cabinets with laminated plastic counter top; stairs. | .017 | .49 | .09 | .58 |
| 8 | Mechanical | 1 lavatory, white, wall hung; 1 water closet, white; 1 bathtub, porcelain enamel steel, white; 1 kitchen sink, stainless steel, single; 1 water heater, gas fired, 30 gal.; gas fired forced air heat. | .061 | 1.28 | .97 | 2.25 |
| 9 | Electrical | 100 Amp. service; romex wiring; incandescent lighting fixtures; switches, receptacles. | .030 | .32 | .66 | .98 |
| 10 | Overhead | Permit and plans. | | .20 | | .20 |
| | | **Total** | | 14.60 | 14.36 | 28.96 |

# Economy Bi-Level

| | Living Area | 2000 S.F. |
|---|---|---|
| | Ground Area | 1000 S.F. |
| | Perimeter | 130 L.F. |
| | Partitions | 265 L.F. |

| | | | MAN-HOURS | COST PER SQUARE FOOT OF LIVING AREA |||
|---|---|---|---|---|---|---|
| | | | | MAT. | LABOR | TOTAL |
| 1 | Site Work | Excavation for lower level, 4' deep. Site preparation for slab. | .029 | .38 | 1.17 | 1.55 |
| 2 | Foundations | Continuous concrete footing 8" deep x 18" wide; 8" thick concrete block foundation wall 4' deep; 4" concrete slab on 4" crushed stone base, trowel finish. | .069 | 1.27 | 1.17 | 2.44 |
| 3 | Framing | 2" x 4" wood studs, 16" O.C.; 1/2" insulation board sheathing; 2" x 4" wood truss roof 24" O.C. with 1/2" plywood sheathing, 4 in 12 pitch; 2" x 8" floor joists 16" O.C. with bridging and 5/8" plywood subfloor. | .107 | 2.48 | 2.55 | 5.03 |
| 4 | Exterior Walls | Horizontal aluminum siding; #15 felt building paper; wood double hung windows; 3 flush solid core wood exterior doors. | .089 | 2.93 | 1.77 | 4.70 |
| 5 | Roofing | 240# asphalt shingles; #15 felt building paper; aluminum flashing; 6" attic insulation. | .024 | .54 | .54 | 1.08 |
| 6 | Interiors | 1/2" drywall, taped and finished, painted with primer and 1 coat; softwood baseboard and trim, painted with primer and 1 coat; rubber backed carpeting 80%, asphalt tile 20%; 27 hollow core wood interior doors. | .213 | 3.84 | 4.74 | 8.58 |
| 7 | Specialties | Kitchen cabinets - base 6 L.F. with laminated plastic counter top, wall 30" high, 8 L.F.; stairs. | .018 | .49 | .09 | .58 |
| 8 | Mechanical | 1 lavatory, white, wall hung; 1 water closet, white; 1 bathtub, porcelain enamel steel, white; 1 kitchen sink, stainless steel, single; 1 water heater, gas fired, 30 gal.; gas fired forced air heat. | .061 | 1.28 | .97 | 2.25 |
| 9 | Electrical | 100 Amp. service; romex wiring; incandescent lighting fixtures; switches, receptacles. | .030 | .32 | .66 | .98 |
| 10 | Overhead | Permit and plans. | | .20 | | .20 |
| | | **Total** | | 13.73 | 13.66 | 27.39 |

# Economy Tri-level

| | Living Area | 2400 S.F. |
|---|---|---|
| | Ground Area | 1650 S.F. |
| | Perimeter | 182 L.F. |
| | Partitions | 240 L.F. |

| | | | MAN-HOURS | COST PER SQUARE FOOT OF LIVING AREA |||
|---|---|---|---|---|---|---|
| | | | | MAT. | LABOR | TOTAL |
| **1** | **Site Work** | Site preparation for slab; excavation for lower level 4' deep; trench 4' deep for foundation wall. | .027 | .32 | 1.18 | 1.50 |
| **2** | **Foundations** | Continuous concrete footing 8" deep x 18" wide; 8" concrete block foundation wall, 4' deep; 4" concrete slab on 4" crushed stone base, trowel finish. | .071 | 1.57 | 1.44 | 3.01 |
| **3** | **Framing** | 2" x 4" wood studs, 16" O.C.; 1/2" insulation board sheathing; 2" x 4" wood truss roof 24" O.C. with 1/2" plywood sheathing, 4 in 12 pitch; 2" x 8" floor joists 16" O.C. with bridging and 5/8" plywood subfloor. | .094 | 2.03 | 1.86 | 3.89 |
| **4** | **Exterior Walls** | Horizontal aluminum siding; #15 felt building paper; wood double hung windows; 3 flush solid core wood exterior doors. | .081 | 2.94 | 2.02 | 4.96 |
| **5** | **Roofing** | 240# asphalt shingles; #15 felt building paper; aluminum flashing; 6" attic insulation. | .032 | .74 | .74 | 1.48 |
| **6** | **Interiors** | 1/2" drywall, taped and finished, painted with primer and 1 coat; softwood baseboard and trim, painted with primer and 1 coat; rubber backed carpeting 80%, asphalt tile 20%; 23 hollow core wood interior doors. | .177 | 3.20 | 3.88 | 7.08 |
| **7** | **Specialties** | Kitchen cabinets - 6 L.F. wall and base cabinets with laminated plastic counter top; stairs. | .014 | .41 | .08 | .49 |
| **8** | **Mechanical** | 1 lavatory, white, wall hung; 1 water closet, white; 1 bathtub, porcelain enamel steel, white; 1 kitchen sink, stainless steel, single; 1 water heater, gas fired, 30 gal; gas fired forced air heat. | .057 | 1.29 | .82 | 2.11 |
| **9** | **Electrical** | 100 Amp. service; romex wiring; incandescent lighting fixtures, switches, receptacles. | .029 | .29 | .62 | .91 |
| **10** | **Overhead** | Permit and plans. | | .17 | | .17 |
| | | **Total** | | 12.96 | 12.64 | 25.60 |

# Average 1 Story

| | | |
|---|---|---|
| Living Area | 1500 S.F. |
| Ground Area | 1500 S.F. |
| Perimeter | 160 L.F. |
| Partitions | 190 L.F. |

| # | Category | Description | Man-Hours | Mat. | Labor | Total |
|---|---|---|---|---|---|---|
| 1 | Site Work | Site preparation for slab; trench 4' deep for foundation wall. | .048 | .51 | 1.74 | 2.25 |
| 2 | Foundations | Continuous concrete footing 8" deep x 18" wide, cast-in-place concrete wall, 8" thick, 4' deep, 2500 psi; waterproofing; draintile; 4" concrete slab on 4" crushed stone base, trowel finish. | .113 | 2.35 | 2.42 | 4.77 |
| 3 | Framing | 2" x 4" wood studs, 16" O.C.; 1/2" plywood sheathing; 2" x 6" rafters 16" O.C. with 1/2" plywood sheathing, 4 in 12 pitch; 2" x 6" ceiling joists 16" O.C.; 1/2" wafer board subfloor on 1" x 2" wood sleepers 16" O.C. | .136 | 2.99 | 3.36 | 6.35 |
| 4 | Exterior Walls | Horizontal beveled wood siding; #15 felt building paper; 3-1/2" batt insulation; 14 wood double hung windows; 3 flush solid core wood exterior doors; storms and screens. | .098 | 4.07 | 2.00 | 6.07 |
| 5 | Roofing | 240# asphalt shingles; #15 felt building paper; aluminum flashing; 6" attic insulation. Aluminum gutters and downspouts. | .047 | 1.09 | 1.08 | 2.17 |
| 6 | Interiors | 1/2" drywall, taped and finished, painted with primer and 1 coat; softwood baseboard and trim, painted with primer and 1 coat; finished hardwood floor 40%, carpet with underlayment 40%, vinyl tile with underlayment 15%, ceramic tile with underlayment 5%; 23 hollow core doors. | .251 | 4.80 | 5.33 | 10.13 |
| 7 | Specialties | Kitchen cabinets - 14 L.F. wall and base cabinets with laminated plastic counter top; medicine cabinet. | .009 | 1.17 | .16 | 1.33 |
| 8 | Mechanical | 1 lavatory, white, wall hung; 1 water closet, white; 1 bathtub with shower, porcelain enamel steel, white; 1 kitchen sink, stainless steel, single; 1 water heater, gas fired, 30 gal.; gas fired forced air heat. | .098 | 1.67 | 1.30 | 2.97 |
| 9 | Electrical | 200 Amp. service; romex wiring; incandescent lighting fixtures, switches, receptacles. | .041 | .48 | .87 | 1.35 |
| 10 | Overhead | Permit and plans. | | .27 | | .27 |
| | | **Total** | | 19.40 | 18.26 | 37.66 |

218

# Average 1½ Story

| | | Living Area | 1800 S.F. |
|---|---|---|---|
| | | Ground Area | 1000 S.F. |
| | | Perimeter | 128 L.F. |
| | | Partitions | 205 L.F. |

| # | Division | Description | MAN-HOURS | MAT. | LABOR | TOTAL |
|---|---|---|---|---|---|---|
| 1 | Site Work | Site preparation for slab; trench 4' deep for foundation wall. | .037 | .43 | 1.33 | 1.76 |
| 2 | Foundations | Continuous concrete footing 8" deep x 18" wide, cast-in-place concrete wall, 8" thick, 4' deep, 2500 psi; waterproofing; draintile; 4" concrete slab on 4" crushed stone base, trowel finish. | .073 | 1.47 | 1.53 | 3.00 |
| 3 | Framing | 2" x 4" wood studs, 16" O.C.; 1/2" plywood sheathing; 2" x 6" rafters 16" O.C. with 1/2" plywood sheathing, 8 in`12 pitch; 2" x 8" floor joists 16" O.C. with bridging and 5/8" plywood subfloor; 1/2" plywood subfloor on 1" x 2" wood sleepers 16" O.C. | .098 | 2.32 | 2.47 | 4.79 |
| 4 | Exterior Walls | Horizontal beveled wood siding; #15 felt building paper; 3-1/2" batt insulation; wood double hung windows; 3 flush solid core wood exterior doors; storms and screens. | .078 | 3.23 | 1.57 | 4.80 |
| 5 | Roofing | 240# asphalt shingles; #15 felt building paper; aluminum flashing; 6" attic insulation. Aluminum gutters and downspouts. | .029 | .62 | .61 | 1.23 |
| 6 | Interiors | 1/2" drywall, taped and finished, painted with primer and 1 coat; softwood baseboard and trim, painted with primer and 1 coat; finished hardwood floor 40%, carpet with underlayment 40%, vinyl tile with underlayment 15%, ceramic tile with underlayment 5%; 25 hollow core doors. | .225 | 3.40 | 3.23 | 6.63 |
| 7 | Specialties | Kitchen cabinets - 14 L.F. wall and base cabinets with laminated plastic counter top; medicine cabinet; stairs. | .022 | 1.24 | .19 | 1.43 |
| 8 | Mechanical | 1 lavatory, white, wall hung; 1 water closet, white; 1 bathtub with shower, porcelain enamel steel, white; 1 kitchen sink, stainless steel, single; 1 water heater, gas fired, 30 gal.; gas fired forced air heat. | .049 | 1.42 | 1.08 | 2.50 |
| 9 | Electrical | 200 Amp. service; romex wiring; incandescent lighting fixtures, switches, receptacles. | .039 | .46 | .85 | 1.31 |
| 10 | Overhead | Permit and plans. | | .22 | | .22 |
| | | **Total** | | 14.81 | 12.86 | 27.67 |

# Average 2 Story

| | | Living Area | 2000 S.F. |
|---|---|---|---|
| | | Ground Area | 1000 S.F. |
| | | Perimeter | 130 L.F. |
| | | Partitions | 265 L.F. |

| | | | MAN-HOURS | COST PER SQUARE FOOT OF LIVING AREA |||
|---|---|---|---|---|---|---|
| | | | | MAT. | LABOR | TOTAL |
| 1 | Site Work | Site preparation for slab; trench 4' deep for foundation wall. | .034 | .38 | 1.20 | 1.58 |
| 2 | Foundations | Continuous concrete footing 8" deep x 18" wide; cast-in-place concrete wall, 8" thick, 4' deep, 2500 psi; 4" concrete slab on 4" crushed stone base, trowel finish. | .066 | 1.33 | 1.39 | 2.72 |
| 3 | Framing | 2" x 4" wood studs, 16" O.C.; 1/2" plywood sheathing; 2" x 6" rafters 16" O.C. with 1/2" plywood sheathing, 4 in 12 pitch; 2" x 6" ceiling joists 16" O.C.; 2" x 8" floor joists 16" O.C. with bridging and 5/8" plywood subfloor; 1/2" plywood subfloor on 1" x 2" wood sleepers 16" O.C. | .131 | 2.95 | 3.27 | 6.22 |
| 4 | Exterior Walls | Horizontal beveled wood siding; #15 felt building paper; 3-1/2" batt insulation; wood double hung windows; 3 flush solid core wood exterior doors; storms and screens. | .111 | 4.50 | 2.20 | 6.70 |
| 5 | Roofing | 240# asphalt shingles; #15 felt building paper; aluminum flashing; 6" attic insulation. Aluminum gutters and downspouts. | .024 | .54 | .54 | 1.08 |
| 6 | Interiors | 1/2" drywall, taped and finished, painted with primer and 1 coat; softwood baseboard and trim, painted with primer and 1 coat; finished hardwood floor 40%, carpet with underlayment 40%, vinyl tile with underlayment 15%, ceramic tile with underlayment 5%; 27 hollow core doors. | .232 | 4.52 | 4.98 | 9.50 |
| 7 | Specialties | Kitchen cabinets - 14 L.F. wall and base cabinets with laminated plastic counter top; medicine cabinet, stairs. | .021 | 1.11 | .17 | 1.28 |
| 8 | Mechanical | 1 lavatory, white, wall hung; 1 water closet, white; 1 bathtub with shower, porcelain enamel steel, white; 1 kitchen sink, stainless steel, single; 1 water heater, gas fired, 30 gal.; gas fired forced air heat. | .060 | 1.28 | .97 | 2.25 |
| 9 | Electrical | 200 Amp. service; romex wiring; incandescent lighting fixtures, switches, receptacles. | .039 | .43 | .81 | 1.24 |
| 10 | Overhead | Permit and plans. | | .20 | | .20 |
| | | **Total** | | 17.24 | 15.53 | 32.77 |

# Average Bi-level

| | | Living Area | 2000 S.F. |
|---|---|---|---|
| | | Ground Area | 1000 S.F. |
| | | Perimeter | 130 L.F. |
| | | Partitions | 180 L.F. |

| | | | MAN-HOURS | COST PER SQUARE FOOT OF LIVING AREA |||
|---|---|---|---|---|---|---|
| | | | | MAT. | LABOR | TOTAL |
| **1** | **Site Work** | Excavation for lower level 4' deep. Site preparation for slab. | .029 | .38 | 1.17 | 1.55 |
| **2** | **Foundations** | Continuous concrete footing 8" deep x 18" wide, cast-in-place concrete wall, 8" thick, 4' deep, 2500 psi; waterproofing; draintile; 4" concrete slab on 4" crushed stone base, trowel finish. | .066 | 1.34 | 1.39 | 2.73 |
| **3** | **Framing** | 2" x 4" wood studs, 16" O.C.; 1/2" plywood sheathing; 2" x 6" rafters 16" O.C. with 1/2" plywood sheathing 4 in 12 pitch; 2" x 6" ceiling joists 16" O.C.; 2" x 8" floor joists 16" O.C. with bridging and 5/8" plywood subfloor; 1/2" plywood subfloor on 1" x 2" wood sleepers 16" O.C. | .118 | 2.86 | 2.99 | 5.85 |
| **4** | **Exterior Walls** | Horizontal beveled wood siding; #15 felt building paper; 3-1/2" batt insulation; wood double hung windows; 3 flush solid core wood exterior doors; storms and screens. | .091 | 3.95 | 1.86 | 5.81 |
| **5** | **Roofing** | 240# asphalt shingles; #15 felt building paper; aluminum flashing; 6" attic insulation. Aluminum gutters and downspouts. | .024 | .48 | .53 | 1.01 |
| **6** | **Interiors** | 1/2" drywall, taped and finished, painted with primer and 1 coat; softwood baseboard and trim, painted with primer and 1 coat; finished hardwood floor 40%, carpet with underlayment 40%, vinyl tile with underlayment 15%, ceramic tile with underlayment 5%; 29 hollow core doors. | .217 | 4.39 | 4.71 | 9.10 |
| **7** | **Specialties** | Kitchen cabinets - 14 L.F. wall and base cabinets with laminated plastic counter top; medicine cabinet; stairs. | .021 | 1.11 | .17 | 1.28 |
| **8** | **Mechanical** | 1 lavatory, white, wall hung; 1 water closet, white; 1 bathtub with shower, porcelain enamel steel, white; 1 kitchen sink, stainless steel, single; 1 water heater, gas fired, 30 gal.; gas fired forced air heat. | .061 | 1.28 | .97 | 2.25 |
| **9** | **Electrical** | 200 Amp. service; romex wiring; incandescent lighting fixtures, switches, receptacles. | .039 | .43 | .81 | 1.24 |
| **10** | **Overhead** | Permit and plans. | | .20 | | .20 |
| | | **Total** | | 16.42 | 14.60 | 31.02 |

# Average Tri-level

| | Living Area | 2400 S.F. |
|---|---|---|
| | Ground Area | 1650 S.F. |
| | Perimeter | 182 L.F. |
| | Partitions | 240 L.F. |

| # | Category | Description | MAN-HOURS | MAT. | LABOR | TOTAL |
|---|---|---|---|---|---|---|
| 1 | **Site Work** | Site preparation for slab; excavation for lower level 4' deep; trench 4' deep for foundation wall. | .029 | .32 | 1.18 | 1.50 |
| 2 | **Foundations** | Continuous concrete footing 8" deep x 18" wide; cast-in-place concrete wall, 8" thick, 4' deep, 2500 psi; waterproofing; draintile; 4" concrete slab on 4" crushed stone base, trowel finish. | .080 | 1.65 | 1.70 | 3.35 |
| 3 | **Framing** | 2" x 4" wood studs, 16" O.C.; 1/2" plywood sheathing; 2" x 6" rafters 16" O.C. with 1/2" plywood sheathing, 4 in 12 pitch; 2" x 6" ceiling joists 16" O.C.; 2" x 8" floor joists 16" O.C. with bridging and 5/8" plywood subfloor; 1/2" plywood subfloor on 1" x 2" wood sleepers 16" O.C. | .124 | 2.68 | 2.99 | 5.67 |
| 4 | **Exterior Walls** | Horizontal beveled wood siding; #15 felt building paper; 3-1/2" batt insulation; wood double hung windows; 3 flush solid core wood exterior doors; storms and screens. | .083 | 3.50 | 1.68 | 5.18 |
| 5 | **Roofing** | 240# asphalt shingles; #15 felt building paper; aluminum flashing; 6" attic insulation. Aluminum gutters and downspouts. | .032 | .75 | .74 | 1.49 |
| 6 | **Interiors** | 1/2" drywall, taped and finished, painted with primer and 1 coat; softwood baseboard and trim, painted with primer and 1 coat; finished hardwood floor 40%, carpet with underlayment 40%, vinyl tile with underlayment 15%, ceramic tile with underlayment 5%; 23 hollow core doors. | .186 | 3.85 | 4.02 | 7.87 |
| 7 | **Specialties** | Kitchen cabinets - 14 L.F. wall and base cabinets with laminated plastic counter top; medicine cabinet. | .012 | .93 | .14 | 1.07 |
| 8 | **Mechanical** | 1 lavatory, white, wall hung; 1 water closet, white; 1 bathtub with shower, porcelain enamel steel, white; 1 kitchen sink, stainless steel, single; 1 water heater, gas fired, 30 gal.; gas fired forced air heat. | .059 | 1.29 | .82 | 2.11 |
| 9 | **Electrical** | 200 Amp. service; romex wiring; incandescent lighting fixtures, switches, receptacles. | .036 | .39 | .74 | 1.13 |
| 10 | **Overhead** | Permit and plans. | | .17 | | .17 |
| | **Total** | | | 15.53 | 14.01 | 29.54 |

# Custom 1 Story

| | | |
|---|---|---|
| Living Area | 2500 S.F. |
| Ground Area | 2500 S.F. |
| Perimeter | 266 L.F. |
| Partitions | 270 L.F. |

| # | Category | Description | Man-Hours | Mat. | Labor | Total |
|---|---|---|---|---|---|---|
| 1 | Site Work | Site preparation for slab; trench 4' deep for foundation wall. | .028 | .31 | 1.10 | 1.41 |
| 2 | Foundations | Continuous concrete footing 8" deep x 18" wide; cast-in-place concrete wall, 8" thick, 4' deep, 3000 psi; waterproofing; draintile; 4" concrete slab on 4" crushed stone base, trowel finish. | .113 | 2.35 | 2.42 | 4.77 |
| 3 | Framing | 2" x 6" wood studs, 16" O.C.; 1/2" plywood sheathing; 2" x 8" rafters 16" O.C. with 5/8" plywood sheathing, 4 in 12 pitch; 2" x 6" ceiling joists 16" O.C.; 5/8" plywood subfloor on 1" x 3" wood sleepers 16" O.C.; #15 felt building paper. | .190 | 3.27 | 3.46 | 6.73 |
| 4 | Exterior Walls | Horizontal beveled wood siding; #15 felt building paper; 6" batt insulation; 18 wood double hung windows; 1 bow/bay window; 3 solid core wood exterior doors; storms and screens. | .085 | 4.27 | 1.71 | 5.98 |
| 5 | Roofing | Wood cedar shingles; #15 felt building paper; copper flashing; 9" attic insulation. Aluminum gutters and downspouts. | .082 | 1.40 | 1.14 | 2.54 |
| 6 | Interiors | 5/8" drywall, skim coat plaster painted with primer and 1 coat; hardwood baseboard and trim, sanded and finished; finished hardwood floor 70%, ceramic tile with underlayment 20%, vinyl tile with underlayment 10%; 33 wood panel interior doors. | .292 | 6.34 | 5.70 | 12.04 |
| 7 | Specialties | Kitchen cabinets - 20 L.F. wall and base cabinets with laminated plastic counter top; 4 L.F. bathroom vanity; medicine cabinet, appliances. | .019 | 2.11 | .38 | 2.49 |
| 8 | Mechanical | 1 kitchen sink, cast iron, double; 1 water heater, gas fired, 50 gal.; gas forced air heat/air conditioning; 1 full bath including: 1 bathtub, color; 1 corner shower; 1 sink, color, built in; 1 water closet, color. 1 lavatory including: 1 sink, color, built in; 1 water closet, color. | .092 | 3.30 | 1.66 | 4.96 |
| 9 | Electrical | 200 Amp. service; romex wiring; fluorescent and incandescent lighting fixtures, switches, receptacles. | .039 | .44 | .84 | 1.28 |
| 10 | Overhead | Permit and plans. | | 2.20 | | 2.20 |
| | | **Total** | | 25.99 | 18.41 | 44.40 |

# Custom 1½ Story

*(Handwritten notes on diagram: 60', 1500 S.F., 25')*

| | Living Area | 2600 S.F. |
|---|---|---|
| | Ground Area | 1500 S.F. |
| | Perimeter | 159 L.F. |
| | Partitions | 275 L.F. |

| | | | MAN-HOURS | COST PER SQUARE FOOT OF LIVING AREA |||
|---|---|---|---|---|---|---|
| | | | | MAT. | LABOR | TOTAL |
| 1 | Site Work | Site preparation for slab; trench 4' deep for foundation wall. | .028 | .29 | 1.00 | 1.29 |
| 2 | Foundations | Continuous concrete footing 8" deep x 18" wide; cast-in-place concrete wall, 8" thick, 4' deep, 3000 psi; waterproofing; draintile; 4" concrete slab on 4" crushed stone base, trowel finish. | .065 | 1.36 | 1.38 | 2.74 |
| 3 | Framing | 2" x 6" wood studs, 16" O.C.; 1/2" plywood sheathing; 2" x 8" rafters 16" O.C. with 5/8" plywood sheathing, 8 in 12 pitch; 2" x 10" floor joists 16" O.C. with bridging and 5/8" plywood subfloor; 5/8" plywood subfloor on 1" x 3" wood sleepers 16" O.C. | .192 | 2.86 | 2.93 | 5.79 |
| 4 | Exterior Walls | Horizontal beveled wood siding; #15 felt building paper; 6" batt insulation; 16 wood double hung windows; 1 bow/bay window; 3 solid core wood exterior doors; storms and screens. | .064 | 3.39 | 1.27 | 4.66 |
| 5 | Roofing | Wood cedar shingles; #15 felt building paper; copper flashing; 9" attic insulation. Aluminum gutters and downspouts. | .048 | .84 | .68 | 1.52 |
| 6 | Interiors | 5/8" drywall, skim coat plaster and finished, painted with primer and 1 coat; hardwood baseboard and trim, sanded and finished; finished hardwood floor 70%, ceramic tile with underlayment 20%, vinyl tile with underlayment 10%; 27 wood panel interior doors. | .259 | 5.49 | 4.94 | 10.43 |
| 7 | Specialties | Kitchen cabinets - 20 L.F. wall and base cabinets with laminated plastic counter top; 4 L.F. bathroom vanity; medicine cabinet; stairs, appliances. | .030 | 2.25 | .41 | 2.66 |
| 8 | Mechanical | 1 kitchen sink, cast iron, double; 1 water heater, gas fired, 75 gal.; gas forced air heat/air conditioning; 1 full bath including: 1 bathtub, color; 1 corner shower; 1 sink, color, built in; 1 water closet, color. 1 lavatory including: 1 sink, color, built in; 1 water closet, color. | .084 | 2.72 | 1.23 | 3.95 |
| 9 | Electrical | 200 Amp. service; romex wiring; fluorescent and incandescent lighting fixtures, switches, receptacles. | .038 | .43 | .82 | 1.25 |
| 10 | Overhead | Permit and plans. | | 2.12 | | 2.12 |
| | | **Total** | | 21.75 | 14.66 | 36.41 |

# Custom 2 Story

| | | | Living Area | 2800 S.F. |
|---|---|---|---|---|
| | | | Ground Area | 1400 S.F. |
| | | | Perimeter | 152 L.F. |
| | | | Partitions | 320 L.F. |

56' × 25' = 1400 s.f.

| | | | MAN-HOURS | COST PER SQUARE FOOT OF LIVING AREA |||
|---|---|---|---|---|---|---|
| | | | | MAT. | LABOR | TOTAL |
| 1 | Site Work | Site preparation for slab; trench 4' deep for foundation wall. | .024 | .27 | .85 | 1.12 |
| 2 | Foundations | Continuous concrete footing 8" deep x 18" wide; cast-in-place concrete wall, 8" thick, 4' deep, 3000 psi; waterproofing; draintile; 4" concrete slab on 4" crushed stone base, trowel finish. | .058 | 1.18 | 1.22 | 2.40 |
| 3 | Framing | 2" x 6" wood studs, 16" O.C.; 1/2" plywood sheathing; 2" x 8" rafters 16" O.C. with 5/8" plywood sheathing, 6 in 12 pitch; 2" x 8" ceiling joists 16" O.C.; 10" floor joists 16" O.C. with bridging and 5/8" plywood subfloor; 5/8" plywood subfloor on 1" x 3" wood sleepers 16" O.C. | .159 | 3.06 | 3.14 | 6.20 |
| 4 | Exterior Walls | Horizontal beveled wood siding; #15 felt building paper; 6" batt insulation; 26 wood double hung windows; 1 bow/bay window; 3 solid core wood exterior doors; storms and screens. | .091 | 4.62 | 1.82 | 6.44 |
| 5 | Roofing | Wood cedar shingles; #15 felt building paper; copper flashing; 9" attic insulation. Aluminum gutters and downspouts. | .042 | .71 | .57 | 1.28 |
| 6 | Interiors | 5/8" drywall, skim coat plaster and finished, painted with primer and 1 coat; hardwood baseboard and trim, sanded and finished; finished hardwood floor 70%, ceramic tile with underlayment 20%, vinyl tile with underlayment 10%; 33 wood panel interior doors. | .271 | 5.96 | 5.31 | 11.27 |
| 7 | Specialties | Kitchen cabinets - 20 L.F. wall and base cabinets with laminated plastic counter top; 4 L.F. bathroom vanity; medicine cabinet; stairs, appliances. | .028 | 2.05 | .37 | 2.42 |
| 8 | Mechanical | 1 kitchen sink, cast iron, double; 1 water heater, gas fired, 75 gal.; gas forced air heat/air conditioning; 1 full bath including: 1 bathtub, color; 1 corner shower; 1 sink, color, built in; 1 water closet, color. 1 lavatory including: 1 sink, color, built in; 1 water closet, color. | .078 | 1.04 | .68 | 1.72 |
| 9 | Electrical | 200 Amp. service; romex wiring; fluorescent and incandescent lighting fixtures, switches, receptacles. | .038 | .42 | .81 | 1.23 |
| 10 | Overhead | Permit and plans. | | 1.97 | | 1.97 |
| | | **Total** | | 21.28 | 14.77 | 36.05 |

225

# Custom Bi-level

| | | Living Area | 2800 S.F. |
|---|---|---|---|
| | | Ground Area | 1400 S.F. |
| | | Perimeter | 152 L.F. |
| | | Partitions | 240 L.F. |

| # | Category | Description | MAN-HOURS | MAT. | LABOR | TOTAL |
|---|---|---|---|---|---|---|
| 1 | Site Work | Site preparation for slab; excavation for lower level 4' deep. | .024 | .27 | .91 | 1.18 |
| 2 | Foundations | Continuous concrete footing 8" deep x 18" wide; cast-in-place concrete wall, 8" thick, 4' deep, 3000 psi; waterproofing; draintile; 4" concrete slab on 4" crushed stone base, trowel finish. | .058 | 1.19 | 1.22 | 2.41 |
| 3 | Framing | 2" x 6" wood studs, 16" O.C.; 1/2" plywood sheathing; 2" x 8" rafters 16" O.C. with 5/8" plywood sheathing, 6 in 12 pitch; 2" x 8" ceiling joists 16" O.C.; 2" x 10" floor joists 16" O.C. with bridging and 5/8" plywood subfloor; 5/8" plywood subfloor on 1" x 3" wood sleepers 16" O.C. | .147 | 2.81 | 2.85 | 5.66 |
| 4 | Exterior Walls | Horizontal beveled wood siding; #15 felt building paper; 6" batt insulation; 26 wood double hung windows; 1 bow/bay window; 3 solid core wood exterior doors; storms and screens. | .079 | 4.21 | 1.58 | 5.79 |
| 5 | Roofing | Wood cedar shingles; #15 felt building paper; copper flashing; 9" attic insulation. Aluminum gutters and downspouts. | .033 | .71 | .57 | 1.28 |
| 6 | Interiors | 5/8" drywall, skim coat plaster and finished, painted with primer and 1 coat; hardwood baseboard and trim, sanded and finished; finished hardwood floor 70%, ceramic tile with underlayment 20%, vinyl tile with underlayment 10%; 34 wood panel interior doors. | .257 | 5.73 | 4.95 | 10.68 |
| 7 | Specialties | Kitchen cabinets - 20 L.F. wall and base cabinets with laminated plastic counter top; 4 L.F. bathroom vanity; medicine cabinet; stairs, appliances. | .028 | 2.09 | .38 | 2.47 |
| 8 | Mechanical | 1 kitchen sink, cast iron, double; 1 water heater, gas fired, 75 gal.; gas forced air heat/air conditioning; 1 full bath including: 1 bathtub, color; 1 corner shower; 1 sink, color, built in; 1 water closet, color. 1 lavatory including: 1 sink, color, built in; 1 water closet, color. | .078 | 2.86 | 1.41 | 4.27 |
| 9 | Electrical | 200 Amp. service; romex wiring; fluorescent and incandescent lighting fixtures, switches, receptacles. | .038 | .42 | .81 | 1.23 |
| 10 | Overhead | Permit and plans. | | 1.97 | | 1.97 |
| | | **Total** | | 22.26 | 14.68 | 36.94 |

# Custom Tri-level

| Living Area | 3000 S.F. |
|---|---|
| Ground Area | 2100 S.F. |
| Perimeter | 197 L.F. |
| Partitions | 285 L.F. |

| | | | MAN-HOURS | COST PER SQUARE FOOT OF LIVING AREA |||
|---|---|---|---|---|---|---|
| | | | | MAT. | LABOR | TOTAL |
| 1 | Site Work | Site preparation for slab; excavation for lower level 4' deep; trench 4' deep for foundation wall. | .023 | .25 | .95 | 1.20 |
| 2 | Foundations | Continuous concrete footing 8" deep x 18" wide; cast-in-place concrete wall, 8" thick, 4' deep, 3000 psi; waterproofing; draintile; 4" concrete slab on 4" crushed stone base, trowel finish. | .073 | 1.52 | 1.56 | 3.08 |
| 3 | Framing | 2" x 6" wood studs, 16" O.C.; 1/2" plywood sheathing; 2" x 8" rafters 16" O.C. with 5/8" plywood sheathing, 6 in 12 pitch; 2" x 8" ceiling joists 16" O.C.; 2" x 10" floor joists 16" O.C. with bridging and 5/8" plywood subfloor; 5/8" plywood subfloor on 1" x 3" wood sleepers 16" O.C. | .162 | 2.96 | 3.05 | 6.01 |
| 4 | Exterior Walls | Horizontal beveled wood siding; #15 felt building paper; 6" batt insulation; 24 wood double hung windows; 1 bow/bay window; 3 solid core wood exterior doors; storms and screens. | .076 | 3.93 | 1.52 | 5.45 |
| 5 | Roofing | Wood cedar shingles; #15 felt building paper; copper flashing; 9" attic insulation. Aluminum gutters and downspouts. | .045 | .98 | .80 | 1.78 |
| 6 | Interiors | 5/8" drywall, skim coat plaster and finished, painted with primer and 1 coat; hardwood baseboard and trim, sanded and finished; finished hardwood floor 70%, ceramic tile with underlayment 20%, vinyl tile with underlayment 10%; 26 wood panel interior doors. | .242 | 5.21 | 4.68 | 9.89 |
| 7 | Specialties | Kitchen cabinets - 20 L.F. wall and base cabinets with laminated plastic counter top; 4 L.F. bathroom vanity; medicine cabinet; stairs, appliances. | .026 | 1.83 | .34 | 2.17 |
| 8 | Mechanical | 1 kitchen sink, cast iron, double; 1 water heater, gas fired, 75 gal.; gas forced air heat/air conditioning; 1 full bath including: 1 bathtub, color; 1 corner shower; 1 sink, color, built in; 1 water closet, color. 1 lavatory including: 1 sink, color, built in; 1 water closet, color. | .073 | 2.67 | 1.32 | 3.99 |
| 9 | Electrical | 200 Amp. service; romex wiring; fluorescent and incandescent lighting fixtures, switches, receptacles. | .036 | .41 | .79 | 1.20 |
| 10 | Overhead | Permit and plans. | | 1.84 | | 1.84 |
| | | **Total** | | 21.60 | 15.01 | 36.61 |

# Luxury 1 Story

| | Living Area | 2500 S.F. |
|---|---|---|
| | Ground Area | 2500 S.F. |
| | Perimeter | 217 L.F. |
| | Partitions | 270 L.F. |

| | | | MAN-HOURS | COST PER SQUARE FOOT OF LIVING AREA | | |
|---|---|---|---|---|---|---|
| | | | | MAT. | LABOR | TOTAL |
| 1 | Site Work | Site preparation for slab; trench 4' deep for foundation wall. | .028 | .31 | 1.14 | 1.45 |
| 2 | Foundations | Continuous concrete footing 8" deep x 18" wide; cast-in-place concrete wall, 12" thick, 4' deep, 3000 psi; waterproofing; draintile; vapor barrier 4" concrete slab on 4" crushed stone base, trowel finish. | .098 | 2.08 | 2.11 | 4.19 |
| 3 | Framing | 2" x 6" wood studs, 16" O.C.; 5/8" plywood sheathing; 2" x 10" rafters 16" O.C. with 5/8" plywood sheathing, 6 in 12 pitch; 2" x 10" ceiling joists 16" O.C.; 5/8 plywood subfloor on 1" x 3" wood sleepers 16" O.C. | .260 | 4.84 | 5.24 | 10.08 |
| 4 | Exterior Walls | Face brick veneer; #15 felt building paper; 6" batt insulation; 18 wood double hung windows; 1 bow/bay window; 3 solid core wood exterior doors; storms and screens. | .204 | 5.40 | 3.68 | 9.08 |
| 5 | Roofing | Wood cedar shingles; #15 felt building paper; copper flashing; 9" attic insulation. Aluminum gutters and downspouts. | .082 | 2.84 | 1.82 | 4.66 |
| 6 | Interiors | 5/8" drywall, thin coat plaster painted with primer and 2 coats; hardwood baseboard and trim, sanded and finished, finished hardwood floor 70%, ceramic tile with underlayment 20%, vinyl tile with underlayment 10%; 33 wood panel interior doors. | .287 | 6.26 | 5.58 | 11.84 |
| 7 | Specialties | Kitchen cabinets - 25 L.F. wall and base cabinets with laminated plastic counter top; 6 L.F. bathroom vanity; medicine cabinet, appliances. | .052 | 2.76 | .97 | 3.73 |
| 8 | Mechanical | 1 kitchen sink, cast iron, double; 1 water heater, gas fired, 75 gal.; gas forced air heat/air conditioning; 1 full bath including: 1 bathtub, color; 1 corner shower; 1 sink, color, built in; 1 water closet, color. 1 lavatory including: 1 sink, color, built in; 1 water closet, color. | .078 | 3.77 | 1.90 | 5.67 |
| 9 | Electrical | 200 Amp. service; romex wiring; fluorescent and incandescent lighting fixtures; intercom, switches, receptacles. | .044 | .47 | .91 | 1.38 |
| 10 | Overhead | Permit and plans. | | 5.20 | | 5.20 |
| | | **Total** | | 33.93 | 23.35 | 57.28 |

# Luxury 1½ Story

| | | Living Area | 2500 S.F. |
|---|---|---|---|
| | | Ground Area | 1500 S.F. |
| | | Perimeter | 155 L.F. |
| | | Partitions | 270 L.F. |

| | | | MAN-HOURS | COST PER SQUARE FOOT OF LIVING AREA |||
|---|---|---|---|---|---|---|
| | | | | MAT. | LABOR | TOTAL |
| 1 | Site Work | Site preparation for slab; trench 4' deep for foundation wall. | .025 | .31 | 1.02 | 1.33 |
| 2 | Foundations | Continuous concrete footing 8" deep x 18" wide; cast-in-place concrete wall, 12" thick, 4' deep, 3000 psi; waterproofing; draintile; vapor barrier; 4" concrete slab on 6" crushed stone base, trowel finish. | .066 | 1.38 | 1.43 | 2.81 |
| 3 | Framing | 2" x 6" wood studs, 16" O.C.; 5/8" plywood sheathing; 2" x 10" rafters 16" O.C. with 5/8" plywood sheathing, 8 in 12 pitch; 2" x 10" ceiling joists 16" O.C.; 2" x 12" floor joists 16" O.C. with bridging and 5/8" plywood subfloor; 5/8" plywood subfloor on 1" x 3" wood sleepers 16" O.C. | .189 | 3.89 | 3.98 | 7.87 |
| 4 | Exterior Walls | Face brick veneer; #15 felt building paper; 6" batt insulation; 18 wood double hung windows; 1 bow/bay window; 3 solid core wood exterior doors; storms and screens. | .174 | 5.00 | 3.03 | 8.03 |
| 5 | Roofing | Wood cedar shingles; #15 felt building paper; copper flashing; 9" attic insulation. Aluminum gutters and downspouts. | .065 | 1.86 | 1.26 | 3.12 |
| 6 | Interiors | 5/8" drywall, thin coat plaster, painted with primer and 2 coats; hardwood baseboard and trim, sanded and finished; finished hardwood floor 70%, ceramic tile with underlayment 20%, vinyl tile with underlayment 10%; 26 wood panel interior doors. | .260 | 5.51 | 4.98 | 10.49 |
| 7 | Specialties | Kitchen cabinets - 25 L.F. wall and base cabinets with laminated plastic counter top; 6 L.F. bathroom vanity; medicine cabinet; stairs, appliances. | .062 | 4.46 | 1.15 | 5.61 |
| 8 | Mechanical | 1 kitchen sink, cast iron, double; 1 water heater, gas fired, 75 gal.; gas forced air heat/air conditioning; 1 full bath including: 1 bathtub, color; 1 corner shower; 1 sink, color, built in; 1 water closet, color. 1 lavatory including: 1 sink, color, built in; 1 water closet, color. | .080 | 3.79 | 1.91 | 5.70 |
| 9 | Electrical | 200 Amp. service; romex wiring; fluorescent and incandescent lighting fixtures; intercom, switches, receptacles. | .044 | .48 | .93 | 1.41 |
| 10 | Overhead | Permit and plans. | | 5.20 | | 5.20 |
| | | **Total** | | 31.88 | 19.69 | 51.57 |

# Luxury 2 Story

| | Living Area | 2800 S.F. |
|---|---|---|
| | Ground Area | 1400 S.F. |
| | Perimeter | 152 L.F. |
| | Partitions | 320 L.F. |

| # | Category | Description | Man-Hours | Mat. | Labor | Total |
|---|---|---|---|---|---|---|
| 1 | Site Work | Site preparation for slab; trench 4' deep for foundation wall. | .024 | .27 | .91 | 1.18 |
| 2 | Foundations | Continuous concrete footing 8" deep x 18" wide; cast-in-place concrete wall, 12" thick, 4' deep, 3000 psi; waterproofing; draintile; vapor barrier; 4" concrete slab on 4" crushed stone base, trowel finish. | .058 | 1.19 | 1.22 | 2.41 |
| 3 | Framing | 2" x 6" wood studs, 16" O.C.; 5/8" plywood sheathing; 2" x 10" rafters 16" O.C. with 5/8" plywood sheathing, 6 in 12 pitch; 2" x 8" ceiling joists 16" O.C.; 2" x 12" floor joists 16" O.C. with bridging and 5/8" plywood subfloor; 5/8" plywood subfloor on 1" x 3" wood sleepers 16" O.C. | .193 | 3.90 | 4.09 | 7.99 |
| 4 | Exterior Walls | Face brick veneer; #15 felt building paper; 6" batt insulation; 26 wood double hung windows; 1 bow/bay window; 3 solid core wood exterior doors; storms and screens. | .247 | 6.39 | 4.48 | 10.87 |
| 5 | Roofing | Wood cedar shingles; #15 felt building paper; copper flashing; 9" attic insulation. Aluminum gutters and downspouts. | .049 | 1.56 | 1.07 | 2.63 |
| 6 | Interiors | 5/8" drywall, thin coat plaster, painted with primer and 2 coats; hardwood baseboard and trim, sanded and finished; finished hardwood floor 70%, ceramic tile with underlayment 20%, vinyl tile with underlayment 10%, 33 wood panel interior doors. | .252 | 5.96 | 5.31 | 11.27 |
| 7 | Specialties | Kitchen cabinets - 25 L.F. wall and base cabinets with laminated plastic counter top; 6 L.F. bathroom vanity; medicine cabinet; stairs, appliances. | .057 | 4.02 | 1.03 | 5.05 |
| 8 | Mechanical | 1 kitchen sink, cast iron, double; 1 water heater, gas fired, 75 gal.; gas forced air heat/air conditioning; 1 full bath including: 1 bathtub, color; 1 corner shower; 1 sink, color, built in; 1 water closet, color. 1 lavatory including: 1 sink, color, built in; 1 water closet, color. | .071 | 3.38 | 1.71 | 5.09 |
| 9 | Electrical | 200 Amp. service; romex wiring; fluorescent and incandescent lighting fixtures; intercom, switches, receptacles. | .042 | .46 | .91 | 1.37 |
| 10 | Overhead | Permit and plans. | | 4.64 | | 4.64 |
| | **Total** | | | 31.77 | 20.73 | 52.50 |

230

# Luxury Bi-level

| | | Living Area | 2800 S.F. |
|---|---|---|---|
| | | Ground Area | 1400 S.F. |
| | | Perimeter | 152 L.F. |
| | | Partitions | 240 L.F. |

| | | | MAN-HOURS | COST PER SQUARE FOOT OF LIVING AREA |||
|---|---|---|---|---|---|---|
| | | | | MAT. | LABOR | TOTAL |
| 1 | Site Work | Site preparation for slab; excavation for lower level 4' deep. | .024 | .27 | .91 | 1.18 |
| 2 | Foundations | Continuous concrete footing 8" deep x 18" wide; cast-in-place concrete wall, 12" thick, 4' deep, 3000 psi; waterproofing; draintile; vapor barrier; 4" concrete slab on 4" crushed stone base, trowel finish. | .058 | 1.18 | 1.22 | 2.40 |
| 3 | Framing | 2" x 6" wood studs, 16" O.C.; 5/8" plywood sheathing; 2" x 10" rafters 16" O.C. with 5/8" plywood sheathing, 6 in 12 pitch; 2" x 8" ceiling joists 16" O.C.; 2" x 12" floor joists 16" O.C. with bridging and 5/8" plywood subfloor; 5/8" plywood subfloor on 1" x 3" wood sleepers 16" O.C. | .232 | 3.40 | 3.42 | 6.82 |
| 4 | Exterior Walls | Face brick veneer, #15 felt building paper; 6" batt insulation; wood double hung windows; 1 bow/bay window; 3 solid core wood exterior doors; storms and screens. | .185 | 5.56 | 3.36 | 8.92 |
| 5 | Roofing | Wood cedar shingles; #15 felt building paper; copper flashing; 9" attic insulation. Aluminum gutters and downspouts. | .042 | 1.43 | .91 | 2.34 |
| 6 | Interiors | 5/8" drywall, thin coat plaster, painted with primer and 2 coats; hardwood baseboard and trim, sanded and finished; finished hardwood floor 70%, ceramic tile with underlayment 20%; vinyl tile with underlayment 10%; 34 wood panel interior doors. | .238 | 5.73 | 4.95 | 10.68 |
| 7 | Specialties | Kitchen cabinets - 25 L.F. wall and base cabinets with laminated plastic counter top; 6 L.F. bathroom vanity; medicine cabinet; stairs. | .056 | 2.50 | .89 | 3.39 |
| 8 | Mechanical | 1 kitchen sink, cast iron, double; 1 water heater, gas fired, 75 gal.; gas forced air heat/air conditioning; 1 full bath including: 1 bathtub, color; 1 corner shower; 1 sink, color, built in; 1 water closet, color. 1 lavatory including: 1 sink, color, built in; 1 water closet, color. | .071 | 1.57 | .97 | 2.54 |
| 9 | Electrical | 200 Amp. service; romex wiring; fluorescent and incandescent lighting fixtures; intercom, switches, receptacles. | .042 | .45 | .89 | 1.34 |
| 10 | Overhead | Permit and plans. | | 4.64 | | 4.64 |
| | | **Total** | | 26.73 | 17.52 | 44.25 |

# Luxury Tri level

| | | | Living Area | 3500 S.F. |
|---|---|---|---|---|
| | | | Ground Area | 2400 S.F. |
| | | | Perimeter | 220 L.F. |
| | | | Partitions | 320 L.F. |

| | | | MAN-HOURS | COST PER SQUARE FOOT OF LIVING AREA |||
|---|---|---|---|---|---|---|
| | | | | MAT. | LABOR | TOTAL |
| 1 | Site Work | Site preparation for slab; excavation for lower level 4' deep; trench 4' deep for foundation wall. | .021 | .22 | .77 | .99 |
| 2 | Foundations | Continuous concrete footing 8" deep x 18" wide; cast-in-place concrete wall, 12" thick, 4' deep, 3000 psi; waterproofing; draintile; vapor barrier; 4" concrete slab on 4" crushed stone base, trowel finish. | .109 | 2.26 | 2.37 | 4.63 |
| 3 | Framing | 2" x 6" wood studs, 16" O.C.; 5/8" plywood sheathing; 2" x 10" rafters 16" O.C. with 5/8" plywood sheathing, 6 in 12 pitch; 2" x 8" ceiling joists 16" O.C.; 2" x 12" floor joists 16" O.C. with bridging and 5/8" plywood subfloor; 5/8" plywood subfloor on 1" x 3" wood sleepers 16" O.C. | .204 | 4.15 | 4.34 | 8.49 |
| 4 | Exterior Walls | Face brick veneer; #15 felt building paper; 6" batt insulation; 26 wood double hung windows; 1 bow/bay window; 3 solid core wood exterior doors; storms and screens. | .181 | 4.49 | 3.31 | 7.80 |
| 5 | Roofing | Wood cedar shingles; #15 felt building paper; copper flashing; 9" attic insulation. Aluminum gutters and downspouts. | .056 | 1.95 | 1.25 | 3.20 |
| 6 | Interiors | 5/8" drywall, thin coat plaster, painted with primer and 2 coats; hardwood baseboard and trim, sanded and finished; finished hardwood floor 70%, ceramic tile with underlayment 20%, vinyl tile with underlayment 10%; 29 wood panel interior doors. | .217 | 4.96 | 4.47 | 9.43 |
| 7 | Specialties | Kitchen cabinets - 25 L.F. wall and base cabinets with laminated plastic counter top; 6 L.F. bathroom vanity; medicine cabinet; stairs, appliances.. | .048 | 2.04 | .72 | 2.76 |
| 8 | Mechanical | 1 kitchen sink, cast iron, double; 1 water heater, gas fired, 75 gal.; gas forced air heat/air conditioning; 1 full bath including: 1 bathtub, color; 1 corner shower; 1 sink, color, built in; 1 water closet, color. 1 lavatory including: 1 sink, color, built in; 1 water closet, color. | .057 | 3.94 | 1.81 | 5.75 |
| 9 | Electrical | 200 Amp. service; romex wiring; fluorescent and incandescent lighting fixtures; intercom, switches, receptacles. | .039 | .41 | .82 | 1.23 |
| 10 | Overhead | Permit and plans. | | 3.72 | | 3.72 |
| | | **Total** | | 28.14 | 19.86 | 48.00 |

# LIGHT COMMERCIAL BUILDING TYPES

### Row House / Townhouse

This type of multi-unit housing usually has a number of attached units made up of inner units and end units. The units are joined by common walls. The inner units have only two exterior walls. The common walls are fireproof. End units have three walls and a common wall. Row House and Townhouse construction generally exits in the four classes listed here: Economy, Average, Custom and Luxury. It is not unusual for condominium developments to be constructed along lines similar to this type of construction, however the builder may incur an additional expense due to more elaborate common wall construction.

### Retail Store

This type of light commercial construction usually has ceiling heights that range from 10' to 12'. Only single-story construction is listed here. Costs are listed separately for four different basic floor areas. Elaborate or decorative exterior wall construction systems are not taken into consideration. The cost of signs and decorative work must be added to the basic costs shown here. Most other elements in this type of construction are standard unless the specific use requires special interior partitions or additional mechanical and electrical work.

### Office Building

This type of light commercial construction usually has a ceiling height of 10'. Only single-story construction is listed in these tables. Costs are listed separately for four different basic floor areas. All costs should be adjusted where necessary for design alternatives and owner's requirements. For example, modern office buildings sometimes require little interior partition work, yet this cost shows up later when more expensive office equipment is required for work stations.

### Warehouse

This type of light commercial construction usually has a ceiling height of 18'. Costs are listed separately for four different basic floor areas. All costs should be adjusted where necessary for design alternatives and owner's requirements.

### Apartment Building

This type of multi-unit housing usually has a number of apartments with common walls. These common walls are not fireproof. It is not unusual for condominium developments to be constructed along lines similar to this type of construction, however the builder may incur an additional expense due to more elaborate common wall construction. There are two basic floor area types described here: 5,000' and 10,000'. It is assumed in both cases that this construction is 2½ story with a ceiling height of 8'.

### Motel

Costs were calculated for a one-story building with 5,000 and 8,000 square feet of floor area; and for a two-story building with 8,000 and 10,000 square feet of floor area. Common walls are not fireproof. It is assumed in both cases that this construction has a ceiling height of 8'. All costs should be adjusted where necessary for design alternatives and owner's requirements.

# Apartment

| | | |
|---|---|---|
| Living Area | 5220 S.F. | |
| Ground Area | 1740 S.F. | |
| Perimeter | 180 L.F. | |
| Partitions | 768 L.F. | |

| | | Man-Hours | COST PER SQUARE FOOT OF LIVING AREA | | |
|---|---|---|---|---|---|
| | | | MAT. | LABOR | TOTAL |
| 1 | **Site Work** — Site preparation for slab; excavation for lower level 4' deep; trench 4' deep for foundation wall; utility trench. | .012 | .08 | .39 | .47 |
| 2 | **Foundations** — Continuous concrete footing 12" deep x 24" wide; 12" thick concrete block foundation wall, 4' deep; 4" concrete slab on 4" gravel base, trowel finish. | .048 | 1.03 | .96 | 1.99 |
| 3 | **Framing** — 2" x 4" wood studs, 16" O.C.; 1/2" plywood sheathing; 2" x 4" wood truss roof 24" O.C. with 1/2" plywood sheathing; 2" x 10" floor joists 16" O.C. with bridging, 1" x 3" furring and 5/8" plywood subfloor. | .108 | 2.55 | 2.53 | 5.08 |
| 4 | **Exterior Walls** — Brick veneer; 27 aluminum sliding windows; 1 metal clad wood door; 3-1/2" batt insulation. | .165 | 2.08 | 3.11 | 5.19 |
| 5 | **Roofing** — 240# asphalt shingles; #15 felt building paper; aluminum flashing; 6" attic insulation. | .016 | .36 | .36 | .72 |
| 6 | **Interiors** — 5/8" drywall, taped and finished, painted with primer and 2 coats; rubber backed carpet 64%, asphalt tile 36%; 18 hollow core wood interior doors, 6 metal, fire rated entry doors. | .183 | 3.70 | 4.10 | 7.80 |
| 7 | **Specialties** — Kitchen cabinets - 72 L.F. wall and base cabinets with laminated plastic counter top; stairs; appliances; bathroom accessories. | .060 | 3.90 | .93 | 4.83 |
| 8 | **Mechanical** — 6 lavatories, white, built-in; 6 water closets, white; 6 bathtubs, cast iron, white; 6 kitchen sinks, stainless steel, single; 6 water heaters, gas fired, 30 gal.; gas fired forced air heat/air conditioning. | .083 | 5.27 | 2.06 | 7.33 |
| 9 | **Electrical** — 200 Amp. service; romex wiring; fluorescent and incandescent lighting; switches, receptacles. | .032 | .43 | .69 | 1.12 |
| 10 | **Overhead** — Permit and plans. | | 4.16 | | 4.16 |
| | **Total** | | 23.56 | 15.13 | 38.69 |

# Apartment

| | | |
|---|---|---|
| Living Area | 10,416 S.F. | |
| Ground Area | 3,472 S.F. | |
| Perimeter | 304 L.F. | |
| Partitions | 1,484 L.F. | |

| | | MAN-HOURS | COST PER SQUARE FOOT OF LIVING AREA |||
|---|---|---|---|---|---|
| | | | MAT. | LABOR | TOTAL |
| 1 | **Site Work** — Site preparation for slab; excavation for lower level 4' deep; trench 4' deep for foundation wall; utility trench. | .008 | .04 | .31 | .35 |
| 2 | **Foundations** — Continuous concrete footing 12" deep x 24" wide; 12" thick concrete block foundation wall, 4' deep; 4" thick concrete slab on 4" gravel base, trowel finish. | .142 | .92 | .84 | 1.76 |
| 3 | **Framing** — 2" x 4" wood studs, 16" O.C.; 1/2" plywood sheathing; 2" x 4" wood truss roof 24" O.C. with 1/2" plywood sheathing; 2" x 10" floor joists, 16" O.C., with bridging, 1" x 3" furring and 5/8" plywood subfloor. | .114 | 2.63 | 2.67 | 5.30 |
| 4 | **Exterior Walls** — Brick veneer; 54 aluminum sliding windows; 2 metal clad wood doors; 3-1/2" batt insulation. | .140 | 1.88 | 2.66 | 4.54 |
| 5 | **Roofing** — 240# asphalt shingles; #15 felt building paper; aluminum flashing; 6" attic insulation. | .016 | .36 | .36 | .72 |
| 6 | **Interiors** — 5/8" drywall, taped and finished, painted with primer and 2 coats; rubber backed carpet 64%, asphalt tile 36%; 36 hollow core wood interior doors; 12 metal, fire rated entry doors. | .176 | 3.61 | 3.97 | 7.58 |
| 7 | **Specialties** — Kitchen cabinets - 144 L.F. wall and base cabinets with laminated plastic counter tops; stairs; appliances; bathroom accessories. | .059 | 3.91 | .93 | 4.84 |
| 8 | **Mechanical** — 12 lavatories, white, built-in; 12 water closets white; 12 bathtubs, cast iron, white; 12 kitchen sinks, stainless steel, single; 12 water heaters, gas fired, 30 gal.; gas fired forced air heat/air conditioning. | .083 | 5.27 | 2.06 | 7.33 |
| 9 | **Electrical** — 400 Amp. service; romex wiring; fluorescent and incandescent lighting fixtures, switches, receptacles. | .032 | .42 | .68 | 1.10 |
| 10 | **Overhead** — Permit and plans. | | 3.11 | | 3.11 |
| | **Total** | | 22.15 | 13.48 | 36.63 |

235

# Motel 1 Story

| | | |
|---|---|---|
| Living Area | 5000 S.F. |
| Ground Area | 5000 S.F. |
| Perimeter | 468 L.F. |
| Partitions | 720 L.F. |

| # | Category | Description | MAN-HOURS | MAT. | LABOR | TOTAL |
|---|---|---|---|---|---|---|
| 1 | Site Work | Site preparation for slab; trench 4' deep for foundation wall; utility trench. | .055 | .08 | 2.05 | 2.13 |
| 2 | Foundations | Continuous concrete footing 8" deep x 18" wide; 12" thick concrete block foundation wall, 4' deep; 4" thick concrete slab on 4" gravel base, trowel finish. | .127 | 2.73 | 2.55 | 5.28 |
| 3 | Framing | 2" x 4" wood studs, 24" O.C.; 2" x 4" wood truss roof, 24" O.C., with 1/2" plywood sheathing. | .069 | 1.79 | 1.69 | 3.48 |
| 4 | Exterior Walls | Concrete block, regular, 8" thick; wire truss strips; poured insulation; 16 aluminum awning windows; 16 metal clad, wood doors and frames; bronze finish window wall system; aluminum entrance door with transom, 3' x 10'. | .188 | 4.03 | 4.34 | 8.37 |
| 5 | Roofing | 240# asphalt shingles; #15 felt building paper; aluminum flashing; 6" attic insulation. | .047 | 1.09 | 1.08 | 2.17 |
| 6 | Interiors | 1/2" drywall, taped and finished, painted with primer and 2 coats; 16 hollow core wood interior doors; 16 wood bi-fold closet doors; 81% carpet; 19% asphalt tile. | .189 | 3.17 | 3.98 | 7.15 |
| 7 | Specialties | Bathroom accessories. | .002 | .27 | .04 | .31 |
| 8 | Mechanical | 16 lavatories, white, built-in; 1 lavatory, white, wall hung; 17 water closets, white; 16 bathtubs, cast iron, white; 16 water heaters, gas fired, 30 gal.; 17 package, thru-wall heater/air conditioner. | .122 | 5.96 | 2.34 | 8.30 |
| 9 | Electrical | 200 Amp. service; EMT conduit; fluorescent lighting fixtures, switches, receptacles. | .037 | .59 | .77 | 1.36 |
| 10 | Overhead | Permit and plans. | | 3.30 | | 3.30 |
| | | **Total** | | 23.01 | 18.84 | 41.85 |

# Motel 1 Story

| Living Area | 8000 S.F. |
|---|---|
| Ground Area | 8000 S.F. |
| Perimeter | 712 L.F. |
| Partitions | 1040 L.F. |

| | | | MAN-HOURS | MAT. | LABOR | TOTAL |
|---|---|---|---|---|---|---|
| 1 | Site Work | Site preparation for slab; trench 4' deep for foundation wall; utility trench. | .050 | .05 | 1.94 | 1.99 |
| 2 | Foundations | Continuous concrete footing 8" deep x 18" wide; 12" thick concrete block foundation wall, 4' deep; 4" thick concrete slab on 4" gravel base, trowel finish. | .123 | 2.64 | 2.46 | 5.10 |
| 3 | Framing | 2" x 4" wood studs, 24" O.C.; 2" x 4" wood truss roof, 24" O.C., with 1/2" plywood sheathing. | .068 | 1.75 | 1.65 | 3.40 |
| 4 | Exterior Walls | Concrete block, regular, 8" thick; wire truss strips; poured insulation; 26 aluminum awning windows; 26 metal clad, wood doors and frames; bronze finish window wall system; aluminum entrance door with transom, 3' x 10'. | .179 | 3.82 | 4.14 | 7.96 |
| 5 | Roofing | 240# asphalt shingles; #15 felt building paper; aluminum flashing; 6" attic insulation. | .047 | 1.09 | 1.08 | 2.17 |
| 6 | Interiors | 1/2" drywall, taped and finished, painted with primer and 2 coats; 27 hollow core wood interior doors; 26 wood bi-fold closet doors; 80% carpet; 20% asphalt tile. | .179 | 3.10 | 3.82 | 6.92 |
| 7 | Specialties | Bathroom accessories. | .002 | .27 | .04 | .31 |
| 8 | Mechanical | 26 lavatories, white, built-in; 1 lavatory, white, wall hung; 27 water closets, white; 26 bathtubs, cast iron, white; 26 water heaters, gas fired, 30 gal.; 27 package, thru-wall heater/air conditioner. | .122 | 5.98 | 2.35 | 8.33 |
| 9 | Electrical | 400 Amp. service; EMT conduit; fluorescent lighting fixtures, switches, receptacles. | .038 | .77 | .80 | 1.57 |
| 10 | Overhead | Permit and plans. | | 3.25 | | 3.25 |
| | | **Total** | | 22.72 | 18.28 | 41.00 |

# Motel 2 Story

| | | |
|---|---|---|
| Living Area | | 8000 S.F. |
| Ground Area | | 4000 S.F. |
| Perimeter | | 384 L.F. |
| Partitions | | 1184 L.F. |

| # | Category | Description | Man-Hours | Mat. | Labor | Total |
|---|---|---|---|---|---|---|
| 1 | Site Work | Site preparation for slab; trench 4' deep for foundation wall; utility trench. | .027 | .05 | 1.02 | 1.07 |
| 2 | Foundations | Continuous concrete footing 8" deep x 18" wide; 12" thick concrete block foundation wall, 4' deep; 4" thick concrete slab on 4" gravel base, trowel finish. | .065 | 1.39 | 1.30 | 2.69 |
| 3 | Framing | 2" x 4" wood studs, 16" O.C.; 2" x 8" roof joists, 16" O.C.; 1/2" exterior plywood deck; 2" x 10" floor joists, 16" O.C.; 3/4" plywood sheathing. | .069 | 2.44 | 2.25 | 4.69 |
| 4 | Exterior Walls | Concrete block, regular, 8" thick; wire truss strips; poured insulation; 27 aluminum awning windows; 27 metal clad, wood doors and frames; bronze finish window wall system; aluminum entrance door with transom, 3' x 10'. | .190 | 3.93 | 4.35 | 8.28 |
| 5 | Roofing | Asphalt built-up roof over 2-1/8" glass fiber insulation; aluminum flashing. | .027 | .66 | .31 | .97 |
| 6 | Interiors | 1/2" drywall, taped and finished, painted with primer and 2 coats; texture spray on concrete deck ceiling; 28 hollow core wood interior doors; 27 wood bi-fold closet doors; 80% carpet; 20% asphalt tile; balcony railings. | .200 | 3.64 | 3.99 | 7.63 |
| 7 | Specialties | Bathroom accessories. | .001 | .28 | .04 | .32 |
| 8 | Mechanical | 27 lavatories, white, built-in; 1 lavatory, white, wall hung; 28 water closets, white; 27 bathtubs, cast iron, white; 27 water heaters, gas fired, 30 gal.; 28 package, thru-wall heater/air conditioner. | .131 | 5.91 | 2.55 | 8.46 |
| 9 | Electrical | 400 Amp. service; EMT conduit; fluorescent lighting fixtures, switches, receptacles. | .039 | .79 | .83 | 1.62 |
| 10 | Overhead | Permit and plans. | | 3.25 | | 3.25 |
| | | **Total** | | 22.34 | 16.64 | 38.98 |

# Motel 2 Story

| | | |
|---|---|---|
| Living Area | 10,368 S.F. |
| Ground Area | 5,000 S.F. |
| Perimeter | 480 L.F. |
| Partitions | 1,528 L.F. |

| # | Category | Description | Man-Hours | Mat. | Labor | Total |
|---|---|---|---|---|---|---|
| 1 | Site Work | Site preparation for slab; trench 4' deep for foundation wall; utility trench. | .026 | .04 | .99 | 1.03 |
| 2 | Foundations | Continuous concrete footing 8" deep x 18" wide; 12" thick concrete block foundation wall, 4' deep; 4" thick concrete slab on 4" gravel base, trowel finish. | .063 | .99 | .91 | 1.90 |
| 3 | Framing | 2" x 4" wood studs, 16" O.C.; 2" x 8" roof joists, 16" O.C.; 1/2" exterior plywood deck; 2" x 10" floor joists, 16" O.C.; 3/4" plywood sheathing. | .075 | 2.55 | 2.39 | 4.94 |
| 4 | Exterior Walls | Concrete block, regular, 8" thick; wire truss strips; poured insulation; 27 aluminum awning windows; 36 metal clad, wood doors and frames; bronze finish window wall system; aluminum entrance door with transom, 3' x 10'. | .180 | 3.75 | 4.18 | 7.93 |
| 5 | Roofing | Asphalt built-up roof over 2-1/8" glass fiber insulation; aluminum flashing. | .027 | .79 | .58 | 1.37 |
| 6 | Interiors | 1/2" drywall, taped and finished, painted with primer and 2 coats; texture spray on concrete deck ceiling; 36 hollow core wood interior doors; 35 wood bi-fold closet doors; 80% carpet; 20% asphalt tile; balcony railings. | .199 | 4.26 | 4.28 | 8.54 |
| 7 | Specialties | Bathroom accessories. | .002 | .28 | .04 | .32 |
| 8 | Mechanical | 35 lavatories, white, built-in; 1 lavatory, white, wall hung; 36 water closets, white; 35 bathtubs, cast iron, white; 35 water heaters, gas fired, 30 gal.; 36 package, thru-wall heater/air conditioner. | .125 | 6.15 | 2.42 | 8.57 |
| 9 | Electrical | 400 Amp. service; EMT conduit; fluorescent lighting fixtures, switches, receptacles. | .037 | .77 | .81 | 1.58 |
| 10 | Overhead | Permit and plans. | | 3.14 | | 3.14 |
| | **Total** | | | 22.72 | 16.60 | 39.32 |

# Office

| | | |
|---|---|---|
| Occupied Area | 2000 S.F. | |
| Ground Area | 2000 S.F. | |
| Perimeter | 180 L.F. | |
| Partitions | 104 L.F. | |

| | | MAN-HOURS | COST PER SQUARE FOOT OF LIVING AREA |||
|---|---|---|---|---|---|
| | | | MAT. | LABOR | TOTAL |
| **1** Site Work | Site preparation for slab; trench 4' deep for foundation wall; utility trench. | .054 | .22 | 1.81 | 2.03 |
| **2** Foundations | Continuous concrete footing 8" deep x 18" wide; 12" thick concrete block foundation wall, 4' deep; 4" thick concrete slab on 4" gravel base, trowel finish. | .123 | 2.65 | 2.48 | 5.13 |
| **3** Framing | Open web steel joists, 24" deep, 12.7#/L.F., 5' O.C., 40' span, 30 PSF superimposed load; 1-1/2" deep, 22 gauge, open type metal deck; 3-5/8" wide 20 ga., non-load bearing metal studs, 24" O.C. | .022 | 1.76 | .61 | 2.37 |
| **4** Exterior Walls | Concrete block, split rib profile, 12" thick; wire truss strips; polystyrene block insert insulation; bronze finish storefront system; aluminum entrance door with transom, 3' x 10'; 1 steel, hollow metal door and frame. | .220 | 5.82 | 4.06 | 9.88 |
| **5** Roofing | Asphalt built-up roof over 2-1/8" glass fiber insulation; aluminum flashing. | .047 | 1.35 | 1.00 | 2.35 |
| **6** Interiors | 5/8" drywall, taped and finished, painted with primer and 2 coats; softwood baseboard, primed and painted; 2' x 4' suspended acoustical ceiling system; 6 hollow core wood interior doors; 85% vinyl tile, 15% concrete floor. | .149 | 5.21 | 2.84 | 8.05 |
| **7** Specialties | Kitchen cabinets - 10 L.F. wall and base with laminated plastic counter top; toilet room accessories. | .004 | .60 | .12 | .72 |
| **8** Mechanical | 2 lavatories, white, wall hung; 2 water closets white; 1 water heater, gas fired, 30 gal.; gas fired, forced air heat/air conditioning; 1 kitchen sink, single, stainless steel. | .063 | 2.90 | 1.13 | 4.03 |
| **9** Electrical | 200 Amp. service; EMT conduit; fluorescent lighting fixtures, switches, receptacles. | .074 | 1.50 | 1.61 | 3.11 |
| **10** Overhead | Permit and plans. | | 3.30 | | 3.30 |
| **Total** | | | 25.31 | 15.66 | 40.97 |

# Office

| | | |
|---|---|---|
| Occupied Area | 3000 S.F. | |
| Ground Area | 3000 S.F. | |
| Perimeter | 220 L.F. | |
| Partitions | 128 L.F. | |

| # | Category | Description | Man-Hours | Mat. | Labor | Total |
|---|---|---|---|---|---|---|
| 1 | **Site Work** | Site preparation for slab; trench 4' deep for foundation wall; utility trench. | .035 | .14 | 1.20 | 1.34 |
| 2 | **Foundations** | Continuous concrete footing 8" deep x 18" wide; 12" thick concrete block foundation wall, 4' deep; 4" thick concrete slab on 4" gravel base, trowel finish; 1 column footing. | .106 | 2.33 | 2.15 | 4.48 |
| 3 | **Framing** | Open web steel joists, 16" deep, 7.8#/L.F., 5' O.C., 25' span, 30 PSF superimposed load; 1-1/2" deep, 22 gauge, open type metal deck; 10" deep wide flange beam; one 4" x 4" steel tube column. | .022 | 1.65 | .66 | 2.31 |
| 4 | **Exterior Walls** | Concrete block, split rib profile, 12" thick; wire truss strips; polystyrene block insert insulation; bronze finish storefront system; aluminum entrance door with transom, 3' x 10'; 1 steel, hollow metal door and frame. | .174 | 4.51 | 3.21 | 7.72 |
| 5 | **Roofing** | Asphalt built-up roof over 2-1/8" glass fiber insulation; aluminum flashing. | .047 | 1.35 | 1.00 | 2.35 |
| 6 | **Interiors** | 5/8" drywall, taped and finished, painted with primer and 2 coats; softwood baseboard, primed and painted; 2' x 4' suspended acoustical ceiling system; 7 hollow core wood interior doors; 90% vinyl tile, 10% concrete floor. | .133 | 4.90 | 2.57 | 7.47 |
| 7 | **Specialties** | Kitchen cabinets - 10 L.F. wall and base with laminated plastic counter top; toilet room accessories. | .005 | .40 | .08 | .48 |
| 8 | **Mechanical** | 2 lavatories, white, wall hung; 2 water closets white; 1 water heater, gas fired, 30 gal.; gas fired, forced air heat/air conditioning; 1 kitchen sink, single, stainless steel. | .056 | 3.52 | 1.13 | 4.65 |
| 9 | **Electrical** | 200 Amp. service; EMT conduit; fluorescent lighting fixtures, switches, receptacles. | .069 | 1.44 | 1.52 | 2.96 |
| 10 | **Overhead** | Permit and plans. | | 3.27 | | 3.27 |
| | **Total** | | | 23.51 | 13.52 | 37.03 |

# Office

| | | Occupied Area | 5000 S.F. |
|---|---|---|---|
| | | Ground Area | 5000 S.F. |
| | | Perimeter | 300 L.F. |
| | | Partitions | 202 L.F. |

| | | | MAN-HOURS | COST PER SQUARE FOOT OF LIVING AREA ||| 
|---|---|---|---|---|---|---|
| | | | | MAT. | LABOR | TOTAL |
| 1 | Site Work | Site preparation for slab; trench 4' deep for foundation wall; utility trench. | .028 | .08 | 1.00 | 1.08 |
| 2 | Foundations | Continuous concrete footing 8" deep x 18" wide; 12" thick concrete block foundation wall, 4' deep; 6" concrete slab on 4" gravel base, trowel finish; 3 column footings. | .096 | 2.43 | 1.97 | 4.40 |
| 3 | Framing | Open web steel joists, 16" deep, 7.8#/L.F., 5' O.C., 25' span, 30 PSF superimposed load; 1-1/2" deep wide flange beam; 3-4"x 4" steel tube columns | .022 | 1.79 | .67 | 2.46 |
| 4 | Exterior Walls | Concrete block, split rib profile, 12" thick; wire truss strips; polystyrene block insert insulation; bronze finish storefront system; aluminum entrance door with transom, 3' x 10'; 1 steel, hollow metal door and frame. | .152 | 4.01 | 2.81 | 6.82 |
| 5 | Roofing | Asphalt built-up roof over 2-1/8" glass fiber insulation; aluminum flashing. | .047 | 1.35 | 1.00 | 2.35 |
| 6 | Interiors | 5/8" drywall, taped and finished, painted with primer and 2 coats; softwood baseboard, primed and painted; 2' x 4' suspended acoustical ceiling system; 10 hollow core wood interior doors; 90% vinyl tile, 10% concrete floor. | .135 | 4.87 | 2.62 | 7.49 |
| 7 | Specialties | Kitchen cabinets - 10 L.F. wall and base with laminated plastic counter top; toilet room accessories. | .005 | .24 | .05 | .29 |
| 8 | Mechanical | 2 lavatories, white, wall hung; 2 water closets white; 1 water heater, gas fired, 30 gal.; gas fired, forced air heat/air conditioning; 1 kitchen sink, single, stainless steel. | .051 | 2.61 | .84 | 3.45 |
| 9 | Electrical | 200 Amp. service; EMT conduit; fluorescent lighting fixtures, switches, receptacles. | .066 | 1.37 | 1.44 | 2.81 |
| 10 | Overhead | Permit and plans. | | 3.26 | | 3.26 |
| | **Total** | | | 22.01 | 12.40 | 34.41 |

# Office

| | | |
|---|---|---|
| Occupied Area | | 10,000 S.F. |
| Ground Area | | 10,000 S.F. |
| Perimeter | | 410 L.F. |
| Partitions | | 425 L.F. |

| | | Description | MAN-HOURS | COST PER SQUARE FOOT OF LIVING AREA | | |
|---|---|---|---|---|---|---|
| | | | | MAT. | LABOR | TOTAL |
| 1 | Site Work | Site preparation for slab; trench 4' deep for foundation wall; utility trench. | .018 | .04 | .64 | .68 |
| 2 | Foundations | Continuous concrete footing 8" deep x 18" wide; 12" thick concrete block foundation wall, 4' deep; 6" concrete slab on 4 column footings. | .076 | 2.04 | 1.60 | 3.64 |
| 3 | Framing | Open web steel joists, 24" deep, 12.7#/L.F., 5' O.C., 40' span, 30 PSF superimposed load; 1-1/2" deep, 22 gauge, open type metal deck; 12" deep wide flange beam; 4- 4" x 4" steel tube columns. | .022 | 1.95 | .63 | 2.58 |
| 4 | Exterior Walls | Concrete block, split rib profile, 12" thick; wire truss strips; steel furring channels 16" O.C.; polystyrene block insert insulation; bronze finish storefront system; aluminum entrance door with transom, 3' x 10'; 3 steel, hollow metal doors and frames. | .127 | 3.40 | 2.42 | 5.82 |
| 5 | Roofing | Asphalt built-up roof over 2-1/8" glass fiber insulation; aluminum flashing. | .047 | 1.35 | 1.00 | 2.35 |
| 6 | Interiors | 5/8" drywall, taped and finished, painted with primer and 2 coats; softwood baseboard, primed and painted; 2' x 4' suspended acoustical ceiling system; 14 hollow core wood interior doors; 84% carpet, 10% vinyl tile, 6% ceramic tile. | .117 | 4.47 | 2.25 | 6.72 |
| 7 | Specialties | Kitchen cabinets - 10 L.F. wall and base with laminated plastic counter top; toilet room accessories. | .003 | .14 | .03 | .17 |
| 8 | Mechanical | 4 lavatories, white, wall hung; 3 water closets, white; 1 water heater, gas fired, 30 gal.; gas fired, forced air heat/air conditioning; 1 kitchen sink, single, stainless steel. | .045 | 2.70 | .67 | 3.37 |
| 9 | Electrical | 600 Amp. service; EMT conduit; fluorescent lighting fixtures, switches, receptacles. | .068 | 1.43 | 1.45 | 2.88 |
| 10 | Overhead | Permit and plans. | | 3.24 | | 3.24 |
| | **Total** | | | 20.76 | 10.69 | 31.45 |

243

# Retail Store

| | | Occupied Area | 2000 S.F. |
|---|---|---|---|
| | | Ground Area | 2000 S.F. |
| | | Perimeter | 180 L.F. |
| | | Partitions | 64 L.F. |

| | | | MAN-HOURS | COST PER SQUARE FOOT OF LIVING AREA | | |
|---|---|---|---|---|---|---|
| | | | | MAT. | LABOR | TOTAL |
| **1** | **Site Work** | Site preparation for slab; trench 4' deep for foundation wall; utility trench. | .037 | .22 | 1.18 | 1.40 |
| **2** | **Foundations** | Continuous concrete footing 8" deep x 18" wide; 12" thick concrete block foundation wall, 4' deep; 4" thick concrete slab on 4" gravel base, trowel finish. | .123 | 2.65 | 2.48 | 5.13 |
| **3** | **Framing** | Open web steel joists, 24" deep, 12.7#/L.F., 5' O.C., 40' span, 30 PSF superimposed load; 1-1/2" deep, 22 gauge, open type metal deck; 3-5/8" wide 20 ga., non-load bearing metal studs, 24" O.C. | .019 | 1.78 | .57 | 2.35 |
| **4** | **Exterior Walls** | Concrete block, split rib profile, 12" thick; wire truss strips; polystyrene block insert insulation; bronze finish storefront system; aluminum entrance door with transom, 3' x 10'; 1 steel, hollow metal door and frame. | .250 | 6.95 | 4.68 | 11.63 |
| **5** | **Roofing** | Asphalt built-up roof over 2-1/8" glass fiber insulation; aluminum flashing. | .047 | 1.35 | 1.00 | 2.35 |
| **6** | **Interiors** | 5/8" drywall, taped and finished, painted with primer and 2 coats; softwood baseboard, primed and painted; 2' x 4' suspended acoustical ceiling system; 3 hollow core wood interior doors; 68% vinyl tile, 32% concrete floor. | .106 | 3.78 | 2.08 | 5.86 |
| **7** | **Specialties** | Toilet room accessories. | .005 | .08 | .03 | .11 |
| **8** | **Mechanical** | 2 lavatories, white, wall hung; 2 water closets white; 1 water heater, gas fired, 30 gal.; gas fired, forced air heat/air conditioning. | .054 | 2.80 | .98 | 3.78 |
| **9** | **Electrical** | 200 Amp. service; EMT conduit; fluorescent lighting fixtures, switches, receptacles. | .059 | 1.23 | 1.27 | 2.50 |
| **10** | **Overhead** | Permit and plans. | | 3.30 | | 3.30 |
| | | **Total** | | 24.14 | 14.27 | 38.41 |

# Retail Store

| | | Occupied Area | 3000 S.F. |
|---|---|---|---|
| | | Ground Area | 3000 S.F. |
| | | Perimeter | 220 L.F. |
| | | Partitions | 74 L.F. |

| # | Category | Description | MAN-HOURS | MAT. | LABOR | TOTAL |
|---|---|---|---|---|---|---|
| 1 | Site Work | Site preparation for slab; trench 4' deep for foundation wall; utility trench. | .035 | .14 | 1.20 | 1.34 |
| 2 | Foundations | Continuous concrete footing 8" deep x 18" wide; 12" thick concrete block foundation wall, 4' deep; 4" thick concrete slab on 4" gravel base, trowel finish; 1 column footing. | .106 | 2.33 | 2.15 | 4.48 |
| 3 | Framing | Open web steel joists, 16" deep, 7.8#/L.F., 5' O.C., 25' span, 30 PSF superimposed load; 1-1/2" deep, 22 gauge, open type metal deck; 10" deep wide flange beam; one 4" x 4" steel tube column. | .019 | 1.40 | .67 | 2.07 |
| 4 | Exterior Walls | Concrete block, split rib profile, 12" thick; wire truss strips; polystyrene block insert insulation; bronze finish storefront system; aluminum entrance door with transom, 3' x 10'; 1 steel, hollow metal door and frame. | .202 | 5.54 | 3.79 | 9.33 |
| 5 | Roofing | Asphalt built-up roof over 2-1/8" glass fiber insulation; aluminum flashing. | .047 | 1.35 | 1.00 | 2.35 |
| 6 | Interiors | 5/8" drywall, taped and finished, painted with primer and 2 coats; softwood baseboard, primed and painted; 2' x 4' suspended acoustical ceiling system; 3 hollow core wood interior doors; 88% vinyl tile, 12% concrete floor. | .119 | 4.08 | 2.37 | 6.45 |
| 7 | Specialties | Toilet room accessories. | .002 | .05 | .02 | .07 |
| 8 | Mechanical | 2 lavatories, white, wall hung; 2 water closets white; 1 water heater, gas fired, 30 gal.; gas fired, forced air heat/air conditioning. | .048 | 3.16 | .91 | 4.07 |
| 9 | Electrical | 200 Amp. service; EMT conduit; fluorescent lighting fixtures, switches, receptacles. | .054 | 1.15 | 1.16 | 2.31 |
| 10 | Overhead | Permit and plans. | | 3.27 | | 3.27 |
| | | **Total** | | 22.47 | 13.27 | 35.74 |

# Retail Store

| | | Occupied Area | 5000 S.F. |
|---|---|---|---|
| | | Ground Area | 5000 S.F. |
| | | Perimeter | 300 L.F. |
| | | Partitions | 78 L.F. |

| # | Section | Description | MAN-HOURS | MAT. | LABOR | TOTAL |
|---|---|---|---|---|---|---|
| 1 | Site Work | Site preparation for slab; trench 4' deep for foundation wall; utility trench. | .028 | .08 | 1.00 | 1.08 |
| 2 | Foundations | Continuous concrete footing 8" deep x 18" wide; 12" thick concrete block foundation wall, 4' deep; 6" concrete slab on 4" gravel base, trowel finish; 3 column footings. | .095 | 2.42 | 1.97 | 4.39 |
| 3 | Framing | Open web steel joists, 16" deep, 7.8#/L.F., 5' O.C., 25' span, 30 PSF superimposed load; 1-1/2" deep, 22 gauge, open type metal deck; 10" deep wide flange beam; 3 - 4" x 4" steel tube columns. | .017 | 1.58 | .58 | 2.16 |
| 4 | Exterior Walls | Concrete block, split rib profile, 12" thick; wire truss strips; polystyrene block insert insulation; bronze finish storefront system; aluminum entrance door with transom, 3' x 10'; 1 steel, hollow metal door and frame. | .157 | 4.11 | 2.90 | 7.01 |
| 5 | Roofing | Asphalt built-up roof over 2-1/8" glass fiber insulation; aluminum flashing. | .047 | 1.35 | 1.00 | 2.35 |
| 6 | Interiors | 5/8" drywall, taped and finished, painted with primer and 2 coats; softwood baseboard, primed and painted; 2' x 4' suspended acoustical ceiling system; 3 hollow core wood interior doors; 68% vinyl tile, 32% concrete floor. | .086 | 3.45 | 1.70 | 5.15 |
| 7 | Specialties | Toilet room accessories. | .001 | .03 | .02 | .05 |
| 8 | Mechanical | 2 lavatories, white, wall hung; 2 water closets white; 1 water heater, gas fired, 30 gal.; gas fired, forced air heat/air conditioning. | .045 | 2.56 | .78 | 3.34 |
| 9 | Electrical | 200 Amp. service; EMT conduit; fluorescent lighting fixtures, switches, receptacles. | .052 | 1.11 | 1.12 | 2.23 |
| 10 | Overhead | Permit and plans. | | 3.26 | | 3.26 |
| | **Total** | | | 19.95 | 11.07 | 31.02 |

# Retail Store

| | | |
|---|---|---|
| Occupied Area | | 10,000 S.F. |
| Ground Area | | 10,000 S.F. |
| Perimeter | | 410 L.F. |
| Partitions | | 110 L.F. |

| | | Description | MAN-HOURS | COST PER SQUARE FOOT OF LIVING AREA ||| 
|---|---|---|---|---|---|---|
| | | | | MAT. | LABOR | TOTAL |
| 1 | Site Work | Site preparation for slab; trench 4' deep for foundation wall; utility trench. | .018 | .04 | .64 | .68 |
| 2 | Foundations | Continuous concrete footing 8" deep x 18" wide; 12" thick concrete block foundation wall, 4' deep; 6" thick concrete slab on 4" gravel base, trowel finish; 4 column footings. | .076 | 2.17 | 1.74 | 3.91 |
| 3 | Framing | Open web steel joists, 24" deep, 12.7#/L.F., 5' O.C., 40' span, 30 PSF superimposed load; 1-1/2" deep, 22 gauge, open type metal deck; 12" deep wide flange beam; 4 - 4" x 4" steel tube columns. | .017 | 1.83 | .53 | 2.36 |
| 4 | Exterior Walls | Concrete block, split rib profile, 12" thick; wire truss strips; steel furring channels 16" O.C.; polystyrene block insert insulation; bronze finish storefront system; aluminum entrance door with transom, 3' x 10'; 3 steel, hollow metal doors and frame. | .121 | 3.04 | 2.27 | 5.31 |
| 5 | Roofing | Asphalt built-up roof over 2-1/8" glass fiber insulation; aluminum flashing. | .047 | 1.35 | 1.00 | 2.35 |
| 6 | Interiors | 5/8" drywall, taped and finished, painted with primer and 2 coats; softwood baseboard, primed and painted; 2' x 4' suspended acoustical ceiling system; 3 hollow core wood interior doors; 84% vinyl tile, 16% concrete floor. | .076 | 3.30 | 1.51 | 4.81 |
| 7 | Specialties | Toilet room accessories. | .001 | .01 | .01 | .02 |
| 8 | Mechanical | 2 lavatories, white, wall hung; 2 water closets white; 1 water heater, gas fired, 30 gal.; gas fired, forced air heat/air conditioning. | .033 | 2.49 | .63 | 3.12 |
| 9 | Electrical | 600 Amp. service; EMT conduit; fluorescent lighting fixtures, switches, receptacles. | .051 | 1.11 | 1.10 | 2.21 |
| 10 | Overhead | Permit and plans. | | 3.24 | | 3.24 |
| | **Total** | | | 18.58 | 9.43 | 28.01 |

247

# Economy Townhouse
## 10 Units/Cluster

| | Living Area | 1200 S.F. |
|---|---|---|
| | Ground Area | 600 S.F. |
| | Perimeter | 100 L.F. |
| | Partitions | 150 L.F. |

| # | Division | Description | Man-Hours | Mat. | Labor | Total |
|---|---|---|---|---|---|---|
| 1 | Site Work | Site preparation for slab; trench 4' deep for foundation wall; utility trench. | .052 | .46 | 1.57 | 2.03 |
| 2 | Foundations | Continuous concrete footing 8" deep x 18" wide; 8" thick concrete block foundation wall, 4' deep; 4" thick concrete slab on 4" gravel base, trowel finish. | .094 | 1.73 | 1.71 | 3.44 |
| 3 | Framing | 2" x 4" wood studs, 16" O.C.; 1/2" insulation board sheathing; 2" x 4" wood truss roof, 24" O.C., with 1/2" plywood sheathing. | .127 | 3.14 | 3.15 | 6.29 |
| 4 | Exterior Walls | Horizontal aluminum siding; #15 felt building paper; 8 wood double hung windows; 2 flush sold core wood exterior doors; 3-1/2" batt insulation. | .067 | 2.00 | 1.38 | 3.38 |
| 5 | Roofing | 240# asphalt shingles; #15 felt building paper; aluminum flashing; 6" attic insulation. | .024 | .54 | .54 | 1.08 |
| 6 | Interiors | 1/2" drywall, 5/8" drywall (party wall), taped and finished, painted with primer and 1 coat; softwood baseboard and trim, painted with primer and 1 coat; rubber backed carpeting, 80%; asphalt tile, 20%; 19 hollow core wood interior doors. | .243 | 3.97 | 5.16 | 9.13 |
| 7 | Specialties | Kitchen cabinets - 6 L.F. wall and base cabinets with laminated plastic counter top; stairs. | .027 | .76 | .53 | 1.29 |
| 8 | Mechanical | 1 lavatory, white, wall hung; 1 water closet, white; 1 bathtub, porcelain enamel steel, white; 1 kitchen sink; stainless steel, single; 1 water heater, gas fired, 30 gal.; gas fired hot air heat. | .086 | 2.09 | 1.62 | 3.71 |
| 9 | Electrical | 100 Amp. service; EMT conduit; incandescent lighting fixtures, switches, receptacles. | .035 | .37 | .76 | 1.13 |
| 10 | Overhead | Permit and plans. | | .35 | | .35 |
| | | **Total** | | 15.41 | 16.42 | 31.83 |

248

# Average Townhouse
# 8 Units/Cluster

| | |
|---|---:|
| Living Area | 1440 S.F. |
| Ground Area | 720 S.F. |
| Perimeter | 112 L.F. |
| Partitions | 180 L.F. |

| # | Category | Description | Man-Hours | Mat. | Labor | Total |
|---|---|---|---:|---:|---:|---:|
| 1 | **Site Work** | Site preparation for slab; trench 4' deep for foundation wall; utility trench. | .045 | .37 | 1.33 | 1.70 |
| 2 | **Foundations** | Continuous concrete footing 8" deep x 18" wide; 12" thick grouted concrete block foundation wall, 4' deep; 4" concrete slab on 4" gravel base, trowel finish. | .096 | 1.97 | 1.88 | 3.85 |
| 3 | **Framing** | 2" x 4" wood studs, 16" O.C.; 1/2" insulation board sheathing; 2" x 4" wood truss roof, 24" O.C., with 1/2" plywood sheathing, 1" x 3" furring 24" O.C., 4 in 12 pitch; 2" x 8" floor joists 16" O.C. with bridging, 1" x 3" furring and 1/2" plywood subfloor. | .121 | 3.01 | 2.97 | 5.98 |
| 4 | **Exterior Walls** | White cedar shingles; #15 felt building paper; 10 wood double hung windows; 1 wood panel and 1 flush sold core wood exterior door; 3-1/2" batt insulation; storms and screens. | .061 | 2.03 | 1.24 | 3.27 |
| 5 | **Roofing** | 240# asphalt shingles; #15 felt building paper; aluminum flashing; 6" attic insulation. Aluminum gutters and downspouts. | .024 | .54 | .54 | 1.08 |
| 6 | **Interiors** | 1/2" drywall, 5/8" drywall (party wall), taped and finished, painted with primer and 1 coat; carpet with underlayment 90%, vinyl tile with underlayment 10%; 18 hollow core doors. | .266 | 4.23 | 5.64 | 9.87 |
| 7 | **Specialties** | Kitchen cabinets - 14 L.F. wall and base cabinets with laminated plastic counter top; medicine cabinet; stairs. | .028 | 1.44 | .55 | 1.99 |
| 8 | **Mechanical** | 1 lavatory, white, wall hung; 1 water closet, white; 1 bathtub with shower, porcelain enamel steel, white; 1 kitchen sink; stainless steel, single; 1 water heater, gas fired, 30 gal.; gas fired hot air heat. | .119 | 2.52 | 2.21 | 4.73 |
| 9 | **Electrical** | 200 Amp. service; EMT conduit; incandescent lighting fixtures, switches, receptacles. | .039 | .34 | .71 | 1.05 |
| 10 | **Overhead** | Permit and plans. | | .45 | | .45 |
| | **Total** | | | 16.90 | 17.07 | 33.97 |

# Custom Townhouse
## 6 Units/Cluster

| | |
|---|---:|
| Living Area | 1800 S.F. |
| Ground Area | 720 S.F. |
| Perimeter | 108 L.F. |
| Partitions | 225 L.F. |

| # | Category | Description | Man-Hours | Mat. | Labor | Total |
|---|---|---|---|---|---|---|
| 1 | **Site Work** | Site preparation for slab; excavation for basement 8' deep; utility trench. | .032 | .31 | 1.10 | 1.41 |
| 2 | **Foundations** | Continuous concrete footing 12" deep x 24" wide; cast-in-place concrete wall, 8" thick x 8' high, 3000 psi; waterproofing; draintile; 4" concrete slab on 4" gravel base, trowel finish. | .100 | 1.93 | 2.07 | 4.00 |
| 3 | **Framing** | 2" x 6" and 2" x 4" (partitions and party walls), wood studs, 16" O.C.; 1/2" plywood sheathing; 2" x 8" rafters, 16" O.C., 1/2" plywood sheathing, 4 in 12 pitch; 2" x 6" ceiling joists, 16" O.C.; 2" x 10" floor joists, 16" O.C.; 1" x 3" furring, 16" O.C. and 5/8" subfloor. | .154 | 3.60 | 3.59 | 7.19 |
| 4 | **Exterior Walls** | Horizontal beveled wood siding; #15 felt building paper; 6" batt insulation; 10 plastic clad, wood double hung windows; 1 bay window; 2 - 6 panel and 1 flush solid core wood exterior doors; 1 - 16' x 7' garage door; storms and screens. | .073 | 3.62 | 1.46 | 5.08 |
| 5 | **Roofing** | Wood cedar shingles; #15 felt building paper; copper flashing; 9" attic insulation. Aluminum gutters and downspouts. | .026 | .96 | .57 | 1.53 |
| 6 | **Interiors** | 1/2" drywall 5/8" drywall (party walls), taped and finished, painted with primer and 1 coat; hardwood baseboard and trim, sanded and finished; finished hardwood floor 70%, ceramic tile floor 20%, vinyl tile 10%; 28 wood panel interior doors. | .325 | 8.67 | 8.82 | 17.49 |
| 7 | **Specialties** | Kitchen - 20 LF wall and base cabinets with laminated plastic counter top; 4 - 4LF bathroom vanities; medicine cabinets; stairs; appliances; fireplace. | .046 | 2.76 | .88 | 3.64 |
| 8 | **Mechanical** | 4 lavatories, color, built-in; 4 water closets, color; 1 bathtub with shower, cast iron, color; 1 shower, steel enamel, stone base; 1 kitchen sink, double, stainless steel 1 water heater, gas fired, 30 gal.; gas fired, forced air heat/air conditioning. | .141 | 4.49 | 2.56 | 7.05 |
| 9 | **Electrical** | 200 Amp. service; EMT conduit; fluorescent and incandescent lighting fixtures, intercom, switches, receptacles. | .062 | .68 | 1.32 | 2.00 |
| 10 | **Overhead** | Permit and plans. | | 3.08 | | 3.08 |
| | **Total** | | | 30.10 | 22.37 | 52.47 |

# Luxury Townhouse
# 4 Units/Cluster

| | | | Living Area | 2000 S.F. |
|---|---|---|---|---|
| | | | Ground Area | 912 S.F. |
| | | | Perimeter | 124 L.F. |
| | | | Partitions | 250 L.F. |

| | | | MAN-HOURS | COST PER SQUARE FOOT OF LIVING AREA |||
|---|---|---|---|---|---|---|
| | | | | MAT. | LABOR | TOTAL |
| **1** | **Site Work** | Site preparation for slab; excavation for basement 8' deep; utility trench. | .029 | .11 | .73 | .84 |
| **2** | **Foundations** | Continuous concrete footing 12" deep x 24" wide; cast-in-place concrete wall, 12" thick x 8' high, 3000 psi; waterproofing; draintile; vapor barrier; 4" concrete slab on 4" gravel base, trowel finish. | .109 | 2.45 | 2.24 | 4.69 |
| **3** | **Framing** | 2" x 6" and 2" x 4" (partitions and party walls). wood studs, 16" O.C.; 1/2" plywood sheathing; 2" x 10" rafters, 16" O.C., 5/8" sheathing, 8 in 12 pitch; 2" x 8" ceiling joists, 16" O.C.; 2" x 12" floor joists, 16" O.C. with bridging, 1" x 3" furring and 5/8" sheathing. | .208 | 4.29 | 4.06 | 8.35 |
| **4** | **Exterior Walls** | Face brick veneer; #15 felt building paper; 6" batt insulation; 14 metal clad, wood double hung windows; 1 bay window; 2 hand carved mahogany and 1 flush solid core wood exterior doors; 1 - 16' x 7' garage door; storms and screens. | .177 | 5.33 | 3.40 | 8.73 |
| **5** | **Roofing** | Wood cedar shingles; #15 felt building paper; copper flashing; 9" attic insulation. Aluminum gutters and downspouts. | .029 | 1.09 | .65 | 1.74 |
| **6** | **Interiors** | 5/8" drywall thin coat plaster, painted with primer and 2 coats; hardwood baseboard and trim, sanded and finished; finished hardwood floor 70%, ceramic tile with underlayment 20%, vinyl tile with underlayment 10%; 33 wood panel interior doors. | .312 | 6.91 | 6.57 | 13.48 |
| **7** | **Specialties** | Kitchen - 25 LF wall and base cabinets with laminated plastic counter top; 4 - 4LF bathroom vanities; medicine cabinets; stairs; range; compactor; dishwasher; disposal; rangehood; fireplace. | .087 | 4.04 | 1.67 | 5.71 |
| **8** | **Mechanical** | 5 lavatories, color, built-in; 4 water closets, color; 2 bathtubs with shower, cast iron, color; 1 shower, steel enamel, stone base; 1 kitchen sink, double, stainless steel; 1 water heater, gas fired, 75 gal.; gas fired, forced air heating/air conditioning.. | .105 | 4.54 | 1.57 | 6.11 |
| **9** | **Electrical** | 200 Amp. service; EMT conduit; fluorescent and incandescent lighting fixtures, intercom, switches, receptacles. | .060 | .65 | 1.27 | 1.92 |
| **10** | **Overhead** | Permit and plans. | | 6.50 | | 6.50 |
| | | **Total** | | 35.91 | 22.16 | 58.07 |

251

# Warehouse

| | | Occupied Area | 2000 S.F. |
|---|---|---|---|
| | | Ground Area | 2000 S.F. |
| | | Perimeter | 180 L.F. |
| | | Partitions | 54 L.F. |

| | | | MAN-HOURS | COST PER SQUARE FOOT OF LIVING AREA |||
|---|---|---|---|---|---|---|
| | | | | MAT. | LABOR | TOTAL |
| 1 | Site Work | Site preparation for slab; trench 4' deep for foundation wall; utility trench. | .037 | .22 | 1.18 | 1.40 |
| 2 | Foundations | Continuous concrete footing 8" deep x 18" wide; 12" thick concrete block foundation wall, 4' deep; 4" thick concrete slab on 4" gravel base, trowel finish. | .126 | 3.01 | 2.54 | 5.55 |
| 3 | Framing | Open web steel joists, 24" deep, 12.7#/L.F., 5' O.C., 40' span, 30 PSF superimposed load; 1-1/2" deep, 22 gauge, open type metal deck; 3-5/8" wide 20 ga., non-load bearing metal studs, 24" O.C. | .018 | 1.77 | .55 | 2.32 |
| 4 | Exterior Walls | Concrete block, regular, 12" thick; wire truss strips; polystyrene block insert insulation; 16 aluminum sliding windows; 2 overhead doors; 2 steel, hollow metal doors and frames. | .224 | 5.54 | 4.13 | 9.67 |
| 5 | Roofing | Asphalt built-up roof over 2-1/8" glass fiber insulation; aluminum flashing; four 2' x 2' skylights. | .052 | 1.76 | 1.10 | 2.86 |
| 6 | Interiors | 5/8" drywall, taped and finished, painted with primer and 2 coats; softwood baseboard, primed and painted; 2' x 4' suspended acoustical ceiling system; 2 steel, hollow metal doors and frames; 5% vinyl tile, 95% concrete floor. | .026 | .74 | .55 | 1.29 |
| 7 | Specialties | Toilet room accessories. | .003 | .04 | .01 | .05 |
| 8 | Mechanical | 1 lavatory, white, wall hung; 1 water closet white; 1 water heater, gas fired, 30 gal. | .014 | .29 | .22 | .51 |
| 9 | Electrical | 200 Amp. service; baseboard heaters; EMT conduit; fluorescent lighting fixtures, switches, receptacles. | .046 | 2.56 | .94 | 3.50 |
| 10 | Overhead | Permit and plans. | | 1.70 | | 1.70 |
| | | **Total** | | 17.62 | 11.22 | 28.84 |

# Warehouse

| | | |
|---|---|---|
| Occupied Area | | 3000 S.F. |
| Ground Area | | 3000 S.F. |
| Perimeter | | 220 L.F. |
| Partitions | | 54 L.F. |

| | | MAN-HOURS | COST PER SQUARE FOOT OF LIVING AREA |||
|---|---|---|---|---|---|
| | | | MAT. | LABOR | TOTAL |
| 1 | **Site Work** — Site preparation for slab; trench 4' deep for foundation wall; utility trench. | .035 | .14 | 1.20 | 1.34 |
| 2 | **Foundations** — Continuous concrete footing 8" deep x 18" wide; 12" thick concrete block foundation wall, 4' deep; 4" concrete slab on 4" gravel base, trowel finish; 1 column footing. | .109 | 2.68 | 2.22 | 4.90 |
| 3 | **Framing** — Open web steel joists, 16" deep, 7.8#/L.F., 5' O.C., 25' span, 30 PSF superimposed load; 1-1/2" deep, 22 gauge, open type metal deck; 10" deep wide flange beam; 1 - 4"x 4" steel tube column. | .017 | 1.79 | .59 | 2.38 |
| 4 | **Exterior Walls** — Concrete block, regular, 12" thick; wire truss strips; polystyrene block insert insulation; 24 aluminum sliding windows; 2 overhead doors; 2 steel, hollow metal doors and frames. | .230 | 4.41 | 4.19 | 8.60 |
| 5 | **Roofing** — Asphalt built-up roof over 2-1/8" glass fiber insulation; aluminum flashing; 6 - 2' x 2' skylights. | .052 | 1.77 | 1.11 | 2.88 |
| 6 | **Interiors** — 5/8" drywall, taped and finished, painted with primer and 2 coats; softwood baseboard, primed and painted; 2' x 4' suspended acoustical ceiling system; 3 hollow core wood interior doors; 68% vinyl tile, 32% concrete floor. | .018 | .49 | .36 | .85 |
| 7 | **Specialties** — Toilet room accessories. | .002 | .03 | .01 | .04 |
| 8 | **Mechanical** — 2 lavatories, white, wall hung, 2 water closets white; 1 water heater, gas fired, 30 gallons. | .009 | .19 | .15 | .34 |
| 9 | **Electrical** — 200 Amp. service; EMT conduit; baseboard heaters; fluorescent lighting fixtures, switches, receptacles. | .034 | 2.41 | .75 | 3.16 |
| 10 | **Overhead** — Permit and plans. | | 1.67 | | 1.67 |
| | **Total** | | 15.58 | 10.58 | 26.16 |

253

# Warehouse

| | | |
|---|---|---|
| Occupied Area | 5000 S.F. |
| Ground Area | 5000 S.F. |
| Perimeter | 300 L.F. |
| Partitions | 54 L.F. |

| # | Category | Description | MAN-HOURS | MAT. | LABOR | TOTAL |
|---|---|---|---|---|---|---|
| 1 | Site Work | Site preparation for slab; trench 4' deep for foundation wall; utility trench. | .028 | .08 | 1.00 | 1.08 |
| 2 | Foundations | Continuous concrete footing 8" deep x 18" wide; 12" thick concrete block foundation wall, 4' deep; 6" concrete slab on 4" gravel base, trowel finish; 3 column footings. | .095 | 2.42 | 1.97 | 4.39 |
| 3 | Framing | Open web steel joists, 16" deep, 7.8#/L.F., 5' O.C., 25' span, 30 PSF superimposed load; 1-1/2" deep, 22 gauge, open type metal deck; 10" wide flange beam; 3 - 4" x 4" steel tube column. | .011 | 1.59 | .57 | 2.16 |
| 4 | Exterior Walls | Concrete block, regular, 12" thick; wire truss strips; polystyrene block insert insulation; 40 aluminum sliding windows; 4 overhead doors; 2 steel, hollow metal doors and frames. | .192 | 4.63 | 3.55 | 8.18 |
| 5 | Roofing | Asphalt built-up roof over 2-1/8" glass fiber insulation; aluminum flashing; ten 2' x 2' skylights. | .052 | 1.77 | 1.11 | 2.88 |
| 6 | Interiors | 5/8" drywall, taped and finished, painted with primer and 2 coats; softwood baseboard, primed and painted; 2' x 4' suspended acoustical ceiling system; 2 steel, hollow metal doors and frames; 5% vinyl tile, 95% concrete floor. | .010 | .29 | .21 | .50 |
| 7 | Specialties | Toilet room accessories. | .001 | .02 | .01 | .03 |
| 8 | Mechanical | 1 lavatory, white, wall hung; 1 water closet white; 1 water heater, gas fired, 30 gal. | .006 | .12 | .09 | .21 |
| 9 | Electrical | 200 Amp. service; baseboard heaters; EMT conduit; fluorescent lighting fixtures, switches, receptacles. | .028 | 2.29 | .59 | 2.88 |
| 10 | Overhead | Permit and plans. | | 1.64 | | 1.64 |
| | | **Total** | | 14.85 | 9.10 | 23.95 |

# Warehouse

| | | Occupied Area | 10,000 S.F. |
|---|---|---|---|
| | | Ground Area | 10,000 S.F. |
| | | Perimeter | 500 L.F. |
| | | Partitions | 54 L.F. |

| | | | MAN-HOURS | COST PER SQUARE FOOT OF LIVING AREA |||
|---|---|---|---|---|---|---|
| | | | | MAT. | LABOR | TOTAL |
| 1 | Site Work | Site preparation for slab; trench 4' deep for foundation wall; utility trench. | .018 | .04 | .64 | .68 |
| 2 | Foundations | Continuous concrete footing 8" deep x 18" wide; 12" thick concrete block foundation wall, 4' deep; 6" concrete slab on 4" gravel base, trowel finish; 4 column footings. | .086 | 2.22 | 1.78 | 4.00 |
| 3 | Framing | Open web steel joists, 24" deep, 12.7#/L.F., 5' O.C., 40' span, 30 PSF superimposed load; 1-1/2" deep, 22 gauge, open type metal deck; 12" deep wide flange beam; four 4" x 4" steel tube column. | .017 | 1.57 | .53 | 2.10 |
| 4 | Exterior Walls | Concrete block, regular, 12" thick; wire truss strips; polystyrene block insert insulation; 38 aluminum sliding windows; 8 overhead doors; 4 steel, hollow metal doors. | .153 | 3.48 | 2.70 | 6.18 |
| 5 | Roofing | Asphalt built-up roof over 2-1/8" glass fiber insulation; aluminum flashing; twenty 2' x 2' skylights. | .052 | 1.37 | 1.00 | 2.37 |
| 6 | Interiors | 5/8" drywall, taped and finished, painted with primer and 2 coats; softwood baseboard, primed and painted; 2' x 4' suspended acoustical ceiling system; 2 steel hollow core metal doors and frames; 5% vinyl tile, 95% concrete floor. | .005 | .14 | .11 | .25 |
| 7 | Specialties | Toilet room accessories. | .001 | .01 | .01 | .02 |
| 8 | Mechanical | 1 lavatory, white, wall hung; 1 water closet white; 1 water heater, gas fired, 30 gal. | .003 | .06 | .04 | .10 |
| 9 | Electrical | 200 Amp. service; baseboard heaters; EMT conduit; fluorescent lighting fixtures, switches, receptacles. | .023 | 2.23 | .48 | 2.71 |
| 10 | Overhead | Permit and plans. | | 1.63 | | 1.63 |
| | | **Total** | | 12.75 | 7.29 | 20.04 |

# HOW TO USE UNIT PRICE PAGES

**NOTE:**
Man-hours are a unit of measure for performing a task. To derive the total man-hours for a line item, the user must multiply the quantity of the project involved times the man-hour figure shown in this column.

**CREW F-1** is shown on page 409 as follows:

| Crew F-1 | Hr. | Daily | Hr. | Daily | Bare Costs | Incl. O&P |
|---|---|---|---|---|---|---|
| 1 Carpenter | $12.35 | $98.80 | $19.55 | $156.40 | $12.35 | $19.55 |
| Power Tools |  | 7.00 |  | 7.70 | .87 | .96 |
| 8 M.H., Daily Totals |  | $105.80 |  | $164.10 | $13.22 | $20.51 |

**BARE COSTS** are developed as follows for line no. 6.1-05-001
**MAT.** is **BARE MATERIAL COST ($.19)**
**LABOR** for Crew F1 = Man-hour Cost **($12.35)** x Man-hour Rate **(.053)** = **$.66**
**EQUIP.** for Crew F1 = Equip. Hour Cost **($.87)** x Man-hour Rate **(.053)** = **$.05**
**TOTAL** = MAT. COST **($.19)** + LABOR COST **($.66)** + EQUIP. COST **($.05)** = **$.90** per linear foot of bracing.
(Note: Equipment and Labor costs are taken from the Crew Tables. See example at the top of this page.)

### Unit
The unit of measure listed here reflects the material being used in the assembly. For example: joists are defined in linear feet (LF).

**TOTAL COSTS INCLUDING O&P** are developed as follows:
**MAT.** is BARE MATERIAL COST + 10% = $.19 + $.02 = **$.21**
**LABOR** for Crew F1 = Man-hour Cost **($19.55)** x Man-hour Rate **(.053)** = **$1.04**
**EQUIP.** for Crew F1 = Equip. Hour Cost **($.96)** x Man-hour Rate **(.053)** = **$.05**
**TOTAL** = MAT. COST **($.21)** + LABOR COST **($1.04)** + EQUIP. COST **($.05)** = **$1.30**
(Note: Equipment and labor costs are taken from the Crew Tables. See example at top of this page.)

### General Meaning
The meaning of this line item is that wood let-in bracing will be installed at a rate of .053 man-hours per linear foot based on a 1" x 6" board installed in a 16" on-center stud wall.

### Line Number Determination
Major UCI subdivision is **6.1**
(Two digits plus decimal point plus last digit)

Major classification within UCI subdivision is **05**
(Two digits)

Item line number is **020**
(Three digits)

Complete line number is
**6.1-05-020**

### IMPORTANT
Prices in this section are listed in two ways. Some are bare costs. Others include the overhead and profit of the subcontractor. In most cases, it is best for a general contractor to add an additional 10% to the figures found in the column titled "TOTAL INCLUDING O&P." For further information, see the introduction to this section.

| 6.1 | Rough Carpentry | CREW | MAN-HOURS | UNIT | MAT. | LABOR | EQUIP. | TOTAL | TOTAL INCL O&P |
|---|---|---|---|---|---|---|---|---|---|
| 02 238 | 2" x 10" | F-1 | .059 | L.F. | .68 | .73 | .05 | 1.46 | 1.95 |
| 240 | 2" x 12" |  | .073 |  | .86 | .91 | .06 | 1.83 | 2.45 |
| 05-001 | BRACING Let-in, with 1" x 6" boards, studs 16" O.C. |  | .053 |  | .19 | .66 | .05 | .90 | 1.30 |
| 020 | Studs at 24" O.C. |  | .035 |  | .19 | .43 | .03 | .65 | .92 |
| 030 | Let-in, "T" shaped, 20 ga. galv. steel, studs at 16" O.C. |  | .014 |  | .53 | .17 | .01 | .71 | .87 |
| 040 | Studs at 24" O.C. |  | .013 |  | .53 | .17 | .01 | .71 | .86 |
| 050 | 16 ga. galv. steel straps, studs at 16" O.C. |  | .013 |  | .53 | .17 | .01 | .71 | .86 |
| 060 | Studs at 24" O.C. |  | .013 |  | .53 | .16 | .01 | .70 | .85 |
| 07-001 | BRIDGING Wood, for joists 16" O.C., 1" x 3" |  | .062 | Pr. | .38 | .76 |  | 1.19 | 1.68 |
| 010 | 2" x 3" bridging |  | .062 |  | .67 | .76 | .05 | 1.48 | 2 |
| 030 | Steel, galvanized, 18 ga., for 2" x 10" joists at 12" O.C. | 1 Carp | .062 |  | .86 | .76 |  | 1.62 | 2.15 |
| 040 | 24" O.C. |  | .057 |  | 1.26 | .71 |  | 1.97 | 2.50 |
| 060 | For 14" joists at 16" O.C. |  | 6.150 |  | 1.26 | 76 |  | 77.26 | 120 |
| 090 | Compression type, 16" O.C., 2" x 8" joists |  | .040 |  | .84 | .49 |  | 1.33 | 1.71 |
| 100 | 2" x 12" joists |  | .040 |  | .90 | .49 |  | 1.39 | 1.77 |
| 10-001 | DOCK BUMPERS Bolts not included, 2" x 6" to 4" x 8", average | F-1 | 26.670 | B.F. | .42 | 332 | 23 | 355.42 | 545 |
| 13-001 | FRAMING, BEAMS & GIRDERS |  |  |  |  |  |  |  |  |
| 002 |  |  |  |  |  |  |  |  |  |
| 100 | Single, 2" x 6" | F-2 | .023 | L.F. | .36 | .28 | .02 | .66 | .86 |
| 102 | 2" x 8" |  | .025 |  | .55 | .31 | .02 | .88 | 1.11 |
| 104 | 2" x 10" |  | .027 |  | .75 | .33 | .02 | 1.10 | 1.37 |
| 106 | 2" x 12" |  | .029 |  | .86 | .35 | .03 | 1.24 | 1.54 |
| 108 | 2" x 14" |  | .032 |  | 1.10 | .39 | .03 | 1.52 | 1.87 |
| 110 | 3" x 6" |  | .029 |  | .85 | .35 | .03 | 1.23 | 1.53 |
| 112 | 3" x 10" |  | .032 |  | 1.40 | .39 | .03 | 1.82 | 2.20 |
| 114 | 3" x 12" |  | .036 |  | 1.68 | .44 | .03 | 2.15 | 2.58 |
| 116 | 3" x 14" |  | .040 |  | 1.96 | .49 | .04 | 2.49 | 2.98 |
| 118 | 4" x 8" | F-3 | .040 |  | 1.50 | .50 | .25 | 2.25 | 2.71 |
| 120 | 4" x 10" |  | .042 |  | 1.87 | .53 | .26 | 2.66 | 3.18 |
| 122 | 4" x 12" |  | .044 |  | 2.24 | .55 | .28 | 3.07 | 3.64 |
| 124 | 4" x 14" |  | .047 |  | 2.60 | .59 | .29 | 3.48 | 4.11 |
| 200 | Double, 2" x 6" | F-2 | .040 |  | .73 | .49 | .04 | 1.26 | 1.62 |
| 202 | 2" x 8" |  | .043 |  | 1.10 | .52 | .04 | 1.66 | 2.09 |
| 204 | 2" x 10" |  | .046 |  | 1.36 | .56 | .04 | 1.96 | 2.43 |
| 206 | 2" x 12" |  | .049 |  | 1.72 | .61 | .04 | 2.37 | 2.90 |
| 208 | 2" x 14" |  | .053 |  | 2.20 | .66 | .05 | 2.91 | 3.51 |
| 210 |  |  |  |  |  |  |  |  |  |
| 300 | Triple, 2" x 6" | F-2 | .058 | L.F. | 1.09 | .72 | .05 | 1.86 | 2.39 |
| 302 | 2" x 8" |  | .064 |  | 1.65 | .79 | .06 | 2.50 | 3.13 |
| 304 | 2" x 10" |  | .071 |  | 2.04 | .88 | .06 | 2.98 | 3.70 |
| 306 | 2" x 12" |  | .080 |  | 2.57 | .99 | .07 | 3.63 | 4.47 |
| 308 | 2" x 14" |  | .084 |  | 3.30 | 1.04 | .07 | 4.41 | 5.35 |
| 310 |  |  |  |  |  |  |  |  |  |

# UNIT PRICE COSTS

Use the data in this section to estimate building costs and to develop totals for custom Residential and Light Commercial systems and projects. All of the unit prices found here adhere to the uniform standards of the construction industry. In these tables, each item listed is completely defined. The figures are representative of the data used to create the first two section of this book. All of these figures and totals are useful in developing specific cost estimates that relate to the systems section of the cost estimating process.

An outline of the major components that go into these tables is shown to the left of this page. When installations are made by one trade and require no crew, the appropriate trade is listed directly in the CREW column. MAN-HOURS are simply a unit of measure to perform a task. TOTAL MAN-HOURS can be found by multiplying MAN-HOURS times the quantities involved.

Develop the bare cost of individual items by adding up the cost of materials, labor and equipment. The TOTAL COST INCLUDING OVERHEAD AND PROFIT (O&P) column includes mark-ups on material and labor for the installing contractor.

Labor costs are for current, state-of-the-art installations. The production rates used to develop this data assume that the worker is a competent tradesman working in a realistic construction environment. Of course, it is also assumed that the work is being done in daylight hours. Equipment costs, if needed, have been listed.

In some cases, construction estimates require a "unit cost approach." The following pages provide the data needed to assemble such itemized estimates. Use this data to make last-minute adjustments in design and to justify any price estimates completed using the other two section of "Residential/Light Commercial Cost Data 1985".

## TABLE OF CONTENTS

| Division | | Page |
|---|---|---|
| 1 | General Requirements | 259 |
| 2 | Site Work | 272 |
| 3 | Concrete | 282 |
| 4 | Moisture Protection | 285 |
| 5 | Metals | 291 |
| 6 | Wood & Plastics | 296 |
| 7 | Moisture | 314 |
| 8 | Doors, Windows & Glass | 330 |
| 9 | Finishes | 352 |
| 10 | Specialties | 366 |
| 11 | Equipment | 372 |
| 12 | Furnishings | 373 |
| 13 | Special Construction | 374 |
| 14 | Conveying Systems | 374 |
| 15 | Mechanical | 375 |
| 16 | Electrical | 394 |

# 1.0 Crews

| | | CREW | MAN-HOURS | UNIT | MAT. | LABOR | EQUIP. | TOTAL | TOTAL INCL O&P |
|---|---|---|---|---|---|---|---|---|---|
| 05-001 | **GENERAL CONDITIONS** | | | | | | | | |
| 025 | Crew A-1 | A-1 | 8 | Day | | 80 | 35.20 | 115.20 | 165.50 |
| 050 | Crew A-2 | A-2 | 24 | | | 242 | 57.15 | 299.15 | 445.25 |
| 075 | Crew A-3 | A-3 | 8 | | | 82.80 | 219.80 | 302.60 | 371.80 |
| 100 | Crew A-4 | A-4 | 24 | | | 292.40 | | 292.40 | 460.80L |
| 125 | Crew A-5 | A-5 | 18 | | | 180.50 | 14.30 | 194.80 | 301.50 |
| 150 | Crew A-6 | A-6 | 16 | | | 183.20 | | 183.20 | 290.40L |
| 175 | Crew A-7 | A-7 | 24 | | | 261.20 | | 261.20 | 415.20L |
| 200 | Crew A-8 | A-8 | 32 | ▼ | | 339.20 | | 339.20 | 540L |
| 10-001 | **SITE WORK** | | | | | | | | |
| 010 | Crew B-1 | B-1 | 24 | Day | | 256 | | 256 | 405.60L |
| 025 | Crew B-2 | B-2 | 40 | | | 416 | | 416 | 659.20L |
| 040 | Crew B-3 | B-3 | 48 | | | 522.80 | 1,143.40 | 1,666.20 | 2,084.15 |
| 055 | Crew B-4 | B-4 | 48 | | | 498.80 | 314.40 | 813.20 | 1,135.05 |
| 070 | Crew B-5 | B-5 | 64 | | | 725.60 | 726.50 | 1,452.10 | 1,949.95 |
| 085 | Crew B-6 | B-6 | 24 | | | 256.40 | 152.20 | 408.60 | 573.80 |
| 100 | Crew B-7 | B-7 | 48 | | | 517.20 | 719.60 | 1,236.80 | 1,611.55 |
| 115 | Crew B-8 | B-8 | 64 | | | 710.80 | 1,527.40 | 2,238.20 | 2,804.95 |
| 130 | Crew B-9 | B-9 | 40 | | | 416 | 141.50 | 557.50 | 814.85 |
| 145 | Crew B-10 | B-10 | 12 | | | 141.20 | | 141.20 | 224.20L |
| 148 | Crew B-10A | B-10A | 12 | | | 141.20 | 72.10 | 213.30 | 303.50 |
| 150 | Crew B-10B | B-10B | 12 | | | 141.20 | 614.40 | 755.60 | 900.05 |
| 152 | Crew B-10C | B-10C | 12 | | | 141.20 | 685.90 | 827.10 | 978.70 |
| 154 | Crew B-10D | B-10D | 12 | | | 141.20 | 693.10 | 834.30 | 986.60 |
| 156 | Crew B-10E | B-10E | 12 | | | 141.20 | 83.85 | 225.05 | 316.40 |
| 158 | Crew B-10F | B-10F | 12 | | | 141.20 | 172.80 | 314 | 414.30 |
| 160 | Crew B-10G | B-10G | 12 | | | 141.20 | 283.20 | 424.40 | 535.70 |
| 162 | Crew B-10H | B-10H | 12 | | | 141.20 | 29.05 | 170.25 | 256.15 |
| 164 | Crew B-10I | B-10I | 12 | | | 141.20 | 59.65 | 200.85 | 289.80 |
| 166 | Crew B-10J | B-10J | 12 | | | 141.20 | 37.90 | 179.10 | 265.90 |
| 168 | Crew B-10K | B-10K | 12 | | | 141.20 | 98.85 | 240.05 | 332.90 |
| 170 | Crew B-10L | B-10L | 12 | | | 141.20 | 198.60 | 339.80 | 442.65 |
| 172 | Crew B-10M | B-10M | 12 | | | 141.20 | 899 | 1,040.20 | 1,213.10 |
| 174 | Crew B-10N | B-10N | 12 | | | 141.20 | 269 | 410.20 | 520.10 |
| 176 | Crew B-10O | B-10O | 12 | | | 141.20 | 410.80 | 552 | 676.10 |
| 178 | Crew B-10P | B-10P | 12 | | | 141.20 | 585 | 726.20 | 867.70 |
| 180 | Crew B-10Q | B-10Q | 12 | | | 141.20 | 985 | 1,126.20 | 1,307.70 |
| 182 | Crew B-10R | B-10R | 12 | | | 141.20 | 199 | 340.20 | 443.10 |
| 184 | Crew B-10S | B-10S | 12 | | | 141.20 | 266.60 | 407.80 | 517.45 |
| 186 | Crew B-10T | B-10T | 12 | | | 141.20 | 437.80 | 579 | 705.80 |
| 188 | Crew B-10U | B-10U | 12 | | | 141.20 | 835 | 976.20 | 1,142.70 |
| 192 | Crew B-10W | B-10W | 12 | | | 141.20 | 335.60 | 476.80 | 593.35 |
| 194 | Crew B-10X | B-10X | 12 | | | 141.20 | 1,142 | 1,283.20 | 1,480.40 |
| 196 | Crew B-10Y | B-10Y | 12 | | | 141.20 | 277 | 418.20 | 528.90 |
| 200 | Crew B-11 | B-11 | 16 | | | 181.20 | | 181.20 | 287.60L |
| 202 | Crew B-11A | B-11A | 16 | | | 181.20 | 614.40 | 795.60 | 963.45 |
| 204 | Crew B-11B | B-11B | 16 | | | 181.20 | 787.25 | 968.45 | 1,153.55 |
| 206 | Crew B-11C | B-11C | 16 | | | 181.20 | 152.20 | 333.40 | 455 |
| 220 | Crew B-11K | B-11K | 16 | | | 181.20 | 415.60 | 596.80 | 744.75 |
| 222 | Crew B-11L | B-11L | 16 | | | 181.20 | 422 | 603.20 | 751.80 |
| 224 | Crew B-11M | B-11M | 16 | | | 181.20 | 228 | 409.20 | 538.40 |
| 250 | Crew B-12 | B-12 | 16 | | | 190 | | 190 | 301.20L |
| 252 | Crew B-12A | B-12A | 16 | | | 190 | 402.20 | 592.20 | 743.60 |
| 254 | Crew B-12B | B-12B | 16 | | | 190 | 560.20 | 750.20 | 917.40 |
| 256 | Crew B-12C | B-12C | 16 | | | 190 | 773 | 963 | 1,151.50 |
| 258 | Crew B-12D | B-12D | 16 | | | 190 | 1,389 | 1,579 | 1,829.10 |
| 260 | Crew B-12E | B-12E | 16 | | | 190 | 255 | 445 | 581.70 |
| 262 | Crew B-12F | B-12F | 16 | | | 190 | 323.20 | 513.20 | 656.70 |
| 264 | Crew B-12G | B-12G | 16 | ▼ | | 190 | 305.60 | 495.60 | 637.35 |

For expanded coverage of these items see *Building Construction Cost Data 1985*

# 1.0 Crews

| | | Crews | CREW | MAN-HOURS | UNIT | MAT. | LABOR | EQUIP. | TOTAL | TOTAL INCL O&P |
|---|---|---|---|---|---|---|---|---|---|---|
| 10 | 266 | Crew B-12H | B-12H | 16 | Day | 190 | 425.05 | | 615.05 | 768.75 |
| | 268 | Crew B-12I | B-12I | 16 | | 190 | 366.50 | | 556.50 | 704.35 |
| | 270 | Crew B-12J | B-12J | 16 | | 190 | 432.20 | | 622.20 | 776.60 |
| | 272 | Crew B-12K | B-12K | 16 | | 190 | 619 | | 809 | 982.10 |
| | 274 | Crew B-12L | B-12L | 16 | | 190 | 317.90 | | 507.90 | 650.90 |
| | 276 | Crew B-12M | B-12M | 16 | | 190 | 422.30 | | 612.30 | 765.70 |
| | 278 | Crew B-12N | B-12N | 16 | | 190 | 495.35 | | 685.35 | 846.10 |
| | 280 | Crew B-12O | B-12O | 16 | | 190 | 601.75 | | 791.75 | 963.10 |
| | 282 | Crew B-12P | B-12P | 16 | | 190 | 512.90 | | 702.90 | 865.40 |
| | 284 | Crew B-12Q | B-12Q | 16 | | 190 | 268 | | 458 | 596 |
| | 286 | Crew B-12R | B-12R | 16 | | 190 | 560.20 | | 750.20 | 917.40 |
| | 288 | Crew B-12S | B-12S | 16 | | 190 | 1,143 | | 1,333 | 1,558.50 |
| | 290 | Crew B-12T | B-12T | 16 | | 190 | 965.20 | | 1,155.20 | 1,362.90 |
| | 292 | Crew B-12V | B-12V | 16 | | 190 | 797.90 | | 987.90 | 1,178.90 |
| | 300 | Crew B-13 | B-13 | 56 | | 606 | 384 | | 990 | 1,382.80 |
| | 315 | Crew B-14 | B-14 | 48 | | 512.40 | 152.20 | | 664.60 | 979.40 |
| | 330 | Crew B-15 | B-15 | 28 | | 306.80 | 1,172.80 | | 1,479.60 | 1,774.30 |
| | 345 | Crew B-16 | B-16 | 32 | | 338.80 | 279.20 | | 618 | 842.70 |
| | 360 | Crew B-17 | B-17 | 32 | | 339.20 | 372 | | 711.20 | 945.60 |
| | 375 | Crew B-18 | B-18 | 24 | | 256 | 38.50 | | 294.50 | 447.95 |
| | 390 | Crew B-19 | B-19 | 64 | | 805.20 | 974.70 | | 1,779.90 | 2,410.15 |
| | 405 | Crew B-20 | B-20 | 24 | | 316.80 | | | 316.80 | 502L |
| | 420 | Crew B-21 | B-21 | 28 | | 368.40 | 87.90 | | 456.30 | 680.50 |
| | 435 | Crew B-22 | B-22 | 30 | | 394.20 | 131.85 | | 526.05 | 769.70 |
| | 450 | Crew B-23 | B-23 | 40 | | 416 | 210.60 | | 626.60 | 890.85 |
| | 465 | Crew B-24 | B-24 | 24 | | 273.60 | | | 273.60 | 429.60L |
| | 480 | Crew B-25 | B-25 | 80 | | 858.40 | 706.80 | | 1,565.20 | 2,138.70 |
| | 495 | Crew B-26 | B-26 | 88 | | 979.60 | 1,529 | | 2,508.60 | 3,241.10 |
| | 510 | Crew B-27 | B-27 | 32 | | 336 | 80.50 | | 416.50 | 620.95 |
| | 525 | Crew B-28 | B-28 | 24 | | 277.60 | | | 277.60 | 439.60L |
| | 540 | Crew B-29 | B-29 | 56 | | 606 | 432.20 | | 1,038.20 | 1,435.80 |
| | 555 | Crew B-30 | B-30 | 24 | | 266.80 | 1,118.60 | | 1,385.40 | 1,651.25 |
| | 570 | Crew B-31 | B-31 | 40 | | 434.80 | 122.65 | | 557.45 | 823.70 |
| | 585 | Crew B-32 | B-32 | 16 | | 202.40 | 594.80 | | 797.20 | 975.90 |
| | 600 | Crew B-33 | B-33 | 14 | | 166.50 | | | 166.50 | 264.40L |
| | 602 | Crew B-33A | B-33A | 14 | | 166.50 | 1,212.95 | | 1,379.45 | 1,598.65 |
| | 604 | Crew B-33B | B-33B | 14 | | 166.50 | 1,243.05 | | 1,409.55 | 1,631.75 |
| | 606 | Crew B-33C | B-33C | 14 | | 166.50 | 1,243.05 | | 1,409.55 | 1,631.75 |
| | 608 | Crew B-33D | B-33D | 14 | | 166.50 | 1,188.75 | | 1,355.25 | 1,572 |
| | 610 | Crew B-33E | B-33E | 14 | | 166.50 | 1,339.75 | | 1,506.25 | 1,738.10 |
| | 612 | Crew B-33F | B-33F | 14 | | 166.50 | 856.75 | | 1,023.25 | 1,206.80 |
| | 614 | Crew B-33G | B-33G | 14 | | 166.50 | 1,172.75 | | 1,339.25 | 1,554.40 |
| | 630 | Crew B-34 | B-34 | 8 | | 82.80 | | | 82.80 | 130L |
| | 632 | Crew B-34A | B-34A | 8 | | 82.80 | 219.80 | | 302.60 | 371.80 |
| | 634 | Crew B-34B | B-34B | 8 | | 82.80 | 279.20 | | 362 | 437.10 |
| | 636 | Crew B-34C | B-34C | 8 | | 82.80 | 379.30 | | 462.10 | 547.20 |
| | 638 | Crew B-34D | B-34D | 8 | | 82.80 | 394.10 | | 476.90 | 563.50 |
| | 650 | Crew B-35 | B-35 | 48 | | 613.52 | 341.35 | | 954.87 | 1,347.50 |
| | 665 | Crew B-36 | B-36 | 32 | | 357.20 | 664.35 | | 1,021.55 | 1,297.20 |
| | 680 | Crew B-37 | B-37 | 48 | | 512.40 | 83.85 | | 596.25 | 904.20 |
| | 695 | Crew B-38 | B-38 | 24 | | 256.40 | 361.40 | | 617.80 | 803.95 |
| | 710 | Crew B-39 | B-39 | 48 | | 512.40 | 141.50 | | 653.90 | 967.65 |
| | 725 | Crew B-40 | B-40 | 64 | | 805.20 | 1,465.60 | | 2,270.80 | 2,950.15 |
| | 740 | Crew B-41 | B-41 | 44 | | 463.50 | 120.65 | | 584.15 | 867.20 |
| | 755 | Crew B-42 | B-42 | 64 | | 712.40 | 917.60 | | 1,630 | 2,152.15 |
| | 770 | Crew B-43 | B-43 | 48 | | 526 | 545.30 | | 1,071.30 | 1,433.45 |
| | 785 | Crew B-44 | B-44 | 64 | | 798.40 | 876.05 | | 1,674.45 | 2,290.85 |
| | 800 | Crew B-45 | B-45 | 16 | | 184 | 240.80 | | 424.80 | 555.70 |
| | 815 | Crew B-46 | B-46 | 48 | | 553.60 | 28.20 | | 581.80 | 946.20 |
| | 830 | Crew B-47 | B-47 | 24 | | 272.40 | 421.70 | | 694.10 | 895.45 |

For expanded coverage of these items see *Building Construction Cost Data 1985*

# 1.0 Crews

| | | CREW | MAN-HOURS | UNIT | MAT. | LABOR | EQUIP. | TOTAL | TOTAL INCL O&P |
|---|---|---|---|---|---|---|---|---|---|
| 845 | Crew B-48 | B-48 | 56 | Day | | 622.40 | 631.80 | 1,254.20 | 1,681.35 |
| 860 | Crew B-49 | B-49 | 88 | | | 1,010.80 | 1,015.80 | 2,026.60 | 2,743.35 |
| 870 | Crew B-50 | B-50 | 112 | | | 1,358.80 | 1,002.90 | 2,361.70 | 3,356.40 |
| 880 | Crew B-51 | B-51 | 48 | | | 498.80 | 57.15 | 555.95 | 852.05 |
| 890 | Crew B-52 | B-52 | 56 | | | 652.20 | 292.50 | 944.70 | 1,356.55 |
| 900 | Crew B-53 | B-53 | 8 | | | 96.40 | 133.40 | 229.80 | 299.55 |
| 902 | Crew B-54 | B-54 | 8 | | | 96.40 | 197.80 | 294.20 | 370.40 |
| 904 | Crew B-55 | B-55 | 16 | | | 162 | 210.60 | 372.60 | 487.25 |
| 906 | Crew B-56 | B-56 | 16 | | | 176.40 | 413.35 | 589.75 | 734.30 |
| 908 | Crew B-57 | B-57 | 48 | | | 542.40 | 747.10 | 1,289.50 | 1,681.40 |
| 910 | Crew B-58 | B-58 | 24 | | | 256.40 | 1,558.45 | 1,814.85 | 2,120.70 |
| 912 | Crew B-59 | B-59 | 8 | | | 82.80 | 433 | 515.80 | 606.30 |
| 914 | Crew B-60 | B-60 | 56 | | | 638.80 | 666.30 | 1,305.10 | 1,745.30 |
| 916 | Crew B-61 | B-61 | 40 | | | 432.40 | 264.65 | 697.05 | 976.30 |
| 918 | Crew B-62 | B-62 | 24 | | | 256.40 | 93.85 | 350.25 | 509.60 |
| 920 | Crew B-63 | B-63 | 40 | | | 416.40 | 93.85 | 510.25 | 763.20 |
| 922 | Crew B-64 | B-64 | 16 | | | 162 | 158.80 | 320.80 | 430.25 |
| 924 | Crew B-65 | B-65 | 16 | | | 162 | 216.75 | 378.75 | 494 |
| 926 | Crew B-66 | B-66 | 8 | ↓ | | 96.40 | 115.30 | 211.70 | 279.60 |
| 15-001 | **CONCRETE** | | | | | | | | |
| 010 | Crew C-1 | C-1 | 32 | Day | | 376.40 | 21 | 397.40 | 619.10 |
| 035 | Crew C-2 | C-2 | 48 | | | 590 | 28 | 618 | 965.20 |
| 060 | Crew C-3 | C-3 | 64 | | | 804.40 | 83.65 | 888.05 | 1,417.25 |
| 085 | Crew C-4 | C-4 | 32 | | | 441.60 | 31.80 | 473.40 | 775.40 |
| 110 | Crew C-5 | C-5 | 56 | | | 738 | 384 | 1,122 | 1,642.40 |
| 135 | Crew C-6 | C-6 | 48 | | | 510.80 | 53 | 563.80 | 863.90 |
| 160 | Crew C-7 | C-7 | 64 | | | 692 | 594.75 | 1,286.75 | 1,747.40 |
| 185 | Crew C-8 | C-8 | 56 | | | 626.80 | 498 | 1,124.80 | 1,533.80 |
| 210 | Crew C-9 | C-9 | 8 | | | 94.80 | 27.10 | 121.90 | 176.20 |
| 235 | Crew C-10 | C-10 | 24 | | | 269.60 | 54.20 | 323.80 | 479.20 |
| 260 | Crew C-11 | C-11 | 72 | | | 950.80 | 1,022 | 1,972.80 | 2,729.40 |
| 285 | Crew C-12 | C-12 | 48 | | | 594.40 | 234.60 | 829 | 1,199.65 |
| 310 | Crew C-13 | C-13 | 24 | | | 311.60 | 56.40 | 368 | 583.25 |
| 335 | Crew C-14 | C-14 | 144 | | | 1,734 | 911.20 | 2,645.20 | 3,781.10 |
| 360 | Crew C-15 | C-15 | 72 | | | 848.40 | 41.10 | 889.50 | 1,391.60 |
| 385 | Crew C-16 | C-16 | 72 | | | 839.60 | 498 | 1,337.60 | 1,890.60 |
| 410 | Crew C-17 | C-17 | 80 | | | 1,040 | | 1,040 | 1,659.20L |
| 415 | Crew C-17A | C-17A | 80 | | | 1,040 | 113.10 | 1,153.10 | 1,783.60 |
| 420 | Crew C-17B | C-17B | 80 | | | 1,040 | 217.50 | 1,257.50 | 1,898.45 |
| 425 | Crew C-17C | C-17C | 80 | | | 1,040 | 330.60 | 1,370.60 | 2,022.90 |
| 430 | Crew C-17D | C-17D | 80 | | | 1,040 | 435.05 | 1,475.05 | 2,137.75 |
| 435 | Crew C-17E | C-17E | 80 | | | 1,040 | 46.65 | 1,086.65 | 1,710.50 |
| 450 | Crew C-18 | C-18 | 9 | | | 92 | 34.90 | 126.90 | 184.20 |
| 475 | Crew C-19 | C-19 | 9 | | | 92 | 54.50 | 146.50 | 205.75 |
| 500 | Crew C-20 | C-20 | 64 | | | 692 | 551 | 1,243 | 1,699.30 |
| 525 | Crew C-21 | C-21 | 64 | | | 692 | 167.40 | 859.40 | 1,277.35 |
| 550 | Crew C-22 | C-22 | 42 | | | 571.75 | 49.90 | 621.65 | 1,011.35 |
| 575 | Crew C-23 | C-23 | 80 | | | 1,028.40 | 790 | 1,818.40 | 2,507.80 |
| 580 | Crew C-23A | C-23A | 80 | | | 1,028.40 | 790 | 1,818.40 | 2,507.80 |
| 600 | Crew C-24 | C-24 | 80 | ↓ | | 1,028.40 | 1,022 | 2,050.40 | 2,763 |
| 20-001 | **MASONRY** | | | | | | | | |
| 010 | Crew D-1 | D-1 | 16 | Day | | 181.20 | | 181.20 | 282.80L |
| 035 | Crew D-2 | D-2 | 44 | | | 512.60 | | 512.60 | 801L |
| 060 | Crew D-3 | D-3 | 42 | | | 487.90 | | 487.90 | 761.90L |
| 085 | Crew D-4 | D-4 | 32 | | | 358 | 146.95 | 504.95 | 722.85 |
| 110 | Crew D-5 | D-5 | 8 | | | 100.80 | 28.30 | 129.10 | 188.30 |
| 135 | Crew D-6 | D-6 | 50 | | | 568.30 | | 568.30 | 887.50L |
| 160 | Crew D-7 | D-7 | 16 | ↓ | | 175.60 | | 175.60 | 270.40L |

For expanded coverage of these items see *Building Construction Cost Data 1985*

# 1.0 Crews

| | | | CREW | MAN-HOURS | UNIT | MAT. | LABOR | EQUIP. | TOTAL | TOTAL INCL O&P |
|---|---|---|---|---|---|---|---|---|---|---|
| 20 | 185 | Crew D-8 | D-8 | 40 | Day | | 463.20 | | 463.20 | 722.80L |
| | 210 | Crew D-9 | D-9 | 48 | | | 543.60 | | 543.60 | 848.40L |
| | 240 | Crew D-10 | D-10 | 40 | | | 481.60 | 219.80 | 701.40 | 996.20 |
| | 265 | Crew D-11 | D-11 | 24 | | | 298 | | 298 | 465.20L |
| | 290 | Crew D-12 | D-12 | 32 | | | 378.40 | | 378.40 | 590.80L |
| | 315 | Crew D-13 | D-13 | 48 | ↓ | | 580.40 | 219.80 | 800.20 | 1,152.60 |
| 25-001 | **METALS** | | | | | | | | | |
| | 010 | Crew E-1 | E-1 | 24 | Day | | 325.20 | 56.40 | 381.60 | 606.85 |
| | 035 | Crew E-2 | E-2 | 56 | | | 738 | 790 | 1,528 | 2,109.40 |
| | 060 | Crew E-3 | E-3 | 24 | | | 335.20 | 104.60 | 439.80 | 689.45 |
| | 085 | Crew E-4 | E-4 | 32 | | | 441.60 | 56.40 | 498 | 818.85 |
| | 110 | Crew E-5 | E-5 | 80 | | | 1,073.20 | 894.60 | 1,967.80 | 2,798.85 |
| | 135 | Crew E-6 | E-6 | 128 | | | 1,717.60 | 991.65 | 2,709.25 | 3,997.60 |
| | 160 | Crew E-7 | E-7 | 80 | | | 1,073.20 | 902.80 | 1,976 | 2,807.90 |
| | 185 | Crew E-8 | E-8 | 104 | | | 1,382.40 | 1,015.60 | 2,398 | 3,449.55 |
| | 210 | Crew E-9 | E-9 | 128 | | | 1,717.60 | 1,120.20 | 2,837.80 | 4,139 |
| | 235 | Crew E-10 | E-10 | 16 | | | 228.80 | 292 | 520.80 | 713.20 |
| | 260 | Crew E-11 | E-11 | 32 | | | 374 | 146.95 | 520.95 | 786.85 |
| | 285 | Crew E-12 | E-12 | 16 | | | 218.80 | 56.40 | 275.20 | 424.45 |
| | 310 | Crew E-13 | E-13 | 12 | | | 170.60 | 56.40 | 227 | 348.05 |
| | 335 | Crew E-14 | E-14 | 8 | | | 122.40 | 56.40 | 178.80 | 271.65 |
| | 360 | Crew E-15 | E-15 | 16 | ↓ | | 197.60 | 26 | 223.60 | 374.20 |
| 30-001 | **WOOD & PLASTICS** | | | | | | | | | |
| | 010 | Crew F-1 | F-1 | 8 | Day | | 98.80 | 7 | 105.80 | 164.10 |
| | 035 | Crew F-2 | F-2 | 16 | | | 197.60 | 14 | 211.60 | 328.20 |
| | 060 | Crew F-3 | F-3 | 40 | | | 498.40 | 248.60 | 747 | 1,062.65 |
| | 085 | Crew F-4 | F-4 | 48 | | | 585.20 | 540.80 | 1,126 | 1,521.70 |
| | 110 | Crew F-5 | F-5 | 32 | | | 411.20 | 14 | 425.20 | 666.60 |
| | 135 | Crew F-6 | F-6 | 40 | | | 460.80 | 248.60 | 709.40 | 1,003.45 |
| | 160 | Crew F-7 | F-7 | 32 | ↓ | | 357.60 | 14 | 371.60 | 581.80 |
| 35-001 | **MOISTURE PROTECTION** | | | | | | | | | |
| | 010 | Crew G-1 | G-1 | 56 | Day | | 620.80 | 97.70 | 718.50 | 1,137.45 |
| | 035 | Crew G-2 | G-2 | 24 | | | 261.60 | 103.75 | 365.35 | 524.10 |
| | 060 | Crew G-3 | G-3 | 32 | | | 380.80 | 22 | 402.80 | 630.55 |
| | 085 | Crew G-4 | G-4 | 24 | | | 256 | 122.40 | 378.40 | 540.20 |
| | 110 | Crew G-5 | G-5 | 40 | ↓ | | 435.20 | 97.70 | 532.90 | 829.45 |
| 40-001 | **DOORS, WINDOWS & GLAZING** | | | | | | | | | |
| | 010 | Crew H-1 | H-1 | 32 | Day | | 412 | | 412 | 676L |
| | 035 | Crew H-2 | H-2 | 24 | " | | 279.20 | | 279.20 | 438L |
| 50-001 | **FINISHES** | | | | | | | | | |
| | 010 | Crew J-1 | J-1 | 40 | Day | | 462 | 35.10 | 497.10 | 759 |
| | 035 | Crew J-2 | J-2 | 48 | | | 561.20 | 35.10 | 596.30 | 912.20 |
| | 060 | Crew J-3 | J-3 | 16 | | | 177.60 | 113.55 | 291.15 | 398.50 |
| | 085 | Crew J-4 | J-4 | 16 | ↓ | | 175.60 | | 175.60 | 270.40L |
| 55-001 | **SPECIALTIES** | | | | | | | | | |
| | 010 | Crew K-1 | K-1 | 16 | Day | | 180.80 | 174.60 | 355.40 | 477.30 |
| | 035 | Crew K-2 | K-2 | 24 | " | | 310.80 | 174.60 | 485.40 | 712.90 |
| 60-001 | **EQUIPMENT** | | | | | | | | | |
| | 010 | Crew L-1 | L-1 | 16 | Day | | 220.40 | | 220.40 | 348.40L |
| | 035 | Crew L-2 | L-2 | 16 | | | 176.80 | | 176.80 | 281.20L |
| | 060 | Crew L-3 | L-3 | 16 | | | 209 | | 209 | 331.40L |
| | 085 | Crew L-4 | L-4 | 24 | | | 279.60 | | 279.60 | 446.40L |
| | 110 | Crew L-5 | L-5 | 56 | | | 757.60 | 384 | 1,141.60 | 1,707.60 |
| | 135 | Crew L-6 | L-6 | 12 | | | 165.40 | | 165.40 | 261.60L |
| | 160 | Crew L-7 | L-7 | 28 | | | 332.60 | | 332.60 | 526.40L |
| | 185 | Crew L-8 | L-8 | 20 | | | 252.80 | | 252.80 | 400.20L |
| | 210 | Crew L-9 | L-9 | 36 | ↓ | | 405.40 | | 405.40 | 656L |

For expanded coverage of these items see *Building Construction Cost Data 1985*

# 1.0 Crews

| | | CREW | MAN-HOURS | UNIT | MAT. | LABOR | EQUIP. | TOTAL | TOTAL INCL O&P |
|---|---|---|---|---|---|---|---|---|---|
| 65-001 | **CONVEYING SYSTEMS** | | | | | | | | |
| 010 | Crew M-1 | M-1 | 32 | Day | | 419.52 | 59 | 478.52 | 732.10 |
| 035 | Crew M-2 | M-2 | 16 | " | | 204 | 14 | 218 | 332.20 |
| 75-001 | **MECHANICAL** | | | | | | | | |
| 010 | Crew Q-1 | Q-1 | 16 | Day | | 198.72 | | 198.72 | 314.80L |
| 035 | Crew Q-2 | Q-2 | 24 | | | 309.12 | | 309.12 | 489.60L |
| 060 | Crew Q-3 | Q-3 | 32 | | | 423.52 | | 423.52 | 670.80L |
| 085 | Crew Q-4 | Q-4 | 32 | | | 423.52 | 18.15 | 441.67 | 690.75 |
| 110 | Crew Q-5 | Q-5 | 16 | | | 199.44 | | 199.44 | 316L |
| 135 | Crew Q-6 | Q-6 | 24 | | | 310.24 | | 310.24 | 491.60L |
| 160 | Crew Q-7 | Q-7 | 32 | | | 425.04 | | 425.04 | 673.60L |
| 185 | Crew Q-8 | Q-8 | 32 | | | 425.04 | 18.15 | 443.19 | 693.55 |
| 210 | Crew Q-9 | Q-9 | 16 | | | 198.72 | | 198.72 | 317.60L |
| 235 | Crew Q-10 | Q-10 | 24 | | | 309.12 | | 309.12 | 494L |
| 260 | Crew Q-11 | Q-11 | 32 | | | 423.52 | | 423.52 | 676.80L |
| 285 | Crew Q-12 | Q-12 | 16 | | | 202.32 | | 202.32 | 322.40L |
| 310 | Crew Q-13 | Q-13 | 32 | | | 431.12 | | 431.12 | 687.20L |
| 335 | Crew Q-14 | Q-14 | 16 | | | 198.72 | | 198.72 | 320.40L |
| 360 | Crew Q-15 | Q-15 | 16 | | | 198.72 | 18.15 | 216.87 | 334.75 |
| 385 | Crew Q-16 | Q-16 | 24 | | | 309.12 | 18.15 | 327.27 | 509.55 |
| 410 | Crew Q-17 | Q-17 | 16 | | | 199.44 | 18.15 | 217.59 | 335.95 |
| 435 | Crew Q-18 | Q-18 | 24 | | | 310.24 | 18.15 | 328.39 | 511.55 |
| 460 | Crew Q-19 | Q-19 | 24 | | | 309.44 | | 309.44 | 489.60L |
| 485 | Crew Q-20 | Q-20 | 20 | | | 253.72 | | 253.72 | 404.40L |
| 510 | Crew Q-21 | Q-21 | 32 | ↓ | | 420.24 | | 420.24 | 665.20L |
| 80-001 | **ELECTRICAL** | | | | | | | | |
| 010 | Crew R-1 | R-1 | 48 | Day | | 600 | | 600 | 950L |
| 035 | Crew R-2 | R-2 | 56 | | | 703.20 | 175.80 | 879 | 1,307 |
| 060 | Crew R-3 | R-3 | 20 | | | 275.60 | 87.90 | 363.50 | 531.70 |
| 085 | Crew R-4 | R-4 | 40 | ↓ | | 551.60 | 56.40 | 608 | 992.45 |
| 90-001 | **TRADE RATES** | | | | | | | | |
| 010 | Skilled average | 1 Skwk | 8 | Day | | 100.80 | | 100.80 | 160.80L |
| 015 | Helpers average | 1 Help | 8 | | | 78 | | 78 | 124.80L |
| 020 | Foremen, inside | 1 Fori | 8 | | | 103.20 | | 103.20 | 164.65L |
| 025 | Foremen, outside | 1 Foro | 8 | | | 110 | | 110 | 175.50L |
| 030 | Common building laborers | 1 Clab | 8 | | | 80 | | 80 | 126.70L |
| 035 | Air tool laborer | 1 Cair | 8 | | | 80.80 | | 80.80 | 127.90L |
| 040 | Asbestos workers | 1 Asbe | 8 | | | 110.40 | | 110.40 | 177.90L |
| 045 | Boilermakers | 1 Boil | 8 | | | 110.80 | | 110.80 | 177.75L |
| 050 | Bricklayers | 1 Bric | 8 | | | 100.80 | | 100.80 | 157.20L |
| 055 | Bricklayer helpers | 1 Brhe | 8 | | | 80.40 | | 80.40 | 125.35L |
| 060 | Carpenters | 1 Carp | 8 | | | 98.80 | | 98.80 | 156.50L |
| 065 | Carpet & linoleum layer | 1 Crpt | 8 | | | 96.80 | | 96.80 | 150.95L |
| 070 | Cement finishers | 1 Cefi | 8 | | | 94.80 | | 94.80 | 146.25L |
| 075 | Electricians | 1 Elec | 8 | | | 110 | | 110 | 173.45L |
| 080 | Elevator constructors | 1 Elev | 8 | | | 110.40 | | 110.40 | 175.60L |
| 085 | Equip. oper. crane or shovel | 1 Eqhv | 8 | | | 103.20 | | 103.20 | 163.75L |
| 090 | Equip. operators med. equip | 1 Eqmd | 8 | | | 101.20 | | 101.20 | 160.55L |
| 095 | Equip. operators lt. equip | 1 Eqlt | 8 | | | 96.40 | | 96.40 | 152.95L |
| 100 | Equip. operators, oilers | 1 Eqol | 8 | | | 86.80 | | 86.80 | 137.70L |
| 105 | Equip. oper., master mechanics | 1 Eqmm | 8 | | | 107.20 | | 107.20 | 170.10L |
| 110 | Glaziers | 1 Glaz | 8 | | | 99.60 | | 99.60 | 155.75L |
| 115 | Lathers | 1 Lath | 8 | | | 99.20 | | 99.20 | 153.35L |
| 120 | Marble setters | 1 Marb | 8 | | | 98.40 | | 98.40 | 153.45L |
| 125 | Millwrights | 1 Mill | 8 | | | 102 | | 102 | 158.25L |
| 130 | Mosaic & terrazzo workers | 1 Mstz | 8 | | | 97.60 | | 97.60 | 150.25L |
| 135 | Mosaic & terrazzo helpers | 1 Mthe | 8 | | | 80 | | 80 | 123.20L |
| 140 | Painters, ordinary | 1 Pord | 8 | ↓ | | 94.80 | | 94.80 | 147.85L |

For expanded coverage of these items see *Building Construction Cost Data 1985*

# 1.0 Crews

| | | | CREW | MAN-HOURS | UNIT | MAT. | LABOR | EQUIP. | TOTAL | TOTAL INCL O&P |
|---|---|---|---|---|---|---|---|---|---|---|
| 90 | 145 | Painters, spray | 1 Pspr | 8 | Day | | 97.20 | | 97.20 | 169.85L |
| | 150 | Painters, structural steel | 1 Psst | 8 | | | 98.80 | | 98.80 | 172.65L |
| | 155 | Paperhangers | 1 Pape | 8 | | | 96.80 | | 96.80 | 150.95L |
| | 160 | Pile drivers | 1 Pile | 8 | | | 99.20 | | 99.20 | 168.95L |
| | 165 | Plasterers | 1 Plas | 8 | | | 98.80 | | 98.80 | 154.10L |
| | 170 | Plasterer helpers | 1 Plah | 8 | | | 82.80 | | 82.80 | 129.10L |
| | 175 | Plumbers | 1 Plum | 8 | | | 110.40 | | 110.40 | 174.95L |
| | 180 | Plumbers helpers | 1 Pluh | 8 | | | 90 | | 90 | 142.65L |
| | 185 | Rodmen (reinforcing) | 1 Rodm | 8 | | | 106.40 | | 106.40 | 178.40L |
| | 190 | Roofers composition | 1 Rofc | 8 | | | 92.80 | | 92.80 | 153.90L |
| | 195 | Roofers, precast | 1 Rofp | 8 | | | 94 | | 94 | 155.90L |
| | 200 | Roofers, tile & slate | 1 Rots | 8 | | | 93.60 | | 93.60 | 155.30L |
| | 205 | Roofer helpers (composition) | 1 Rohe | 8 | | | 70.40 | | 70.40 | 116.70L |
| | 210 | Sheet metal workers | 1 Shee | 8 | | | 110.40 | | 110.40 | 176.40L |
| | 215 | Sprinkler installers | 1 Spri | 8 | | | 112.40 | | 112.40 | 179.10L |
| | 220 | Steamfitters or pipefitters | 1 Stpi | 8 | | | 110.80 | | 110.80 | 175.60L |
| | 225 | Stone masons | 1 Ston | 8 | | | 99.60 | | 99.60 | 155.35L |
| | 230 | Structural steel workers | 1 Sswk | 8 | | | 106.40 | | 106.40 | 182.25L |
| | 235 | Tile layers (floor) | 1 Tilf | 8 | | | 97.20 | | 97.20 | 149.50L |
| | 240 | Tile layer helpers | 1 Tilh | 8 | | | 78.40 | | 78.40 | 120.65L |
| | 245 | Truck drivers, light | 1 Trlt | 8 | | | 82 | | 82 | 128.90L |
| | 250 | Truck drivers, heavy | 1 Trhv | 8 | | | 82.80 | | 82.80 | 130.15L |
| | 255 | Welders, structural steel | 1 Sswl | 8 | | | 106.40 | | 106.40 | 182.25L |
| | 260 | Wreckers | 1 Wrck | 8 | | | 80 | | 80 | 135.60L |

# 1.1 Overhead

| | | CREW | MAN-HOURS | UNIT | MAT. | LABOR | EQUIP. | TOTAL | TOTAL INCL O&P |
|---|---|---|---|---|---|---|---|---|---|
| 02-001 | **ARCHITECTURAL FEES** New construction, minimum | | | | | | | | |
| 002 | For work to $10,000 | | | Project | | | | | 15% |
| 004 | To $25,000 | | | | | | | | 13% |
| 006 | To $100,000 | | | | | | | | 10% |
| 04-001 | **CLEANING UP** After job completion, minimum | | | | | | | | .30% |
| 004 | Maximum | | | | | | | | 1% |
| 09-001 | **CONSTRUCTION MANAGEMENT FEES** | | | | | | | | |
| 006 | For work to $10,000 | | | Project | | | | | 10% |
| 007 | To $25,000 | | | | | | | | 9% |
| 009 | To $100,000 | | | | | | | | 6% |
| 11-001 | **CONTINGENCIES** Allowance to add at conceptual stage | | | | | | | | 20% |
| 015 | Final working drawing stage | | | | | | | | 2% |
| 18-001 | **INSURANCE** Builders risk, standard, minimum | | | Job Cost | | | | | .10% |
| 005 | Maximum | | | | | | | | .50% |
| 020 | All-risk type, minimum | | | | | | | | .12% |
| 025 | Maximum | | | | | | | | .68% |
| 040 | Contractor's equipment floater, minimum | | | Value | | | | | .90% |
| 045 | Maximum | | | " | | | | | 1.60% |
| 060 | Public liability, average | | | Job Cost | | | | | .82% |
| 081 | Workers compensation & employer's liability | | | Payroll | | | | | |
| 200 | Range of 36 trades in 50 states, excl. wrecking, min. | | | | | 1.04% | | | |
| 210 | Average | | | | | 9% | | | |
| 220 | Maximum | | | | | 54.13% | | | |
| 36-001 | **PERMITS** Rule of thumb, most cities, minimum | | | Job Cost | | | | | .50% |
| 010 | Maximum | | | " | | | | | 2% |
| 40-001 | **RENDERINGS** Water color, matted, 20" x 30", eye level, | | | | | | | | |
| 005 | Average | | | Ea. | 500 | | | 500 | 550M |

For expanded coverage of these items see *Building Construction Cost Data 1985*

## 1.1 Overhead

| | | CREW | MAN-HOURS | UNIT | MAT. | LABOR | EQUIP. | TOTAL | TOTAL INCL O&P |
|---|---|---|---|---|---|---|---|---|---|
| 44-001 | **SCAFFOLDING SPECIALTIES** | | | | | | | | |
| 010 | Building exterior, 1 to 5 stories | 3 Carp | 1.430 | C.S.F. | 9.60 | 17.65 | | 27.25 | 39 |
| 48-001 | **SMALL TOOLS** As % of contractor's work, minimum | | | Total | | | | | .50% |
| 010 | Maximum | | | " | | | | | 2% |
| 50-001 | **SURVEYING** Conventional, topographical, minimum | | | | | | | | |
| 030 | Lot location and lines, minimum | | | Acre | | | | | 310 |
| 032 | Average | | | | | | | | 500 |
| 040 | Maximum | | | ↓ | | | | | 815 |
| 110 | Crew for building layout, 2 man crew | | | Day | | | | | 410 |
| 120 | 3 man crew | | | " | | | | | 580 |
| 54-001 | **TARPAULINS** Cotton duck, 10 oz. to 13.13 oz. per S.Y., minimum | | | S.F. | .25 | | | .25 | .27M |
| 005 | Maximum | | | | .42 | | | .42 | .46M |
| 020 | Reinforced polyethylene 3 mils thick, white | | | | .04 | | | .04 | .04M |
| 030 | 4 mils thick, white, clear or black | | | ↓ | .05 | | | .05 | .05M |
| 56-001 | **TAXES** Sales tax, State, County & City, average | | | % | 4 | | | | |
| 005 | Maximum | | | | 8 | | | | |
| 020 | Social Security, on first $40,000 of wages | | | | | 7.05 | | | |
| 030 | Unemployment, Ma., combined Federal and State, minimum | | | | | 2.30 | | | |
| 035 | Average | | | | | 5.50 | | | |
| 040 | Maximum | | | ↓ | | 6.50 | | | |

## 1.5 Contractor Equipment

| | | UNIT | HOURLY OPER. COST | RENT PER DAY | RENT PER WEEK | RENT PER MONTH | CREWS EQUIPMENT COST |
|---|---|---|---|---|---|---|---|
| 05-001 | **EARTHWORK EQUIPMENT RENTAL** Without operators unless | | | | | | |
| 002 | noted by* | | | | | | |
| 004 | Aggregate Spreader, push type 8' to 12' wide | Ea. | .62 | 77 | 225 | 675 | 49.95 |
| 005 | Augers for truck or trailer mounting, vertical drilling | | | | | | |
| 006 | 4" to 36" diam., 54 H.P., gas, 10' spindle travel | Ea. | 6.65 | 150 | 455 | 1,450 | 144.20 |
| 007 | 14' spindle travel | | 6.95 | 180 | 540 | 1,625 | 163.60 |
| 008 | Auger, horizontal boring machine, 12" to 36" diameter, 45 H.P. | | 6.15 | 590 | 1,750 | 5,300 | 399.20 |
| 009 | 12" to 48" diameter, 65 H.P. | | 8.40 | 620 | 2,050 | 6,050 | 477.20 |
| 010 | Backhoe, diesel hydraulic, crawler mounted, 1/2 C.Y. capacity | | 6.25 | 480* | 1,025 | 3,325 | 255 |
| 012 | 5/8 C.Y. capacity | | 7.25 | 505* | 1,050 | 3,400 | 268 |
| 014 | 3/4 C.Y. capacity | | 10.40 | 560* | 1,200 | 3,875 | 323.20 |
| 015 | 1 C.Y. capacity | | 13.40 | 810* | 1,475 | 4,300 | 402.20 |
| 020 | 1-1/2 C.Y. capacity | | 16.90 | 960* | 2,125 | 6,175 | 560.20 |
| 030 | 2 C.Y. capacity | | 26 | 1,150* | 2,825 | 8,550 | 773 |
| 032 | 2-1/2 C.Y. capacity | | 41 | 1,450* | 4,075 | 11,500 | 1,143 |
| 034 | 3-1/2 C.Y. capacity | | 48 | 1,750* | 5,025 | 15,000 | 1,389 |
| 035 | Gradall type, truck mounted, 3 ton @ 15' radius, 5/8 C.Y. | | 17.15 | 850* | 1,475 | 4,450 | 432.20 |
| 037 | 1 C.Y. capacity | | 23 | 1,025* | 2,175 | 6,500 | 619 |
| 040 | Backhoe-loader, wheel type, 40 to 45 H.P., 5/8 C.Y. capacity | | 4.66 | 325* | 390 | 1,200 | 115.30 |
| 045 | 45 H.P. to 60 H.P., 3/4 C.Y. capacity | | 5.65 | 395* | 535 | 1,650 | 152.20 |
| 046 | 80 H.P., 1-1/4 C.Y. capacity | | 8.50 | 450* | 800 | 2,250 | 228 |
| 050 | Brush chipper, gas engine, 6" cutter head, 35 H.P. | | 3.08 | 120 | 385 | 1,125 | 101.65 |
| 055 | 12" cutter head, 130 H.P. | | 6.20 | 160 | 425 | 1,250 | 134.60 |
| 060 | 15" cutter head, 165 H.P. | | 7.95 | 195 | 480 | 1,475 | 159.60 |
| 075 | Bucket, clamshell, general purpose, 3/8 C.Y. | | .40 | 33 | 100 | 290 | 23.20 |
| 080 | 1/2 C.Y. | | .55 | 38 | 120 | 360 | 28.40 |
| 085 | 3/4 C.Y. | | .62 | 50 | 145 | 440 | 33.95 |
| 090 | 1 C.Y. | | .73 | 52 | 160 | 480 | 37.85 |
| 095 | 1-1/2 C.Y. | | 1.05 | 65 | 200 | 600 | 48.40 |
| 100 | 2 C.Y. | | 1.18 | 82 | 270 | 820 | 63.45 |
| 101 | Bucket, dragline, medium duty, 1/2 C.Y. | ↓ | .35 | 25 | 65 | 190 | 15.80 |
| 102 | 3/4 C.Y. | | .41 | 29 | 87 | 270 | 20.70 |

For expanded coverage of these items see *Building Construction Cost Data 1985*

## 1.5 Contractor Equipment

| | | Description | UNIT | HOURLY OPER. COST | RENT PER DAY | RENT PER WEEK | RENT PER MONTH | CREWS EQUIPMENT COST |
|---|---|---|---|---|---|---|---|---|
| 05 | 103 | 1 C.Y. | Ea. | .47 | 35 | 105 | 310 | 24.75 |
| | 104 | 1-1/2 C.Y. | | .54 | 43 | 130 | 375 | 30.30 |
| | 105 | 2 C.Y. | | .67 | 56 | 170 | 505 | 39.35 |
| | 107 | 3 C.Y. | | .99 | 88 | 250 | 730 | 57.90 |
| | 120 | Compactor, roller, 2 drum, 2000 lb., operator walking | | 2.39 | 75 | 265 | 800 | 72.10 |
| | 125 | Rammer compactor, gas, 1000 lb. blow | | .97 | 50 | 145 | 445 | 36.75 |
| | 130 | Vibratory plate, gas, 13" plate, 1000 lb. blow | | .52 | 40 | 115 | 345 | 27.15 |
| | 135 | 24" plate, 5000 lb. blow | | .69 | 49 | 165 | 480 | 38.50 |
| | 137 | Curb builder, 14 H.P., gas, single screw | | 2.82 | 87 | 260 | 790 | 74.55 |
| | 139 | Double screw | ↓ | 2.56 | 100 | 300 | 915 | 80.50 |
| | 175 | Extractor, piling, see lines 250 to 275 | | | | | | |
| | 180 | | | | | | | |
| | 186 | Grader, self-propelled, 25,000 lb. | Ea. | 11.80 | 405 | 1,220 | 4,325 | 338.40 |
| | 191 | 30,000 lb. | | 16.50 | 485 | 1,450 | 5,100 | 422 |
| | 192 | 40,000 lb. | | 22 | 635 | 1,900 | 6,300 | 556 |
| | 193 | 55,000 lb. | | 28 | 950 | 2,850 | 9,100 | 794 |
| | 195 | Hammer, pavement demo., hyd., gas, self-prop., 1000 to 1250 lb. | | 6.15 | 260 | 800 | 2,400 | 209.20 |
| | 200 | 1300 to 1500 lb. | | 6.60 | 285 | 850 | 2,550 | 222.80 |
| | 205 | Pile driving hammer, steam or air, 4150 ft.-lb. @ 225 BPM | | 3.80 | 125 | 380 | 1,125 | 106.40 |
| | 210 | 8750 ft.-lb. @ 145 BPM | | 5.10 | 195 | 585 | 1,600 | 157.80 |
| | 215 | 15,000 ft.-lb. @ 60 BPM | | 5.90 | 280 | 750 | 1,950 | 197.20 |
| | 220 | 24,450 ft.-lb. @ 111 BPM | ↓ | 7.20 | 355 | 1,100 | 2,825 | 277.60 |
| | 225 | Leads, 15,000 ft.-lb. hammers | L.F. | | | | 11 | .70 |
| | 230 | 24,450 ft.-lb. hammers and heavier | " | | | | 12 | .80 |
| | 235 | Diesel type hammer, 22,400 ft.-lb. | Ea. | 8.35 | 425 | 1,250 | 2,950 | 316.80 |
| | 240 | 41,300 ft.-lb. | | 12.50 | 610 | 2,100 | 4,900 | 520 |
| | 245 | 141,000 ft.-lb. | | 21 | 1,050 | 3,125 | 9,450 | 793 |
| | 250 | Vib. elec. hammer/ext., 200 KW diesel generator, 34 H.P. | | 13.40 | 570 | 1,725 | 5,150 | 452.20 |
| | 255 | 80 H.P. | | 18.15 | 1,050 | 3,125 | 9,400 | 770.20 |
| | 260 | 150 H.P. | | 21 | 1,350 | 4,075 | 10,000 | 983 |
| | 270 | Extractor, steam or air, 700 ft.-lb. | | .90 | 150 | 425 | 1,100 | 92.20 |
| | 275 | 1000 ft.-lb. | | 1.39 | 230 | 620 | 1,475 | 135.10 |
| | 300 | Roller, tandem, gas, 3 to 5 ton | | 3.23 | 100 | 290 | 890 | 83.85 |
| | 305 | Diesel, 8 to 12 ton | | 6.60 | 180 | 600 | 1,550 | 172.80 |
| | 310 | Towed type, vibratory, gas 12.5 H.P., 2 ton | | 2.19 | 92 | 270 | 810 | 71.50 |
| | 315 | Sheepsfoot, double 60" x 60" | | 2.34 | 100 | 300 | 900 | 78.70 |
| | 320 | Pneumatic tire diesel roller, 12 ton | | 8.40 | 145 | 450 | 1,350 | 157.20 |
| | 325 | 21 to 25 ton | | 13.10 | 320 | 950 | 2,900 | 294.80 |
| | 330 | Sheepsfoot roller, self-propelled, 4 wheel, 130 H.P. | | 10.65 | 350 | 990 | 3,050 | 283.20 |
| | 332 | 300 H.P. | | 15.95 | 435 | 1,300 | 3,950 | 387.60 |
| | 335 | Vibratory steel drum & pneumatic tire, diesel, 18,000 lb. | | 12 | 305 | 905 | 2,725 | 277 |
| | 340 | 29,000 lb. | | 15 | 365 | 1,100 | 3,275 | 340 |
| | 345 | Scrapers, towed type, 7 to 9 C.Y. capacity | | 2.65 | 115 | 340 | 1,025 | 89.20 |
| | 350 | 12 to 17 C.Y. capacity | | 4.41 | 140 | 420 | 1,250 | 119.30 |
| | 355 | Self-propelled, 4 x 4 drive, 2 engine, 14 C.Y. capacity | | 38 | 1,375* | 3,300 | 10,250 | 964 |
| | 360 | 1 engine, 24 C.Y. capacity | | 45 | 1,525* | 3,775 | 11,200 | 1,115 |
| | 365 | Self-loading, 11 C.Y. capacity | | 34 | 860* | 1,800 | 5,350 | 632 |
| | 370 | 22 C.Y. capacity | ↓ | 51 | 1,200* | 2,700 | 8,200 | 948 |
| | 385 | Shovels, see Cranes division 1.5-20 | | | | | | |
| | 386 | Shovel front attachment, mechanical, 1/2 C.Y. | Ea. | .59 | 59 | 180 | 525 | 40.70 |
| | 387 | 3/4 C.Y. | | 2.56 | 94 | 280 | 830 | 76.50 |
| | 388 | 1 C.Y. | | 2.77 | 145 | 430 | 1,275 | 108.15 |
| | 389 | 1-1/2 C.Y. | | 3.02 | 165 | 475 | 1,450 | 119.15 |
| | 391 | 3 C.Y. | | 5.65 | 290 | 900 | 2,675 | 225.20 |
| | 411 | Tractor, crawler, with bulldozer, torque converter, diesel 75 H.P. | | 6.95 | 425 * | 715 | 2,150 | 198.60 |
| | 415 | 105 H.P. | | 10.70 | 560 * | 1,250 | 3,825 | 335.60 |
| | 420 | 140 H.P. | | 14.25 | 660 * | 1,550 | 4,625 | 424 |
| | 426 | 200 H.P. | | 19.30 | 850 * | 2,300 | 7,500 | 614.40 |
| | 431 | 300 H.P. | | 27 | 1,125* | 3,415 | 9,775 | 899 |
| | 436 | 410 H.P. | ↓ | 39 | 1,375* | 4,150 | 12,700 | 1,142 |

For expanded coverage of these items see *Building Construction Cost Data 1985*

## 1.5 Contractor Equipment

| | | UNIT | HOURLY OPER. COST | RENT PER DAY | RENT PER WEEK | RENT PER MONTH | CREWS EQUIPMENT COST |
|---|---|---|---|---|---|---|---|
| 05 438 | 700 H.P. | Ea. | 68 | 2,150* | 6,950 | 20,800 | 1,934 |
| 440 | Loader, crawler, torque converter, diesel, 1-1/2 C.Y., 80 H.P. | | 10.50 | 495* | 925 | 2,750 | 269 |
| 445 | 1-1/2 to 1-3/4 C.Y., 95 H.P. | | 12.45 | 530* | 1,025 | 3,175 | 304.60 |
| 451 | 1-3/4 to 2-1/4 C.Y., 130 H.P. | | 16.35 | 640* | 1,400 | 4,275 | 410.80 |
| 453 | 2-1/2 to 3-1/4 C.Y., 190 H.P. | | 25 | 805* | 1,925 | 5,775 | 585 |
| 456 | 4-1/2 to 5 C.Y., 275 H.P. | | 35 | 1,200* | 3,525 | 10,700 | 985 |
| 461 | Tractor loader, wheel, torque conv 4 x 4, 1 to 1-1/4 C.Y., 65 H.P. | | 8.25 | 415* | 665 | 2,125 | 199 |
| 462 | 1-1/2 to 1-3/4 C.Y., 80 H.P. | | 10.70 | 485* | 905 | 2,725 | 266.60 |
| 465 | 1-3/4 to 2 C.Y., 100 H.P. | | 12.55 | 540* | 1,100 | 3,300 | 320.40 |
| 471 | 2-1/2 to 3-1/2 C.Y., 130 H.P. | | 17.85 | 640* | 1,475 | 4,425 | 437.80 |
| 473 | 3 to 4-1/2 C.Y., 170 H.P. | | 22 | 750* | 1,800 | 4,850 | 536 |
| 476 | 5-1/4 to 5-3/4 C.Y., 270 H.P. | | 35 | 1,025* | 2,775 | 9,175 | 835 |
| 481 | 7 to 8 C.Y., 375 H.P. | | 59 | 1,450* | 4,000 | 10,400 | 1,272 |
| 487 | 12-1/2 C.Y., 690 H.P. | | 96 | 2,600* | 7,950 | 24,000 | 2,358 |
| 488 | Wheeled, skid steer, 10 C.F., 30 H.P. gas | | 3.48 | 87 | 330 | 815 | 93.85 |
| 489 | 1 C.Y., 78 H.P., diesel | | 4.82 | 150 | 500 | 1,450 | 138.55 |
| 490 | Trencher, chain, boom type, gas, operator walking, 12 H.P. | | 8.05 | 98 | 345 | 1,050 | 133.40 |
| 491 | Operator riding, 40 H.P. | | 12.85 | 165 | 475 | 1,375 | 197.80 |
| 500 | Wheel type, gas, 4' deep, 12" wide | | 11.25 | 440* | 730 | 2,200 | 236 |
| 510 | Diesel, 6' deep, 20" wide | | 15 | 610* | 1,450 | 4,400 | 410 |
| 515 | Ladder type, gas, 5' deep, 8" wide | | 8.55 | 485* | 950 | 2,775 | 258.40 |
| 520 | Diesel, 8' deep, 16" wide | | 10.70 | 650* | 1,650 | 5,150 | 415.60 |
| 525 | Truck, dump, tandem, 12 ton payload | | 13.10 | 350* | 575 | 1,725 | 219.80 |
| 530 | Three axle dump, 16 ton payload | | 16.15 | 435* | 750 | 2,275 | 279.20 |
| 535 | Dump trailer only, rear dump, 16-1/2 C.Y. | | 3.91 | 80 | 240 | 715 | 79.30 |
| 540 | 20 C.Y. | | 4.76 | 92 | 280 | 845 | 94.10 |
| 545 | Flatbed, single axle, 1-1/2 ton rating | | 3.77 | 46 | 135 | 415 | 57.15 |
| 550 | 3 ton rating | | 4.30 | 58 | 160 | 505 | 66.40 |
| 555 | Off highway rear dump, 25 ton capacity | | 25 | 840* | 2,200 | 6,750 | 640 |
| 560 | 35 ton capacity | ↓ | 34 | 1,150* | 3,150 | 9,475 | 902 |
| 10-001 | **CONCRETE EQUIPMENT RENTAL** | | | | | | |
| 010 | without operators unless noted by * | | | | | | |
| 020 | Bucket, concrete lightweight, 1/2 C.Y. | Ea. | .12 | 20 | 50 | 125 | 10.95 |
| 030 | 1 C.Y. | | .12 | 29 | 70 | 190 | 14.95 |
| 040 | 1-1/2 C.Y. | | .19 | 23 | 75 | 220 | 16.50 |
| 050 | 2 C.Y. | | .19 | 31 | 105 | 315 | 22.50 |
| 060 | Cart, concrete, operator walking, 10 C.F. | | .74 | 48 | 145 | 390 | 34.90 |
| 070 | Operator riding, 18 C.F. | | 1.31 | 69 | 220 | 620 | 54.50 |
| 080 | Conveyer for concrete, portable, gas, 16" wide, 26' long | | 1.37 | 75 | 205 | 595 | 51.95 |
| 090 | 46' long | | 1.59 | 135 | 425 | 1,250 | 97.70 |
| 100 | 56' long | | 1.80 | 170 | 500 | 1,450 | 114.40 |
| 110 | Core drill, electric, 2-1/2 H.P., 1" to 8" bit diameter | | .58 | 45 | 115 | 360 | 27.65 |
| 120 | Finisher, concrete floor, gas, riding trowel, 48" diameter | | 1.60 | 90 | 250 | 610 | 62.80 |
| 130 | Gas, manual, 3 blade, 36" trowel | | .38 | 30 | 90 | 265 | 21.05 |
| 140 | 4 blade, 48" trowel | | .51 | 38 | 115 | 315 | 27.10 |
| 150 | Float, hand-operated (Bull float) 48" wide | | .12 | 9 | 26 | 45 | 6.15 |
| 160 | Grinder, concrete and terrazzo, electric, floor | | .88 | 37 | 110 | 325 | 29.05 |
| 170 | Wall grinder | | .29 | 17 | 43 | 180 | 10.90 |
| 180 | Mixer, powered, mortar and concrete, gas, 6 C.F., 18 H.P. | | 2.01 | 33 | 95 | 260 | 35.10 |
| 190 | 10 C.F., 25 H.P. | | 2.93 | 36 | 125 | 350 | 48.45 |
| 200 | 16 C.F. | | 3.06 | 90 | 300 | 885 | 84.50 |
| 210 | Concrete, stationary, tilt drum, 2 C.Y. | | 5.80 | 250 | 765 | 2,300 | 199.40 |
| 212 | Pump, concrete, truck mounted, 4" line, 80' boom | | 18.50 | 770* | 1,750 | 5,150 | 498 |
| 214 | 5" line, 110' boom | | 28 | 830* | 1,950 | 5,875 | 614 |
| 216 | Mud jack, 47 C.F. per hr. | | 1.10 | 27 | 80 | 240 | 24.80 |
| 218 | 225 C.F. per hr. | | 2.97 | 140 | 400 | 1,175 | 103.75 |
| 260 | Saw, concrete, manual, gas, 18 H.P. | | 1.65 | 37 | 110 | 335 | 35.20 |
| 265 | Self-propelled, gas, 30 H.P. | | 2.67 | 55 | 165 | 495 | 54.35 |
| 270 | Vibrators, concrete, electric, 60 cycle, 2 H.P. | | .23 | 20 | 65 | 190 | 14.85 |
| 280 | 3 H.P. | ↓ | .29 | 26 | 75 | 225 | 17.30 |

For expanded coverage of these items see *Building Construction Cost Data 1985*

## 1.5 Contractor Equipment

| | | | UNIT | HOURLY OPER. COST | RENT PER DAY | RENT PER WEEK | RENT PER MONTH | CREWS EQUIPMENT COST |
|---|---|---|---|---|---|---|---|---|
| 10 | 290 | Gas engine, 3 H.P. | Ea. | .66 | 21 | 65 | 190 | 18.30 |
| | 300 | 5 H.P. | " | .94 | 33 | 95 | 265 | 26.50 |
| | 15-001 | **GENERAL EQUIPMENT RENTAL** | | | | | | |
| | 010 | Without operators unless noted by* | | | | | | |
| | 015 | Aerial lift, scissor type, to 15' high, 1000 lb. capacity | Ea. | .15 | 77 | 248 | 745 | 50.80 |
| | 016 | To 25' high, 2000 lb. capacity | | .18 | 115 | 365 | 1,100 | 74.45 |
| | 017 | Telescoping boom to 40' high, 750 lb. capacity | | .21 | 240 | 700 | 2,250 | 141.70 |
| | 018 | 2000 lb. capacity | | .24 | 300 | 1,000 | 3,000 | 201.90 |
| | 019 | To 60' high, 750 lb. capacity | | .27 | 540 | 1,250 | 3,300 | 252.15 |
| | 020 | Air compressor, portable, gas engine, 60 C.F.M. | | 3.54 | 34 | 95 | 290 | 47.30 |
| | 030 | 160 C.F.M. | | 4.28 | 51 | 155 | 460 | 65.25 |
| | 040 | Diesel engine, rotary screw, 250 C.F.M. | | 6.90 | 83 | 255 | 745 | 106.20 |
| | 050 | 365 C.F.M. | | 10.35 | 105 | 330 | 935 | 148.80 |
| | 060 | 600 C.F.M. | | 17.15 | 160 | 495 | 1,400 | 236.20 |
| | 070 | 750 C.F.M. | | 21 | 205 | 605 | 1,800 | 289 |
| | 080 | For silenced models, small sizes, add | | 3% | 5% | 5% | | |
| | 090 | Large sizes, add | ▼ | 4% | 7% | 7% | | |
| | 092 | Air tools and accessories | | | | | | |
| | 093 | Breaker, pavement, 60 lb. | Ea. | .43 | 15 | 42 | 125 | 11.85 |
| | 094 | 80 lb. | | .56 | 17 | 50 | 145 | 14.50 |
| | 095 | Drills, hand (jackhammer) 65 lb. | | .57 | 16 | 55 | 170 | 15.55 |
| | 096 | Wagon, swing boom, 4" drifter | | 7.35 | 185 | 550 | 1,600 | 168.80 |
| | 097 | 5" drifter | | 7.65 | 210 | 810 | 2,350 | 223.20 |
| | 098 | Dust control per drill | | .12 | 10 | 25 | 75 | 5.95 |
| | 099 | Hammer, chipping, 12 lb. | | .15 | 13 | 35 | 100 | 8.20 |
| | 100 | Hose, air with couplings, 50' long, 3/4" diameter | | .10 | 4 | 12 | 25 | 3.20 |
| | 110 | 1" diameter | | .10 | 5.50 | 14 | 35 | 3.60 |
| | 120 | 1-1/2" diameter | | .10 | 9.35 | 25 | 72 | 5.80 |
| | 130 | 2" diameter | | .11 | 12 | 35 | 100 | 7.90 |
| | 140 | 2-1/2" diameter | | .12 | 12 | 37 | 105 | 8.35 |
| | 141 | 3" diameter | | .12 | 12.30 | 37 | 110 | 8.35 |
| | 145 | Drill, steel, 7/8" x 2' | | | 3.50 | 10 | 16 | 2 |
| | 146 | 7/8" x 6' | | | 5 | 11 | 23 | 2.20 |
| | 152 | Moil points | | .67 | 2.50 | 8 | 12 | 6.95 |
| | 153 | 20 Sheeting driver for 60 lb. breaker | | .08 | 8.10 | 19 | 53 | 4.45 |
| | 154 | For 125 lb. breaker | | .08 | 8.10 | 21 | 64 | 4.85 |
| | 155 | Spade | | .13 | 4.50 | 11 | 33 | 3.25 |
| | 156 | Tamper, single, 35 lb. | | .43 | 15 | 45 | 135 | 12.45 |
| | 157 | Triple, 140 lb. | | 1.41 | 30 | 95 | 285 | 30.30 |
| | 158 | Wrenches, impact, air powered up to 3/4" bolt | | .08 | 19 | 58 | 170 | 12.25 |
| | 159 | Up to 1-1/4" bolt | | .09 | 26 | 76 | 215 | 15.90 |
| | 160 | Barricades, barrels, reflectorized 1 to 50 barrels | | | .48 | 1.42 | 4.30 | .30 |
| | 161 | 100 to 200 barrels | | | .40 | 1.19 | 3.52 | .20 |
| | 162 | Barrels with flashers, 1 to 50 barrels | | | .80 | 2.38 | 7.10 | .45 |
| | 163 | 100 to 200 barrels | | | .73 | 2.15 | 6.45 | .40 |
| | 164 | Barrels with steady burn type C lights | | | .87 | 2.44 | 7.90 | .50 |
| | 165 | Illuminated board, trailer mounted, with generator | | 1.08 | 81 | 310 | 785 | 70.65 |
| | 167 | Portable, stock, with flashers, 1 to 6 units | | | .56 | 1.64 | 4.80 | .30 |
| | 168 | 25 to 50 units | | | .50 | 1.36 | 4.10 | .25 |
| | 170 | Carts, brick, hand powered, 1000 lb. capacity | | .96 | 14.10 | 41 | 120 | 15.90 |
| | 180 | Gas engine, 1500 lb., 7-1/2 ft. lift | | 1.18 | 70 | 205 | 615 | 50.45 |
| | 183 | Distributor, asphalt, trailer mtd, 2000 gal., 38 H.P. diesel | | 4.80 | 260 | 855 | 2,550 | 209.40 |
| | 184 | 3000 gal., 38 H.P. diesel | | 5.35 | 300 | 990 | 2,625 | 240.80 |
| | 185 | Drill, rotary hammer, electric, 1-1/2" diameter | | .12 | 32 | 95 | 220 | 19.95 |
| | 186 | Carbide bit for above | | | 10 | 30 | 45 | 6 |
| | 187 | Emulsion sprayer, 65 gal., 5 H.P. engine | | 1.01 | 17 | 50 | 170 | 18.10 |
| | 188 | 200 gal., 5 H.P. engine | ▼ | 1.50 | 34 | 100 | 335 | 32 |
| | 190 | Fencing, see division 1.1-58 and 2.7-09 | | | | | | |
| | 192 | Floodlight, mercury, vapor or quartz, on tripod | | | | | | |
| | 193 | 1000 watt | Ea. | | 20 | 60 | 180 | 196 |

For expanded coverage of these items see *Building Construction Cost Data 1985*

# 1.5 Contractor Equipment

| | | UNIT | HOURLY OPER. COST | RENT PER DAY | RENT PER WEEK | RENT PER MONTH | CREWS EQUIPMENT COST |
|---|---|---|---|---|---|---|---|
| 15 194 | 2000 watt | Ea. | 35 | 100 | 300 | | 340 |
| 195 | Floodlights, trailer mounted with generator, 1-300 watt light | | 3.75 | 11 | 33 | 100 | 36.60 |
| 200 | 4-300 watt lights | | 4.44 | 23 | 70 | 220 | 49.50 |
| 202 | Forklift, wheeled, for brick, 18', 3000 lb. | | 6.10 | 145 | 470 | 1,375 | 142.80 |
| 204 | 28', 4000 lb. | | 7.50 | 185 | 545 | 1,650 | 169 |
| 210 | Generator, electric, gas engine, 1.5 KW to 3 KW | | .50 | 22 | 67 | 195 | 17.40 |
| 220 | 5 KW | | .82 | 30 | 125 | 315 | 31.55 |
| 230 | 10 KW | | 1.93 | 55 | 180 | 540 | 51.45 |
| 240 | 25 KW | | 3.64 | 84 | 300 | 780 | 89.10 |
| 250 | Diesel engine, 20 KW | | 3.25 | 59 | 190 | 575 | 64 |
| 260 | 50 KW | | 3.50 | 130 | 330 | 985 | 94 |
| 270 | 100 KW | | 8.10 | 205 | 520 | 1,500 | 168.80 |
| 280 | 250 KW | | 14.40 | 285 | 860 | 2,575 | 287.20 |
| 285 | Hammer, hydraulic, for mounting on boom, to 500 ft.-lb. | | 2.25 | 180 | 540 | 1,600 | 126 |
| 286 | 500 to 1200 ft.-lb. | | 2.55 | 260 | 775 | 2,300 | 175.40 |
| 290 | Heaters, space, oil or electric, 50 MBH | | .99 | 14 | 40 | 125 | 15.90 |
| 300 | 100 MBH | | 1.24 | 20 | 56 | 160 | 21.10 |
| 310 | 300 MBH | | 1.70 | 32 | 95 | 280 | 32.60 |
| 315 | 500 MBH | | 2.15 | 46 | 135 | 405 | 44.20 |
| 320 | Hose, water, suction with coupling, 20' long, 2" diameter | | | 7 | 17 | 45 | 3.40 |
| 321 | 3" diameter | | | 9 | 23 | 65 | 4.60 |
| 322 | 4" diameter | | | 12 | 33 | 100 | 6.60 |
| 323 | 6" diameter | | | 25 | 65 | 170 | 13 |
| 324 | 8" diameter | | | 34 | 83 | 255 | 16.60 |
| 325 | Discharge hose with coupling, 50' long, 2" diameter | | | 7 | 17 | 45 | 3.40 |
| 326 | 3" diameter | | | 9 | 23 | 60 | 4.60 |
| 327 | 4" diameter | | | 12 | 35 | 90 | 7 |
| 328 | 6" diameter | | | 24 | 62 | 160 | 12.40 |
| 329 | 8" diameter | | | 40 | 87 | 180 | 17.40 |
| 330 | Ladders, extension type, 16' to 36' long | | | 9 | 27 | 65 | 5.40 |
| 340 | 40' to 60' long | ↓ | | 18 | 56 | 165 | 11.20 |
| 341 | Level, laser type, for pipe laying, self leveling | | | | | | |
| 343 | Manual leveling | Ea. | | 125 | 375 | 650 | 75 |
| 344 | Rotary beacon with rod and sensor | | | 140 | 425 | 795 | 85 |
| 346 | Builders level with tripod and rod | | | 15 | 120 | 144 | 24 |
| 350 | Light towers, towable, with generator, 1000 watt | | 1.28 | 78 | 200 | 610 | 50.25 |
| 360 | 2000 watt | | 1.93 | 110 | 335 | 1,000 | 82.45 |
| 370 | Mixer, powered, plaster and mortar, 6 C.F., 7 H.P. | | .91 | 30 | 90 | 255 | 25.30 |
| 380 | 10 C.F., 9 H.P. | | 1.21 | 35 | 120 | 330 | 33.70 |
| 390 | Paint sprayers complete, 8 CFM | | | 31 | 85 | 255 | 17 |
| 400 | 17 CFM | | | 39 | 130 | 330 | 26 |
| 402 | Pavers, bituminous, rubber tires, 8' wide, 52 H.P., gas | | 11.75 | 385 | 1,275 | 3,750 | 349 |
| 403 | 8' wide, 64 H.P., diesel | | 10.90 | 395 | 1,425 | 4,200 | 372.20 |
| 405 | Crawler, 10' wide, 78 H.P., gas | | 19.25 | 650 | 1,900 | 5,700 | 534 |
| 406 | 10' wide, 87 H.P., diesel | | 15.35 | 765 | 2,550 | 7,500 | 632.80 |
| 407 | Concrete paver, 12' to 24' wide, 250 H.P. | | 30 | 1,300 | 4,050 | 12,500 | 1,050 |
| 408 | Placer-spreader-trimmer, 24' wide, 300 H.P. | | 34 | 1,375 | 4,175 | 12,600 | 1,107 |
| 410 | Pump, centrifugal gas pump, 1-1/2"-4 MGPH | | .42 | 16 | 48 | 135 | 12.95 |
| 420 | 2"-8 MGPH | | .46 | 22 | 64 | 190 | 16.50 |
| 430 | 3"-15 MGPH | | 1.01 | 27 | 80 | 225 | 24.10 |
| 440 | 6"-90 MGPH | | 1.63 | 80 | 240 | 700 | 61.05 |
| 450 | Submersible electric pump, 1-1/4"-55 GPM | | .14 | 18 | 50 | 150 | 11.10 |
| 460 | 1-1/2"-83 GPM | | .22 | 22 | 67 | 185 | 15.15 |
| 470 | 2"-120 GPM | | .34 | 31 | 85 | 240 | 19.70 |
| 480 | 3"-300 GPM | | .60 | 41 | 115 | 350 | 27.80 |
| 490 | 4"-560 GPM | | 1.60 | 51 | 150 | 425 | 42.80 |
| 500 | 6"-1590 GPM | | 2.05 | 135 | 400 | 1,200 | 96.40 |
| 510 | Diaphragm pump, gas, single, 1-1/2" diameter | | .44 | 14 | 40 | 121 | 11.50 |
| 520 | 2" diameter | | .58 | 23 | 71 | 200 | 18.85 |
| 530 | 3" diameter | ↓ | .91 | 29 | 99 | 245 | 27.10 |

For expanded coverage of these items see *Building Construction Cost Data 1985*

## 1.5 Contractor Equipment

| | | UNIT | HOURLY OPER. COST | RENT PER DAY | RENT PER WEEK | RENT PER MONTH | CREWS EQUIPMENT COST |
|---|---|---|---|---|---|---|---|
| 15 540 | Double, 4" diameter | Ea. | 1.13 | 49 | 150 | 440 | 39.05 |
| 550 | Trash pump, self-priming, gas, 2" diameter | ↓ | .43 | 23 | 68 | 210 | 17.05 |
| 560 | Diesel, 4" diameter | | 1.23 | 58 | 180 | 550 | 45.85 |
| 565 | Diesel, 6" diameter | ↓ | 2.60 | 89 | 280 | 815 | 76.80 |
| 566 | Rollers, see division 1.5-05 | | | | | | |
| 570 | Salamanders, L.P. gas fired, 100,000 B.T.U. | Ea. | .50 | 14 | 35 | 110 | 11 |
| 572 | Sandblaster, portable, open top, 3 C.F. capacity | | 1.22 | 40 | 115 | 300 | 32.75 |
| 573 | 6 C.F. capacity | | 2.52 | 50 | 145 | 420 | 49.15 |
| 574 | Accessories for above | | | 13 | 40 | 120 | 8 |
| 580 | Saw, chain, gas engine, 18" long | | .50 | 26 | 93 | 260 | 22.60 |
| 590 | 36" long | | .65 | 31 | 115 | 325 | 28.20 |
| 595 | 60" long | | 1.10 | 38 | 145 | 400 | 37.80 |
| 600 | Masonry, table mounted, 14" diameter, 5 H.P. | | .79 | 31 | 110 | 220 | 28.30 |
| 610 | Circular, hand held, electric, 7" diameter | | .15 | 11 | 29 | 85 | 7 |
| 620 | 12" diameter | | .16 | 20 | 72 | 185 | 15.70 |
| 630 | Steam cleaner, 100 gallons per hour | | .29 | 34 | 105 | 305 | 23.30 |
| 631 | 200 gallons per hour | ↓ | .75 | 44 | 130 | 375 | 32 |
| 635 | Torch, cutting, acetylene-oxygen, 150' hose | | | | | | |
| 636 | Hourly operating cost includes tips and gas | Ea. | 5.10 | 13 | 37 | 130 | 48.20 |
| 641 | Toilet, portable chemical | | | | 22 | 70 | 4.40 |
| 642 | Recycle flush type | | | | 27 | 85 | 5.40 |
| 643 | Toilet, fresh water flush, garden hose, | | | | | 130 | 8.65 |
| 644 | Hoisted, non-flush, for high rise | | | | | 70 | 4.65 |
| 645 | Toilet trailers, minimum | | | | | 60 | 4 |
| 646 | Maximum | ↓ | | | | 95 | 6.30 |
| 647 | Trailer, office, see division 1.1-58-420 | | | | | | |
| 650 | Trailers, platform, flush deck, 2 axle, 25 ton capacity | Ea. | 1 | 82 | 285 | 1,000 | 65 |
| 660 | 40 ton capacity | | 1.20 | 120 | 420 | 1,475 | 93.60 |
| 670 | 3 axle, 50 ton capacity | | 2.30 | 150 | 525 | 1,825 | 123.40 |
| 680 | 75 ton capacity | ↓ | 2.90 | 220 | 770 | 2,700 | 177.20 |
| 686 | | | | | | | |
| 690 | Water tank, engine driven discharge, 5000 gallons | Ea. | 9.15 | 235 | 695 | 2,075 | 212.20 |
| 700 | 10,000 gallons | | 10.95 | 340 | 1,025 | 3,000 | 292.60 |
| 702 | Transit with tripod | ↓ | | 20 | 75 | 225 | 15 |
| 703 | | | | | | | |
| 705 | Trench box, 8,000 lbs. 8' x 16' | Ea. | 1.30 | 185 | 410 | 1,175 | 92.40 |
| 707 | 12,000 lbs., 10' x 20' | | 1.60 | 265 | 625 | 1,750 | 137.80 |
| 710 | Truck, pickup, 3/4 ton, 2 wheel drive | | 4.90 | 37 | 120 | 375 | 63.20 |
| 720 | 4 wheel drive | | 5.05 | 48 | 175 | 520 | 75.40 |
| 730 | Tractor, 4 x 2, 30 ton capacity, 195 H.P. | | 9.10 | 210 | 740 | 2,525 | 220.80 |
| 741 | 250 H.P. | | 13.50 | 235 | 820 | 2,900 | 272 |
| 750 | 6 x 2, 40 ton capacity, 240 H.P. | | 12.50 | 290 | 1,000 | 3,500 | 300 |
| 760 | 6 x 4, 45 ton capacity, 240 H.P. | | 14.15 | 295 | 1,025 | 3,050 | 318.20 |
| 770 | Welder, electric, 200 amp | | .52 | 16 | 50 | 145 | 14.15 |
| 780 | 300 amp | | .67 | 20 | 64 | 175 | 18.15 |
| 790 | Gas engine, 200 amp | | 3.08 | 35 | 100 | 300 | 44.65 |
| 800 | 300 amp | | 4.30 | 37 | 110 | 320 | 56.40 |
| 810 | Wheelbarrow, any size | | | 8 | 22 | 52 | 4.40 |
| 820 | Wrecking ball, 4000 lb. | ↓ | | 22 | 70 | 210 | 14 |
| 20-001 | **LIFTING & HOISTING EQUIPMENT RENTAL** | | | | | | |
| 010 | without operators unless noted by* | | | | | | |
| 020 | Crane, climbing, 106 ft. jib, 6000 lb. capacity, 410 FPM | Ea. | 12.60 | | | 7,750 | 617.45 |
| 030 | 101 ft. jib, 10,250 lb. capacity, 270 FPM | " | 16.60 | | | 10,000 | 799.45 |
| 040 | Tower, static, 130' high, 106 ft. jib, | | | | | | |
| 050 | 6200 lb. capacity at 400 FPM | Ea. | 12.60 | | | 8,300 | 654.10 |
| 060 | crawler, cable, 1/2 C.Y., 15 tons at 12' radius | ↓ | 12.15 | 660* | 900 | 2,700 | 277.20 |
| 070 | 3/4 C.Y., 20 tons at 12' radius | | 12.60 | 810* | 1,225 | 3,800 | 345.80 |
| 080 | 1 C.Y., 25 tons at 12' radius | | 13.40 | 895* | 1,400 | 4,375 | 387.20 |
| 090 | 1-1/2 C.Y., 40 tons at 12' radius | ↓ | 15.95 | 935* | 1,775 | 5,625 | 482.60 |

For expanded coverage of these items see *Building Construction Cost Data 1985*

## 1.5 Contractor Equipment

| | | | UNIT | HOURLY OPER. COST | RENT PER DAY | RENT PER WEEK | RENT PER MONTH | CREWS EQUIPMENT COST |
|---|---|---|---|---|---|---|---|---|
| 20 | 100 | 2 C.Y., 50 tons at 12' radius | Ea. | 19.95 | 1,225* | 2,400 | 6,275 | 639.60 |
| | 110 | 3 C.Y., 75 tons at 12' radius | | 25 | 1,375* | 2,700 | 7,100 | 740 |
| | 120 | 100 ton capacity, standard boom | | 32 | 1,075 | 2,925 | 7,950 | 841 |
| | 130 | 165 ton capacity, standard boom | | 46 | 1,250 | 4,100 | 12,200 | 1,188 |
| | 140 | 200 ton capacity, 150' boom | | 52 | 1,750 | 5,425 | 16,200 | 1,501 |
| | 150 | 450' boom | | 53 | 1,950 | 6,100 | 18,400 | 1,644 |
| | 160 | Truck mounted, cable operated, 6 x 4, 20 tons at 10' radius | | 9.80 | 780* | 1,125 | 3,375 | 303.40 |
| | 170 | 25 tons at 10' radius | | 11.90 | 940* | 1,475 | 4,375 | 390.20 |
| | 180 | 8 x 4, 30 tons at 10' radius | | 14.60 | 1,000* | 1,600 | 4,850 | 436.80 |
| | 190 | 40 tons at 12' radius | | 14.75 | 1,050* | 1,825 | 5,500 | 483 |
| | 200 | 8 x 4, 60 tons at 15' radius | | 19.40 | 1,100* | 2,000 | 6,075 | 555.20 |
| | 210 | 90 tons at 15' radius | | 25 | 1,425* | 2,950 | 8,875 | 790 |
| | 220 | 115 tons at 15' radius | | 28 | 1,575* | 3,350 | 10,100 | 894 |
| | 230 | 150 tons at 18' radius | | 34 | 1,700* | 3,750 | 11,200 | 1,022 |
| | 240 | Truck mounted, hydraulic, 12 ton capacity | | 8.70 | 695* | 825 | 2,500 | 234.60 |
| | 250 | 25 ton capacity | | 13 | 880* | 1,400 | 4,375 | 384 |
| | 255 | 33 ton capacity | | 13.50 | 925* | 1,475 | 4,625 | 403 |
| | 260 | 55 ton capacity | | 18.35 | 1,075* | 1,900 | 5,900 | 526.80 |
| | 270 | 80 ton capacity | | 22 | 1,425* | 3,300 | 10,700 | 836 |
| | 280 | Self-propelled, 4 x 4, with telescoping boom, 5 ton | | 5.35 | 535* | 665 | 1,900 | 175.80 |
| | 290 | 12-1/2 ton capacity | | 9.35 | 440* | 725 | 2,075 | 219.80 |
| | 300 | 15 ton capacity | | 9.50 | 480* | 875 | 2,725 | 251 |
| | 310 | 25 ton capacity | | 11.75 | 770* | 1,425 | 4,250 | 379 |
| | 320 | Derricks, guy, 20 ton capacity, 60' boom, 75' mast | | 3.80 | 135 | 410 | 1,225 | 112.40 |
| | 330 | 100' boom, 115' mast | | 4.32 | 175 | 515 | 1,525 | 137.55 |
| | 340 | Stiffleg, 20 ton capacity, 70' boom, 37' mast | | 3.14 | 180 | 535 | 1,625 | 132.10 |
| | 350 | 100' boom, 47' mast | | 3.67 | 220 | 660 | 2,000 | 161.35 |
| | 360 | Hoists, chain type, overhead, manual, 3/4 ton | | | 9 | 28 | 83 | 5.60 |
| | 390 | 10 ton | | | 18 | 59 | 160 | 11.80 |
| | 400 | Hoist and tower, 4000 lb. capacity, portable, 40' high | | 3 | | | 1,050 | 94 |
| | 410 | For each added 10' section, add | | | | | 52 | 3.45 |
| | 420 | Hoist and single tubular tower, 5000 lb., 100' high | | 3.69 | | | 1,650 | 139.50 |
| | 430 | For each added 6'-6" section, add | | | | | 44 | 2.90 |
| | 440 | Hoist and double tubular tower, 5000 lb., 100' high | | 3.14 | | | 1,975 | 156.80 |
| | 450 | For each added 6'-6" section, add | | | | | 86 | 5.70 |
| | 455 | Hoist and tower, mast type, 6000 lb., 100' high | | 5.05 | | | 1,575 | 145.40 |
| | 457 | For each added 10' section, add | | | | | 59 | 3.90 |
| | 460 | Hoist and tower, personnel, electric, 2000 lb., 100' @ 125 FPM | | 3.75 | | | 2,325 | 185 |
| | 470 | 3000 lb., 100' @ 200 FPM | | 4.23 | | | 2,550 | 203.85 |
| | 480 | 3000 lb., 150' @ 300 FPM | | 4.66 | | | 2,650 | 213.95 |
| | 490 | 4000 lb., 100' @ 300 FPM | | 4.93 | | | 3,625 | 281.10 |
| | 500 | 6000 lb., 100' @ 275 FPM | ▼ | 5.85 | | | 3,725 | 295.10 |
| | 510 | For added heights up to 500', add | L.F. | | | | 6 | .40 |
| | 520 | Jacks, hydraulic, 20 ton | Ea. | | 10 | 29 | 70 | 5.80 |
| | 550 | 100 ton | " | | 15 | 53 | 150 | 10.60 |
| | 600 | Jacks, hydraulic, climbing with 50' jackrods | | | | | | |
| | 601 | and control consoles, minimum 3 mo. rental | | | | | | |
| | 610 | 30 ton capacity | Ea. | | | | 700 | 46.65 |
| | 615 | For each added 10' jackrod section, add | | | | | 17.30 | 1.15 |
| | 630 | 50 ton capacity | | | | | 1,300 | 86.65 |
| | 635 | For each added 10' jackrod section, add | | | | | 24 | 1.60 |
| | 650 | 125 ton capacity | | | | | 3,675 | 245 |
| | 655 | For each added 10' jackrod section, add | | | | | 180 | 12 |
| | 660 | Cable jack, 10 ton capacity with 200' cable | | | | | 635 | 42.30 |
| | 665 | For each added 50' of cable, add | ▼ | | | | 37 | 2.45 |

For expanded coverage of these items see *Building Construction Cost Data 1985*

## 1.5 Contractor Equipment

| | | UNIT | HOURLY OPER. COST | RENT 1ST MO. | RENT 2ND MO. | RENT 3RD MO. | |
|---|---|---|---|---|---|---|---|
| 25-001 | **WELLPOINT EQUIPMENT RENTAL** See also division 2.3-55 | | | | | | |
| 020 | | | | | | | |
| 010 | Combination jetting & wellpoint pump, 60 H.P. diesel | Ea. | 2.50 | 2,125 | 1,475 | 1,200 | |
| 020 | High pressure gas jet pump, 200 H.P., 300 psi | " | 6.95 | 1,550 | 1,400 | 1,100 | |
| 030 | Discharge pipe, 8" diameter | L.F. | | 3.51 | 1.46 | 1.13 | |
| 035 | 12" diameter | | | 4.60 | 2.27 | 1.73 | |
| 040 | Header pipe, flows up to 150 G.P.M., 4" diameter | | | 1.89 | 1.08 | .81 | |
| 050 | 400 G.P.M., 6" diameter | | | 2.92 | 1.18 | .86 | |
| 060 | 800 G.P.M., 8" diameter | | | 4.60 | 2.27 | 1.51 | |
| 070 | 1500 G.P.M., 10" diameter | | | 4.76 | 2.43 | 2.16 | |
| 080 | 2500 G.P.M., 12" diameter | | | 5.30 | 4.21 | 2.33 | |
| 090 | 4500 G.P.M., 16" diameter | | | 9.20 | 5.72 | 4 | |
| 095 | For quick coupling aluminum and plastic pipe, add | | | 10 | 10 | 10 | |
| 110 | Wellpoint, 25' long, with fittings & riser pipe, 1-1/2" or 2" diameter | Ea. | | 25 | 17.85 | 14.05 | |
| 120 | Wellpoint pump, diesel powered, 4" diameter, 20 H.P. | | 2.45 | 1,750 | 1,200 | 760 | |
| 130 | 6" diameter, 30 H.P. | | 3.59 | 2,000 | 1,300 | 1,200 | |
| 140 | 8" suction, 40 H.P. | | 3.92 | 2,350 | 1,400 | 1,275 | |
| 150 | 10" suction, 75 H.P. | | 4.72 | 2,600 | 1,600 | 1,450 | |
| 160 | 12" suction, 100 H.P. | | 5.85 | 3,550 | 2,350 | 2,150 | |
| 170 | 12" suction, 175 H.P. | | 6.15 | 4,200 | 2,350 | 2,150 | |

## 2.1 Exploration & Clearing

| | | CREW | MAN-HOURS | UNIT | MAT. | LABOR | EQUIP. | TOTAL | TOTAL INCL O&P |
|---|---|---|---|---|---|---|---|---|---|
| 05-001 | **BORINGS** Initial field stake out and determination of elevations | A-6 | 16 | Day | | 185 | | 185 | 290L |
| 010 | Drawings showing boring details | 1 Skwk | 8 | Total | | 100 | | 100 | 160L |
| 020 | Report and recommendations from P.E. | " | 17.780 | " | | 225 | | 225 | 355L |
| 021 | | | | | | | | | |
| 030 | Mobilization and demobilization, minimum | B-55 | 3.330 | Total | | 34 | 44 | 78 | 100 |
| 035 | For over 100 miles, per added mile | | .036 | Mile | | .36 | .47 | .83 | 1.08 |
| 060 | Auger holes in earth, no samples, 2-1/2" diameter | | .204 | L.F. | | 2.06 | 2.68 | 4.74 | 6.20 |
| 080 | Cased borings in earth, with samples, 2-1/2" diameter | | .288 | " | | 2.92 | 3.79 | 6.71 | 8.80 |
| 140 | Drill rig and crew with light duty rig | | 16 | Day | | 160 | 210 | 370 | 485 |
| 10-001 | **CLEAR AND GRUB** Light, trees to 6" diam., cut & chip | B-7 | 60 | Acre | | 645 | 900 | 1,545 | 2,025 |
| 015 | Grub stumps and remove | B-30 | 12 | | | 135 | 560 | 695 | 825 |
| 020 | Medium, trees to 10" diam., cut & chip | B-7 | 68.570 | | | 740 | 1,025 | 1,765 | 2,300 |
| 025 | Grub stumps and remove | B-30 | 16 | | | 180 | 745 | 925 | 1,100 |
| 030 | Heavy, trees to 16" diam., cut & chip | B-7 | 80 | | | 860 | 1,200 | 2,060 | 2,675 |
| 035 | Grub stumps and remove | B-30 | 20 | | | 220 | 930 | 1,150 | 1,375 |
| 040 | If burning is allowed, reduce cut & chip | | | | | | | | 33% |
| 15-001 | **CLEARING** Brush with brush saw & rake | 1 Clab | .014 | S.Y. | | .14 | | .14 | .22L |
| 010 | By hand | " | .029 | | | .29 | | .29 | .45L |
| 030 | With dozer, ball and chain, light clearing | B-11A | .004 | | | .05 | .17 | .22 | .26 |
| 040 | Medium clearing | " | 25 | Acre | | 285 | 960 | 1,245 | 1,500 |
| 32-001 | **DUMP CHARGES** Typical urban city, fees only | | | | | | | | |
| 010 | Building construction materials | | | C.Y. | 3.18 | | | | 3.49M |
| 020 | Demolition lumber, trees, brush | | | | 5.45 | | | | 6M |
| 030 | Rubbish only | | | | 3.18 | | | | 3.49M |
| 100 | Reclamation station, usual charge | | | Ton | 14.55 | | | | 16M |
| 45-001 | **SITE DEMOLITION** Abandon catch basin or manhole | B-6 | 3.430 | Ea. | | 37 | 22 | 59 | 82 |
| 002 | Remove existing catch basin or manhole | | 6 | | | 64 | 38 | 102 | 145 |
| 003 | Catch basin or manhole frames & covers stored | | 1.710 | | | 18.30 | 10.85 | 29.15 | 41 |
| 004 | Remove and reset | | 3.430 | | | 37 | 22 | 59 | 82 |
| 005 | Concrete foundations, plain | B-5 | 1.940 | C.Y. | | 22 | 22 | 44 | 59 |
| 008 | Reinforced | " | 2.560 | " | | 29 | 29 | 58 | 78 |

## 2.1 Exploration & Clearing

| | | CREW | MAN-HOURS | UNIT | MAT. | LABOR | EQUIP. | TOTAL | TOTAL INCL O&P |
|---|---|---|---|---|---|---|---|---|---|
| 010 | Concrete walls | B-5 | 1.880 | C.Y. | | 22 | 21 | 43 | 57 |
| 020 | Elevated slabs | | 2.460 | | | 28 | 28 | 56 | 75 |
| 030 | Slab on grade, plain | | 1.420 | | | 15.85 | 16.15 | 32 | 43 |
| 035 | Mesh reinforcing | ↓ | 1.940 | ↓ | | 22 | 22 | 44 | 59 |
| 039 | For congested sites or small quantities, add up to | | | | | | | | 200% |
| 040 | Add for disposal, on site | B-11A | .069 | C.Y. | | .78 | 2.65 | 3.43 | 4.15 |
| 045 | Offsite, to 5 miles | B-34D | .105 | " | | 1.10 | 5.20 | 6.30 | 7.40 |
| 060 | Fencing, barbed wire, 3 strand | 2 Clab | .037 | L.F. | | .37 | | .37 | .59L |
| 065 | 5 strand | | .057 | | | .57 | | .57 | .91L |
| 070 | Chain link, remove only | | .052 | | | .52 | | .52 | .82L |
| 075 | Remove and reset | ↓ | .320 | ↓ | | 3.20 | | 3.20 | 5.05L |
| 100 | Masonry walls, block or tile, solid | B-5 | .036 | C.F. | | .40 | .40 | .80 | 1.08 |
| 110 | Cavity | | .029 | | | .33 | .33 | .66 | .89 |
| 120 | Brick, solid | | .071 | | | .81 | .81 | 1.62 | 2.17 |
| 130 | With block | | .057 | | | .64 | .64 | 1.28 | 1.73 |
| 140 | Stone, with mortar | | .071 | | | .81 | .81 | 1.62 | 2.17 |
| 150 | Dry set | ↓ | .043 | ↓ | | .48 | .48 | .96 | 1.30 |

## 2.3 Earthwork

| | | CREW | MAN-HOURS | UNIT | MAT. | LABOR | EQUIP. | TOTAL | TOTAL INCL O&P |
|---|---|---|---|---|---|---|---|---|---|
| 03-001 | **BACKFILL** By hand, no compaction, light soil | 1 Clab | .571 | C.Y. | | 5.70 | | 5.70 | 9.05L |
| 010 | Heavy soil | | .667 | | | 6.65 | | 6.65 | 10.55L |
| 030 | compaction in 6" layers, hand tamp, add | ↓ | .388 | | | 3.88 | | 3.88 | 6.15L |
| 050 | Air tamp, add | B-9 | .211 | | | 2.19 | .74 | 2.93 | 4.29 |
| 060 | Vibrating plate, add | A-1 | .133 | | | 1.33 | .59 | 1.92 | 2.76 |
| 080 | Compaction in 12" layers, hand tamp, add | 1 Clab | .235 | | | 2.35 | | 2.35 | 3.73L |
| 130 | Dozer backfilling, bulk, up to 300' haul, no compaction | B-10B | .010 | | | .12 | .51 | .63 | .75 |
| 140 | Air tamped | B-11B | .067 | ↓ | | .76 | 3.28 | 4.04 | 4.81 |
| 05-001 | **BORROW** Buy and load at pit, haul 2 miles to site, place | | | | | | | | |
| 002 | and spread, with 180 H.P. dozer, no compaction | | | | | | | | |
| 010 | Bank run gravel | B-15 | .047 | C.Y. | 3 | .52 | 1.95 | 5.47 | 6.25 |
| 020 | Common borrow | | .047 | | 1.50 | .52 | 1.95 | 3.97 | 4.61 |
| 030 | Crushed stone, 1-1/2" | | .047 | | 7 | .52 | 1.95 | 9.47 | 10.65 |
| 032 | 3/4" | | .047 | | 7 | .52 | 1.95 | 9.47 | 10.65 |
| 034 | 1/2" | | .047 | | 8 | .52 | 1.95 | 10.47 | 11.75 |
| 036 | 3/8" | | .047 | | 9 | .52 | 1.95 | 11.47 | 12.85 |
| 040 | Sand, washed, concrete or masonry | | .047 | | 6.50 | .52 | 1.95 | 8.97 | 10.10 |
| 050 | Dead or bank sand | ↓ | .047 | ↓ | 3 | .52 | 1.95 | 5.47 | 6.25 |
| 08-001 | **COMPACTION** Rolling with road roller, 5 tons | B-10E | 1.500 | Hr. | | 17.65 | 10.50 | 28.15 | 40 |
| 005 | Air tamp 8" lifts common fill | B-9 | .160 | C.Y. | | 1.66 | .57 | 2.23 | 3.26 |
| 006 | Select fill | " | .133 | | | 1.39 | .47 | 1.86 | 2.72 |
| 060 | Vibratory plate, 8" lifts, common fill | A-1 | .107 | | | 1.07 | .47 | 1.54 | 2.21 |
| 070 | Select fill | " | .089 | ↓ | | .89 | .39 | 1.28 | 1.84 |
| 16-001 | **EXCAVATING, BULK** Medium earth piled or truck | | | | | | | | |
| 002 | loaded, no trucks or haul included | | | | | | | | |
| 020 | Backhoe, hydraulic, crawler mtd., 1 C.Y. cap.=45 C.Y./hr. | B-12A | .044 | C.Y. | | .53 | 1.12 | 1.65 | 2.07 |
| 031 | Wheel mounted, 1/2 C.Y. cap. = 20 C.Y./hr. | B-12E | .100 | | | 1.19 | 1.59 | 2.78 | 3.64 |
| 120 | Front end loader, track mtd., 1-1/2 C.Y. cap. = 70 C.Y./hr. | B-10N | .021 | | | .25 | .48 | .73 | .93 |
| 150 | Wheel mounted, 3/4 C.Y. cap. = 45 C.Y./hr. | B-10R | .033 | | | .39 | .55 | .94 | 1.23 |
| 17-001 | **EXCAVATING, STRUCTURAL** Hand, pits to 6' deep, sandy soil | 1 Clab | 1 | | | 10 | | 10 | 15.85L |
| 010 | Heavy soil or clay | | 2 | | | 20 | | 20 | 32L |
| 110 | Hand loading trucks from stock pile, sandy soil | | .667 | | | 6.65 | | 6.65 | 10.55L |
| 130 | Heavy soil or clay | ↓ | 1 | ↓ | | 10 | | 10 | 15.85L |

For expanded coverage of these items see *Means' Site Work Cost Data 1985*

## 2.3 Earthwork

| | | | CREW | MAN-HOURS | UNIT | MAT. | LABOR | EQUIP. | TOTAL | TOTAL INCL O&P |
|---|---|---|---|---|---|---|---|---|---|---|
| 17 | 150 | For wet or muck hand excavation, add to above | | | C.Y. | | | | | 50% |
| | 18-001 | **EXCAVATING, TRENCH** or continuous footing, 4' wide | | | | | | | | |
| | 003 | 3' deep, 3/8 C.Y. tractor backhoe = 245 L.F./day | B-11C | .107 | C.Y. | | 1.21 | 1.01 | 2.22 | 3.03 |
| | 004 | 1/2 C.Y. tractor backhoe = 325 L.F./day | B-11M | .080 | | | .91 | 1.14 | 2.05 | 2.69 |
| | 005 | 4' deep, 3/8 C.Y. tractor backhoe = 170 L.F./day | B-11C | .107 | | | 1.21 | 1.01 | 2.22 | 3.03 |
| | 006 | 1/2 C.Y. tractor backhoe = 225 L.F./day | B-11M | .080 | | | .91 | 1.14 | 2.05 | 2.69 |
| | 007 | 5' deep, 3/8 C.Y. tractor backhoe = 125 L.F./day | B-11C | .107 | | | 1.21 | 1.01 | 2.22 | 3.03 |
| | 008 | 1/2 C.Y. tractor backhoe = 165 L.F./day | B-11M | .080 | | | .91 | 1.14 | 2.05 | 2.69 |
| | 009 | 6' deep, 1/2 C.Y. tractor backhoe = 130 L.F./day | " | .080 | | | .91 | 1.14 | 2.05 | 2.69 |
| | 010 | 5/8 C.Y. hydraulic backhoe = 160 L.F./day | B-12Q | .064 | | | .76 | 1.07 | 1.83 | 2.38 |
| | 020 | 8' deep, 1/2 C.Y. tractor backhoe = 85 L.F./day | B-11M | .080 | | | .91 | 1.14 | 2.05 | 2.69 |
| | 030 | 1/2 C.Y. Gradall = 85 L.F./day | B-12J | .080 | | | .95 | 2.16 | 3.11 | 3.88 |
| | 140 | By hand with pick and shovel to 6' deep, light soil | 1 Clab | 1 | | | 10 | | 10 | 15.85L |
| | 150 | Heavy soil | " | 2 | | | 20 | | 20 | 32L |
| | 19-001 | **EXCAVATING, UTILITY TRENCH** Sandy clay soil | | | | | | | | |
| | 005 | Trenching with chain trencher, 12 H.P., operator walking | | | | | | | | |
| | 010 | 4" wide trench, 12" deep | B-53 | .010 | L.F. | | .12 | .17 | .29 | .37 |
| | 100 | Backfill by hand including compaction, add | | | | | | | | |
| | 105 | 4" wide trench, 12" deep | A-1 | .010 | L.F. | | .10 | .04 | .14 | .21 |
| | 20-001 | **FILL** Spread dumped material, no compaction, by dozer | B-10B | .012 | C.Y. | | .14 | .61 | .75 | .90 |
| | 010 | By hand | 1 Clab | .667 | " | | 6.65 | | 6.65 | 10.55L |
| | 050 | Gravel fill, compacted, under floor slabs, 3" deep | B-14 | .005 | S.F. | .06 | .05 | .02 | .13 | .16 |
| | 060 | 6" deep (gravel at $4.05 ton) | | .005 | | .12 | .05 | .02 | .19 | .24 |
| | 070 | 9" deep | | .007 | | .18 | .07 | .02 | .27 | .33 |
| | 080 | 12" deep | | .008 | | .24 | .08 | .03 | .35 | .43 |
| | 100 | Alternate pricing method, 3" deep | | .533 | C.Y. | 6.50 | 5.71 | 1.69 | 13.90 | 18.05 |
| | 110 | 6" deep | | .300 | | 6.50 | 3.20 | .95 | 10.65 | 13.25 |
| | 120 | 9" deep | | .240 | | 6.50 | 2.56 | .76 | 9.82 | 12.05 |
| | 130 | 12" deep | | .218 | | 6.50 | 2.33 | .69 | 9.52 | 11.60 |
| | 22-001 | **GRADING** Site excav. & fill, not incl. mobilization, demobilization or | | | | | | | | |
| | 002 | compaction. Includes 1/4 push dozer per scraper. | | | | | | | | |
| | 010 | Dozer 300' haul, 75 H.P., = 20 C.Y./hr. | B-10L | .075 | C.Y. | | .88 | 1.24 | 2.12 | 2.77 |
| | 240 | Hand grading, finish | 1 Clab | .012 | S.F. | | .12 | | .12 | .19L |
| | 250 | Rough | " | .062 | S.Y. | | .62 | | .62 | .97L |
| | 30-001 | **HAULING** Earth 6 C.Y. dump truck 1/4 mile round trip, 5.0 loads/hr. | B-34A | .033 | C.Y. | | .35 | .92 | 1.27 | 1.55 |
| | 020 | 4 mile round trip, 1.8 loads/hr. | " | .094 | | | .97 | 2.59 | 3.56 | 4.37 |
| | 030 | C.Y. dump truck, 1 mile round trip, 2.7 loads/hr. | B-34B | .031 | | | .32 | 1.07 | 1.39 | 1.68 |
| | 050 | 4 mile round trip, 1.6 loads/hr. | " | .053 | | | .55 | 1.86 | 2.41 | 2.91 |
| | 35-001 | **MOBILIZATION AND DEMOBILIZATION** Dozer, 105 H.P. | B-34C | 1.290 | Ea. | | 13.35 | 61 | 74.35 | 88 |
| | 090 | Shovel, backhoe or dragline, 3/4 C.Y. | | 1.860 | | | 19.25 | 88 | 107.25 | 125 |
| | 120 | Tractor shovel or front end loader, 1 C.Y. | | 1.290 | | | 13.35 | 61 | 74.35 | 88 |
| | 50-001 | **TERMITE PRETREATMENT** | 1 Skwk | .005 | S.F.FLR | .10 | .07 | | .17 | .22 |
| | 040 | Insecticides for termite control, minimum | | | Gal. | 10 | | | 10 | 11M |
| | 050 | Maximum | | | " | 15 | | | 15 | 16.50M |

## 2.4 Caissons & Piling

| | | CREW | MAN-HOURS | UNIT | MAT. | LABOR | EQUIP. | TOTAL | TOTAL INCL O&P |
|---|---|---|---|---|---|---|---|---|---|
| 15-001 | **PILES, CONCRETE** 200 piles, 60' long | | | | | | | | |
| 002 | unless specified otherwise, not incl. pile caps or mobilization | | | | | | | | |
| 080 | Cast in place friction pile, 50' long, fluted, | | | | | | | | |
| 081 | tapered steel, 4000 psi concrete, no reinforcing | | | | | | | | |
| 090 | 12" diameter, 7 ga. | B-19 | .107 | V.L.F. | 10.65 | 1.35 | 1.62 | 13.62 | 15.75 |
| 130 | End bearing, fluted, constant diameter, | | | | | | | | |
| 132 | 4000 psi concrete, no reinforcing | | | | | | | | |
| 134 | 12" diameter, 7 ga. | B-19 | .107 | V.L.F. | 10.95 | 1.35 | 1.62 | 13.92 | 16.05 |
| 310 | Precast, prestressed, 40' long, 10" thick, square | | .091 | | 5.35 | 1.15 | 1.39 | 7.89 | 9.35 |
| 320 | 12" thick, square | | .094 | | 7 | 1.19 | 1.43 | 9.62 | 11.25 |
| 340 | 14" thick, square | | .107 | | 10.70 | 1.35 | 1.62 | 13.67 | 15.80 |
| 400 | 18" thick, square | ↓ | .123 | ↓ | 15 | 1.55 | 1.87 | 18.42 | 21 |
| 25-001 | **PILES, WOOD** Untreated, friction or end bearing, not including | | | | | | | | |
| 005 | mobilization or demobilization | | | | | | | | |
| 010 | Up to 30' long, 12" butts, 8" points | B-19 | .102 | V.L.F. | 3.30 | 1.29 | 1.56 | 6.15 | 7.50 |
| 080 | Treated piles, 12 lb. creosote per C.F., | | | | | | | | |
| 081 | friction or end bearing, ASTM class B | | | | | | | | |
| 100 | Up to 30' long, 12" butts, 8" points | B-19 | .102 | V.L.F. | 5.40 | 1.29 | 1.56 | 8.25 | 9.80 |
| 30-001 | **PILING SPECIAL COSTS** Concrete pile caps | | | | | | | | |
| 050 | Cutoffs, concrete piles, plain | 1 Pile | 1.450 | Ea. | | 18.05 | | 18.05 | 31L |
| 060 | With steel thin shell | | .211 | | | 2.61 | | 2.61 | 4.45L |
| 070 | Steel pile or "H" piles | | .421 | | | 5.20 | | 5.20 | 8.90L |
| 080 | Wood piles | ↓ | .211 | ↓ | | 2.61 | | 2.61 | 4.45L |
| 100 | Testing, any type piles, test load is twice the design load | | | | | | | | |
| 105 | 50 ton design load, 100 ton test | B-19 | 136 | Ea. | 2,500 | 1,700 | 2,075 | 6,275 | 7,875 |
| 36-001 | **MOBILIZATION** Set up & remove air compressor, 600 C.F.M. | A-5 | 5.450 | Ea. | | 54.67 | 4.33 | 59 | 91 |
| 015 | 1200 C.F.M. | " | 8.180 | | | 82.50 | 6.50 | 89 | 135 |
| 020 | Crane, with pile leads and pile hammer, 75 ton | B-8 | 128 | ↓ | | 1,425 | 3,050 | 4,475 | 5,600 |

## 2.5 Site Drainage & Utilities

| | | CREW | MAN-HOURS | UNIT | MAT. | LABOR | EQUIP. | TOTAL | TOTAL INCL O&P |
|---|---|---|---|---|---|---|---|---|---|
| 02-001 | **CATCH BASINS OR MANHOLES** Including footing & excavation, | | | | | | | | |
| 002 | not including frame and cover | | | | | | | | |
| 005 | Brick, 4' inside diameter, 4' deep | D-1 | 16 | Ea. | 300 | 180 | | 480 | 615 |
| 111 | Precast 4', I.D., riser @ $27.75 per V.L.F., 4' deep | B-6 | 5.850 | | 210 | 63 | 37 | 310 | 370 |
| 160 | Frames and covers, 24" square, 500 lb. | " | 3.080 | ↓ | 140 | 32.50 | 19.50 | 192 | 230 |
| 08-001 | **GAS SERVICE AND DISTRIBUTION** Not including excavation | | | | | | | | |
| 005 | or backfill | | | | | | | | |
| 010 | Polyethylene, 60 psi, coils, 5/8" diameter, SDR 7 | B-20 | .053 | L.F. | .12 | .70 | | .82 | 1.25 |
| 015 | 1-1/4" diameter, SDR 10 | " | .060 | " | .51 | .79 | | 1.30 | 1.82 |
| 050 | Steel, schedule 40, plain end, tar coated & wrapped | | | | | | | | |
| 055 | 1" diameter | Q-4 | .107 | L.F. | 2.20 | 1.41 | .06 | 3.67 | 4.72 |
| 060 | 2" diameter | | .114 | | 3.50 | 1.52 | .06 | 5.08 | 6.30 |
| 065 | 3" diameter | ↓ | .123 | ↓ | 6.30 | 1.63 | .07 | 8 | 9.60 |
| 10-001 | **FILL** In trench, crushed bank run | B-6 | .092 | C.Y. | 5.75 | .98 | .59 | 7.32 | 8.55 |
| 010 | Screened 3/4" to 1/2" | " | .092 | " | 7 | .98 | .59 | 8.57 | 9.90 |
| 27-001 | **PIPE, DRAINAGE AND SEWAGE** Excavation & backfill | | | | | | | | |
| 010 | Asbestos Cement, class 2400, 6" diameter | B-20 | .073 | L.F. | 3 | .96 | | 3.96 | 4.82 |
| 012 | 8" diameter | B-21 | .074 | | 3.90 | .97 | .23 | 5.10 | 6.10 |
| 014 | 10" diameter | " | .085 | | 5.25 | 1.11 | .27 | 6.63 | 7.85 |
| 090 | Bituminous fiber, plain, 2" diameter | 2 Plum | .040 | | .84 | .55 | | 1.39 | 1.80 |
| 094 | 4" diameter | " | .042 | ↓ | 1.25 | .58 | | 1.83 | 2.30 |

For expanded coverage of these items see *Means' Site Work Cost Data 1985*

# 2.5 Site Drainage & Utilities

| | | | CREW | MAN-HOURS | UNIT | MAT. | LABOR | EQUIP. | TOTAL | TOTAL INCL O&P |
|---|---|---|---|---|---|---|---|---|---|---|
| 27 | 098 | 6" diameter | 2 Plum | .047 | L.F. | 2.50 | .65 | | 3.15 | 3.78 |
| | 100 | 8" diameter | " | .053 | " | 5.95 | .74 | | 6.69 | 7.70 |
| | 130 | Concrete, non-reinforced, extra strength, B&S or T&G joints | | | | | | | | |
| | 132 | 6" diameter | B-20 | .091 | L.F. | 2.25 | 1.20 | | 3.45 | 4.37 |
| | 134 | 8" diameter | B-21 | .093 | | 2.45 | 1.23 | .29 | 3.97 | 4.96 |
| | 136 | 10" diameter | " | .108 | ↓ | 2.80 | 1.42 | .34 | 4.56 | 5.70 |
| | 350 | Corrugated Metal Pipe, galvanized or aluminum, bituminous | | | | | | | | |
| | 351 | coated with paved invert, 20' to 30' lengths | | | | | | | | |
| | 355 | 8" diameter 16 ga. | B-20 | .073 | L.F. | 4.35 | .96 | | 5.31 | 6.30 |
| | 605 | Polyvinyl chloride pipe, 13' lengths, S.D.R. 35, 4" diameter | | .064 | | .65 | .84 | | 1.49 | 2.05 |
| | 615 | 6" diameter | | .069 | | 1.25 | .91 | | 2.16 | 2.81 |
| | 625 | 8" diameter | ↓ | .072 | | 2.20 | .95 | | 3.15 | 3.92 |
| | 630 | 10" diameter | B-21 | .085 | | 3.60 | 1.11 | .27 | 4.98 | 6 |
| | 650 | Vitrified clay sewer pipe, premium joint, C-200, 4" diameter | B-20 | .091 | | 1.60 | 1.20 | | 2.80 | 3.65 |
| | 655 | 4' and 5' lengths, 6" diameter | " | .120 | | 2.50 | 1.58 | | 4.08 | 5.25 |
| | 660 | 8" diameter | B-21 | .140 | | 3.50 | 1.84 | .44 | 5.78 | 7.25 |
| | 665 | 10" diameter | " | .147 | ↓ | 5 | 1.94 | .46 | 7.40 | 9.10 |
| 29-001 | | **SUBDRAINAGE** Foundation underdrains | | | | | | | | |
| | 005 | Asbestos Cement, class 4000 underdrain, perforated | | | | | | | | |
| | 010 | 4" diameter | B-20 | .062 | L.F. | 1.80 | .81 | | 2.61 | 3.27 |
| | 015 | 6" diameter | " | .063 | | 2.85 | .83 | | 3.68 | 4.46 |
| | 060 | Bituminous fiber, perforated underdrain, 3" diameter | 2 Plum | .040 | | 1.05 | .55 | | 1.60 | 2.03 |
| | 065 | 4" diameter | | .042 | | 1.20 | .58 | | 1.78 | 2.24 |
| | 070 | 5" diameter | | .044 | | 2.10 | .61 | | 2.71 | 3.28 |
| | 075 | 6" diameter | ↓ | .047 | | 2.60 | .65 | | 3.25 | 3.89 |
| | 140 | Porous wall concrete underdrain, std. strength, 4" diameter | B-20 | .072 | | 1.30 | .95 | | 2.25 | 2.93 |
| | 145 | 6" diameter | " | .076 | | 1.40 | 1.01 | | 2.41 | 3.13 |
| | 150 | 8" diameter | B-21 | .090 | | 2.25 | 1.19 | .28 | 3.72 | 4.67 |
| | 200 | Vitrified clay, perforated, 2' lengths (C-211), 4" diameter | B-20 | .060 | | 1.80 | .79 | | 2.59 | 3.24 |
| | 205 | 6" diameter | | .076 | | 3.05 | 1.01 | | 4.06 | 4.95 |
| | 210 | 8" diameter | ↓ | .083 | ↓ | 5.25 | 1.09 | | 6.34 | 7.50 |
| 30-001 | | **PIPING, WATER DISTRIBUTION SYSTEMS** Pipe laid in trench, | | | | | | | | |
| | 002 | excavation and backfill not included | | | | | | | | |
| | 140 | Ductile Iron pipe, class 250 water piping, 18' lengths | | | | | | | | |
| | 141 | Mechanical joint, 4" diameter | B-20 | .167 | L.F. | 5 | 2.20 | | 7.20 | 9 |
| | 170 | Copper tubing, type K, 20' joints, 1-1/2" diameter | | .067 | | 2.55 | .88 | | 3.43 | 4.20 |
| | 171 | 2" diameter | ↓ | .076 | ↓ | 3.98 | 1.01 | | 4.99 | 5.95 |
| | 175 | Gate valves with boxes, cast iron | | | | | | | | |
| | 178 | 4" diameter | B-20 | 4 | Ea. | 250 | 53 | | 303 | 360 |
| | 240 | Polyvinyl chloride pipe, class 150, S.D.R.-18, 4" diameter | | .120 | L.F. | 2.93 | 1.58 | | 4.51 | 5.75 |
| | 245 | 6" diameter | ↓ | .133 | | 5.15 | 1.76 | | 6.91 | 8.45 |
| | 250 | 8" diameter | B-21 | .175 | | 7.90 | 2.30 | .55 | 10.75 | 12.95 |
| | 255 | 10" diameter | | .200 | | 11.70 | 2.63 | .63 | 14.96 | 17.75 |
| | 260 | 12" diameter | ↓ | .280 | | 15.50 | 3.68 | .88 | 20.06 | 24 |
| | 265 | Class 160, S.D.R.-26, 1-1/2" diameter | B-20 | .080 | | .40 | 1.06 | | 1.46 | 2.11 |
| | 270 | 2" diameter | | .096 | | .52 | 1.27 | | 1.79 | 2.58 |
| | 275 | 2-1/2" diameter | | .096 | | .75 | 1.27 | | 2.02 | 2.83 |
| | 280 | 3" diameter | | .120 | | 1.08 | 1.58 | | 2.66 | 3.70 |
| | 285 | 4" diameter | ↓ | .120 | ↓ | 1.76 | 1.58 | | 3.34 | 4.45 |
| 32-001 | | **SEPTIC TANKS** Not incl. excav. or piping, precast, 1,000 gallon | B-21 | 3.500 | Ea. | 275 | 46 | 11 | 332 | 390 |
| | 010 | 2,000 gallon | | 5.600 | | 530 | 73.40 | 17.60 | 621 | 720 |
| | 060 | Fiberglass, 1,000 gallon | | 4.670 | | 385 | 61.35 | 14.65 | 461 | 535 |
| | 070 | 1,500 gallon | ↓ | 7 | | 490 | 93 | 22 | 605 | 710 |
| | 100 | Distribution boxes, concrete, 5 outlets | 2 Plum | 1 | | 25 | 13.80 | | 38.80 | 49 |
| | 110 | 12 outlets | " | 2 | | 175 | 28 | | 203 | 235 |
| | 142 | Leaching pit, 6', complete | | | ↓ | | | | | 550 |
| | 220 | Excavation for septic tank, 3/4 C.Y. backhoe | B-12F | .110 | C.Y. | | 1.31 | 2.23 | 3.54 | 4.53 |
| | 240 | 4' trench for disposal field, 3/4 C.Y. backhoe | " | .048 | L.F. | | .57 | .96 | 1.53 | 1.96 |
| | 260 | Gravel fill, run of the bank | B-6 | .160 | C.Y. | 3 | 1.71 | 1.01 | 5.72 | 7.15 |

For expanded coverage of these items see *Means' Site Work Cost Data 1985*

## 2.5 Site Drainage & Utilities

| | | CREW | MAN-HOURS | UNIT | MAT. | LABOR | EQUIP. | TOTAL | TOTAL INCL O&P |
|---|---|---|---|---|---|---|---|---|---|
| 32 280 | Crushed stone, 3/4" | B-6 | .160 | C.Y. | 7 | 1.71 | 1.01 | 9.72 | 11.55 |
| 38-001 | **UTILITY VAULTS** Precast concrete, 6" thick | | | | | | | | |
| 005 | 5' x 10' x 6' high, I.D. | B-13 | 28 | Ea. | 1,300 | 305 | 190 | 1,795 | 2,125 |
| 035 | Hand hole, precast concrete, 1-1/2" thick | | | | | | | | |
| 040 | 1'-4" x 2'-4" x 1'-3", I.D., light duty | B-1 | 6 | Ea. | 130 | 64 | | 194 | 245 |
| 045 | 4'-6" x 5'-10" x 2'-7", O.D., heavy duty | B-6 | 8 | " | 285 | 84 | 51 | 420 | 505 |
| 40-001 | **WELLS** Domestic water, drilled and cased, including casing | | | | | | | | |
| 010 | 4" to 6" diameter | B-23 | .250 | V.L.F. | 7 | 2.60 | 1.32 | 10.92 | 13.25 |
| 150 | Pumps, installed in wells to 100' deep, 4" submersible | B-21 | 15.560 | Ea. | 350 | 206 | 49 | 605 | 765 |
| 152 | 3/4 H.P., 144 to 1110 GPH | | 16.970 | | 400 | 222 | 53 | 675 | 850 |
| 160 | To 180' deep, 1 H.P., 102 to 1356 GPH | ↓ | 25.450 | ↓ | 450 | 335 | 80 | 865 | 1,125 |

## 2.6 Roads & Walks

| | | CREW | MAN-HOURS | UNIT | MAT. | LABOR | EQUIP. | TOTAL | TOTAL INCL O&P |
|---|---|---|---|---|---|---|---|---|---|
| 02-001 | **ASPHALT BLOCKS** Premolded, 6"x12" x 1-1/4", w/bed & neopr. adhesive | D-1 | .119 | S.F. | 2.31 | 1.34 | | 3.65 | 4.64 |
| 010 | 3" thick | | .123 | | 3 | 1.39 | | 4.39 | 5.50 |
| 030 | Hexagonal tile, 8" wide, 1-1/4" thick | | .119 | | 2.20 | 1.34 | | 3.54 | 4.51 |
| 040 | 2" thick | | .123 | | 2.80 | 1.39 | | 4.19 | 5.25 |
| 050 | Square, 8" x 8", 1-1/4" thick | | .119 | | 2.20 | 1.34 | | 3.54 | 4.51 |
| 060 | 2" thick | ↓ | .123 | | 2.94 | 1.39 | | 4.33 | 5.40 |
| 05-001 | **BASE** Prepare and roll sub-base, small areas | B-32 | .003 | ↓ | | .03 | .10 | .13 | .16 |
| 07-001 | **BASE COURSE** Crushed 3/4" stone @ $10.50 per ton, delivered, 3" deep | B-36 | .001 | S.F. | .18 | .01 | .02 | .21 | .23 |
| 010 | 6" deep | | .001 | | .37 | .02 | .02 | .41 | .46 |
| 020 | 9" deep | | .002 | | .56 | .02 | .03 | .61 | .68 |
| 030 | 12" deep | ↓ | .002 | ↓ | .74 | .02 | .04 | .80 | .89 |
| 035 | Bank run gravel @ $4.50 per ton, delivered, spread to sub-grade | | | | | | | | |
| 037 | 6" deep | B-10B | | S.F. | .15 | .01 | .02 | .18 | .20 |
| 039 | 9" deep | | .001 | | .23 | .01 | .03 | .27 | .30 |
| 040 | 12" deep | ↓ | .001 | ↓ | .30 | .01 | .04 | .35 | .39 |
| 11-001 | **PAVING**, Bituminous | | | | | | | | |
| 002 | 6" stone base, 2" binder course, 1" topping | B-25 | .008 | S.F. | .83 | .08 | .07 | .98 | 1.12 |
| 030 | Binder course, 1-1/2" thick | | .002 | | .22 | .02 | .02 | .26 | .29 |
| 040 | 2" thick | | .004 | | .30 | .05 | .03 | .38 | .43 |
| 050 | 3" thick | | .005 | | .44 | .06 | .05 | .55 | .63 |
| 060 | 4" thick | | .007 | | .59 | .07 | .07 | .73 | .85 |
| 080 | Sand finish course, 3/4" thick | | .002 | | .12 | .02 | .01 | .15 | .17 |
| 090 | 1" thick | ↓ | .002 | | .17 | .03 | .02 | .22 | .25 |
| 15-001 | **BRICK PAVING** 4" x 8" x 1-1/2", without joints (4.5 brick/S.F.) | D-1 | .145 | | 1.75 | 1.65 | | 3.40 | 4.50 |
| 010 | Grouted, 3/8" joint (3.9 brick/S.F.) | | .178 | | 1.55 | 2.01 | | 3.56 | 4.85 |
| 020 | 4" x 8" x 2-1/4", without joints (4.5 bricks/S.F.) | | .145 | | 1.88 | 1.65 | | 3.53 | 4.64 |
| 030 | Grouted, 3/8" joint (3.9 brick/S.F.) | ↓ | .178 | ↓ | 1.61 | 2.01 | | 3.62 | 4.91 |
| 20-001 | **CONCRETE PAVING** With mesh, not incl. base, joints or | | | | | | | | |
| 002 | finish, 4500 psi concrete, 6" thick | B-26 | .005 | S.F. | 1.21 | .06 | .08 | 1.35 | 1.51 |
| 010 | 8" thick | " | .007 | | 1.54 | .08 | .11 | 1.73 | 1.93 |
| 070 | Finishing, broom finish, add to above | 2 Cefi | .013 | ↓ | | .16 | | .16 | .24L |
| 101 | | | | | | | | | |
| 22-001 | **CURBS** Bituminous, plain, 8" wide, 6" high, 50 L.F. per ton | B-27 | .032 | L.F. | .55 | .34 | .08 | .97 | 1.23 |
| 015 | Bituminous berm, 12" w., 3"to 6"h. 35lf / ton, before pvmt. | | .046 | | .79 | .48 | .12 | 1.39 | 1.76 |
| 020 | 12" w. 1-1/2"to 4" H.,60 L.F. per ton, laid with pavement | ↓ | .030 | | .46 | .32 | .08 | .86 | 1.10 |
| 030 | Concrete, 6" x 18", cast in place, straight | C-2 | .096 | ↓ | 2.75 | 1.18 | .06 | 3.99 | 4.96 |

For expanded coverage of these items see *Means' Site Work Cost Data 1985*

## 2.6 Roads & Walks

| | | CREW | MAN-HOURS | UNIT | MAT. | LABOR | EQUIP. | TOTAL | TOTAL INCL O&P |
|---|---|---|---|---|---|---|---|---|---|
| 040 | 6" x 18" radius | C-2 | .107 | L.F. | 2.95 | 1.31 | .06 | 4.32 | 5.40 |
| 045 | Precast, 6" x 18", straight | B-29 | .160 | | 4.50 | 1.74 | 1.23 | 7.47 | 9.05 |
| 060 | 6" x 18" radius | | .172 | | 6.75 | 1.86 | 1.33 | 9.94 | 11.85 |
| 100 | Granite, split face, straight, 5" x 16" | | .112 | | 9.25 | 1.22 | .86 | 11.33 | 13.05 |
| 110 | 6" x 18" (see also division 4.4-15) | | .124 | | 10.75 | 1.35 | .96 | 13.06 | 15 |
| 130 | Radius curbing, 6" x 18", over 10' radius | | .215 | | 13 | 2.33 | 1.66 | 16.99 | 19.80 |
| 140 | Corners, 2' radius | | .700 | Ea. | 52 | 7.60 | 5.40 | 65 | 75 |
| 160 | Edging, 4-1/2" x 12", straight | | .187 | L.F. | 4.20 | 2.02 | 1.44 | 7.66 | 9.40 |
| 180 | Curb inlets, (guttermouth) straight | | 1.370 | Ea. | 105 | 14.45 | 10.55 | 130 | 150 |
| 37-001 | **SEALCOATING** 2 coat tar pitch emulsion, over 10,000 S.Y. | B-36 | .003 | S.F. | .03 | .03 | .07 | .13 | .16 |
| 010 | Under 1000 S.Y. | B-1 | .023 | S.Y. | .31 | .24 | | .55 | .73 |
| 40-001 | **SIDEWALKS** Bituminous, no base included, 2" thick | B-37 | .008 | S.F. | .34 | .09 | .01 | .44 | .53 |
| 010 | 2-1/2" thick | " | .010 | | .42 | .11 | .02 | .55 | .66 |
| 015 | brick on 4" thick sand bed laid flat, 4.5 per S.F. | D-1 | .145 | | 1.88 | 1.65 | | 3.53 | 4.64 |
| 020 | Laid on edge, 7.2 per S.F. (see also 2.6-15) | " | .229 | | 2.92 | 2.59 | | 5.51 | 7.25 |
| 031 | broomed finish, no base, 4" thick | B-24 | .040 | | .79 | .46 | | 1.25 | 1.59 |
| 035 | 5" thick | | .044 | | .95 | .50 | | 1.45 | 1.83 |
| 040 | 6" thick | | .047 | | 1.10 | .54 | | 1.64 | 2.05 |
| 045 | For bank run gravel base, 4" thick, add | B-6 | .010 | | .10 | .10 | .06 | .26 | .34 |
| 052 | 8" thick, add | " | .015 | | .20 | .16 | .10 | .46 | .58 |
| 100 | Crushed stone, 1" thick, white marble | 2 Clab | .009 | | .43 | .09 | | .52 | .62 |
| 105 | Bluestone | " | .009 | | .09 | .09 | | .18 | .25 |
| 110 | Flagging, bluestone, irregular 1" thick, | D-1 | .198 | | 1.80 | 2.24 | | 4.04 | 5.45 |
| 115 | Snapped rectangular, 1" thick | | .174 | | 2.25 | 1.97 | | 4.22 | 5.55 |
| 120 | Snapped random rectangular, 1-1/2" thick | | .188 | | 3.25 | 2.13 | | 5.38 | 6.90 |
| 125 | 2" thick | | .193 | | 3.50 | 2.18 | | 5.68 | 7.25 |
| 130 | Slate, natural cleft, irregular, 3/4" thick | | .174 | | 1.30 | 1.97 | | 3.27 | 4.50 |
| 135 | Random rectangular, gauged, 1/2" thick | | .152 | | 3.05 | 1.73 | | 4.78 | 6.05 |
| 140 | Random rectangular, butt joint, gauged, 1/4" thick | | .107 | | 3.30 | 1.21 | | 4.51 | 5.50 |
| 145 | For sand rubbed finish, add | | | | 2.06 | | | 2.06 | 2.26M |
| 155 | Granite blocks, 3-1/2" x 3-1/2" x 3-1/2" | D-1 | .174 | | 4.25 | 1.97 | | 6.22 | 7.75 |
| 170 | Redwood, prefabricated, 4' x 4' sections | F-2 | .051 | | 2.28 | .63 | .04 | 2.95 | 3.55 |
| 175 | Redwood planks, 1" thick, on sleepers | " | .067 | | 1.50 | .82 | .06 | 2.38 | 3.02 |
| 45-001 | **STEPS** Including excavation, borrow and concrete base, where applicable | | | | | | | | |
| 010 | Bricks | D-1 | .696 | LF Riser | 18 | 7.90 | | 25.90 | 32 |
| 020 | Railroad ties | 2 Carp | .229 | | 11.60 | 2.82 | | 14.42 | 17.25 |
| 030 | Bluestone treads, 12" x 2" or 12" x 1-1/2" | D-1 | .246 | | 16.30 | 2.79 | | 19.09 | 22 |
| 050 | Concrete, cast in place, see division 3.3-14-680 | | | | | | | | |
| 060 | Precast concrete, see division 3.4-50 | | | | | | | | |
| 47-001 | **TERRACES** Compared to sidewalks, deduct | | | S.F. | | | | | 10% |

## 2.7 Site Improvements

| | | CREW | MAN-HOURS | UNIT | MAT. | LABOR | EQUIP. | TOTAL | TOTAL INCL O&P |
|---|---|---|---|---|---|---|---|---|---|
| 08-001 | **CHAIN LINK FENCE,** 9 ga. wire, | | | | | | | | |
| 002 | 1-5/8" post 10'O.C., 1-3/8" top rail, 2" corner post galv. steel 3' high | B-1 | .130 | L.F. | 2.50 | 1.38 | | 3.88 | 4.94 |
| 005 | 4' high | | .141 | | 2.90 | 1.51 | | 4.41 | 5.60 |
| 010 | 6' high | | .209 | | 3.75 | 2.23 | | 5.98 | 7.65 |
| 015 | Add for gate 3' wide, 1-3/8" frame 3' high | | 1.040 | Ea. | 25 | 11.15 | | 36.15 | 45 |
| 017 | 4' high | | 1.200 | | 33 | 12.80 | | 45.80 | 57 |
| 019 | 6' high | | 1.600 | | 50 | 17.05 | | 67.05 | 82 |
| 020 | Add for gate 4' wide, 1-3/8" frame 3' high | | 1.090 | | 33 | 11.65 | | 44.65 | 55 |
| 022 | 4' high | | 1.330 | | 44 | 14.20 | | 58.20 | 71 |
| 024 | 6' high | | 2 | | 66 | 21 | | 87 | 105 |

For expanded coverage of these items see *Means' Site Work Cost Data 1985*

## 2.7 Site Improvements

| | | CREW | MAN-HOURS | UNIT | MAT. | LABOR | EQUIP. | TOTAL | TOTAL INCL O&P |
|---|---|---|---|---|---|---|---|---|---|
| 08 035 | Aluminum, 9 ga. wire, 3' high | B-1 | .130 | L.F. | 3 | 1.38 | | 4.38 | 5.50 |
| 038 | 4' high | | .141 | | 3.50 | 1.51 | | 5.01 | 6.25 |
| 040 | 6' high | | .209 | ↓ | 4.50 | 2.23 | | 6.73 | 8.50 |
| 045 | Add for gate 3' wide, 1-3/8" frame 3' high | | 1.040 | Ea. | 30 | 11.15 | | 41.15 | 51 |
| 047 | 4' high | | 1.200 | | 40 | 12.80 | | 52.80 | 64 |
| 049 | 6' high | | 1.600 | | 60 | 17.05 | | 77.05 | 93 |
| 050 | Add for gate 4' wide, 1-3/8" frame 3' high | | 1.090 | | 40 | 11.65 | | 51.65 | 62 |
| 052 | 4' high | | 1.330 | | 53 | 14.20 | | 67.20 | 81 |
| 054 | 6' high | | 2 | ↓ | 80 | 21 | | 101 | 120 |
| 062 | Vinyl covered 9 ga. wire, 3' high | | .130 | L.F. | 2.80 | 1.38 | | 4.18 | 5.25 |
| 064 | 4' high | | .141 | | 3.20 | 1.51 | | 4.71 | 5.90 |
| 066 | 6' high | | .209 | ↓ | 4 | 2.23 | | 6.23 | 7.95 |
| 072 | Add for gate 3' wide, 1-3/8" frame 3' high | | 1.040 | Ea. | 33 | 11.15 | | 44.15 | 54 |
| 074 | 4' high | | 1.200 | | 44 | 12.80 | | 56.80 | 69 |
| 076 | 6' high | | 1.600 | | 67 | 17.05 | | 84.05 | 100 |
| 078 | Add for gate 4' wide, 1-3/8" frame 3' high | | 1.090 | | 44 | 11.65 | | 55.65 | 67 |
| 080 | 4' high | | 1.330 | | 59 | 14.20 | | 73.20 | 87 |
| 082 | 6' high | ↓ | 2 | ↓ | 88 | 21 | | 109 | 130 |
| 086 | Tennis courts, 11 ga. wire, 2 1/2" post 10' O.C., 1-5/8" top rail | | | | | | | | |
| 090 | 2-1/2" corner post, 10' high | B-1 | .253 | L.F. | 6.90 | 2.69 | | 9.59 | 11.85 |
| 092 | 12' high | | .300 | " | 8.90 | 3.20 | | 12.10 | 14.85 |
| 100 | Add for gate 3' wide, 1-5/8" frame 10' high | | 2.400 | Ea. | 90 | 26 | | 116 | 140 |
| 104 | Aluminized, 11 ga. wire 10' high | | .253 | L.F. | 8 | 2.69 | | 10.69 | 13.05 |
| 110 | 12' high | | .300 | " | 10.50 | 3.20 | | 13.70 | 16.60 |
| 114 | Add for gate 3' wide, 1-5/8" frame, 10' high | | 2.400 | Ea. | 110 | 26 | | 136 | 160 |
| 125 | Vinyl covered 11 ga. wire, 10' high | | .253 | L.F. | 8.90 | 2.69 | | 11.59 | 14.05 |
| 130 | 12' high | | .300 | " | 11 | 3.20 | | 14.20 | 17.15 |
| 140 | Add for gate 3' wide, 1-3/8" frame, 10' high | ↓ | 2.400 | Ea. | 120 | 26 | | 146 | 175 |
| 13-001 | **WOOD FENCE**, Basket weave, 3/8" x 4" boards, 2" x 4" | | | | | | | | |
| 002 | stringers on spreaders, 4" x 4" posts | | | | | | | | |
| 005 | No. 1 cedar, 6' high | B-1 | .150 | L.F. | 12.75 | 1.60 | | 14.35 | 16.55 |
| 007 | Treated pine, 6' high | | .160 | | 6.50 | 1.71 | | 8.21 | 9.85 |
| 009 | Vertical weave 6' high | ↓ | .166 | ↓ | 7.25 | 1.77 | | 9.02 | 10.75 |
| 020 | Board fence, 1" x 4" boards, 2" x 4" rails, 4" x 4" post | | | | | | | | |
| 022 | Preservative treated, 2 rail, 3' high | B-1 | .166 | L.F. | 4.15 | 1.77 | | 5.92 | 7.35 |
| 024 | 4' high | | .178 | | 4.50 | 1.90 | | 6.40 | 7.95 |
| 026 | 3 rail, 5' high | | .185 | | 5.60 | 1.97 | | 7.57 | 9.30 |
| 030 | 6' high | | .192 | | 6.10 | 2.05 | | 8.15 | 9.95 |
| 032 | No. 2 grade western cedar, 2 rail, 3' high | | .166 | | 4.40 | 1.77 | | 6.17 | 7.65 |
| 034 | 4' high | | .178 | | 8.35 | 1.90 | | 10.25 | 12.20 |
| 036 | 3 rail, 5' high | | .185 | | 10.75 | 1.97 | | 12.72 | 14.95 |
| 040 | 6' high | | .192 | | 13 | 2.05 | | 15.05 | 17.55 |
| 042 | No. 1 grade cedar, 2 rail, 3' high | | .166 | | 10.50 | 1.77 | | 12.27 | 14.35 |
| 044 | 4' high | | .178 | | 12 | 1.90 | | 13.90 | 16.20 |
| 046 | 3 rail, 5' high | | .185 | | 13.25 | 1.97 | | 15.22 | 17.70 |
| 050 | 6' high | ↓ | .192 | ↓ | 14.50 | 2.05 | | 16.55 | 19.20 |
| 054 | Shadow box, 1" x 6" board, 2" x 4" rail, 4" x 4" post | | | | | | | | |
| 056 | Pine, pressure treated, 3 rail, 6' high | B-1 | .160 | L.F. | 6 | 1.71 | | 7.71 | 9.30 |
| 060 | Gate, 3 1/2 ft. wide | | .171 | | 40 | 1.83 | | 41.83 | 47 |
| 062 | No. 1 cedar, 3 rail, 4' high | | .178 | | 11.40 | 1.90 | | 13.30 | 15.55 |
| 064 | 6' high | | .185 | | 12.75 | 1.97 | | 14.72 | 17.15 |
| 086 | Open rail fence, split rails, 2 rail 3' high, no. 1 cedar | | .150 | | 2.85 | 1.60 | | 4.45 | 5.65 |
| 087 | No. 2 cedar | | .150 | | 1.50 | 1.60 | | 3.10 | 4.19 |
| 088 | 3 rail, 4' high, no. 1 cedar | | .160 | | 3.75 | 1.71 | | 5.46 | 6.85 |
| 089 | No. 2 cedar | | .160 | | 1.75 | 1.71 | | 3.46 | 4.63 |
| 092 | Rustic rails, 2 rail 3' high, no. 1 cedar | | .150 | | 2.50 | 1.60 | | 4.10 | 5.30 |
| 093 | No. 2 cedar | | .150 | | 1.60 | 1.60 | | 3.20 | 4.30 |
| 094 | 3 rail, 4' high | | .160 | | 3.10 | 1.71 | | 4.81 | 6.10 |
| 095 | No. 2 cedar | ↓ | .160 | ↓ | 2.10 | 1.71 | | 3.81 | 5 |
| 096 | Picket fence, gothic, pressure treated pine | | | | | | | | |

For expanded coverage of these items see *Means' Site Work Cost Data 1985*

## 2.7 Site Improvements

| | | Crew | Man-Hours | Unit | Mat. | Labor | Equip. | Total | Total Incl O&P |
|---|---|---|---|---|---|---|---|---|---|
| 13 100 | 2 rail, 3' high | B-1 | .171 | L.F. | 2.75 | 1.83 | | 4.58 | 5.90 |
| 102 | 3 rail, 4' high | | .185 | " | 3.20 | 1.97 | | 5.17 | 6.65 |
| 104 | Gate, 3 1/2 wide | | .185 | Ea. | 29 | 1.97 | | 30.97 | 35 |
| 106 | No. 2 cedar, 2 rail, 3' high | | .171 | L.F. | 3.20 | 1.83 | | 5.03 | 6.40 |
| 110 | 3 rail, 4' high | | .185 | | 3.50 | 1.97 | | 5.47 | 6.95 |
| 112 | Gate, 3 1/2' wide | | .185 | | 34 | 1.97 | | 35.97 | 41 |
| 114 | No. 1 cedar, 2 rail 3' high | | .171 | | 7.25 | 1.83 | | 9.08 | 10.85 |
| 116 | 3 rail, 4' high | | .185 | | 8.25 | 1.97 | | 10.22 | 12.20 |
| 120 | Rustic picket, molded pine, 2 rail, 3' high | | .171 | | 2.35 | 1.83 | | 4.18 | 5.50 |
| 122 | No. 1 cedar, 2 rail, 3' high | | .171 | | 4.80 | 1.83 | | 6.63 | 8.20 |
| 124 | Stockade fence, no. 1 cedar, 3 1/4" rails, 6' high | | .150 | | 9.75 | 1.60 | | 11.35 | 13.25 |
| 126 | 8' high | | .155 | | 15.25 | 1.65 | | 16.90 | 19.40 |
| 130 | No. 2 cedar, treated wood rails, 6' high | | .150 | | 4 | 1.60 | | 5.60 | 6.95 |
| 132 | Gate, 3 1/2' wide | | .155 | | 33 | 1.65 | | 34.65 | 39 |
| 136 | Treated pine, treated rails, 6' high | | .960 | Ea. | 6.25 | 10.25 | | 16.50 | 23 |
| 140 | 8' high | | .960 | | 8.25 | 10.25 | | 18.50 | 25 |
| 27-001 | **PLANTERS** Concrete, sandblasted, precast, 48" diameter, 24" high | 2 Clab | 1.070 | | 400 | 10.65 | | 410.65 | 455 |
| 030 | Fiberglass, circular, 36" diameter, 24" high | " | 1.070 | | 450 | 10.65 | | 460.65 | 510 |
| 37-001 | **SPRINKLER SYSTEM** For lawns | | | | | | | | |
| 080 | Residential system, custom, 1" supply | B-20 | .009 | S.F.Grnd | .19 | .12 | | .31 | .40 |
| 090 | 1-1/2" supply | " | .010 | " | .17 | .14 | | .31 | .40 |
| 40-001 | **STONE WALLS** Including excavation, concrete footing and | | | | | | | | |
| 002 | stone 3' below grade. Price is exposed face area. | | | | | | | | |
| 020 | Decorative random stone, to 6' high, 1'-6" thick, dry set | D-1 | .457 | S.F. | 5.95 | 5.20 | | 11.15 | 14.65 |
| 030 | Mortar set | | .400 | | 7 | 4.53 | | 11.53 | 14.75 |
| 050 | Cut stone, to 6' high, 1'-6" thick, dry set | | .457 | | 9.50 | 5.20 | | 14.70 | 18.55 |
| 060 | Mortar set | | .400 | | 10 | 4.53 | | 14.53 | 18.05 |
| 080 | Retaining wall, random stone, 6' to 10' high, 2' thick, dry set | | .356 | | 7.40 | 4.03 | | 11.43 | 14.40 |
| 090 | Mortar set | | .320 | | 8.80 | 3.62 | | 12.42 | 15.35 |
| 110 | Cut stone, 6' to 10' high, 2' thick, dry set | | .356 | | 12.10 | 4.03 | | 16.13 | 19.60 |
| 120 | Mortar set | | .320 | | 12.80 | 3.62 | | 16.42 | 19.75 |

## 2.8 Lawns & Planting

| | | Crew | Man-Hours | Unit | Mat. | Labor | Equip. | Total | Total Incl O&P |
|---|---|---|---|---|---|---|---|---|---|
| 05-001 | **EDGING** Redwood, untreated, 1" x 4" | F-2 | .032 | L.F. | .64 | .39 | .03 | 1.06 | 1.36 |
| 010 | 2" x 4" | | .048 | | .63 | .60 | .04 | 1.27 | 1.69 |
| 020 | Steel edge strips, 1/4" x 5" including stakes | | .048 | | 2.22 | .60 | .04 | 2.86 | 3.44 |
| 030 | 3/16" x 4" | | .048 | | 1.65 | .60 | .04 | 2.29 | 2.81 |
| 050 | Brick edging, set on edge, 3 brick per L.F. | D-1 | .119 | | 1.27 | 1.34 | | 2.61 | 3.49 |
| 060 | Set flat, 1-1/2 brick per L.F. | " | .043 | | .64 | .49 | | 1.13 | 1.47 |
| 07-001 | **EROSION CONTROL** Jute mesh, 100 S.Y. per roll, 4' wide, stapled | B-1 | .010 | S.Y. | .58 | .10 | | .68 | .80 |
| 010 | Plastic netting, stapled, 2" x 1" mesh, 20 mil | | .010 | | .30 | .10 | | .40 | .49 |
| 020 | Polypropylene mesh, stapled, 6.5 oz/S.Y. | | .010 | | 1.54 | .10 | | 1.64 | 1.86 |
| 030 | Tobacco netting, #2, stapled | | .010 | | .02 | .10 | | .12 | .18 |
| 10-001 | **GROUND COVER** Plants, pachysandra, in prepared beds | | 2.400 | C | 10.50 | 26 | | 36.50 | 52 |
| 020 | Vinca minor | | 2.400 | " | 41 | 26 | | 67 | 86 |
| 060 | Stone chips, in 50 lb. bags, Georgia marble | | .046 | Bag | 2.15 | .49 | | 2.64 | 3.15 |
| 070 | Onyx gemstone | | .092 | | 8.50 | .98 | | 9.48 | 10.90 |
| 080 | Quartz | | .092 | | 3.70 | .98 | | 4.68 | 5.65 |
| 090 | Pea gravel, truckload lots | | .857 | C.Y. | 13.05 | 9.15 | | 22.20 | 29 |
| 15-001 | **HEDGE PLANTS** Barberry, 2' to 3' high | | .185 | Ea. | 22 | 1.97 | | 23.97 | 27 |
| 020 | Privet, 18" to 24" high | | .137 | | 2.25 | 1.46 | | 3.71 | 4.79 |
| 040 | Boxwood, 18" to 20" high | | .300 | | 21 | 3.20 | | 24.20 | 28 |
| 060 | Ilex, 15" to 18" high | | .462 | | 13.25 | 4.92 | | 18.17 | 22 |

For expanded coverage of these items see *Means' Site Work Cost Data 1985*

## 2.8 Lawns & Planting

| | | CREW | MAN-HOURS | UNIT | MAT. | LABOR | EQUIP. | TOTAL | TOTAL INCL O&P |
|---|---|---|---|---|---|---|---|---|---|
| 25-001 | **LOAM OR TOPSOIL** Remove and stockpile on site | | | | | | | | |
| 070 | Furnish and place, truck dumped @ $9.00 per C.Y., 4" deep | B-10S | .001 | S.F. | .12 | .01 | .02 | .15 | .18 |
| 080 | 6" deep | " | .002 | " | .18 | .02 | .04 | .24 | .27 |
| 090 | Fine grading and seeding, incl. lime, fertilizer & seed, | | | | | | | | |
| 100 | With equipment | | | S.F. | .15 | .07 | | .22 | .17 |
| 35-001 | **PLANT BED** Preparation, 18" deep, by machine | A-1 | .002 | S.F. | .53 | .02 | .01 | .56 | .62 |
| 010 | By hand | 2 Clab | .052 | " | .53 | .52 | | 1.05 | 1.40 |
| 40-001 | **PLANTING** Moving shrubs on site, 12" ball | B-1 | .857 | Ea. | | 9.15 | | 9.15 | 14.50L |
| 010 | 24" ball | " | 1.090 | | | 11.65 | | 11.65 | 18.45L |
| 030 | Moving trees on site, 36" ball | B-6 | 6.400 | | | 69 | 41 | 110 | 155 |
| 040 | 60" ball | " | 24 | | | 260 | 150 | 410 | 575 |
| 45-001 | **SEEDING** Mechanical seeding, $1.85/lb., 215 lb./acre | A-1 | 25.810 | Acre | 400 | 255 | 115 | 770 | 975 |
| 010 | $1.85/lb., 44 lb./M.S.Y. | | .001 | S.F. | .02 | .01 | | .03 | .03 |
| 040 | Fertilizer applied 35 lbs. per M.S.F. | | .040 | M.S.F. | 5.25 | .40 | .18 | 5.83 | 6.60 |
| 060 | Limestone applied 50 lbs. per M.S.F. | | .040 | " | 2 | .40 | .18 | 2.58 | 3.03 |
| 50-001 | **SODDING** In East, 1 inch deep, incl. fine grade, on level ground | B-14 | .005 | S.F. | .16 | .05 | .02 | .23 | .28 |
| 020 | On slopes | | .007 | | .17 | .07 | .02 | .26 | .32 |
| 120 | in Midwest on level ground, prepared area, over 400 S.Y. | | .006 | | .10 | .07 | .02 | .19 | .24 |
| 123 | 100 S.Y. area | | .007 | | .14 | .07 | .02 | .23 | .29 |
| 126 | 50 S.Y. area | | .007 | | .21 | .08 | .02 | .31 | .38 |
| 130 | On slopes, 400 S.Y. area | | .007 | | .11 | .08 | .02 | .21 | .27 |
| 170 | Polyurethane with ceramic chips for median strip, minimum | | | | .57 | | | .57 | .62M |
| 180 | Maximum | | | | .78 | | | .78 | .85M |
| 65-001 | **TREES AND SHRUBS** Guying trees, 2" to 4" diameter | B-6 | 1.500 | Ea. | 22 | 16.50 | 9.50 | 48 | 60 |
| 010 | 6" to 10" diameter | B-3 | 6 | " | 80 | 65 | 145 | 290 | 350 |
| 030 | Trees in place, balled & burlapped | | | | | | | | |
| 050 | Pines, black, 2-1/2' to 3' high | B-1 | .480 | Ea. | 14.50 | 5.10 | | 19.60 | 24 |
| 070 | Yews, spreading, 12" to 15" spread | | .400 | | 9 | 4.27 | | 13.27 | 16.65 |
| 080 | Upright, 2' to 2.5' high | | .421 | | 25 | 4.49 | | 29.49 | 35 |
| 100 | Junipers, prostrate 15" to 18" spread | | .300 | | 14 | 3.20 | | 17.20 | 20 |
| 110 | Pfitzer, 2' to 2.5' spread | | .436 | | 21 | 4.65 | | 25.65 | 30 |
| 120 | Upright junipers, 4-1/2' to 5' high | B-17 | .582 | | 27 | 6.20 | 6.75 | 39.95 | 47 |
| 140 | Spruce, 4' to 5' high | | .427 | | 35 | 4.54 | 4.96 | 44.50 | 51 |
| 160 | Birch, 6' to 8' high, 3 stems | | 1.600 | | 36 | 17.40 | 18.60 | 72 | 87 |
| 180 | Flowering crab, 6' to 8' high | | 1.600 | | 35 | 17.40 | 18.60 | 71 | 86 |
| 200 | Hawthorn, 8' to 10' high | | 1.600 | | 65 | 17.40 | 18.60 | 101 | 120 |
| 220 | Maple, Norway, 8' to 10' high, 1-1/2" to 1-3/4" caliper | | 3.200 | | 37 | 34 | 37 | 108 | 135 |
| 240 | Oak, 2-1/2" to 3" caliper | | 10.670 | | 100 | 110 | 125 | 335 | 425 |
| 260 | Willow, 6' to 8' high | | 1.280 | | 22 | 13.10 | 14.90 | 50 | 62 |
| 280 | Rhododendron, 18" to 24" high | B-1 | .500 | | 13.50 | 5.35 | | 18.85 | 23 |
| 290 | 3' to 4' high | B-17 | .427 | | 37 | 4.54 | 4.96 | 46.50 | 53 |
| 310 | Japanese holly, 12" to 15" high | B-1 | .338 | | 5.30 | 3.61 | | 8.91 | 11.55 |
| 320 | 18" to 24" high | | .511 | | 9.55 | 5.45 | | 15 | 19.15 |
| 340 | American holly, 15" to 18" high | | .250 | | 13 | 2.67 | | 15.67 | 18.55 |
| 350 | 5' to 6' high | B-17 | .582 | | 55 | 6.20 | 6.75 | 67.95 | 78 |
| 370 | Canadian hemlock, 2-1/2' to 3' high | B-1 | .667 | | 24 | 7.10 | | 31.10 | 38 |
| 380 | 5' to 6' high | B-17 | .970 | | 58 | 10.75 | 11.25 | 80 | 92 |
| 400 | Douglas fir, 3' to 4' high | | .427 | | 15.90 | 4.54 | 4.96 | 25.40 | 30 |
| 410 | 6' to 7' high | | 1.280 | | 58 | 13.10 | 14.90 | 86 | 100 |
| 430 | Arborvitae, globe, 12" to 15" high | B-1 | .267 | | 7.50 | 2.84 | | 10.34 | 12.75 |
| 440 | Pyramidal, 4' to 5' high | B-17 | .970 | | 21 | 10.75 | 11.25 | 43 | 52 |
| 460 | White pine, 4' to 5' high | | .427 | | 32 | 4.54 | 4.96 | 41.50 | 48 |
| 470 | 7' to 8' high | | 1.600 | | 59 | 17.40 | 18.60 | 95 | 110 |
| 490 | Juniper, andorra, 18" to 24" spread | B-1 | .300 | | 14.30 | 3.20 | | 17.50 | 21 |
| 500 | Blue pfitzer, 2' to 2.5' spread | | .545 | | 21 | 5.80 | | 26.80 | 32 |
| 520 | Euonymus, compacta, 15" to 18" high | | .300 | | 7.50 | 3.20 | | 10.70 | 13.30 |
| 530 | 4' to 5' high | B-17 | .640 | | 34 | 6.75 | 7.45 | 48.20 | 56 |
| 550 | Dogwood, pink, 5' to 6' high | | 1.070 | | 30 | 11.60 | 12.40 | 54 | 65 |
| 560 | 1-1/2" caliper | | 2.290 | | 85 | 24 | 27 | 136 | 160 |

For expanded coverage of these items see *Means' Site Work Cost Data 1985*

## 2.8 Lawns & Planting

| | | | CREW | MAN-HOURS | UNIT | MAT. | LABOR | EQUIP. | TOTAL | TOTAL INCL O&P |
|---|---|---|---|---|---|---|---|---|---|---|
| 65 | 580 | Dogwood, white, 4' to 5' high | B-17 | .427 | Ea. | 19.10 | 4.54 | 4.96 | 28.60 | 34 |
| | 590 | 1-1/4" caliper | " | 1.780 | | 64 | 19 | 21 | 104 | 125 |
| | 610 | Shrubs in place, forsythia, 2' to 3' high | B-1 | .400 | | 6.35 | 4.27 | | 10.62 | 13.75 |
| | 620 | Weigelia, 3' to 4' high | | .343 | | 17 | 3.66 | | 20.66 | 24 |
| | 630 | Deutzia, 12" to 15" high | | .250 | | 6.35 | 2.67 | | 9.02 | 11.20 |
| | 640 | Spirea, 3' to 4' high | | .343 | | 19 | 3.66 | | 22.66 | 27 |
| | 650 | Honeysuckle, 3' to 4' high | | .400 | | 8.50 | 4.27 | | 12.77 | 16.10 |
| | 660 | Flowering almond, 2' to 3' high | ↓ | .667 | ↓ | 7.50 | 7.10 | | 14.60 | 19.50 |

## 3.1 Formwork

| | | CREW | MAN-HOURS | UNIT | MAT. | LABOR | EQUIP. | TOTAL | TOTAL INCL O&P |
|---|---|---|---|---|---|---|---|---|---|
| 45-001 | **FORMS IN PLACE, FOOTINGS** Continuous wall, 1 use | C-1 | .085 | S.F.C.A. | .76 | 1 | .06 | 1.82 | 2.49 |
| 015 | 4 use | " | .066 | " | .23 | .78 | .04 | 1.05 | 1.53 |
| 150 | Keyway, 4 uses, tapered wood, 2" x 4" | 1 Carp | .015 | L.F. | .07 | .19 | | .26 | .37 |
| 155 | 2" x 6" | " | .016 | " | .09 | .20 | | .29 | .41 |
| 500 | Spread footings, 1 use | C-1 | .105 | S.F.C.A. | .87 | 1.23 | .07 | 2.17 | 2.99 |
| 515 | 4 use | " | .077 | " | .32 | .91 | .05 | 1.28 | 1.84 |
| 55-001 | **FORMS IN PLACE, SLAB ON GRADE** | | | | | | | | |
| 100 | Bulkhead forms with keyway, 1 use, 2 piece | C-1 | .063 | L.F. | .34 | .74 | .04 | 1.12 | 1.59 |
| 200 | Curb forms, wood, 6" to 12" high, on grade, 1 use | | .149 | S.F.C.A. | 1.10 | 1.75 | .10 | 2.95 | 4.09 |
| 215 | 4 use | | .116 | " | .40 | 1.37 | .08 | 1.85 | 2.69 |
| 300 | Edge forms, to 6" high, 4 use, on grade | | .053 | L.F. | .16 | .62 | .04 | .82 | 1.21 |
| 305 | 7" to 12" high, 4 use, on grade | | .074 | S.F.C.A. | .51 | .86 | .05 | 1.42 | 1.98 |
| 400 | For slab blockouts, 1 use to 12" high | ↓ | .160 | L.F. | .42 | 1.88 | .11 | 2.41 | 3.56 |
| 65-001 | **FORMS IN PLACE, WALLS** | | | | | | | | |
| 010 | Box out for wall openings, to 16" thick, to 10 S.F. | C-2 | 2 | Ea. | 12.20 | 24.83 | 1.17 | 38.20 | 54 |
| 015 | Over 10 S.F. (use perimeter) | " | .171 | L.F. | 1.15 | 2.11 | .10 | 3.36 | 4.71 |
| 025 | Brick shelf, 4" wide, add to wall forms, use wall area | | | | | | | | |
| 026 | above shelf, 1 use | C-2 | .200 | S.F.C.A. | 1.20 | 2.46 | .12 | 3.78 | 5.35 |
| 035 | 4 use | | .160 | " | .47 | 1.97 | .09 | 2.53 | 3.73 |
| 050 | Bulkhead forms for walls, with keyway, 1 use, 2 piece | | .181 | L.F. | 1.40 | 2.22 | .11 | 3.73 | 5.20 |
| 055 | 3 piece | | .274 | " | 1.82 | 3.37 | .16 | 5.35 | 7.50 |
| 200 | Job built plyform wall forms, to 8' high, 1 use | | .130 | S.F.C.A. | 1.27 | 1.59 | .08 | 2.94 | 4.01 |
| 215 | 4 use | | .095 | | .50 | 1.16 | .06 | 1.72 | 2.46 |
| 240 | Over 8' to 16' high, 1 use | | .171 | | 1.37 | 2.11 | .10 | 3.58 | 4.95 |
| 255 | 4 use | | .122 | | .57 | 1.49 | .07 | 2.13 | 3.07 |
| 300 | For architectural finish, add | | .026 | | .30 | .32 | .02 | .64 | .86 |
| 780 | Modular prefabricated plywood, to 8' high, 1 use per month | | .053 | | .74 | .65 | .03 | 1.42 | 1.87 |
| 786 | 4 use per month | | .053 | | .25 | .65 | .03 | .93 | 1.34 |
| 800 | To 16' high, 1 use per month | | .087 | | .91 | 1.07 | .05 | 2.03 | 2.76 |
| 806 | 4 use per month | ↓ | .087 | ↓ | .37 | 1.07 | .05 | 1.49 | 2.16 |

## 3.2 Reinforcing Steel

| | | CREW | MAN-HOURS | UNIT | MAT. | LABOR | EQUIP. | TOTAL | TOTAL INCL O&P |
|---|---|---|---|---|---|---|---|---|---|
| 04-001 | **REINFORCING IN PLACE** A615 Grade 60 | | | | | | | | |
| 050 | Footings, #4 to #7 | 4 Rodm | .008 | Lb. | .24 | .10 | | .34 | .43 |
| 055 | #8 to #14 | | .004 | | .24 | .06 | | .30 | .36 |
| 070 | Walls, #3 to #7 | ↓ | .005 | ↓ | .25 | .07 | | .32 | .39 |

For expanded coverage of these items see Means' *Concrete & Masonry Cost Data 1985*

## 3.2 Reinforcing Steel

| | | CREW | MAN-HOURS | UNIT | MAT. | LABOR | EQUIP. | TOTAL | TOTAL INCL O&P |
|---|---|---|---|---|---|---|---|---|---|
| 04 075 | #8 to #14 | 4 Rodm | .004 | Lb. | .24 | .05 | | .29 | .35 |
| 240 | Dowels, 2 feet long, deformed, #3 bar | 2 Rodm | .114 | Ea. | .71 | 1.52 | | 2.23 | 3.33 |
| 241 | #4 bar | | .128 | | .85 | 1.70 | | 2.55 | 3.79 |
| 242 | #5 bar | | .145 | | 1.03 | 1.93 | | 2.96 | 4.38 |
| 243 | #6 bar | ↓ | .152 | ↓ | 1.29 | 2.03 | | 3.32 | 4.82 |
| 06-001 | WELDED WIRE FABRIC Rolls, 6 x 6 = #10/10 (W1.4/W1.4) 21 lb. | 2 Rodm | .005 | C.S.F. | .08 | .06 | | .14 | .19 |
| 030 | 6 x 6 = #6/6 (W2.9/W2.9) 42 lb. per C.S.F. | | .006 | S.F. | .15 | .07 | | .22 | .29 |
| 050 | 4 x 4 = #10/10 (W1.4/W1.4) 31 lb. per C.S.F. | | .005 | | .10 | .07 | | .17 | .23 |
| 090 | 2 x 2 = #12 galv. for gunite reinforcing | ↓ | .025 | ↓ | .16 | .33 | | .49 | .72 |
| 095 | Material prices for above include 10% lap | | | | | | | | |

## 3.3 Cast in Place Concrete

| | | CREW | MAN-HOURS | UNIT | MAT. | LABOR | EQUIP. | TOTAL | TOTAL INCL O&P |
|---|---|---|---|---|---|---|---|---|---|
| 12-001 | CONCRETE, READY MIX Regular weight, 2000 psi | | | C.Y. | 43 | | | 43 | 47M |
| 010 | 2500 psi | | | | 44 | | | 44 | 48M |
| 015 | 3000 psi | | | | 46.15 | | | 46.15 | 51M |
| 020 | 3500 psi | | | | 47 | | | 47 | 52M |
| 025 | 3750 psi | | | | 48.60 | | | 48.60 | 53M |
| 030 | 4000 psi | | | | 49.25 | | | 49.25 | 54M |
| 035 | 4500 psi | | | | 51.10 | | | 51.10 | 56M |
| 040 | 5000 psi | | | | 52.05 | | | 52.05 | 57M |
| 100 | For high early strength cement, add | | | | 10 | | | | |
| 200 | For all lightweight aggregate, add | | | ↓ | 50 | | | | |
| 14-001 | CONCRETE IN PLACE Including forms (4 uses), reinforcing | | | | | | | | |
| 005 | steel, including finishing unless otherwise indicated | | | | | | | | |
| 050 | Chimney foundations, minimum | C-17A | 3 | C.Y. | 77.50 | 38.76 | 4.24 | 120.50 | 150 |
| 051 | Maximum | " | 4.060 | " | 88.90 | 53.25 | 5.75 | 147.90 | 190 |
| 170 | Curbs, formed in place, 6" x 18", straight, | C-15 | .180 | L.F. | 2.75 | 2.12 | .10 | 4.97 | 6.50 |
| 175 | Curb and gutter | " | .424 | " | 4.45 | 5.01 | .24 | 9.70 | 13.10 |
| 380 | Footings, spread under 1 C.Y. | C-17B | 2.230 | C.Y. | 57 | 28.95 | 6.05 | 92 | 115 |
| 385 | Over 5 C.Y. | C-17C | 1.080 | | 64 | 14.08 | 4.47 | 82.55 | 98 |
| 390 | Footings, strip, 18" x 9", plain | C-17B | 2.740 | | 58 | 35.55 | 7.45 | 101 | 130 |
| 395 | 36" x 12", reinforced | | 1.560 | | 53 | 19.77 | 4.23 | 77 | 95 |
| 400 | Foundation mat, under 10 C.Y. | | 2.480 | | 98 | 32.25 | 6.75 | 137 | 165 |
| 405 | Over 20 C.Y. | ↓ | 1.690 | ↓ | 87 | 22.41 | 4.59 | 114 | 135 |
| 475 | Ground slab, incl. troweled finish, not incl. forms | | | | | | | | |
| 476 | or reinforcing, over 10,000 S.F., 4" thick slab | C-8 | .016 | S.F. | .62 | .18 | .14 | .94 | 1.12 |
| 482 | 6" thick slab | " | .016 | " | .91 | .17 | .14 | 1.22 | 1.43 |
| 620 | Retaining walls, gravity, 4' high (see also division 2.7-32) | C-17B | 4.190 | C.Y. | 70 | 54.60 | 11.40 | 136 | 175 |
| 16-001 | CURING With burlap, 4 uses assumed, 7.5 oz. | 2 Clab | .003 | S.F. | .02 | .03 | | .05 | .07 |
| 010 | 12 oz. | | .003 | C.S.F. | .03 | .03 | | .06 | .08 |
| 020 | With waterproof curing paper, 2 ply, reinforced | | .002 | " | .04 | .02 | | .06 | .07 |
| 030 | With sprayed membrane curing compound | ↓ | .002 | S.F. | .02 | .02 | | .04 | .05 |
| 26-001 | FINISHING FLOORS Monolithic, screed finish | 1 Cefi | .009 | S.F. | | .11 | | .11 | .16L |
| 005 | Darby finish | " | .011 | | | .13 | | .13 | .19L |
| 010 | Float finish | C-9 | .011 | | | .13 | .04 | .17 | .24 |
| 015 | Broom finish | | .012 | | | .14 | .04 | .18 | .26 |
| 020 | Steel trowel finish, for resilient tile | | .013 | | | .16 | .04 | .20 | .28 |
| 025 | For finish floor | ↓ | .015 | | | .17 | .05 | .22 | .32 |
| 160 | Exposed local aggregate finish, minimum | 1 Cefi | .013 | | .06 | .15 | | .21 | .30 |
| 165 | Maximum | | .017 | | .18 | .20 | | .38 | .51 |
| 28-001 | FINISHING WALLS Break ties and patch voids | | .015 | | .01 | .18 | | .19 | .28 |
| 005 | Burlap rub with grout | ↓ | .018 | ↓ | .05 | .21 | | .26 | .38 |

For expanded coverage of these items see Means' *Concrete & Masonry Cost Data 1985*

## 3.3 Cast in Place Concrete

| | | | CREW | MAN-HOURS | UNIT | MAT. | LABOR | EQUIP. | TOTAL | TOTAL INCL O&P |
|---|---|---|---|---|---|---|---|---|---|---|
| 28 | 030 | Bush hammer, green concrete | 1 Cefi | .047 | S.F. | .01 | .56 | | .57 | .87 |
| | 035 | Cured concrete | | .073 | | .02 | .86 | | .88 | 1.35 |
| 30-001 | | FLOOR PATCHING 1/4" thick, small areas, regular | | .047 | | .70 | .56 | | 1.26 | 1.63 |
| | 010 | Epoxy | ↓ | .080 | ↓ | 2.75 | .95 | | 3.70 | 4.49 |
| 38-001 | | PLACING CONCRETE and vibrating, including labor & equipment | | | | | | | | |
| | 140 | Elevated slabs, less than 6" thick, pumped | C-20 | .582 | C.Y. | | 6.30 | 5 | 11.30 | 15.45 |
| | 145 | With crane and bucket | C-7 | .674 | | | 7.30 | 6.25 | 13.55 | 18.40 |
| | 150 | 6" to 10" thick, pumped | C-20 | .492 | | | 5.30 | 4.24 | 9.54 | 13.05 |
| | 155 | With crane and bucket | C-7 | .582 | | | 6.30 | 5.40 | 11.70 | 15.90 |
| | 160 | Slabs over 10" thick, pumped | C-20 | .427 | | | 4.61 | 3.67 | 8.28 | 11.35 |
| | 165 | With crane and bucket | C-7 | .492 | | | 5.30 | 4.58 | 9.88 | 13.45 |
| | 190 | Footings, continuous, shallow, direct chute | C-6 | .400 | | | 4.26 | .44 | 4.70 | 7.20 |
| | 195 | Pumped | C-20 | .640 | | | 6.90 | 5.50 | 12.40 | 17 |
| | 200 | With crane and bucket | C-7 | .711 | | | 7.70 | 6.60 | 14.30 | 19.40 |
| | 240 | Footings, spread, under 1 C.Y., direct chute | C-6 | .873 | | | 9.30 | .96 | 10.26 | 15.70 |
| | 260 | Spread footings, over 5 C.Y., direct chute | | .436 | | | 4.64 | .48 | 5.12 | 7.85 |
| | 290 | Foundation mats, over 20 C.Y., direct chute | | .137 | | | 1.46 | .15 | 1.61 | 2.47 |
| | 430 | Slab on grade, 4" thick, direct chute | ↓ | .436 | | | 4.64 | .48 | 5.12 | 7.85 |
| | 435 | Pumped | C-20 | .533 | | | 5.75 | 4.59 | 10.34 | 14.15 |
| | 440 | With crane and bucket | C-7 | .582 | | | 6.30 | 5.40 | 11.70 | 15.90 |
| | 490 | Walls, 8" thick, direct chute | C-6 | .533 | | | 5.70 | .59 | 6.29 | 9.60 |
| | 495 | Pumped | C-20 | .753 | | | 8.15 | 6.50 | 14.65 | 20 |
| | 500 | With crane and bucket | C-7 | .800 | | | 8.65 | 7.45 | 16.10 | 22 |
| | 505 | 12" thick, direct chute | C-6 | .480 | | | 5.10 | .53 | 5.63 | 8.65 |
| | 510 | Pumped | C-20 | .674 | | | 7.30 | 5.80 | 13.10 | 17.90 |
| | 520 | With crane and bucket | C-7 | .711 | ↓ | | 7.70 | 6.60 | 14.30 | 19.40 |
| | 560 | Wheeled concrete dumping, add to placing costs above | | | | | | | | |
| | 561 | Walking cart, 50' haul, add | A-1 | .258 | C.Y. | | 2.58 | 1.14 | 3.72 | 5.35 |
| | 562 | 150' haul, add | | .296 | | | 2.96 | 1.30 | 4.26 | 6.15 |
| | 570 | 250' haul, add | ↓ | .348 | | | 3.48 | 1.53 | 5.01 | 7.20 |
| | 580 | Riding cart, 50' haul, add | B-9 | .125 | | | 1.30 | .44 | 1.74 | 2.55 |
| | 581 | 150' haul, add | | .174 | | | 1.81 | .62 | 2.43 | 3.54 |
| | 590 | 250' haul, add | ↓ | .200 | ↓ | | 2.08 | .71 | 2.79 | 4.07 |
| 54-001 | | WINTER PROTECTION for heated ready mix, add, minimum | | | C.Y. | 2.50 | | | 2.50 | 2.75M |
| | 005 | Maximum | | | " | 2.75 | | | 2.75 | 3.02M |
| | 010 | Protecting concrete and temporary heat, add, minimum | 2 Clab | .003 | S.F. | .05 | .03 | | .08 | .10 |
| | 016 | Maximum | " | .008 | " | .38 | .08 | | .46 | .54 |

## 3.4 Precast Concrete

| | | | CREW | MAN-HOURS | UNIT | MAT. | LABOR | EQUIP. | TOTAL | TOTAL INCL O&P |
|---|---|---|---|---|---|---|---|---|---|---|
| 50-001 | | STAIRS concrete treads on steel stringers, 3' wide | C-12 | .640 | Riser | 40 | 7.92 | 3.13 | 51.05 | 60 |
| | 030 | Front entrance, 5' wide with 48" platform, 2 risers | | 3 | Flight | 220 | 37.35 | 14.65 | 272 | 315 |
| | 035 | 5 risers | | 4 | | 260 | 49.45 | 19.55 | 329 | 385 |
| | 050 | 6' wide, 2 risers | ↓ | 3.200 | | 240 | 39.35 | 15.65 | 295 | 345 |
| | 120 | Basement entrance stairs, steel bulkhead doors, minimum | B-51 | 2.180 | | 380 | 22.40 | 2.60 | 405 | 455 |
| | 125 | Maximum | " | 4.360 | ↓ | 500 | 45.80 | 5.20 | 551 | 625 |

For expanded coverage of these items see Means' *Concrete & Masonry Cost Data 1985*

# 4.1 Mortar & Masonry Accessories

| | | CREW | MAN-HOURS | UNIT | MAT. | LABOR | EQUIP. | TOTAL | TOTAL INCL O&P |
|---|---|---|---|---|---|---|---|---|---|
| 06-001 | **ANCHOR BOLTS** Hooked type with nut, 1/2" diam., 8" long | 1 Bric | .040 | Ea. | .45 | .50 | | .95 | 1.28 |
| 003 | 12" long | | .042 | | .55 | .53 | | 1.08 | 1.43 |
| 006 | 3/4" diameter, 8" long | | .050 | | 1.20 | .63 | | 1.83 | 2.30 |
| 007 | 12" long | ↓ | .053 | ↓ | 1.45 | .67 | | 2.12 | 2.64 |
| 10-001 | **CEMENT** Gypsum 80 lb. bag, T.L. lots | | | Bag | 8.75 | | | 8.75 | 9.60M |
| 005 | L.T.L. lots | | | | 9.05 | | | 9.05 | 9.95M |
| 010 | Masonry, 70 lb. bag, T.L. lots | | | | 4.50 | | | 4.50 | 4.95M |
| 015 | L.T.L. lots | | | | 4.80 | | | 4.80 | 5.30M |
| 020 | White masonry cement, 70 lb. bag, T.L. lots | | | | 13 | | | 13 | 14.30M |
| 025 | L.T.L. lots | | | ↓ | 13.75 | | | 13.75 | 15.10M |
| 25-001 | **FIREPLACE ACCESSORIES** Chimney screens, galv., 13" x 13" flue | 1 Bric | 1 | Ea. | 32 | 12.60 | | 44.60 | 55 |
| 005 | Galv., 24" x 24" flue | | 1.600 | | 90 | 20 | | 110 | 130 |
| 020 | Stainless steel, 13" x 13" flue | | 1 | | 185 | 12.60 | | 197.60 | 225 |
| 025 | 20" x 20" flue | | 1.600 | | 295 | 20 | | 315 | 355 |
| 040 | Cleanout doors and frames, cast iron, 8" x 8" | | .667 | | 12 | 8.40 | | 20.40 | 26 |
| 045 | 12" x 12" | | .800 | | 24 | 10.10 | | 34.10 | 42 |
| 050 | 18" x 24" | | 1 | | 70 | 12.60 | | 82.60 | 97 |
| 055 | Cast iron frame, steel door, 24" x 30" | | 1.600 | | 160 | 20 | | 180 | 205 |
| 080 | Damper, rotary control, steel, 30" opening | | 1.330 | | 40 | 16.80 | | 56.80 | 70 |
| 085 | Cast iron, 30" opening | | 1.330 | | 45 | 16.80 | | 61.80 | 76 |
| 120 | Steel plate, poker control, 60" opening | | 1 | | 125 | 12.60 | | 137.60 | 155 |
| 125 | 84" opening | | 1.600 | | 240 | 20 | | 260 | 295 |
| 140 | "Universal" type, chain operated, 32" x 20" opening | | 1 | | 90 | 12.60 | | 102.60 | 120 |
| 145 | 48" x 24" opening | | 1.600 | | 160 | 20 | | 180 | 205 |
| 160 | Dutch Oven door and frame, cast iron, 12" x 15" opening | | .615 | | 50 | 7.75 | | 57.75 | 67 |
| 165 | Copper plated, 12" x 15" opening | | .615 | | 85 | 7.75 | | 92.75 | 105 |
| 180 | Fireplace forms with registers, 25" opening | | 2.670 | | 295 | 34 | | 329 | 375 |
| 190 | 34" opening | | 3.200 | | 320 | 40 | | 360 | 415 |
| 200 | 48" opening | | 4 | | 650 | 50 | | 700 | 795 |
| 210 | 72" opening | | 5.330 | | 1,025 | 67 | | 1,092 | 1,225 |
| 240 | Squirrel and bird screens, galvanized, 8" x 8" flue | | .500 | | 28 | 6.30 | | 34.30 | 41 |
| 245 | 13" x 13" flue | ↓ | 1 | ↓ | 32 | 12.60 | | 44.60 | 55 |
| 30-001 | **GROUTING** Bond beams and lintels, 8" deep, pumped, not incl. block, | | | | | | | | |
| 020 | Concrete block cores, solid, 4" thick, by hand, .067 C.F./S.F. | D-3 | .035 | S.F. | .14 | .41 | | .55 | .79 |
| 025 | 8" thick, pumped .258 C.F. per S.F. | D-4 | .038 | | .56 | .42 | .17 | 1.15 | 1.47 |
| 030 | 10" thick, .340 C.F. per S.F. | | .039 | | .74 | .43 | .18 | 1.35 | 1.69 |
| 035 | 12" thick, .422 C.F. per S.F. | ↓ | .040 | ↓ | .92 | .45 | .18 | 1.55 | 1.92 |
| 35-001 | **JOINT REINFORCING** Steel bars, placed horizontal, #3 & #4 bars | | | | | | | | |
| 005 | Placed vertical, #3 & #4 bars | 1 Bric | .010 | Lb. | .28 | .13 | | .41 | .50 |
| 006 | #5 & #6 bars | | .012 | " | .28 | .16 | | .44 | .55 |
| 020 | Wire strips, regular truss, to 6" wide | | .267 | C.L.F. | 12.50 | 3.36 | | 15.86 | 19 |
| 025 | 12" wide | | .400 | | 15.50 | 5.05 | | 20.55 | 25 |
| 040 | Cavity wall with drip section to 6" wide | | .267 | | 13 | 3.36 | | 16.36 | 19.55 |
| 045 | 12" wide | ↓ | .400 | ↓ | 16 | 5.05 | | 21.05 | 25 |
| 45-001 | **LINTELS** Steel angles, minimum | 1 Bric | .008 | Lb. | .38 | .10 | | .48 | .58 |
| 005 | Maximum, (see also division 5.4-500) | | .016 | " | .54 | .20 | | .74 | .91 |
| 020 | Steel angles, 3-1/2" x 3", 1/4" thick, 2'-6" long | | .160 | Ea. | 5.15 | 2.02 | | 7.17 | 8.80 |
| 025 | 4'-6" long | | .178 | | 9.25 | 2.24 | | 11.49 | 13.65 |
| 040 | 4" x 3-1/2", 1/4" thick, 5'-0" long | | .200 | | 11.80 | 2.52 | | 14.32 | 16.90 |
| 045 | 9'-0" long | ↓ | .229 | ↓ | 21 | 2.88 | | 23.88 | 28 |
| 080 | Precast concrete, 4" x 8", stock units to 5' long | D-1 | .091 | L.F. | 3.55 | 1.04 | | 4.59 | 5.50 |
| 085 | To 12' long | D-4 | .168 | | 4.35 | 1.89 | .77 | 7.01 | 8.60 |
| 100 | 6" wide, 8" high, solid, stock units to 5' long | | .173 | | 5.55 | 1.94 | .79 | 8.28 | 10 |
| 105 | To 12' long | | .168 | ↓ | 6.35 | 1.89 | .77 | 9.01 | 10.80 |
| 70-001 | **WALL TIES** To brick veneer, galv., corrugated, 7/8" x 7", 24 gauge | 1 Bric | .008 | Ea. | .03 | .10 | | .13 | .18 |
| 005 | 16 gauge | | .008 | | .10 | .10 | | .20 | .26 |
| 060 | Cavity wall, 6" long, Z type, galvanized, 1/8" diameter | | .008 | | .15 | .10 | | .25 | .31 |
| 065 | 3/16" diameter | ↓ | .008 | ↓ | .06 | .10 | | .16 | .22 |

For expanded coverage of these items see Means' Concrete & Masonry Cost Data 1985

## 4.1 Mortar & Masonry Accessories

| | | | CREW | MAN-HOURS | UNIT | MAT. | LABOR | EQUIP. | TOTAL | TOTAL INCL O&P |
|---|---|---|---|---|---|---|---|---|---|---|
| 70 | 080 | 8" long, Z type, 3/16" diameter, galvanized | 1 Bric | .008 | Ea. | .07 | .10 | | .17 | .23 |
| | 085 | Copperweld | | .008 | | .11 | .10 | | .21 | .27 |
| | 150 | Stone anchors, galv., U or Z shaped, 6" long, 1/8" x 1" | | .008 | | .43 | .10 | | .53 | .62 |
| | 155 | 1/4" x 1" | | .008 | | .88 | .10 | | .98 | 1.12 |

## 4.2 Brick Masonry

| | | | CREW | MAN-HOURS | UNIT | MAT. | LABOR | EQUIP. | TOTAL | TOTAL INCL O&P |
|---|---|---|---|---|---|---|---|---|---|---|
| 06-001 | | CHIMNEY Standard bricks @ $208 per M, 16" x 16" with one 8" x 8" flue | D-1 | .889 | V.L.F. | 9.20 | 10.05 | | 19.25 | 26 |
| | 005 | 16" x 20" with one 8" x 12" flue | | 1 | | 11.30 | 11.35 | | 22.65 | 30 |
| | 010 | 16" x 24" with two 8" x 8" flues | | 1.140 | | 13.60 | 12.95 | | 26.55 | 35 |
| | 015 | 20" x 20" with one 12" x 12" flue | | 1.140 | | 13.35 | 12.95 | | 26.30 | 35 |
| | 020 | 20" x 24" with two 8" x 12" flues | | 1.330 | | 17.60 | 15.10 | | 32.70 | 43 |
| | 025 | 20" x 32" with two 12" x 12" flues | | 1.600 | | 22 | 18.10 | | 40.10 | 52 |
| 09-001 | | COLUMNS Standard bricks @ $208 per M, 8" x 8", 9 brick | | .286 | | 2.15 | 3.24 | | 5.39 | 7.40 |
| | 010 | 12" x 8", 13.5 brick | | .432 | | 3.25 | 4.90 | | 8.15 | 11.20 |
| | 020 | 12" x 12", 20.3 brick | | .640 | | 4.90 | 7.25 | | 12.15 | 16.70 |
| | 030 | 16" x 12", 27 brick | | .842 | | 6.50 | 9.55 | | 16.05 | 22 |
| | 040 | 16" x 16", 36 brick | | 1.140 | | 8.65 | 12.95 | | 21.60 | 30 |
| | 050 | 20" x 16", 45 brick | | 1.450 | | 10.80 | 16.45 | | 27.25 | 38 |
| | 060 | 20" x 20", 56.3 brick | | 1.780 | | 13.55 | 20 | | 33.55 | 46 |
| 10-001 | | COMMON BRICK Standard size, material only, minimum | | | M | 180 | | | 180 | 200M |
| | 005 | Average | | | " | 200 | | | 200 | 220M |
| 12-001 | | COPING For 12" wall, stock units, aluminum | D-1 | .200 | L.F. | 5.50 | 2.27 | | 7.77 | 9.60 |
| | 005 | Precast concrete, stock units, 6" wide | | .160 | | 3.50 | 1.81 | | 5.31 | 6.70 |
| | 010 | 10" wide | | .178 | | 4.70 | 2.01 | | 6.71 | 8.30 |
| | 015 | 14" wide | | .200 | | 5.10 | 2.27 | | 7.37 | 9.15 |
| | 030 | Limestone for 12" wall, 4" thick | | .178 | | 8.25 | 2.01 | | 10.26 | 12.20 |
| | 035 | 6" thick | | .200 | | 11.50 | 2.27 | | 13.77 | 16.20 |
| | 050 | Marble to 4" thick, no wash, 9" wide | | .178 | | 14.85 | 2.01 | | 16.86 | 19.50 |
| | 055 | 12" wide | | .200 | | 19.80 | 2.27 | | 22.07 | 25 |
| | 070 | Terra cotta, 9" wide | | .178 | | 2.90 | 2.01 | | 4.91 | 6.35 |
| | 075 | 12" wide | | .200 | | 4.75 | 2.27 | | 7.02 | 8.75 |
| 15-001 | | CORNICES Brick cornice on existing building | | | | | | | | |
| | 011 | Standard bricks @ $215 per M, minimum | D-1 | .533 | S.F.Face | 3.35 | 6.05 | | 9.40 | 13.10 |
| | 015 | Maximum | " | .696 | " | 3.35 | 7.90 | | 11.25 | 16 |
| 22-001 | | FIREPLACE | | | | | | | | |
| | 010 | Brick fireplace, not incl. foundations or chimneys | | | | | | | | |
| | 011 | 30" x 24" opening, plain brickwork | D-1 | 40 | Ea. | 225 | 455 | | 680 | 955 |
| | 020 | Fireplace box only (110 brick) | | 8 | | 95 | 91 | | 186 | 245 |
| | 030 | Elaborate brickwork and details | | 80 | | 420 | 905 | | 1,325 | 1,875 |
| | 040 | For hearth, add | | 8 | | 60 | 91 | | 151 | 205 |
| | 060 | Plain brickwork, incl. metal circulator | | 32 | | 405 | 360 | | 765 | 1,000 |
| | 080 | Face brick only, standard size, 8" x 2-2/3" x 4" | | 53.330 | M | 215 | 605 | | 820 | 1,175 |
| | 090 | Stone fireplace, fieldstone, add | | | S.F.Face | | | | | 9 |
| | 100 | Cut stone, add | | | " | | | | | 8.75 |
| 23-001 | | FIREPLACE, PREFABRICATED Free standing or wall hung | | | | | | | | |
| | 010 | with hood & screen, minimum | F-1 | 6.150 | Ea. | 485 | 75.60 | 5.40 | 566 | 660 |
| | 015 | Average | | 8 | | 915 | 98 | 7 | 1,020 | 1,175 |
| | 020 | Maximum | | 8.890 | | 3,500 | 112.20 | 7.80 | 3,620 | 4,025 |
| | 025 | For chimney heights over 8'-6", 7" diameter, add | | .267 | V.L.F. | 18 | 3.30 | .23 | 21.53 | 25 |
| | 030 | 10" diameter, add | | .267 | | 26 | 3.30 | .23 | 29.53 | 34 |
| | 035 | 12" diameter, add | | .267 | | 36 | 3.30 | .23 | 39.53 | 45 |
| | 040 | 14" diameter, add | | .267 | | 45 | 3.30 | .23 | 48.53 | 55 |

For expanded coverage of these items see Means' *Concrete & Masonry Cost Data 1985*

## 4.2 Brick Masonry

| | | | CREW | MAN-HOURS | UNIT | MAT. | LABOR | EQUIP. | TOTAL | TOTAL INCL O&P |
|---|---|---|---|---|---|---|---|---|---|---|
| 23 | 045 | Simulated brick chimney top, 3' high, 16" x 16" | F-1 | .800 | Ea. | 120 | 9.90 | .70 | 130.60 | 150 |
| | 050 | 24" x 24" | " | 1.140 | | 195 | 14.10 | 1 | 210.10 | 240 |
| | 055 | Woodburning stoves, cast iron, minimum | F-2 | 12.310 | | 175 | 154.25 | 10.75 | 340 | 445 |
| | 060 | Average | ↓ | 16 | | 455 | 196 | 14 | 665 | 830 |
| | 065 | Maximum | | 20 | | 1,175 | 247.50 | 17.50 | 1,440 | 1,700 |
| | 070 | For gas log lighter, add | | | ↓ | 28 | | | 28 | 31M |
| | 075 | Simulated logs, gas fired, 40,000 BTU, 2' long, minimum | F-1 | 1.140 | Set | 95 | 14.10 | 1 | 110.10 | 130 |
| | 080 | Maximum | | 1.330 | | 160 | 16.48 | 1.17 | 177.65 | 205 |
| | 085 | Electric, 11,500 BTU, 1'-6" long, minimum | | 1.140 | | 75 | 14.10 | 1 | 90.10 | 105 |
| | 090 | Maximum | ↓ | 1.330 | ↓ | 120 | 16.48 | 1.17 | 137.65 | 160 |
| | 300 | Fireplace accessories, screens, minimum | 1 Carp | .800 | Ea. | 40 | 9.90 | | 49.90 | 60 |
| | 305 | Average | | 1 | | 200 | 12.35 | | 212.35 | 240 |
| | 310 | Maximum | ↓ | 2 | ↓ | 550 | 25 | | 575 | 645 |
| 28-001 | | FLOORING Acid proof shales, red, 8" x 3-3/4" x 1-1/4" thick | D-7 | 37.210 | M | 375 | 410 | | 785 | 1,050 |
| | 005 | 2-1/4" thick | D-1 | 40 | | 400 | 455 | | 855 | 1,150 |
| | 020 | Acid proof clay brick, 8" x 3-3/4" x 2-1/4" thick | | 40 | | 400 | 455 | | 855 | 1,150 |
| | 025 | 9" x 4-1/2" x 3" thick | ↓ | 43.240 | ↓ | 670 | 490 | | 1,160 | 1,500 |
| | 026 | Cast ceramic, pressed, 4" x 8" x 1/2", unglazed | D-7 | .160 | S.F. | 3.55 | 1.76 | | 5.31 | 6.60 |
| | 027 | Glazed | | .160 | | 4.35 | 1.76 | | 6.11 | 7.50 |
| | 028 | Hand molded flooring, 4" x 8" x 3/4", unglazed | | .168 | | 4.30 | 1.85 | | 6.15 | 7.60 |
| | 029 | Glazed | | .168 | | 5.25 | 1.85 | | 7.10 | 8.60 |
| | 030 | 8" hexagonal, 3/4" thick, unglazed | | .188 | | 4.80 | 2.07 | | 6.87 | 8.45 |
| | 031 | Glazed | ↓ | .188 | | 6.75 | 2.07 | | 8.82 | 10.60 |
| | 045 | Acid proof joints | D-1 | .246 | | 3.75 | 2.79 | | 6.54 | 8.50 |
| | 050 | Pavers, 8" x 4", 1" to 1-1/4" thick, red | D-7 | .168 | | 1.80 | 1.85 | | 3.65 | 4.83 |
| | 051 | Ironspot | " | .168 | | 2.60 | 1.85 | | 4.45 | 5.70 |
| | 054 | 1-3/8" to 1-3/4" thick, red | D-1 | .168 | | 1.85 | 1.91 | | 3.76 | 5 |
| | 056 | Ironspot | | .168 | | 2.80 | 1.91 | | 4.71 | 6.05 |
| | 058 | 2-1/4" thick, red | | .178 | | 1.90 | 2.01 | | 3.91 | 5.25 |
| | 059 | Ironspot | | .178 | | 3 | 2.01 | | 5.01 | 6.45 |
| | 060 | Sidewalk or patios, on sand bed, laid flat, no mortar, 4.5 per S.F. | | .145 | | 1.75 | 1.65 | | 3.40 | 4.50 |
| | 065 | Laid on edge, 7 per S.F. | ↓ | .229 | | 2.75 | 2.59 | | 5.34 | 7.05 |
| | 080 | For basket weave pattern, add | | | | | 15% | | | |
| | 085 | For herringbone pattern, add | | | | | 20% | | | |
| | 086 | For acid-resistant joints, add | D-1 | .008 | | .40 | .09 | | .49 | .57 |
| | 087 | For epoxy joints, add | | .027 | | 1 | .30 | | 1.30 | 1.57 |
| | 088 | For Furan underlayment, add | ↓ | .027 | | 1.40 | .30 | | 1.70 | 2.01 |
| | 089 | For waxed surface, steam cleaned, add | D-5 | .008 | ↓ | .05 | .10 | .03 | .18 | .24 |
| 34-001 | | LINTELS See division 4.1-45 & 5.4-50 | | | | | | | | |
| 56-001 | | VENEER | | | | | | | | |
| | 105 | Std., sel. common, 8" x 2-2/3" x 4" $200 per M(6.75 per S.F.) | D-2 | .191 | S.F. | 1.60 | 2.23 | | 3.83 | 5.25 |
| | 150 | Standard 8" x 2-2/3" x 4", running bond, red face, @ $215 per M | | .200 | | 1.70 | 2.33 | | 4.03 | 5.50 |
| | 155 | Buff or gray face, brick at $250 per M (6.75 per S.F.) | | .200 | | 2 | 2.33 | | 4.33 | 5.85 |
| | 160 | Full header every 6th course (7.88 per S.F.) | | .238 | | 2.45 | 2.77 | | 5.22 | 7 |
| | 165 | English, full header every 2nd course (10.13 per S.F.) | | .314 | | 3.20 | 3.66 | | 6.86 | 9.25 |
| | 170 | Flemish, alter. header every course (9.00 per S.F.) | | .293 | | 2.80 | 3.42 | | 6.22 | 8.40 |
| | 180 | Flemish, alt. header every 6th course (7.13/S.F.) | | .215 | | 2.20 | 2.50 | | 4.70 | 6.35 |
| | 182 | Full headers throughout (13.50 per S.F.) | | .419 | | 4.20 | 4.88 | | 9.08 | 12.25 |
| | 184 | Rowlock course (13.50 per S.F.) | | .440 | | 4.20 | 5.15 | | 9.35 | 12.65 |
| | 186 | Rowlock stretcher (4.50 per S.F.) | | .142 | | 1.40 | 1.65 | | 3.05 | 4.12 |
| | 188 | Soldier course (6.75 per S.F.) | | .220 | | 2.10 | 2.56 | | 4.66 | 6.30 |
| | 190 | Sailor course (4.50 per S.F.) | | .152 | | 1.40 | 1.77 | | 3.17 | 4.30 |
| | 195 | Glazed face, brick at $750 per M, running bond | | .210 | | 5.55 | 2.44 | | 7.99 | 9.90 |
| | 197 | Full header every 6th course (7.88 per S.F.) | | .259 | | 6.45 | 3.02 | | 9.47 | 11.80 |
| | 200 | Jumbo 12" x 4" x 6" running bond, @ $900 per M (3.00 per S.F.) | | .101 | | 3.15 | 1.18 | | 4.33 | 5.30 |
| | 205 | Norman, 12" x 2-2/3" x 4" running bond, @ $385/M (4.50 / S.F.) | | .138 | | 2.12 | 1.60 | | 3.72 | 4.84 |
| | 210 | Norwegian 12" x 3-1/5" x 4" at $450 / M (3.75 / S.F.) | ↓ | .117 | ↓ | 2.05 | 1.37 | | 3.42 | 4.39 |

For expanded coverage of these items see Means' *Concrete & Masonry Cost Data 1985*

## 4.2 Brick Masonry

| | | Description | CREW | MAN-HOURS | UNIT | MAT. | LABOR | EQUIP. | TOTAL | TOTAL INCL O&P |
|---|---|---|---|---|---|---|---|---|---|---|
| | 220 | Economy 8" x 4" x 4" $370 / M (4.50 / S.F.) | D-2 | .142 | S.F. | 2 | 1.65 | | 3.65 | 4.78 |
| | 230 | Engineer 8" x 3-1/5" x 4" at $255 per M (5.63 per S.F.) | | .169 | | 1.80 | 1.97 | | 3.77 | 5.05 |
| | 240 | Roman 12" x 2" x 4" at $460 per M (6.00 per S.F.) | | .176 | | 3.21 | 2.05 | | 5.26 | 6.75 |
| | 250 | SCR 12" x 2-2/3" x 6" at $545 per M (4.50 per S.F.) | | .142 | | 2.95 | 1.65 | | 4.60 | 5.85 |
| | 260 | Utility 12" x 4" x 4" at $650 per M (3.00 per S.F.) | ↓ | .098 | | 2.30 | 1.14 | | 3.44 | 4.31 |
| | 290 | For interior veneer construction, add | | | ↓ | | 15% | | | |
| 60-001 | | **WALLS** | | | | | | | | |
| | 080 | Common, 8" x 2-2/3" x 4" at $178 per M, 4" wall, as face brick | D-2 | .205 | S.F. | 1.55 | 2.38 | | 3.93 | 5.45 |
| | 085 | 4" thick, as back-up | " | .183 | | 1.55 | 2.14 | | 3.69 | 5.05 |
| | 090 | 8" thick wall, 13.50 bricks per S.F. | D-3 | .311 | | 3.10 | 3.61 | | 6.71 | 9.05 |
| | 100 | 12" thick wall, 20.25 bricks per S.F. | | .442 | | 4.65 | 5.15 | | 9.80 | 13.15 |
| | 105 | 16" thick wall, 27.00 bricks per S.F. | ↓ | .560 | | 6.20 | 6.50 | | 12.70 | 17 |
| | 120 | Reinf., straight hard, 8" x 2-2/3" x 4" at $200 per M, 4" wall | D-2 | .215 | | 1.75 | 2.50 | | 4.25 | 5.85 |
| | 125 | 8" thick wall, 13.50 bricks per S.F. | D-3 | .323 | | 3.35 | 3.75 | | 7.10 | 9.55 |
| | 130 | 12" thick wall, 20.25 bricks per S.F. | | .467 | | 5.15 | 5.40 | | 10.55 | 14.15 |
| | 135 | 16" thick wall, 27.00 bricks per S.F. | ↓ | .600 | | 6.90 | 6.95 | | 13.85 | 18.45 |
| 70-001 | | **WINDOW SILL** Bluestone, natural cleft, 12" wide, 1-1/2" thick | D-1 | .188 | | 5.80 | 2.13 | | 7.93 | 9.70 |
| | 005 | 2" thick | | .188 | ↓ | 6.20 | 2.13 | | 8.33 | 10.15 |
| | 010 | Cut stone, 5" x 8" plain | | .188 | L.F. | 6.50 | 2.13 | | 8.63 | 10.50 |
| | 020 | Face brick on edge, brick @ $215 per M | | .200 | | 1.35 | 2.27 | | 3.62 | 5 |
| | 040 | Marble, 12" wide, 1" thick | | .188 | | 9.55 | 2.13 | | 11.68 | 13.85 |
| | 060 | Precast concrete, stock sections, 6" wide | | .188 | | 3.50 | 2.13 | | 5.63 | 7.20 |
| | 065 | 10" wide | | .188 | | 4.60 | 2.13 | | 6.73 | 8.40 |
| | 070 | 14" wide | | .188 | | 5 | 2.13 | | 7.13 | 8.85 |
| | 090 | Slate, colored, unfading, 12" wide, 1" thick | | .188 | | 7.40 | 2.13 | | 9.53 | 11.45 |
| | 095 | 2" thick | | .188 | | 11 | 2.13 | | 13.13 | 15.45 |
| | 120 | Stainless steel, stock | | .188 | | 7.90 | 2.13 | | 10.03 | 12 |
| | 125 | Custom | ↓ | .188 | ↓ | 11.70 | 2.13 | | 13.83 | 16.20 |

## 4.3 Block & Tile Masonry

| | | Description | CREW | MAN-HOURS | UNIT | MAT. | LABOR | EQUIP. | TOTAL | TOTAL INCL O&P |
|---|---|---|---|---|---|---|---|---|---|---|
| 07-001 | | **CONCRETE BLOCK, BACK-UP,** incl. scaffolding | | | | | | | | |
| | 002 | Sand aggregate, tooled joint 1 side | | | | | | | | |
| | 100 | Reinforced, alternate courses, 4" thick | D-8 | .092 | S.F. | .89 | 1.06 | | 1.95 | 2.64 |
| | 110 | 6" thick | | .096 | | 1.07 | 1.12 | | 2.19 | 2.92 |
| | 115 | 8" thick | | .101 | | 1.27 | 1.17 | | 2.44 | 3.23 |
| | 120 | 10" thick | ↓ | .104 | | 1.80 | 1.20 | | 3 | 3.86 |
| | 125 | 12" thick | D-9 | .132 | ↓ | 1.86 | 1.49 | | 3.35 | 4.37 |
| 17-001 | | **CONCRETE BLOCK, DECORATIVE** Incl. scaffolding | | | | | | | | |
| | 002 | Embossed, simulated brick face, not reinforced | | | | | | | | |
| | 010 | 8" x 16" units, 4" thick | D-8 | .100 | S.F. | 1.25 | 1.16 | | 2.41 | 3.18 |
| | 020 | 8" thick | | .118 | ↓ | 1.52 | 1.36 | | 2.88 | 3.80 |
| | 025 | 12" thick | ↓ | .133 | ↓ | 2.25 | 1.54 | | 3.79 | 4.88 |
| | 040 | Embossed both sides | | | | | | | | |
| | 050 | 8" thick | D-8 | .133 | S.F. | 2.05 | 1.54 | | 3.59 | 4.66 |
| | 055 | 12" thick | " | .145 | " | 2.60 | 1.68 | | 4.28 | 5.50 |
| | 100 | Fluted high strength | | | | | | | | |
| | 110 | Flutes 1 side, 8" x 16" x 4" thick | D-8 | .116 | S.F. | 1.48 | 1.34 | | 2.82 | 3.72 |
| | 115 | Flutes 2 sides, 8" x 16" x 4" thick | | .119 | | 1.78 | 1.38 | | 3.16 | 4.12 |
| | 120 | 8" thick | ↓ | .133 | | 2.63 | 1.54 | | 4.17 | 5.30 |
| | 125 | For special colors, add | | | ↓ | .22 | | | .22 | .24M |
| | 140 | Deep grooved, smooth face | | | | | | | | |
| | 145 | 8" x 16" x 4" thick | D-8 | .116 | S.F. | 1.35 | 1.34 | | 2.69 | 3.58 |
| | 150 | 8" thick | " | .133 | " | 2.07 | 1.54 | | 3.61 | 4.69 |

For expanded coverage of these items see Means' *Concrete & Masonry Cost Data 1985*

## 4.3 Block & Tile Masonry

| | | CREW | MAN-HOURS | UNIT | MAT. | LABOR | EQUIP. | TOTAL | TOTAL INCL O&P |
|---|---|---|---|---|---|---|---|---|---|
| 17 200 | Formbloc, incl. inserts & reinforcing | | | | | | | | |
| 210 | 8" x 16" x 8" thick | D-8 | .116 | S.F. | 2.30 | 1.34 | | 3.64 | 4.63 |
| 215 | 12" thick | " | .129 | " | 2.55 | 1.49 | | 4.04 | 5.15 |
| 250 | Ground face | | | | | | | | |
| 260 | 8" x 16" x 4" thick | D-8 | .116 | S.F. | 2.28 | 1.34 | | 3.62 | 4.60 |
| 265 | 6" thick | ↓ | .129 | | 2.75 | 1.49 | | 4.24 | 5.35 |
| 270 | 8" thick | ↓ | .138 | | 3.24 | 1.60 | | 4.84 | 6.05 |
| 275 | 12" thick | D-9 | .181 | | 4.16 | 2.05 | | 6.21 | 7.80 |
| 290 | For special colors, add, minimum | | | | 5 | | | | |
| 295 | Maximum | | | ↓ | 15 | | | | |
| 400 | Slump block | | | | | | | | |
| 410 | 4" face height x 16" x 4" thick | D-1 | .097 | S.F. | 1.32 | 1.10 | | 2.42 | 3.17 |
| 415 | 6" thick | | .100 | | 1.68 | 1.13 | | 2.81 | 3.62 |
| 420 | 8" thick | | .103 | | 2.15 | 1.17 | | 3.32 | 4.19 |
| 425 | 10" thick | | .114 | | 3.28 | 1.29 | | 4.57 | 5.65 |
| 430 | 12" thick | | .123 | | 3.46 | 1.39 | | 4.85 | 6 |
| 440 | 6" face height x 16" x 6" thick | | .103 | | 1.34 | 1.17 | | 2.51 | 3.30 |
| 445 | 8" thick | | .107 | | 1.68 | 1.21 | | 2.89 | 3.73 |
| 450 | 10" thick | | .123 | | 2.52 | 1.39 | | 3.91 | 4.95 |
| 455 | 12" thick | ↓ | .133 | ↓ | 2.63 | 1.51 | | 4.14 | 5.25 |
| 500 | Split rib profile units, 1" deep ribs, 8 ribs | | | | | | | | |
| 510 | 8" x 16" x 4" thick | D-8 | .116 | S.F. | 1.60 | 1.34 | | 2.94 | 3.86 |
| 515 | 6" thick | | .123 | | 1.97 | 1.43 | | 3.40 | 4.39 |
| 520 | 8" thick | ↓ | .131 | | 2.52 | 1.52 | | 4.04 | 5.15 |
| 525 | 12" thick | D-9 | .175 | ↓ | 2.91 | 1.98 | | 4.89 | 6.30 |
| 535 | | | | | | | | | |
| 540 | For special deeper colors, 4" thick, add | | | S.F. | .15 | | | .15 | .16M |
| 545 | 12" thick, add | | | | .32 | | | .32 | .35M |
| 560 | For white, 4" thick, add | | | | .68 | | | .68 | .74M |
| 565 | 6" thick, add | | | | .90 | | | .90 | .99M |
| 570 | 8" thick, add | | | | 1.17 | | | 1.17 | 1.28M |
| 575 | 12" thick, add | | | ↓ | 1.40 | | | 1.40 | 1.54M |
| 600 | Split face or scored split face | | | | | | | | |
| 610 | 8" x 16" x 4" thick | D-8 | .114 | S.F. | 1.46 | 1.32 | | 2.78 | 3.67 |
| 615 | 6" thick | | .127 | | 1.88 | 1.47 | | 3.35 | 4.36 |
| 620 | 8" thick | ↓ | .136 | | 2.13 | 1.57 | | 3.70 | 4.79 |
| 625 | 12" thick | D-9 | .178 | ↓ | 2.58 | 2.01 | | 4.59 | 6 |
| 635 | | | | | | | | | |
| 640 | For special deeper colors, 4" thick, add | | | S.F. | .21 | | | .21 | .23M |
| 645 | 6" thick, add | | | | .30 | | | .30 | .33M |
| 650 | 8" thick, add | | | | .36 | | | .36 | .39M |
| 655 | 12" thick, add | | | | .45 | | | .45 | .49M |
| 665 | For white, 4" thick, add | | | | .57 | | | .57 | .62M |
| 670 | 6" thick, add | | | | .97 | | | .97 | 1.06M |
| 675 | 8" thick, add | | | | 1.02 | | | 1.02 | 1.12M |
| 680 | 12" thick, add | | | ↓ | 1.31 | | | 1.31 | 1.44M |
| 700 | Scored ground face, 2 to 5 scores | | | | | | | | |
| 710 | 8" x 16" x 4" thick | D-8 | .116 | S.F. | 2.40 | 1.34 | | 3.74 | 4.74 |
| 715 | 6" thick | | .129 | | 2.82 | 1.49 | | 4.31 | 5.45 |
| 720 | 8" thick | ↓ | .138 | | 3.36 | 1.60 | | 4.96 | 6.20 |
| 725 | 12" thick | D-9 | .181 | ↓ | 4.28 | 2.05 | | 6.33 | 7.90 |
| 800 | Hexagonal face profile units, 8" x 16" units | | | | | | | | |
| 810 | 4" thick, hollow | D-8 | .116 | S.F. | 1.32 | 1.34 | | 2.66 | 3.55 |
| 820 | Solid | | .116 | | 1.74 | 1.34 | | 3.08 | 4.01 |
| 830 | 6" thick, hollow | | .129 | | 1.70 | 1.49 | | 3.19 | 4.20 |
| 835 | 8" thick, hollow | | .138 | ↓ | 2.42 | 1.60 | | 4.02 | 5.15 |
| 850 | For stacked bond, add | | 9.520 | M.S.F. | | 110 | | 110 | 170L |
| 855 | For high rise construction, add per story | ↓ | .590 | " | | 6.85 | | 6.85 | 10.65L |
| 860 | For scored block, per score, per face, add | | | Ea. | .05 | | | .05 | .05M |
| 865 | For honed or ground face, per face, add | | | " | .13 | | | .13 | .14M |

For expanded coverage of these items see Means' *Concrete & Masonry Cost Data 1985*

## 4.3 Block & Tile Masonry

| | | | CREW | MAN-HOURS | UNIT | MAT. | LABOR | EQUIP. | TOTAL | TOTAL INCL O&P |
|---|---|---|---|---|---|---|---|---|---|---|
| 17 | 870 | For honed or ground end, per end, add | | | Ea. | .95 | | | .95 | 1.04M |
| | 875 | For bullnose block, add | | | " | 1.06 | | | 1.06 | 1.16M |
| | 880 | For special colors, add | | | S.F. | 15% | | | | |
| | 27-001 | **CONCRETE BLOCK FOUNDATION WALL** Incl. scaffolding | | | | | | | | |
| | 005 | Sand aggregate, trowel cut joints, not reinf., parged 1/2" thick | | | | | | | | |
| | 020 | Regular, 8" x 16" x 6" thick | D-8 | .089 | S.F. | 1.07 | 1.03 | | 2.10 | 2.78 |
| | 025 | 8" thick | | .093 | | 1.30 | 1.08 | | 2.38 | 3.11 |
| | 030 | 10" thick | ↓ | .095 | | 1.80 | 1.10 | | 2.90 | 3.70 |
| | 035 | 12" thick | D-9 | .122 | | 1.86 | 1.38 | | 3.24 | 4.19 |
| | 050 | Solid, 8" x 16" block, 6" thick | D-8 | .091 | | 1.38 | 1.05 | | 2.43 | 3.16 |
| | 055 | 8" thick | " | .096 | | 1.71 | 1.12 | | 2.83 | 3.62 |
| | 060 | 12" thick | D-9 | .126 | ↓ | 2.58 | 1.43 | | 4.01 | 5.05 |

## 4.4 Stonework

| | | | CREW | MAN-HOURS | UNIT | MAT. | LABOR | EQUIP. | TOTAL | TOTAL INCL O&P |
|---|---|---|---|---|---|---|---|---|---|---|
| | 15-001 | **GRANITE** Cut to size | | | | | | | | |
| | 005 | Veneer, polished face, 3/4" to 1-1/2" thick | | | | | | | | |
| | 015 | Low price, gray, light gray, etc. | D-10 | .308 | S.F. | 10 | 3.71 | 1.69 | 15.40 | 18.65 |
| | 022 | High price, red, black, etc. | " | .308 | " | 26 | 3.71 | 1.69 | 31.40 | 36 |
| | 030 | 1-1/2" to 2-1/2" thick, veneer | | | | | | | | |
| | 035 | Low price, gray, light gray, etc. | D-10 | .308 | S.F. | 12 | 3.71 | 1.69 | 17.40 | 21 |
| | 055 | High price, red, black, etc. | " | .308 | " | 16 | 3.71 | 1.69 | 21.40 | 25 |
| | 070 | 2-1/2" to 4" thick, veneer | | | | | | | | |
| | 075 | Low price, gray, light gray, etc. | D-10 | .364 | S.F. | 20 | 4.40 | 2 | 26.40 | 31 |
| | 095 | High price, red, black, etc. | " | .364 | | 24 | 4.40 | 2 | 30.40 | 35 |
| | 100 | Deduct from polished for bush hammered finish, deduct | | | | 5% | | | | |
| | 105 | Coarse rubbed finish, deduct | | | | 10% | | | | |
| | 110 | Honed finish, deduct | | | | 5% | | | | |
| | 115 | Thermal finish, deduct | | | ↓ | 18% | | | | |
| | 245 | For radius under 5', add | | | L.F. | 100% | | | | |
| | 250 | Steps, copings, etc., finished on more than one surface | | | | | | | | |
| | 255 | Minimum | D-10 | .800 | C.F. | 50 | 9.65 | 4.40 | 64.05 | 75 |
| | 260 | Maximum | " | .800 | " | 65 | 9.65 | 4.40 | 79.05 | 91 |
| | 280 | Pavers, 4" x 4" x 4" blocks, split face and joints | | | | | | | | |
| | 285 | Minimum | D-11 | .300 | S.F. | 6.25 | 3.73 | | 9.98 | 12.70 |
| | 290 | Maximum | " | .300 | " | 11 | 3.73 | | 14.73 | 17.90 |
| | 320 | | | | | | | | | |
| | 350 | Curbing, city street type, 6" x 18", split face, | | | | | | | | |
| | 351 | sawn top, radius nosing, 4' to 7' lengths | D-10 | .533 | L.F. | 10.25 | 6.42 | 2.93 | 19.60 | 25 |
| | 360 | Highway type, 5" x 16", split face, | | | | | | | | |
| | 361 | sawn top, 4' to 7' lengths | D-10 | .133 | L.F. | 8.75 | 1.61 | .73 | 11.09 | 12.95 |
| | 400 | Soffits, 2" thick, minimum | D-13 | 1.370 | S.F. | 25 | 16.70 | 6.30 | 48 | 60 |
| | 410 | Maximum | | 1.370 | | 40 | 16.70 | 6.30 | 63 | 77 |
| | 420 | 4" thick, minimum | | 1.370 | | 30 | 16.70 | 6.30 | 53 | 66 |
| | 430 | Maximum | ↓ | 1.370 | ↓ | 50 | 16.70 | 6.30 | 73 | 88 |
| | 25-001 | **LIMESTONE**, cut to size | | | | | | | | |
| | 002 | Veneer facing panels | | | | | | | | |
| | 010 | Sawn finish, 2" thick, to 3' x 5' panels | D-10 | .308 | S.F. | 8 | 3.71 | 1.69 | 13.40 | 16.45 |
| | 015 | Smooth finish, 2" thick, to 3' x 5' panels | | .308 | | 9.50 | 3.71 | 1.69 | 14.90 | 18.10 |
| | 030 | 3" thick, to 4' x 9' panels | | .178 | | 10.25 | 2.14 | .98 | 13.37 | 15.70 |
| | 035 | 4" thick, to 5' x 11' panels | ↓ | .145 | | 12 | 1.75 | .80 | 14.55 | 16.80 |
| | 050 | Texture finish, light stick, 4-1/2" thick, 5' x 12' | D-4 | .107 | | 9 | 1.19 | .49 | 10.68 | 12.30 |
| | 075 | 5" thick, to 5' x 14' panels | D-10 | .145 | ↓ | 13.50 | 1.75 | .80 | 16.05 | 18.45 |

For expanded coverage of these items see Means' *Concrete & Masonry Cost Data 1985*

## 4.4 Stonework

| | | | CREW | MAN-HOURS | UNIT | MAT. | LABOR | EQUIP. | TOTAL | TOTAL INCL O&P |
|---|---|---|---|---|---|---|---|---|---|---|
| 25 | 100 | Medium ribbed, textured finish, 4-1/2" thick, to 5' x 12' | D-10 | .145 | S.F. | 9 | 1.75 | .80 | 11.55 | 13.50 |
| | 105 | 5" thick, to 5' x 14' panels | | .145 | | 13.50 | 1.75 | .80 | 16.05 | 18.45 |
| | 120 | Deep ribbed, textured finish, 4-1/2" thick, to 5' x 10' | | .145 | | 9 | 1.75 | .80 | 11.55 | 13.50 |
| | 125 | 5" thick, to 5' x 14' panels | | .145 | | 13.50 | 1.75 | .80 | 16.05 | 18.45 |
| | 140 | Sugar cube, textured finish, 4-1/2" thick, to 5' x 12' | | .145 | | 9 | 1.75 | .80 | 11.55 | 13.50 |
| | 145 | 5" thick, to 5' x 14' panels | | .145 | | 13.50 | 1.75 | .80 | 16.05 | 18.45 |
| | 200 | Coping, smooth finish, top & 2 sides | | 1.330 | C.F. | 45 | 15.65 | 7.35 | 68 | 83 |
| | 205 | | | | | | | | | |
| | 210 | Sills, lintels, jambs, smooth finish, average | D-10 | 2 | C.F. | 45 | 24 | 11 | 80 | 99 |
| | 215 | Detailed | | 2 | " | 60 | 24 | 11 | 95 | 115 |
| | 230 | Steps, extra hard, 14" wide, 6" rise | | .800 | L.F. | 25 | 9.65 | 4.40 | 39.05 | 47 |
| 30-001 | | MARBLE Base, 3/4" thick, polished, group A, 4" thick | | | | 7.15 | | | 7.15 | 7.85M |
| | 100 | Facing, polished finish, cut to size, 3/4" to 7/8" thick | | | | | | | | |
| | 105 | Average | D-10 | .308 | S.F. | 12.50 | 3.71 | 1.69 | 17.90 | 21 |
| | 110 | Maximum | " | .308 | " | 30 | 3.71 | 1.69 | 35.40 | 41 |
| | 220 | Window sills, 6" x 2" thick | D-1 | .188 | L.F. | 8 | 2.13 | | 10.13 | 12.15 |
| | 250 | Flooring, polished tiles, 12" x 12" x 3/8" thick | | | | | | | | |
| | 251 | Thin set, average | D-11 | .267 | S.F. | 5.30 | 3.31 | | 8.61 | 11 |
| | 260 | Maximum | | .267 | | 15 | 3.31 | | 18.31 | 22 |
| | 270 | Mortar bed, average | | .369 | | 5 | 4.58 | | 9.58 | 12.65 |
| | 274 | Maximum | | .369 | | 15 | 4.58 | | 19.58 | 24 |
| | 278 | Travertine, 1-1/4" thick, average | D-10 | .308 | | 12.50 | 3.71 | 1.69 | 17.90 | 21 |
| | 279 | Maximum | " | .308 | | 16.50 | 3.71 | 1.69 | 21.90 | 26 |
| | 350 | Thresholds, 3' long, 7/8" thick, 4" to 5" wide, plain | D-12 | 1.330 | Ea. | 8 | 15.75 | | 23.75 | 33 |
| | 355 | Beveled | | 1.330 | " | 9 | 15.75 | | 24.75 | 35 |
| | 370 | Window stools, polished, 7/8" thick, 5" wide | | .376 | L.F. | 8 | 4.45 | | 12.45 | 15.75 |
| 40-001 | | ROUGH STONE wall in mortar, under 18" thick | D-1 | .291 | C.F. | 8.75 | 3.29 | | 12.04 | 14.75 |
| | 015 | Over 18" thick | D-12 | .508 | " | 7.75 | 6 | | 13.75 | 17.90 |
| 45-001 | | SANDSTONE OR BROWNSTONE | | | | | | | | |
| | 010 | Sawed face veneer, 2-1/2" thick, to 2' x 4' panels | D-10 | | S.F. | 8.80 | | | 8.80 | 9.68M |
| | 015 | 4" thick, to 3'-6" x 8' panels | | | | 8.40 | | | 8.40 | 9.24M |
| | 030 | Split face, random sizes | | | | 5.25 | | | 5.25 | 5.77M |
| | 035 | Cut stone trim (sandstone) | | | | | | | | |
| | 036 | Ribbon stone, 4" thick, 5' pieces | D-8 | .333 | L.F. | 65 | 3.86 | | 68.86 | 78 |
| | 037 | Cove stone, 4" thick, 5' pieces | | .381 | | 55 | 4.41 | | 59.41 | 67 |
| | 038 | Cornice stone, 10" to 12" wide | | .444 | | 85 | 5.15 | | 90.15 | 100 |
| | 039 | Band stone, 4" thick, 5' pieces | | .276 | | 45 | 3.19 | | 48.19 | 54 |
| | 041 | Window and door trim, 3" to 4" wide | | .250 | | 30 | 2.90 | | 32.90 | 38 |
| | 042 | Key stone, 18" long | | .667 | Ea. | 22 | 7.70 | | 29.70 | 36 |
| 53-001 | | SLATE Pennsylvania, blue gray to gray black; Vermont, | | | | | | | | |
| | 350 | Stair treads, sand finish, 1" thick x 12" wide | | | | | | | | |
| | 360 | 3 L.F. to 6 L.F. | D-10 | .333 | L.F. | 12.95 | 4.02 | 1.83 | 18.80 | 23 |
| | 370 | Ribbon, sand finish, 1" thick x 12" wide | | | | | | | | |
| | 375 | To 6 L.F. | D-10 | .333 | L.F. | 8 | 4.02 | 1.83 | 13.85 | 17.10 |

## 5.1 Structural Metals

| | | | CREW | MAN-HOURS | UNIT | MAT. | LABOR | EQUIP. | TOTAL | TOTAL INCL O&P |
|---|---|---|---|---|---|---|---|---|---|---|
| 26-001 | | COLUMNS | | | | | | | | |
| | 080 | Steel, concrete filled, extra strong pipe, 3-1/2" diameter | E-2 | .085 | L.F. | 8.75 | 1.12 | 1.20 | 11.07 | 12.80 |
| | 083 | 4" diameter | | .072 | | 9.50 | .95 | 1.01 | 11.46 | 13.15 |
| | 085 | 4-1/2" diameter | | .062 | | 10 | .82 | .88 | 11.70 | 13.35 |

For expanded coverage of these items see *Building Construction Cost Data 1985*

## 5.1 Structural Metals

| | | | CREW | MAN-HOURS | UNIT | MAT. | LABOR | EQUIP. | TOTAL | TOTAL INCL O&P |
|---|---|---|---|---|---|---|---|---|---|---|
| 26 | 089 | 5" diameter | E-2 | .055 | L.F. | 12 | .73 | .77 | 13.50 | 15.25 |
| | 090 | 5-1/2" diameter | | .049 | | 12 | .65 | .69 | 13.34 | 15.05 |
| | 093 | 6-5/8" diameter | | .046 | | 17 | .61 | .65 | 18.26 | 20 |
| | 100 | Lightweight units, 3-1/2" diameter | | .072 | | 2.75 | .95 | 1.01 | 4.71 | 5.75 |
| | 105 | 4" diameter | ↓ | .062 | ↓ | 3 | .82 | .88 | 4.70 | 5.65 |
| | 110 | For galvanizing, add | | | Lb. | .25 | | | .25 | .27M |
| | 130 | For web ties, angles, etc., add per added lb. | 1 Sswk | .008 | | .65 | .11 | | .76 | .91 |
| | 150 | Steel pipe, extra strong, no concrete, 3" to 5" O.D. | E-2 | .004 | | .60 | .06 | .06 | .72 | .82 |
| | 160 | 6" to 12" O.D. | " | .001 | ↓ | .60 | .02 | .01 | .63 | .70 |
| | 210 | Square structural tubing, 4" x 4" x 1/4" | | | L.F. | | | | .50 | .55 |
| | 240 | Rectangular structural tubing, 5" to 6" wide, light section | E-2 | .005 | Lb. | .82 | .07 | .07 | .96 | 1.09 |
| | 270 | 12" x 8" x 1/2" | | .002 | | .61 | .02 | .03 | .66 | .75 |
| | 280 | Heavy section | ↓ | .001 | ↓ | .60 | .02 | .01 | .63 | .70 |
| | 800 | Lally columns, to 8', 2-3/4" diameter | F-2 | .667 | Ea. | 16.50 | 8.22 | .58 | 25.30 | 32 |
| | 808 | 4" diameter | " | .800 | " | 19 | 9.90 | .70 | 29.60 | 37 |
| 50-001 | | **STRUCTURAL STEEL PROJECTS** Bolted, unless mentioned otherwise | | | | | | | | |
| | 020 | Apts., nursing homes, etc., steel bearing, 1 to 2 stories | E-5 | 7.770 | Ton | 770 | 103 | 87 | 960 | 1,125 |
| | 070 | Offices, hospitals, etc., steel bearing, 1 to 2 stories | | 7.770 | | 750 | 103 | 87 | 940 | 1,100 |
| | 310 | Roof trusses, minimum | | 6.150 | Ton | 800 | 81 | 69 | 950 | 1,100 |
| | 320 | Maximum | ↓ | 9.640 | " | 1,300 | 125 | 110 | 1,535 | 1,775 |
| 52-001 | | **STRUCTURAL STEEL** Bolted, including fabrication | | | | | | | | |
| | 005 | Beams, 6 WF 9 | E-2 | .078 | L.F. | 4.50 | 1.02 | 1.10 | 6.62 | 7.90 |
| | 010 | 8 WF 10 | | .078 | | 5 | 1.02 | 1.10 | 7.12 | 8.45 |
| | 020 | Columns, 6 WF 15.5 | | .104 | | 7.75 | 1.37 | 1.46 | 10.58 | 12.45 |
| | 025 | 8 WF 31 | ↓ | .104 | ↓ | 15.50 | 1.37 | 1.46 | 18.33 | 21 |
| | 080 | Not including trucking | | | | | | | | |

## 5.2 Metal Joists & Decks

| | | | CREW | MAN-HOURS | UNIT | MAT. | LABOR | EQUIP. | TOTAL | TOTAL INCL O&P |
|---|---|---|---|---|---|---|---|---|---|---|
| 30-001 | | **METAL DECKING** Steel floor panels, over 15,000 S.F. | | | | | | | | |
| | 210 | Open type, galv., 1-1/2" deep, 22 ga., under 50 square | E-4 | .007 | S.F. | .70 | .10 | .01 | .81 | .95 |
| | 260 | 20 ga., under 50 square | | .008 | | .80 | .12 | .01 | .93 | 1.09 |
| | 290 | 18 ga., under 50 square | ↓ | .008 | | 1.10 | .12 | .01 | 1.23 | 1.43 |
| | 370 | | | | | | | | | |
| | 610 | Slab form, steel 28 gauge, 9/16" deep, uncoated | E-1 | .006 | | .35 | .09 | .01 | .45 | .54 |
| | 620 | Galvanized | | .006 | | .38 | .09 | .01 | .48 | .57 |
| | 630 | 24 gauge, 1-5/16" deep, uncoated | | .006 | | .50 | .09 | .01 | .60 | .71 |
| | 640 | Galvanized | | .006 | | .55 | .09 | .01 | .65 | .76 |
| | 650 | 22 gauge, 1-5/16" deep, uncoated | | .006 | | .62 | .08 | .02 | .72 | .85 |
| | 660 | Galvanized | ↓ | .006 | ↓ | .72 | .08 | .02 | .82 | .96 |
| 42-001 | | **OPEN WEB JOISTS** | | | | | | | | |
| | 002 | | | | | | | | | |
| | 100 | Bar joists installed, no material included, 15' span | E-2 | .053 | L.F. | | .71 | .75 | 1.46 | 2.01 |
| | 110 | 20' span | | .043 | | | .57 | .61 | 1.18 | 1.62 |
| | 120 | 25' span | | .037 | | | .49 | .53 | 1.02 | 1.41 |
| | 130 | 30' span | | .034 | | | .45 | .48 | .93 | 1.28 |
| | 140 | 35' span | | .032 | | | .42 | .45 | .87 | 1.21 |
| | 150 | 40' span | | .035 | | | .47 | .49 | .96 | 1.32 |
| | 160 | 45' span | | .034 | | | .45 | .47 | .92 | 1.27 |
| | 170 | 50' span | ↓ | .032 | ↓ | | .42 | .45 | .87 | 1.21 |
| | 300 | Add per pound for material, incl. bridging | | | Lb. | .30 | | | .30 | .33M |

For expanded coverage of these items see *Building Construction Cost Data 1985*

## 5.4 Misc. & Ornamental Metals

| | | CREW | MAN-HOURS | UNIT | MAT. | LABOR | EQUIP. | TOTAL | TOTAL INCL O&P |
|---|---|---|---|---|---|---|---|---|---|
| 46-001 | **LAMP POSTS** Only, 6' high, stock units, aluminum | 1 Carp | .500 | Ea. | 22 | 6.20 | | 28.20 | 34 |
| 010 | Mild steel, plain | " | .500 | " | 16 | 6.20 | | 22.20 | 27 |
| 50-001 | **LINTELS** Plain steel angles, under 500 lb. (see also division 4.1-450) | 1 Bric | .016 | Lb. | .40 | .20 | | .60 | .75 |
| 010 | 500 to 1000 lb. | " | .013 | " | .34 | .17 | | .51 | .64 |
| 58-001 | **RAILINGS, COMMERCIAL** Aluminum pipe rail, anodized | E-4 | .164 | L.F. | 35 | 2.26 | .29 | 37.55 | 43 |
| 090 | Aluminum balcony rail, 1-1/2" posts, with pickets | | .164 | | 25 | 2.26 | .29 | 27.55 | 32 |
| 100 | With expanded metal panels | | .164 | | 40 | 2.26 | .29 | 42.55 | 48 |
| 110 | Aluminum faced wood panel inserts | | .164 | | 50 | 2.26 | .29 | 52.55 | 59 |
| 120 | Hammered wire glass panel inserts | | .164 | | 50 | 2.26 | .29 | 52.55 | 59 |
| 130 | Porcelain enamel panel inserts | | .164 | | 50 | 2.26 | .29 | 52.55 | 59 |
| 140 | For anodizing finish all above, add | | | | 2 | | | 2 | 2.20M |
| 160 | Mild steel, ornamental rounded top rail | E-4 | .164 | | 25 | 2.26 | .29 | 27.55 | 32 |
| 170 | As above but flat top rail | | .164 | | 22 | 2.26 | .29 | 24.55 | 28 |
| 180 | As above but pitch down stairs | | .183 | | 25 | 2.53 | .32 | 27.85 | 32 |
| 190 | Residential, stock units, mild steel, deluxe | | .102 | | 10 | 1.40 | .18 | 11.58 | 13.60 |
| 191 | Economy | | .102 | | 8 | 1.40 | .18 | 9.58 | 11.40 |
| 240 | Steel pipe, welded, 1-1/2" round, painted | | .160 | | 20 | 2.21 | .28 | 22.49 | 26 |
| 250 | Galvanized | | .160 | | 30 | 2.21 | .28 | 32.49 | 37 |
| 270 | Steel pipe with 4" x 1/4" toe plate, painted | | .160 | | 22 | 2.21 | .28 | 24.49 | 28 |
| 280 | Galvanized | | .160 | | 32 | 2.21 | .28 | 34.49 | 39 |
| 300 | For stainless steel, add to painted | | | | 300% | | | | |
| 320 | For curved rails, add | | | | 30% | 30% | | | |
| 64-001 | **STAIR** | | | | | | | | |
| 170 | Pre-erected, steel pan tread, 3'-6" wide, with flat bar rail | E-4 | .376 | Riser | 100 | 5.19 | .66 | 105.85 | 120 |
| 180 | With picket rail | | 1.880 | | 135 | 25.68 | 3.32 | 164 | 195 |
| 181 | Spiral aluminum, 5'-0" diameter, stock units | | .711 | | 135 | 9.80 | 1.25 | 146.05 | 165 |
| 182 | Custom units | | .711 | | 200 | 9.80 | 1.25 | 211.05 | 240 |
| 190 | Spiral, cast iron, 4'-0" diameter, ornamental, minimum | | .711 | | 125 | 9.80 | 1.25 | 136.05 | 155 |
| 192 | Maximum | | 1.280 | | 185 | 17.64 | 2.26 | 204.90 | 235 |
| 72-001 | **WEATHERVANES** Residential types, minimum | 1 Carp | 1 | Ea. | 15 | 12.35 | | 27.35 | 36 |
| 010 | Maximum | " | 4 | " | 550 | 49 | | 599 | 685 |

## 5.8 Expansion Control & Fasteners

| | | CREW | MAN-HOURS | UNIT | MAT. | LABOR | EQUIP. | TOTAL | TOTAL INCL O&P |
|---|---|---|---|---|---|---|---|---|---|
| 02-001 | **BOLTS & HEX NUTS** Steel, A307 | | | | | | | | |
| 010 | 1/4" diameter, 1/2" long | | | Ea. | .04 | | | .04 | .04M |
| 020 | 1" long | | | | .05 | | | .05 | .05M |
| 030 | 2" long | | | | .06 | | | .06 | .06M |
| 040 | 3" long | | | | .09 | | | .09 | .09M |
| 050 | 4" long | | | | .10 | | | .10 | .11M |
| 060 | 3/8" diameter, 1" long | | | | .10 | | | .10 | .11M |
| 070 | 2" long | | | | .14 | | | .14 | .15M |
| 080 | 3" long | | | | .18 | | | .18 | .19M |
| 090 | 4" long | | | | .22 | | | .22 | .24M |
| 100 | 5" long | | | | .28 | | | .28 | .30M |
| 110 | 1/2" diameter, 1-1/2" long | | | | .23 | | | .23 | .25M |
| 120 | 2" long | | | | .26 | | | .26 | .28M |
| 130 | 4" long | | | | .32 | | | .32 | .35M |
| 140 | 6" long | | | | .58 | | | .58 | .63M |
| 150 | 8" long | | | | .75 | | | .75 | .82M |
| 160 | 5/8" diameter, 1-1/2" long | | | | .42 | | | .42 | .46M |
| 170 | 2" long | | | | .46 | | | .46 | .50M |
| 180 | 4" long | | | | .57 | | | .57 | .62M |
| 190 | 6" long | | | | .67 | | | .67 | .73M |

For expanded coverage of these items see *Building Construction Cost Data 1985*

## 5.8 Expansion Control & Fasteners

| | | | CREW | MAN-HOURS | UNIT | MAT. | LABOR | EQUIP. | TOTAL | TOTAL INCL O&P |
|---|---|---|---|---|---|---|---|---|---|---|
| 02 | 200 | 8" long | | | Ea. | 1.25 | | | 1.25 | 1.37M |
| | 210 | 10" long | | | | 1.43 | | | 1.43 | 1.57M |
| | 220 | 3/4" diameter, 2" long | | | | .68 | | | .68 | .74M |
| | 230 | 4" long | | | | 1 | | | 1 | 1.10M |
| | 240 | 6" long | | | | 1.30 | | | 1.30 | 1.43M |
| | 250 | 8" long | | | | 1.60 | | | 1.60 | 1.76M |
| | 260 | 10" long | | | | 1.90 | | | 1.90 | 2.09M |
| | 270 | 12" long | | | | 2.15 | | | 2.15 | 2.36M |
| | 280 | 1" diameter, 3" long | | | | 1.85 | | | 1.85 | 2.03M |
| | 290 | 6" long | | | | 2.50 | | | 2.50 | 2.75M |
| | 300 | 12" long | | | | 4.60 | | | 4.60 | 5.05M |
| | 310 | For galvanized, add | | | | 20% | | | | |
| | 320 | For stainless, add | | | | 150% | | | | |
| 03-001 | | **DRILLING** And layout for anchors, per | | | | | | | | |
| | 005 | inch of depth, concrete or brick walls. | | | | | | | | |
| | 100 | For ceiling installations add | | | | | 40% | | | |
| | 110 | Drilling & layout for drywall or plaster walls | | | | | | | | |
| | 120 | Holes, 1/4" diameter | 1 Carp | .053 | Ea. | .05 | .66 | | .71 | 1.10 |
| | 130 | 3/8" diameter | | .057 | | .05 | .71 | | .76 | 1.17 |
| | 140 | 1/2" diameter | | .062 | | .05 | .76 | | .81 | 1.26 |
| | 150 | 3/4" diameter | | .067 | | .05 | .82 | | .87 | 1.36 |
| | 160 | 1" diameter | | .073 | | .06 | .90 | | .96 | 1.49 |
| | 170 | 1-1/4" diameter | | .080 | | .07 | .99 | | 1.06 | 1.64 |
| | 180 | 1-1/2" diameter | | .089 | | .08 | 1.10 | | 1.18 | 1.83 |
| | 190 | For ceiling installations add | | | | | 40% | | | |
| 12-001 | | **EXPANSION ANCHORS** Bolts & shields | | | | | | | | |
| | 010 | Bolt anchors for concrete, brick or stone, no layout and drilling | | | | | | | | |
| | 020 | Expansion shields, zinc, 1/4" diameter, 1" long, single | 1 Carp | .089 | Ea. | .40 | 1.10 | | 1.50 | 2.18 |
| | 030 | 1-3/8" long, double | | .094 | | .45 | 1.16 | | 1.61 | 2.34 |
| | 050 | 2" long, double | | .100 | | .82 | 1.24 | | 2.06 | 2.86 |
| | 070 | 2-1/2" long, double | | .107 | | 1.06 | 1.32 | | 2.38 | 3.25 |
| | 090 | 3" long, double | | .114 | | 1.65 | 1.41 | | 3.06 | 4.05 |
| | 110 | 4" long, double | | .123 | | 3.05 | 1.52 | | 4.57 | 5.75 |
| | 130 | 1" diameter, 6" long, double | | .133 | | 8.80 | 1.65 | | 10.45 | 12.30 |
| | 210 | Hollow wall anchors for gypsum board, | | | | | | | | |
| | 220 | plaster, tile or wall board | | | | | | | | |
| | 250 | 3/16" diameter, short | | | Ea. | .28 | | | .28 | .30M |
| | 300 | Toggle bolts, bright steel, 1/8" diameter, 2" long | 1 Carp | .094 | | .16 | 1.16 | | 1.32 | 2.02 |
| | 310 | 4" long | | .100 | | .21 | 1.24 | | 1.45 | 2.19 |
| | 320 | 3/16" diameter, 3" long | | .100 | | .19 | 1.24 | | 1.43 | 2.17 |
| | 330 | 6" long | | .107 | | .30 | 1.32 | | 1.62 | 2.42 |
| | 340 | 1/4" diameter, 3" long | | .107 | | .22 | 1.32 | | 1.54 | 2.33 |
| | 350 | 6" long | | .114 | | .33 | 1.41 | | 1.74 | 2.60 |
| | 360 | 3/8" diameter, 3" long | | .114 | | .52 | 1.41 | | 1.93 | 2.81 |
| | 370 | 6" long | | .133 | | .72 | 1.65 | | 2.37 | 3.40 |
| | 380 | 1/2" diameter, 4" long | | .133 | | 1.50 | 1.65 | | 3.15 | 4.26 |
| | 390 | 6" long | | .160 | | 1.96 | 1.98 | | 3.94 | 5.30 |
| | 400 | Nailing anchors | | | | | | | | |
| | 410 | Nylon anchor, stainless nail, 1/4" diameter, 1" long | | | C | 13 | | | 13 | 14.30M |
| | 420 | 1-1/2" long | | | | 16 | | | 16 | 17.60M |
| | 430 | 2" long | | | | 27 | | | 27 | 30M |
| | 440 | Zamac anchor, stainless nail, 1/4" diameter, 1" long | | | | 22 | | | 22 | 24M |
| | 450 | 1-1/2" long | | | | 25 | | | 25 | 28M |
| | 460 | 2" long | | | | 35 | | | 35 | 39M |
| | 570 | Lag screw, 1/4" diameter, short | | | Ea. | .21 | | | .21 | .23M |
| | 580 | Long | | | | .24 | | | .24 | .26M |
| | 590 | 3/8" diameter, short | | | | .37 | | | .37 | .40M |
| | 600 | Long | | | | .42 | | | .42 | .46M |
| | 610 | 1/2" diameter, short | | | | .73 | | | .73 | .80M |

For expanded coverage of these items see *Building Construction Cost Data 1985*

## 5.8 Expansion Control & Fasteners

| | | CREW | MAN-HOURS | UNIT | MAT. | LABOR | EQUIP. | TOTAL | TOTAL INCL O&P |
|---|---|---|---|---|---|---|---|---|---|
| 12 620 | Long | | | Ea. | .86 | | | .86 | .94M |
| 630 | 3/4" diameter, short | | | | 1.17 | | | 1.17 | 1.28M |
| 640 | Long | | | | 1.50 | | | 1.50 | 1.65M |
| 660 | Lead, #6 & #8, 3/4" long | | | | .08 | | | .08 | .08M |
| 670 | #10 & #12, 1-1/2" long | | | | .12 | | | .12 | .13M |
| 680 | #16 & #18, 1-1/2" long | | | | .16 | | | .16 | .17M |
| 690 | Plastic, #6 & #8, 3/4" long | | | | .02 | | | .02 | .02M |
| 700 | #8 & #10, 7/8" long | | | | .03 | | | .03 | .03M |
| 710 | #10 & #12, 1" long | | | | .03 | | | .03 | .03M |
| 720 | #14 & #16, 1-1/2" long | | | | .04 | | | .04 | .04M |
| 15-001 | **HIGH-STRENGTH BOLTS** Structural steel, see division 5.1-50-520 | | | | | | | | |
| 17-001 | **LAG SCREWS** 1/4" diameter, steel, 2" long | 1 Carp | .057 | Ea. | .12 | .71 | | .83 | 1.25 |
| 010 | 3/8" diameter, 3" long | | .076 | | .24 | .94 | | 1.18 | 1.75 |
| 020 | 1/2" diameter, 3" long | | .084 | | .38 | 1.04 | | 1.42 | 2.07 |
| 030 | 5/8" diameter, 3" long | | .094 | | .75 | 1.16 | | 1.91 | 2.67 |

| | | UNIT | BARE MAT. ONLY PLAIN | GALV. | ALUM. |
|---|---|---|---|---|---|
| 25-001 | **NAILS** Prices of material only, copper | Lb. | 3.25 | | |
| 040 | Stainless steel | | 4.40 | | |
| 060 | Common 3d to 20d | | .39 | .48 | 1.85 |
| 070 | 30d to 60d | | .38 | .47 | 1.85 |
| 080 | Annular or spiral thread, 4d to 60d | | .62 | .74 | |
| 100 | Finish nails, 4d to 10d | | .41 | .49 | 1.95 |
| 120 | Drywall nails | | .80 | .96 | |
| 140 | Flooring nails, hardened steel, 2d to 10d | | .70 | .84 | |
| 160 | Masonry nails, hardened steel, 3/4" to 3" long | | .70 | .84 | |
| 180 | Concrete nails, hardened steel | | .70 | .84 | |
| 200 | Roofing nails, threaded | | | .62 | 2.55 |
| 210 | Threaded, with washers | | | .80 | |
| 230 | Compressed lead head, threaded | | | .91 | |
| 240 | Screw-down | | | .96 | |
| 260 | Siding nails, plain shank | | | .55 | 1.55 |
| 270 | Threaded | | | .60 | |
| 290 | Add to prices above for cement coating | | .04 | | |
| 310 | Zinc or tin plating | | .10 | | |
| 27-001 | **RIVETS** 1/2" grip length | | | | |
| 002 | | C | | | |
| 010 | Aluminum rivet & mandrel, 1/8" diameter | | | | 3.02 |
| 020 | 3/8" diameter | | | | 4.65 |
| 030 | Aluminum rivet, steel mandrel, 1/8" diameter | | | | 3.20 |
| 040 | 3/8" diameter | | | | 5.12 |
| 050 | Copper rivet, steel mandrel, 1/8" diameter | | 4 | | |
| 060 | Monel rivet, steel mandrel, 1/8" diameter | | 13.70 | | |
| 070 | 3/8" diameter | | 25 | | |
| 080 | Stainless rivet & mandrel, 1/8" diameter | | 8.50 | | |
| 090 | 3/8" diameter | | 19.55 | | |
| 100 | Stainless rivet, steel mandrel, 1/8" diameter | | 7.20 | | |
| 110 | 3/8" diameter | | 12.05 | | |
| 120 | Steel rivet and mandrel, 1/8" diameter | | 3.05 | | |
| 130 | 3/8" diameter | | 5.05 | | |
| 140 | Hand riveting tool, minimum | Ea. | 28 | | |
| 150 | Maximum | | 275 | | |
| 160 | Power riveting tool, minimum | | 200 | | |
| 170 | Maximum | | 775 | | |

| | | CREW | MAN-HOURS | UNIT | MAT. | LABOR | EQUIP. | TOTAL | TOTAL INCL O&P |
|---|---|---|---|---|---|---|---|---|---|
| 35-001 | **TIMBER CONNECTORS** Add up cost of each part for total | | | | | | | | |
| 002 | cost of connection | | | | | | | | |

For expanded coverage of these items see *Building Construction Cost Data 1985*

## 5.8 Expansion Control & Fasteners

| | | | CREW | MAN-HOURS | UNIT | MAT. | LABOR | EQUIP. | TOTAL | TOTAL INCL O&P |
|---|---|---|---|---|---|---|---|---|---|---|
| 35 | 020 | Bolts, machine, sq. hd. with nut & washer, 1/2" diameter, 4" long | 1 Carp | .057 | Ea. | .65 | .71 | | 1.36 | 1.83 |
| | 030 | 7-1/2" long | | .062 | " | .80 | .76 | | 1.56 | 2.08 |
| | 080 | Drilling bolt holes in timber, 1/2" diameter | | .018 | Inch | | .22 | | .22 | .35L |
| | 090 | 1" diameter | | .023 | " | | .28 | | .28 | .45L |
| | 110 | Framing anchors, 2 or 3 dimensional, 10 gauge, no nails incl. | | .046 | Ea. | .35 | .56 | | .91 | 1.28 |
| | 130 | Joist and beam hangers, 18 ga. galv., for 2" x 4" joist | | .046 | | .35 | .56 | | .91 | 1.28 |
| | 140 | 2" x 6" to 2" x 10" joist | | .048 | | .75 | .60 | | 1.35 | 1.77 |
| | 160 | 16 ga. galv., 3" x 6" to 3" x 10" joist | | .050 | | 1.05 | .62 | | 1.67 | 2.13 |
| | 170 | 3" x 10" to 3" x 14" joist | | .050 | | 1.40 | .62 | | 2.02 | 2.52 |
| | 180 | 4" x 6" to 4" x 10" joist | | .052 | | 1.05 | .64 | | 1.69 | 2.16 |
| | 190 | 4" x 10" to 4" x 14" joist | | .052 | | 1.50 | .64 | | 2.14 | 2.66 |
| | 200 | Two 2" x 6" to two 2" x 10" joist | | .053 | | 1 | .66 | | 1.66 | 2.14 |
| | 210 | Two 2" x 10" to two 2" x 14" joist | | .053 | | 1.40 | .66 | | 2.06 | 2.58 |
| | 230 | 3/16" thick for 6" x 8" joist | | .055 | | 3.50 | .68 | | 4.18 | 4.93 |
| | 240 | 6" x 10" joist | | .057 | | 3.75 | .71 | | 4.46 | 5.25 |
| | 250 | 6" x 12" joist | | .059 | | 6.30 | .73 | | 7.03 | 8.10 |
| | 270 | 1/4" thick, 6" x 14" joist | | .062 | | 6.50 | .76 | | 7.26 | 8.35 |
| | 280 | Joist anchors, 1/4" x 1-1/4" x 18" | | .057 | | 2.50 | .71 | | 3.21 | 3.87 |
| | 290 | Plywood clips, extruded aluminum H clip, for 3/4" panels | | | | .05 | | | .05 | .05M |
| | 300 | Galvanized 18 ga. back-up clip | | | | .04 | | | .04 | .04M |
| | 320 | Post framing, 16 ga. galv. for 4" x 4" base, 2 piece | 1 Carp | .062 | | 3.65 | .76 | | 4.41 | 5.20 |
| | 330 | Cap | | .062 | | 2.30 | .76 | | 3.06 | 3.73 |
| | 350 | Rafter anchors, 18 ga. galv., 1-1/2" wide, 5-1/4" long | | .055 | | .33 | .68 | | 1.01 | 1.44 |
| | 360 | 10-3/4" long | | .055 | | .60 | .68 | | 1.28 | 1.74 |
| | 380 | Shear plates, 2-5/8" diameter | | .067 | | .95 | .82 | | 1.77 | 2.35 |
| | 390 | 4" diameter | | .070 | | 2 | .86 | | 2.86 | 3.56 |
| | 400 | Sill anchors, (embedded in concrete or block), 18-5/8" long | | .070 | | .50 | .86 | | 1.36 | 1.91 |
| | 410 | Spike grids, 4" x 4", flat or curved | | .067 | | 4.35 | .82 | | 5.17 | 6.10 |
| | 440 | Split rings, 2-1/2" diameter | | .067 | | .60 | .82 | | 1.42 | 1.96 |
| | 450 | 4" diameter | | .073 | | 1 | .90 | | 1.90 | 2.52 |
| | 470 | Strap ties, 14 ga., 1-3/8" wide, 12" long | | .044 | | .60 | .55 | | 1.15 | 1.53 |
| | 480 | 24" long | | .050 | | 1 | .62 | | 1.62 | 2.08 |
| | 500 | Toothed rings, 2-5/8" or 4" diameter | | .089 | | .70 | 1.10 | | 1.80 | 2.51 |
| | 520 | Truss plates, toothed, 18 gauge, for 36' span | | .471 | Truss | 6 | 5.80 | | 11.80 | 15.80 |
| | 540 | Washers, 2" x 2" x 1/8" | | | Ea. | .14 | | | .14 | .15M |
| | 550 | 3" x 3" x 3/16" | | | | .40 | | | .40 | .44M |
| 50-001 | | WOOD SCREWS #9, 1" long, steel | | | | .02 | | | .02 | .02M |
| | 030 | #12, 2" long | | | | .05 | | | .05 | .05M |
| | 060 | #14, 3" long | | | C | .09 | | | .09 | .09M |

## 6

### 6.1 Rough Carpentry

| | | | CREW | MAN-HOURS | UNIT | MAT. | LABOR | EQUIP. | TOTAL | TOTAL INCL O&P |
|---|---|---|---|---|---|---|---|---|---|---|
| 02-001 | | BLOCKING | | | | | | | | |
| 02-001 | | | | | | | | | | |
| | 195 | Miscellaneous, to wood construction | | | | | | | | |
| | 200 | 2" x 4" | F-1 | .032 | L.F. | .23 | .39 | .03 | .65 | .91 |
| | 205 | 2" x 6" | | .036 | | .36 | .45 | .03 | .84 | 1.14 |
| | 210 | 2" x 8" | | .040 | | .50 | .49 | .04 | 1.03 | 1.37 |
| | 215 | 2" x 10" | | .045 | | .68 | .55 | .04 | 1.27 | 1.67 |
| | 220 | 2" x 12" | | .053 | | .86 | .65 | .05 | 1.56 | 2.03 |
| | 230 | To steel construction | | | | | | | | |
| | 232 | 2" x 4" | F-1 | .038 | L.F. | .23 | .48 | .03 | .74 | 1.04 |
| | 234 | 2" x 4" | | .044 | | .36 | .55 | .04 | .95 | 1.31 |
| | 236 | 2" x 8" | | .051 | | .50 | .63 | .04 | 1.17 | 1.59 |

For expanded coverage of these items see *Building Construction Cost Data 1985*

## 6.1 Rough Carpentry

| | | CREW | MAN-HOURS | UNIT | MAT. | LABOR | EQUIP. | TOTAL | TOTAL INCL O&P |
|---|---|---|---|---|---|---|---|---|---|
| 02 238 | 2" x 10" | F-1 | .059 | L.F. | .68 | .73 | .05 | 1.46 | 1.95 |
| 240 | 2" x 12" | | .073 | | .86 | .91 | .06 | 1.83 | 2.45 |
| 05-001 | **BRACING** Let-in, with 1" x 6" boards, studs 16" O.C. | | .053 | | .19 | .66 | .05 | .90 | 1.30 |
| 020 | Studs at 24" O.C. | | .035 | | .19 | .43 | .03 | .65 | .92 |
| 030 | Let-in, "T" shaped, 20 ga. galv. steel, studs at 16" O.C. | | .014 | | .53 | .17 | .01 | .71 | .87 |
| 040 | Studs at 24" O.C. | | .013 | | .53 | .17 | .01 | .71 | .86 |
| 050 | 16 ga. galv. steel straps, studs at 16" O.C. | | .013 | | .53 | .17 | .01 | .71 | .86 |
| 060 | Studs at 24" O.C. | | .013 | | .53 | .16 | .01 | .70 | .85 |
| 07-001 | **BRIDGING** Wood, for joists 16" O.C., 1" x 3" | | .062 | Pr. | .38 | .76 | .05 | 1.19 | 1.68 |
| 010 | 2" x 3" bridging | | .062 | | .67 | .76 | .05 | 1.48 | 2 |
| 030 | Steel, galvanized, 18 ga., for 2" x 10" joists at 12" O.C. | 1 Carp | .062 | | .86 | .76 | | 1.62 | 2.15 |
| 040 | 24" O.C. | | .057 | | 1.26 | .71 | | 1.97 | 2.50 |
| 060 | For 14" joists at 16" O.C. | | 6.150 | | 1.26 | 76 | | 77.26 | 120 |
| 090 | Compression type, 16" O.C., 2" x 8" joists | | .040 | | .84 | .49 | | 1.33 | 1.71 |
| 100 | 2" x 12" joists | | .040 | | .90 | .49 | | 1.39 | 1.77 |
| 10-001 | **DOCK BUMPERS** Bolts not included, 2" x 6" to 4" x 8", average | F-1 | 26.670 | B.F. | .42 | 332 | 23 | 355.42 | 545 |
| 13-001 | **FRAMING, BEAMS & GIRDERS** | | | | | | | | |
| 002 | | | | | | | | | |
| 100 | Single, 2" x 6" | F-2 | .023 | L.F. | .36 | .28 | .02 | .66 | .86 |
| 102 | 2" x 8" | | .025 | | .55 | .31 | .02 | .88 | 1.11 |
| 104 | 2" x 10" | | .027 | | .75 | .33 | .02 | 1.10 | 1.37 |
| 106 | 2" x 12" | | .029 | | .86 | .35 | .03 | 1.24 | 1.54 |
| 108 | 2" x 14" | | .032 | | 1.10 | .39 | .03 | 1.52 | 1.87 |
| 110 | 3" x 6" | | .029 | | .85 | .35 | .03 | 1.23 | 1.53 |
| 112 | 3" x 10" | | .032 | | 1.40 | .39 | .03 | 1.82 | 2.20 |
| 114 | 3" x 12" | | .036 | | 1.68 | .44 | .03 | 2.15 | 2.58 |
| 116 | 3" x 14" | | .040 | | 1.96 | .49 | .04 | 2.49 | 2.98 |
| 118 | 4" x 8" | F-3 | .040 | | 1.50 | .50 | .25 | 2.25 | 2.71 |
| 120 | 4" x 10" | | .042 | | 1.87 | .53 | .26 | 2.66 | 3.18 |
| 122 | 4" x 12" | | .044 | | 2.24 | .55 | .28 | 3.07 | 3.64 |
| 124 | 4" x 14" | | .047 | | 2.60 | .59 | .29 | 3.48 | 4.11 |
| 200 | Double, 2" x 6" | F-2 | .040 | | .73 | .49 | .04 | 1.26 | 1.62 |
| 202 | 2" x 8" | | .043 | | 1.10 | .52 | .04 | 1.66 | 2.09 |
| 204 | 2" x 10" | | .046 | | 1.36 | .56 | .04 | 1.96 | 2.43 |
| 206 | 2" x 12" | | .049 | | 1.72 | .61 | .04 | 2.37 | 2.90 |
| 208 | 2" x 14" | | .053 | | 2.20 | .66 | .05 | 2.91 | 3.51 |
| 210 | | | | | | | | | |
| 300 | Triple, 2" x 6" | F-2 | .058 | L.F. | 1.09 | .72 | .05 | 1.86 | 2.39 |
| 302 | 2" x 8" | | .064 | | 1.65 | .79 | .06 | 2.50 | 3.13 |
| 304 | 2" x 10" | | .071 | | 2.04 | .88 | .06 | 2.98 | 3.70 |
| 306 | 2" x 12" | | .080 | | 2.57 | .99 | .07 | 3.63 | 4.47 |
| 308 | 2" x 14" | | .084 | | 3.30 | 1.04 | .07 | 4.41 | 5.35 |
| 310 | | | | | | | | | |
| 16-001 | **FRAMING, CEILINGS** | | | | | | | | |
| 002 | | | | | | | | | |
| 600 | Suspended, 2" x 3" | F-2 | .016 | L.F. | .18 | .20 | .01 | .39 | .53 |
| 605 | 2" x 4" | | .018 | | .23 | .22 | .02 | .47 | .62 |
| 610 | 2" x 6" | | .020 | | .36 | .24 | .02 | .62 | .81 |
| 615 | 2" x 8" | | .025 | | .50 | .31 | .02 | .83 | 1.05 |
| 19-001 | **FRAMING, JOISTS** | | | | | | | | |
| 200 | Joists, 2" x 4" | F-2 | .013 | L.F. | .23 | .16 | .01 | .40 | .52 |
| 210 | 2" x 6" | | .013 | | .36 | .16 | .01 | .53 | .66 |
| 215 | 2" x 8" | | .015 | | .50 | .18 | .01 | .69 | .85 |
| 220 | 2" x 10" | | .018 | | .68 | .22 | .02 | .92 | 1.11 |
| 225 | 2" x 12" | | .018 | | .86 | .22 | .02 | 1.10 | 1.32 |
| 230 | 2" x 14" | | .021 | | 1.10 | .25 | .02 | 1.37 | 1.64 |
| 235 | 3" x 6" | | .017 | | .84 | .21 | .02 | 1.07 | 1.28 |

297

## 6.1 Rough Carpentry

| | | | CREW | MAN-HOURS | UNIT | MAT. | LABOR | EQUIP. | TOTAL | TOTAL INCL O&P |
|---|---|---|---|---|---|---|---|---|---|---|
| 19 | 240 | 3" x 10" | F-2 | .021 | L.F. | 1.40 | .25 | .02 | 1.67 | 1.96 |
| | 245 | 3" x 12" | | .027 | | 1.68 | .33 | .02 | 2.03 | 2.40 |
| | 250 | 4" x 6" | | .020 | | 1.12 | .24 | .02 | 1.38 | 1.64 |
| | 255 | 4" x 10" | | .027 | | 1.87 | .33 | .02 | 2.22 | 2.60 |
| | 260 | 4" x 12" | | .036 | | 2.24 | .44 | .03 | 2.71 | 3.19 |
| | 261 | 2" x 8" | | .025 | | .50 | .31 | .02 | .83 | 1.06 |
| | 261 | 2" x 10" | | .030 | | .68 | .37 | .03 | 1.08 | 1.36 |
| | 262 | 2" x 12" | ↓ | .035 | ↓ | .86 | .44 | .03 | 1.33 | 1.67 |
| 22-001 | | **FRAMING, MISCELLANEOUS** | | | | | | | | |
| | 002 | | | | | | | | | |
| | 200 | Firestops, 2" x 4" | F-2 | .021 | L.F. | .23 | .25 | .02 | .50 | .67 |
| | 210 | 2" x 6" | | .027 | | .36 | .33 | .02 | .71 | .94 |
| | 510 | 2" x 6" | | .021 | | .64 | .26 | .02 | .92 | 1.14 |
| | 512 | 2" x 8" | | .023 | | .90 | .28 | .02 | 1.20 | 1.46 |
| | 520 | Steel construction, 2" x 4" | | .021 | | .36 | .26 | .02 | .64 | .83 |
| | 522 | 2" x 6" | | .023 | | .64 | .28 | .02 | .94 | 1.17 |
| | 524 | 2" x 8" | | .025 | | .90 | .31 | .02 | 1.23 | 1.49 |
| | 700 | Rough bucks, treated, for doors or windows, 2" x 6" | | .040 | | .64 | .49 | .04 | 1.17 | 1.52 |
| | 710 | 2" x 8" | | .042 | | .90 | .52 | .04 | 1.46 | 1.85 |
| | 800 | Stair stringers, 2" x 10" | | .123 | | .68 | 1.52 | .11 | 2.31 | 3.27 |
| | 810 | 2" x 12" | | .123 | | .86 | 1.52 | .11 | 2.49 | 3.47 |
| | 815 | 3" x 10" | | .128 | | 1.40 | 1.58 | .11 | 3.09 | 4.17 |
| | 820 | 3" x 12" | ↓ | .128 | ↓ | 1.68 | 1.58 | .11 | 3.37 | 4.47 |
| 25-001 | | **FRAMING, COLUMNS** | | | | | | | | |
| | 002 | | | | | | | | | |
| | 010 | 4" x 4" | F-2 | .041 | L.F. | .75 | .50 | .04 | 1.29 | 1.67 |
| | 015 | 4" x 6" | | .058 | | 1.12 | .72 | .05 | 1.89 | 2.43 |
| | 020 | 4" x 8" | | .073 | | 1.50 | .90 | .06 | 2.46 | 3.14 |
| | 025 | 6" x 6" | | .074 | | 1.68 | .91 | .07 | 2.66 | 3.37 |
| | 030 | 6" x 8" | | .091 | | 2.24 | 1.13 | .08 | 3.45 | 4.34 |
| | 035 | 6" x 10" | ↓ | .107 | ↓ | 2.80 | 1.32 | .09 | 4.21 | 5.25 |
| | 037 | | | | | | | | | |
| 28-001 | | **FRAMING, ROOFS** | | | | | | | | |
| | 200 | Fascia boards, 2" x 8" | F-2 | .071 | L.F. | .50 | .88 | .06 | 1.44 | 2.01 |
| | 210 | 2" x 10" | | .089 | | .68 | 1.10 | .08 | 1.86 | 2.57 |
| | 500 | Rafters, ordinary, to 4 in 12 pitch, 2" x 6", ordinary | | .016 | | .36 | .20 | .01 | .57 | .72 |
| | 502 | On steep roofs | | .020 | | .36 | .24 | .02 | .62 | .81 |
| | 504 | On dormers or complex roofs | | .027 | | .36 | .34 | .02 | .72 | .95 |
| | 506 | 2" x 8", ordinary | | .017 | | .50 | .21 | .01 | .72 | .90 |
| | 508 | On steep roofs | | .021 | | .50 | .26 | .02 | .78 | .99 |
| | 510 | On dormers or complex roofs | | .030 | | .50 | .36 | .03 | .89 | 1.16 |
| | 512 | 2" x 10", ordinary | | .025 | | .68 | .32 | .02 | 1.02 | 1.27 |
| | 514 | On steep roofs | | .032 | | .68 | .40 | .03 | 1.11 | 1.41 |
| | 516 | On dormers or complex roofs | | .038 | | .68 | .47 | .03 | 1.18 | 1.52 |
| | 518 | 2" x 12", ordinary | | .028 | | .86 | .35 | .02 | 1.23 | 1.52 |
| | 520 | On steep roofs | | .035 | | .86 | .44 | .03 | 1.33 | 1.67 |
| | 522 | On dormers or complex roofs | | .041 | | .86 | .50 | .04 | 1.40 | 1.78 |
| | 530 | Hip and valley rafters, 2" x 6", ordinary | | .021 | | .36 | .26 | .02 | .64 | .83 |
| | 532 | On steep roofs | | .027 | | .36 | .34 | .02 | .72 | .96 |
| | 534 | On dormers or complex roofs | | .031 | | .36 | .38 | .03 | .77 | 1.04 |
| | 536 | 2" x 8", ordinary | | .022 | | .50 | .27 | .02 | .79 | 1.01 |
| | 538 | On steep roofs | | .029 | | .50 | .36 | .03 | .89 | 1.15 |
| | 540 | On dormers or complex roofs | | .034 | | .50 | .42 | .03 | .95 | 1.25 |
| | 542 | 2" x 10", ordinary | | .028 | | .68 | .35 | .02 | 1.05 | 1.32 |
| | 544 | On steep roofs | | .036 | | .68 | .45 | .03 | 1.16 | 1.49 |
| | 546 | On dormers or complex roofs | ↓ | .042 | ↓ | .68 | .52 | .04 | 1.24 | 1.61 |
| | 546 | | | | | | | | | |
| | 548 | Hip and valley rafters, 2" x 12", ordinary | F-2 | .030 | L.F. | .86 | .37 | .03 | 1.26 | 1.57 |

For expanded coverage of these items see *Building Construction Cost Data 1985*

## 6.1 Rough Carpentry

| | | | CREW | MAN-HOURS | UNIT | MAT. | LABOR | EQUIP. | TOTAL | TOTAL INCL O&P |
|---|---|---|---|---|---|---|---|---|---|---|
| 28 | 550 | On steep roofs | F-2 | .039 | L.F. | .86 | .49 | .03 | 1.38 | 1.75 |
| | 552 | On dormers or complex roofs | | .045 | | .86 | .56 | .04 | 1.46 | 1.87 |
| | 554 | Hip and valley jacks, 2" x 6", ordinary | | .027 | | .36 | .33 | .02 | .71 | .94 |
| | 556 | On steep roofs | | .034 | | .36 | .42 | .03 | .81 | 1.09 |
| | 558 | On dormers or complex roofs | | .039 | | .36 | .49 | .03 | .88 | 1.20 |
| | 560 | 2" x 8", ordinary | | .033 | | .50 | .40 | .03 | .93 | 1.22 |
| | 562 | On steep roofs | | .042 | | .50 | .51 | .04 | 1.05 | 1.40 |
| | 564 | On dormers or complex roofs | | .048 | | .50 | .59 | .04 | 1.13 | 1.53 |
| | 566 | 2" x 10", ordinary | | .036 | | .68 | .44 | .03 | 1.15 | 1.48 |
| | 568 | On steep roofs | | .046 | | .68 | .56 | .04 | 1.28 | 1.69 |
| | 570 | On dormers or complex roofs | | .052 | | .68 | .64 | .05 | 1.37 | 1.82 |
| | 572 | 2" x 12", ordinary | | .043 | | .86 | .52 | .04 | 1.42 | 1.82 |
| | 574 | On steep roofs | | .054 | | .86 | .67 | .05 | 1.58 | 2.06 |
| | 576 | On dormers or complex roofs | | .063 | | .86 | .78 | .05 | 1.69 | 2.23 |
| | 578 | Rafter tie, 1" x 4", #3 | | .020 | | .13 | .24 | .02 | .39 | .55 |
| | 580 | Ridge board, #2 or better, 1" x 6" | | .027 | | .20 | .33 | .02 | .55 | .77 |
| | 582 | 1" x 8" | | .029 | | .26 | .35 | .03 | .64 | .88 |
| | 584 | 1" x 10" | | .032 | | .34 | .39 | .03 | .76 | 1.03 |
| | 586 | 2" x 6" | | .032 | | .36 | .39 | .03 | .78 | 1.05 |
| | 588 | 2" x 8" | | .036 | | .50 | .44 | .03 | .97 | 1.28 |
| | 590 | 2" x 10" | | .040 | | .68 | .49 | .04 | 1.21 | 1.57 |
| | 592 | Roof cants, split, 4" x 4" | | .025 | | .80 | .31 | .02 | 1.13 | 1.38 |
| | 594 | 6" x 6" | | .027 | | 1.55 | .33 | .02 | 1.90 | 2.25 |
| | 596 | Roof curbs, untreated, 2" x 6" | | .031 | | .36 | .38 | .03 | .77 | 1.03 |
| | 598 | 2" x 12" | | .040 | | .86 | .49 | .04 | 1.39 | 1.77 |
| | 600 | Sister rafters, 2" x 6" | | .020 | | .36 | .24 | .02 | .62 | .81 |
| | 602 | 2" x 8" | | .025 | | .50 | .31 | .02 | .83 | 1.06 |
| | 604 | 2" x 10" | | .030 | | .68 | .37 | .03 | 1.08 | 1.36 |
| | 606 | 2" x 12" | | .035 | | .86 | .44 | .03 | 1.33 | 1.67 |
| 31-001 | | **FRAMING, SILLS** | | | | | | | | |
| | 181 | | | | | | | | | |
| | 200 | Ledgers, nailed, 2" x 4" | F-2 | .021 | L.F. | .23 | .26 | .02 | .51 | .69 |
| | 205 | 2" x 6" | | .027 | | .36 | .33 | .02 | .71 | .94 |
| | 210 | Bolted, not including bolts, 3" x 6" | | .049 | | .85 | .61 | .04 | 1.50 | 1.94 |
| | 215 | 3" x 12" | | .069 | | 1.68 | .85 | .06 | 2.59 | 3.26 |
| | 260 | Mud sills, redwood, construction grade, 2" x 4" | | .018 | | .64 | .22 | .02 | .88 | 1.07 |
| | 262 | 2" x 6" | | .021 | | .95 | .25 | .02 | 1.22 | 1.47 |
| | 400 | Sills, 2" x 4" | | .027 | | .23 | .33 | .02 | .58 | .80 |
| | 405 | 2" x 6" | | .029 | | .36 | .35 | .03 | .74 | .99 |
| | 408 | 2" x 8" | | .032 | | .50 | .39 | .03 | .92 | 1.21 |
| | 410 | 2" x 10" | | .036 | | .68 | .44 | .03 | 1.15 | 1.48 |
| | 412 | 2" x 12" | | .040 | | .86 | .49 | .04 | 1.39 | 1.77 |
| | 420 | Treated, 2" x 4" | | .029 | | .36 | .35 | .03 | .74 | .99 |
| | 422 | 2" x 6" | | .032 | | .54 | .39 | .03 | .96 | 1.25 |
| | 424 | 2" x 8" | | .036 | | .72 | .44 | .03 | 1.19 | 1.52 |
| | 426 | 2" x 10" | | .040 | | 1.05 | .49 | .04 | 1.58 | 1.98 |
| | 428 | 2" x 12" | | .046 | | 1.20 | .56 | .04 | 1.80 | 2.26 |
| | 440 | 4" x 4" | | .036 | | .89 | .44 | .03 | 1.36 | 1.71 |
| | 442 | 4" x 6" | | .046 | | 1.32 | .56 | .04 | 1.92 | 2.39 |
| | 446 | 4" x 8" | | .053 | | 1.78 | .66 | .05 | 2.49 | 3.05 |
| | 448 | 4" x 10" | | .062 | | 2.25 | .76 | .05 | 3.06 | 3.74 |
| 33-001 | | **PARTITIONS** Metal studs with runners, partitions 10' high | | | | | | | | |
| | 002 | L.B., Lt. ga., structural, 24" O.C. 20 ga. galv., 2-1/2" wide | F-2 | .016 | S.F. | .36 | .20 | .01 | .57 | .72 |
| | 004 | 3-5/8" wide | | .017 | | .38 | .21 | .01 | .60 | .76 |
| | 006 | 4" wide | | .018 | | .42 | .22 | .02 | .66 | .83 |
| | 008 | 6" wide | | .019 | | .47 | .23 | .02 | .72 | .90 |
| | 010 | 18 ga., 2-1/2" wide | | .016 | | .73 | .21 | .01 | .95 | 1.14 |
| | 012 | 3-5/8" wide | | .017 | | .77 | .21 | .02 | 1 | 1.20 |
| | 014 | 4" wide | | .018 | | .84 | .22 | .02 | 1.08 | 1.30 |

For expanded coverage of these items see *Building Construction Cost Data 1985*

## 6.1 Rough Carpentry

| | | | CREW | MAN-HOURS | UNIT | MAT. | LABOR | EQUIP. | TOTAL | TOTAL INCL O&P |
|---|---|---|---|---|---|---|---|---|---|---|
| 33 | 016 | 6" wide | F-2 | .019 | S.F. | .96 | .23 | .02 | 1.21 | 1.45 |
| | 018 | 16 ga., 2-1/2" wide | | .017 | | .87 | .21 | .01 | 1.09 | 1.30 |
| | 020 | 3-5/8" wide | | .018 | | .87 | .22 | .02 | 1.11 | 1.32 |
| | 022 | 4" wide | | .019 | | .97 | .23 | .02 | 1.22 | 1.45 |
| | 024 | 6" wide | | .019 | | 1.15 | .23 | .02 | 1.40 | 1.66 |
| | 026 | 8" wide | | .020 | | 1.46 | .24 | .02 | 1.72 | 2.02 |
| | 030 | Nonload bearing, 24" O.C., 25 ga., galv. 1-5/8" wide | | .016 | | .16 | .20 | .01 | .37 | .50 |
| | 032 | 2-1/2" wide | | .016 | | .18 | .20 | .01 | .39 | .53 |
| | 034 | 3-5/8" wide | | .017 | | .21 | .21 | .01 | .43 | .57 |
| | 036 | 4" wide | | .018 | | .25 | .22 | .02 | .49 | .64 |
| | 038 | 6" wide | | .018 | | .31 | .22 | .02 | .55 | .71 |
| | 040 | 20 ga., 2-1/2" wide | | .016 | | .22 | .20 | .01 | .43 | .57 |
| | 042 | 3-5/8" wide | | .017 | | .25 | .21 | .01 | .47 | .62 |
| | 044 | 4" wide | | .018 | | .28 | .22 | .02 | .52 | .67 |
| | 046 | 6" wide | ↓ | .018 | ↓ | .33 | .22 | .02 | .57 | .74 |
| | 048 | | | | | | | | | |
| | 101 | Wood stud with single bottom plate and | | | | | | | | |
| | 102 | double top plate, no waste, std. & better lumber | | | | | | | | |
| | 104 | 2" x 4" studs, 8 ft. high studs, 16" O.C. | F-2 | .160 | L.F. | 2.55 | 1.98 | .14 | 4.67 | 6.10 |
| | 106 | 24" O.C. | | .128 | | 1.92 | 1.58 | .11 | 3.61 | 4.74 |
| | 108 | 10 ft. high, studs 16" O.C. | | .160 | | 3.02 | 1.98 | .14 | 5.14 | 6.60 |
| | 110 | 24" O.C. | | .128 | | 2.26 | 1.58 | .11 | 3.95 | 5.10 |
| | 112 | 12 ft. high, studs 16" O.C. | | .200 | | 3.68 | 2.47 | .18 | 6.33 | 8.15 |
| | 114 | 24" O.C. | | .160 | | 2.72 | 1.98 | .14 | 4.84 | 6.25 |
| | 116 | 2" x 6" studs, 8 ft. high, studs 16" O.C. | | .178 | | 3.97 | 2.19 | .16 | 6.32 | 8 |
| | 118 | 24" O.C. | | .139 | | 2.99 | 1.72 | .12 | 4.83 | 6.15 |
| | 120 | 10 ft. high, studs 16" O.C. | | .178 | | 4.73 | 2.19 | .16 | 7.08 | 8.85 |
| | 122 | 24" O.C. | | .139 | | 3.52 | 1.72 | .12 | 5.36 | 6.75 |
| | 124 | 12 ft. high, studs 16" O.C. | | .229 | | 5.40 | 2.82 | .20 | 8.42 | 10.65 |
| | 126 | 24" O.C. | | .178 | | 3.97 | 2.19 | .16 | 6.32 | 8 |
| | 128 | For horizontal blocking, 2" x 4", add | | .027 | | .23 | .33 | .02 | .58 | .80 |
| | 130 | 2" x 6", add | | .027 | | .36 | .33 | .02 | .71 | .94 |
| | 134 | For openings, add | ↓ | .064 | ↓ | | .79 | .06 | .85 | 1.31 |
| | 138 | For headers, add | | | B.F. | .39 | | | .39 | .42M |
| 34-001 | | FRAMING, SLEEPERS | | | | | | | | |
| | 002 | | | | | | | | | |
| | 010 | On concrete, treated, 1" x 2" | F-2 | .007 | L.F. | .11 | .08 | .01 | .20 | .26 |
| | 015 | 1" x 3" | | .008 | | .16 | .10 | .01 | .27 | .34 |
| | 020 | 2" x 4" | | .011 | | .33 | .13 | .01 | .47 | .58 |
| | 025 | 2" x 6" | ↓ | .012 | ↓ | .49 | .15 | .01 | .65 | .79 |
| | 027 | | | | | | | | | |
| | 100 | Canopy or soffit framing, 1" x 4" | F-2 | .018 | L.F. | .13 | .22 | .02 | .37 | .51 |
| | 102 | 1" x 6" | | .019 | | .20 | .23 | .02 | .45 | .61 |
| | 104 | 1" x 8" | | .021 | | .26 | .26 | .02 | .54 | .72 |
| | 110 | 2" x 4" | | .026 | | .23 | .32 | .02 | .57 | .78 |
| | 112 | 2" x 6" | | .029 | | .36 | .35 | .03 | .74 | .98 |
| | 114 | 2" x 8" | | .032 | | .50 | .39 | .03 | .92 | 1.21 |
| | 120 | 3" x 4" | | .032 | | .56 | .39 | .03 | .98 | 1.27 |
| | 122 | 3" x 6" | | .040 | | .85 | .49 | .04 | 1.38 | 1.76 |
| | 124 | 3" x 10" | ↓ | .053 | ↓ | 1.40 | .66 | .05 | 2.11 | 2.63 |
| | 125 | | | | | | | | | |
| 40-001 | | FRAMING, WALLS | | | | | | | | |
| | 200 | Headers over openings, 2" x 6" | F-2 | .044 | L.F. | .36 | .55 | .04 | .95 | 1.31 |
| | 205 | 2" x 8" | | .047 | | .50 | .58 | .04 | 1.12 | 1.52 |
| | 210 | Headers over openings, 2" x 10" | | .050 | | .68 | .62 | .04 | 1.34 | 1.77 |
| | 215 | 2" x 12" | | .053 | | .86 | .66 | .05 | 1.57 | 2.04 |
| | 220 | 4" x 12" | | .084 | | 2.24 | 1.04 | .07 | 3.35 | 4.19 |
| | 225 | 6" x 12" | ↓ | .114 | ↓ | 3.36 | 1.41 | .10 | 4.87 | 6.05 |

For expanded coverage of these items see *Building Construction Cost Data 1985*

## 6.1 Rough Carpentry

| | | Crew | Man-Hours | Unit | Mat. | Labor | Equip. | Total | Total Incl O&P |
|---|---|---|---|---|---|---|---|---|---|
| 500 | Plates, untreated, 2" x 3" | F-2 | .019 | L.F. | .18 | .23 | .02 | .43 | .58 |
| 502 | 2" x 4" | | .020 | | .23 | .24 | .02 | .49 | .66 |
| 504 | 2" x 6" | | .021 | | .36 | .26 | .02 | .64 | .83 |
| 506 | Treated, 2" x 3" | | .021 | | .28 | .26 | .02 | .56 | .75 |
| 508 | 2" x 4" | | .023 | | .36 | .28 | .02 | .66 | .86 |
| 510 | 2" x 6" | | .025 | | .54 | .31 | .02 | .87 | 1.10 |
| 512 | Studs, 8' high wall, 2" x 3" | | .013 | | .18 | .17 | .01 | .36 | .47 |
| 514 | 2" x 4" | | .011 | | .23 | .14 | .01 | .38 | .49 |
| 516 | 2" x 6" | | .016 | | .36 | .20 | .01 | .57 | .72 |
| 518 | 3" x 4" | | .020 | | .56 | .24 | .02 | .82 | 1.03 |
| 520 | Installed on second story, 2" x 3" | | .014 | | .18 | .17 | .01 | .36 | .48 |
| 522 | 2" x 4" | | .016 | | .23 | .20 | .01 | .44 | .58 |
| 524 | 2" x 6" | | .018 | | .36 | .22 | .02 | .60 | .76 |
| 526 | 3" x 4" | | .020 | | .56 | .24 | .02 | .82 | 1.03 |
| 528 | Installed on dormer or gable, 2" x 3" | | .015 | | .18 | .19 | .01 | .38 | .51 |
| 530 | 2" x 4" | | .018 | | .23 | .21 | .02 | .46 | .62 |
| 532 | 2" x 6" | | .020 | | .36 | .24 | .02 | .62 | .81 |
| 534 | 3" x 4" | | .023 | | .55 | .28 | .02 | .85 | 1.07 |
| 536 | 6' high wall, 2" x 3" | | .016 | | .18 | .21 | .01 | .40 | .54 |
| 538 | 2" x 4" | | .019 | | .23 | .23 | .02 | .48 | .64 |
| 540 | 2" x 6" | | .022 | | .36 | .27 | .02 | .65 | .84 |
| 542 | 3" x 4" | | .027 | | .55 | .33 | .02 | .90 | 1.15 |
| 544 | Installed on second story, 2" x 3" | | .017 | | .18 | .21 | .01 | .40 | .54 |
| 546 | 2" x 4" | | .020 | | .23 | .24 | .02 | .49 | .66 |
| 548 | 2" x 6" | | .023 | | .36 | .28 | .02 | .66 | .86 |
| 550 | 3" x 4" | | .029 | | .55 | .35 | .03 | .93 | 1.20 |
| 552 | Installed on dormer or gable, 2" x 3" | | .019 | | .18 | .23 | .02 | .43 | .58 |
| 554 | 2" x 4" | | .022 | | .23 | .27 | .02 | .52 | .71 |
| 556 | 2" x 6" | | .026 | | .36 | .32 | .02 | .70 | .93 |
| 558 | 3" x 4" | | .033 | | .55 | .41 | .03 | .99 | 1.29 |
| 560 | 3' high wall, 2" x 3" | | .022 | | .18 | .27 | .02 | .47 | .64 |
| 562 | 2" x 4" | | .025 | | .23 | .31 | .02 | .56 | .77 |
| 564 | 2" x 6" | | .029 | | .36 | .35 | .03 | .74 | .99 |
| 566 | 3" x 4" | | .036 | | .55 | .45 | .03 | 1.03 | 1.35 |
| 568 | Installed on second story, 2" x 3" | | .023 | | .18 | .28 | .02 | .48 | .67 |
| 570 | 2" x 4" | | .026 | | .23 | .33 | .02 | .58 | .79 |
| 572 | 2" x 6" | | .031 | | .36 | .38 | .03 | .77 | 1.03 |
| 574 | 3" x 4" | | .037 | | .55 | .46 | .03 | 1.04 | 1.37 |
| 576 | Installed on dormer or gable, 2" x 3" | | .026 | | .18 | .32 | .02 | .52 | .72 |
| 578 | 2" x 4" | | .029 | | .23 | .36 | .03 | .62 | .86 |
| 580 | 2" x 6" | | .034 | | .36 | .43 | .03 | .82 | 1.10 |
| 582 | 3" x 4" | | .042 | | .55 | .52 | .04 | 1.11 | 1.47 |
| 825 | For second story & above, add | | | | | 5% | | | |
| 830 | Dormer & gable, add | | | | | 15% | | | |
| 43-001 | FRAMING, HEAVY Mill timber, beams, single 6" x 10" | F-2 | .073 | | 2.80 | .90 | .06 | 3.76 | 4.57 |
| 010 | Single 8" x 16" | | .139 | | 6.10 | 1.72 | .12 | 7.94 | 9.55 |
| 020 | Built from 2" lumber, multiple 2" x 14" | | .018 | B.F. | .47 | .22 | .02 | .71 | .88 |
| 021 | Built from 3" lumber, multiple 3" x 6" | | .023 | | .56 | .28 | .02 | .86 | 1.08 |
| 022 | Multiple 3" x 8" | | .020 | | .56 | .24 | .02 | .82 | 1.03 |
| 023 | Multiple 3" x 10" | | .018 | | .56 | .22 | .02 | .80 | .98 |
| 024 | Multiple 3" x 12" | | .016 | | .56 | .20 | .01 | .77 | .94 |
| 025 | Built from 4" lumber, multiple 4" x 6" | | .020 | | .56 | .24 | .02 | .82 | 1.03 |
| 026 | Multiple 4" x 8" | | .018 | | .56 | .22 | .02 | .80 | .98 |
| 027 | Multiple 4" x 10" | | .016 | | .56 | .20 | .01 | .77 | .94 |
| 028 | Multiple 4" x 12" | | .015 | | .56 | .18 | .01 | .75 | .91 |
| 029 | Columns, structural grade, 1500f, 4" x 4" | | .036 | | .73 | .44 | .03 | 1.20 | 1.53 |
| 030 | 6" x 6" | | .071 | L.F. | 1.53 | .88 | .06 | 2.47 | 3.14 |
| 040 | 8" x 8" | | .067 | | 2.93 | .82 | .06 | 3.81 | 4.59 |
| 050 | 10" x 10" | F-2 | .178 | | 4.58 | 2.19 | .16 | 6.93 | 8.70 |
| 060 | 12" x 12" | | .229 | | 6.60 | 2.82 | .20 | 9.62 | 11.95 |

For expanded coverage of these items see *Building Construction Cost Data 1985*

## 6.1 Rough Carpentry

| | | Crew | Man-Hours | Unit | Mat. | Labor | Equip. | Total | Total Incl O&P |
|---|---|---|---|---|---|---|---|---|---|
| 43 061 080 | Floor planks, 2" thick, T & G, 2" x 6" | F-2 | .015 | B.F. | .52 | .19 | .01 | .72 | .88 |
| 090 | 2" x 10" | | .015 | | .55 | .18 | .01 | .74 | .90 |
| 110 | 3" thick, 3" x 6" | | .015 | | .73 | .19 | .01 | .93 | 1.12 |
| 120 | 3" x 10" | | .015 | | .74 | .18 | .01 | .93 | 1.11 |
| 140 | Girders, structural grade, 12" x 12" | | .020 | | .58 | .24 | .02 | .84 | 1.05 |
| 150 | 10" x 16" | ↓ | .016 | ↓ | .58 | .20 | .01 | .79 | .97 |
| 205 | Roof planks, see division 6.1-620 | | | | | | | | |
| 230 | Roof purlins, 4" thick, structural grade | F-2 | .015 | B.F. | .55 | .19 | .01 | .75 | .92 |
| 250 | Roof trusses, add timber connectors, division 5.8-35 | | .036 | " | .55 | .44 | .03 | 1.02 | 1.33 |
| 50-001 | **FURRING** Wood strips, on walls, 1" x 2", on wood | ↓ | .029 | L.F. | .07 | .35 | .03 | .45 | .67 |
| 030 | On masonry | 1 Carp | .016 | | .07 | .20 | | .27 | .39 |
| 040 | On concrete | | .031 | | .08 | .38 | | .46 | .69 |
| 060 | 1" x 3", wood strips, on walls, on wood | | .015 | | .10 | .18 | | .28 | .39 |
| 070 | On masonry | | .016 | | .11 | .20 | | .31 | .44 |
| 080 | On concrete | | .031 | | .08 | .38 | | .46 | .69 |
| 085 | 1" x 3", wood strips, on ceilings, on wood | | .023 | | .09 | .28 | | .37 | .55 |
| 090 | On masonry | | .025 | | .07 | .31 | | .38 | .57 |
| 095 | On concrete | ↓ | .038 | ↓ | .10 | .47 | | .57 | .86 |
| 54-001 | **GROUNDS** For casework, 1" x 2" wood strips, on wood | 1 Carp | .024 | L.F. | .07 | .30 | | .37 | .55 |
| 010 | On masonry | | .028 | | .07 | .35 | | .42 | .63 |
| 020 | On concrete | | .032 | | .08 | .40 | | .48 | .71 |
| 040 | For plaster, 3/4" deep, on wood | | .018 | | .07 | .22 | | .29 | .42 |
| 050 | On masonry | | .036 | | .07 | .44 | | .51 | .77 |
| 060 | On concrete | | .046 | | .09 | .56 | | .65 | .99 |
| 070 | On metal lath | ↓ | .040 | ↓ | .09 | .49 | | .58 | .88 |
| 55-001 | **INSULATION** See division 7.2 | | | | | | | | |

| | | | Unit | 20 MBF | 5 MBF | 1 MBF | | |
|---|---|---|---|---|---|---|---|---|
| 57-001 | **LUMBER TREATMENT** Creosoted 8 lbs. per C.F., add | | | M.B.F. | 160 | 170 | 330 | | |
| 020 | For every added 2#/C.F., add per increment | | | | 21 | 21 | 42 | | |
| 040 | Fire retardant, wet | | | | 170 | 180 | 350 | | |
| 050 | KDAT | | | | 190 | 200 | 390 | | |
| 070 | Salt treated, water borne, .40 lb. retention | | | | 110 | 125 | 235 | | |
| 080 | Oil borne, 8 lb. retention | | | | 135 | 145 | 280 | | |
| 100 | Kiln dried lumber, 1" & 2" thick, soft woods | | | | 80 | 85 | 165 | | |
| 110 | Hard woods | | | | 85 | 90 | 175 | | |
| 150 | For small size 1" stock, add | | | | 10 | 10 | 20 | | |
| 170 | For full size rough lumber, add | | | ↓ | 20% | 20% | 20% | | |

| | | Crew | Man-Hours | Unit | Mat. | Labor | Equip. | Total | Total Incl O&P |
|---|---|---|---|---|---|---|---|---|---|
| 58-001 002 | **PARTITIONS** Wood stud with single bottom plate and double top plate, no waste, std. & better lumber | | | | | | | | |
| 018 | 2" x 4" studs, 8 ft. high, studs 12" O.C. | F-2 | .200 | L.F. | 3.35 | 2.47 | .18 | 6 | 7.80 |
| 020 | 16" O.C. | | .160 | | 2.55 | 1.98 | .14 | 4.67 | 6.10 |
| 030 | 24" O.C. | | .128 | | 1.92 | 1.58 | .11 | 3.61 | 4.74 |
| 038 | 10 ft. high, studs 12" O.C. | | .200 | | 3.97 | 2.47 | .18 | 6.62 | 8.45 |
| 040 | 16" O.C. | | .160 | | 3.02 | 1.98 | .14 | 5.14 | 6.60 |
| 050 | 24" O.C. | | .128 | | 2.28 | 1.58 | .11 | 3.97 | 5.15 |
| 058 | 12 ft. high, studs 12" O.C. | | .246 | | 4.83 | 3.04 | .22 | 8.09 | 10.35 |
| 060 | 16" O.C. | | .200 | | 3.69 | 2.47 | .18 | 6.34 | 8.15 |
| 070 | 24" O.C. | | .160 | | 2.73 | 1.98 | .14 | 4.85 | 6.30 |
| 078 | 2" x 6" studs, 8 ft. high, studs 12" O.C. | | .229 | | 4.86 | 2.82 | .20 | 7.88 | 10.05 |
| 080 | 16" O.C. | | .178 | | 3.67 | 2.19 | .16 | 6.02 | 7.70 |
| 090 | 24" O.C. | | .139 | | 2.77 | 1.72 | .12 | 4.61 | 5.90 |
| 098 | 10 ft. high, studs 12" O.C. | | .229 | | 5.75 | 2.82 | .20 | 8.77 | 11 |
| 100 | 16" O.C. | ↓ | .178 | ↓ | 4.38 | 2.19 | .16 | 6.73 | 8.45 |

For expanded coverage of these items see *Building Construction Cost Data 1985*

## 6.1 Rough Carpentry

| | | CREW | MAN-HOURS | UNIT | MAT. | LABOR | EQUIP. | TOTAL | TOTAL INCL O&P |
|---|---|---|---|---|---|---|---|---|---|
| 110 | 24" O.C. | F-2 | .139 | L.F. | 3.25 | 1.72 | .12 | 5.09 | 6.45 |
| 118 | 12 ft. high, studs 12" O.C. | | .291 | | 6.60 | 3.60 | .25 | 10.45 | 13.25 |
| 120 | 16" O.C. | | .229 | | 5 | 2.82 | .20 | 8.02 | 10.20 |
| 130 | 24" O.C. | | .178 | | 3.67 | 2.19 | .16 | 6.02 | 7.70 |
| 140 | For horizontal blocking, 2" x 4", add | | .027 | | .23 | .33 | .02 | .58 | .80 |
| 150 | 2" x 6", add | ↓ | .027 | ↓ | .36 | .33 | .02 | .71 | .94 |
| 160 | For openings, add | | | | | | | | |
| 170 | Headers for above openings, material only, add | | | B.F. | .40 | | | .40 | .44M |

| | | UNIT | 20 MSF | 5 MSF | 1 MSF |
|---|---|---|---|---|---|
| 60-001 | PLYWOOD TREATMENT Fire retardant, 1/4" thick | M.S.F. | 160 | 185 | 225 |
| 003 | 3/8" thick | | 160 | 185 | 225 |
| 005 | 1/2" thick | | 160 | 185 | 225 |
| 007 | 5/8" thick | | 180 | 195 | 240 |
| 010 | 3/4" thick | | 190 | 220 | 265 |
| 020 | For KDAT, add | | 40 | 40 | 40 |
| 050 | Salt treated water borne, .25 lb., wet, 1/4" thick | | 80 | 100 | 145 |
| 053 | 3/8" thick | | 85 | 105 | 155 |
| 055 | 1/2" thick | ↓ | 100 | 120 | 160 |
| 057 | 5/8" thick | | | | |
| 060 | 3/4" thick | M.S.F. | 125 | 145 | 200 |
| 080 | For KDAT add | | 47 | 47 | 47 |
| 090 | For .40 lb., per C.F. retention, add | | 32 | 32 | 32 |
| 100 | For certification stamp, add | ↓ | 30 | 30 | 30 |

| | | CREW | MAN-HOURS | UNIT | MAT. | LABOR | EQUIP. | TOTAL | TOTAL INCL O&P |
|---|---|---|---|---|---|---|---|---|---|
| 62-001 | ROOF DECKS | | | | | | | | |
| 020 | For cementitious decks, see division 3.5 | | | | | | | | |
| 040 | Cedar planks, 3.65 B.F. per S.F., 3" thick | F-2 | .050 | S.F. | 3.26 | .62 | .04 | 3.92 | 4.61 |
| 050 | 4.65 B.F. per S.F., 4" thick | | .064 | | 4.15 | .79 | .06 | 5 | 5.90 |
| 070 | Douglas fir, 3" thick | | .050 | | 2.45 | .62 | .04 | 3.11 | 3.72 |
| 080 | 4" thick | | .064 | | 3.12 | .79 | .06 | 3.97 | 4.74 |
| 100 | Hemlock, 3" thick | | .050 | | 2.36 | .62 | .04 | 3.02 | 3.62 |
| 110 | 4" thick | | .064 | | 3.03 | .79 | .06 | 3.88 | 4.65 |
| 130 | Western white spruce, 3" thick | | .050 | | 2.45 | .62 | .04 | 3.11 | 3.72 |
| 140 | 4" thick | ↓ | .064 | ↓ | 3.12 | .79 | .06 | 3.97 | 4.74 |
| 66-001 | ROOF TRUSSES For timber connector trusses, see div. 5.8-35 | | | | | | | | |
| | Flat wood, 2" x 4" metal plate connected, 24" O.C., 3/12 slope | | | | | | | | |
| 500 | 1' overhang, 12' span | F-5 | .582 | Ea. | 25 | 7.50 | .25 | 32.75 | 40 |
| 503 | Gable end | " | .914 | | 31 | 11.75 | .40 | 43.15 | 53 |
| 505 | 1' overhang, 20' span | F-6 | .645 | | 33 | 7.44 | 4.01 | 44.45 | 52 |
| 506 | Gable end | | .690 | | 40 | 7.96 | 4.29 | 52.25 | 61 |
| 510 | 24' span | | .667 | | 41 | 7.66 | 4.14 | 52.80 | 62 |
| 510 | Gable end | | .714 | | 50 | 8.21 | 4.44 | 62.65 | 73 |
| 515 | 26' span | | .702 | | 44 | 8.09 | 4.36 | 56.45 | 66 |
| 515 | Gable end | | .784 | | 56 | 9.03 | 4.87 | 69.90 | 81 |
| 520 | 28' span | | .755 | | 53 | 8.71 | 4.69 | 66.40 | 77 |
| 520 | Gable end | | .816 | | 58 | 9.45 | 5.05 | 72.50 | 84 |
| 524 | 30' span | | .784 | | 59 | 9.03 | 4.87 | 72.90 | 85 |
| 524 | Gable end | | .851 | | 70 | 9.80 | 5.30 | 85.10 | 98 |
| 525 | 32' span | | .800 | | 64 | 9.23 | 4.97 | 78.20 | 90 |
| 527 | Gable end | | .870 | | 78 | 10 | 5.40 | 93.40 | 110 |
| 528 | 34' span | | .833 | | 68 | 9.60 | 5.20 | 82.80 | 96 |
| 529 | Gable end | | .952 | | 82 | 11 | 5.90 | 98.90 | 115 |
| 532 | Gable end | | 1 | | 84 | 11.55 | 6.20 | 101.75 | 115 |
| 535 | 8/12 pitch, 1' overhang, 20' span | | .702 | | 40 | 8.09 | 4.36 | 52.45 | 62 |
| 540 | 24' span | | .727 | | 45 | 8.38 | 4.52 | 57.90 | 68 |
| 545 | 26' span | | .769 | | 50 | 8.87 | 4.78 | 63.65 | 74 |
| 550 | 28' span | | .816 | | 61 | 9.45 | 5.05 | 75.50 | 88 |
| 555 | 32' span | ↓ | .889 | ↓ | 67 | 10.25 | 5.50 | 82.75 | 96 |

For expanded coverage of these items see *Building Construction Cost Data 1985*

## 6.1 Rough Carpentry

| | | | CREW | MAN-HOURS | UNIT | MAT. | LABOR | EQUIP. | TOTAL | TOTAL INCL O&P |
|---|---|---|---|---|---|---|---|---|---|---|
| 66 | 560 | 36' span | F-6 | .976 | Ea. | 78 | 11.25 | 6.05 | 95.30 | 110 |
| | 68-001 | **ROUGH HARDWARE** Average % of carpentry material, minimum | | | | .50% | | | | |
| | 020 | Maximum | | | | 1.50% | | | | |
| | 70-001 | **SHEATHING** Plywood on roof, CDX | | | | | | | | |
| | 003 | 5/16" thick | F-2 | .010 | S.F. | .26 | .12 | .01 | .39 | .49 |
| | 005 | 3/8" thick | | .010 | | .29 | .13 | .01 | .43 | .53 |
| | 010 | 1/2" thick | | .011 | | .37 | .14 | .01 | .52 | .64 |
| | 020 | 5/8" thick | | .012 | | .41 | .15 | .01 | .57 | .70 |
| | 030 | 3/4" thick | | .013 | | .46 | .17 | .01 | .64 | .78 |
| | 050 | Plywood on walls with exterior standard, 3/8" thick | | .013 | | .29 | .17 | .01 | .47 | .59 |
| | 060 | 1/2" thick | ↓ | .014 | ↓ | .37 | .18 | .01 | .56 | .70 |
| | 070 | 5/8" thick | F-2 | .015 | S.F. | .41 | .19 | .01 | .61 | .76 |
| | 080 | 3/4" thick | " | .016 | | .46 | .21 | .01 | .68 | .84 |
| | 100 | For shear wall construction, add | | | | | 20% | | | |
| | 120 | For structural 1 exterior plywood, add | | | | 10% | | | | |
| | 140 | With boards, on roof 1" x 6" boards, laid horizontal | F-2 | .022 | | .70 | .27 | .02 | .99 | 1.22 |
| | 150 | Laid diagonal | | .025 | | .70 | .31 | .02 | 1.03 | 1.27 |
| | 170 | 1" x 8" boards, laid horizontal | | .018 | | .70 | .22 | .02 | .94 | 1.15 |
| | 180 | Laid diagonal | ↓ | .022 | | .70 | .27 | .02 | .99 | 1.22 |
| | 200 | For steep roofs, add | | | | | 40% | | | |
| | 220 | For dormers, hips and valleys, add | | | | 5% | 50% | | | |
| | 240 | Boards on walls, 1" x 6" boards, laid regular | F-2 | .025 | | .70 | .31 | .02 | 1.03 | 1.27 |
| | 250 | Laid diagonal | | .027 | | .70 | .34 | .02 | 1.06 | 1.33 |
| | 270 | 1" x 8" boards, laid regular | | .021 | | .70 | .26 | .02 | .98 | 1.20 |
| | 280 | Laid diagonal | | .025 | | .70 | .31 | .02 | 1.03 | 1.27 |
| | 285 | Gypsum, weatherproof, 1/2" thick | | .015 | | .20 | .19 | .01 | .40 | .53 |
| | 290 | Sealed, 4/10" thick | | .015 | | .18 | .18 | .01 | .37 | .50 |
| | 300 | Wood fiber, regular, no vapor barrier, 1/2" thick | | .013 | | .36 | .17 | .01 | .54 | .67 |
| | 310 | 5/8" thick | | .013 | | .47 | .17 | .01 | .65 | .79 |
| | 330 | No vapor barrier, in colors, 1/2" thick | | .013 | | .47 | .17 | .01 | .65 | .79 |
| | 340 | 5/8" thick | | .013 | | .58 | .17 | .01 | .76 | .91 |
| | 360 | With vapor barrier one side, white, 1/2" thick | | .013 | | .46 | .17 | .01 | .64 | .78 |
| | 370 | Vapor barrier 2 sides | | .013 | | .70 | .17 | .01 | .88 | 1.04 |
| | 380 | Asphalt impregnated, 25/32" thick | | .013 | | .25 | .17 | .01 | .43 | .55 |
| | 385 | Intermediate, 1/2" thick | ↓ | .013 | ↓ | .21 | .17 | .01 | .39 | .50 |
| | 78-001 | **STRUCTURAL JOISTS** Fabricated "I" joists with wood flanges, | | | | | | | | |
| | 010 | Plywood webs, incl. bridging & blocking, panels 24" O.C. | | | | | | | | |
| | 120 | 15' to 24' span, 50 psf live load | F-5 | .013 | S.F. | 1.03 | .17 | .01 | 1.21 | 1.41 |
| | 130 | 55 psf live load | | .014 | | 1.07 | .18 | .01 | 1.26 | 1.47 |
| | 140 | 24' to 30' span, 45 psf live load | | .012 | | 1.18 | .15 | .01 | 1.34 | 1.55 |
| | 150 | 55 psf live load | ↓ | .013 | | 1.28 | .17 | .01 | 1.46 | 1.69 |
| | 160 | Tubular steel open webs, 45 psf, 24" O.C., 40' span | F-3 | .006 | | 1.14 | .08 | .04 | 1.26 | 1.42 |
| | 170 | 55' span | | .008 | | 1.10 | .10 | .05 | 1.25 | 1.42 |
| | 180 | 70' span | | .004 | | 1.60 | .05 | .03 | 1.68 | 1.87 |
| | 190 | 85 psf live load, 26' span | ↓ | .017 | | 1.44 | .21 | .11 | 1.76 | 2.05 |
| | 82-001 | **SUBFLOOR** Plywood, CDX, 1/2" thick | F-2 | .011 | | .37 | .13 | .01 | .51 | .63 |
| | 010 | 5/8" thick | | .012 | | .41 | .15 | .01 | .57 | .69 |
| | 020 | 3/4" thick | | .013 | | .46 | .16 | .01 | .63 | .77 |
| | 030 | 1-1/8" thick, 2-4-1 including underlayment | | .015 | | 1.05 | .19 | .01 | 1.25 | 1.47 |
| | 050 | With boards, 1" x 10" S4S, laid regular | | .015 | | .71 | .18 | .01 | .90 | 1.08 |
| | 060 | Laid diagonal | | .018 | | .71 | .22 | .02 | .95 | 1.15 |
| | 080 | 1" x 8" S4S, laid regular | | .016 | | .70 | .20 | .01 | .91 | 1.10 |
| | 090 | Laid diagonal | | .019 | | .70 | .23 | .02 | .95 | 1.16 |
| | 110 | Wood fiber, T&G, 2' x 8' planks, 1" thick | | .016 | | .95 | .20 | .01 | 1.16 | 1.37 |
| | 120 | 1-3/8" thick | | .018 | | 1.25 | .22 | .02 | 1.49 | 1.74 |
| | 86-001 | **UNDERLAYMENT** Plywood, underlayment grade, 3/8" thick | | .011 | | .34 | .13 | .01 | .48 | .59 |
| | 010 | 1/2" thick | | .011 | | .39 | .14 | .01 | .54 | .66 |
| | 020 | 5/8" thick | | .011 | | .49 | .14 | .01 | .64 | .77 |
| | 030 | 3/4" thick | ↓ | .012 | ↓ | .68 | .15 | .01 | .84 | 1 |

For expanded coverage of these items see *Building Construction Cost Data 1985*

## 6.1 Rough Carpentry

| | | CREW | MAN-HOURS | UNIT | MAT. | LABOR | EQUIP. | TOTAL | TOTAL INCL O&P |
|---|---|---|---|---|---|---|---|---|---|
| 050 | Particle board, 3/8" thick | F-2 | .011 | S.F. | .21 | .13 | .01 | .35 | .45 |
| 060 | 1/2" thick | | .011 | | .22 | .14 | .01 | .37 | .47 |
| 080 | 5/8" thick | | .011 | | .25 | .14 | .01 | .40 | .51 |
| 090 | 3/4" thick | | .012 | | .31 | .15 | .01 | .47 | .59 |
| 110 | Hardboard, underlayment grade, 4' x 4', .215" thick | ↓ | .011 | ↓ | .23 | .13 | .01 | .37 | .47 |

## 6.2 Finish Carpentry

| | | CREW | MAN-HOURS | UNIT | MAT. | LABOR | EQUIP. | TOTAL | TOTAL INCL O&P |
|---|---|---|---|---|---|---|---|---|---|
| 03-001 | BEAMS, DECORATIVE Rough sawn cedar, non-load bearing, 4" x 4" | 2 Carp | .089 | L.F. | 1.32 | 1.10 | | 2.42 | 3.19 |
| 010 | 4" x 6" | | .094 | | 2.10 | 1.16 | | 3.26 | 4.15 |
| 030 | 4" x 10" | | .107 | | 3.62 | 1.32 | | 4.94 | 6.05 |
| 040 | 4" x 12" | | .114 | | 4.34 | 1.41 | | 5.75 | 7 |
| 050 | 8" x 8" | | .123 | ↓ | 5.80 | 1.52 | | 7.32 | 8.80 |
| 090 | Connector plates, steel, with bolts, straight | | .213 | Ea. | 12 | 2.63 | | 14.63 | 17.35 |
| 100 | Tee | ↓ | .320 | " | 18.50 | 3.95 | | 22.45 | 27 |
| 06-001 | CABINETS Corner china cabinets, stock pine, | | | | | | | | |
| 002 | 80" high, unfinished, minimum | 2 Carp | 2.420 | Ea. | 225 | 30 | | 255 | 295 |
| 010 | Maximum | " | 3.640 | " | 365 | 45 | | 410 | 475 |
| 070 | Kitchen base cabinets, hardwood, not incl. counter tops, | | | | | | | | |
| 071 | 24" deep, 35" high, prefinished | | | | | | | | |
| 080 | One top drawer, one door below, 12" wide | 2 Carp | .645 | Ea. | 88 | 7.95 | | 95.95 | 110 |
| 082 | 15" wide | | .667 | | 93 | 8.25 | | 101.25 | 115 |
| 084 | 18" wide | | .687 | | 100 | 8.50 | | 108.50 | 125 |
| 086 | 21" wide | | .705 | | 105 | 8.70 | | 113.70 | 130 |
| 088 | 24" wide | ↓ | .717 | ↓ | 110 | 8.85 | | 118.85 | 135 |
| 089 | | | | | | | | | |
| 100 | Four drawers, 12" wide | 2 Carp | .645 | Ea. | 135 | 7.95 | | 142.95 | 160 |
| 102 | 15" wide | | .667 | | 135 | 8.25 | | 143.25 | 160 |
| 104 | 18" wide | | .687 | | 140 | 8.50 | | 148.50 | 165 |
| 106 | 24" wide | | .717 | | 160 | 8.85 | | 168.85 | 190 |
| 120 | Two top drawers, two doors below, 27" wide | | .727 | | 135 | 9 | | 144 | 165 |
| 122 | 30" wide | | .748 | | 145 | 9.25 | | 154.25 | 175 |
| 124 | 33" wide | | .766 | | 155 | 9.45 | | 164.45 | 185 |
| 126 | 36" wide | | .788 | | 160 | 9.75 | | 169.75 | 190 |
| 128 | 42" wide | | .808 | | 170 | 10 | | 180 | 205 |
| 130 | 48" wide | | .847 | | 175 | 10.45 | | 185.45 | 210 |
| 150 | Range or sink base, two doors below, 30" wide | | .748 | | 120 | 9.25 | | 129.25 | 145 |
| 152 | 33" wide | | .766 | | 125 | 9.45 | | 134.45 | 150 |
| 154 | 36" wide | | .788 | | 135 | 9.75 | | 144.75 | 165 |
| 156 | 42" wide | | .808 | | 145 | 10 | | 155 | 175 |
| 158 | 48" wide | ↓ | .847 | | 150 | 10.45 | | 160.45 | 180 |
| 180 | For sink front units, deduct | | | | 37 | | | 37 | 41M |
| 200 | Corner base cabinets, 36" wide, standard | 2 Carp | .889 | | 140 | 11 | | 151 | 170 |
| 210 | Lazy Susan with revolving door | " | .970 | ↓ | 175 | 12 | | 187 | 210 |
| 400 | Kitchen wall cabinets, hardwood, 12" deep with two doors | | | | | | | | |
| 405 | 12" high, 30" wide | 2 Carp | .645 | Ea. | 68 | 7.95 | | 75.95 | 87 |
| 410 | 36" wide | | .667 | | 71 | 8.25 | | 79.25 | 91 |
| 440 | 15" high, 30" wide | | .667 | | 70 | 8.25 | | 78.25 | 90 |
| 442 | 33" wide | | .687 | | 73 | 8.50 | | 81.50 | 94 |
| 444 | 36" wide | | .705 | | 75 | 8.70 | | 83.70 | 96 |
| 444 | 42" wide | | .705 | | 78 | 8.70 | | 86.70 | 100 |
| 470 | 24" high, 30" wide | | .687 | | 88 | 8.50 | | 96.50 | 110 |
| 472 | 36" wide | ↓ | .705 | | 98 | 8.70 | | 106.70 | 120 |

For expanded coverage of these items see *Building Construction Cost Data 1985*

## 6.2 Finish Carpentry

| | | CREW | MAN-HOURS | UNIT | MAT. | LABOR | EQUIP. | TOTAL | TOTAL INCL O&P |
|---|---|---|---|---|---|---|---|---|---|
| 06 474 | 42" wide | 2 Carp | .717 | Ea. | 105 | 8.85 | | 113.85 | 130 |
| 500 | 30" high, one door, 12" wide | | .727 | | 61 | 9 | | 70 | 81 |
| 502 | 15" wide | | .748 | | 65 | 9.25 | | 74.25 | 86 |
| 504 | 18" wide | | .766 | | 72 | 9.45 | | 81.45 | 94 |
| 506 | 24" wide | | .788 | | 78 | 9.75 | | 87.75 | 100 |
| 530 | Two doors, 27" wide | | .808 | | 94 | 10 | | 104 | 120 |
| 532 | 30" wide | | .829 | | 100 | 10.25 | | 110.25 | 125 |
| 534 | 36" wide | | .851 | | 110 | 10.50 | | 120.50 | 140 |
| 536 | 42" wide | | .865 | | 120 | 10.70 | | 130.70 | 150 |
| 538 | 48" wide | | .870 | | 135 | 10.75 | | 145.75 | 165 |
| 600 | Corner wall, 30" high, 24" wide | | .889 | | 78 | 11 | | 89 | 105 |
| 605 | 30" wide | | .930 | | 87 | 11.50 | | 98.50 | 115 |
| 610 | 36" wide | | .970 | | 95 | 12 | | 107 | 125 |
| 650 | Revolving Lazy Susan | | 1.050 | | 115 | 13 | | 128 | 145 |
| 700 | Broom cabinet, 84" high, 24" deep, 18" wide | | 1.600 | | 200 | 19.75 | | 219.75 | 250 |
| 750 | Oven cabinets, 84" high, 24" deep, 27" wide | | 2 | | 210 | 25 | | 235 | 270 |
| 775 | Valance board trim | ▼ | .040 | ▼ | 5.80 | .50 | | 6.30 | 7.15 |
| 776 | | | | | | | | | |
| 900 | For deluxe models of all cabinets, add to above | | | Ea. | 40% | | | | |
| 950 | For custom built in place, add to above | | | " | 25% | 10% | | | |
| 955 | Rule of thumb, kitchen cabinets not including | | | | | | | | |
| 956 | appliances & counter top, minimum | 2 Carp | .533 | L.F. | 90 | 6.60 | | 96.60 | 110 |
| 960 | Maximum | " | .640 | " | 125 | 7.90 | | 132.90 | 150 |
| 961 | For metal cabinets, see division 12.1-150 | | | | | | | | |
| 09-001 | **COLUMNS** For base plates, see division 5.4-10 | | | | | | | | |
| 005 | Aluminum, round colonial, 6" diameter | 2 Carp | .200 | V.L.F. | 6.75 | 2.47 | | 9.22 | 11.35 |
| 010 | 8" diameter | | .257 | | 8.75 | 3.17 | | 11.92 | 14.65 |
| 020 | 10" diameter | | .182 | | 12.90 | 2.25 | | 15.15 | 17.75 |
| 025 | Fir, stock units, hollow round, 6" diameter | | .200 | | 8.90 | 2.47 | | 11.37 | 13.70 |
| 030 | 8" diameter | | .200 | | 11.20 | 2.47 | | 13.67 | 16.25 |
| 035 | 10" diameter | | .229 | | 13.90 | 2.82 | | 16.72 | 19.75 |
| 040 | Solid turned, to 8' high, 3-1/2" diameter | | .200 | | 3.90 | 2.47 | | 6.37 | 8.20 |
| 050 | 4-1/2" diameter | | .213 | | 6.15 | 2.63 | | 8.78 | 10.95 |
| 060 | 5-1/2" diameter | | .229 | | 8.10 | 2.82 | | 10.92 | 13.40 |
| 080 | Square columns, built-up, 5" x 5" | | .246 | | 5.40 | 3.04 | | 8.44 | 10.75 |
| 090 | Solid, 3-1/2" x 3-1/2" | | .123 | | 4.10 | 1.52 | | 5.62 | 6.90 |
| 160 | Pine, tapered, T & G, 10' to 14' high, 12" diam., minimum | | .160 | | 32 | 1.98 | | 33.98 | 38 |
| 170 | Maximum | | .246 | | 69 | 3.04 | | 72.04 | 81 |
| 190 | 12' to 16' high, 14" diameter, minimum | | .160 | | 50 | 1.98 | | 51.98 | 58 |
| 200 | Maximum | | .246 | | 79 | 3.04 | | 82.04 | 92 |
| 220 | 12' to 20' high, 18" diameter, minimum | | .246 | | 68 | 3.04 | | 71.04 | 80 |
| 230 | Maximum | | .320 | | 100 | 3.95 | | 103.95 | 115 |
| 250 | 20' high, 20" diameter, minimum | | .400 | | 80 | 4.94 | | 84.94 | 96 |
| 260 | Maximum | ▼ | .457 | ▼ | 115 | 5.65 | | 120.65 | 135 |
| 280 | For flat pilasters, deduct | | | ▼ | 33% | | | | |
| 300 | For splitting into halves, add | | | Ea. | 45 | | | 45 | 50M |
| 400 | Rough sawn cedar posts, 4" x 4" | 2 Carp | .064 | V.L.F. | 1.08 | .79 | | 1.87 | 2.44 |
| 410 | 4" x 6" | | .068 | | 2.10 | .84 | | 2.94 | 3.64 |
| 420 | 6" x 6" | | .073 | | 3.26 | .90 | | 4.16 | 5 |
| 430 | 8" x 8" | ▼ | .080 | ▼ | 4.80 | .99 | | 5.79 | 6.85 |
| 12-001 | **COUNTER TOP** Stock, plastic lam., 25" wide with backsplash, minimum | 1 Carp | .267 | L.F. | 5.45 | 3.29 | | 8.74 | 11.20 |
| 010 | Maximum (Also see 6.4-800) | | .320 | | 12.45 | 3.95 | | 16.40 | 19.95 |
| 030 | Custom plastic, 7/8" thick, aluminum molding, no splash | | .267 | | 12.75 | 3.29 | | 16.04 | 19.25 |
| 040 | Cove splash | | .267 | | 16.90 | 3.29 | | 20.19 | 24 |
| 060 | 1-1/4" thick, no splash | | .286 | | 14.80 | 3.53 | | 18.33 | 22 |
| 070 | Square splash | | .286 | | 18.95 | 3.53 | | 22.48 | 26 |
| 090 | Square edge, plastic face, 7/8" thick, no splash | | .267 | | 16.35 | 3.29 | | 19.64 | 23 |
| 100 | With splash | ▼ | .267 | | 21 | 3.29 | | 24.29 | 28 |
| 120 | For stainless channel edge, 7/8" thick, add | | | | 1.50 | | | 1.50 | 1.65M |
| 130 | 1-1/4" thick, add | | | ▼ | 1.90 | | | 1.90 | 2.09M |

For expanded coverage of these items see *Building Construction Cost Data 1985*

# 6.2 Finish Carpentry

| | | Description | CREW | MAN-HOURS | UNIT | MAT. | LABOR | EQUIP. | TOTAL | TOTAL INCL O&P |
|---|---|---|---|---|---|---|---|---|---|---|
| 12 | 150 | For solid color suede finish, add | | | L.F. | 1.32 | | | 1.32 | 1.45M |
| | 170 | For end splash, add | | | Ea. | 10.45 | | | 10.45 | 11.50M |
| | 190 | For cut outs, standard, add, minimum | 1 Carp | .250 | | 1 | 3.09 | | 4.09 | 6 |
| | 200 | Maximum | | 1 | ↓ | 1 | 12.35 | | 13.35 | 21 |
| | 210 | Postformed, including backsplash and front edge | | .267 | L.F. | 8 | 3.29 | | 11.29 | 14 |
| | 211 | Mitred, add | | .667 | Ea. | | 8.25 | | 8.25 | 13.05L |
| | 220 | Built-in place, 25" wide, plastic laminate | | .320 | L.F. | 7.50 | 3.95 | | 11.45 | 14.50 |
| | 230 | Ceramic tile mosaic | ↓ | .320 | | 18.40 | 3.95 | | 22.35 | 27 |
| | 250 | Marble, stock, with splash, 1/2" thick, minimum | 1 Bric | .471 | | 21 | 5.95 | | 26.95 | 32 |
| | 270 | 3/4" thick, maximum | " | .615 | | 66 | 7.75 | | 73.75 | 85 |
| | 290 | Maple, solid, laminated, 1-1/2" thick, no splash | 1 Carp | .286 | ↓ | 26 | 3.53 | | 29.53 | 34 |
| | 300 | With square splash | | .286 | | 30 | 3.53 | | 33.53 | 39 |
| | 320 | Stainless steel | | .333 | S.F. | 40 | 4.12 | | 44.12 | 51 |
| | 340 | Cutting block for all counters, 16" x 20" x 1" | | 1 | Ea. | 32 | 12.35 | | 44.35 | 55 |
| | 340 | Installed in counter top, self-edged | | 2 | " | 22 | 25 | | 47 | 63 |
| | 360 | Table tops, plastic laminate, square edge, 7/8" thick | | .178 | S.F. | 6 | 2.20 | | 8.20 | 10.10 |
| | 370 | 1-1/8" thick | ↓ | .200 | " | 6.75 | 2.47 | | 9.22 | 11.35 |
| 15-001 | | **CUPOLA** Stock units, 1" redwood, 2' x 2' x 2', aluminum roof | 1 Carp | 2 | Ea. | 110 | 25 | | 135 | 160 |
| | 010 | Copper roof | | 2 | | 210 | 25 | | 235 | 270 |
| | 030 | 3' x 3' base, 3' high, aluminum roof | | 2.290 | | 210 | 28 | | 238 | 275 |
| | 040 | Copper roof | | 2.290 | | 275 | 28 | | 303 | 345 |
| | 060 | 31" x 31" base, 51" high, aluminum roof | | 2.290 | | 300 | 28 | | 328 | 375 |
| | 070 | Copper roof | | 2.290 | | 420 | 28 | | 448 | 505 |
| | 090 | Hexagonal, 31" wide, 37" high, copper roof | | 2.670 | | 345 | 33 | | 378 | 430 |
| | 100 | 35" wide, 43" high, copper roof | | 2.670 | | 465 | 33 | | 498 | 565 |
| | 120 | For deluxe stock units, add to above | | | | 30 | | | | |
| | 140 | For custom built units, add to above | | | | 50 | 50 | | | |
| | 160 | Fiberglass, 5'-0" square base, 9'-3" high | F-3 | 6.670 | | 2,400 | 84 | 41 | 2,525 | 2,825 |
| | 170 | 6'-0" square base, 8'-0" high | " | 8 | ↓ | 2,600 | 100 | 50 | 2,750 | 3,075 |
| 17-001 | | **DOORS AND FRAMES** See division 8.1 & 8.2 | | | | | | | | |
| 20-001 | | **FIREPLACE MANTEL BEAMS** Rough texture wood, 4" x 8" | 1 Carp | .222 | L.F. | 3.25 | 2.74 | | 5.99 | 7.90 |
| | 010 | 4" x 10" | | .229 | " | 4.05 | 2.82 | | 6.87 | 8.95 |
| | 030 | Laminated hardwood, 2-1/4" x 10-1/2" wide, 6' long | | 1.600 | Ea. | 85 | 19.75 | | 104.75 | 125 |
| | 040 | 8' long | | 1.600 | " | 110 | 19.75 | | 129.75 | 150 |
| | 060 | Brackets for above, rough sawn | | .667 | Pr. | 7.55 | 8.25 | | 15.80 | 21 |
| | 070 | Laminated | | .667 | " | 11.20 | 8.25 | | 19.45 | 25 |
| 22-001 | | **FIREPLACE MANTELS** 6" molding, 6' x 3'-6" opening, minimum | | 1.600 | Opng. | 70 | 19.75 | | 89.75 | 110 |
| | 010 | Maximum | | 1.600 | | 105 | 19.75 | | 124.75 | 145 |
| | 030 | Prefabricated pine, colonial type, stock, deluxe | | 4 | | 485 | 49 | | 534 | 610 |
| | 040 | Economy | ↓ | 2.670 | ↓ | 180 | 33 | | 213 | 250 |
| 25-001 | | **FLOORING, WOOD** See division 9.6-40 | | | | | | | | |
| 27-001 | | **GRILLES** And panels, hardwood, sanded | | | | | | | | |
| | 002 | 2' x 4' to 4' x 8', custom designs, unfinished, minimum | 1 Carp | .211 | S.F. | 9.40 | 2.60 | | 12 | 14.45 |
| | 005 | Average | | .267 | | 14.60 | 3.29 | | 17.89 | 21 |
| | 010 | Maximum | | .421 | | 22 | 5.20 | | 27.20 | 32 |
| | 030 | As above, but prefinished, minimum | | .211 | | 12.25 | 2.60 | | 14.85 | 17.60 |
| | 040 | Maximum | ↓ | .421 | ↓ | 29 | 5.20 | | 34.20 | 40 |
| 30-001 | | **HARDWARE** Finish, see division 6.4 & 8.7 | | | | | | | | |
| | 010 | Rough, see division 5.8 | | | | | | | | |
| 32-001 | | **LOUVERS** Redwood, 2'-0" opening, full circle | 1 Carp | .500 | Ea. | 70 | 6.20 | | 76.20 | 87 |
| | 010 | Half circle, 2'-0" diameter | | .500 | | 60 | 6.20 | | 66.20 | 76 |
| | 020 | Octagonal | | .500 | | 55 | 6.20 | | 61.20 | 70 |
| | 030 | Triangular, 5/12 pitch, 5'-0" at base | ↓ | .500 | ↓ | 73 | 6.20 | | 79.20 | 90 |
| 36-001 | | **MILLWORK,** Rule of thumb: Milled material cost | | | | | | | | |
| | 002 | equals three times cost of lumber | | | | | | | | |
| | 100 | Typical finish hardwood milled material | | | | | | | | |
| | 102 | 1" x 12", custom birch | | | L.F. | 1.85 | | | 1.85 | 2.03M |

For expanded coverage of these items see *Building Construction Cost Data 1985*

## 6.2 Finish Carpentry

| | | | CREW | MAN-HOURS | UNIT | MAT. | LABOR | EQUIP. | TOTAL | TOTAL INCL O&P |
|---|---|---|---|---|---|---|---|---|---|---|
| 36 | 104 | Cedar | | | L.F. | 2.45 | | | 2.45 | 2.69M |
| | 106 | Oak | | | | 1.98 | | | 1.98 | 2.17M |
| | 108 | Redwood | | | | 1.89 | | | 1.89 | 2.07M |
| | 110 | Southern yellow pine | | | | 1.35 | | | 1.35 | 1.48M |
| | 112 | Sugar pine | | | | 1.82 | | | 1.82 | 2M |
| | 114 | Teak | | | | 6.50 | | | 6.50 | 7.15M |
| | 116 | Walnut | | | | 3.60 | | | 3.60 | 3.96M |
| | 118 | White pine | | | ↓ | 1.55 | | | 1.55 | 1.70M |
| 37-001 | | **MOLDINGS, BASE** | | | | | | | | |
| | 050 | Base, stock pine, 9/16" x 3-1/2" | 1 Carp | .033 | L.F. | .58 | .41 | | .99 | 1.29 |
| | 055 | 9/16" x 4-1/2" | | .040 | | .77 | .49 | | 1.26 | 1.63 |
| | 056 | Base shoe, oak, 3/4" x 1" | ↓ | .033 | ↓ | .86 | .41 | | 1.27 | 1.60 |
| 41-001 | | **MOLDINGS, CASINGS** | | | | | | | | |
| | 002 | | | | | | | | | |
| | 009 | Apron, stock pine, 5/8" x 2" | 1 Carp | .032 | L.F. | .38 | .40 | | .78 | 1.04 |
| | 011 | 5/8" x 3-1/2" | | .036 | | .98 | .45 | | 1.43 | 1.79 |
| | 030 | Band, stock pine, 11/16" x 1-1/8" | | .030 | | .23 | .37 | | .60 | .83 |
| | 035 | 11/16" x 1-3/4" | | .032 | | .35 | .40 | | .75 | 1.01 |
| | 070 | Casing, stock pine, 11/16" x 2-1/2" | | .033 | | .44 | .41 | | .85 | 1.14 |
| | 075 | 11/16" x 3-1/2" | ↓ | .037 | ↓ | .77 | .46 | | 1.23 | 1.57 |
| 43-001 | | **MOLDINGS, CEILINGS** | | | | | | | | |
| | 002 | | | | | | | | | |
| | 060 | Bed, stock pine, 9/16" x 1-3/4" | 1 Carp | .030 | L.F. | .31 | .37 | | .68 | .92 |
| | 065 | 9/16" x 2" | | .033 | | .42 | .41 | | .83 | 1.11 |
| | 120 | Cornice molding, stock pine, 9/16" x 1-3/4" | | .024 | | .32 | .30 | | .62 | .83 |
| | 130 | 9/16" x 2-1/4" | | .027 | | .41 | .33 | | .74 | .97 |
| | 240 | Cove scotia, stock pine, 9/16" x 1-3/4" | | .030 | | .32 | .37 | | .69 | .93 |
| | 250 | 11/16" x 2-3/4" | | .031 | | .74 | .39 | | 1.13 | 1.43 |
| | 260 | Crown, stock pine, 9/16" x 3-5/8" | | .032 | | .65 | .40 | | 1.05 | 1.34 |
| | 270 | 11/16" x 4-5/8" | ↓ | .036 | ↓ | 1.36 | .45 | | 1.81 | 2.21 |
| 46-001 | | **MOLDINGS, EXTERIOR** | | | | | | | | |
| | 150 | Cornice, boards, pine, 1" x 2" | 1 Carp | .024 | L.F. | .16 | .30 | | .46 | .65 |
| | 160 | 1" x 4" | | .032 | | .32 | .40 | | .72 | .98 |
| | 170 | 1" x 6" | | .040 | | .54 | .49 | | 1.03 | 1.38 |
| | 180 | 1" x 8" | | .040 | | .74 | .49 | | 1.23 | 1.60 |
| | 190 | 1" x 10" | | .044 | | .89 | .55 | | 1.44 | 1.85 |
| | 200 | 1" x 12" | | .044 | | 1.65 | .55 | | 2.20 | 2.68 |
| | 220 | Three piece, built-up, pine, minimum | | .100 | | .65 | 1.24 | | 1.89 | 2.67 |
| | 230 | Maximum | | .123 | | 3.10 | 1.52 | | 4.62 | 5.80 |
| | 300 | Trim, exterior, sterling pine, corner board, 1" x 4" | | .040 | | .30 | .49 | | .79 | 1.11 |
| | 310 | 1" x 6" (see also 6.2-370) | | .040 | | .46 | .49 | | .95 | 1.29 |
| | 320 | 2" x 6" | | .048 | | .87 | .60 | | 1.47 | 1.91 |
| | 330 | 2" x 8" | | .048 | | 1.16 | .60 | | 1.76 | 2.22 |
| | 335 | Fascia, 1" x 6" | | .032 | | .55 | .40 | | .95 | 1.23 |
| | 337 | 1" x 8" | | .036 | | .61 | .44 | | 1.05 | 1.37 |
| | 340 | Moldings, back band | | .032 | | .33 | .40 | | .73 | .99 |
| | 350 | Casing | | .032 | | .58 | .40 | | .98 | 1.26 |
| | 360 | Crown | | .032 | | .83 | .40 | | 1.23 | 1.54 |
| | 370 | Porch rail with balusters | | .364 | | 4.50 | 4.49 | | 8.99 | 12.05 |
| | 380 | Screen | | .020 | | .12 | .25 | | .37 | .53 |
| | 410 | Verge board, sterling pine, 1" x 4" | | .040 | | .31 | .49 | | .80 | 1.12 |
| | 420 | 1" x 6" | | .040 | | .46 | .49 | | .95 | 1.29 |
| | 430 | 2" x 6" | | .048 | | .87 | .60 | | 1.47 | 1.91 |
| | 440 | 2" x 8" | ↓ | .048 | | 1.16 | .60 | | 1.76 | 2.22 |
| | 470 | For redwood trim, add | | | ↓ | 150% | | | | |
| 49-001 | | **MOLDINGS, TRIM** | | | | | | | | |
| | 020 | Astragal, stock pine, 11/16" x 1-3/4" | 1 Carp | .031 | L.F. | .43 | .39 | | .82 | 1.09 |
| | 025 | 1-5/16" x 2-3/16" | | .033 | | .81 | .41 | | 1.22 | 1.54 |
| | 080 | Chair rail, stock pine, 5/8" x 2-1/2" | ↓ | .030 | ↓ | .58 | .37 | | .95 | 1.22 |

For expanded coverage of these items see *Building Construction Cost Data 1985*

# 6.2 Finish Carpentry

| | | CREW | MAN-HOURS | UNIT | MAT. | LABOR | EQUIP. | TOTAL | TOTAL INCL O&P |
|---|---|---|---|---|---|---|---|---|---|
| 50 090 | 5/8" x 3-1/2" | 1 Carp | .033 | L.F. | .95 | .41 | | 1.36 | 1.70 |
| 100 | Closet pole, stock pine, 1-1/8" diameter | | .040 | | .57 | .49 | | 1.06 | 1.41 |
| 110 | Fir, 1-5/8" diameter | | .040 | | .62 | .49 | | 1.11 | 1.46 |
| 330 | Half round, stock pine, 1/4" x 1/2" | | .030 | | .07 | .37 | | .44 | .66 |
| 335 | 1/2" x 1" | ▼ | .031 | ▼ | .20 | .39 | | .59 | .83 |
| 340 | Handrail, fir, single piece, stock, hardware not included | | | | | | | | |
| 345 | 1-1/2" x 1-3/4" | 1 Carp | .100 | L.F. | .62 | 1.24 | | 1.86 | 2.64 |
| 347 | Pine, 1-1/2" x 1-3/4" | | .100 | | .63 | 1.24 | | 1.87 | 2.65 |
| 350 | 1-1/2" x 2-1/2" | | .105 | | .94 | 1.30 | | 2.24 | 3.09 |
| 360 | Lattice, stock pine, 1/4" x 1-1/8" | | .030 | | .12 | .37 | | .49 | .71 |
| 370 | 1/4" x 1-3/4" | | .032 | | .17 | .40 | | .57 | .81 |
| 380 | Miscellaneous, custom, pine or cedar, 1" x 1" | | .030 | | .11 | .37 | | .48 | .70 |
| 390 | Nominal 1" x 3" | | .033 | | .31 | .41 | | .72 | .99 |
| 410 | Birch or oak, custom, nominal 1" x 1" | | .033 | | .15 | .41 | | .56 | .82 |
| 420 | Nominal 1" x 3" | | .037 | | .45 | .46 | | .91 | 1.22 |
| 440 | Walnut, custom, nominal 1" x 1" | | .037 | | .26 | .46 | | .72 | 1.01 |
| 450 | Nominal 1" x 3" | | .040 | | .74 | .49 | | 1.23 | 1.60 |
| 470 | Teak, custom, nominal 1" x 1" | | .037 | | .60 | .46 | | 1.06 | 1.39 |
| 480 | Nominal 1" x 3" | | .040 | | 1.70 | .49 | | 2.19 | 2.65 |
| 490 | Quarter round, stock pine, 1/4" x 1/4" | | .029 | | .07 | .36 | | .43 | .65 |
| 495 | 3/4" x 3/4" | | .031 | | .17 | .39 | | .56 | .80 |
| 560 | Wainscot moldings, 1-1/8" x 9/16", 2 ft. high, minimum | | .105 | | 5.30 | 1.30 | | 6.60 | 7.90 |
| 570 | Maximum | ▼ | .123 | ▼ | 10.50 | 1.52 | | 12.02 | 13.95 |
| 51-001 002 | **MOLDINGS, WINDOW AND DOOR** | | | | | | | | |
| 280 | Door moldings, stock, decorative, 1-1/8" wide, plain | 1 Carp | .471 | Set | 17.40 | 5.80 | | 23.20 | 28 |
| 290 | Detailed | " | .471 | " | 30 | 5.80 | | 35.80 | 42 |
| 310 | Door trim, interior, including headers, | | | | | | | | |
| 315 | stops and casings, 2 sides, pine, 2-1/2" wide | 1 Carp | 1.360 | Opng. | 18 | 16.75 | | 34.75 | 46 |
| 317 | 4-1/2" wide | | 1.510 | " | 23 | 18.65 | | 41.65 | 55 |
| 320 | Glass beads, stock pine, 1/4" x 11/16" | | .028 | L.F. | .14 | .35 | | .49 | .70 |
| 325 | 3/8" x 1/2" | | .029 | | .17 | .36 | | .53 | .76 |
| 327 | 3/8" x 7/8" | | .030 | | .20 | .37 | | .57 | .80 |
| 485 | Parting bead, stock pine, 3/8" x 3/4" | | .029 | | .13 | .36 | | .49 | .71 |
| 487 | 1/2" x 3/4" | | .031 | | .19 | .39 | | .58 | .82 |
| 500 | Stool caps, stock pine, 11/16" x 3-1/2" | | .040 | | .70 | .49 | | 1.19 | 1.55 |
| 510 | 1-1/16" x 3-1/4" | | .053 | ▼ | 1.79 | .66 | | 2.45 | 3.01 |
| 530 | Threshold, oak, 3' long, inside, 5/8" x 3-5/8" | | .250 | Ea. | 3.94 | 3.09 | | 7.03 | 9.20 |
| 540 | Outside, 1-1/2" x 7-5/8" | ▼ | .500 | " | 14.70 | 6.20 | | 20.90 | 26 |
| 590 | Window trim sets, including casings, header, stops, | | | | | | | | |
| 591 | stool and apron, 2-1/2" wide, minimum | 1 Carp | .615 | Opng. | 9 | 7.60 | | 16.60 | 22 |
| 595 | Average | | .800 | | 11.50 | 9.90 | | 21.40 | 28 |
| 600 | Maximum | ▼ | 1.330 | ▼ | 14 | 16.45 | | 30.45 | 41 |
| 53-001 | **PANELING, BOARDS** | | | | | | | | |
| 640 | Wood board paneling, 3/4" thick, knotty pine | F-2 | .053 | S.F. | .73 | .66 | .05 | 1.44 | 1.90 |
| 650 | Rough sawn cedar | | .053 | | 1.42 | .66 | .05 | 2.13 | 2.66 |
| 670 | Redwood, clear, 1" x 4" boards | | .053 | | 2.60 | .66 | .05 | 3.31 | 3.95 |
| 690 | Aromatic cedar, closet lining, boards | ▼ | .058 | ▼ | 1.52 | .72 | .05 | 2.29 | 2.87 |
| 56-001 | **PANELING, HARDBOARD** | | | | | | | | |
| 002 | Not incl. furring or trim, hardboard, tempered, 1/8" thick | F-2 | .032 | S.F. | .22 | .39 | .03 | .64 | .90 |
| 010 | 1/4" thick | | .032 | | .34 | .39 | .03 | .76 | 1.03 |
| 030 | Tempered pegboard, 1/8" thick | | .032 | | .22 | .39 | .03 | .64 | .90 |
| 040 | 1/4" thick | | .032 | | .35 | .39 | .03 | .77 | 1.04 |
| 060 | Untempered hardboard, natural finish, 1/8" thick | | .032 | | .19 | .39 | .03 | .61 | .87 |
| 070 | 1/4" thick | | .032 | | .25 | .39 | .03 | .67 | .93 |
| 090 | Untempered pegboard, 1/8" thick | | .032 | | .20 | .39 | .03 | .62 | .88 |
| 100 | 1/4" thick | | .032 | | .33 | .39 | .03 | .75 | 1.02 |
| 120 | Plastic faced hardboard, 1/8" thick | ▼ | .032 | ▼ | .38 | .39 | .03 | .80 | 1.07 |

For expanded coverage of these items see *Building Construction Cost Data 1985*

## 6.2 Finish Carpentry

| | | CREW | MAN-HOURS | UNIT | MAT. | LABOR | EQUIP. | TOTAL | TOTAL INCL O&P |
|---|---|---|---|---|---|---|---|---|---|
| 56 130 | 1/4" thick | F-2 | .032 | S.F. | .51 | .39 | .03 | .93 | 1.22 |
| 150 | Plastic faced pegboard, 1/8" thick | | .032 | | .36 | .39 | .03 | .78 | 1.05 |
| 160 | 1/4" thick | | .032 | | .45 | .39 | .03 | .87 | 1.15 |
| 180 | Wood grained, plain or grooved, 1/4" thick, minimum | | .032 | | .32 | .39 | .03 | .74 | 1.01 |
| 190 | Maximum | | .038 | ↓ | .60 | .47 | .03 | 1.10 | 1.43 |
| 210 | Moldings for hardboard, wood or aluminum, minimum | | .032 | L.F. | .22 | .39 | .03 | .64 | .90 |
| 220 | Maximum | ↓ | .038 | " | .60 | .47 | .03 | 1.10 | 1.43 |
| 58-001 | **PANELING, PLYWOOD** | | | | | | | | |
| 240 | Plywood, prefinished, 1/4" thick, 4' x 8' sheets | | | | | | | | |
| 241 | with vertical grooves. Birch faced, minimum | F-2 | .032 | S.F. | .50 | .39 | .03 | .92 | 1.21 |
| 242 | Average | | .038 | | .78 | .47 | .03 | 1.28 | 1.64 |
| 243 | Maximum | | .046 | | 1.10 | .56 | .04 | 1.70 | 2.15 |
| 260 | Mahogany, African | | .040 | | 1.25 | .49 | .04 | 1.78 | 2.20 |
| 270 | Philippine (Lauan) | | .032 | | .33 | .39 | .03 | .75 | 1.02 |
| 290 | Oak or Cherry, minimum | | .032 | | 1.15 | .39 | .03 | 1.57 | 1.92 |
| 300 | Maximum | | .040 | | 2.15 | .49 | .04 | 2.68 | 3.19 |
| 320 | Rosewood | | .050 | | 8 | .62 | .04 | 8.66 | 9.85 |
| 340 | Teak | | .040 | | 2.10 | .49 | .04 | 2.63 | 3.13 |
| 360 | Chestnut | | .043 | | 3 | .52 | .04 | 3.56 | 4.18 |
| 380 | Pecan | | .040 | | 1.20 | .49 | .04 | 1.73 | 2.14 |
| 390 | Walnut, minimum | | .032 | | 1.80 | .39 | .03 | 2.22 | 2.64 |
| 395 | Maximum | | .040 | | 3 | .49 | .04 | 3.53 | 4.12 |
| 400 | Plywood, prefinished, 3/4" thick, stock grades, minimum | | .050 | | .75 | .62 | .04 | 1.41 | 1.85 |
| 410 | Maximum | | .071 | | 3.50 | .88 | .06 | 4.44 | 5.30 |
| 430 | Architectural grade, minimum | | .071 | | 2.50 | .88 | .06 | 3.44 | 4.22 |
| 440 | Maximum | | .100 | | 3.80 | 1.23 | .09 | 5.12 | 6.25 |
| 460 | Plywood, unfin. "A" face, birch, V.C., 1/2" thick, natural | | .036 | | 1.23 | .44 | .03 | 1.70 | 2.08 |
| 470 | Select | | .036 | | 1.35 | .44 | .03 | 1.82 | 2.21 |
| 490 | Veneer core, 3/4" thick, natural | | .050 | | 1.24 | .62 | .04 | 1.90 | 2.39 |
| 500 | Select | | .050 | | 1.46 | .62 | .04 | 2.12 | 2.63 |
| 520 | Lumber core, 3/4" thick, natural | | .050 | | 1.70 | .62 | .04 | 2.36 | 2.90 |
| 550 | Plywood, unfinished, knotty pine, 1/4" thick, A2 grade | | .036 | | 1.25 | .44 | .03 | 1.72 | 2.10 |
| 560 | A3 grade | | .036 | | 1.32 | .44 | .03 | 1.79 | 2.18 |
| 580 | 3/4" thick, veneer core, A2 grade | | .050 | | 1.56 | .62 | .04 | 2.22 | 2.74 |
| 590 | A3 grade | | .050 | | 1.25 | .62 | .04 | 1.91 | 2.40 |
| 610 | Aromatic cedar, 1/4" thick, plywood | | .040 | | 1.20 | .49 | .04 | 1.73 | 2.14 |
| 620 | 1/4" thick, particle board | ↓ | .040 | ↓ | .62 | .49 | .04 | 1.15 | 1.50 |
| 61-001 | **RAILING** Custom design, architectural grade, hardwood, minimum | 1 Carp | .211 | L.F. | 10.50 | 2.60 | | 13.10 | 15.65 |
| 010 | Maximum | | .267 | | 23 | 3.29 | | 26.29 | 31 |
| 030 | Stock interior railing with spindles 6" O.C., 4' long | | .200 | | 21 | 2.47 | | 23.47 | 27 |
| 040 | 8' long | ↓ | .167 | | 19.70 | 2.06 | | 21.76 | 25 |
| 64-001 | **SHELVING** Pine, clear grade, no edge band, 1" x 8" | F-1 | .070 | | .63 | .86 | .06 | 1.55 | 2.12 |
| 010 | 1" x 10" | | .073 | | .79 | .90 | .06 | 1.75 | 2.36 |
| 020 | 1" x 12" | ↓ | .076 | | .98 | .94 | .07 | 1.99 | 2.64 |
| 040 | For lumber edge band, by hand, add | | | | 1.20 | | | 1.20 | 1.32M |
| 042 | By machine, add | | | | .74 | | | .74 | .81M |
| 060 | Plywood, 3/4" thick with lumber edge, 12" wide | F-1 | .107 | | .94 | 1.32 | .09 | 2.35 | 3.22 |
| 070 | 24" wide | | .114 | ↓ | 1.78 | 1.41 | .10 | 3.29 | 4.30 |
| 090 | Bookcase, pine, clear grade, 8" shelves, 12" O.C. | | .114 | S.F.Face | 3.48 | 1.41 | .10 | 4.99 | 6.15 |
| 100 | 12" wide shelves | | .123 | " | 4.16 | 1.52 | .11 | 5.79 | 7.10 |
| 120 | Adjustable closet rod and shelf, 12" wide, 3' long | | .400 | Ea. | 6.60 | 4.95 | .35 | 11.90 | 15.45 |
| 130 | 8' long | | .533 | " | 15.40 | 6.58 | .47 | 22.45 | 28 |
| 150 | Prefinished shelves with supports, stock, 8" wide | | .107 | L.F. | 4.90 | 1.32 | .09 | 6.31 | 7.60 |
| 160 | 10" wide | ↓ | .114 | | 5.50 | 1.41 | .10 | 7.01 | 8.40 |
| 180 | Custom, high quality dadoed pine shelving units, minimum | | | S.F. Face | | | | 22 | 27 |
| 190 | Maximum | | | " | | | | 30 | 38 |
| 65-001 | **SHUTTERS, EXTERIOR** Aluminum, louvered, 1'-4" wide, 3'-0" long | 1 Carp | .800 | Pr. | 22 | 9.90 | | 31.90 | 40 |
| 020 | 4'-0" long | " | .800 | " | 25 | 9.90 | | 34.90 | 43 |

For expanded coverage of these items see *Building Construction Cost Data 1985*

## 6.2 Finish Carpentry

| | | | CREW | MAN-HOURS | UNIT | MAT. | LABOR | EQUIP. | TOTAL | TOTAL INCL O&P |
|---|---|---|---|---|---|---|---|---|---|---|
| 65 | 030 | 5'-4" long | 1 Carp | .800 | Pr. | 29 | 9.90 | | 38.90 | 48 |
| | 040 | 6'-8" long | | .889 | | 38 | 11 | | 49 | 59 |
| | 100 | Pine, louvered, primed, each 1'-2" wide, 3'-3" long | | .800 | | 27 | 9.90 | | 36.90 | 45 |
| | 110 | 4'-7" long | | .800 | | 36 | 9.90 | | 45.90 | 55 |
| | 125 | Each 1'-4" wide, 3'-0" long | | .800 | | 27 | 9.90 | | 36.90 | 45 |
| | 135 | 5'-3" long | | .800 | | 39 | 9.90 | | 48.90 | 59 |
| | 150 | Each 1'-6" wide, 3'-3" long | | .800 | | 30 | 9.90 | | 39.90 | 49 |
| | 160 | 4'-7" long | | .800 | | 42 | 9.90 | | 51.90 | 62 |
| | 162 | Hemlock, louvered, 1'-2" wide, 5'-7" long | | .800 | | 44 | 9.90 | | 53.90 | 64 |
| | 163 | Each 1'-4" wide, 2'-2" long | | .800 | | 26 | 9.90 | | 35.90 | 44 |
| | 164 | 3'-0" long | | .800 | | 27 | 9.90 | | 36.90 | 45 |
| | 165 | 3'-3" long | | .800 | | 28 | 9.90 | | 37.90 | 46 |
| | 166 | 3'-11" long | | .800 | | 31 | 9.90 | | 40.90 | 50 |
| | 167 | 4'-3" long | | .800 | | 31 | 9.90 | | 40.90 | 50 |
| | 168 | 5'-3" long | | .800 | | 38 | 9.90 | | 47.90 | 57 |
| | 169 | 5'-11" long | | .800 | | 44 | 9.90 | | 53.90 | 64 |
| | 170 | Door blinds, 6'-9" long, each 1'-3" wide | | .889 | | 51 | 11 | | 62 | 73 |
| | 171 | 1'-6" wide | | .889 | | 53 | 11 | | 64 | 76 |
| | 172 | Hemlock, solid raised panel, each 1'-4" wide, 3'-3" long | | .800 | | 42 | 9.90 | | 51.90 | 62 |
| | 173 | 3'-11" long | | .800 | | 50 | 9.90 | | 59.90 | 71 |
| | 174 | 4'-3" long | | .800 | | 53 | 9.90 | | 62.90 | 74 |
| | 175 | 4'-7" long | | .800 | | 57 | 9.90 | | 66.90 | 78 |
| | 176 | 4'-11" long | | .800 | | 62 | 9.90 | | 71.90 | 84 |
| | 177 | 5'-11" long | | .800 | | 74 | 9.90 | | 83.90 | 97 |
| | 180 | Door blinds, 6'-9" long, each 1'-3" wide | | .889 | | 82 | 11 | | 93 | 110 |
| | 190 | 1'-6" wide | | .889 | | 84 | 11 | | 95 | 110 |
| | 250 | Polystyrene, solid raised panel, each 1'-4" wide, 3'-3" long | | .800 | | 33 | 9.90 | | 42.90 | 52 |
| | 260 | 3'-11" long | | .800 | | 41 | 9.90 | | 50.90 | 61 |
| | 270 | 4'-7" long | | .800 | | 45 | 9.90 | | 54.90 | 65 |
| | 280 | 5'-3" long | | .800 | | 48 | 9.90 | | 57.90 | 68 |
| | 290 | 6'-8" long | ▼ | .889 | | 68 | 11 | | 79 | 92 |
| | 350 | For brick walls, add | | | | 5.10 | | | 5.10 | 5.60M |
| | 450 | Polystyrene, louvered, each 1'-2" wide, 3'-3" long | 1 Carp | .800 | | 28 | 9.90 | | 37.90 | 46 |
| | 460 | 4'-7" long | | .800 | | 36 | 9.90 | | 45.90 | 55 |
| | 475 | 5'-3" long | | .800 | | 40 | 9.90 | | 49.90 | 60 |
| | 485 | 6'-8" long | | .889 | | 64 | 11 | | 75 | 88 |
| | 600 | Vinyl, louvered, each 1'-2" x 4'-7" long | | .800 | | 34 | 9.90 | | 43.90 | 53 |
| | 620 | Each 1'-4" x 6'-8" long | ▼ | .889 | ▼ | 60 | 11 | | 71 | 83 |
| 67-001 | | SIDING, BOARDS | | | | | | | | |
| | 320 | Wood, cedar bevel, short lengths, A grade, 1/2" x 6" | 1 Carp | .032 | S.F. | 1.25 | .40 | | 1.65 | 2 |
| | 330 | 1/2" x 8" | | .029 | | 1.14 | .36 | | 1.50 | 1.82 |
| | 350 | 3/4" x 10", clear grade, 3' to 16' | | .027 | | 1.30 | .33 | | 1.63 | 1.95 |
| | 360 | "B" grade | | .027 | | 1.28 | .33 | | 1.61 | 1.93 |
| | 380 | Cedar, rough sawn, 1" x 4", B & Btr., natural | | .033 | | 1.53 | .41 | | 1.94 | 2.34 |
| | 390 | Stained | | .033 | | 1.58 | .41 | | 1.99 | 2.39 |
| | 410 | 1" x 12", board & batten, #3 & Btr., natural | | .031 | | .98 | .38 | | 1.36 | 1.68 |
| | 420 | Stained | | .031 | | 1.03 | .38 | | 1.41 | 1.73 |
| | 440 | 1" x 8" channel siding, #3 & Btr., natural | | .032 | | 1.07 | .40 | | 1.47 | 1.80 |
| | 450 | Stained | | .032 | | 1.12 | .40 | | 1.52 | 1.86 |
| | 470 | Redwood, clear, beveled, vertical grain, 1/2" x 4" | | .040 | | 1.60 | .49 | | 2.09 | 2.54 |
| | 480 | 1/2" x 8" | | .032 | | 1.29 | .40 | | 1.69 | 2.04 |
| | 500 | 3/4" x 10" | | .027 | | 1.57 | .33 | | 1.90 | 2.25 |
| | 520 | Channel siding, 1" x 10", clear | ▼ | .028 | | 1.02 | .35 | | 1.37 | 1.67 |
| | 525 | Redwood, T&G boards, clear, 1" x 4" | F-2 | .053 | | 2.30 | .66 | .05 | 3.01 | 3.62 |
| | 527 | 1" x 8" | " | .043 | | 2 | .52 | .04 | 2.56 | 3.08 |
| | 540 | White pine, rough sawn, 1" x 8", natural | 1 Carp | .029 | | .47 | .36 | | .83 | 1.09 |
| | 550 | Stained | " | .029 | ▼ | .52 | .36 | | .88 | 1.14 |
| 70-001 | | SIDING, SHEETS | | | | | | | | |
| | 002 | Siding, hardboard, 7/16" thick, prime painted, lap, | | | | | | | | |

For expanded coverage of these items see *Building Construction Cost Data 1985*

## 6.2 Finish Carpentry

| | | | CREW | MAN-HOURS | UNIT | MAT. | LABOR | EQUIP. | TOTAL | TOTAL INCL O&P |
|---|---|---|---|---|---|---|---|---|---|---|
| 70 | 003 | plain or grooved finish | F-2 | .021 | S.F. | .38 | .26 | .02 | .66 | .86 |
| | 010 | Board finish, 7/16" thick, lap or grooved, primed | | .021 | | .60 | .26 | .02 | .88 | 1.10 |
| | 020 | Stained | | .021 | | .65 | .26 | .02 | .93 | 1.15 |
| | 070 | Particle board, overlaid, 3/8" thick | | .021 | | .57 | .26 | .02 | .85 | 1.06 |
| | 090 | Plywood, medium density overlaid, 3/8" thick | | .021 | | .57 | .26 | .02 | .85 | 1.06 |
| | 100 | 1/2" thick | | .023 | | .73 | .28 | .02 | 1.03 | 1.27 |
| | 110 | 3/4" thick | | .025 | | .91 | .31 | .02 | 1.24 | 1.51 |
| | 160 | Texture 1-11, cedar, 5/8" thick, natural | | .024 | | 1.30 | .29 | .02 | 1.61 | 1.92 |
| | 170 | Factory stained | | .024 | | 1.35 | .29 | .02 | 1.66 | 1.97 |
| | 190 | Texture 1-11, fir, 5/8" thick, natural | | .024 | | .55 | .29 | .02 | .86 | 1.09 |
| | 200 | Factory stained | | .024 | | .60 | .29 | .02 | .91 | 1.15 |
| | 205 | Texture 1-11, S.Y.P., 5/8" thick, natural | | .024 | | .53 | .29 | .02 | .84 | 1.07 |
| | 210 | Factory stained | | .024 | | .58 | .29 | .02 | .89 | 1.12 |
| | 220 | Rough sawn cedar, 3/8" thick, natural | | .024 | | .97 | .29 | .02 | 1.28 | 1.55 |
| | 230 | Factory stained | | .024 | | 1.03 | .29 | .02 | 1.34 | 1.62 |
| | 250 | Rough sawn fir, 3/8" thick, natural | | .024 | | .38 | .29 | .02 | .69 | .90 |
| | 260 | Factory stained | | .024 | | .43 | .29 | .02 | .74 | .96 |
| | 280 | Redwood, textured siding, 5/8" thick | | .024 | | 1.33 | .29 | .02 | 1.64 | 1.95 |
| | 300 | Polyvinyl chloride coated, 3/8" thick | ↓ | .021 | ↓ | .75 | .26 | .02 | 1.03 | 1.26 |
| 73-001 | | SOFFITS Wood fiber, no vapor barrier, 15/32" thick | F-2 | .030 | S.F. | .38 | .37 | .03 | .78 | 1.04 |
| | 010 | 5/8" thick | | .030 | | .49 | .37 | .03 | .89 | 1.16 |
| | 030 | As above, 5/8" thick, with factory finish | | .030 | | .56 | .37 | .03 | .96 | 1.24 |
| | 050 | Hardboard, 3/8" thick, slotted | | .030 | | .60 | .37 | .03 | 1 | 1.29 |
| | 100 | Exterior AC plywood, 1/4" thick | | .038 | | .38 | .47 | .03 | .88 | 1.20 |
| | 110 | 1/2" thick | ↓ | .038 | ↓ | .54 | .47 | .03 | 1.04 | 1.38 |
| 76-001 | | STAIR PARTS Balusters, turned, 30" high, pine, minimum | 1 Carp | .286 | Ea. | 3.60 | 3.53 | | 7.13 | 9.55 |
| | 010 | Maximum | | .308 | | 4.50 | 3.80 | | 8.30 | 10.95 |
| | 030 | 30" high birch balusters, minimum | | .286 | | 4.60 | 3.53 | | 8.13 | 10.65 |
| | 040 | Maximum | | .308 | | 4.90 | 3.80 | | 8.70 | 11.40 |
| | 060 | 42" high, pine balusters, minimum | | .296 | | 4 | 3.66 | | 7.66 | 10.20 |
| | 070 | Maximum | | .320 | | 4.89 | 3.95 | | 8.84 | 11.65 |
| | 090 | 42" high birch balusters, minimum | | .296 | | 5.85 | 3.66 | | 9.51 | 12.25 |
| | 100 | Maximum | | .320 | ↓ | 6.25 | 3.95 | | 10.20 | 13.15 |
| | 105 | Baluster, stock pine, 1-1/16" x 1-1/16" | | .033 | L.F. | .44 | .41 | | .85 | 1.14 |
| | 110 | 1-5/8" x 1-5/8" | | .036 | " | .89 | .45 | | 1.34 | 1.69 |
| | 120 | Newels, 3-1/4" wide, starting, minimum | | 1.140 | Ea. | 31 | 14.10 | | 45.10 | 56 |
| | 130 | Maximum | | 1.330 | | 150 | 16.45 | | 166.45 | 190 |
| | 150 | Landing, minimum | | 1.600 | | 36 | 19.75 | | 55.75 | 71 |
| | 160 | Maximum | | 2 | ↓ | 145 | 25 | | 170 | 200 |
| | 180 | Railings, oak, built-up, minimum | | .133 | L.F. | 3.75 | 1.65 | | 5.40 | 6.75 |
| | 190 | Maximum | | .145 | | 7.70 | 1.80 | | 9.50 | 11.30 |
| | 210 | Add for sub rail | ↓ | .073 | ↓ | 1.60 | .90 | | 2.50 | 3.18 |
| | 211 | | | | | | | | | |
| | 230 | Risers, Beech, 3/4" x 7-1/2" high | 1 Carp | .125 | L.F. | 3.50 | 1.54 | | 5.04 | 6.30 |
| | 240 | Fir, 3/4" x 7-1/2" high | | .125 | | .95 | 1.54 | | 2.49 | 3.49 |
| | 260 | Oak, 3/4" x 7-1/2" high | | .125 | | 2.95 | 1.54 | | 4.49 | 5.70 |
| | 280 | Pine, 3/4" x 7-1/2" high | | .121 | | .95 | 1.50 | | 2.45 | 3.42 |
| | 285 | Skirt board, pine, 1" x 10" | | .145 | | 1.24 | 1.80 | | 3.04 | 4.21 |
| | 290 | 1" x 12" | | .154 | ↓ | 1.52 | 1.90 | | 3.42 | 4.68 |
| | 300 | Treads, oak, 1-1/16" x 9-1/2" wide, 3' long | | .444 | Ea. | 14.15 | 5.50 | | 19.65 | 24 |
| | 310 | 4' long | | .471 | | 19.70 | 5.80 | | 25.50 | 31 |
| | 330 | 1-1/16" x 11-1/2" wide, 3' long | | .444 | | 16.80 | 5.50 | | 22.30 | 27 |
| | 340 | 6' long | ↓ | .571 | ↓ | 35 | 7.05 | | 42.05 | 50 |
| | 360 | Beech treads, add | | | | 40% | | | | |
| | 380 | For mitered return nosings, add | | | L.F. | 1.50 | | | | 1.65M |
| 79-001 | | STAIRS, PREFABRICATED | | | | | | | | |
| | 010 | Box stairs, prefabricated 3'-0" wide | | | | | | | | |
| | 011 | Oak treads, no handrails, 2 ft. high | 2 Carp | 3.200 | Flight | 105 | 40 | | 145 | 180 |
| | 020 | 4 ft. high | " | 4 | " | 180 | 49 | | 229 | 275 |

For expanded coverage of these items see *Building Construction Cost Data 1985*

## 6.2 Finish Carpentry

| | | CREW | MAN-HOURS | UNIT | MAT. | LABOR | EQUIP. | TOTAL | TOTAL INCL O&P |
|---|---|---|---|---|---|---|---|---|---|
| 030 | 6 ft. high | 2 Carp | 4.570 | Flight | 280 | 56 | | 336 | 395 |
| 040 | 8 ft. high | | 5.330 | | 340 | 66 | | 406 | 480 |
| 060 | With pine treads for carpet, 2 ft. high | | 3.200 | | 65 | 40 | | 105 | 135 |
| 070 | 4 ft. high | | 4 | | 110 | 49 | | 159 | 200 |
| 080 | 6 ft. high | | 4.570 | | 165 | 56 | | 221 | 270 |
| 090 | 8 ft. high | | 5.330 | | 200 | 66 | | 266 | 325 |
| 110 | For 4' wide stairs, add | | | | 10% | | | | |
| 150 | Prefabricated stair rail with balusters, 5 risers | 2 Carp | 1.070 | Ea. | 99 | 13.15 | | 112.15 | 130 |
| 160 | | | | | | | | | |
| 170 | Basement stairs, prefabricated, soft wood, | | | | | | | | |
| 171 | open risers, 3' wide, 8' high | 2 Carp | 4 | Flight | 115 | 49 | | 164 | 205 |
| 190 | Open stairs, prefabricated prefinished poplar, metal stringers, | | | | | | | | |
| 191 | treads 3'-6" wide, no railings | | | | | | | | |
| 200 | 3 ft. high | 2 Carp | 3.200 | Flight | 285 | 40 | | 325 | 375 |
| 210 | 4 ft. high | | 4 | | 355 | 49 | | 404 | 470 |
| 220 | 6 ft. high | | 4.570 | | 620 | 56 | | 676 | 770 |
| 230 | 8 ft. high | | 5.330 | | 910 | 66 | | 976 | 1,100 |
| 250 | For prefab. 3 piece wood railings & balusters, add for | | | | | | | | |
| 260 | 3 ft. high stairs | 2 Carp | 1.070 | Ea. | 98 | 13.15 | | 111.15 | 130 |
| 270 | 4 ft. high stairs | | 1.140 | | 120 | 14.10 | | 134.10 | 155 |
| 280 | 6 ft. high stairs | | 1.230 | | 180 | 15.20 | | 195.20 | 220 |
| 290 | 8 ft. high stairs | | 1.330 | | 245 | 16.45 | | 261.45 | 295 |
| 310 | For 3'-6" x 3'-6" platform, add | | 4 | | 82 | 49 | | 131 | 170 |
| 330 | Curved stairways, 3'-3" wide, prefabricated, oak, unfinished, | | | | | | | | |
| 331 | incl. curved balustrade system, open one side | | | | | | | | |
| 340 | 9' high | 2 Carp | 22.860 | Flight | 4,100 | 280 | | 4,380 | 4,950 |
| 350 | 10' high | | 22.860 | | 4,500 | 280 | | 4,780 | 5,400 |
| 370 | Open two sides, 9' high | | 32 | | 6,600 | 395 | | 6,995 | 7,875 |
| 380 | 10' high | | 32 | | 7,300 | 395 | | 7,695 | 8,650 |
| 400 | Residential, wood, oak treads, prefabricated | | 10.670 | | 620 | 130 | | 750 | 890 |
| 420 | Built-in place | | 36.360 | | 735 | 450 | | 1,185 | 1,525 |
| 440 | Spiral, oak, 4'-6" diameter, unfinished, prefabricated, | | | | | | | | |
| 450 | incl. railing, 9' high | 2 Carp | 10.670 | Flight | 2,675 | 130 | | 2,805 | 3,150 |
| 88-001 | **VANITIES** | | | | | | | | |
| 002 | | | | | | | | | |
| 800 | Vanity bases, 2 doors, 30" high, 21" deep, 24" wide | 2 Carp | 1.45 | Ea. | 97 | 36 | | 133 | 146 |
| 805 | 30" wide | | 1.63 | | 105 | 40 | | 145 | 179 |
| 810 | 36" wide | | 2.00 | | 140 | 49 | | 189 | 232 |
| 815 | 48" wide | | 2.42 | | 180 | 60 | | 240 | 293 |
| 900 | For deluxe models of all cabinets, add to above | | | | 40% | | | | |
| 950 | For custom built in place, add to above | | | | 25% | 10% | | | |

## 6.3 Laminated Construction

| | | CREW | MAN-HOURS | UNIT | MAT. | LABOR | EQUIP. | TOTAL | TOTAL INCL O&P |
|---|---|---|---|---|---|---|---|---|---|
| 10-001 | **LAMINATED FRAMING** Not including decking | | | | | | | | |
| 002 | | | | | | | | | |
| 020 | Straight roof beams, 20' clear span, 8' O.C. | F-3 | .016 | S.F.Flr. | 1.90 | .19 | .10 | 2.19 | 2.51 |
| 030 | 16" O.C. | | .013 | | 1.28 | .15 | .08 | 1.51 | 1.74 |
| 050 | 40' clear span, beams 8' O.C. | | .013 | | 3.58 | .15 | .08 | 3.81 | 4.27 |
| 060 | Beams 16' O.C. | | .010 | | 2.70 | .13 | .06 | 2.89 | 3.25 |
| 080 | 60' clear span, 8' O.C. | F-4 | .017 | | 5.70 | .20 | .19 | 6.09 | 6.80 |
| 090 | Beams 16' O.C. | " | .013 | | 4.32 | .15 | .14 | 4.61 | 5.15 |
| 110 | Tudor arches, 30' to 40' clear span, frames 8' O.C. | F-3 | .024 | | 5 | .29 | .15 | 5.44 | 6.15 |
| 120 | Frames 16' O.C. | " | .018 | | 3.40 | .22 | .11 | 3.73 | 4.21 |

For expanded coverage of these items see *Building Construction Cost Data 1985*

## 6.3 Laminated Construction

| | | | CREW | MAN-HOURS | UNIT | MAT. | LABOR | EQUIP. | TOTAL | TOTAL INCL O&P |
|---|---|---|---|---|---|---|---|---|---|---|
| 10 | 140 | 50' to 60' clear span, frames 8' O.C. | F-4 | .022 | S.F.Flr. | 4.92 | .26 | .25 | 5.43 | 6.10 |
| | 150 | Frames 16' O.C. | | .018 | | 4 | .23 | .20 | 4.43 | 4.98 |
| | 170 | Radial arches, 60' clear span, 8' O.C. | | .025 | | 5.10 | .31 | .28 | 5.69 | 6.40 |
| | 180 | 16' O.C. | | .017 | | 3.40 | .20 | .19 | 3.79 | 4.27 |
| | 200 | 100' clear span, frames @ 8' O.C. | | .030 | | 3.87 | .36 | .34 | 4.57 | 5.20 |
| | 210 | 16' O.C. | | .020 | | 4 | .24 | .23 | 4.47 | 5.05 |
| | 230 | 120' clear span, 8' O.C. | | .033 | | 6 | .40 | .38 | 6.78 | 7.65 |
| | 240 | Frames 16' O.C. | ↓ | .025 | | 5.40 | .31 | .28 | 5.99 | 6.75 |
| | 260 | Bowstring trusses, 20' O.C., 40' clear span | F-3 | .017 | | 2.30 | .21 | .10 | 2.61 | 2.97 |
| | 270 | 60' clear span | F-4 | .013 | | 2.16 | .16 | .15 | 2.47 | 2.80 |
| | 280 | 100' clear span | | .012 | | 3.40 | .14 | .14 | 3.68 | 4.12 |
| | 290 | 120' clear span | ↓ | .013 | | 3.70 | .16 | .15 | 4.01 | 4.49 |
| | 310 | For premium appearance, add to S.F. prices | | | | 5% | | | | |
| | 330 | For industrial type, deduct | | | | 15% | | | | |
| | 350 | For stain and varnish, add | | | | 5% | | | | |
| | 390 | For 3/4" laminations, add to straight | | | | 25% | | | | |
| | 410 | Add to curved | | | ↓ | 15% | | | | |
| | 430 | Alternate pricing method: (use nominal footage of | | | | | | | | |
| | 431 | components). Straight beams, camber less than 6" | F-3 | 11.430 | M.B.F. | 1,400 | 144 | 71 | 1,615 | 1,850 |
| | 440 | Columns, including hardware | | 20 | | 1,550 | 250 | 125 | 1,925 | 2,225 |
| | 460 | Curved members, radius over 32 ft. | | 16 | | 1,600 | 201 | 99 | 1,900 | 2,175 |
| | 470 | Radius 10 ft. to 31 ft. | ↓ | 13.330 | | 1,750 | 167 | 83 | 2,000 | 2,275 |
| | 490 | For complicated shapes, add maximum | | | | 100% | | | | |
| | 510 | For pressure treating, add to straight | | | | 35% | | | | |
| | 520 | Add to curved | | | ↓ | 45% | | | | |
| 15-001 | | **LAMINATED BEAMS** (Fb 2400 psi) | | | | | | | | |
| | 005 | 3" x 18" | F-3 | .083 | L.F. | 13.75 | 1.04 | .52 | 15.31 | 17.35 |
| | 010 | 5" x 12" | | .089 | | 14 | 1.11 | .55 | 15.66 | 17.75 |
| | 015 | 5" x 15" | | .111 | | 17 | 1.39 | .69 | 19.08 | 22 |
| | 020 | 5" x 18" | | .138 | | 20 | 1.72 | .86 | 22.58 | 26 |
| | 025 | 5" x 24" | | .182 | | 28 | 2.27 | 1.13 | 31.40 | 36 |
| | 030 | 7" x 12" | | .125 | | 17.50 | 1.55 | .78 | 19.83 | 23 |
| | 035 | 7" x 15" | | .154 | | 21 | 1.91 | .96 | 23.87 | 27 |
| | 040 | 7" x 18" | | .190 | | 25 | 2.38 | 1.18 | 28.56 | 33 |
| | 045 | 7" x 24" | ↓ | .250 | | 38 | 3.12 | 1.55 | 42.67 | 48 |
| | 050 | For premium appearance, add to S.F. prices | | | | 5% | | | | |
| | 055 | For industrial type, deduct | | | | 15% | | | | |
| | 060 | For stain and varnish, add | | | | 5% | | | | |
| | 065 | For 3/4" laminations, add | | | ↓ | 25% | | | | |
| 20-001 | | **LAMINATED ROOF DECK** Pine or hemlock, 3" thick | F-2 | .038 | S.F. | 4.80 | .47 | .03 | 5.30 | 6.05 |
| | 010 | 4" thick | | .049 | | 5.85 | .61 | .04 | 6.50 | 7.45 |
| | 030 | Cedar, 3" thick | | .038 | | 5.80 | .47 | .03 | 6.30 | 7.15 |
| | 040 | 4" thick | | .049 | | 6.90 | .61 | .04 | 7.55 | 8.60 |
| | 060 | Fir, 3" thick | | .038 | | 4.80 | .47 | .03 | 5.30 | 6.05 |
| | 070 | 4" thick | ↓ | .049 | ↓ | 5.80 | .61 | .04 | 6.45 | 7.40 |

## 7.1 Waterproofing

| | | | CREW | MAN-HOURS | UNIT | MAT. | LABOR | EQUIP. | TOTAL | TOTAL INCL O&P |
|---|---|---|---|---|---|---|---|---|---|---|
| 10-001 | | **BITUMINOUS ASPHALT COATING** For foundation | | | | | | | | |
| | 003 | Brushed on, below grade, 1 coat | 1 Rofc | .012 | S.F. | .08 | .14 | | .22 | .32 |
| | 010 | 2 coat | | .016 | | .15 | .19 | | .34 | .47 |
| | 030 | Sprayed on, below grade, 1 coat, 25.6 S.F./gal. | | .010 | | .12 | .11 | | .23 | .32 |
| | 040 | 2 coat, 20.5 S.F./gal. | | .016 | | .15 | .19 | | .34 | .47 |
| | 060 | Troweled on, asphalt with fibers, 1/16" thick | ↓ | .016 | ↓ | .50 | .19 | | .69 | .86 |

For expanded coverage of these items see *Building Construction Cost Data 1985*

# 7.1 Waterproofing

| | | | CREW | MAN-HOURS | UNIT | MAT. | LABOR | EQUIP. | TOTAL | TOTAL INCL O&P |
|---|---|---|---|---|---|---|---|---|---|---|
| 10 | 070 | 1/8" thick | 1 Rofc | .020 | S.F. | 1 | .23 | | 1.23 | 1.48 |
| | 100 | 1/2" thick | " | .023 | " | 4.30 | .27 | | 4.57 | 5.15 |
| | 340 | Glass fibered roof & patching cement, 5 gallon | | | Gal. | 12.50 | | | 12.50 | 13.75M |
| | 345 | Reinforcing glass membrane, 450 S.F./roll | | | Ea. | 72 | | | 72 | 79M |
| | 350 | Neoprene roof coating, 5 gal., 2 gal./sq. | | | Gal. | 24 | | | 24 | 26M |
| | 370 | Roof patch & flashing cement, 5 gallon | | | | 35 | | | 35 | 39M |
| | 600 | Roof resaturant, glass fibered, 3 gal./sq. | | | | 11 | | | 11 | 12.10M |
| | 620 | Mineral rubber, 3 gal./sq. | | | | 10 | | | 10 | 11M |
| 15-001 | | **BUILDING PAPER** Aluminum and kraft laminated, foil 1 side | 1 Carp | .004 | S.F. | .03 | .05 | | .08 | .12 |
| | 010 | Foil 2 sides | | .004 | | .06 | .05 | | .11 | .15 |
| | 030 | Asphalt, two ply, #30, for subfloors | | .002 | | .07 | .03 | | .10 | .12 |
| | 040 | Asphalt felt sheathing paper, #15 | | .002 | | .03 | .03 | | .06 | .08 |
| | 060 | Polyethylene vapor barrier, standard, .002" thick | | .002 | | .03 | .03 | | .06 | .08 |
| | 070 | .004" thick | | .002 | | .05 | .03 | | .08 | .10 |
| | 090 | .006" thick | | .002 | | .05 | .03 | | .08 | .10 |
| | 100 | .008" thick | | .002 | | .05 | .03 | | .08 | .10 |
| | 120 | .010" thick | | .002 | | .06 | .03 | | .09 | .11 |
| | 150 | Red rosin paper, 5 sq. rolls, 4 lbs. per square | | .002 | | .02 | .03 | | .05 | .06 |
| | 160 | 5 lbs. per square | | .002 | | .02 | .03 | | .05 | .06 |
| | 180 | Reinf. waterproof, .002" polyethylene backing, 1 side | | .002 | | .06 | .03 | | .09 | .11 |
| | 190 | 2 sides | ↓ | .002 | ↓ | .08 | .03 | | .11 | .13 |
| 20-001 | | **CAULKING AND SEALANTS** | | | | | | | | |
| | 002 | Acoustical sealant, elastomeric | | | Gal. | 18.25 | | | 18.25 | 20M |
| | 010 | Caulking compound, oil base, bulk | | | | 13.35 | | | 13.35 | 14.70M |
| | 020 | Brilliant white color | | | | 10.10 | | | 10.10 | 11.10M |
| | 030 | Aluminum pigment and other colors | | | ↓ | 12.95 | | | 12.95 | 14.25M |
| | 050 | Bulk, in place, 1/4" x 1/2", 154 L.F./gal. | 1 Bric | .031 | L.F. | .07 | .39 | | .46 | .68 |
| | 060 | 1/2" x 1/2", 77 L.F./gal. | | .032 | | .13 | .40 | | .53 | .77 |
| | 080 | 3/4" x 3/4", 34 L.F./gal. | | .035 | | .30 | .44 | | .74 | 1.01 |
| | 090 | 3/4" x 1", 26 L.F./gal. | | .040 | | .39 | .50 | | .89 | 1.22 |
| | 100 | 1" x 1", 19 L.F./gal. | ↓ | .044 | ↓ | .53 | .56 | | 1.09 | 1.46 |
| | 110 | Acrylic based, bulk | | | Gal. | 15.25 | | | 15.25 | 16.75M |
| | 120 | Cartridges | | | | 22 | | | 22 | 24M |
| | 140 | Butyl based, bulk | | | | 10.25 | | | 10.25 | 11.25M |
| | 150 | Cartridges | | | ↓ | 14.30 | | | 14.30 | 15.75M |
| | 170 | Bulk, in place 1/4" x 1/2", 154 L.F./gal. | 1 Bric | .035 | L.F. | .22 | .44 | | .66 | .93 |
| | 180 | 1/2" x 1/2", 77 L.F./gal. | " | .044 | " | .41 | .56 | | .97 | 1.32 |
| | 185 | Hypalon, bulk | | | Gal. | 34 | | | 34 | 37M |
| | 190 | Cartridges | | | | 45 | | | 45 | 50M |
| | 200 | Latex based, bulk | | | | 13.65 | | | 13.65 | 15M |
| | 210 | Cartridges | | | ↓ | 22 | | | 22 | 24M |
| | 220 | Bulk in place, 1/4" x 1/2", 154 L.F./gal. | 1 Bric | .035 | L.F. | .09 | .44 | | .53 | .78 |
| | 230 | Polysulfide compounds, 1 component, bulk | | | Gal. | 45 | | | 45 | 50M |
| | 240 | Cartridges | | | " | 39 | | | 39 | 43M |
| | 260 | 1 or 2 component, in place, 1/4" x 1/4", 308 L.F./gal. | 1 Bric | .055 | L.F. | .15 | .70 | | .85 | 1.25 |
| | 270 | 1/2" x 1/4", 154 L.F./gal. | | .059 | | .30 | .75 | | 1.05 | 1.49 |
| | 290 | 3/4" x 3/8", 68 L.F./gal. | | .062 | | .69 | .78 | | 1.47 | 1.97 |
| | 300 | 1" x 1/2", 38 L.F./gal. | ↓ | .062 | ↓ | .92 | .78 | | 1.70 | 2.22 |
| | 320 | Polyurethane, 1 or 2 component, bulk | | | Gal. | 31 | | | 31 | 34M |
| | 330 | Cartridges | | | " | 46 | | | 46 | 51M |
| | 350 | 1 or 2 component, in place, 1/4" x 1/4", 308 L.F./gal. | 1 Bric | .053 | L.F. | .05 | .67 | | .72 | 1.10 |
| | 360 | 1/2" x 1/4", 154 L.F./gal. | | .055 | | .11 | .70 | | .81 | 1.21 |
| | 380 | 3/4" x 3/8", 68 L.F./gal. | | .062 | | .23 | .78 | | 1.01 | 1.46 |
| | 390 | 1" x 1/2", 38 L.F./gal. | ↓ | .073 | ↓ | .41 | .92 | | 1.33 | 1.88 |
| | 410 | Silicone rubber, bulk | | | Gal. | 72 | | | 72 | 79M |
| | 420 | Cartridges | | | " | 75 | | | 75 | 83M |
| 70-001 | | **SILICONE OR STEARATE** Sprayed on masonry, 1 coat | 1 Rofc | .002 | S.F. | .25 | .02 | | .27 | .31 |
| | 010 | 2 coats | " | .004 | " | .42 | .05 | | .47 | .54 |

For expanded coverage of these items see *Building Construction Cost Data 1985*

315

## 7.2 Insulation

| | | CREW | MAN-HOURS | UNIT | MAT. | LABOR | EQUIP. | TOTAL | TOTAL INCL O&P |
|---|---|---|---|---|---|---|---|---|---|
| 02-001 | **BLOWN-IN INSULATION** Ceilings, with open access | | | | | | | | |
| 002 | Cellulose, 3-1/2" thick, R-11 | G-4 | .008 | S.F. | .12 | .09 | .04 | .25 | .31 |
| 003 | 5-3/16" thick, R-19 | | .013 | | .18 | .14 | .06 | .38 | .48 |
| 005 | 6-1/2" thick, R-22 | | .016 | | .23 | .17 | .08 | .48 | .61 |
| 100 | Fiberglass, 5" thick, R-11 | | .013 | | .11 | .14 | .06 | .31 | .41 |
| 105 | 6" thick, R-13 | | .016 | | .14 | .17 | .08 | .39 | .51 |
| 110 | 8-1/2" thick, R-19 | | .022 | | .17 | .23 | .11 | .51 | .68 |
| 200 | Mineral wool, 4" thick, R-11 | | .011 | | .17 | .11 | .06 | .34 | .43 |
| 205 | 6" thick, R-13 | | .016 | | .27 | .17 | .08 | .52 | .66 |
| 210 | 9" thick, R-19 | ↓ | .024 | ↓ | .46 | .26 | .12 | .84 | 1.05 |
| 17-001 | **FLOOR INSULATION, NONRIGID** Including | | | | | | | | |
| 002 | spring type wire fasteners | | | | | | | | |
| 200 | Fiberglass, blankets or batts, paper or foil backing | | | | | | | | |
| 210 | 1 side, 3-1/2" thick, R11 | 1 Carp | .011 | S.F. | .20 | .14 | | .34 | .44 |
| 215 | 6" thick, R19 | | .013 | | .31 | .16 | | .47 | .60 |
| 220 | 8-1/2" thick, R30 | ↓ | .015 | ↓ | .45 | .18 | | .63 | .78 |
| 25-001 | **MASONRY INSULATION** Vermiculite or perlite, poured | | | | | | | | |
| 010 | In cores of concrete block, 4" thick wall | D-1 | .003 | | .40 | .04 | | .44 | .50 |
| 070 | Foamed in place, urethane in 2-5/8" cavity | G-2 | .023 | S.F. | .15 | .25 | .10 | .50 | .67 |
| 080 | For each 1" added thickness, add | " | .010 | | .05 | .11 | .04 | .20 | .28 |
| 30-001 | **PERIMETER INSULATION** Asphalt impregnated cork, 1/2" thick, R1.12 | 1 Carp | .012 | | .98 | .14 | | 1.12 | 1.31 |
| 010 | 1" thick, R2.24 | | .012 | | 2.18 | .14 | | 2.32 | 2.63 |
| 060 | Polystyrene, molded bead board, 1" thick, R4 | | .012 | | .21 | .14 | | .35 | .46 |
| 070 | 2" thick, R8 | | .012 | ↓ | .39 | .14 | | .53 | .66 |
| 35-001 | **POURED INSULATION** Cellulose fiber, R3.8 per inch | | .040 | C.F. | .41 | .49 | | .90 | 1.23 |
| 008 | Fiberglass wool, R4 per inch | | .040 | | .27 | .49 | | .76 | 1.08 |
| 010 | Mineral wool, R3 per inch | | .040 | | .73 | .49 | | 1.22 | 1.59 |
| 030 | Polystyrene, R4 per inch | | .040 | | 1.35 | .49 | | 1.84 | 2.27 |
| 040 | Vermiculite or perlite, R2.7 per inch | ↓ | .040 | ↓ | 1.39 | .49 | | 1.88 | 2.31 |
| 40-001 | **REFLECTIVE** Aluminum foil on 40 lb. kraft, foil 1 side, R9 | 1 Carp | .004 | S.F. | .03 | .05 | | .08 | .12 |
| 010 | Multilayered with air spaces, 2 ply, R14 | | .004 | | .12 | .05 | | .17 | .21 |
| 050 | 3 ply, R17 | | .005 | | .16 | .07 | | .23 | .28 |
| 060 | 5 ply, R22 | ↓ | .005 | ↓ | .24 | .07 | | .31 | .37 |
| 50-001 | **ROOF DECK INSULATION** | | | | | | | | |
| 003 | Fiberboard, mineral, 1" thick, R2.78 | 1 Rofc | .010 | S.F. | .26 | .12 | | .38 | .48 |
| 008 | 1-1/2" thick, R4 | | .010 | | .36 | .12 | | .48 | .59 |
| 010 | 2" thick, R5.26 | ↓ | .010 | ↓ | .50 | .12 | | .62 | .74 |
| 030 | Fiberglass, in 3' x 4' or 4' x 8' sheets | | | | | | | | |
| 040 | 15/16" thick, R3.3 | 1 Rofc | .008 | S.F. | .35 | .09 | | .44 | .54 |
| 046 | 1-1/16" thick, R3.8 | | .008 | | .39 | .09 | | .48 | .58 |
| 060 | 1-5/16" thick, R5.3 | | .008 | | .52 | .09 | | .61 | .73 |
| 065 | 1-5/8" thick, R5.7 | | .008 | | .60 | .09 | | .69 | .81 |
| 070 | 1-7/8" thick, R7.7 | | .008 | | .64 | .09 | | .73 | .86 |
| 080 | 2-1/4" thick, R8 | ↓ | .010 | ↓ | .69 | .12 | | .81 | .95 |
| 090 | Fiberglass and urethane composite, 3' x 4' sheets | | | | | | | | |
| 100 | 1-11/16" thick, R11.1 | 1 Rofc | .008 | S.F. | .55 | .09 | | .64 | .76 |
| 120 | 2" thick, R14.3 | | .010 | | .65 | .12 | | .77 | .91 |
| 130 | 2-5/8" thick, R18.2 | ↓ | .010 | ↓ | .81 | .12 | | .93 | 1.08 |
| 150 | Foamglass, 2' x 4' sheets, rectangular | | | | | | | | |
| 151 | 1-1/2" thick R 3.95 | 1 Rofc | .010 | S.F. | 1.42 | .12 | | 1.54 | 1.75 |
| 152 | 2" thick R 5.26 | | .010 | | 1.88 | .12 | | 2 | 2.26 |
| 153 | 3" thick R7.89 | | .011 | | 2.21 | .13 | | 2.34 | 2.65 |
| 154 | 4" thick R 10.53 | ↓ | .011 | ↓ | 3.86 | .13 | | 3.99 | 4.47 |
| 160 | Tapered 1/16", 1/8" or 1/4" per foot | | | B.F. | .96 | | | .96 | 1.05M |
| 165 | Perlite, 2' x 4' sheets | 1 Rofc | .010 | S.F. | .20 | .12 | | .32 | .41 |
| 165 | 3/4" thick, R2.08 | | .010 | | .20 | .12 | | .32 | .41 |
| 166 | 1" thick, R2.78 | | .010 | | .27 | .12 | | .39 | .49 |
| 167 | 1-1/2" thick, R4.17 | | .010 | | .40 | .12 | | .52 | .63 |
| 168 | 2" thick, R5.26 | ↓ | .011 | ↓ | .54 | .13 | | .67 | .81 |

For expanded coverage of these items see *Building Construction Cost Data 1985*

# 7.2 Insulation

| | | | MAN-HOURS | UNIT | BARE COSTS MAT. | LABOR | EQUIP. | TOTAL | TOTAL INCL O&P |
|---|---|---|---|---|---|---|---|---|---|
| 50 | 170 | Perlite/urethane composite | | | | | | | |
| | 171 | 1-1/4" thick, R5.88 | 1 Rofc .008 | S.F. | .60 | .09 | | .69 | .81 |
| | 172 | 1-1/2" thick, R7.2 | .008 | | .63 | .09 | | .72 | .85 |
| | 173 | 1-3/4" thick, R10 | .008 | | .67 | .09 | | .76 | .89 |
| | 174 | 2" thick, R12.5 | .010 | | .70 | .12 | | .82 | .96 |
| | 175 | 2-1/2" thick, R14.3 | .011 | | .75 | .12 | | .87 | 1.03 |
| | 176 | 3" thick, R20 | .011 | | .91 | .13 | | 1.04 | 1.22 |
| | 180 | Phenolic foam, 2' x 4' sheets | | | | | | | |
| | 181 | 1-3/16" thick, R10 | 1 Rofc .008 | S.F. | .65 | .09 | | .74 | .87 |
| | 182 | 1-3/8" thick, R11.1 | .008 | | .76 | .09 | | .85 | .99 |
| | 183 | 1-3/4" thick, R14.3 | .008 | | .96 | .09 | | 1.05 | 1.21 |
| | 184 | 2" thick, R16.7 | .010 | | 1.10 | .12 | | 1.22 | 1.40 |
| | 185 | 2-1/2" thick, R20 | .010 | | 1.38 | .12 | | 1.50 | 1.71 |
| | 186 | 3" thick, R25 | .010 | | 1.65 | .12 | | 1.77 | 2.01 |
| | 190 | Polystyrene | | | | | | | |
| | 191 | Extruded, 2.3#/C.F., 1" thick, R5.26 | 1 Rofc .005 | S.F. | .28 | .06 | | .34 | .41 |
| | 192 | 2" thick, R10 | .006 | | .55 | .07 | | .62 | .73 |
| | 193 | 3" thick, R15 | .008 | | .84 | .09 | | .93 | 1.08 |
| | 201 | Expanded bead board, 1" thick, R3.57 | .005 | | .15 | .06 | | .21 | .27 |
| | 210 | 2" thick, R7.14 | .006 | | .30 | .07 | | .37 | .45 |
| | 220 | Urethane, felt both sides | | | | | | | |
| | 221 | 1" thick, R6.7 | 1 Rofc .008 | S.F. | .41 | .09 | | .50 | .60 |
| | 222 | 1-1/2" thick, R11.11 | .008 | | .50 | .09 | | .59 | .70 |
| | 223 | 2" thick, R14.3 | .010 | | .61 | .12 | | .73 | .86 |
| | 224 | 2-1/2" thick, R20 | .010 | | .71 | .12 | | .83 | .97 |
| | 225 | 3" thick, R25 | .010 | | .88 | .12 | | 1 | 1.16 |
| | 230 | Urethane and gypsum board composite | | | | | | | |
| | 231 | 1-5/8" thick, R7.7 | 1 Rofc .008 | S.F. | .72 | .09 | | .81 | .95 |
| | 232 | 2" thick, R10 | .010 | | .94 | .12 | | 1.06 | 1.23 |
| | 233 | 2-1/2" thick, R14.3 | .010 | | 1.05 | .12 | | 1.17 | 1.35 |
| | 234 | 3" thick, R18.2 | .010 | | 1.09 | .12 | | 1.21 | 1.39 |
| 77 | 001 | **VENTS, ONE-WAY** For insulated decks, 1 per M.S.F., plastic, minimum | 1 Rofc .200 | Ea. | 4.98 | 2.32 | | 7.30 | 9.35 |
| | 010 | Maximum | .400 | | 55 | 4.64 | | 59.64 | 68 |
| | 030 | Aluminum | .267 | | 13.10 | 3.09 | | 16.19 | 19.55 |
| | 050 | Copper | .267 | | 15.45 | 3.09 | | 18.54 | 22 |
| | 080 | Fiber board baffles, 12" wide for 16" O.C. rafter spacing | 1 Carp .080 | | .45 | .99 | | 1.44 | 2.06 |
| | 090 | For 24" O.C. rafter spacing | " .080 | | .67 | .99 | | 1.66 | 2.30 |
| | 095 | For louvers and vents see division 7.6 - 42 | | | | | | | |
| 82 | 001 | **WALL INSULATION RIGID** | | | | | | | |
| | 002 | | | | | | | | |
| | 004 | Fiberglass, 1.5#/C.F., unfaced | | | | | | | |
| | 006 | 1-1/2" thick, R6.2 | 1 Carp .008 | S.F. | .25 | .10 | | .35 | .44 |
| | 008 | 2" thick, R8.3 | .009 | | .37 | .11 | | .48 | .58 |
| | 012 | 3" thick, R12.4 | .010 | | .56 | .12 | | .68 | .81 |
| | 037 | 1" thick, 3#/C.F. R4.3 unfaced | .008 | | .40 | .10 | | .50 | .60 |
| | 039 | 1-1/2" thick, R6.5 | .009 | | .60 | .11 | | .71 | .83 |
| | 040 | 2" thick, R8.7 | .009 | | .80 | .12 | | .92 | 1.06 |
| | 042 | 2-1/2" thick, R10.9 | .010 | | 1 | .12 | | 1.12 | 1.30 |
| | 044 | 3" thick, R13 | .011 | | 1.20 | .13 | | 1.33 | 1.53 |
| | 052 | 1" thick, 3#/C.F. R4.3 foil faced | .008 | | .74 | .10 | | .84 | .98 |
| | 054 | 1-1/2" thick, R6.5 | .009 | | .77 | .11 | | .88 | 1.02 |
| | 056 | 2" thick, R8.7 | .009 | | 1.14 | .12 | | 1.26 | 1.44 |
| | 058 | 2-1/2" thick, R10.9 | .010 | | 1.28 | .12 | | 1.40 | 1.60 |
| | 060 | 3" thick, R13 | .011 | | 1.53 | .13 | | 1.66 | 1.89 |
| | 067 | 1" thick, 6#/C.F. R4.3 unfaced | .009 | | .70 | .11 | | .81 | .94 |
| | 069 | 1-1/2" thick, R6.5 | .009 | | 1.05 | .12 | | 1.17 | 1.34 |
| | 070 | 2" thick, R8.7 | .010 | | 1.40 | .12 | | 1.52 | 1.74 |
| | 072 | 2-1/2" thick, R10.9 | .011 | | 1.75 | .13 | | 1.88 | 2.13 |

For expanded coverage of these items see *Building Construction Cost Data 1985*

## 7.2 Insulation

| | | | CREW | MAN-HOURS | UNIT | MAT. | LABOR | EQUIP. | TOTAL | TOTAL INCL O&P |
|---|---|---|---|---|---|---|---|---|---|---|
| 82 | 074 | 3" thick, R13 | 1 Carp | .011 | S.F. | 2.54 | .14 | | 2.68 | 3.02 |
| | 082 | 1" thick, 6#/C.F. R4.3 foil faced | | .009 | | 1.05 | .11 | | 1.16 | 1.33 |
| | 084 | 1-1/2" thick, R6.5 | | .009 | | 1.32 | .12 | | 1.44 | 1.64 |
| | 086 | 2" thick, R8.7 | | .010 | | 1.75 | .12 | | 1.87 | 2.12 |
| | 088 | 2-1/2" thick, R10.9 | | .011 | | 2.18 | .13 | | 2.31 | 2.61 |
| | 090 | 3" thick, R13 | | .011 | | 2.49 | .14 | | 2.63 | 2.96 |
| | 150 | Foamglass, 1-1/2" thick, R2.64 | | .011 | | 1.42 | .13 | | 1.55 | 1.77 |
| | 155 | 2" thick, R5.26 | | .011 | | 1.92 | .14 | | 2.06 | 2.34 |
| | 170 | Perlite, 1" thick, R2.77 | | .011 | | .28 | .13 | | .41 | .52 |
| | 175 | 2" thick, R5.55 | | .011 | | .50 | .14 | | .64 | .77 |
| | 190 | Polystyrene, extruded, blue, 2.2#/C.F.,3/4" thick, R4 | | .010 | | .36 | .13 | | .49 | .60 |
| | 194 | 1-1/2" thick, R8.1 | | .011 | | .70 | .14 | | .84 | .99 |
| | 196 | 2" thick, R10.8 | | .011 | | .78 | .14 | | .92 | 1.08 |
| | 210 | Molded bead board, white, 1" thick, R3.85 | | .011 | | .15 | .13 | | .28 | .37 |
| | 212 | 1-1/2" thick, R5.6 | | .011 | | .26 | .14 | | .40 | .50 |
| | 214 | 2" thick, R7.7 | | .011 | | .31 | .14 | | .45 | .56 |
| | 235 | Sheathing, insulating foil faced fiberboard, 3/8" thick | ↓ | .013 | ↓ | .18 | .16 | | .34 | .45 |
| | 240 | | | | | | | | | |
| | 250 | Urethane, no paper backing 1/2" thick, R2.9 | 1 Carp | .010 | S.F. | .21 | .13 | | .34 | .43 |
| | 252 | 1" thick, R5.8 | | .011 | | .41 | .13 | | .54 | .66 |
| | 254 | 1-1/2" thick, R8.7 | | .011 | | .60 | .14 | | .74 | .88 |
| | 256 | 2" thick, R11.7 | | .011 | | .72 | .14 | | .86 | 1.02 |
| | 270 | Fire resistant 1/2" thick, R2.9 | | .010 | | .26 | .13 | | .39 | .49 |
| | 272 | 1" thick, R5.8 | | .011 | | .50 | .13 | | .63 | .76 |
| | 274 | 1-1/2" thick, R8.7 | | .011 | | .76 | .14 | | .90 | 1.05 |
| | 276 | 2" thick, R11.7 | ↓ | .011 | ↓ | .90 | .14 | | 1.04 | 1.21 |
| 85-001 | | **WALL OR CEILING INSULATION, NON-RIGID** | | | | | | | | |
| | 004 | Fiberglass, kraft faced, batts or blankets | | | | | | | | |
| | 006 | 3-1/2" thick, R11, 11" wide | 1 Carp | .005 | S.F. | .18 | .06 | | .24 | .30 |
| | 014 | 6" thick, R19, 11" wide | | .006 | | .30 | .07 | | .37 | .45 |
| | 020 | 9" thick, R30, 15" wide | | .006 | | .48 | .07 | | .55 | .64 |
| | 024 | 12" thick, R38, 15" wide | ↓ | .006 | ↓ | .65 | .07 | | .72 | .83 |
| | 040 | Fiberglass, foil faced, batts or blankets | | | | | | | | |
| | 042 | 3-1/2" thick, R11, 15" wide | 1 Carp | .005 | S.F. | .20 | .06 | | .26 | .32 |
| | 046 | 6" thick, R19, 15" wide | | .005 | | .32 | .06 | | .38 | .45 |
| | 050 | 9" thick, R30, 15" wide | ↓ | .006 | ↓ | .52 | .07 | | .59 | .69 |
| | 080 | Fiberglass, unfaced, batts or blankets | | | | | | | | |
| | 082 | 3-1/2" thick, R11, 15" wide | 1 Carp | .005 | S.F. | .16 | .06 | | .22 | .27 |
| | 086 | 6" thick, R19, 15" wide | | .006 | | .29 | .07 | | .36 | .43 |
| | 090 | 9" thick, R30, 15" wide | | .007 | | .46 | .09 | | .55 | .64 |
| | 094 | 12" thick, R38, 15" wide | ↓ | .007 | ↓ | .62 | .09 | | .71 | .82 |
| | 130 | Mineral fiber batts, kraft faced | | | | | | | | |
| | 132 | 3-1/2" thick, R13 | 1 Carp | .005 | S.F. | .48 | .06 | | .54 | .63 |
| | 134 | 6" thick, R19 | | .005 | | .58 | .06 | | .64 | .74 |
| | 138 | 10" thick, R30 | ↓ | .006 | | .92 | .07 | | .99 | 1.13 |
| | 190 | For foil backing 2 sides, add | | | ↓ | .06 | | | .06 | .06M |

## 7.3 Shingles

| | | | CREW | MAN-HOURS | UNIT | MAT. | LABOR | EQUIP. | TOTAL | TOTAL INCL O&P |
|---|---|---|---|---|---|---|---|---|---|---|
| 05-001 | | **ALUMINUM** Shingles, mill finish,.020" thick | 1 Carp | 3.480 | Sq. | 115 | 43 | | 158 | 195 |
| | 010 | .030" thick | " | 3.480 | | 146 | 43 | | 189 | 230 |
| | 030 | For colors, anodized finish, add | | | | 32 | | | 32 | 35M |
| | 040 | For bonderized finish, add | | | ↓ | 64 | | | 64 | 70M |
| | 060 | Ridge cap, .020" thick | 1 Carp | .047 | L.F. | 3.98 | .58 | | 4.56 | 5.30 |
| | 070 | .030" thick | " | .047 | " | 5.70 | .58 | | 6.28 | 7.20 |

For expanded coverage of these items see *Building Construction Cost Data 1985*

## 7.3 Shingles

| | | | CREW | MAN-HOURS | UNIT | MAT. | LABOR | EQUIP. | TOTAL | TOTAL INCL O&P |
|---|---|---|---|---|---|---|---|---|---|---|
| 05 | 090 | Valley section for above, .020" thick | 1 Carp | .047 | L.F. | 1.46 | .58 | | 2.04 | 2.53 |
| | 100 | .030" thick | " | .047 | " | 1.57 | .58 | | 2.15 | 2.65 |
| | 120 | For 1" factory applied polystyrene insulation, add | | | Sq. | 22.50 | | | 22.50 | 25M |
| 07-001 | | **ALUMINUM** Tiles, .019" thick, mission tile | 1 Carp | 3.480 | Sq. | 226 | 43 | | 269 | 315 |
| | 020 | Spanish tiles | | 2.670 | | 180 | 33 | | 213 | 250 |
| 10-001 | | **ASBESTOS** Mineral fiber strip shingles, 14" x 30", 325 lb. per square | | 2 | | 100 | 25 | | 125 | 150 |
| | 010 | 12" x 24", 167 lb. per square | | 2.290 | | 82 | 28 | | 110 | 135 |
| | 020 | Shakes, 9.35" x 16", 500 lb. per square (siding) | | 3.640 | | 147 | 45 | | 192 | 235 |
| | 030 | Hip & ridge shingles, 5-3/8" x 14" | | .080 | L.F. | 285 | .99 | | 285.99 | 315 |
| | 040 | Hexagonal shape, 16" x 16" | | 2.670 | Sq. | 112 | 33 | | 145 | 175 |
| | 050 | Square, 16" x 16" | | 2.670 | | 101 | 33 | | 134 | 165 |
| | 200 | For steep roofs, add | | | | | 50% | | | |
| 15-001 | | **ASPHALT SHINGLES** | | | | | | | | |
| | 010 | Standard strip shingles | | | | | | | | |
| | 015 | Inorganic, class A, 210-235 lb./square, 3 bundles/square | 1 Rofc | 1.450 | Sq. | 28 | 16.85 | | 44.85 | 59 |
| | 020 | Organic, class C, 235-240 lb./square, 3 bundles/square | " | 1.600 | " | 28 | 18.55 | | 46.55 | 62 |
| | 025 | Standard, laminated multi-layered shingles | | | | | | | | |
| | 030 | Class A, 240-260 lb./square, 3 bundles/square | 1 Rofc | 1.780 | Sq. | 51 | 21 | | 72 | 90 |
| | 035 | Class C, 260-300 lb./square, 4 bundles/square | " | 2 | " | 48 | 23 | | 71 | 91 |
| | 040 | Premium, laminated multi-layered shingles | | | | | | | | |
| | 045 | Class A, 260-300 lb./square, 4 bundles/square | 1 Rofc | 2.290 | Sq. | 84 | 27 | | 111 | 135 |
| | 050 | Class C, 300-385 lb./square, 5 bundles/square | " | 2.670 | " | 80 | 31 | | 111 | 140 |
| | 070 | Hip and ridge roll | | .020 | L.F. | .55 | .23 | | .78 | .99 |
| | 090 | Ridge shingles | | .024 | " | .65 | .28 | | .93 | 1.18 |
| | 100 | For steep roofs, add | | | | | 50% | | | |
| 20-001 | | **CLAY TILE** 8-1/4" x 11" exposure, colors, 730 lb. per sq. | | | | | | | | |
| | 020 | Lanai tile or Classic tile | 1 Rots | 4.850 | Sq. | 290 | 57 | | 347 | 415 |
| | 030 | Americana, most colors | | 4.850 | | 777 | 57 | | 834 | 950 |
| | 035 | Green, gray or brown | | 4.850 | | 777 | 57 | | 834 | 950 |
| | 040 | Blue | | 4.850 | | 749 | 57 | | 806 | 920 |
| | 060 | Spanish tile, 900 lb. per sq., red | | 4.440 | | 236 | 52 | | 288 | 345 |
| | 080 | Buff, green, gray, brown | | 4.440 | | 410 | 52 | | 462 | 535 |
| | 090 | Blue | | 4.440 | | 769 | 52 | | 821 | 930 |
| | 110 | Mission tile, 1220 lb. per sq., machine scored finish, red | | 6.960 | | 466 | 81 | | 547 | 650 |
| | 170 | French tile, 935 lb. per sq., smooth finish, red | | 5.930 | | 910 | 69 | | 979 | 1,125 |
| | 175 | Blue or green | | 5.930 | | 935 | 69 | | 1,004 | 1,150 |
| | 180 | Norman tile, 1600 lb. per sq. | | 8 | | 1,022 | 94 | | 1,116 | 1,275 |
| | 220 | Williamsburg tile, 950 lb. per sq., aged cedar | | 5.930 | | 289 | 69 | | 358 | 435 |
| | 225 | Gray or green | | 5.930 | | 289 | 69 | | 358 | 435 |
| | 235 | Ridge shingles, clay tile | | .040 | L.F. | 3.50 | .47 | | 3.97 | 4.63 |
| | 300 | For steep roofs, add to above | | | | | | | | |
| 25-001 | | **CONCRETE TILE** Including installation of accessories | 1 Rots | 5.930 | Sq. | 90 | 69 | | 159 | 215 |
| | 015 | Custom blues | " | 5.930 | " | 270 | 69 | | 339 | 410 |
| 35-001 | | **SLATE** Including felt underlay & nails, Buckingham, Virginia, black | | | | | | | | |
| | 010 | 3/16" thick | 1 Rots | 4.570 | Sq. | 570 | 53 | | 623 | 715 |
| | 020 | 1/4" thick | | 4.570 | | 570 | 53 | | 623 | 715 |
| | 090 | Pennsylvania black, Bangor, #1 clear | | 4.570 | | 570 | 53 | | 623 | 715 |
| | 120 | Vermont, unfading colors, green, mottled green | | 4.570 | | 560 | 53 | | 613 | 705 |
| | 130 | Semi-weathering green & gray | | 4.570 | | 570 | 53 | | 623 | 715 |
| | 140 | Purple | | 4.570 | | 630 | 53 | | 683 | 780 |
| | 150 | Black or gray | | 4.570 | | 630 | 53 | | 683 | 780 |
| | 151 | For steep roofs, add to above | | | | | 50% | | | |
| | 152 | | | | | | | | | |
| | 270 | Ridge shingles, slate | 1 Rots | .040 | L.F. | 3.60 | .47 | | 4.07 | 4.74 |
| 40-001 | | **STEEL** Shingles, galvanized, 26 gauge | 1 Rots | 3.640 | Sq. | 42 | 43 | | 85 | 115 |
| | 020 | 24 gauge | " | 3.640 | " | 44 | 43 | | 87 | 120 |

For expanded coverage of these items see *Building Construction Cost Data 1985*

## 7.3 Shingles

| | | | CREW | MAN-HOURS | UNIT | MAT. | LABOR | EQUIP. | TOTAL | TOTAL INCL O&P |
|---|---|---|---|---|---|---|---|---|---|---|
| 40 | 030 | For colored galvanized shingles, add | | | Sq. | 36 | | | 36 | 40M |
| | 050 | For 1" factory applied polystyrene insulation, add | | | | 24 | | | 24 | 26M |
| 45-001 | | WOOD 16" No. 1 red cedar shingles, 5X, 5" exposure, on roof | 1 Carp | 3.200 | | 133 | 40 | | 173 | 210 |
| | 001 | 5" exposure, 16" #1 red cedar fire proof | | 3.200 | | 196 | 40 | | 236 | 278 |
| | 020 | 7-1/2" exposure, on walls | | 3.900 | | 88 | 48 | | 136 | 175 |
| | 020 | With fire retardant shingles | | 3.900 | | 180 | 48 | | 228 | 275 |
| | 030 | 18" No. 1 red cedar perfections, 5-1/2" exposure, on roof | | 2.910 | | 110 | 36 | | 146 | 180 |
| | 030 | 5-1/2" exposure, 18", #1 red cedar fire proof | | 2.910 | | 216 | 36 | | 252 | 295 |
| | 050 | 7-1/2" exposure, on walls | | 3.560 | | 110 | 44 | | 154 | 190 |
| | 050 | With fire retardant shingles | | 3.560 | | 199 | 44 | | 243 | 290 |
| | 060 | Resquared, and rebutted, 5-1/2" exposure, on roof | | 2.670 | | 111 | 33 | | 144 | 175 |
| | 060 | 5-1/2" exposure, 18", fireproof resquared | | 2.670 | | 201 | 33 | | 234 | 275 |
| | 090 | 7-1/2" exposure, on walls | | 3.270 | | 102 | 40 | | 142 | 175 |
| | 090 | With fire retardant shingles | | 3.270 | | 189 | 40 | | 229 | 270 |
| | 110 | Hand-split red cedar shakes, on walls, 24" long, 10" exposure | | 3.200 | | 109 | 40 | | 149 | 180 |
| | 110 | With fire retardant shakes | | 3.200 | | 211 | 40 | | 251 | 295 |
| | 120 | 18" long, 8-1/2" exposure | | 4 | | 84 | 49 | | 133 | 170 |
| | 120 | With fire retardant shakes | | 4 | | 255 | 49 | | 304 | 360 |
| | 200 | White cedar shingles, 16" long, extras, 5" exposure, on roof | | 3.330 | | 80 | 41 | | 121 | 155 |
| | 210 | 7-1/2" exposure, on walls | | 4 | | 70 | 49 | | 119 | 155 |
| | 215 | "B" grade, 5" exposure on walls | | 3.330 | | 57.40 | 41 | | 98.40 | 130 |
| | 230 | For #15 organic felt underlayment on roof, 1 layer, add | | .125 | | 2.62 | 1.54 | | 4.16 | 5.35 |
| | 240 | 2 layers, add | | .250 | | 5.24 | 3.09 | | 8.33 | 10.65 |
| | 250 | For plastic-coated steel foil underlayment, class B roofs, add | | .125 | | 43.55 | 1.54 | | 45.09 | 50 |
| | 260 | For steep roofs, add to above | | | | | 50% | | | |
| | 300 | Ridge shakes or shingle wood | 1 Carp | .029 | L.F. | 1.06 | .35 | | 1.41 | 1.72 |

## 7.4 Roofing & Siding

| | | | CREW | MAN-HOURS | UNIT | MAT. | LABOR | EQUIP. | TOTAL | TOTAL INCL O&P |
|---|---|---|---|---|---|---|---|---|---|---|
| 06-001 | | ALUMINUM SIDING .019" thick, on steel construction, natural | | | | | | | | |
| | 604 | .024 thick smooth white single 8" wide | F-2 | .036 | S.F. | .90 | .45 | .03 | 1.38 | 1.73 |
| | 606 | Double 4" pattern 8" wide | | .036 | | .95 | .45 | .03 | 1.43 | 1.78 |
| | 608 | 5" pattern 10" wide | | .032 | | 1 | .40 | .03 | 1.43 | 1.76 |
| | 612 | Embossed white single 8" wide | | .036 | | .95 | .45 | .03 | 1.43 | 1.78 |
| | 614 | Double 4" pattern 8" wide | | .036 | | .98 | .45 | .03 | 1.46 | 1.82 |
| | 616 | 5" pattern 10" wide | | .032 | | 1.05 | .40 | .03 | 1.48 | 1.82 |
| | 632 | .019 thick insulated backed smooth white single 8" wide | | .048 | | .94 | .59 | .04 | 1.57 | 2.01 |
| | 634 | Double 4" pattern 8" wide | | .048 | | .97 | .59 | .04 | 1.60 | 2.05 |
| | 636 | 5" pattern 10" wide | | .043 | | 1 | .53 | .04 | 1.57 | 1.99 |
| | 640 | Embossed white single 8" wide | | .048 | | .95 | .59 | .04 | 1.58 | 2.02 |
| | 642 | Double 4" pattern 8" wide | | .048 | | .98 | .59 | .04 | 1.61 | 2.06 |
| | 644 | 5" pattern 10" wide | | .043 | | 1 | .53 | .04 | 1.57 | 1.99 |
| | 650 | Shake finish 10" wide white | | .032 | | 1.20 | .40 | .03 | 1.63 | 1.98 |
| | 660 | Vertical pattern 12" wide white | | .031 | | .95 | .38 | .03 | 1.36 | 1.68 |
| | 664 | For colors add | | | | .05 | | | .05 | .05M |
| | 670 | Accessories white | | | | | | | | |
| | 672 | Starter strip 2-1/8" | F-2 | .026 | L.F. | .23 | .33 | .02 | .58 | .79 |
| | 674 | Sill trim | | .036 | | .25 | .44 | .03 | .72 | 1 |
| | 676 | Inside corner | | .026 | | .71 | .33 | .02 | 1.06 | 1.32 |
| | 678 | Outside corner post | | .026 | | 1.10 | .33 | .02 | 1.45 | 1.75 |
| | 680 | Door & window trim | | .036 | | .26 | .45 | .03 | .74 | 1.03 |
| | 682 | For colors add | | | | .03 | | | .03 | .03M |
| | 690 | Soffit & fascia 1' overhang solid (see 7.6-20 and 7.6-54) | F-2 | .145 | | 2.35 | 1.79 | .13 | 4.27 | 5.55 |
| | 692 | Vented | | .145 | | 2.40 | 1.79 | .13 | 4.32 | 5.60 |
| | 694 | 2' overhang solid | | .160 | | 4.70 | 1.98 | .14 | 6.82 | 8.45 |

For expanded coverage of these items see *Building Construction Cost Data 1985*

## 7.4 Roofing & Siding

| | | CREW | MAN-HOURS | UNIT | MAT. | LABOR | EQUIP. | TOTAL | TOTAL INCL O&P |
|---|---|---|---|---|---|---|---|---|---|
| 06 696 | Vented | F-2 | .160 | L.F. | 4.85 | 1.98 | .14 | 6.97 | 8.60 |
| 12-001 | **ASPHALT** Coated felt, #30, 2 sq. per roll, not mopped | 1 Rofc | .138 | Sq. | 9.98 | 1.60 | | 11.58 | 13.65 |
| 020 | #15, 4 sq. per roll, plain or perforated, not mopped | | .138 | | 9.10 | 1.60 | | 10.70 | 12.65 |
| 025 | Perforated | | .138 | | 3.55 | 1.60 | | 5.15 | 6.55 |
| 030 | Roll roofing, smooth, #55 | | .533 | | 11.15 | 6.18 | | 17.33 | 22.52 |
| 050 | #90 | | .533 | | 13.97 | 6.20 | | 20.17 | 26 |
| 052 | Mineralized | | .533 | | 15.80 | 6.20 | | 22 | 28 |
| 054 | D.C. (Double coverage), 19" selvage edge | ↓ | .800 | ↓ | 29.10 | 9.30 | | 38.40 | 47 |
| 058 | Adhesive (lap cement) | | | Gal. | 3.78 | | | 3.78 | 4.15M |
| 15-001 | **BUILT-UP ROOFING** | | | | | | | | |
| 012 | Asphalt flood coat with gravel/slag surfacing, not including | | | | | | | | |
| 014 | Insulation, flashing or wood nailers | | | | | | | | |
| 020 | Asbestos base sheet, 3 plies #15 asbestos felt, mopped | G-1 | 2.550 | Sq. | 53 | 28.56 | 4.44 | 86 | 110 |
| 035 | On nailable decks | | 2.670 | | 81 | 29.35 | 4.65 | 115 | 145 |
| 050 | 4 plies #15 asbestos felt, mopped | | 2.800 | | 87.72 | 31.11 | 4.89 | 123.72 | 155 |
| 055 | On nailable decks | | 2.950 | | 74.18 | 32.85 | 5.15 | 112.18 | 140 |
| 070 | Coated glass base sheet, 2 plies glass (type IV), mopped | | 2.550 | | 42 | 28.56 | 4.44 | 75 | 98 |
| 085 | 3 plies glass, mopped | | 2.800 | | 50 | 31.11 | 4.89 | 86 | 110 |
| 095 | On nailable decks | | 2.950 | | 50 | 32.85 | 5.15 | 88 | 115 |
| 100 | 3 plies glass fiber felt (type IV), mopped | | 2.550 | | 38 | 28.56 | 4.44 | 71 | 94 |
| 105 | On nailable decks | | 2.670 | | 38 | 29.35 | 4.65 | 72 | 96 |
| 110 | 4 plies glass fiber felt (type IV), mopped | | 2.800 | | 47 | 31.11 | 4.89 | 83 | 110 |
| 115 | On nailable decks | | 2.950 | | 47 | 32.85 | 5.15 | 85 | 110 |
| 120 | Organic base sheet, 3 plies #15 organic felt, mopped | | 2.800 | | 33 | 31.11 | 4.89 | 69 | 93 |
| 125 | On nailable decks | | 2.950 | | 33 | 32.85 | 5.15 | 71 | 96 |
| 130 | 4 plies #15 organic felt, mopped | ↓ | 2.550 | ↓ | 32 | 28.56 | 4.44 | 65 | 87 |
| 200 | Asphalt flood coat, smooth surface | | | | | | | | |
| 220 | Asbestos base sheet & 3 plies #15 asbestos felt, mopped | G-1 | 2.330 | Sq. | 48 | 25.93 | 4.07 | 78 | 100 |
| 240 | On nailable decks | | 2.430 | | 48 | 26.75 | 4.25 | 79 | 100 |
| 260 | 4 plies #15 asbestos felt, mopped | | 2.330 | | 43 | 25.93 | 4.07 | 73 | 95 |
| 270 | On nailable decks | ↓ | 2.430 | ↓ | 43 | 26.75 | 4.25 | 74 | 97 |
| 290 | Coated glass fiber base sheet, mopped, and 2 plies of | | | | | | | | |
| 291 | glass fiber felt (type IV) | G-1 | 2.240 | Sq. | 40 | 25.09 | 3.91 | 69 | 90 |
| 310 | On nailable decks | | 2.330 | | 40 | 25.93 | 4.07 | 70 | 91 |
| 320 | 3 plies, mopped | | 2.430 | | 48 | 26.75 | 4.25 | 79 | 100 |
| 330 | On nailable decks | | 2.550 | | 48 | 28.56 | 4.44 | 81 | 105 |
| 350 | 3 plies glass fiber felt (type IV), mopped | | 2.240 | | 36 | 25.09 | 3.91 | 65 | 85 |
| 360 | On nailable decks | | 2.330 | | 36 | 25.93 | 4.07 | 66 | 87 |
| 380 | 4 plies glass fiber felt (type IV), mopped | | 2.430 | | 45 | 26.75 | 4.25 | 76 | 99 |
| 390 | On nailable decks | | 2.550 | | 45 | 28.56 | 4.44 | 78 | 100 |
| 400 | Organic base sheet & 3 plies #15 organic felt, mopped | | 2.430 | | 31 | 26.75 | 4.25 | 62 | 84 |
| 420 | On nailable decks | | 2.550 | | 31 | 28.56 | 4.44 | 64 | 86 |
| 430 | 4 plies #15 organic felt, mopped | ↓ | 2.430 | ↓ | 30 | 26.75 | 4.25 | 61 | 82 |
| 450 | Coal tar pitch with gravel/slag surfacing | | | | | | | | |
| 460 | 4 plies #15 asbestos felt, mopped | G-1 | 2.550 | Sq. | 95 | 28.56 | 4.44 | 128 | 155 |
| 480 | 3 plies glass fiber felt (type IV), mopped | " | 2.800 | " | 68 | 31.11 | 4.89 | 104 | 130 |
| 500 | Coated glass fiber base sheet, and 2 plies of | | | | | | | | |
| 501 | glass fiber felt, type IV, mopped | G-1 | 2.800 | Sq. | 72 | 31.11 | 4.89 | 108 | 135 |
| 530 | On nailable decks | | 2.950 | | 72 | 32.85 | 5.15 | 110 | 140 |
| 540 | On wood decks | | 2.950 | | 72 | 32.85 | 5.15 | 110 | 140 |
| 560 | 4 plies glass fiber felt (type IV), mopped | | 2.550 | | 84 | 28.56 | 4.44 | 117 | 145 |
| 580 | On nailable decks | | 2.670 | | 84 | 29.35 | 4.65 | 118 | 145 |
| 590 | On wood decks | | 2.670 | | 84 | 29.35 | 4.65 | 118 | 145 |
| 600 | 4 plies #15 organic felt, mopped | ↓ | 2.550 | ↓ | 79 | 28.56 | 4.44 | 112 | 140 |
| 660 | Asphalt mineral surface, roll roofing | | | | | | | | |
| 670 | 1 ply #15 organic felt, 2 plies mineral surfaced | | | | | | | | |
| 680 | selvage, edge roofing, lap 19", nailed & mopped | G-1 | 2.070 | Sq. | 38 | 23.38 | 3.62 | 65 | 84 |
| 700 | 3 plies glass fiber felt (type IV), 1 ply mineral surfaced | | | | | | | | |
| 710 | selvage, edge roofing, lapped 18", mopped | G-1 | 2.240 | Sq. | 58 | 25.09 | 3.91 | 87 | 110 |

For expanded coverage of these items see *Building Construction Cost Data 1985*

## 7.4 Roofing & Siding

| | | | CREW | MAN-HOURS | UNIT | MAT. | LABOR | EQUIP. | TOTAL | TOTAL INCL O&P |
|---|---|---|---|---|---|---|---|---|---|---|
| 15 | 740 | Coated glass fiber base sheet, 2 plies of glass fiber | | | | | | | | |
| | 750 | felt (type IV), 1 ply mineral surfaced selvage | | | | | | | | |
| | 760 | edge roofing, lapped 18", mopped | G-1 | 2.240 | Sq. | 62 | 25.09 | 3.91 | 91 | 115 |
| | 770 | On nailable decks | " | 2.330 | " | 62 | 25.93 | 4.07 | 92 | 115 |
| | 780 | 3 plies glass fiber felt (type III), 1 ply mineral surfaced | | | | | | | | |
| | 790 | selvage, edge roofing, lapped 18", mopped | G-1 | 2.240 | Sq. | 56 | 25.09 | 3.91 | 85 | 105 |
| | 791 | Cold applied, 3-ply system (components listed below) | G-5 | .800 | " | | 8.70 | 1.95 | 10.65 | 16.60 |
| | 792 | Spunbond poly.fabric,1.35 oz./S.Y., 36" wide, 10.8 Sq./roll | | | Ea. | 60 | | | 60 | 66M |
| | 793 | 49" wide, 14.6 Sq./roll | | | | 81 | | | 81 | 89M |
| | 794 | 2.10 oz./S.Y., 36" wide, 10.8 Sq./roll | | | | 94 | | | 94 | 105M |
| | 795 | 49" wide, 14.6 Sq./roll | | | ↓ | 129 | | | 129 | 140M |
| | 796 | Base & finish coat, 3 gal./Sq., 5 gal./can | | | Gal. | 5 | | | 5 | 5.50M |
| | 797 | Coating, ceramic granules, 1/2 Sq./bag | | | Ea. | 9.40 | | | 9.40 | 10.35M |
| | 798 | Aluminum, 2 gal./Sq. | | | Gal. | 8.50 | | | 8.50 | 9.35M |
| | 799 | Emulsion, fibered or non-fibered, 4 gal./Sq. | | | " | 3.80 | | | 3.80 | 4.18M |
| | 800 | Roof covering | | | Sq. | 6.45 | | | 6.45 | 7.10M |
| | 810 | For 15 year bond | | | | 12 | | | 12 | 13.20M |
| | 820 | For 20 year bond | | | ↓ | 25 | | | 25 | 28M |
| | 880 | Maintenance agreement,entire system, 5 years, add 1-5% total cost | | | Job | | | | | |
| | 900 | Bond charge, asphalt felt roofs, 10 year bond | | | Sq. | 11.50 | | | 11.50 | 12.65M |
| | 910 | New work | 1 Skwk | .364 | | | 4.58 | | 4.58 | 7.30L |
| | 920 | Reroofing | " | .800 | ↓ | | 10.10 | | 10.10 | 16.10L |
| | 940 | Roof samples-laboratory testing | | | | | | | | |
| | 950 | Minimum labor/ equipment charge | | | Ea. | | | | | 175 |
| | 960 | Maximum | | | " | | | | | 300 |
| | 965 | Non-destructive inspection, includes inspection & written report | | | | | | | | |
| | 966 | Capacitance meter, minimum | | | S.F. | | | | | .08 |
| | 967 | Maximum | | | | | | | | .25 |
| | 970 | Nuclear moisture meter, minimum with $500.00 guaranty | | | | | | | | .03 |
| | 971 | Maximum | | | | | | | | .10 |
| | 975 | Infra-red, hand-held, minimum with $500.00 guaranty | | | | | | | | .03 |
| | 976 | Maximum | | | ↓ | | | | | .10 |
| | 980 | Reduce daily outputs by 25% for complicated, or | | | | | | | | |
| | 981 | Unusual shaped roofs and those with many penetrations | | | | | | | | |
| 18-001 | | CANTS 4" x 4" treated timber, cut diagonally | 1 Rofc | .025 | L.F. | 1.08 | .29 | | 1.37 | 1.66 |
| | 010 | Foamglass | | .025 | | .44 | .29 | | .73 | .96 |
| | 030 | Mineral or fiber, trapezoidal, 1"x 4" x 48" | | .025 | | .23 | .29 | | .52 | .73 |
| | 040 | 1-1/2" x 5-5/8" x 48" | | .025 | ↓ | .28 | .29 | | .57 | .78 |
| 33-001 | | FELT Asphalt asbestos, #15, no mopping | | .138 | Sq. | 2.63 | 1.60 | | 4.23 | 5.55 |
| | 020 | #43 | | .138 | | 5.23 | 1.60 | | 6.83 | 8.40 |
| | 030 | Base sheet, #45 | | .138 | | 5.20 | 1.60 | | 6.80 | 8.35 |
| | 040 | #50 | | .138 | | 6.50 | 1.60 | | 8.10 | 9.80 |
| | 050 | Cap, mineral surfaced | | .138 | | 31.30 | 1.60 | | 32.90 | 37 |
| | 060 | Flashing membrane, #65 | | .500 | | 7.80 | 5.80 | | 13.60 | 18.20 |
| | 080 | Coal tar asbestos, #15, no mopping | | .138 | | 10.40 | 1.60 | | 12 | 14.10 |
| | 090 | Asphalt felt, #15, 4 sq. per roll, no mopping | | .138 | | 2.62 | 1.60 | | 4.22 | 5.55 |
| | 110 | #30, 2 sq. per roll | | .138 | | 5.24 | 1.60 | | 6.84 | 8.40 |
| | 120 | Double coated, #30 | | .138 | | 6.27 | 1.60 | | 7.87 | 9.55 |
| | 140 | #40 | | .138 | | 7.55 | 1.60 | | 9.15 | 10.95 |
| | 150 | Tarred felt, organic, #15, 4 sq. rolls | | .138 | | 4.75 | 1.60 | | 6.35 | 7.90 |
| | 155 | #30, 2 sq. roll | | .138 | | 10.15 | 1.60 | | 11.75 | 13.80 |
| | 170 | Add for mopping above felts, per ply, asphalt, 20 lbs. per sq. | | .286 | | 2 | 3.31 | | 5.31 | 7.70 |
| | 180 | Coal tar mopping, 30 lbs. per sq. | ↓ | .286 | | 6.40 | 3.31 | | 9.71 | 12.55 |
| | 190 | Flood coat, with asphalt (60 lbs. per sq.) | 2 Rofc | .982 | | 6 | 11.40 | | 17.40 | 25 |
| | 200 | With coal tar (75 lbs. per sq.) | " | .982 | ↓ | 15.95 | 11.40 | | 27.35 | 36 |
| 52-001 | | SINGLE-PLY MEMBRANES | | | | | | | | |
| | 002 | | | | | | | | | |
| | 020 | Chlorinated polyethylene(CPE), 40 mils, 0.31 P.S.F. | | | | | | | | |
| | 030 | Partially adhered with mechanical fasteners | G-5 | .008 | S.F. | 1.47 | .08 | .02 | 1.57 | 1.78 |

For expanded coverage of these items see *Building Construction Cost Data 1985*

## 7.4 Roofing & Siding

| | | | CREW | MAN-HOURS | UNIT | MAT. | LABOR | EQUIP. | TOTAL | TOTAL INCL O&P |
|---|---|---|---|---|---|---|---|---|---|---|
| 52 | 080 | Chlorosulfonated polyethylene-hypalon (CSPE), 35 mils, 0.25 PSF | | | | | | | | |
| | 090 | Fully adhered with neoprene latex | G-5 | .011 | S.F. | 1.28 | .12 | .03 | 1.43 | 1.64 |
| | 100 | 45 mils, 0.29 P.S.F. | | | | | | | | |
| | 110 | Loose-laid & ballasted with stone (10 P.S.F.) | G-5 | .006 | S.F. | 1.44 | .07 | .01 | 1.52 | 1.70 |
| | 120 | Partially adhered with fastening strips | | .008 | | 1.61 | .08 | .02 | 1.71 | 1.93 |
| | 130 | Plates with adhesive attachment | | .008 | | 1.75 | .08 | .02 | 1.85 | 2.09 |
| | 200 | Elastomer modified asphalt, 150 mils, 1.47 P.S.F. | | | | | | | | |
| | 240 | Partially adhered-torched | G-5 | .016 | S.F. | 2.55 | .17 | .04 | 2.76 | 3.14 |
| | 250 | Hot mopped with asphalt | | .016 | | 2.60 | .17 | .04 | 2.81 | 3.19 |
| | 270 | Fully adhered-torched | | .020 | | 2.90 | .22 | .05 | 3.17 | 3.60 |
| | 280 | Hot asphalt attachment | | .020 | | 2.95 | .22 | .05 | 3.22 | 3.66 |
| | 290 | Adhesive attachment | | .020 | | 2.96 | .22 | .05 | 3.23 | 3.67 |
| | 350 | Ethylene propylene diene monomer (EPDM), 45 mils, 0.28 P.S.F. | | | | | | | | |
| | 360 | Loose-laid & ballasted with stone (10 P.S.F.) | G-5 | .006 | S.F. | 1.05 | .07 | .01 | 1.13 | 1.27 |
| | 370 | Partially adhered with batten strips | | .008 | | 1.20 | .08 | .02 | 1.30 | 1.48 |
| | 380 | Fully adhered with adhesive | | .011 | | 1.45 | .12 | .03 | 1.60 | 1.83 |
| | 400 | 55 mils, 0.40 P.S.F. | | | | | | | | |
| | 410 | Loose-laid & ballasted with stone (10 P.S.F.) | G-5 | .006 | S.F. | .60 | .07 | .01 | .68 | .78 |
| | 420 | Partially adhered to plates @ 4' O.C. with adhesive | | .008 | | 1.95 | .08 | .02 | 2.05 | 2.31 |
| | 430 | Fully adhered with adhesive | | .011 | | .80 | .12 | .03 | .95 | 1.12 |
| | 450 | 60 mils, 0.35 P.S.F. | | | | | | | | |
| | 460 | Loose-laid & ballasted with stone (10 P.S.F.) | G-5 | .006 | S.F. | 1.29 | .07 | .01 | 1.37 | 1.54 |
| | 470 | Partially adhered with bar anchors | | .008 | | 1.45 | .08 | .02 | 1.55 | 1.76 |
| | 480 | Fully adhered with adhesive | | .011 | | 1.67 | .12 | .03 | 1.82 | 2.07 |
| | 485 | Vulcanizing tape for membrane, 2" x 50' roll | | | Ea. | 34 | | | 34 | 37M |
| | 490 | Batten strips, 10' sections | | | | 2.50 | | | 2.50 | 2.75M |
| | 491 | Vulcanizing tape for batten strips, 4" x 50' roll | | | | 76 | | | 76 | 84M |
| | 493 | Plate anchors | | | M | 81 | | | 81 | 89M |
| | 497 | Adhesive for fully adhered systems, 60 S.F./gal. | | | Gal. | 15 | | | 15 | 16.50M |
| | 500 | Modified bitumen | | | | | | | | |
| | 530 | 120 mils, 0.92 P.S.F., fully adhered with solvent | G-5 | .014 | S.F. | 1.79 | .16 | .03 | 1.98 | 2.27 |
| | 540 | 150 mils, 0.82 P.S.F. | | | | | | | | |
| | 550 | Loose-laid & ballasted with gravel (4 P.S.F.) | G-5 | .013 | S.F. | .80 | .14 | .03 | .97 | 1.14 |
| | 560 | Partially adhered with torch welding | | .016 | | .70 | .17 | .04 | .91 | 1.10 |
| | 570 | Fully adhered with torch welding | | .020 | | .70 | .22 | .05 | .97 | 1.18 |
| | 580 | Hot asphalt attachment | | .020 | | .75 | .22 | .05 | 1.02 | 1.24 |
| | 581 | | | | | | | | | |
| | 600 | 160 mils, 0.78 to 1.2 P.S.F., with asphalt emulsion coating | | | | | | | | |
| | 610 | Loose-laid & ballasted with stone/gravel (10 P.S.F.) | G-5 | .013 | S.F. | 2.25 | .14 | .03 | 2.42 | 2.73 |
| | 620 | Partially adhered with torch welding | | .016 | | 2.48 | .17 | .04 | 2.69 | 3.06 |
| | 630 | Fully adhered with torch welding | | .020 | | 2.73 | .22 | .05 | 3 | 3.42 |
| | 640 | Hot asphalt attachment | | .020 | | 2.78 | .22 | .05 | 3.05 | 3.47 |
| | 641 | | | | | | | | | |
| | 680 | Neoprene, 60 mils, 0.45 P.S.F. | | | | | | | | |
| | 700 | Partially adhered with mechanical fasteners | G-5 | .008 | S.F. | 1.65 | .08 | .02 | 1.75 | 1.98 |
| | 710 | Fully adhered with contact adhesive | " | .011 | " | 1.40 | .12 | .03 | 1.55 | 1.78 |
| | 711 | Uncured neoprene, 60 mils, for flashing | | | | | | | | |
| | 750 | Polyisobutylene (PIB), 100 mils, 0.57 P.S.F. | | | | | | | | |
| | 760 | Loose-laid & ballasted with stone/gravel (10 P.S.F.) | G-5 | .006 | S.F. | 1.40 | .07 | .01 | 1.48 | 1.66 |
| | 770 | Partially adhered with adhesive | | .008 | | 1.30 | .08 | .02 | 1.40 | 1.59 |
| | 780 | Hot asphalt attachment | | .008 | | 2.45 | .08 | .02 | 2.55 | 2.86 |
| | 790 | Fully adhered with contact cement | | .011 | | 1.40 | .12 | .03 | 1.55 | 1.78 |
| | 791 | | | | | | | | | |
| | 800 | Polymer Modified Bitumen, 160 mils, 0.88 P.S.F. | | | | | | | | |
| | 810 | Partially adhered with torch welding | G-5 | .016 | S.F. | 2.54 | .17 | .04 | 2.75 | 3.13 |
| | 815 | Fully adhered with self contained bitumen | " | .020 | " | 2.72 | .22 | .05 | 2.99 | 3.41 |
| | 816 | | | | | | | | | |
| | 820 | Polyvinyl chloride (P.V.C.) | | | | | | | | |
| | 825 | 45 mils, 0.30 P.S.F. | | | | | | | | |
| | 830 | Loose-laid & ballasted with stone/gravel (10 P.S.F.) | G-5 | .006 | S.F. | 1.45 | .07 | .01 | 1.53 | 1.71 |

For expanded coverage of these items see *Building Construction Cost Data 1985*

## 7.4 Roofing & Siding

| | | | CREW | MAN-HOURS | UNIT | MAT. | LABOR | EQUIP. | TOTAL | TOTAL INCL O&P |
|---|---|---|---|---|---|---|---|---|---|---|
| 52 | 835 | Partially adhered with mechanical fasteners | G-5 | .008 | S.F. | 1.96 | .08 | .02 | 2.06 | 2.32 |
| | 840 | 48 mils, 0.33 to 0.38 P.S.F. | | | | | | | | |
| | 845 | Loose-laid & ballasted with stone/gravel(10 P.S.F.) | G-5 | .006 | S.F. | .70 | .07 | .01 | .78 | .89 |
| | 850 | Partially adhered with mechanical & solvent weld | | .008 | | 1.06 | .08 | .02 | 1.16 | 1.33 |
| | 855 | Fully adhered with cold emulsion | | .011 | | 1.18 | .12 | .03 | 1.33 | 1.53 |
| | 860 | Hot asphalt attachment | | .011 | | 1.35 | .12 | .03 | 1.50 | 1.72 |
| | 865 | 60 mils, 0.40#, partially adhered with PVC coated strips | ↓ | .008 | | 2.50 | .08 | .02 | 2.60 | 2.91 |
| | 868 | Uncured neoprene, 60 mils, for flashing | 1 Rofc | .013 | | 1.02 | .15 | | 1.17 | 1.38 |
| | 869 | Separator sheet | G-5 | .010 | ↓ | .03 | .11 | .02 | .16 | .24 |
| | 870 | Reinforced PVC, 48 mils, 0.33 P.S.F. | | | | | | | | |
| | 875 | Loose-laid & ballasted with stone/gravel (12 P.S.F.) | G-5 | .006 | S.F. | 1.89 | .07 | .01 | 1.97 | 2.20 |
| | 880 | Partially adhered with mechanical fasteners | | .008 | | 2.38 | .08 | .02 | 2.48 | 2.78 |
| | 885 | Fully adhered with adhesive | ↓ | .011 | ↓ | 2.60 | .12 | .03 | 2.75 | 3.10 |
| 60-001 | | **STEEL SIDING** Beveled, vinyl coated, 8" wide | 1 Carp | .030 | S.F. | .58 | .37 | | .95 | 1.23 |
| | 005 | 10" wide | " | .030 | | .52 | .37 | | .89 | 1.15 |
| | 008 | Galv., corrugated or ribbed, on steel frame, 29 gauge | G-3 | .041 | | .51 | .49 | .03 | 1.03 | 1.37 |
| | 010 | 26 gauge | | .041 | | .69 | .49 | .03 | 1.21 | 1.57 |
| | 030 | 24 gauge | | .041 | | .95 | .49 | .03 | 1.47 | 1.86 |
| | 040 | 22 gauge | | .041 | | 1.11 | .49 | .03 | 1.63 | 2.03 |
| | 060 | 20 gauge | | .041 | | 1.27 | .49 | .03 | 1.79 | 2.21 |
| | 070 | Colored, corrugated or ribbed, on steel, 10 yr. finish, 29 gauge | | .041 | | .60 | .49 | .03 | 1.12 | 1.47 |
| | 090 | 26 gauge | | .041 | | .86 | .49 | .03 | 1.38 | 1.76 |
| | 100 | 24 gauge | ↓ | .041 | ↓ | 1.30 | .49 | .03 | 1.82 | 2.24 |
| 69-001 | | **VINYL SIDING** Solid PVC panels, 8" to 10" wide, plain | | | | .65 | | | .65 | .71M |
| | 200 | Smooth, white, single, 8" wide | F-2 | .032 | S.F. | .55 | .40 | .03 | .98 | 1.27 |
| | 202 | 10" wide | | .029 | | .57 | .35 | .03 | .95 | 1.22 |
| | 210 | Double 4" pattern, 8" wide | | .032 | | .60 | .40 | .03 | 1.03 | 1.32 |
| | 212 | 5" pattern, 10" wide | | .029 | | .62 | .35 | .03 | 1 | 1.28 |
| | 220 | Embossed, white, single, 8" wide | | .032 | | .60 | .40 | .03 | 1.03 | 1.32 |
| | 222 | 10 " wide | | .029 | | .62 | .35 | .03 | 1 | 1.28 |
| | 230 | Double 4" pattern, 8" wide | | .032 | | .64 | .40 | .03 | 1.07 | 1.37 |
| | 232 | 5" pattern, 10" wide | | .029 | | .66 | .35 | .03 | 1.04 | 1.32 |
| | 240 | Shake finish, 10" wide, white | | .029 | | .65 | .35 | .03 | 1.03 | 1.31 |
| | 260 | Vertical pattern, double 5", 10" wide, white | ↓ | .029 | ↓ | .58 | .35 | .03 | .96 | 1.23 |
| | 262 | | | | | | | | | |
| | 270 | For colors, add | | | S.F. | .05 | | | .05 | .05M |
| | 272 | For insulated backer, add | F-2 | .008 | " | .08 | .10 | .01 | .19 | .25 |
| | 300 | Accessories, starter, strip | | .023 | L.F. | .19 | .28 | .02 | .49 | .68 |
| | 310 | "J" channel, 1/2" | | .023 | | .16 | .28 | .02 | .46 | .64 |
| | 312 | 5/8" | | .023 | | .18 | .28 | .02 | .48 | .67 |
| | 314 | 3/4" | | .023 | | .19 | .28 | .02 | .49 | .68 |
| | 316 | 1" | | .023 | | .21 | .29 | .02 | .52 | .71 |
| | 318 | 1-1/8" | | .023 | | .22 | .29 | .02 | .53 | .72 |
| | 319 | 1-1/4" | | .024 | | .23 | .29 | .02 | .54 | .74 |
| | 320 | Under sill trim | | .032 | | .18 | .39 | .03 | .60 | .85 |
| | 330 | Outside corner post, 3" face, pocket 5/8" | | .023 | | .72 | .28 | .02 | 1.02 | 1.26 |
| | 332 | 7/8" | | .023 | | .74 | .29 | .02 | 1.05 | 1.29 |
| | 334 | 1-1/4" | | .024 | | .76 | .29 | .02 | 1.07 | 1.32 |
| | 340 | Inside corner post, pocket 5/8" | | .023 | | .72 | .28 | .02 | 1.02 | 1.26 |
| | 342 | 7/8" | | .023 | | .74 | .29 | .02 | 1.05 | 1.29 |
| | 344 | 1-1/4" | | .024 | | .76 | .29 | .02 | 1.07 | 1.32 |
| | 350 | Door & window trim, 2-1/2" face, pocket 5/8" | | .031 | | .33 | .38 | .03 | .74 | 1.01 |
| | 352 | 7/8" | | .032 | | .36 | .39 | .03 | .78 | 1.05 |
| | 354 | 1-1/4" | | .033 | | .38 | .40 | .03 | .81 | 1.09 |
| | 360 | Soffit & fascia, 1' overhang, solid | | .133 | | 1.31 | 1.64 | .12 | 3.07 | 4.18 |
| | 362 | Vented | | .133 | | 1.35 | 1.64 | .12 | 3.11 | 4.22 |
| | 370 | 2' overhang, solid | | .145 | | 2.62 | 1.79 | .13 | 4.54 | 5.85 |
| | 372 | Vented | ↓ | .145 | ↓ | 2.70 | 1.79 | .13 | 4.62 | 5.95 |

For expanded coverage of these items see *Building Construction Cost Data 1985*

## 7.6 Sheet Metal Work

| | | CREW | MAN-HOURS | UNIT | MAT. | LABOR | EQUIP. | TOTAL | TOTAL INCL O&P |
|---|---|---|---|---|---|---|---|---|---|
| 05-001 | **COPPER ROOFING** Batten seam, over 10 squares, 16 oz., 130 lb. per sq. | 1 Shee | 7.270 | Sq. | 175 | 100 | | 275 | 355 |
| 020 | 18 oz., 145 lb. per sq. | | 8 | | 190 | 110 | | 300 | 385 |
| 040 | Standing seam, over 10 squares, 16 oz., 125 lb. per sq. | | 6.150 | | 170 | 85 | | 255 | 325 |
| 060 | 18 oz., 140 lb. per sq. | | 6.670 | | 185 | 92 | | 277 | 350 |
| 090 | Flat seam, over 10 squares, 16 oz., 115 lb. per sq. | ▼ | 6.670 | | 160 | 92 | | 252 | 325 |
| 120 | For abnormal conditions or small areas, add | | | | 25% | 100% | | | |
| 130 | For lead-coated copper, add | | | ▼ | 18% | | | | |
| 10-001 | **DOWNSPOUTS** Aluminum 2" x 3", .020" thick, embossed | 1 Shee | .042 | L.F. | .50 | .58 | | 1.08 | 1.48 |
| 010 | Enameled | | .042 | | .55 | .58 | | 1.13 | 1.53 |
| 030 | Enameled, .024" thick, 2" x 3" | | .042 | | .71 | .58 | | 1.29 | 1.71 |
| 040 | 3" x 4" | | .057 | | 1.20 | .79 | | 1.99 | 2.58 |
| 060 | Round, corrugated aluminum, 3" diameter, .020" thick | | .042 | | .95 | .58 | | 1.53 | 1.97 |
| 070 | 4" diameter, .025" thick | | .057 | ▼ | 1.50 | .79 | | 2.29 | 2.91 |
| 090 | Wire strainer, round, 2" diameter | | .052 | Ea. | .80 | .71 | | 1.51 | 2.02 |
| 100 | 4" diameter | | .052 | | 1.03 | .71 | | 1.74 | 2.27 |
| 120 | Rectangular, perforated, 2" x 3" | | .055 | | 1.38 | .76 | | 2.14 | 2.73 |
| 130 | 3" x 4" | | .055 | ▼ | 2.24 | .76 | | 3 | 3.68 |
| 150 | Copper, round, 16 oz., stock, 2" diameter | | .042 | L.F. | 2.41 | .58 | | 2.99 | 3.58 |
| 160 | 3" diameter | | .042 | | 3.10 | .58 | | 3.68 | 4.34 |
| 180 | 4" diameter | | .055 | | 4.10 | .76 | | 4.86 | 5.75 |
| 190 | 5" diameter | | .062 | | 4.36 | .85 | | 5.21 | 6.15 |
| 210 | Rectangular, corrugated copper, stock, 2" x 3" | | .042 | | 3.35 | .58 | | 3.93 | 4.61 |
| 220 | 3" x 4" | | .055 | | 4.48 | .76 | | 5.24 | 6.15 |
| 240 | Rectangular, plain copper, stock, 2" x 3" | | .042 | | 3.38 | .58 | | 3.96 | 4.65 |
| 250 | 3" x 4" | | .055 | ▼ | 4.12 | .76 | | 4.88 | 5.75 |
| 270 | Wire strainers, rectangular, 2" x 3" | | .055 | Ea. | 1.80 | .76 | | 2.56 | 3.20 |
| 280 | 3" x 4" | | .055 | | 2.83 | .76 | | 3.59 | 4.33 |
| 300 | Round, 2" diameter | | .055 | | 1.69 | .76 | | 2.45 | 3.08 |
| 310 | 3" diameter | | .055 | | 2.47 | .76 | | 3.23 | 3.93 |
| 330 | 4" diameter | | .055 | | 3.81 | .76 | | 4.57 | 5.40 |
| 340 | 5" diameter | | .070 | ▼ | 5.30 | .96 | | 6.26 | 7.35 |
| 360 | Lead-coated copper, round, stock, 2" diameter | | .042 | L.F. | 4.32 | .58 | | 4.90 | 5.70 |
| 370 | 3" diameter | | .042 | | 5.30 | .58 | | 5.88 | 6.75 |
| 390 | 4" diameter | | .055 | | 6.45 | .76 | | 7.21 | 8.30 |
| 430 | Rectangular, corrugated, stock, 2" x 3" | | .042 | | 5.75 | .58 | | 6.33 | 7.25 |
| 450 | Plain, stock, 2" x 3" | | .042 | | 6.80 | .58 | | 7.38 | 8.40 |
| 460 | 3" x 4" | | .055 | | 8.55 | .76 | | 9.31 | 10.60 |
| 480 | Steel, galvanized, round, corrugated, 2" or 3" diam., 28 gauge | | .042 | | .45 | .58 | | 1.03 | 1.42 |
| 490 | 4" diameter, 28 gauge | | .055 | | .57 | .76 | | 1.33 | 1.84 |
| 570 | Rectangular, corrugated, 28 gauge, 2" x 3" | | .042 | | .46 | .58 | | 1.04 | 1.43 |
| 580 | 3" x 4" | | .055 | | 1.31 | .76 | | 2.07 | 2.66 |
| 600 | Rectangular, plain, 28 gauge, galvanized, 2" x 3" | | .042 | | .42 | .58 | | 1 | 1.39 |
| 610 | 3" x 4" | | .055 | | 1.27 | .76 | | 2.03 | 2.61 |
| 630 | Epoxy painted, 24 gauge, corrugated, 2" x 3" | | .042 | | .87 | .58 | | 1.45 | 1.89 |
| 640 | 3" x 4" | | .055 | ▼ | 1.12 | .76 | | 1.88 | 2.45 |
| 660 | Wire strainers, rectangular, 2" x 3" | | .055 | Ea. | 1.39 | .76 | | 2.15 | 2.75 |
| 670 | 3" x 4" | | .055 | | 2.27 | .76 | | 3.03 | 3.71 |
| 690 | Round strainers, 2" or 3" diameter | | .055 | | .94 | .76 | | 1.70 | 2.25 |
| 700 | 4" diameter | | .055 | | 1.03 | .76 | | 1.79 | 2.35 |
| 720 | 5" diameter | | .055 | | 1.08 | .76 | | 1.84 | 2.40 |
| 730 | 6" diameter | ▼ | .070 | ▼ | 1.30 | .96 | | 2.26 | 2.96 |
| 12-001 | **DRIP EDGE** Aluminum, .016" thick, 5" girth, mill finish | 1 Carp | .020 | L.F. | .13 | .25 | | .38 | .53 |
| 010 | White finish | | .020 | | .28 | .25 | | .53 | .70 |
| 020 | 8" girth | | .020 | | .20 | .25 | | .45 | .61 |
| 030 | 28" girth | | .080 | | .90 | .99 | | 1.89 | 2.55 |
| 040 | Galvanized, 5" girth | | .020 | | .15 | .25 | | .40 | .56 |
| 050 | 8" girth | ▼ | .020 | ▼ | .22 | .25 | | .47 | .63 |
| 13-001 | **ELBOWS** Aluminum, 2" x 3", embossed | 1 Shee | .080 | Ea. | 1.75 | 1.10 | | 2.85 | 3.69 |
| 010 | Enameled | " | .080 | " | 1.75 | 1.10 | | 2.85 | 3.69 |

For expanded coverage of these items see *Building Construction Cost Data 1985*

## 7.6 Sheet Metal Work

| | | | CREW | MAN-HOURS | UNIT | MAT. | LABOR | EQUIP. | TOTAL | TOTAL INCL O&P |
|---|---|---|---|---|---|---|---|---|---|---|
| 13 | 020 | 3" x 4", .025" thick, embossed | 1 Shee | .080 | Ea. | 2.05 | 1.10 | | 3.15 | 4.02 |
| | 030 | Enameled | | .080 | | 2.05 | 1.10 | | 3.15 | 4.02 |
| | 040 | Round corrugated, 3", embossed, .020" thick | | .080 | | 1.70 | 1.10 | | 2.80 | 3.63 |
| | 050 | 4", .025" thick | | .080 | | 1.85 | 1.10 | | 2.95 | 3.80 |
| | 060 | Copper, 16 oz. round, corrugated, 2" diameter | | .080 | | 6 | 1.10 | | 7.10 | 8.35 |
| | 070 | 3" diameter | | .080 | | 7.40 | 1.10 | | 8.50 | 9.90 |
| | 080 | 4" diameter | | .080 | | 9.80 | 1.10 | | 10.90 | 12.55 |
| | 100 | 2" x 3" corrugated | | .080 | | 8.05 | 1.10 | | 9.15 | 10.60 |
| | 110 | 3" x 4" corrugated | ▼ | .080 | ▼ | 10.75 | 1.10 | | 11.85 | 13.60 |
| 20-001 | | **FASCIA** Aluminum, reverse board and batten, | | | | | | | | |
| | 010 | .032" thick, colored, no furring included | 1 Shee | .055 | S.F. | 2.25 | .76 | | 3.01 | 3.69 |
| | 030 | Steel, galv. and enameled, stock, no furring, long panels | | .055 | | 1.95 | .76 | | 2.71 | 3.36 |
| | 060 | Short panels | | .070 | | 3.05 | .96 | | 4.01 | 4.89 |
| 25-001 | | **FLASHING** Aluminum, mill finish, .013" thick | | .055 | | .27 | .76 | | 1.03 | 1.51 |
| | 003 | .016" thick | | .055 | | .31 | .76 | | 1.07 | 1.56 |
| | 006 | .019" thick | | .055 | | .63 | .76 | | 1.39 | 1.91 |
| | 010 | .032" thick | | .055 | | .77 | .76 | | 1.53 | 2.06 |
| | 020 | .040" thick | | .055 | | 1.30 | .76 | | 2.06 | 2.65 |
| | 030 | .050" thick | ▼ | .055 | | 1.60 | .76 | | 2.36 | 2.98 |
| | 040 | Painted finish, add | | | | .16 | | | .16 | .17M |
| | 050 | Fabric-backed 2 sides, .004" thick | 1 Shee | .024 | | .37 | .33 | | .70 | .94 |
| | 070 | .016" thick | | .024 | | .91 | .33 | | 1.24 | 1.54 |
| | 075 | Mastic-backed, self adhesive | | .017 | | 1.95 | .24 | | 2.19 | 2.53 |
| | 080 | Mastic-coated 2 sides, .004" thick | | .024 | | .44 | .33 | | .77 | 1.02 |
| | 100 | .005" thick | | .024 | | .55 | .33 | | .88 | 1.14 |
| | 110 | .016" thick | ▼ | .024 | ▼ | .97 | .33 | | 1.30 | 1.60 |
| | 130 | Asphalt flashing cement, 5 gallon | | | Gal. | 13.40 | | | 13.40 | 14.75M |
| | 160 | Copper, 16 oz., sheets, under 6000 lbs. | 1 Shee | .070 | S.F. | 1.11 | .96 | | 2.07 | 2.75 |
| | 170 | Over 6000 lbs. | | .052 | | 1.11 | .71 | | 1.82 | 2.36 |
| | 190 | 20 oz. sheets, under 6000 lbs. | | .073 | | 1.40 | 1 | | 2.40 | 3.14 |
| | 200 | Over 6000 lbs. | | .055 | | 1.40 | .76 | | 2.16 | 2.76 |
| | 220 | 24 oz. sheets, under 6000 lbs. | | .076 | | 1.67 | 1.05 | | 2.72 | 3.52 |
| | 250 | 32 oz. sheets, under 6000 lbs. | | .080 | | 2.22 | 1.10 | | 3.32 | 4.21 |
| | 280 | Copper, paperbacked 1 side, 2 oz. | | .024 | | .75 | .33 | | 1.08 | 1.36 |
| | 290 | 3 oz. | | .024 | | 1.10 | .33 | | 1.43 | 1.74 |
| | 310 | Paperbacked 2 sides, copper, 2 oz. | | .024 | | .95 | .33 | | 1.28 | 1.58 |
| | 315 | 3 oz. | | .024 | | 1.15 | .33 | | 1.48 | 1.80 |
| | 320 | 5 oz. | | .024 | | 1.95 | .33 | | 2.28 | 2.68 |
| | 340 | Mastic-backed 2 sides, copper, 2 oz. | | .024 | | .95 | .33 | | 1.28 | 1.58 |
| | 350 | 3 oz. | | .024 | | 1.26 | .33 | | 1.59 | 1.92 |
| | 370 | 5 oz. | | .024 | | 2.10 | .33 | | 2.43 | 2.84 |
| | 380 | Fabric-backed 2 sides, copper, 2 oz. | | .024 | | 1.10 | .33 | | 1.43 | 1.74 |
| | 400 | 3 oz. | | .024 | | 1.37 | .33 | | 1.70 | 2.04 |
| | 410 | 5 oz. | | .024 | | 2.15 | .33 | | 2.48 | 2.90 |
| | 430 | Copper-clad stainless steel, .015" thick, under 500 lbs. | | .070 | | 2.10 | .96 | | 3.06 | 3.84 |
| | 460 | .018" thick, under 500 lbs. | | .080 | | 2.70 | 1.10 | | 3.80 | 4.73 |
| | 465 | CPE Clad Metal | | .028 | | 1.85 | .39 | | 2.24 | 2.65 |
| | 471 | CSPE Bonded | | .028 | | 1.61 | .39 | | 2 | 2.39 |
| | 475 | EPDM Cured | | .028 | | 1.10 | .39 | | 1.49 | 1.83 |
| | 480 | Uncured | ▼ | .028 | ▼ | .90 | .39 | | 1.29 | 1.61 |
| | 490 | Fabric, asphalt-saturated cotton, specification grade | 1 Rofc | .229 | S.Y. | 1.21 | 2.65 | | 3.86 | 5.75 |
| | 500 | Utility grade | | .025 | | .76 | .29 | | 1.05 | 1.32 |
| | 520 | Open-mesh fabric, saturated, 40 oz. per S.Y. | | .025 | | 1.55 | .29 | | 1.84 | 2.19 |
| | 530 | Close-mesh fabric, saturated, 17 oz. per S.Y. | | .025 | | 1.55 | .29 | | 1.84 | 2.19 |
| | 550 | Fiberglass, resin-coated | | .025 | | 1.42 | .29 | | 1.71 | 2.05 |
| | 560 | Asphalt-coated, 40 oz. per S.Y. | ▼ | .025 | ▼ | 2.12 | .29 | | 2.41 | 2.82 |
| | 565 | Hypalon Clad Metal | 1 Shee | .028 | S.F. | 2.56 | .39 | | 2.95 | 3.43 |
| | 580 | Lead, 2.5 lb. per S.F., up to 12" wide | 1 Rofc | .059 | | 2.20 | .69 | | 2.89 | 3.56 |
| | 590 | Over 12" wide | " | .059 | ▼ | 2.95 | .69 | | 3.64 | 4.39 |

For expanded coverage of these items see *Building Construction Cost Data 1985*

## 7.6 Sheet Metal Work

| | | | CREW | MAN-HOURS | UNIT | MAT. | LABOR | EQUIP. | TOTAL | TOTAL INCL O&P |
|---|---|---|---|---|---|---|---|---|---|---|
| 25 | 610 | Lead-coated copper, fabric-backed, 2 oz. | 1 Shee | .024 | S.F. | .92 | .33 | | 1.25 | 1.55 |
| | 620 | 5 oz. | | .024 | | 1.35 | .33 | | 1.68 | 2.02 |
| | 640 | Mastic-backed 2 sides, 2 oz. | | .024 | | .59 | .33 | | .92 | 1.18 |
| | 650 | 5 oz. | | .024 | | 1.02 | .33 | | 1.35 | 1.66 |
| | 670 | Paperbacked 1 side, 2 oz. | | .024 | | .57 | .33 | | .90 | 1.16 |
| | 680 | 3 oz. | | .024 | | .83 | .33 | | 1.16 | 1.45 |
| | 700 | Paperbacked 2 sides, 2 oz. | | .024 | | .64 | .33 | | .97 | 1.24 |
| | 710 | 5 oz. | | .024 | | 1.28 | .33 | | 1.61 | 1.94 |
| | 715 | Neoprene, 60 mil | | .028 | | 1.94 | .39 | | 2.33 | 2.75 |
| | 716 | Self-curing | | .028 | | 1.95 | .39 | | 2.34 | 2.76 |
| | 717 | Uncured | | .028 | | 1.78 | .39 | | 2.17 | 2.58 |
| | 730 | Polyvinyl chloride, black, .010" thick | 1 Rofc | .028 | | .47 | .33 | | .80 | 1.06 |
| | 740 | .020" thick | | .028 | | .85 | .33 | | 1.18 | 1.48 |
| | 760 | .030" thick | | .028 | | 1.28 | .33 | | 1.61 | 1.95 |
| | 770 | .056" thick | | .028 | | 2.38 | .33 | | 2.71 | 3.16 |
| | 790 | Black or white for exposed roofs, .060" thick | | .028 | | 2.50 | .33 | | 2.83 | 3.29 |
| | 800 | Asbestos-backed for parking decks, .045" thick | | .028 | | .80 | .33 | | 1.13 | 1.42 |
| | 805 | PVC (19 mils) coated galv. steel (24 mils), 4' x 8' sheets | | .033 | | 1.20 | .39 | | 1.59 | 1.96 |
| | 806 | PVC tape, 5" x 45 mils, for joint covers, 100 L.F./roll | | | Ea. | 80 | | | 80 | 88M |
| | 810 | Rubber, butyl, 1/32" thick | 1 Rofc | .028 | S.F. | .59 | .33 | | .92 | 1.19 |
| | 820 | 1/16" thick | | .028 | | .86 | .33 | | 1.19 | 1.49 |
| | 830 | Neoprene, cured, 1/16" thick | | .028 | | 1.35 | .33 | | 1.68 | 2.03 |
| | 840 | 1/8" thick | | .028 | | 2.75 | .33 | | 3.08 | 3.57 |
| | 850 | Shower pan, bituminous membrane, 7 oz. | 1 Shee | .052 | | 1.40 | .71 | | 2.11 | 2.68 |
| | 855 | 3 ply copper and fabric, 3 oz. | | .052 | | 1.70 | .71 | | 2.41 | 3.01 |
| | 860 | 7 oz. | | .052 | | 3.65 | .71 | | 4.36 | 5.15 |
| | 865 | Copper, 16 oz. | | .080 | | 2.75 | 1.10 | | 3.85 | 4.79 |
| | 870 | Lead on copper and fabric, 5 oz. | | .052 | | 1.40 | .71 | | 2.11 | 2.68 |
| | 880 | 7 oz. | | .052 | | 2.75 | .71 | | 3.46 | 4.16 |
| | 890 | Stainless steel sheets, 32 ga., .010" thick | | .052 | | 1.70 | .71 | | 2.41 | 3.01 |
| | 900 | 28 ga., .015" thick | | .052 | | 2.10 | .71 | | 2.81 | 3.45 |
| | 910 | 26 ga., .018" thick | | .052 | | 2.60 | .71 | | 3.31 | 4 |
| | 920 | 24 ga., .025" thick | | .052 | | 3.20 | .71 | | 3.91 | 4.66 |
| | 929 | For mechanically keyed flashing, add | | | | 50% | | | | |
| | 930 | Stainless steel, paperbacked 2 sides, .005" thick | 1 Shee | .024 | | 1.05 | .33 | | 1.38 | 1.69 |
| | 932 | Steel sheets, galvanized, 20 gauge | | .062 | | .50 | .85 | | 1.35 | 1.91 |
| | 934 | 30 gauge | | .050 | | .18 | .69 | | .87 | 1.30 |
| | 940 | Terne coated stainless steel, .015" thick, 28 ga. | | .052 | | 2.10 | .71 | | 2.81 | 3.45 |
| | 950 | .018" thick, 26 ga. | | .052 | | 2.15 | .71 | | 2.86 | 3.50 |
| | 960 | Zinc and copper alloy, .020" thick | | .052 | | 1.05 | .71 | | 1.76 | 2.29 |
| | 970 | .027" thick | | .052 | | 1.35 | .71 | | 2.06 | 2.62 |
| | 980 | .032" thick | | .052 | | 1.60 | .71 | | 2.31 | 2.90 |
| | 990 | .040" thick | | .052 | | 2 | .71 | | 2.71 | 3.34 |
| 30-001 | | GRAVEL STOP Aluminum, .050" thick, 4" height, mill finish | 1 Shee | .055 | L.F. | 3.65 | .76 | | 4.41 | 5.25 |
| | 008 | Duranodic finish | | .055 | | 4.30 | .76 | | 5.06 | 5.95 |
| | 010 | Painted | | .055 | | 3.80 | .76 | | 4.56 | 5.40 |
| | 120 | Copper, 16 oz., 3" face height | | .055 | | 3.55 | .76 | | 4.31 | 5.10 |
| | 135 | Galv. steel, 24 ga., 4" leg, plain, with continuous cleat, 4" face | | .055 | | 1.80 | .76 | | 2.56 | 3.20 |
| | 150 | Polyvinyl chloride, 6" face height | | .059 | | 2.97 | .82 | | 3.79 | 4.57 |
| | 180 | Stainless steel, 24 ga., 6" face height | | .059 | | 5.90 | .82 | | 6.72 | 7.80 |
| 33-001 | | GUTTERS Aluminum, stock units, 5" box, .027" thick, plain | 1 Shee | .067 | L.F. | .95 | .92 | | 1.87 | 2.52 |
| | 002 | Inside corner | | .320 | Ea. | 2.45 | 4.42 | | 6.87 | 9.75 |
| | 003 | Outside corner | | .320 | " | 2.45 | 4.42 | | 6.87 | 9.75 |
| | 010 | Enameled | | .067 | L.F. | .90 | .92 | | 1.82 | 2.46 |
| | 011 | Inside corner | | .320 | Ea. | 2.50 | 4.42 | | 6.92 | 9.80 |
| | 012 | Outside corner | | .320 | " | 2.50 | 4.42 | | 6.92 | 9.80 |
| | 030 | 5" box type, .032" thick, plain | | .067 | L.F. | 1.15 | .92 | | 2.07 | 2.74 |
| | 031 | Inside corner | | .320 | Ea. | 2.60 | 4.42 | | 7.02 | 9.90 |

For expanded coverage of these items see *Building Construction Cost Data 1985*

## 7.6 Sheet Metal Work

| | | | CREW | MAN-HOURS | UNIT | MAT. | LABOR | EQUIP. | TOTAL | TOTAL INCL O&P |
|---|---|---|---|---|---|---|---|---|---|---|
| 33 | 032 | Outside corner | 1 Shee | .320 | Ea. | 2.60 | 4.42 | | 7.02 | 9.90 |
| | 040 | Enameled | | .067 | L.F. | 1.10 | .92 | | 2.02 | 2.68 |
| | 041 | Inside corner | | .320 | Ea. | 2.65 | 4.42 | | 7.07 | 9.95 |
| | 042 | Outside corner | | .320 | " | 2.65 | 4.42 | | 7.07 | 9.95 |
| | 060 | 5" x 6" combination fascia & gutter, .032" thick, enameled | | .133 | L.F. | 2.50 | 1.84 | | 4.34 | 5.70 |
| | 070 | Copper, half round, 16 oz., stock units, 4" wide | | .067 | | 3.75 | .92 | | 4.67 | 5.60 |
| | 090 | 5" wide | | .067 | | 4 | .92 | | 4.92 | 5.85 |
| | 100 | 6" wide | | .070 | | 4.90 | .96 | | 5.86 | 6.90 |
| | 120 | K type copper gutter, stock, 4" wide | | .067 | | 4.89 | .92 | | 5.81 | 6.85 |
| | 130 | 5" wide | | .067 | | 5.85 | .92 | | 6.77 | 7.90 |
| | 150 | Lead coated copper, half round, stock, 4" wide | | .067 | | 3.50 | .92 | | 4.42 | 5.30 |
| | 160 | 6" wide | | .070 | | 4.22 | .96 | | 5.18 | 6.20 |
| | 180 | K type leadcoated copper, stock, 4" wide | | .067 | | 3.55 | .92 | | 4.47 | 5.40 |
| | 190 | 5" wide | | .067 | | 4.37 | .92 | | 5.29 | 6.30 |
| | 210 | Stainless steel, half round or box, stock, 4" wide | | .067 | | 4.37 | .92 | | 5.29 | 6.30 |
| | 220 | 5" wide | | .067 | | 4.63 | .92 | | 5.55 | 6.55 |
| | 240 | Steel, galv., half round or box, 28 ga., 5" wide, plain | | .067 | | .65 | .92 | | 1.57 | 2.19 |
| | 250 | Enameled | | .067 | | .70 | .92 | | 1.62 | 2.24 |
| | 270 | 26 ga. galvanized steel, stock, 5" wide | | .067 | | .78 | .92 | | 1.70 | 2.33 |
| | 280 | 6" wide | ↓ | .067 | | .85 | .92 | | 1.77 | 2.41 |
| | 300 | Vinyl, O.G., 4" wide | 1 Carp | .073 | | 1.18 | .90 | | 2.08 | 2.72 |
| | 310 | 5" wide | | .073 | | 1.33 | .90 | | 2.23 | 2.89 |
| | 320 | 4" half round, stock units | ↓ | .073 | ↓ | .92 | .90 | | 1.82 | 2.43 |
| | 325 | Joint connectors | | | Ea. | 2.05 | | | 2.05 | 2.25M |
| | 330 | Wood, clear treated cedar, fir or hemlock, 3" x 4" | 1 Carp | .080 | L.F. | 3.65 | .99 | | 4.64 | 5.60 |
| | 340 | 4" x 5" | | .080 | | 4.75 | .99 | | 5.74 | 6.80 |
| 36- | 001 | GUTTER GUARD 6" wide strip, aluminum mesh | | .016 | | .30 | .20 | | .50 | .64 |
| | 010 | Vinyl mesh | ↓ | .016 | ↓ | .15 | .20 | | .35 | .48 |
| 39- | 001 | LEAD ROOFING 3 lb. per S.F., batten seam | 1 Shee | 20 | Sq. | 440 | 275 | | 715 | 925 |
| | 010 | Flat seam | " | 18.180 | " | 395 | 250 | | 645 | 835 |
| 42- | 001 | LOUVERS Aluminum with screen, residential, 8" x 8" | 1 Carp | .211 | Ea. | 3.61 | 2.60 | | 6.21 | 8.10 |
| | 010 | 12" x 12" | | .211 | | 5.05 | 2.60 | | 7.65 | 9.65 |
| | 020 | 12" x 18" | | .229 | | 7.80 | 2.82 | | 10.62 | 13.05 |
| | 025 | 14" x 24" | | .267 | | 9.85 | 3.29 | | 13.14 | 16.05 |
| | 030 | 18" x 24" | | .296 | | 11.85 | 3.66 | | 15.51 | 18.85 |
| | 050 | 30" x 24" | | .333 | | 25 | 4.12 | | 29.12 | 34 |
| | 070 | Triangle, adjustable, small | | .400 | | 12.60 | 4.94 | | 17.54 | 22 |
| | 080 | Large | | .533 | | 13.30 | 6.60 | | 19.90 | 25 |
| | 210 | Midget, aluminum, 3/4" deep, 1" diameter | | .094 | | .43 | 1.16 | | 1.59 | 2.31 |
| | 215 | 3" diameter | | .133 | | 1.14 | 1.65 | | 2.79 | 3.86 |
| | 220 | 4" diameter | | .160 | | 1.43 | 1.98 | | 3.41 | 4.70 |
| | 225 | 6" diameter | ↓ | .267 | ↓ | 2.01 | 3.29 | | 5.30 | 7.45 |
| | 230 | Ridge vent strip, mill finish | 1 Shee | .052 | L.F. | 1.56 | .71 | | 2.27 | 2.85 |
| | 240 | Under eaves vent, aluminum, mill finish, 16" x 4" | 1 Carp | .107 | Ea. | .88 | 1.32 | | 2.20 | 3.05 |
| | 250 | 16" x 8" | | .107 | | 1.08 | 1.32 | | 2.40 | 3.27 |
| | 300 | Vinyl wall louvers, 1-1/2" deep, 8" x 8" | | 1.330 | | 2.32 | 16.45 | | 18.77 | 29 |
| | 302 | Area to 104 S.F. | | 1.140 | | 3.20 | 14.10 | | 17.30 | 26 |
| | 304 | To 85 S.F. | | 1 | | 3 | 12.35 | | 15.35 | 23 |
| | 700 | Vinyl wall louvers, 1-1/2" deep, 8" x 8" | | .211 | | 2.25 | 2.60 | | 4.85 | 6.60 |
| | 702 | 12" x 12" | | .211 | | 2.75 | 2.60 | | 5.35 | 7.15 |
| | 708 | 12" x 18" | | .229 | | 3.50 | 2.83 | | 6.33 | 8.33 |
| | 720 | 14" x 24" | ↓ | .267 | ↓ | 5.00 | 3.30 | | 8.30 | 10.73 |
| 48- | 001 | MONEL ROOFING Batten seam, over 10 squares, .018" thick | 1 Shee | 6.670 | Sq. | 470 | 92 | | 562 | 665 |
| | 010 | .021" thick | | 6.960 | | 535 | 96 | | 631 | 740 |
| | 030 | Standing seam, .018" thick | | 5.930 | | 455 | 82 | | 537 | 630 |
| | 040 | .021" thick | | 6.150 | | 525 | 85 | | 610 | 715 |
| | 060 | Flat seam, .018" thick | | 6.150 | | 450 | 85 | | 535 | 630 |
| | 070 | .021" thick | ↓ | 6.670 | ↓ | 515 | 92 | | 607 | 715 |
| 51- | 001 | REGLET Aluminum, .025" thick, in concrete parapet | 1 Carp | .036 | L.F. | 1.25 | .44 | | 1.69 | 2.07 |
| | 010 | Copper, 10 oz. | " | .036 | " | 1.18 | .44 | | 1.62 | 1.99 |

For expanded coverage of these items see *Building Construction Cost Data 1985*

## 7.6 Sheet Metal Work

| | | | CREW | MAN-HOURS | UNIT | MAT. | LABOR | EQUIP. | TOTAL | TOTAL INCL O&P |
|---|---|---|---|---|---|---|---|---|---|---|
| 51 | 030 | 16 oz. | 1 Carp | .036 | L.F. | 1.69 | .44 | | 2.13 | 2.55 |
| | 040 | Galvanized steel, 24 gauge | | .036 | | .42 | .44 | | .86 | 1.16 |
| | 060 | Stainless steel, .020" thick | | .036 | | 1.75 | .44 | | 2.19 | 2.62 |
| | 070 | Zinc and copper alloy, 20 oz. | ↓ | .036 | | 1.39 | .44 | | 1.83 | 2.22 |
| | 090 | Counter flashing for above, 12" wide, .032" aluminum | 1 Shee | .053 | | 1.49 | .74 | | 2.23 | 2.82 |
| | 100 | Copper, 10 oz. | | .053 | | 1.39 | .74 | | 2.13 | 2.71 |
| | 120 | 16 oz. | | .053 | | 1.64 | .74 | | 2.38 | 2.98 |
| | 130 | Galvanized steel, .020" thick | | .053 | | .59 | .74 | | 1.33 | 1.83 |
| | 150 | Stainless steel, .020" thick | | .053 | | 2 | .74 | | 2.74 | 3.38 |
| | 160 | Zinc and copper alloy, 20 oz. | ↓ | .053 | ↓ | 2.57 | .74 | | 3.31 | 4 |
| 54-001 | | SOFFIT Aluminum, residential, stock units, .020" thick | 1 Carp | .038 | S.F. | .70 | .47 | | 1.17 | 1.52 |
| | 010 | Baked enamel on steel, 16 or 18 gauge | | .076 | | 5.05 | .94 | | 5.99 | 7.05 |
| | 030 | Polyvinyl chloride, white, solid | | .035 | | .76 | .43 | | 1.19 | 1.52 |
| | 040 | Perforated | ↓ | .035 | | .79 | .43 | | 1.22 | 1.55 |
| | 050 | For colors, add | | | ↓ | .05 | | | .05 | .05M |
| 60-001 | | TERMITE Shields, zinc, 10" wide, .012" thick | 1 Carp | .023 | L.F. | 1 | .28 | | 1.28 | 1.55 |
| | 010 | .020" thick | " | .023 | " | .70 | .28 | | .98 | 1.22 |
| 63-001 | | ZINC Copper alloy roofing, batten seam, .020" thick | 1 Shee | 6.670 | Sq. | 360 | 92 | | 452 | 545 |
| | 010 | .027" thick | | 6.960 | | 445 | 96 | | 541 | 645 |
| | 030 | .032" thick | | 7.270 | | 525 | 100 | | 625 | 740 |
| | 040 | .040" thick | ↓ | 7.620 | | 670 | 105 | | 775 | 905 |
| | 060 | For standing seam construction, deduct | | | | 2% | | | | |
| | 070 | For flat seam construction, deduct | | | ↓ | 3% | | | | |

## 7.8 Roof Accessories

| | | | CREW | MAN-HOURS | UNIT | MAT. | LABOR | EQUIP. | TOTAL | TOTAL INCL O&P |
|---|---|---|---|---|---|---|---|---|---|---|
| 10-001 | | CEILING HATCHES 2'-6" x 2'-6", single leaf, steel frame & cover | G-3 | 2.910 | Ea. | 252 | 35 | 2 | 289 | 335 |
| | 010 | Aluminum cover | | 2.910 | | 284 | 35 | 2 | 321 | 370 |
| | 030 | 2'-6" x 3'-0", single leaf, steel frame & steel cover | | 2.910 | | 288 | 35 | 2 | 325 | 375 |
| | 040 | Aluminum cover | ↓ | 2.910 | ↓ | 326 | 35 | 2 | 363 | 415 |
| 20-001 | | ROOF HATCHES With curb, 1" fiberglass insulation, 2'-6" x 3'-0" | | | | | | | | |
| | 050 | Aluminum curb and cover | G-3 | 3.200 | Ea. | 361 | 37.80 | 2.20 | 401 | 460 |
| | 052 | Galvanized steel | | 3.200 | | 340 | 37.80 | 2.20 | 380 | 435 |
| | 054 | Plain steel, primed | | 3.200 | | 304 | 37.80 | 2.20 | 344 | 395 |
| | 060 | 2'-6" x 4'-6", aluminum curb & cover | | 3.560 | | 517 | 42.56 | 2.44 | 562 | 640 |
| | 080 | Galvanized steel | | 3.560 | | 489 | 42.56 | 2.44 | 534 | 610 |
| | 090 | Plain steel, primed | | 3.560 | | 430 | 42.56 | 2.44 | 475 | 545 |
| | 120 | 2'-6" x 8'-0", aluminum curb and cover | | 4.850 | | 897 | 57.67 | 3.33 | 958 | 1,075 |
| | 140 | Galvanized steel | | 4.850 | | 850 | 57.67 | 3.33 | 911 | 1,025 |
| | 150 | Plain steel, primed | ↓ | 4.850 | | 742 | 57.67 | 3.33 | 803 | 910 |
| | 180 | For plexiglass panels, add to above | | | | 197 | | | 197 | 215M |
| | 200 | For galv. curb and alum. cover, deduct from aluminum | | | ↓ | 14 | | | 14 | 15.40M |
| 40-001 | | SKYLIGHT Plastic roof domes, flush or curb mounted, ten or | | | | | | | | |
| | 010 | more units, curb not included, "L" frames | | | | | | | | |
| | 030 | Nominal size under 10 S.F., double | G-3 | .246 | S.F. | 15.75 | 2.93 | .17 | 18.85 | 22 |
| | 040 | Single | | .200 | | 12.60 | 2.38 | .14 | 15.12 | 17.80 |
| | 060 | 10 S.F. to 20 S.F., double | | .102 | | 14.70 | 1.21 | .07 | 15.98 | 18.15 |
| | 070 | Single | | .081 | | 10.50 | .96 | .06 | 11.52 | 13.15 |
| | 090 | 20 S.F. to 30 S.F., double | | .081 | | 12.60 | .96 | .06 | 13.62 | 15.45 |
| | 100 | Single | | .069 | | 8.40 | .82 | .05 | 9.27 | 10.60 |
| | 120 | 30 S.F. to 65 S.F., double | | .069 | | 9.45 | .82 | .05 | 10.32 | 11.75 |
| | 130 | Single | ↓ | .052 | | 6.30 | .62 | .04 | 6.96 | 7.95 |
| | 150 | For insulated 4" curbs, double, add | | | | 15% | | | | |
| | 160 | Single, add | | | ↓ | 30% | | | | |

For expanded coverage of these items see *Building Construction Cost Data 1985*

## 7.8 Roof Accessories

| | | | CREW | MAN-HOURS | UNIT | MAT. | LABOR | EQUIP. | TOTAL | TOTAL INCL O&P |
|---|---|---|---|---|---|---|---|---|---|---|
| 40 | 180 | For integral insulated 9" curbs, double, add | | | S.F. | 30% | | | | |
| | 190 | Single, add | | | " | 45% | | | | |
| | 212 | Ventilating insulated plexiglass dome with | | | | | | | | |
| | 213 | curb mounting, 36" x 36" | G-3 | 2.670 | Ea. | 349 | 32.17 | 1.83 | 383 | 435 |
| | 215 | 52" x 52" | | 2.670 | | 467 | 32.17 | 1.83 | 501 | 565 |
| | 216 | 28" x 52" | | 3.200 | | 386 | 37.80 | 2.20 | 426 | 490 |
| | 217 | 36" x 52" | ↓ | 3.200 | | 410 | 37.80 | 2.20 | 450 | 515 |
| | 218 | For electric opening system, add | | | | 210 | | | 210 | 230M |
| | 221 | Operating skylight, with thermopane glass, 24" x 48" | G-3 | 3.200 | | 424 | 37.80 | 2.20 | 464 | 530 |
| | 222 | 32" x 48" | | 3.560 | ↓ | 415 | 42.56 | 2.44 | 460 | 525 |
| | 240 | Sandwich panels, fiberglass, for walls, 1-9/16" thick, to 250 S.F. | | .160 | S.F. | 11.20 | 1.90 | .11 | 13.21 | 15.45 |
| | 250 | 250 S.F. and up | | .121 | | 9 | 1.44 | .08 | 10.52 | 12.30 |
| | 270 | As above, but for roofs, 2-3/4" thick, to 250 S.F. | | .108 | | 16.65 | 1.30 | .07 | 18.02 | 20 |
| | 280 | 250 S.F. and up | ↓ | .097 | ↓ | 13.35 | 1.15 | .07 | 14.57 | 16.60 |

## 8.1 Metal Doors

| | | | CREW | MAN-HOURS | UNIT | MAT. | LABOR | EQUIP. | TOTAL | TOTAL INCL O&P |
|---|---|---|---|---|---|---|---|---|---|---|
| 10-001 | | STEEL FRAMES, KNOCK DOWN 18 ga., up to 5-3/4" deep | | | | | | | | |
| | 002 | 6'-8" high, 3'-0" wide, single | F-2 | 1 | Ea. | 42 | 12.37 | .88 | 55.25 | 67 |
| | 004 | 6'-0" wide, double | | 1.140 | | 49 | 14.10 | 1 | 64.10 | 77 |
| | 010 | 7'-0" high, 3'-0" wide, single | | 1 | | 45 | 12.37 | .88 | 58.25 | 70 |
| | 014 | 6'-0" wide, double | ↓ | 1.140 | ↓ | 54 | 14.10 | 1 | 69.10 | 83 |
| | 015 | | | | | | | | | |
| | 280 | 18 ga. drywall, up to 4-7/8" deep, 7'-0" high, 3'-0" wide, single | F-2 | 1 | Ea. | 50 | 12.37 | .88 | 63.25 | 76 |
| | 284 | 6'-0" wide, double | | 1.140 | | 59 | 14.10 | 1 | 74.10 | 88 |
| | 360 | 16 ga., up to 5-3/4" deep, 7'-0" high, 4'-0" wide, single | | 1.070 | | 60 | 13.17 | .93 | 74.10 | 88 |
| | 364 | 8'-0" wide, double | | 1.330 | | 75 | 16.48 | 1.17 | 92.65 | 110 |
| | 370 | 8'-0" high, 4'-0" wide, single | | 1.070 | | 66 | 13.17 | .93 | 80.10 | 94 |
| | 374 | 8'-0" wide, double | | 1.330 | | 81 | 16.48 | 1.17 | 98.65 | 115 |
| | 400 | 6-3/4" deep, 7'-0" high, 4'-0" wide, single | | 1.070 | | 65 | 13.17 | .93 | 79.10 | 93 |
| | 404 | 8'-0" wide, double | | 1.330 | | 80 | 16.48 | 1.17 | 97.65 | 115 |
| | 410 | 8'-0" high, 4'-0" wide, single | | 1.070 | | 71 | 13.17 | .93 | 85.10 | 100 |
| | 414 | 8'-0" wide, double | | 1.330 | | 86 | 16.48 | 1.17 | 103.65 | 120 |
| | 440 | 8-3/4" deep, 7'-0" high, 4'-0" wide, single | | 1.070 | | 74 | 13.17 | .93 | 88.10 | 105 |
| | 444 | 8'-0" wide, double | | 1.330 | | 86 | 16.48 | 1.17 | 103.65 | 120 |
| | 450 | 8'-0" high, 4'-0" wide, single | | 1.070 | | 80 | 13.17 | .93 | 94.10 | 110 |
| | 454 | 8'-0" wide, double | | 1.330 | | 93 | 16.48 | 1.17 | 110.65 | 130 |
| | 480 | 16 ga. drywall, up to 3-7/8" deep, 7'-0" high, 3'-0" wide, single | | 1 | | 60 | 12.37 | .88 | 73.25 | 87 |
| | 484 | 6'-0" wide, double | ↓ | 1.140 | | 70 | 14.10 | 1 | 85.10 | 100 |
| | 490 | For welded frames, add | | | ↓ | 10 | | | 10 | 11M |
| | 490 | | | | | | | | | |
| | 540 | 16 ga. "B" label, up to 5-3/4" deep, 7'-0" high, 4'-0" wide, single | F-2 | 1.070 | Ea. | 66 | 13.17 | .93 | 80.10 | 94 |
| | 544 | 8'-0" wide, double | | 1.330 | | 76 | 16.48 | 1.17 | 93.65 | 110 |
| | 580 | 6-3/4" deep, 7'-0" high, 4'-0" wide, single | | 1.070 | | 72 | 13.17 | .93 | 86.10 | 100 |
| | 584 | 8'-0" wide, double | | 1.330 | ↓ | 84 | 16.48 | 1.17 | 101.65 | 120 |
| | 620 | 8-3/4" deep, 7'-0" high, 4'-0" wide, single | | 1.070 | | 77 | 13.17 | .93 | 91.10 | 105 |
| | 624 | 8'-0" wide, double | ↓ | 1.330 | ↓ | 92 | 16.48 | 1.17 | 109.65 | 130 |
| | 630 | For "A" label use same price as "B" label | | | | | | | | |
| | 640 | For baked enamel finish, add | | | Ea. | 40% | 90% | | | |
| | 650 | For galvanizing, add | | | ↓ | 10% | | | | |
| | 660 | For porcelain enamel finish, add | | | | 100% | 150% | | | |
| | 790 | Transom lite frames, fixed, add | F-2 | .103 | S.F. | 6 | 1.28 | .09 | 7.37 | 8.70 |
| | 800 | Movable, add | " | .123 | " | 7.25 | 1.52 | .11 | 8.88 | 10.50 |
| 21-001 | | COMMERCIAL STEEL DOORS Flush, full panel, hollow core, 1-3/8" thick | | | | | | | | |
| | 002 | 20 ga., 2'-0" x 6'-8" | F-2 | .800 | Ea. | 98 | 9.90 | .70 | 108.60 | 125 |

For expanded coverage of these items see *Building Construction Cost Data 1985*

## 8.1 Metal Doors

| | | | MAN- | | \multicolumn{4}{c|}{BARE COSTS} | TOTAL |
|---|---|---|---|---|---|---|---|---|---|
| | | CREW | HOURS | UNIT | MAT. | LABOR | EQUIP. | TOTAL | INCL O&P |
| 21 | 004 | 2'-6" x 6'-8" | F-2 | .889 | Ea. | 104 | 10.97 | .78 | 115.75 | 135 |
| | 006 | 3'-0" x 6'-8" | | .941 | | 110 | 11.63 | .82 | 122.45 | 140 |
| | 010 | 3'-0" x 7'-0" | ↓ | .941 | | 120 | 11.63 | .82 | 132.45 | 150 |
| | 012 | For vision lite, add | | | | 47 | | | 47 | 52M |
| | 014 | For narrow lite, add | | | | 55 | | | 55 | 61M |
| | 016 | For bottom louver, add | | | | 105 | | | 105 | 115M |
| | 023 | For baked enamel finish, add | | | | 40% | 90% | | | |
| | 026 | For galvanizing, add | | | | 30% | | | | |
| | 029 | For porcelain enamel finish, add | | | ↓ | 100% | 150% | | | |
| | 030 | | | | | | | | | |
| | 032 | Half glass, 20 ga., 2'-0" x 6'-8" | F-2 | .800 | Ea. | 135 | 9.90 | .70 | 145.60 | 165 |
| | 034 | 2'-6" x 6'-8" | | .889 | | 140 | 10.97 | .78 | 151.75 | 170 |
| | 036 | 3'-0" x 6'-8" | | .941 | | 150 | 11.63 | .82 | 162.45 | 185 |
| | 040 | 3'-0" x 7'-0" | | .941 | | 155 | 11.63 | .82 | 167.45 | 190 |
| | 102 | Hollow core, 1-3/4" thick, full panel, 20 ga., 2'-6" x 6'-8" | | .889 | | 110 | 10.97 | .78 | 121.75 | 140 |
| | 104 | 3'-0" x 6'-8" | | .941 | | 115 | 11.63 | .82 | 127.45 | 145 |
| | 106 | 3'-0" x 7'-0" | | .941 | | 125 | 11.63 | .82 | 137.45 | 155 |
| | 108 | 4'-0" x 7'-0" | | 1.070 | | 165 | 13.17 | .93 | 179.10 | 205 |
| | 110 | 4'-0" x 8'-0" | | 1.230 | | 200 | 15.22 | 1.08 | 216.30 | 245 |
| | 112 | 18 ga., 2'-6" x 6'-8" | | .941 | | 130 | 11.63 | .82 | 142.45 | 160 |
| | 114 | 3'-0" x 6'-8" | | 1 | | 140 | 12.37 | .88 | 153.25 | 175 |
| | 116 | 3'-0" x 7'-0" | | 1 | | 150 | 12.37 | .88 | 163.25 | 185 |
| | 118 | 4'-0" x 7'-0" | | 1.140 | | 185 | 14.10 | 1 | 200.10 | 225 |
| | 120 | 4'-0" x 8'-0" | | 1.140 | | 215 | 14.10 | 1 | 230.10 | 260 |
| | 122 | Half glass, 20 ga., 2'-6" x 6'-8" | | .800 | | 145 | 9.90 | .70 | 155.60 | 175 |
| | 124 | 3'-0" x 6'-8" | | .889 | | 155 | 10.97 | .78 | 166.75 | 190 |
| | 126 | 3'-0" x 7'-0" | | .889 | | 165 | 10.97 | .78 | 176.75 | 200 |
| | 128 | 4'-0" x 7'-0" | | 1 | | 205 | 12.37 | .88 | 218.25 | 245 |
| | 130 | 4'-0" x 8'-0" | | 1.230 | | 235 | 15.22 | 1.08 | 251.30 | 285 |
| | 132 | 18 ga., 2'-6" x 6'-8" | | .889 | | 175 | 10.97 | .78 | 186.75 | 210 |
| | 134 | 3'-0" x 6'-8" | | .941 | | 165 | 11.63 | .82 | 177.45 | 200 |
| | 136 | 3'-0" x 7'-0" | | .941 | | 200 | 11.63 | .82 | 212.45 | 240 |
| | 138 | 4'-0" x 7'-0" | | 1.070 | | 230 | 13.17 | .93 | 244.10 | 275 |
| | 140 | 4'-0" x 8'-0" | | 1.140 | | 260 | 14.10 | 1 | 275.10 | 310 |
| | 172 | Composite, 1-3/4" thick, full panel, 18 ga., 3'-0" x 6'-8" | | 1.070 | | 155 | 13.17 | .93 | 169.10 | 190 |
| | 174 | 2'-6" x 7'-0" | | 1 | | 145 | 12.37 | .88 | 158.25 | 180 |
| | 176 | 3'-0" x 7'-0" | | 1.070 | | 165 | 13.17 | .93 | 179.10 | 205 |
| | 180 | 4'-0" x 8'-0" | | 1.230 | | 235 | 15.22 | 1.08 | 251.30 | 285 |
| | 182 | Half glass, 18 ga., 3'-0" x 6'-8" | | 1 | | 200 | 12.37 | .88 | 213.25 | 240 |
| | 184 | 2'-6" x 7'-0" | | .941 | | 185 | 11.63 | .82 | 197.45 | 225 |
| | 186 | 3'-0" x 7'-0" | | 1 | | 215 | 12.37 | .88 | 228.25 | 255 |
| | 190 | 4'-0" x 8'-0" | ↓ | 1.140 | ↓ | 285 | 14.10 | 1 | 300.10 | 335 |
| 23 | 001 | **FIRE DOOR** Steel, flush, "B" label, 90 minute, hollow | | | | | | | | |
| | 002 | Full panel, 20 ga., 2'-0" x 6'-8" | F-2 | .800 | Ea. | 115 | 9.90 | .70 | 125.60 | 145 |
| | 004 | 2'-6" x 6'-8" | | .889 | | 120 | 10.97 | .78 | 131.75 | 150 |
| | 006 | 3'-0" x 6'-8" | | .941 | | 135 | 11.63 | .82 | 147.45 | 170 |
| | 008 | 3'-0" x 7'-0" | | .941 | | 140 | 11.63 | .82 | 152.45 | 175 |
| | 014 | 18 ga., 3'-0" x 6'-8" | | 1 | | 165 | 12.37 | .88 | 178.25 | 200 |
| | 016 | 2'-6" x 7'-0" | | .941 | | 155 | 11.63 | .82 | 167.45 | 190 |
| | 018 | 3'-0" x 7'-0" | | 1 | | 170 | 12.37 | .88 | 183.25 | 210 |
| | 020 | 4'-0" x 7'-0" | ↓ | 1.070 | ↓ | 200 | 13.17 | .93 | 214.10 | 240 |
| | 020 | | | | | | | | | |
| | 022 | For "A" label, 3 hour, 18 ga., use same price as "B" label | | | | | | | | |
| | 024 | For vision lite, add | | | Ea. | 65 | | | 65 | 72M |
| | 052 | Flush, "B" label 90 min., composite, 20 ga., 2'-0" x 6'-8" | F-2 | .889 | | 130 | 10.97 | .78 | 141.75 | 160 |
| | 054 | 2'-6" x 6'-8" | | .941 | | 140 | 11.63 | .82 | 152.45 | 175 |
| | 056 | 3'-0" x 6'-8" | | 1 | | 150 | 12.37 | .88 | 163.25 | 185 |
| | 058 | 3'-0" x 7'-0" | | 1 | | 165 | 12.37 | .88 | 178.25 | 200 |
| | 064 | Flush, "A" label 3 hour, composite, 18 ga., 3'-0" x 6'-8" | | 1.070 | | 185 | 13.17 | .93 | 199.10 | 225 |
| | 066 | 2'-6" x 7'-0" | ↓ | 1 | ↓ | 175 | 12.37 | .88 | 188.25 | 215 |

For expanded coverage of these items see *Building Construction Cost Data 1985*

## 8.1 Metal Doors

| | | | CREW | MAN-HOURS | UNIT | MAT. | LABOR | EQUIP. | TOTAL | TOTAL INCL O&P |
|---|---|---|---|---|---|---|---|---|---|---|
| 23 | 068 | 3'-0" x 7'-0" | F-2 | 1.070 | Ea. | 200 | 13.17 | .93 | 214.10 | 240 |
| | 070 | 4'-0" x 7'-0" | | 1.140 | | 240 | 14.10 | 1 | 255.10 | 285 |
| 24-001 | | RESIDENTIAL DOOR Steel, 24 ga., embossed, full panel, 2'-8" x 6'-8" | | 1 | | 105 | 12.37 | .88 | 118.25 | 135 |
| | 004 | 3'-0" x 6'-8" | | 1.070 | | 115 | 13.17 | .93 | 129.10 | 150 |
| | 006 | 3'-0" x 7'-0" | | 1.070 | | 125 | 13.17 | .93 | 139.10 | 160 |
| | 022 | Half glass, 2'-8" x 6'-8" | | .941 | | 150 | 11.63 | .82 | 162.45 | 185 |
| | 024 | 3'-0" x 6'-8" | | 1 | | 160 | 12.37 | .88 | 173.25 | 195 |
| | 026 | 3'-0" x 7'-0" | | 1 | | 170 | 12.37 | .88 | 183.25 | 210 |
| | 072 | Raised plastic face, full panel, 2'-8" x 6'-8" | | 1 | | 115 | 12.37 | .88 | 128.25 | 145 |
| | 074 | 3'-0" x 6'-8" | | 1.070 | | 125 | 13.17 | .93 | 139.10 | 160 |
| | 076 | 3'-0" x 7'-0" | | 1.070 | | 145 | 13.17 | .93 | 159.10 | 180 |
| | 082 | Half glass, 2'-8" x 6'-8" | | .941 | | 150 | 11.63 | .82 | 162.45 | 185 |
| | 084 | 3'-0" x 6'-8" | | 1 | | 160 | 12.37 | .88 | 173.25 | 195 |
| | 086 | 3'-0" x 7'-0" | | 1 | | 180 | 12.37 | .88 | 193.25 | 220 |
| | 132 | Flush face, full panel, 2'-6" x 6'-8" | | 1 | | 85 | 12.37 | .88 | 98.25 | 115 |
| | 134 | 3'-0" x 6'-8" | | 1.070 | | 97 | 13.17 | .93 | 111.10 | 130 |
| | 136 | 3'-0" x 7'-0" | | 1.070 | | 115 | 13.17 | .93 | 129.10 | 150 |
| | 142 | Half glass, 2'-8" x 6'-8" | | .941 | | 120 | 11.63 | .82 | 132.45 | 150 |
| | 144 | 3'-0" x 6'-8" | | 1 | | 130 | 12.37 | .88 | 143.25 | 165 |
| | 146 | 3'-0" x 7'-0" | | 1 | | 140 | 12.37 | .88 | 153.25 | 175 |
| | 230 | Interior, residential, closet, bi-fold, 6'-8" x 2'-0" wide | | 1 | | 38 | 12.37 | .88 | 51.25 | 62 |
| | 233 | 3'-0" wide | | 1 | | 45 | 12.37 | .88 | 58.25 | 70 |
| | 236 | 4'-0" wide | | 1.070 | | 60 | 13.17 | .93 | 74.10 | 88 |
| | 240 | 5'-0" wide | | 1.140 | | 70 | 14.10 | 1 | 85.10 | 100 |
| | 242 | 6'-0" wide | | 1.230 | | 78 | 15.22 | 1.08 | 94.30 | 110 |
| 30-001 | | ALUMINUM FRAMES Entrance, 3' x 7' opening, clear finish | 2 Sswk | 2.290 | Opng. | 125 | 30 | | 155 | 190 |
| | 006 | Black finish | | 2.290 | Ea. | 175 | 30 | | 205 | 245 |
| | 010 | Bronze finish | | 2.290 | Opng. | 150 | 30 | | 180 | 215 |
| | 020 | 3'-6" x 7'-0", mill finish | | 2.290 | Ea. | 140 | 30 | | 170 | 205 |
| | 022 | Bronze finish | | 2.290 | | 170 | 30 | | 200 | 240 |
| | 024 | Black finish | | 2.290 | | 185 | 30 | | 215 | 255 |
| | 052 | Bronze finish | | 2.670 | | 155 | 35 | | 190 | 230 |
| | 054 | Black finish | | 2.670 | | 195 | 35 | | 230 | 275 |
| | 060 | 7'-0" x 7'-0", mill finish | | 2.670 | | 155 | 35 | | 190 | 230 |
| | 062 | Bronze finish | | 2.670 | | 185 | 35 | | 220 | 265 |
| | 064 | Black finish | | 2.670 | | 200 | 35 | | 235 | 280 |
| | 200 | Transoms, 3'-0" x 3'-0", mill finish | | .200 | | 60 | 2.66 | | 62.66 | 71 |
| | 202 | Bronze finish | | .200 | | 70 | 2.66 | | 72.66 | 82 |
| | 204 | Black finish | | .200 | | 75 | 2.66 | | 77.66 | 87 |
| | 220 | 3'-6" x 3'-0", mill finish | | .200 | | 62 | 2.66 | | 64.66 | 73 |
| | 222 | Bronze finish | | .200 | | 75 | 2.66 | | 77.66 | 87 |
| | 224 | Black finish | | .200 | | 80 | 2.66 | | 82.66 | 93 |
| | 250 | 6'-0" x 3'-0", mill finish | | .246 | | 71 | 3.27 | | 74.27 | 84 |
| | 252 | Bronze finish | | .246 | | 74 | 3.27 | | 77.27 | 87 |
| | 254 | Black finish | | .246 | | 91 | 3.27 | | 94.27 | 105 |
| | 270 | 7'-0" x 3'-0", mill finish | | .246 | | 77 | 3.27 | | 80.27 | 90 |
| | 272 | Bronze finish | | .246 | | 90 | 3.27 | | 93.27 | 105 |
| | 274 | Black finish | | .246 | | 97 | 3.27 | | 100.27 | 110 |
| 32-001 | | ALUMINUM DOORS Commercial entrance | | | | | | | | |
| | 100 | Narrow style, no glazing, 3'-0" x 7'-0", mill finish | F-2 | 5.330 | Ea. | 245 | 66.33 | 4.67 | 316 | 380 |
| | 105 | Bronze finish | | 5.330 | | 290 | 66.33 | 4.67 | 361 | 430 |
| | 110 | Black finish | | 5.330 | | 310 | 66.33 | 4.67 | 381 | 450 |
| | 150 | 3-6" x 7'-0", mill finish | | 5.330 | | 275 | 66.33 | 4.67 | 346 | 410 |
| | 155 | Bronze finish | | 5.330 | | 325 | 66.33 | 4.67 | 396 | 465 |
| | 160 | Black finish | | 5.330 | | 350 | 66.33 | 4.67 | 421 | 495 |
| 40-001 | | ALUMINUM DOORS & FRAMES Entrance, narrow stile, including | | | | | | | | |
| | 002 | hardware & closer, clear finish, not incl. glass, 3' x 7' opening | 2 Sswk | 8 | Ea. | 505 | 105 | | 610 | 740 |

For expanded coverage of these items see *Building Construction Cost Data 1985*

## 8.1 Metal Doors

| | | CREW | MAN-HOURS | UNIT | MAT. | LABOR | EQUIP. | TOTAL | TOTAL INCL O&P |
|---|---|---|---|---|---|---|---|---|---|
| 010 | 3' x 10' opening, 3' high transom | 2 Sswk | 8.890 | Ea. | 585 | 120 | | 705 | 845 |
| 020 | 3'-6" x 10' opening, 3' high transom | | 8.890 | " | 610 | 120 | | 730 | 875 |
| 030 | 6' x 7' opening | | 12.310 | Pr. | 790 | 165 | | 955 | 1,150 |
| 040 | 6' x 10' opening, 3' high transom | ↓ | 14.550 | " | 990 | 195 | | 1,185 | 1,425 |
| 100 | Add to above for wide stile doors | | | Leaf | 30% | | | | |
| 110 | Full vision doors, with 1/2" glass, add | | | | 55% | | | | |
| 120 | Non-standard size, add | | | | 35% | | | | |
| 130 | Light bronze finish, add | | | | 15% | | | | |
| 140 | Dark bronze finish, add | | | | 18% | | | | |
| 150 | Black finish, add | | | ↓ | 27% | | | | |
| 160 | Concealed panic device, add | | | | 365 | | | 365 | 400M |
| 170 | Electric striker release, add | | | Opng. | 260 | | | 260 | 285M |
| 180 | Floor check, add | F-2 | 5.330 | Leaf | 235 | 66.33 | 4.67 | 306 | 370 |
| 190 | Concealed closer, add | " | 12.310 | " | 150 | 154.25 | 10.75 | 315 | 415 |
| 85-001 | **STORM DOORS & FRAMES** Aluminum, residential, | | | | | | | | |
| 002 | combination storm and screen | | | | | | | | |
| 040 | Anodized, 6'-8" x 2'-6" wide | F-2 | 1.070 | Ea. | 80 | 13.17 | .93 | 94.10 | 110 |
| 042 | 2'-8" wide | | 1.140 | | 85 | 14.10 | 1 | 100.10 | 115 |
| 044 | 3'-0" wide | ↓ | .022 | | 90 | .27 | .02 | 90.29 | 99 |
| 050 | For 7'-0" door, add | | | | 5 | | | | |
| 100 | Mill finish, 6'-8" x 2'-6" wide | F-2 | 1.070 | | 70 | 13.17 | .93 | 84.10 | 99 |
| 102 | 2'-8" wide | | 1.140 | | 76 | 14.10 | 1 | 91.10 | 105 |
| 104 | 3'-0" wide | ↓ | 1.140 | ↓ | 78 | 14.10 | 1 | 93.10 | 110 |
| 110 | For 7'-0" door, add | | | | 5 | | | | |
| 150 | White painted, 6'-8" x 2'-6" wide | F-2 | 1.070 | Ea. | 80 | 13.17 | .93 | 94.10 | 110 |
| 152 | 2'-8" wide | | 1.140 | | 85 | 14.10 | 1 | 100.10 | 115 |
| 154 | 3'-0" wide | ↓ | 1.140 | | 88 | 14.10 | 1 | 103.10 | 120 |
| 160 | For 7'-0" door, add | | | ↓ | 5% | | | | |
| 200 | Wood door & screen, see division 8.2-32 | | | | | | | | |
| 202 | | | | | | | | | |

## 8.2 Wood & Plastic Doors

| | | CREW | MAN-HOURS | UNIT | MAT. | LABOR | EQUIP. | TOTAL | TOTAL INCL O&P |
|---|---|---|---|---|---|---|---|---|---|
| 12-001 | **WOOD FRAMES** | | | | | | | | |
| 040 | Exterior frame, incl. ext. trim, pine, 5/4 x 4-9/16" deep | F-2 | .043 | L.F. | 2.46 | .52 | .04 | 3.02 | 3.58 |
| 042 | 5-3/16" deep | | .043 | | 2.76 | .52 | .04 | 3.32 | 3.91 |
| 044 | 6-9/16" deep | | .043 | | 3.14 | .52 | .04 | 3.70 | 4.33 |
| 060 | Oak, 5/4 x 4-9/16" deep | | .046 | | 3.10 | .56 | .04 | 3.70 | 4.35 |
| 062 | 5-3/16" deep | | .046 | | 3.38 | .56 | .04 | 3.98 | 4.66 |
| 064 | 6-9/16" deep | | .046 | | 3.95 | .56 | .04 | 4.55 | 5.30 |
| 080 | Walnut, 5/4 x 4-9/16" deep | | .046 | | 4.57 | .56 | .04 | 5.17 | 5.95 |
| 082 | 5-3/16" deep | | .046 | | 5.25 | .56 | .04 | 5.85 | 6.70 |
| 084 | 6-9/16" deep | | .046 | | 5.85 | .56 | .04 | 6.45 | 7.35 |
| 100 | Sills, 8/4 x 8" deep, oak, no horns | | .160 | | 5.05 | 1.98 | .14 | 7.17 | 8.85 |
| 102 | 2" horns | | .160 | | 5.35 | 1.98 | .14 | 7.47 | 9.15 |
| 104 | 3" horns | | .160 | | 6.25 | 1.98 | .14 | 8.37 | 10.15 |
| 110 | 8/4 x 10" deep, oak, no horns | | .178 | | 6.45 | 2.19 | .16 | 8.80 | 10.75 |
| 112 | 2" horns | | .178 | | 6.90 | 2.19 | .16 | 9.25 | 11.25 |
| 114 | 3" horns | ↓ | .178 | ↓ | 7.55 | 2.19 | .16 | 9.90 | 11.95 |
| 120 | For casing, see division 6.2-41 | | | | | | | | |
| 122 | | | | | | | | | |
| 200 | Exterior, colonial, frame & trim, 3' opng., in-swing, minimum | F-2 | .727 | Ea. | 160 | 8.96 | .64 | 169.60 | 190 |
| 202 | Maximum | | .800 | | 375 | 9.90 | .70 | 385.60 | 430 |
| 210 | 5'-4" opening, in-swing, minimum | | .941 | ↓ | 270 | 11.63 | .82 | 282.45 | 315 |
| 212 | Maximum | ↓ | 1.070 | | 560 | 13.17 | .93 | 574.10 | 640 |

For expanded coverage of these items see *Building Construction Cost Data 1985*

## 8.2 Wood & Plastic Doors

| | | | CREW | MAN-HOURS | UNIT | MAT. | LABOR | EQUIP. | TOTAL | TOTAL INCL O&P |
|---|---|---|---|---|---|---|---|---|---|---|
| 12 | 214 | Out-swing, minimum | F-2 | .941 | Ea. | 290 | 11.63 | .82 | 302.45 | 340 |
| | 216 | Maximum | | 1.070 | | 575 | 13.17 | .93 | 589.10 | 655 |
| | 240 | 6'-0" opening, in-swing, minimum | | 1 | | 280 | 12.37 | .88 | 293.25 | 330 |
| | 242 | Maximum | | 1.600 | | 590 | 19.60 | 1.40 | 611 | 680 |
| | 246 | Out-swing, minimum | | 1 | | 295 | 12.37 | .88 | 308.25 | 345 |
| | 248 | Maximum | | 1.600 | | 605 | 19.60 | 1.40 | 626 | 700 |
| | 260 | For two sidelights, add, minimum | | .533 | Opng. | 135 | 6.58 | .47 | 142.05 | 160 |
| | 262 | Maximum | | .800 | " | 245 | 9.90 | .70 | 255.60 | 285 |
| | 270 | Custom birch frame, 3'-0" opening | | 1 | Ea. | 95 | 12.37 | .88 | 108.25 | 125 |
| | 275 | 6'-0" opening | | 1 | " | 125 | 12.37 | .88 | 138.25 | 160 |
| | 300 | Interior frame, pine, 11/16" x 3-5/8" deep | | .043 | L.F. | 1.02 | .52 | .04 | 1.58 | 2 |
| | 302 | 4-9/16" deep | | .043 | | 1.28 | .52 | .04 | 1.84 | 2.28 |
| | 304 | 5-3/16" deep | | .043 | | 1.53 | .52 | .04 | 2.09 | 2.56 |
| | 320 | Oak, 11/16" x 3-5/8" deep | | .046 | | 1.25 | .56 | .04 | 1.85 | 2.31 |
| | 322 | 4-9/16" deep | | .046 | | 1.49 | .56 | .04 | 2.09 | 2.58 |
| | 324 | 5-3/16" deep | | .046 | | 1.79 | .56 | .04 | 2.39 | 2.91 |
| | 340 | Walnut, 11/16" x 3-5/8" deep | | .046 | | 1.39 | .56 | .04 | 1.99 | 2.47 |
| | 342 | 4-9/16" deep | | .046 | | 1.82 | .56 | .04 | 2.42 | 2.94 |
| | 344 | 5-3/16" deep | | .046 | | 2.34 | .56 | .04 | 2.94 | 3.51 |
| | 360 | Pocket door frame | | 1 | Ea. | 58 | 12.37 | .88 | 71.25 | 84 |
| | 380 | Threshold, oak, 5/8" x 3-5/8" deep | | .080 | L.F. | 1.09 | .99 | .07 | 2.15 | 2.84 |
| | 382 | 4-5/8" deep | | .084 | | 1.45 | 1.04 | .07 | 2.56 | 3.32 |
| | 384 | 5-5/8" deep | | .089 | | 1.85 | 1.10 | .08 | 3.03 | 3.86 |
| | 400 | For casing see division 6.2-370 | | | | | | | | |
| 22-001 | | **WOOD DOORS, DECORATOR** | | | | | | | | |
| | 400 | Hand carved door, mahogany, simple design | | | | | | | | |
| | 402 | 1-3/4" x 7'-0" x 3'-0" wide | F-2 | 1.140 | Ea. | 310 | 14.10 | 1 | 325.10 | 365 |
| | 404 | 3'-6" wide | | 1.230 | | 405 | 15.22 | 1.08 | 421.30 | 470 |
| | 420 | Rosewood, 1-3/4" x 7'-0" x 3'-0" wide | | 1.140 | | 405 | 14.10 | 1 | 420.10 | 470 |
| | 422 | 3'-6" wide | | 1.230 | | 550 | 15.22 | 1.08 | 566.30 | 630 |
| | 428 | For 6'-8" high door, deduct from 7'-0" door | | | | 10% | | | | |
| | 432 | For detailed design, add | | | | 50% | | | | |
| | 434 | For hand carved back, add | | | | 20% | | | | |
| | 436 | For ornate mahogany door, 2-1/4" thick, add | | | | 20% | | | | |
| | 438 | For ornate rosewood door, 2-1/4" thick, add | | | | 20% | | | | |
| | 440 | For custom finish, add | | | | 84 | | | 84 | 92M |
| | 460 | Side panel, mahogany, simple design, 7'-0" x 1'-0" wide | F-2 | .762 | | 63 | 9.43 | .67 | 73.10 | 85 |
| | 462 | 1'-2" wide | | .800 | | 68 | 9.90 | .70 | 78.60 | 91 |
| | 464 | 1'-4" wide | | .842 | | 74 | 10.41 | .74 | 85.15 | 99 |
| | 480 | Rosewood, simple design 7'-0" x 1'-0" wide | | .762 | | 105 | 9.43 | .67 | 115.10 | 130 |
| | 482 | 1'-2" wide | | .800 | | 115 | 9.90 | .70 | 125.60 | 145 |
| | 484 | 1'-4" wide | | .842 | | 145 | 10.41 | .74 | 156.15 | 175 |
| | 490 | For detailed design, add | | | | 50% | | | | |
| | 492 | For hand carved back, add | | | | 20% | | | | |
| | 652 | Interior cafe doors, 2'-6" opening, stock, panel pine | F-2 | 1 | | 76 | 12.37 | .88 | 89.25 | 105 |
| | 654 | 3'-0" opening | " | 1 | | 84 | 12.37 | .88 | 97.25 | 115 |
| | 654 | Custom hardwood or louvered pine | | | | | | | | |
| | 656 | 2-'6" opening | F-2 | 1 | Ea. | 65 | 12.37 | .88 | 78.25 | 92 |
| | 800 | 3'-0" opening | " | 1 | " | 75 | 12.37 | .88 | 88.25 | 105 |
| | 880 | Pre-hung doors, see division 8.2-50 | | | | | | | | |
| 25-001 | | **WOOD DOORS, PANELED** Interior, six panel, hollow core, 1-3/4" thick | | | | | | | | |
| | 004 | Molded hardboard, 2'-0" x 6'-8" | F-2 | .941 | Ea. | 33 | 11.63 | .82 | 45.45 | 56 |
| | 006 | 2'-6" x 6'-8" | | .941 | | 35 | 11.63 | .82 | 47.45 | 58 |
| | 008 | 3'-0" x 6'-8" | | .941 | | 38 | 11.63 | .82 | 50.45 | 61 |
| | 014 | Embossed print, molded hardboard, 2'-0" x 6'-8" | | .941 | | 51 | 11.63 | .82 | 63.45 | 75 |
| | 016 | 2'-6" x 6'-8" | | .941 | | 56 | 11.63 | .82 | 68.45 | 81 |
| | 018 | 3'-0" x 6'-8" | | .941 | | 61 | 11.63 | .82 | 73.45 | 86 |
| | 054 | Six panel, solid, 1-3/8" thick, pine, 2'-0" x 6'-8" | | 1.070 | | 110 | 13.17 | .93 | 124.10 | 145 |
| | 056 | 2'-6" x 6'-8" | | 1.140 | | 120 | 14.10 | 1 | 135.10 | 155 |
| | 058 | 3'-0" x 6'-8" | | 1.230 | | 140 | 15.22 | 1.08 | 156.30 | 180 |

For expanded coverage of these items see *Building Construction Cost Data 1985*

## 8.2 Wood & Plastic Doors

| | | | MAN- | | \multicolumn{4}{c|}{BARE COSTS} | TOTAL |
|---|---|---|---|---|---|---|---|---|---|
| | | CREW | HOURS | UNIT | MAT. | LABOR | EQUIP. | TOTAL | INCL O&P |
| 25 | 102 | Two panel, bored rail, solid, 1-3/8" thick, pine, 1'-6" x 6'-8" | F-2 | 1 | Ea. | 130 | 12.37 | .88 | 143.25 | 165 |
| | 104 | 2'-0" x 6'-8" | | 1.070 | | 190 | 13.17 | .93 | 204.10 | 230 |
| | 106 | 2'-6" x 6'-8" | | 1.140 | | 225 | 14.10 | 1 | 240.10 | 270 |
| | 134 | Two panel, solid, 1-3/8" thick, fir, 2'-0" x 6'-8" | | 1.070 | | 115 | 13.17 | .93 | 129.10 | 150 |
| | 136 | 2'-6" x 6'-8" | | 1.140 | | 120 | 14.10 | 1 | 135.10 | 155 |
| | 138 | 3'-0" x 6'-8" | | 1.230 | | 130 | 15.22 | 1.08 | 146.30 | 170 |
| | 174 | Five panel, solid, 1-3/8" thick, fir, 2'-0" x 6'-8" | | 1.070 | | 125 | 13.17 | .93 | 139.10 | 160 |
| | 176 | 2'-6" x 6'-8" | | 1.140 | | 130 | 14.10 | 1 | 145.10 | 165 |
| | 178 | 3'-0" x 6'-8" | ↓ | 1.230 | ↓ | 135 | 15.22 | 1.08 | 151.30 | 175 |
| 32-001 | | **WOOD DOORS, RESIDENTIAL** | | | | | | | | |
| | 020 | Exterior, combination storm & screen, pine | | | | | | | | |
| | 022 | Cross buck, 6'-9" x 2'-6" wide | F-2 | 1.450 | Ea. | 125 | 17.98 | 1.27 | 144.25 | 165 |
| | 026 | 2'-8" wide | | 1.600 | | 125 | 19.60 | 1.40 | 146 | 170 |
| | 028 | 3'-0" wide | | 1.780 | | 130 | 22.44 | 1.56 | 154 | 180 |
| | 030 | 7'-1" x 3'-0" wide | | 1.780 | | 135 | 22.44 | 1.56 | 159 | 185 |
| | 040 | Full lite, 6'-9" x 2'-6" wide | | 1.450 | | 115 | 17.98 | 1.27 | 134.25 | 155 |
| | 042 | 2'-8" wide | | 1.600 | | 115 | 19.60 | 1.40 | 136 | 160 |
| | 044 | 3'-0" wide | | 1.780 | | 120 | 22.44 | 1.56 | 144 | 170 |
| | 050 | 7'-1" x 3'-0" wide | | 1.780 | | 125 | 22.44 | 1.56 | 149 | 175 |
| | 070 | Dutch door, pine, 1-3/4" x 6'-8" x 2'-8" wide, minimum | | 1.330 | | 175 | 16.48 | 1.17 | 192.65 | 220 |
| | 072 | Maximum | | 1.600 | | 215 | 19.60 | 1.40 | 236 | 270 |
| | 080 | 3'-0" wide, minimum | | 1.330 | | 185 | 16.48 | 1.17 | 202.65 | 230 |
| | 082 | Maximum | | 1.600 | | 230 | 19.60 | 1.40 | 251 | 285 |
| | 100 | Entrance door, colonial, 1-3/4" x 6'-8" x 2'-8" wide | | 1 | | 145 | 12.37 | .88 | 158.25 | 180 |
| | 102 | 6 panel pine, 3'-0" wide | | 1.070 | | 160 | 13.17 | .93 | 174.10 | 200 |
| | 110 | 8 panel pine, 2'-8" wide | | 1 | | 185 | 12.37 | .88 | 198.25 | 225 |
| | 112 | 3'-0" wide | ↓ | 1.070 | | 200 | 13.17 | .93 | 214.10 | 240 |
| | 120 | For tempered safety glass lites, add | | | ↓ | 16 | | | 16 | 17.60M |
| | 122 | | | | | | | | | |
| | 130 | Flush, birch, solid core, 1-3/4" x 6'-8" x 2'-8" wide | F-2 | 1 | Ea. | 99 | 12.37 | .88 | 112.25 | 130 |
| | 132 | 3'-0" wide | | 1.070 | | 100 | 13.17 | .93 | 114.10 | 130 |
| | 134 | 7'-0" x 2'-8" wide | | 1 | | 105 | 12.37 | .88 | 118.25 | 135 |
| | 136 | 3'-0" wide | | 1.070 | | 110 | 13.17 | .93 | 124.10 | 145 |
| | 170 | Hand carved door, mahogany 2'-8" x 6'-8" | | 1.070 | | 140 | 13.17 | .93 | 154.10 | 175 |
| | 172 | 3'-0" x 6'-8" | | 1.070 | | 160 | 13.17 | .93 | 174.10 | 200 |
| | 174 | Rosewood, 2'-8" x 6'-8" | | 1.070 | | 340 | 13.17 | .93 | 354.10 | 395 |
| | 176 | 3'-0" x 6'-8" | ↓ | 1.070 | ↓ | 370 | 13.17 | .93 | 384.10 | 430 |
| | 270 | Interior, closet, bi-folding, with hardware, no frame or trim incl. | | | | | | | | |
| | 272 | Flush, birch, 6'-6" or 6'-8" x 2'-6" wide | F-2 | 1.230 | Ea. | 46 | 15.22 | 1.08 | 62.30 | 76 |
| | 274 | 3'-0" wide | | 1.230 | | 49 | 15.22 | 1.08 | 65.30 | 79 |
| | 276 | 4'-0" wide | | 1.330 | | 77 | 16.48 | 1.17 | 94.65 | 110 |
| | 278 | 5'-0" wide | | 1.450 | | 82 | 17.98 | 1.27 | 101.25 | 120 |
| | 280 | 6'-0" wide | | 1.600 | | 89 | 19.60 | 1.40 | 110 | 130 |
| | 300 | Raised panel pine, 6'-6" or 6'-8" x 2'-6" wide | | 1.230 | | 120 | 15.22 | 1.08 | 136.30 | 155 |
| | 302 | 3'-0" wide | | 1.230 | | 130 | 15.22 | 1.08 | 146.30 | 170 |
| | 304 | 4'-0" wide | | 1.330 | | 210 | 16.48 | 1.17 | 227.65 | 260 |
| | 306 | 5'-0" wide | | 1.450 | | 240 | 17.98 | 1.27 | 259.25 | 295 |
| | 308 | 6'-0" wide | | 1.600 | | 260 | 19.60 | 1.40 | 281 | 320 |
| | 320 | Louvered, pine, 6'-6" or 6'-8" x 2'-6" wide | | 1.230 | | 82 | 15.22 | 1.08 | 98.30 | 115 |
| | 322 | 3'-0" wide | | 1.230 | | 88 | 15.22 | 1.08 | 104.30 | 120 |
| | 324 | 4'-0" wide | | 1.330 | | 150 | 16.48 | 1.17 | 167.65 | 190 |
| | 326 | 5'-0" wide | | 1.450 | | 160 | 17.98 | 1.27 | 179.25 | 205 |
| | 328 | 6'-0" wide | ↓ | 1.600 | ↓ | 170 | 19.60 | 1.40 | 191 | 220 |
| | 440 | Bi-passing closet, incl. hardware, no frame or trim incl., | | | | | | | | |
| | 442 | Flush, lauan, 6'-8" x 4'-0" wide | F-2 | 1.330 | Opng. | 62 | 16.48 | 1.17 | 79.65 | 96 |
| | 444 | 5'-0" wide | | 1.450 | | 72 | 17.98 | 1.27 | 91.25 | 110 |
| | 446 | 6'-0" wide | | 1.600 | | 81 | 19.60 | 1.40 | 102 | 120 |
| | 460 | Flush, birch, 6'-8" x 4'-0" wide | | 1.330 | | 79 | 16.48 | 1.17 | 96.65 | 115 |
| | 462 | 5'-0" wide | ↓ | 1.450 | ↓ | 92 | 17.98 | 1.27 | 111.25 | 130 |

For expanded coverage of these items see *Building Construction Cost Data 1985*

## 8.2 Wood & Plastic Doors

| | | | CREW | MAN-HOURS | UNIT | MAT. | LABOR | EQUIP. | TOTAL | TOTAL INCL O&P |
|---|---|---|---|---|---|---|---|---|---|---|
| 32 | 464 | 6'-0" wide | F-2 | 1.600 | Opng. | 105 | 19.60 | 1.40 | 126 | 150 |
| | 480 | Louvered, pine, 6'-8" x 4'-0" wide | | 1.330 | | 110 | 16.48 | 1.17 | 127.65 | 150 |
| | 482 | 5'-0" wide | | 1.450 | | 125 | 17.98 | 1.27 | 144.25 | 165 |
| | 484 | 6'-0" wide | | 1.600 | | 140 | 19.60 | 1.40 | 161 | 185 |
| | 500 | Paneled, pine, 6'-8" x 4'-0" wide | | 1.330 | | 215 | 16.48 | 1.17 | 232.65 | 265 |
| | 502 | 5'-0" wide | | 1.450 | | 235 | 17.98 | 1.27 | 254.25 | 290 |
| | 504 | 6'-0" wide | ▼ | 1.600 | ▼ | 260 | 19.60 | 1.40 | 281 | 320 |
| | 505 | | | | | | | | | |
| | 610 | Folding accordian, closet, not including frame | | | | | | | | |
| | 612 | Vinyl, 2 layer, stock (see also division 10.1-45) | F-2 | .040 | S.F. | 2.42 | .49 | .04 | 2.95 | 3.48 |
| | 614 | Woven mahogany and vinyl, stock | | .040 | | 3.10 | .49 | .04 | 3.63 | 4.23 |
| | 616 | Wood slats with vinyl overlay, stock | | .040 | | 5.25 | .49 | .04 | 5.78 | 6.60 |
| | 618 | Economy vinyl, stock | | .040 | | 1.10 | .49 | .04 | 1.63 | 2.03 |
| | 620 | Rigid PVC | ▼ | .040 | ▼ | 1.49 | .49 | .04 | 2.02 | 2.46 |
| | 622 | For custom folding, add to above | | | Ea. | 25% | | | | |
| | 623 | | | | | | | | | |
| | 740 | Passage doors, flush, no frame included | | | | | | | | |
| | 742 | Lauan, hollow core, 1-3/8" x 6'-8" x 1'-6" wide | F-2 | .842 | Ea. | 16 | 10.41 | .74 | 27.15 | 35 |
| | 744 | 2'-0" wide | | .889 | | 16 | 10.97 | .78 | 27.75 | 36 |
| | 746 | 2'-6" wide | | .889 | | 18 | 10.97 | .78 | 29.75 | 38 |
| | 748 | 2'-8" wide | | .889 | | 19 | 10.97 | .78 | 30.75 | 39 |
| | 750 | 3'-0" wide | | .941 | | 20 | 11.63 | .82 | 32.45 | 41 |
| | 770 | Birch, hollow core, 1-3/8" x 6'-8" x 1'-6" wide | | .842 | | 20 | 10.41 | .74 | 31.15 | 39 |
| | 772 | 2'-0" wide | | .889 | | 23 | 10.97 | .78 | 34.75 | 44 |
| | 774 | 2'-6" wide | | .889 | | 26 | 10.97 | .78 | 37.75 | 47 |
| | 776 | 2'-8" wide | | .889 | | 27 | 10.97 | .78 | 38.75 | 48 |
| | 778 | 3'-0" wide | | .941 | | 29 | 11.63 | .82 | 41.45 | 51 |
| | 800 | Pine louvered, 1-3/8" x 6'-8" x 1'-6" wide | | .842 | | 45 | 10.41 | .74 | 56.15 | 67 |
| | 802 | 2'-0" wide | | .889 | | 55 | 10.97 | .78 | 66.75 | 79 |
| | 804 | 2'-6" wide | | .889 | | 72 | 10.97 | .78 | 83.75 | 97 |
| | 806 | 2'-8" wide | | .889 | | 74 | 10.97 | .78 | 85.75 | 100 |
| | 808 | 3'-0" wide | | .941 | | 78 | 11.63 | .82 | 90.45 | 105 |
| | 830 | Pine paneled, 1-3/8" x 6'-8" x 1'-6" wide | | .842 | | 65 | 10.41 | .74 | 76.15 | 89 |
| | 832 | 2'-0" wide | | .889 | | 85 | 10.97 | .78 | 96.75 | 110 |
| | 834 | 2'-6" wide | ▼ | .889 | ▼ | 90 | 10.97 | .78 | 101.75 | 115 |
| | 836 | 2'-8" wide | F-2 | .889 | Ea. | 94 | 10.97 | .78 | 105.75 | 120 |
| | 838 | 3'-0" wide | " | .941 | " | 105 | 11.63 | .82 | 117.45 | 135 |
| | 840 | | | | | | | | | |
| | 855 | For over 20 doors, deduct | | | | 15% | | | | |
| | 50-001 | **PRE-HUNG DOORS** | | | | | | | | |
| | 030 | Exterior, wood, combination storm & screen, 6'-9" x 2'-6" wide | F-2 | 1.070 | Ea. | 145 | 13.17 | .93 | 159.10 | 180 |
| | 032 | 2'-8" wide | | 1.070 | | 145 | 13.17 | .93 | 159.10 | 180 |
| | 034 | 3'-0" wide | ▼ | 1.070 | | 150 | 13.17 | .93 | 164.10 | 185 |
| | 036 | For 7'-0" high door, add | | | ▼ | 7% | | | | |
| | 037 | For aluminum storm doors, see division 8.1-85 | | | | | | | | |
| | 160 | Entrance door, flush, birch, solid core | | | | | | | | |
| | 162 | 4-5/8" solid jamb, 1-3/4" x 6'-8" x 2'-8" wide | F-2 | 1 | Ea. | 160 | 12.37 | .88 | 173.25 | 195 |
| | 162 | 7' door, 4-5/8" jamb | | 1 | | 175 | 12.37 | .88 | 188.25 | 215 |
| | 162 | 6'-8" door, 5-5/8" jamb | | 1 | | 165 | 12.37 | .88 | 178.25 | 200 |
| | 162 | 7' door, 5-5/8" jamb | | 1 | | 180 | 12.37 | .88 | 193.25 | 220 |
| | 164 | 4-5/8" solid jamb, 1/3/4" x 6'-8" x 3'-0" wide | | 1 | | 165 | 12.37 | .88 | 178.25 | 200 |
| | 164 | 7' door, 4-5/8" jamb | | 1 | | 180 | 12.37 | .88 | 193.25 | 220 |
| | 164 | 6'-8" door, 5-5/8" jamb | | 1 | | 170 | 12.37 | .88 | 183.25 | 210 |
| | 164 | 7' door, 5-5/8" jamb | ▼ | 1 | ▼ | 190 | 12.37 | .88 | 203.25 | 230 |
| | 200 | Entrance door, colonial, 6 panel pine | | | | | | | | |
| | 202 | 4-5/8" solid jamb, 1-3/4" x 6'-8" x 2'-8" wide | F-2 | 1 | Ea. | 265 | 12.37 | .88 | 278.25 | 310 |
| | 202 | 7' door, 4-5/8" jamb | | 1 | | 290 | 12.37 | .88 | 303.25 | 340 |
| | 202 | 6'-8" door 5-5/8" jamb | ▼ | 1 | ▼ | 260 | 12.37 | .88 | 273.25 | 305 |

For expanded coverage of these items see *Building Construction Cost Data 1985*

## 8.2 Wood & Plastic Doors

| | | CREW | MAN-HOURS | UNIT | MAT. | LABOR | EQUIP. | TOTAL | TOTAL INCL O&P |
|---|---|---|---|---|---|---|---|---|---|
| 202 | 7' door, 5-5/8" jamb | F-2 | 1 | Ea. | 270 | 12.37 | .88 | 283.25 | 320 |
| 204 | 4-5/8" solid jamb, 1-3/4" x 6'-8" x 3'-0" wide | | 1 | | 275 | 12.37 | .88 | 288.25 | 325 |
| 204 | 7' door, 4-5/8" jamb | | 1 | | 300 | 12.37 | .88 | 313.25 | 350 |
| 204 | 6'-8" door, 5-5/8" jamb | | 1 | | 285 | 12.37 | .88 | 298.25 | 335 |
| 204 | 7' door, 5-5/8" jamb | ↓ | 1 | | 310 | 12.37 | .88 | 323.25 | 360 |
| 206 | | | | | | | | | |
| 220 | For 5-5/8" solid jamb, add | | | | 10.50 | | | 10.50 | 11.55M |
| 230 | French door, 6'-8" x 6'-0" wide, 1/2" insul. glass and grille | F-2 | 2.290 | ↓ | 605 | 28 | 2 | 635 | 710 |
| 299 | | | | | | | | | |
| 400 | Interior, passage door, 4-5/8" solid jamb | | | | | | | | |
| 440 | Lauan, flush, solid core, 1-3/8" x 6'-8" x 2'-6" wide | F-2 | .800 | Ea. | 65 | 9.90 | .70 | 75.60 | 88 |
| 440 | 7' door, 4-5/8" jamb | | .800 | | 71 | 9.90 | .70 | 81.60 | 95 |
| 440 | 6'-8" door, 5-5/8" jamb | | .800 | | 75 | 9.90 | .70 | 85.60 | 99 |
| 440 | 7' door, 5-5/8" jamb | | .800 | | 81 | 9.90 | .70 | 91.60 | 105 |
| 442 | 2'-8" wide | | .800 | | 68 | 9.90 | .70 | 78.60 | 91 |
| 442 | 7' door, 4-5/8" jamb | | .800 | | 75 | 9.90 | .70 | 85.60 | 99 |
| 442 | 6'-8" door, 5-5/8" jamb | | .800 | | 78 | 9.90 | .70 | 88.60 | 100 |
| 442 | 7' door, 5-5/8" jamb | | .800 | | 85 | 9.90 | .70 | 95.60 | 110 |
| 444 | 3'-0" wide | | .800 | | 74 | 9.90 | .70 | 84.60 | 98 |
| 444 | 7' door, 4-5/8" jamb | | .800 | | 80 | 9.90 | .70 | 90.60 | 105 |
| 444 | 6'-8" door, 5-5/8" jamb | | .800 | | 83 | 9.90 | .70 | 93.60 | 110 |
| 444 | 7' door, 5-5/8" jamb | | .800 | | 90 | 9.90 | .70 | 100.60 | 115 |
| 460 | Hollow core, 1-3/8" x 6'-8" x 2'-6" wide | | .800 | | 49 | 9.90 | .70 | 59.60 | 70 |
| 460 | 7' door, 4-5/8" jamb | | .800 | | 54 | 9.90 | .70 | 64.60 | 76 |
| 460 | 6'-8" door, 5-5/8" jamb | | .800 | | 59 | 9.90 | .70 | 69.60 | 81 |
| 460 | 7' door, 5-5/8" jamb | | .800 | | 64 | 9.90 | .70 | 74.60 | 87 |
| 462 | 2'-8" wide | | .800 | | 50 | 9.90 | .70 | 60.60 | 71 |
| 462 | 7' door, 4-5/8" jamb | | .800 | | 55 | 9.90 | .70 | 65.60 | 77 |
| 462 | 6'-8" door, 5-5/8" jamb | | .800 | | 60 | 9.90 | .70 | 70.60 | 82 |
| 462 | 7' door, 5-5/8" jamb | | .800 | | 65 | 9.90 | .70 | 75.60 | 88 |
| 464 | 3'-0" wide | | .800 | | 53 | 9.90 | .70 | 63.60 | 75 |
| 464 | 7' door, 4-5/8" jamb | | .800 | | 57 | 9.90 | .70 | 67.60 | 79 |
| 464 | 6'-8" door, 5-5/8" jamb | | .800 | | 62 | 9.90 | .70 | 72.60 | 85 |
| 464 | 7' door, 5-5/8" jamb | | .800 | | 73 | 9.90 | .70 | 83.60 | 97 |
| 470 | | | | | | | | | |
| 500 | Birch, flush, solid core, 1-3/8" x 6'-8" x 2'-6" wide | F-2 | .800 | | 130 | 9.90 | .70 | 140.60 | 160 |
| 500 | 7' door, 4-5/8" jamb | | .800 | | 120 | 9.90 | .70 | 130.60 | 150 |
| 500 | 6'-8" door, 5-5/8" jamb | | .800 | | 120 | 9.90 | .70 | 130.60 | 150 |
| 500 | 7' door, 5-5/8" jamb | | .800 | | 130 | 9.90 | .70 | 140.60 | 160 |
| 502 | 2'-8" wide | | .800 | | 110 | 9.90 | .70 | 120.60 | 135 |
| 502 | 7' door, 4-5/8" jamb | | .800 | | 120 | 9.90 | .70 | 130.60 | 150 |
| 502 | 6'-8" door, 5-5/8" jamb | | .800 | | 120 | 9.90 | .70 | 130.60 | 150 |
| 502 | 7' door 5-5/8" jamb | | .800 | | 130 | 9.90 | .70 | 140.60 | 160 |
| 504 | 3'-0" wide | | .800 | | 135 | 9.90 | .70 | 145.60 | 164 |
| 504 | 7' door, 4-5/8" jamb | | .800 | | 135 | 9.90 | .70 | 145.60 | 165 |
| 504 | 6'-8" door, 5-5/8" jamb | | .800 | | 135 | 9.90 | .70 | 145.60 | 165 |
| 504 | 7' door, 5-5/8" jamb | | .800 | | 145 | 9.90 | .70 | 155.60 | 175 |
| 520 | Hollow core, 1-3/8" x 6'-8" x 2'-6" wide | | .800 | | 63 | 9.90 | .70 | 73.60 | 86 |
| 520 | 7' door, 4-5/8" jamb | | .800 | | 69 | 9.90 | .70 | 79.60 | 92 |
| 520 | 6'-8" door, 5-5/8" jamb | | .800 | | 73 | 9.90 | .70 | 83.60 | 97 |
| 520 | 5' door, 5-5/8" jamb | | .800 | | 83 | 9.90 | .70 | 93.60 | 110 |
| 522 | 2'-8" wide | | .800 | | 65 | 9.90 | .70 | 75.60 | 88 |
| 522 | 7' door, 4-5/8" jamb | | .800 | | 71 | 9.90 | .70 | 81.60 | 95 |
| 522 | 6'-8" door, 5-5/8" jamb | | .800 | | 75 | 9.90 | .70 | 85.60 | 99 |
| 522 | 7' door, 5-5/8" jamb | | .800 | | 81 | 9.90 | .70 | 91.60 | 104 |
| 524 | 3'-0" wide | | .800 | | 67 | 9.90 | .70 | 77.60 | 90 |
| 524 | 7' door, 4-5/8" jamb | | .800 | | 73 | 9.90 | .70 | 83.60 | 97 |
| 524 | 6'-8" door, 5-5/8" jamb | | .800 | | 67 | 9.90 | .70 | 77.60 | 90 |
| 524 | 7' door, 5-5/8" jamb | ↓ | .800 | ↓ | 83 | 9.90 | .70 | 93.60 | 110 |
| 528 | | | | | | | | | |

For expanded coverage of these items see *Building Construction Cost Data 1985*

## 8.2 Wood & Plastic Doors

| | | | CREW | MAN-HOURS | UNIT | MAT. | LABOR | EQUIP. | TOTAL | TOTAL INCL O&P |
|---|---|---|---|---|---|---|---|---|---|---|
| 50 | 550 | Pine louvered, 1-3/8" x 6'-8" x 2'-6" wide | F-2 | .800 | Ea. | 140 | 9.90 | .70 | 150.60 | 170 |
| | 550 | 5-5/8" jamb | | .800 | | 150 | 9.90 | .70 | 160.60 | 180 |
| | 552 | 2'-8" wide | | .800 | | 145 | 9.90 | .70 | 155.60 | 175 |
| | 552 | 5-5/8" jamb | | .800 | | 155 | 9.90 | .70 | 165.60 | 185 |
| | 554 | 3'-0" wide | | .842 | | 150 | 9.90 | .70 | 160.60 | 179 |
| | 554 | 5-5/8" jamb | ↓ | .800 | ↓ | 160 | 9.90 | .70 | 170.60 | 190 |
| | 560 | | | | | | | | | |
| | 600 | Paneled, 1-3/8" x 6'-8" x 2'-6" wide | F-2 | .800 | Ea. | 155 | 9.90 | .70 | 165.60 | 185 |
| | 600 | 5-5/8" jamb | | .800 | | 165 | 9.90 | .70 | 175.60 | 200 |
| | 602 | 2'-8" wide | | .800 | | 155 | 9.90 | .70 | 165.60 | 185 |
| | 602 | 5-5/8" jamb | | .800 | | 170 | 9.90 | .70 | 180.60 | 205 |
| | 604 | 3'-0" wide | | .800 | | 160 | 9.90 | .70 | 170.60 | 190 |
| | 604 | 5-5/8" jamb | ↓ | .800 | ↓ | 175 | 9.90 | .70 | 185.60 | 210 |
| | 620 | | | | | | | | | |
| | 650 | For 5-5/8" solid jamb, add | | | Ea. | 5.25 | | | 5.25 | 5.75M |
| | 652 | For split jamb, deduct | | | " | 5.25 | | | 5.25 | 5.75M |

## 8.3 Special Doors

| | | | CREW | MAN-HOURS | UNIT | MAT. | LABOR | EQUIP. | TOTAL | TOTAL INCL O&P |
|---|---|---|---|---|---|---|---|---|---|---|
| 03-001 | | ACOUSTICAL Incl. framed seals, 3' x 7', wood, 27 STC rating | F-2 | 10.670 | Ea. | 375 | 130.65 | 9.35 | 515 | 630 |
| | 010 | Steel, 40 STC rating | | 10.670 | | 1,050 | 130.65 | 9.35 | 1,190 | 1,375 |
| | 020 | 45 STC rating | | 10.670 | | 1,200 | 130.65 | 9.35 | 1,340 | 1,550 |
| | 030 | 48 STC rating | | 10.670 | | 1,400 | 130.65 | 9.35 | 1,540 | 1,750 |
| | 040 | 53 STC rating | ↓ | 10.670 | ↓ | 1,500 | 130.65 | 9.35 | 1,640 | 1,875 |
| 04-001 | | BULKHEAD CELLAR DOORS Steel, not incl. sides, minimum | 1 Carp | 1.450 | Ea. | 160 | 17.95 | | 177.95 | 205 |
| | 010 | Maximum | | 1.570 | | 177 | 19.35 | | 196.35 | 225 |
| | 050 | With sides and foundation plates, minimum | | 1.700 | | 170 | 21 | | 191 | 220 |
| | 060 | Maximum | ↓ | 1.860 | ↓ | 195 | 23 | | 218 | 250 |
| 09-001 | | COUNTER DOORS 4' high roll-up, 6' long, galv. steel or aluminum | 2 Carp | 8 | Opng. | 570 | 99 | | 669 | 785 |
| | 030 | Galvanized steel, UL label | | 8.890 | | 710 | 110 | | 820 | 955 |
| | 060 | Stainless steel, 4' high roll-up, 6' long | | 8 | | 855 | 99 | | 954 | 1,100 |
| | 070 | 10' long | ↓ | 8.890 | ↓ | 1,050 | 110 | | 1,160 | 1,325 |
| 13-001 | | DOUBLE ACTING With vision panel, incl. frame, closer & hardware | | | | | | | | |
| | 100 | .063" aluminum, 7'-2" high, 3'-4" wide | 2 Carp | 3.810 | Pr. | 530 | 47 | | 577 | 660 |
| | 105 | 6'-4" wide | " | 4 | " | 955 | 49 | | 1,004 | 1,125 |
| | 200 | Solid core wood, 1-3/4" thick, metal frame, stainless steel | | | | | | | | |
| | 201 | base plate, 7' high opening, 4' wide | 2 Carp | 4 | Pr. | 625 | 49 | | 674 | 765 |
| | 205 | 7' wide | " | 4.210 | " | 1,075 | 52 | | 1,127 | 1,275 |
| 15-001 | | FLOOR, COMMERCIAL Aluminum tile, steel frame, single leaf, 2'x2' opng. | 2 Sswk | 4.570 | Opng. | 300 | 61 | | 361 | 435 |
| | 005 | 3'-6" x 3'-6" opening | | 4.570 | | 520 | 61 | | 581 | 675 |
| | 050 | Double leaf, 4' x 4' opening | | 5.330 | | 720 | 71 | | 791 | 915 |
| | 055 | 5' x 5' opening | | 5.330 | | 1,025 | 71 | | 1,096 | 1,250 |
| 18-001 | | FLOOR, INDUSTRIAL Steel 300 psf L.L., single leaf, 2' x 2' opening | | 2.670 | | 320 | 35 | | 355 | 415 |
| | 005 | 3' x 3' opening | | 2.910 | | 450 | 39 | | 489 | 560 |
| | 030 | Double leaf, 4' x 4' opening | | 3.200 | | 680 | 43 | | 723 | 820 |
| | 035 | 5' x 5' opening | ↓ | 3.560 | ↓ | 905 | 47 | | 952 | 1,075 |
| 22-001 | | GLASS, SLIDING Wood, 5/8" tempered insul. glass, 6' wide, premium | 2 Carp | 4 | Ea. | 680 | 49 | | 729 | 825 |
| | 010 | Economy | | 4 | | 350 | 49 | | 399 | 465 |
| | 015 | 8' wide, wood, premium | | 5.330 | | 800 | 66 | | 866 | 985 |
| | 020 | Economy | | 5.330 | | 600 | 66 | | 666 | 765 |
| | 025 | 12' wide, wood, premium | | 6.400 | | 1,250 | 79 | | 1,329 | 1,500 |
| | 030 | Economy | ↓ | 6.400 | ↓ | 830 | 79 | | 909 | 1,050 |
| | 035 | Aluminum sliding, 5/8" tempered insulated glass, 6' wide | | | | | | | | |
| | 040 | Premium | 2 Carp | 4 | Ea. | 310 | 49 | | 359 | 420 |

## 8.3 Special Doors

| | | CREW | MAN-HOURS | UNIT | MAT. | LABOR | EQUIP. | TOTAL | TOTAL INCL O&P |
|---|---|---|---|---|---|---|---|---|---|
| 22 045 | Economy | 2 Carp | 4 | Ea. | 260 | 49 | | 309 | 365 |
| 046 | | | | | | | | | |
| 050 | 8' wide, premium | 2 Carp | 5.330 | Ea. | 365 | 66 | | 431 | 505 |
| 055 | Economy | | 5.330 | | 310 | 66 | | 376 | 445 |
| 060 | 12' wide, premium | | 6.400 | | 415 | 79 | | 494 | 580 |
| 065 | Economy | ↓ | 6.400 | ↓ | 365 | 79 | | 444 | 525 |
| 100 | Replacement doors, wood | | | | | | | | |
| 105 | 6' wide, premium | 2 Carp | 4 | Ea. | 520 | 49 | | 569 | 650 |
| 24-001 | **GLASS, SWING** Tempered, 1/2" thick, incl. hardware, 3' x 7' opening | 2 Glaz | 8 | Opng. | 1,500 | 100 | | 1,600 | 1,800 |
| 010 | 6' x 7' opening | " | 11.430 | " | 3,000 | 140 | | 3,140 | 3,525 |
| 31-001 | **MALL FRONT** 2 fixed end panels with remaining panels sliding, | | | | | | | | |
| 070 | incl. automatic oper. see division 8.4-30 | | | | | | | | |
| 33-001 | **KALAMEIN** Interior, flush type, 3' x 7' | 2 Carp | 3.720 | Opng. | 210 | 46 | | 256 | 305 |
| 39-001 | **OVERHEAD, COMMERCIAL** Frames not included | | | | | | | | |
| 100 | Stock, sectional, heavy duty, wood, 1-3/4" thick, 8' x 8' high | 2 Carp | 8 | Ea. | 380 | 99 | | 479 | 575 |
| 110 | 10' x 10' high | | 8.890 | | 530 | 110 | | 640 | 755 |
| 120 | 12' x 12' high | | 10.670 | | 760 | 130 | | 890 | 1,050 |
| 130 | Chain hoist, 14' x 14' high | | 12.310 | | 1,200 | 150 | | 1,350 | 1,550 |
| 140 | 12' x 16' high | | 16 | | 1,225 | 200 | | 1,425 | 1,650 |
| 150 | 20' x 8' high | | 20 | | 1,025 | 245 | | 1,270 | 1,525 |
| 160 | 20' x 16' high | | 26.670 | | 2,350 | 330 | | 2,680 | 3,100 |
| 180 | Center mullion openings, 8' high | | 4 | | 280 | 49 | | 329 | 385 |
| 190 | 20' high | ↓ | 8 | | 525 | 99 | | 624 | 735 |
| 210 | For medium duty custom doors, deduct | | | | 5% | 5% | | | |
| 215 | For medium duty stock doors, deduct | | | ↓ | 20% | 5% | | | |
| 216 | | | | | | | | | |
| 230 | Fiberglass and aluminum, heavy duty, sectional, 12' x 12' high | 2 Carp | 10.670 | Ea. | 860 | 130 | | 990 | 1,150 |
| 245 | Chain hoist, 20' x 20' high | | 32 | | 2,100 | 395 | | 2,495 | 2,925 |
| 260 | Steel, 24 ga. sectional, manual, 8' x 8' high | | 8 | | 315 | 99 | | 414 | 505 |
| 265 | 10' x 10' high | | 8.890 | | 475 | 110 | | 585 | 695 |
| 270 | 12' x 12' high | | 10.670 | | 685 | 130 | | 815 | 960 |
| 280 | Chain hoist, 20' x 14' high | ↓ | 22.860 | ↓ | 1,575 | 280 | | 1,855 | 2,175 |
| 285 | For 1-1/4" rigid insulation and 26 ga. galv. | | | | | | | | |
| 286 | back panel, add | | | S.F. | 2 | | | 2 | 2.20M |
| 290 | For electric trolley operator, to 14' x 14', add | 1 Carp | 4 | Ea. | 500 | 49 | | 549 | 630 |
| 295 | Over 14' x 14', add | " | 8 | " | 600 | 99 | | 699 | 815 |
| 42-001 | **RESIDENTIAL GARAGE DOORS** Including hardware, no frame | | | | | | | | |
| 002 | | | | | | | | | |
| 005 | Hinged, wood, custom, double door, 9' x 7' | 2 Carp | 5.330 | Ea. | 200 | 66 | | 266 | 325 |
| 007 | 16' x 7' | | 8 | | 265 | 99 | | 364 | 450 |
| 020 | Overhead, sectional, incl. hardware, fiberglass, 9' x 7', standard | | 2 | | 285 | 25 | | 310 | 355 |
| 022 | Deluxe | | 2 | | 315 | 25 | | 340 | 385 |
| 030 | 16' x 7', standard | | 2.670 | | 475 | 33 | | 508 | 575 |
| 032 | Deluxe | | 2.670 | | 525 | 33 | | 558 | 630 |
| 050 | Hardboard, 9' x 7', standard | | 2 | | 185 | 25 | | 210 | 245 |
| 052 | Deluxe | | 2 | | 255 | 25 | | 280 | 320 |
| 060 | 16' x 7', standard | | 2.670 | | 275 | 33 | | 308 | 355 |
| 062 | Deluxe | | 2.670 | | 520 | 33 | | 553 | 625 |
| 070 | Metal, 9' x 7', standard | | 2 | | 230 | 25 | | 255 | 290 |
| 072 | Deluxe | | 2 | | 365 | 25 | | 390 | 440 |
| 080 | 16' x 7', standard | | 2.670 | | 335 | 33 | | 368 | 420 |
| 082 | Deluxe | | 2.670 | | 695 | 33 | | 728 | 815 |
| 090 | Wood, 9' x 7', standard | | 2 | | 200 | 25 | | 225 | 260 |
| 092 | Deluxe | | 2 | | 600 | 25 | | 625 | 700 |
| 100 | 16' x 7', standard | | 2.670 | | 400 | 33 | | 433 | 490 |
| 102 | Deluxe | ↓ | 2.670 | | 775 | 33 | | 808 | 905 |
| 180 | Door hardware only, sectional | 1 Carp | 2 | ↓ | 40 | 25 | | 65 | 83 |
| 182 | One side only | | | | 25 | | | 25 | 28M |

For expanded coverage of these items see *Building Construction Cost Data 1985*

339

## 8.3 Special Doors

| | | | CREW | MAN-HOURS | UNIT | MAT. | LABOR | EQUIP. | TOTAL | TOTAL INCL O&P |
|---|---|---|---|---|---|---|---|---|---|---|
| 42 | 300 | Swing-up, including hardware, fiberglass, 9' x 7', standard | 2 Carp | 2 | Ea. | 185 | 25 | | 210 | 245 |
| | 302 | Deluxe | | 2 | | 285 | 25 | | 310 | 355 |
| | 310 | 16' x 7' high, standard | | 2.670 | | 320 | 33 | | 353 | 405 |
| | 312 | Deluxe | | 2.670 | | 385 | 33 | | 418 | 475 |
| | 320 | Hardboard, 9' x 7', standard | | 2 | | 210 | 25 | | 235 | 270 |
| | 322 | Deluxe | | 2 | | 275 | 25 | | 300 | 340 |
| | 330 | 16' x 7', standard | | 2.670 | | 325 | 33 | | 358 | 410 |
| | 332 | Deluxe | | 2.670 | | 425 | 33 | | 458 | 520 |
| | 340 | Metal, 9' x 7', standard | | 2 | | 205 | 25 | | 230 | 265 |
| | 342 | Deluxe | | 2 | | 335 | 25 | | 360 | 410 |
| | 350 | 16' x 7', standard | | 2.670 | | 360 | 33 | | 393 | 450 |
| | 352 | Deluxe | | 2.670 | | 600 | 33 | | 633 | 710 |
| | 360 | Wood, 9' x 7', standard | | 2 | | 155 | 25 | | 180 | 210 |
| | 362 | Deluxe | | 2 | | 275 | 25 | | 300 | 340 |
| | 370 | 16' x 7', standard | | 2.670 | | 300 | 33 | | 333 | 380 |
| | 372 | Deluxe | ▼ | 2.670 | | 435 | 33 | | 468 | 530 |
| | 390 | Door hardware only, swing up | 1 Carp | 2 | | 40 | 25 | | 65 | 83 |
| | 392 | One side only | | 1.140 | | 23 | 14.10 | | 37.10 | 48 |
| | 400 | For electric operator, economy, add | | 1 | | 160 | 12.35 | | 172.35 | 195 |
| | 410 | Deluxe, including remote control, add | ▼ | 1 | ▼ | 240 | 12.35 | | 252.35 | 285 |
| | 450 | For electronic control, 1 transmitter, add to operator | | | Total | 40 | | | 40 | 44M |
| | 460 | 2 transmitters, add to operator | | | " | 75 | | | 75 | 83M |
| | 600 | Replace section, on sectional door, fiberglass, 9' x 7' | 1 Carp | 2 | Ea. | 81 | 25 | | 106 | 130 |
| | 602 | 16' x 7' | | 2.290 | | 145 | 28 | | 173 | 205 |
| | 620 | Hardboard, 9' x 7' | | 2 | | 94 | 25 | | 119 | 145 |
| | 622 | 16' x 7' | | 2.290 | | 165 | 28 | | 193 | 225 |
| | 630 | Metal, 9' x 7' | | 2 | | 110 | 25 | | 135 | 160 |
| | 632 | 16' x 7' | | 2.290 | | 185 | 28 | | 213 | 250 |
| | 650 | Wood, 9' x 7' | | 2 | | 105 | 25 | | 130 | 155 |
| | 652 | 16' x 7' | ▼ | 2.290 | | 175 | 28 | | 203 | 235 |
| 45-001 | | ROLLING SERVICE DOORS Steel, manual, 20 ga., 8' x 8' high, standard | 2 Sswk | 10 | | 600 | 135 | | 735 | 890 |
| | 010 | 10' x 10' high | | 11.430 | | 750 | 150 | | 900 | 1,075 |
| | 012 | 8' x 8' high, class A fire door | | 11.430 | | 1,100 | 150 | | 1,250 | 1,475 |
| | 013 | 12' x 12' high, standard | | 16 | | 1,900 | 215 | | 2,115 | 2,450 |
| | 014 | 12' x 12' high, class A fire door | | 20 | | 2,400 | 265 | | 2,665 | 3,100 |
| | 016 | 10' x 20' high, standard | | 32 | | 1,350 | 425 | | 1,775 | 2,225 |
| | 018 | 10' x 20' high, class A fire door | ▼ | 40 | ▼ | 2,350 | 530 | | 2,880 | 3,500 |
| | 300 | For 18 ga. doors, add | | | S.F. | .75 | | | .75 | .82M |
| | 330 | For enamel finish, add | | | " | .52 | | | .52 | .57M |
| | 360 | For safety edge bottom bar, pneumatic, add | | | L.F. | 10 | | | 10 | 11M |
| | 370 | Electric, add | | | | 14 | | | 14 | 15.40M |
| | 400 | For weatherstripping, extruded rubber, jambs, add | | | | 7 | | | 7 | 7.70M |
| | 410 | Hood, add | | | | 9 | | | 9 | 9.90M |
| | 420 | Sill, add | | | ▼ | 3.50 | | | 3.50 | 3.85M |
| | 450 | Motor operators, to 14' x 14' opening | 2 Sswk | 3.200 | Ea. | 400 | 43 | | 443 | 515 |
| | 470 | For fire door, fusible link, add | | | " | 100 | | | 100 | 110M |
| 51-001 | | ROLL UP GRILLE Aluminum, manual operated, mill finish | 2 Sswk | .195 | S.F. | 14.05 | 2.60 | | 16.65 | 19.90 |
| | 010 | Bronze anodized | | .195 | " | 17.50 | 2.60 | | 20.10 | 24 |
| | 040 | Steel, manual operated, 10' x 10' high | | 16 | Opng. | 850 | 215 | | 1,065 | 1,300 |
| | 050 | 15' x 8' high | ▼ | 20 | | 1,150 | 265 | | 1,415 | 1,725 |
| | 100 | For safety edge bottom bar, add | | | | 210 | | | 210 | 230M |
| | 110 | For motor operation, add | 2 Sswk | 3.200 | ▼ | 650 | 43 | | 693 | 790 |
| 52-001 | | SECURITY GATES See division 10.1-72 | | | | | | | | |
| 54-001 | | SHOCK ABSORBING Rigid, no frame, insulated, 1-13/16" thick, 5' x 7' | 2 Sswk | 8.420 | Opng. | 1,350 | 110 | | 1,460 | 1,675 |
| | 010 | 8' x 8' | | 8.890 | | 1,775 | 120 | | 1,895 | 2,150 |
| | 050 | Flexible, frame not incl., 5' x 7' opening, economy | | 8 | | 1,150 | 105 | | 1,255 | 1,450 |
| | 060 | Deluxe | | 8.420 | | 1,700 | 110 | | 1,810 | 2,050 |
| | 100 | 8' x 8' opening, economy | | 8 | | 1,800 | 105 | | 1,905 | 2,150 |
| | 110 | Deluxe | ▼ | 8.420 | ▼ | 2,100 | 110 | | 2,210 | 2,500 |

For expanded coverage of these items see *Building Construction Cost Data 1985*

## 8.3 Special Doors

| | | CREW | MAN-HOURS | UNIT | MAT. | LABOR | EQUIP. | TOTAL | TOTAL INCL O&P |
|---|---|---|---|---|---|---|---|---|---|
| 66-001 | TIN CLAD 3 ply, 6' x 7', double sliding, manual | 2 Carp | 16 | Opng. | 1,840 | 200 | | 2,040 | 2,325 |
| 100 | For electric operator, add | 1 Elec | 4 | " | 1,300 | 55 | | 1,355 | 1,525 |

## 8.4 Entrances & Storefronts

| | | CREW | MAN-HOURS | UNIT | MAT. | LABOR | EQUIP. | TOTAL | TOTAL INCL O&P |
|---|---|---|---|---|---|---|---|---|---|
| 10-001 | BALANCED DOORS Incl. hdwre & frame, alum. & glass, 3' x 7', economy | 2 Sswk | 17.780 | Ea. | 2,200 | 235 | | 2,435 | 2,825 |
| 015 | Premium | | 22.860 | | 3,570 | 305 | | 3,875 | 4,450 |
| 050 | Stainless steel and glass, 3' x 7', economy | | 17.780 | | 3,475 | 235 | | 3,710 | 4,225 |
| 060 | Premium | ↓ | 22.860 | ↓ | 5,775 | 305 | | 6,080 | 6,875 |
| 25-001 | SLIDING ENTRANCE 12' x 7'-6" opening, 5' x 7' door, two way traffic, | | | | | | | | |
| 002 | mat activated, panic pushout, incl. operator & hardware, | | | | | | | | |
| 003 | not incl. glass or glazing | 2 Glaz | 22.860 | Opng. | 6,275 | 285 | | 6,560 | 7,350 |
| 30-001 | SLIDING PANEL Mall fronts, aluminum & glass, 15' x 9' high | 2 Glaz | 12.310 | Opng. | 1,400 | 155 | | 1,555 | 1,775 |
| 010 | 24' x 9' high | | 22.860 | | 2,275 | 285 | | 2,560 | 2,950 |
| 020 | 48' x 9' high, with fixed panels | ↓ | 17.780 | ↓ | 4,575 | 220 | | 4,795 | 5,375 |
| 050 | For bronze finish, add | | | | 15% | | | | |
| 35-001 | STAINLESS STEEL and glass entrance unit, narrow stiles | | | | | | | | |
| 002 | 3' x 7' opening, including hardware, minimum | 2 Sswk | 10 | Opng. | 1,200 | 135 | | 1,335 | 1,550 |
| 005 | Average | | 11.430 | | 2,000 | 150 | | 2,150 | 2,450 |
| 010 | Maximum | ↓ | 13.330 | | 3,200 | 175 | | 3,375 | 3,825 |
| 100 | For solid bronze entrance units, statuary finish, add | | | | 60% | | | | |
| 110 | Without statuary finish, add | | | ↓ | 40% | | | | |
| 40-001 | STOREFRONT SYSTEMS Aluminum frame, clear 3/8" plate glass, | | | | | | | | |
| 002 | incl. 3' x 7' door with hardware (400 sq. ft. max. wall) | | | | | | | | |
| 050 | Wall height to 12' high, commercial grade | 2 Glaz | .107 | S.F. | 10.30 | 1.33 | | 11.63 | 13.40 |
| 060 | Institutional grade | | .123 | | 11.95 | 1.53 | | 13.48 | 15.55 |
| 070 | Monumental grade | | .139 | | 16.75 | 1.73 | | 18.48 | 21 |
| 100 | 6' x 7' door with hardware, commerical grade | | .119 | | 13 | 1.48 | | 14.48 | 16.60 |
| 110 | Institutional grade | | .139 | | 15.75 | 1.73 | | 17.48 | 20 |
| 120 | Monumental grade | ↓ | .160 | | 21.50 | 1.99 | | 23.49 | 27 |
| 150 | For bronze anodized finish, add | | | | 15% | | | | |
| 160 | For black anodized finish, add | | | | 20% | | | | |
| 170 | For stainless steel framing, add to monumental | | | | 75% | | | | |
| 200 | For no 3' x 7' door and hardware, deduct | | | ↓ | 3.18 | | | 3.18 | 3.49M |
| 250 | For no 6' x 7' door and hardware, deduct | | | | | | | | |
| 60-001 | SWING DOORS Aluminum entrance, 6' x 7', incl. hardware & operator | 2 Sswk | 22.860 | Opng. | 4,500 | 305 | | 4,805 | 5,475 |
| 002 | For anodized finish, add | | | " | 380 | | | 380 | 420M |

## 8.5 Metal Windows

| | | CREW | MAN-HOURS | UNIT | MAT. | LABOR | EQUIP. | TOTAL | TOTAL INCL O&P |
|---|---|---|---|---|---|---|---|---|---|
| 11-001 | ALUMINUM SASH Stock #2 grd. glazing and trim not included, casement | 2 Sswk | .080 | S.F. | 10.15 | 1.06 | | 11.21 | 13 |
| 005 | Double hung | | .080 | | 6.05 | 1.06 | | 7.11 | 8.50 |
| 010 | Fixed casement | | .080 | | 4.70 | 1.06 | | 5.76 | 7 |
| 015 | Picture window | | .080 | | 5.35 | 1.06 | | 6.41 | 7.70 |
| 020 | Projected window | | .080 | | 12 | 1.06 | | 13.06 | 15 |
| 025 | Single hung | | .080 | | 5.60 | 1.06 | | 6.66 | 8 |
| 030 | Sliding | ↓ | .080 | ↓ | 7.70 | 1.06 | | 8.76 | 10.30 |
| 100 | Mullions for above, tubular | | .080 | L.F. | 1.50 | 1.06 | | 2.56 | 3.47 |

For expanded coverage of these items see *Building Construction Cost Data 1985*

## 8.5 Metal Windows

| | | | CREW | MAN-HOURS | UNIT | MAT. | LABOR | EQUIP. | TOTAL | TOTAL INCL O&P |
|---|---|---|---|---|---|---|---|---|---|---|
| 11 | 200 | Double glazing for above, add | 2 Sswk | .080 | S.F. | 5 | 1.06 | | 6.06 | 7.30 |
| | 210 | Triple glazing for above, add | " | .188 | " | 6 | 2.50 | | 8.50 | 10.90 |
| 20-001 | | ALUMINUM WINDOWS Including frame and glazing, grade 2 | | | | | | | | |
| | 100 | Stock units, casement, 3'-1" x 3'-2" opening | 2 Sswk | 1.600 | Ea. | 155 | 21 | | 176 | 205 |
| | 104 | Insulating glass | " | 1.600 | " | 145 | 21 | | 166 | 195 |
| | 105 | Add for storms | | | | 30 | | | 30 | 33M |
| | 160 | Projected, with screen, 3'-1" x 3'-2" opening | 2 Sswk | 1.600 | Ea. | 105 | 21 | | 126 | 150 |
| | 165 | Insulating glass | " | 1.600 | | 105 | 21 | | 126 | 150 |
| | 170 | Add for storms | | | | 30 | | | 30 | 33M |
| | 200 | 4'-5" x 5'-3" opening | 2 Sswk | 2 | | 145 | 27 | | 172 | 205 |
| | 205 | Insulating glass | " | 2 | | 165 | 27 | | 192 | 225 |
| | 210 | Add for storms | | | | 44 | | | 44 | 48M |
| | 250 | Enamel finish windows, 3'-1" x 3'-2" | 2 Sswk | 1.600 | | 93 | 21 | | 114 | 140 |
| | 255 | Insulating glass | | 1.600 | | 120 | 21 | | 141 | 170 |
| | 260 | 4'-5" x 5'-3" | | 2 | | 165 | 27 | | 192 | 225 |
| | 270 | Insulating glass | | 2 | | 180 | 27 | | 207 | 245 |
| | 300 | Single hung, 2' x 3' opening, enameled, standard glazed | | 1.600 | | 70 | 21 | | 91 | 115 |
| | 310 | Insulating glass | | 1.600 | | 115 | 21 | | 136 | 165 |
| | 330 | 2'-8" x 6'-8" opening, standard glazed | | 2 | | 150 | 27 | | 177 | 210 |
| | 340 | Insulating glass | | 2 | | 190 | 27 | | 217 | 255 |
| | 370 | 3'-4" x 5'-0" opening, standard glazed | | 1.780 | | 97 | 24 | | 121 | 145 |
| | 380 | Insulating glass | | 1.780 | | 130 | 24 | | 154 | 185 |
| | 400 | Sliding aluminum, 3' x 2' opening, standard glazed | | 1.600 | | 46 | 21 | | 67 | 87 |
| | 410 | Insulating glass | | 1.600 | | 77 | 21 | | 98 | 120 |
| | 430 | 5' x 3' opening, standard glazed | | 1.780 | | 80 | 24 | | 104 | 130 |
| | 440 | Insulating glass | | 1.780 | | 130 | 24 | | 154 | 185 |
| | 460 | 8' x 4' opening, standard glazed | | 2.670 | | 115 | 35 | | 150 | 185 |
| | 470 | Insulating glass | | 2.670 | | 230 | 35 | | 265 | 315 |
| | 500 | 9' x 5' opening, standard glazed | | 4 | | 180 | 53 | | 233 | 290 |
| | 510 | Insulating glass | | 4 | | 305 | 53 | | 358 | 425 |
| | 550 | Sliding, with thermal barrier and screen, 6' x 4', 2 track | | 2 | | 245 | 27 | | 272 | 315 |
| | 570 | 4 track | | 2 | | 310 | 27 | | 337 | 385 |
| | 600 | For above units with bronze finish, add | | | | 8% | | | | |
| | 620 | For installation in concrete openings, add | | | | | 80% | | | |
| | 640 | | | | | | | | | |
| 30-001 | | JALOUSIES Aluminum incl. glazing & screens, stock, 1'-7" x 3'-2" | 2 Sswk | 1.600 | Ea. | 57 | 21 | | 78 | 99 |
| | 010 | 2'-3" x 4'-0" | | 1.600 | | 105 | 21 | | 126 | 150 |
| | 020 | 3'-1" x 2'-0" | | 1.600 | | 98 | 21 | | 119 | 145 |
| | 030 | 3'-1" x 5'-3" | | 1.600 | | 135 | 21 | | 156 | 185 |
| | 100 | Mullions for above, 2'-0" long | | .200 | | 6.20 | 2.66 | | 8.86 | 11.40 |
| | 110 | 5'-3" long | | .200 | | 11 | 2.66 | | 13.66 | 16.65 |
| 40-001 | | LOUVERS See division 4.1-80, 6.2-32 & 7.6-42 | | | | | | | | |
| 50-001 | | SCREENS For metal sash, aluminum or bronze mesh, flat screen | 2 Sswk | .013 | S.F. | 2 | .17 | | 2.17 | 2.50 |
| | 050 | Wicket screen, inside window | | .016 | | 3 | .21 | | 3.21 | 3.66 |
| | 080 | Security screen, aluminum frame with stainless steel cloth | | .013 | | 12 | .17 | | 12.17 | 13.50 |
| | 090 | Steel grate, painted, on steel frame | | .010 | | 6.50 | .13 | | 6.63 | 7.40 |
| | 100 | For solar louvers, add | | .100 | | 12.50 | 1.33 | | 13.83 | 16.05 |
| 61-001 | | STEEL SASH Stock, glazing and trim not included | | | | | | | | |
| | 010 | Casement, 100% vented | 2 Sswk | .080 | S.F. | 15.90 | 1.06 | | 16.96 | 19.30 |
| | 020 | 50% vented | | .080 | | 15.90 | 1.06 | | 16.96 | 19.30 |
| | 030 | Fixed | | .080 | | 8.50 | 1.06 | | 9.56 | 11.15 |
| | 100 | Projected, commercial, 40% vented | | .080 | | 15.90 | 1.06 | | 16.96 | 19.30 |
| | 110 | Intermediate, 50% vented | | .080 | | 17.50 | 1.06 | | 18.56 | 21 |
| | 150 | Industrial, horizontally pivoted | | .080 | | 13.25 | 1.06 | | 14.31 | 16.40 |
| | 160 | Fixed | | .080 | | 11.60 | 1.06 | | 12.66 | 14.60 |
| | 200 | Industrial security sash | | .080 | | 21 | 1.06 | | 22.06 | 25 |
| | 210 | Fixed | | .080 | | 16.95 | 1.06 | | 18.01 | 20 |

For expanded coverage of these items see *Building Construction Cost Data 1985*

## 8.5 Metal Windows

| | | Description | CREW | MAN-HOURS | UNIT | MAT. | LABOR | EQUIP. | TOTAL | TOTAL INCL O&P |
|---|---|---|---|---|---|---|---|---|---|---|
| 61 | 250 | Picture window | 2 Sswk | .080 | S.F. | 7.40 | 1.06 | | 8.46 | 9.95 |
| | 300 | Double hung | | .080 | " | 20 | 1.06 | | 21.06 | 24 |
| | 500 | Mullions for above, open interior face | | .067 | L.F. | 3.71 | .89 | | 4.60 | 5.60 |
| | 510 | With interior cover | | .067 | " | 7.40 | .89 | | 8.29 | 9.65 |
| | 600 | Double glazing for above, add | | .080 | S.F. | 5 | 1.06 | | 6.06 | 7.30 |
| | 610 | Triple glazing for above, add | ↓ | .188 | " | 6 | 2.50 | | 8.50 | 10.90 |
| 70-001 | | **STEEL WINDOWS** Stock, including frame, trim and insulating glass | | | | | | | | |
| | 100 | Stock units, double hung, 2'-8" x 4'-6" opening | 2 Sswk | 1.330 | Ea. | 110 | 17.75 | | 127.75 | 150 |
| | 110 | 2'-4" x 3'-9" opening | | 1.330 | | 95 | 17.75 | | 112.75 | 135 |
| | 150 | Commercial projected, 3'-9" x 5'-5" opening | | 1.600 | | 200 | 21 | | 221 | 255 |
| | 160 | 6'-9" x 4'-1" opening | | 2.290 | | 240 | 30 | | 270 | 315 |
| | 200 | Intermediate projected, 2'-9" x 4'-1" opening | | 1.330 | | 105 | 17.75 | | 122.75 | 145 |
| | 210 | 4'-1" x 5'-5" opening | ↓ | 1.600 | ↓ | 160 | 21 | | 181 | 210 |
| 85-001 | | **STORM WINDOWS** Aluminum, residential | | | | | | | | |
| | 002 | | | | | | | | | |
| | 030 | Basement, mill finish, incl. fiberglass screen | | | | | | | | |
| | 032 | 1'-10" x 1'-0" high | F-2 | .533 | Ea. | 10.30 | 6.58 | .47 | 17.35 | 22 |
| | 034 | 2'-9" x 1'-6" high | | .533 | | 15.45 | 6.58 | .47 | 22.50 | 28 |
| | 036 | 3'-4" x 2'-0" high | ↓ | .533 | ↓ | 18.55 | 6.58 | .47 | 25.60 | 31 |
| | 160 | Double-hung, combination, storm & screen | | | | | | | | |
| | 170 | Custom, anodized, 2'-0" x 3'-5" high | F-2 | .533 | Ea. | 39 | 6.58 | .47 | 46.05 | 54 |
| | 172 | 2'-6" x 5'-0" high | | .571 | | 42 | 7.05 | .50 | 49.55 | 58 |
| | 174 | 4'-0" x 6'-0" high | | .640 | | 53 | 7.89 | .56 | 61.45 | 71 |
| | 180 | White painted, 2'-0" x 3'-5" high | | .533 | | 37 | 6.58 | .47 | 44.05 | 52 |
| | 182 | 2'-6" x 5'-0" high | | .571 | | 39 | 7.05 | .50 | 46.55 | 55 |
| | 184 | 4'-0" x 6'-0" high | | .640 | | 50 | 7.89 | .56 | 58.45 | 68 |
| | 200 | Average quality, anodized, 2'-0" x 3'-5" high | | .533 | | 33 | 6.58 | .47 | 40.05 | 47 |
| | 202 | 2'-6" x 5'-0" high | | .571 | | 38 | 7.05 | .50 | 45.55 | 54 |
| | 204 | 4'-0" x 6'-0" high | | .640 | | 47 | 7.89 | .56 | 55.45 | 65 |
| | 240 | White painted, 2'-0" x 3'-5" high | | .533 | | 32 | 6.58 | .47 | 39.05 | 46 |
| | 242 | 2'-6" x 5'-0" high | | .571 | | 38 | 7.05 | .50 | 45.55 | 54 |
| | 244 | 4'-0" x 6'-0" high | | .640 | | 45 | 7.89 | .56 | 53.45 | 63 |
| | 260 | Mill finish, 2'-0" x 3'-5" high | | .533 | | 28 | 6.58 | .47 | 35.05 | 42 |
| | 262 | 2'-6" x 5'-0" high | | .571 | | 33 | 7.05 | .50 | 40.55 | 48 |
| | 264 | 4'-0" x 6-8" high | ↓ | .640 | ↓ | 41 | 7.89 | .56 | 49.45 | 58 |
| | 400 | Picture window, storm, 1 lite, white or bronze finish | | | | | | | | |
| | 402 | 4'-6" x 4'-6" high | F-2 | .640 | Ea. | 70 | 7.89 | .56 | 78.45 | 90 |
| | 404 | 5'-8" x 4'-6" high | | .800 | | 82 | 9.90 | .70 | 92.60 | 105 |
| | 440 | Mill finish, 4'-6" x 4'-6" high | | .640 | | 64 | 7.89 | .56 | 72.45 | 84 |
| | 442 | 5'-8" x 4'-6" high | ↓ | .800 | ↓ | 73 | 9.90 | .70 | 83.60 | 97 |
| | 460 | 3 lite, white or bronze finish | | | | | | | | |
| | 462 | 4'-6" x 4'-6" high | F-2 | .640 | Ea. | 82 | 7.89 | .56 | 90.45 | 105 |
| | 464 | 5'-8" x 4'-6" high | | .800 | | 97 | 9.90 | .70 | 107.60 | 125 |
| | 480 | Mill finish, 4'-6" x 4'-6" high | | .640 | | 73 | 7.89 | .56 | 81.45 | 93 |
| | 482 | 5'-8" x 4'-6" high | ↓ | .800 | | 84 | 9.90 | .70 | 94.60 | 110 |
| | 500 | Sliding glass door, storm 6' x 6'-8", standard | 1 Glaz | 4 | | 195 | 50 | | 245 | 290 |
| | 500 | Sliding glass door, storm window, 6'-0" x 6'-8", fixed | | 5 | | 135 | 62 | | 197 | 245 |
| | 510 | Economy | | 4 | | 135 | 50 | | 185 | 225 |
| | 510 | Operable | ↓ | 3.810 | ↓ | 195 | 47 | | 242 | 290 |
| | 600 | Sliding window storm, 2 lite, white or bronze finish | | | | | | | | |
| | 602 | 3'-4" x 2'-7" high | F-2 | .571 | Ea. | 36 | 7.05 | .50 | 43.55 | 51 |
| | 604 | 4'-4" x 3'-3" high | | .640 | | 43 | 7.89 | .56 | 51.45 | 60 |
| | 606 | 5'-4" x 6'-0" high | ↓ | .800 | ↓ | 61 | 9.90 | .70 | 71.60 | 84 |
| | 640 | 3 lite, white or bronze finish | | | | | | | | |
| | 642 | 4'-4" x 3'-3" high | F-2 | .640 | Ea. | 55 | 7.89 | .56 | 63.45 | 74 |
| | 644 | 5'-4" x 6'-0" high | | .800 | | 84 | 9.90 | .70 | 94.60 | 110 |
| | 646 | 6'-0" x 6'-0" high | | .889 | | 88 | 10.97 | .78 | 99.75 | 115 |
| | 680 | Mill finish, 4'-4" x 3'-3" high | | .640 | | 49 | 7.89 | .56 | 57.45 | 67 |
| | 682 | 5'-4" x 6'-0" high | ↓ | .800 | ↓ | 74 | 9.90 | .70 | 84.60 | 98 |

For expanded coverage of these items see *Building Construction Cost Data 1985*

## 8.5 Metal Windows

| | | | CREW | MAN-HOURS | UNIT | MAT. | LABOR | EQUIP. | TOTAL | TOTAL INCL O&P |
|---|---|---|---|---|---|---|---|---|---|---|
| 85 | 684 | 6'-0" x 6-0" high | F-2 | .889 | Ea. | 78 | 10.97 | .78 | 89.75 | 105 |
| | 900 | Magnetic interior storm window | | | | | | | | |
| | 910 | 3/16" plate glass | 1 Glaz | .075 | S.F. | 3.35 | .93 | | 4.28 | 5.15 |

## 8.6 Wood Windows

| | | CREW | MAN-HOURS | UNIT | MAT. | LABOR | EQUIP. | TOTAL | TOTAL INCL O&P |
|---|---|---|---|---|---|---|---|---|---|
| 10-001 | AWNING WINDOW Including frame, screen, and exterior trim | | | | | | | | |
| 002 | | | | | | | | | |
| 010 | Average quality, builders model, 34" x 22", standard glazed | 1 Carp | .800 | Ea. | 58 | 9.90 | | 67.90 | 79 |
| 020 | Insulating glass | | .800 | | 76 | 9.90 | | 85.90 | 99 |
| 030 | 40" x 28", standard glazed | | .889 | | 71 | 11 | | 82 | 95 |
| 040 | Insulating glass | | .889 | | 99 | 11 | | 110 | 125 |
| 050 | 48" x 36", standard glazed | | 1 | | 88 | 12.35 | | 100.35 | 115 |
| 060 | Insulating glass | | 1 | | 120 | 12.35 | | 132.35 | 150 |
| 100 | Plastic clad, premium, insulating glass, 34" x 22" | | .800 | | 115 | 9.90 | | 124.90 | 140 |
| 110 | 40" x 22" | | .800 | | 125 | 9.90 | | 134.90 | 155 |
| 120 | 36" x 28" | | .889 | | 140 | 11 | | 151 | 170 |
| 130 | 40" x 28" | | .889 | | 145 | 11 | | 156 | 175 |
| 140 | 48" x 28" | | 1 | | 150 | 12.35 | | 162.35 | 185 |
| 150 | 48" x 36" | | 1 | | 180 | 12.35 | | 192.35 | 220 |
| 200 | Metal clad, deluxe, insulating glass, 34" x 22" | | .800 | | 150 | 9.90 | | 159.90 | 180 |
| 210 | 40" x 22" | | .800 | | 155 | 9.90 | | 164.90 | 185 |
| 220 | 36" x 28" | | .889 | | 165 | 11 | | 176 | 200 |
| 230 | 40" x 28" | | .889 | | 175 | 11 | | 186 | 210 |
| 240 | 48" x 28" | | 1 | | 190 | 12.35 | | 202.35 | 230 |
| 250 | 48" x 36" | ↓ | 1 | ↓ | 225 | 12.35 | | 237.35 | 265 |
| 20-001 | BOW-BAY WINDOW Including frame, screen and exterior trim, | | | | | | | | |
| 002 | end panels operable | | | | | | | | |
| 100 | Awning type, builders model, 8'-0" x 5'-0" high, standard glazed, 4 | 2 Carp | 1.600 | Ea. | 425 | 19.75 | | 444.75 | 500 |
| 105 | Insulating glass | | 1.600 | | 490 | 19.75 | | 509.75 | 570 |
| 110 | 12'-0" x 6'-0" high, standard glazed | | 2.670 | | 735 | 33 | | 768 | 860 |
| 120 | Insulating glass, 6 panels | | 2.670 | | 805 | 33 | | 838 | 940 |
| 130 | Plastic clad, premium, insulating glass, 6'-0" x 4'-0" | | 1.600 | | 640 | 19.75 | | 659.75 | 735 |
| 134 | 9'-0" x 4'-0" | | 2 | | 850 | 25 | | 875 | 975 |
| 138 | 10'-0" x 5'-0" | | 2.290 | | 1,100 | 28 | | 1,128 | 1,250 |
| 142 | 12'-0" x 6'-0" | | 2.670 | | 1,350 | 33 | | 1,383 | 1,525 |
| 160 | Metal clad, deluxe, insul. glass, 6'-0" x 4'-0" high, 3 panels | | 1.600 | | 805 | 19.75 | | 824.75 | 915 |
| 164 | 9'-0" x 4'-0" high, 4 panels | | 2 | | 1,080 | 25 | | 1,105 | 1,225 |
| 168 | 10'-0" x 5'-0" high, 5 panels | | 2.290 | | 1,425 | 28 | | 1,453 | 1,600 |
| 172 | 12'-0" x 6'-0" high, 6 panels | | 2.670 | | 1,750 | 33 | | 1,783 | 1,975 |
| 200 | Casement type, bldrs. model, 8'-0" x 5'-0" high, standard glazed, 4 | | 1.600 | | 390 | 19.75 | | 409.75 | 460 |
| 205 | Insulating glass | | 1.600 | | 495 | 19.75 | | 514.75 | 575 |
| 210 | 12'-0" x 6'-0" high, 6 panels, standard glazed | | 2.670 | | 580 | 33 | | 613 | 690 |
| 220 | Insulating glass | | 2.670 | | 775 | 33 | | 808 | 905 |
| 230 | Plastic clad, premium, insulating glass, 8'-0" x 5'-0" | | 1.600 | | 745 | 19.75 | | 764.75 | 850 |
| 234 | 10'-0" x 5'-0" | | 2 | | 985 | 25 | | 1,010 | 1,125 |
| 238 | 10'-0" x 6'-0" | | 2.290 | | 1,100 | 28 | | 1,128 | 1,250 |
| 242 | 12'-0" x 6'-0" | | 2.670 | | 1,250 | 33 | | 1,283 | 1,425 |
| 260 | Metal clad, deluxe, insul. glass, 8'-0" x 5'-0" high, 4 panels | | 1.600 | | 955 | 19.75 | | 974.75 | 1,075 |
| 264 | 10'-0" x 5'-0" high, 5 panels | | 2 | | 1,225 | 25 | | 1,250 | 1,375 |
| 268 | 10'-0" x 6'-0" high, 5 panels | | 2.290 | | 1,350 | 28 | | 1,378 | 1,525 |
| 272 | 12'-0" x 6'-0" high, 6 panels | | 2.670 | | 1,650 | 33 | | 1,683 | 1,875 |
| 300 | Double hung type, bldrs. model, 8'-0" x 4'-0" high, std. glazed, 4 | | 1.600 | | 575 | 19.75 | | 594.75 | 665 |
| 305 | Insulating glass | ↓ | 1.600 | ↓ | 640 | 19.75 | | 659.75 | 735 |

For expanded coverage of these items see *Building Construction Cost Data 1985*

## 8.6 Wood Windows

| | | | MAN- | | \multicolumn{4}{c|}{BARE COSTS} | TOTAL |
|---|---|---|---|---|---|---|---|---|---|
| | | CREW | HOURS | UNIT | MAT. | LABOR | EQUIP. | TOTAL | INCL O&P |
| 20 310 | 9'-0" x 5'-0" high, 4 panels, standard glazed | 2 Carp | 2.670 | Ea. | 625 | 33 | | 658 | 740 |
| 320 | Insulating glass | | 2.670 | | 725 | 33 | | 758 | 850 |
| 330 | Plastic clad, premium, insulating glass, 7'-0" x 4'-0" | | 1.600 | | 750 | 19.75 | | 769.75 | 855 |
| 334 | 8'-0" x 4'-0" | | 2 | | 850 | 25 | | 875 | 975 |
| 338 | 8'-0" x 5'-0" | | 2.290 | | 800 | 28 | | 828 | 925 |
| 342 | 9'-0" x 5'-0" | | 2.670 | | 940 | 33 | | 973 | 1,075 |
| 360 | Metal clad, deluxe, insul. glass, 7'-0" x 4'-0" high, 3 panels | | 1.600 | | 995 | 19.75 | | 1,014.75 | 1,125 |
| 364 | 8'-0" x 4'-0" high, 4 panels | | 2 | | 1,025 | 25 | | 1,050 | 1,175 |
| 368 | 8'-0" x 5'-0" high, 4 panels | | 2.290 | | 1,125 | 28 | | 1,153 | 1,275 |
| 372 | 9'-0" x 5'-0" high, 4 panels | | 2.670 | | 1,225 | 33 | | 1,258 | 1,400 |
| 700 | Drip cap, premolded vinyl, 8' long | | .533 | | 50 | 6.60 | | 56.60 | 65 |
| 704 | 12' long | ▼ | .615 | ▼ | 56 | 7.60 | | 63.60 | 74 |
| 25-001 | **WEATHERSTRIPPING** See division 8.7-70 | | | | | | | | |
| 31-001 | **CASEMENT WINDOW** Including frame, screen, and exterior trim | | | | | | | | |
| 002 | | | | | | | | | |
| 010 | Average quality, bldrs. model, 2'-0" x 3'-0" high, standard glazed | 1 Carp | .800 | Ea. | 80 | 9.90 | | 89.90 | 105 |
| 015 | Insulating glass | | .800 | | 100 | 9.90 | | 109.90 | 125 |
| 020 | 2'-0" x 4'-6" high, standard glazed | | .889 | | 105 | 11 | | 116 | 135 |
| 025 | Insulating glass | | .889 | | 135 | 11 | | 146 | 165 |
| 030 | 2'-0" x 6'-0" high, standard glazed | | 1 | | 120 | 12.35 | | 132.35 | 150 |
| 035 | Insulating glass | | 1 | | 160 | 12.35 | | 172.35 | 195 |
| 200 | Metal clad, deluxe, insulating glass, 2'-0" x 3'-0" high | | .800 | | 150 | 9.90 | | 159.90 | 180 |
| 204 | 2'-0" x 4'-0" high | | .889 | | 180 | 11 | | 191 | 215 |
| 208 | 2'-0" x 5'-0" high | | 1 | | 205 | 12.35 | | 217.35 | 245 |
| 212 | 2'-0" x 6'-0" high | ▼ | 1 | ▼ | 245 | 12.35 | | 257.35 | 290 |
| 220 | For multiple leaf units, deduct for stationary sash | | | | | | | | |
| 221 | 2' high | | | Ea. | 17 | | | 17 | 18.70M |
| 230 | 4'-6" high | | | | 20 | | | 20 | 22M |
| 240 | 6' high | | | | 24 | | | 24 | 26M |
| 300 | For installation, add per leaf | | | ▼ | | 15% | | | |
| 40-001 | **DOUBLE HUNG** Including frame, screen, and exterior trim | | | | | | | | |
| 002 | | | | | | | | | |
| 010 | Average quality, bldrs. model, 2'-0" x 3'-0" high, standard glazed | 1 Carp | .800 | Ea. | 61 | 9.90 | | 70.90 | 83 |
| 015 | Insulating glass | | .800 | | 91 | 9.90 | | 100.90 | 115 |
| 020 | 3'-0" x 4'-0" high, standard glazed | | .889 | | 82 | 11 | | 93 | 110 |
| 025 | Insulating glass | | .889 | | 130 | 11 | | 141 | 160 |
| 030 | 4'-0" x 4'-6" high, standard glazed | | 1 | | 99 | 12.35 | | 111.35 | 130 |
| 035 | Insulating glass | | 1 | | 160 | 12.35 | | 172.35 | 195 |
| 100 | Plastic clad, premium, insulating glass, 2'-6" x 3'-0" | | .800 | | 135 | 9.90 | | 144.90 | 165 |
| 110 | 3'-0" x 3'-6" | | .800 | | 150 | 9.90 | | 159.90 | 180 |
| 120 | 3'-0" x 4'-0" | | .889 | | 160 | 11 | | 171 | 195 |
| 130 | 3'-0" x 4'-6" | | .889 | | 175 | 11 | | 186 | 210 |
| 140 | 3'-0" x 5'-0" | | 1 | | 195 | 12.35 | | 207.35 | 235 |
| 150 | 3'-6" x 6'-0" | | 1 | | 220 | 12.35 | | 232.35 | 260 |
| 200 | Metal clad, deluxe, insulating glass, 2'-6" x 3'-0" high | | .800 | | 170 | 9.90 | | 179.90 | 205 |
| 210 | 3'-0" x 3'-6" high | | .800 | | 195 | 9.90 | | 204.90 | 230 |
| 220 | 3'-0" x 4'-0" high | | .889 | | 210 | 11 | | 221 | 250 |
| 230 | 3'-0" x 4'-6" high | | .889 | | 225 | 11 | | 236 | 265 |
| 240 | 3'-0" x 5'-0" high | | 1 | | 250 | 12.35 | | 262.35 | 295 |
| 250 | 3'-6" x 6'-0" high | ▼ | 1 | ▼ | 280 | 12.35 | | 292.35 | 330 |
| 50-001 | **PICTURE WINDOW** Including frame, screen, and exterior trim | | | | | | | | |
| 002 | | | | | | | | | |
| 010 | Average quality, bldrs. model, 4'-0" x 4'-0" high, standard glazed | 2 Carp | 1.330 | Ea. | 155 | 16.45 | | 171.45 | 195 |
| 015 | Insulating glass | | 1.330 | | 180 | 16.45 | | 196.45 | 225 |
| 020 | 4'-0" x 4'-6" high, standard glazed | | 1.450 | | 170 | 17.95 | | 187.95 | 215 |
| 025 | Insulating glass | | 1.450 | | 195 | 17.95 | | 212.95 | 245 |
| 030 | 5'-0" x 4'-0" high, standard glazed | | 1.450 | | 185 | 17.95 | | 202.95 | 230 |
| 035 | Insulating glass | ▼ | 1.450 | ▼ | 220 | 17.95 | | 237.95 | 270 |

For expanded coverage of these items see *Building Construction Cost Data 1985*

## 8.6 Wood Windows

| | | | MAN- | | BARE COSTS | | | | TOTAL |
|---|---|---|---|---|---|---|---|---|---|
| | | CREW | HOURS | UNIT | MAT. | LABOR | EQUIP. | TOTAL | INCL O&P |
| 50 040 | 6'-0" x 4'-6" high, standard glazed | 2 Carp | 1.600 | Ea. | 205 | 19.75 | | 224.75 | 255 |
| 045 | Insulating glass | | 1.600 | | 245 | 19.75 | | 264.75 | 300 |
| 100 | Plastic clad, premium, insulating glass, 4'-0" x 4'-0" | | 1.330 | | 190 | 16.45 | | 206.45 | 235 |
| 110 | 4'-6" x 6'-6" | | 1.450 | | 270 | 17.95 | | 287.95 | 325 |
| 120 | 5'-6" x 6'-6" | | 1.600 | | 410 | 19.75 | | 429.75 | 480 |
| 130 | 6'-6" x 6'-6" | | 1.600 | | 485 | 19.75 | | 504.75 | 565 |
| 200 | Metal clad, deluxe, insulating glass, 4'-0" x 4'-0" high | | 1.330 | | 240 | 16.45 | | 256.45 | 290 |
| 210 | 4'-6" x 6'-6" high | | 1.450 | | 345 | 17.95 | | 362.95 | 410 |
| 220 | 5'-6" x 6'-6" high | | 1.600 | | 520 | 19.75 | | 539.75 | 605 |
| 230 | 6'-6" x 6'-6" high | ↓ | 1.600 | ↓ | 615 | 19.75 | | 634.75 | 710 |
| 61-001 | **SLIDING WINDOW** Including frame, screen, and exterior trim | | | | | | | | |
| 002 | | | | | | | | | |
| 010 | Average quality, bldrs. model, 3'-0" x 2'-0" high, standard glazed | 1 Carp | .800 | Ea. | 62 | 9.90 | | 71.90 | 84 |
| 012 | Insulating glass | | .800 | | 85 | 9.90 | | 94.90 | 110 |
| 020 | 4'-0" x 3'-6" high, standard glazed | | .889 | | 84 | 11 | | 95 | 110 |
| 022 | Insulating glass | | .889 | | 120 | 11 | | 131 | 150 |
| 030 | 6'-0" x 5'-0" high, standard glazed | | 1 | | 140 | 12.35 | | 152.35 | 175 |
| 032 | Insulating glass | | 1 | | 195 | 12.35 | | 207.35 | 235 |
| 200 | Metal clad, deluxe, insulating glass, 3'-0" x 3'-0" high | | .800 | | 215 | 9.90 | | 224.90 | 250 |
| 205 | 4'-0" x 3'-6" high | | .889 | | 265 | 11 | | 276 | 310 |
| 210 | 5'-0" x 4'-0" high | | .889 | | 320 | 11 | | 331 | 370 |
| 215 | 6'-0" x 5'-0" high | ↓ | 1 | ↓ | 455 | 12.35 | | 467.35 | 520 |
| 65-001 | **WINDOW GRILLE OR MUNTIN** Snap-in type | | | | | | | | |
| 002 | Colonial or diamond pattern | | | | | | | | |
| 200 | Wood, awning window, glass size 28" x 16" high | 1 Carp | .267 | Ea. | 9.70 | 3.29 | | 12.99 | 15.90 |
| 206 | 44" x 24" high | | .286 | | 13.20 | 3.53 | | 16.73 | 20 |
| 210 | Casement, glass size, 20" x 36" high | | .267 | | 19.80 | 3.29 | | 23.09 | 27 |
| 218 | 20" x 56" high | | .286 | ↓ | 44 | 3.53 | | 47.53 | 54 |
| 220 | Double hung, glass size, 16" x 24" high | | .333 | Set | 21 | 4.12 | | 25.12 | 30 |
| 228 | 32" x 32" high | | .364 | " | 39 | 4.49 | | 43.49 | 50 |
| 250 | Picture, glass size, 48" x 48" high | | .267 | Ea. | 40 | 3.29 | | 43.29 | 49 |
| 258 | 60" x 68" high | | .286 | " | 65 | 3.53 | | 68.53 | 77 |
| 260 | Sliding, glass size, 14" x 36" high | | .333 | Set | 24 | 4.12 | | 28.12 | 33 |
| 268 | 36" x 36" high | ↓ | .364 | " | 40 | 4.49 | | 44.49 | 51 |
| 70-001 | **WOOD SASH** Including glazing but not including trim | | | | | | | | |
| 002 | | | | | | | | | |
| 005 | Custom, 5'-0" x 4'-0", 1" dbl. glazed, 3/16" thick lites | 2 Carp | 5 | Ea. | 220 | 62 | | 282 | 340 |
| 010 | 1/4" thick lites | | 3.200 | | 255 | 40 | | 295 | 345 |
| 020 | 1" thick, triple glazed | | 3.200 | | 300 | 40 | | 340 | 395 |
| 030 | 7'-0" x 4'-6" high, 1" double glazed, 3/16" thick lites | | 3.720 | | 300 | 46 | | 346 | 405 |
| 040 | 1/4" thick lites | | 3.720 | | 340 | 46 | | 386 | 445 |
| 050 | 1" thick, triple glazed | | 3.720 | | 380 | 46 | | 426 | 490 |
| 060 | 8'-6" x 5'-0" high, 1" double glazed, 3/16" thick lites | | 4.570 | | 400 | 56 | | 456 | 530 |
| 070 | 1/4" thick lites | | 4.570 | | 440 | 56 | | 496 | 575 |
| 080 | 1" thick, triple glazed | ↓ | 4.570 | ↓ | 485 | 56 | | 541 | 625 |
| 090 | Window frames only, based on perimeter length | | | L.F. | 1.30 | | | 1.30 | 1.43M |
| 091 | | | | | | | | | |
| 300 | Replacement sash, double hung, double glazing, window to 12 S.F. | 1 Carp | .125 | S.F. | 8.25 | 1.54 | | 9.79 | 11.50 |
| 310 | 12 S.F. to 20 S.F. | | .085 | | 7.60 | 1.05 | | 8.65 | 10 |
| 320 | 20 S.F. and over | ↓ | .075 | | 7 | .93 | | 7.93 | 9.20 |
| 380 | Triple glazing for above, add | | | ↓ | 1.60 | | | 1.60 | 1.76M |
| 700 | Sash, single lite, 2'-0" x 2'-0" high | 1 Carp | .400 | Ea. | 35 | 4.94 | | 39.94 | 46 |
| 705 | 2'-6" x 2'-0" high | | .421 | | 40 | 5.20 | | 45.20 | 52 |
| 710 | 2'-6" x 2'-6" high | | .444 | | 42 | 5.50 | | 47.50 | 55 |
| 715 | 3'-0" x 2'-0" high | ↓ | .471 | ↓ | 45 | 5.80 | | 50.80 | 59 |
| 80-001 | **WOOD SCREENS** Over 3 S.F., 3/4" frames | 2 Carp | .043 | S.F. | 1.70 | .53 | | 2.23 | 2.70 |
| 010 | 1-1/8" frames | " | .043 | " | 1.90 | .53 | | 2.43 | 2.92 |

For expanded coverage of these items see *Building Construction Cost Data 1985*

## 8.7 Finish Hardware Specialties

| | | CREW | MAN-HOURS | UNIT | MAT. | LABOR | EQUIP. | TOTAL | TOTAL INCL O&P |
|---|---|---|---|---|---|---|---|---|---|
| 01-001 | **AVERAGE** Percentage for hardware, total job cost, minimum | | | | | | | .60% | .60% |
| 005 | Maximum | | | | | | | 3% | 3% |
| 050 | Total hardware for building, average distribution | | | | 85% | 15% | | | |
| 100 | Door hardware, apartment, interior | | | Door | 66 | | | 66 | 73M |
| 150 | Hospital bedroom, minimum | | | ↓ | 90 | | | 90 | 99M |
| 200 | Maximum | | | ↓ | 425 | | | 425 | 470M |
| 210 | Pocket door | | | Ea. | 79 | | | 79 | 87M |
| 225 | School, single exterior, incl. lever, not incl. panic device | | | Door | 265 | | | 265 | 290M |
| 250 | Single interior, regular use, no lever included | | | " | 120 | | | 120 | 130M |
| 255 | Including handicap lever | | | | | | | | |
| 260 | Heavy use, incl. lever and closer | | | Door | 220 | | | 220 | 240M |
| 285 | Stairway, single interior | | | " | 145 | | | 145 | 160M |
| 310 | Double exterior, with panic device | | | Pr. | 630 | | | 630 | 695M |
| 330 | | | | | | | | | |
| 360 | Toilet, public, single interior | | | Door | 91 | | | 91 | 100M |
| 05-001 | **AUTOMATIC OPENERS** Swing doors, single | 2 Carp | 20 | Ea. | 1,600 | 245 | | 1,845 | 2,150 |
| 010 | Single operating pair | | 32 | Pr. | 2,950 | 395 | | 3,345 | 3,875 |
| 040 | For double simultaneous doors, one way, add | | 13.330 | | 200 | 165 | | 365 | 480 |
| 050 | Two way, add | | 17.780 | ↓ | 285 | 220 | | 505 | 660 |
| 100 | Sliding doors, 3' wide, including track & hanger, single | | 26.670 | Opng. | 2,900 | 330 | | 3,230 | 3,700 |
| 130 | Bi-parting | | 32 | | 3,500 | 395 | | 3,895 | 4,475 |
| 145 | Activating carpet, single door, one way, add | | 7.270 | | 500 | 90 | | 590 | 690 |
| 155 | Two way, add | | 12.310 | ↓ | 775 | 150 | | 925 | 1,100 |
| 175 | Handicap opener, button operating | ↓ | 2 | Ea. | 850 | 25 | | 875 | 975 |
| 13-001 | **DETECTION SYSTEMS** See division 16.8-15 | | | | | | | | |
| 14-001 | **DOOR ACCESSORIES** | | | | | | | | |
| 002 | | | | | | | | | |
| 100 | Knockers, brass, standard | 1 Carp | .800 | Ea. | 52 | 9.90 | | 61.90 | 73 |
| 110 | Deluxe | | .800 | | 114 | 9.90 | | 123.90 | 140 |
| 400 | Security chain, standard | | .444 | | 10 | 5.50 | | 15.50 | 19.70 |
| 410 | Deluxe | | .444 | | 25 | 5.50 | | 30.50 | 36 |
| 15-001 | **DOOR CLOSER** Rack and pinion | | 1.230 | | 46 | 15.20 | | 61.20 | 75 |
| 002 | Adjustable backcheck, 3 way mount, all sizes, regular arm | | 1.330 | | 55 | 16.45 | | 71.45 | 87 |
| 004 | Hold open arm | | 1.330 | | 61 | 16.45 | | 77.45 | 93 |
| 010 | Fusible link | | 1.230 | | 52 | 15.20 | | 67.20 | 81 |
| 020 | Non sized, regular arm | | 1.330 | | 60 | 16.45 | | 76.45 | 92 |
| 024 | Hold open arm | | 1.330 | | 67 | 16.45 | | 83.45 | 100 |
| 040 | 4 way mount, non sized, regular arm | | 1.330 | | 78 | 16.45 | | 94.45 | 110 |
| 044 | Hold open arm | | 1.330 | | 85 | 16.45 | | 101.45 | 120 |
| 20-001 | **DOORSTOPS** Holder and bumper, floor or wall | | .333 | | 16 | 4.12 | | 20.12 | 24 |
| 130 | Wall bumper | | .333 | | 3 | 4.12 | | 7.12 | 9.80 |
| 160 | Floor bumper, 1" high | | .333 | | 2.50 | 4.12 | | 6.62 | 9.25 |
| 190 | Plunger type, door mounted | ↓ | .333 | ↓ | 12 | 4.12 | | 16.12 | 19.70 |
| 33-001 | **HINGES** Full mortise, average frequency, steel base, 4-1/2"x4-1/2", USP | | | Pr. | 16.65 | | | 16.65 | 18.30M |
| 010 | 5" x 5", USP | | | | 27 | | | 27 | 30M |
| 020 | " x 6", USP | | | | 56 | | | 56 | 62M |
| 040 | Brass base, 4-1/2" x 4-1/2", US10 | | | | 36 | | | 36 | 40M |
| 050 | 5" x 5", US10 | | | | 46 | | | 46 | 51M |
| 060 | 6" x 6", US10 | | | | 83 | | | 83 | 91M |
| 080 | Stainless steel base, 4-1/2" x 4-1/2", US32 | | | ↓ | 57 | | | 57 | 63M |
| 090 | For non removable pin, add | | | Ea. | 2.75 | | | 2.75 | 3.02M |
| 091 | For floating pin, driven tips, add | | | | 3.38 | | | 3.38 | 3.71M |
| 093 | For hospital type tip on pin, add | | | | 9.65 | | | 9.65 | 10.60M |
| 094 | For steeple type tip on pin, add | | | ↓ | 8.85 | | | 8.85 | 9.75M |
| 100 | Full mortise, high frequency, steel base, 4-1/2" x 4-1/2", USP | | | Pr. | 45 | | | 45 | 50M |
| 110 | 5" x 5", USP | | | | 52 | | | 52 | 57M |
| 120 | 6" x 6", USP | | | ↓ | 96 | | | 96 | 105M |

For expanded coverage of these items see *Building Construction Cost Data 1985*

## 8.7 Finish Hardware Specialties

| | | | Crew | Man-Hours | Unit | Mat. | Labor | Equip. | Total | Total Incl O&P |
|---|---|---|---|---|---|---|---|---|---|---|
| 33 | 140 | Brass base, 4-1/2" x 4-1/2", US10 | | | Pr. | 68 | | | 68 | 75M |
| | 150 | 5" x 5", US10 | | | | 80 | | | 80 | 88M |
| | 160 | 6" x 6", US10 | | | | 125 | | | 125 | 140M |
| | 180 | Stainless steel base, 4-1/2" x 4-1/2", US32 | | | ↓ | 105 | | | 105 | 115M |
| | 193 | For hospital type tip on pin, add | | | Ea. | 11.45 | | | 11.45 | 12.60M |
| | 200 | Full mortise low frequency, steel base, 4-1/2" x 4-1/2", USP | | | Pr. | 7.02 | | | 7.02 | 7.70M |
| | 210 | 5" x 5", USP | | | | 18.95 | | | 18.95 | 21M |
| | 220 | 6" x 6", USP | | | | 36 | | | 36 | 40M |
| | 240 | Brass base, 4-1/2" x 4-1/2", US10 | | | | 28 | | | 28 | 31M |
| | 250 | 5" x 5", US10 | | | | 42 | | | 42 | 46M |
| | 280 | Stainless steel base, 4-1/2" x 4-1/2", US32 | | | ↓ | 44 | | | 44 | 48M |
| 35-001 | | KICK PLATE 6" high, for 3' door, aluminum | 1 Carp | .533 | Ea. | 12.50 | 6.60 | | 19.10 | 24 |
| | 050 | Bronze | | .533 | | 33 | 6.60 | | 39.60 | 47 |
| 40-001 | | LOCKSET Heavy duty cylindrical, passage doors | | .800 | | 68 | 9.90 | | 77.90 | 90 |
| | 002 | Non-keyed, passage | | .667 | | 21 | 8.25 | | 29.25 | 36 |
| | 010 | Privacy | | .667 | | 29 | 8.25 | | 37.25 | 45 |
| | 040 | Keyed, single cylinder function | | .800 | | 47 | 9.90 | | 56.90 | 67 |
| | 042 | Hotel | | 1 | | 55 | 12.35 | | 67.35 | 80 |
| | 060 | Bedroom, bathroom and inner office doors | | .800 | | 87 | 9.90 | | 96.90 | 110 |
| | 090 | Apartment, office and corridor doors | | .800 | | 110 | 9.90 | | 119.90 | 135 |
| | 140 | Keyed, single cylinder function | | .800 | | 115 | 9.90 | | 124.90 | 140 |
| | 170 | Residential, interior door, minimum | | .500 | | 7.30 | 6.20 | | 13.50 | 17.80 |
| | 172 | Maximum | | 1 | | 23 | 12.35 | | 35.35 | 45 |
| | 180 | Exterior, minimum | | .571 | | 15 | 7.05 | | 22.05 | 28 |
| | 182 | Maximum | | 1 | | 80 | 12.35 | | 92.35 | 110 |
| 42-001 | | ENTRANCE LOCKS Cylinder, grip handle, deadlocking latch | | .889 | | 80 | 11 | | 91 | 105 |
| | 002 | Deadbolt | | 1 | | 93 | 12.35 | | 105.35 | 120 |
| | 010 | Push and pull plate, dead bolt | ↓ | 1 | | 90 | 12.35 | | 102.35 | 120 |
| | 090 | For handicapped lever, add | | | | 60 | | | 60 | 66M |
| 45-001 | | PANIC DEVICE For rim locks, single door, exit only | 1 Carp | 2.670 | ↓ | 200 | 33 | | 233 | 270 |
| | 100 | Mortise, bar, exit only | | 4 | Pr. | 250 | 49 | | 299 | 355 |
| | 300 | Mortise, bar, exit only | | 2.670 | Ea. | 260 | 33 | | 293 | 340 |
| | 400 | Double doors, exit only | | 4 | Pr. | 500 | 49 | | 549 | 630 |
| | 450 | Exit & entrance | ↓ | 4 | " | 525 | 49 | | 574 | 655 |
| 50-001 | | PUSH-PULL Push plate, pull plate, aluminum | 1 Carp | .667 | Ea. | 25 | 8.25 | | 33.25 | 41 |
| | 050 | Bronze | | .667 | | 41 | 8.25 | | 49.25 | 58 |
| | 150 | Pull handle and push bar, aluminum | | .727 | | 73 | 9 | | 82 | 95 |
| | 200 | Bronze | | .800 | | 93 | 9.90 | | 102.90 | 120 |
| | 400 | Door pull, designer style, cast aluminum, minimum | | .667 | | 40 | 8.25 | | 48.25 | 57 |
| | 500 | Maximum | | 1 | | 175 | 12.35 | | 187.35 | 210 |
| 60-001 | | THRESHOLD 3' long door saddles, aluminum, minimum | | .400 | | 19.75 | 4.94 | | 24.69 | 30 |
| | 010 | Maximum | | .667 | | 62 | 8.25 | | 70.25 | 81 |
| | 050 | Bronze, minimum | | .400 | | 36 | 4.94 | | 40.94 | 47 |
| | 060 | Maximum | | .667 | | 105 | 8.25 | | 113.25 | 130 |
| | 070 | Rubber, 1/2" thick, 5-1/2" wide | | .400 | | 21 | 4.94 | | 25.94 | 31 |
| | 080 | 2-3/4" wide | | .400 | ↓ | 11.45 | 4.94 | | 16.39 | 20 |
| 70-001 | | WEATHERSTRIPPING Window, double hung, 3' x 5', zinc | | 1.110 | Opng. | 13 | 13.70 | | 26.70 | 36 |
| | 010 | Bronze | | 1.110 | | 25 | 13.70 | | 38.70 | 49 |
| | 020 | Vinyl V strip | | 1.140 | | 2.50 | 14.10 | | 16.60 | 25 |
| | 050 | As above but heavy duty, zinc | | 1.740 | | 16.10 | 21 | | 37.10 | 52 |
| | 060 | Bronze | | 1.740 | | 30 | 21 | | 51 | 67 |
| | 100 | Doors, wood frame, interlocking, for 3' x 7' door, zinc | | 2.670 | | 18.20 | 33 | | 51.20 | 72 |
| | 110 | Bronze | | 2.670 | | 30 | 33 | | 63 | 85 |
| | 120 | Vinyl V strip | | 1.140 | | 6.05 | 14.10 | | 20.15 | 29 |
| | 130 | 6' x 7' opening, zinc | | 4 | | 22 | 49 | | 71 | 100 |
| | 140 | Bronze | ↓ | 4 | ↓ | 36 | 49 | | 85 | 120 |
| | 170 | Wood frame, spring type, bronze | | | | | | | | |
| | 180 | 3' x 7' door | 1 Carp | 1.050 | Opng. | 8.85 | 13 | | 21.85 | 30 |

For expanded coverage of these items see *Building Construction Cost Data 1985*

## 8.7 Finish Hardware Specialties

| | | CREW | MAN-HOURS | UNIT | MAT. | LABOR | EQUIP. | TOTAL | TOTAL INCL O&P |
|---|---|---|---|---|---|---|---|---|---|
| 70 190 | 6' x 7' door | 1 Carp | 1.140 | Opng. | 12.50 | 14.10 | | 26.60 | 36 |
| 192 | Felt, 3' x 7' door | | .571 | | 1.50 | 7.05 | | 8.55 | 12.85 |
| 193 | 6' x 7' door | | .615 | | 1.75 | 7.60 | | 9.35 | 13.95 |
| 195 | Rubber, 3' x 7' door | | 1.050 | | 3.65 | 13 | | 16.65 | 25 |
| 196 | 6' x 7' door | ↓ | 1.140 | ↓ | 4.15 | 14.10 | | 18.25 | 27 |
| 220 | Metal frame, spring type, bronze | | | | | | | | |
| 230 | 3' x 7' door | 1 Carp | 2.860 | Opng. | 35 | 35 | | 70 | 94 |
| 240 | 6' x 7' door | " | 3.480 | | 45 | 43 | | 88 | 120 |
| 250 | For stainless steel, spring type, add | | | ↓ | 100% | | | | |
| 251 | | | | | | | | | |
| 270 | Metal frame, extruded sections, 3' x 7' door, aluminum | 1 Carp | 4 | Opng. | 25 | 49 | | 74 | 105 |
| 280 | Bronze | | 4 | | 66 | 49 | | 115 | 150 |
| 310 | Metal frame, extruded sections, 6' x 7' door, aluminum | ↓ | 6.670 | | 29 | 82 | | 111 | 160 |
| 320 | Bronze | ↓ | 6.670 | ↓ | 66 | 82 | | 148 | 205 |
| 350 | Threshold weatherstripping | | | | | | | | |
| 360 | | | | | | | | | |
| 365 | Door sweep, flush mounted, aluminum | 1 Carp | .250 | Ea. | 10.90 | 3.09 | | 13.99 | 16.90 |
| 370 | Vinyl | | .320 | " | 3.60 | 3.95 | | 7.55 | 10.20 |
| 400 | Astragal for double doors, aluminum | | 2 | Opng. | 39 | 25 | | 64 | 82 |
| 410 | Bronze | ↓ | 2 | " | 88 | 25 | | 113 | 135 |
| 80-001 | **WINDOW HARDWARE** | | | | | | | | |
| 100 | Handles, surface mounted, aluminum | 1 Carp | .333 | Ea. | .88 | 4.12 | | 5 | 7.50 |
| 102 | Brass | | .333 | | .99 | 4.12 | | 5.11 | 7.60 |
| 104 | Chrome | | .333 | | .90 | 4.12 | | 5.02 | 7.50 |
| 150 | Recessed, aluminum | | .667 | | .78 | 8.25 | | 9.03 | 13.90 |
| 152 | Brass | | .667 | | .89 | 8.25 | | 9.14 | 14 |
| 154 | Chrome | | .667 | | .81 | 8.25 | | 9.06 | 13.95 |
| 200 | Latches, aluminum | | .400 | | 1.05 | 4.94 | | 5.99 | 9 |
| 202 | Brass | | .400 | | 1.15 | 4.94 | | 6.09 | 9.10 |
| 204 | Chrome | ↓ | .400 | ↓ | 1.10 | 4.94 | | 6.04 | 9.05 |

## 8.8 Glass & Glazing

| | | CREW | MAN-HOURS | UNIT | MAT. | LABOR | EQUIP. | TOTAL | TOTAL INCL O&P |
|---|---|---|---|---|---|---|---|---|---|
| 12-001 | **FULL VISION** Window system with 3/4" glass mullions, 10' high | H-2 | .200 | S.F. | 11.50 | 2.33 | | 13.83 | 16.30 |
| 010 | 10' to 20' high, minimum | | .240 | | 14.50 | 2.79 | | 17.29 | 20 |
| 015 | Average | | .267 | | 18.50 | 3.10 | | 21.60 | 25 |
| 020 | Maximum | ↓ | .343 | ↓ | 24 | 3.99 | | 27.99 | 33 |
| 18-001 | **GLAZING VARIABLES** | | | | | | | | |
| 050 | For high rise glazing, from exterior, add per S.F. per story | | | Story | | .06 | | .06 | .08L |
| 060 | For glass replacement, add | | | S.F. | | 100% | | | |
| 070 | For gasket settings, add | | | L.F. | 1.80 | | | 1.80 | 1.98M |
| 080 | For concrete reglet settings, add | | | S.F. | 20 | 25 | | | |
| 090 | For sloped glazing, add | | | " | | 25 | | | |
| 200 | Fabrication, polished edges, 1/4" thick | | | Inch | .50 | | | .50 | .55M |
| 210 | 1/2" thick | | | | .70 | | | .70 | .77M |
| 250 | Mitered edges, 1/4" thick | | | | 1 | | | 1 | 1.10M |
| 260 | 1/2" thick | | | ↓ | 1.39 | | | 1.39 | 1.52M |
| 21-001 | **INSULATING GLASS UNITS** 2 lites 1/8" float, 1/2" thick unit, under 15 S.F. | | | | | | | | |
| 010 | Tinted | 2 Glaz | .168 | S.F. | 5.30 | 2.10 | | 7.40 | 9.10 |
| 020 | Double glazed, 5/8" thick unit, 3/16" float, 15-30 S.F., clear | | .178 | | 5.05 | 2.21 | | 7.26 | 9 |
| 040 | 1" thick, double glazed, 1/4" float, 30 to 70 S.F., clear | | .213 | | 6.30 | 2.66 | | 8.96 | 11.10 |
| 050 | Tinted | | .213 | | 7.45 | 2.66 | | 10.11 | 12.35 |
| 200 | Both lites, light & heat reflective | | .188 | | 9.75 | 2.34 | | 12.09 | 14.40 |
| 250 | Heat reflective, film inside, 1" thick unit, clear | | .188 | | 9.25 | 2.34 | | 11.59 | 13.85 |
| 260 | Tinted | ↓ | .188 | ↓ | 9.70 | 2.34 | | 12.04 | 14.35 |

For expanded coverage of these items see *Building Construction Cost Data 1985*

## 8.8 Glass & Glazing

| | | | CREW | MAN-HOURS | UNIT | MAT. | LABOR | EQUIP. | TOTAL | TOTAL INCL O&P |
|---|---|---|---|---|---|---|---|---|---|---|
| 21 | 300 | Film on weatherside, clear, 1/2" thick unit | 2 Glaz | .168 | S.F. | 6.70 | 2.10 | | 8.80 | 10.65 |
| | 310 | 5/8" thick unit | | .178 | | 7.85 | 2.21 | | 10.06 | 12.10 |
| | 320 | 1" thick unit | ↓ | .188 | ↓ | 8.95 | 2.34 | | 11.29 | 13.50 |
| 36-001 | | **FLOAT GLASS** 3/16" thick, clear, plain | 2 Glaz | .123 | S.F. | .90 | 1.53 | | 2.43 | 3.39 |
| | 010 | Tinted | | .123 | | 1.20 | 1.53 | | 2.73 | 3.72 |
| | 020 | Tempered, clear | | .123 | | 2.89 | 1.53 | | 4.42 | 5.60 |
| | 030 | Tempered, tinted | | .123 | | 3.75 | 1.53 | | 5.28 | 6.50 |
| | 060 | 1/4" thick, clear, plain | | .133 | | 1.25 | 1.66 | | 2.91 | 3.97 |
| | 070 | Tinted | | .133 | | 1.65 | 1.66 | | 3.31 | 4.41 |
| | 080 | Tempered, clear | | .133 | | 2.91 | 1.66 | | 4.57 | 5.80 |
| | 090 | Tempered, tinted | | .133 | | 3.80 | 1.66 | | 5.46 | 6.80 |
| | 120 | 5/16" thick, clear, plain | | .160 | | 2.48 | 1.99 | | 4.47 | 5.85 |
| | 130 | Tempered, clear | | .160 | | 7.90 | 1.99 | | 9.89 | 11.80 |
| | 160 | 3/8" thick, clear, plain | | .213 | | 2.55 | 2.66 | | 5.21 | 6.95 |
| | 170 | Tinted | | .213 | | 3.90 | 2.66 | | 6.56 | 8.45 |
| | 180 | Tempered, clear | | .213 | | 8.40 | 2.66 | | 11.06 | 13.40 |
| | 190 | Tempered, tinted | | .213 | | 10 | 2.66 | | 12.66 | 15.15 |
| | 220 | 1/2" thick, clear, plain | | .291 | | 3.97 | 3.62 | | 7.59 | 10.05 |
| | 230 | Tinted | | .291 | | 6.05 | 3.62 | | 9.67 | 12.30 |
| | 240 | Tempered, clear | | .291 | | 12.35 | 3.62 | | 15.97 | 19.25 |
| | 250 | Tempered, tinted | | .291 | | 14.20 | 3.62 | | 17.82 | 21 |
| | 280 | 5/8" thick, clear, plain | | .356 | | 4.72 | 4.43 | | 9.15 | 12.10 |
| | 290 | Tempered, clear | | .356 | | 14.75 | 4.43 | | 19.18 | 23 |
| | 320 | 3/4" thick, clear, plain | | .457 | | 5.60 | 5.70 | | 11.30 | 15.05 |
| | 330 | Tempered, clear | | .457 | | 21 | 5.70 | | 26.70 | 32 |
| | 360 | 1" thick, clear, plain | ↓ | .533 | ↓ | 13.30 | 6.65 | | 19.95 | 25 |
| 42-001 | | **POLYCARBONATE** Clear, masked, cut sheets, 1/8" thick | 2 Glaz | .094 | S.F. | 1.90 | 1.17 | | 3.07 | 3.92 |
| | 050 | 3/16" thick | | .097 | | 3 | 1.21 | | 4.21 | 5.20 |
| | 100 | 1/4" thick | | .103 | | 3.50 | 1.29 | | 4.79 | 5.85 |
| | 150 | 3/8" thick | | .107 | | 6.50 | 1.33 | | 7.83 | 9.25 |
| 60-001 | | **WINDOW GLASS** Clear float, stops, putty bed, 1/8" thick | | .033 | | 1.88 | .42 | | 2.30 | 2.72 |
| | 050 | 3/16" thick, clear | | .033 | | 1.95 | .42 | | 2.37 | 2.79 |
| | 060 | Tinted | | .033 | | 2.20 | .42 | | 2.62 | 3.07 |
| | 070 | Tempered | ↓ | .033 | ↓ | 3.75 | .42 | | 4.17 | 4.77 |
| 50-001 | | **TUBE FRAMING** For window walls and store fronts, aluminum, stock | | | | | | | | |
| | 002 | | | | | | | | | |
| | 005 | Plain tube frame, mill finish, 1-3/4" x 1-3/4" | 2 Glaz | .178 | L.F. | 3.74 | 2.21 | | 5.95 | 7.60 |
| | 010 | 1-3/4" x 3" | | .178 | | 4.31 | 2.21 | | 6.52 | 8.20 |
| | 015 | 1-3/4" x 4" | | .178 | | 5.10 | 2.21 | | 7.31 | 9.05 |
| | 020 | 1-3/4" x 4-1/2" | | .178 | | 5.45 | 2.21 | | 7.66 | 9.45 |
| | 025 | 2" x 6" | | .178 | | 7.20 | 2.21 | | 9.41 | 11.40 |
| | 030 | 3" x 3" | | .178 | | 6.70 | 2.21 | | 8.91 | 10.85 |
| | 035 | 4" x 4" | | .178 | | 7.35 | 2.21 | | 9.56 | 11.55 |
| | 040 | 4-1/2" x 4-1/2" | | .178 | | 9.45 | 2.21 | | 11.66 | 13.85 |
| | 045 | Glass bead | | .067 | | 1.20 | .83 | | 2.03 | 2.62 |
| | 100 | Flush tube frame, mill finish, 1/4" glass, 1-3/4" x 4", open header | | .200 | | 3.78 | 2.49 | | 6.27 | 8.05 |
| | 100 | Flush tube frame, bronze fin., 1/4" glass, 1-3/4"x4", open header | | .200 | | 3.78 | 2.49 | | 6.27 | 8.05 |
| | 100 | Flush tube frame, black fin., 1/4" glass, 1-3/4"x4", open header | | .200 | | 4.10 | 2.49 | | 6.59 | 8.40 |
| | 105 | Open sill | | .200 | | 3.83 | 2.49 | | 6.32 | 8.10 |
| | 105 | Open sill, bronze finish | | .200 | | 3.88 | 2.49 | | 6.37 | 8.15 |
| | 105 | Open sill, black finish | | .200 | | 4.13 | 2.49 | | 6.62 | 8.45 |
| | 110 | Closed back header | ↓ | .200 | | 5.05 | 2.49 | | 7.54 | 9.45 |
| | 115 | Closed back sill | 3 Glaz | .300 | | 5.40 | 3.74 | | 9.14 | 11.80 |
| | 120 | Vertical mullion, one piece | 2 Glaz | .200 | | 5.20 | 2.49 | | 7.69 | 9.60 |
| | 120 | Vertical mullion, one piece, bronze finish | | .200 | | 5.20 | 2.49 | | 7.69 | 9.60 |
| | 120 | Vertical mullion, One piece, black finish | | .200 | | 5.60 | 2.49 | | 8.09 | 10.05 |
| | 125 | Two piece | | .200 | | 5.75 | 2.49 | | 8.24 | 10.20 |
| | 130 | 90° or 180° vertical corner post | ↓ | .200 | ↓ | 9.35 | 2.49 | | 11.84 | 14.20 |

For expanded coverage of these items see *Building Construction Cost Data 1985*

## 8.8 Glass & Glazing

| | | Description | CREW | MAN-HOURS | UNIT | MAT. | LABOR | EQUIP. | TOTAL | TOTAL INCL O&P |
|---|---|---|---|---|---|---|---|---|---|---|
| 50 | 135 | | | | | | | | | |
| | 140 | 1-3/4" x 4-1/2", open header | 2 Glaz | .200 | L.F. | 4.40 | 2.49 | | 6.89 | 8.75 |
| | 140 | 1-3/4" x 4-1/2", bronze finish | | .200 | | 4.40 | 2.49 | | 6.89 | 8.75 |
| | 140 | 1-3/4" x 4-1/2", black finish | | .200 | | 4.73 | 2.49 | | 7.22 | 9.10 |
| | 145 | Open sill | | .200 | | 4.51 | 2.49 | | 7 | 8.85 |
| | 145 | Open sill, bronze finish | | .200 | | 4.54 | 2.49 | | 7.03 | 8.90 |
| | 145 | Open sill, black finish | | .200 | | 4.90 | 2.49 | | 7.39 | 9.30 |
| | 150 | Closed back header | | .200 | | 5.35 | 2.49 | | 7.84 | 9.80 |
| | 155 | Closed back sill | | .200 | | 6.75 | 2.49 | | 9.24 | 11.30 |
| | 160 | Vertical mullion, one piece | | .200 | | 5.40 | 2.49 | | 7.89 | 9.85 |
| | 160 | Vertical mullion, one piece, bronze finish | | .200 | | 5.45 | 2.49 | | 7.94 | 9.90 |
| | 160 | Vertical mullion, one piece, black finish | | .200 | | 5.85 | 2.49 | | 8.34 | 10.35 |
| | 165 | Two piece | | .200 | | 6.65 | 2.49 | | 9.14 | 11.20 |
| | 170 | 90° or 180° vertical corner post | ▼ | .200 | ▼ | 9.90 | 2.49 | | 12.39 | 14.80 |
| | 175 | | | | | | | | | |
| | 200 | Flush tube frame, mill fin. for ins. glass, 2" x 4-1/2", open header | 2 Glaz | .213 | L.F. | 5.30 | 2.66 | | 7.96 | 10 |
| | 200 | Flush tube frame, bronze, insulating glass, 2"x4-1/2", open header | | .213 | | 5.30 | 2.66 | | 7.96 | 10 |
| | 200 | Flush tube frame, black, insulating glass, 2"x4-1/2", open header | | .213 | | 5.75 | 2.66 | | 8.41 | 10.50 |
| | 205 | Open sill, mill finish | | .213 | | 5.45 | 2.66 | | 8.11 | 10.15 |
| | 205 | Open sill, bronze finish | | .213 | | 5.45 | 2.66 | | 8.11 | 10.15 |
| | 205 | Open sill, black finish | | .213 | | 5.85 | 2.66 | | 8.51 | 10.60 |
| | 210 | Closed back header | | .213 | | 7.15 | 2.66 | | 9.81 | 12 |
| | 215 | Closed back sill | | .213 | | 7.45 | 2.66 | | 10.11 | 12.35 |
| | 220 | Vertical mullion, one piece, mill finish | | .229 | | 6.80 | 2.85 | | 9.65 | 11.95 |
| | 220 | Vertical mullion, one piece, bronze finish | | .213 | | 6.90 | 2.66 | | 9.56 | 11.75 |
| | 220 | Vertical mullion, one piece, black finish | | .213 | | 7.40 | 2.66 | | 10.06 | 12.30 |
| | 225 | Two piece | | .235 | | 7.60 | 2.93 | | 10.53 | 12.95 |
| | 230 | 90° or 180° vertical corner post | ▼ | .229 | ▼ | 12.60 | 2.85 | | 15.45 | 18.30 |
| | 235 | | | | | | | | | |
| | 500 | Flush tube frame, mill fin., thermal brk., 2-1/4"x 4-1/2" | 2 Glaz | .216 | L.F. | 7.85 | 2.69 | | 10.54 | 12.85 |
| | 500 | Flush tube frame, bronze, thermal brk., 2-1/4"x4", open header | | .213 | | 7.85 | 2.66 | | 10.51 | 12.80 |
| | 500 | Flush tube frame, black, thermal brk., 2-1/4"x4-1/2", open header | | .213 | | 8.50 | 2.66 | | 11.16 | 13.50 |
| | 505 | Open sill | | .213 | | 7.90 | 2.66 | | 10.56 | 12.85 |
| | 510 | Vertical mullion, one piece | | .232 | | 8.95 | 2.89 | | 11.84 | 14.35 |
| | 510 | Vertical mullion, one piece, bronze finish | | .213 | | 6.20 | 2.66 | | 8.86 | 10.95 |
| | 510 | Vertical mullion, one piece, black finish | | .213 | | 6.65 | 2.66 | | 9.31 | 11.45 |
| | 515 | Two piece | | .246 | | 9.80 | 3.06 | | 12.86 | 15.55 |
| | 520 | 90° or 180° vertical corner post | ▼ | .232 | ▼ | 16.15 | 2.89 | | 19.04 | 22 |
| | 525 | | | | | | | | | |
| | 530 | Door stop (snap in) | 2 Glaz | .042 | L.F. | 1.20 | .52 | | 1.72 | 2.14 |
| | 535 | | | | | | | | | |
| | 700 | For joints, 90°, clip type, add | | | Ea. | 3.15 | | | 3.15 | 3.46M |
| | 705 | Screw spline joint, mill finish, add | | | | 2.63 | | | 2.63 | 2.89M |
| | 705 | Screw spline joint, bronze finish, add | | | | 7.10 | | | 7.10 | 7.80M |
| | 705 | Screw spline joint, black finish, add | | | | 7.65 | | | 7.65 | 8.40M |
| | 710 | For joint other than 90°, add | | | ▼ | 6.30 | | | 6.30 | 6.95M |
| | 715 | | | | | | | | | |
| | 800 | For bronze finish, add | | | L.F. | 18% | | | | |
| | 805 | For stainless steel, add | | | | 75% | | | | |
| | 810 | For monumental grade, add | | | | 50% | | | | |
| | 815 | For steel stiffener, add | 2 Glaz | .080 | ▼ | 3.60 | 1 | | 4.60 | 5.50 |
| | 820 | For 2 to 5 stories, add per story | | | Story | | 5% | | | |

For expanded coverage of these items see *Building Construction Cost Data 1985*

## 9.1 Lath & Plaster

| | | CREW | MAN-HOURS | UNIT | MAT. | LABOR | EQUIP. | TOTAL | TOTAL INCL O&P |
|---|---|---|---|---|---|---|---|---|---|
| 15-001 | **FURRING** Beams & columns, 3/4" galvanized channels, | | | | | | | | |
| 003 | 12" O.C. | 1 Lath | .052 | S.F. | .27 | .64 | | .91 | 1.29 |
| 005 | 16" O.C. | | .047 | | .23 | .58 | | .81 | 1.16 |
| 007 | 24" O.C. | | .043 | | .17 | .54 | | .71 | 1.02 |
| 010 | Ceilings, on steel, 3/4" channels, galvanized, 12" O.C. | | .038 | | .27 | .47 | | .74 | 1.03 |
| 030 | 16" O.C. | | .028 | | .23 | .34 | | .57 | .78 |
| 040 | 24" O.C. | | .019 | | .17 | .24 | | .41 | .55 |
| 060 | 1-1/2" channels, galvanized, 12" O.C. | | .042 | | .34 | .52 | | .86 | 1.18 |
| 070 | 16" O.C. | | .031 | | .29 | .38 | | .67 | .91 |
| 090 | 24" O.C. | | .021 | | .22 | .25 | | .47 | .64 |
| 100 | Walls, galvanized, 3/4" channels, 12" O.C. | | .034 | | .27 | .42 | | .69 | .95 |
| 120 | 16" O.C. | | .030 | | .23 | .37 | | .60 | .83 |
| 130 | 24" O.C. | | .023 | | .17 | .28 | | .45 | .63 |
| 150 | 1-1/2" channels, galvanized, 12" O.C., | | .038 | | .34 | .47 | | .81 | 1.10 |
| 160 | 16" O.C. | | .033 | | .29 | .41 | | .70 | .96 |
| 180 | 24" O.C. | | .026 | | .22 | .33 | | .55 | .74 |
| 800 | Suspended ceilings, including carriers | | | | | | | | |
| 820 | 1-1/2" carriers, 24" O.C. | | | | | | | | |
| 830 | 3/4" channels, 16" O.C. | 1 Lath | .048 | S.F. | .42 | .60 | | 1.02 | 1.39 |
| 832 | 24" OC | | .040 | | .37 | .50 | | .87 | 1.17 |
| 840 | 1-1/2" channels, 16" O.C. | | .052 | | .48 | .64 | | 1.12 | 1.52 |
| 842 | 24" OC | | .042 | | .41 | .52 | | .93 | 1.26 |
| 860 | 2" carriers, 24" OC | | | | | | | | |
| 870 | 3/4" channels, 16" OC | 1 Lath | .052 | S.F. | .49 | .64 | | 1.13 | 1.53 |
| 872 | 24" OC | | .042 | | .44 | .52 | | .96 | 1.29 |
| 880 | 1-1/2" channels, 16" OC | | .055 | | .55 | .68 | | 1.23 | 1.66 |
| 882 | 24" OC | | .044 | | .48 | .55 | | 1.03 | 1.38 |
| 20-001 | **GYPSUM LATH** Plain or perforated, nailed, 3/8" thick | 1 Lath | .010 | S.F. | .22 | .13 | | .35 | .44 |
| 010 | 1/2" thick, nailed | | .011 | | .24 | .14 | | .38 | .48 |
| 030 | Clipped to steel studs, 3/8" thick | | .012 | | .25 | .15 | | .40 | .50 |
| 040 | 1/2" thick | | .013 | | .26 | .16 | | .42 | .53 |
| 060 | Firestop gypsum base, to steel studs, 1/2" thick | | .013 | | .31 | .16 | | .47 | .58 |
| 070 | 5/8" thick | | .014 | | .33 | .17 | | .50 | .63 |
| 090 | Moisture resistant, 4' x 8' sheets, 1/2" thick | | .012 | | .33 | .15 | | .48 | .59 |
| 100 | 5/8" thick | | .013 | | .38 | .16 | | .54 | .66 |
| 120 | Laminated, 1" thick, to steel studs | | .014 | | .52 | .17 | | .69 | .83 |
| 130 | For foil facing, add to above | | | | .06 | | | .06 | .06M |
| 150 | For ceiling installations, add | 1 Lath | .040 | | | .50 | | .50 | .77L |
| 160 | For columns and beams, add | " | .040 | | | .50 | | .50 | .77L |
| 21-001 | **GYPSUM PLASTER** 80# bag, less than 1 ton | | | Bag | 9.35 | | | 9.35 | 10.30M |
| 002 | | | | | | | | | |
| 030 | 2 coats, no lath included, on walls | J-1 | .042 | S.F. | .29 | .49 | .04 | .82 | 1.12 |
| 040 | On ceilings | | .048 | | .29 | .56 | .04 | .89 | 1.24 |
| 090 | 3 coats, no lath included, on walls | | .051 | | .40 | .59 | .04 | 1.03 | 1.41 |
| 100 | On ceilings | | .057 | | .40 | .66 | .05 | 1.11 | 1.52 |
| 160 | For irregular or curved surfaces, add | | | | 30% | | | | 30% |
| 180 | For columns and beams, add | | | | 50% | | | | 50% |
| 25-001 | **METAL LATH** Diamond, expanded, 2.5 lb. per S.Y., painted | | | | | | | | |
| 360 | 2.5 lb. diamond painted, on wood framing, on walls | 1 Lath | .010 | S.F. | .18 | .13 | | .31 | .40 |
| 370 | On ceilings | | .012 | | .18 | .15 | | .33 | .43 |
| 420 | 3.4 lb. diamond painted, wired to steel framing, on walls | | .012 | | .22 | .15 | | .37 | .47 |
| 430 | On ceilings | | .015 | | .22 | .18 | | .40 | .53 |
| 510 | Rib lath, painted, wired to steel, on walls, 2.75 lb. | | .012 | | .25 | .15 | | .40 | .50 |
| 520 | 3.4 lb. | | .013 | | .26 | .16 | | .42 | .53 |
| 570 | Suspended ceiling system, incl. 3.4 lb. diamond lath, painted | | .059 | | .61 | .73 | | 1.34 | 1.81 |
| 580 | Galvanized | | .059 | | .65 | .73 | | 1.38 | 1.85 |
| 35-001 | **PERLITE OR VERMICULITE PLASTER** Under 200 bags | | | | | | | | |
| 030 | 2 coats, no lath included, on walls | J-1 | .048 | S.F. | .29 | .56 | .04 | .89 | 1.23 |

For expanded coverage of these items see *Building Construction Cost Data 1985*

## 9.1 Lath & Plaster

| | | | CREW | MAN-HOURS | UNIT | MAT. | LABOR | EQUIP. | TOTAL | TOTAL INCL O&P |
|---|---|---|---|---|---|---|---|---|---|---|
| 35 | 040 | On ceilings | J-1 | .056 | S.F. | .29 | .65 | .05 | .99 | 1.39 |
| | 090 | 3 coats, no lath included, on walls | | .060 | | .45 | .70 | .05 | 1.20 | 1.64 |
| | 100 | On ceilings | ▼ | .071 | | .45 | .82 | .06 | 1.33 | 1.84 |
| | 170 | For irregular or curved surfaces, add to above | | | | 30% | | | | 30% |
| | 180 | For columns and beams, add to above | | | | 50% | | | | 50% |
| | 190 | For soffits, add to ceiling prices | | | ▼ | | 40% | | | |
| 50-001 | | SPRAYED Mineral fiber or cementitious for fireproofing, | | | | | | | | |
| | 005 | not incl. tamping or canvas protection | | | | | | | | |
| | 010 | 1" thick, on flat plate steel | G-2 | .008 | S.F. | .31 | .09 | .03 | .43 | .52 |
| | 020 | Flat decking | | .010 | | .31 | .11 | .04 | .46 | .56 |
| | 040 | Beams | | .016 | | .31 | .17 | .07 | .55 | .69 |
| | 050 | Corrugated or fluted decks | | .019 | | .31 | .21 | .08 | .60 | .76 |
| | 070 | Columns, 1-1/8" thick | | .022 | | .37 | .24 | .09 | .70 | .88 |
| | 080 | 2-3/16" thick | ▼ | .034 | | .62 | .37 | .15 | 1.14 | 1.43 |
| | 085 | For tamping, add | | | | | 10% | | | |
| | 090 | For canvas protection, add | G-2 | .005 | | .02 | .05 | .02 | .09 | .13 |
| 55-001 | | STUCCO 3 coats 1" thick, float finish, on frame construction | J-2 | .102 | | .68 | 1.20 | .07 | 1.95 | 2.69 |
| | 010 | On masonry construction | J-1 | .081 | | .20 | .93 | .07 | 1.20 | 1.75 |
| | 015 | 2 coats, 3/4" thick, float finish, no lath incl. | " | .041 | | .22 | .47 | .04 | .73 | 1.02 |
| | 030 | For trowel finish, add | 1 Plas | .005 | | | .06 | | .06 | .10L |
| | 060 | For coloring and special finish, add, minimum | | | | .01 | | | .01 | .01M |
| | 070 | Maximum | | | | .07 | | | .07 | .07M |
| | 100 | Exterior plaster, with bonding agent, 1 coat, on walls | J-1 | .019 | | .25 | .21 | .02 | .48 | .63 |
| | 120 | Ceilings | | .022 | | .25 | .26 | .02 | .53 | .70 |
| | 130 | Beams | | .044 | | .25 | .51 | .04 | .80 | 1.12 |
| | 150 | Columns | ▼ | .037 | | .25 | .43 | .03 | .71 | .98 |
| | 160 | Mesh, painted, nailed to wood, 1.8 lb. | 1 Lath | .015 | | .22 | .18 | | .40 | .53 |
| | 180 | 3.6 lb. | | .016 | | .32 | .20 | | .52 | .66 |
| | 190 | Wired to steel, painted, 1.8 lb. | | .017 | | .24 | .21 | | .45 | .59 |
| | 210 | 3.6 lb. | | .018 | | .29 | .22 | | .51 | .66 |
| | 220 | Clinton cloth, on wood | | .015 | | .39 | .18 | | .57 | .71 |
| | 240 | On steel | ▼ | .017 | | .39 | .21 | | .60 | .75 |
| 57-001 | | TILE OR TERRAZZO BASE Scratch coat only | J-1 | .015 | | .04 | .17 | .01 | .22 | .33 |
| | 050 | Scratch and brown coat only | " | .039 | ▼ | .10 | .45 | .03 | .58 | .84 |

## 9.2 Drywall

| | | | CREW | MAN-HOURS | UNIT | MAT. | LABOR | EQUIP. | TOTAL | TOTAL INCL O&P |
|---|---|---|---|---|---|---|---|---|---|---|
| 02-001 | | ACCESSORIES, DRYWALL Casing bead, galvanized steel | 1 Carp | .028 | L.F. | .10 | .34 | | .44 | .65 |
| | 010 | Vinyl | | .028 | | .09 | .34 | | .43 | .64 |
| | 030 | Corner bead, galvanized steel, 1" x 1" | | .028 | | .08 | .34 | | .42 | .63 |
| | 040 | 1-1/4" x 1-1/4" | | .028 | | .10 | .34 | | .44 | .65 |
| | 060 | Vinyl corner bead | | .028 | | .07 | .34 | | .41 | .62 |
| | 070 | Door casing, vinyl, for 2" wall systems | | .032 | | .23 | .40 | | .63 | .88 |
| | 090 | Furring channel, galv. steel, 7/8" deep, standard | | .031 | | .17 | .38 | | .55 | .79 |
| | 100 | Resilient | | .031 | | .18 | .38 | | .56 | .80 |
| | 110 | J bead, galvanized steel, 1/2" wide | | .027 | | .12 | .33 | | .45 | .65 |
| | 112 | 5/8" wide | | .027 | | .13 | .33 | | .46 | .66 |
| | 150 | Z bar, galvanized steel, 1-1/2" wide | ▼ | .031 | ▼ | .21 | .38 | | .59 | .83 |
| 07-001 | | DRYWALL Gypsum plasterboard, nailed or screwed to studs, | | | | | | | | |
| | 010 | unless otherwise noted | | | | | | | | |
| | 015 | 3/8" thick, on walls, standard, no finish included | 2 Carp | .008 | S.F. | .18 | .10 | | .28 | .35 |
| | 020 | On ceiling | | .011 | | .18 | .14 | | .32 | .42 |
| | 025 | On beams, columns, or soffits | | .021 | | .18 | .26 | | .44 | .62 |
| | 030 | 1/2" thick, on walls, standard, no finish included | ▼ | .009 | ▼ | .21 | .11 | | .32 | .40 |

For expanded coverage of these items see *Building Construction Cost Data 1985*

## 9.2 Drywall

| | | | CREW | MAN-HOURS | UNIT | MAT. | LABOR | EQUIP. | TOTAL | TOTAL INCL O&P |
|---|---|---|---|---|---|---|---|---|---|---|
| 07 | 035 | Taped and finished | 2 Carp | .018 | S.F. | .23 | .22 | | .45 | .60 |
| | 040 | Fire resistant, no finish included | | .009 | | .24 | .11 | | .35 | .44 |
| | 045 | Taped and finished | | .018 | | .26 | .22 | | .48 | .63 |
| | 050 | Water resistant, no finish included | | .009 | | .27 | .11 | | .38 | .47 |
| | 055 | Taped and finished | | .018 | | .29 | .22 | | .51 | .67 |
| | 060 | Prefinished, vinyl, clipped to studs | ↓ | .015 | ↓ | .64 | .18 | | .82 | .99 |
| | 065 | | | | | | | | | |
| | 100 | On ceilings, standard, no finish included | 2 Carp | .013 | S.F. | .21 | .16 | | .37 | .49 |
| | 105 | Taped and finished | | .020 | | .23 | .25 | | .48 | .64 |
| | 110 | Fire resistant, no finish included | | .013 | | .24 | .16 | | .40 | .52 |
| | 115 | Taped and finished | | .020 | | .26 | .25 | | .51 | .68 |
| | 120 | Water resistant, no finish included | | .013 | | .27 | .16 | | .43 | .56 |
| | 125 | Taped and finished | | .020 | | .29 | .25 | | .54 | .71 |
| | 150 | On beams, columns, or soffits, standard, no finish included | | .024 | | .21 | .29 | | .50 | .69 |
| | 155 | Taped and finished | | .034 | | .23 | .42 | | .65 | .91 |
| | 160 | Fire resistant, no finish included | | .024 | | .24 | .29 | | .53 | .73 |
| | 165 | Taped and finished | | .034 | | .26 | .42 | | .68 | .94 |
| | 170 | Water resistant, no finish included | | .024 | | .27 | .29 | | .56 | .76 |
| | 175 | Taped and finished | | .034 | | .29 | .42 | | .71 | .98 |
| | 200 | 5/8" thick, on walls, standard, no finish included | | .009 | | .23 | .12 | | .35 | .44 |
| | 205 | Taped and finished | | .019 | | .25 | .23 | | .48 | .64 |
| | 210 | Fire resistant, no finish included | | .009 | | .26 | .12 | | .38 | .47 |
| | 215 | Taped and finished | | .019 | | .28 | .23 | | .51 | .68 |
| | 220 | Water resistant, no finish included | | .009 | | .30 | .12 | | .42 | .51 |
| | 225 | Taped and finished | | .019 | | .32 | .23 | | .55 | .72 |
| | 230 | Prefinished, vinyl, clipped to studs | ↓ | .015 | ↓ | .67 | .19 | | .86 | 1.04 |
| | 235 | | | | | | | | | |
| | 300 | On ceilings, standard, no finish included | 2 Carp | .015 | S.F. | .23 | .18 | | .41 | .54 |
| | 305 | Taped and finished | | .021 | | .25 | .26 | | .51 | .69 |
| | 310 | Fire resistant, no finish included | | .015 | | .26 | .18 | | .44 | .57 |
| | 315 | Taped and finished | | .021 | | .28 | .26 | | .54 | .73 |
| | 320 | Water resistant, no finish included | | .015 | | .30 | .18 | | .48 | .61 |
| | 325 | Taped and finished | | .021 | | .32 | .26 | | .58 | .77 |
| | 350 | On beams, columns, or soffits, standard, no finish included | | .025 | | .23 | .30 | | .53 | .73 |
| | 355 | Taped and finished | | .036 | | .25 | .44 | | .69 | .97 |
| | 360 | Fire resistant, no finish included | | .025 | | .26 | .30 | | .56 | .77 |
| | 365 | Taped and finished | | .036 | | .28 | .44 | | .72 | 1 |
| | 370 | Water resistant, no finish included | | .025 | | .30 | .30 | | .60 | .81 |
| | 375 | Taped and finished | | .036 | | .32 | .44 | | .76 | 1.05 |
| | 400 | Fireproofing, beams or columns, 2 layers, 1/2" thick | | .048 | | .38 | .60 | | .98 | 1.37 |
| | 405 | 5/8" thick | | .053 | | .46 | .66 | | 1.12 | 1.55 |
| | 410 | 3 layers, 1/2" thick | | .071 | | .57 | .88 | | 1.45 | 2.02 |
| | 415 | 5/8" thick | | .076 | | .69 | .94 | | 1.63 | 2.25 |
| | 460 | Blueboard, 1/2" thick, standard | | .009 | | .23 | .11 | | .34 | .43 |
| | 465 | Fireproof | | .009 | | .25 | .11 | | .36 | .45 |
| | 470 | 5/8" thick, fireproof | ↓ | .009 | | .26 | .12 | | .38 | .47 |
| | 515 | For thin coat plaster instead of taping, add | J-1 | .013 | ↓ | .07 | .16 | .01 | .24 | .33 |
| | 517 | Thin coat plaster, in 50 lb. bags | | | Bag | 7.10 | | | 7.10 | 7.80M |
| | 520 | For high ceilings, over 8' high, add | 2 Carp | .005 | S.F. | .10 | .06 | | .16 | .21 |
| | 525 | For prime coat (residential construction), add | | .007 | | .05 | .08 | | .13 | .19 |
| | 527 | For textured spray, add | | .011 | | .09 | .14 | | .23 | .31 |
| | 530 | For over 3 stories high, add per story | | .003 | ↓ | .05 | .03 | | .08 | .11 |
| | 535 | For finishing corners, inside or outside, add | ↓ | .015 | L.F. | .04 | .18 | | .22 | .33 |
| | 550 | For acoustical sealant, add per bead | 1 Carp | .016 | " | .02 | .20 | | .22 | .33 |
| | 555 | Sealant, 1 quart tube | | | Ea. | 5.90 | | | 5.90 | 6.50M |
| | 560 | Sound deadening board, 1/4" gypsum | 2 Carp | .009 | S.F. | .16 | .11 | | .27 | .35 |
| | 565 | 1/2" wood fiber | " | .009 | " | .19 | .11 | | .30 | .38 |

354  For expanded coverage of these items see *Building Construction Cost Data 1985*

# 9.3 Tile & Terrazzo

| | | CREW | MAN-HOURS | UNIT | MAT. | LABOR | EQUIP. | TOTAL | TOTAL INCL O&P |
|---|---|---|---|---|---|---|---|---|---|
| 05-001 | **CERAMIC TILE** Base | | | | | | | | |
| 060 | Cove base, 4-1/4" x 4-1/4" high, mud set | D-7 | .176 | L.F. | 1.53 | 1.93 | | 3.46 | 4.65 |
| 070 | Thin set | | .125 | | 1.42 | 1.37 | | 2.79 | 3.67 |
| 090 | 6" x 4-1/4" high, mud set | | .160 | | 1.52 | 1.76 | | 3.28 | 4.38 |
| 100 | Thin set | | .117 | | 1.37 | 1.28 | | 2.65 | 3.48 |
| 120 | Sanitary cove base, 6" x 4-1/4" high, mud set | | .172 | | 1.66 | 1.89 | | 3.55 | 4.73 |
| 130 | Thin set | | .129 | | 1.61 | 1.42 | | 3.03 | 3.95 |
| 150 | 6" x 6" high, mud set | | .190 | | 1.75 | 2.09 | | 3.84 | 5.15 |
| 160 | Thin set | | .137 | | 1.65 | 1.50 | | 3.15 | 4.13 |
| 240 | Bullnose trim, 4-1/4" x 4-1/4", mud set | | .195 | | 1.59 | 2.14 | | 3.73 | 5.05 |
| 250 | Thin set | | .125 | | 1.48 | 1.37 | | 2.85 | 3.74 |
| 270 | 6" x 4-1/4" bullnose trim, mud set | | .190 | | 1.62 | 2.09 | | 3.71 | 5 |
| 280 | Thin set | | .129 | ↓ | 1.55 | 1.42 | | 2.97 | 3.89 |
| 300 | Floors, natural clay, random or uniform, thin set, color group 1 | | .087 | S.F. | 1.67 | .96 | | 2.63 | 3.31 |
| 310 | Color group 2 | | .087 | | 1.79 | .96 | | 2.75 | 3.45 |
| 330 | Porcelain type, 1 color, color group 2, 1" x 1" | | .087 | | 1.88 | .96 | | 2.84 | 3.55 |
| 340 | 2" x 2" or 2" x 1", thin set | ↓ | .087 | | 2.09 | .96 | | 3.05 | 3.78 |
| 360 | For random blend, 2 colors, add | | | | .10 | | | .10 | .11M |
| 370 | 4 colors, add | | | | .20 | | | .20 | .22M |
| 430 | Specialty tile, 3" x 6" x 1/2", decorator finish | D-7 | .087 | ↓ | 4.17 | .96 | | 5.13 | 6.05 |
| 431 | | | | | | | | | |
| 450 | Add for epoxy grout, 1/16" joint, 1" x 1" tile | D-7 | .017 | S.F. | .61 | .18 | | .79 | .95 |
| 460 | 2" x 2" tile | " | .017 | " | .47 | .18 | | .65 | .80 |
| 480 | Pregrouted sheets, walls, 4-1/4" x 4-1/4", 6" x 4-1/4" | | | | | | | | |
| 481 | and 8-1/2" x 4-1/4", 4 S.F. sheets, silicone grout | D-7 | .067 | S.F. | 2.01 | .73 | | 2.74 | 3.34 |
| 510 | Floors, unglazed, 2 S.F. sheets, | | | | | | | | |
| 511 | urethane adhesive | D-7 | .089 | S.F. | 3.44 | .98 | | 4.42 | 5.30 |
| 540 | Walls, interior, thin set, 4-1/4" x 4-1/4" tile | | .089 | | 1.53 | .98 | | 2.51 | 3.19 |
| 550 | 6" x 4-1/4" tile | | .084 | | 1.65 | .92 | | 2.57 | 3.24 |
| 570 | 8-1/2" x 4-1/4" tile | | .084 | | 2 | .92 | | 2.92 | 3.62 |
| 580 | 6" x 6" tile | | .080 | ↓ | 1.73 | .88 | | 2.61 | 3.26 |
| 600 | Decorated wall tile, 4-1/4" x 4-1/4", minimum | | .018 | Ea. | .86 | .20 | | 1.06 | 1.26 |
| 610 | Maximum | | .028 | " | 11.10 | .30 | | 11.40 | 12.70 |
| 660 | Crystalline glazed, 4-1/4" x 4-1/4", mud set, plain | | .160 | S.F. | 1.85 | 1.76 | | 3.61 | 4.74 |
| 670 | 4-1/4" x 4-1/4", scored tile | | .160 | | 2.05 | 1.76 | | 3.81 | 4.96 |
| 690 | 1-3/8" squares | | .172 | | 4.34 | 1.89 | | 6.23 | 7.70 |
| 700 | For epoxy grout, 1/16" joints, 4-1/4" tile, add | | .017 | | .40 | .18 | | .58 | .72 |
| 720 | For tile set in dry mortar, add | | .009 | | | .10 | | .10 | .16L |
| 730 | For tile set in portland cement mortar, add | ↓ | .055 | ↓ | | .61 | | .61 | .93L |
| 20-001 | **MARBLE** Thin gauge tile, 12" x 6", 9/32", White Carara | D-7 | .250 | S.F. | 4.23 | 2.74 | | 6.97 | 8.90 |
| 010 | Filled Travertine | | .250 | | 4.75 | 2.74 | | 7.49 | 9.45 |
| 020 | Synthetic tiles, 12" x 12" x 5/8", thin set, floors | | .250 | | 5.55 | 2.74 | | 8.29 | 10.35 |
| 030 | On walls (see also division 4.4-30) | ↓ | .291 | ↓ | 5.55 | 3.19 | | 8.74 | 11 |
| 25-001 | **METAL TILE** Cove base, standard colors, 4-1/4" square | 1 Carp | .053 | L.F. | 1 | .66 | | 1.66 | 2.14 |
| 020 | 4-1/8" x 8-1/2" | | .040 | " | 1 | .49 | | 1.49 | 1.88 |
| 040 | Walls, aluminum, 4-1/4" square, thin set, plain | | .100 | S.F. | 1.69 | 1.24 | | 2.93 | 3.82 |
| 050 | Epoxy enameled | | .100 | | 1.96 | 1.24 | | 3.20 | 4.11 |
| 070 | Leather on aluminum, colors | | .123 | | 16.90 | 1.52 | | 18.42 | 21 |
| 080 | Stainless steel | | .100 | | 4.41 | 1.24 | | 5.65 | 6.80 |
| 100 | Suede on aluminum | ↓ | .123 | | 16.90 | 1.52 | | 18.42 | 21 |
| 110 | For sizes other than 4-1/4" x 4-1/4", add | | | | .13 | | | .13 | .14M |
| 30-001 | **PLASTIC TILE** Walls, 4-1/4" x 4-1/4", .050" thick | 1 Carp | .064 | | .95 | .79 | | 1.74 | 2.30 |
| 010 | .110" thick | " | .064 | ↓ | 1.16 | .79 | | 1.95 | 2.53 |
| 35-001 | **QUARRY TILE** Base, cove or sanitary, 2" or 5" high, mud set | | | | | | | | |
| 010 | 1/2" thick | D-7 | .145 | L.F. | 1.74 | 1.60 | | 3.34 | 4.37 |
| 030 | Bullnose trim, red, mud set, 6" x 6" x 1/2" thick | | .133 | | 1.68 | 1.46 | | 3.14 | 4.10 |
| 040 | 4" x 4" x 1/2" thick | | .145 | | 1.43 | 1.60 | | 3.03 | 4.03 |
| 060 | 4" x 8" x 1/2" thick, using 8" as edge | ↓ | .123 | ↓ | 1.87 | 1.35 | | 3.22 | 4.14 |
| 061 | | | | | | | | | |

For expanded coverage of these items see *Building Construction Cost Data 1985*

## 9.3 Tile & Terrazzo

| | | | CREW | MAN-HOURS | UNIT | MAT. | LABOR | EQUIP. | TOTAL | TOTAL INCL O&P |
|---|---|---|---|---|---|---|---|---|---|---|
| 35 | 070 | Floors, mud set, 1000 S.F. lots, red, 4" x 4" x 1/2" thick | D-7 | .133 | S.F. | 1.96 | 1.46 | | 3.42 | 4.41 |
| | 090 | 6" x 6" x 1/2" thick | | .114 | | 1.72 | 1.25 | | 2.97 | 3.82 |
| | 100 | 4" x 8" x 1/2" thick | ↓ | .123 | | 1.80 | 1.35 | | 3.15 | 4.06 |
| | 130 | For waxed coating, add | | | | .27 | | | .27 | .29M |
| | 150 | For colors other than green, add | | | | .21 | | | .21 | .23M |
| | 160 | For abrasive surface, add | | | | .27 | | | .27 | .29M |
| | 180 | Brown tile, imported, 6" x 6" x 7/8" | D-7 | .133 | | 3.81 | 1.46 | | 5.27 | 6.45 |
| | 190 | 9" x 9" x 1-1/4" | | .145 | | 5.20 | 1.60 | | 6.80 | 8.20 |
| | 210 | For thin set mortar application, deduct | ↓ | .023 | ↓ | | .25 | | .25 | .39L |
| | 250 | | | | | | | | | |
| | 270 | Stair tread & riser, 6" x 6" x 3/4", plain | D-7 | .320 | S.F. | 4.19 | 3.51 | | 7.70 | 10 |
| | 280 | Abrasive | | .320 | | 4.64 | 3.51 | | 8.15 | 10.50 |
| | 300 | Wainscot, 6" x 6" x 1/2", thin set, red | | .152 | | 1.43 | 1.67 | | 3.10 | 4.15 |
| | 310 | Colors other than green | | .152 | ↓ | 1.48 | 1.67 | | 3.15 | 4.20 |
| | 330 | Window sill, 6" wide, 3/4" thick | | .178 | L.F. | 2.95 | 1.95 | | 4.90 | 6.25 |
| | 340 | Corners | | .178 | Ea. | 2.20 | 1.95 | | 4.15 | 5.40 |
| 37 | 001 | SLATE TILE Vermont, 6" x 6" x 1/4" thick, thin set | ↓ | .089 | S.F. | 1.68 | .98 | | 2.66 | 3.35 |
| | 002 | | | | | | | | | |
| 40 | 001 | TERRAZZO, CAST IN PLACE Cove base, 6" high | 1 Mstz | .348 | L.F. | .92 | 4.24 | | 5.16 | 7.55 |
| | 010 | Curb, 6" high and 6" wide | " | .533 | | 1.64 | 6.50 | | 8.14 | 11.80 |
| | 030 | Divider strip for floors, 12 ga., 1-1/4" deep, zinc | | | | .71 | | | .71 | .78M |
| | 040 | Brass | | | | 1.93 | | | 1.93 | 2.12M |
| | 060 | Solid 1/4" thick, 1-1/4" deep, zinc | | | | 2.46 | | | 2.46 | 2.70M |
| | 120 | For thin set floors, 16 ga., 1/2" x 1/4", zinc | | | ↓ | .37 | | | .37 | .40M |
| | 150 | Floor, bonded to concrete, 1-3/4" thick, gray cement | J-3 | .123 | S.F. | 1.72 | 1.37 | .87 | 3.96 | 4.96 |
| | 160 | White cement | | .123 | | 2.41 | 1.37 | .87 | 4.65 | 5.70 |
| | 180 | Not bonded, 3" total thickness, gray cement | | .154 | | 2.11 | 1.71 | 1.09 | 4.91 | 6.15 |
| | 190 | White cement | ↓ | .154 | ↓ | 2.80 | 1.71 | 1.09 | 5.60 | 6.90 |
| 45 | 001 | TERRAZZO, PRECAST Base, 6" high, straight | 1 Mstz | .067 | L.F. | 5.20 | .81 | | 6.01 | 6.95 |
| | 010 | Cove | | .067 | | 5.40 | .81 | | 6.21 | 7.20 |
| | 030 | 8" high base, straight | | .073 | | 5.60 | .89 | | 6.49 | 7.55 |
| | 040 | Cove | ↓ | .073 | | 5.60 | .89 | | 6.49 | 7.55 |
| | 060 | For white cement, add | | | | .16 | | | .16 | .17M |
| | 070 | For 16 ga. zinc toe strip, add | | | | .42 | | | .42 | .46M |
| | 090 | Curbs, 4" x 4" high | 1 Mstz | .145 | | 10.70 | 1.77 | | 12.47 | 14.50 |
| | 100 | 8" x 8" high | " | .178 | ↓ | 12.70 | 2.17 | | 14.87 | 17.30 |
| | 120 | Floor tiles, non-slip, 1" thick, 12" x 12" | D-1 | .267 | S.F. | 5.55 | 3.02 | | 8.57 | 10.80 |
| | 130 | 1-1/4" thick, 12" x 12" | | .267 | | 6.55 | 3.02 | | 9.57 | 11.90 |
| | 150 | 16" x 16" | | .291 | | 6.15 | 3.29 | | 9.44 | 11.90 |
| | 160 | 1-1/2" thick, 16" x 16" | ↓ | .320 | | 7 | 3.62 | | 10.62 | 13.35 |
| | 480 | Wainscot, 12" x 12" x 1" tiles | 1 Mstz | .229 | ↓ | 5.55 | 2.79 | | 8.34 | 10.40 |
| | 490 | 16" x 16" x 1-1/2" tiles | " | .267 | ↓ | 6.80 | 3.25 | | 10.05 | 12.50 |

## 9.5 Acoustical Treatment

| | | | CREW | MAN-HOURS | UNIT | MAT. | LABOR | EQUIP. | TOTAL | TOTAL INCL O&P |
|---|---|---|---|---|---|---|---|---|---|---|
| 10 | 001 | CEILING TILE Stapled, cemented or installed on suspension | | | | | | | | |
| | 010 | system, 12" x 12" or 12" x 24", not including furring | | | | | | | | |
| | 060 | Mineral fiber, plastic coated, 5/8" thick | 1 Carp | .020 | S.F. | .62 | .25 | | .87 | 1.07 |
| | 070 | 3/4" thick | | .020 | | .80 | .25 | | 1.05 | 1.27 |
| | 090 | Fire rated, 3/4" thick, plain faced | | .020 | | .89 | .25 | | 1.14 | 1.37 |
| | 100 | Plastic coated face | | .020 | | .96 | .25 | | 1.21 | 1.45 |
| | 120 | Aluminum faced, 5/8" thick, plain | ↓ | .020 | ↓ | 1.98 | .25 | | 2.23 | 2.57 |
| | 121 | | | | | | | | | |
| | 330 | For flameproofing, add | | | S.F. | .06 | | | .06 | .06M |
| | 340 | For sculptured 3 dimensional, add | | | " | .18 | | | .18 | .19M |

For expanded coverage of these items see *Building Construction Cost Data 1985*

# 9.5 Acoustical Treatment

| | | | MAN- | | | BARE COSTS | | | TOTAL |
|---|---|---|---|---|---|---|---|---|---|
| | | CREW | HOURS | UNIT | MAT. | LABOR | EQUIP. | TOTAL | INCL O&P |
| 10 | 390 | For ceiling primer, add | | | S.F. | .08 | | | .08 | .08M |
| | 400 | For ceiling cement, add | | | " | .23 | | | .23 | .25M |
| 15-001 | | **SUSPENDED ACOUSTIC CEILING BOARDS** Not including | | | | | | | | |
| | 010 | suspension system | | | | | | | | |
| | 030 | Fiberglass boards, film faced, 2' x 2' or 2' x 4', 5/8" thick | 1 Carp | .012 | S.F. | .31 | .15 | | .46 | .57 |
| | 040 | 3/4" thick | | .016 | | .41 | .20 | | .61 | .76 |
| | 050 | 3" thick, thermal, R11 | | .016 | | .74 | .20 | | .94 | 1.13 |
| | 060 | Glass cloth faced fiberglass, 3/4" thick | | .016 | | .75 | .20 | | .95 | 1.14 |
| | 070 | 1" thick | | .016 | | .85 | .20 | | 1.05 | 1.25 |
| | 082 | 1-1/2" thick, nubby face | | .016 | | 1.17 | .20 | | 1.37 | 1.60 |
| | 090 | Mineral fiber boards, 5/8" thick, aluminum faced, 24" x 24" | | .013 | | .68 | .16 | | .84 | 1.01 |
| | 093 | 24" x 48" | | .012 | | .68 | .15 | | .83 | .99 |
| | 096 | Standard face | | .012 | | .33 | .15 | | .48 | .59 |
| | 100 | Plastic coated face | | .020 | | .36 | .25 | | .61 | .79 |
| | 120 | Mineral fiber, 2 hour rating, 5/8" thick | | .012 | | .40 | .15 | | .55 | .67 |
| | 130 | Mirror faced panels, 15/16" thick | ↓ | .016 | ↓ | 4.30 | .20 | | 4.50 | 5.05 |
| | 150 | Air distributing ceilings, fire rated | | | | | | | | |
| | 160 | 5/8" thick, water felted board | 1 Carp | .020 | S.F. | .47 | .25 | | .72 | .91 |
| | 190 | Eggcrate, acrylic, 1/2" x 1/2" x 1/2" cubes | | .016 | | 1.80 | .20 | | 2 | 2.29 |
| | 210 | Polystyrene eggcrate, 3/8" x 3/8" x 1/2" cubes | | .016 | | .88 | .20 | | 1.08 | 1.28 |
| | 220 | 1/2" x 1/2" x 1/2" cubes | | .016 | | .91 | .20 | | 1.11 | 1.31 |
| | 240 | Luminous panels, prismatic, acrylic | | .020 | | .73 | .25 | | .98 | 1.19 |
| | 250 | Polystyrene | | .020 | | .42 | .25 | | .67 | .85 |
| | 270 | Flat or ribbed, acrylic | | .020 | | 1.17 | .25 | | 1.42 | 1.68 |
| | 280 | Polystyrene | | .020 | | .68 | .25 | | .93 | 1.14 |
| | 300 | Drop pan, white, acrylic | | .020 | | 2.78 | .25 | | 3.03 | 3.45 |
| | 310 | Polystyrene | | .020 | | 2.25 | .25 | | 2.50 | 2.87 |
| | 360 | Perforated aluminum sheets, .024" thick, corrugated, painted | | .016 | | 1.30 | .20 | | 1.50 | 1.74 |
| | 370 | Plain | | .016 | | .92 | .20 | | 1.12 | 1.32 |
| | 375 | Wood fiber in cementitious binder, 2' x 2' or 4', painted, 1" thick | | .013 | | 1.20 | .16 | | 1.36 | 1.58 |
| | 376 | 2" thick | | .015 | | 2.06 | .18 | | 2.24 | 2.55 |
| | 377 | 2-1/2" thick | | .016 | | 2.36 | .20 | | 2.56 | 2.91 |
| | 378 | 3" thick | ↓ | .018 | ↓ | 2.62 | .22 | | 2.84 | 3.23 |
| 20-001 | | **SUSPENSION SYSTEMS** For boards and tile listed above | | | | | | | | |
| | 005 | Class A suspension system, T bar, 2' x 4' grid | 1 Carp | .010 | S.F. | .36 | .12 | | .48 | .59 |
| | 030 | 2' x 2' grid | | .012 | | .38 | .15 | | .53 | .66 |
| | 040 | Concealed Z bar suspension system, 12" module | | .015 | | .46 | .19 | | .65 | .81 |
| | 060 | 1-1/2" carrier channels, 4' O.C., add | | .017 | | .11 | .21 | | .32 | .45 |
| | 065 | 1-1/2" x 3-1/2" channels | ↓ | .017 | ↓ | .33 | .21 | | .54 | .70 |
| | 070 | Carrier channels for ceilings with | | | | | | | | |
| | 090 | recessed lighting fixtures, add | 1 Carp | .017 | S.F. | .25 | .21 | | .46 | .62 |
| | 500 | Wire hangers, #12 wire | " | .027 | Ea. | .25 | .33 | | .58 | .80 |
| 25-001 | | **SUSPENDED CEILINGS, COMPLETE** Including standard | | | | | | | | |
| | 010 | suspension system but not incl. 1-1/2" carrier channels | | | | | | | | |
| | 060 | Ceiling board system, 2' x 4', plain faced, supermarkets | 1 Carp | .016 | S.F. | .66 | .20 | | .86 | 1.04 |
| | 070 | Offices | | .021 | | .71 | .26 | | .97 | 1.19 |
| | 180 | Tile, Z bar suspension, 5/8" mineral fiber tile | | .034 | | 1.09 | .42 | | 1.51 | 1.86 |
| | 190 | 3/4" mineral fiber tile | ↓ | .034 | ↓ | 1.20 | .42 | | 1.62 | 1.99 |
| 30-001 | | **SOUND ABSORBING PANELS** Perforated steel facing, painted with | | | | | | | | |
| | 010 | fiberglass or mineral filler, no backs, 2-1/4" thick, modular | | | | | | | | |
| | 020 | space units, ceiling or wall hung, white or colored | 1 Carp | .040 | S.F. | 4.14 | .49 | | 4.63 | 5.35 |
| | 030 | Fiberboard sound deadening panels, 1/2" thick | " | .013 | " | .13 | .16 | | .29 | .40 |
| | 050 | Fiberglass panels, 4' x 8' x 1" thick, with | | | | | | | | |
| | 060 | glass cloth face for walls, cemented | 1 Carp | .052 | S.F. | 1.36 | .64 | | 2 | 2.51 |
| | 070 | 1-1/2" thick, dacron covered, inner aluminum frame, | | | | | | | | |
| | 071 | wall mounted | 1 Carp | .013 | S.F. | 6.15 | .16 | | 6.31 | 7.05 |
| 40-001 | | **SOUND ATTENUATION** Blanket, 1" thick | ↓ | .009 | ↓ | .27 | .11 | | .38 | .47 |
| | 050 | 1-1/2" thick | | .009 | | .33 | .11 | | .44 | .53 |

For expanded coverage of these items see *Building Construction Cost Data 1985*

## 9.5 Acoustical Treatment

| | | CREW | MAN-HOURS | UNIT | MAT. | LABOR | EQUIP. | TOTAL | TOTAL INCL O&P |
|---|---|---|---|---|---|---|---|---|---|
| 40 100 | 2" thick | 1 Carp | .009 | S.F. | .52 | .11 | | .63 | .74 |
| 150 | 3" thick | " | .009 | " | .68 | .11 | | .79 | .92 |

## 9.6 Flooring

| | | CREW | MAN-HOURS | UNIT | MAT. | LABOR | EQUIP. | TOTAL | TOTAL INCL O&P |
|---|---|---|---|---|---|---|---|---|---|
| 05-001 | **CARPET** Commercial grades, cemented | | | | | | | | |
| 070 | Acrylic, 26 oz., light to medium traffic | 1 Tilf | .028 | S.F. | 1.16 | .34 | | 1.50 | 1.79 |
| 090 | 28 oz., medium traffic | | .028 | | 1.43 | .34 | | 1.77 | 2.09 |
| 110 | 35 oz., medium to heavy traffic | | .028 | | 1.58 | .34 | | 1.92 | 2.25 |
| 210 | Nylon, non anti-static, 15 oz., light traffic | | .028 | | .93 | .34 | | 1.27 | 1.54 |
| 280 | Nylon, with anti-static, 17 oz., light to medium traffic | | .028 | | 1.10 | .34 | | 1.44 | 1.73 |
| 290 | 20 oz., medium traffic | | .028 | | 1.17 | .34 | | 1.51 | 1.80 |
| 300 | 22 oz., medium traffic | | .028 | | 1.22 | .34 | | 1.56 | 1.86 |
| 310 | 24 oz., medium to heavy traffic | | .028 | | 1.32 | .34 | | 1.66 | 1.97 |
| 320 | 26 oz., medium to heavy traffic | | .028 | | 1.41 | .34 | | 1.75 | 2.07 |
| 330 | 28 oz., heavy traffic | | .028 | | 1.51 | .34 | | 1.85 | 2.18 |
| 340 | Needle bonded, 20 oz., no padding | | .017 | | .94 | .21 | | 1.15 | 1.35 |
| 350 | Polypropylene, 15 oz., light traffic | | .017 | | .60 | .21 | | .81 | .98 |
| 365 | 22 oz., medium traffic | | .022 | | .84 | .27 | | 1.11 | 1.34 |
| 380 | Scrim installed, nylon sponge back carpet, 20 oz. | | .028 | | 2.01 | .34 | | 2.35 | 2.73 |
| 385 | 60 oz. | | .032 | | 3.44 | .39 | | 3.83 | 4.38 |
| 400 | Tile, foam-backed, needle punch | | .017 | | .73 | .21 | | .94 | 1.13 |
| 410 | Tufted loop or shag | | .017 | | 1.30 | .21 | | 1.51 | 1.76 |
| 450 | Wool, 36 oz., medium to heavy traffic | | .028 | | 2.72 | .34 | | 3.06 | 3.51 |
| 470 | Sponge back, wool, 36 oz., medium to heavy traffic | | .017 | | 2.68 | .21 | | 2.89 | 3.27 |
| 490 | 42 oz., heavy traffic | ▼ | .028 | | 3 | .34 | | 3.34 | 3.82 |
| 550 | For stretched and edge fastened, deduct | | | ▼ | .05 | | | .05 | .05M |
| 560 | For bound carpet baseboard, add | 1 Tilf | .027 | L.F. | .50 | .32 | | .82 | 1.05 |
| 561 | For stairs, not incl. price of carpet, add | | .267 | Riser | | 3.24 | | 3.24 | 4.98L |
| 900 | Padding, sponge rubber cushion, minimum | | .011 | S.F. | .15 | .14 | | .29 | .37 |
| 910 | Maximum | | .011 | | .37 | .14 | | .51 | .61 |
| 920 | Felt, 32 oz. to 56 oz., minimum | | .011 | | .18 | .14 | | .32 | .41 |
| 930 | Maximum | | .012 | | .42 | .14 | | .56 | .68 |
| 940 | Bonded urethane, 3/8" thick, minimum | | .010 | | .15 | .13 | | .28 | .36 |
| 950 | Maximum | | .012 | | .33 | .14 | | .47 | .58 |
| 960 | Prime urethane, 1/4" thick, minimum | | .010 | | .15 | .13 | | .28 | .36 |
| 970 | Maximum | ▼ | .012 | ▼ | .52 | .14 | | .66 | .79 |
| 10-001 | **COMPOSITION FLOORING** Acrylic, 1/4" thick | | | | | | | | |
| 060 | Epoxy, with colored quartz chips, broadcast, minimum | C-6 | .071 | S.F. | 1.04 | .76 | .08 | 1.88 | 2.42 |
| 070 | Maximum | | .098 | | 1.62 | 1.04 | .11 | 2.77 | 3.55 |
| 090 | Trowelled, minimum | | .086 | | 1.31 | .92 | .09 | 2.32 | 2.98 |
| 100 | Maximum | ▼ | .100 | ▼ | 2.09 | 1.06 | .11 | 3.26 | 4.10 |
| 120 | Heavy duty epoxy topping, 1/4" thick, | | | | | | | | |
| 130 | 500 to 1,000 S.F. | C-6 | .114 | S.F. | 2.40 | 1.21 | .13 | 3.74 | 4.70 |
| 150 | 1,000 to 2,000 S.F. | | .107 | | 2.10 | 1.13 | .12 | 3.35 | 4.23 |
| 160 | Over 10,000 S.F. | | .100 | | 2.10 | 1.06 | .11 | 3.27 | 4.11 |
| 180 | Epoxy terrazzo, 1/4" thick, chemical resistant, minimum | | .128 | | 3.46 | 1.36 | .14 | 4.96 | 6.10 |
| 190 | Maximum | ▼ | .171 | ▼ | 5.75 | 1.82 | .19 | 7.76 | 9.40 |
| 17-001 | **MASONRY FLOORS** See division 4.2-28 | | | | | | | | |
| 20-001 | **RESILIENT** Asphalt tile, on concrete, 1/8" thick | | | | | | | | |
| 005 | Color group B | 1 Tilf | .015 | S.F. | .62 | .18 | | .80 | .96 |
| 010 | Color group C & D | " | .015 | | .67 | .18 | | .85 | 1.01 |
| 030 | For wood subfloor, add to above for felt underlayment | | | ▼ | .07 | | | .07 | .07M |

For expanded coverage of these items see *Building Construction Cost Data 1985*

## 9.6 Flooring

| | | Description | CREW | MAN-HOURS | UNIT | MAT. | LABOR | EQUIP. | TOTAL | TOTAL INCL O&P |
|---|---|---|---|---|---|---|---|---|---|---|
| 20 | 080 | Base, cove, rubber or vinyl, .080" thick | | | | | | | | |
| | 110 | Standard colors, 2-1/2" high | 1 Tilf | .027 | L.F. | .36 | .32 | | .68 | .89 |
| | 115 | 4" high | | .027 | | .47 | .32 | | .79 | 1.02 |
| | 120 | 6" high | | .027 | | .57 | .32 | | .89 | 1.13 |
| | 145 | 1/8" thick, standard colors, 2-1/2" high | | .027 | | .36 | .32 | | .68 | .89 |
| | 150 | 4" high | | .027 | | .52 | .32 | | .84 | 1.07 |
| | 155 | 6" high | | .027 | | .57 | .32 | | .89 | 1.13 |
| | 160 | Corners, 2-1/2" high | | .027 | Ea. | .57 | .32 | | .89 | 1.13 |
| | 163 | 4" high | | .027 | | .60 | .32 | | .92 | 1.16 |
| | 166 | 6" high | | .027 | | .73 | .32 | | 1.05 | 1.30 |
| | 170 | Conductive flooring, rubber tile, 1/8" thick | | .025 | S.F. | 2.72 | .30 | | 3.02 | 3.45 |
| | 180 | Homogeneous vinyl tile, 1/8" thick | | .025 | | 3.66 | .30 | | 3.96 | 4.49 |
| | 220 | Cork tile, standard finish, 1/8" thick | | .025 | | 1.15 | .30 | | 1.45 | 1.73 |
| | 225 | 3/16" thick | | .025 | | 1.31 | .30 | | 1.61 | 1.90 |
| | 230 | 5/16" thick | | .025 | | 1.46 | .30 | | 1.76 | 2.07 |
| | 235 | 1/2" thick | | .025 | | 1.83 | .30 | | 2.13 | 2.47 |
| | 250 | Urethane finish, 1/8" thick | | .025 | | 1.41 | .30 | | 1.71 | 2.01 |
| | 255 | 3/16" thick | | .025 | | 1.62 | .30 | | 1.92 | 2.24 |
| | 260 | 5/16" thick | | .025 | | 1.83 | .30 | | 2.13 | 2.47 |
| | 265 | 1/2" thick | | .025 | | 2.20 | .30 | | 2.50 | 2.88 |
| | 370 | Polyethylene, in rolls, no base incl., landscape surfaces | | .029 | | 1.56 | .35 | | 1.91 | 2.26 |
| | 380 | Nylon action surface, 1/8" thick | | .029 | | 1.74 | .35 | | 2.09 | 2.46 |
| | 390 | 1/4" thick | | .029 | | 2.50 | .35 | | 2.85 | 3.29 |
| | 400 | 3/8" thick | | .029 | | 3.44 | .35 | | 3.79 | 4.33 |
| | 590 | Rubber, sheet goods, 36" wide, 1/8" thick | | .034 | | 2.10 | .41 | | 2.51 | 2.95 |
| | 595 | 3/16" thick | | .036 | | 2.31 | .44 | | 2.75 | 3.22 |
| | 600 | 1/4" thick | | .040 | | 2.67 | .49 | | 3.16 | 3.68 |
| | 605 | Tile, marbleized colors, 12" x 12", 1/8" thick | | .016 | | 1.46 | .20 | | 1.66 | 1.91 |
| | 610 | 3/16" thick | | .016 | | 1.99 | .20 | | 2.19 | 2.50 |
| | 615 | 1/4" thick | | .016 | | 2.15 | .20 | | 2.35 | 2.67 |
| | 630 | Special tile, plain colors, 1/8" thick | | .016 | | 2.20 | .20 | | 2.40 | 2.73 |
| | 635 | 3/16" thick | | .016 | | 2.93 | .20 | | 3.13 | 3.53 |
| | 640 | 1/4" thick | | .016 | | 3.95 | .20 | | 4.15 | 4.65 |
| | 650 | | | | | | | | | |
| | 700 | Vinyl composition tile, 12" x 12", 1/16" thick | 1 Tilf | .015 | S.F. | .53 | .18 | | .71 | .86 |
| | 705 | Embossed | | .015 | | .67 | .18 | | .85 | 1.01 |
| | 710 | Marbleized | | .015 | | .67 | .18 | | .85 | 1.01 |
| | 715 | Plain | | .015 | | .73 | .18 | | .91 | 1.08 |
| | 720 | 1/32" thick, embossed | | .015 | | .78 | .18 | | .96 | 1.13 |
| | 725 | Marbleized | | .015 | | .84 | .18 | | 1.02 | 1.20 |
| | 730 | Plain | | .015 | | .99 | .18 | | 1.17 | 1.37 |
| | 731 | | | | | | | | | |
| | 735 | 1/8" thick, marbleized | 1 Tilf | .015 | S.F. | .85 | .18 | | 1.03 | 1.21 |
| | 740 | Plain | | .015 | | 1.05 | .18 | | 1.23 | 1.43 |
| | 750 | Vinyl tile, 12" x 12", .050" thick, minimum | | .019 | | 1.23 | .23 | | 1.46 | 1.70 |
| | 755 | Maximum | | .023 | | 4.18 | .28 | | 4.46 | 5.05 |
| | 760 | 1/8" thick, minimum | | .019 | | 1.81 | .23 | | 2.04 | 2.34 |
| | 765 | Solid colors | | .023 | | 2.20 | .28 | | 2.48 | 2.85 |
| | 770 | Marbleized or Travertine pattern | | .021 | | 2.98 | .25 | | 3.23 | 3.66 |
| | 775 | Florentine pattern | | .021 | | 4.02 | .25 | | 4.27 | 4.81 |
| | 780 | Maximum | | .021 | | 7.50 | .25 | | 7.75 | 8.65 |
| | 781 | | | | | | | | | |
| | 800 | Vinyl sheet goods, backed, .070" thick, minimum | 1 Tilf | .012 | S.F. | .84 | .15 | | .99 | 1.15 |
| | 805 | Maximum | | .025 | | 1.97 | .30 | | 2.27 | 2.63 |
| | 810 | .093" thick, minimum | | .012 | | 1.10 | .15 | | 1.25 | 1.44 |
| | 815 | Maximum | | .025 | | 2.46 | .30 | | 2.76 | 3.17 |
| | 820 | .125" thick, minimum | | .012 | | 1.31 | .15 | | 1.46 | 1.67 |
| | 825 | Maximum | | .025 | | 3.15 | .30 | | 3.45 | 3.93 |
| | 830 | .250" thick, minimum | | .012 | | 1.78 | .15 | | 1.93 | 2.19 |
| | 835 | Maximum | | .025 | | 4.56 | .30 | | 4.86 | 5.50 |

For expanded coverage of these items see *Building Construction Cost Data 1985*

## 9.6 Flooring

| | | | CREW | MAN-HOURS | UNIT | MAT. | LABOR | EQUIP. | TOTAL | TOTAL INCL O&P |
|---|---|---|---|---|---|---|---|---|---|---|
| 20 | 870 | Adhesive cement, 1 gallon does 100 to 300 S.F. | | | Gal. | 12.60 | | | 12.60 | 13.85M |
| | 880 | Asphalt primer, 1 gallon per 300 S.F. | | | | 5.20 | | | 5.20 | 5.70M |
| | 890 | Emulsion, 1 gallon per 140 S.F. | | | | 2.35 | | | 2.35 | 2.58M |
| | 895 | Latex underlayment | | | | 15 | | | 15 | 16.50M |
| 35-001 | | TILE AND TERRAZZO See division 9.3 | | | | | | | | |
| 40-001 | | WOOD Fir, vertical grain, 1" x 4", not incl. finish, B & better | 1 Carp | .031 | S.F. | 1.71 | .39 | | 2.10 | 2.49 |
| | 010 | C grade & better | | .031 | | 1.57 | .39 | | 1.96 | 2.34 |
| | 030 | Flat grain, 1" x 4", not incl. finish, B & better | | .031 | | 1.45 | .39 | | 1.84 | 2.21 |
| | 040 | C & better | | .031 | | 1.35 | .39 | | 1.74 | 2.10 |
| | 460 | Oak, white or red, 25/32" x 2-1/4", not incl. finish | | | | | | | | |
| | 470 | Clear quartered | 1 Carp | .047 | S.F. | 1.63 | .58 | | 2.21 | 2.71 |
| | 490 | Clear/select, 2-1/4" wide | | .047 | | 1.45 | .58 | | 2.03 | 2.52 |
| | 500 | #1 common | | .043 | | 1.15 | .53 | | 1.68 | 2.11 |
| | 520 | Parquetry, standard, 5/16" thick, not incl. finish, oak, minimum | | .050 | | 1 | .62 | | 1.62 | 2.08 |
| | 530 | Maximum | | .080 | | 5.20 | .99 | | 6.19 | 7.30 |
| | 550 | Teak, minimum | | .050 | | 2.44 | .62 | | 3.06 | 3.66 |
| | 560 | Maximum | | .080 | | 5.15 | .99 | | 6.14 | 7.25 |
| | 565 | 13/16" thick, select grade oak, minimum | | .050 | | 8.50 | .62 | | 9.12 | 10.35 |
| | 570 | Maximum | | .080 | | 10.65 | .99 | | 11.64 | 13.30 |
| | 580 | Custom parquetry, including finish, minimum | | .080 | | 10.80 | .99 | | 11.79 | 13.45 |
| | 590 | Maximum | | .160 | | 15 | 1.98 | | 16.98 | 19.65 |
| | 610 | Prefinished white oak, prime grade, 2-1/4" wide | | .047 | | 1.68 | .58 | | 2.26 | 2.77 |
| | 620 | 3-1/4" wide | | .043 | | 1.68 | .53 | | 2.21 | 2.69 |
| | 640 | Ranch plank | | .055 | | 2.87 | .68 | | 3.55 | 4.24 |
| | 650 | Hardwood blocks, 9" x 9", 25/32" thick | | .050 | | 3.06 | .62 | | 3.68 | 4.34 |
| | 670 | Parquetry, 5/16" thick, oak, minimum | | .050 | | 1 | .62 | | 1.62 | 2.08 |
| | 680 | Maximum | | .080 | | 1.78 | .99 | | 2.77 | 3.52 |
| | 700 | Walnut or teak, parquetry, minimum | | .050 | | 3.15 | .62 | | 3.77 | 4.44 |
| | 710 | Maximum | | .080 | | 5.70 | .99 | | 6.69 | 7.85 |
| | 720 | Acrylic wood parquet blocks, 12" x 12" x 5/16", | | | | | | | | |
| | 721 | irradiated, set in epoxy | 1 Carp | .050 | S.F. | 3.78 | .62 | | 4.40 | 5.15 |
| | 740 | Yellow pine, 3/4" x 3-1/8", T & G, C & better, not incl. finish | " | .040 | " | 1.33 | .49 | | 1.82 | 2.25 |
| | 741 | | | | | | | | | |
| | 780 | Sanding and finishing, fill, shellac, wax | 1 Carp | .027 | S.F. | .36 | .33 | | .69 | .93 |
| | 790 | Subfloor and underlayment, see division 6.1-82 & 86 | | | | | | | | |
| 45-001 | | WOOD BLOCK FLOORING End grain flooring, creosoted, 2" thick | 1 Carp | .027 | S.F. | 1.82 | .33 | | 2.15 | 2.53 |
| | 040 | Natural finish, 1" thick | | .029 | | 2.03 | .36 | | 2.39 | 2.80 |
| | 060 | 1-1/2" thick | | .031 | | 2.70 | .39 | | 3.09 | 3.58 |
| | 070 | 2" thick, | | .107 | | 2.96 | 1.32 | | 4.28 | 5.35 |

## 9.8 Painting & Wall Covering

| | | | CREW | MAN-HOURS | UNIT | MAT. | LABOR | EQUIP. | TOTAL | TOTAL INCL O&P |
|---|---|---|---|---|---|---|---|---|---|---|
| 05-001 | | CORNER GUARDS Rubber, 3" wide, standard | 1 Pord | .059 | L.F. | 2.08 | .70 | | 2.78 | 3.38 |
| | 010 | Bullnose (see also division 5.4-14) | " | .059 | " | 3.74 | .70 | | 4.44 | 5.20 |
| | 030 | 1/4" thick, 2-3/4" wide, rubber | 1 Pord | .059 | L.F. | 2.28 | .70 | | 2.98 | 3.60 |
| | 040 | Vinyl, 5/16" thick, 2-1/2" wide | " | .059 | " | 2.38 | .70 | | 3.08 | 3.71 |
| 16-001 | | CABINETS AND CASEWORK | | | | | | | | |
| | 002 | Labor cost includes protection of adjacent items not painted | | | | | | | | |
| | 100 | Primer coat, oil base, brushwork | 1 Pord | .020 | S.F. | .03 | .24 | | .27 | .40 |
| | 200 | Paint, oil base, brushwork, 1 coat | | .021 | | .04 | .25 | | .29 | .43 |
| | 250 | 2 coats | | .040 | | .08 | .47 | | .55 | .83 |
| | 300 | Stain, brushwork, wipe off | | .022 | | .03 | .26 | | .29 | .44 |

For expanded coverage of these items see *Building Construction Cost Data 1985*

## 9.8 Painting & Wall Covering

| | | | CREW | MAN-HOURS | UNIT | MAT. | LABOR | EQUIP. | TOTAL | TOTAL INCL O&P |
|---|---|---|---|---|---|---|---|---|---|---|
| 16 | 400 | Shellac, 1 coat, brushwork | 1 Pord | .021 | S.F. | .03 | .25 | | .28 | .42 |
| | 450 | Varnish, 3 coats, brushwork | " | .034 | | .09 | .40 | | .49 | .73 |
| | 500 | For latex paint, deduct | | | ↓ | 10% | | | | |
| 17-001 | | **DOORS AND WINDOWS** | | | | | | | | |
| | 002 | Labor cost includes protection of adjacent items not painted | | | | | | | | |
| | 050 | Flush door and frame, per side, oil base, primer coat, brushwork | 1 Pord | .571 | Ea. | 1.02 | 6.75 | | 7.77 | 11.70 |
| | 100 | Paint, 1 coat | | .320 | | 1.31 | 3.79 | | 5.10 | 7.35 |
| | 120 | 2 coats | | .640 | | 2.55 | 7.60 | | 10.15 | 14.65 |
| | 140 | Stain, brushwork, wipe off | | .800 | | .98 | 9.50 | | 10.48 | 15.85 |
| | 160 | Shellac, 1 coat, brushwork | | .667 | | .89 | 7.90 | | 8.79 | 13.30 |
| | 180 | Varnish, 3 coats, brushwork | | 1.600 | | 2.70 | 18.95 | | 21.65 | 33 |
| | 200 | Panel door and frame, per side, oil base, primer coat, brushwork | | .615 | | 1.12 | 7.30 | | 8.42 | 12.60 |
| | 220 | Paint, 1 coat | | .667 | | 1.41 | 7.90 | | 9.31 | 13.85 |
| | 240 | 2 coats | | 1.330 | | 2.81 | 15.80 | | 18.61 | 28 |
| | 260 | Stain, brushwork, wipeoff | | 1.330 | | 1.08 | 15.80 | | 16.88 | 26 |
| | 280 | Shellac, 1 coat, brushwork | | 1 | | .98 | 11.85 | | 12.83 | 19.55 |
| | 300 | Varnish, 3 coats, brushwork | ↓ | 2.670 | ↓ | 2.95 | 32 | | 34.95 | 53 |
| | 440 | Windows, including frame and trim, per side | | | | | | | | |
| | 460 | Colonial type, 2' x 3', oil base, primer coat, brushwork | 1 Pord | .333 | Ea. | .82 | 3.95 | | 4.77 | 7.05 |
| | 580 | Paint, 1 coat | | .364 | | 1 | 4.31 | | 5.31 | 7.80 |
| | 600 | 2 coats | | .615 | | 1.95 | 7.30 | | 9.25 | 13.50 |
| | 620 | 3' x 5' opening, primer coat, brushwork | | .533 | | 1.02 | 6.30 | | 7.32 | 11 |
| | 640 | Paint, 1 coat | | .615 | | 1.26 | 7.30 | | 8.56 | 12.75 |
| | 660 | 2 coats | | 1 | | 2.43 | 11.85 | | 14.28 | 21 |
| | 680 | 4' x 8' opening, primer coat, brushwork | | .667 | | 1.22 | 7.90 | | 9.12 | 13.65 |
| | 700 | Paint, 1 coat | | .800 | | 1.51 | 9.50 | | 11.01 | 16.45 |
| | 720 | 2 coats | | 1.330 | | 2.94 | 15.80 | | 18.74 | 28 |
| | 800 | Single lite type, 2' x 3', oil base, primer coat, brushwork | | .200 | | .78 | 2.37 | | 3.15 | 4.55 |
| | 820 | Paint, 1 coat | | .216 | | .97 | 2.56 | | 3.53 | 5.05 |
| | 840 | 2 coats | | .400 | | 1.87 | 4.74 | | 6.61 | 9.45 |
| | 860 | 3' x 5' opening, primer coat, brushwork | | .296 | | .97 | 3.51 | | 4.48 | 6.55 |
| | 880 | Paint, 1 coat | | .320 | | 1.23 | 3.79 | | 5.02 | 7.25 |
| | 900 | 2 coats | | .571 | | 2.39 | 6.75 | | 9.14 | 13.20 |
| | 920 | 4' x 8' opening, primer coat, brushwork | | .571 | | 2.20 | 6.75 | | 8.95 | 13 |
| | 940 | Paint, 1 coat | | .615 | | 2.45 | 7.30 | | 9.75 | 14.05 |
| | 960 | 2 coats | ↓ | 1 | | 3.79 | 11.85 | | 15.64 | 23 |
| | 980 | For latex paint deduct | | | ↓ | 10% | | | | |
| 18-001 | | **MISCELLANEOUS** | | | | | | | | |
| | 002 | Labor cost includes protection of adjacent items not painted | | | | | | | | |
| | 070 | Fence, chain link, per side, oil base, primer coat, brushwork | 2 Pord | .013 | S.F. | .03 | .16 | | .19 | .28 |
| | 100 | Spray | | .010 | | .03 | .12 | | .15 | .22 |
| | 120 | Paint 1 coat, brushwork | | .014 | | .04 | .16 | | .20 | .30 |
| | 140 | Spray | | .010 | | .04 | .12 | | .16 | .23 |
| | 160 | Picket, wood, primer coat, brushwork | | .039 | | .05 | .46 | | .51 | .78 |
| | 180 | Spray | | .031 | | .05 | .36 | | .41 | .62 |
| | 200 | Paint 1 coat, brushwork | | .040 | | .06 | .47 | | .53 | .81 |
| | 220 | Spray | | .031 | | .06 | .36 | | .42 | .63 |
| | 240 | Floors, concrete or wood, oil base, primer or sealer coat, brushwork | | .007 | | .03 | .09 | | .12 | .17 |
| | 245 | Roller | | .006 | | .03 | .08 | | .11 | .15 |
| | 260 | Spray | | .006 | | .03 | .07 | | .10 | .15 |
| | 265 | Paint 1 coat, brushwork | | .008 | | .04 | .09 | | .13 | .19 |
| | 280 | Roller | | .007 | | .04 | .08 | | .12 | .17 |
| | 285 | Spray | | .006 | | .04 | .07 | | .11 | .16 |
| | 300 | Stain, wood floor, brushwork | | .007 | | .03 | .09 | | .12 | .17 |
| | 320 | Roller | | .006 | | .03 | .08 | | .11 | .15 |
| | 325 | Spray | | .006 | | .03 | .07 | | .10 | .15 |
| | 340 | Varnish, wood floor, brushwork | | .008 | | .04 | .09 | | .13 | .19 |
| | 345 | Roller | | .007 | | .04 | .08 | | .12 | .17 |
| | 360 | Spray | ↓ | .006 | ↓ | .04 | .08 | | .12 | .16 |

For expanded coverage of these items see *Building Construction Cost Data 1985*

## 9.8 Painting & Wall Covering

| | | | MAN- | | BARE COSTS ||| | TOTAL |
|---|---|---|---|---|---|---|---|---|---|
| | | CREW | HOURS | UNIT | MAT. | LABOR | EQUIP. | TOTAL | INCL O&P |
| 18 | 365 | For dust proofing or anti skid, see division 3.3-26 | | | | | | | | |
| | 380 | Grilles, per side, oil base, primer coat, brushwork | 1 Pord | .020 | Ea. | .03 | .24 | | .27 | .40 |
| | 385 | Spray | | .016 | | .03 | .19 | | .22 | .33 |
| | 388 | Paint 1 coat, brushwork | | .021 | | .04 | .25 | | .29 | .43 |
| | 390 | Spray | | .016 | | .04 | .19 | | .23 | .34 |
| | 392 | Paint 2 coats, brushwork | | .040 | | .06 | .47 | | .53 | .81 |
| | 394 | Spray | ▼ | .032 | ▼ | .06 | .38 | | .44 | .66 |
| | 420 | Gutters and downspouts, oil base, primer coat, brushwork | 2 Pord | .025 | L.F. | .06 | .29 | | .35 | .52 |
| | 425 | Paint 1 coat, brushwork | | .027 | | .07 | .32 | | .39 | .57 |
| | 430 | Paint 2 coats, brushwork | | .049 | | .13 | .58 | | .71 | 1.05 |
| | 500 | Pipe, to 4" diameter, primer or sealer coat, oil base, brushwork | | .020 | | .05 | .24 | | .29 | .42 |
| | 510 | Spray | | .015 | | .05 | .17 | | .22 | .32 |
| | 520 | Paint 1 coat, brushwork | | .021 | | .06 | .25 | | .31 | .46 |
| | 530 | Spray | | .015 | | .06 | .17 | | .23 | .33 |
| | 535 | Paint 2 coats, brushwork | | .040 | | .11 | .47 | | .58 | .86 |
| | 540 | Spray | | .029 | | .11 | .34 | | .45 | .66 |
| | 545 | To 8" diameter, primer or sealer coat, brushwork | | .040 | | .09 | .47 | | .56 | .84 |
| | 550 | Spray | | .025 | | .09 | .29 | | .38 | .55 |
| | 555 | Paint 1 coat, brushwork | | .046 | | .10 | .54 | | .64 | .95 |
| | 560 | Spray | | .025 | | .10 | .29 | | .39 | .56 |
| | 565 | Paint 2 coats, brushwork | | .080 | | .19 | .95 | | 1.14 | 1.69 |
| | 570 | Spray | | .043 | | .19 | .51 | | .70 | 1 |
| | 575 | To 12" diameter, primer or sealer coat, brushwork | | .067 | | .14 | .79 | | .93 | 1.39 |
| | 580 | Spray | | .050 | | .14 | .59 | | .73 | 1.08 |
| | 585 | Paint 1 coat, brushwork | | .073 | | .15 | .86 | | 1.01 | 1.51 |
| | 600 | Spray | | .050 | | .15 | .59 | | .74 | 1.09 |
| | 620 | Paint 2 coats, brushwork | | .133 | | .29 | 1.58 | | 1.87 | 2.78 |
| | 625 | Spray | | .100 | | .29 | 1.19 | | 1.48 | 2.17 |
| | 630 | To 16" diameter, primer or sealer coat, brushwork | | .083 | | .16 | .99 | | 1.15 | 1.72 |
| | 635 | Spray | | .067 | | .16 | .79 | | .95 | 1.41 |
| | 640 | Paint 1 coat, brushwork | | .089 | | .17 | 1.05 | | 1.22 | 1.83 |
| | 645 | Spray | | .067 | | .17 | .79 | | .96 | 1.42 |
| | 650 | Paint 2 coats, brushwork | | .160 | | .33 | 1.90 | | 2.23 | 3.32 |
| | 655 | Spray | ▼ | .123 | ▼ | .33 | 1.46 | | 1.79 | 2.64 |
| | 700 | Trim, wood, incl. puttying, under 6" wide | | | | | | | | |
| | 720 | Primer coat, oil base, brushwork | 1 Pord | .009 | L.F. | .02 | .11 | | .13 | .19 |
| | 725 | Paint, 1 coat, brushwork | | .009 | | .03 | .11 | | .14 | .20 |
| | 740 | 2 coats | | .016 | | .05 | .19 | | .24 | .35 |
| | 745 | 3 coats | | .027 | | .08 | .32 | | .40 | .58 |
| | 750 | Over 6" wide, primer coat, brushwork | | .013 | | .03 | .16 | | .19 | .28 |
| | 755 | Paint, 1 coat, brushwork | | .018 | | .04 | .21 | | .25 | .37 |
| | 760 | 2 coats | | .027 | | .07 | .32 | | .39 | .57 |
| | 765 | 3 coats | | .042 | ▼ | .11 | .50 | | .61 | .90 |
| | 800 | Cornice, simple design, primer coat, oil base, brushwork | | .029 | S.F. | .03 | .34 | | .37 | .57 |
| | 825 | Paint, 1 coat, brushwork | | .032 | | .04 | .38 | | .42 | .64 |
| | 830 | 2 coats | | .050 | | .07 | .59 | | .66 | 1 |
| | 835 | Ornate design, primer coat | | .053 | | .03 | .63 | | .66 | 1.02 |
| | 840 | Paint, 1 coat | | .057 | | .04 | .68 | | .72 | 1.10 |
| | 845 | 2 coats | | .089 | | .07 | 1.05 | | 1.12 | 1.72 |
| | 860 | Balustrades, per side, primer coat, oil base, brushwork | | .027 | | .03 | .32 | | .35 | .53 |
| | 865 | Paint, 1 coat | | .028 | | .04 | .33 | | .37 | .56 |
| | 870 | 2 coats | | .047 | | .07 | .56 | | .63 | .95 |
| | 890 | Trusses and exposed wood frames, primer coat, oil base, brushwork | | .010 | | .03 | .12 | | .15 | .22 |
| | 895 | Spray | | .007 | | .03 | .08 | | .11 | .16 |
| | 900 | Paint 1 coat, brushwork | | 0.11 | | .04 | .13 | | .17 | .24 |
| | 920 | Spray | | .007 | | .04 | .08 | | .12 | .17 |
| | 922 | Paint 2 coats, brushwork | | .016 | | .07 | .19 | | .26 | .37 |
| | 924 | Spray | | .013 | | .07 | .16 | | .23 | .32 |
| | 926 | Stain, brushwork, wipe off | | .013 | | .03 | .16 | | .19 | .28 |
| | 928 | Varnish, 3 coats, brushwork | ▼ | .029 | ▼ | .15 | .34 | | .49 | .70 |

For expanded coverage of these items see *Building Construction Cost Data 1985*

## 9.8 Painting & Wall Covering

| | | | CREW | MAN-HOURS | UNIT | MAT. | LABOR | EQUIP. | TOTAL | TOTAL INCL O&P |
|---|---|---|---|---|---|---|---|---|---|---|
| 18 | 935 | For latex paint, deduct | | | S.F. | 10% | | | | |
| 19-001 | | **SIDING** Exterior | | | | | | | | |
| | 002 | Labor cost includes protection of adjacent items not painted | | | | | | | | |
| | 010 | Steel siding, oil base, primer or sealer coat, brushwork | 2 Pord | .009 | S.F. | .05 | .11 | | .16 | .23 |
| | 050 | Spray | | .005 | | .05 | .06 | | .11 | .15 |
| | 080 | Paint 2 coats, brushwork | | .012 | | .08 | .14 | | .22 | .31 |
| | 100 | Spray | | .006 | | .08 | .07 | | .15 | .20 |
| | 120 | Stucco, rough, oil base, paint 2 coats, brushwork | | .012 | | .10 | .15 | | .25 | .34 |
| | 140 | Roller | | .008 | | .10 | .09 | | .19 | .26 |
| | 160 | Spray | | .006 | | .10 | .07 | | .17 | .22 |
| | 180 | Texture 1-11 or clapboard, oil base, primer coat, brushwork | | .006 | | .06 | .07 | | .13 | .18 |
| | 200 | Spray | | .004 | | .06 | .05 | | .11 | .14 |
| | 210 | Paint 1 coat, brushwork | | .006 | | .05 | .08 | | .13 | .17 |
| | 220 | Spray | | .004 | | .05 | .05 | | .10 | .13 |
| | 240 | Paint 2 coats, brushwork | | .013 | | .09 | .15 | | .24 | .34 |
| | 260 | Spray | | .008 | | .09 | .09 | | .18 | .24 |
| | 300 | Stain 1 coat, brushwork | | .006 | | .07 | .07 | | .14 | .19 |
| | 320 | Spray | | .004 | | .07 | .05 | | .12 | .15 |
| | 340 | Stain 2 coats, brushwork | | .013 | | .12 | .15 | | .27 | .37 |
| | 400 | Spray | | .008 | | .12 | .09 | | .21 | .27 |
| | 420 | Wood shingles, oil base primer coat, brushwork | | .006 | | .07 | .08 | | .15 | .20 |
| | 440 | Spray | | .004 | | .07 | .05 | | .12 | .15 |
| | 460 | Paint 1 coat, brushwork | | .007 | | .06 | .08 | | .14 | .19 |
| | 480 | Spray | | .004 | | .06 | .05 | | .11 | .14 |
| | 500 | Paint 2 coats, brushwork | | .014 | | .10 | .16 | | .26 | .37 |
| | 520 | Spray | | .008 | | .10 | .09 | | .19 | .25 |
| | 580 | Stain 1 coat, brushwork | | .006 | | .08 | .08 | | .16 | .21 |
| | 600 | Spray | | .004 | | .08 | .05 | | .13 | .16 |
| | 650 | Stain 2 coats, brushwork | | .014 | | .13 | .16 | | .29 | .40 |
| | 700 | Spray | ↓ | .008 | | .13 | .09 | | .22 | .28 |
| | 800 | For latex paint, deduct | | | ↓ | 10% | | | | |
| 21-001 | | **WALL AND CEILINGS** | | | | | | | | |
| | 002 | Labor cost includes protection of adjacent items not painted | | | | | | | | |
| | 010 | Concrete, dry wall or plaster, oil base, primer or sealer coat | | | | | | | | |
| | 020 | Smooth finish, brushwork | 1 Pord | .004 | S.F. | .04 | .05 | | .09 | .12 |
| | 024 | Roller | | .004 | | .04 | .04 | | .08 | .11 |
| | 028 | Spray | | .002 | | .04 | .02 | | .06 | .07 |
| | 030 | Sand finish, brushwork | | .005 | | .05 | .06 | | .11 | .14 |
| | 034 | Roller | | .004 | | .05 | .05 | | .10 | .13 |
| | 038 | Spray | | .002 | | .05 | .03 | | .08 | .09 |
| | 040 | Paint 1 coat, smooth finish, brushwork | | .004 | | .05 | .05 | | .10 | .14 |
| | 044 | Roller | | .004 | | .05 | .05 | | .10 | .13 |
| | 048 | Spray | | .002 | | .05 | .03 | | .08 | .09 |
| | 050 | Sand finish, brushwork | | .005 | | .06 | .06 | | .12 | .16 |
| | 054 | Roller | | .004 | | .06 | .05 | | .11 | .14 |
| | 058 | Spray | | .002 | | .06 | .03 | | .09 | .11 |
| | 080 | Paint 2 coats, smooth finish, brushwork | | .008 | | .09 | .10 | | .19 | .25 |
| | 084 | Roller | | .007 | | .09 | .08 | | .17 | .23 |
| | 088 | Spray | | .004 | | .09 | .04 | | .13 | .16 |
| | 090 | Sand finish, brushwork | | .010 | | .11 | .11 | | .22 | .30 |
| | 094 | Roller | | .008 | | .11 | .09 | | .20 | .26 |
| | 098 | Spray | | .004 | | .11 | .04 | | .15 | .19 |
| | 120 | Paint 3 coats, smooth finish, brushwork | | .012 | | .13 | .14 | | .27 | .36 |
| | 124 | Roller | | .010 | | .13 | .12 | | .25 | .33 |
| | 128 | Spray | | .005 | | .13 | .06 | | .19 | .24 |
| | 130 | Sand finish, brushwork | | .014 | | .16 | .17 | | .33 | .44 |
| | 134 | Roller | | .011 | | .16 | .13 | | .29 | .38 |
| | 138 | Spray | | .005 | | .16 | .06 | | .22 | .27 |
| | 160 | Glaze coating, 5 coats, spray, clear | ↓ | .009 | ↓ | .50 | .11 | | .61 | .71 |

For expanded coverage of these items see *Building Construction Cost Data 1985*

## 9.8 Painting & Wall Covering

| | | | CREW | MAN-HOURS | UNIT | MAT. | LABOR | EQUIP. | TOTAL | TOTAL INCL O&P |
|---|---|---|---|---|---|---|---|---|---|---|
| 21 | 164 | Multicolor | 1 Pord | .009 | S.F. | .60 | .11 | | .71 | .82 |
| | 170 | For latex paint, deduct | | | | | | | | |
| | 180 | For ceiling installations, add | | | S.F. | | 25% | | | |
| | 190 | | | | | | | | | |
| | 200 | Masonry or concrete block, oil base, primer or sealer coat | | | | | | | | |
| | 210 | Smooth finish, brushwork | 1 Pord | .005 | S.F. | .06 | .05 | | .11 | .15 |
| | 218 | Spray | | .002 | | .06 | .03 | | .09 | .11 |
| | 220 | Sand finish, brushwork | | .006 | | .07 | .07 | | .14 | .18 |
| | 228 | Spray | | .002 | | .07 | .03 | | .10 | .12 |
| | 240 | Paint 1 coat, smooth finish, brushwork | | .005 | | .06 | .06 | | .12 | .16 |
| | 248 | Spray | | .002 | | .06 | .03 | | .09 | .11 |
| | 250 | Sand finish, brushwork | | .006 | | .07 | .07 | | .14 | .18 |
| | 258 | Spray | | .002 | | .07 | .03 | | .10 | .12 |
| | 280 | Paint 2 coats, smooth finish, brushwork | | .010 | | .10 | .11 | | .21 | .29 |
| | 288 | Spray | | .004 | | .10 | .04 | | .14 | .18 |
| | 290 | Sand finish, brushwork | | .011 | | .11 | .13 | | .24 | .32 |
| | 298 | Spray | | .004 | | .11 | .04 | | .15 | .19 |
| | 320 | Paint 3 coats, smooth finish, brushwork | | .013 | | .15 | .16 | | .31 | .41 |
| | 328 | Spray | | .005 | | .15 | .06 | | .21 | .26 |
| | 330 | Sand finish, brushwork | | .023 | | .16 | .27 | | .43 | .60 |
| | 338 | Spray | | .005 | | .16 | .06 | | .22 | .27 |
| | 360 | Glaze coating, 5 coats, spray, clear | | .009 | | .50 | .11 | | .61 | .71 |
| | 362 | Multicolor | | .009 | | .60 | .11 | | .71 | .82 |
| | 400 | Block filler, 1 coat, brushwork | | .006 | | .08 | .07 | | .15 | .20 |
| | 410 | Silicone, water repellent, 2 coats, spray | ↓ | .009 | ↓ | .04 | .11 | | .15 | .21 |
| | 412 | For latex paint, deduct | | | | 10% | | | | |
| 22-001 | | REMOVAL Existing lead paint, by chemicals, | | | | | | | | |
| | 002 | refinish with 2 coats of paint | | | | | | | | |
| | 005 | Baseboard, to 6" wide | 1 Pord | .042 | L.F. | .13 | .50 | | .63 | .92 |
| | 007 | To 12" wide | | .053 | " | .26 | .63 | | .89 | 1.27 |
| | 020 | Balustrades, one side | ↓ | .089 | S.F. | .29 | 1.05 | | 1.34 | 1.96 |
| | 022 | | | | | | | | | |
| | 140 | Cabinets, simple design | 1 Pord | .094 | S.F. | .13 | 1.12 | | 1.25 | 1.88 |
| | 142 | Ornate design | | .200 | | .15 | 2.37 | | 2.52 | 3.86 |
| | 160 | Cornice, simple design | | .123 | | .13 | 1.46 | | 1.59 | 2.42 |
| | 162 | Ornate design | | .229 | | .15 | 2.71 | | 2.86 | 4.39 |
| | 280 | Doors, one side, flush | | .064 | | .13 | .76 | | .89 | 1.33 |
| | 282 | Two panel | | .073 | | .13 | .86 | | .99 | 1.49 |
| | 284 | Four panel | | .084 | ↓ | .15 | 1 | | 1.15 | 1.72 |
| | 288 | For trim, one side, add | | .040 | L.F. | .13 | .47 | | .60 | .88 |
| | 300 | Fence, picket, one side | ↓ | .100 | S.F. | .14 | 1.19 | | 1.33 | 2 |
| | 302 | | | | | | | | | |
| | 320 | Grilles, one side, simple design | 1 Pord | .084 | S.F. | .15 | 1 | | 1.15 | 1.72 |
| | 322 | Ornate design | | .178 | " | .18 | 2.11 | | 2.29 | 3.48 |
| | 440 | Pipes, to 4" diameter | | .040 | L.F. | .13 | .47 | | .60 | .88 |
| | 442 | To 8" diameter | | .089 | | .26 | 1.05 | | 1.31 | 1.93 |
| | 444 | To 12" diameter | | .123 | ↓ | .40 | 1.46 | | 1.86 | 2.71 |
| | 446 | To 16" diameter | | .178 | | .53 | 2.11 | | 2.64 | 3.87 |
| | 450 | For hangers, add | ↓ | .080 | Ea. | .20 | .95 | | 1.15 | 1.70 |
| | 451 | | | | | | | | | |
| | 480 | Siding | 1 Pord | .047 | S.F. | .13 | .56 | | .69 | 1.01 |
| | 500 | Trusses, open | | .107 | S.F.Face | .13 | 1.26 | | 1.39 | 2.11 |
| | 620 | Windows, one side only, double hung, 1/1 light, 24" x 48" high | | .667 | Ea. | 1.06 | 7.90 | | 8.96 | 13.50 |
| | 622 | 30" x 60" high | | .889 | | 1.65 | 10.55 | | 12.20 | 18.25 |
| | 624 | 36" x 72" high | | 1 | | 2.38 | 11.85 | | 14.23 | 21 |
| | 628 | 40" x 80" high | | 1.330 | | 3.30 | 15.80 | | 19.10 | 28 |
| | 640 | Colonial window, 6/6 light, 24" x 48" high | | 1.140 | | 1.32 | 13.55 | | 14.87 | 23 |
| | 642 | 30" x 60" high | | 1.600 | | 1.91 | 18.95 | | 20.86 | 32 |
| | 644 | 36" x 72" high | ↓ | 2 | ↓ | 2.64 | 24 | | 26.64 | 40 |
| | 648 | 40" x 80" high | | 2.290 | | 3.56 | 27 | | 30.56 | 46 |

For expanded coverage of these items see *Building Construction Cost Data 1985*

# 9.8 Painting & Wall Covering

| | | CREW | MAN-HOURS | UNIT | MAT. | LABOR | EQUIP. | TOTAL | TOTAL INCL O&P |
|---|---|---|---|---|---|---|---|---|---|
| 22 660 | 8/8 light, 24" x 48" high | 1 Pord | 1.330 | Ea. | 1.58 | 15.80 | | 17.38 | 26 |
| 662 | 40" x 80" high | | 2.670 | | 3.96 | 32 | | 35.96 | 54 |
| 680 | 12/12 light, 24" x 48" high | | 1.600 | | 1.98 | 18.95 | | 20.93 | 32 |
| 682 | 40" x 80" high | | 3.200 | | 4.29 | 38 | | 42.29 | 64 |
| 684 | For frame & trim, add | | .053 | L.F. | .13 | .63 | | .76 | 1.13 |
| 25-001 | **SANDING** And puttying interior trim, compared to | | | | | | | | |
| 010 | painting 1 coat, on quality work | | | SF or LF | | 100% | | | |
| 030 | Medium work | | | | | 50% | | | |
| 040 | Industrial grade | | | | | 25% | | | |
| 27-001 | **SCRAPE AFTER FIRE DAMAGE** | | | | | | | | |
| 002 | | | | | | | | | |
| 005 | Boards, 1" x 4" | 1 Pord | .055 | L.F. | | .65 | | .65 | 1.02L |
| 006 | 1" x 6" | | .073 | | | .86 | | .86 | 1.34L |
| 007 | 1" x 8" | | .100 | | | 1.19 | | 1.19 | 1.85L |
| 008 | 1" x 10" | | .123 | | | 1.46 | | 1.46 | 2.27L |
| 050 | Framing, 2" x 4" | | .073 | | | .86 | | .86 | 1.34L |
| 051 | 2" x 6" | | .089 | | | 1.05 | | 1.05 | 1.64L |
| 052 | 2" x 8" | | .114 | | | 1.35 | | 1.35 | 2.11L |
| 053 | 2" x 10" | | .133 | | | 1.58 | | 1.58 | 2.46L |
| 054 | 2" x 12" | | .160 | | | 1.90 | | 1.90 | 2.96L |
| 055 | | | | | | | | | |
| 100 | Heavy framing, 3" x 4" | 1 Pord | .073 | L.F. | | .86 | | .86 | 1.34L |
| 101 | 4" x 4" | | .089 | | | 1.05 | | 1.05 | 1.64L |
| 102 | 4" x 6" | | .107 | | | 1.26 | | 1.26 | 1.97L |
| 103 | 4" x 8" | | .133 | | | 1.58 | | 1.58 | 2.46L |
| 104 | 4" x 10" | | .160 | | | 1.90 | | 1.90 | 2.96L |
| 106 | 4" x 12" | | .178 | | | 2.11 | | 2.11 | 3.29L |
| 290 | For sealing, minimum | | .010 | S.F. | | .11 | | .11 | .18L |
| 292 | Maximum | | .017 | " | | .21 | | .21 | .32L |
| 300 | For sandblasting, see division 4.5-12 | | | | | | | | |
| 302 | | | | | | | | | |
| 30-001 | **VARNISH** 1 coat + sealer, on wood trim, no sanding included | 1 Pord | .009 | S.F. | .07 | .11 | | .18 | .24 |
| 010 | Hardwood floors, 2 coat, no sanding included | | .010 | | .07 | .12 | | .19 | .26 |
| 35-001 | **WALL COATINGS** Acrylic glazed coatings, minimum | | .015 | | .22 | .18 | | .40 | .52 |
| 010 | Maximum | | .026 | | .43 | .31 | | .74 | .96 |
| 030 | Epoxy coatings, minimum | | .015 | | .24 | .18 | | .42 | .55 |
| 040 | Maximum | | .047 | | .93 | .56 | | 1.49 | 1.89 |
| 060 | Exposed aggregate, troweled on, 1/16" to 1/4", minimum | | .034 | | .59 | .40 | | .99 | 1.28 |
| 070 | Maximum (epoxy or polyacrylate) | | .062 | | 1.16 | .73 | | 1.89 | 2.41 |
| 090 | 1/2" to 5/8" aggregate, minimum | | .062 | | 1.05 | .73 | | 1.78 | 2.29 |
| 100 | Maximum | | .100 | | 1.80 | 1.19 | | 2.99 | 3.83 |
| 120 | 1" aggregate size, minimum | | .089 | | 1.48 | 1.05 | | 2.53 | 3.27 |
| 130 | Maximum | | .145 | | 2.30 | 1.72 | | 4.02 | 5.20 |
| 150 | Exposed aggregate, sprayed on, 1/8" aggregate, minimum | | .027 | | .46 | .32 | | .78 | 1.01 |
| 160 | Maximum | | .055 | | .96 | .65 | | 1.61 | 2.08 |
| 180 | High build epoxy, 50 mil, minimum | | .021 | | .36 | .24 | | .60 | .78 |
| 190 | Maximum | | .084 | | 1.36 | 1 | | 2.36 | 3.05 |
| 210 | Laminated epoxy with fiberglass, minimum | | .027 | | .46 | .32 | | .78 | 1.01 |
| 220 | Maximum | | .055 | | .87 | .65 | | 1.52 | 1.98 |
| 240 | Sprayed perlite or vermiculite, 1/16" thick, minimum | | .003 | | .05 | .03 | | .08 | .11 |
| 250 | Maximum | | .013 | | .33 | .15 | | .48 | .59 |
| 270 | Vinyl plastic wall coating, minimum | | .011 | | .19 | .13 | | .32 | .41 |
| 280 | Maximum | | .033 | | .69 | .40 | | 1.09 | 1.38 |
| 300 | Urethane on smooth surface, 2 coat, minimum | | .007 | | .11 | .08 | | .19 | .25 |
| 310 | Maximum | | .012 | | .23 | .14 | | .37 | .48 |
| 330 | 3 coat, minimum | | .010 | | .17 | .11 | | .28 | .36 |
| 340 | Maximum | | .017 | | .36 | .20 | | .56 | .71 |
| 360 | Ceramic-like glazed coating, cementitious, minimum | | .018 | | .28 | .22 | | .50 | .64 |
| 370 | Maximum | | .023 | | .36 | .27 | | .63 | .82 |

For expanded coverage of these items see *Building Construction Cost Data 1985*

## 9.8 Painting & Wall Covering

| | | | CREW | MAN-HOURS | UNIT | MAT. | LABOR | EQUIP. | TOTAL | TOTAL INCL O&P |
|---|---|---|---|---|---|---|---|---|---|---|
| 35 | 390 | Resin base, minimum | 1 Pord | .013 | S.F. | .20 | .15 | | .35 | .45 |
| | 400 | Maximum | " | .024 | " | .33 | .29 | | .62 | .81 |
| 40-001 | | WALL COVERING | | | | | | | | |
| | 005 | Aluminum foil | 1 Pape | .029 | S.F. | .67 | .35 | | 1.02 | 1.29 |
| | 010 | Copper sheets, .025" thick, vinyl backing | | .033 | | 3.01 | .40 | | 3.41 | 3.94 |
| | 030 | Phenolic backing | | .033 | | 4.03 | .40 | | 4.43 | 5.05 |
| | 060 | Cork tiles, light or dark, 12" x 12" x 3/16" | | .033 | | 1.18 | .40 | | 1.58 | 1.93 |
| | 070 | 5/16" thick | | .033 | | 1.47 | .40 | | 1.87 | 2.25 |
| | 090 | 1/4" basket weave | | .033 | | 3.26 | .40 | | 3.66 | 4.22 |
| | 100 | 1/2" natural, non-directional pattern | | .033 | | 2.95 | .40 | | 3.35 | 3.87 |
| | 120 | Granular surface, 12" x 36", 1/2" thick | | .021 | | .62 | .25 | | .87 | 1.07 |
| | 130 | 1" thick | | .021 | | .88 | .25 | | 1.13 | 1.36 |
| | 150 | Polyurethane coated, 12" x 12" x 3/16" thick | | .033 | | 1.14 | .40 | | 1.54 | 1.88 |
| | 160 | 5/16" thick | | .033 | | 1.42 | .40 | | 1.82 | 2.19 |
| | 180 | Cork wallpaper, paperbacked, natural | | .017 | | .98 | .20 | | 1.18 | 1.39 |
| | 190 | Colors | | .017 | | 1.08 | .20 | | 1.28 | 1.50 |
| | 210 | Flexible wood veneer, 1/32" thick, plain woods | | .080 | | 1.79 | .97 | | 2.76 | 3.48 |
| | 220 | Exotic woods | ↓ | .080 | ↓ | 4.45 | .97 | | 5.42 | 6.40 |
| | 240 | Gypsum-based, fabric-backed, fire | | | | | | | | |
| | 250 | resistant for masonry walls, minimum | 1 Pape | .010 | S.F. | .54 | .12 | | .66 | .78 |
| | 260 | Average | | .011 | | .61 | .13 | | .74 | .88 |
| | 270 | Maximum | ↓ | .013 | | .69 | .15 | | .84 | .99 |
| | 275 | Acrylic, modified, semi-rigid PVC, .028" thick | 2 Carp | .048 | | .67 | .60 | | 1.27 | 1.69 |
| | 280 | .040" thick | " | .050 | | .88 | .62 | | 1.50 | 1.95 |
| | 300 | Vinyl wall covering, fabric-backed, lightweight | 1 Pape | .013 | | .40 | .15 | | .55 | .68 |
| | 330 | Medium weight | | .017 | | .60 | .20 | | .80 | .97 |
| | 340 | Heavy weight | ↓ | .018 | ↓ | .80 | .22 | | 1.02 | 1.23 |
| | 360 | Adhesive, 5 gal. lots | | | Gal. | 4.68 | | | 4.68 | 5.15M |
| | 370 | Wallpaper at $8.00 per double roll, average workmanship | 1 Pape | .013 | S.F. | .16 | .15 | | .31 | .41 |
| | 390 | Paper at $17 per double roll, average workmanship | | .015 | | .33 | .18 | | .51 | .65 |
| | 400 | Paper at $40 per double roll, quality workmanship | ↓ | .018 | ↓ | .71 | .22 | | .93 | 1.13 |
| | 401 | | | | | | | | | |
| | 410 | Linen wall covering, paper backed | | | | | | | | |
| | 415 | Flame treatment, minimum | 1 Pape | .020 | S.F. | | .24 | | .24 | .38L |
| | 418 | Maximum | | .023 | | | .28 | | .28 | .43L |
| | 420 | Grass cloths with lining paper, minimum | | .020 | | .55 | .24 | | .79 | .98 |
| | 430 | Maximum | ↓ | .023 | ↓ | 1.63 | .28 | | 1.91 | 2.22 |
| 45-001 | | WALLGUARD See division 5.4 | | | | | | | | |
| | 050 | Neoprene with aluminum fastening strip, 1-1/2" x 2" | 1 Carp | .073 | L.F. | 3.35 | .90 | | 4.25 | 5.10 |
| | 100 | Trolley rail, PVC, clipped to wall, 5" high | | .043 | | 2.70 | .53 | | 3.23 | 3.82 |
| | 105 | 8" high | | .044 | ↓ | 3.55 | .55 | | 4.10 | 4.77 |
| | 120 | Vinyl acrylic bed aligner and bumper, 37" long | | .800 | Ea. | 30 | 9.90 | | 39.90 | 49 |
| | 130 | 43" long | ↓ | .889 | " | 33 | 11 | | 44 | 54 |

## 10.1 Specialties

| | | | CREW | MAN-HOURS | UNIT | MAT. | LABOR | EQUIP. | TOTAL | TOTAL INCL O&P |
|---|---|---|---|---|---|---|---|---|---|---|
| 02-001 | | BATHROOM ACCESSORIES | | | | | | | | |
| | 020 | Curtain rod, stainless steel, 5' long, 1" diameter | 1 Carp | .123 | Ea. | 12.85 | 1.52 | | 14.37 | 16.55 |
| | 030 | 1-1/4" diameter | " | .133 | " | 9.65 | 1.65 | | 11.30 | 13.20 |
| | 050 | Dispenser units, combined soap & towel dispensers, | | | | | | | | |
| | 051 | mirror and shelf, flush mounted | 1 Carp | .800 | Ea. | 195 | 9.90 | | 204.90 | 230 |
| | 060 | Towel dispenser and waste receptacle, | | | | | | | | |
| | 061 | flush mounted | 1 Carp | .800 | Ea. | 280 | 9.90 | | 289.90 | 325 |
| | 080 | Grab bar, straight, 1" diameter, stainless steel, 12" long | " | .333 | " | 16.05 | 4.12 | | 20.17 | 24 |

For expanded coverage of these items see *Building Construction Cost Data 1985*

## 10.1 Specialties

| | | | CREW | MAN-HOURS | UNIT | MAT. | LABOR | EQUIP. | TOTAL | TOTAL INCL O&P |
|---|---|---|---|---|---|---|---|---|---|---|
| 02 | 110 | 36" long | 1 Carp | .400 | Ea. | 21 | 4.94 | | 25.94 | 31 |
| | 200 | For 1-1/4" diameter bars, add | | | | 20% | | | | |
| | 210 | For 1-1/2" diameter bars, add | | | | 30% | | | | |
| | 300 | Mirror with stainless steel, 3/4" square frame, 18" x 24" | 1 Carp | .400 | | 56 | 4.94 | | 60.94 | 69 |
| | 310 | 36" x 24" | | .533 | | 98 | 6.60 | | 104.60 | 120 |
| | 330 | 72" x 24" | | 1.330 | | 180 | 16.45 | | 196.45 | 225 |
| | 350 | Mirror with 5" stainless steel shelf, 3/4" sq. frame, 18" x 24" | | .400 | | 73 | 4.94 | | 77.94 | 88 |
| | 360 | 36" x 24" | | .533 | | 135 | 6.60 | | 141.60 | 160 |
| | 380 | 72" x 24" | | 1.330 | | 265 | 16.45 | | 281.45 | 320 |
| | 420 | Napkin/tampon dispenser, surface mounted | | .533 | | 245 | 6.60 | | 251.60 | 280 |
| | 430 | Robe hook, single, regular | | .222 | | 10.70 | 2.74 | | 13.44 | 16.10 |
| | 440 | Heavy duty, concealed mounting | | .222 | | 13.90 | 2.74 | | 16.64 | 19.65 |
| | 460 | Soap dispenser, chrome, surface mounted, liquid | | .400 | | 47 | 4.94 | | 51.94 | 60 |
| | 500 | Recessed stainless steel, liquid | | .800 | | 54 | 9.90 | | 63.90 | 75 |
| | 560 | Shelf, stainless steel, 5" wide, 18 ga., 24" long | | .333 | | 20 | 4.12 | | 24.12 | 29 |
| | 610 | Toilet tissue dispenser, surface mounted, S.S., single roll | | .267 | | 17.10 | 3.29 | | 20.39 | 24 |
| | 620 | Double roll | | .333 | | 28 | 4.12 | | 32.12 | 37 |
| | 640 | Towel bar, stainless steel, 18" long | | .348 | | 17.10 | 4.30 | | 21.40 | 26 |
| | 650 | 30" long | | .381 | | 18 | 4.70 | | 22.70 | 27 |
| | 670 | Towel dispenser, stainless steel, surface mounted | | .500 | | 41 | 6.20 | | 47.20 | 55 |
| | 680 | Flush mounted, recessed | | .800 | | 75 | 9.90 | | 84.90 | 98 |
| | 740 | Tumbler holder, tumbler only | | .267 | | 9.65 | 3.29 | | 12.94 | 15.85 |
| | 750 | Soap, tumbler & toothbrush | | .267 | | 16.05 | 3.29 | | 19.34 | 23 |
| | 770 | Wall urn ash receiver, recessed, 14" long | | .667 | | 75 | 8.25 | | 83.25 | 96 |
| | 800 | Waste receptacles, stainless steel, with top, 13 gallon | | .800 | | 130 | 9.90 | | 139.90 | 160 |
| 05-001 | | **BULLETIN BOARD** Cork sheets, unbacked, no frame, 1/8" thick | | | | | | | | |
| | 002 | | | | | | | | | |
| | 212 | Prefabricated, 1/4" cork, 4' x 4' with aluminum frame | 2 Carp | 1.140 | Ea. | 100 | 14.10 | | 114.10 | 130 |
| | 214 | Aluminum frame with glass door | | 1.140 | | 465 | 14.10 | | 479.10 | 535 |
| | 230 | Prefabricated, sliding glass, enclosed, 3' x 4' | | 1 | | 415 | 12.35 | | 427.35 | 475 |
| | 260 | 6' x 4' | | 2.290 | | 575 | 28 | | 603 | 675 |
| 07-001 | | **CANOPIES** Wall hung, aluminum, prefinished, 8' x 10' | K-2 | 18.460 | | 725 | 240 | 135 | 1,100 | 1,350 |
| | 030 | 8' x 20' | " | 21.820 | | 850 | 280 | 160 | 1,290 | 1,575 |
| | 230 | Aluminum entrance canopies, flat soffit | | | | | | | | |
| | 250 | 3'-6" x 4'-0", clear anodized | 2 Carp | 4 | Ea. | 300 | 49 | | 349 | 410 |
| | 470 | Canvas awnings, including canvas, frame & lettering | | | | | | | | |
| | 500 | Minimum | 2 Carp | .160 | S.F. | 5 | 1.98 | | 6.98 | 8.65 |
| | 530 | Average | | .178 | | 10 | 2.20 | | 12.20 | 14.50 |
| | 550 | Maximum | | .200 | | 15 | 2.47 | | 17.47 | 20 |
| 08-001 | | **CANOPIES, RESIDENTIAL** Prefabricated | | | | | | | | |
| | 002 | | | | | | | | | |
| | 050 | Carport, free standing, aluminum, 16' x 8' | 2 Carp | 5.330 | Ea. | 300 | 66 | | 366 | 435 |
| | 052 | 20' x 10' | " | 8 | | 545 | 99 | | 644 | 755 |
| | 100 | Door canopies, aluminum, 36" projection, 4' wide | 1 Carp | 1 | | 140 | 12.35 | | 152.35 | 175 |
| | 102 | 5' wide | | 1.330 | | 175 | 16.45 | | 191.45 | 220 |
| | 104 | 6' wide | | 1.600 | | 200 | 19.75 | | 219.75 | 250 |
| | 106 | 8' wide | | 2 | | 240 | 25 | | 265 | 305 |
| | 108 | 10' wide | | 2.670 | | 295 | 33 | | 328 | 375 |
| | 110 | | | | | | | | | |
| | 120 | 48" projection, 4' wide | 1 Carp | 1 | Ea. | 190 | 12.35 | | 202.35 | 230 |
| | 122 | 5' wide | | 1.330 | | 235 | 16.45 | | 251.45 | 285 |
| | 124 | 6' wide | | 1.600 | | 270 | 19.75 | | 289.75 | 330 |
| | 126 | 8' wide | | 2 | | 325 | 25 | | 350 | 395 |
| | 128 | 10' wide | | 2.670 | | 395 | 33 | | 428 | 485 |
| | 130 | | | | | | | | | |
| | 200 | Patio cover, aluminum, 16' x 8' | 2 Carp | 5.330 | Ea. | 265 | 66 | | 331 | 395 |
| | 204 | 20' x 10' | " | 8 | | 385 | 99 | | 484 | 580 |
| | 300 | Window awnings, aluminum, window 3' high, 4' wide | 1 Carp | .800 | | 73 | 9.90 | | 82.90 | 96 |
| | 302 | 6' wide | " | 1 | | 105 | 12.35 | | 117.35 | 135 |

For expanded coverage of these items see *Building Construction Cost Data 1985*

## 10.1 Specialties

| | | CREW | MAN-HOURS | UNIT | MAT. | LABOR | EQUIP. | TOTAL | TOTAL INCL O&P |
|---|---|---|---|---|---|---|---|---|---|
| 08 304 | 9' wide | 1 Carp | 1.600 | Ea. | 140 | 19.75 | | 159.75 | 185 |
| 306 | 12' wide | | 2.670 | | 185 | 33 | | 218 | 255 |
| 310 | Window, 4' high, 4' wide | | .800 | | 120 | 9.90 | | 129.90 | 150 |
| 312 | 6' wide | | 1 | | 155 | 12.35 | | 167.35 | 190 |
| 314 | 9' wide | | 1.600 | | 210 | 19.75 | | 229.75 | 260 |
| 316 | 12' wide | | 2.670 | | 275 | 33 | | 308 | 355 |
| 320 | Window, 6' high, 4' wide | | .800 | | 185 | 9.90 | | 194.90 | 220 |
| 322 | 6' wide | | 1 | | 245 | 12.35 | | 257.35 | 290 |
| 324 | 9' wide | | 1.600 | | 310 | 19.75 | | 329.75 | 370 |
| 326 | 12' wide | | 2.670 | | 420 | 33 | | 453 | 515 |
| 340 | Roll-up aluminum, 2'-6" wide | | .571 | | 54 | 7.05 | | 61.05 | 71 |
| 342 | 3' wide | | .667 | | 71 | 8.25 | | 79.25 | 91 |
| 344 | 4' wide | | .800 | | 86 | 9.90 | | 95.90 | 110 |
| 346 | 6' wide | | 1 | | 110 | 12.35 | | 122.35 | 140 |
| 348 | 9' wide | | 1.600 | | 145 | 19.75 | | 164.75 | 190 |
| 350 | 12' wide | | 2.670 | | 180 | 33 | | 213 | 250 |
| 360 | Window awnings, canvas, 24" drop, 3' wide | | .800 | | 26 | 9.90 | | 35.90 | 44 |
| 362 | 4' wide | | .800 | | 30 | 9.90 | | 39.90 | 49 |
| 370 | 30" drop, 3' wide | | .800 | | 29 | 9.90 | | 38.90 | 48 |
| 372 | 4' wide | | .800 | | 32 | 9.90 | | 41.90 | 51 |
| 374 | 5' wide | | .889 | | 42 | 11 | | 53 | 64 |
| 376 | 6' wide | | 1 | | 46 | 12.35 | | 58.35 | 70 |
| 378 | 8' wide | | 1.330 | | 57 | 16.45 | | 73.45 | 89 |
| 380 | 10' wide | | 2 | | 68 | 25 | | 93 | 115 |
| 10-001 | **CHALKBOARD** Cement asbestos, no frame, economy | 2 Carp | .060 | S.F. | 2.45 | .75 | | 3.20 | 3.88 |
| 010 | Deluxe | " | .060 | " | 2.80 | .75 | | 3.55 | 4.26 |
| 080 | Porcelain enamel, 24 ga. steel, 4' high, with aluminum | | | | | | | | |
| 090 | trim, alum. foil backing, with core materials as follows: | | | | | | | | |
| 210 | Slate, 3/8" thick, frame not included, to 4' wide | 2 Carp | .070 | S.F. | 7.55 | .86 | | 8.41 | 9.65 |
| 220 | To 4'-6" wide | | .070 | " | 8.30 | .86 | | 9.16 | 10.50 |
| 300 | Frame for chalkboards, aluminum, chalk tray | | .055 | L.F. | 3.60 | .68 | | 4.28 | 5.05 |
| 310 | Trim | | .042 | " | 1.05 | .51 | | 1.56 | 1.97 |
| 12-001 | **CHUTES** Linen or refuse, incl. sprinklers | | | | | | | | |
| 005 | Aluminized steel, 16 ga., 18" diameter | 2 Shee | 4.570 | Floor | 525 | 63 | | 588 | 680 |
| 010 | 24" diameter | | 5 | | 595 | 69 | | 664 | 765 |
| 040 | Galvanized steel, 16 ga., 18" diameter | | 4.570 | | 485 | 63 | | 548 | 635 |
| 050 | 24" diameter | | 5 | | 535 | 69 | | 604 | 700 |
| 080 | Stainless steel, 18" diameter | | 4.570 | | 695 | 63 | | 758 | 865 |
| 090 | 24" diameter | | 5 | | 830 | 69 | | 899 | 1,025 |
| 200 | Mail chutes, aluminum & glass, 14-1/4" wide, 4-5/8" deep | | 4 | | 500 | 55 | | 555 | 640 |
| 210 | 8-5/8" deep | | 4.210 | | 600 | 58 | | 658 | 755 |
| 260 | Lobby collection boxes, aluminum | | 3.560 | Ea. | 755 | 49 | | 804 | 910 |
| 270 | Bronze or stainless | | 3.560 | " | 1,050 | 49 | | 1,099 | 1,225 |
| 20-001 | **DIRECTORY BOARDS** Plastic, glass covered, 30" x 20" | 2 Carp | 5.330 | Ea. | 210 | 66 | | 276 | 335 |
| 010 | 36" x 48" | | 8 | | 450 | 99 | | 549 | 650 |
| 090 | Outdoor, weatherproof, black plastic, 36" x 24" | | 8 | | 420 | 99 | | 519 | 620 |
| 100 | 36" x 36" | | 10.670 | | 470 | 130 | | 600 | 725 |
| 22-001 | **DISAPPEARING STAIRWAY** No trim included | | | | | | | | |
| 002 | 8'-0" ceiling | 2 Carp | 4 | Ea. | 925 | 49 | | 974 | 1,100 |
| 003 | 9'-0" ceiling | | 4 | | 950 | 49 | | 999 | 1,125 |
| 004 | 10'-0" ceiling | | 5.330 | | 1,000 | 66 | | 1,066 | 1,200 |
| 005 | 11'-0" ceiling | | 5.330 | | 1,175 | 66 | | 1,241 | 1,400 |
| 006 | 12'-0" ceiling | | 5.330 | | 1,225 | 66 | | 1,291 | 1,450 |
| 010 | Custom grade, pine, 8'-6" ceiling, minimum | 1 Carp | 2 | | 62 | 25 | | 87 | 105 |
| 015 | Average | | 2.290 | | 145 | 28 | | 173 | 205 |
| 020 | Maximum | | 2.670 | | 225 | 33 | | 258 | 300 |
| 050 | Custom grade, pine, heavy duty, pivoted, 8'-6" ceiling pine | | 2.670 | | 270 | 33 | | 303 | 350 |
| 060 | 16'-0" ceiling | | 4 | | 470 | 49 | | 519 | 595 |
| 080 | Economy folding, pine, 8'-6" ceiling | | 2 | | 54 | 25 | | 79 | 99 |

For expanded coverage of these items see *Building Construction Cost Data 1985*

# 10.1 Specialties

| | | | CREW | MAN-HOURS | UNIT | MAT. | LABOR | EQUIP. | TOTAL | TOTAL INCL O&P |
|---|---|---|---|---|---|---|---|---|---|---|
| 22 | 090 | 9'-6" ceiling | 1 Carp | 2 | Ea. | 70 | 25 | | 95 | 115 |
| | 100 | Fire escape, galvanized steel, 8'-0" to 10'-4" ceiling | 2 Carp | 16 | | 1,000 | 200 | | 1,200 | 1,425 |
| | 101 | 10'-6" to 13'-6" ceiling | | 16 | | 1,250 | 200 | | 1,450 | 1,700 |
| | 110 | Automatic electric, aluminum, floor to floor height, 8' to 9' | ↓ | 16 | ↓ | 4,200 | 200 | | 4,400 | 4,925 |
| 37-001 | | LOCKERS Steel, baked enamel, 60" or 72", single tier, minimum | 1 Shee | .571 | Opng. | 52 | 7.90 | | 59.90 | 70 |
| | 010 | Maximum | | .667 | | 95 | 9.20 | | 104.20 | 120 |
| | 030 | 2 tier, 60" or 72" total height, minimum | | .308 | | 34 | 4.25 | | 38.25 | 44 |
| | 040 | Maximum | ↓ | .400 | ↓ | 49 | 5.50 | | 54.50 | 63 |
| | 240 | Teacher and pupil wardrobes, enameled | | | | | | | | |
| | 250 | 22" x 15" x 61" high, minimum | 1 Shee | .800 | Ea. | 90 | 11.05 | | 101.05 | 115 |
| | 255 | Average | | .889 | | 150 | 12.25 | | 162.25 | 185 |
| | 270 | Maximum | ↓ | 1 | | 155 | 13.80 | | 168.80 | 195 |
| | 360 | For hanger rods, add | | | | 1.20 | | | 1.20 | 1.32M |
| | 370 | For stainless steel lockers, add | | | | 100% | | | | |
| 40-001 | | MAIL Boxes, horizontal, key lock, 5"H x 6"W x 15"D, alum., rear loading | 1 Carp | .235 | | 50 | 2.91 | | 52.91 | 60 |
| | 010 | Front loading | | .235 | | 54 | 2.91 | | 56.91 | 64 |
| | 020 | Double, 5"H x 12"W x 15"D, rear loading | | .308 | | 92 | 3.80 | | 95.80 | 105 |
| | 030 | Front loading | | .308 | | 100 | 3.80 | | 103.80 | 115 |
| | 050 | Quadruple, 10"H x 12"W x 15"D, rear loading | | .400 | | 160 | 4.94 | | 164.94 | 185 |
| | 060 | Front loading | | .400 | | 170 | 4.94 | | 174.94 | 195 |
| | 160 | Vault type, horizontal, for apartments, 4" x 5" | | .235 | | 50 | 2.91 | | 52.91 | 60 |
| | 170 | Alphabetical directories, 50 names | | .800 | | 115 | 9.90 | | 124.90 | 140 |
| | 190 | Letter slot, residential | | .400 | | 27 | 4.94 | | 31.94 | 38 |
| | 200 | Post office type | | 1 | | 105 | 12.35 | | 117.35 | 135 |
| 42-001 | | MEDICINE CABINETS With mirror, stock, 16" x 22", unlighted | | .571 | | 41 | 7.05 | | 48.05 | 56 |
| | 010 | Lighted | | 1.330 | | 80 | 16.45 | | 96.45 | 115 |
| | 030 | Sliding mirror doors, 34" x 21", unlighted | | 1.140 | | 63 | 14.10 | | 77.10 | 92 |
| | 040 | Lighted | | 1.600 | | 130 | 19.75 | | 149.75 | 175 |
| | 060 | Center mirror, 2 end cabinets, unlighted, 48" long | | 1.140 | | 165 | 14.10 | | 179.10 | 205 |
| | 070 | 72" long | ↓ | 1.600 | | 195 | 19.75 | | 214.75 | 245 |
| | 090 | For lighting, 48" long, add | 1 Elec | 2.290 | | 47 | 31 | | 78 | 100 |
| | 100 | 72" long, add | " | 2.670 | | 70 | 37 | | 107 | 135 |
| | 120 | Hotel cabinets, stainless, with lower shelf, unlighted | 1 Carp | .800 | | 140 | 9.90 | | 149.90 | 170 |
| | 130 | Lighted | " | 1.600 | ↓ | 195 | 19.75 | | 214.75 | 245 |
| 45-001 | | PARTITIONS, FOLDING ACCORDION See also division 8.2-32 | | | | | | | | |
| | 010 | Vinyl covered, over 150 S.F., frame not included | | | | | | | | |
| | 030 | Residential, 1.25 lb. per S.F., 8 ft. maximum height | 2 Carp | .053 | S.F. | 6.50 | .66 | | 7.16 | 8.20 |
| | 040 | Commercial, 1.75 lb. per S.F., 8 ft. maximum height | | .071 | | 6.65 | .88 | | 7.53 | 8.70 |
| | 090 | Acoustical, 3 lb. per S.F., 17 ft. maximum height | | .160 | | 7.90 | 1.98 | | 9.88 | 11.80 |
| | 120 | 5 lb. per S.F., 27 ft. maximum height | | .168 | | 11.35 | 2.08 | | 13.43 | 15.80 |
| | 150 | Vinyl clad wood or steel, electric operation, 5.0 psf | | .100 | | 17.85 | 1.24 | | 19.09 | 22 |
| | 190 | Wood, non-acoustic, birch or mahogany, to 10 ft. high | ↓ | .053 | ↓ | 10.50 | .66 | | 11.16 | 12.60 |
| 47-001 | | PARTITIONS, FOLDING LEAF Acoustic, wood | | | | | | | | |
| | 010 | Vinyl faced, to 18' high, 6 psf, minimum | 2 Carp | .267 | S.F. | 24 | 3.29 | | 27.29 | 32 |
| | 015 | Average | | .356 | | 25 | 4.39 | | 29.39 | 34 |
| | 020 | Maximum | | .533 | | 26 | 6.60 | | 32.60 | 39 |
| | 040 | Formica or hardwood finish, minimum | | .267 | | 24 | 3.29 | | 27.29 | 32 |
| | 050 | Maximum | | .533 | | 25 | 6.60 | | 31.60 | 38 |
| | 060 | Wood, low acoustical type, 4.5 psf, to 12' high | ↓ | .320 | ↓ | 14.50 | 3.95 | | 18.45 | 22 |
| 52-001 | | PARTITIONS, MOVABLE OFFICE Demountable, add for doors | | | | | | | | |
| | 010 | Do not deduct door openings from total L.F. | | | | | | | | |
| | 090 | Asbestos cement, 1-3/4" thick, prefinished, low walls | 2 Carp | .200 | L.F. | 29 | 2.47 | | 31.47 | 36 |
| | 100 | Full height | " | .400 | | 36 | 4.94 | | 40.94 | 47 |
| | 120 | Economy grade, 1-3/4" thick, deduct | | | | 20% | | | | |
| | 130 | High quality, 4" thick, add | | | | 100% | | | | |
| | 340 | Metal, to 9'-6" high, enameled steel, no glass | 2 Carp | .400 | | 62 | 4.94 | | 66.94 | 76 |
| | 350 | Steel frame, all glass | | .400 | | 71 | 4.94 | | 75.94 | 86 |
| | 370 | Vinyl covered, no glass | | .400 | | 72 | 4.94 | | 76.94 | 87 |
| | 380 | Steel frame with 52% glass | ↓ | .400 | ↓ | 64 | 4.94 | | 68.94 | 78 |

For expanded coverage of these items see *Building Construction Cost Data 1985*

## 10.1 Specialties

| | | | CREW | MAN-HOURS | UNIT | MAT. | LABOR | EQUIP. | TOTAL | TOTAL INCL O&P |
|---|---|---|---|---|---|---|---|---|---|---|
| 52 | 400 | Free standing, 4'-8" high, steel with glass | 2 Carp | .160 | L.F. | 38 | 1.98 | | 39.98 | 45 |
| | 410 | Acoustical | | .160 | | 51 | 1.98 | | 52.98 | 59 |
| | 430 | Low rails, 3'-3" high, enameled steel | | .160 | | 30 | 1.98 | | 31.98 | 36 |
| | 440 | Vinyl covered | ↓ | .160 | ↓ | 38 | 1.98 | | 39.98 | 45 |
| | 550 | For acoustical partitions, add, minimum | | | S.F. | .40 | | | .40 | .44M |
| | 555 | Maximum | | | " | .55 | | | .55 | .60M |
| | 570 | For doors, not incl. hardware, hollow metal door, add | 2 Carp | 3.720 | Ea. | 135 | 46 | | 181 | 220 |
| | 580 | Hardwood door, add | | 4.710 | | 97 | 58 | | 155 | 200 |
| | 600 | Hardware for doors, not incl. closers, keyed | | .552 | | 65 | 6.80 | | 71.80 | 82 |
| | 610 | Non-keyed | ↓ | .552 | ↓ | 58 | 6.80 | | 64.80 | 75 |
| 56-001 | | **PARTITIONS, PORTABLE** Divider walls, free standing, fiber core | | | | | | | | |
| | 002 | | | | | | | | | |
| | 150 | 5 ft. high, burlap face, straight | 2 Carp | .107 | L.F. | 47 | 1.32 | | 48.32 | 54 |
| | 160 | Curved | | .107 | | 67 | 1.32 | | 68.32 | 76 |
| | 180 | Carpeted face, straight | | .107 | | 65 | 1.32 | | 66.32 | 74 |
| | 190 | Curved | | .107 | | 92 | 1.32 | | 93.32 | 105 |
| | 220 | Plastic laminated face, straight | | .107 | | 42 | 1.32 | | 43.32 | 48 |
| | 230 | Curved | ↓ | .107 | ↓ | 60 | 1.32 | | 61.32 | 68 |
| 57-001 | | **PARTITIONS, SHOWER** Economy, painted steel, steel | | | | | | | | |
| | 010 | base, no door or plumbing included | 2 Shee | 3.200 | Ea. | 115 | 44 | | 159 | 195 |
| | 030 | Square, 32" x 32", stock, with receptor & door, fiberglass | ↓ | 3.200 | | 255 | 44 | | 299 | 350 |
| | 060 | Galvanized and painted steel | ↓ | 3.200 | ↓ | 175 | 44 | | 219 | 265 |
| | 070 | Shower stall, double wall, incl. receptor but not including | | | | | | | | |
| | 080 | door or plumbing, enameled steel | 2 Shee | 3.200 | Ea. | 570 | 44 | | 614 | 700 |
| | 150 | Circular fiberglass, 36" diameter, no plumbing included | | 4 | | 225 | 55 | | 280 | 335 |
| | 170 | One piece, 36" diameter, less door | | 4 | | 255 | 55 | | 310 | 370 |
| | 180 | With door | ↓ | 4.570 | ↓ | 410 | 63 | | 473 | 550 |
| | 190 | | | | | | | | | |
| | 410 | Doors, economy plastic, 24" wide | 1 Shee | .889 | Ea. | 63 | 12.25 | | 75.25 | 89 |
| | 420 | Tempered glass door, economy | | 1 | | 94 | 13.80 | | 107.80 | 125 |
| | 470 | Deluxe, tempered glass, chrome on brass frame, minimum | | 1.600 | | 265 | 22 | | 287 | 325 |
| | 500 | Maximum | ↓ | 1.600 | ↓ | 490 | 22 | | 512 | 575 |
| 60-001 | | **PARTITIONS, TOILET** For ceiling framing, see division 5.4-68 | | | | | | | | |
| | 010 | Cubicles, ceiling hung, marble | 2 Marb | 8 | Ea. | 450 | 98 | | 548 | 650 |
| | 020 | Painted metal | 2 Carp | 4 | | 200 | 49 | | 249 | 300 |
| | 030 | Plastic laminate on particle board | | 4 | | 295 | 49 | | 344 | 405 |
| | 040 | Porcelain enamel | | 4 | | 605 | 49 | | 654 | 745 |
| | 050 | Stainless steel | ↓ | 4 | | 555 | 49 | | 604 | 690 |
| | 060 | For handicap units, add | | | ↓ | 66 | | | 66 | 73M |
| | 070 | | | | | | | | | |
| | 080 | Floor & ceiling anchored, marble | 2 Marb | 6.400 | Ea. | 465 | 79 | | 544 | 635 |
| | 100 | Painted metal | 2 Carp | 3.200 | | 200 | 40 | | 240 | 285 |
| | 110 | Plastic laminate on particle board | | 3.200 | | 350 | 40 | | 390 | 450 |
| | 120 | Porcelain enamel | | 3.200 | | 615 | 40 | | 655 | 740 |
| | 130 | Stainless steel | ↓ | 3.200 | | 555 | 40 | | 595 | 675 |
| | 140 | For handicap units, add | | | | 66 | | | 66 | 73M |
| | 160 | Floor mounted, marble | 2 Marb | 5.330 | | 400 | 66 | | 466 | 540 |
| | 170 | Painted metal | 2 Carp | 2.290 | | 180 | 28 | | 208 | 245 |
| | 180 | Plastic laminate on particle board | | 2.290 | | 170 | 28 | | 198 | 230 |
| | 190 | Porcelain enamel | | 2.290 | | 605 | 28 | | 633 | 710 |
| | 200 | Stainless steel | ↓ | 2.290 | | 555 | 28 | | 583 | 655 |
| | 210 | For handicap units, add | | | | 62 | | | 62 | 68M |
| | 220 | For juvenile units, deduct | | | ↓ | 20 | | | 20 | 22M |
| | 230 | | | | | | | | | |
| | 240 | Floor mounted, headrail braced, marble | 2 Marb | 5.330 | Ea. | 425 | 66 | | 491 | 570 |
| | 250 | Painted metal | 2 Carp | 2.670 | | 190 | 33 | | 223 | 260 |
| | 260 | Plastic laminate on particle board | | 2.670 | | 180 | 33 | | 213 | 250 |
| | 270 | Porcelain enamel | | 2.670 | | 600 | 33 | | 633 | 710 |
| | 280 | Stainless steel | ↓ | 2.670 | | 545 | 33 | | 578 | 650 |
| | 290 | For handicap units, add | | | ↓ | 66 | | | 66 | 73M |

For expanded coverage of these items see *Building Construction Cost Data 1985*

## 10.1 Specialties

| | | | Crew | Man-Hours | Unit | Mat. | Labor | Equip. | Total | Total Incl O&P |
|---|---|---|---|---|---|---|---|---|---|---|
| 60 | 300 | Wall hung partitions, painted metal | 2 Carp | 2.290 | Ea. | 370 | 28 | | 398 | 450 |
| | 310 | Plastic laminate | | 2.290 | | 430 | 28 | | 458 | 520 |
| | 320 | Porcelain enamel | | 2.290 | | 600 | 28 | | 628 | 705 |
| | 330 | Stainless steel | ↓ | 2.290 | | 650 | 28 | | 678 | 760 |
| | 340 | For handicap units, add | | | ↓ | 66 | | | 66 | 73M |
| | 400 | Screens, entrance, floor mounted, 54" high | | | | | | | | |
| | 410 | Marble | D-1 | .457 | L.F. | 72 | 5.20 | | 77.20 | 87 |
| | 420 | Painted metal | 2 Carp | .267 | | 46 | 3.29 | | 49.29 | 56 |
| | 430 | Plastic laminate on particle board | | .267 | | 56 | 3.29 | | 59.29 | 67 |
| | 440 | Porcelain enamel | | .267 | | 110 | 3.29 | | 113.29 | 125 |
| | 450 | Stainless steel | ↓ | .267 | ↓ | 97 | 3.29 | | 100.29 | 110 |
| | 460 | Urinal screen, 18" wide, ceiling braced, marble | D-1 | 2.670 | Ea. | 215 | 30 | | 245 | 285 |
| | 470 | Painted metal | 2 Carp | 2 | | 110 | 25 | | 135 | 160 |
| | 480 | Plastic laminate on particle board | | 2 | | 145 | 25 | | 170 | 200 |
| | 490 | Porcelain enamel | | 2 | | 210 | 25 | | 235 | 270 |
| | 500 | Stainless steel | ↓ | 2 | ↓ | 210 | 25 | | 235 | 270 |
| | 505 | | | | | | | | | |
| | 510 | Floor mounted, head rail braced | | | | | | | | |
| | 520 | Marble | D-1 | 2.670 | Ea. | 190 | 30 | | 220 | 255 |
| | 530 | Painted metal | 2 Carp | 2 | | 115 | 25 | | 140 | 165 |
| | 540 | Plastic laminate on particle board | | 2 | | 135 | 25 | | 160 | 190 |
| | 550 | Porcelain enamel | | 2 | | 210 | 25 | | 235 | 270 |
| | 560 | Stainless steel | ↓ | 2 | | 210 | 25 | | 235 | 270 |
| | 570 | Pilaster, flush, marble | D-1 | 1.780 | | 235 | 20 | | 255 | 290 |
| | 580 | Painted metal | 2 Carp | 1.600 | | 60 | 19.75 | | 79.75 | 97 |
| | 590 | Plastic laminate on particle board | | 1.600 | | 75 | 19.75 | | 94.75 | 115 |
| | 600 | Porcelain enamel | | 1.600 | | 150 | 19.75 | | 169.75 | 195 |
| | 610 | Stainless steel | ↓ | 1.600 | ↓ | 150 | 19.75 | | 169.75 | 195 |
| | 615 | | | | | | | | | |
| | 620 | Post braced, marble | D-1 | 1.780 | Ea. | 230 | 20 | | 250 | 285 |
| | 630 | Painted metal | 2 Carp | 1.600 | | 60 | 19.75 | | 79.75 | 97 |
| | 640 | Plastic laminate on particle board | | 1.600 | | 75 | 19.75 | | 94.75 | 115 |
| | 650 | Porcelain enamel | | 1.600 | | 150 | 19.75 | | 169.75 | 195 |
| | 660 | Stainless steel | ↓ | 1.600 | ↓ | 150 | 19.75 | | 169.75 | 195 |
| | 670 | Wall hung, bracket supported | | | | | | | | |
| | 680 | Painted metal | 2 Carp | 1.600 | Ea. | 92 | 19.75 | | 111.75 | 135 |
| | 690 | Plastic laminate on particle board | | 1.600 | | 87 | 19.75 | | 106.75 | 125 |
| | 700 | Porcelain enamel | | 1.600 | | 180 | 19.75 | | 199.75 | 230 |
| | 710 | Stainless steel | ↓ | 1.600 | ↓ | 170 | 19.75 | | 189.75 | 220 |
| | 720 | | | | | | | | | |
| | 740 | Flange supported, painted metal | 2 Carp | 1.600 | Ea. | 185 | 19.75 | | 204.75 | 235 |
| | 750 | Plastic laminate on particle board | | 1.600 | | 105 | 19.75 | | 124.75 | 145 |
| | 760 | Porcelain enamel | | 1.600 | | 220 | 19.75 | | 239.75 | 275 |
| | 770 | Stainless steel | | 1.600 | | 210 | 19.75 | | 229.75 | 260 |
| | 780 | Wedge type, painted metal | | 1.600 | | 125 | 19.75 | | 144.75 | 170 |
| | 790 | Plastic laminate on ptcl. board | | 1.600 | | 90 | 19.75 | | 109.75 | 130 |
| | 800 | Porcelain enamel | | 1.600 | | 195 | 19.75 | | 214.75 | 245 |
| | 810 | Stainless steel | ↓ | 1.600 | ↓ | 195 | 19.75 | | 214.75 | 245 |
| 67-001 | | **PROJECTION SCREENS** Wall or ceiling hung, glass beaded | | | | | | | | |
| | 010 | Manually operated, economy | 2 Carp | .032 | S.F. | 3.25 | .40 | | 3.65 | 4.20 |
| | 030 | Intermediate | | .036 | | 5.15 | .44 | | 5.59 | 6.35 |
| | 040 | Deluxe | ↓ | .040 | ↓ | 7.50 | .49 | | 7.99 | 9.05 |
| 75-001 | | **SHELVING** Metal, industrial, cross-braced, 3' wide, 12" deep | | | | | | | | |
| | 010 | 24" deep | 1 Sswk | .024 | SF Shlf | 1.95 | .32 | | 2.27 | 2.70 |
| | 220 | Wide span, 1600 lb. capacity per shelf, 6' wide, 30" deep | | .021 | | 3.90 | .28 | | 4.18 | 4.77 |
| | 240 | 42" deep | ↓ | .018 | ↓ | 3.50 | .24 | | 3.74 | 4.26 |
| | 300 | Residential, adjustable metal, 12" deep | | | | | | | | |
| | 310 | 23" to 37" wide | 1 Carp | .500 | Ea. | 5.90 | 6.20 | | 12.10 | 16.25 |
| | 320 | 47" to 61" wide | | .500 | | 8.55 | 6.20 | | 14.75 | 19.20 |
| | 330 | 71" to 88" wide | ↓ | .500 | ↓ | 11.75 | 6.20 | | 17.95 | 23 |

For expanded coverage of these items see *Building Construction Cost Data 1985*

## 10.1 Specialties

| | | | CREW | MAN-HOURS | UNIT | MAT. | LABOR | EQUIP. | TOTAL | TOTAL INCL O&P |
|---|---|---|---|---|---|---|---|---|---|---|
| 75 | 340 | 95" to 109" | 1 Carp | .500 | Ea. | 13.90 | 6.20 | | 20.10 | 25 |
| | 350 | For closet rod, add | | | | 22% | 2% | | | |
| | 360 | For 14" shelves, add | | | ↓ | 12% | 5% | | | |
| 77-001 | | SIGNS Letters, individual, 2" high, cast aluminum | | | | | | | | |
| | 390 | Plaques, 20" x 30", for up to 300 letters, cast aluminum | 2 Carp | 4 | Ea. | 650 | 49 | | 699 | 795 |
| | 400 | Cast bronze | | 4 | | 720 | 49 | | 769 | 870 |
| | 420 | 30" x 40", cast aluminum | | 5.330 | | 1,250 | 66 | | 1,316 | 1,475 |
| | 430 | Cast bronze | | 5.330 | | 1,375 | 66 | | 1,441 | 1,625 |
| | 510 | Acrylic exit signs, 15" x 6", surface mounted, minimum | | .533 | | 8.20 | 6.60 | | 14.80 | 19.45 |
| | 520 | Maximum | | .800 | ↓ | 26 | 9.90 | | 35.90 | 44 |
| | 570 | Plexiglass, exterior, illuminated, single face | | .160 | S.F. | 44 | 1.98 | | 45.98 | 52 |
| | 580 | Double face | | .213 | | 57 | 2.63 | | 59.63 | 67 |
| | 640 | Painted plywood (MDO), over 4' x 8' | | .133 | | 6.50 | 1.65 | | 8.15 | 9.75 |
| | 660 | Under 4' x 8' | ↓ | .160 | ↓ | 7.30 | 1.98 | | 9.28 | 11.15 |
| | 670 | For metal edge moldings, add | | | L.F. | 2.65 | | | 2.65 | 2.91M |
| 80-001 | | TELEPHONE ENCLOSURE Desk-top type | 2 Carp | 2 | Ea. | 425 | 25 | | 450 | 505 |
| | 030 | Shelf type, wall hung, minimum | | 3.200 | | 265 | 40 | | 305 | 355 |
| | 040 | Maximum | ↓ | 3.200 | | 925 | 40 | | 965 | 1,075 |
| | 050 | | | | | | | | | |
| | 060 | Booth type, painted steel, indoor or outdoor, minimum | 2 Carp | 10.670 | Ea. | 1,425 | 130 | | 1,555 | 1,775 |
| | 070 | Maximum (stainless steel) | | 10.670 | | 3,625 | 130 | | 3,755 | 4,200 |
| | 190 | Outdoor, drive-up type, wall mounted | ↓ | 4 | | 495 | 49 | | 544 | 625 |
| | 200 | Post mounted, stainless steel posts | ↓ | 5.330 | ↓ | 875 | 66 | | 941 | 1,075 |

## 11.1 Architectural Equipment

| | | | CREW | MAN-HOURS | UNIT | MAT. | LABOR | EQUIP. | TOTAL | TOTAL INCL O&P |
|---|---|---|---|---|---|---|---|---|---|---|
| 03-001 | | APPLIANCES Cooking range, 30" free standing, 1 oven, minimum | 2 Clab | 1.600 | Ea. | 320 | 16 | | 336 | 375 |
| | 005 | Maximum | " | 4 | | 1,200 | 40 | | 1,240 | 1,375 |
| | 035 | Built-in, 30" wide, 1 oven, minimum | 2 Carp | 4 | | 480 | 49 | | 529 | 605 |
| | 040 | Maximum | " | 8 | | 800 | 99 | | 899 | 1,025 |
| | 090 | Counter top cook tops, 4 burner, standard, minimum | 1 Elec | 1.330 | | 200 | 18.35 | | 218.35 | 250 |
| | 095 | Maximum | | 2.670 | | 430 | 37 | | 467 | 530 |
| | 125 | Microwave oven, minimum | | 2 | | 220 | 28 | | 248 | 285 |
| | 130 | Maximum | ↓ | 4 | | 1,000 | 55 | | 1,055 | 1,175 |
| | 150 | Combination range, refrigerator and sink, 30" wide, minimum | L-1 | 8 | | 640 | 110 | | 750 | 880 |
| | 155 | Maximum | " | 16 | ↓ | 1,180 | 220 | | 1,400 | 1,650 |
| | 164 | Combination range, refrigerator, sink, microwave | | | | | | | | |
| | 166 | oven and ice maker | L-1 | 20 | Ea. | 3,175 | 275 | | 3,450 | 3,925 |
| | 175 | Compactor, residential size, 4 to 1 compaction, minimum | 1 Carp | 1.600 | | 340 | 19.75 | | 359.75 | 405 |
| | 180 | Maximum | " | 2.670 | | 490 | 33 | | 523 | 590 |
| | 275 | Dishwasher, built-in, 2 cycles, minimum | L-1 | 4 | | 270 | 55 | | 325 | 385 |
| | 280 | Maximum | | 8 | | 360 | 110 | | 470 | 570 |
| | 330 | Garbage disposer, sink type, minimum | | 3.200 | | 50 | 44 | | 94 | 125 |
| | 335 | Maximum | ↓ | 5.330 | | 180 | 73 | | 253 | 315 |
| | 415 | Hood for range, 2 speed, vented, 30" wide, minimum | L-3 | 3.200 | | 60 | 42 | | 102 | 130 |
| | 420 | Maximum | | 5.330 | | 230 | 70 | | 300 | 365 |
| | 430 | 42" wide, minimum | | 3.200 | | 155 | 42 | | 197 | 235 |
| | 435 | Maximum | ↓ | 5.330 | | 270 | 70 | | 340 | 405 |
| | 450 | For ventless hood, 2 speed, add | | | | 12 | | | 12 | 13.20M |
| | 465 | For vented 1 speed, deduct from maximum | | | | 25 | | | 25 | 28M |
| | 538 | Oven, built in, standard | 1 Elec | 4 | | 400 | 55 | | 455 | 525 |
| | 539 | Deluxe | " | 4 | | 680 | 55 | | 735 | 835 |
| | 550 | Refrigerator, no frost, 10 C.F. to 12 C.F. minimum | 2 Clab | 1.600 | | 350 | 16 | | 366 | 410 |
| | 560 | Maximum | " | 2.670 | ↓ | 580 | 27 | | 607 | 680 |

For expanded coverage of these items see *Building Construction Cost Data 1985*

## 11.1 Architectural Equipment

| | | CREW | MAN-HOURS | UNIT | MAT. | LABOR | EQUIP. | TOTAL | TOTAL INCL O&P |
|---|---|---|---|---|---|---|---|---|---|
| 03 640 | Sump pump cellar drainer, 1/3 H.P., minimum | 1 Plum | 2.670 | Ea. | 75 | 37 | | 112 | 140 |
| 645 | Maximum | " | 4 | | 200 | 55 | | 255 | 305 |
| 735 | Water softener, automatic, to 30 grains per gallon | 2 Plum | 3.200 | | 330 | 44 | | 374 | 435 |
| 740 | To 75 grains per gallon | " | 4 | | 650 | 55 | | 705 | 800 |
| 46-001 | SAUNA Prefabricated, incl. heater & controls, 7' high, 6' x 4' | L-7 | 12.730 | | 2,300 | 150 | | 2,450 | 2,775 |
| 170 | Door only, with tempered insulated glass window | 2 Carp | 4.710 | | 150 | 58 | | 208 | 255 |
| 180 | Prehung, incl. jambs, pulls & hardware | " | 1.330 | | 190 | 16.45 | | 206.45 | 235 |
| 250 | Heaters only (incl. above), wall mounted, to 200 C.F. | | | | 325 | | | 325 | 360M |
| 58-001 | STEAM BATH Heater, timer & head, single, to 140 C.F. | 1 Plum | 6.670 | | 725 | 92 | | 817 | 945 |
| 050 | To 300 C.F. | " | 7.270 | | 800 | 100 | | 900 | 1,050 |
| 270 | Conversion unit for residential tub | ↓ | | | 700 | | | 700 | 770M |
| 61-001 | VACUUM CLEANING Central, 3 valve, residential | 1 Skwk | 8.890 | Total | 540 | 110 | | 650 | 775 |
| 040 | 5 valve system | | 16 | | 575 | 200 | | 775 | 955 |
| 060 | 7 valve system | ↓ | 20 | | 750 | 250 | | 1,000 | 1,225 |
| 080 | 9 valve system | 1 Sswk | 26.670 | ↓ | 860 | 355 | | 1,215 | 1,550 |
| 67-001 | WINE VAULT Redwood, air conditioned, walk-in type | | | | | | | | |
| 002 | 6'-8" high, incl. racks, 2' x 4' for 156 bottles | 2 Carp | 8 | Ea. | 3,000 | 99 | | 3,099 | 3,450 |
| 020 | 4' x 6' for 614 bottles | | 10.670 | | 5,000 | 130 | | 5,130 | 5,700 |
| 040 | 6' x 12' for 1940 bottles | ↓ | 16 | ↓ | 9,500 | 200 | | 9,700 | 10,800 |
| 060 | Portable chillers, reach-in, 71"high x 27"wide x 23"deep | | | | | | | | |
| 065 | One temperature, 235 bottles | | | Ea. | 900 | | | 900 | 990M |
| 070 | Three temperature, 200 bottles | | | | 1,200 | | | 1,200 | 1,325M |
| 075 | 31" x 23" x 23", 60 bottles | | | ↓ | 600 | | | 600 | 660M |

## 12.1 Furnishings

| | | CREW | MAN-HOURS | UNIT | MAT. | LABOR | EQUIP. | TOTAL | TOTAL INCL O&P |
|---|---|---|---|---|---|---|---|---|---|
| 05-001 | BLINDS, EXTERIOR Aluminum, louvered, 1'-4" wide, 3'-0" long | 1 Carp | .800 | Pr. | 24 | 9.90 | | 33.90 | 42 |
| 020 | 4'-0" long | | .800 | | 28 | 9.90 | | 37.90 | 46 |
| 030 | 5'-4" long | | .800 | | 31 | 9.90 | | 40.90 | 50 |
| 040 | 6'-8" long | | .889 | | 41 | 11 | | 52 | 62 |
| 100 | Pine, louvered, primed, each 1'-2" wide, 3'-3" long | | .800 | | 36 | 9.90 | | 45.90 | 55 |
| 110 | 4'-7" long | | .800 | | 49 | 9.90 | | 58.90 | 70 |
| 125 | Each 1'-4" wide, 3'-0" long | | .800 | | 38 | 9.90 | | 47.90 | 57 |
| 135 | 5'-3" long | | .800 | | 44 | 9.90 | | 53.90 | 64 |
| 150 | Each 1'-6"wide, 3'-3" long | | .800 | | 41 | 9.90 | | 50.90 | 61 |
| 160 | 4'-7" long | | .800 | | 55 | 9.90 | | 64.90 | 76 |
| 250 | Polystyrene, solid raised panel, each 3'-3" wide, 3'-0" long | | .800 | | 52 | 9.90 | | 61.90 | 73 |
| 260 | 3'-11" long | | .800 | | 61 | 9.90 | | 70.90 | 83 |
| 270 | 4'-7" long | | .800 | | 66 | 9.90 | | 75.90 | 88 |
| 280 | 5'-3" long | | .800 | | 73 | 9.90 | | 82.90 | 96 |
| 290 | 6'-8" long | ↓ | .889 | ↓ | 97 | 11 | | 108 | 125 |
| 350 | | | | | | | | | |
| 450 | Polystyrene, louvered, 1'-2" wide, 3'-3" long | 1 Carp | .800 | Pair | 21 | 9.90 | | 30.90 | 39 |
| 460 | 4'-7" long | | .800 | Pr. | 28 | 9.90 | | 37.90 | 46 |
| 475 | 5'-3" long | | .800 | | 32 | 9.90 | | 41.90 | 51 |
| 485 | 6'-8" long | | .889 | | 42 | 11 | | 53 | 64 |
| 600 | Vinyl, louvered, each 1'-2" x 4'-7" long | | .800 | | 27 | 9.90 | | 36.90 | 45 |
| 620 | Each 1'-4" x 6'-8" long | ↓ | .889 | ↓ | 41 | 11 | | 52 | 62 |

For expanded coverage of these items see *Building Construction Cost Data 1985*

## 13.1 Special Construction

| | | CREW | MAN-HOURS | UNIT | MAT. | LABOR | EQUIP. | TOTAL | TOTAL INCL O&P |
|---|---|---|---|---|---|---|---|---|---|
| 37-001 | **GREENHOUSE** Shell only, stock units, not incl. 2' stub walls, foundation, | | | | | | | | |
| 002 | floors, heat or compartments | | | | | | | | |
| 030 | Residential type, free standing, 8'-6" long x 7'-6" wide | 2 Carp | .271 | S.F.Flr. | 33 | 3.35 | | 36.35 | 42 |
| 040 | 10'-6" wide | | .188 | | 26 | 2.32 | | 28.32 | 32 |
| 060 | 13'-6" wide | | .148 | | 23 | 1.83 | | 24.83 | 28 |
| 070 | 17'-0" wide | | .100 | | 26 | 1.24 | | 27.24 | 31 |
| 090 | Lean-to type, 3'-10" wide | | .471 | | 30 | 5.80 | | 35.80 | 42 |
| 100 | 6'-10" wide | | .276 | | 23 | 3.41 | | 26.41 | 31 |
| 110 | Wall mounted, to existing window, 3' x 3' | 1 Carp | 2 | Ea. | 320 | 25 | | 345 | 390 |
| 112 | 4' x 5' | " | 2.670 | " | 475 | 33 | | 508 | 575 |
| 120 | Deluxe quality, free standing, 7'-6" wide | 2 Carp | .291 | S.F.Flr. | 66 | 3.59 | | 69.59 | 78 |
| 122 | 10'-6" wide | | .198 | | 61 | 2.44 | | 63.44 | 71 |
| 124 | 13'-6" wide | | .154 | | 57 | 1.90 | | 58.90 | 66 |
| 126 | 17'-0" wide | | .107 | | 49 | 1.32 | | 50.32 | 56 |
| 140 | Lean-to type, 3'-10" wide | | .516 | | 76 | 6.35 | | 82.35 | 94 |
| 142 | 6'-10" wide | | .291 | | 71 | 3.59 | | 74.59 | 84 |
| 144 | 8'-0" wide | | .165 | | 67 | 2.04 | | 69.04 | 77 |
| 70-001 | **SWIMMING POOL ENCLOSURE** Translucent, free standing, | | | | | | | | |
| 002 | not including foundations, heat or light | | | | | | | | |
| 020 | Economy, minimum | 2 Carp | .080 | S.F. Hor | 5.60 | .99 | | 6.59 | 7.70 |
| 030 | Maximum | | .160 | | 13.40 | 1.98 | | 15.38 | 17.85 |
| 040 | Deluxe, minimum | | .160 | | 19.25 | 1.98 | | 21.23 | 24 |
| 060 | Maximum | | .229 | | 200 | 2.82 | | 202.82 | 225 |
| 75-001 | **SWIMMING POOLS** Residential in-ground, vinyl lined, concrete sides | | | | | | | | |
| 002 | Sides including equipment, sand bottom | | | S.F.Surf | | | | | 16 |
| 010 | metal or polystyrene sides | | | | | | | | 12 |
| 020 | Add for vermiculite bottom | | | | | | | | 1.65 |
| 050 | Gunite bottom and sides, white plaster finish | | | | | | | | |
| 060 | 350 S.F. | | | S.F.Surf | | | | | 25 |
| 075 | 800 S.F. | | | " | | | | | 16 |

## 14.1 Conveying Systems

| | | CREW | MAN-HOURS | UNIT | MAT. | LABOR | EQUIP. | TOTAL | TOTAL INCL O&P |
|---|---|---|---|---|---|---|---|---|---|
| 10-001 | **CORRESPONDENCE LIFT** 1 floor 2 stop, 25 lb. capacity, electric | 2 Elev | 80 | Ea. | 3,400 | 1,100 | | 4,500 | 5,500 |
| 15-001 | **DUMBWAITERS** 2 stop, electric, minimum | 2 Elev | 123 | Ea. | 1,750 | 1,700 | | 3,450 | 4,625 |
| 010 | Maximum | | 145 | | 5,000 | 2,000 | | 7,000 | 8,700 |
| 030 | Hand, minimum | | 69.570 | | 600 | 960 | | 1,560 | 2,175 |
| 040 | Maximum | | 84.210 | | 1,025 | 1,150 | | 2,175 | 2,975 |
| 060 | For each additional stop, electric, add | | 29.630 | Stop | 750 | 410 | | 1,160 | 1,475 |
| 070 | Hand, add | | 26.670 | " | 545 | 370 | | 915 | 1,175 |
| 20-001 | **ELEVATORS** For multi-story buildings, housing project, minimum | | | | | | | | |
| 700 | Residential, cab type, 1 floor, 2 stop, minimum | 2 Elev | 80 | Ea. | 5,200 | 1,100 | | 6,300 | 7,475 |
| 710 | Maximum | | 160 | | 9,300 | 2,200 | | 11,500 | 13,700 |
| 720 | 2 floor, 3 stop, minimum | | 133 | | 6,300 | 1,850 | | 8,150 | 9,850 |
| 730 | Maximum | | 267 | | 15,700 | 3,675 | | 19,375 | 23,100 |
| 770 | Stair climber (chair lift), single seat, minimum | | 16 | | 1,900 | 220 | | 2,120 | 2,450 |
| 780 | Maximum | | 80 | | 2,500 | 1,100 | | 3,600 | 4,500 |

For expanded coverage of these items see *Building Construction Cost Data 1985*

## 15.1 Pipe & Fittings

| | | CREW | MAN-HOURS | UNIT | MAT. | LABOR | EQUIP. | TOTAL | TOTAL INCL O&P |
|---|---|---|---|---|---|---|---|---|---|
| 04-001 | **BACKFLOW PREVENTER** Includes gate valves, | | | | | | | | |
| 002 | and four test cocks, corrosion resistant, automatic operation | | | | | | | | |
| 410 | Threaded | | | | | | | | |
| 412 | 3/4" pipe size | 1 Plum | .500 | Ea. | 150 | 6.90 | | 156.90 | 175 |
| 414 | 1" pipe size | " | .571 | " | 180 | 7.90 | | 187.90 | 210 |
| 07-001 | **CLEANOUTS** | | | | | | | | |
| 008 | Round or square, scoriated nickel bronze top | | | | | | | | |
| 010 | 2" pipe size | 1 Plum | .800 | Ea. | 34 | 11.05 | | 45.05 | 55 |
| 014 | 4" pipe size | " | 1.330 | " | 48 | 18.40 | | 66.40 | 82 |
| 098 | Round top, recessed for terrazzo | | | | | | | | |
| 100 | 2" pipe size | 1 Plum | .889 | Ea. | 54 | 12.25 | | 66.25 | 79 |
| 110 | 4" pipe size | " | 2 | " | 70 | 28 | | 98 | 120 |
| 10-001 | **CLEANOUT TEE** Cast iron with countersunk plug | | | | | | | | |
| 022 | 3" pipe size | 1 Plum | 2.220 | Ea. | 15.45 | 31 | | 46.45 | 66 |
| 024 | 4" pipe size | " | 2.420 | ↓ | 24 | 33 | | 57 | 79 |
| 050 | For round smooth access cover, add | | | | 20% | | | | |
| 13-001 | **CONNECTORS** Flexible, corrugated, 7/8" O.D., 1/2" I.D. | | | | | | | | |
| 005 | Gas, seamless brass, steel fittings | | | | | | | | |
| 020 | 12" long | 1 Plum | .250 | Ea. | 4.60 | 3.45 | | 8.05 | 10.55 |
| 022 | 18" long | | .250 | | 5.70 | 3.45 | | 9.15 | 11.75 |
| 024 | 24" long | | .250 | | 6.80 | 3.45 | | 10.25 | 12.95 |
| 026 | 30" long | ↓ | .250 | ↓ | 7.30 | 3.45 | | 10.75 | 13.50 |
| 16-001 | **DRAINS** | | | | | | | | |
| 014 | Cornice, C.I., 45° or 90° outlet | | | | | | | | |
| 020 | 3" and 4" pipe size | Q-1 | 1.330 | Ea. | 40 | 16.55 | | 56.55 | 70 |
| 026 | For galvanized body, add | | | | 7.70 | | | 7.70 | 8.45M |
| 028 | For polished bronze dome, add | | | ↓ | 6.60 | | | 6.60 | 7.25M |
| 200 | Floor, medium duty, C.I., deep flange, 7" top | | | | | | | | |
| 204 | 2" and 3" pipe size | Q-1 | 1.330 | Ea. | 24 | 16.55 | | 40.55 | 53 |
| 208 | For galvanized body, add | | | ↓ | 10.30 | | | 10.30 | 11.35M |
| 212 | For polished bronze top, add | | | | 13.40 | | | 13.40 | 14.75M |
| 250 | Heavy duty, cleanout & trap w/bucket, CI, 15" top | | | | | | | | |
| 254 | 2", 3", and 4" pipe size | Q-1 | 2.670 | Ea. | 1,075 | 33 | | 1,108 | 1,225 |
| 256 | For galvanized body, add | | | | 175 | | | 175 | 195M |
| 258 | For polished bronze top, add | | | ↓ | 110 | | | 110 | 120M |
| 386 | Roof, flat metal deck, C.I. body, 10" aluminum dome | | | | | | | | |
| 389 | 3" pipe size | Q-1 | 1.140 | Ea. | 86 | 14.20 | | 100.20 | 115 |
| 22-001 | **FAUCETS/FITTINGS** | | | | | | | | |
| 015 | Bath, faucets, diverter spout combination, sweat | 1 Plum | 1 | Ea. | 37 | 13.80 | | 50.80 | 63 |
| 020 | For integral stops, IPS unions, add | | | | 18 | | | 18 | 19.80M |
| 050 | Drain, central lift, 1-1/2" IPS male | 1 Plum | .400 | | 23 | 5.50 | | 28.50 | 34 |
| 060 | Trip lever, 1-1/2" IPS male | | .400 | | 22 | 5.50 | | 27.50 | 33 |
| 100 | Kitchen sink faucets, top mount, cast spout | ↓ | .800 | | 27 | 11.05 | | 38.05 | 47 |
| 110 | For spray, add | | | | 8 | 10% | | | |
| 200 | Laundry faucets, shelf type, I.P.S. or copper unions | 1 Plum | .667 | ↓ | 25 | 9.20 | | 34.20 | 42 |
| 202 | | | | | | | | | |
| 210 | Lavatory faucet, centerset, without drain | 1 Plum | .800 | Ea. | 19 | 11.05 | | 30.05 | 38 |
| 220 | For pop-up drain, add | | | | 8.95 | 15% | | | |
| 280 | Self-closing, center set | 1 Plum | .800 | | 53 | 11.05 | | 64.05 | 76 |
| 300 | Service sink faucet, cast spout, pail hook, hose end | | .571 | | 38 | 7.90 | | 45.90 | 54 |
| 400 | Shower by-pass valve with union | | .444 | | 29 | 6.15 | | 35.15 | 42 |
| 420 | Shower thermostatic mixing valve, concealed | ↓ | 1 | | 140 | 13.80 | | 153.80 | 175 |
| 430 | For inlet strainer, check, and stops, add | | | | 40 | 5% | | | |
| 500 | Sillcock, compact, brass, I.P.S. or copper to hose | 1 Plum | .333 | ↓ | 4.07 | 4.60 | | 8.67 | 11.75 |
| 34-001 | **PIPE, BRASS** Plain end, | | | | | | | | |
| 002 | Regular weight | | | | | | | | |

For expanded coverage of these items see *Means' Mechanical Cost Data 1985*

## 15.1 Pipe & Fittings

| | | | CREW | MAN-HOURS | UNIT | MAT. | LABOR | EQUIP. | TOTAL | TOTAL INCL O&P |
|---|---|---|---|---|---|---|---|---|---|---|
| 34 | 112 | 1/2" diameter | 1 Plum | .167 | L.F. | 3.86 | 2.30 | | 6.16 | 7.90 |
| | 114 | 3/4" diameter | ↓ | .174 | | 5.20 | 2.40 | | 7.60 | 9.50 |
| | 116 | 1" diameter | ↓ | .186 | | 7.32 | 2.57 | | 9.89 | 12.10 |
| | 118 | 1-1/4" diameter | Q-1 | .222 | | 10.90 | 2.76 | | 13.66 | 16.35 |
| | 120 | 1-1/2" diameter | | .246 | | 12.70 | 3.06 | | 15.76 | 18.80 |
| | 122 | 2" diameter | ↓ | .302 | ↓ | 16.52 | 3.75 | | 20.27 | 24 |
| 35-001 | | **PIPE, BRASS, FITTINGS**, Rough bronze, threaded | | | | | | | | |
| | 100 | Standard wt., 90° Elbow | | | | | | | | |
| | 110 | 1/2" | 1 Plum | .667 | Ea. | 1.76 | 9.20 | | 10.96 | 16.50 |
| | 112 | 3/4" | | .727 | | 2.40 | 10.05 | | 12.45 | 18.55 |
| | 114 | 1" | ↓ | .800 | | 3.72 | 11.05 | | 14.77 | 22 |
| | 116 | 1-1/4" | Q-1 | .941 | | 6 | 11.70 | | 17.70 | 25 |
| | 118 | 1-1/2" | " | 1 | | 7.50 | 12.40 | | 19.90 | 28 |
| | 150 | 45° Elbow, 1/8" | 1 Plum | .615 | | 1.30 | 8.50 | | 9.80 | 14.90 |
| | 158 | 1/2" | | .667 | | 1.63 | 9.20 | | 10.83 | 16.35 |
| | 160 | 3/4" | | .727 | | 2.40 | 10.05 | | 12.45 | 18.55 |
| | 162 | 1" | ↓ | .800 | | 4.10 | 11.05 | | 15.15 | 22 |
| | 164 | 1-1/4" | Q-1 | .941 | | 6.55 | 11.70 | | 18.25 | 26 |
| | 166 | 1-1/2" | " | 1 | | 8.20 | 12.40 | | 20.60 | 29 |
| | 200 | Tee, 1/8" | 1 Plum | .889 | | 1.65 | 12.25 | | 13.90 | 21 |
| | 208 | 1/2" | | 1 | | 2.06 | 13.80 | | 15.86 | 24 |
| | 210 | 3/4" | | 1.140 | | 2.94 | 15.75 | | 18.69 | 28 |
| | 212 | 1" | ↓ | 1.330 | | 5.30 | 18.40 | | 23.70 | 35 |
| | 214 | 1-1/4" | Q-1 | 1.450 | | 7.50 | 18.05 | | 25.55 | 37 |
| | 216 | 1-1/2" | " | 1.600 | | 10.30 | 19.85 | | 30.15 | 43 |
| | 250 | Coupling, 1/8" | 1 Plum | .500 | | 1.19 | 6.90 | | 8.09 | 12.25 |
| | 258 | 1/2" | | .533 | | 1.48 | 7.35 | | 8.83 | 13.30 |
| | 260 | 3/4" | | .571 | | 1.98 | 7.90 | | 9.88 | 14.70 |
| | 262 | 1" | ↓ | .615 | | 3 | 8.50 | | 11.50 | 16.75 |
| | 264 | 1-1/4" | Q-1 | .727 | | 4.69 | 9.05 | | 13.74 | 19.45 |
| | 266 | 1-1/2" | " | .800 | ↓ | 6.40 | 9.95 | | 16.35 | 23 |
| 37-001 | | **PIPE, CAST IRON** Soil, on hangers 5' O.C. | | | | | | | | |
| | 002 | Single hub, service wt., lead & oakum joints 10' O.C. | | | | | | | | |
| | 212 | 2" diameter | Q-1 | .320 | L.F. | 2.57 | 3.97 | | 6.54 | 9.10 |
| | 214 | 3" diameter | | .333 | | 3.67 | 4.14 | | 7.81 | 10.60 |
| | 216 | 4" diameter | ↓ | .364 | | 4.85 | 4.52 | | 9.37 | 12.50 |
| | 218 | 5" diameter | Q-2 | .387 | | 6.35 | 4.99 | | 11.34 | 14.90 |
| | 220 | 6" diameter | " | .407 | ↓ | 9.27 | 5.25 | | 14.52 | 18.50 |
| | 400 | No hub, couplings 10' O.C. | | | | | | | | |
| | 410 | 1-1/2" diameter | Q-1 | .250 | L.F. | 2.68 | 3.11 | | 5.79 | 7.85 |
| | 412 | 2" diameter | | .267 | | 2.74 | 3.31 | | 6.05 | 8.25 |
| | 414 | 3" diameter | | .276 | | 3.67 | 3.43 | | 7.10 | 9.45 |
| | 416 | 4" diameter | ↓ | .296 | | 4.70 | 3.68 | | 8.38 | 11 |
| | 418 | 5" diameter | Q-2 | .333 | ↓ | 7 | 4.29 | | 11.29 | 14.50 |
| 38-001 | | **PIPE, CAST IRON, FITTINGS**, Soil | | | | | | | | |
| | 004 | Hub and spigot, service weight | | | | | | | | |
| | 008 | 1/4 Bend, 2" | Q-1 | 1 | Ea. | 3.01 | 12.40 | | 15.41 | 23 |
| | 012 | 3" | | 1.140 | | 5.60 | 14.20 | | 19.80 | 29 |
| | 014 | 4" | ↓ | 1.230 | | 8.20 | 15.30 | | 23.50 | 33 |
| | 016 | 5" | Q-2 | 1.330 | | 10.75 | 17.15 | | 27.90 | 39 |
| | 018 | 6" | " | 1.410 | | 13.90 | 18.20 | | 32.10 | 44 |
| | 034 | 1/8 Bend, 2" | Q-1 | 1 | | 2.27 | 12.40 | | 14.67 | 22 |
| | 035 | 3" | | 1.140 | | 4.48 | 14.20 | | 18.68 | 27 |
| | 036 | 4" | ↓ | 1.230 | | 6.65 | 15.30 | | 21.95 | 32 |
| | 038 | 5" | Q-2 | 1.330 | | 9 | 17.15 | | 26.15 | 37 |
| | 040 | 6" | " | 1.410 | | 11.05 | 18.20 | | 29.25 | 41 |
| | 050 | Sanitary Tee, 2" | Q-1 | 1.600 | | 5.20 | 19.85 | | 25.05 | 37 |
| | 054 | 3" | " | 1.780 | ↓ | 9.75 | 22 | | 31.75 | 46 |

For expanded coverage of these items see *Means' Mechanical Cost Data 1985*

## 15.1 Pipe & Fittings

| | | CREW | MAN-HOURS | UNIT | MAT. | LABOR | EQUIP. | TOTAL | TOTAL INCL O&P |
|---|---|---|---|---|---|---|---|---|---|
| 062 | 4" | Q-1 | 2 | Ea. | 11.65 | 25 | | 36.65 | 52 |
| 070 | 5" | Q-2 | 2 | | 21 | 26 | | 47 | 64 |
| 080 | 6" | " | 2.180 | ↓ | 25 | 28 | | 53 | 72 |
| 599 | No hub | | | | | | | | |
| 600 | Cplg. & labor required at joints not incl. in fitting | | | | | | | | |
| 601 | price. Add 1 per joint for installed price | | | | | | | | |
| 602 | 1/4 Bend, 1-1/2" | | | Ea. | 2.58 | | | 2.58 | 2.83M |
| 606 | 2" | | | | 2.64 | | | 2.64 | 2.90M |
| 608 | 3" | | | | 3.43 | | | 3.43 | 3.77M |
| 612 | 4" | | | | 5.10 | | | 5.10 | 5.60M |
| 620 | 1/8 Bend, 1-1/2" | | | | 1.95 | | | 1.95 | 2.14M |
| 624 | 2" | | | | 2 | | | 2 | 2.20M |
| 626 | 3" | | | | 2.85 | | | 2.85 | 3.13M |
| 628 | 4" | | | | 3.75 | | | 3.75 | 4.12M |
| 640 | Sanitary Tee, 1-1/2" | | | | 3.22 | | | 3.22 | 3.54M |
| 646 | 2" | | | | 3.59 | | | 3.59 | 3.94M |
| 652 | 3" | | | | 4.48 | | | 4.48 | 4.92M |
| 660 | 4" | | | ↓ | 6.90 | | | 6.90 | 7.60M |
| 800 | Coupling, standard (by CISPI Mfrs.) | | | | | | | | |
| 802 | 1-1/2" | Q-1 | .571 | Ea. | 2.20 | 7.10 | | 9.30 | 13.65 |
| 804 | 2" | | .615 | | 2.20 | 7.65 | | 9.85 | 14.55 |
| 808 | 3" | | .667 | | 2.60 | 8.30 | | 10.90 | 16 |
| 812 | 4" | ↓ | .727 | ↓ | 3 | 9.05 | | 12.05 | 17.60 |
| 40-001 | **PIPE, COPPER** 50/50 solder joints, | | | | | | | | |
| 002 | Type K tubing, couplings & clevis hangers 10' O.C. | | | | | | | | |
| 110 | 1/4" diameter | 1 Plum | .145 | L.F. | .49 | 2.01 | | 2.50 | 3.72 |
| 118 | 3/4" diameter | | .190 | | 1.33 | 2.63 | | 3.96 | 5.65 |
| 120 | 1" diameter | | .235 | | 1.80 | 3.25 | | 5.05 | 7.15 |
| 126 | 2" diameter | ↓ | .364 | ↓ | 4.34 | 5 | | 9.34 | 12.75 |
| 200 | Type L tubing, couplings & hangers 10' O.C. | | | | | | | | |
| 210 | 1/4" diameter | 1 Plum | .138 | L.F. | .43 | 1.90 | | 2.33 | 3.49 |
| 212 | 3/8" diameter | | .143 | | .59 | 1.97 | | 2.56 | 3.77 |
| 214 | 1/2" diameter | | .151 | | .69 | 2.08 | | 2.77 | 4.06 |
| 216 | 5/8" diameter | | .163 | | .96 | 2.25 | | 3.21 | 4.63 |
| 218 | 3/4" diameter | | .182 | | 1 | 2.51 | | 3.51 | 5.10 |
| 220 | 1" diameter | | .222 | | 1.47 | 3.07 | | 4.54 | 6.50 |
| 222 | 1-1/4" diameter | | .276 | | 1.90 | 3.81 | | 5.71 | 8.10 |
| 224 | 1-1/2" diameter | | .296 | | 2.40 | 4.09 | | 6.49 | 9.10 |
| 226 | 2" diameter | ↓ | .348 | | 3.61 | 4.80 | | 8.41 | 11.60 |
| 228 | 2-1/2" diameter | Q-1 | .485 | | 5.24 | 6 | | 11.24 | 15.30 |
| 230 | 3" diameter | " | .533 | | 6.91 | 6.60 | | 13.51 | 18.10 |
| 241 | For other than full hard temper, add | | | | 25% | | | | |
| 258 | For 95/5 solder, add | | | | | 6% | | | |
| 259 | For silver solder, add | | | ↓ | | 15% | | | |
| 400 | Type DWV tubing, couplings & hangers 10' O.C. | | | | | | | | |
| 410 | 1-1/4" diameter | 1 Plum | .267 | L.F. | 1.54 | 3.68 | | 5.22 | 7.55 |
| 412 | 1-1/2" diameter | | .286 | | 1.89 | 3.94 | | 5.83 | 8.35 |
| 414 | 2" diameter | ↓ | .333 | | 2.44 | 4.60 | | 7.04 | 9.95 |
| 416 | 3" diameter | Q-1 | .500 | | 4.03 | 6.20 | | 10.23 | 14.25 |
| 418 | 4" diameter | " | .667 | ↓ | 6.93 | 8.30 | | 15.23 | 21 |
| 41-001 | **PIPE, COPPER, FITTINGS,** Wrought unless otherwise noted | | | | | | | | |
| 004 | Solder joints, copper x copper | | | | | | | | |
| 010 | 90° Elbow, 1/2" | 1 Plum | .400 | Ea. | .13 | 5.50 | | 5.63 | 8.90 |
| 012 | 3/4" | | .421 | | .30 | 5.80 | | 6.10 | 9.55 |
| 025 | 45° Elbow, 1/4" | | .364 | | .94 | 5 | | 5.94 | 9 |
| 027 | 3/8" | | .364 | | .85 | 5 | | 5.85 | 8.90 |
| 028 | 1/2" | | .400 | | .30 | 5.50 | | 5.80 | 9.10 |
| 029 | 5/8" | ↓ | .421 | ↓ | 1.39 | 5.80 | | 7.19 | 10.75 |

For expanded coverage of these items see *Means' Mechanical Cost Data 1985*

## 15.1 Pipe & Fittings

| | | | CREW | MAN-HOURS | UNIT | MAT. | LABOR | EQUIP. | TOTAL | TOTAL INCL O&P |
|---|---|---|---|---|---|---|---|---|---|---|
|41| 030 | 3/4" | 1 Plum | .421 | Ea. | .52 | 5.80 | | 6.32 | 9.80 |
| | 031 | 1" | | .500 | | 1.19 | 6.90 | | 8.09 | 12.25 |
| | 032 | 1-1/4" | | .533 | | 1.77 | 7.35 | | 9.12 | 13.60 |
| | 033 | 1-1/2" | | .615 | | 2.13 | 8.50 | | 10.63 | 15.80 |
| | 034 | 2" | ▼ | .727 | | 3.56 | 10.05 | | 13.61 | 19.80 |
| | 035 | 2-1/2" | Q-1 | 1.230 | | 7.55 | 15.30 | | 22.85 | 33 |
| | 036 | 3" | " | 1.230 | | 9.65 | 15.30 | | 24.95 | 35 |
| | 045 | Tee, 1/4" | 1 Plum | .571 | | 1.19 | 7.90 | | 9.09 | 13.80 |
| | 047 | 3/8" | | .571 | | 1.07 | 7.90 | | 8.97 | 13.65 |
| | 048 | 1/2" | | .615 | | .22 | 8.50 | | 8.72 | 13.70 |
| | 049 | 5/8" | | .667 | | 1.63 | 9.20 | | 10.83 | 16.35 |
| | 050 | 3/4" | | .667 | | .55 | 9.20 | | 9.75 | 15.20 |
| | 051 | 1" | | .800 | | 2.15 | 11.05 | | 13.20 | 19.85 |
| | 052 | 1-1/4" | | .889 | | 3.12 | 12.25 | | 15.37 | 23 |
| | 053 | 1-1/2" | | 1 | | 4.07 | 13.80 | | 17.87 | 26 |
| | 054 | 2" | ▼ | 1.140 | | 6.60 | 15.75 | | 22.35 | 32 |
| | 055 | 2-1/2" | Q-1 | 2 | | 13.70 | 25 | | 38.70 | 54 |
| | 056 | 3" | " | 2.290 | | 18.85 | 28 | | 46.85 | 66 |
| | 065 | Coupling, 1/4" | 1 Plum | .333 | | .07 | 4.60 | | 4.67 | 7.35 |
| | 067 | 3/8" | | .333 | | .35 | 4.60 | | 4.95 | 7.70 |
| | 068 | 1/2" | | .364 | | .12 | 5 | | 5.12 | 8.10 |
| | 069 | 5/8" | | .381 | | .44 | 5.25 | | 5.69 | 8.80 |
| | 070 | 3/4" | | .381 | | .21 | 5.25 | | 5.46 | 8.55 |
| | 071 | 1" | | .444 | | .84 | 6.15 | | 6.99 | 10.65 |
| | 072 | 1-1/4" | | .471 | | .95 | 6.50 | | 7.45 | 11.35 |
| | 073 | 1-1/2" | | .533 | | 1.30 | 7.35 | | 8.65 | 13.10 |
| | 074 | 2" | ▼ | .615 | | 1.94 | 8.50 | | 10.44 | 15.60 |
| | 075 | 2-1/2" | Q-1 | 1.070 | | 3.40 | 13.25 | | 16.65 | 25 |
| | 076 | 3" | " | 1.230 | ▼ | 4.70 | 15.30 | | 20 | 29 |
| | 200 | DWV, solder joints, copper x copper | | | | | | | | |
| | 203 | 90° Elbow, 1-1/4" | 1 Plum | .615 | Ea. | 1.58 | 8.50 | | 10.08 | 15.20 |
| | 205 | 1-1/2" | | .667 | | 1.58 | 9.20 | | 10.78 | 16.30 |
| | 207 | 2" | ▼ | .800 | | 3.05 | 11.05 | | 14.10 | 21 |
| | 209 | 3" | Q-1 | 1.600 | | 7.35 | 19.85 | | 27.20 | 40 |
| | 210 | 4" | " | 1.780 | | 24 | 22 | | 46 | 61 |
| | 225 | Tee, Sanitary, 1-1/4" | 1 Plum | .889 | | 2.63 | 12.25 | | 14.88 | 22 |
| | 227 | 1-1/2" | | 1 | | 2.42 | 13.80 | | 16.22 | 25 |
| | 229 | 2" | ▼ | 1.140 | | 4.49 | 15.75 | | 20.24 | 30 |
| | 231 | 3" | Q-1 | 2.290 | | 10.20 | 28 | | 38.20 | 56 |
| | 233 | 4" | " | 2.670 | | 31 | 33 | | 64 | 87 |
| | 240 | Coupling, 1-1/4" | 1 Plum | .571 | | .60 | 7.90 | | 8.50 | 13.15 |
| | 242 | 1-1/2" | | .615 | | .91 | 8.50 | | 9.41 | 14.45 |
| | 244 | 2" | ▼ | .727 | | 1.24 | 10.05 | | 11.29 | 17.25 |
| | 246 | 3" | Q-1 | 1.450 | | 2.08 | 18.05 | | 20.13 | 31 |
| | 248 | 4" | " | 1.600 | ▼ | 5.30 | 19.85 | | 25.15 | 37 |
|15| 49-001 | **PIPE, PLASTIC** | | | | | | | | |
| | 002 | Fiberglass reinforced, couplings 10' O.C., hangers 3 per 10' | | | | | | | | |
| | 024 | 2" diameter | Q-1 | .254 | L.F. | 5.98 | 3.15 | | 9.13 | 11.55 |
| | 028 | 4" diameter | | .308 | | 9.90 | 3.82 | | 13.72 | 16.95 |
| | 030 | 6" diameter | ▼ | .372 | ▼ | 15.11 | 4.62 | | 19.73 | 24 |
| | 180 | PVC, couplings 10' O.C., hangers 3 per 10' | | | | | | | | |
| | 182 | Schedule 40 | | | | | | | | |
| | 186 | 1/2" diameter | 1 Plum | .170 | L.F. | .65 | 2.35 | | 3 | 4.44 |
| | 187 | 3/4" diameter | | .222 | | .71 | 3.07 | | 3.78 | 5.65 |
| | 188 | 1" diameter | | .235 | | .88 | 3.25 | | 4.13 | 6.10 |
| | 189 | 1-1/4" diameter | | .267 | | .98 | 3.68 | | 4.66 | 6.90 |
| | 190 | 1-1/2" diameter | ▼ | .296 | | 1.01 | 4.09 | | 5.10 | 7.60 |
| | 191 | 2" diameter | Q-1 | .348 | | 1.21 | 4.32 | | 5.53 | 8.15 |
| | 192 | 2-1/2" diameter | " | .372 | ▼ | 2.04 | 4.62 | | 6.66 | 9.55 |

For expanded coverage of these items see *Means' Mechanical Cost Data 1985*

## 15.1 Pipe & Fittings

| | | CREW | MAN-HOURS | UNIT | MAT. | LABOR | EQUIP. | TOTAL | TOTAL INCL O&P |
|---|---|---|---|---|---|---|---|---|---|
| 193 | 3" diameter | Q-1 | .400 | L.F. | 2.54 | 4.97 | | 7.51 | 10.65 |
| 194 | 4" diameter | | .471 | | 2.98 | 5.85 | | 8.83 | 12.55 |
| 195 | 5" diameter | | .500 | | 5.07 | 6.20 | | 11.27 | 15.40 |
| 196 | 6" diameter | ↓ | .552 | ↓ | 6.30 | 6.85 | | 13.15 | 17.80 |
| 410 | DWV type, schedule 40, couplings 10' O.C., hangers 3 per 10' | | | | | | | | |
| 412 | ABS | | | | | | | | |
| 414 | 1-1/4" diameter | 1 Plum | .267 | L.F. | .89 | 3.68 | | 4.57 | 6.80 |
| 415 | 1-1/2" diameter | " | .296 | | .90 | 4.09 | | 4.99 | 7.45 |
| 416 | 2" diameter | Q-1 | .348 | ↓ | 1.04 | 4.32 | | 5.36 | 8 |
| 440 | PVC | | | | | | | | |
| 441 | 1-1/4" diameter | 1 Plum | .267 | L.F. | 1.04 | 3.68 | | 4.72 | 7 |
| 442 | 1-1/2" diameter | " | .267 | | 1.08 | 3.68 | | 4.76 | 7 |
| 446 | 2" diameter | Q-1 | .348 | | 1.28 | 4.32 | | 5.60 | 8.25 |
| 447 | 3" diameter | | .400 | | 2.28 | 4.97 | | 7.25 | 10.40 |
| 448 | 4" diameter | ↓ | .471 | ↓ | 3.15 | 5.85 | | 9 | 12.70 |
| 536 | CPVC, couplings 10' O.C., hangers 3 per 10' | | | | | | | | |
| 538 | Schedule 40 | | | | | | | | |
| 540 | 1/4" diameter | 1 Plum | .157 | L.F. | 1.23 | 2.16 | | 3.39 | 4.78 |
| 545 | 3/8" diameter | | .163 | | 1.23 | 2.25 | | 3.48 | 4.92 |
| 546 | 1/2" diameter | | .170 | | 1.29 | 2.35 | | 3.64 | 5.15 |
| 547 | 3/4" diameter | | .222 | | 1.56 | 3.07 | | 4.63 | 6.60 |
| 548 | 1" diameter | | .235 | | 2.08 | 3.25 | | 5.33 | 7.45 |
| 549 | 1-1/4" diameter | | .267 | | 2.62 | 3.68 | | 6.30 | 8.70 |
| 550 | 1-1/2" diameter | ↓ | .296 | | 3.02 | 4.09 | | 7.11 | 9.80 |
| 551 | 2" diameter | Q-1 | .348 | | 3.75 | 4.32 | | 8.07 | 10.95 |
| 552 | 2-1/2" diameter | | .372 | | 6.56 | 4.62 | | 11.18 | 14.55 |
| 553 | 3" diameter | | .400 | | 7.70 | 4.97 | | 12.67 | 16.35 |
| 554 | 4" diameter | ↓ | .333 | ↓ | 10.61 | 4.14 | | 14.75 | 18.25 |
| 50-001 | PIPE, PLASTIC, FITTINGS | | | | | | | | |
| 270 | PVC (white), schedule 40, socket joints | | | | | | | | |
| 276 | 90° Elbow, 1/2" | 1 Plum | .364 | Ea. | .14 | 5 | | 5.14 | 8.10 |
| 277 | 3/4" | | .381 | | .16 | 5.25 | | 5.41 | 8.50 |
| 278 | 1" | | .444 | | .28 | 6.15 | | 6.43 | 10.05 |
| 279 | 1-1/4" | | .471 | | .47 | 6.50 | | 6.97 | 10.80 |
| 280 | 1-1/2" | ↓ | .500 | | .49 | 6.90 | | 7.39 | 11.45 |
| 281 | 2" | Q-1 | .667 | | .77 | 8.30 | | 9.07 | 13.95 |
| 282 | 2-1/2" | | .727 | | 2.39 | 9.05 | | 11.44 | 16.95 |
| 283 | 3" | | .941 | | 2.37 | 11.70 | | 14.07 | 21 |
| 284 | 4" | | 1.140 | | 5.10 | 14.20 | | 19.30 | 28 |
| 285 | 5" | | 1.330 | | 16.30 | 16.55 | | 32.85 | 44 |
| 286 | 6" | ↓ | 2 | | 16.30 | 25 | | 41.30 | 57 |
| 318 | Tee, 1/2" | 1 Plum | .571 | | .17 | 7.90 | | 8.07 | 12.70 |
| 319 | 3/4" | | .615 | | .19 | 8.50 | | 8.69 | 13.65 |
| 320 | 1" | | .667 | | .35 | 9.20 | | 9.55 | 14.95 |
| 321 | 1-1/4" | | .727 | | .54 | 10.05 | | 10.59 | 16.50 |
| 322 | 1-1/2" | ↓ | .800 | | .68 | 11.05 | | 11.73 | 18.25 |
| 323 | 2" | Q-1 | .941 | | .93 | 11.70 | | 12.63 | 19.55 |
| 324 | 2-1/2" | | 1.140 | | 3.38 | 14.20 | | 17.58 | 26 |
| 325 | 3" | | 1.450 | | 4.57 | 18.05 | | 22.62 | 34 |
| 326 | 4" | | 1.780 | | 7.60 | 22 | | 29.60 | 43 |
| 327 | 5" | | 2 | | 26 | 25 | | 51 | 68 |
| 328 | 6" | ↓ | 3.200 | | 26 | 40 | | 66 | 92 |
| 338 | Coupling, 1/2" | 1 Plum | .364 | | .08 | 5 | | 5.08 | 8.05 |
| 339 | 3/4" | | .381 | | .12 | 5.25 | | 5.37 | 8.45 |
| 340 | 1" | | .444 | | .20 | 6.15 | | 6.35 | 9.95 |
| 341 | 1-1/4" | | .471 | | .26 | 6.50 | | 6.76 | 10.60 |
| 342 | 1-1/2" | ↓ | .500 | | .28 | 6.90 | | 7.18 | 11.25 |
| 343 | 2" | Q-1 | .571 | | .46 | 7.10 | | 7.56 | 11.75 |
| 344 | 2-1/2" | | .800 | ↓ | 1.04 | 9.95 | | 10.99 | 16.90 |
| 345 | 3" | ↓ | .842 | ↓ | 1.63 | 10.45 | | 12.08 | 18.35 |

For expanded coverage of these items see *Means' Mechanical Cost Data 1985*

## 15.1 Pipe & Fittings

| | | | CREW | MAN-HOURS | UNIT | MAT. | LABOR | EQUIP. | TOTAL | TOTAL INCL O&P |
|---|---|---|---|---|---|---|---|---|---|---|
| 50 | 346 | 4" | Q-1 | 1 | Ea. | 2.31 | 12.40 | | 14.71 | 22 |
| | 347 | 5" | ↓ | 1.140 | ↓ | 7.45 | 14.20 | | 21.65 | 31 |
| | 348 | 6" | ↓ | 1.330 | ↓ | 7.45 | 16.55 | | 24 | 34 |
| | 450 | DWV, ABS, non pressure, socket joints | | | | | | | | |
| | 454 | 1/4 Bend, 1-1/4" | 1 Plum | .471 | Ea. | .62 | 6.50 | | 7.12 | 10.95 |
| | 456 | 1-1/2" | " | .500 | | .46 | 6.90 | | 7.36 | 11.45 |
| | 457 | 2" | Q-1 | .571 | ↓ | .62 | 7.10 | | 7.72 | 11.90 |
| | 480 | Tee, sanitary | | | | | | | | |
| | 482 | 1-1/4" | 1 Plum | .727 | Ea. | .72 | 10.05 | | 10.77 | 16.70 |
| | 483 | 1-1/2" | " | .800 | | .72 | 11.05 | | 11.77 | 18.30 |
| | 484 | 2" | Q-1 | .941 | ↓ | 1.06 | 11.70 | | 12.76 | 19.70 |
| | 500 | PVC, schedule 40, socket joints | | | | | | | | |
| | 504 | 1/4 Bend, 1-1/4" diameter | 1 Plum | .471 | Ea. | 1.40 | 6.50 | | 7.90 | 11.85 |
| | 506 | 1-1/2" | " | .500 | | .86 | 6.90 | | 7.76 | 11.90 |
| | 507 | 2" | Q-1 | .571 | | 1.16 | 7.10 | | 8.26 | 12.50 |
| | 508 | 3" | | .941 | | 3.71 | 11.70 | | 15.41 | 23 |
| | 509 | 4" | | 1.140 | | 6.80 | 14.20 | | 21 | 30 |
| | 510 | 6" | ↓ | 2 | | 46 | 25 | | 71 | 90 |
| | 525 | Tee, sanitary 1-1/4" | 1 Plum | .727 | | 1.86 | 10.05 | | 11.91 | 17.95 |
| | 527 | 1-1/2" | " | .800 | | 1.34 | 11.05 | | 12.39 | 18.95 |
| | 528 | 2" | Q-1 | .941 | | 2 | 11.70 | | 13.70 | 21 |
| | 529 | 3" | | 1.450 | | 4.84 | 18.05 | | 22.89 | 34 |
| | 530 | 4" | ↓ | 1.780 | ↓ | 9.45 | 22 | | 31.45 | 45 |
| | 550 | CPVC, schedule 80, socket or threaded joints | | | | | | | | |
| | 554 | 90° Elbow, 1/4" | 1 Plum | .400 | Ea. | 4.69 | 5.50 | | 10.19 | 13.90 |
| | 556 | 1/2" | | .444 | | 1.54 | 6.15 | | 7.69 | 11.40 |
| | 557 | 3/4" | | .471 | | 2.40 | 6.50 | | 8.90 | 12.95 |
| | 558 | 1" | | .533 | | 3.83 | 7.35 | | 11.18 | 15.90 |
| | 559 | 1-1/4" | | .571 | | 8.85 | 7.90 | | 16.75 | 22 |
| | 560 | 1-1/2" | ↓ | .615 | | 7.30 | 8.50 | | 15.80 | 21 |
| | 561 | 2" | Q-1 | .727 | | 11.15 | 9.05 | | 20.20 | 27 |
| | 562 | 2-1/2" | | .889 | | 27 | 11.05 | | 38.05 | 47 |
| | 563 | 3" | ↓ | 1.140 | | 29 | 14.20 | | 43.20 | 54 |
| | 600 | Coupling, 1/4" | 1 Plum | .400 | | 5.45 | 5.50 | | 10.95 | 14.75 |
| | 602 | 1/2" | | .444 | | 2.03 | 6.15 | | 8.18 | 11.95 |
| | 603 | 3/4" | | .471 | | 2.78 | 6.50 | | 9.28 | 13.35 |
| | 604 | 1" | | .533 | | 3.71 | 7.35 | | 11.06 | 15.75 |
| | 605 | 1-1/4" | | .571 | | 6.05 | 7.90 | | 13.95 | 19.15 |
| | 606 | 1-1/2" | ↓ | .615 | | 7.10 | 8.50 | | 15.60 | 21 |
| | 607 | 2" | Q-1 | .727 | | 8.20 | 9.05 | | 17.25 | 23 |
| | 608 | 2-1/2" | | .889 | | 19.80 | 11.05 | | 30.85 | 39 |
| | 609 | 3" | ↓ | 1.140 | ↓ | 20 | 14.20 | | 34.20 | 44 |
| | 55-001 | **PIPE, STEEL** | | | | | | | | |
| | 005 | Schedule 40, threaded, with couplings, and clevis type | | | | | | | | |
| | 006 | hangers sized for covering, 10' O.C. | | | | | | | | |
| | 054 | Black, 1/4" diameter | 1 Plum | .121 | L.F. | 1.03 | 1.67 | | 2.70 | 3.78 |
| | 057 | 3/4" diameter | | .131 | | 1.12 | 1.81 | | 2.93 | 4.10 |
| | 058 | 1" diameter | ↓ | .151 | | 1.63 | 2.08 | | 3.71 | 5.10 |
| | 059 | 1-1/4" diameter | Q-1 | .180 | | 2.04 | 2.23 | | 4.27 | 5.80 |
| | 060 | 1-1/2" diameter | | .200 | | 2.39 | 2.48 | | 4.87 | 6.55 |
| | 061 | 2" diameter | | .250 | | 3.27 | 3.11 | | 6.38 | 8.50 |
| | 062 | 2-1/2" diameter | | .320 | | 5.31 | 3.97 | | 9.28 | 12.15 |
| | 063 | 3" diameter | | .372 | | 6.86 | 4.62 | | 11.48 | 14.85 |
| | 064 | 3-1/2" diameter | | .400 | | 9.13 | 4.97 | | 14.10 | 17.90 |
| | 065 | 4" diameter | ↓ | .444 | | 10.42 | 5.50 | | 15.92 | 20 |
| | 067 | 6" diameter | Q-2 | .774 | | 28.64 | 9.95 | | 38.59 | 47 |
| | 068 | 8" diameter | " | .889 | ↓ | 40.59 | 11.45 | | 52.04 | 63 |
| | 080 | Pipe nipple std blk 2"lg 1/2" | 1 Plum | .421 | Ea. | .61 | 5.80 | | 6.41 | 9.90 |
| | 082 | Pipe nipple std blk 2"lg 1" | " | .533 | | 1.01 | 7.35 | | 8.36 | 12.80 |
| | 086 | Pipe nipple std blk 2"lg 2-1/2" | Q-1 | 1 | ↓ | 6.10 | 12.40 | | 18.50 | 26 |

## 15.1 Pipe & Fittings

| | | | CREW | MAN-HOURS | UNIT | MAT. | LABOR | EQUIP. | TOTAL | TOTAL INCL O&P |
|---|---|---|---|---|---|---|---|---|---|---|
| 55 | 129 | Galvanized | | | | | | | | |
| | 130 | 3/8" diameter | 1 Plum | .123 | L.F. | 1.45 | 1.70 | | 3.15 | 4.29 |
| | 131 | 1/2" diameter | | .127 | | 1.14 | 1.75 | | 2.89 | 4.03 |
| | 132 | 3/4" diameter | | .131 | | 1.30 | 1.81 | | 3.11 | 4.30 |
| | 133 | 1" diameter | ↓ | .151 | | 1.87 | 2.08 | | 3.95 | 5.35 |
| | 134 | 1-1/4" diameter | Q-1 | .180 | | 2.35 | 2.23 | | 4.58 | 6.10 |
| | 135 | 1-1/2" diameter | | .200 | | 2.75 | 2.48 | | 5.23 | 6.95 |
| | 136 | 2" diameter | | .250 | | 3.77 | 3.11 | | 6.88 | 9.05 |
| | 137 | 2-1/2" diameter | | .320 | | 6.04 | 3.97 | | 10.01 | 12.95 |
| | 138 | 3" diameter | | .372 | | 7.80 | 4.62 | | 12.42 | 15.90 |
| | 140 | 4" diameter | ↓ | .444 | | 11.82 | 5.50 | | 17.32 | 22 |
| | 142 | 6" diameter | Q-2 | .774 | ↓ | 33.14 | 9.95 | | 43.09 | 52 |
| 56 | 001 | **PIPE, STEEL, FITTINGS**, Threaded | | | | | | | | |
| | 002 | Cast Iron, | | | | | | | | |
| | 004 | Standard weight, black | | | | | | | | |
| | 006 | 90° Elbow, straight | | | | | | | | |
| | 010 | 3/4" | 1 Plum | .571 | Ea. | 1.13 | 7.90 | | 9.03 | 13.75 |
| | 011 | 1" | " | .615 | | 1.39 | 8.50 | | 9.89 | 15 |
| | 013 | 1-1/2" | Q-1 | .800 | | 2.71 | 9.95 | | 12.66 | 18.70 |
| | 014 | 2" | | .889 | | 4.19 | 11.05 | | 15.24 | 22 |
| | 015 | 2-1/2" | | 1.140 | | 7.95 | 14.20 | | 22.15 | 31 |
| | 016 | 3" | | 1.600 | | 12 | 19.85 | | 31.85 | 45 |
| | 018 | 4" | ↓ | 2.670 | | 22 | 33 | | 55 | 77 |
| | 020 | 6" | Q-2 | 2.670 | ↓ | 48 | 34 | | 82 | 105 |
| | 050 | Tee, straight | | | | | | | | |
| | 051 | 1/4" | 1 Plum | .800 | Ea. | 1.11 | 11.05 | | 12.16 | 18.70 |
| | 054 | 3/4" | | .889 | | 1.64 | 12.25 | | 13.89 | 21 |
| | 055 | 1" | ↓ | 1 | | 1.80 | 13.80 | | 15.60 | 24 |
| | 057 | 1-1/2" | Q-1 | 1.230 | | 3.75 | 15.30 | | 19.05 | 28 |
| | 058 | 2" | | 1.450 | | 5.65 | 18.05 | | 23.70 | 35 |
| | 059 | 2-1/2" | | 1.780 | | 10.10 | 22 | | 32.10 | 46 |
| | 060 | 3" | | 2.670 | | 15.25 | 33 | | 48.25 | 69 |
| | 062 | 4" | | 2.670 | | 29 | 33 | | 62 | 84 |
| | 064 | 6" | ↓ | 4 | | 65 | 50 | | 115 | 150 |
| | 065 | 8" | Q-2 | 4.800 | ↓ | 135 | 62 | | 197 | 245 |
| | 070 | Standard weight, galvanized | | | | | | | | |
| | 072 | 90° Elbow, straight | | | | | | | | |
| | 076 | 3/4" | 1 Plum | .571 | Ea. | 1.74 | 7.90 | | 9.64 | 14.40 |
| | 077 | 1" | " | .615 | | 2.04 | 8.50 | | 10.54 | 15.70 |
| | 079 | 1-1/2" | Q-1 | .800 | | 4.31 | 9.95 | | 14.26 | 20 |
| | 080 | 2" | | .889 | | 6.40 | 11.05 | | 17.45 | 25 |
| | 081 | 2-1/2" | | 1.140 | | 12.30 | 14.20 | | 26.50 | 36 |
| | 082 | 3" | | 1.600 | | 18.75 | 19.85 | | 38.60 | 52 |
| | 084 | 4" | ↓ | 2.670 | | 34 | 33 | | 67 | 90 |
| | 086 | 6" | Q-2 | 2.670 | ↓ | 75 | 34 | | 109 | 135 |
| | 110 | Tee, straight | | | | | | | | |
| | 114 | 3/4" | 1 Plum | .889 | Ea. | 2.38 | 12.25 | | 14.63 | 22 |
| | 115 | 1" | " | 1 | | 2.67 | 13.80 | | 16.47 | 25 |
| | 117 | 1-1/2" | Q-1 | 1.230 | | 5.85 | 15.30 | | 21.15 | 31 |
| | 118 | 2" | | 1.450 | | 7.70 | 18.05 | | 25.75 | 37 |
| | 119 | 2-1/2" | | 1.780 | | 15.70 | 22 | | 37.70 | 52 |
| | 120 | 3" | | 2.670 | | 24 | 33 | | 57 | 79 |
| | 122 | 4" | ↓ | 4 | | 46 | 50 | | 96 | 130 |
| | 124 | 6" | Q-2 | 4.800 | ↓ | 100 | 62 | | 162 | 210 |
| | 500 | Malleable iron, 150 lb | | | | | | | | |
| | 502 | Black | | | | | | | | |
| | 504 | 90° Elbow, straight | | | | | | | | |
| | 509 | 3/4" | 1 Plum | .571 | Ea. | .65 | 7.90 | | 8.55 | 13.20 |
| | 510 | 1" | " | .615 | | 1.21 | 8.50 | | 9.71 | 14.80 |
| | 512 | 1-1/2" | Q-1 | .800 | ↓ | 2.61 | 9.95 | | 12.56 | 18.60 |

For expanded coverage of these items see *Means' Mechanical Cost Data 1985*

## 15.1 Pipe & Fittings

| | | | CREW | MAN-HOURS | UNIT | MAT. | LABOR | EQUIP. | TOTAL | TOTAL INCL O&P |
|---|---|---|---|---|---|---|---|---|---|---|
| 56 | 513 | 2" | Q-1 | .889 | Ea. | 3.80 | 11.05 | | 14.85 | 22 |
| | 514 | 2-1/2" | | 1.140 | | 9.25 | 14.20 | | 23.45 | 33 |
| | 515 | 3" | | 1.600 | | 14.40 | 19.85 | | 34.25 | 47 |
| | 517 | 4" | ↓ | 2.670 | | 24 | 33 | | 57 | 79 |
| | 519 | 6" | Q-2 | 3.430 | ↓ | 71 | 44 | | 115 | 150 |
| | 545 | Tee, straight | | | | | | | | |
| | 550 | 3/4" | 1 Plum | .889 | Ea. | 1.01 | 12.25 | | 13.26 | 21 |
| | 551 | 1" | " | 1 | | 1.87 | 13.80 | | 15.67 | 24 |
| | 552 | 1-1/4" | Q-1 | 1.140 | | 3.04 | 14.20 | | 17.24 | 26 |
| | 553 | 1-1/2" | | 1.230 | | 3.74 | 15.30 | | 19.04 | 28 |
| | 554 | 2" | | 1.450 | | 5.50 | 18.05 | | 23.55 | 35 |
| | 555 | 2-1/2" | | 1.780 | | 12.25 | 22 | | 34.25 | 48 |
| | 556 | 3" | | 2.670 | | 16.05 | 33 | | 49.05 | 70 |
| | 557 | 3-1/2" | | 3.200 | | 29 | 40 | | 69 | 95 |
| | 558 | 4" | ↓ | 4 | | 32 | 50 | | 82 | 115 |
| | 560 | 6" | Q-2 | 6 | ↓ | 92 | 77 | | 169 | 225 |
| | 565 | Coupling, straight | | | | | | | | |
| | 570 | 3/4" | 1 Plum | .500 | Ea. | .92 | 6.90 | | 7.82 | 11.95 |
| | 571 | 1" | " | .533 | | 1.39 | 7.35 | | 8.74 | 13.20 |
| | 573 | 1-1/2" | Q-1 | .667 | | 2.21 | 8.30 | | 10.51 | 15.55 |
| | 574 | 2" | | .762 | | 3.20 | 9.45 | | 12.65 | 18.50 |
| | 575 | 2-1/2" | | 1 | | 7.10 | 12.40 | | 19.50 | 27 |
| | 576 | 3" | | 1.330 | | 9.75 | 16.55 | | 26.30 | 37 |
| | 578 | 4" | ↓ | 2.290 | | 18.10 | 28 | | 46.10 | 65 |
| | 580 | 6" | Q-2 | 3 | | 46 | 39 | | 85 | 110 |
| | 600 | For galvanized elbows, tees, and couplings add | | | ↓ | 23% | | | | |
| 61-001 | | PIPE, GROOVED-JOINT STEEL FITTINGS & VALVES | | | | | | | | |
| | 002 | Pipe includes coupling & clevis type hanger 10' O.C. | | | | | | | | |
| | 100 | Schedule 40, black | | | | | | | | |
| | 104 | 3/4" diameter | 1 Plum | .030 | L.F. | 1.44 | .42 | | 1.86 | 2.24 |
| | 105 | 1" diameter | | .040 | | 1.75 | .55 | | 2.30 | 2.80 |
| | 106 | 1-1/4" diameter | | .050 | | 2.22 | .69 | | 2.91 | 3.54 |
| | 107 | 1-1/2" diameter | | .050 | | 2.56 | .69 | | 3.25 | 3.91 |
| | 108 | 2" diameter | ↓ | .059 | | 3.23 | .82 | | 4.05 | 4.85 |
| | 109 | 2-1/2" diameter | Q-1 | .070 | | 4.78 | .86 | | 5.64 | 6.65 |
| | 110 | 3" diameter | | .080 | | 6.01 | .99 | | 7 | 8.20 |
| | 111 | 4" diameter | | .089 | | 8.68 | 1.10 | | 9.78 | 11.30 |
| | 112 | 5" diameter | ↓ | .100 | ↓ | 22.36 | 1.24 | | 23.60 | 27 |
| | 400 | Elbow, 90° or 45°, black steel | | | | | | | | |
| | 404 | 1" diameter | 1 Plum | .160 | Ea. | 4.17 | 2.21 | | 6.38 | 8.10 |
| | 405 | 1-1/4" diameter | | .200 | | 5.55 | 2.76 | | 8.31 | 10.50 |
| | 406 | 1-1/2" diameter | | .242 | | 5.95 | 3.35 | | 9.30 | 11.85 |
| | 407 | 2" diameter | ↓ | .320 | | 5.95 | 4.42 | | 10.37 | 13.55 |
| | 408 | 2-1/2" diameter | Q-1 | .400 | | 8.05 | 4.97 | | 13.02 | 16.75 |
| | 410 | 4" diameter | | .640 | | 15.75 | 7.95 | | 23.70 | 30 |
| | 411 | 5" diameter | ↓ | .800 | | 38 | 9.95 | | 47.95 | 58 |
| | 416 | For galvanized elbows, add | | | ↓ | 15% | | | | |
| | 469 | Tee, black steel | | | | | | | | |
| | 470 | 3/4" diameter | 1 Plum | .211 | Ea. | 5.90 | 2.91 | | 8.81 | 11.10 |
| | 474 | 1" diameter | | .242 | | 5.90 | 3.35 | | 9.25 | 11.80 |
| | 475 | 1-1/4" diameter | | .296 | | 6.90 | 4.09 | | 10.99 | 14.05 |
| | 476 | 1-1/2" diameter | | .364 | | 7.40 | 5 | | 12.40 | 16.10 |
| | 477 | 2" diameter | ↓ | .471 | | 9.20 | 6.50 | | 15.70 | 20 |
| | 478 | 2-1/2" diameter | Q-1 | .593 | | 12.50 | 7.35 | | 19.85 | 25 |
| | 480 | 4" diameter | | .941 | | 27 | 11.70 | | 38.70 | 48 |
| | 481 | 5" diameter | ↓ | 1.230 | | 63 | 15.30 | | 78.30 | 94 |
| | 490 | For galvanized tees, add | | | ↓ | 15% | | | | |
| 67-001 | | SHOCK ABSORBERS | | | | | | | | |
| | 050 | 3/4" male I.P.S. For 1 to 11 fixtures | 1 Plum | .667 | Ea. | 25 | 9.20 | | 34.20 | 42 |

For expanded coverage of these items see *Means' Mechanical Cost Data 1985*

## 15.1 Pipe & Fittings

| | | CREW | MAN-HOURS | UNIT | MAT. | LABOR | EQUIP. | TOTAL | TOTAL INCL O&P |
|---|---|---|---|---|---|---|---|---|---|
| 67 060 | 1" male I.P.S., For 12 to 32 fixtures | 1 Plum | 1 | Ea. | 51 | 13.80 | | 64.80 | 78 |
| 70-001 | **SUPPORTS/CARRIERS** For plumbing fixtures | | | | | | | | |
| 060 | Plate type with studs, top back plate | 1 Plum | 1.140 | Ea. | 10.55 | 15.75 | | 26.30 | 37 |
| 300 | Lavatory, concealed arm | | | | | | | | |
| 305 | Floor mounted, single | | | | | | | | |
| 310 | High back fixture | 1 Plum | 1.330 | Ea. | 51 | 18.40 | | 69.40 | 85 |
| 320 | Flat slab fixture | " | 1.330 | " | 60 | 18.40 | | 78.40 | 95 |
| 820 | Water closet, residential | | | | | | | | |
| 822 | Vertical centerline, floor mount | | | | | | | | |
| 824 | Single, 3" caulk, 2" or 3" vent | 1 Plum | 1.330 | Ea. | 56 | 18.40 | | 74.40 | 91 |
| 826 | 4" caulk, 2" or 4" vent | " | 1.330 | " | 75 | 18.40 | | 93.40 | 110 |
| 73-001 | **TRAPS** | | | | | | | | |
| 003 | Cast iron, service weight | | | | | | | | |
| 005 | Long P trap, 2" pipe size | | | | | | | | |
| 110 | 12" long | Q-1 | 1 | Ea. | 8.55 | 12.40 | | 20.95 | 29 |
| 300 | P trap, 2" pipe size | | 1 | | 6.20 | 12.40 | | 18.60 | 27 |
| 304 | 3" pipe size | ↓ | 1.140 | | 7.75 | 14.20 | | 21.95 | 31 |
| 380 | Drum trap, 4" x 5", 1-1/2" tapping | Q-2 | 1.410 | ↓ | 9.40 | 18.20 | | 27.60 | 39 |
| 470 | Copper, drainage, drum trap | | | | | | | | |
| 484 | 3" x 6" swivel, 1-1/2" pipe size | 1 Plum | .500 | Ea. | 14.65 | 6.90 | | 21.55 | 27 |
| 510 | P trap, standard pattern | | | | | | | | |
| 520 | 1-1/4" pipe size | 1 Plum | .444 | Ea. | 8.70 | 6.15 | | 14.85 | 19.30 |
| 524 | 1-1/2" pipe size | | .471 | | 7.55 | 6.50 | | 14.05 | 18.60 |
| 526 | 2" pipe size | | .533 | | 15.25 | 7.35 | | 22.60 | 28 |
| 528 | 3" pipe size | ↓ | .727 | ↓ | 38 | 10.05 | | 48.05 | 58 |
| 671 | ABS DWV P trap, solvent weld joint | | | | | | | | |
| 672 | 1-1/2" pipe size | 1 Plum | .444 | Ea. | 1.44 | 6.15 | | 7.59 | 11.30 |
| 673 | 2" pipe size | | .471 | | 2.60 | 6.50 | | 9.10 | 13.15 |
| 674 | 3" pipe size | | .533 | | 9.75 | 7.35 | | 17.10 | 22 |
| 675 | 4" pipe size | | .571 | | 19.95 | 7.90 | | 27.85 | 34 |
| 676 | PP DWV mech. jnt., dilution trap, 1-1/2" pipe size | | .500 | | 61 | 6.90 | | 67.90 | 78 |
| 677 | P trap, 1-1/2" pipe size | | .471 | | 10 | 6.50 | | 16.50 | 21 |
| 678 | 2" pipe size | | .500 | | 11 | 6.90 | | 17.90 | 23 |
| 679 | 3" pipe size | | .571 | | 21 | 7.90 | | 28.90 | 36 |
| 680 | 4" pipe size | | .615 | | 37 | 8.50 | | 45.50 | 54 |
| 681 | Running trap, 1-1/2" pipe size | | .500 | | 12 | 6.90 | | 18.90 | 24 |
| 682 | 2" pipe size | | .533 | | 18 | 7.35 | | 25.35 | 31 |
| 683 | S trap, 1-1/2" pipe size | | .500 | | 11.50 | 6.90 | | 18.40 | 24 |
| 684 | 2" pipe size | | .533 | | 17.80 | 7.35 | | 25.15 | 31 |
| 685 | Universal trap, 1-1/2" pipe size | | .571 | | 9.65 | 7.90 | | 17.55 | 23 |
| 686 | PVC DWV hub x hub, basin trap, 1-1/4" pipe size | | .444 | | 3 | 6.15 | | 9.15 | 13 |
| 687 | Sink P trap, 1-1/2" pipe size | | .444 | | 2 | 6.15 | | 8.15 | 11.90 |
| 688 | Tubular S trap, 1-1/2" pipe size | | .471 | | 2.66 | 6.50 | | 9.16 | 13.20 |
| 689 | PVC sch. 40 DWV, drum trap, 3/4" pipe size | | .444 | | 2.85 | 6.15 | | 9 | 12.85 |
| 690 | 1-1/2" pipe size | | .500 | | 3.11 | 6.90 | | 10.01 | 14.35 |
| 691 | P trap, 1-1/2" pipe size | | .444 | | .88 | 6.15 | | 7.03 | 10.70 |
| 692 | 2" pipe size | | .471 | | 1.67 | 6.50 | | 8.17 | 12.15 |
| 693 | 3" pipe size | | .533 | | 5.10 | 7.35 | | 12.45 | 17.25 |
| 694 | 4" pipe size | | .571 | | 8.65 | 7.90 | | 16.55 | 22 |
| 695 | P trap w/clean out, 1-1/2" pipe size | | .444 | | 1.45 | 6.15 | | 7.60 | 11.30 |
| 696 | 2" pipe size | | .471 | | 2.44 | 6.50 | | 8.94 | 13 |
| 697 | P trap adjustable, 1-1/2" pipe size | | .471 | | 1.19 | 6.50 | | 7.69 | 11.60 |
| 698 | P trap adj. w/union & cleanout, 1-1/2" pipe size | | .500 | | 1.76 | 6.90 | | 8.66 | 12.85 |
| 699 | P trap fixture, 1-1/2" pipe size | ↓ | .444 | ↓ | 1.97 | 6.15 | | 8.12 | 11.90 |
| 76-001 | **VACUUM BREAKERS** Hot or cold water | | | | | | | | |
| 103 | Anti-siphon, brass | | | | | | | | |
| 106 | 1/2" size | 1 Plum | .333 | Ea. | 10.65 | 4.60 | | 15.25 | 19 |
| 108 | 3/4" size | " | .400 | " | 12.60 | 5.50 | | 18.10 | 23 |

For expanded coverage of these items see *Means' Mechanical Cost Data 1985*

## 15.1 Pipe & Fittings

| | | CREW | MAN-HOURS | UNIT | MAT. | LABOR | EQUIP. | TOTAL | TOTAL INCL O&P |
|---|---|---|---|---|---|---|---|---|---|
| 110 | 1" size | 1 Plum | .421 | Ea. | 19.70 | 5.80 | | 25.50 | 31 |
| 80-001 | **VALVES, BRONZE** | | | | | | | | |
| 102 | Angle, 150 lb., rising stem, threaded | | | | | | | | |
| 107 | 3/4" size | 1 Plum | .400 | Ea. | 26 | 5.50 | | 31.50 | 37 |
| 108 | 1" size | " | .421 | " | 37 | 5.80 | | 42.80 | 50 |
| 175 | Check, swing, class 150, regrinding disc, threaded | | | | | | | | |
| 186 | 3/4" size | 1 Plum | .400 | Ea. | 12.45 | 5.50 | | 17.95 | 22 |
| 187 | 1" size | " | .421 | " | 16.30 | 5.80 | | 22.10 | 27 |
| 285 | Gate, NRS, soldered, 300 psi | | | | | | | | |
| 294 | 3/4" size | 1 Plum | .400 | Ea. | 13.55 | 5.50 | | 19.05 | 24 |
| 295 | 1" size | " | .421 | " | 16.40 | 5.80 | | 22.20 | 27 |
| 560 | Relief, pressure & temperature, self-closing, ASME | | | | | | | | |
| 565 | 1" size | 1 Plum | .333 | Ea. | 42 | 4.60 | | 46.60 | 53 |
| 566 | 1-1/4" size | " | .400 | " | 89 | 5.50 | | 94.50 | 105 |
| 640 | Pressure, water, ASME, threaded | | | | | | | | |
| 644 | 3/4" size | 1 Plum | .286 | Ea. | 18.75 | 3.94 | | 22.69 | 27 |
| 645 | 1" size | " | .333 | " | 39 | 4.60 | | 43.60 | 50 |
| 690 | Reducing, water pressure | | | | | | | | |
| 694 | 1/2" size | 1 Plum | .333 | Ea. | 36 | 4.60 | | 40.60 | 47 |
| 696 | 1" size | " | .421 | " | 68 | 5.80 | | 73.80 | 84 |
| 835 | Tempering, water, sweat connections | | | | | | | | |
| 840 | 1/2" size | 1 Plum | .333 | Ea. | 17.35 | 4.60 | | 21.95 | 26 |
| 844 | 3/4" size | " | .400 | " | 20 | 5.50 | | 25.50 | 31 |
| 865 | Threaded connections | | | | | | | | |
| 870 | 1/2" size | 1 Plum | .333 | Ea. | 20 | 4.60 | | 24.60 | 29 |
| 874 | 3/4" size | " | .400 | " | 25 | 5.50 | | 30.50 | 36 |
| 97-001 | **WATER SUPPLY METERS** | | | | | | | | |
| 200 | Domestic/commercial, bronze | | | | | | | | |
| 202 | Threaded | | | | | | | | |
| 206 | 5/8" diameter, to 20 GPM | 1 Plum | .500 | Ea. | 36 | 6.90 | | 42.90 | 51 |
| 208 | 3/4" diameter, to 30 GPM | ↓ | .571 | ↓ | 45 | 7.90 | | 52.90 | 62 |
| 210 | 1" diameter, to 50 GPM | | .667 | | 75 | 9.20 | | 84.20 | 97 |

## 15.2 Plumbing Fixtures

| | | CREW | MAN-HOURS | UNIT | MAT. | LABOR | EQUIP. | TOTAL | TOTAL INCL O&P |
|---|---|---|---|---|---|---|---|---|---|
| 04-001 | **BATHS** | | | | | | | | |
| 010 | Tubs, recessed porcelain enamel on cast iron, with trim | | | | | | | | |
| 018 | 48" x 42" | Q-1 | 4 | Ea. | 680 | 50 | | 730 | 825 |
| 022 | 72" x 36" | | 5.330 | | 705 | 66 | | 771 | 880 |
| 030 | Mat bottom, 4' long | | 2.910 | | 585 | 36 | | 621 | 700 |
| 038 | 5' long | | 3.640 | | 245 | 45 | | 290 | 340 |
| 048 | Above floor drain, 5' long | | 4 | | 425 | 50 | | 475 | 545 |
| 056 | Corner 48" x 44" | | 4 | | 785 | 50 | | 835 | 940 |
| 200 | Enameled formed steel, 4'-6" long | | 2.760 | | 175 | 34 | | 209 | 245 |
| 220 | 5' long | ↓ | 2.910 | ↓ | 165 | 36 | | 201 | 240 |
| 460 | Module tub & showerwall surround, molded fiberglass | | | | | | | | |
| 461 | 5' long x 34" wide x 76" high | Q-1 | 4 | Ea. | 470 | 50 | | 520 | 595 |
| 600 | Whirlpool, bath with vented overflow, molded fiberglass | | | | | | | | |
| 700 | Redwood tub system | | | | | | | | |
| 705 | 4' diameter x 4' deep | Q-1 | 16 | Ea. | 1,250 | 200 | | 1,450 | 1,700 |
| 715 | 6' diameter x 4' deep | | 20 | | 1,500 | 250 | | 1,750 | 2,050 |
| 720 | 8' diameter x 4' deep | ↓ | 20 | ↓ | 1,750 | 250 | | 2,000 | 2,325 |
| 960 | Rough-in, supply, waste and vent, for all above tubs, add | | 7.270 | | 77.26 | 90 | | 167.26 | 230 |

For expanded coverage of these items see *Means' Mechanical Cost Data 1985*

## 15.2 Plumbing Fixtures

| | | CREW | MAN-HOURS | UNIT | MAT. | LABOR | EQUIP. | TOTAL | TOTAL INCL O&P |
|---|---|---|---|---|---|---|---|---|---|
| 32-001 | **LAVATORIES** With trim, white unless noted otherwise | | | | | | | | |
| 050 | Vanity top, porcelain enamel on cast iron | | | | | | | | |
| 060 | 20" x 18" | Q-1 | 2.500 | Ea. | 100 | 31 | | 131 | 160 |
| 064 | 26" x 18" oval | | 2.500 | | 130 | 31 | | 161 | 190 |
| 072 | 18" round | ↓ | 2.500 | | 92 | 31 | | 123 | 150 |
| 086 | For color, add | | | | 25% | | | | |
| 100 | Cultured marble, 19" x 17", single bowl | Q-1 | 2.500 | | 65 | 31 | | 96 | 120 |
| 112 | 25" x 22", single bowl | | 2.500 | | 78 | 31 | | 109 | 135 |
| 116 | 37" x 22", single bowl | ↓ | 2.500 | ↓ | 95 | 31 | | 126 | 155 |
| 156 | | | | | | | | | |
| 190 | Stainless steel, self-rimming, 25" x 22", single bowl, ledge | Q-1 | 2.500 | Ea. | 125 | 31 | | 156 | 185 |
| 196 | 17" x 22", single bowl | | 2.500 | | 115 | 31 | | 146 | 175 |
| 260 | Steel, enameled, 20" x 17", single bowl | | 2.500 | | 60 | 31 | | 91 | 115 |
| 290 | Vitreous china, 20" x 16", single bowl | | 2.500 | | 150 | 31 | | 181 | 215 |
| 320 | 22" x 13", single bowl | | 2.500 | | 100 | 31 | | 131 | 160 |
| 358 | Rough-in, supply, waste and vent for all above lavatories | ↓ | 5.520 | ↓ | 59.49 | 69 | | 128.49 | 175 |
| 400 | Wall hung | | | | | | | | |
| 404 | Porcelain enamel on cast iron, 16" x 14", single bowl | Q-1 | 2 | Ea. | 165 | 25 | | 190 | 220 |
| 418 | 20" x 18", single bowl | " | 2 | | 115 | 25 | | 140 | 165 |
| 458 | For color, add | | | | 30% | | | | |
| 600 | Vitreous china, 18" x 15", single bowl with backsplash | Q-1 | 2 | | 120 | 25 | | 145 | 170 |
| 606 | 19" x 17", single bowl | | 2 | | 72 | 25 | | 97 | 120 |
| 696 | Rough-in, supply, waste and vent for above lavatories | ↓ | 5.520 | ↓ | 98.49 | 69 | | 167.49 | 215 |
| 36-001 | **LAUNDRY SINKS** With trim | | | | | | | | |
| 002 | Porcelain enamel on cast iron, black iron frame | | | | | | | | |
| 005 | 24" x 20", single compartment | Q-1 | 2.670 | Ea. | 160 | 33 | | 193 | 230 |
| 010 | 24" x 23", single compartment | " | 2.670 | " | 185 | 33 | | 218 | 255 |
| 300 | Plastic, on wall hanger or legs | | | | | | | | |
| 302 | 18" x 23", single compartment | Q-1 | 2.460 | Ea. | 51 | 31 | | 82 | 105 |
| 310 | 20" x 24", single compartment | | 2.460 | | 70 | 31 | | 101 | 125 |
| 320 | 36" x 23", double compartment | | 2.910 | | 75 | 36 | | 111 | 140 |
| 330 | 40" x 24", double compartment | | 2.910 | | 120 | 36 | | 156 | 190 |
| 500 | Stainless steel, counter top, 22" x 17" single compartment | | 2.670 | | 135 | 33 | | 168 | 200 |
| 510 | 19" x 22", single compartment | | 2.670 | | 145 | 33 | | 178 | 210 |
| 520 | 33" x 22", double compartment | | 3.200 | | 220 | 40 | | 260 | 305 |
| 960 | Rough-in, supply, waste and vent, for all laundry sinks | ↓ | 8.420 | ↓ | 65.97 | 105 | | 170.97 | 240 |
| 41-001 | **PUMPS, CIRCULATING** Heated or chilled water application | | | | | | | | |
| 060 | Bronze, sweat connections, 1/40 HP, in line | | | | | | | | |
| 064 | 3/4" size | Q-1 | 1 | Ea. | 69 | 12.40 | | 81.40 | 96 |
| 066 | 1" size | " | 1.140 | " | 185 | 14.20 | | 199.20 | 225 |
| 100 | Flange connection, 3/4" to 1-1/2" size | | | | | | | | |
| 104 | 1/12 HP | Q-1 | 2.670 | Ea. | 180 | 33 | | 213 | 250 |
| 106 | 1/8 HP | | 2.670 | | 275 | 33 | | 308 | 355 |
| 110 | 1/3 HP | | 2.670 | | 290 | 33 | | 323 | 370 |
| 114 | 2" size, 1/6 HP | | 3.200 | | 355 | 40 | | 395 | 455 |
| 118 | 2-1/2" size, 1/4 HP | | 3.200 | | 535 | 40 | | 575 | 650 |
| 122 | 3" size, 1/4 HP | ↓ | 4 | ↓ | 560 | 50 | | 610 | 695 |
| 54-001 | **PUMPS, SUBMERSIBLE** Dewatering | | | | | | | | |
| 700 | Sump pump, 10' head, automatic | | | | | | | | |
| 710 | Bronze, 22 GPM, 1/4 HP, 1-1/4" discharge | 1 Plum | 1.330 | Ea. | 200 | 18.40 | | 218.40 | 250 |
| 750 | Cast iron, 23 GPM, 1/4 HP, 1-1/4" discharge | " | 1.330 | " | 94 | 18.40 | | 112.40 | 135 |
| 56-001 | **SHOWERS** | | | | | | | | |
| 003 | Stall, galvanized steel, baked enamel receptor, with door | | | | | | | | |
| 005 | 30" x 30" square | Q-1 | 7.270 | Ea. | 155 | 90 | | 245 | 315 |
| 010 | 32" x 32" square | | 7.270 | | 175 | 90 | | 265 | 335 |
| 020 | 36" x 36" square | ↓ | 7.270 | ↓ | 170 | 90 | | 260 | 330 |
| 150 | Stall, with door and trim | | | | | | | | |

For expanded coverage of these items see *Means' Mechanical Cost Data 1985*

## 15.2 Plumbing Fixtures

| | | | CREW | MAN-HOURS | UNIT | MAT. | LABOR | EQUIP. | TOTAL | TOTAL INCL O&P |
|---|---|---|---|---|---|---|---|---|---|---|
| 56 | 152 | 32" square | Q-1 | 8 | Ea. | 260 | 99 | | 359 | 445 |
| | 154 | Terrazzo receptor, 32" square | | 8 | | 460 | 99 | | 559 | 665 |
| | 156 | 36" square | | 8.890 | | 490 | 110 | | 600 | 715 |
| | 158 | 36" corner angle | | 8.890 | | 490 | 110 | | 600 | 715 |
| | 300 | Fiberglass, one piece, with 3 walls, 32" x 32" square | | 6.670 | | 210 | 83 | | 293 | 360 |
| | 310 | 36" x 36" square | | 6.670 | | 240 | 83 | | 323 | 395 |
| | 496 | Rough-in, supply, waste and vent for above showers | ↓ | 5.710 | ↓ | 59.86 | 71 | | 130.86 | 180 |
| | 497 | | | | | | | | | |
| 60-001 | | **SINKS** With faucets and drain | | | | | | | | |
| | 200 | Kitchen, counter top, P.E. on C.I., 24" x 21" single bowl | Q-1 | 5 | Ea. | 110 | 62 | | 172 | 220 |
| | 210 | 30" x 21" single bowl | | 5 | | 130 | 62 | | 192 | 240 |
| | 220 | 32" x 21" double bowl | | 6.150 | | 145 | 76 | | 221 | 280 |
| | 300 | Stainless steel, self rimming, 19" x 18" single bowl | | 5 | | 200 | 62 | | 262 | 320 |
| | 310 | 25" x 22", single bowl | | 5 | | 220 | 62 | | 282 | 340 |
| | 320 | 33" x 22" double bowl | | 6.150 | | 235 | 76 | | 311 | 380 |
| | 330 | 43" x 22" double bowl | | 6.150 | | 260 | 76 | | 336 | 405 |
| | 400 | Steel, enameled, with ledge, 24" x 21" single bowl | | 5 | | 59 | 62 | | 121 | 165 |
| | 410 | 32" x 21" double bowl | ↓ | 6.150 | | 77 | 76 | | 153 | 205 |
| | 496 | For color sinks except stainless steel, add | | | | 10% | | | | |
| | 498 | For rough-in, supply, waste and vent, counter top sinks | Q-1 | 8.420 | ↓ | 74.75 | 105 | | 179.75 | 250 |
| | 500 | Kitchen, raised deck, P.E. on C.I. | | | | | | | | |
| | 510 | 32" x 21", dual level, double bowl | Q-1 | 10 | Ea. | 220 | 125 | | 345 | 440 |
| | 579 | For rough-in, supply, waste & vent, sinks | | 8.420 | | 77.82 | 105 | | 182.82 | 250 |
| | 665 | Service, floor, corner, P.E. on C.I., 28" x 28" | ↓ | 4 | | 315 | 50 | | 365 | 425 |
| | 675 | Vinyl coated rim guard, add | | | | 35 | | | 35 | 39M |
| | 677 | For brass bar type rim guard, add | | | | 41 | | | 41 | 45M |
| | 679 | For rough-in, supply, waste & vent, floor service sinks | Q-1 | 15.240 | ↓ | 59.25 | 190 | | 249.25 | 365 |
| 80-001 | | **WATER CLOSETS** | | | | | | | | |
| | 015 | Tank type, vitreous china, including seat, supply pipe with stop | | | | | | | | |
| | 020 | Wall hung, one piece | Q-1 | 3.020 | Ea. | 425 | 37 | | 462 | 525 |
| | 020 | In color | | 3.020 | | 575 | 37 | | 612 | 690 |
| | 040 | Two piece, close coupled | | 3.020 | | 260 | 37 | | 297 | 345 |
| | 040 | In color | | 3.020 | | 340 | 37 | | 377 | 435 |
| | 096 | For rough-in, supply, waste, vent and carrier | | 6.400 | | 133.62 | 79 | | 212.62 | 275 |
| | 100 | Floor mounted, one piece | | 3.020 | | 370 | 37 | | 407 | 465 |
| | 102 | One piece, low profile | | 3.020 | | 570 | 37 | | 607 | 685 |
| | 110 | Two piece, close coupled, water saver | ↓ | 3.020 | | 105 | 37 | | 142 | 175 |
| | 196 | For color, add | | | | 30% | | | | |
| | 198 | For rough-in, supply, waste and vent | Q-1 | 5.710 | ↓ | 80.62 | 71 | | 151.62 | 200 |
| | 300 | Bowl only, with flush valve, seat | | | | | | | | |
| | 310 | Wall hung | Q-1 | 2.760 | Ea. | 220 | 34 | | 254 | 295 |
| | 320 | For rough-in, supply, waste and vent, single WC | | 6.400 | | 141.56 | 79 | | 220.56 | 280 |
| | 330 | Floor mounted | ↓ | 2.760 | ↓ | 195 | 34 | | 229 | 270 |

## 15.3 Plumbing Appliances

| | | | CREW | MAN-HOURS | UNIT | MAT. | LABOR | EQUIP. | TOTAL | TOTAL INCL O&P |
|---|---|---|---|---|---|---|---|---|---|---|
| 50-001 | | **WATER HEATERS** | | | | | | | | |
| | 100 | Residential, electric, glass lined tank, 10 gal., single element | 1 Plum | 3.478 | Ea. | 105 | 48 | | 153 | 185 |
| | 106 | 30 gallon, double element | | 3.636 | | 135 | 51 | | 186 | 220 |
| | 108 | 40 gallon, double element | | 4 | | 140 | 56 | | 196 | 235 |
| | 110 | 52 gallon, double element | | 4 | | 165 | 56 | | 221 | 260 |
| | 112 | 66 gallon, double element | | 4.44 | | 225 | 62 | | 287 | 335 |
| | 114 | 80 gallon, double element | | 5 500 | | 265 | 70 | | 335 | 395 |
| | 200 | Gas fired, glass lined tank, vent not incl., 20 gallon | ↓ | 3.809 | ↓ | 140 | 53 | | 193 | 230 |

For expanded coverage of these items see *Means' Mechanical Cost Data 1985*

## 15.3 Plumbing Appliances

| | | CREW | MAN-HOURS | UNIT | MAT. | LABOR | EQUIP. | TOTAL | TOTAL INCL O&P |
|---|---|---|---|---|---|---|---|---|---|
| 50 204 | 30 gallon | 1 Plum | 4 | Ea. | 150 | 56 | | 206 | 245 |
| 210 | 75 gallon | | 5.333 | | 400 | 75 | | 475 | 550 |
| 300 | Oil fired, glass lined tank, vent not included, 30 gallon | | 4 | | 675 | 55 | | 730 | 830 |
| 304 | 50 gallon | ↓ | 4.440 | ↓ | 890 | 61 | | 951 | 1,075 |

## 15.4 Fire Extinguishing Systems

| | | CREW | MAN-HOURS | UNIT | MAT. | LABOR | EQUIP. | TOTAL | TOTAL INCL O&P |
|---|---|---|---|---|---|---|---|---|---|
| 20-001 | FIRE EXTINGUISHERS | | | | | | | | |
| 012 | CO2, portable with swivel horn, 5 lb. | | | Ea. | 60 | | | 60 | 66M |
| 014 | With hose and "H" horn, 10 lb. | | | " | 90 | | | 90 | 99M |
| 100 | Dry chemical, pressurized | | | | | | | | |
| 104 | Standard type, portable, painted, 2-1/2 lb. | | | Ea. | 15.75 | | | 15.75 | 17.30M |
| 108 | 10 lb. | | | | 42 | | | 42 | 46M |
| 110 | 20 lb. | | | | 62 | | | 62 | 68M |
| 200 | ABC all purpose type, portable, 2-1/2 lb. | | | | 17.35 | | | 17.35 | 19.10M |
| 208 | 9-1/2 lb. | | | | 41 | | | 41 | 45M |
| 212 | 30 lb. | | | ↓ | 85 | | | 85 | 94M |

## 15.5 Heating

| | | CREW | MAN-HOURS | UNIT | MAT. | LABOR | EQUIP. | TOTAL | TOTAL INCL O&P |
|---|---|---|---|---|---|---|---|---|---|
| 06-001 | BOILERS, ELECTRIC, ASME Standard controls and trim | | | | | | | | |
| 100 | Steam, 6 KW, 20.5 MBH | Q-19 | 20 | Ea. | 2,450 | 260 | | 2,710 | 3,100 |
| 116 | 60 KW, 205 MBH | | 24 | | 3,650 | 310 | | 3,960 | 4,500 |
| 200 | Hot water, 12 KW, 41 MBH | | 18.460 | | 2,350 | 240 | | 2,590 | 2,950 |
| 204 | 24 KW, 82 MBH | | 20 | | 2,525 | 260 | | 2,785 | 3,175 |
| 206 | 30 KW, 103 MBH | | 20 | | 2,550 | 260 | | 2,810 | 3,225 |
| 207 | 36 KW, 123 MBH | ↓ | 20 | ↓ | 2,650 | 260 | | 2,910 | 3,325 |
| 08-001 | BOILERS, GAS FIRED Natural or propane, standard controls | | | | | | | | |
| 100 | Cast iron, with insulated jacket | | | | | | | | |
| 300 | Hot water, gross output, 80 MBH | Q-7 | 21.920 | Ea. | 810 | 290 | | 1,100 | 1,350 |
| 302 | 100 MBH | " | 23.700 | " | 990 | 315 | | 1,305 | 1,600 |
| 400 | Steel, insulating jacket | | | | | | | | |
| 600 | Hot water, including burner & one zone valve, gross output | | | | | | | | |
| 601 | 51.2 MBH | Q-6 | 10.670 | Ea. | 945 | 140 | | 1,085 | 1,250 |
| 602 | 72 MBH | | 12 | | 1,100 | 155 | | 1,255 | 1,450 |
| 604 | 86 MBH | | 12.630 | | 1,150 | 165 | | 1,315 | 1,525 |
| 606 | 101 MBH | | 13.330 | | 1,275 | 170 | | 1,445 | 1,675 |
| 608 | 132 MBH | | 14.120 | | 1,475 | 180 | | 1,655 | 1,900 |
| 610 | 150 MBH | | 16 | | 1,700 | 205 | | 1,905 | 2,200 |
| 611 | 186 MBH | | 17.140 | | 2,050 | 220 | | 2,270 | 2,600 |
| 614 | 292 MBH | ↓ | 20 | ↓ | 2,925 | 260 | | 3,185 | 3,625 |
| 615 700 | For tankless water heater on smaller gas units, add | | | Ea. | 10% | | | | |
| 705 706 | For additional zone valves up to 312 MBH add | | | " | 64 | | | 64 | 70M |
| 10-001 | BOILERS, OIL FIRED Standard controls, flame retention burner | | | | | | | | |
| 100 | Cast iron, with insulated flush jacket | | | | | | | | |
| 200 | Steam, gross output, 109 MBH | Q-7 | 26.670 | Ea. | 1,100 | 355 | | 1,455 | 1,775 |
| 206 | 207 MBH | " | 35.560 | " | 1,575 | 470 | | 2,045 | 2,475 |

For expanded coverage of these items see *Means' Mechanical Cost Data 1985*

## 15.5 Heating

| | | CREW | MAN-HOURS | UNIT | MAT. | LABOR | EQUIP. | TOTAL | TOTAL INCL O&P |
|---|---|---|---|---|---|---|---|---|---|
| 10 218 | 940 MBH | Q-7 | 76.190 | Ea. | 5,550 | 1,000 | | 6,550 | 7,700 |
| 300 | Hot water, same price as steam | | | | | | | | |
| 700 | Hot water, gross output, 103 MBH | Q-6 | 12.630 | Ea. | 1,300 | 165 | | 1,465 | 1,700 |
| 702 | 122 MBH | | 13.330 | | 1,325 | 170 | | 1,495 | 1,725 |
| 706 | 168 MBH | | 16 | | 1,675 | 205 | | 1,880 | 2,175 |
| 708 | 225 MBH | ↓ | 17.140 | ↓ | 1,975 | 220 | | 2,195 | 2,525 |
| 12-001 | **BOILERS, GAS/OIL** Combination with burners and controls | | | | | | | | |
| 100 | Cast Iron with insulated jacket | | | | | | | | |
| 200 | Steam, gross output, 720 MBH | Q-7 | 80 | Ea. | 6,625 | 1,075 | | 7,700 | 8,975 |
| 300 | Hot water, gross output, 584 MBH | " | 59.260 | " | 7,900 | 785 | | 8,685 | 9,925 |
| 400 | Steel, insulated jacket, skid base, tubeless | | | | | | | | |
| 450 | Steam, 150 psi gross output, 335 MBH, 10 BHP | Q-6 | 36.920 | Ea. | 6,575 | 475 | | 7,050 | 8,000 |
| 19-001 | **BURNERS** | | | | | | | | |
| 099 | Residential, conversion, gas fired, LP or natural | | | | | | | | |
| 100 | Gun type, atmospheric input 35 to 180 MBH | | | Ea. | 140 | | | 140 | 155M |
| 102 | 50 to 240 MBH | | | | 160 | | | 160 | 175M |
| 104 | 200 to 400 MBH | | | ↓ | 300 | | | 300 | 330M |
| 300 | Flame retention oil fired assembly, input | | | | | | | | |
| 302 | .50 to 2.25 GPH | | | Ea. | 225 | | | 225 | 250M |
| 304 | 2.0 to 5.0 GPH | | | | 255 | | | 255 | 280M |
| 306 | 3.0 to 7.0 GPH | | | | 390 | | | 390 | 430M |
| 308 | 6.0 to 12.0 GPH | | | | 615 | | | 615 | 675M |
| 320 | For mounting flange or pedestal, add | | | ↓ | 7.50 | | | 7.50 | 8.25M |
| 29-001 | **DUCT HEATERS** Electric, slip-in. Includes blast coil, controls, fused | | | | | | | | |
| 002 | transformer, air flow switch, thermal cutouts, contactors | | | | | | | | |
| 010 | 120 volt, 1 phase, 0.5 KW, 8" wide x 8" high | Q-20 | 1.430 | Ea. | 200 | 18.10 | | 218.10 | 250 |
| 012 | 1 KW, 10" wide x 10" high | | 1.670 | | 205 | 21 | | 226 | 260 |
| 016 | 5 KW, 16" wide x 16" high | | 2 | | 240 | 25 | | 265 | 305 |
| 200 | 208 or 240 volt, 3 phase, 1.5 KW, 8" wide x 8" high | | 1.670 | | 230 | 21 | | 251 | 285 |
| 212 | 50 KW, 60" wide x 30" high | | 3.330 | | 880 | 42 | | 922 | 1,025 |
| 216 | 100 KW, 72" wide x 30" high | ↓ | 5 | ↓ | 1,625 | 63 | | 1,688 | 1,900 |
| 35-001 | **FURNACES** Hot air heating, blowers, standard controls, | | | | | | | | |
| 002 | not including gas, oil or flue piping. | | | | | | | | |
| 100 | Electric, UL listed, heat staging, 240 volt | | | | | | | | |
| 102 | 30 MBH | Q-20 | 5 | Ea. | 260 | 63 | | 323 | 385 |
| 108 | 76 MBH | | 5.560 | | 505 | 70 | | 575 | 670 |
| 109 | 85.3 MBH | | 5.410 | | 420 | 69 | | 489 | 570 |
| 110 | 91 MBH | ↓ | 5.880 | ↓ | 525 | 75 | | 600 | 695 |
| 300 | Gas, AGA certified, direct drive models | | | | | | | | |
| 302 | 42 MBH output | Q-9 | 4 | Ea. | 320 | 50 | | 370 | 430 |
| 304 | 63 MBH output | | 4.210 | | 335 | 52 | | 387 | 450 |
| 306 | 79 MBH output | | 4.440 | | 350 | 55 | | 405 | 475 |
| 308 | 84 MBH output | | 4.710 | | 380 | 58 | | 438 | 510 |
| 310 | 105 MBH output | | 5 | | 425 | 62 | | 487 | 565 |
| 312 | 126 MBH output | | 5.330 | | 550 | 66 | | 616 | 710 |
| 313 | 160 MBH output | | 5.710 | | 910 | 71 | | 981 | 1,125 |
| 314 | 200 MBH output | | 6.150 | | 1,275 | 76 | | 1,351 | 1,525 |
| 400 | For starter plenum, add | ↓ | 1 | ↓ | 45 | 12.40 | | 57.40 | 69 |
| 600 | Oil, UL listed, atomizing gun type burner | | | | | | | | |
| 602 | 55 MBH output | Q-9 | 4.440 | Ea. | 585 | 55 | | 640 | 730 |
| 603 | 84 MBH output | | 4.570 | | 600 | 57 | | 657 | 750 |
| 604 | 99 MBH output | | 4.710 | | 650 | 58 | | 708 | 810 |
| 606 | 125 MBH output | | 5 | | 740 | 62 | | 802 | 915 |
| 608 | 152 MBH output | | 5.330 | | 935 | 66 | | 1,001 | 1,125 |
| 610 | 200 MBH output | ↓ | 6.150 | ↓ | 1,175 | 76 | | 1,251 | 1,425 |
| 37-001 | **FURNACES, COMBINATION SYSTEMS** Heating, cooling, electric air | | | | | | | | |
| 002 | cleaner, humidification, dehumidification. | | | | | | | | |
| 200 | Gas fired, 80 MBH heat output, 24 MBH cooling | Q-9 | 13.330 | Ea. | 2,275 | 165 | | 2,440 | 2,775 |
| 202 | 80 MBH heat output, 36 MBH cooling | " | 13.330 | " | 2,450 | 165 | | 2,615 | 2,950 |

For expanded coverage of these items see *Means' Mechanical Cost Data 1985*

## 15.5 Heating

| | | CREW | MAN-HOURS | UNIT | MAT. | LABOR | EQUIP. | TOTAL | TOTAL INCL O&P |
|---|---|---|---|---|---|---|---|---|---|
| 37 204 | 100 MBH heat output, 29 MBH cooling | Q-9 | 16 | Ea. | 2,475 | 200 | | 2,675 | 3,050 |
| 206 | 100 MBH heat output, 36 MBH cooling | | 16 | | 2,600 | 200 | | 2,800 | 3,175 |
| 208 | 100 MBH heat output, 47 MBH cooling | ↓ | 17.780 | | 2,875 | 220 | | 3,095 | 3,525 |
| 210 | 120 MBH heat output, 29 MBH cooling | Q-10 | 18.460 | | 2,600 | 240 | | 2,840 | 3,250 |
| 212 | 120 MBH heat output, 42 MBH cooling | | 18.460 | | 2,950 | 240 | | 3,190 | 3,625 |
| 214 | 120 MBH heat output, 47 MBH cooling | | 20 | | 2,975 | 260 | | 3,235 | 3,675 |
| 216 | 120 MBH heat output, 55 MBH cooling | | 21.820 | | 3,150 | 280 | | 3,430 | 3,925 |
| 218 | 144 MBH heat output, 42 MBH cooling | | 20 | | 3,000 | 260 | | 3,260 | 3,700 |
| 220 | 144 MBH heat output, 47 MBH cooling | | 20 | | 3,025 | 260 | | 3,285 | 3,750 |
| 222 | 144 MBH heat output, 58 MBH cooling | | 24 | | 3,175 | 310 | | 3,485 | 3,975 |
| 225 | 144 MBH heat output, 60 MBH cooling | ↓ | 34.290 | | 3,225 | 440 | | 3,665 | 4,250 |
| 300 | Oil fired, 84 MBH heat output, 24 MBH cooling | Q-9 | 13.330 | | 2,425 | 165 | | 2,590 | 2,925 |
| 302 | 84 MBH heat output, 36 MBH cooling | | 13.330 | | 2,600 | 165 | | 2,765 | 3,125 |
| 304 | 95.2 MBH heat output, 29 MBH cooling | | 16 | | 2,625 | 200 | | 2,825 | 3,200 |
| 306 | 95.2 MBH heat output, 36 MBH cooling | ↓ | 16 | | 2,725 | 200 | | 2,925 | 3,325 |
| 328 | 184.8 MBH heat output, 60 MBH cooling | Q-10 | 24 | ↓ | 3,375 | 310 | | 3,685 | 4,200 |
| 350 | For precharged tubing with connection, add | | | | | | | | |
| 352 | 15 feet | | | Ea. | 83 | | | 83 | 91M |
| 354 | 25 feet | | | | 110 | | | 110 | 120M |
| 356 | 35 feet | | | ↓ | 140 | | | 140 | 155M |
| 51-001 | **HYDRONIC HEATING** Terminal units, not including pipe to main supply | | | | | | | | |
| 100 | Radiation | | | | | | | | |
| 115 | Fin tube, wall hung, 14" slope top cover, with damper | | | | | | | | |
| 120 | 1-1/4" copper tube, 4-1/4" alum. fin | Q-5 | 1 | L.F. | 23 | 12.45 | | 35.45 | 45 |
| 130 | 2" steel tube, 4-1/4" steel fin | | 1.230 | | 23 | 15.35 | | 38.35 | 50 |
| 131 | Baseboard, pkgd, 1/2" copper tube, alum. fin, 7" h. | | .615 | | 5.60 | 7.65 | | 13.25 | 18.30 |
| 132 | 3/4" copper tube, alum. fin, 7" high | | .667 | | 5.85 | 8.30 | | 14.15 | 19.60 |
| 134 | 1" copper tube, alum. fin, 8-7/8" high | | .667 | | 10.25 | 8.30 | | 18.55 | 24 |
| 136 | 1-1/4" copper tube, alum. fin, 8-7/8" high | ↓ | .667 | ↓ | 14.40 | 8.30 | | 22.70 | 29 |
| 300 | Radiators, cast iron | | | | | | | | |
| 310 | Free standing or wall hung, 6 tube, 25" high | Q-5 | .167 | Section | 16.60 | 2.08 | | 18.68 | 22 |
| 395 | Unit heaters, propeller, 1 speed, 200° EWT | | | | | | | | |
| 400 | Horizontal, 14.7 MBH | Q-5 | 1.330 | Ea. | 190 | 16.60 | | 206.60 | 235 |
| 406 | 44.8 MBH | | 2 | | 260 | 25 | | 285 | 325 |
| 424 | 292.5 MBH | ↓ | 8 | | 710 | 100 | | 810 | 940 |
| 430 | For vertical diffuser, add | | | | 76 | | | 76 | 84M |
| 500 | Vertical flow, 52.4 MBH | Q-5 | 1.450 | | 260 | 18.15 | | 278.15 | 315 |
| 508 | 140 MBH | " | 4 | | 415 | 50 | | 465 | 535 |
| 516 | 408 MBH | Q-6 | 13.330 | | 1,050 | 170 | | 1,220 | 1,425 |
| 518 | 520 MBH | " | 17.140 | ↓ | 1,250 | 220 | | 1,470 | 1,725 |
| 54-001 | **HUMIDIFIERS** | | | | | | | | |
| 003 | Centrifugal atomizing | | | | | | | | |
| 010 | 10 lb. per hour | Q-5 | 1.600 | Ea. | 960 | 19.95 | | 979.95 | 1,100 |
| 052 | Steam, room or duct, filter, regulators, automatic controls, 220Volt | | | | | | | | |
| 056 | 17 lb. per hour | Q-5 | 3.200 | Ea. | 1,900 | 40 | | 1,940 | 2,150 |
| 058 | 30 lb. per hour | | 4 | | 1,925 | 50 | | 1,975 | 2,200 |
| 062 | 90 lb. per hour | ↓ | 5.330 | ↓ | 2,850 | 66 | | 2,916 | 3,250 |
| 65-001 | **INSULATION** | | | | | | | | |
| 010 | Rule of thumb, as a percentage of total mechanical costs | | | | | | | 10% | |
| 290 | Domestic water heater wrap kit | | | | | | | | |
| 292 | 1-1/2" with vinyl jacket, 20-60 gal. | 1 Plum | 1 | Ea. | 21 | 13.80 | | 34.80 | 45 |
| 300 | Ductwork | | | | | | | | |
| 302 | Blanket type, fiberglass, flexible | | | | | | | | |
| 303 | Fire resistant liner, black coating one side | | | | | | | | |
| 305 | 1/2" thick, 2 lb. density | Q-14 | .042 | S.F. | .25 | .52 | | .77 | 1.12 |
| 306 | 1" thick, 1-1/2 lb. density | " | .046 | " | .31 | .57 | | .88 | 1.26 |
| 314 | FRK vapor barrier wrap, .75 lb. density | | | | | | | | |
| 316 | 1" thick | Q-14 | .046 | S.F. | .18 | .57 | | .75 | 1.11 |
| 317 | 1-1/2" thick | " | .050 | " | .21 | .62 | | .83 | 1.23 |

For expanded coverage of these items see *Means' Mechanical Cost Data 1985*

## 15.5 Heating

| | | | CREW | MAN-HOURS | UNIT | MAT. | LABOR | EQUIP. | TOTAL | TOTAL INCL O&P |
|---|---|---|---|---|---|---|---|---|---|---|
| 65 | 318 | 2" thick | Q-14 | .053 | S.F. | .27 | .66 | | .93 | 1.37 |
| | 320 | 4" thickness | " | .062 | " | .55 | .76 | | 1.31 | 1.84 |
| | 349 | Board type, fiberglass, 3 lb. density | | | | | | | | |
| | 350 | Fire resistant, black pigmented, 1 side | | | | | | | | |
| | 352 | 1" thick | Q-14 | .107 | S.F. | .66 | 1.32 | | 1.98 | 2.86 |
| | 354 | 1-1/2" thick | " | .123 | " | .69 | 1.53 | | 2.22 | 3.22 |
| | 400 | Pipe covering | | | | | | | | |
| | 660 | Fiberglass, with all service jacket | | | | | | | | |
| | 684 | 1" wall, 1/2" iron pipe size | Q-14 | .067 | L.F. | .75 | .83 | | 1.58 | 2.16 |
| | 686 | 3/4" iron pipe size | | .070 | | .86 | .86 | | 1.72 | 2.34 |
| | 687 | 1" iron pipe size | | .073 | | .91 | .90 | | 1.81 | 2.46 |
| | 690 | 2" iron pipe size | ↓ | .080 | ↓ | 1.22 | .99 | | 2.21 | 2.94 |
| | 786 | Rubber tubing, flexible closed cell foam | | | | | | | | |
| | 810 | 1/2" wall, 1/4" iron pipe size | 1 Asbe | .089 | L.F. | .25 | 1.23 | | 1.48 | 2.25 |
| | 812 | 3/8" iron pipe size | | .089 | | .28 | 1.23 | | 1.51 | 2.28 |
| | 813 | 1/2" iron pipe size | | .089 | | .31 | 1.23 | | 1.54 | 2.32 |
| | 814 | 3/4" iron pipe size | | .089 | | .34 | 1.23 | | 1.57 | 2.35 |
| | 815 | 1" iron pipe size | | .089 | | .38 | 1.23 | | 1.61 | 2.39 |
| | 817 | 1-1/2" iron pipe size | | .089 | | .53 | 1.23 | | 1.76 | 2.56 |
| | 818 | 2" iron pipe size | | .089 | | .64 | 1.23 | | 1.87 | 2.68 |
| | 820 | 3" iron pipe size | | .089 | | .88 | 1.23 | | 2.11 | 2.94 |
| | 822 | 4" iron pipe size | | .100 | | 1.32 | 1.38 | | 2.70 | 3.68 |
| | 830 | 3/4" wall, 1/4" iron pipe size | | .089 | | .37 | 1.23 | | 1.60 | 2.38 |
| | 833 | 1/2" iron pipe size | | .089 | | .51 | 1.23 | | 1.74 | 2.54 |
| | 834 | 3/4" iron pipe size | | .089 | | .62 | 1.23 | | 1.85 | 2.66 |
| | 835 | 1" iron pipe size | | .089 | | .71 | 1.23 | | 1.94 | 2.76 |
| | 838 | 2" iron pipe size | ↓ | .089 | ↓ | 1.27 | 1.23 | | 2.50 | 3.37 |
| 80 | 001 | **SPACE HEATERS** Cabinet, grilles, fan, controls, burner, | | | | | | | | |
| | 002 | thermostat, no piping. For flue see division 15.5-92 | | | | | | | | |
| | 100 | Gas fired, floor mounted | | | | | | | | |
| | 110 | 60 MBH output | Q-5 | 1.600 | Ea. | 340 | 19.95 | | 359.95 | 405 |
| | 118 | 180 MBH output | | 2.670 | | 520 | 33 | | 553 | 625 |
| | 200 | Suspension mounted, propeller fan, 36 MBH output | | 2 | | 315 | 25 | | 340 | 385 |
| | 204 | 60 MBH output | | 2.290 | | 360 | 28 | | 388 | 440 |
| | 210 | 120 MBH output | | 3.200 | | 495 | 40 | | 535 | 610 |
| | 224 | 320 MBH output | ↓ | 8 | | 1,100 | 100 | | 1,200 | 1,375 |
| | 250 | For powered venter and adapter, add | | | | 155 | | | 155 | 170M |
| | 500 | Wall furnace, 17 MBH output | Q-5 | 2.670 | | 185 | 33 | | 218 | 255 |
| | 502 | 24 MBH output | | 3.200 | | 230 | 40 | | 270 | 315 |
| | 504 | 49 MBH output | ↓ | 4 | ↓ | 325 | 50 | | 375 | 435 |
| 89 | 001 | **TANKS** | | | | | | | | |
| | 002 | Fiberglass, underground, U.L. listed, not including | | | | | | | | |
| | 003 | manway or hold-down strap | | | | | | | | |
| | 010 | 1000 gallon capacity | Q-6 | 12 | Ea. | 1,400 | 155 | | 1,555 | 1,775 |
| | 014 | 2000 gallon capacity | Q-7 | 16 | | 2,250 | 215 | | 2,465 | 2,800 |
| | 050 | For manway, fittings and hold-downs, add | | | | 20% | 15% | | | |
| | 200 | Steel, liquid expansion, painted, 8 gallon capacity | Q-5 | .800 | | 44 | 9.95 | | 53.95 | 64 |
| | 204 | 15 gallon capacity | | .941 | | 53 | 11.75 | | 64.75 | 77 |
| | 208 | 24 gallon capacity | | 1.140 | | 67 | 14.25 | | 81.25 | 96 |
| | 212 | 40 gallon capacity | | 1.600 | | 105 | 19.95 | | 124.95 | 145 |
| | 300 | Steel ASME expansion, rubber diaphragm, 19 gallon capacity | | .941 | | 515 | 11.75 | | 526.75 | 585 |
| | 302 | 31 gallon capacity | | 1.330 | | 665 | 16.60 | | 681.60 | 760 |
| | 304 | 61 gallon capacity | | 1.780 | | 1,025 | 22 | | 1,047 | 1,175 |
| | 308 | 119 gallon capacity | ↓ | 2 | ↓ | 1,350 | 25 | | 1,375 | 1,525 |
| | 400 | Steel, storage, above ground, including supports, coating, | | | | | | | | |
| | 402 | fittings, not including mat, pumps or piping | | | | | | | | |
| | 406 | 550 gallon capacity | Q-5 | 4 | Ea. | 585 | 50 | | 635 | 725 |
| | 408 | 1000 gallon capacity | Q-7 | 8 | " | 815 | 105 | | 920 | 1,075 |

## 15.5 Heating

| | | | CREW | MAN-HOURS | UNIT | MAT. | LABOR | EQUIP. | TOTAL | TOTAL INCL O&P |
|---|---|---|---|---|---|---|---|---|---|---|
| 89 | 412 | 2000 gallon capacity | Q-7 | 10.670 | Ea. | 1,150 | 140 | | 1,290 | 1,500 |
| | 500 | Steel underground, coated, set in place, incl. hold-down bars. | | | | | | | | |
| | 550 | Excavation, pad, pumps and piping not included | | | | | | | | |
| | 552 | 1000 gallon capacity, 7 gauge shell | Q-7 | 8 | Ea. | 725 | 105 | | 830 | 965 |
| | 554 | 5000 gallon capacity, 1/4" thick shell | " | 32 | " | 2,625 | 425 | | 3,050 | 3,550 |
| 92-001 | | **VENT CHIMNEY** Prefab metal, U.L. listed | | | | | | | | |
| | 002 | Gas, double wall, galvanized steel | | | | | | | | |
| | 008 | 3" diameter | Q-9 | .222 | V.L.F. | 2 | 2.76 | | 4.76 | 6.60 |
| | 010 | 4" diameter | | .235 | | 2.43 | 2.92 | | 5.35 | 7.35 |
| | 012 | 5" diameter | | .250 | | 2.88 | 3.11 | | 5.99 | 8.15 |
| | 014 | 6" diameter | | .267 | | 3.38 | 3.31 | | 6.69 | 9 |
| | 016 | 7" diameter | | .286 | | 4.59 | 3.55 | | 8.14 | 10.70 |
| | 018 | 8" diameter | | .308 | | 5.10 | 3.82 | | 8.92 | 11.70 |
| | 020 | 10" diameter | | .333 | | 10.80 | 4.14 | | 14.94 | 18.50 |
| | 300 | All fuel, double wall, stainless steel, 6" diameter | | .267 | | 13.95 | 3.31 | | 17.26 | 21 |
| | 302 | 7" diameter | | .286 | | 17.25 | 3.55 | | 20.80 | 25 |
| | 500 | Vent damper bi-metal 6" flue | | 1 | Ea. | 49 | 12.40 | | 61.40 | 74 |
| | 510 | Gas, auto., electric | | 2 | " | 92 | 25 | | 117 | 140 |
| | 800 | All fuel, double wall, stainless steel fittings | | | | | | | | |
| | 801 | Roof support 6" diameter | Q-9 | .533 | Ea. | 35 | 6.60 | | 41.60 | 49 |
| | 802 | 7" diameter | | .571 | | 39 | 7.10 | | 46.10 | 54 |
| | 803 | 8" diameter | | .615 | | 42 | 7.65 | | 49.65 | 58 |
| | 804 | 10" diameter | | .667 | | 54 | 8.30 | | 62.30 | 73 |
| | 805 | 12" diameter | | .727 | | 65 | 9.05 | | 74.05 | 86 |
| | 806 | 14" diameter | | .762 | | 81 | 9.45 | | 90.45 | 105 |
| | 810 | Elbow 15°, 6" diameter | | .533 | | 30 | 6.60 | | 36.60 | 44 |
| | 812 | 7" diameter | | .571 | | 34 | 7.10 | | 41.10 | 49 |
| | 830 | Insulated tee with insulated tee cap, 6" diameter | | .533 | | 56 | 6.60 | | 62.60 | 72 |
| | 834 | 7" diameter | | .571 | | 73 | 7.10 | | 80.10 | 92 |
| | 850 | Joist shield, 6" diameter | | .533 | | 16.45 | 6.60 | | 23.05 | 29 |
| | 851 | 7" diameter | | .571 | | 17.95 | 7.10 | | 25.05 | 31 |
| | 852 | 8" diameter | | .615 | | 18.75 | 7.65 | | 26.40 | 33 |
| | 853 | 10" diameter | | .667 | | 22 | 8.30 | | 30.30 | 37 |
| | 854 | 12" diameter | | .727 | | 32 | 9.05 | | 41.05 | 50 |
| | 855 | 14" diameter | | .762 | | 33 | 9.45 | | 42.45 | 51 |
| | 860 | Round top, 6" diameter | | .533 | | 20 | 6.60 | | 26.60 | 33 |
| | 862 | 7" diameter | | .571 | | 26 | 7.10 | | 33.10 | 40 |
| | 864 | 8" diameter | | .615 | | 36 | 7.65 | | 43.65 | 52 |
| | 866 | 10" diameter | | .667 | | 62 | 8.30 | | 70.30 | 81 |
| | 868 | 12" diameter | | .727 | | 88 | 9.05 | | 97.05 | 110 |
| | 870 | 14" diameter | | .762 | | 115 | 9.45 | | 124.45 | 140 |
| | 880 | Adjustable roof flashing, 6" diameter | | .533 | | 63 | 6.60 | | 69.60 | 80 |
| | 882 | 7" diameter | | .571 | | 71 | 7.10 | | 78.10 | 89 |

## 15.7 Air Conditioning & Ventilating

| | | | CREW | MAN-HOURS | UNIT | MAT. | LABOR | EQUIP. | TOTAL | TOTAL INCL O&P |
|---|---|---|---|---|---|---|---|---|---|---|
| 10-001 | | **AIR FILTERS** | | | | | | | | |
| | 005 | Activated charcoal type, full flow | | | MCFM | 400 | | | 400 | 440M |
| | 200 | Electronic air cleaner, self-contained | | | | | | | | |
| | 215 | 500 CFM | 1 Shee | 3.480 | Ea. | 520 | 48 | | 568 | 650 |
| | 220 | 1000 CFM | | 3.640 | | 715 | 50 | | 765 | 865 |
| | 225 | 1200 CFM | | 3.810 | | 1,050 | 53 | | 1,103 | 1,250 |
| | 230 | 2500 CFM | | 4 | | 1,200 | 55 | | 1,255 | 1,400 |
| | 295 | Mechanical media filtration units | | | | | | | | |

For expanded coverage of these items see *Means' Mechanical Cost Data 1985*

## 15.7 Air Conditioning & Ventilating

| | | | CREW | MAN-HOURS | UNIT | MAT. | LABOR | EQUIP. | TOTAL | TOTAL INCL O&P |
|---|---|---|---|---|---|---|---|---|---|---|
| 10 | 300 | High efficiency type, with frame, non-supported | | | MCFM | 35 | | | 35 | 39M |
| | 310 | Supported type | | | | 47 | | | 47 | 52M |
| | 400 | Medium efficiency, extended surface | | | | 5.25 | | | 5.25 | 5.75M |
| | 450 | Permanent washable | | | | 20 | | | 20 | 22M |
| | 500 | Renewable disposable roll | | | | 70 | | | 70 | 77M |
| | 550 | Throwaway glass or paper media type | | | Ea. | 1.87 | | | 1.87 | 2.05M |
| 40-001 | | **CONTROLS COMPONENTS** | | | | | | | | |
| | 503 | Thermostats, 1 set back, manual | 1 Shee | 1 | Ea. | 40 | 13.80 | | 53.80 | 66 |
| | 504 | 1 set back, electric, timed | | 1 | | 63 | 13.80 | | 76.80 | 91 |
| | 505 | 2 set back, electric, timed | ↓ | 1 | ↓ | 64 | 13.80 | | 77.80 | 92 |
| 52-001 | | **DIFFUSERS** Aluminum, opposed blade damper unless noted | | | | | | | | |
| | 010 | Ceiling, linear, also for sidewall | | | | | | | | |
| | 012 | 2" wide | 1 Shee | .250 | L.F. | 14.40 | 3.45 | | 17.85 | 21 |
| | 016 | 4" wide | | .308 | " | 18.55 | 4.25 | | 22.80 | 27 |
| | 050 | Perforated, 24" x 24", panel size 6" x 6" | | .500 | Ea. | 41 | 6.90 | | 47.90 | 56 |
| | 052 | 8" x 8" | | .533 | | 43 | 7.35 | | 50.35 | 59 |
| | 100 | Rectangular, 1 to 4 way blow, 6" x 6" | | .500 | | 25 | 6.90 | | 31.90 | 39 |
| | 102 | 12" x 6" | | .533 | | 32 | 7.35 | | 39.35 | 47 |
| | 104 | 12" x 9" | | .571 | | 39 | 7.90 | | 46.90 | 56 |
| | 106 | 12" x 12" | | .667 | | 44 | 9.20 | | 53.20 | 63 |
| | 110 | 24" x 12" | | .800 | | 73 | 11.05 | | 84.05 | 98 |
| | 118 | 24" x 24" | | 1.140 | | 130 | 15.75 | | 145.75 | 170 |
| | 150 | Round, butterfly damper, 6" diameter | | .444 | | 10.70 | 6.15 | | 16.85 | 22 |
| | 152 | 8" diameter | | .500 | | 11.80 | 6.90 | | 18.70 | 24 |
| | 200 | T bar mounting, 24" x 24" lay-in frame, 6" x 6" | | .500 | | 43 | 6.90 | | 49.90 | 58 |
| | 202 | 9" x 9" | | .571 | | 49 | 7.90 | | 56.90 | 67 |
| | 204 | 12" x 12" | | .667 | | 63 | 9.20 | | 72.20 | 84 |
| | 206 | 15" x 15" | | .727 | | 82 | 10.05 | | 92.05 | 105 |
| | 208 | 18" x 18" | ↓ | .800 | | 81 | 11.05 | | 92.05 | 105 |
| | 600 | For steel diffusers instead of aluminum, deduct | | | ↓ | 10% | | | | |
| 55-001 | | **GRILLES** | | | | | | | | |
| | 002 | Aluminum | | | | | | | | |
| | 100 | Air return, 6" x 6" | 1 Shee | .308 | Ea. | 5.45 | 4.25 | | 9.70 | 12.80 |
| | 102 | 10" x 6" | | .333 | | 5.85 | 4.60 | | 10.45 | 13.80 |
| | 108 | 16" x 8" | | .364 | | 8.40 | 5 | | 13.40 | 17.25 |
| | 110 | 12" x 12" | | .364 | | 8.40 | 5 | | 13.40 | 17.25 |
| | 112 | 24" x 12" | | .444 | | 15.30 | 6.15 | | 21.45 | 27 |
| | 118 | 16" x 16" | ↓ | .364 | ↓ | 12.65 | 5 | | 17.65 | 22 |
| 58-001 | | **REGISTERS** | | | | | | | | |
| | 098 | Air supply | | | | | | | | |
| | 300 | Baseboard, hand adj. damper, enameled steel | | | | | | | | |
| | 302 | 10" x 6" | 1 Shee | .333 | Ea. | 5.10 | 4.60 | | 9.70 | 12.95 |
| | 304 | 12" x 5" | | .364 | | 12 | 5 | | 17 | 21 |
| | 306 | 12" x 6" | ↓ | .348 | ↓ | 5.45 | 4.80 | | 10.25 | 13.65 |
| | 400 | Floor, toe operated damper, enameled steel | | | | | | | | |
| | 402 | 4" x 6" | 1 Shee | .250 | Ea. | 16.65 | 3.45 | | 20.10 | 24 |
| | 404 | 4" x 12" | " | .308 | " | 9.90 | 4.25 | | 14.15 | 17.65 |
| 61-001 | | **DUCT ACCESSORIES** | | | | | | | | |
| | 005 | Air extractors, 12" x 4" | 1 Shee | .333 | Ea. | 7.15 | 4.60 | | 11.75 | 15.20 |
| | 010 | 8" x 6" | | .364 | | 6.45 | 5 | | 11.45 | 15.10 |
| | 020 | 20" x 8" | | .500 | | 15.40 | 6.90 | | 22.30 | 28 |
| | 300 | Fire damper, curtain type, vertical, 8" x 4" | | .333 | | 18.15 | 4.60 | | 22.75 | 27 |
| | 302 | 12" x 4" | | .364 | | 18.15 | 5 | | 23.15 | 28 |
| | 324 | 16" x 14" | | .444 | | 19.80 | 6.15 | | 25.95 | 32 |
| | 600 | Multi-blade dampers, opposed blade, 12" x 12" | | .381 | | 7.65 | 5.25 | | 12.90 | 16.80 |
| | 602 | 12" x 18" | | .444 | | 10.60 | 6.15 | | 16.75 | 21 |
| | 608 | 24" x 24" | ↓ | 1 | ↓ | 25 | 13.80 | | 38.80 | 50 |
| | 750 | Variable volume modulating motorized damper | | | | | | | | |
| | 752 | 12" x 12" | 1 Shee | .667 | Ea. | 92 | 9.20 | | 101.20 | 115 |

For expanded coverage of these items see *Means' Mechanical Cost Data 1985*

## 15.7 Air Conditioning & Ventilating

| | | | CREW | MAN-HOURS | UNIT | MAT. | LABOR | EQUIP. | TOTAL | TOTAL INCL O&P |
|---|---|---|---|---|---|---|---|---|---|---|
| 61 | 756 | 24" x 12" | 1 Shee | 1 | Ea. | 110 | 13.80 | | 123.80 | 145 |
| | 770 | For thermostat, add | " | 1 | " | 37 | 13.80 | | 50.80 | 63 |
| | 800 | Volume control, dampers | | | | | | | | |
| | 810 | 8" x 8" | 1 Shee | .333 | Ea. | 14.75 | 4.60 | | 19.35 | 24 |
| | 816 | 18" x 12" | " | .444 | " | 26 | 6.15 | | 32.15 | 38 |
| | 900 | Silencers, noise control for air flow, duct | | | MCFM | 36 | | | 36 | 40M |
| | 920 | Plenums, measured by panel surface | | | S.F. | 8 | | | 8 | 8.80M |
| 64-001 | | **DUCTWORK** | | | | | | | | |
| | 002 | Fabricated rectangular, includes fittings, joints, supports, | | | | | | | | |
| | 003 | allowance for flexible connections, no insulation | | | | | | | | |
| | 010 | Aluminum, alloy 3003-H14, under 300 lb. | Q-10 | .300 | Lb. | 2.08 | 3.86 | | 5.94 | 8.45 |
| | 011 | 300 to 500 lb. | | .280 | | 1.81 | 3.60 | | 5.41 | 7.20 |
| | 012 | 500 to 1000 lb. | | .253 | | 1.70 | 3.25 | | 4.95 | 7.05 |
| | 014 | 1000 to 2000 lb. | | .200 | | 1.54 | 2.58 | | 4.12 | 5.80 |
| | 050 | Galvanized steel, under 400 lb. | | .102 | | 1.19 | 1.32 | | 2.51 | 3.41 |
| | 052 | 400 to 1000 lb. | | .094 | | .76 | 1.21 | | 1.97 | 2.77 |
| | 054 | 1000 to 2000 lb. | ↓ | .091 | ↓ | .58 | 1.17 | | 1.75 | 2.50 |
| | 130 | Flexible, vinyl coated spring steel or aluminum, | | | | | | | | |
| | 140 | pressure to 10" (WG) UL-181 | | | | | | | | |
| | 150 | Non-insulated, 3" diameter | Q-9 | .040 | L.F. | .30 | .50 | | .80 | 1.12 |
| | 154 | 5" diameter | | .050 | | .38 | .62 | | 1 | 1.41 |
| | 156 | 6" diameter | | .057 | | .41 | .71 | | 1.12 | 1.59 |
| | 158 | 7" diameter | | .067 | | .48 | .83 | | 1.31 | 1.85 |
| | 190 | Insulated, 4" diameter | | .047 | | .99 | .58 | | 1.57 | 2.02 |
| | 192 | 5" diameter | | .053 | | 1.15 | .66 | | 1.81 | 2.32 |
| | 194 | 6" diameter | | .062 | | 1.39 | .76 | | 2.15 | 2.75 |
| | 196 | 7" diameter | | .073 | | 1.51 | .90 | | 2.41 | 3.10 |
| | 198 | 8" diameter | | .089 | | 1.68 | 1.10 | | 2.78 | 3.61 |
| | 204 | 12" diameter | ↓ | .160 | ↓ | 2.40 | 1.99 | | 4.39 | 5.80 |
| 70-001 | | **FANS** | | | | | | | | |
| | 800 | Ventilation, residential | | | | | | | | |
| | 802 | Attic, roof type | | | | | | | | |
| | 803 | Aluminum dome, damper & curb | | | | | | | | |
| | 804 | 6" diameter, 300 CFM | 1 Elec | .500 | Ea. | 120 | 6.90 | | 126.90 | 145 |
| | 805 | 7" diameter, 450 CFM | | .533 | | 140 | 7.35 | | 147.35 | 165 |
| | 806 | 9" diameter, 900 CFM | | .571 | | 250 | 7.85 | | 257.85 | 285 |
| | 808 | 12" diameter, 1000 CFM (gravity) | | .800 | | 180 | 11 | | 191 | 215 |
| | 809 | 16" diameter, 1500 CFM (gravity) | | .889 | | 220 | 12.20 | | 232.20 | 260 |
| | 810 | 20" diameter, 2500 CFM (gravity) | | 1 | | 270 | 13.75 | | 283.75 | 320 |
| | 811 | 26" diameter, 4000 CFM (gravity) | | 1.140 | | 340 | 15.70 | | 355.70 | 400 |
| | 812 | 32" diameter, 6500 CFM (gravity) | | 1.330 | | 465 | 18.35 | | 483.35 | 540 |
| | 813 | 38" diameter, 8000 CFM (gravity) | | 1.600 | | 640 | 22 | | 662 | 740 |
| | 814 | 50" diameter, 13,000 CFM (gravity) | ↓ | 2 | ↓ | 1,025 | 28 | | 1,053 | 1,175 |
| | 816 | Plastic, ABS dome | | | | | | | | |
| | 818 | 930 CFM | 1 Elec | .571 | Ea. | 66 | 7.85 | | 73.85 | 85 |
| | 820 | 1600 CFM | " | .667 | " | 92 | 9.15 | | 101.15 | 115 |
| | 824 | Attic, wall type, with shutter, one speed | | | | | | | | |
| | 825 | 12" diameter, 1000 CFM | 1 Elec | .571 | Ea. | 94 | 7.85 | | 101.85 | 115 |
| | 826 | 14" diameter, 1500 CFM | | .667 | | 110 | 9.15 | | 119.15 | 135 |
| | 827 | 16" diameter, 2000 CFM | ↓ | .889 | ↓ | 135 | 12.20 | | 147.20 | 170 |
| | 829 | Whole house, wall type, with shutter, one speed | | | | | | | | |
| | 830 | 30" diameter, 4800 CFM | 1 Elec | 1.140 | Ea. | 265 | 15.70 | | 280.70 | 315 |
| | 831 | 36" diameter, 7000 CFM | | 1.330 | | 310 | 18.35 | | 328.35 | 370 |
| | 832 | 42" diameter, 10,000 CFM | | 1.600 | | 370 | 22 | | 392 | 440 |
| | 833 | 48" diameter, 16,000 CFM | ↓ | 2 | ↓ | 445 | 28 | | 473 | 535 |
| | 834 | For two speed, add | | | | 10 | | | 10 | 11M |
| | 835 | Whole house, lay-down type, with shutter, one speed | | | | | | | | |
| | 836 | 30" diameter, 4500 CFM | 1 Elec | 1 | Ea. | 275 | 13.75 | | 288.75 | 325 |
| | 837 | 36" diameter, 6500 CFM | " | 1.140 | " | 325 | 15.70 | | 340.70 | 380 |

For expanded coverage of these items see *Means' Mechanical Cost Data 1985*

## 15.7 Air Conditioning & Ventilating

| | | CREW | MAN-HOURS | UNIT | MAT. | LABOR | EQUIP. | TOTAL | TOTAL INCL O&P |
|---|---|---|---|---|---|---|---|---|---|
| 70 838 | 42" diameter, 9000 CFM | 1 Elec | 1.330 | Ea. | 380 | 18.35 | | 398.35 | 445 |
| 839 | 48" diameter, 12,000 CFM | " | 1.600 | | 455 | 22 | | 477 | 535 |
| 844 | For two speed, add | | | | 10 | | | 10 | 11M |
| 845 | For 12 hour timer switch, add | 1 Elec | .250 | ↓ | 15 | 3.44 | | 18.44 | 22 |
| 73-001 | **FAN COIL AIR CONDITIONING** Cabinet mounted, filters | | | | | | | | |
| 010 | Chilled water, 1/2 ton cooling | Q-5 | 2 | Ea. | 720 | 25 | | 745 | 830 |
| 012 | 1 ton cooling | | 2.670 | | 810 | 33 | | 843 | 945 |
| 018 | 3 ton cooling | | 4 | | 1,475 | 50 | | 1,525 | 1,700 |
| 100 | Direct expansion, air cooled condensing, 5 ton cooling | ↓ | 5.330 | ↓ | 525 | 66 | | 591 | 685 |
| 76-001 | **HEAT PUMPS** | | | | | | | | |
| 100 | Air to air, split system, not including curbs or pads | | | | | | | | |
| 102 | 2 ton cooling, 8.5 MBH heat @ 0°F | Q-5 | 13.330 | Ea. | 1,250 | 165 | | 1,415 | 1,650 |
| 150 | Single package, not including curbs, pads, or plenums | | | | | | | | |
| 152 | 2 ton cooling, 6.5 MBH heat @ 0°F | Q-5 | 10.670 | Ea. | 1,200 | 135 | | 1,335 | 1,525 |
| 200 | Water source to air, single package | | | | | | | | |
| 210 | 1 ton cooling, 13 MBH heat @ 75°F | Q-5 | 8 | Ea. | 845 | 100 | | 945 | 1,100 |
| 500 | Water heater, 12,000 BTU | " | 4 | " | 560 | 50 | | 610 | 695 |
| 82-001 | **PACKAGED TERMINAL AIR CONDITIONER** Cabinet, wall sleeve, | | | | | | | | |
| 010 | louver, electric heat, thermostat, manual changeover, 208 V | | | | | | | | |
| 020 | 6000 BTUH cooling, 8800 BTU heat | Q-5 | 2.670 | Ea. | 630 | 33 | | 663 | 745 |
| 022 | 9,000 BTUH cooling, 13,900 BTU heat | | 3.200 | | 655 | 40 | | 695 | 785 |
| 024 | 12,000 BTUH cooling, 13,900 BTU heat | | 4 | | 700 | 50 | | 750 | 850 |
| 026 | 15,000 BTUH cooling, 13,900 BTU heat | ↓ | 5.330 | ↓ | 730 | 66 | | 796 | 910 |
| 85-001 | **ROOF TOP AIR CONDITIONERS** Standard controls, curb | | | | | | | | |
| 100 | Single zone, electric cool, gas heat | | | | | | | | |
| 114 | 5 ton cooling, 112 MBH heating | Q-5 | 28.570 | Ea. | 3,350 | 355 | | 3,705 | 4,250 |
| 116 | 10 ton cooling, 200 MBH heating | Q-6 | 52.170 | | 6,475 | 675 | | 7,150 | 8,200 |
| 118 | 15 ton cooling, 270 MBH heating | " | 77.420 | | 8,625 | 1,000 | | 9,625 | 11,100 |
| 122 | 30 ton cooling, 540 MBH heating | Q-7 | 145 | ↓ | 17,800 | 1,925 | | 19,725 | 22,600 |
| 88-001 | **SELF-CONTAINED SINGLE PACKAGE** | | | | | | | | |
| 010 | Air cooled, for free blow or duct, not including remote condenser | | | | | | | | |
| 020 | 3 ton cooling | Q-5 | 16 | Ea. | 2,450 | 200 | | 2,650 | 3,000 |
| 021 | 4 ton cooling | " | 17.780 | " | 2,675 | 220 | | 2,895 | 3,300 |
| 100 | Water cooled for free blow or duct, not including tower | | | | | | | | |
| 110 | 3 ton cooling | Q-6 | 24 | Ea. | 2,125 | 310 | | 2,435 | 2,825 |
| 96-001 | **VENTILATORS** Base, damper & bird screen, CFM in 5 MPH wind | | | | | | | | |
| 052 | 8" neck diameter, 215 CFM | Q-9 | 1.140 | Ea. | 36 | 14.20 | | 50.20 | 62 |

## 16.0 Raceways

| | | CREW | MAN-HOURS | UNIT | MAT. | LABOR | EQUIP. | TOTAL | TOTAL INCL O&P |
|---|---|---|---|---|---|---|---|---|---|
| 20-001 002 | **CONDUIT TO 15' HIGH** Includes 2 terminations, 2 elbows and 10 beam clamps per 100 L.F. | | | | | | | | |
| 250 | Steel, intermediate conduit (IMC), 1/2" diameter | 1 Elec | .080 | L.F. | .43 | 1.10 | | 1.53 | 2.21 |
| 253 | 3/4" diameter | | .089 | | .52 | 1.22 | | 1.74 | 2.50 |
| 255 | 1" diameter | | .114 | | .75 | 1.57 | | 2.32 | 3.30 |
| 257 | 1-1/4" diameter | | .123 | | .99 | 1.69 | | 2.68 | 3.76 |
| 260 | 1-1/2" diameter | | .133 | | 1.18 | 1.83 | | 3.01 | 4.19 |
| 263 | 2" diameter | | .160 | | 1.60 | 2.20 | | 3.80 | 5.25 |
| 500 | Electric metallic tubing (EMT), 1/2" diameter | | .047 | | .22 | .65 | | .87 | 1.26 |
| 502 | 3/4" diameter | | .062 | | .32 | .85 | | 1.17 | 1.69 |
| 504 | 1" diameter | | .070 | | .53 | .96 | | 1.49 | 2.09 |
| 506 | 1-1/4" diameter | | .080 | | .81 | 1.10 | | 1.91 | 2.63 |
| 508 | 1-1/2" diameter | | .089 | | 1.02 | 1.22 | | 2.24 | 3.05 |
| 550 | Unicouple, 2-1/2" diameter | ↓ | .114 | ↓ | 2.50 | 1.57 | | 4.07 | 5.25 |

For expanded coverage of these items see *Means' Electrical Cost Data 1985*

# 16.0 Raceways

| | | | CREW | MAN-HOURS | UNIT | MAT. | LABOR | EQUIP. | TOTAL | TOTAL INCL O&P |
|---|---|---|---|---|---|---|---|---|---|---|
| 20 | 554 | 3" diameter | 1 Elec | .133 | L.F. | 3 | 1.83 | | 4.83 | 6.20 |
| 30-001 | | **CONDUIT IN CONCRETE SLAB** Including terminations, | | | | | | | | |
| | 002 | fittings | | | | | | | | |
| | 323 | PVC, schedule 40, 1/2" diameter | 1 Elec | .030 | L.F. | .16 | .41 | | .57 | .82 |
| | 325 | 3/4" diameter | | .035 | | .21 | .48 | | .69 | .99 |
| | 327 | 1" diameter | | .040 | | .32 | .55 | | .87 | 1.22 |
| | 330 | 1-1/4" diameter | | .047 | | .43 | .65 | | 1.08 | 1.49 |
| | 333 | 1-1/2" diameter | | .057 | | .53 | .79 | | 1.32 | 1.82 |
| | 335 | 2" diameter | | .067 | | .69 | .92 | | 1.61 | 2.20 |
| | 435 | Rigid galvanized steel, 1/2" diameter | | .040 | | .51 | .55 | | 1.06 | 1.43 |
| | 440 | 3/4" diameter | | .047 | | .61 | .65 | | 1.26 | 1.69 |
| | 445 | 1" diameter | | .062 | | .89 | .85 | | 1.74 | 2.31 |
| | 450 | 1-1/4" diameter | | .073 | | 1.18 | 1 | | 2.18 | 2.87 |
| | 460 | 1-1/2" diameter | | .080 | | 1.40 | 1.10 | | 2.50 | 3.27 |
| | 480 | 2" diameter | | .089 | | 1.92 | 1.22 | | 3.14 | 4.04 |
| 50-001 | | **CONDUIT IN TRENCH** Includes terminations and fittings | | | | | | | | |
| | 020 | Rigid galvanized steel, 2" diameter | 1 Elec | .053 | L.F. | 1.90 | .73 | | 2.63 | 3.25 |
| | 040 | 2-1/2" diameter | | .080 | | 3.10 | 1.10 | | 4.20 | 5.15 |
| | 060 | 3" diameter | | .100 | | 4.25 | 1.38 | | 5.63 | 6.85 |
| | 080 | 3-1/2" diameter | | .114 | | 5.60 | 1.57 | | 7.17 | 8.65 |
| 55-001 | | **CONDUIT FITTINGS** | | | | | | | | |
| | 228 | LB fittings, with cover, 1/2" diameter | 1 Elec | .500 | Ea. | 3.35 | 6.90 | | 10.25 | 14.55 |
| | 229 | 3/4" diameter | | .615 | | 3.50 | 8.45 | | 11.95 | 17.20 |
| | 230 | 1" diameter | | .727 | | 5 | 10 | | 15 | 21 |
| | 233 | 1-1/4" diameter | | 1 | | 7.30 | 13.75 | | 21.05 | 30 |
| | 235 | 1-1/2" diameter | | 1.330 | | 9.10 | 18.35 | | 27.45 | 39 |
| | 237 | 2" diameter | | 1.600 | | 15.30 | 22 | | 37.30 | 52 |
| | 238 | 2-1/2" diameter | | 2 | | 37 | 28 | | 65 | 84 |
| | 239 | 3" diameter | | 2.290 | | 47 | 31 | | 78 | 100 |
| | 240 | 3-1/2" diameter | | 2.670 | | 74 | 37 | | 111 | 140 |
| | 241 | 4" diameter | | 3.200 | | 82 | 44 | | 126 | 160 |
| | 528 | Service entrance cap, 1/2" diameter | | .500 | | 2.15 | 6.90 | | 9.05 | 13.20 |
| | 530 | 3/4" diameter | | .615 | | 2.85 | 8.45 | | 11.30 | 16.50 |
| | 532 | 1" diameter | | .800 | | 3.20 | 11 | | 14.20 | 21 |
| | 534 | 1-1/4" diameter | | 1 | | 3.65 | 13.75 | | 17.40 | 26 |
| | 536 | 1-1/2" diameter | | 1.230 | | 5.55 | 16.90 | | 22.45 | 33 |
| | 538 | 2" diameter | | 1.450 | | 9.20 | 20 | | 29.20 | 42 |
| | 540 | 2-1/2" diameter | | 2 | | 28 | 28 | | 56 | 74 |
| | 542 | 3" diameter | | 2.350 | | 47 | 32 | | 79 | 105 |
| | 544 | 3-1/2" diameter | | 2.670 | | 63 | 37 | | 100 | 125 |
| | 546 | 4" diameter | | 2.960 | | 77 | 41 | | 118 | 150 |
| 60-001 | | **FLEXIBLE METALLIC CONDUIT** | | | | | | | | |
| | 005 | Greenfield, 3/8" diameter | 1 Elec | .040 | L.F. | .18 | .55 | | .73 | 1.07 |
| | 010 | 1/2" diameter | | .040 | | .25 | .55 | | .80 | 1.14 |
| | 020 | 3/4" diameter | | .050 | | .32 | .69 | | 1.01 | 1.44 |
| | 025 | 1" diameter | | .080 | | .65 | 1.10 | | 1.75 | 2.45 |
| | 030 | 1-1/4" diameter | | .114 | | .83 | 1.57 | | 2.40 | 3.39 |
| | 035 | 1-1/2" diameter | | .160 | | 1.07 | 2.20 | | 3.27 | 4.65 |
| | 037 | 2" diameter | | .200 | | 1.40 | 2.75 | | 4.15 | 5.90 |
| | 038 | 2-1/2" diameter | | .267 | | 1.62 | 3.67 | | 5.29 | 7.55 |
| | 039 | 3" diameter | | .320 | | 2.02 | 4.40 | | 6.42 | 9.15 |
| | 040 | 3-1/2" diameter | | .400 | | 3.65 | 5.50 | | 9.15 | 12.70 |
| | 041 | 4" diameter | | .533 | | 5.25 | 7.35 | | 12.60 | 17.35 |
| 90-001 | | **WIREMOLD RACEWAY** | | | | | | | | |
| | 010 | No. 500 | 1 Elec | .080 | L.F. | .42 | 1.10 | | 1.52 | 2.20 |
| | 020 | No. 1000 | | .089 | | .70 | 1.22 | | 1.92 | 2.70 |
| | 040 | No. 1500, small pancake | | .089 | | .65 | 1.22 | | 1.87 | 2.64 |

For expanded coverage of these items see *Means' Electrical Cost Data 1985*

## 16.0 Raceways

| | | | CREW | MAN-HOURS | UNIT | MAT. | LABOR | EQUIP. | TOTAL | TOTAL INCL O&P |
|---|---|---|---|---|---|---|---|---|---|---|
| 90 | 060 | No. 2000, base & cover | 1 Elec | .089 | L.F. | .70 | 1.22 | | 1.92 | 2.70 |
| | 080 | No. 3000, base & cover | | .107 | | 1.37 | 1.47 | | 2.84 | 3.82 |
| | 100 | No. 4000, base & cover | | .123 | | 2.55 | 1.69 | | 4.24 | 5.45 |
| | 120 | No. 6000, base & cover | | .160 | ↓ | 3.80 | 2.20 | | 6 | 7.65 |
| | 240 | Fittings, elbows, No. 500 | | .200 | Ea. | .95 | 2.75 | | 3.70 | 5.40 |
| | 280 | Elbow cover, No. 2000 | | .200 | | 1.05 | 2.75 | | 3.80 | 5.50 |
| | 300 | Switch box, No. 500 | | .500 | | 2.50 | 6.90 | | 9.40 | 13.60 |
| | 340 | Telephone outlet, No. 1500 | | .500 | | 4.25 | 6.90 | | 11.15 | 15.50 |
| | 360 | Junction box, No. 1500 | ↓ | .500 | ↓ | 3.35 | 6.90 | | 10.25 | 14.55 |
| | 380 | Plugmold wired sections, No. 2000 | | | | | | | | |
| | 400 | 1 circuit, 6 outlets, 3 ft. long | 1 Elec | 1 | Ea. | 12.45 | 13.75 | | 26.20 | 35 |
| | 410 | 2 circuits, 8 outlets, 6 ft. long | | 1.510 | | 20 | 21 | | 41 | 55 |
| | 420 | Tele-power poles, aluminum, 4 outlets | ↓ | 2.960 | ↓ | 58 | 41 | | 99 | 130 |
| 95-001 | | **WIREWAY** | | | | | | | | |
| | 010 | Screw cover with fittings and supports, 2-1/2" x 2-1/2" | 1 Elec | .178 | L.F. | 4.10 | 2.44 | | 6.54 | 8.35 |
| | 020 | 4" x 4" | | .200 | | 4.70 | 2.75 | | 7.45 | 9.50 |
| | 040 | 6" x 6" | | .267 | | 7.70 | 3.67 | | 11.37 | 14.25 |
| | 060 | 8" x 8" | ↓ | .400 | ↓ | 13 | 5.50 | | 18.50 | 23 |

## 16.1 Conductors & Grounding

| | | | CREW | MAN-HOURS | UNIT | MAT. | LABOR | EQUIP. | TOTAL | TOTAL INCL O&P |
|---|---|---|---|---|---|---|---|---|---|---|
| 10-001 | | **WIRE** | | | | | | | | |
| | 002 | 600 volt, type THW, copper, solid, #14 | 1 Elec | .006 | L.F. | .03 | .08 | | .11 | .17 |
| | 003 | #12 | | .007 | | .04 | .10 | | .14 | .20 |
| | 004 | #10 | | .008 | | .06 | .11 | | .17 | .24 |
| | 016 | Stranded #6 | | .012 | | .15 | .17 | | .32 | .43 |
| | 018 | #4 | | .002 | | .23 | .02 | | .25 | .29 |
| | 020 | #3 | | .016 | | .29 | .22 | | .51 | .67 |
| | 022 | #2 | | .018 | | .35 | .24 | | .59 | .77 |
| | 024 | #1 | | .020 | | .46 | .28 | | .74 | .94 |
| | 026 | 1/0 | | .024 | | .54 | .33 | | .87 | 1.12 |
| | 028 | 2/0 | | .028 | | .65 | .38 | | 1.03 | 1.31 |
| | 030 | 3/0 | | .032 | | .79 | .44 | | 1.23 | 1.56 |
| | 035 | 4/0 | ↓ | .036 | ↓ | .98 | .50 | | 1.48 | 1.87 |
| 20-001 | | **ARMORED CABLE** | | | | | | | | |
| | 005 | 600 volt, copper (BX), #14, 2 wire | 1 Elec | .033 | L.F. | .24 | .46 | | .70 | .99 |
| | 010 | 3 wire | | .040 | | .30 | .55 | | .85 | 1.20 |
| | 015 | #12, 2 wire | | .038 | | .28 | .52 | | .80 | 1.13 |
| | 020 | 3 wire | | .044 | | .38 | .61 | | .99 | 1.38 |
| | 025 | #10, 2 wire | | .044 | | .46 | .61 | | 1.07 | 1.47 |
| | 030 | 3 wire | | .053 | | .60 | .73 | | 1.33 | 1.82 |
| | 035 | #8, 3 wire | ↓ | .067 | ↓ | 1 | .92 | | 1.92 | 2.55 |
| 70-001 | | **NON-METALLIC SHEATHED CABLE** 600 volt | | | | | | | | |
| | 010 | Copper with ground wire, (Romex) | | | | | | | | |
| | 015 | #14, 2 wire | 1 Elec | .032 | L.F. | .07 | .44 | | .51 | .77 |
| | 020 | 3 wire | | .035 | | .13 | .48 | | .61 | .90 |
| | 025 | #12, 2 wire | | .036 | | .10 | .50 | | .60 | .90 |
| | 030 | 3 wire | | .040 | | .19 | .55 | | .74 | 1.08 |
| | 035 | #10, 2 wire | | .040 | | .18 | .55 | | .73 | 1.07 |
| | 040 | 3 wire | | .057 | | .28 | .79 | | 1.07 | 1.55 |
| | 045 | #8, 3 wire | | .062 | | .60 | .85 | | 1.45 | 1.99 |
| | 050 | #6, 3 wire | ↓ | .067 | ↓ | .84 | .92 | | 1.76 | 2.37 |

For expanded coverage of these items see *Means' Electrical Cost Data 1985*

## 16.1 Conductors & Grounding

| | | CREW | MAN-HOURS | UNIT | MAT. | LABOR | EQUIP. | TOTAL | TOTAL INCL O&P |
|---|---|---|---|---|---|---|---|---|---|
| 70 055 | SE type SER aluminum cable, 3 RHW and | | | | | | | | |
| 060 | 1 bare neutral, 3 #8 & 1 #8 | 1 Elec | .053 | L.F. | .36 | .73 | | 1.09 | 1.55 |
| 065 | 3 #6 & 1 #6 | | .062 | | .46 | .85 | | 1.31 | 1.84 |
| 070 | 3 #4 & 1 #6 | | .073 | | .53 | 1 | | 1.53 | 2.16 |
| 075 | 3 #2 & 1 #4 | | .080 | | .76 | 1.10 | | 1.86 | 2.57 |
| 080 | 3 #1/0 & 1 #2 | | .089 | | 1.15 | 1.22 | | 2.37 | 3.19 |
| 085 | 3 #2/0 & 1 #1 | | .100 | | 1.30 | 1.38 | | 2.68 | 3.60 |
| 090 | 3 #4/0 & 1 #2/0 | | .114 | | 1.90 | 1.57 | | 3.47 | 4.57 |
| 240 | SEU service entrance cable, copper 2 conductors, #8 + #8 neut. | | .053 | | .53 | .73 | | 1.26 | 1.74 |
| 260 | #6 + #8 neutral | | .062 | | .66 | .85 | | 1.51 | 2.06 |
| 280 | #6 + #6 neutral | | .062 | | .71 | .85 | | 1.56 | 2.12 |
| 300 | #4 + #6 neutral | | .073 | | .94 | 1 | | 1.94 | 2.61 |
| 320 | #4 + #4 neutral | | .073 | | 1.05 | 1 | | 2.05 | 2.73 |
| 340 | #3 + #5 neutral | | .076 | | 1.15 | 1.05 | | 2.20 | 2.92 |
| 650 | Service entrance cap for copper SEU | | | | | | | | |
| 660 | 100 amp | 1 Elec | .667 | Ea. | 3.20 | 9.15 | | 12.35 | 17.95 |
| 670 | 150 amp | | .800 | | 7.50 | 11 | | 18.50 | 26 |
| 680 | 200 amp | | 1 | | 10 | 13.75 | | 23.75 | 33 |
| 80-001 | GROUNDING | | | | | | | | |
| 003 | Rod, copper clad, 8' long, 1/2" diameter | 1 Elec | 1.510 | Ea. | 15 | 21 | | 36 | 49 |
| 005 | 3/4" diameter | | 1.510 | | 15.70 | 21 | | 36.70 | 50 |
| 008 | 10' long, 1/2" diameter | | 1.670 | | 9.85 | 23 | | 32.85 | 47 |
| 010 | 3/4" diameter | | 1.820 | | 19.35 | 25 | | 44.35 | 61 |
| 026 | Wire, ground, bare armored, #8-1 conductor | | .040 | L.F. | .41 | .55 | | .96 | 1.32 |
| 027 | #6-1 conductor | | .044 | | .50 | .61 | | 1.11 | 1.51 |
| 040 | Bare copper, #6 wire | | .008 | | .16 | .11 | | .27 | .35 |
| 060 | #2 | | .016 | | .39 | .22 | | .61 | .78 |
| 180 | Water pipe ground clamps, heavy duty | | | | | | | | |
| 200 | Bronze, 1/2" to 1" diameter | 1 Elec | 1 | Ea. | 4.60 | 13.75 | | 18.35 | 27 |
| 280 | Brazed connections, #6 wire | | .667 | | 3.80 | 9.15 | | 12.95 | 18.65 |
| 300 | #2 wire | | .800 | | 3.80 | 11 | | 14.80 | 22 |

## 16.2 Boxes & Wiring Devices

| | | CREW | MAN-HOURS | UNIT | MAT. | LABOR | EQUIP. | TOTAL | TOTAL INCL O&P |
|---|---|---|---|---|---|---|---|---|---|
| 10-001 | PULL BOXES & CABINETS | | | | | | | | |
| 010 | Sheet metal, pull box, NEMA 1, type SC, 6"W x 6"H x 4"D | 1 Elec | 1 | Ea. | 5 | 13.75 | | 18.75 | 27 |
| 020 | 8"W x 8"H x 4"D | | 1 | | 7 | 13.75 | | 20.75 | 29 |
| 030 | 10"W x 12"H x 6"D | | 1.510 | | 12 | 21 | | 33 | 46 |
| 080 | 12"W x 16"H x 6"D | | 2 | | 22 | 28 | | 50 | 68 |
| 100 | 20"W x 20"H x 6"D | | 2.220 | | 37 | 31 | | 68 | 89 |
| 700 | Cabinets, current transformer | | | | | | | | |
| 705 | Single door, 24"H x 24"W x 10"D | 1 Elec | 5 | Ea. | 62 | 69 | | 131 | 175 |
| 710 | 30"H x 24"W x 10"D | | 6.150 | | 76 | 85 | | 161 | 215 |
| 715 | 36"H x 24"W x 10"D | | 7.270 | | 83 | 100 | | 183 | 250 |
| 720 | 30"H x 30"W x 10"D | | 8 | | 115 | 110 | | 225 | 300 |
| 725 | 36"H x 30"W x 10"D | | 8.890 | | 135 | 120 | | 255 | 340 |
| 730 | 36"H x 36"W x 10"D | | 10 | | 150 | 140 | | 290 | 380 |
| 750 | Double door, 48"H x 36"W x 10"D | | 13.330 | | 255 | 185 | | 440 | 570 |
| 755 | 24"H x 24"W x 12"D | | 8 | | 125 | 110 | | 235 | 310 |
| 20-001 | OUTLET BOXES | | | | | | | | |
| 002 | Pressed steel, octagon, 4" | 1 Elec | .444 | Ea. | 1 | 6.10 | | 7.10 | 10.75 |
| 010 | Extension | | .200 | | 1 | 2.75 | | 3.75 | 5.45 |
| 015 | Square 4" | | .444 | | 1 | 6.10 | | 7.10 | 10.75 |

For expanded coverage of these items see *Means' Electrical Cost Data 1985*

## 16.2 Boxes & Wiring Devices

| | | | CREW | MAN-HOURS | UNIT | MAT. | LABOR | EQUIP. | TOTAL | TOTAL INCL O&P |
|---|---|---|---|---|---|---|---|---|---|---|
| 20 | 020 | Extension | 1 Elec | .200 | Ea. | 1.29 | 2.75 | | 4.04 | 5.75 |
| | 025 | Covers, blank | | .125 | | .38 | 1.72 | | 2.10 | 3.13 |
| | 030 | Plaster rings | | .125 | | .59 | 1.72 | | 2.31 | 3.36 |
| | 065 | Switchbox | | .333 | | .87 | 4.58 | | 5.45 | 8.20 |
| | 110 | Concrete, floor, 1 gang | ↓ | 1.670 | ↓ | 34 | 23 | | 57 | 74 |
| 25-001 | | **OUTLET BOXES, PLASTIC** | | | | | | | | |
| | 005 | 4", round, with 2 mounting nails | 1 Elec | .348 | Ea. | .80 | 4.78 | | 5.58 | 8.40 |
| | 010 | Bar hanger mounted | | .348 | | 1.50 | 4.78 | | 6.28 | 9.20 |
| | 020 | Square with 2 mounting nails | | .348 | | .92 | 4.78 | | 5.70 | 8.55 |
| | 030 | Plaster ring | | .125 | | .44 | 1.72 | | 2.16 | 3.19 |
| | 040 | Switch box with 2 mounting nails, 1 gang | | .296 | | .48 | 4.07 | | 4.55 | 6.95 |
| | 050 | 2 gang | | .348 | | 1.32 | 4.78 | | 6.10 | 9 |
| | 060 | 3 gang | ↓ | .444 | ↓ | 2 | 6.10 | | 8.10 | 11.85 |
| 30-001 | | **WIRING DEVICES** | | | | | | | | |
| | 020 | Toggle switch, quiet type, single pole, 15 amp | 1 Elec | .200 | Ea. | 4.25 | 2.75 | | 7 | 9 |
| | 060 | 3 way, 15 amp | | .348 | | 6.25 | 4.78 | | 11.03 | 14.40 |
| | 090 | 4 way, 15 amp | ↓ | .533 | ↓ | 17.65 | 7.35 | | 25 | 31 |
| | 165 | Dimmer switch, 120 volt, incandescent, 600 watt, 1 pole | | | | | | | | |
| | 220 | Receptacle, duplex, 120 V grounded, 15 amp | 1 Elec | .200 | Ea. | 3.30 | 2.75 | | 6.05 | 7.95 |
| | 230 | 20 amp | | .296 | | 5.40 | 4.07 | | 9.47 | 12.35 |
| | 240 | Dryer, 30 amp | | .533 | | 4 | 7.35 | | 11.35 | 15.95 |
| | 250 | Range, 50 amp | | .727 | | 4.20 | 10 | | 14.20 | 20 |
| | 260 | Wall plates, stainless steel, 1 gang | | .100 | | 2.15 | 1.38 | | 3.53 | 4.53 |
| | 280 | 2 gang | | .151 | | 4.45 | 2.08 | | 6.53 | 8.15 |
| | 320 | Lampholder, keyless | | .308 | | 1.55 | 4.23 | | 5.78 | 8.40 |
| | 340 | Pullchain with receptacle | ↓ | .364 | ↓ | 3.95 | 5 | | 8.95 | 12.25 |

## 16.3 Starters, Boards & Switches

| | | | CREW | MAN-HOURS | UNIT | MAT. | LABOR | EQUIP. | TOTAL | TOTAL INCL O&P |
|---|---|---|---|---|---|---|---|---|---|---|
| 10-001 | | **CIRCUIT BREAKERS** | | | | | | | | |
| | 010 | Enclosed (NEMA 1), 600 volt, 3 pole, 30 amp | 1 Elec | 2.500 | Ea. | 185 | 34 | | 219 | 260 |
| | 020 | 60 amp | | 2.860 | | 185 | 39 | | 224 | 265 |
| | 040 | 100 amp | | 3.480 | | 220 | 48 | | 268 | 315 |
| | 060 | 225 amp | | 5.330 | | 475 | 73 | | 548 | 640 |
| | 070 | 400 amp | ↓ | 10 | ↓ | 835 | 140 | | 975 | 1,125 |
| 23-001 | | **LOAD CENTERS** | | | | | | | | |
| | 010 | 3 wire, 120/240V, 1 phase, including 1 pole plug-in breakers | | | | | | | | |
| | 020 | 100 amp main lugs, indoor, 8 circuits | 1 Elec | 5.710 | Ea. | 63 | 79 | | 142 | 195 |
| | 030 | 12 circuits | | 6.670 | | 86 | 92 | | 178 | 240 |
| | 040 | Rainproof, 8 circuits | | 5.710 | | 72 | 79 | | 151 | 205 |
| | 050 | 12 circuits | | 6.670 | | 95 | 92 | | 187 | 250 |
| | 060 | 200 amp main lugs, indoor, 16 circuits | | 8.890 | | 165 | 120 | | 285 | 375 |
| | 070 | 20 circuits | | 10.670 | | 195 | 145 | | 340 | 445 |
| | 080 | 24 circuits | | 12.310 | | 215 | 170 | | 385 | 505 |
| | 120 | Rainproof, 16 circuits | | 8.890 | | 200 | 120 | | 320 | 415 |
| | 130 | 20 circuits | | 10.670 | | 225 | 145 | | 370 | 480 |
| | 140 | 24 circuits | ↓ | 12.310 | ↓ | 250 | 170 | | 420 | 540 |
| 24-001 | | **METER CENTERS AND SOCKETS** | | | | | | | | |
| | 010 | Sockets, single position, 4 terminal, 100 amp | 1 Elec | 2.500 | Ea. | 18 | 34 | | 52 | 74 |
| | 020 | 150 amp | | 3.480 | | 21 | 48 | | 69 | 99 |
| | 030 | 200 amp | | 4.210 | | 27 | 58 | | 85 | 120 |
| | 040 | 20 amp | | 2.500 | | 31 | 34 | | 65 | 88 |
| | 050 | Double position, 4 terminal, 100 amp | ↓ | 2.860 | ↓ | 58 | 39 | | 97 | 125 |

For expanded coverage of these items see *Means' Electrical Cost Data 1985*

## 16.3 Starters, Boards & Switches

| | | CREW | MAN-HOURS | UNIT | MAT. | LABOR | EQUIP. | TOTAL | TOTAL INCL O&P |
|---|---|---|---|---|---|---|---|---|---|
| 24 060 | 150 amp | 1 Elec | 3.810 | Ea. | 74 | 52 | | 126 | 165 |
| 070 | 200 amp | " | 4.710 | " | 110 | 65 | | 175 | 225 |
| 27-001 | **MOTOR CONNECTIONS** | | | | | | | | |
| 002 | Flexible conduit and fittings, 1 HP motor | 1 Elec | 1 | Ea. | 3 | 13.75 | | 16.75 | 25 |
| 50-001 | **PANELBOARDS** (Including breakers) | | | | | | | | |
| 005 | NQOB, w/20 amp 1 pole bolt-on circuit breakers | | | | | | | | |
| 060 | 4 wire, 120/208 volts, 100 amp main lugs, 12 circuits | 1 Elec | 8 | Ea. | 300 | 110 | | 410 | 505 |
| 065 | 16 circuits | | 10.670 | | 345 | 145 | | 490 | 610 |
| 070 | 20 circuits | | 12.310 | | 395 | 170 | | 565 | 700 |
| 075 | 24 circuits | | 13.330 | | 445 | 185 | | 630 | 780 |
| 080 | 30 circuits | | 15.090 | | 515 | 210 | | 725 | 895 |
| 085 | 225 amp main lugs, 32 circuits | | 17.780 | | 560 | 245 | | 805 | 1,000 |
| 090 | 34 circuits | | 19.050 | | 580 | 260 | | 840 | 1,050 |
| 095 | 36 circuits | | 20 | | 605 | 275 | | 880 | 1,100 |
| 100 | 42 circuits | ↓ | 23.530 | ↓ | 680 | 325 | | 1,005 | 1,250 |
| 160 | NQOB panel, w/20 amp, 1 pole, circuit breakers | | | | | | | | |
| 200 | 4 wire, 120/208 volts with main circuit breaker | | | | | | | | |
| 205 | 100 amp main, 24 circuits | 1 Elec | 17.020 | Ea. | 580 | 235 | | 815 | 1,000 |
| 210 | 30 circuits | | 20 | | 650 | 275 | | 925 | 1,150 |
| 220 | 225 amp main, 32 circuits | | 22.220 | | 1,050 | 305 | | 1,355 | 1,625 |
| 225 | 42 circuits | | 28.570 | | 1,175 | 395 | | 1,570 | 1,900 |
| 230 | 400 amp main, 42 circuits | ↓ | 33.330 | ↓ | 1,650 | 460 | | 2,110 | 2,550 |
| 53-001 | **PANELBOARD CIRCUIT BREAKERS** | | | | | | | | |
| 005 | Bolt-on, 10,000 amp I.C., 120 volt, 1 pole | | | | | | | | |
| 010 | 15 to 50 amp | 1 Elec | .800 | Ea. | 6.95 | 11 | | 17.95 | 25 |
| 020 | 60 amp | | 1 | | 6.95 | 13.75 | | 20.70 | 29 |
| 030 | 70 amp | ↓ | 1 | ↓ | 13.55 | 13.75 | | 27.30 | 37 |
| 035 | 240 volt, 2 pole | | | | | | | | |
| 040 | 15 to 50 amp | 1 Elec | 1 | Ea. | 15.40 | 13.75 | | 29.15 | 39 |
| 050 | 60 amp | | 1.070 | | 15.40 | 14.65 | | 30.05 | 40 |
| 060 | 80 to 100 amp | | 1.600 | | 41 | 22 | | 63 | 80 |
| 070 | 3 pole, bolt-on, 15 to 60 amp | | 1.290 | | 50 | 17.75 | | 67.75 | 83 |
| 080 | 70 amp | | 1.600 | | 64 | 22 | | 86 | 105 |
| 090 | 80 to 100 amp | | 2.220 | | 73 | 31 | | 104 | 130 |
| 100 | 22,000 amp I.C., 240 volt, 2 pole, 70 to 225 amp | | 2.960 | | 205 | 41 | | 246 | 290 |
| 110 | 3 pole, 70 to 225 amp | ↓ | 3.480 | ↓ | 315 | 48 | | 363 | 420 |
| 55-001 | **SAFETY SWITCHES** | | | | | | | | |
| 010 | General duty, 240 volt, 3 pole, fused, 30 amp | 1 Elec | 2.500 | Ea. | 34 | 34 | | 68 | 92 |
| 020 | 60 amp | | 3.480 | | 59 | 48 | | 107 | 140 |
| 030 | 100 amp | | 4.210 | | 100 | 58 | | 158 | 200 |
| 040 | 200 amp | | 6.150 | | 220 | 85 | | 305 | 375 |
| 050 | 400 amp | ↓ | 8.890 | ↓ | 480 | 120 | | 600 | 720 |
| 56-001 | **TIME SWITCHES** | | | | | | | | |
| 010 | Single pole, single throw, 24 hour dial | 1 Elec | 2 | Ea. | 39 | 28 | | 67 | 86 |
| 020 | 24 hour dial with reserve power | | 2.220 | | 195 | 31 | | 226 | 265 |
| 030 | Astronomic dial | | 2.220 | | 64 | 31 | | 95 | 120 |
| 040 | Astronomic dial with reserve power | | 2.420 | | 200 | 33 | | 233 | 275 |
| 050 | 7 day calendar dial | | 2.420 | | 64 | 33 | | 97 | 125 |
| 060 | 7 day calendar dial with reserve power | | 2.500 | | 175 | 34 | | 209 | 245 |
| 070 | Photo cell 2000 watt | ↓ | 1 | ↓ | 14 | 13.75 | | 27.75 | 37 |

For expanded coverage of these items see *Means' Electrical Cost Data 1985*

## 16.5 Power Systems & Capacitors

| | CREW | MAN-HOURS | UNIT | MAT. | LABOR | EQUIP. | TOTAL | TOTAL INCL O&P |
|---|---|---|---|---|---|---|---|---|
| 10-001 **GENERATOR SET** | | | | | | | | |
| 002   Gas or gasoline operated, includes battery, | | | | | | | | |
| 005     charger, muffler & transfer switch | | | | | | | | |
| 020       3 phase, 4 wire, 277/480 volt, 7.5 KW | R-3 | 24.100 | Ea. | 4,975 | 335 | 105 | 5,415 | 6,125 |

## 16.6 Lighting

| | CREW | MAN-HOURS | UNIT | MAT. | LABOR | EQUIP. | TOTAL | TOTAL INCL O&P |
|---|---|---|---|---|---|---|---|---|
| 10-001 **INTERIOR LIGHTING FIXTURES** Including lamps, mounting | | | | | | | | |
| 003   hardware and connections | | | | | | | | |
| 010   Fluorescent, C.W. lamps, ceiling, recess mounted in grid, RS | | | | | | | | |
| 013     grid ceiling mount | | | | | | | | |
| 020       Acrylic lens, 1'W x 4'L, two 40 watt | 1 Elec | 1.400 | Ea. | 35 | 19.30 | | 54.30 | 69 |
| 030       2'W x 2'L, two U40 watt | ↓ | 1.400 | ↓ | 44 | 19.30 | | 63.30 | 79 |
| 060       2'W x 4'L, four 40 watt | ↓ | 1.700 | ↓ | 46 | 23 | | 69 | 88 |
| 100   Surface mounted, RS | | | | | | | | |
| 103     Acrylic lens with hinged & latched door frame | | | | | | | | |
| 110       1'W x 4'L, two 40 watt | 1 Elec | 1.140 | Ea. | 36 | 15.70 | | 51.70 | 64 |
| 120       2'W x 2'L, two U40 watt | | 1.140 | | 57 | 15.70 | | 72.70 | 87 |
| 150       2'W x 4'L, four 40 watt | ↓ | 1.510 | ↓ | 60 | 21 | | 81 | 99 |
| 210 | | | | | | | | |
| 213   Strip fixture, surface mounted | | | | | | | | |
| 220     4' long, one 40 watt RS | 1 Elec | .941 | Ea. | 17 | 12.95 | | 29.95 | 39 |
| 230     4' long, two 40 watt RS | | 1 | | 18 | 13.75 | | 31.75 | 41 |
| 260     8' long, one 75 watt, SL | | 1.190 | | 30 | 16.40 | | 46.40 | 59 |
| 270     8' long, two 75 watt, SL | ↓ | 1.290 | ↓ | 35 | 17.75 | | 52.75 | 66 |
| 358   Mercury vapor, integral ballast, ceiling, recess mounted, | | | | | | | | |
| 359     prismatic glass lens, floating door | | | | | | | | |
| 360       'W x 2'L, 250 watt DX lamp | 1 Elec | 2.500 | Ea. | 210 | 34 | | 244 | 285 |
| 370       2'W x 2'L, 400 watt DX lamp | | 2.760 | | 220 | 38 | | 258 | 300 |
| 380     Surface mtd., prismatic lens, 2'W x 2'L, 250 watt DX lamp | | 2.960 | | 200 | 41 | | 241 | 285 |
| 390       2'W x 2'L, 400 watt DX lamp | ↓ | 3.330 | ↓ | 210 | 46 | | 256 | 305 |
| 400   High bay, aluminum reflector | | | | | | | | |
| 403     Single unit, 400 watt DX lamp | 1 Elec | 3.480 | Ea. | 190 | 48 | | 238 | 285 |
| 410     Single unit, 1000 watt DX lamp | | 4 | | 340 | 55 | | 395 | 460 |
| 420     Twin unit, two 400 watt DX lamps | | 5 | | 380 | 69 | | 449 | 525 |
| 421   Low bay, aluminum reflector, 250W DX lamp | ↓ | 2.500 | ↓ | 235 | 34 | | 269 | 315 |
| 422   Metal halide, integral ballast, ceiling, recess mounted | | | | | | | | |
| 423     prismatic glass lens, floating door | | | | | | | | |
| 424       2'W x 2'L, 250 watt | 1 Elec | 2.500 | Ea. | 220 | 34 | | 254 | 295 |
| 425       2'W x 2'L, 400 watt | | 2.760 | | 260 | 38 | | 298 | 345 |
| 426     Surface mounted, 2'W x 2'L, 250 watt | | 2.960 | | 210 | 41 | | 251 | 295 |
| 427       2'W x 2'L, 400 watt | ↓ | 3.330 | ↓ | 250 | 46 | | 296 | 345 |
| 428   High bay, aluminum reflector, | | | | | | | | |
| 429     Single unit, 400 watt | 1 Elec | 3.480 | Ea. | 220 | 48 | | 268 | 315 |
| 430     Single unit, 1000 watt | | 4 | | 410 | 55 | | 465 | 540 |
| 431     Twin unit, 400 watt | | 5 | | 440 | 69 | | 509 | 590 |
| 432   Low bay, aluminum reflector, 250W DX lamp | ↓ | 2.500 | ↓ | 275 | 34 | | 309 | 355 |
| 445   Incandescent, ceiling, recess mtd., round alzak reflector, prewired | | | | | | | | |
| 447     100 watt | 1 Elec | 1 | Ea. | 36 | 13.75 | | 49.75 | 61 |
| 448     150 watt | " | 1 | " | 37 | 13.75 | | 50.75 | 62 |
| 520   Ceiling, surface mounted, opal glass drum | | | | | | | | |
| 530     8", one 60 watt | 1 Elec | .800 | Ea. | 21 | 11 | | 32 | 40 |
| 540     10", two 60 watt lamps | | 1 | | 26 | 13.75 | | 39.75 | 50 |
| 550     12", four 60 watt lamps | | 1.190 | | 52 | 16.40 | | 68.40 | 83 |
| 690   Mirror light, fluorescent, RS, acrylic enclosure, two 40 watt | ↓ | 1 | ↓ | 37 | 13.75 | | 50.75 | 62 |

For expanded coverage of these items see *Means' Electrical Cost Data 1985*

## 16.6 Lighting

| | | | CREW | MAN-HOURS | UNIT | MAT. | LABOR | EQUIP. | TOTAL | TOTAL INCL O&P |
|---|---|---|---|---|---|---|---|---|---|---|
| 10 | 691 | One 40 watt | 1 Elec | 1 | Ea. | 32 | 13.75 | | 45.75 | 57 |
| | 692 | One 20 watt | " | .667 | " | 31 | 9.15 | | 40.15 | 49 |
| 50-001 | | EXTERIOR FIXTURES With lamps | | | | | | | | |
| | 040 | Quartz, 500 watt | 1 Elec | 1.510 | Ea. | 65 | 21 | | 86 | 105 |
| | 080 | Wall pack, mercury vapor, 175 watt | | 2 | | 185 | 28 | | 213 | 245 |
| | 100 | 250 watt | | 2 | | 205 | 28 | | 233 | 270 |
| | 110 | Low pressure sodium, 35 watt | | 2 | | 155 | 28 | | 183 | 215 |
| | 115 | 55 watt | ↓ | 2 | ↓ | 235 | 28 | | 263 | 300 |
| 75-001 | | LAMPS | | | | | | | | |
| | 008 | Fluorescent, rapid start, cool white, 2' long, 20 watt | 1 Elec | .080 | Ea. | 3 | 1.10 | | 4.10 | 5.05 |
| | 010 | 4' long, 40 watt | | .089 | | 1.70 | 1.22 | | 2.92 | 3.80 |
| | 100 | Metal halide, mogul base, 175 watt | | .267 | | 32 | 3.67 | | 35.67 | 41 |
| | 110 | 250 watt | | .267 | | 37 | 3.67 | | 40.67 | 46 |
| | 135 | Sodium high pressure, 70 watt | | .267 | | 44 | 3.67 | | 47.67 | 54 |
| | 137 | 150 watt | ↓ | .267 | ↓ | 47 | 3.67 | | 50.67 | 57 |
| 80-001 | | TRACK LIGHTING | | | | | | | | |
| | 010 | 8' section | 1 Elec | 1.510 | Ea. | 37 | 21 | | 58 | 73 |
| | 030 | 3 circuits, 4' section | | 1.190 | | 24 | 16.40 | | 40.40 | 52 |
| | 040 | 8' section | ↓ | 1.510 | ↓ | 39 | 21 | | 60 | 76 |
| 90-001 | | RESIDENTIAL FIXTURES | | | | | | | | |
| | 020 | Pendant globe with shade, 150 watt | 1 Elec | .400 | Ea. | 50 | 5.50 | | 55.50 | 64 |
| | 040 | Fluorescent, interior, surface, circline, 32 watt & 40 watt | | .400 | | 34 | 5.50 | | 39.50 | 46 |
| | 050 | 2' x 2', two U 40 watt | | 1 | | 56 | 13.75 | | 69.75 | 83 |
| | 070 | Shallow under cabinet, two 20 watt | | .500 | | 38 | 6.90 | | 44.90 | 53 |
| | 090 | Wall mounted, 4'L, one 40 watt, with baffle | | .800 | | 32 | 11 | | 43 | 53 |
| | 200 | Incandescent, exterior lantern, wall mounted, 60 watt | | .500 | | 22 | 6.90 | | 28.90 | 35 |
| | 210 | Post light, 150W, with 7' post | | 2 | | 42 | 28 | | 70 | 90 |
| | 250 | Lamp holder, weatherproof with 150W PAR | | .500 | | 13 | 6.90 | | 19.90 | 25 |
| | 255 | With reflector and guard | | .667 | | 22 | 9.15 | | 31.15 | 39 |
| | 260 | Interior pendent, globe with shade, 150 watt | ↓ | .400 | ↓ | 50 | 5.50 | | 55.50 | 64 |

## 16.7 Lighting Utilities

| | | | CREW | MAN-HOURS | UNIT | MAT. | LABOR | EQUIP. | TOTAL | TOTAL INCL O&P |
|---|---|---|---|---|---|---|---|---|---|---|
| 01-001 | | ELECTRIC & TELEPHONE SITEWORK Not including excavation, backfill | | | | | | | | |
| | 020 | and cast in place concrete | | | | | | | | |
| | 420 | Underground duct, banks ready for concrete fill, minimum of 1-1/2 | | | | | | | | |
| | 440 | between ducts. For wire & cable see division 16.1 | | | | | | | | |
| | 460 | PVC, type EB, 2 @ 2" diameter | 1 Elec | .067 | L.F. | .56 | .91 | | 1.48 | 2.06 |
| | 480 | 4 @ 2" diameter | | .133 | | 1.12 | 1.83 | | 2.95 | 4.12 |
| | 560 | 4 @ 4" diameter | | .200 | | 2.60 | 2.75 | | 5.35 | 7.20 |
| | 620 | Rigid galvanized steel, 2 @ 2" diameter | | .089 | | 3.80 | 1.22 | | 5.02 | 6.10 |
| | 640 | 4 @ 2" diameter | | .178 | | 7.60 | 2.44 | | 10.04 | 12.20 |
| | 740 | 4 @ 4" diameter | ↓ | .471 | ↓ | 26. | 6.45 | | 32.45 | 39 |

For expanded coverage of these items see *Means' Electrical Cost Data 1985*

## 16.8 Special Systems

| | | CREW | MAN-HOURS | UNIT | MAT. | LABOR | EQUIP. | TOTAL | TOTAL INCL O&P |
|---|---|---|---|---|---|---|---|---|---|
| 15-001 | **DETECTION SYSTEMS** | | | | | | | | |
| 010 | Burglar alarm, battery operated, mechanical trigger | 1 Elec | 2 | Ea. | 125 | 28 | | 153 | 180 |
| 020 | Electrical trigger | | 2 | | 155 | 28 | | 183 | 215 |
| 040 | For outside key control, add | | 1 | | 21 | 13.75 | | 34.75 | 45 |
| 060 | For remote signaling circuitry, add | | 1 | | 37 | 13.75 | | 50.75 | 62 |
| 120 | Door switches, hinge switch | | 1.510 | | 38 | 21 | | 59 | 75 |
| 140 | Magnetic switch | | 1.510 | | 56 | 21 | | 77 | 94 |
| 280 | Ultrasonic motion detector, 12 volt | | 3.480 | | 190 | 48 | | 238 | 285 |
| 300 | Infrared photoelectric detector | | 3.480 | | 190 | 48 | | 238 | 285 |
| 320 | Passive infrared detector | | 3.480 | | 210 | 48 | | 258 | 305 |
| 342 | Switchmats, 30" x 5 ft. | | 1.510 | | 35 | 21 | | 56 | 71 |
| 344 | 25 ft. | | 2 | | 125 | 28 | | 153 | 180 |
| 346 | Police connect panel | | 2 | | 160 | 28 | | 188 | 220 |
| 348 | Telephone dialer | | 1.510 | | 315 | 21 | | 336 | 380 |
| 350 | Alarm bell | | 2 | | 29 | 28 | | 57 | 75 |
| 352 | Siren | | 2 | | 69 | 28 | | 97 | 120 |
| 520 | Smoke detector, ceiling type | | 1.290 | | 64 | 17.75 | | 81.75 | 98 |
| 560 | Light and horn | | 1.510 | | 75 | 21 | | 96 | 115 |
| 580 | Fire alarm horn | | 1.190 | | 21 | 16.40 | | 37.40 | 49 |
| 600 | Door holder, electro-magnetic | | 2 | | 48 | 28 | | 76 | 96 |
| 620 | Combination holder and closer | | 2.500 | | 320 | 34 | | 354 | 405 |
| 660 | Drill switch | | 1 | | 32 | 13.75 | | 45.75 | 57 |
| 680 | Master box | | 2.960 | | 1,150 | 41 | | 1,191 | 1,325 |
| 780 | Remote annunciator, 8 zone lamp | | 4.440 | | 160 | 61 | | 221 | 270 |
| 800 | 12 zone lamp | | 6.150 | | 210 | 85 | | 295 | 365 |
| 820 | 16 zone lamp | | 7.270 | | 265 | 100 | | 365 | 450 |
| 840 | Standpipe or sprinkler alarm, alarm device | | 1 | | 80 | 13.75 | | 93.75 | 110 |
| 860 | Actuating device | ↓ | 1 | ↓ | 215 | 13.75 | | 228.75 | 260 |
| 30-001 | **DOORBELL SYSTEM** Incl. transformer, button & signal | | | | | | | | |
| 100 | Door chimes, 2 notes, minimum | 1 Elec | .500 | Ea. | 13 | 6.90 | | 19.90 | 25 |
| 102 | Maximum | | .667 | | 33 | 9.15 | | 42.15 | 51 |
| 110 | Tube type, 3 tube system | | .667 | | 43 | 9.15 | | 52.15 | 62 |
| 118 | 4 tube system | | .800 | | 110 | 11 | | 121 | 140 |
| 190 | For transformer & button, minimum add | | 1.600 | | 9 | 22 | | 31 | 45 |
| 196 | Maximum, add | | 1.780 | | 33 | 24 | | 57 | 75 |
| 300 | For push button only, minimum | | .333 | | 9 | 4.58 | | 13.58 | 17.15 |
| 310 | Maximum | ↓ | .400 | ↓ | 17 | 5.50 | | 22.50 | 27 |
| 33-001 | **ELECTRIC HEATING** | | | | | | | | |
| 130 | Baseboard heaters, 2' long, 375 watt | 1 Elec | 1 | Ea. | 28 | 13.75 | | 41.75 | 52 |
| 140 | 3' long, 500 watt | | 1 | | 35 | 13.75 | | 48.75 | 60 |
| 160 | 4' long, 750 watt | | 1.190 | | 43 | 16.40 | | 59.40 | 73 |
| 180 | 5' long, 935 watt | | 1.400 | | 57 | 19.30 | | 76.30 | 93 |
| 200 | 6' long, 1125 watt | | 1.600 | | 64 | 22 | | 86 | 105 |
| 240 | 8' long, 1500 watt | | 2 | | 86 | 28 | | 114 | 140 |
| 280 | 10' long, 1875 watt | ↓ | 2.420 | ↓ | 98 | 33 | | 131 | 160 |
| 295 | Wall heaters with fan, 120 to 277 volt | | | | | | | | |
| 297 | surface mounted, residential, 750 watt | 1 Elec | 1.140 | Ea. | 45 | 15.70 | | 60.70 | 74 |
| 298 | 1000 watt | | 1.140 | | 48 | 15.70 | | 63.70 | 78 |
| 299 | 1250 watt | | 1.330 | | 65 | 18.35 | | 83.35 | 100 |
| 300 | 1500 watt | | 2 | | 68 | 28 | | 96 | 120 |
| 301 | 2000 watt | | 1.600 | | 69 | 22 | | 91 | 110 |
| 305 | 2500 watt | | 2 | | 105 | 28 | | 133 | 160 |
| 307 | 4000 watt | | 2.290 | | 105 | 31 | | 136 | 165 |
| 360 | Thermostats, integral | | .500 | | 15 | 6.90 | | 21.90 | 27 |
| 380 | Line voltage, 1 pole | ↓ | 1 | ↓ | 14 | 13.75 | | 27.75 | 37 |
| 40-001 | **SOUND SYSTEM** | | | | | | | | |
| 360 | House telephone, talking station | 1 Elec | 5 | Ea. | 195 | 69 | | 264 | 325 |
| 380 | Press to talk, release to listen | " | 1.510 | | 46 | 21 | | 67 | 83 |
| 400 | System-on button | | | ↓ | 20 | | | 20 | 22M |

For expanded coverage of these items see *Means' Electrical Cost Data 1985*

# 16.8 Special Systems

| | | CREW | MAN-HOURS | UNIT | BARE COSTS MAT. | LABOR | EQUIP. | TOTAL | TOTAL INCL O&P |
|---|---|---|---|---|---|---|---|---|---|
| 420 | Door release | 1 Elec | 2 | Ea. | 45 | 28 | | 73 | 93 |
| 440 | Combination speaker and microphone | | 1 | | 78 | 13.75 | | 91.75 | 105 |
| 460 | Termination box | | 2.500 | | 25 | 34 | | 59 | 82 |
| 480 | Amplifier or power supply | | 1.510 | | 315 | 21 | | 336 | 380 |
| 500 | Vestibule door unit | | .500 | | 55 | 6.90 | | 61.90 | 71 |
| 520 | Strip cabinet | | .296 | | 110 | 4.07 | | 114.07 | 125 |
| 540 | Directory | ↓ | .500 | ↓ | 30 | 6.90 | | 36.90 | 44 |
| 50-001 | **T.V. SYSTEMS** | | | | | | | | |
| 500 | T.V. Antenna only, minimum | 1 Elec | 1.330 | Ea. | 25 | 18.35 | | 43.35 | 56 |
| 510 | Maximum | " | 2 | " | 110 | 28 | | 138 | 165 |
| 81-001 | **WIRING, RESIDENTIAL** 20' average runs | | | | | | | | |
| 018 | Air conditioning receptacle | | | | | | | | |
| 020 | Using non-metallic sheathed cable | 1 Elec | .800 | Ea. | 6 | 11 | | 17 | 24 |
| 024 | BX cable | | .964 | | 10 | 13.25 | | 23.25 | 32 |
| 026 | EMT conduit | | 1.190 | | 10 | 16.40 | | 26.40 | 37 |
| 028 | Aluminum conduit | | 1.600 | | 17.40 | 22 | | 39.40 | 54 |
| 030 | Galvanized steel conduit | ↓ | 1.700 | ↓ | 16.60 | 23 | | 39.60 | 55 |
| 038 | Disposal wiring | | | | | | | | |
| 040 | Using non-metallic sheathed cable | 1 Elec | .889 | Ea. | 5 | 12.20 | | 17.20 | 25 |
| 044 | BX cable | | 1.070 | | 8 | 14.65 | | 22.65 | 32 |
| 046 | EMT conduit | | 1.330 | | 8 | 18.35 | | 26.35 | 38 |
| 048 | Aluminum conduit | | 1.780 | | 16 | 24 | | 40 | 56 |
| 050 | Galvanized steel conduit | ↓ | 1.900 | ↓ | 15 | 26 | | 41 | 58 |
| 058 | Dryer circuits | | | | | | | | |
| 060 | Using non-metallic sheathed cable | 1 Elec | 1.450 | Ea. | 13 | 20 | | 33 | 46 |
| 062 | BX cable | | 1.740 | | 20 | 24 | | 44 | 60 |
| 064 | EMT conduit | | 2.160 | | 15 | 30 | | 45 | 63 |
| 068 | Aluminum conduit | | 2.860 | | 23 | 39 | | 62 | 87 |
| 070 | Galvanized steel conduit | ↓ | 3.080 | ↓ | 22 | 42 | | 64 | 91 |
| 078 | Duplex receptacles | | | | | | | | |
| 080 | Using non-metallic sheathed cable | 1 Elec | .615 | Ea. | 5.50 | 8.45 | | 13.95 | 19.40 |
| 084 | BX cable | | .741 | | 9 | 10.20 | | 19.20 | 26 |
| 086 | EMT conduit | | .920 | | 9 | 12.65 | | 21.65 | 30 |
| 088 | Aluminum conduit | | 1.230 | | 17 | 16.90 | | 33.90 | 45 |
| 090 | Galvanized steel conduit | ↓ | 1.310 | ↓ | 16 | 18.05 | | 34.05 | 46 |
| 098 | Exhaust fan wiring | | | | | | | | |
| 100 | Using non-metallic sheathed cable | 1 Elec | .800 | Ea. | 5 | 11 | | 16 | 23 |
| 102 | BX cable | | .964 | | 8 | 13.25 | | 21.25 | 30 |
| 106 | EMT conduit | | 1.190 | | 8 | 16.40 | | 24.40 | 35 |
| 108 | Aluminum conduit | | 1.600 | | 16 | 22 | | 38 | 52 |
| 110 | Galvanized steel conduit | ↓ | 1.700 | ↓ | 15 | 23 | | 38 | 53 |
| 118 | Fire alarm smoke detector & horn | | | | | | | | |
| 120 | Using non-metallic sheathed cable | 1 Elec | .800 | Ea. | 36 | 11 | | 47 | 57 |
| 122 | BX cable | | .964 | | 39 | 13.25 | | 52.25 | 64 |
| 124 | EMT conduit | | 1.190 | | 39 | 16.40 | | 55.40 | 69 |
| 128 | Aluminum conduit | | 1.600 | | 46 | 22 | | 68 | 85 |
| 130 | Galvanized steel conduit | ↓ | 1.700 | ↓ | 45 | 23 | | 68 | 86 |
| 158 | Furnace circuit and switch | | | | | | | | |
| 160 | Using non-metallic sheathed cable | 1 Elec | 1.330 | Ea. | 7.10 | 18.35 | | 25.45 | 37 |
| 162 | BX cable | | 1.600 | | 11 | 22 | | 33 | 47 |
| 164 | EMT conduit | | 2 | | 11 | 28 | | 39 | 55 |
| 168 | Aluminum conduit | | 2.670 | | 18.60 | 37 | | 55.60 | 78 |
| 169 | Galvanized steel conduit | ↓ | 2.860 | ↓ | 17.60 | 39 | | 56.60 | 81 |
| 170 | Ground fault receptacle | | | | | | | | |
| 172 | Using non-metallic sheathed cable | 1 Elec | 1 | Ea. | 36 | 13.75 | | 49.75 | 61 |
| 174 | BX cable | | 1.210 | | 41 | 16.65 | | 57.65 | 71 |
| 176 | EMT conduit | | 1.480 | | 41 | 20 | | 61 | 77 |
| 177 | Aluminum conduit | ↓ | 2 | ↓ | 48 | 28 | | 76 | 96 |

For expanded coverage of these items see *Means' Electrical Cost Data 1985*

## 16.8 Special Systems

| | | | CREW | MAN-HOURS | UNIT | MAT. | LABOR | EQUIP. | TOTAL | TOTAL INCL O&P |
|---|---|---|---|---|---|---|---|---|---|---|
| 81 | 178 | Galvanized steel conduit | 1 Elec | 2.110 | Ea. | 47 | 29 | | 76 | 97 |
| | 179 | Heater circuits | | | | | | | | |
| | 180 | Using non-metallic sheathed cable | 1 Elec | 1 | Ea. | 5 | 13.75 | | 18.75 | 27 |
| | 182 | BX cable | | 1.210 | | 8 | 16.65 | | 24.65 | 35 |
| | 184 | EMT conduit | | 1.480 | | 8 | 20 | | 28 | 41 |
| | 186 | Aluminum conduit | | 2 | | 16 | 28 | | 44 | 61 |
| | 188 | Galvanized steel conduit | ↓ | 2.110 | ↓ | 15 | 29 | | 44 | 62 |
| | 198 | Intercom, 8 stations | | | | | | | | |
| | 200 | Using non-metallic sound cable | 1 Elec | 11.430 | Total | 260 | 155 | | 415 | 535 |
| | 202 | BX cable | | 13.790 | | 315 | 190 | | 505 | 645 |
| | 204 | EMT conduit | | 17.020 | | 380 | 235 | | 615 | 785 |
| | 206 | Aluminum conduit | | 22.860 | | 595 | 315 | | 910 | 1,150 |
| | 208 | Galvanized steel conduit | ↓ | 24.240 | ↓ | 560 | 335 | | 895 | 1,150 |
| | 220 | Light fixtures, average | | .602 | Ea. | 20 | 8.25 | | 28.25 | 35 |
| | 238 | Lighting wiring | | | | | | | | |
| | 240 | Using non-metallic sheathed cable | 1 Elec | .500 | Ea. | 5 | 6.90 | | 11.90 | 16.35 |
| | 242 | BX cable | | .602 | | 9 | 8.25 | | 17.25 | 23 |
| | 244 | EMT conduit | | .748 | | 9 | 10.30 | | 19.30 | 26 |
| | 246 | Aluminum conduit | | 1 | | 17 | 13.75 | | 30.75 | 40 |
| | 248 | Galvanized steel conduit | ↓ | 1.070 | ↓ | 16 | 14.65 | | 30.65 | 41 |
| | 258 | Range circuits | | | | | | | | |
| | 260 | Using non-metallic sheathed cable | 1 Elec | 2 | Ea. | 30 | 28 | | 58 | 76 |
| | 262 | BX cable | | 2.420 | | 42 | 33 | | 75 | 99 |
| | 264 | EMT conduit | | 2.960 | | 26 | 41 | | 67 | 93 |
| | 266 | Aluminum conduit | | 4 | | 34 | 55 | | 89 | 125 |
| | 268 | Galvanized steel conduit | | 4.210 | | 33 | 58 | | 91 | 130 |
| | 280 | Service and panel, 100 amp | | 10 | | 220 | 140 | | 360 | 460 |
| | 300 | 200 amp | ↓ | 16 | ↓ | 420 | 220 | | 640 | 810 |
| | 318 | Switch, single pole | | | | | | | | |
| | 320 | Using non-metallic sheathed cable | 1 Elec | .500 | Ea. | 5.40 | 6.90 | | 12.30 | 16.80 |
| | 322 | BX cable | | .602 | | 9 | 8.25 | | 17.25 | 23 |
| | 324 | EMT conduit | | .748 | | 9 | 10.30 | | 19.30 | 26 |
| | 326 | Aluminum conduit | | 1 | | 17 | 13.75 | | 30.75 | 40 |
| | 328 | Galvanized steel conduit | ↓ | 1.070 | ↓ | 16 | 14.65 | | 30.65 | 41 |
| | 339 | Switch, 3-way | | | | | | | | |
| | 340 | Using non-metallic sheathed cable | 1 Elec | .667 | Ea. | 7 | 9.15 | | 16.15 | 22 |
| | 342 | BX cable | | .800 | | 12 | 11 | | 23 | 31 |
| | 343 | EMT conduit | | 1 | | 12 | 13.75 | | 25.75 | 35 |
| | 344 | Aluminum conduit | | 1.330 | | 18 | 18.35 | | 36.35 | 49 |
| | 345 | Galvanized steel conduit | ↓ | 1.380 | ↓ | 17 | 18.95 | | 35.95 | 49 |
| | 358 | Water heater circuit | | | | | | | | |
| | 360 | Using non-metallic sheathed cable | 1 Elec | 1.600 | Ea. | 7 | 22 | | 29 | 42 |
| | 362 | BX cable | | 1.900 | | 12 | 26 | | 38 | 55 |
| | 364 | EMT conduit | | 2.350 | | 12 | 32 | | 44 | 64 |
| | 366 | Aluminum conduit | | 3.200 | | 18 | 44 | | 62 | 89 |
| | 368 | Galvanized steel conduit | ↓ | 3.330 | ↓ | 17 | 46 | | 63 | 91 |
| | 378 | Weatherproof receptacle with ground fault breaker at panel | | | | | | | | |
| | 380 | Using non-metallic sheathed cable | 1 Elec | 1.330 | Ea. | 51 | 18.35 | | 69.35 | 85 |
| | 382 | BX cable | | 1.600 | | 55 | 22 | | 77 | 95 |
| | 384 | EMT conduit | | 2 | | 55 | 28 | | 83 | 105 |
| | 386 | Aluminum conduit | | 2.670 | | 62 | 37 | | 99 | 125 |
| | 388 | Galvanized steel conduit | ↓ | 2.860 | ↓ | 60 | 39 | | 99 | 130 |

For expanded coverage of these items see *Means' Electrical Cost Data 1985*

# CREWS

| Crew No. | Bare Costs | | Incl. Subs O & P | | Cost Per Man-hour | |
|---|---|---|---|---|---|---|
| **Crew A-1** | Hr. | Daily | Hr. | Daily | Bare Costs | Incl. O&P |
| 1 Building Laborer | $10.00 | $80.00 | $15.85 | $126.80 | $10.00 | $15.85 |
| 1 Gas Eng. Power Tool | | 35.20 | | 38.70 | 4.40 | 4.83 |
| 8 M.H., Daily Totals | | $115.20 | | $165.50 | $14.40 | $20.68 |
| **Crew A-5** | Hr. | Daily | Hr. | Daily | Bare Costs | Incl. O&P |
| 2 Building Laborers | $10.00 | $160.00 | $15.85 | $253.60 | $10.02 | $15.87 |
| .25 Truck Driver (light) | 10.25 | 20.50 | 16.10 | 32.20 | | |
| .25 Light Truck, 1.5 Ton | | 14.30 | | 15.70 | .79 | .87 |
| 18 M.H., Daily Totals | | $194.80 | | $301.50 | $10.81 | $16.74 |
| **Crew A-6** | Hr. | Daily | Hr. | Daily | Bare Costs | Incl. O&P |
| 1 Chief Of Party | $12.05 | $96.40 | $19.10 | $152.80 | $11.45 | $18.15 |
| 1 Instrument Man | 10.85 | 86.80 | 17.20 | 137.60 | | |
| 16 M.H., Daily Totals | | $183.20 | | $290.40 | $11.45 | $18.15 |
| **Crew A-7** | Hr. | Daily | Hr. | Daily | Bare Costs | Incl. O&P |
| 1 Chief Of Party | $12.05 | $96.40 | $19.10 | $152.80 | $10.88 | $17.30 |
| 1 Instrument Man | 10.85 | 86.80 | 17.20 | 137.60 | | |
| 1 Rodman/Chainman | 9.75 | 78.00 | 15.60 | 124.80 | | |
| 24 M.H., Daily Totals | | $261.20 | | $415.20 | $10.88 | $17.30 |
| **Crew A-8** | Hr. | Daily | Hr. | Daily | Bare Costs | Incl. O&P |
| 1 Chief Of Party | $12.05 | $96.40 | $19.10 | $152.80 | $10.60 | $16.87 |
| 1 Instrument Man | 10.85 | 86.80 | 17.20 | 137.60 | | |
| 2 Rodmen/Chainmen | 9.75 | 156.00 | 15.60 | 249.60 | | |
| 32 M.H., Daily Totals | | $339.20 | | $540.00 | $10.60 | $16.87 |
| **Crew B-1** | Hr. | Daily | Hr. | Daily | Bare Costs | Incl. O&P |
| 1 Labor Foreman (outside) | $12.00 | $96.00 | $19.00 | $152.00 | $10.66 | $16.90 |
| 2 Building Laborers | 10.00 | 160.00 | 15.85 | 253.60 | | |
| 24 M.H., Daily Totals | | $256.00 | | $405.60 | $10.66 | $16.90 |
| **Crew B-3** | Hr. | Daily | Hr. | Daily | Bare Costs | Incl. O&P |
| 1 Labor Foreman (outside) | $12.00 | $96.00 | $19.00 | $152.00 | $10.89 | $17.21 |
| 2 Building Laborers | 10.00 | 160.00 | 15.85 | 253.60 | | |
| 1 Equip. Oper. (med.) | 12.65 | 101.20 | 20.10 | 160.80 | | |
| 2 Truck Drivers (heavy) | 10.35 | 165.60 | 16.25 | 260.00 | | |
| F.E. Loader, T.M., 2.5 C.Y. | | 585.00 | | 643.50 | | |
| 2 Dump Trucks, 16 Ton | | 558.40 | | 614.25 | 23.82 | 26.20 |
| 48 M.H., Daily Totals | | $1666.20 | | $2084.15 | $34.71 | $43.41 |
| **Crew B-5** | Hr. | Daily | Hr. | Daily | Bare Costs | Incl. O&P |
| 1 Labor Foreman (outside) | $12.00 | $96.00 | $19.00 | $152.00 | $11.33 | $17.98 |
| 4 Building Laborers | 10.00 | 320.00 | 15.85 | 507.20 | | |
| 2 Equip. Oper. (med.) | 12.65 | 202.40 | 20.10 | 321.60 | | |
| 1 Mechanic | 13.40 | 107.20 | 21.25 | 170.00 | | |
| 1 Air Compr., 250 C.F.M. | | 106.20 | | 116.80 | | |
| Air Tools & Accessories | | 23.70 | | 26.05 | | |
| 2-50 Ft. Air Hoses, 1.5" Dia. | | 11.60 | | 12.75 | | |
| F.E. Loader, T.M., 2.5 C.Y. | | 585.00 | | 643.50 | 11.35 | 12.48 |
| 64 M.H., Daily Totals | | $1452.10 | | $1949.90 | $22.68 | $30.46 |
| **Crew B-6** | Hr. | Daily | Hr. | Daily | Bare Costs | Incl. O&P |
| 2 Building Laborers | $10.00 | $160.00 | $15.85 | $253.60 | $10.68 | $16.93 |
| 1 Equip. Oper. (light) | 12.05 | 96.40 | 19.10 | 152.80 | | |
| 1 Backhoe Loader, 48 H.P. | | 152.20 | | 167.40 | 6.34 | 6.97 |
| 24 M.H., Daily Totals | | $408.60 | | $573.80 | $17.02 | $23.90 |

| Crew No. | Bare Costs | | Incl. Subs O & P | | Cost Per Man-hour | |
|---|---|---|---|---|---|---|
| **Crew B-7** | Hr. | Daily | Hr. | Daily | Bare Costs | Incl. O&P |
| 1 Labor Foreman (outside) | $12.00 | $96.00 | $19.00 | $152.00 | $10.77 | $17.08 |
| 4 Building Laborers | 10.00 | 320.00 | 15.85 | 507.20 | | |
| 1 Equip. Oper. (med.) | 12.65 | 101.20 | 20.10 | 160.80 | | |
| 1 Chipping Machine | | 134.60 | | 148.05 | | |
| F.E. Loader, T.M., 2.5 C.Y. | | 585.00 | | 643.50 | 14.99 | 16.49 |
| 48 M.H., Daily Totals | | $1236.80 | | $1611.55 | $25.76 | $33.57 |
| **Crew B-8** | Hr. | Daily | Hr. | Daily | Bare Costs | Incl. O&P |
| 1 Labor Foreman (outside) | $12.00 | $96.00 | $19.00 | $152.00 | $11.10 | $17.57 |
| 2 Building Laborers | 10.00 | 160.00 | 15.85 | 253.60 | | |
| 2 Equip. Oper. (med.) | 12.65 | 202.40 | 20.10 | 321.60 | | |
| 1 Equip. Oper. Oiler | 10.85 | 86.80 | 17.20 | 137.60 | | |
| 2 Truck Drivers (heavy) | 10.35 | 165.60 | 16.25 | 260.00 | | |
| 1 Hyd. Crane, 25 Ton | | 384.00 | | 422.40 | | |
| F.E. Loader, T.M., 2.5 C.Y. | | 585.00 | | 643.50 | | |
| 2 Dump Trucks, 16 Ton | | 558.40 | | 614.25 | 23.86 | 26.25 |
| 64 M.H., Daily Totals | | $2238.20 | | $2804.95 | $34.96 | $43.82 |
| **Crew B-9** | Hr. | Daily | Hr. | Daily | Bare Costs | Incl. O&P |
| 1 Labor Foreman (outside) | $12.00 | $96.00 | $19.00 | $152.00 | $10.40 | $16.48 |
| 4 Building Laborers | 10.00 | 320.00 | 15.85 | 507.20 | | |
| 1 Air Compr., 250 C.F.M. | | 106.20 | | 116.80 | | |
| Air Tools & Accessories | | 23.70 | | 26.05 | | |
| 2-50 Ft. Air Hoses, 1.5" Dia. | | 11.60 | | 12.75 | 3.53 | 3.89 |
| 40 M.H., Daily Totals | | $557.50 | | $814.80 | $13.93 | $20.37 |
| **Crew B-10B** | Hr. | Daily | Hr. | Daily | Bare Costs | Incl. O&P |
| 1 Equip. Oper. (med.) | $12.65 | $101.20 | $20.10 | $160.80 | $11.76 | $18.68 |
| .5 Building Laborer | 10.00 | 40.00 | 15.85 | 63.40 | | |
| 1 Dozer, 200 H.P. | | 614.40 | | 675.85 | 51.20 | 56.32 |
| 12 M.H., Daily Totals | | $755.60 | | $900.05 | $62.96 | $75.00 |
| **Crew B-10E** | Hr. | Daily | Hr. | Daily | Bare Costs | Incl. O&P |
| 1 Equip. Oper. (med.) | $12.65 | $101.20 | $20.10 | $160.80 | $11.76 | $18.68 |
| .5 Building Laborer | 10.00 | 40.00 | 15.85 | 63.40 | | |
| 1 Tandem Roller, 5 Ton | | 83.85 | | 92.25 | 6.98 | 7.68 |
| 12 M.H., Daily Totals | | $225.05 | | $316.45 | $18.74 | $26.36 |
| **Crew B-10L** | Hr. | Daily | Hr. | Daily | Bare Costs | Incl. O&P |
| 1 Equip. Oper. (med.) | $12.65 | $101.20 | $20.10 | $160.80 | $11.76 | $18.68 |
| .5 Building Laborer | 10.00 | 40.00 | 15.85 | 63.40 | | |
| 1 Dozer, 75 H.P. | | 198.60 | | 218.45 | 16.55 | 18.20 |
| 12 M.H., Daily Totals | | $339.80 | | $442.65 | $28.31 | $36.88 |
| **Crew B-10N** | Hr. | Daily | Hr. | Daily | Bare Costs | Incl. O&P |
| 1 Equip. Oper. (med.) | $12.65 | $101.20 | $20.10 | $160.80 | $11.76 | $18.68 |
| .5 Building Laborer | 10.00 | 40.00 | 15.85 | 63.40 | | |
| F.E. Loader, T.M., 1.5 C.Y. | | 269.00 | | 295.90 | 22.41 | 24.65 |
| 12 M.H., Daily Totals | | $410.20 | | $520.10 | $34.17 | $43.33 |
| **Crew B-10R** | Hr. | Daily | Hr. | Daily | Bare Costs | Incl. O&P |
| 1 Equip. Oper. (med.) | $12.65 | $101.20 | $20.10 | $160.80 | $11.76 | $18.68 |
| .5 Building Laborer | 10.00 | 40.00 | 15.85 | 63.40 | | |
| F.E. Loader, W.M., 1 C.Y. | | 199.00 | | 218.90 | 16.58 | 18.24 |
| 12 M.H., Daily Totals | | $340.20 | | $443.10 | $28.34 | $36.92 |
| **Crew B-10S** | Hr. | Daily | Hr. | Daily | Bare Costs | Incl. O&P |
| 1 Equip. Oper. (med.) | $12.65 | $101.20 | $20.10 | $160.80 | $11.76 | $18.68 |
| .5 Building Laborer | 10.00 | 40.00 | 15.85 | 63.40 | | |
| F.E. Loader, W.M., 1.5 C.Y. | | 266.60 | | 293.25 | 22.21 | 24.43 |
| 12 M.H., Daily Totals | | $407.80 | | $517.45 | $33.97 | $43.11 |

# CREWS

| Crew No. | | Bare Costs | | Incl. Subs O & P | | Cost Per Man-hour | |
|---|---|---|---|---|---|---|---|
| **Crew B-11A** | Hr. | Daily | Hr. | Daily | Bare Costs | Incl. O&P |
| 1 Equipment Oper. (med.) | $12.65 | $101.20 | $20.10 | $160.80 | $11.32 | $17.97 |
| 1 Building Laborer | 10.00 | 80.00 | 15.85 | 126.80 | | |
| 1 Dozer, 200 H.P. | | 614.40 | | 675.85 | 38.40 | 42.24 |
| 16 M.H., Daily Totals | | $795.60 | | $963.45 | $49.72 | $60.21 |
| **Crew B-11B** | Hr. | Daily | Hr. | Daily | Bare Costs | Incl. O&P |
| 1 Equipment Oper. (med.) | $12.65 | $101.20 | $20.10 | $160.80 | $11.32 | $17.97 |
| 1 Building Laborer | 10.00 | 80.00 | 15.85 | 126.80 | | |
| 1 Dozer, 200 H.P. | | 614.40 | | 675.85 | | |
| 1 Air Powered Tamper | | 12.45 | | 13.70 | | |
| 1 Air Compr. 365 C.F.M. | | 148.80 | | 163.70 | | |
| 2-50 Ft. Air Hoses, 1.5" Dia. | | 11.60 | | 12.75 | 49.20 | 54.12 |
| 16 M.H., Daily Totals | | $968.45 | | $1153.60 | $60.52 | $72.09 |
| **Crew B-11C** | Hr. | Daily | Hr. | Daily | Bare Costs | Incl. O&P |
| 1 Equipment Oper. (med.) | $12.65 | $101.20 | $20.10 | $160.80 | $11.32 | $17.97 |
| 1 Building Laborer | 10.00 | 80.00 | 15.85 | 126.80 | | |
| 1 Backhoe Loader, 48 H.P. | | 152.20 | | 167.40 | 9.51 | 10.46 |
| 16 M.H., Daily Totals | | $333.40 | | $455.00 | $20.83 | $28.43 |
| **Crew B-11M** | Hr. | Daily | Hr. | Daily | Bare Costs | Incl. O&P |
| 1 Equipment Oper. (med.) | $12.65 | $101.20 | $20.10 | $160.80 | $11.32 | $17.97 |
| 1 Building Laborer | 10.00 | 80.00 | 15.85 | 126.80 | | |
| 1 Backhoe Loader, 80 H.P. | | 228.00 | | 250.80 | 14.25 | 15.67 |
| 16 M.H., Daily Totals | | $409.20 | | $538.40 | $25.57 | $33.64 |
| **Crew B-12A** | Hr. | Daily | Hr. | Daily | Bare Costs | Incl. O&P |
| 1 Equip. Oper. (crane) | $12.90 | $103.20 | $20.45 | $163.60 | $11.87 | $18.82 |
| 1 Equip. Oper. Oiler | 10.85 | 86.80 | 17.20 | 137.60 | | |
| 1 Hyd. Excavator, 1 C.Y. | | 402.20 | | 442.40 | 25.13 | 27.65 |
| 16 M.H., Daily Totals | | $592.20 | | $743.60 | $37.00 | $46.47 |
| **Crew B-12E** | Hr. | Daily | Hr. | Daily | Bare Costs | Incl. O&P |
| 1 Equip. Oper. (crane) | $12.90 | $103.20 | $20.45 | $163.60 | $11.87 | $18.82 |
| 1 Equip. Oper. Oiler | 10.85 | 86.80 | 17.20 | 137.60 | | |
| 1 Hyd. Excavator, .5 C.Y. | | 255.00 | | 280.50 | 15.93 | 17.53 |
| 16 M.H., Daily Totals | | $445.00 | | $581.70 | $27.80 | $36.35 |
| **Crew B-12F** | Hr. | Daily | Hr. | Daily | Bare Costs | Incl. O&P |
| 1 Equip. Oper. (crane) | $12.90 | $103.20 | $20.45 | $163.60 | $11.87 | $18.82 |
| 1 Equip. Oper. Oiler | 10.85 | 86.80 | 17.20 | 137.60 | | |
| 1 Hyd. Excavator, .75 C.Y. | | 323.20 | | 355.50 | 20.20 | 22.21 |
| 16 M.H., Daily Totals | | $513.20 | | $656.70 | $32.07 | $41.03 |
| **Crew B-12J** | Hr. | Daily | Hr. | Daily | Bare Costs | Incl. O&P |
| 1 Equip. Oper. (crane) | $12.90 | $103.20 | $20.45 | $163.60 | $11.87 | $18.82 |
| 1 Equip. Oper. Oiler | 10.85 | 86.80 | 17.20 | 137.60 | | |
| 1 Gradall, 3 Ton, .5 C.Y. | | 432.20 | | 475.40 | 27.01 | 29.71 |
| 16 M.H., Daily Totals | | $622.20 | | $776.60 | $38.88 | $48.53 |
| **Crew B-12Q** | Hr. | Daily | Hr. | Daily | Bare Costs | Incl. O&P |
| 1 Equip. Oper. (crane) | $12.90 | $103.20 | $20.45 | $163.60 | $11.87 | $18.82 |
| 1 Equip. Oper. Oiler | 10.85 | 86.80 | 17.20 | 137.60 | | |
| 1 Hyd. Excavator, 5/8 C.Y. | | 268.00 | | 294.80 | 16.75 | 18.42 |
| 16 M.H., Daily Totals | | $458.00 | | $596.20 | $28.62 | $37.24 |
| **Crew B-12R** | Hr. | Daily | Hr. | Daily | Bare Costs | Incl. O&P |
| 1 Equip. Oper. (crane) | $12.90 | $103.20 | $20.45 | $163.60 | $11.87 | $18.82 |
| 1 Equip. Oper. Oiler | 10.85 | 86.80 | 17.20 | 137.60 | | |
| 1 Hyd. Excavator, 1.5 C.Y. | | 560.20 | | 616.20 | 35.01 | 38.51 |
| 16 M.H., Daily Totals | | $750.20 | | $917.40 | $46.88 | $57.33 |

| Crew No. | | Bare Costs | | Incl. Subs O & P | | Cost Per Man-hour | |
|---|---|---|---|---|---|---|---|
| **Crew B-13** | Hr. | Daily | Hr. | Daily | Bare Costs | Incl. O&P |
| 1 Labor Foreman (outside) | $12.00 | $96.00 | $19.00 | $152.00 | $10.82 | $17.15 |
| 4 Building Laborers | 10.00 | 320.00 | 15.85 | 507.20 | | |
| 1 Equip. Oper. (crane) | 12.90 | 103.20 | 20.45 | 163.60 | | |
| 1 Equip. Oper. Oiler | 10.85 | 86.80 | 17.20 | 137.60 | | |
| 1 Hyd. Crane, 25 Ton | | 384.00 | | 422.40 | 6.85 | 7.54 |
| 56 M.H., Daily Totals | | $990.00 | | $1382.80 | $17.67 | $24.69 |
| **Crew B-14** | Hr. | Daily | Hr. | Daily | Bare Costs | Incl. O&P |
| 1 Labor Foreman (outside) | $12.00 | $96.00 | $19.00 | $152.00 | $10.67 | $16.91 |
| 4 Building Laborers | 10.00 | 320.00 | 15.85 | 507.20 | | |
| 1 Equip. Oper. (light) | 12.05 | 96.40 | 19.10 | 152.80 | | |
| 1 Backhoe Loader, 48 H.P. | | 152.20 | | 167.40 | 3.17 | 3.48 |
| 48 M.H., Daily Totals | | $664.60 | | $979.40 | $13.84 | $20.39 |
| **Crew B-15** | Hr. | Daily | Hr. | Daily | Bare Costs | Incl. O&P |
| 1 Equipment Oper. (med.) | $12.65 | $101.20 | $20.10 | $160.80 | $10.95 | $17.29 |
| .5 Building Laborer | 10.00 | 40.00 | 15.85 | 63.40 | | |
| 2 Truck Drivers (heavy) | 10.35 | 165.60 | 16.25 | 260.00 | | |
| 2 Dump Trucks, 16 Ton | | 558.40 | | 614.25 | | |
| 1 Dozer, 200 H.P. | | 614.40 | | 675.85 | 41.88 | 46.07 |
| 28 M.H., Daily Totals | | $1479.60 | | $1774.30 | $52.83 | $63.36 |
| **Crew B-16** | Hr. | Daily | Hr. | Daily | Bare Costs | Incl. O&P |
| 1 Labor Foreman (outside) | $12.00 | $96.00 | $19.00 | $152.00 | $10.66 | $16.90 |
| 2 Building Laborers | 10.00 | 160.00 | 15.85 | 253.60 | | |
| 24 M.H., Daily Totals | | $256.00 | | $405.60 | $10.66 | $16.90 |
| **Crew B-17** | Hr. | Daily | Hr. | Daily | Bare Costs | Incl. O&P |
| 2 Building Laborers | $10.00 | $160.00 | $15.85 | $253.60 | $10.60 | $16.76 |
| 1 Equip. Oper. (light) | 12.05 | 96.40 | 19.10 | 152.80 | | |
| 1 Truck Driver (heavy) | 10.35 | 82.80 | 16.25 | 130.00 | | |
| 1 Backhoe Loader, 48 H.P. | | 152.20 | | 167.40 | | |
| 1 Dump Truck, 12 Ton | | 219.80 | | 241.80 | 11.62 | 12.78 |
| 32 M.H., Daily Totals | | $711.20 | | $945.60 | $22.22 | $29.54 |
| **Crew B-19** | Hr. | Daily | Hr. | Daily | Bare Costs | Incl. O&P |
| 1 Pile Driver Foreman | $14.40 | $115.20 | $24.55 | $196.40 | $12.58 | $20.90 |
| 4 Pile Drivers | 12.40 | 396.80 | 21.15 | 676.80 | | |
| 2 Equip. Oper. (crane) | 12.90 | 206.40 | 20.45 | 327.20 | | |
| 1 Equip. Oper. Oiler | 10.85 | 86.80 | 17.20 | 137.60 | | |
| 1 Crane, 40 Ton & Access. | | 482.60 | | 530.85 | | |
| 60 L.F. Leads, 15K Ft. Lbs. | | 42.00 | | 46.20 | | |
| 1 Hammer, 15K Ft. Lbs. | | 197.20 | | 216.90 | | |
| 1 Air Compr., 600 C.F.M. | | 236.20 | | 259.80 | | |
| 2-50 Ft. Air Hoses, 3" Dia. | | 16.70 | | 18.35 | 15.22 | 16.75 |
| 64 M.H., Daily Totals | | $1779.90 | | $2410.10 | $27.80 | $37.65 |
| **Crew B-20** | Hr. | Daily | Hr. | Daily | Bare Costs | Incl. O&P |
| 1 Plumber Foreman (out) | $15.80 | $126.40 | $25.05 | $200.40 | $13.20 | $20.91 |
| 1 Plumber | 13.80 | 110.40 | 21.85 | 174.80 | | |
| 1 Building Laborer | 10.00 | 80.00 | 15.85 | 126.80 | | |
| 24 M.H., Daily Totals | | $316.80 | | $502.00 | $13.20 | $20.91 |
| **Crew B-21** | Hr. | Daily | Hr. | Daily | Bare Costs | Incl. O&P |
| 1 Plumber Foreman (out) | $15.80 | $126.40 | $25.05 | $200.40 | $13.15 | $20.85 |
| 1 Plumber | 13.80 | 110.40 | 21.85 | 174.80 | | |
| 1 Building Laborer | 10.00 | 80.00 | 15.85 | 126.80 | | |
| .5 Equip. Oper. (crane) | 12.90 | 51.60 | 20.45 | 81.80 | | |
| .5 S.P. Crane, 5 Ton | | 87.90 | | 96.70 | 3.13 | 3.45 |
| 28 M.H., Daily Totals | | $456.30 | | $680.50 | $16.28 | $24.30 |

# CREWS

| Crew B-23 | Hr. | Daily | Hr. | Daily | Bare Costs | Incl. O&P |
|---|---|---|---|---|---|---|
| 1 Labor Foreman (outside) | $12.00 | $96.00 | $19.00 | $152.00 | $10.40 | $16.48 |
| 4 Building Laborers | 10.00 | 320.00 | 15.85 | 507.20 | | |
| Truck & Drill Rig | | 210.60 | | 231.65 | 5.26 | 5.79 |
| 40 M.H., Daily Totals | | $626.60 | | $890.85 | $15.66 | $22.27 |

| Crew B-24 | Hr. | Daily | Hr. | Daily | Bare Costs | Incl. O&P |
|---|---|---|---|---|---|---|
| 1 Cement Finisher | $11.85 | $94.80 | $18.30 | $146.40 | $11.40 | $17.90 |
| 1 Building Laborer | 10.00 | 80.00 | 15.85 | 126.80 | | |
| 1 Carpenter | 12.35 | 98.80 | 19.55 | 156.40 | | |
| 24 M.H., Daily Totals | | $273.60 | | $429.60 | $11.40 | $17.90 |

| Crew B-25 | Hr. | Daily | Hr. | Daily | Bare Costs | Incl. O&P |
|---|---|---|---|---|---|---|
| 1 Labor Foreman (outside) | $12.00 | $96.00 | $19.00 | $152.00 | $10.73 | $17.01 |
| 7 Building Laborers | 10.00 | 560.00 | 15.85 | 887.60 | | |
| 2 Equip. Oper. (med.) | 12.65 | 202.40 | 20.10 | 321.60 | | |
| 1 Paving Machine | | 534.00 | | 587.40 | | |
| 1 Tandem Roller, 10 Ton | | 172.80 | | 190.10 | 8.83 | 9.71 |
| 80 M.H., Daily Totals | | $1565.20 | | $2138.70 | $19.56 | $26.72 |

| Crew B-26 | Hr. | Daily | Hr. | Daily | Bare Costs | Incl. O&P |
|---|---|---|---|---|---|---|
| 1 Labor Foreman (outside) | $12.00 | $96.00 | $19.00 | $152.00 | $11.13 | $17.71 |
| 6 Building Laborers | 10.00 | 480.00 | 15.85 | 760.80 | | |
| 2 Equip. Oper. (med.) | 12.65 | 202.40 | 20.10 | 321.60 | | |
| 1 Rodman (reinf.) | 13.30 | 106.40 | 22.30 | 178.40 | | |
| 1 Cement Finisher | 11.85 | 94.80 | 18.30 | 146.40 | | |
| 1 Grader, 30,000 Lbs. | | 422.00 | | 464.20 | | |
| 1 Paving Mach. & Equip. | | 1107.00 | | 1217.70 | 17.37 | 19.11 |
| 88 M.H., Daily Totals | | $2508.60 | | $3241.10 | $28.50 | $36.82 |

| Crew B-27 | Hr. | Daily | Hr. | Daily | Bare Costs | Incl. O&P |
|---|---|---|---|---|---|---|
| 1 Labor Foreman (outside) | $12.00 | $96.00 | $19.00 | $152.00 | $10.50 | $16.63 |
| 3 Building Laborers | 10.00 | 240.00 | 15.85 | 380.40 | | |
| 1 Berm Machine | | 80.50 | | 88.55 | 2.51 | 2.76 |
| 32 M.H., Daily Totals | | $416.50 | | $620.95 | $13.01 | $19.39 |

| Crew B-29 | Hr. | Daily | Hr. | Daily | Bare Costs | Incl. O&P |
|---|---|---|---|---|---|---|
| 1 Labor Foreman (outside) | $12.00 | $96.00 | $19.00 | $152.00 | $10.82 | $17.15 |
| 4 Building Laborers | 10.00 | 320.00 | 15.85 | 507.20 | | |
| 1 Equip. Oper. (crane) | 12.90 | 103.20 | 20.45 | 163.60 | | |
| 1 Equip. Oper. Oiler | 10.85 | 86.80 | 17.20 | 137.60 | | |
| 1 Gradall, 3 Ton, 1/2 C.Y. | | 432.20 | | 475.40 | 7.71 | 8.48 |
| 56 M.H., Daily Totals | | $1038.20 | | $1435.80 | $18.53 | $25.63 |

| Crew B-30 | Hr. | Daily | Hr. | Daily | Bare Costs | Incl. O&P |
|---|---|---|---|---|---|---|
| 1 Equip. Oper. (med.) | $12.65 | $101.20 | $20.10 | $160.80 | $11.11 | $17.53 |
| 2 Truck Drivers (heavy) | 10.35 | 165.60 | 16.25 | 260.00 | | |
| 1 Hyd. Excavator, 1.5 C.Y. | | 560.20 | | 616.20 | | |
| 2 Dump Trucks, 16 Ton | | 558.40 | | 614.25 | 46.60 | 51.26 |
| 24 M.H., Daily Totals | | $1385.40 | | $1651.25 | $57.71 | $68.79 |

| Crew B-32 | Hr. | Daily | Hr. | Daily | Bare Costs | Incl. O&P |
|---|---|---|---|---|---|---|
| 2 Equip. Oper. (med.) | $12.65 | $202.40 | $20.10 | $321.60 | $12.65 | $20.10 |
| 1 Grader, 30,000 lbs. | | 422.00 | | 464.20 | | |
| 1 Tandem Roller, 10 Ton | | 172.80 | | 190.10 | 37.17 | 40.89 |
| 16 M.H., Daily Totals | | $797.20 | | $975.90 | $49.82 | $60.99 |

| Crew B-34A | Hr. | Daily | Hr. | Daily | Bare Costs | Incl. O&P |
|---|---|---|---|---|---|---|
| 1 Truck Driver (heavy) | $10.35 | $82.80 | $16.25 | $130.00 | $10.35 | $16.25 |
| 1 Dump Truck, 12 Ton | | 219.80 | | 241.80 | 27.47 | 30.22 |
| 8 M.H., Daily Totals | | $302.60 | | $371.80 | $37.82 | $46.47 |

| Crew B-34B | Hr. | Daily | Hr. | Daily | Bare Costs | Incl. O&P |
|---|---|---|---|---|---|---|
| 1 Truck Driver (heavy) | $10.35 | $82.80 | $16.25 | $130.00 | $10.35 | $16.25 |
| 1 Dump Truck, 16 Ton | | 279.20 | | 307.10 | 34.90 | 38.38 |
| 8 M.H., Daily Totals | | $362.00 | | $437.10 | $45.25 | $54.63 |

| Crew B-34C | Hr. | Daily | Hr. | Daily | Bare Costs | Incl. O&P |
|---|---|---|---|---|---|---|
| 1 Truck Driver (heavy) | $10.35 | $82.80 | $16.25 | $130.00 | $10.35 | $16.25 |
| 1 Truck Tractor, 40 Ton | | 300.00 | | 330.00 | | |
| 1 Dump Trailer, 16.5 C.Y. | | 79.30 | | 87.25 | 47.41 | 52.15 |
| 8 M.H., Daily Totals | | $462.10 | | $547.25 | $57.76 | $68.40 |

| Crew B-34D | Hr. | Daily | Hr. | Daily | Bare Costs | Incl. O&P |
|---|---|---|---|---|---|---|
| 1 Truck Driver (heavy) | $10.35 | $82.80 | $16.25 | $130.00 | $10.35 | $16.25 |
| 1 Truck Tractor, 40 Ton | | 300.00 | | 330.00 | | |
| 1 Dump Trailer, 20 C.Y. | | 94.10 | | 103.50 | 49.26 | 54.18 |
| 8 M.H., Daily Totals | | $476.90 | | $563.50 | $59.61 | $70.43 |

| Crew B-36 | Hr. | Daily | Hr. | Daily | Bare Costs | Incl. O&P |
|---|---|---|---|---|---|---|
| 1 Labor Foreman (outside) | $12.00 | $96.00 | $19.00 | $152.00 | $11.16 | $17.70 |
| 2 Building Laborers | 10.00 | 160.00 | 15.85 | 253.60 | | |
| 1 Equip. Oper. (med.) | 12.65 | 101.20 | 20.10 | 160.80 | | |
| 1 Dozer, 200 H.P. | | 614.40 | | 675.85 | | |
| 1 Aggregate Spreader | | 49.95 | | 54.95 | 20.76 | 22.83 |
| 32 M.H., Daily Totals | | $1021.55 | | $1297.20 | $31.92 | $40.53 |

| Crew B-37 | Hr. | Daily | Hr. | Daily | Bare Costs | Incl. O&P |
|---|---|---|---|---|---|---|
| 1 Labor Foreman (outside) | $12.00 | $96.00 | $19.00 | $152.00 | $10.67 | $16.91 |
| 4 Building Laborers | 10.00 | 320.00 | 15.85 | 507.20 | | |
| 1 Equip. Oper. (light) | 12.05 | 96.40 | 19.10 | 152.80 | | |
| 1 Tandem Roller, 5 Ton | | 83.85 | | 92.25 | 1.74 | 1.92 |
| 48 M.H., Daily Totals | | $596.25 | | $904.25 | $12.41 | $18.83 |

| Crew B-51 | Hr. | Daily | Hr. | Daily | Bare Costs | Incl. O&P |
|---|---|---|---|---|---|---|
| 1 Labor Foreman (outside) | $12.00 | $96.00 | $19.00 | $152.00 | $10.39 | $16.44 |
| 4 Building Laborers | 10.00 | 320.00 | 15.85 | 507.20 | | |
| 1 Truck Driver (heavy) | 10.35 | 82.80 | 16.25 | 130.00 | | |
| 1 Light Truck, 1.5 Ton | | 57.15 | | 62.85 | 1.19 | 1.30 |
| 48 M.H., Daily Totals | | $555.95 | | $852.05 | $11.58 | $17.74 |

| Crew B-53 | Hr. | Daily | Hr. | Daily | Bare Costs | Incl. O&P |
|---|---|---|---|---|---|---|
| 1 Equip. Oper. (light) | $12.05 | $96.40 | $19.10 | $152.80 | $12.05 | $19.10 |
| 1 Trencher, Chain, 12 H.P. | | 133.40 | | 146.75 | 16.67 | 18.34 |
| 8 M.H., Daily Totals | | $229.80 | | $299.55 | $28.72 | $37.44 |

| Crew B-55 | Hr. | Daily | Hr. | Daily | Bare Costs | Incl. O&P |
|---|---|---|---|---|---|---|
| 1 Building Laborer | $10.00 | $80.00 | $15.85 | $126.80 | $10.12 | $15.97 |
| 1 Truck Driver (light) | 10.25 | 82.00 | 16.10 | 128.80 | | |
| 1 Flatbed Truck w/Auger | | 144.20 | | 158.60 | | |
| 1 Truck, 3 Ton | | 66.40 | | 73.05 | 13.16 | 14.47 |
| 16 M.H., Daily Totals | | $372.60 | | $487.25 | $23.28 | $30.44 |

| Crew C-1 | Hr. | Daily | Hr. | Daily | Bare Costs | Incl. O&P |
|---|---|---|---|---|---|---|
| 3 Carpenters | $12.35 | $296.40 | $19.55 | $469.20 | $11.76 | $18.62 |
| 1 Building Laborer | 10.00 | 80.00 | 15.85 | 126.80 | | |
| Power Tools | | 21.00 | | 23.10 | .65 | .72 |
| 32 M.H., Daily Totals | | $397.40 | | $619.10 | $12.41 | $19.34 |

# CREWS

| Crew No. | Bare Costs | | Incl. Subs O & P | | Cost Per Man-hour | |
|---|---|---|---|---|---|---|
| **Crew C-2** | Hr. | Daily | Hr. | Daily | Bare Costs | Incl. O&P |
| 1 Carpenter Foreman (out) | $14.35 | $114.80 | $22.75 | $182.00 | $12.29 | $19.46 |
| 4 Carpenters | 12.35 | 395.20 | 19.55 | 625.60 | | |
| 1 Building Laborer | 10.00 | 80.00 | 15.85 | 126.80 | | |
| Power Tools | | 28.00 | | 30.80 | .58 | .64 |
| 48 M.H., Daily Totals | | $618.00 | | $965.20 | $12.87 | $20.10 |
| **Crew C-6** | Hr. | Daily | Hr. | Daily | Bare Costs | Incl. O&P |
| 1 Labor Foreman (outside) | $12.00 | $96.00 | $19.00 | $152.00 | $10.64 | $16.78 |
| 4 Building Laborers | 10.00 | 320.00 | 15.85 | 507.20 | | |
| 1 Cement Finisher | 11.85 | 94.80 | 18.30 | 146.40 | | |
| 2 Gas Engine Vibrators | | 53.00 | | 58.30 | 1.10 | 1.21 |
| 48 M.H., Daily Totals | | $563.80 | | $863.90 | $11.74 | $17.99 |
| **Crew C-7** | Hr. | Daily | Hr. | Daily | Bare Costs | Incl. O&P |
| 1 Labor Foreman (outside) | $12.00 | $96.00 | $19.00 | $152.00 | $10.81 | $17.08 |
| 5 Building Laborers | 10.00 | 400.00 | 15.85 | 634.00 | | |
| 1 Cement Finisher | 11.85 | 94.80 | 18.30 | 146.40 | | |
| 1 Equip. Oper. (med.) | 12.65 | 101.20 | 20.10 | 160.80 | | |
| 2 Gas Engine Vibrators | | 53.00 | | 58.30 | | |
| 1 Concrete Bucket, 1 C.Y. | | 14.95 | | 16.45 | | |
| 1 Hyd. Crane, 55 Ton | | 526.80 | | 579.50 | 9.29 | 10.22 |
| 64 M.H., Daily Totals | | $1286.75 | | $1747.45 | $20.10 | $27.30 |
| **Crew C-8** | Hr. | Daily | Hr. | Daily | Bare Costs | Incl. O&P |
| 1 Labor Foreman (outside) | $12.00 | $96.00 | $19.00 | $152.00 | $11.19 | $17.60 |
| 3 Building Laborers | 10.00 | 240.00 | 15.85 | 380.40 | | |
| 2 Cement Finishers | 11.85 | 189.60 | 18.30 | 292.80 | | |
| 1 Equip. Oper. (med.) | 12.65 | 101.20 | 20.10 | 160.80 | | |
| 1 Concrete Pump (small) | | 498.00 | | 547.80 | 8.89 | 9.78 |
| 56 M.H., Daily Totals | | $1124.80 | | $1533.80 | $20.08 | $27.38 |
| **Crew C-9** | Hr. | Daily | Hr. | Daily | Bare Costs | Incl. O&P |
| 1 Cement Finisher | $11.85 | $94.80 | $18.30 | $146.40 | $11.85 | $18.30 |
| 1 Gas Finishing Mach. | | 27.10 | | 29.80 | 3.38 | 3.72 |
| 8 M.H., Daily Totals | | $121.90 | | $176.20 | $15.23 | $22.02 |
| **Crew C-12** | Hr. | Daily | Hr. | Daily | Bare Costs | Incl. O&P |
| 1 Carpenter Foreman (out) | $14.35 | $114.80 | $22.75 | $182.00 | $12.38 | $19.61 |
| 3 Carpenters | 12.35 | 296.40 | 19.55 | 469.20 | | |
| 1 Building Laborer | 10.00 | 80.00 | 15.85 | 126.80 | | |
| 1 Equip. Oper. (crane) | 12.90 | 103.20 | 20.45 | 163.60 | | |
| 1 Hyd. Crane, 12 Ton | | 234.60 | | 258.05 | 4.88 | 5.37 |
| 48 M.H., Daily Totals | | $829.00 | | $1199.65 | $17.26 | $24.98 |
| **Crew C-15** | Hr. | Daily | Hr. | Daily | Bare Costs | Incl. O&P |
| 1 Carpenter Foreman (out) | $14.35 | $114.80 | $22.75 | $182.00 | $11.78 | $18.70 |
| 2 Carpenters | 12.35 | 197.60 | 19.55 | 312.80 | | |
| 3 Building Laborers | 10.00 | 240.00 | 15.85 | 380.40 | | |
| 2 Cement Finishers | 11.85 | 189.60 | 18.30 | 292.80 | | |
| 1 Rodman (reinf.) | 13.30 | 106.40 | 22.30 | 178.40 | | |
| Power Tools | | 14.00 | | 15.40 | | |
| 1 Gas Finishing Mach. | | 27.10 | | 29.80 | .57 | .62 |
| 72 M.H., Daily Totals | | $889.50 | | $1391.60 | $12.35 | $19.32 |
| **Crew C-17A** | Hr. | Daily | Hr. | Daily | Bare Costs | Incl. O&P |
| 2 Skilled Worker Foremen | $14.60 | $233.60 | $23.30 | $372.80 | $13.00 | $20.74 |
| 8 Skilled Workers | 12.60 | 806.40 | 20.10 | 1286.40 | | |
| .125 Crane, 80 Ton, & Tools | | 113.10 | | 124.45 | 1.41 | 1.55 |
| 80 M.H., Daily Totals | | $1153.10 | | $1783.65 | $14.41 | $22.29 |

| Crew No. | Bare Costs | | Incl. Subs O & P | | Cost Per Man-hour | |
|---|---|---|---|---|---|---|
| **Crew C-17B** | Hr. | Daily | Hr. | Daily | Bare Costs | Incl. O&P |
| 2 Skilled Worker Foremen | $14.60 | $233.60 | $23.30 | $372.80 | $13.00 | $20.74 |
| 8 Skilled Workers | 12.60 | 806.40 | 20.10 | 1286.40 | | |
| .25 Crane, 80 Ton & Tools | | 217.55 | | 239.30 | 2.71 | 2.99 |
| 80 M.H., Daily Totals | | $1257.55 | | $1898.50 | $15.71 | $23.73 |
| **Crew C-17C** | Hr. | Daily | Hr. | Daily | Bare Costs | Incl. O&P |
| 2 Skilled Worker Foremen | $14.60 | $233.60 | $23.30 | $372.80 | $13.00 | $20.74 |
| 8 Skilled Workers | 12.60 | 806.40 | 20.10 | 1286.40 | | |
| .375 Crane, 80 Ton & Tools | | 330.65 | | 363.75 | 4.13 | 4.54 |
| 80 M.H., Daily Totals | | $1370.65 | | $2022.95 | $17.13 | $25.28 |
| **Crew C-18** | Hr. | Daily | Hr. | Daily | Bare Costs | Incl. O&P |
| .125 Labor Foreman (out) | $12.00 | $12.00 | $19.00 | $19.00 | $10.22 | $16.20 |
| 1 Building Laborer | 10.00 | 80.00 | 15.85 | 126.80 | | |
| 1 Concrete Cart, 10 C.F. | | 34.90 | | 38.40 | 3.87 | 4.26 |
| 9 M.H., Daily Totals | | $126.90 | | $184.20 | $14.09 | $20.46 |
| **Crew C-19** | Hr. | Daily | Hr. | Daily | Bare Costs | Incl. O&P |
| .125 Labor Foreman (out) | $12.00 | $12.00 | $19.00 | $19.00 | $10.22 | $16.20 |
| 1 Building Laborer | 10.00 | 80.00 | 15.85 | 126.80 | | |
| 1 Concrete Cart, 18 C.F. | | 54.50 | | 59.95 | 6.05 | 6.66 |
| 9 M.H., Daily Totals | | $146.50 | | $205.75 | $16.27 | $22.86 |
| **Crew C-20** | Hr. | Daily | Hr. | Daily | Bare Costs | Incl. O&P |
| 1 Labor Foreman (outside) | $12.00 | $96.00 | $19.00 | $152.00 | $10.81 | $17.08 |
| 5 Building Laborers | 10.00 | 400.00 | 15.85 | 634.00 | | |
| 1 Cement Finisher | 11.85 | 94.80 | 18.30 | 146.40 | | |
| 1 Equip. Oper. (med.) | 12.65 | 101.20 | 20.10 | 160.80 | | |
| 2 Gas Engine Vibrators | | 53.00 | | 58.30 | | |
| 1 Concrete Pump (small) | | 498.00 | | 547.80 | 8.60 | 9.47 |
| 64 M.H., Daily Totals | | $1243.00 | | $1699.30 | $19.41 | $26.55 |
| **Crew D-1** | Hr. | Daily | Hr. | Daily | Bare Costs | Incl. O&P |
| 1 Bricklayer | $12.60 | $100.80 | $19.65 | $157.20 | $11.32 | $17.67 |
| 1 Bricklayer Helper | 10.05 | 80.40 | 15.70 | 125.60 | | |
| 16 M.H., Daily Totals | | $181.20 | | $282.80 | $11.32 | $17.67 |
| **Crew D-2** | Hr. | Daily | Hr. | Daily | Bare Costs | Incl. O&P |
| 3 Bricklayers | $12.60 | $302.40 | $19.65 | $471.60 | $11.65 | $18.20 |
| 2 Bricklayer Helpers | 10.05 | 160.80 | 15.70 | 251.20 | | |
| .5 Carpenter | 12.35 | 49.40 | 19.55 | 78.20 | | |
| 44 M.H., Daily Totals | | $512.60 | | $801.00 | $11.65 | $18.20 |
| **Crew D-3** | Hr. | Daily | Hr. | Daily | Bare Costs | Incl. O&P |
| 3 Bricklayers | $12.60 | $302.40 | $19.65 | $471.60 | $11.61 | $18.14 |
| 2 Bricklayer Helpers | 10.05 | 160.80 | 15.70 | 251.20 | | |
| .25 Carpenter | 12.35 | 24.70 | 19.55 | 39.10 | | |
| 42 M.H., Daily Totals | | $487.90 | | $761.90 | $11.61 | $18.14 |
| **Crew D-4** | Hr. | Daily | Hr. | Daily | Bare Costs | Incl. O&P |
| 1 Bricklayer | $12.60 | $100.80 | $19.65 | $157.20 | $11.18 | $17.53 |
| 2 Bricklayer Helpers | 10.05 | 160.80 | 15.70 | 251.20 | | |
| 1 Equip. Oper. (light) | 12.05 | 96.40 | 19.10 | 152.80 | | |
| Power Equipment | | 146.95 | | 161.65 | 4.59 | 5.05 |
| 32 M.H., Daily Totals | | $504.95 | | $722.85 | $15.77 | $22.58 |
| **Crew D-5** | Hr. | Daily | Hr. | Daily | Bare Costs | Incl. O&P |
| 1 Bricklayer | $12.60 | $100.80 | $19.65 | $157.20 | $12.60 | $19.65 |
| 1 Power Tool | | 28.30 | | 31.15 | 3.53 | 3.89 |
| 8 M.H., Daily Totals | | $129.10 | | $188.35 | $16.13 | $23.54 |

# CREWS

| Crew No. | Bare Costs Hr. | Daily | Incl. Subs O & P Hr. | Daily | Cost Per Man-hour Bare Costs | Incl. O&P |
|---|---|---|---|---|---|---|
| **Crew D-6** | Hr. | Daily | Hr. | Daily | Bare Costs | Incl. O&P |
| 3 Bricklayers | $12.60 | $302.40 | $19.65 | $471.60 | $11.36 | $17.75 |
| 3 Bricklayer Helpers | 10.05 | 241.20 | 15.70 | 376.80 | | |
| .25 Carpenter | 12.35 | 24.70 | 19.55 | 39.10 | | |
| 50 M.H., Daily Totals | | $568.30 | | $887.50 | $11.36 | $17.75 |
| **Crew D-7** | Hr. | Daily | Hr. | Daily | Bare Costs | Incl. O&P |
| 1 Tile Layer | $12.15 | $97.20 | $18.70 | $149.60 | $10.97 | $16.90 |
| 1 Tile Layer Helper | 9.80 | 78.40 | 15.10 | 120.80 | | |
| 16 M.H., Daily Totals | | $175.60 | | $270.40 | $10.97 | $16.90 |
| **Crew D-8** | Hr. | Daily | Hr. | Daily | Bare Costs | Incl. O&P |
| 3 Bricklayers | $12.60 | $302.40 | $19.65 | $471.60 | $11.58 | $18.07 |
| 2 Bricklayer Helpers | 10.05 | 160.80 | 15.70 | 251.20 | | |
| 40 M.H., Daily Totals | | $463.20 | | $722.80 | $11.58 | $18.07 |
| **Crew D-10** | Hr. | Daily | Hr. | Daily | Bare Costs | Incl. O&P |
| 1 Bricklayer Foreman | $14.60 | $116.80 | $22.80 | $182.40 | $12.04 | $18.86 |
| 1 Bricklayer | 12.60 | 100.80 | 19.65 | 157.20 | | |
| 2 Bricklayer Helpers | 10.05 | 160.80 | 15.70 | 251.20 | | |
| 1 Equip. Oper. (crane) | 12.90 | 103.20 | 20.45 | 163.60 | | |
| 1 Truck Crane, 12.5 Ton | | 219.80 | | 241.80 | 5.49 | 6.04 |
| 40 M.H., Daily Totals | | $701.40 | | $996.20 | $17.53 | $24.90 |
| **Crew D-11** | Hr. | Daily | Hr. | Daily | Bare Costs | Incl. O&P |
| 1 Bricklayer Foreman | $14.60 | $116.80 | $22.80 | $182.40 | $12.41 | $19.38 |
| 1 Bricklayer | 12.60 | 100.80 | 19.65 | 157.20 | | |
| 1 Bricklayer Helper | 10.05 | 80.40 | 15.70 | 125.60 | | |
| 24 M.H., Daily Totals | | $298.00 | | $465.20 | $12.41 | $19.38 |
| **Crew D-12** | Hr. | Daily | Hr. | Daily | Bare Costs | Incl. O&P |
| 1 Bricklayer Foreman | $14.60 | $116.80 | $22.80 | $182.40 | $11.82 | $18.46 |
| 1 Bricklayer | 12.60 | 100.80 | 19.65 | 157.20 | | |
| 2 Bricklayer Helpers | 10.05 | 160.80 | 15.70 | 251.20 | | |
| 32 M.H., Daily Totals | | $378.40 | | $590.80 | $11.82 | $18.46 |
| **Crew E-1** | Hr. | Daily | Hr. | Daily | Bare Costs | Incl. O&P |
| 1 Welder Foreman | $15.30 | $122.40 | $26.20 | $209.60 | $13.55 | $22.70 |
| 1 Welder | 13.30 | 106.40 | 22.80 | 182.40 | | |
| 1 Equip. Oper. (light) | 12.05 | 96.40 | 19.10 | 152.80 | | |
| 1 Gas Welding Machine | | 56.40 | | 62.05 | 2.35 | 2.58 |
| 24 M.H., Daily Totals | | $381.60 | | $606.85 | $15.90 | $25.28 |
| **Crew E-2** | Hr. | Daily | Hr. | Daily | Bare Costs | Incl. O&P |
| 1 Struc. Steel Foreman | $15.30 | $122.40 | $26.20 | $209.60 | $13.17 | $22.15 |
| 4 Struc. Steel Workers | 13.30 | 425.60 | 22.80 | 729.60 | | |
| 1 Equip. Oper. (crane) | 12.90 | 103.20 | 20.45 | 163.60 | | |
| 1 Equip. Oper. Oiler | 10.85 | 86.80 | 17.20 | 137.60 | | |
| 1 Crane, 90 Ton | | 790.00 | | 869.00 | 14.10 | 15.51 |
| 56 M.H., Daily Totals | | $1528.00 | | $2109.40 | $27.27 | $37.66 |
| **Crew E-4** | Hr. | Daily | Hr. | Daily | Bare Costs | Incl. O&P |
| 1 Struc. Steel Foreman | $15.30 | $122.40 | $26.20 | $209.60 | $13.80 | $23.65 |
| 3 Struc. Steel Workers | 13.30 | 319.20 | 22.80 | 547.20 | | |
| 1 Gas Welding Machine | | 56.40 | | 62.05 | 1.76 | 1.93 |
| 32 M.H., Daily Totals | | $498.00 | | $818.85 | $15.56 | $25.58 |

| Crew No. | Bare Costs Hr. | Daily | Incl. Subs O & P Hr. | Daily | Cost Per Man-hour Bare Costs | Incl. O&P |
|---|---|---|---|---|---|---|
| **Crew F-1** | Hr. | Daily | Hr. | Daily | Bare Costs | Incl. O&P |
| 1 Carpenter | $12.35 | $98.80 | $19.55 | $156.40 | $12.35 | $19.55 |
| Power Tools | | 7.00 | | 7.70 | .87 | .96 |
| 8 M.H., Daily Totals | | $105.80 | | $164.10 | $13.22 | $20.51 |
| **Crew F-2** | Hr. | Daily | Hr. | Daily | Bare Costs | Incl. O&P |
| 2 Carpenters | $12.35 | $197.60 | $19.55 | $312.80 | $12.35 | $19.55 |
| Power Tools | | 14.00 | | 15.40 | .87 | .96 |
| 16 M.H., Daily Totals | | $211.60 | | $328.20 | $13.22 | $20.51 |
| **Crew F-3** | Hr. | Daily | Hr. | Daily | Bare Costs | Incl. O&P |
| 4 Carpenters | $12.35 | $395.20 | $19.55 | $625.60 | $12.46 | $19.73 |
| 1 Equip. Oper. (crane) | 12.90 | 103.20 | 20.45 | 163.60 | | |
| 1 Hyd. Crane, 12 Ton | | 234.60 | | 258.05 | | |
| Power Tools | | 14.00 | | 15.40 | 6.21 | 6.83 |
| 40 M.H., Daily Totals | | $747.00 | | $1062.65 | $18.67 | $26.56 |
| **Crew F-4** | Hr. | Daily | Hr. | Daily | Bare Costs | Incl. O&P |
| 4 Carpenters | $12.35 | $395.20 | $19.55 | $625.60 | $12.19 | $19.30 |
| 1 Equip. Oper. (crane) | 12.90 | 103.20 | 20.45 | 163.60 | | |
| 1 Equip. Oper. Oiler | 10.85 | 86.80 | 17.20 | 137.60 | | |
| 1 Hyd. Crane, 55 Ton | | 526.80 | | 579.50 | | |
| Power Tools | | 14.00 | | 15.40 | 11.26 | 12.39 |
| 48 M.H., Daily Totals | | $1126.00 | | $1521.70 | $23.45 | $31.69 |
| **Crew F-5** | Hr. | Daily | Hr. | Daily | Bare Costs | Incl. O&P |
| 1 Carpenter Foreman | $14.35 | $114.80 | $22.75 | $182.00 | $12.85 | $20.35 |
| 3 Carpenters | 12.35 | 296.40 | 19.55 | 469.20 | | |
| Power Tools | | 14.00 | | 15.40 | .43 | .48 |
| 32 M.H., Daily Totals | | $425.20 | | $666.60 | $13.28 | $20.83 |
| **Crew F-6** | Hr. | Daily | Hr. | Daily | Bare Costs | Incl. O&P |
| 2 Carpenters | $12.35 | $197.60 | $19.55 | $312.80 | $11.52 | $18.25 |
| 2 Building Laborers | 10.00 | 160.00 | 15.85 | 253.60 | | |
| 1 Equip. Oper. (crane) | 12.90 | 103.20 | 20.45 | 163.60 | | |
| 1 Hyd. Crane, 12 Ton | | 234.60 | | 258.05 | | |
| Power Tools | | 14.00 | | 15.40 | 6.21 | 6.83 |
| 40 M.H., Daily Totals | | $709.40 | | $1003.45 | $17.73 | $25.08 |
| **Crew G-1** | Hr. | Daily | Hr. | Daily | Bare Costs | Incl. O&P |
| 1 Roofer Foreman | $13.60 | $108.80 | $22.55 | $180.40 | $11.08 | $18.39 |
| 4 Roofers, Composition | 11.60 | 371.20 | 19.25 | 616.00 | | |
| 2 Roofer Helpers | 8.80 | 140.80 | 14.60 | 233.60 | | |
| Application Equipment | | 97.70 | | 107.45 | 1.74 | 1.91 |
| 56 M.H., Daily Totals | | $718.50 | | $1137.45 | $12.82 | $20.30 |
| **Crew G-2** | Hr. | Daily | Hr. | Daily | Bare Costs | Incl. O&P |
| 1 Plasterer | $12.35 | $98.80 | $19.25 | $154.00 | $10.90 | $17.08 |
| 1 Plasterer Helper | 10.35 | 82.80 | 16.15 | 129.20 | | |
| 1 Building Laborer | 10.00 | 80.00 | 15.85 | 126.80 | | |
| Grouting Equipment | | 103.75 | | 114.15 | 4.32 | 4.75 |
| 24 M.H., Daily Totals | | $365.35 | | $524.15 | $15.22 | $21.83 |
| **Crew G-3** | Hr. | Daily | Hr. | Daily | Bare Costs | Incl. O&P |
| 2 Sheet Metal Workers | $13.80 | $220.80 | $22.05 | $352.80 | $11.90 | $18.95 |
| 2 Building Laborers | 10.00 | 160.00 | 15.85 | 253.60 | | |
| Power Tools | | 22.00 | | 24.20 | .68 | .75 |
| 32 M.H., Daily Totals | | $402.80 | | $630.60 | $12.58 | $19.70 |

# CREWS

| Crew No. | Bare Costs Hr. | Bare Costs Daily | Incl. Subs O&P Hr. | Incl. Subs O&P Daily | Cost Per Man-hour Bare Costs | Cost Per Man-hour Incl. O&P |
|---|---|---|---|---|---|---|
| **Crew G-4** | Hr. | Daily | Hr. | Daily | Bare Costs | Incl. O&P |
| 1 Labor Foreman (outside) | $12.00 | $96.00 | $19.00 | $152.00 | $10.66 | $16.90 |
| 2 Building Laborers | 10.00 | 160.00 | 15.85 | 253.60 | | |
| 1 Light Truck, 1.5 Ton | | 57.15 | | 62.85 | | |
| 1 Air Compr., 160 C.F.M. | | 65.25 | | 71.80 | 5.10 | 5.61 |
| 24 M.H., Daily Totals | | $378.40 | | $540.25 | $15.76 | $22.51 |
| **Crew G-5** | Hr. | Daily | Hr. | Daily | Bare Costs | Incl. O&P |
| 1 Roofer Foreman | $13.60 | $108.80 | $22.55 | $180.40 | $10.88 | $18.05 |
| 2 Roofers, Composition | 11.60 | 185.60 | 19.25 | 308.00 | | |
| 2 Roofer Helpers | 8.80 | 140.80 | 14.60 | 233.60 | | |
| Application Equipment | | 97.70 | | 107.45 | 2.44 | 2.68 |
| 40 M.H., Daily Totals | | $532.90 | | $829.45 | $13.32 | $20.73 |
| **Crew H-2** | Hr. | Daily | Hr. | Daily | Bare Costs | Incl. O&P |
| 2 Glaziers | $12.45 | $199.20 | $19.45 | $311.20 | $11.63 | $18.25 |
| 1 Building Laborer | 10.00 | 80.00 | 15.85 | 126.80 | | |
| 24 M.H., Daily Totals | | $279.20 | | $438.00 | $11.63 | $18.25 |
| **Crew J-1** | Hr. | Daily | Hr. | Daily | Bare Costs | Incl. O&P |
| 3 Plasterers | $12.35 | $296.40 | $19.25 | $462.00 | $11.55 | $18.01 |
| 2 Plasterer Helpers | 10.35 | 165.60 | 16.15 | 258.40 | | |
| 1 Mixing Machine, 6 C.F. | | 35.10 | | 38.60 | .87 | .96 |
| 40 M.H., Daily Totals | | $497.10 | | $759.00 | $12.42 | $18.97 |
| **Crew J-2** | Hr. | Daily | Hr. | Daily | Bare Costs | Incl. O&P |
| 3 Plasterers | $12.35 | $296.40 | $19.25 | $462.00 | $11.69 | $18.20 |
| 2 Plasterer Helpers | 10.35 | 165.60 | 16.15 | 258.40 | | |
| 1 Lather | 12.40 | 99.20 | 19.15 | 153.20 | | |
| 1 Mixing Machine, 6 C.F. | | 35.10 | | 38.60 | .73 | .80 |
| 48 M.H., Daily Totals | | $596.30 | | $912.20 | $12.42 | $19.00 |
| **Crew J-3** | Hr. | Daily | Hr. | Daily | Bare Costs | Incl. O&P |
| 1 Terrazzo Worker | $12.20 | $97.60 | $18.80 | $150.40 | $11.10 | $17.10 |
| 1 Terrazzo Helper | 10.00 | 80.00 | 15.40 | 123.20 | | |
| 1 Mixing Mach. & Grinder | | 113.55 | | 124.90 | 7.09 | 7.80 |
| 16 M.H., Daily Totals | | $291.15 | | $398.50 | $18.19 | $24.90 |
| **Crew K-2** | Hr. | Daily | Hr. | Daily | Bare Costs | Incl. O&P |
| 1 Struc. Steel Foreman | $15.30 | $122.40 | $26.20 | $209.60 | $12.95 | $21.70 |
| 1 Struc. Steel Worker | 13.30 | 106.40 | 22.80 | 182.40 | | |
| 1 Truck Driver (light) | 10.25 | 82.00 | 16.10 | 128.80 | | |
| 1 Truck w/Power Equip. | | 174.65 | | 192.10 | 7.27 | 8.00 |
| 24 M.H., Daily Totals | | $485.45 | | $712.90 | $20.22 | $29.70 |
| **Crew L-1** | Hr. | Daily | Hr. | Daily | Bare Costs | Incl. O&P |
| 1 Electrician | $13.75 | $110.00 | $21.70 | $173.60 | $13.77 | $21.77 |
| 1 Plumber | 13.80 | 110.40 | 21.85 | 174.80 | | |
| 16 M.H., Daily Totals | | $220.40 | | $348.40 | $13.77 | $21.77 |
| **Crew L-3** | Hr. | Daily | Hr. | Daily | Bare Costs | Incl. O&P |
| 1 Carpenter | $12.35 | $98.80 | $19.55 | $156.40 | $13.06 | $20.71 |
| .5 Electrician | 13.75 | 55.00 | 21.70 | 86.80 | | |
| .5 Sheet Metal Worker | 13.80 | 55.20 | 22.05 | 88.20 | | |
| 16 M.H., Daily Totals | | $209.00 | | $331.40 | $13.06 | $20.71 |
| **Crew L-7** | Hr. | Daily | Hr. | Daily | Bare Costs | Incl. O&P |
| 2 Carpenters | $12.35 | $197.60 | $19.55 | $312.80 | $11.87 | $18.80 |
| 1 Building Laborer | 10.00 | 80.00 | 15.85 | 126.80 | | |
| .5 Electrician | 13.75 | 55.00 | 21.70 | 86.80 | | |
| 28 M.H., Daily Totals | | $332.60 | | $526.40 | $11.87 | $18.80 |

| Crew No. | Bare Costs Hr. | Bare Costs Daily | Incl. Subs O&P Hr. | Incl. Subs O&P Daily | Cost Per Man-hour Bare Costs | Cost Per Man-hour Incl. O&P |
|---|---|---|---|---|---|---|
| **Crew Q-1** | Hr. | Daily | Hr. | Daily | Bare Costs | Incl. O&P |
| 1 Plumber | $13.80 | $110.40 | $21.85 | $174.80 | $12.42 | $19.67 |
| 1 Plumber Apprentice | 11.04 | 88.32 | 17.50 | 140.00 | | |
| 16 M.H., Daily Totals | | $198.72 | | $314.80 | $12.42 | $19.67 |
| **Crew Q-2** | Hr. | Daily | Hr. | Daily | Bare Costs | Incl. O&P |
| 2 Plumbers | $13.80 | $220.80 | $21.85 | $349.60 | $12.88 | $20.40 |
| 1 Plumber Apprentice | 11.04 | 88.32 | 17.50 | 140.00 | | |
| 24 M.H., Daily Totals | | $309.12 | | $489.60 | $12.88 | $20.40 |
| **Crew Q-4** | Hr. | Daily | Hr. | Daily | Bare Costs | Incl. O&P |
| 1 Plumber Foreman (ins) | $14.30 | $114.40 | $22.65 | $181.20 | $13.23 | $20.96 |
| 1 Plumber | 13.80 | 110.40 | 21.85 | 174.80 | | |
| 1 Welder (plumber) | 13.80 | 110.40 | 21.85 | 174.80 | | |
| 1 Plumber Apprentice | 11.04 | 88.32 | 17.50 | 140.00 | | |
| 1 Electric Welding Mach. | | 18.15 | | 19.95 | .56 | .62 |
| 32 M.H., Daily Totals | | $441.67 | | $690.75 | $13.79 | $21.58 |
| **Crew Q-5** | Hr. | Daily | Hr. | Daily | Bare Costs | Incl. O&P |
| 1 Steamfitter | $13.85 | $110.80 | $21.95 | $175.60 | $12.46 | $19.75 |
| 1 Steamfitter Apprentice | 11.08 | 88.64 | 17.55 | 140.40 | | |
| 16 M.H., Daily Totals | | $199.44 | | $316.00 | $12.46 | $19.75 |
| **Crew Q-6** | Hr. | Daily | Hr. | Daily | Bare Costs | Incl. O&P |
| 2 Steamfitters | $13.85 | $221.60 | $21.95 | $351.20 | $12.92 | $20.48 |
| 1 Steamfitter Apprentice | 11.08 | 88.64 | 17.55 | 140.40 | | |
| 24 M.H., Daily Totals | | $310.24 | | $491.60 | $12.92 | $20.48 |
| **Crew Q-7** | Hr. | Daily | Hr. | Daily | Bare Costs | Incl. O&P |
| 1 Steamfitter Foreman (ins) | $14.35 | $114.80 | $22.75 | $182.00 | $13.28 | $21.05 |
| 2 Steamfitters | 13.85 | 221.60 | 21.95 | 351.20 | | |
| 1 Steamfitter Apprentice | 11.08 | 88.64 | 17.55 | 140.40 | | |
| 32 M.H., Daily Totals | | $425.04 | | $673.60 | $13.28 | $21.05 |
| **Crew Q-9** | Hr. | Daily | Hr. | Daily | Bare Costs | Incl. O&P |
| 1 Sheet Metal Worker | $13.80 | $110.40 | $22.05 | $176.40 | $12.42 | $19.85 |
| 1 Sheet Metal Apprentice | 11.04 | 88.32 | 17.65 | 141.20 | | |
| 16 M.H., Daily Totals | | $198.72 | | $317.60 | $12.42 | $19.85 |
| **Crew Q-10** | Hr. | Daily | Hr. | Daily | Bare Costs | Incl. O&P |
| 2 Sheet Metal Workers | $13.80 | $220.80 | $22.05 | $352.80 | $12.88 | $20.58 |
| 1 Sheet Metal Apprentice | 11.04 | 88.32 | 17.65 | 141.20 | | |
| 24 M.H., Daily Totals | | $309.12 | | $494.00 | $12.88 | $20.58 |
| **Crew Q-14** | Hr. | Daily | Hr. | Daily | Bare Costs | Incl. O&P |
| 1 Asbestos Worker | $13.80 | $110.40 | $22.25 | $178.00 | $12.42 | $20.02 |
| 1 Asbestos Apprentice | 11.04 | 88.32 | 17.80 | 142.40 | | |
| 16 M.H., Daily Totals | | $198.72 | | $320.40 | $12.42 | $20.02 |
| **Crew Q-19** | Hr. | Daily | Hr. | Daily | Bare Costs | Incl. O&P |
| 1 Steamfitter | $13.85 | $110.80 | $21.95 | $175.60 | $12.89 | $20.40 |
| 1 Steamfitter Apprentice | 11.08 | 88.64 | 17.55 | 140.40 | | |
| 1 Electrician | 13.75 | 110.00 | 21.70 | 173.60 | | |
| 24 M.H., Daily Totals | | $309.44 | | $489.60 | $12.89 | $20.40 |

# CREWS

| Crew No. | Bare Costs | | Incl. Subs O & P | | Cost Per Man-hour | |
|---|---|---|---|---|---|---|
| **Crew Q-20** | Hr. | Daily | Hr. | Daily | Bare Costs | Incl. O&P |
| 1 Sheet Metal Worker | $13.80 | $110.40 | $22.05 | $176.40 | $12.68 | $20.22 |
| 1 Sheet Metal Apprentice | 11.04 | 88.32 | 17.65 | 141.20 | | |
| .5 Electrician | 13.75 | 55.00 | 21.70 | 86.80 | | |
| 20 M.H., Daily Totals | | $253.72 | | $404.40 | $12.68 | $20.22 |

| Crew No. | Bare Costs | | Incl. Subs O & P | | Cost Per Man-hour | |
|---|---|---|---|---|---|---|
| **Crew R-3** | Hr. | Daily | Hr. | Daily | Bare Costs | Incl. O&P |
| 1 Electrician Foreman | $14.25 | $114.00 | $22.45 | $179.60 | $13.78 | $21.75 |
| 1 Electrician | 13.75 | 110.00 | 21.70 | 173.60 | | |
| .5 Equip. Oper. (crane) | 12.90 | 51.60 | 20.45 | 81.80 | | |
| .5 S.P. Crane, 5 Ton | | 87.90 | | 96.70 | 4.39 | 4.83 |
| 20 M.H., Daily Totals | | $363.50 | | $531.70 | $18.17 | $26.58 |

# TRADES

| Trade | Bare Costs Hr. | Daily | Incl. Subs O & P Hr. | Daily | Cost Per Man-hour Bare Costs | Incl. O&P |
|---|---|---|---|---|---|---|
| **Skwk** | Hr. | Daily | Hr. | Daily | Bare Costs | Incl. O&P |
| 1 Skilled Worker | $12.60 | $100.80 | $20.10 | $160.80 | $12.60 | $20.10 |
| 8 M.H., Daily Total | | $100.80 | | $160.80 | $12.60 | $20.10 |
| **Clab** | Hr. | Daily | Hr. | Daily | Bare Costs | Incl. O&P |
| 1 Building Laborer | $10.00 | $80.00 | $15.85 | $126.80 | $10.00 | $15.85 |
| 8 M.H., Daily Total | | $80.00 | | $126.80 | $10.00 | $15.85 |
| **Asbe** | Hr. | Daily | Hr. | Daily | Bare Costs | Incl. O&P |
| 1 Asbestos Worker | $13.80 | $110.40 | $22.25 | $178.00 | $13.80 | $22.25 |
| 8 M.H., Daily Total | | $110.40 | | $178.00 | $13.80 | $22.25 |
| **Boil** | Hr. | Daily | Hr. | Daily | Bare Costs | Incl. O&P |
| 1 Boilermaker | $13.85 | $110.80 | $22.25 | $178.00 | $13.85 | $22.25 |
| 8 M.H., Daily Total | | $110.80 | | $178.00 | $13.85 | $22.25 |
| **Bric** | Hr. | Daily | Hr. | Daily | Bare Costs | Incl. O&P |
| 1 Bricklayer | $12.60 | $100.80 | $19.65 | $157.20 | $12.60 | $19.65 |
| 8 M.H., Daily Total | | $100.80 | | $157.20 | $12.60 | $19.65 |
| **Brhe** | Hr. | Daily | Hr. | Daily | Bare Costs | Incl. O&P |
| 1 Bricklayer Helper | $10.05 | $80.40 | $15.70 | $125.60 | $10.05 | $15.70 |
| 8 M.H., Daily Total | | $80.40 | | $125.60 | $10.05 | $15.70 |
| **Carp** | Hr. | Daily | Hr. | Daily | Bare Costs | Incl. O&P |
| 1 Carpenter | $12.35 | $98.80 | $19.55 | $156.40 | $12.35 | $19.55 |
| 8 M.H., Daily Total | | $98.80 | | $156.40 | $12.35 | $19.55 |
| **Cefi** | Hr. | Daily | Hr. | Daily | Bare Costs | Incl. O&P |
| 1 Cement Finisher | $11.85 | $94.80 | $18.30 | $146.40 | $11.85 | $18.30 |
| 8 M.H., Daily Total | | $94.80 | | $146.40 | $11.85 | $18.30 |
| **Elec** | Hr. | Daily | Hr. | Daily | Bare Costs | Incl. O&P |
| 1 Electrician | $13.75 | $110.00 | $21.70 | $173.60 | $13.75 | $21.70 |
| 8 M.H., Daily Total | | $110.00 | | $173.60 | $13.75 | $21.70 |
| **Elev** | Hr. | Daily | Hr. | Daily | Bare Costs | Incl. O&P |
| 1 Elevator Constructor | $13.80 | $110.40 | $21.95 | $175.60 | $13.80 | $21.95 |
| 8 M.H., Daily Total | | $110.40 | | $175.60 | $13.80 | $21.95 |
| **Eqhv** | Hr. | Daily | Hr. | Daily | Bare Costs | Incl. O&P |
| 1 Equipment Oper. (crane) | $12.90 | $103.20 | $20.45 | $163.60 | $12.90 | $20.45 |
| 8 M.H., Daily Total | | $103.20 | | $163.60 | $12.90 | $20.45 |
| **Eqmd** | Hr. | Daily | Hr. | Daily | Bare Costs | Incl. O&P |
| 1 Equipment Oper. (med.) | $12.65 | $101.20 | $20.10 | $160.80 | $12.65 | $20.10 |
| 8 M.H., Daily Total | | $101.20 | | $160.80 | $12.65 | $20.10 |
| **Eqlt** | Hr. | Daily | Hr. | Daily | Bare Costs | Incl. O&P |
| 1 Equipment Oper. (light) | $12.05 | $96.40 | $19.10 | $152.80 | $12.05 | $19.10 |
| 8 M.H., Daily Total | | $96.40 | | $152.80 | $12.05 | $19.10 |
| **Eqol** | Hr. | Daily | Hr. | Daily | Bare Costs | Incl. O&P |
| 1 Equipment Oper. Oiler | $10.85 | $86.80 | $17.20 | $137.60 | $10.85 | $17.20 |
| 8 M.H., Daily Total | | $86.80 | | $137.60 | $10.85 | $17.20 |
| **Eqmm** | Hr. | Daily | Hr. | Daily | Bare Costs | Incl. O&P |
| 1 Master Mechanic | $13.40 | $107.20 | $21.25 | $170.00 | $13.40 | $21.25 |
| 8 M.H., Daily Total | | $107.20 | | $170.00 | $13.40 | $21.25 |
| **Glaz** | Hr. | Daily | Hr. | Daily | Bare Costs | Incl. O&P |
| 1 Glazier | $12.45 | $99.60 | $19.45 | $155.60 | $12.45 | $19.45 |
| 8 M.H., Daily Total | | $99.60 | | $155.60 | $12.45 | $19.45 |
| **Lath** | Hr. | Daily | Hr. | Daily | Bare Costs | Incl. O&P |
| 1 Lather | $12.40 | $99.20 | $19.15 | $153.20 | $12.40 | $19.15 |
| 8 M.H., Daily Total | | $99.20 | | $153.20 | $12.40 | $19.15 |
| **Marb** | Hr. | Daily | Hr. | Daily | Bare Costs | Incl. O&P |
| 1 Marble Setter | $12.30 | $98.40 | $19.20 | $153.60 | $12.30 | $19.20 |
| 8 M.H., Daily Total | | $98.40 | | $153.60 | $12.30 | $19.20 |
| **Mill** | Hr. | Daily | Hr. | Daily | Bare Costs | Incl. O&P |
| 1 Millwright | $12.75 | $102.00 | $19.80 | $158.40 | $12.75 | $19.80 |
| 8 M.H., Daily Total | | $102.00 | | $158.40 | $12.75 | $19.80 |
| **Mstz** | Hr. | Daily | Hr. | Daily | Bare Costs | Incl. O&P |
| 1 Mosaic/Terrazzo Worker | $12.20 | $97.60 | $18.80 | $150.40 | $12.20 | $18.80 |
| 8 M.H., Daily Total | | $97.60 | | $150.40 | $12.20 | $18.80 |
| **Pord** | Hr. | Daily | Hr. | Daily | Bare Costs | Incl. O&P |
| 1 Painter (Ordinary) | $11.85 | $94.80 | $18.50 | $148.00 | $11.85 | $18.50 |
| 8 M.H., Daily Total | | $94.80 | | $148.00 | $11.85 | $18.50 |

| Trade | Bare Costs Hr. | Daily | Incl. Subs O & P Hr. | Daily | Cost Per Man-hour Bare Costs | Incl. O&P |
|---|---|---|---|---|---|---|
| **Psst** | Hr. | Daily | Hr. | Daily | Bare Costs | Incl. O&P |
| 1 Painter (Structural Steel) | $12.35 | $98.80 | $21.60 | $172.80 | $12.35 | $21.60 |
| 8 M.H., Daily Total | | $98.80 | | $172.80 | $12.35 | $21.60 |
| **Pape** | Hr. | Daily | Hr. | Daily | Bare Costs | Incl. O&P |
| 1 Paper Hanger | $12.10 | $96.80 | $18.90 | $151.20 | $12.10 | $18.90 |
| 8 M.H., Daily Total | | $96.80 | | $151.20 | $12.10 | $18.90 |
| **Pile** | Hr. | Daily | Hr. | Daily | Bare Costs | Incl. O&P |
| 1 Pile Driver | $12.40 | $99.20 | $21.15 | $169.20 | $12.40 | $21.15 |
| 8 M.H., Daily Total | | $99.20 | | $169.20 | $12.40 | $21.15 |
| **Plas** | Hr. | Daily | Hr. | Daily | Bare Costs | Incl. O&P |
| 1 Plasterer | $12.35 | $98.80 | $19.25 | $154.00 | $12.35 | $19.25 |
| 8 M.H., Daily Total | | $98.80 | | $154.00 | $12.35 | $19.25 |
| **Plah** | Hr. | Daily | Hr. | Daily | Bare Costs | Incl. O&P |
| 1 Plasterer Helper | $10.35 | $82.80 | $16.15 | $129.20 | $10.35 | $16.15 |
| 8 M.H., Daily Total | | $82.80 | | $129.20 | $10.35 | $16.15 |
| **Plum** | Hr. | Daily | Hr. | Daily | Bare Costs | Incl. O&P |
| 1 Plumber | $13.80 | $110.40 | $21.85 | $174.80 | $13.80 | $21.85 |
| 8 M.H., Daily Total | | $110.40 | | $174.80 | $13.80 | $21.85 |
| **Rodm** | Hr. | Daily | Hr. | Daily | Bare Costs | Incl. O&P |
| 1 Rodmen, (Reinf.) | $13.30 | $106.40 | $22.30 | $178.40 | $13.30 | $22.30 |
| 8 M.H., Daily Total | | $106.40 | | $178.40 | $13.30 | $22.30 |
| **Rofc** | Hr. | Daily | Hr. | Daily | Bare Costs | Incl. O&P |
| 1 Roofer (Composition) | $11.60 | $92.80 | $19.25 | $154.00 | $11.60 | $19.25 |
| 8 M.H., Daily Total | | $92.80 | | $154.00 | $11.60 | $19.25 |
| **Rots** | Hr. | Daily | Hr. | Daily | Bare Costs | Incl. O&P |
| 1 Roofer (Tile & Slate) | $11.70 | $93.60 | $19.40 | $155.20 | $11.70 | $19.40 |
| 8 M.H., Daily Total | | $93.60 | | $155.20 | $11.70 | $19.40 |
| **Rohe** | Hr. | Daily | Hr. | Daily | Bare Costs | Incl. O&P |
| 1 Roofer Helper | $8.80 | $70.40 | $14.60 | $116.80 | $8.80 | $14.60 |
| 8 M.H., Daily Total | | $70.40 | | $116.80 | $8.80 | $14.60 |
| **Shee** | Hr. | Daily | Hr. | Daily | Bare Costs | Incl. O&P |
| 1 Sheet Metal Worker | $13.80 | $110.40 | $22.05 | $176.40 | $13.80 | $22.05 |
| 8 M.H., Daily Total | | $110.40 | | $176.40 | $13.80 | $22.05 |
| **Spri** | Hr. | Daily | Hr. | Daily | Bare Costs | Incl. O&P |
| 1 Sprinkler Installer | $14.05 | $112.40 | $22.40 | $179.20 | $14.05 | $22.40 |
| 8 M.H., Daily Total | | $112.40 | | $179.20 | $14.05 | $22.40 |
| **Stpi** | Hr. | Daily | Hr. | Daily | Bare Costs | Incl. O&P |
| 1 Steamfitter/Pipefitter | $13.85 | $110.80 | $21.95 | $175.60 | $13.85 | $21.95 |
| 8 M.H., Daily Total | | $110.80 | | $175.60 | $13.85 | $21.95 |
| **Ston** | Hr. | Daily | Hr. | Daily | Bare Costs | Incl. O&P |
| 1 Stone Mason | $12.45 | $99.60 | $19.40 | $155.20 | $12.45 | $19.40 |
| 8 M.H., Daily Total | | $99.60 | | $155.20 | $12.45 | $19.40 |
| **Sswk** | Hr. | Daily | Hr. | Daily | Bare Costs | Incl. O&P |
| 1 Struc. Steel Worker | $13.30 | $106.40 | $22.80 | $182.40 | $13.30 | $22.80 |
| 8 M.H., Daily Total | | $106.40 | | $182.40 | $13.30 | $22.80 |
| **Tilf** | Hr. | Daily | Hr. | Daily | Bare Costs | Incl. O&P |
| 1 Tile Layer (Floor) | $12.15 | $97.20 | $18.70 | $149.60 | $12.15 | $18.70 |
| 8 M.H., Daily Total | | $97.20 | | $149.60 | $12.15 | $18.70 |
| **Tilh** | Hr. | Daily | Hr. | Daily | Bare Costs | Incl. O&P |
| 1 Tile Layer Helper | $9.80 | $78.40 | $15.10 | $120.80 | $9.80 | $15.10 |
| 8 M.H., Daily Total | | $78.40 | | $120.80 | $9.80 | $15.10 |
| **Trlt** | Hr. | Daily | Hr. | Daily | Bare Costs | Incl. O&P |
| 1 Truck Driver (light) | $10.25 | $82.00 | $16.10 | $128.80 | $10.25 | $16.10 |
| 8 M.H., Daily Total | | $82.00 | | $128.80 | $10.25 | $16.10 |
| **Trhv** | Hr. | Daily | Hr. | Daily | Bare Costs | Incl. O&P |
| 1 Truck Driver (heavy) | $10.35 | $82.80 | $16.25 | $130.00 | $10.35 | $16.25 |
| 8 M.H., Daily Total | | $82.80 | | $130.00 | $10.35 | $16.25 |
| **Sswl** | Hr. | Daily | Hr. | Daily | Bare Costs | Incl. O&P |
| 1 Welder (Struc. Steel) | $13.30 | $106.40 | $22.80 | $182.40 | $13.30 | $22.80 |
| 8 M.H., Daily Total | | $106.40 | | $182.40 | $13.30 | $22.80 |

# LOCATION FACTORS

Costs shown in "Residential Cost Data" are based on National Averages for materials and installation. To adjust these costs to a specific location, simply multiply the base cost by the residential factor for that city. The data is arranged alphabetically by state and postal zip code numbers. For a city not listed, use the factor for a nearby city with similar economic characteristics.

| STATE/ZIP | CITY | RES. | COMM. |
|---|---|---|---|
| **ALABAMA** | | | |
| 350-352 | Birmingham | .84 | .85 |
| 354 | Tuscaloosa | .85 | .82 |
| 355 | Jasper | .84 | .85 |
| 356 | Decatur | .84 | .86 |
| 357, 358 | Huntsville | .85 | .88 |
| 359 | Gadsden | .84 | .84 |
| 360, 361 | Montgomery | .88 | .86 |
| 362 | Anniston | .84 | .84 |
| 363 | Dothan | .88 | .83 |
| 364 | Evergreen | .88 | .83 |
| 365, 366 | Mobile | .93 | .87 |
| 367 | Selma | .88 | .83 |
| 368 | Phenix City | .88 | .83 |
| 369 | Butler | .88 | .83 |
| **ALASKA** | | | |
| 995, 996 | Anchorage | 1.32 | 1.34 |
| 997 | Fairbanks | 1.32 | 1.35 |
| 998 | Juneau | 1.32 | 1.33 |
| 999 | Ketchikan | 1.32 | 1.34 |
| **ARIZONA** | | | |
| 850 | | | |
| 852, 853 | Phoenix | .96 | .98 |
| 855 | Globe | .96 | .94 |
| 856, 857 | Tucson | .95 | .96 |
| 859 | Show Low | .96 | .94 |
| 860 | Flagstaff | .96 | .95 |
| 863 | Prescott | .96 | .95 |
| 864 | Kingman | .96 | .95 |
| 865 | Chambers | .96 | .95 |
| **ARKANSAS** | | | |
| 716 | Pine Bluff | .87 | .87 |
| 717 | Camden | .87 | .87 |
| 718 | Texarkana | .87 | .86 |
| 719 | Hot Springs | .87 | .87 |
| 720-722 | Little Rock | .87 | .87 |
| 723 | West Memphis | .90 | .90 |
| 724 | Jonesboro | .90 | .90 |
| 725 | Batesville | .87 | .87 |
| 726 | Harrison | .87 | .87 |
| 727 | Fayetteville | .88 | .85 |
| 728 | Russellville | .88 | .85 |
| 729 | Fort Smith | .88 | .85 |
| **CALIFORNIA** | | | |
| 900-918 | Los Angeles | 1.11 | 1.11 |
| 920, 921 | San Diego | 1.11 | 1.10 |
| 922 | Palm Springs | 1.14 | 1.11 |
| 923, 924 | San Bernardino | 1.14 | 1.11 |
| 925 | Riverside | 1.14 | 1.11 |
| 926, 927 | Santa Ana | 1.14 | 1.11 |
| 928 | Anaheim | 1.14 | 1.12 |
| 930 | Ventura | 1.13 | 1.12 |
| 931 | Santa Barbara | 1.15 | 1.13 |
| 932, 933 | Bakersfield | 1.12 | 1.07 |
| 934 | San Luis Obispo | 1.15 | 1.13 |
| 935 | Mojave | 1.12 | 1.09 |
| 936, 937 | Fresno | 1.12 | 1.09 |
| 939 | Salinas | 1.12 | 1.12 |
| 940, 941 | San Francisco | 1.20 | 1.24 |
| 943 | Palo Alto | 1.20 | 1.20 |
| 944 | San Mateo | 1.20 | 1.24 |
| 945, 946 | Oakland | 1.20 | 1.24 |
| 947 | Berkeley | 1.20 | 1.24 |
| 948 | Richmond | 1.20 | 1.24 |
| 948 | San Rafael | 1.20 | 1.15 |
| 950, 951 | San Jose | 1.20 | 1.20 |
| 952, 953 | Stockton | 1.14 | 1.11 |
| 954 | Santa Rosa | 1.15 | 1.18 |

| STATE/ZIP | CITY | RES. | COMM. |
|---|---|---|---|
| **CALIFORNIA** (Cont.) | | | |
| 955 | Eureka | 1.15 | 1.15 |
| 956-958 | Sacramento | 1.15 | 1.15 |
| 959 | Marysville | 1.15 | 1.15 |
| 960 | Redding | 1.15 | 1.15 |
| 961 | Susanville | 1.15 | 1.15 |
| **COLORADO** | | | |
| 800-802 | Denver | 1.00 | .96 |
| 803 | Boulder | 1.00 | .96 |
| 804 | Golden | 1.00 | .96 |
| 805 | Fort Collins | .98 | .93 |
| 806 | Greeley | .98 | .93 |
| 807 | Fort Morgan | .98 | .93 |
| 808, 809 | Colorado Springs | .96 | .94 |
| 810 | Pueblo | .97 | .95 |
| 811 | Alamosa | .97 | .95 |
| 812 | Salida | .96 | .94 |
| 813 | Durango | .97 | .95 |
| 814 | Montrose | .96 | .94 |
| 815 | Grand Junction | 1.00 | .96 |
| 816 | Glenwood Springs | 1.00 | .96 |
| **CONNECTICUT** | | | |
| 060, 061 | Hartford | .98 | .99 |
| 062 | Willimantic | .98 | .99 |
| 063 | New London | .98 | .99 |
| 064, 065 | New Haven | .98 | .99 |
| 066 | Bridgeport | .97 | 1.00 |
| 067 | Waterbury | .97 | .97 |
| 068, 069 | Stamford | .97 | 1.01 |
| **DELAWARE** | | | |
| 197, 198 | Wilmington | 1.00 | 1.01 |
| 199 | Dover | 1.00 | 1.01 |
| **DISTRICT OF COLUMBIA** | | | |
| 200-205 | Washington | .94 | .96 |
| **FLORIDA** | | | |
| 320, 322 | Jacksonville | .88 | .87 |
| 323 | Tallahassee | .81 | .83 |
| 324 | Panama City | .81 | .83 |
| 325 | Pensacola | .81 | .88 |
| 326 | Gainesville | .88 | .85 |
| 327, 328 | Orlando | .87 | .85 |
| 329 | Melbourne | .87 | .85 |
| 330, 331 | Miami | .92 | .94 |
| 333 | Fort Lauderdale | .91 | .92 |
| 334 | W. Palm Beach | .92 | .92 |
| 335, 336 | Tampa | .89 | .93 |
| 337 | St. Petersburg | .89 | .93 |
| 338 | Lakeland | .87 | .89 |
| 339 | Fort Myers | .89 | .90 |
| **GEORGIA** | | | |
| 300-303 | Atlanta | .84 | .89 |
| 304 | Statesboro | .85 | .87 |
| 305 | Gainesville | .84 | .87 |
| 306 | Athens | .84 | .87 |
| 307 | Dalton | .89 | .88 |
| 308, 309 | Augusta | .84 | .86 |
| 310, 312 | Macon | .84 | .84 |
| 313, 314 | Savannah | .85 | .87 |
| 315 | Waycross | .83 | .82 |
| 316 | Valdosta | .83 | .82 |
| 317 | Albany | .83 | .83 |
| 318, 319 | Columbus | .82 | .84 |
| **HAWAII** | | | |
| 967, 968 | Honolulu | 1.16 | 1.12 |

# LOCATION FACTORS

| STATE/ZIP | CITY | RES. | COMM. |
|---|---|---|---|
| **IDAHO** | | | |
| 832 | Pocatello | .96 | .95 |
| 833 | Twin Falls | .96 | .95 |
| 834 | Idaho Falls | .96 | .95 |
| 835 | Lewiston | 1.04 | 1.04 |
| 836, 837 | Boise | .96 | .96 |
| 838 | Coeur D'Alene | 1.04 | 1.04 |
| **ILLINOIS** | | | |
| 600-606 | Chicago | 1.01 | .99 |
| 609 | Kankakee | 1.01 | 1.01 |
| 610, 611 | Rockford | 1.01 | 1.01 |
| 612 | Rock Island | 1.01 | .94 |
| 613 | La Salle | 1.01 | .96 |
| 614 | Galesburg | 1.01 | .96 |
| 615, 616 | Peoria | 1.01 | .96 |
| 617 | Bloomington | 1.01 | .96 |
| 618, 619 | Champaign | 1.00 | .96 |
| 620, 622 | East St. Louis | .99 | .99 |
| 623 | Quincy | .99 | .96 |
| 624 | Effingham | 1.00 | .97 |
| 625-627 | Springfield | .99 | .98 |
| 628 | Centralia | .99 | .99 |
| 629 | Carbondale | .99 | .99 |
| **INDIANA** | | | |
| 460-462 | Indianapolis | 1.00 | .98 |
| 463, 464 | Gary | 1.06 | 1.03 |
| 465, 466 | South Bend | .99 | .98 |
| 467, 468 | Fort Wayne | .94 | .95 |
| 469 | Kokomo | .95 | .95 |
| 470 | Lawrenceburg | 1.01 | .98 |
| 471 | New Albany | .96 | .92 |
| 472 | Columbus | 1.00 | .98 |
| 473 | Muncie | .95 | .94 |
| 474 | Bloomington | 1.00 | .98 |
| 475 | Washington | .98 | .99 |
| 476, 477 | Evansville | .98 | .99 |
| 478 | Terre Haute | .98 | .96 |
| 479 | Lafayette | .95 | .95 |
| **IOWA** | | | |
| 500-503 | Des Moines | .96 | .94 |
| 504 | Mason City | .94 | .88 |
| 505 | Fort Dodge | .94 | .88 |
| 506, 507 | Waterloo | .94 | .88 |
| 508 | Creston | .96 | .91 |
| 510, 511 | Sioux City | .88 | .86 |
| 512 | Sibley | .88 | .86 |
| 513 | Spencer | .88 | .86 |
| 514 | Carroll | .96 | .91 |
| 515 | Council Bluffs | .96 | .91 |
| 516 | Shenandoah | .96 | .91 |
| 520 | Dubuque | .96 | .86 |
| 521 | Decorah | .96 | .86 |
| 522-524 | Cedar Rapids | .97 | .87 |
| 525 | Ottumwa | .97 | .87 |
| 526 | Burlington | 1.01 | .95 |
| 527, 528 | Davenport | 1.01 | .98 |
| **KANSAS** | | | |
| 660-662 | Kansas City | 1.00 | .98 |
| 664-666 | Topeka | .93 | .93 |
| 667 | Fort Scott | 1.00 | .98 |
| 668 | Emporia | .93 | .91 |
| 669 | Belleville | .93 | .88 |
| 670-672 | Wichita | .90 | .92 |
| 673 | Independence | .90 | .90 |
| 674 | Salina | .90 | .88 |
| 675 | Hutchinson | .90 | .88 |
| 676 | Hays | .90 | .88 |
| 677 | Colby | .90 | .88 |
| 678 | Dodge City | .90 | .90 |
| 679 | Liberal | .90 | .90 |

| STATE/ZIP | CITY | RES. | COMM. |
|---|---|---|---|
| **KENTUCKY** | | | |
| 400-402 | Louisville | .96 | .95 |
| 403-405 | Lexington | .95 | .93 |
| 406 | Frankfort | .95 | .91 |
| 407-409 | Corbin | .95 | .91 |
| 410 | Covington | 1.01 | .98 |
| 411, 412 | Ashland | .95 | .97 |
| 413, 414 | Campton | .95 | .91 |
| 415, 416 | Pikeville | .95 | .97 |
| 417, 418 | Hazard | .95 | .91 |
| 420 | Paducah | .96 | .90 |
| 421, 422 | Bowling Green | .96 | .90 |
| 423 | Owensboro | .96 | .94 |
| 424 | Henderson | .96 | .94 |
| 425, 426 | Somerset | .95 | .91 |
| 427 | Elizabethtown | .96 | .92 |
| **LOUISIANA** | | | |
| 700, 701 | New Orleans | .93 | .93 |
| 703 | Thibodaux | .93 | .93 |
| 704 | Hammond | .94 | .94 |
| 705 | Lafayette | .94 | .90 |
| 706 | Lake Charles | .93 | .93 |
| 707, 708 | Baton Rouge | .94 | .94 |
| 710, 711 | Shreveport | .89 | .90 |
| 712 | Monroe | .89 | .86 |
| 713, 714 | Alexandria | .89 | .87 |
| **MAINE** | | | |
| 039 | Kittery | .91 | .90 |
| 040, 041 | Portland | .88 | .88 |
| 042 | Lewiston | .88 | .88 |
| 043 | Augusta | .88 | .87 |
| 044 | Bangor | .88 | .87 |
| 045 | Bath | .88 | .87 |
| 046 | Machias | .88 | .87 |
| 047 | Houlton | .88 | .87 |
| 048 | Rockland | .88 | .87 |
| 049 | Waterville | .88 | .87 |
| **MARYLAND** | | | |
| 206 | Waldorf | .94 | .93 |
| 207, 208 | College Park | .94 | .93 |
| 209 | Silver Spring | .94 | .92 |
| 210-212 | Baltimore | .92 | .92 |
| 214 | Annapolis | .92 | .93 |
| 215 | Cumberland | .92 | .93 |
| 216 | Easton | .92 | .93 |
| 217 | Hagerstown | .92 | .91 |
| 218 | Salisbury | .92 | .93 |
| 219 | Elkton | .92 | .92 |
| **MASSACHUSETTS** | | | |
| 010, 011 | Springfield | .98 | .97 |
| 012 | Pittsfield | .96 | .96 |
| 013 | Greenfield | .98 | .97 |
| 014 | Fitchburg | 1.04 | 1.00 |
| 015, 016 | Worcester | 1.04 | 1.00 |
| 017 | Framingham | 1.07 | 1.08 |
| 018 | Lowell | 1.04 | 1.03 |
| 019 | Lynn | 1.07 | 1.08 |
| 020-022 | Boston | 1.07 | 1.08 |
| 023, 024 | Brockton | .99 | 1.01 |
| 025 | Buzzards Bay | .99 | 1.01 |
| 026 | Hyannis | .99 | 1.01 |
| 027 | New Bedford | .99 | 1.00 |
| **MICHIGAN** | | | |
| 480 | Royal Oak | 1.03 | 1.02 |
| 480-482 | Detroit | 1.03 | 1.02 |
| 484, 485 | Flint | .97 | .99 |
| 486, 487 | Saginaw | .96 | .99 |
| 488, 489 | Lansing | 1.01 | 1.06 |
| 490, 491 | Kalamazoo | .98 | .96 |
| 492 | Jackson | 1.01 | .97 |
| 493-495 | Grand Rapids | .96 | .94 |
| 496 | Traverse City | .96 | .92 |
| 497 | Gaylord | .95 | .95 |
| 498, 499 | Iron Mountain | .95 | .92 |

# LOCATION FACTORS

| STATE/ZIP | CITY | RES. | COMM. |
|---|---|---|---|
| **MINNESOTA** | | | |
| 550, 551 | St. Paul | 1.01 | 1.00 |
| 553, 554 | Minneapolis | 1.01 | .98 |
| 556-558 | Duluth | .95 | .97 |
| 559 | Rochester | .98 | .97 |
| 560 | Mankato | 1.01 | 1.00 |
| 561 | Windom | 1.01 | 1.00 |
| 562 | Willmar | 1.01 | 1.00 |
| 563 | St. Cloud | 1.01 | .94 |
| 564 | Brainerd | 1.01 | .94 |
| 565 | Detroit Lakes | .85 | .92 |
| 566 | Bemidji | .85 | .92 |
| 567 | Thief River Falls | .85 | .92 |
| **MISSISSIPPI** | | | |
| 386 | Clarksdale | .86 | .82 |
| 387 | Greenville | .86 | .82 |
| 388 | Tupelo | .86 | .80 |
| 389 | Greenwood | .86 | .82 |
| 390-392 | Jackson | .86 | .86 |
| 393 | Meridian | .86 | .80 |
| 394 | Laurel | .89 | .85 |
| 395 | Biloxi | .89 | .85 |
| 396 | Mc Comb | .86 | .83 |
| 397 | Columbus | .86 | .80 |
| **MISSOURI** | | | |
| 630, 631 | St. Louis | .99 | .99 |
| 633 | Bowling Green | .99 | .99 |
| 634 | Hannibal | .99 | .90 |
| 635 | Kirksville | .89 | .92 |
| 636 | Flat River | .99 | .99 |
| 637 | Cape Girardeau | .99 | .99 |
| 638 | Sikeston | .99 | .99 |
| 639 | Poplar Bluff | .99 | .99 |
| 640, 641 | Kansas City | 1.00 | .98 |
| 644, 645 | Saint Joseph | .89 | .99 |
| 646 | Chillicothe | .89 | .92 |
| 647 | Harrisonville | 1.00 | .98 |
| 648 | Joplin | .88 | .90 |
| 650, 652 | Columbia | .99 | .90 |
| 651 | Jefferson City | .99 | .91 |
| 653 | Sedalia | .99 | .91 |
| 654, 655 | Rolla | .99 | .91 |
| 656-658 | Springfield | .88 | .90 |
| **MONTANA** | | | |
| 590, 591 | Billings | .96 | .95 |
| 592 | Wolf Point | .96 | .94 |
| 593 | Miles City | .96 | .94 |
| 594 | Great Falls | .96 | .95 |
| 595 | Havre | .96 | .94 |
| 596 | Helena | .96 | .94 |
| 597 | Butte | .96 | .92 |
| 598 | Missoula | .96 | .92 |
| 599 | Kalispell | .96 | .94 |
| **NEBRASKA** | | | |
| 680, 681 | Omaha | .92 | .95 |
| 683-685 | Lincoln | .92 | .91 |
| 686 | Columbus | .92 | .93 |
| 687 | Norfolk | .92 | .93 |
| 688 | Grand Island | .92 | .85 |
| 689 | Hastings | .92 | .85 |
| 690 | Mc Cook | .92 | .85 |
| 691 | North Platte | .92 | .85 |
| 692 | Valentine | .92 | .85 |
| 693 | Alliance | .91 | .85 |
| **NEVADA** | | | |
| 890, 891 | Las Vegas | 1.07 | 1.08 |
| 893 | Ely | 1.07 | 1.06 |
| 894, 895 | Reno | 1.05 | 1.06 |
| 897 | Carson City | 1.05 | 1.06 |
| 898 | Elko | 1.05 | 1.06 |

| STATE/ZIP | CITY | RES. | COMM. |
|---|---|---|---|
| **NEW HAMPSHIRE** | | | |
| 030, 031 | Manchester | .91 | .92 |
| 032, 033 | Concord | .91 | .92 |
| 034 | Keene | .91 | .92 |
| 035 | Littleton | .91 | .92 |
| 036 | Charlestown | .91 | .92 |
| 037 | Claremont | .91 | .92 |
| 038 | Portsmouth | .91 | .90 |
| **NEW JERSEY** | | | |
| 070, 071 | Newark | 1.06 | 1.04 |
| 072 | Elizabeth | 1.06 | 1.04 |
| 073 | Jersey City | 1.06 | 1.05 |
| 074, 075 | Paterson | 1.06 | 1.06 |
| 076 | Hackensack | 1.06 | 1.06 |
| 077 | Long Branch | 1.06 | 1.04 |
| 078 | Dover | 1.06 | 1.04 |
| 079 | Summit | 1.06 | 1.04 |
| 080, 083 | Vineland | 1.06 | 1.02 |
| 081 | Camden | 1.06 | 1.03 |
| 082, 084 | Atlantic City | 1.06 | 1.03 |
| 085, 086 | Trenton | 1.06 | 1.05 |
| 087 | Point Pleasant | 1.06 | 1.04 |
| 088, 089 | New Brunswick | 1.06 | 1.04 |
| **NEW MEXICO** | | | |
| 870, 871 | Albuquerque | .91 | .93 |
| 873 | Gallup | .91 | .94 |
| 874 | Farmington | .91 | .94 |
| 875 | Santa Fe | .91 | .94 |
| 877 | Las Vegas | .91 | .94 |
| 878 | Socorro | .91 | .93 |
| 879 | Truth/Conseq. | .91 | .91 |
| 880 | Las Cruces | .91 | .91 |
| 881 | Clovis | .91 | .91 |
| 882 | Roswell | .91 | .91 |
| 883 | Carrizozo | .91 | .91 |
| 884 | Tucumcari | .91 | .94 |
| **NEW YORK** | | | |
| 100 | Manhattan | 1.14 | 1.14 |
| 103 | Staten Island | 1.14 | 1.14 |
| 104 | Bronx | 1.14 | 1.14 |
| 105 | Mount Vernon | 1.14 | 1.08 |
| 106 | White Plains | 1.14 | 1.08 |
| 107 | Yonkers | 1.11 | 1.08 |
| 108 | New Rochelle | 1.11 | 1.08 |
| 109 | Suffern | 1.11 | 1.08 |
| 110 | Queens | 1.14 | 1.14 |
| 111 | Long Island City | 1.14 | 1.14 |
| 112 | Brooklyn | 1.14 | 1.14 |
| 113 | Flushing | 1.14 | 1.14 |
| 114 | Jamaica | 1.14 | 1.14 |
| 115, 117, 118 | Hicksville | 1.14 | 1.14 |
| 116 | Far Rockaway | 1.14 | 1.14 |
| 119 | Riverhead | 1.14 | 1.14 |
| 120-122 | Albany | .96 | .95 |
| 123 | Schenectady | .97 | .96 |
| 124 | Kingston | 1.11 | 1.08 |
| 125, 126 | Poughkeepsie | 1.11 | 1.08 |
| 127 | Monticello | 1.11 | 1.08 |
| 128 | Glens Falls | .98 | .96 |
| 129 | Plattsburgh | .98 | .96 |
| 130-132 | Syracuse | .96 | .95 |
| 133-135 | Utica | .92 | .93 |
| 136 | Watertown | .92 | .93 |
| 137-139 | Binghamton | .92 | .91 |
| 140-142 | Buffalo | 1.07 | 1.04 |
| 143 | Niagara Falls | 1.07 | 1.04 |
| 144-146 | Rochester | .99 | 1.00 |
| 147 | Jamestown | 1.07 | 1.04 |
| 148, 149 | Elmira | .92 | .89 |

# LOCATION FACTORS

| STATE/ZIP | CITY | RES. | COMM. |
|---|---|---|---|
| **NORTH CAROLINA** | | | |
| 270, 272-274 | Greensboro | .80 | .81 |
| 271 | Winston-Salem | .81 | .82 |
| 275, 276 | Raleigh | .81 | .81 |
| 277 | Durham | .80 | .81 |
| 278 | Rocky Mount | .81 | .81 |
| 279 | Elizabeth City | .85 | .85 |
| 280-282 | Charlotte | .81 | .82 |
| 283 | Fayetteville | .81 | .81 |
| 284 | Wilmington | .81 | .83 |
| 285 | Kingston | .81 | .81 |
| 286 | Hickory | .81 | .82 |
| 287, 288 | Asheville | .81 | .83 |
| 289 | Murphy | .81 | .82 |
| **NORTH DAKOTA** | | | |
| 580, 581 | Fargo | .85 | .90 |
| 582 | Grand Forks | .85 | .90 |
| 583 | Devils Lake | .85 | .90 |
| 584 | Jamestown | .85 | .90 |
| 585 | Bismarck | .85 | .89 |
| 586 | Dickinson | .85 | .89 |
| 587 | Minot | .85 | .89 |
| 588 | Williston | .85 | .89 |
| **OHIO** | | | |
| 430-432 | Columbus | .97 | .97 |
| 433 | Marion | .97 | .99 |
| 434-436 | Toledo | 1.07 | 1.04 |
| 437, 438 | Zanesville | .97 | .97 |
| 439 | Steubenville | 1.00 | 1.00 |
| 440, 441 | Cleveland | 1.17 | 1.10 |
| 442, 443 | Akron | 1.05 | 1.03 |
| 444, 445 | Youngstown | 1.02 | 1.00 |
| 446, 447 | Canton | 1.00 | .99 |
| 448, 449 | Mansfield | 1.00 | .99 |
| 450-452 | Cincinnati | 1.01 | .98 |
| 453, 454 | Dayton | .98 | .98 |
| 455 | Springfield | .97 | .97 |
| 456 | Chillicothe | 1.01 | .98 |
| 457 | Athens | .97 | .97 |
| 458 | Lima | 1.07 | .97 |
| **OKLAHOMA** | | | |
| 730, 731 | Oklahoma City | .92 | .94 |
| 734 | Ardmore | .92 | .90 |
| 735 | Lawton | .92 | .90 |
| 736 | Clinton | .92 | .94 |
| 737 | Enid | .92 | .92 |
| 738 | Woodward | .92 | .92 |
| 739 | Guymon | .92 | .92 |
| 740, 741 | Tulsa | .95 | .92 |
| 743 | Miami | .95 | .92 |
| 744 | Muskogee | .95 | .92 |
| 745 | McAlester | .92 | .94 |
| 746 | Ponca City | .95 | .92 |
| 747 | Durant | .92 | .94 |
| 748 | Shawnee | .92 | .94 |
| 749 | Poteau | .88 | .85 |
| **OREGON** | | | |
| 970-972 | Portland | 1.08 | 1.08 |
| 973 | Salem | 1.08 | 1.03 |
| 974 | Eugene | 1.07 | 1.05 |
| 975 | Medford | 1.07 | 1.05 |
| 976 | Klamath Falls | 1.07 | 1.05 |
| 977 | Bend | 1.07 | 1.05 |
| 978 | Pendleton | .96 | .93 |
| 979 | Vale | .96 | .93 |
| **PENNSYLVANIA** | | | |
| 150-152 | Pittsburgh | 1.04 | 1.02 |
| 153 | Washington | 1.04 | 1.02 |
| 154 | Uniontown | 1.04 | 1.02 |
| 155 | Bedford | 1.04 | .98 |
| 156 | Greensburg | 1.04 | 1.02 |
| 157 | Indiana | 1.04 | .98 |
| 158 | Du Bois | 1.04 | .98 |
| 159 | Johnstown | 1.04 | .98 |

| STATE/ZIP | CITY | RES. | COMM. |
|---|---|---|---|
| **PENNSYLVANIA (Cont.)** | | | |
| 160 | Butler | 1.04 | 1.02 |
| 161 | New Castle | 1.04 | 1.02 |
| 162 | Kittanning | 1.04 | 1.02 |
| 163 | Oil City | .97 | 1.02 |
| 164, 165 | Erie | .97 | .98 |
| 166 | Altoona | 1.04 | .98 |
| 167 | Bradford | .97 | .94 |
| 168 | State College | .94 | .94 |
| 169 | Wellsboro | .94 | .94 |
| 170, 171 | Harrisburg | .94 | .93 |
| 172 | Chambersburg | .94 | .93 |
| 173, 174 | York | .94 | .91 |
| 175, 176 | Lancaster | .94 | .91 |
| 177 | Williamsport | .94 | .94 |
| 178 | Sunbury | .94 | .94 |
| 179 | Pottsville | .94 | .93 |
| 180 | Lehigh Valley | .99 | .98 |
| 181 | Allentown | .99 | .98 |
| 182 | Hazleton | .99 | .98 |
| 183 | Stroudsburg | .99 | .98 |
| 184, 185 | Scranton | .94 | .97 |
| 186, 187 | Wilkes Barre | .94 | .97 |
| 188 | Montrose | .94 | .97 |
| 189 | Doylestown | .94 | 1.02 |
| 190, 191 | Philadelphia | 1.05 | 1.02 |
| 193 | Westchester | 1.05 | 1.02 |
| 194 | Norristown | 1.05 | 1.02 |
| 195, 196 | Reading | .92 | .93 |
| **RHODE ISLAND** | | | |
| 028, 029 | Providence | .96 | .98 |
| **SOUTH CAROLINA** | | | |
| 290-292 | Columbia | .79 | .84 |
| 293 | Spartanburg | .79 | .84 |
| 294 | Charleston | .81 | .84 |
| 295 | Florence | .79 | .80 |
| 296 | Greenville | .79 | .81 |
| 297 | Rock Hill | .79 | .81 |
| 298 | Aiken | .79 | .81 |
| 299 | Beaufort | .81 | .81 |
| **SOUTH DAKOTA** | | | |
| 570, 571 | Sioux Falls | .90 | .86 |
| 572 | Watertown | .90 | .85 |
| 573 | Mitchell | .90 | .85 |
| 574 | Aberdeen | .90 | .85 |
| 575 | Pierre | .90 | .85 |
| 576 | Mobridge | .91 | .85 |
| 577 | Rapid City | .91 | .85 |
| **TENNESSEE** | | | |
| 370-372 | Nashville | .83 | .83 |
| 373, 374 | Chattanooga | .89 | .88 |
| 376 | Johnson City | .86 | .85 |
| 377-379 | Knoxville | .86 | .86 |
| 380, 381 | Memphis | .90 | .90 |
| 382 | Mc Kenzie | .83 | .83 |
| 383 | Jackson | .90 | .90 |
| 384 | Columbia | .83 | .83 |
| 385 | Cookeville | .83 | .83 |
| **TEXAS** | | | |
| 750 | Mc Kinney | .96 | .90 |
| 751 | Waxahackie | .94 | .95 |
| 752, 753 | Dallas | .96 | .94 |
| 754 | Greenville | .96 | .90 |
| 755 | Texarkana | .96 | .86 |
| 756 | Longview | .96 | .86 |
| 757 | Tyler | .96 | .86 |
| 758 | Palestine | .86 | .86 |
| 759 | Lufkin | .86 | .86 |
| 760, 761 | Forth Worth | .94 | .95 |
| 762 | Denton | .96 | .90 |
| 763 | Wichita Falls | .89 | .90 |
| 764 | Eastland | .94 | .95 |

# LOCATION FACTORS

| STATE/ZIP | CITY | RES. | COMM. |
|---|---|---|---|
| **TEXAS** (Cont.) | | | |
| 765 | Temple | .91 | .84 |
| 766, 767 | Waco | .86 | .86 |
| 768 | Brownwood | .87 | .87 |
| 769 | San Angelo | .87 | .83 |
| 770-772 | Houston | .92 | .96 |
| 773 | Huntsville | .92 | .92 |
| 774 | Wharton | .92 | .96 |
| 775 | Galveston | .92 | .95 |
| 776, 777 | Beaumont | .95 | .98 |
| 778 | Bryan | .91 | .92 |
| 779 | Victoria | .85 | .86 |
| 780-782 | San Antonio | .86 | .89 |
| 783, 784 | Corpus Christi | .85 | .86 |
| 785 | Mc Allen | .85 | .84 |
| 786, 787 | Austin | .91 | .90 |
| 788 | Del Rio | .86 | .88 |
| 789 | Giddings | .91 | .92 |
| 790, 791 | Amarillo | .91 | .91 |
| 792 | Childress | .88 | .90 |
| 793, 794 | Lubbock | .88 | .89 |
| 795, 796 | Abilene | .87 | .87 |
| 797 | Midland | .86 | .88 |
| 798, 799 | El Paso | .85 | .85 |
| **UTAH** | | | |
| 840, 841 | Salt Lake City | .94 | .96 |
| 843, 844 | Ogden | .95 | .92 |
| 845 | Price | .95 | .94 |
| 846 | Provo | .95 | .94 |
| **VERMONT** | | | |
| 050 | White River Jct. | .89 | .88 |
| 051 | Bellows Falls | .89 | .88 |
| 052 | Bennington | .89 | .88 |
| 053 | Brattleboro | .89 | .88 |
| 054 | Burlington | .89 | .89 |
| 056 | Montpelier | .89 | .89 |
| 057 | Rutland | .89 | .88 |
| 058 | St. Johnsbury | .89 | .89 |
| 059 | Guildhall | .89 | .89 |
| **VIRGINIA** | | | |
| 220, 221 | Fairfax | .94 | .96 |
| 222 | Arlington | .94 | .96 |
| 223 | Alexandria | .94 | .96 |
| 224, 225 | Fredericksburg | .94 | .96 |
| 226 | Winchester | .94 | .96 |
| 227 | Culpeper | .94 | .96 |
| 228 | Harrisonburg | .94 | .96 |
| 229 | Charlottesville | .88 | .87 |
| 230-232 | Richmond | .88 | .86 |
| 233-235 | Norfolk | .85 | .85 |
| 236 | Newport News | .86 | .85 |
| 237 | Portsmouth | .85 | .85 |
| 238 | Petersburg | .88 | .86 |
| 239 | Farmville | .88 | .86 |
| 240, 241 | Roanoke | .86 | .83 |
| 242 | Bristol | .86 | .79 |
| 243 | Pulaski | .86 | .83 |
| 244 | Staunton | .88 | .87 |
| 245 | Lynchburg | .86 | .82 |
| 246 | Grundy | .99 | .99 |
| **WASHINGTON** | | | |
| 980, 981 | Seattle | .99 | 1.05 |
| 982 | Everett | .99 | 1.05 |
| 983, 984 | Tacoma | 1.06 | 1.06 |
| 985 | Olympia | 1.06 | 1.05 |
| 986 | Vancouver | 1.08 | 1.04 |
| 988 | Wenatchee | .99 | 1.04 |
| 989 | Yakima | 1.06 | 1.04 |
| 990-992 | Spokane | 1.04 | 1.06 |
| 993 | Richland | 1.04 | 1.04 |
| 994 | Clarkston | 1.04 | 1.04 |

| STATE/ZIP | CITY | RES. | COMM. |
|---|---|---|---|
| **WEST VIRGINIA** | | | |
| 247, 248 | Bluefield | .99 | .99 |
| 249 | Lewisburg | .99 | .99 |
| 250-253 | Charleston | .99 | .99 |
| 254 | Martinsburg | .94 | .96 |
| 255-257 | Huntington | .97 | .99 |
| 258, 259 | Beckley | .99 | .99 |
| 260 | Wheeling | .97 | .98 |
| 261 | Parkersburg | .97 | .98 |
| 262 | Buckhannon | .99 | .96 |
| 263, 264 | Clarksburg | .99 | .96 |
| 265 | Morgantown | .99 | .96 |
| 266 | Gassaway | .99 | .99 |
| 267 | Romney | .92 | .93 |
| 268 | Petersburg | .99 | .96 |
| **WISCONSIN** | | | |
| 530-532 | Milwaukee | .96 | .95 |
| 534 | Racine | .99 | .93 |
| 535, 537 | Madison | .95 | .94 |
| 538 | Lancaster | .95 | .92 |
| 539 | Portage | .95 | .92 |
| 540 | New Richmond | 1.01 | .92 |
| 541-543 | Green Bay | .95 | .92 |
| 544 | Wausau | .95 | .91 |
| 545 | Rhinelander | .95 | .91 |
| 546 | La Crosse | .95 | .92 |
| 547 | Eau Claire | 1.01 | .92 |
| 548 | Superior | 1.01 | .95 |
| 549 | Oshkosh | .95 | .92 |
| **WYOMING** | | | |
| 820 | Cheyenne | .98 | .98 |
| 821 | Yellowstone Nat'l. Park | .98 | .93 |
| 822 | Wheatland | .98 | .93 |
| 823 | Rawlins | .98 | .93 |
| 824 | Worland | .98 | .93 |
| 825 | Riverton | .98 | .92 |
| 826 | Casper | .98 | .92 |
| 827 | Newcastle | .98 | .92 |
| 828 | Sheridan | .98 | .93 |
| 829-831 | Rock Springs | .98 | .93 |
| **CANADIAN FACTORS** (reflect Canadian currency) | | | |
| **ALBERTA** | | | |
| | Calgary | 1.03 | 1.01 |
| | Edmonton | 1.04 | 1.01 |
| **BRITISH COLUMBIA** | | | |
| | Vancouver | 1.06 | 1.05 |
| **MANITOBA** | | | |
| | Winnipeg | .99 | .99 |
| **NEW BRUNSWICK** | | | |
| | Saint John | .92 | .90 |
| | Moncton | .92 | .90 |
| **NEWFOUNDLAND** | | | |
| | St. John's | .93 | .91 |
| **NOVA SCOTIA** | | | |
| | Halifax | .91 | .90 |
| **ONTARIO** | | | |
| | Hamilton | 1.05 | 1.01 |
| | London | 1.03 | 1.00 |
| | Ottawa | 1.04 | .99 |
| | Sudbury | 1.02 | .99 |
| | Toronto | 1.06 | 1.05 |
| **PRINCE EDWARD ISLAND** | | | |
| | Charlottetown | .92 | .90 |
| **QUEBEC** | | | |
| | Montreal | 1.05 | .97 |
| | Quebec | 1.04 | .96 |
| **SASKATCHEWAN** | | | |
| | Regina | 1.00 | .99 |
| | Saskatoon | 1.00 | .99 |

# Abbreviations

| | | | | | |
|---|---|---|---|---|---|
| A | Area Square Feet; Ampere | Calc | Calculated | D.H. | Double Hung |
| ABS | Acrylonitrile Butadiene Styrene; Asbestos Bonded Steel | Cap. | Capacity | DHW | Domestic Hot Water |
| A.C. | Alternating Current; Air Conditioning; Asbestos Cement | Carp. | Carpenter | Diag. | Diagonal |
| | | C.B. | Circuit Breaker | Diam. | Diameter |
| | | C.C.F. | Hundred Cubic Feet | Distrib. | Distribution |
| | | cd | Candela | Dk. | Deck |
| A.C.I. | American Concrete Institute | cd/sf | Candela per Square Foot | D.L. | Dead Load; Diesel |
| Addit. | Additional | CD | Grade of Plywood Face & Back | Do. | Ditto |
| Adj. | Adjustable | CDX | Plywood, grade C&D, exterior glue | Dp. | Depth |
| af | Audio-frequency | Cefi. | Cement Finisher | D.P.S.T. | Double Pole, Single Throw |
| A.G.A. | American Gas Association | Cem. | Cement | Dr. | Driver |
| Agg. | Aggregate | CF | Hundred Feet | Drink. | Drinking |
| A.H. | Ampere Hours | C.F. | Cubic Feet | D.S. | Double Strength |
| A hr | Ampere-hour | CFM | Cubic Feet per Minute | D.S.A. | Double Strength A Grade |
| A.I.A. | American Institute of Architects | c.g. | Center of Gravity | D.S.B. | Double Strength B Grade |
| AIC | Ampere Interrupting Capacity | CHW | Commercial Hot Water | Dty. | Duty |
| Allow. | Allowance | C.I. | Cast Iron | DWV | Drain Waste Vent |
| alt. | Altitude | C.I.P. | Cast in Place | DX | Deluxe White, Direct Expansion |
| Alum. | Aluminum | Circ. | Circuit | dyn | Dyne |
| a.m. | Ante Meridiem | C.L. | Carload Lot | e | Eccentricity |
| Amp. | Ampere | Clab. | Common Laborer | E | Equipment Only; East |
| Approx. | Approximate | C.L.F. | Hundred Linear Feet | Ea. | Each |
| Apt. | Apartment | CLF | Current Limiting Fuse | Econ. | Economy |
| Asb. | Asbestos | CLP | Cross Linked Polyethylene | EDP | Electronic Data Processing |
| A.S.B.C. | American Standard Building Code | cm | Centimeter | E.D.R. | Equiv. Direct Radiation |
| Asbe. | Asbestos Worker | CMP | Corr. Metal Pipe | Eq. | Equation |
| A.S.H.R.A.E. | American Society of Heating, Refrig. & AC Engineers | C.M.U. | Concrete Masonry Unit | Elec. | Electrician; Electrical |
| | | Col. | Column | Elev. | Elevator; Elevating |
| A.S.M.E. | American Society of Mechanical Engineers | $CO_2$ | Carbon Dioxide | EMT | Electrical Metallic Conduit; Thin Wall Conduit |
| | | Comb. | Combination | | |
| A.S.T.M. | American Society for Testing and Materials | Compr. | Compressor | Eng. | Engine |
| | | Conc. | Concrete | EPDM | Ethylene Propylene Diene Monomer |
| Attchmt. | Attachment | Cont. | Continuous; Continued | | |
| Avg. | Average | | | Eqhv. | Equip. Oper., heavy |
| Bbl. | Barrel | Corr. | Corrugated | Eqlt. | Equip. Oper., light |
| B.&B. | Grade B and Better; Balled & Burlapped | Cos | Cosine | Eqmd. | Equip. Oper., medium |
| | | Cot | Cotangent | Eqmm. | Equip. Oper., Master Mechanic |
| B.&S. | Bell and Spigot | Cov. | Cover | Eqol. | Equip. Oper., oilers |
| B.&W. | Black and White | CPA | Control Point Adjustment | Equip. | Equipment |
| b.c.c. | Body-centered Cubic | Cplg. | Coupling | ERW | Electric Resistant Weld |
| B.F. | Board Feet | C.P.M. | Critical Path Method | Est. | Estimated |
| Bg. Cem. | Bag of Cement | CPVC | Chlorinated Polyvinyl Chloride | esu | Electrostatic Units |
| BHP | Brake Horse Power | C. Pr. | Hundred Pair | E.W. | Each Way |
| B.I. | Black Iron | CRC | Cold Rolled Channel | EWT | Entering Water Temperature |
| Bit.; Bitum. | Bituminous | Creos. | Creosote | Excav. | Excavation |
| Bk. | Backed | Crpt. | Carpet & Linoleum Layer | Exp. | Expansion |
| Bkrs. | Breakers | CRT | Cathode-ray Tube | Ext. | Exterior |
| Bldg. | Building | CS | Carbon Steel | Extru. | Extrusion |
| Blk. | Block | Csc | Cosecant | f. | Fiber stress |
| Bm. | Beam | C.S.F. | Hundred Square Feet | F | Fahrenheit; Female; Fill |
| Boil. | Boilermaker | C.S.I. | Construction Specification Institute | Fab. | Fabricated |
| | | | | FBGS | Fiberglass |
| B.P.M. | Blows per Minute | C.T. | Current Transformer | F.C. | Footcandles |
| BR | Bedroom | CTS | Copper Tube Size | f.c.c. | Face-centered Cubic |
| Brg. | Bearing | Cu | Cubic | f'c. | Compressive Stress in Concrete; Extreme Compressive Stress |
| Brhe. | Bricklayer Helper | Cu. Ft. | Cubic Foot | | |
| Bric. | Bricklayer | cw | Continuous Wave | F.E. | Front End |
| Brk. | Brick | C.W. | Cool White | FEP | Fluorinated Ethylene Propylene (Teflon) |
| Brng. | Bearing | Cwt. | 100 Pounds | | |
| Brs. | Brass | C.W.X. | Cool White Deluxe | F.G. | Flat Grain |
| Brz. | Bronze | C.Y. | Cubic Yard (27 cubic feet) | F.H.A. | Federal Housing Administration |
| Bsn. | Basin | C.Y./Hr. | Cubic Yard per Hour | Fig. | Figure |
| Btr. | Better | Cyl. | Cylinder | Fin. | Finished |
| BTU | British Thermal Unit | d | Penny (nail size) | Fixt. | Fixture |
| BTUH | BTU per Hour | D | Deep; Depth; Discharge | Fl. Oz. | Fluid Ounces |
| BX | Interlocked Armored Cable | Dis.; Disch. | Discharge | Flr. | Floor |
| c | Conductivity | | | F.M. | Frequency Modulation; Factory Mutual |
| C | Hundred; Centigrade | Db. | Decibel | | |
| | | Dbl. | Double | | |
| C/C | Center to Center | DC | Direct Current | Fmg. | Framing |
| Cab. | Cabinet | Demob. | Demobilization | Fndtn. | Foundation |
| Cair. | Air Tool Laborer | d.f.u. | Drainage Fixture Units | Fori. | Foreman, inside |
| | | | | Foro. | Foreman, outside |

# Abbreviations

| | | | | | |
|---|---|---|---|---|---|
| Fount. | Fountain | I.P.S. | Iron Pipe Size | M.C.M. | Thousand Circular Mils |
| FPM | Feet per Minute | I.P.T. | Iron Pipe Threaded | M.C.P. | Motor Circuit Protector |
| FPT | Female Pipe Thread | J | Joule | MD | Medium Duty |
| Fr. | Frame | J.I.C. | Joint Industrial Council | M.D.O. | Medium Density Overlaid |
| F.R. | Fire Rating | K. | Thousand; Thousand Pounds | Med. | Medium |
| FRK | Foil Reinforced Kraft | K.D.A.T. | Kiln Dried After Treatment | MF | Thousand Feet |
| FRP | Fiberglass Reinforced Plastic | kg | Kilogram | M.F.B.M. | Thousand Feet Board Measure |
| FS | Forged Steel | kG | Kilogauss | Mfg. | Manufacturing |
| FSC | Cast Body; Cast Switch Box | kgf | Kilogram force | Mfrs. | Manufacturers |
| Ft. | Foot; Feet | kHz | Kilohertz | mg | Milligram |
| Ftng. | Fitting | Kip. | 1000 Pounds | MGD | Million Gallons per Day |
| Ftg. | Footing | KJ | Kiljoule | MGPH | Thousand Gallons per Hour |
| Ft. Lb. | Foot Pound | K.L. | Effective Length Factor | MH | Manhole; Metal Halide; Man-Hour |
| Furn. | Furniture | Km | Kilometer | MHz | Megahertz |
| FVNR | Full Voltage Non Reversing | K.L.F. | Kips per Linear Foot | Mi. | Mile |
| FXM | Female by Male | K.S.F. | Kips per Square Foot | MI | Malleable Iron; Mineral Insulated |
| Fy. | Minimum Yield Stress of Steel | K.S.I. | Kips per Square Inch | mm | Millimeter |
| g | Gram | K.V. | Kilo Volt | Mill. | Millwright |
| G | Gauss | K.V.A. | Kilo Volt Ampere | Min. | Minimum |
| Ga. | Gauge | K.V.A.R. | Kilovar (Reactance) | Misc. | Miscellaneous |
| Gal. | Gallon | KW | Kilo Watt | ml | Milliliter |
| Gal./Min. | Gallon Per Minute | KWh | Kilowatt-hour | M.L.F. | Thousand Linear Feet |
| Galv. | Galvanized | L | Labor Only; Length; Long | Mo. | Month |
| Gen. | General | Lab. | Labor | Mobil. | Mobilization |
| Glaz. | Glazier | lat | Latitude | Mog. | Mogul Base |
| GPD | Gallons per Day | Lath. | Lather | MPH | Miles per Hour |
| GPH | Gallons per Hour | Lav. | Lavatory | MPT | Male Pipe Thread |
| GPM | Gallons per Minute | lb.; # | Pound | MRT | Mile Round Trip |
| GR | Grade | L.B. | Load Bearing; L Conduit Body | ms | millisecond |
| Gran. | Granular | L. & E. | Labor & Equipment | M.S.F. | Thousand Square Feet |
| Grnd. | Ground | lb./hr. | Pounds per Hour | Mstz. | Mosaic & Terrazzo Worker |
| H | High; High Strength Bar Joist; | lb./L.F. | Pounds per Linear Foot | M.S.Y. | Thousand Square Yards |
| | Henry | lbf/sq in. | Pound-force per Square Inch | Mtd. | Mounted |
| H.C. | High Capacity | L.C.L. | Less than Carload Lot | Mthe. | Mosaic & Terrazzo Helper |
| H.D. | Heavy Duty; High Density | Ld. | Load | Mtng. | Mounting |
| H.D.O. | High Density Overlaid | L.F. | Linear Foot | Mult. | Multi; Multiply |
| Hdr. | Header | Lg. | Long; Length; Large | MV | Megavolt |
| Hdwe. | Hardware | L. & H. | Light and Heat | MW | Megawatt |
| Help. | Helper average | L.H. | Long Span High Strength Bar Joist | MXM | Male by Male |
| HEPA | High Efficiency Particulate Air Filter | L.J. | Long Span Standard Strength | MYD | Thousand yards |
| Hg | Mercury | | Bar Joist | N | Natural; North |
| H.O. | High Output | L.L. | Live Load | nA | nanoampere |
| Horiz. | Horizontal | L.L.D. | Lamp Lumen Depreciation | NA | Not Available; Not Applicable |
| H.P. | Horsepower; High Pressure | lm | Lumen | N.B.C. | National Building Code |
| H.P.F. | High Power Factor | lm/sf | Lumen per Square Foot | NC | Normally Closed |
| Hr. | Hour | lm/W | Lumen Per Watt | N.E.M.A. | National Electrical |
| Hrs./Day | Hours Per Day | L.O.A. | Length Over All | | Manufacturers Association |
| HSC | High Short Circuit | log | Logarithm | NEHB | Bolted Circuit Breaker to 600V. |
| Ht. | Height | L.P. | Liquefied Petroleum; | N.L.B. | Non-Load-Bearing |
| Htg. | Heating | | Low Pressure | nm | nanometer |
| Htrs. | Heaters | L.P.F. | Low Power Factor | No. | Number |
| HVAC | Heating, Ventilating & | Lt. | Light | NO | Normally Open |
| | Air Conditioning | Lt. Ga. | Light Gauge | N.O.C. | Not Otherwise Classified |
| Hvy. | Heavy | L.T.L. | Less than Truckload Lot | Nose. | Nosing |
| HW | Hot Water | Lt. Wt. | Lightweight | N.P.T. | National Pipe Thread |
| Hyd.; | | L.V. | Low Voltage | NQOB | Bolted Circuit Breaker to 240V. |
| Hydr. | Hydraulic | M | Thousand; Material; Male; | N.R.C. | Noise Reduction Coefficient |
| Hz. | Hertz (cycles) | | Medium Wall Copper | N.R.S. | Non Rising Stem |
| I. | Moment of Inertia | m/hr | Manhour | ns | nanosecond |
| I.C. | Interrupting Capacity | mA | Milliampere | nW | nanowatt |
| ID | Inside Diameter | Mach. | Machine | OB | Opposing Blade |
| I.D. | Inside Dimension; | Mag. Str. | Magnetic Starter | OC | On Center |
| | Identification | Maint. | Maintenance | OD | Outside Diameter |
| I.F. | Inside Frosted | Marb. | Marble Setter | O.D. | Outside Dimension |
| I.M.C. | Intermediate Metal Conduit | Mat. | Material | ODS | Overhead Distribution System |
| In. | Inch | Mat'l. | Material | O & P | Overhead and Profit |
| Incan. | Incandescent | Max. | Maximum | Oper. | Operator |
| Incl. | Included; Including | MBF | Thousand Board Feet | Opng. | Opening |
| Int. | Interior | MBH | Thousand BTU's per hr. | Orna. | Ornamental |
| Inst. | Installation | M.C.F. | Thousand Cubic Feet | O.S.&Y. | Outside Stem and Yoke |
| Insul. | Insulation | M.C.F.M. | Thousand Cubic Feet | Ovhd | Overhead |
| I.P. | Iron Pipe | | per Minute | Oz. | Ounce |

# Abbreviations

| | | | | | |
|---|---|---|---|---|---|
| P. | Pole; Applied Load; Projection | S. | Suction; Single Entrance; South | T.S. | Trigger Start |
| p. | Page | | | Tr. | Trade |
| Pape. | Paperhanger | Scaf. | Scaffold | Transf. | Transformer |
| PAR | Weatherproof Reflector | Sch.; Sched. | Schedule | Trhv. | Truck Driver, Heavy |
| Pc. | Piece | | | Trlr. | Trailer |
| P.C. | Portland Cement; Power Connector | S.C.R. | Modular Brick | Trlt. | Truck Driver, Light |
| | | S.D.R. | Standard Dimension Ratio | TV | Television |
| P.C.F. | Pounds per Cubic Foot | S.E. | Surfaced Edge | T.W. | Thermoplastic Water Resistant Wire |
| P.E. | Professional Engineer; Porcelain Enamel; Polyethylene; Plain End | S.E.R.; S.E.U. | Service Entrance Cable | UCI | Uniform Construction Index |
| | | S.F. | Square Foot | UF | Underground Feeder |
| Perf. | Perforated | S.F.C.A. | Square Foot Contact Area | U.H.F. | Ultra High Frequency |
| Ph. | Phase | S.F.G. | Square Foot of Ground | U.L. | Underwriters Laboratory |
| P.I. | Pressure Injected | S.F. Hor. | Square Foot Horizontal | Unfin. | Unfinished |
| Pile. | Pile Driver | S.F.R. | Square Feet of Radiation | URD | Underground Residential Distribution |
| Pkg. | Package | S.F.Shlf. | Square Foot of Shelf | | |
| Pl. | Plate | S4S | Surface 4 Sides | V | Volt |
| Plah. | Plasterer Helper | Shee. | Sheet Metal Worker | VA | Volt/amp |
| Plas. | Plasterer | Sin | Sine | V.A.T. | Vinyl Asbestos Tile |
| Pluh. | Plumbers Helper | Skwk. | Skilled Worker | VAV | Variable Air Volume |
| Plum. | Plumber | SL | Saran Lined | Vent. | Ventilating |
| Ply. | Plywood | S.L. | Slimline | Vert. | Vertical |
| p.m. | Post Meridiem | Sldr. | Solder | V.G. | Vertical Grain |
| Pord. | Painter, Ordinary | S.N. | Solid Neutral | V.H.F. | Very High Frequency |
| pp | Pages | S.P. | Static Pressure; Single Pole; Self-Propelled | VHO | Very High Output |
| PP; PPL | Polypropylene | | | Vib. | Vibrating |
| P.P.M. | Parts Per Million | Spri. | Sprinkler Installer | V.L.F. | Vertical Linear Foot |
| Pr. | Pair | Sq. | Square; 100 square feet | Vol. | Volume |
| Prefab. | Prefabricated | S.P.D.T. | Single Pole, Double Throw | W | Wire; Watt; Wide; West |
| Prefin. | Prefinished | S.P.S.T. | Single Pole, Single Throw | w/ | With |
| Prop. | Propelled | SPT | Standard Pipe Thread | W.C. | Water Column; Water Closet |
| PSF; psf | Pounds per Square Foot | Sq. Hd. | Square Head | W.F. | Wide Flange |
| PSI; psi | Pounds per Square Inch | S.S. | Single Strength; Stainless Steel | W.G. | Water Gauge |
| PSIG | Pounds per Square Inch Gauge | S.S.B. | Single Strength B Grade | Wldg. | Welding |
| PSP | Plastic Sewer Pipe | Sswk. | Structural Steel Worker | Wrck. | Wrecker |
| Pspr. | Painter, Spray | Sswl. | Structural Steel Welder | W.S.P. | Water, Steam, Petroleum |
| Psst. | Painter, Structural Steel | St.; Stl. | Steel | WT, Wt. | Weight |
| P.T. | Potential Transformer | S.T.C. | Sound Transmission Coefficient | WWF | Welded Wire Fabric |
| P. & T. | Pressure & Temperature | Std. | Standard | XFMR | Transformer |
| Ptd. | Painted | STP | Standard Temperature & Pressure | XHD | Extra Heavy Duty |
| Ptns. | Partitions | Stpi. | Steamfitter; Pipefitter | Y | Wye |
| Pu | Ultimate Load | Str. | Strength; Starter; Straight | yd | Yard |
| PVC | Polyvinyl Chloride | Strd. | Stranded | yr | Year |
| Pvmt. | Pavement | Struct. | Structural | Δ | Delta |
| Pwr. | Power | Sty. | Story | % | Percent |
| Q | Quantity Heat Flow | Subj. | Subject | ~ | Approximately |
| Quan.; Qty. | Quantity | Subs. | Subcontractors | ∅ | Phase |
| Q.C. | Quick Coupling | Surf. | Surface | @ | At |
| r | Radius of Gyration | Sw. | Switch | # | Pound; Number |
| R | Resistance | Swbd. | Switchboard | < | Less Than |
| R.C.P. | Reinforced Concrete Pipe | S.Y. | Square Yard | > | Greater Than |
| Rect. | Rectangle | Syn. | Synthetic | | |
| Reg. | Regular | Sys. | System | | |
| Reinf. | Reinforced | t. | Thickness | | |
| Req'd. | Required | T | Temperature; Ton | | |
| Resi | Residential | Tan | Tangent | | |
| Rgh. | Rough | T.C. | Terra Cotta | | |
| R.H.W. | Rubber, Heat & Water Resistant; Residential Hot Water | T.D. | Temperature Difference | | |
| | | TFE | Tetrafluoroethylene (Teflon) | | |
| rms | Root Mean Square | T. & G. | Tongue & Groove; Tar & Gravel | | |
| Rnd. | Round | | | | |
| Rodm. | Rodman | Th.; Thk. | Thick | | |
| Rofc. | Roofer, Composition | Thn. | Thin | | |
| Rofp. | Roofer, Precast | Thrded | Threaded | | |
| Rohe. | Roofer Helpers (Composition) | Tilf. | Tile Layer, Floor | | |
| Rots. | Roofer, Tile & Slate | Tilh. | Tile Layer Helper | | |
| R.O.W. | Right of Way | THW | Insulated Strand Wire | | |
| RPM | Revolutions per Minute | THWN; THHN | Nylon Jacketed Wire | | |
| R.R. | Direct Burial Feeder Conduit | | | | |
| R.S. | Rapid Start | T.L. | Truckload | | |
| RT | Round Trip | Tot. | Total | | |

# INDEX

## A

| | |
|---|---|
| Abandon catch basin | 272 |
| ABC extinguisher | 387 |
| Abrasive floor tile | 356 |
| tread | 356 |
| ABS DWV pipe | 379 |
| Access door basement | 284 |
| door ceiling | 329 |
| door duct | 392 |
| door floor | 338 |
| door roof | 329 |
| Accessory bath | 151 |
| bathroom | 366 |
| door | 347 |
| drainage | 326 |
| drywall | 353 |
| duct | 392 |
| Accordion door | 336 |
| Acidproof floor | 287 |
| Acoustical ceiling | 356 |
| door | 338, 369 |
| folding partition | 369 |
| panel | 357 |
| partition | 369 |
| sealant | 315, 354 |
| wallboard | 354 |
| Acrylic carpet | 142, 358 |
| caulking | 315 |
| ceiling | 357 |
| sign | 372 |
| wall coating | 365 |
| wallcovering | 366 |
| wood block | 360 |
| Adhesive cement | 360 |
| floor | 359 |
| roof | 321 |
| wallpaper | 366 |
| Adjustable shelf | 371 |
| Admixture masonry | 285 |
| Aerial lift | 268 |
| Aggregate exposed | 283 |
| spreader | 265 |
| stone | 280 |
| Air cleaner electronic | 391 |
| compressor | 268 |
| conditioning | 186, 188 |
| conditioning wine vault | 373 |
| extractor | 392 |
| hose | 268 |
| spade | 268 |
| tool | 268 |
| vent roof | 317 |
| Air-conditioner cooling & heating | 394 |
| packaged terminal | 394 |
| receptacle wiring | 403 |
| rooftop | 394 |
| self-contained | 394 |
| thru-wall | 394 |
| Air-conditioning direct expansion | 394 |
| fan-coil | 394 |
| ventilating | 393-394 |
| Air-filter | 391 |
| roll type | 392 |
| Air-return grille | 392 |
| Air-supply register | 392 |
| Alarm burglar | 402 |
| All fuel vent-chimney | 391 |
| Aluminum ceiling tile | 356 |
| chain link | 279 |
| coping | 286 |
| diffuser perforated | 392 |
| door | 332, 338-339 |
| door frame | 332 |
| downspout | 325 |
| ductwork | 393 |
| flashing | 121, 123, 326 |
| foil | 315-316, 366 |
| gravel stop | 327 |
| grille | 392 |
| gutter | 327 |
| nail | 295 |
| pipe | 276 |
| push pull plate | 348 |
| rail | 293 |
| reglet | 328 |
| rivet | 295 |
| service entrance cable | 397 |
| sheet metal | 325 |
| shingle | 318 |
| siding | 70, 320 |
| siding accessories | 320 |
| sliding door | 338 |
| stair | 293 |
| storefront | 341 |
| storm door | 333 |
| tile | 319, 355 |
| tube frame | 350 |
| window | 342 |
| Anchor expansion | 294 |
| hollow wall | 294 |
| joist | 296 |
| nailing | 294 |
| stone | 286 |
| Angle valve | 384 |
| Antenna system | 403 |
| TV | 403 |
| Apartment call system | 402 |
| Appliance | 150, 372 |
| receptacle | 403 |
| residential | 372 |
| wiring | 403 |
| Apron wood | 308 |
| Arch laminated | 313 |
| radial | 314 |
| Architectural equipment | 372 |
| fee | 264 |
| Armored cable | 396 |
| Asbestos base sheet | 321 |
| cement partition | 369 |
| cement pipe | 275-276 |
| felt | 320-322 |
| shingle | 319 |
| Ash receiver | 367 |
| Asphalt block | 277 |
| block floor | 277 |
| coating | 314 |
| curb | 277 |
| distributor | 268 |
| driveway | 14 |
| felt | 321-322 |
| flashing | 326 |
| flood coat | 321 |
| floor tile | 358 |
| paper | 315 |
| paver | 269 |
| primer | 360 |
| sheathing | 304 |
| shingle | 319 |
| sidewalk | 12, 278 |
| tile | 143 |
| Astragal | 349 |
| molding | 308 |
| Atomizing humidifier | 389 |
| Atrium | 156 |
| Attic stair | 368 |
| ventilation fan | 393 |
| Auger | 265 |
| hole | 272 |
| Automatic opener | 347 |
| Awning | 367 |
| canvas | 368 |
| window | 78, 344 |

## B

| | |
|---|---|
| Back-up block | 288 |
| Backfill | 273-274 |
| trench | 274 |
| Backflow preventer | 375 |
| Backhoe | 273 |
| excavation | 274 |
| rental | 265 |
| Backsplash counter top | 306 |
| Baffle roof | 317 |
| Baked enamel doors | 331 |
| enamel frame | 330 |
| Balanced door | 341 |
| Balcony rail | 293 |
| Ball wrecking | 270 |
| Ballast roof | 323 |
| Balled & burlapped tree | 281 |
| Baluster | 145, 312 |
| birch | 145 |
| pine | 145 |
| Balustrades painting | 362 |
| Band molding | 308 |
| Bankrun gravel | 273 |
| Bar grab | 366 |
| panic | 348 |
| towel | 367 |
| zee | 353 |
| Barrel | 268 |
| Barricade | 268 |
| Base cabinet | 305 |
| carpet | 358 |
| column | 296 |
| course | 277 |
| cove | 359 |
| metal | 355 |
| quarry tile | 355 |

421

# INDEX

| | |
|---|---|
| resilient | 359 |
| road | 277 |
| sheet | 321-322 |
| sink | 305 |
| terrazzo | 356 |
| vanity | 313 |
| wood | 308 |
| Baseboard heater electric | 402 |
| heating | 197, 389 |
| register | 392 |
| Basement stair | 145, 284 |
| Basketweave fence | 279 |
| Bath | 384 |
| accessory | 151 |
| communal | 384 |
| faucet | 375 |
| steam | 373 |
| whirlpool | 384 |
| Bathroom | 162, 164, 166, 168, 170, 172, 174, 176, 178 |
| accessory | 366 |
| fixture | 384-386 |
| Batt insulation | 72 |
| Bay window | 82, 344 |
| Bead board | 317 |
| board insulation | 317 |
| casing | 353 |
| corner | 353 |
| parting | 309 |
| Beam bondcrete | 353 |
| ceiling | 305 |
| drywall | 353-354 |
| fireproofing | 353 |
| hanger | 296 |
| laminated | 313-314 |
| mantel | 307 |
| plaster | 352-353 |
| steel | 292 |
| wood | 301 |
| Beams wood | 297 |
| Bed molding | 308 |
| plant | 281 |
| Beech tread | 312 |
| Bell & spigot pipe | 376 |
| Berm paving | 277 |
| Bevel siding | 311 |
| Bi-fold door | 138, 332, 335 |
| Bi-passing closet doors | 335 |
| door | 138 |
| Birch door | 335-336 |
| molding | 309 |
| paneling | 310 |
| stair | 312 |
| wood frame | 334 |
| Bituminous block | 277 |
| coating | 314 |
| curb | 277 |
| fiber pipe | 276 |
| paver | 269 |
| sidewalk | 278 |
| Blackboard | 368 |
| Blanket insulation | 318, 389 |
| sound attenuation | 357 |
| Blind | 99, 373 |

| | |
|---|---|
| exterior | 310-311 |
| Block asphalt | 277 |
| asphalt floor | 277 |
| back-up | 288 |
| bituminous | 277 |
| concrete | 288 |
| concrete foundation | 290 |
| decorative concrete | 288-289 |
| filler | 364 |
| floor | 360 |
| floor wood | 360 |
| grooved | 288 |
| ground face | 289 |
| profile | 289 |
| scored | 289 |
| slump | 289 |
| wall | 62 |
| wall foundation | 24 |
| Blocking | 296 |
| wood | 303 |
| Blockout slab | 282 |
| Blown-in cellulose | 316 |
| fiberglass | 316 |
| insulation | 316 |
| Blueboard | 354 |
| Bluestone sidewalk | 278 |
| sill | 288 |
| step | 278 |
| Board and batten siding | 311 |
| bead | 317 |
| bulletin | 367 |
| ceiling | 357 |
| directory | 368 |
| fence | 279 |
| insulation | 317 |
| paneling | 309 |
| sheathing | 304 |
| siding | 66 |
| valance | 306 |
| verge | 308 |
| Boiler electric | 387 |
| electric steam | 387 |
| gas-fired | 387 |
| gas/oil combination | 388 |
| hot-water | 387-388 |
| oil-fired | 387 |
| steam | 387-388 |
| Bolt | 293 |
| dead | 348 |
| expansion | 294 |
| toggle | 294 |
| Bolt-on circuit breaker | 399 |
| Bond roof | 322 |
| Bondcrete | 353 |
| Bookcase | 310 |
| Boom lift | 268 |
| Booth telephone | 372 |
| Boring | 272 |
| Borrow | 273 |
| Bottle storage | 373 |
| Bow window | 82, 344 |
| Bowstring truss | 314 |
| Box distribution | 276 |
| mail | 369 |

| | |
|---|---|
| out | 282 |
| pull | 397 |
| stair | 312 |
| termination | 403 |
| Boxes & wiring devices | 397 |
| Bracing let-in | 297 |
| Brass hinge | 347-348 |
| pipe | 375 |
| pipe-fitting | 376 |
| screw | 296 |
| Brazed wire to grounding | 397 |
| Breaker circuit | 398 |
| vacuum | 383 |
| Brick | 287 |
| anchor | 285 |
| cart | 268 |
| driveway | 14 |
| edging | 280 |
| forklift | 269 |
| molding | 308 |
| paving | 277 |
| removal | 273 |
| shelf | 282 |
| sidewalk | 12, 278 |
| sill | 288 |
| step | 278 |
| veneer | 64 |
| wall | 64, 288 |
| Bridging | 297 |
| Bronze plaque | 372 |
| push-pull plate | 348 |
| Broom cabinet | 306 |
| finish | 283 |
| Brown coat | 353 |
| Brownstone | 291 |
| Brush clearing | 272 |
| cutter | 265 |
| Buck rough | 298 |
| Bucket concrete | 267 |
| crane | 265 |
| Buggy concrete | 267, 284 |
| Builder risk insurance | 264 |
| Building directory | 403 |
| excavation | 6, 8 |
| greenhouse | 374 |
| hardware | 347 |
| paper | 315 |
| permit | 264 |
| prefabricated | 374 |
| Built-up roof | 321 |
| roofing | 120 |
| Bulkhead door | 338 |
| formwork | 282 |
| Bulldozer | 266, 273 |
| Bulletin board | 367 |
| Bullnose block | 290 |
| Bumper dock | 297 |
| wall | 347, 366 |
| Burglar alarm | 402 |
| Burlap curing | 283 |
| rubbing | 283 |
| Burner gas conversion | 388 |
| gun type | 388 |
| oil | 388 |

422

# INDEX

residential . . . . . . . . . . . . . . . . . . 388
Bush hammer concrete . . . . . . . . 284
Butyl caulking . . . . . . . . . . . . . . . . 315
      flashing . . . . . . . . . . . . . . . . . . 327
Bypassing closet door . . . . . 138, 335

## C

Cabinet . . . . . . . . . . . . . . . . . . . . . 148
    broom . . . . . . . . . . . . . . . . . . . 306
    current transformer . . . . . . . . . 397
    electrical . . . . . . . . . . . . . . . . . 397
    hotel . . . . . . . . . . . . . . . . . . . . 369
    kitchen . . . . . . . . . . . . . . . . . . 305
    medicine . . . . . . . . . . . . . . . . . 369
    oven . . . . . . . . . . . . . . . . . . . . 306
    painting . . . . . . . . . . . . . . . . . 360
    shower . . . . . . . . . . . . . . . . . . 370
    stain . . . . . . . . . . . . . . . . . . . . 360
    strip . . . . . . . . . . . . . . . . . . . . 403
    transformer . . . . . . . . . . . . . . 397
    varnish . . . . . . . . . . . . . . . . . . 361
Cable armored . . . . . . . . . . . . . . . 396
    electric . . . . . . . . . . . . . . . . . . 396
    jack . . . . . . . . . . . . . . . . . . . . . 271
    sheathed nonmetallic . . . . . . . 396
    sheathed romex . . . . . . . . . . . 396
Cafe door . . . . . . . . . . . . . . . . . . . 334
Call system apartment . . . . . . . . . 402
Canopy . . . . . . . . . . . . . . . . . . . . . 367
    door . . . . . . . . . . . . . . . . . . . . 367
    entrance . . . . . . . . . . . . . . . . . 367
    framing . . . . . . . . . . . . . . . . . 300
Cant roof . . . . . . . . . . . . . . . 299, 322
Canvas awning . . . . . . . . . . 367-368
Cap service entrance . . . . . . . . . . 397
Carbon dioxide extinguisher . . . . 387
Carpentry crew . . . . . . . . . . . . . . 262
Carpet . . . . . . . . . . . . . . . . . 142, 358
    base . . . . . . . . . . . . . . . . . . . . 358
    floor . . . . . . . . . . . . . . . . . . . . 358
    padding . . . . . . . . . . . . . . . . . 358
    tile . . . . . . . . . . . . . . . . . . . . . 358
Carrier ceiling . . . . . . . . . . . . . . . 352
    channel . . . . . . . . . . . . . . . . . 357
    fixture . . . . . . . . . . . . . . . . . . 383
Cart brick . . . . . . . . . . . . . . . . . . . 268
    concrete . . . . . . . . . . . . . 267, 284
Cased boring . . . . . . . . . . . . . . . . 272
Casement window . . . . . . 76, 344-345
Casework ground . . . . . . . . . . . . 302
    painting . . . . . . . . . . . . . . . . . 360
    stain . . . . . . . . . . . . . . . . . . . . 360
    varnish . . . . . . . . . . . . . . . . . . 361
Casing bead . . . . . . . . . . . . . . . . . 353
    door . . . . . . . . . . . . . . . . . . . . 353
    wood . . . . . . . . . . . . . . . . . . . 308
Cast in place concrete . . . . . . . . . 283
    in place terrazzo . . . . . . . . . . . 356
    iron damper . . . . . . . . . . . . . . 285
    iron stair . . . . . . . . . . . . . . . . 293
    stone floor . . . . . . . . . . . . . . . 287
Cast-iron drain . . . . . . . . . . . . . . . 375
    pipe . . . . . . . . . . . . . . . . . . . . 376

pipe-fitting . . . . . . . . . . . . 376, 381
radiator . . . . . . . . . . . . . . . . . . 389
trap . . . . . . . . . . . . . . . . . . . . . 383
Catch basin . . . . . . . . . . . . . . . . . 275
    basin removal . . . . . . . . . . . . 272
Caulking . . . . . . . . . . . . . . . . . . . 315
    sealant . . . . . . . . . . . . . . . . . . 315
Cavity wall reinforcing . . . . . . . . 285
Cedar closet . . . . . . . . . . . . . . . . . 310
    fence . . . . . . . . . . . . . . . . 19, 279
    molding . . . . . . . . . . . . . . . . . 308
    paneling . . . . . . . . . . . . . . . . . 309
    post . . . . . . . . . . . . . . . . . . . . 306
    roof deck . . . . . . . . . . . . . . . . 314
    roof plank . . . . . . . . . . . . . . . 303
    shake siding . . . . . . . . . . . . . . . 68
    shingle . . . . . . . . . . . . . . . . . . 320
    siding . . . . . . . . . . . . . . . . . . . 311
    stair . . . . . . . . . . . . . . . . . . . . 313
Ceiling . . . . . . . . . . . . . . . . . 128, 132
    acoustical . . . . . . . . . . . . . . . 356
    beam . . . . . . . . . . . . . . . . . . . 305
    board . . . . . . . . . . . . . . . . . . . 357
    bondcrete . . . . . . . . . . . . . . . . 353
    carrier . . . . . . . . . . . . . . . . . . 352
    diffuser . . . . . . . . . . . . . . . . . 392
    drill . . . . . . . . . . . . . . . . . . . . 294
    drywall . . . . . . . . . . . . . . . . . . 354
    eggcrate . . . . . . . . . . . . . . . . . 357
    framing . . . . . . . . . . . . . . . . . 297
    furring . . . . . . . . . . . . . . 302, 352
    hatch . . . . . . . . . . . . . . . . . . . 329
    insulation . . . . . . . . . . . . . . . . 316
    lath . . . . . . . . . . . . . . . . . . . . 352
    luminous . . . . . . . . . . . . 357, 401
    molding . . . . . . . . . . . . . . . . . 308
    painting . . . . . . . . . . . . . 363-364
    plaster . . . . . . . . . . . . . . . 352-353
    stair . . . . . . . . . . . . . . . . . . . . 369
    suspended . . . . . . . . 134, 352, 357
    suspension system . . . . . . . . . 357
    tile . . . . . . . . . . . . . . . . . . . . . 357
Cellar door . . . . . . . . . . . . . . . . . 338
Cellulose blown-in . . . . . . . . . . . 316
    insulation . . . . . . . . . . . . . . . . 316
Cement adhesive . . . . . . . . . . . . . 360
    flashing . . . . . . . . . . . . . . . . . 315
    gypsum . . . . . . . . . . . . . . . . . 285
    masonry . . . . . . . . . . . . . . . . . 285
    masonry unit . . . . . . . . . 288-290
    mortar . . . . . . . . . . . . . . . . . . 355
Centrifugal pump . . . . . . . . . . . . 269
Ceramic coating . . . . . . . . . . . . . 365
    tile . . . . . . . . . . . . . . . . . 143, 355
    tile counter top . . . . . . . . . . . 307
    tile floor . . . . . . . . . . . . . . . . 355
Chain hoist . . . . . . . . . . . . . . . . . 271
    hoist door . . . . . . . . . . . . . . . 339
    link fence . . . . . . . . . . 18, 278-279
    link fence removal . . . . . . . . . 273
    saw . . . . . . . . . . . . . . . . . . . . 270
    trencher . . . . . . . . . . . . . . . . . 267
Chair molding . . . . . . . . . . . . . . . 308
Chalk tray . . . . . . . . . . . . . . . . . . 368

Chalkboard . . . . . . . . . . . . . . . . . 368
    frame . . . . . . . . . . . . . . . . . . 368
Channel carrier . . . . . . . . . . . . . . 357
    furring . . . . . . . . . . . . . . 352-353
    siding . . . . . . . . . . . . . . . . . . . 311
Charge dump . . . . . . . . . . . . . . . . 272
Chemical dry extinguisher . . . . . . 387
    toilet . . . . . . . . . . . . . . . . . . . 270
Chiller portable . . . . . . . . . . . . . . 373
Chime door . . . . . . . . . . . . . . . . . 402
Chimney brick . . . . . . . . . . . . . . . 286
    foundation . . . . . . . . . . . . . . . 283
    prefabricated . . . . . . . . . . . . . 286
    screen . . . . . . . . . . . . . . . . . . 285
    simulated brick . . . . . . . . . . . 287
    vent . . . . . . . . . . . . . . . . . . . . 391
    vent fitting . . . . . . . . . . . . . . . 391
China cabinet . . . . . . . . . . . . . . . 305
Chipper brush . . . . . . . . . . . . . . . 265
Chipping hammer . . . . . . . . . . . . 268
Chute linen . . . . . . . . . . . . . . . . . 368
    mail . . . . . . . . . . . . . . . . . . . . 368
    refuse . . . . . . . . . . . . . . . . . . 368
Circline fixture . . . . . . . . . . . . . . 401
Circuit breaker bolt-on . . . . . . . . 399
    electric dryer . . . . . . . . . . . . . 403
    wiring . . . . . . . . . . . . . . . . . . 198
    wiring heater . . . . . . . . . . . . . 404
    wiring water-heater . . . . . . . . 404
Circuit-breaker . . . . . . . . . . . . . . . 398
Circular saw . . . . . . . . . . . . . . . . 270
Circulating pump . . . . . . . . . . . . . 385
Clamp water pipe ground . . . . . . 397
Clamshell bucket . . . . . . . . . . . . . 265
Clapboard painting . . . . . . . . . . . 363
Clay pipe . . . . . . . . . . . . . . . . . . . 276
    roofing tile . . . . . . . . . . . . . . 319
    tile coping . . . . . . . . . . . . . . . 286
Cleaner steam . . . . . . . . . . . . . . . 270
Cleaning up . . . . . . . . . . . . . . . . . 264
Cleanout door . . . . . . . . . . . . . . . 285
    pipe . . . . . . . . . . . . . . . . . . . . 375
    tee . . . . . . . . . . . . . . . . . . . . . 375
Clear & grub . . . . . . . . . . . . . . . . 272
Clearing brush . . . . . . . . . . . . . . . 272
Climbing crane . . . . . . . . . . . . . . 270
    jack . . . . . . . . . . . . . . . . . . . . 271
Clinton cloth . . . . . . . . . . . . . . . . 353
Clip plywood . . . . . . . . . . . . . . . . 296
Closer door . . . . . . . . . . . . . . . . . 347
Closet . . . . . . . . . . . . . . . . . . . . . 369
    cedar . . . . . . . . . . . . . . . . . . . 310
    door . . . . . . . . . . 138, 332, 335-336
    pole . . . . . . . . . . . . . . . . . . . . 309
    rod . . . . . . . . . . . . . . . . . . . . . 310
Cloth clinton . . . . . . . . . . . . . . . . 353
Coal tar pitch . . . . . . . . . . . . . . . . 321
Coat brown . . . . . . . . . . . . . . . . . 353
    glaze . . . . . . . . . . . . . . . . . . . 363
    scratch . . . . . . . . . . . . . . . . . . 353
Coating bituminous . . . . . . . . . . . 314
    ceramic . . . . . . . . . . . . . . . . . 365
    epoxy . . . . . . . . . . . . . . . . . . . 365
    flood . . . . . . . . . . . . . . . . . . . 322

423

# INDEX

| | | |
|---|---|---|
| silicone .................... 315 | painting .................... 363 | wall covering ................ 366 |
| wall ..................... 365 | paver ..................... 269 | wire...................... 396 |
| waterproofing ............ 315 | paving .................... 277 | Core drill .................... 267 |
| Coffee maker ................ 366 | pile ...................... 275 | Cork floor ................... 359 |
| Cold roofing ................. 322 | pipe ..................... 276 | tile...................... 359 |
| Colonial door............. 335-336 | place .................... 37 | wall tile ................. 366 |
| wood frame ............. 333 | placing .................. 284 | Corner base cabinet ........... 305 |
| Column ................ 291, 306 | protection ............... 284 | bead .................... 353 |
| base .................... 296 | pump .................... 267 | guard ................... 360 |
| bondcrete ............... 353 | ready mix ............... 283 | wall cabinet ............. 306 |
| brick ................... 286 | reinforcing .............. 282 | Cornice ..................... 286 |
| drywall .................. 354 | saw ..................... 267 | drain .................... 375 |
| fireproofing ............. 353 | septic tank ............... 276 | molding ................. 308 |
| lally .................... 291 | shingle................... 319 | painting ................. 362 |
| laminated wood ......... 314 | sidewalk ................. 12 | Correspondence lift........... 374 |
| lath .................... 352 | sill ...................... 288 | Corrugated metal pipe ...... 276 |
| pipe .................... 292 | slab ..................... 30 | roof tile ................. 319 |
| plaster ............... 352-353 | spreader ................. 269 | siding ................... 324 |
| wood ................ 301, 306 | trowel ................... 267 | Counter door ................ 338 |
| Columns wood ............... 298 | utility vault .............. 277 | flashing ................. 329 |
| Combination storm door ... 335-336 | vibrator ................. 267 | top ..................... 306-307 |
| Commercial door ....... 331, 338 | wheeling ................. 284 | top sink ................. 386 |
| folding partition .......... 369 | Conductive floor............. 359 | Coupling PVC .............. 379 |
| steel window ............ 342 | Conductor & grounding ...... 397 | Course base ................. 277 |
| Common brick .............. 286 | wire...................... 396 | Cove base .................. 359 |
| nail .................... 295 | Conduit & fitting flexible ...... 399 | base ceramic tile.......... 355 |
| rafter roof framing ........ 40 | electrical ................. 394 | base terrazzo ............. 356 |
| Communal bath ............. 384 | fitting ................... 395 | molding ................. 308 |
| Compact fill ................ 274 | flexible metallic........... 395 | scotia molding............ 308 |
| Compaction ................. 273 | greenfield ................ 395 | Cover ground ............... 280 |
| soil .................... 273 | in-slab .................. 395 | Covering wall................ 366 |
| Compactor ................. 150 | in-slab PVC .............. 395 | CPE roof ................... 322 |
| earth ................... 266 | in-trench electrical......... 395 | CPVC pipe ................. 379 |
| residential ............... 372 | in-trench steel ............ 395 | Crane ....................... 270 |
| Component sound system ..... 402 | intermediate steel ......... 394 | bucket ................... 265 |
| Composite door ............. 331 | rigid in-slab ............. 395 | climbing ................. 270 |
| Composition floor ........... 358 | Connection motor ........... 399 | crawler................... 270 |
| Compressor air .............. 268 | Connector flexible ........... 375 | hydraulic................. 271 |
| Concrete block .......... 288-290 | joist..................... 296 | material handling ......... 270 |
| block decorative ........ 288-289 | timber .................. 295 | tower ................... 270 |
| block foundation ......... 290 | Contractor equipment......... 265 | Creosote floor ................ 360 |
| block insulation ........... 316 | pump .................... 269 | lumber .................. 302 |
| block painting ............ 364 | Control-component............ 392 | Crew carpentry ............... 262 |
| block wall ............... 62 | Conveyor ................... 267 | concrete ................. 261 |
| bucket .................. 267 | Cooking equipment........... 372 | conveying system......... 263 |
| buggy................... 284 | range .................... 372 | electrical ................. 263 |
| bush hammer............. 284 | Cooling ................ 186, 188 | equipment ............... 262 |
| cart .................. 267, 284 | & heating air-conditioner ..... 394 | finishing ................. 262 |
| cast in place ............. 283 | electric furnace ............ 388 | general condition ......... 259 |
| conveyer ................ 267 | Coping................... 286, 291 | glazing................... 262 |
| coping .................. 286 | clay tile .................. 286 | masonry ................. 261 |
| crew .................... 261 | terra cotta ............... 286 | mechanical system ........ 263 |
| curb ................. 277, 283 | Copper cable ............... 396 | metal ................... 262 |
| curing .................. 283 | downspout ............... 325 | moisture protection ....... 262 |
| driveway ................ 14 | drum trap ............... 383 | site work ............. 259-260 |
| filled column ............. 291 | DWV tubing .............. 377 | specialty ................. 262 |
| finish ................ 37, 277 | flashing ................. 326 | survey ................... 265 |
| float .................... 267 | gravel stop ............... 327 | Crown molding .............. 308 |
| footing ............... 283-284 | gutter ................... 328 | Crushed stone sidewalk ...... 278 |
| foundation ............ 26, 283 | pipe ................ 276, 377 | Cubicle shower .............. 386 |
| furring .................. 302 | reglet ................... 329 | toilet.................... 370 |
| lintel .................... 285 | rivet ..................... 295 | Cupola ..................... 307 |
| mixer ................... 267 | roof .................... 325 | Curb asphalt ................ 277 |
| nail ..................... 295 | tubing ................... 276 | bituminous ............... 277 |

424

# INDEX

concrete .................. 277, 283
edging ........................ 278
formwork ...................... 282
granite .................. 278, 290
highway ....................... 290
inlet .......................... 278
precast ....................... 278
roof ........................... 299
terrazzo ...................... 356
Curing concrete ................. 283
paper .......................... 315
Current transformer cabinet .... 397
Curtain rod .................... 366
type fire-damper .............. 392
Curved stair ................... 313
stairway ...................... 145
Cut stone trim ................. 291
Cutoff pile .................... 275
Cutout counter ................. 307
Cutter brush ................... 265
Cutting block .................. 307
torch .......................... 270
Cylinder lockset ............... 348

## D

Damper ......................... 392
fireplace ..................... 285
multi-blade ................... 392
Darby finish ................... 283
Dead bolt ...................... 348
Deck metal ..................... 292
roof ..................... 292, 303-304
wood ................... 158, 303, 314
Decking ..................... 37, 53
Decorative block ............... 288
Demolition hammer .............. 266
Derrick ........................ 271
guyed ......................... 271
stiffleg ...................... 271
Detection system ............... 402
Detector infra-red ............. 402
motion ........................ 402
smoke ......................... 402
wiring smoke .................. 403
Device wiring .................. 398
Dewater ........................ 272
Diaphragm pump ................. 269
Diesel hammer .................. 266
Diffuser ceiling ............... 392
linear ........................ 392
opposed blade damper .......... 392
perforated aluminum ........... 392
rectangular ................... 392
steel ......................... 392
T-bar mount ................... 392
Dimmer switch .................. 398
Direct expansion
air-conditioning .............. 394
Directory board ................ 368
building ...................... 403
Disappearing stair ............. 368
Discharge hose ................. 269
Dishwasher ..................... 372

Dispenser napkin ............... 367
soap ..................... 366-367
toilet tissue ................. 366
towel ..................... 366-367
Disposer garbage .......... 150, 372
Distribution box ............... 276
Distributor asphalt ............ 268
Divider strip terrazzo ......... 356
Dock bumper .................... 297
Door accessory ................. 347
accordion ..................... 336
acoustical ................ 338, 369
aluminum ...................... 332
balanced ...................... 341
bifolding ..................... 332
blind ......................... 311
bulkhead ...................... 338
cafe .......................... 334
canopy ........................ 367
casing ........................ 353
ceiling ....................... 329
chime ......................... 402
cleanout ...................... 285
closer ........................ 347
closet ........................ 332
commercial .................... 331
composite ..................... 331
counter ....................... 338
dutch ......................... 335
dutch oven .................... 285
entrance ........... 333, 335, 341
exterior ....................... 86
fire ..................... 331, 402
flexible ...................... 340
floor ......................... 338
frame .............. 309, 330, 333
French ........................ 337
garage ......................... 94
glass ..................... 339, 341
hardware ...................... 347
industrial .................... 340
kick plate .................... 348
knocker ....................... 347
labeled ....................... 331
metal ......................... 339
molding ....................... 309
moulded ....................... 334
opener .................... 340, 347
overhead ...................... 339
painting ...................... 361
panel ......................... 334
paneled ....................... 332
partition ..................... 370
passage ....................... 337
pre-hung ...................... 336
pull .......................... 348
release ....................... 403
residential .......... 332-333, 335
rolling ....................... 340
roof .......................... 329
rough buck .................... 298
sauna ......................... 373
shower ........................ 370
sill .............. 309, 333, 348

sliding .................... 88-89
special ....................... 338
stain ......................... 361
steel ..................... 330-332
stop ...................... 347, 351
storm ..................... 98, 333
switch burglar alarm .......... 402
system commercial ............. 140
threshold ..................... 334
varnish ....................... 361
weatherstrip .................. 348
wood ..................... 136, 138
Door-bell system ............... 402
Dormer framing ............. 54, 56
gable ......................... 301
roofing ................... 114, 116
Double acting door ............. 338
hung window ........ 74, 344-345
Dowel & cap .................... 283
Downspout .................. 325-326
aluminum ...................... 325
copper ........................ 325
elbow ......................... 325
painting ...................... 362
steel ......................... 325
strainer ...................... 325
Dozer ................. 266, 273-274
Dragline bucket ................ 265
Drain .......................... 375
cast-iron ..................... 375
floor ......................... 375
Drainage accessory ............. 326
site .......................... 275
trap .......................... 383
Drill ceiling .................. 294
core .......................... 267
drywall ....................... 294
hammer ........................ 268
plaster ....................... 294
rig ........................... 272
steel ......................... 268
wagon ......................... 268
wood .......................... 296
Drinking fountain support ...... 383
Drip edge ...................... 325
Driver sheeting ................ 268
Driveway ........................ 14
Drop pan ceiling ............... 357
Drum trap ...................... 383
trap copper ................... 383
Dry kiln ....................... 303
Dryer circuit electric ......... 403
wiring ........................ 198
Drywall ............... 126, 128, 353
accessory ..................... 353
drill ......................... 294
frame ......................... 330
gypsum ........................ 353
nail .......................... 295
painting ...................... 363
prefinished ................... 354
Duck tarpaulin ................. 265
Duct access door ............... 392
accessory ..................... 392

425

# INDEX

    flexible insulated ............ 393
    flexible noninsulated ......... 393
    heater ....................... 388
    humidifier ................... 389
    HVAC ........................ 393
    insulation ................... 389
    mechanical ................... 393
    silencer ..................... 393
    underground .................. 401
Ductile iron pipe ................ 276
Ductwork ........................ 393
    aluminum ..................... 393
    fabricated ................... 393
    galvanized ................... 393
    rectangular .................. 393
    rigid ........................ 393
    vinyl-coated flexible ........ 393
Dumbwaiter ...................... 374
Dump charge ..................... 272
    truck ........................ 267
Duplex receptacle ......... 198, 403
Dutch door ................. 86, 335
    oven door .................... 285
DWV pipe ABS ................... 379
    pipe-fitting ................. 380
    PVC pipe ..................... 379
    tubing copper ................ 377

### E

Earth compactor ................. 266
    vibrator ..................... 273
Earthwork .................... 273-274
    equipment ............... 265, 273
    equipment rental ............. 265
Edge drip ....................... 325
    formwork ..................... 282
Edging curb ..................... 278
    landscape .................... 280
Eggcrate ceiling ................ 357
Elbow downspout ................. 325
    PVC .......................... 379
Electric baseboard heater ....... 402
    boiler ....................... 387
    cable ........................ 396
    fixture .................. 400-401
    furnace ...................... 388
    generator .................... 269
    generator set ................ 400
    heat ......................... 388
    heating ...................... 197
    lamp ......................... 401
    metallic tubing .............. 394
    service .................. 196, 398
    stair ........................ 369
    switch ....................... 399
    unit heater .................. 402
    utilities .................... 401
    water-heater ................. 386
    wire ......................... 396
Electrical cabinet .............. 397
    conduit ...................... 394
    conduit greenfield ........... 395
    crew ......................... 263

    sitework ..................... 401
Electronic air cleaner .......... 391
Elevated slab ................... 284
Elevator ........................ 374
    construction ................. 271
Embossed print door ............. 334
EMT ............................. 394
Emulsion adhesive ............... 360
    sprayer ...................... 268
Enclosure swimming pool ......... 374
    telephone .................... 372
Entrance cable service .......... 397
    canopy ....................... 367
    door ........ 86, 333, 335-336, 341
    frame ........................ 333
    lock ......................... 348
    screen ....................... 371
    sliding ...................... 341
    system commercial ............. 90
    weather cap .................. 395
EPDM roof ....................... 323
Epoxy coating ................... 365
    floor ........................ 358
    grout ........................ 355
    terrazzo ..................... 358
    wall coating ................. 365
Equipment crew .................. 262
    earthwork ............... 265, 273
    insurance .................... 264
    rental ....................... 270
Erosion control ................. 280
Excavation ...................... 273
    backhoe ...................... 274
    footing ........................ 6
    foundation ..................... 8
    hand ..................... 273-274
    heavy equipment .............. 274
    septic tank .................. 276
    structural ................... 273
    trench ....................... 274
    utility ....................... 10
Exhaust hood .................... 372
    vent ......................... 394
Expansion anchor ................ 294
    bolt ......................... 294
    shield ....................... 294
    tank ......................... 390
Exposed aggregate ............... 283
    aggregate coating ............ 365
Exterior blind .............. 310-311
    door .......................... 86
    door frame ................... 333
    fixture mercury-vapor ........ 401
    light ........................ 199
    light fixture ................ 401
    molding ...................... 308
    plaster ...................... 353
    pre-hung door ................ 336
    residential door ............. 335
    wall framing .................. 38
    wood frame ................... 333
Extinguisher fire ............... 387
Extractor ....................... 266
    air .......................... 392

### F

Fabric flashing ................. 326
Fabricated ductwork ............. 393
Facing stone .................... 291
Fan ............................. 393
    attic ventilation ............ 393
    house ventilation ............ 393
    wiring ....................... 198
Fan-coil air-conditioning ....... 394
Fascia aluminum ................. 320
    board ........................ 298
    metal ........................ 326
    vinyl ........................ 324
    wood ......................... 308
Fastener anchor bolt ............ 285
    timber ....................... 295
    wood ......................... 296
Faucet & fitting ................ 375
    bath ......................... 375
Fee architectural ............... 264
    management ................... 264
Felt ............................ 322
    asbestos ................ 320, 322
    asphalt ................. 321-322
    carpet ....................... 358
    fiberglass ................... 321
    tarred ....................... 322
    underlayment ................. 358
Fence basketweave ............... 279
    board ........................ 279
    chain link .................... 18
    metal .................... 278-279
    painting ..................... 361
    removal ...................... 273
    residential .................. 278
    wood ......................... 279
Fertilizer ...................... 281
Fiberboard roof deck ............ 316
Fiberglass blown-in ............. 316
    ceiling board ................ 357
    column ....................... 306
    cupola ....................... 307
    door ......................... 339
    felt ......................... 321
    insulation ....... 72, 317-318, 389
    panel ........................ 357
    pipe-covering ................ 390
    planter ...................... 280
    roof deck .................... 316
    septic tank .................. 276
    shower stall ................. 370
    tank ......................... 390
    wall lamination .............. 365
    wool ......................... 316
Field septic ..................... 16
Fieldstone ...................... 291
Fill ........................ 273-274
    gravel ............... 273-275, 277
Filled column concrete .......... 291
Filler block .................... 364
Filter air ...................... 391
    mechanical media ............. 391
Fin-tube radiation .............. 389
Finish concrete ................. 277

# INDEX

| | | |
|---|---|---|
| floor .............................. 365 | counter .......................... 329 | screen ........................... 285 |
| hardware ...................... 347 | fabric ............................ 326 | Fluorescent fixture ......... 400-401 |
| nail ............................. 295 | membrane ...................... 322 | lamp ............................ 401 |
| steel trowel ................. 283 | PVC .............................. 327 | light ............................ 199 |
| Finisher floor ..................... 267 | stainless ........................ 327 | Flush door ........................ 335 |
| Finishing crew .................... 262 | valley ........................... 319 | tube framing ................. 350 |
| floor ............................. 360 | vent-chimney .................. 391 | Foam glass roof deck ......... 316 |
| Fir column ........................ 306 | Flat wood truss .................. 303 | insulation .................. 316-317 |
| floor ............................. 360 | Flatbed truck ..................... 267 | phenolic ....................... 317 |
| molding ........................ 309 | Flemish bond ..................... 287 | Foamglass insulation ......... 318 |
| roof deck ..................... 314 | Flexible conduit & fitting ..... 399 | Foil aluminum ........ 315-316, 366 |
| roof plank .................... 303 | connector .................... 375 | back insulation ............. 318 |
| Fire call pullbox ................. 402 | door ............................ 340 | faced fiberboard ........... 318 |
| damage repair .............. 365 | ductwork vinyl-coated ..... 393 | Folding accordion door ..... 336 |
| door ..................... 331, 402 | insulated duct .............. 393 | accordion partition ......... 369 |
| door frame ................... 330 | metallic conduit ............ 395 | partition ...................... 369 |
| escape disappearing ....... 369 | Float concrete .................. 267 | Footing ............................ 22 |
| extinguisher ................. 387 | finish .......................... 283 | concrete .................. 283-284 |
| extinguishing system ..... 387 | Floater equipment .............. 264 | excavation .................... 6 |
| horn ........................... 402 | Floating pin ....................... 347 | reinforcing ................... 282 |
| resistant drywall ........... 354 | Flood coating .................... 322 | spread ........................ 282 |
| retardant lumber ........... 302 | Floodlight .................. 268-269 | Forklift ............................ 269 |
| retardant plywood ......... 303 | Floor .............................. 287 | brick ........................... 269 |
| Fire-damper curtain type ..... 392 | acid proof .................... 287 | Form fireplace .................. 285 |
| Fireplace accessory ........... 285 | adhesive ...................... 359 | slab ............................ 292 |
| box ............................. 286 | asphalt block ................ 277 | Formbloc ......................... 289 |
| damper ........................ 285 | brick ........................... 287 | Formwork plywood ............ 282 |
| form ........................... 285 | bumper ........................ 347 | Foundation ........................ 22 |
| mantel ........................ 307 | carpet .......................... 358 | chimney ...................... 283 |
| masonry ................ 152, 286 | ceramic tile .................. 355 | concrete .................. 26, 283 |
| prefabricated .......... 154, 286 | cleanout ....................... 375 | concrete block .............. 290 |
| Fireproofing plaster ........... 353 | composition .................. 358 | excavation .................... 8 |
| sprayed ....................... 353 | concrete ....................... 30 | masonry ....................... 24 |
| Firestop gypsum ................ 352 | conductive .................... 359 | mat ....................... 283-284 |
| wood ........................... 298 | door ............................ 338 | removal ........................ 272 |
| Fitting conduit .................. 395 | drain ............................ 375 | wall ............................ 290 |
| pipe ............................ 375 | epoxy .......................... 358 | wood ........................... 28 |
| vent-chimney ................ 391 | finish .......................... 365 | Frame chalkboard ............. 368 |
| Fitting-pipe see pipe-fitting .. 376 | finisher ........................ 267 | door ............... 309, 330, 333 |
| Fixed window ..................... 84 | flagging ....................... 278 | drywall ........................ 330 |
| Fixture bathroom .......... 384-386 | framing ..................... 34, 36 | entrance ...................... 333 |
| carrier ......................... 383 | insulation ..................... 316 | labeled ........................ 330 |
| electric ........................ 400 | marble ......................... 291 | metal .......................... 330 |
| fluorescent .............. 400-401 | nail ............................. 295 | steel ........................... 330 |
| halide-metal ................. 400 | painting ....................... 361 | welded ........................ 330 |
| incandescent ................ 400 | plank .......................... 302 | window ........................ 346 |
| interior light ................. 400 | plywood ....................... 304 | wood ........................... 333 |
| lantern ........................ 401 | polyethylene ................. 359 | Framing ............................ 52 |
| mercury-vapor ............... 400 | quarry tile .................... 356 | anchor ......................... 296 |
| metal-halide ................. 400 | register ....................... 392 | canopy ........................ 300 |
| mirror light ................... 400 | rubber ......................... 359 | dormer ..................... 54, 56 |
| plumbing ................. 385-386 | stain ........................... 361 | floor ............................ 34 |
| residential ................... 401 | stone .......................... 278 | laminated .................... 313 |
| sodium-low-pressure ...... 401 | subfloor ....................... 304 | partition ....................... 58 |
| wiring .......................... 404 | terrazzo ....................... 356 | roof ........... 40, 42, 44, 46, 48, 50 |
| Fixtures light .................... 199 | tile terrazzo .................. 356 | steel ........................... 292 |
| Flagging ..................... 278, 287 | underlayment ................ 304 | timber ......................... 301 |
| slate ........................... 278 | varnish ........................ 361 | tube ............................ 350 |
| Flashing .......................... 326 | vinyl ........................... 359 | window wall .................. 350 |
| aluminum ................ 121, 123 | wood ..................... 143, 360 | wood ..................... 297, 301 |
| asphalt ........................ 326 | Flooring .......................... 143 | French door ..................... 337 |
| butyl ........................... 327 | Flue chimney metal ............ 391 | Friction pile ..................... 275 |
| cement ........................ 315 | lining .......................... 286 | Front end loader ............... 273 |
| copper ........................ 326 | prefab metal ................. 391 | shovel ......................... 266 |

427

# INDEX

Full vision door ................ 333
    vision glass ................ 349
Furnace electric ................ 388
    gas-fired ................ 388
    heating & cooling ................ 388
    hot-air ................ 388
    oil-fired ................ 388-389
    wall ................ 390
    wiring ................ 198, 403
Furring ................ 53
    ceiling ................ 37, 302, 352
    channel ................ 353
    metal ................ 352
    wall ................ 302, 352
Fusible link closer ................ 347

## G

Gable dormer ................ 301
    dormer framing ................ 54
    dormer roofing ................ 114
    roof framing ................ 40
Galvanized ductwork ................ 393
Gambrel roof framing ................ 46
Garage door ................ 94, 339
Garbage disposer ................ 150, 372
Gas connector ................ 375
    conversion burner ................ 388
    fired space heater ................ 390
    heating ................ 186, 190
    operated generator ................ 400
    pipe ................ 275
    service and distribution ................ 275
    vent ................ 391
Gas-fired boiler ................ 387
    furnace ................ 388
    space heater ................ 390
Gate metal ................ 278-279
    valve ................ 276
General condition ................ 264
    condition crew ................ 259
Generator construction ................ 269
    emergency ................ 400
    gas operated ................ 400
    set ................ 400
Girder wood ................ 297, 302
Glass ................ 350
    bead ................ 350
    bead molding ................ 309
    door ................ 333, 339, 341
    door shower ................ 370
    full vision ................ 349
    heat reflective ................ 349
    mirror ................ 367
    plate ................ 350
    tempered ................ 350
    tinted ................ 350
    window ................ 350
Glaze coat ................ 363
Glazed brick ................ 287
    wall coating ................ 365
Glazing crew ................ 262
    polycarbonate ................ 350
Glued laminated ................ 313-314

Grab bar ................ 366
Gradall ................ 265
Grader motorized ................ 266
Grading site ................ 274
Granite ................ 290
    building ................ 290
    curb ................ 278, 290
    paver ................ 290
    sidewalk ................ 278
Grass ................ 281
    cloth wallpaper ................ 366
    sprinkler ................ 280
Grating sewer ................ 275
Gravel bankrun ................ 273
    fill ................ 273-275, 277
    stop ................ 327
Greenfield conduit ................ 395
Greenhouse ................ 156, 374
Grid spike ................ 296
Grille air-return ................ 392
    aluminum ................ 392
    decorative wood ................ 307
    painting ................ 362
    roll up ................ 340
    window ................ 346
Grinder terrazzo ................ 267
Grooved block ................ 288
Grooved-joint pipe ................ 382
    pipe-fitting ................ 382
Ground ................ 302
    clamp water pipe ................ 397
    cover ................ 280
    face block ................ 289
    fault ................ 198
    fault receptacle wiring ................ 403
    rod ................ 397
    wire ................ 397
Grounding ................ 397
    & conductor ................ 397
    wire brazed ................ 397
Grout ................ 285
    concrete block ................ 285
    epoxy ................ 355
    wall ................ 285
Guard corner ................ 360
    gutter ................ 328
Gunite pool ................ 157
    reinforcing ................ 283
Gutter ................ 328
    aluminum ................ 327
    copper ................ 328
    guard ................ 328
    painting ................ 362
    stainless ................ 328
    steel ................ 328
    strainer ................ 328
    vinyl ................ 328
    wood ................ 328
Guyed derrick ................ 271
    tree ................ 281
Gypsum board ................ 353
    cement ................ 285
    drywall ................ 353
    fabric wallcovering ................ 366

firestop ................ 352
lath ................ 352
plaster ................ 352
sheathing ................ 304
weatherproof ................ 304

## H

Half round molding ................ 309
Halide-metal fixture ................ 400
Hammer bush ................ 284
    chipping ................ 268
    demolition ................ 266
    drill ................ 268
    hydraulic ................ 269
    pile ................ 266
Hand carved door ................ 334
    clearing ................ 272
    excavation ................ 273-274
    hole ................ 277
    split shake ................ 320
Hand-split shake ................ 68
Handicap opener ................ 347
Handicapped lever ................ 348
Handrail metal ................ 293
    oak ................ 145
    wood ................ 309-310
Hanger joist ................ 296
Hardboard cabinet ................ 305
    overhead door ................ 339
    siding ................ 311
    soffit ................ 312
    tempered ................ 309
    underlayment ................ 305
Hardware door ................ 347, 370
    finish ................ 347
    rough ................ 304
    window ................ 347
Hardwood floor ................ 360
    grille ................ 307
Hatch ceiling ................ 329
    roof ................ 329
Hauling truck ................ 274
Header pipe ................ 272
    wood ................ 300
Hearth ................ 286
Heat electric ................ 388
    pump ................ 394
    reflective glass ................ 349
Heater circuit wiring ................ 404
    contractor ................ 269
    duct ................ 388
    electric baseboard ................ 402
    electric wall ................ 402
    floor mounted space ................ 390
    sauna ................ 373
    water ................ 387
    wiring ................ 198
Heating ................ 186-193, 387-391
    baseboard ................ 389
    electric ................ 197
    hot-air ................ 388
    hot-water ................ 387, 389
    hydronic ................ 387

# INDEX

| | |
|---|---|
| Heavy framing | 301 |
| Hedge | 280 |
| Hex nut | 293 |
| Hexagonal face block | 289 |
| High bay lighting | 400 |
|    intensity discharge lamp | 401 |
|    intensity discharge lighting | 400 |
|    rise glazing | 349 |
|    strength block | 288 |
| Highway curb | 290 |
|    paver | 278 |
| Hinge brass | 347-348 |
|    residential | 348 |
|    steel | 347 |
| Hip rafter | 298 |
|    roof framing | 44 |
| Hoist and tower | 271 |
|    contractor | 271 |
|    lift equipment | 270 |
|    personnel | 271 |
| Hollow core door | 336 |
|    metal | 330 |
|    metal door | 330-331 |
| Honed block | 289 |
| Hood range | 372 |
| Hook robe | 367 |
| Horizontal aluminum siding | 320 |
|    auger | 265 |
|    vinyl siding | 324 |
| Horn fire | 402 |
| Hose air | 268 |
|    discharge | 269 |
|    suction | 269 |
|    water | 269 |
| Hospital door hardware | 347 |
|    tip pin | 347-348 |
| Hot tub | 384 |
| Hot-air furnace | 388 |
|    heating | 186, 188, 388 |
| Hot-water boiler | 387-388 |
|    heating | 190, 387, 389 |
| Hotel cabinet | 369 |
|    furnishing | 311 |
|    lockset | 348 |
| House telephone | 402 |
|    ventilation fan | 393 |
| Humidifier duct | 389 |
|    room | 389 |
| HVAC duct | 393 |
| Hydraulic crane | 271 |
|    hammer | 269 |
|    jack | 271 |
| Hydronic heating | 387 |
| Hypalon caulking | 315 |
|    roof | 323 |

## I

| | |
|---|---|
| Illuminated sign | 372 |
| Impact wrench | 268 |
| In-slab conduit | 395 |
| Incandescent fixture | 400-401 |
|    light | 199 |
| Industrial door | 338 |
|    steel window | 342 |
| Infra-red detector | 402 |
| Inlet curb | 278 |
| Insecticide | 274 |
| Insulation | 72, 316, 390 |
|    blanket | 389 |
|    blown-in | 316 |
|    board | 121, 123 |
|    building | 316 |
|    cavity wall | 316 |
|    cellulose | 316 |
|    duct | 389 |
|    fiberglass | 389 |
|    foam | 316-317 |
|    masonry | 316 |
|    perlite | 316 |
|    pipe | 390 |
|    roof deck | 316 |
|    shingle | 319 |
|    vapor barrier | 315 |
|    water-heater | 389 |
| Insurance | 264 |
|    builder risk | 264 |
|    equipment | 264 |
|    public liability | 264 |
| Interior door | 136, 138 |
|    door frame | 334 |
|    light | 199 |
|    light fixture | 400 |
|    partition framing | 58 |
|    pre-hung door | 337 |
|    residential door | 335 |
|    wood frame | 334 |
| Intermediate conduit | 394 |
| Ironspot brick | 287 |

## J

| | |
|---|---|
| Jack cable | 271 |
|    hydraulic | 271 |
|    mud | 267 |
|    rafter | 299 |
| Jackhammer | 268 |
| Jalousie | 342 |
| Jetting pump | 272 |
| Joint reinforcing | 285 |
| Joist | 37 |
|    anchor | 296 |
|    connector | 296 |
|    hanger | 296 |
|    wood | 297, 304 |
| Jute mesh | 280 |

## K

| | |
|---|---|
| Kalamein door | 339 |
| Keyway | 282 |
| Kick plate door | 348 |
| Kiln dried lumber | 302 |
|    dry | 303 |
| Kitchen | 148 |
|    cabinet | 305 |
|    sink | 386 |
|    sink faucet | 375 |
| Knocker door | 347 |

## L

| | |
|---|---|
| Labeled door | 331 |
|    frame | 330 |
| Labor rate major trade | 263-264 |
| Ladder | 269 |
| Lag screw | 294-295 |
| Lally column | 291 |
| Laminated beam | 313-314 |
|    counter top | 307 |
|    counter top plastic | 307 |
|    epoxy & fiberglass | 365 |
|    framing | 313 |
|    glued | 314 |
|    gypsum lath | 352 |
|    roof deck | 314 |
|    wood | 313-314 |
| Lamp fluorescent | 401 |
|    high intensity discharge | 401 |
|    metal-halide | 401 |
|    post | 293 |
| Lampholder | 398 |
| Landing newel | 312 |
| Landscape | 280-281 |
|    edging | 280 |
|    surface | 359 |
| Lantern fixture | 401 |
| Laser | 269 |
| Latch set | 348 |
| Latex caulking | 315 |
|    underlayment | 360 |
| Lath gypsum | 352 |
|    metal | 352 |
| Lattice molding | 309 |
| Lauan door | 336 |
| Laundry faucet | 375 |
|    tray | 385 |
| Lavatory | 162, 164 |
|    faucet | 375 |
|    support | 383 |
|    vanity top | 385 |
| Lazy susan | 305 |
| Lead coated downspout | 325 |
|    coated flashing | 327 |
|    flashing | 326 |
|    paint removal | 364 |
|    pile | 266 |
|    roof | 328 |
| Lean-to type greenhouse | 374 |
| Leather tile | 355 |
| Ledger | 299 |
| Leeching field | 16 |
|    pit | 276 |
| Let-in bracing | 297 |
| Letter slot | 369 |
| Lever handicapped | 348 |
| Lexan | 350 |
| Lift aerial | 268 |
|    correspondence | 374 |
| Light fixture exterior | 401 |
|    fixture interior | 400 |
|    fixture wiring | 404 |

# INDEX

| | |
|---|---|
| fixtures | 199 |
| portable | 268 |
| post | 401 |
| tower | 269 |
| Lighting | 198, 400-401 |
| high intensity discharge | 400 |
| incandescent | 400 |
| mercury-vapor | 400 |
| metal-halide | 400 |
| outdoor | 268 |
| strip | 400 |
| track | 401 |
| wiring | 404 |
| Limestone | 281, 290-291 |
| coping | 286 |
| Linear diffuser | 392 |
| Linen chute | 368 |
| wallcovering | 366 |
| Lintel | 285, 291, 293 |
| Load bearing stud | 299 |
| Load-center indoor | 398 |
| plug-in breaker | 398 |
| rainproof | 398 |
| Loader front end | 273 |
| tractor | 267 |
| Loam | 281 |
| Lobby collection box | 368 |
| Lock entrance | 348 |
| Locker | 369 |
| Lockset cylinder | 348 |
| hotel | 348 |
| Log simulated | 287 |
| Louver | 328 |
| redwood | 307 |
| ventilation | 307 |
| wood | 307 |
| Louvered blind | 310 |
| door | 136, 138, 336, 338 |
| Lumber | 297 |
| core paneling | 310 |
| creosote | 302 |
| kiln dried | 302 |
| treatment | 302 |
| Luminous ceiling | 357, 401 |
| panel | 357 |

## M

| | |
|---|---|
| Machine welding | 270 |
| Mahogany door | 334 |
| Mail box | 369 |
| box call system | 402 |
| chute | 368 |
| slot | 369 |
| Major trade labor rate | 263-264 |
| Mall front | 341 |
| Malleable iron pipe-fitting | 381 |
| Management fee | 264 |
| Manhole removal | 272 |
| Mansard roof framing | 48 |
| Mantel beam | 307 |
| fireplace | 307 |
| Maple counter top | 307 |
| Marble base | 291 |

| | |
|---|---|
| coping | 286 |
| counter top | 307 |
| floor | 291 |
| screen | 371 |
| sill | 288 |
| synthetic | 355 |
| tile | 355 |
| toilet partition | 370 |
| Masonry brick | 286-287 |
| cement | 285 |
| cornice | 286 |
| crew | 261 |
| fireplace | 152, 286 |
| flashing | 326 |
| foundation | 24 |
| furring | 302 |
| insulation | 316 |
| nail | 295 |
| painting | 364 |
| reinforcing | 285 |
| removal | 273 |
| saw | 270 |
| sidewalk | 278 |
| sill | 288 |
| step | 278 |
| wall | 62, 64, 280, 288 |
| wall tie | 285 |
| Mat foundation | 283-284 |
| Material hoist | 271 |
| Mechanical duct | 393 |
| media filter | 391 |
| system crew | 263 |
| Median strip | 281 |
| Medicine cabinet | 369 |
| Melting snow | 402 |
| Membrane | 121, 123 |
| curing | 283 |
| flashing | 322 |
| roof | 323 |
| Mercury vapor fixture | 199, 400 |
| Exterior fixture | 401 |
| Mesh reinforcing | 283 |
| wire | 283 |
| Metal base | 355 |
| butt frame | 330 |
| chalkboard | 368 |
| clad window | 74-85 |
| crew | 262 |
| deck | 292 |
| door | 333, 339 |
| door frame | 332 |
| door residential | 332 |
| fascia | 326 |
| fence | 278-279 |
| fireplace | 154 |
| flue chimney | 391 |
| frame | 330 |
| furring | 352 |
| gate | 278-279 |
| halide fixture | 199 |
| hollow | 330 |
| lath | 352 |
| locker | 369 |
| overhead door | 339 |

| | |
|---|---|
| partition | 299, 369 |
| pipe | 276 |
| railing | 18, 293 |
| screen | 342, 371 |
| sheet | 325 |
| shelf | 371 |
| shingle | 319 |
| siding | 70 |
| soffit | 329 |
| stud | 299 |
| threshold | 348 |
| tile | 355 |
| toilet partition | 370 |
| truss | 292 |
| tube frame | 350 |
| window | 342-343 |
| Metal-halide fixture | 400 |
| lamp | 401 |
| lighting | 400 |
| Metallic foil | 316 |
| Meter socket | 398 |
| water supply | 384 |
| water supply domestic | 384 |
| Microwave oven | 372 |
| Mill construction | 301 |
| Millwork | 307 |
| Mineral fiber ceiling | 356-357 |
| fiber insulation | 318 |
| roof | 321 |
| shingle | 319 |
| wool blown-in | 316 |
| Mirror | 367 |
| ceiling boards | 357 |
| light fixture | 400 |
| Mixer concrete | 267 |
| mortar | 267, 269 |
| plaster | 269 |
| Mobilization | 274 |
| Modulating damper motorized | 392 |
| Module tub/shower | 384 |
| Moil point | 268 |
| Moisture protection crew | 262 |
| Molding | 308 |
| base | 308 |
| brick | 308 |
| ceiling | 308 |
| exterior | 308 |
| hardboard | 310 |
| wood | 312 |
| Monel rivet | 295 |
| roof | 328 |
| Monolithic finish | 283 |
| Mop roof | 322 |
| Mortar | 285 |
| and masonry accessory | 285 |
| cement | 355 |
| mixer | 267, 269 |
| thinset | 356 |
| Motion detector | 402 |
| Motor connection | 399 |
| generator | 269 |
| Motorized modulating damper | 392 |
| Moulded door | 334 |
| Mounting board plywood | 304 |

430

# INDEX

| | |
|---|---|
| Movable office partition | 369 |
| Movie screen | 371 |
| Moving shrub | 281 |
| tree | 281 |
| Mud jack | 267 |
| Mullion vertical | 351 |
| Multi-blade damper | 392 |
| Muntin window | 346 |

## N

| | |
|---|---|
| Nail | 295 |
| Nailers steel | 298 |
| wood | 298 |
| Napkin dispenser | 367 |
| Neoprene flashing | 327 |
| roof | 323 |
| Newel | 145, 312 |
| No-hub pipe | 376 |
| pipe-fitting | 377 |
| Non-removable pin | 347 |
| Nut hex | 293 |
| Nylon carpet | 142, 358 |

## O

| | |
|---|---|
| Oak door frame | 333 |
| floor | 143, 360 |
| molding | 309 |
| paneling | 310 |
| stair tread | 312 |
| threshold | 334 |
| Off highway truck | 267 |
| Office partition | 369 |
| Oil burner | 388 |
| heater temporary | 269 |
| heating | 188, 190 |
| water-heater | 387 |
| Oil-fired boiler | 387 |
| furnace | 388-389 |
| Open rail fence | 279 |
| Opener automatic | 347 |
| door | 340, 347 |
| handicap | 347 |
| Operating cost | 265 |
| cost equipment | 265 |
| Outdoor lighting | 268 |
| Outlet-box plastic | 398 |
| steel | 397 |
| Oven cabinet | 306 |
| microwave | 372 |
| Overhead door | 94, 339 |
| door system | 92 |

## P

| | |
|---|---|
| P trap | 383 |
| P&T relief valve | 384 |
| Packaged terminal air-conditioner | 394 |
| Padding | 142 |
| carpet | 358 |
| Paint removal | 364 |
| Painting balustrades | 362 |
| cabinet | 360 |
| casework | 360 |
| ceiling | 363-364 |
| clapboard | 363 |
| concrete | 363 |
| concrete block | 364 |
| concrete floor | 361 |
| cornice | 362 |
| door | 361 |
| downspout | 362 |
| drywall | 354, 363 |
| fence | 361 |
| floor | 361 |
| grille | 362 |
| gutter | 362 |
| masonry | 364 |
| pipe | 362 |
| plaster | 363 |
| siding | 363 |
| sprayer | 269 |
| steel siding | 363 |
| stucco | 363 |
| texture 1-11 | 363 |
| trim | 362 |
| truss | 362 |
| wall | 363-364 |
| window | 361 |
| wood floor | 361 |
| wood shingle | 363 |
| Pan form stair | 293 |
| shower | 327 |
| Panel acoustical | 357 |
| door | 334-335 |
| luminous | 357 |
| wall | 290 |
| Panelboard circuit breaker | 399 |
| Paneled door | 136, 332 |
| pine door | 336 |
| Paneling | 309-310 |
| board | 309 |
| plywood | 310 |
| wood | 309 |
| Panic bar | 348 |
| device door hardware | 347 |
| Paper building | 315 |
| curing | 283 |
| sheathing | 315 |
| Paperhanging | 366 |
| Paperholder | 367 |
| Parquet floor | 143, 360 |
| Particle board siding | 312 |
| board underlayment | 305 |
| Parting bead | 309 |
| Partition | 126, 130, 299, 370 |
| acoustical | 369 |
| door | 370 |
| folding | 369 |
| folding leaf | 369 |
| framing | 58 |
| metal | 369 |
| office | 369 |
| portable | 370 |
| shower | 370 |
| toilet | 370-371 |
| wood frame | 302 |
| Passage door | 136, 336-337 |
| lock set | 348 |
| Patch roof | 315 |
| Patching floor concrete | 284 |
| Patio | 278, 287 |
| cover | 367 |
| Paver bituminous | 269 |
| concrete | 269 |
| floor | 287 |
| highway | 278 |
| Paving berm | 277 |
| brick | 277 |
| concrete | 277 |
| stone | 277 |
| Peastone | 280 |
| Pegboard | 309 |
| Perforated ceiling | 357 |
| Perimeter insulation | 316 |
| Perlite insulation | 316, 318 |
| plaster | 352 |
| sprayed | 365 |
| Permit building | 264 |
| Personnel hoist | 271 |
| Phenolic foam | 317 |
| Phone booth | 372 |
| PIB roof | 323 |
| Picket fence | 280 |
| Pickup truck | 270 |
| Picture window | 84, 345 |
| Pier brick | 286 |
| Pilaster toilet partition | 371 |
| wood column | 306 |
| Pile cast in place | 275 |
| concrete | 275 |
| cutoff | 275 |
| driver | 266 |
| hammer | 266 |
| testing | 275 |
| wood | 275 |
| Pin floating | 347 |
| non-removable | 347 |
| Pine column | 306 |
| door | 335 |
| door frame | 333 |
| fireplace mantel | 307 |
| floor | 360 |
| molding | 308 |
| roof deck | 314 |
| shelving | 310 |
| siding | 311 |
| stair | 312 |
| stair tread | 313 |
| Pipe | 276 |
| & fittings | 375-384 |
| asbestos cement | 275 |
| brass | 375 |
| cast-iron | 376 |
| cleanout | 375 |
| column | 292 |
| concrete | 276 |
| copper | 377 |
| CPVC | 379 |
| drainage and sewage | 275 |

431

# INDEX

| | | | | | |
|---|---|---|---|---|---|
| DWV ABS | 379 | wall tile | 355 | Postal specialty | 369 |
| DWV PVC | 379 | Plate glass | 350 | Poured insulation | 72 |
| fitting | 375 | shear | 296 | Power capacitors | 400 |
| gas | 275 | wall switch | 398 | wiring | 398 |
| grooved-joint | 382 | Platform trailer | 270 | Pre-hung door | 336 |
| header | 272 | Plating zinc | 295 | Precast catch basin | 275 |
| insulation | 390 | Plenum silencer | 393 | coping | 286 |
| no-hub | 376 | Plexiglass sign | 372 | curb | 278 |
| painting | 362 | Plugmold raceway | 396 | lintel | 285 |
| plastic | 378-379 | Plumbing | 375 | manhole | 275 |
| polyethylene | 275 | fixture | 385-386 | sill | 288 |
| quick coupling | 272 | Plywood clip | 296 | stair | 284 |
| rail | 293 | floor | 304 | terrazzo | 356 |
| sewer | 275 | formwork | 282 | Prefabricated building | 374 |
| single hub | 376 | joist | 304 | fireplace | 154 |
| soil | 376 | mounting board | 304 | Prefinished drywall | 354 |
| steel | 275 | paneling | 310 | floor | 143, 360 |
| Pipe-covering | 390 | sheathing | 304 | Pressure pipe | 276 |
| fiberglass | 390 | shelving | 310 | Primer asphalt | 360 |
| Pipe-fitting brass | 376 | siding | 312 | Profile block | 289 |
| cast-iron | 376-377, 381 | sign | 372 | Projected window | 342-343 |
| copper | 377 | soffit | 312 | Projection screen | 371 |
| DWV | 380 | subfloor | 304 | Protection slope | 280 |
| grooved-joint | 382 | treatment | 303 | Pull box | 397 |
| malleable iron | 381 | underlayment | 304 | Pump circulating | 385 |
| no-hub | 377 | Pocket door frame | 334 | concrete | 267 |
| plastic | 379-380 | Point moil | 268 | diaphragm | 269 |
| soil | 376 | Pole closet | 309 | heat | 394 |
| steel | 381 | tele-power | 396 | jetting | 272 |
| Piping hydrant | 276 | Police connect panel | 402 | rental | 269 |
| Pit leeching | 276 | Polycarbonate glazing | 350 | submersible | 269, 385 |
| Pitch coal tar | 321 | Polyethylene film | 315 | sump | 373, 385 |
| emulsion tar | 278 | floor | 359 | trash | 270 |
| Placing concrete | 284 | pipe | 275 | water | 269, 272, 277 |
| Plain tube framing | 350 | tarpaulin | 265 | wellpoint | 272 |
| Plank floor | 302 | Polypropylene carpet | 358 | Purlin roof | 302 |
| roof | 303 | Polystyrene bead board | 318 | Push button lock | 348 |
| sheathing | 304 | blind | 311 | pull plate | 348 |
| Plant bed | 281 | ceiling | 357 | Puttying | 365 |
| Planter | 280 | ceiling panel | 357 | PVC conduit in-slab | 395 |
| Planting | 281 | insulation | 316-317 | coupling | 379 |
| Plaque | 372 | molded | 318 | DWV pipe | 379 |
| Plaster | 130, 132 | Polysulfide caulking | 315 | elbow | 379 |
| board | 126, 128 | Polyurethane caulking | 315 | flashing | 327 |
| ceiling | 353 | varnish | 365 | gravel stop | 327 |
| drill | 294 | Polyvinyl chloride pipe | 276 | pipe | 276, 378 |
| ground | 302 | chloride roof | 323 | siding | 312, 324 |
| gypsum | 352 | soffit | 329 | tee | 379 |
| mixer | 269 | Pool swimming | 157, 374 | | |
| painting | 363 | Porcelain enamel doors | 331 | **Q** | |
| perlite | 352 | enamel frame | 330 | Quarry tile | 355 |
| wall | 352 | tile | 355 | Quarter round molding | 309 |
| Plasterboard | 353-354 | Porch molding | 308 | | |
| Plastic blind | 99 | Porous wall pipe | 276 | **R** | |
| clad window | 74-85 | Portable air compressor | 268 | Raceway | 394 |
| faced hardboard | 309 | chiller | 373 | plugmold | 396 |
| laminated counter top | 307 | heater | 269 | wiremold | 396 |
| outlet-box | 398 | light | 268 | wireway | 396 |
| pipe | 276, 378-379 | partition | 370 | Radial arch | 314 |
| pipe-fitting | 379-380 | Post cap | 296 | Radiation fin-tube | 389 |
| siding | 70 | cedar | 306 | Radiator cast-iron | 389 |
| skylight | 329 | lamp | 293 | Rafter | 298 |
| trap | 383 | light | 401 | | |
| tubing | 275-276 | wood | 298 | | |

# INDEX

| | |
|---|---|
| anchor | 296 |
| roof framing | 40 |
| Rail balcony | 293 |
| pipe | 293 |
| trolley | 366 |
| Railing metal | 18, 293 |
| wood | 309-310, 312-313 |
| Railroad tie step | 278 |
| Ranch plank floor | 360 |
| Range | 150 |
| circuit | 198 |
| cooking | 372 |
| hood | 150, 372 |
| receptacle | 398 |
| wiring | 404 |
| Ready mix concrete | 283 |
| Receptacle | 198 |
| appliance | 403 |
| duplex | 403 |
| range | 398 |
| waste | 367 |
| wiring weatherproof | 404 |
| Recessed light | 199 |
| Rectangular diffuser | 392 |
| ductwork | 393 |
| Redwood cupola | 307 |
| louver | 307 |
| paneling | 309 |
| siding | 66, 311 |
| wine vault | 373 |
| Reflective insulation | 316 |
| Refrigeration residential | 372 |
| Refuse chute | 368 |
| Register air-supply | 392 |
| baseboard | 392 |
| Reglet | 328 |
| Reinforcing concrete | 282 |
| masonry | 285 |
| Release door | 403 |
| Relief valve P&T | 384 |
| valve pressure | 384 |
| Removal catch basin | 272 |
| chain link fence | 273 |
| fence | 273 |
| masonry | 273 |
| paint | 364 |
| stone | 273 |
| tree | 272 |
| Rental equipment | 270 |
| equipment rate | 265 |
| generator | 269 |
| Repair fire damage | 365 |
| Repellent water | 364 |
| Replacement sash | 346 |
| sliding door | 339 |
| Resaurant roof | 315 |
| Residential appliance | 372 |
| burner | 388 |
| closet door | 332 |
| door | 332-335 |
| elevator | 374 |
| fence | 278 |
| fixture | 401 |
| folding partition | 369 |

| | |
|---|---|
| greenhouse | 374 |
| hinge | 348 |
| lock | 348 |
| overhead door | 339 |
| rail | 293 |
| refrigeration | 372 |
| stair | 313 |
| storm door | 333 |
| ventilation | 393 |
| water-heater | 386 |
| wiring | 403-404 |
| Resilient floor | 358 |
| flooring | 143 |
| Resquared shingle | 320 |
| Restoration window | 364 |
| Retaining wall | 280, 283 |
| Ridge board | 299 |
| cap | 318 |
| shingle | 319-320 |
| shingle clay | 319 |
| shingle slate | 319 |
| vent | 328 |
| Rig drill | 272 |
| Rigid conduit in-trench | 395 |
| in-slab conduit | 395 |
| insulation | 72, 316-317 |
| Ring split | 296 |
| toothed | 296 |
| Riser beech | 145 |
| oak | 145 |
| wood stair | 312 |
| Rivet | 295 |
| tool | 295 |
| Road scraper | 266 |
| Robe hook | 367 |
| Rod closet | 310 |
| curtain | 366 |
| ground | 397 |
| Roll roof | 321 |
| type air-filter | 392 |
| up grille | 340 |
| Roller earth | 266 |
| Rolling door | 340-341 |
| service door | 340 |
| Romex copper | 396 |
| Roof adhesive | 321 |
| baffle | 317 |
| ballast | 323 |
| beam | 313 |
| bond | 322 |
| built-up | 321 |
| cant | 299, 322 |
| clay tile | 319 |
| copper | 325 |
| CPE | 322 |
| CSPE | 323 |
| deck | 292, 303 |
| deck insulation | 316 |
| deck laminated | 314 |
| deck wood | 314 |
| EPDM | 323 |
| framing | 40, 42, 44, 46, 48, 50 |
| hatch | 329 |
| lead | 328 |

| | |
|---|---|
| maintenance agreement | 322 |
| mineral | 321 |
| modified asphalt | 323 |
| modified bitumen | 323 |
| monel | 328 |
| mop | 322 |
| nail | 295 |
| neoprene | 323 |
| patch | 315 |
| PIB | 323 |
| polyvinyl chloride | 323-324 |
| purlin | 302 |
| rafter | 298 |
| reinforced PVC | 324 |
| resaurant | 315 |
| roll | 321 |
| sample tests | 322 |
| sheathing | 304 |
| shingle | 318 |
| single-ply | 322 |
| skylight | 329 |
| slate | 319 |
| tile | 319 |
| truss | 292, 302-303, 314 |
| ventilator | 317 |
| zinc | 329 |
| Roofing | 104, 106, 108, 110, 112 |
| built-up | 120 |
| cold | 322 |
| gable end | 104 |
| gambrel | 108 |
| hip roof | 106 |
| mansard | 110 |
| shed | 112 |
| Rooftop air-conditioner | 394 |
| Room humidifier | 389 |
| Rosewood door | 334 |
| Rosin paper | 315 |
| Rotary hammer drill | 268 |
| Rough buck | 298 |
| grade | 274 |
| hardware | 304 |
| Rough-in sink counter top | 386 |
| sink raised deck | 386 |
| sink service floor | 386 |
| tub | 384 |
| Round diffuser | 392 |
| Rubber base | 359 |
| corner guard | 360 |
| floor | 359 |
| floor tile | 359 |
| sheet | 359 |
| threshold | 348 |
| tired roller | 266 |

### S

| | |
|---|---|
| Safety switch | 399 |
| Salamander | 270 |
| Sales tax | 265 |
| Salt treatment lumber | 302 |
| Sand fill | 273 |
| Sandblasting equipment | 270 |
| Sanding | 365 |

433

# INDEX

| | |
|---|---|
| floor | 360 |
| Sandstone | 291 |
| flagging | 278 |
| Sandwich panel skylight | 330 |
| Sanitary cove base | 355 |
| Sash replacement | 346 |
| wood | 346 |
| Sauna | 373 |
| door | 373 |
| Saw | 270 |
| chain | 270 |
| circular | 270 |
| concrete | 267 |
| masonry | 270 |
| Scaffold | 265 |
| Scaffolding specialties | 265 |
| School door hardware | 347 |
| wardrobe | 369 |
| Scored block | 289 |
| Scrape after damage | 365 |
| Scraper | 266 |
| Scratch coat | 353 |
| Screen chimney | 285 |
| entrance | 371 |
| metal | 342 |
| molding | 308 |
| projection | 371 |
| security | 342 |
| squirrel and bird | 285 |
| urinal | 371 |
| window | 342, 346 |
| wood | 346 |
| Screw brass | 296 |
| lag | 294-295 |
| spline joint | 351 |
| Sealant | 314 |
| acoustical | 354 |
| caulking | 315 |
| Sealcoat | 278 |
| Sectional door | 339-340 |
| overhead door | 339 |
| Security sash | 342 |
| screen | 342 |
| Seeding | 281 |
| Self-propelled crane | 271 |
| Self-contained air-conditioner | 394 |
| Septic system | 16 |
| tank | 276 |
| Service electric | 196, 398 |
| entrance cable aluminum | 397 |
| entrance cap | 397 |
| sink | 386 |
| sink faucet | 375 |
| wiring | 404 |
| Sewer grating | 275 |
| pipe | 275 |
| Shadow box fence | 279 |
| Shake siding | 68 |
| wood | 320 |
| Shear plate | 296 |
| wall | 304 |
| Sheathed nonmetallic cable | 396 |
| Sheathing | 304 |
| asphalt | 304 |

| | |
|---|---|
| gypsum | 304 |
| insulation | 318 |
| paper | 315 |
| plywood | 304 |
| roof | 304 |
| Shed dormer framing | 56 |
| dormer roofing | 116 |
| roof framing | 50 |
| Sheepsfoot roller | 266 |
| Sheet base | 321-322 |
| floor | 358 |
| metal | 325 |
| rock | 126, 128 |
| Sheeting driver | 268 |
| Sheetrock | 354 |
| Shelf adjustable | 371 |
| brick | 282 |
| metal | 371 |
| Shellac door | 361 |
| Shelving wood | 310 |
| Shield expansion | 294 |
| termite | 329 |
| Shingle | 104-113, 318 |
| asbestos | 319 |
| asphalt | 319 |
| concrete | 319 |
| metal | 319 |
| ridge | 320 |
| roof | 318 |
| siding | 68 |
| stain | 363 |
| wood | 320 |
| Shock absorbing door | 340 |
| Shovel front | 266 |
| Shower by-pass valve | 375 |
| door | 370 |
| pan | 327 |
| partition | 370 |
| stall | 370, 385 |
| Shrub | 280, 282 |
| moving | 281 |
| Shutter | 99, 373 |
| Sidelight | 334 |
| Sidewalk | 12, 287 |
| asphalt | 278 |
| Siding aluminum | 320 |
| hardboard | 311 |
| metal | 70 |
| nail | 295 |
| painting | 363 |
| redwood | 311 |
| shingle | 68 |
| stain | 363 |
| steel | 324 |
| vinyl | 324 |
| wood | 66, 311 |
| Sign | 372 |
| Silencer duct | 393 |
| plenum | 393 |
| Silicone caulking | 315 |
| coating | 315 |
| Sill | 291 |
| door | 309, 333, 348 |
| masonry | 288 |

| | |
|---|---|
| quarry tile | 356 |
| Sillcock | 375 |
| Sills | 299 |
| Simulated log | 287 |
| Single hub pipe | 376 |
| hung window | 342 |
| zone rooftop unit | 394 |
| Single-ply roof | 322 |
| Sink | 150 |
| base | 305 |
| counter top | 386 |
| counter top rough-in | 386 |
| laundry | 385 |
| lavatory | 385 |
| raised deck rough-in | 386 |
| service | 386 |
| service floor rough-in | 386 |
| Siren | 402 |
| Site drainage | 275 |
| grading | 274 |
| removal | 272 |
| work crew | 259-260 |
| Sitework | 6 |
| electrical | 401 |
| Skim coat | 126, 128 |
| Skirtboard | 312 |
| Skylight | 118, 329 |
| roof | 329 |
| Skywindow | 118 |
| Slab blockout | 282 |
| concrete | 30, 284 |
| elevated | 284 |
| form | 292 |
| on grade | 282 |
| removal | 273 |
| Slate | 291 |
| chalkboard | 368 |
| flagging | 278 |
| roof | 319 |
| shingle | 319 |
| sidewalk | 278 |
| sill | 288 |
| stair | 291 |
| tile | 356 |
| Sleeper | 300 |
| wood | 143 |
| Sliding door | 88-89 |
| entrance | 341 |
| glass board | 367 |
| glass door | 338 |
| mirror | 369 |
| panel door | 341 |
| window | 80, 342, 346 |
| Slop sink | 385-386 |
| Slope protection | 280 |
| Slot letter | 369 |
| Slump block | 289 |
| Small tool | 265 |
| Smoke detector | 402 |
| detector wiring | 403 |
| vent-chimney | 391 |
| Snow melting | 402 |
| Soap dispenser | 366-367 |
| holder | 367 |

434

# INDEX

| | |
|---|---|
| Socket meter | 398 |
| Sodding | 281 |
| Sodium-low-pressure fixture | 401 |
| Soffit | 300 |
|    aluminum | 320 |
|    drywall | 354 |
|    metal | 329 |
|    plaster | 353 |
|    plywood | 312 |
|    vinyl | 324 |
|    wood | 312 |
| Softener water | 373 |
| Soil compaction | 273 |
|    compactor | 266 |
|    pipe | 376 |
|    tamping | 273 |
| Soldier course | 287 |
| Solid wood door | 334 |
| Sound attenuation | 357 |
|    system component | 402 |
| Space heater floor mounted | 390 |
|    heater rental | 269 |
| Spade air | 268 |
| Spanish roof tile | 319 |
| Special door | 338 |
|    stairway | 145 |
|    systems | 402, 404 |
| Specialties scaffolding | 265 |
| Specialty crew | 262 |
| Spike grid | 296 |
| Spiral stair | 293, 313 |
| Split rib block | 289 |
|    ring | 296 |
| Sprayed coating | 314 |
|    fireproofing | 353 |
| Sprayer emulsion | 268 |
| Spread footing | 282-283 |
| Spreader concrete | 269 |
| Square Foot Costs | 213-255 |
| Stacked bond block | 289 |
| Stain cabinet | 360 |
|    casework | 360 |
|    door | 361 |
|    floor | 361 |
|    shingle | 363 |
|    siding | 363 |
|    truss | 362 |
| Staining lumber | 314 |
|    siding | 312 |
| Stainless flashing | 327 |
|    gutter | 328 |
|    reglet | 329 |
|    screen | 371 |
|    steel bolt | 294 |
|    steel gravel stop | 327 |
|    steel hinge | 347-348 |
|    steel rivet | 295 |
|    steel shelf | 367 |
|    steel storefront | 341 |
|    tile | 355 |
| Stair | 312 |
|    basement | 145, 284, 313 |
|    ceiling | 369 |
|    climber | 374 |

| | |
|---|---|
|    disappearing | 368 |
|    electric | 369 |
|    pan form | 293 |
|    precast | 284 |
|    railroad tie | 278 |
|    residential | 313 |
|    slate | 291 |
|    spiral | 313 |
|    stringer | 298 |
|    tread | 145 |
|    tread quarry tile | 356 |
|    tread wood | 312 |
|    wood | 312-313 |
| Stairway | 144 |
|    curved | 145 |
|    door hardware | 347 |
|    special | 145 |
| Stall shower | 370, 385 |
|    toilet | 370 |
| Starters boards & switches | 399 |
| Starting newel | 312 |
| Steam bath | 373 |
|    bath, residential | 373 |
|    boiler | 387-388 |
|    boiler electric | 387 |
|    cleaner | 270 |
| Stearate coating | 315 |
| Steel bolt | 293 |
|    bridging | 297 |
|    chain link | 278 |
|    conduit in-slab | 395 |
|    conduit in-trench | 395 |
|    conduit intermediate | 394 |
|    diffuser | 392 |
|    door | 330-332, 338-339 |
|    downspout | 325 |
|    drill | 268 |
|    edging | 280 |
|    fireplace | 154 |
|    flashing | 327 |
|    frame | 330 |
|    furring | 352 |
|    gravel stop | 327 |
|    gutter | 328 |
|    hinge | 347 |
|    lintel | 285, 293 |
|    nailers | 298 |
|    pipe | 275 |
|    pipe-fitting | 381 |
|    railing | 293 |
|    shingle | 319 |
|    siding | 324 |
|    sill | 288 |
|    structural | 292 |
|    window | 343 |
| Steeple tip pin | 347 |
| Step masonry | 278 |
|    stone | 291 |
| Stiffleg derrick | 271 |
| Stockade fence | 280 |
| Stone aggregate | 280 |
|    anchor | 286 |
|    fill | 273 |
|    floor | 278, 291 |

| | |
|---|---|
|    ground cover | 280 |
|    paver | 290 |
|    paving | 277 |
|    removal | 273 |
|    step | 291 |
|    threshold | 288 |
|    trim cut | 291 |
|    veneer | 290-291 |
|    wall | 64, 280, 291 |
| Stool cap | 309 |
|    window | 288, 291 |
| Stop door | 309 |
|    gravel | 327 |
| Storage bottle | 373 |
| Storefront aluminum | 341 |
|    system | 100 |
| Storm door | 98, 333 |
|    window | 98, 343 |
| Stove woodburning | 287 |
| Strainer downspout | 325 |
|    gutter | 328 |
|    roof | 325 |
|    wire | 325 |
| Strap tie | 296 |
| Stringer | 145 |
|    stair | 298 |
| Strip cabinet | 403 |
|    footing | 283 |
|    lighting | 400 |
|    shingle | 319 |
| Structural excavation | 273 |
|    steel | 292 |
| Stucco | 132, 353 |
|    interior | 130 |
|    painting | 363 |
|    wall | 62 |
| Stud metal | 299 |
|    partition | 302 |
|    wall | 38, 302 |
|    wood | 300 |
| Subdrainage pipe | 276 |
| Subfloor | 304 |
|    plywood | 143, 304 |
| Submersible pump | 269, 277, 385 |
| Suction hose | 269 |
| Sump pump | 373, 385 |
| Sunroom | 156 |
| Support drinking fountain | 383 |
|    lavatory | 383 |
| Surface landscape | 359 |
| Survey crew | 265 |
| Suspended ceiling | 134, 352, 357 |
| Suspension system ceiling | 357 |
| Swimming pool | 157, 374 |
|    pool enclosure | 374 |
| Swing check valve | 384 |
|    door | 341 |
| Swing-up door | 94 |
|    overhead door | 340 |
| Switch | 398 |
|    box plastic | 398 |
|    dimmer | 398 |
|    electric | 399 |
|    general duty | 399 |

# INDEX

safety . . . . . . . . . . . . . . . . . . . . . . 399
time . . . . . . . . . . . . . . . . . . . . . . . 399
toggle . . . . . . . . . . . . . . . . . . . . . . 398
wiring . . . . . . . . . . . . . . . . . 198, 404
Switchbox . . . . . . . . . . . . . . . . . . . . . 398
Synthetic marble . . . . . . . . . . . . . . 355
Syphon ventilator rotary . . . . . . . 394
System antenna . . . . . . . . . . . . . . 403
fire extinguishing . . . . . . . . . . . 387
square foot . . . . . . . . . . . . . 213-255

## T

T & G siding . . . . . . . . . . . . . . . . . . . 66
T-bar mount diffuser . . . . . . . . . . 392
Table top . . . . . . . . . . . . . . . . . . . . . 307
Tamper . . . . . . . . . . . . . . . . . . . . . . 268
Tamping soil . . . . . . . . . . . . . . . . . 273
Tandem roller . . . . . . . . . . . . . . . . 266
Tank expansion . . . . . . . . . . . . . . 390
fiberglass . . . . . . . . . . . . . . . . . . 390
oil & gas . . . . . . . . . . . . . . . . . . 390
septic . . . . . . . . . . . . . . . . . . 16, 276
Tar pitch emulsion . . . . . . . . . . . . 278
Tarpaulin . . . . . . . . . . . . . . . . . . . . 265
duck . . . . . . . . . . . . . . . . . . . . . . 265
polyethylene . . . . . . . . . . . . . . . 265
Tarred felt . . . . . . . . . . . . . . . . . . . 322
Tax . . . . . . . . . . . . . . . . . . . . . . . . . 265
sales . . . . . . . . . . . . . . . . . . . . . 265
social security . . . . . . . . . . . . . 265
unemployment . . . . . . . . . . . . . 265
Teak floor . . . . . . . . . . . . . . . 143, 360
molding . . . . . . . . . . . . . . . . . . . 309
paneling . . . . . . . . . . . . . . . . . . 310
Tee cleanout . . . . . . . . . . . . . . . . . 375
pipe . . . . . . . . . . . . . . . . . . . . . . 382
PVC . . . . . . . . . . . . . . . . . . . . . . 379
Tele-power pole . . . . . . . . . . . . . . 396
Telephone booth . . . . . . . . . . . . . 372
enclosure . . . . . . . . . . . . . . . . . 372
house . . . . . . . . . . . . . . . . . . . . 402
Temperature relief valve . . . . . . . 384
Tempered glass . . . . . . . . . . . . . . 350
hardboard . . . . . . . . . . . . . . . . . 309
Tempering valve . . . . . . . . . . . . . . 384
Temporary oil heater . . . . . . . . . . 269
toilet . . . . . . . . . . . . . . . . . . . . . 270
Tennis court fence . . . . . . . . . 18, 279
Terminal air-conditioner
packaged . . . . . . . . . . . . . . . . . 394
Termination box . . . . . . . . . . . . . . 403
Termite shield . . . . . . . . . . . . . . . . 329
Terne coated flashing . . . . . . . . . 327
Terra cotta coping . . . . . . . . . . . . 286
cotta pipe . . . . . . . . . . . . . . . . . 276
Terrace . . . . . . . . . . . . . . . . . . . . . 278
Terrazzo base . . . . . . . . . . . . . . . . 356
epoxy . . . . . . . . . . . . . . . . . . . . 358
floor . . . . . . . . . . . . . . . . . . . . . . 356
precast . . . . . . . . . . . . . . . . . . . 356
wainscot . . . . . . . . . . . . . . . . . . 356
Testing pile . . . . . . . . . . . . . . . . . . 275
Texture 1-11 painting . . . . . . . . . . 363

Thermostat integral . . . . . . . . . . . 402
Thin coat plaster . . . . . . . . . . . . . 354
set ceramic tile . . . . . . . . . . . . 355
Thincoat . . . . . . . . . . . . . . . . 126, 128
Thinset mortar . . . . . . . . . . . . . . . 356
Threshold . . . . . . . . . . . . . . . . . . . 348
door . . . . . . . . . . . . . . . . . . . . . . 334
stone . . . . . . . . . . . . . . . . . 288, 291
wood . . . . . . . . . . . . . . . . . . . . . 309
Thru-wall air-conditioner . . . . . . . 394
Tie masonry . . . . . . . . . . . . . . . . . 285
rafter . . . . . . . . . . . . . . . . . . . . . 299
strap . . . . . . . . . . . . . . . . . . . . . 296
Tile . . . . . . . . . . . . . . . . . . . . . . . . . 359
aluminum . . . . . . . . . . . . . . . . . 319
carpet . . . . . . . . . . . . . . . . . . . . 358
ceiling . . . . . . . . . . . . . . . . 356-357
ceramic . . . . . . . . . . . . . . . . . . . 355
cork wall . . . . . . . . . . . . . . . . . . 366
marble . . . . . . . . . . . . . . . . . . . . 355
metal . . . . . . . . . . . . . . . . . . . . . 355
plastic . . . . . . . . . . . . . . . . . . . . 355
porcelain . . . . . . . . . . . . . . . . . . 355
quarry . . . . . . . . . . . . . . . . . . . . 355
roof . . . . . . . . . . . . . . . . . . . . . . 319
slate . . . . . . . . . . . . . . . . . . . . . 356
vinyl asbestos . . . . . . . . . . . . . 359
wall . . . . . . . . . . . . . . . . . . . . . . 355
Timber connector . . . . . . . . . . . . . 295
fastener . . . . . . . . . . . . . . . . . . 295
framing . . . . . . . . . . . . . . . . . . . 301
laminated . . . . . . . . . . . . . . . . . 313
roof deck . . . . . . . . . . . . . . . . . 303
Time switch . . . . . . . . . . . . . . . . . . 399
Timer switch ventilator . . . . . . . . 394
Tin clad door . . . . . . . . . . . . . . . . 341
Tinted glass . . . . . . . . . . . . . . . . . 350
Toe plate . . . . . . . . . . . . . . . . . . . . 293
Toggle bolt . . . . . . . . . . . . . . . . . . 294
switch . . . . . . . . . . . . . . . . . . . . 398
Toilet accessory . . . . . . . . . . . . . . 366
bowl . . . . . . . . . . . . . . . . . . . . . . 386
chemical . . . . . . . . . . . . . . . . . . 270
door hardware . . . . . . . . . . . . . 347
partition . . . . . . . . . . . . . . . 370-371
room system . . . . . . . . . . . . . . 184
stall . . . . . . . . . . . . . . . . . . . . . . 370
temporary . . . . . . . . . . . . . . . . . 270
tissue dispenser . . . . . . . . 366-367
trailer . . . . . . . . . . . . . . . . . . . . . 270
Tongue and groove siding . . . . . . 66
Tool air . . . . . . . . . . . . . . . . . . . . . 268
rivet . . . . . . . . . . . . . . . . . . . . . . 295
small . . . . . . . . . . . . . . . . . . . . . 265
Toothed ring . . . . . . . . . . . . . . . . . 296
Top counter . . . . . . . . . . . . . 306-307
dressing . . . . . . . . . . . . . . . . . . 281
table . . . . . . . . . . . . . . . . . . . . . 307
Topping epoxy . . . . . . . . . . . . . . . 358
Torch cutting . . . . . . . . . . . . . . . . 270
Towel bar . . . . . . . . . . . . . . . . . . . 367
dispenser . . . . . . . . . . . . . 366-367
Tower crane . . . . . . . . . . . . . . . . . 270
hoist . . . . . . . . . . . . . . . . . . . . . 271

light . . . . . . . . . . . . . . . . . . . . . . 269
Track lighting . . . . . . . . . . . . . . . . 401
Tractor . . . . . . . . . . . . . . . . . . 266, 274
loader . . . . . . . . . . . . . . . . . . . . 267
truck . . . . . . . . . . . . . . . . . . . . . 270
Trailer platform . . . . . . . . . . . . . . 270
toilet . . . . . . . . . . . . . . . . . . . . . 270
truck . . . . . . . . . . . . . . . . . 267, 270
Transformer cabinet . . . . . . . . . . 397
Transit . . . . . . . . . . . . . . . . . . . . . . 270
Transom lite frame . . . . . . . . . . . 330
Trap cast-iron . . . . . . . . . . . . . . . . 383
drainage . . . . . . . . . . . . . . . . . . 383
plastic . . . . . . . . . . . . . . . . . . . . 383
Trash pump . . . . . . . . . . . . . . . . . 270
Travertine . . . . . . . . . . . . . . . . . . . 291
Tray laundry . . . . . . . . . . . . . . . . . 385
Tread abrasive . . . . . . . . . . . . . . . 356
beech . . . . . . . . . . . . . . . . . . . . 145
oak . . . . . . . . . . . . . . . . . . . . . . 145
stone . . . . . . . . . . . . . . . . . . . . . 291
wood . . . . . . . . . . . . . . . . . . . . . 312
Treatment plywood . . . . . . . . . . . 303
wood . . . . . . . . . . . . . . . . . . . . . 302
Tree . . . . . . . . . . . . . . . . . . . . . . . . 281
guyed . . . . . . . . . . . . . . . . . . . . 281
moving . . . . . . . . . . . . . . . . . . . 281
removal . . . . . . . . . . . . . . . . . . . 272
Trench backfill . . . . . . . . . . . . . . . 274
box . . . . . . . . . . . . . . . . . . . . . . 270
excavation . . . . . . . . . . . . . . . . 274
fill . . . . . . . . . . . . . . . . . . . . . . . 275
Trencher . . . . . . . . . . . . . . . . . . . . 267
chain . . . . . . . . . . . . . . . . . . . . . 267
Trenching . . . . . . . . . . . . . . . . . . . . 10
Trim exterior . . . . . . . . . . . . . . . . . 308
painting . . . . . . . . . . . . . . . . . . . 362
tile . . . . . . . . . . . . . . . . . . . . . . . 355
wood . . . . . . . . . . . . . . . . . . . . . 308
Trolley rail . . . . . . . . . . . . . . . . . . . 366
Trowel concrete . . . . . . . . . . . . . . 267
finish . . . . . . . . . . . . . . . . . . . . . 283
Troweled coating . . . . . . . . . . . . . 314
Truck dump . . . . . . . . . . . . . . . . . 267
flatbed . . . . . . . . . . . . . . . . . . . . 267
hauling . . . . . . . . . . . . . . . . . . . 274
loading . . . . . . . . . . . . . . . . . . . 273
mounted crane . . . . . . . . . . . . 271
off highway . . . . . . . . . . . . . . . . 267
pickup . . . . . . . . . . . . . . . . . . . . 270
rental . . . . . . . . . . . . . . . . . . . . . 267
tractor . . . . . . . . . . . . . . . . . . . . 270
trailer . . . . . . . . . . . . . . . . . 267, 270
Truss bowstring . . . . . . . . . . . . . . 314
flat wood . . . . . . . . . . . . . . . . . . 303
metal . . . . . . . . . . . . . . . . . . . . . 292
painting . . . . . . . . . . . . . . . . . . . 362
plate . . . . . . . . . . . . . . . . . . . . . 296
roof . . . . . . . . . . . . . . . 302-303, 314
roof framing . . . . . . . . . . . . . . . . 42
stain . . . . . . . . . . . . . . . . . . . . . 362
varnish . . . . . . . . . . . . . . . . . . . 362
Tub bath . . . . . . . . . . . . . . . . . . . . 384
hot . . . . . . . . . . . . . . . . . . . . . . . 384

# INDEX

| | |
|---|---|
| redwood | 384 |
| rough-in | 384 |
| Tub/shower module | 384 |
| Tube framing | 100, 350 |
| Tubing copper | 276, 377 |
| electric metallic | 394 |
| plastic | 275-276 |
| Tubular steel joist | 304 |
| Tumbler holder | 367 |
| Turned column | 306 |
| TV antenna | 403 |

## U

| | |
|---|---|
| Under eave vent | 328 |
| Underground duct | 401 |
| tank | 390 |
| Underlayment | 304 |
| felt | 358 |
| latex | 360 |
| Unemployment tax | 265 |
| Unit heater electric | 402 |
| Urethane fire resistant | 318 |
| insulation | 318 |
| roof deck | 316 |
| wall coating | 365 |
| Urinal screen | 371 |
| Utilities electric | 401 |
| Utility | 10 |
| excavation | 10 |
| sitework | 401 |
| trench | 274 |
| vault | 277 |

## V

| | |
|---|---|
| Vacuum breaker | 383 |
| cleaning | 373 |
| Valance board | 306 |
| Valley flashing | 319 |
| rafter | 299 |
| Valve | 384 |
| angle | 384 |
| gate | 276 |
| pressure relief | 384 |
| shower by-pass | 375 |
| swing check | 384 |
| tempering | 384 |
| water pressure | 384 |
| Vanity base | 313 |
| top lavatory | 385 |
| Vapor barrier | 315 |
| barrier sheathing | 304 |
| Variable volume damper | 392 |
| Varnish cabinet | 361 |
| casework | 361 |
| door | 361 |
| floor | 361, 365 |
| polyurethane | 365 |
| truss | 362 |
| Vault utility | 277 |
| wine | 373 |
| Veneer brick | 64 |
| core paneling | 310 |

| | |
|---|---|
| granite | 290 |
| stone | 290-291 |
| Vent exhaust | 394 |
| metal chimney | 391 |
| ridge | 328 |
| ridge strip | 328 |
| Vent-chimney | 391 |
| all fuel | 391 |
| fitting | 391 |
| flashing | 391 |
| Ventilating | |
| air-conditioning | 393-394 |
| Ventilation louver | 307 |
| residential | 393 |
| Ventilator roof | 317 |
| rotary syphon | 394 |
| timer switch | 394 |
| Verge board | 308 |
| Vermiculite insulation | 316 |
| Vertical aluminum siding | 320 |
| vinyl siding | 324 |
| Vibrator concrete | 267 |
| earth | 266, 273 |
| Vibratory equipment | 266 |
| Vinyl asbestos tile | 359 |
| blind | 99, 311 |
| chain link | 279 |
| composition floor | 359 |
| corner guard | 360 |
| door trim | 324 |
| faced wallboard | 354 |
| floor | 359 |
| flooring | 143 |
| gutter | 328 |
| pool | 157 |
| shutter | 99 |
| siding | 70, 324 |
| siding accessories | 324 |
| wall coating | 365 |
| wallpaper | 366 |
| window trim | 324 |
| Vitrified clay pipe | 276 |
| Volume control damper | 393 |

## W

| | |
|---|---|
| Wagon drill | 268 |
| Wainscot molding | 309 |
| quarry tile | 356 |
| terrazzo | 356 |
| Walk | 278 |
| Wall brick | 288 |
| bumper | 347, 366 |
| cabinets | 305 |
| canopy | 367 |
| ceramic tile | 355 |
| coating | 365 |
| concrete | 284 |
| covering | 366 |
| drywall | 353 |
| finish | 283 |
| formwork | 282 |
| foundation | 290 |
| framing | 38 |

| | |
|---|---|
| furnace | 390 |
| furring | 302, 352 |
| grout | 285 |
| heater electric | 402 |
| insulation | 317 |
| interior | 126, 130 |
| lath | 352 |
| masonry | 62, 64, 280, 288 |
| painting | 363-364 |
| panel | 290 |
| paneling | 310 |
| plaster | 352 |
| plate | 398 |
| reinforcing | 282 |
| retaining | 280, 283 |
| shear | 304 |
| sheathing | 304 |
| siding | 312 |
| stucco | 353 |
| stud | 302 |
| switch plate | 398 |
| tile | 355 |
| tile cork | 366 |
| window | 100 |
| Wallcovering acrylic | 366 |
| Wallguard | 366 |
| Wallpaper | 366 |
| Walnut door frame | 333 |
| floor | 360 |
| Wardrobe | 369 |
| school | 369 |
| Wash bowl | 385 |
| Washer | 296 |
| Waste receptacle | 367 |
| Water distribution system | 276 |
| heating wiring | 198 |
| hose | 269 |
| pipe ground clamp | 397 |
| pressure relief valve | 384 |
| pressure valve | 384 |
| pump | 269, 272, 277, 373, 385 |
| repellent | 364 |
| softener | 373 |
| supply domestic meter | 384 |
| supply meter | 384 |
| tempering valve | 384 |
| trailer | 270 |
| well | 277 |
| Water-closet | 386 |
| support | 383 |
| Water-heater circuit wiring | 404 |
| electric | 386 |
| insulation | 389 |
| oil | 387 |
| wrap kit | 389 |
| Waterproofing coating | 315 |
| Watertank sprayer | 270 |
| Weather cap entrance | 395 |
| strip window | 348 |
| Weatherproof receptacle | 198 |
| receptacle wiring | 404 |
| Weatherstrip | 348 |
| door | 348 |
| Weatherstripping | 340 |

437

# INDEX

| | |
|---|---|
| Weathervane | 293 |
| Welded frame | 330 |
|     wire mesh | 283 |
| Welding | 292 |
|     machine | 270 |
| Well | 277 |
| Wellpoint equipment | 271 |
|     header pipe | 272 |
|     pump | 272 |
| Wheelbarrow | 270 |
| Whirlpool bath | 384 |
| Window | 74–85 |
|     aluminum | 342 |
|     blind | 311 |
|     casement | 344 |
|     casing | 308 |
|     double hung | 74, 345 |
|     frame | 346 |
|     glass | 350 |
|     grille | 346 |
|     hardware | 347 |
|     metal | 342 |
|     muntin | 346 |
|     painting | 361 |
|     picture | 345 |
|     restoration | 364 |
|     screen | 342, 346 |
|     sill | 288 |
|     sill marble | 291 |
|     sill quarry tile | 356 |
|     sliding | 346 |
|     steel | 343 |
|     stool | 288, 291 |
|     storm | 98, 343 |
|     trim set | 309 |
|     wall | 100 |
|     wall framing | 350 |
|     weather strip | 348 |
|     wood | 344–346 |
| Wine vault | 373 |
| Winter protection | 284 |
| Wire copper | 396 |
|     electric | 396 |
|     ground | 397 |
|     mesh | 283, 353 |
|     strainer | 325 |
|     THW | 396 |
|     to grounding brazed | 397 |
| Wiremold raceway | 396 |
| Wireway raceway | 396 |
| Wiring | 198 |
|     appliance | 403 |
|     device | 398 |
|     furnace | 403 |
|     light fixture | 404 |
|     lighting | 404 |
|     power | 398 |
|     range | 404 |
|     residential | 403–404 |
|     service | 404 |
|     switch | 404 |
| Wood base | 308 |
|     beams | 297 |
|     blind | 99, 311 |
| block floor | 360 |
| blocking | 121, 123, 303 |
| canopy | 300 |
| casing | 308 |
| column | 301, 306 |
| columns | 298 |
| cupola | 307 |
| deck | 158, 303, 314 |
| door | 136, 138, 335 |
| fascia | 308 |
| fastener | 296 |
| fence | 279 |
| fiber ceiling | 357 |
| fiber sheathing | 304 |
| fiber soffit | 312 |
| fiber subfloor | 304 |
| fiber underlayment | 305 |
| floor | 34, 143, 360 |
| folding partition | 369 |
| foundation | 28 |
| frame | 333 |
| framing | 297 |
| furring | 302 |
| girders | 297 |
| gutter | 328 |
| handrail | 309 |
| joist | 297, 304 |
| laminated | 313–314 |
| louver | 307 |
| molding | 312 |
| nailers | 298 |
| overhead door | 339 |
| panel door | 334 |
| paneling | 309 |
| partition | 300, 302 |
| pile | 275 |
| railing | 310, 312 |
| roof deck | 303 |
| sash | 346 |
| screen | 346 |
| shake | 320 |
| sheathing | 304 |
| shelving | 310 |
| shingle | 320 |
| shutter | 99 |
| sidewalk | 278 |
| siding | 66, 311 |
| sill | 299 |
| sliding door | 338 |
| soffit | 312 |
| stair | 312–313 |
| storm door | 98, 336 |
| storm window | 98 |
| stud | 300 |
| subfloor | 304 |
| threshold | 309 |
| tread | 312 |
| treatment | 302 |
| trim | 308 |
| truss | 303 |
| veneer wallpaper | 366 |
| wall framing | 38 |
| window | 344–346 |
| Woodburning stove | 287 |
| Wool carpet | 142 |
|     fiberglass | 316 |
| Wrecking ball | 270 |
| Wrench impact | 268 |

### Y

| | |
|---|---|
| Yellow pine floor | 360 |

### Z

| | |
|---|---|
| Z bar suspension | 357 |
| Zee bar | 353 |
| Zinc divider strip | 356 |
|     flashing | 327 |
|     plating | 295 |
|     roof | 329 |
|     terrazzo strip | 356 |
|     weatherstrip | 348 |

# NOTES

# 1985—Means Seminar Program—1985

*During the upcoming year, R.S. Means offers you a series of 2-day seminars oriented to a wide range of construction-related topics. Conducted throughout the United States, scheduled seminars thoroughly cover the subjects of cost estimating, scheduling, management, and computerized systems for construction. R.S. Means is proud to present these seminars, each specifically designed to give you the expertise you need to be successful in your profession. All seminars include comprehensive workbooks plus the latest innovations in computerized estimating and scheduling.*

## Repair & Remodeling Estimating
(two days) #104

Here are the specialized techniques you need to approach repair and remodeling projects with new skill and confidence. This seminar sorts out and discusses practical solutions to the difficult and varied cost estimating problems associated with the reuse and conversion of existing structures. A wealth of eye-opening procedures is presented, including information about computerized cost estimating as applied to repair and remodeling projects. Based on optimum use of the Means cost manual *Repair & Remodeling Cost Data*. A "must" for new estimators.

## Unit Price Estimating
(two days) #100

For those with estimating experience, this course shows you how today's advanced estimating techniques and cost information sources (such as computerized estimating systems and cost data banks) can be used to develop unit price estimates for projects of any size. You'll get down-to-earth know-how and easy-to-apply directions on organizing information effectively, using plans efficiently, ensuring error-free results, and maximizing your use of cost data resources, including *Building Construction Cost Data*.

## Square Foot Cost Estimating
(two days) #106

For estimators with experience, here's the expert guidance you've asked for. We'll help you develop the practical know-how for producing rapid, reliable square foot cost estimates using only minimal design details. You'll perfect your skills for "before plans" estimating, replacement cost estimating, and assessing buildings of all types. In square foot costing methods, participants receive clear-cut explanations and suggestions for dealing with several types of construction projects at various stages of completion. Centered around techniques used in *Means Systems Costs*, *Means Square Foot Costs*, and *Square Foot Estimating*.

## Scheduling and Project Management
(two days) #102

This seminar helps you successfully establish project priorities, develop realistic schedules, and apply today's advanced management techniques to your construction projects. Included are the "network" approach, Precedence, and Critical Path Methods. Special emphasis is placed on cost interface with various monitoring setups such as computer-based control systems. Through this seminar you'll perfect your scheduling and management skills, ensuring completion of your projects *on time* and *within budget*. Includes hands-on application of *Means Scheduling Manual* and *Building Construction Cost Data*.

## Computers in Construction
(two days) #112

Prepared for those interested in using computers for construction activities, this seminar describes and demonstrates the latest available hardware and software products for construction cost estimating, project scheduling, and accounting applications. Actual demonstrations of the Means line of construction-oriented software — as well as other construction and business software — highlight this Seminar. Optimum use is made of *Building Construction Cost Data*.

### 1985 SPRING SEMINARS

| Location | Seminar No. & Name | Dates (1985) |
|---|---|---|
| **TAMPA** Bay Harbor Inn 7700 Courtney Campbell Causeway Tampa, FL 33607 (813) 885-2541 | 10453TA Repair & Remodeling Est. 10653TA Square Foot Cost Est. 10053TA Unit Price Estimating 10253TA Sched. & Proj. Mgmt. 11253TA Computers in Constr. | Mar. 18 & 19 Mar. 18 & 19 Mar. 20 & 21 Mar. 20 & 21 Mar. 20 & 21 |
| **BEVERLY HILLS** Ramada Inn 1150 S. Beverly Drive Beverly Hills, CA 90035 (213) 553-6561 | 10453BH Repair & Remodeling Est. 10653BH Square Foot Cost Est. 10053BH Unit Price Estimating 10253BH Sched. & Proj. Mgmt. 11253BH Computers in Constr. | Mar. 25 & 26 Mar. 25 & 26 Mar. 27 & 28 Mar. 27 & 28 Mar. 27 & 28 |
| **BOSTON** Natick Hilton Inn Route 9, Speen Street Natick, MA 01760 (617) 653-5000 | 10454BO Repair & Remodeling Est. 10654BO Square Foot Cost Est. 10054BO Unit Price Estimating 10254BO Sched. & Proj. Mgmt. 11254BO Computers in Constr. | Apr. 1 & 2 Apr. 1 & 2 Apr. 3 & 4 Apr. 3 & 4 Apr. 3 & 4 |
| **DENVER** Holiday Inn 1475 S. Colorado Blvd. Denver, CO 80222 (303) 757-7731 | 10454DE Repair & Remodeling Est. 10654DE Square Foot Cost Est. 10054DE Unit Price Estimating 10254DE Sched. & Proj. Mgmt. 11254DE Computers in Constr. | Apr. 15 & 16 Apr. 15 & 16 Apr. 17 & 18 Apr. 17 & 18 Apr. 17 & 18 |
| **WASHINGTON, DC** Howard Johnson's-Airport 2650 Jefferson Davis Arlington, VA 22202 (703) 684-7200 | 10454AR Repair & Remodeling Est. 10654AR Square Foot Cost Est. 10054AR Unit Price Estimating 10254AR Sched. & Proj. Mgmt. 11254AR Computers in Constr. | Apr. 29 & 30 Apr. 29 & 30 May 1 & 2 May 1 & 2 May 1 & 2 |
| **PHOENIX** Holiday Inn Metro Center Holidome 2532 W. Peoria Ave. Phoenix, AZ 85029 (602) 943-2341 | 10455PX Repair & Remodeling Est. 10655PX Square Foot Cost Est. 10055PX Unit Price Estimating 10255PX Sched. & Proj. Mgmt. 11255PX Computers in Constr. | May 20 & 21 May 20 & 21 May 22 & 23 May 22 & 23 May 22 & 23 |
| **NEW YORK** Westchester Marriott 670 White Plains Rd. Tarrytown, NY 10591 (914) 631-2200 | 10456TT Repair & Remodeling Est. 10656TT Square Foot Cost Est. 10056TT Unit Price Estimating 10256TT Sched. & Proj. Mgmt 11256TT Computers in Constr. | Jun 3 & 4 Jun 3 & 4 Jun 5 & 6 Jun 5 & 6 Jun 5 & 6 |
| **ATLANTA** Omni International Hotel One Omni International Plaza Atlanta, GA 30303 (404) 659-0000 | 10456AT Repair & Remodeling Est. 10656AT Square Foot Cost Est. 10056AT Unit Price Estimating 10256AT Sched. & Proj. Mgmt. 11256AT Computers in Constr. | Jun 24 & 25 Jun 24 & 25 Jun 26 & 27 Jun 26 & 27 Jun 26 & 27 |

### 1985 FALL SEMINARS

| Location | Seminar No. & Name | Dates (1985) |
|---|---|---|
| **HYANNIS** Sheraton Regal Inn Rt. 132 & Bearse Way Hyannis, MA 02601 (617) 771-3000 | 10459HY Repair & Remodeling Est. 10659HY Square Foot Cost Est. 10059HY Unit Price Estimating 10259HY Sched. & Proj. Mgmt. 11259HY Computers in Constr. | Sept. 9 & 10 Sept. 9 & 10 Sept. 11 & 12 Sept. 11 & 12 Sept. 11 & 12 |
| **SAN FRANCISCO** San Franciscan Hotel 1231 Market Street at Civic Center San Francisco, CA 94103 (415) 626-8000 | 1045ASF Repair & Remodeling Est. 1065ASF Square Foot Cost Est. 1005ASF Unit Price Estimating 1025ASF Sched. & Proj. Mgmt. 1125ASF Computers in Constr. | Oct. 7 & 8 Oct. 7 & 8 Oct. 9 & 10 Oct. 9 & 10 Oct. 9 & 10 |
| **WASHINGTON, DC** Howard Johnson's-Airport 2650 Jefferson Davis Arlington, VA 22202 (703) 684-7200 | 1045BAR Repair & Remodeling Est. 1065BAR Square Foot Cost Est. 1005BAR Unit Price Estimating 1025BAR Sched. & Proj. Mgmt. 1125BAR Computers in Constr. | Nov. 4 & 5 Nov. 4 & 5 Nov. 6 & 7 Nov. 6 & 7 Nov. 6 & 7 |
| **DALLAS** Amfac Hotel & Resort P.O. Box 61025 Dallas/Ft. Worth Airport Dallas, TX 75261 (214) 453-8400 | 1045CDA Repair & Remodeling Est. 1065CDA Square Foot Cost Est. 1005CDA Unit Price Estimating 1025CDA Sched. & Proj. Mgmt. 1125CDA Computers in Constr. | Dec. 2 & 3 Dec. 2 & 3 Dec. 4 & 5 Dec. 4 & 5 Dec. 4 & 5 |

# 1985 — Seminar Registration Form and Information — 1985

## Registration Fees
- One 2-day seminar, $550 per person
- One 2-day seminar, two or more people registering at the same time, $500 per person
- Two 2-day seminars, $900 per person
- Two 2-day seminars, two or more people registering at the same time, $815 per person

Full payment must be received at least 15 days prior to seminar dates in order to confirm attendance (U.S. Funds).

## How to Register
Complete the registration form and return with your full fee (or minimum deposit of $220/person for one 2-day seminar, or minimum deposit of $360/person for two 2-day seminars) to:

Seminar Division
R.S. Means Company, Inc.
100 Construction Plaza
P.O. Box 800
Kingston, MA 02364-0800

*Class sizes are limited so please return your registration form as soon as possible.*

## Refunds
Advance payments are fully refundable up to 15 days prior to the seminar dates. After that time, a 2-day seminar cancellation is subject to a service charge of $160/person and two 2-day seminar cancellations $270/person.

## Tax Deduction
A federal income tax deduction is allowed for expenses (including registration fees, travel, meals and lodging) for your attendance at a Means Seminar (see Treas. Reg. 1.162-5).

## AACE Approved Courses:
The R.S. Means Construction Estimating and Management Seminars described and offered to you here have each been approved for 14 hours (1.4 CEU) of credit by the American Association of Cost Engineers (AACE, Inc.) Certification Board toward meeting the continuing education requirements for re-certification as a Certified Cost Engineer/Certified Cost Consultant.

## Coffee/Cola Breaks
Your registration includes the cost of a coffee break in the morning and an afternoon cola break. These informal segments will allow you to discuss topics of mutual interest with other members of the seminar.

## Late Telephone Registrations
If you wish to register for any seminar with less than two weeks remaining before the seminar date, please call our Kingston office (617) 747-1270 and ask for the Seminar Registrar. Arrangements may be made if space is available.

## Daily Course Schedule
The first day of each seminar session begins at 8:30 A.M. and ends at 4:30 P.M. The second day is 8:00 - 4:00. Participants are urged to bring a hand-held calculator since many actual problems will be worked out in each session.

## Transportation/Hotel Accommodations
For your convenience, you can obtain assistance in making transportation and hotel arrangements by calling Waterfront Travel in Boston, MA at 1-800-343-6524, and notifying them of your planned attendance at the Means Seminar of your choice. Reservations for lodging can be made through the Waterfront Travel Service. Airline reservations, hotel accommodations and car rental arrangements will be made for you by Waterfront to fit your schedule. We suggest that you contact Waterfront Travel as soon as you return your registration form, in order to reserve the lodging of your choice. Reservations should be received by Waterfront Travel Service one month prior to the seminar date.

## In-House Presentation
All R.S. Means seminars are available for in-house presentation and can be tailored to meet individual client/company needs. If you are interested in having a program conducted at your location, just contact Dan Braman, P.E., Director of Sales, R.S. Means Co., Inc. 156 Construction Plaza, P.O. Box 800, Kingston, MA 02364-0800. Telephone: (617) 747-1270.

---

# Registration Form

Please register the following people for the Means Construction Seminars as shown here. Full payment or deposit is enclosed, and we understand we must make our own hotel reservations if overnight stays are necessary.

☐ Full payment of $ _____ enclosed.
☐ Deposit of $ _____ enclosed.
   Balance due $ _____
   (U.S. Funds)

Firm Name _____
Address _____
City/State/Zip _____
Telephone Number _____

☐ Charge our registration(s) to:
___ American Express  ___ Visa  ___ MasterCard
Account No. _____ Exp. Date _____
_____
CARDHOLDER'S SIGNATURE

**NAME OF REGISTRANT(S)**          **SEMINAR NAME AND NUMBER**          **CITY**          **DATES**
(To appear on certificate of completion)

*Please mail check to:* **R.S. MEANS CO., INC., 156 Construction Plaza, P.O. Box 800, Kingston, MA 02364-0800**

# GALAXY
## Automated Quantity Takeoff and Pricing System

*Modern technology can save you money by cutting takeoff time at least 50%*

Quantity takeoffs don't have to be difficult—not when you can take advantage of modern technology. GALAXY Automated Takeoff and Pricing System lets you determine the detailed cost information you need with the aid of a computer.

You can reduce the time it takes to do an estimate by as much as 50 percent, and more for many projects, thanks to GALAXY's electronic digitizer table. Working directly from a set of plans, an estimator easily feeds the information into the computer, which then makes the proper calculations to determine length, height, width, volume, area or numeric count—all in a fraction of the time it normally would take.

Using a special hand-held cursor, specific points are selected from the plans and "read" into the computer. The estimator chooses the items in convenient order and is able to quickly determine the quantities as desired. Designed for use with the STAR General Estimating Program, GALAXY provides you with access to your own files of known costs as well as any of the R.S. Means standard cost information files of up-to-date unit price information.

Best of all, GALAXY is easy to operate. Any construction professional can learn to run this system after attending the training workshop. It's all part of The Solution by R.S. Means Company, Inc., the people you depend on for your construction cost data.

## GALAXY Main Benefits

- Saves valuable time during takeoffs
- Easy to operate
- Increases productivity
- Reduces need for additional manpower
- Streamlines the estimating process
- Produces detailed estimates
- Makes changes easily
- Customizes estimates to specific requirements

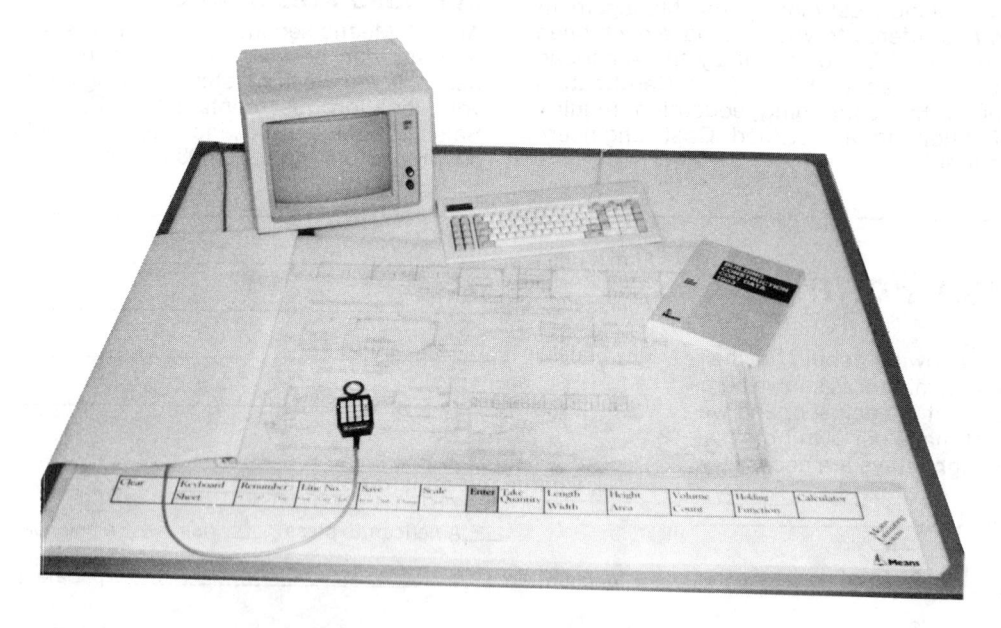

"Takeoffs" right from drawings.

## Detail of Digitizer Menu

| Clear | Keyboard | Renumber | | | Line No. | | | Save | | | Scale | |
|-------|----------|----------|----|----|----------|------|-----|------|-----|--------|-------|------|
|       | Sheet    | #1       | #2 | Tag| Cost     | User | Job | Item | Sub | Change | Arch. | Eng. |

| Take Quantity | Length | Height | Volume | Holding | Calculator |
|---------------|--------|--------|--------|---------|------------|
|               | Width  | Area   | Count  | Function|            |

# STaR
## General Estimating Program

**Helps develop fast and accurate estimates and manage vital cost data**

## A computerized estimating service from the people you can depend on

Now you can use the speed and dependability of a computer to simplify your estimating process. STAR General Estimating Program is easy to use, reduces errors, provides more accurate estimates and enables you to increase productivity and improve profit margins.

Using known construction costs, which reflect your marketplace, STAR helps you to quickly develop your data file, which is organized in the UCI-16 division format. This file then gives you, the estimator, access to necessary information, such as item description, crew make-up, daily productivity, unit of measure and unit prices for material, labor, equipment and total.

In addition STAR can also access any of the Means standard cost files.

With STAR, you can store and update estimates with ease and make modifications with the touch of a button. Changes are no problem because the computer does all the work.

More importantly, STAR enables you to save time—and money! An estimator can now evaluate more jobs, which means more work, increased productivity and higher profits. It's all part of The Solution by R.S. Means Company, Inc., the people you can depend on for your construction cost data.

### STaR Main Benefits

- Develop your own cost file—quickly
- Provides fast, accurate estimates
- Easy to operate
- Provides up-to-date unit prices
- Saves hours of tedious calculating
- Customizes estimates to specific requirements
- Two methods—daily output and man-hour analysis
- Helps in scheduling and planning
- Evaluated overhead and profit

**Unburdened Job Report Summary**
Here you see a summary report of the entire estimate by division, including your evaluation of overhead and profit.

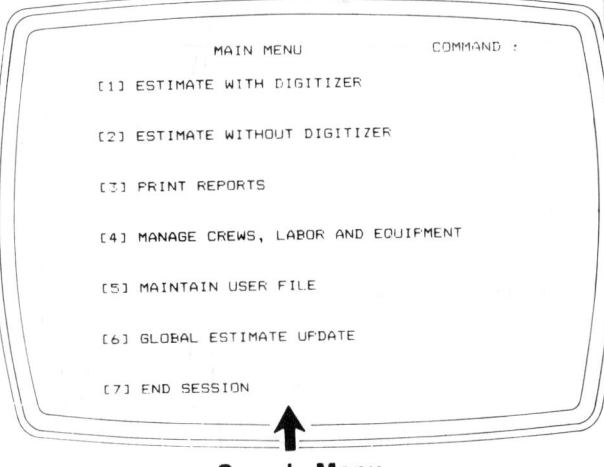

**Sample Menu**
This menu shows you the seven different directions you can go in the program.

# UNIT PRICE COST FILES

*Specific construction cost files help you create detailed estimates*

Rapidly changing costs in the construction industry are no longer a problem thanks to a series of specific building information files from R.S. Means Company, Inc. The Unit Price Cost Files provide you with up-to-date cost data in several different contractor and subcontractor construction categories that help to eliminate much of the uncertainty of creating detailed estimates and bids.

These cost data files, which can be purchased together or separately, contain more than 20,000 thoroughly-researched unit prices for labor, materials and installation. They can be used in conjunction with the STAR General Estimating Program and the GALAXY Automated Quantity Takeoff and Pricing System to determine construction costs in specific building applications, including mechanical, electrical, sitework and more.

Each construction task is defined by providing productivity and crew make-up assumptions as well as material costs. These can be easily adjusted to reflect your market place needs.

The Unit Price Cost Files provide the estimator with quick access to the cost figures he needs to know in the UCI format. National average union labor rates and equipment rental rates are included. It saves time and money in the preparation of estimates and gives you the confidence of knowing that your bid is as accurate as it can be.

R.S. Means also provides you with a valuable annual update service for these important Unit Price Cost Files. Offered on a subscription basis, each priced item is checked and adjusted as needed before the files are sent to you each year.

The Unit Price Cost Files can be changed and modified to your estimate without sacrificing accuracy. It's all part of The Solution by R.S. Means Company, Inc., the people you depend on for all your construction cost data.

## UNIT PRICE COST FILES

- Building Construction
- Repair and Remodeling
- Concrete and Masonry
- Residential/Light Commercial
- Mechanical
- Electrical
- Sitework
- Interiors

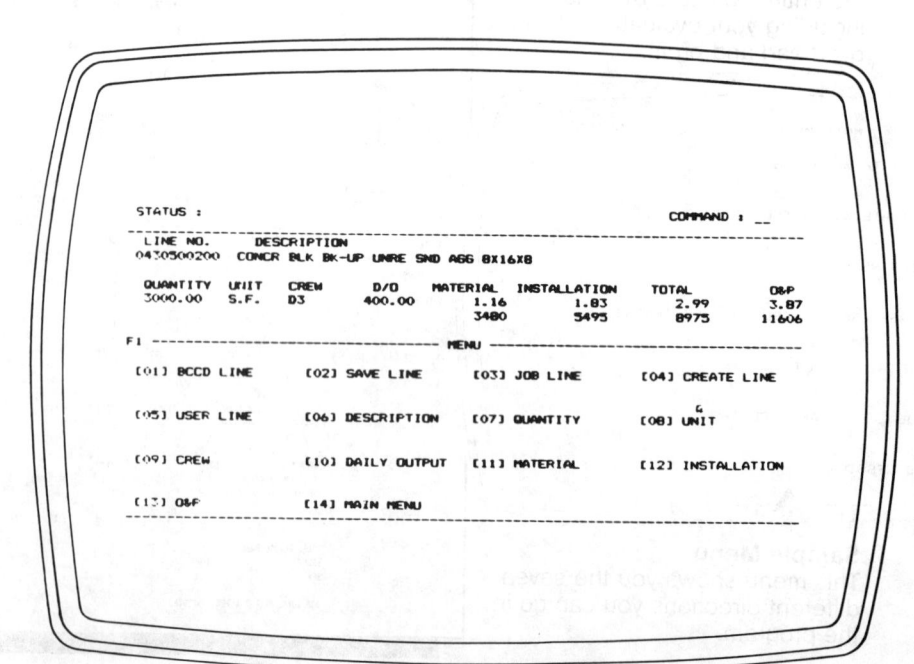

**Screen showing typical line item.**

You can add, delete, or change any item when used with STAR or GALAXY. Fourteen options are available.

# COMET
## Report Writer Sorting Program

*Designed for the construction and facilities professional with a growing need for customized reports*

An estimate must be competitive and precise in order to be successful. It must also look professional. A low bid isn't always going to win the job unless all the aspects of the project are carefully explained and presented in a clear manner. A sloppy report just doesn't cut it in today's business world.

COMET Report Writer is designed to help you create winning proposals in a convenient form to meet the needs of a necessary format. COMET is easy to use and will help you produce reports to your specific requirements.

This program enables you to calculate, accumulate and print individual items from estimating data in the STAR and GALAXY programs. You can customize the report to your needs and have the computer print out only the things you're looking for. Your report will be streamlined and neither you nor your client will waste valuable time sifting through volumes of material.

Virtually every job requires a report containing different information. COMET will let you devise a basic report that can be easily tailored to meet an individual situation. It's all part of The Solution by R.S. Means Company, Inc., the people you depend on for all your construction cost data.

### COMET Main Benefits

- Easy to use
- Ideal for prescribed reports
- Uses any number of variables
- Determines layout of detailed lines
- Customizes to your specifications
- Enhances STAR and GALAXY
- Helps you sell
- Lets the computer work for you

---

# MEANS Software – Information Request

☐ I'm considering an immediate purchase
☐ I plan to buy in _____ months
☐ Want information for my file

I'd like to know more about:
☐ Unit Price Estimating (STAR)
☐ Electronic "Takeoff" (GALAXY)
☐ Cost Data Files
☐ Report Writer (COMET)
☐ Accounting Packages

**Call (617) 747-1270
Software Sales
Or mail this card today!**

Name _____
Title _____
Company _____
Address _____
City/State/Zip _____
Telephone _____

My business is:
☐ General Contractor
☐ Subcontractor–Type _____
☐ Architect
☐ Engineer
☐ Facilities Management
☐ Other _____
☐ Dollar Volume _____

I have _____ estimators.

Takeoff involves _____ % of their estimating time.

I have a _____ computer.

**R.S. Means Co., Inc.**
100 Construction Plaza • Kingston, MA 02364

# NEW...
## Boating Cost Guide
*1st Annual Edition*

**Now for the first time ever... authoritative cost data for boats up to 65 feet, Sail and Power.**

Over 300 pages, illustrated

Over 11,000 actual cost items covering virtually every expense associated with boat maintenance, repair and improvement

The **BOATING COST GUIDE** gives you **all the cost facts** you need. This includes **both** the material or equipment cost **and** installation expenses. These costs, where appropriate, are shown three ways... the **minimum, average, or maximum** you might expect to pay.

- make realistic estimates for boat repairs
- discover cost-saving options
- make better planning decisions
- budget accurately for improvements
- compare price quotations
- check bills for work done

**Book No. 50185 .......... $29.95/copy**

---

# Labor Rates for the Construction Industry 1985

Over 300 pages

**12th Annual Edition**

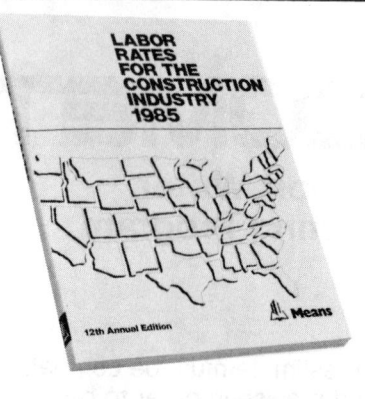

**CITY**

**STATE**

**NATIONAL**

- detailed wage rates by trade for over 300 U.S. and Canadian cities
- forty-six construction trades listed by local union number in each city
- base hourly wage rates plus fringe benefit package costs gathered from reliable sources
- dependable estimates for the trade wage rates not reported at press time
- effective dates for newly negotiated union contracts for both 1985 and 1986
- factors for comparing each trade rate by city, state, and national averages
- historical 1983-1984 wage rates also included for comparison purposes
- each city chart is alphabetically arranged with handy visual flip tabs for quick reference

**Book No. 60285 .................. $34.95/copy**

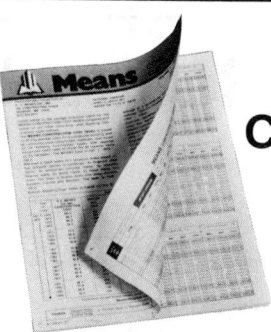

---

# Means Construction Cost Indexes 1985

**$47.95 per year**

(Individual and back issues available at $15.95 per copy)

**The index service providing updated cost adjustment factors.**

Whether updating construction cost estimates based on information in the Means cost manuals, or those from other sources, the construction cost index service is the efficient way to assure 90-day cost accuracy.

Published quarterly (January, April, July, October), this handy report provides cost adjustment factors for the preparation of more precise estimates no matter how late in the year. It's also the ideal method for making continuous cost revisions on on-going projects as the year progresses.

The report comes in four unique sections:
- Breakdowns for 209 major cities
- National averages for 30 key cities
- Five large city averages
- Historical construction cost indexes

**Book No. 61185 .................... $47.95/year**

---

# Means Historical Cost Indexes 1985

**$15.95 per copy**

(20-page booklet)

**Your latest estimating resource for historical cost information.**

This fifth annual edition of *Means Historical Cost Indexes* is designed especially for professionals with estimating activities involving construction costs not only in various *cities* but in various *years*. The booklet indexes historical construction costs for 209 U.S. and Canadian cities from 1940 through 1985.

These historical indexes are the result of careful monitoring of specific material quantities, man-hours, building trades, equipment rental, and installation. This extensive coverage can be depended upon for functional and reliable assistance in all of your historical cost estimating activities. *Means Historical Cost Indexes* includes detailed explanations of uses, with complete examples.

**Book No. 61485 .................. $15.95/copy**

Concise, authoritative, highly acclaimed...

# ESTIMATING FOR THE GENERAL CONTRACTOR

by Paul J. Cook

**First Edition** — over 225 pages, illustrated

**The first book prepared specifically to help contractors and estimators in medium-sized companies estimate more efficiently and accurately**

Light on theory, heavy on practical estimating methods and ideas, here's powerful help for the contractor/estimator who wants to evaluate and polish every facet of their estimating procedure.

No matter how much estimating experience you've had, no matter what your particular construction specialty, there's something of importance for you in this extraordinary volume.

Prepared by one of the industry's most successful estimating consultants, this guide is designed to give your estimating a refreshing breath of new ideas... insights... better techniques. It's a book you'll use over and over again, providing the "extra" dimension you need to make your estimates more reliable and usable than ever before.

Hardcover Book No. 67160 .... $35.95/copy
Softcover Book No. 67170 ..... $30.95/copy

# Square Foot Estimating

by F. William Horsley and Billy Joe Cox

**First Edition** — over 300 pages, illustrated

**A new generation of techniques for conceptual and design stage cost planning.**

*Square Foot Estimating* gives you the expertise to transform conceptual and design-stage estimating into a powerful tool for new business.

You'll see how to maximize the use of your time and effort to achieve new heights of creativity in designs, materials, and construction approaches... all within given budget limitations!

This fast-moving new book begins with a clear look at four estimating techniques used in design stage costing. Each is evaluated by degree of accuracy in relation to time and cost.

It spotlights the use of daring new departures from traditional thinking about square foot costs. It shows how one method — systems estimating — achieves remarkable reliability, typically in a day's work.

The editors demonstrate how building systems can be tried in an infinite number of combinations to quickly determine the direction to take for budgets, codes, loads, insulation, fire-proofing, acoustics, energy use, and special requirements.

Hardcover Book No. 67145 .... $36.95/copy
Softcover Book No. 67146 ..... $31.95/copy

...planned, written and produced to solve — with remarkable speed — the scheduling problems of the eighties...

# Means Scheduling Manual

Now with special new emphasis on using microcomputers and programs for scheduling and project controls.

Tightly written in one easy-to-use spiral-bound volume, here are today's most advanced methods for scheduling and managing building construction projects in the intermediate size range. This is a masterfully-prepared construction management tool equal to today's — and tomorrow's — burgeoning scheduling problems.

*Means Scheduling Manual* filled with the kind of sound practical advice and guidance you've always wanted in a scheduling book. This is a scheduling aid that's truly one step ahead of any other you've ever seen.

**$32.95 per copy** — 2nd edition, over 200 pages, illustrated

**NEWLY REVISED SECOND EDITION...**

Book No. 67152 .............. $32.95/copy

# Means Systems Costs 1985

*The Means illustrated systems cost manual...preferred by more construction professionals for its superb organization and price reliability*

This indispensible tool provides you with thousands of detailed systems costs, accompanied by hundreds of drawings, explanations, component breakdowns and tables -- all within one easy-to-use format.

You'll find separate design aids to assist you in the conceptual phase of estimating any job. Tables of unit prices allow you to cost-out virtually any combination of building systems in logical, step-by-step sequence, customizing your project as more information becomes available.

Now uses the UNIFORMAT numbering system as a guideline for consistent estimating and reporting. It enables you to price and compare various framing and envelope systems with ease.

Use the manual's complete examples, suggestions, and standardized format to compare past and future jobs and build your own cost data file.

**$38.95 per copy** — 10th annual edition, over 525 pages, illustrations

Book No. 60385 .............. $38.95/copy

# MEANS SQUARE FOOT COSTS 1985

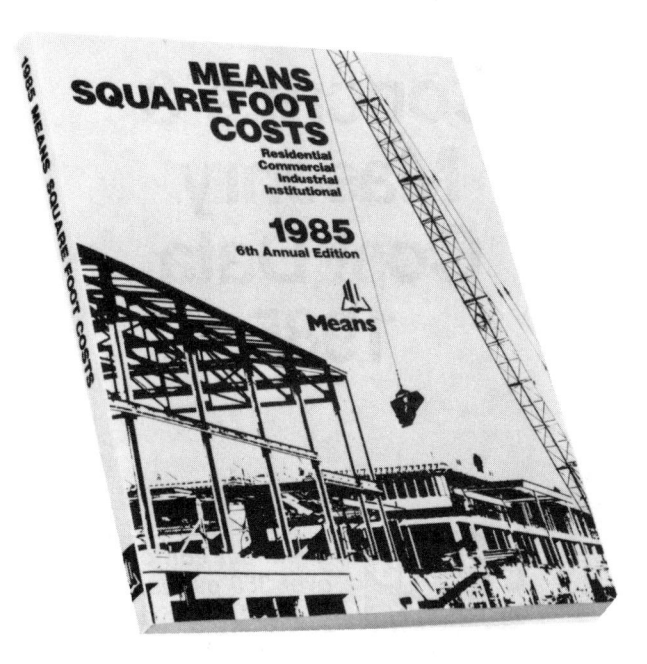

**$39.95 per copy**

6th annual edition with over 425 pages, photographs, illustrations

*Residential • Commercial*
*Industrial • Institutional*

### Here's what you'll find packed into the 1985 edition:

- over 6,000 costs for typical buildings and attachments... residential, commercial, industrial, institutional
- reliable costs for use in 1985
- four classes of residential quality: economy, average, custom, luxury
- over 350 illustrations of unit-in-place components and costs, fully detailed for systems style estimates
- worked-out examples and clear instructions
- cost adjustment factors for all construction-related variables
- over 6,500 unit-in-place costs illustrated for detailed budgets
- factors for local cost differences by Zip Code
- depreciation guidelines
- samples of convenient work sheets for preparing the estimate
- nearly 150 photographs of typical building types and qualities

**Designed for effective use at all levels of building cost budgeting, estimating and valuation... the planning guide and replacement cost resource for architects, design/build contractors, real estate developers, corporate planners, appraisers and assessors.**

Useful for anyone who needs rapid budget cost estimates in the office, with a client, or in the field, this manual provides clear descriptions, photographs, and illustrations of hundreds of residential, commercial, industrial, and institutional buildings. Corresponding costs by square foot are conveniently located on the same page or adjoining pages, minimizing effort for "on-the-spot" estimates.

Convenient tables provide calculation directions and costs for attached structures and special kinds of additions or adjustments.

For those preparing more detailed, itemized estimates, costs are broken down into "systems style" component specifications and unit costs. This fully illustrated unit-in-place section contains over 6,500 costs simplifying construction component identification and pricing.

You can use this new manual to achieve any degree of cost precision you might need.

---

Book No. 60585 .......................... $39.95/copy

# Concrete & Masonry Cost Data 1985

**$36.95 per copy**  3rd annual edition over 350 pages, illustrated

## Full-scope concrete/masonry cost coverage

- thousands of new unit prices
- the most popular systems prices
- in-depth labor cost data
- advice, assistance, ideas

**a "must" resource for:**
general contractors
concrete subcontractors
masonry subcontractors
architects
builders
estimators
engineers
foundation contractors
designers
carpentry subcontractors
governmental agencies

**The first all-inclusive estimating manual devoted to the broad range of today's advanced concrete and masonry methods and materials**

Prices for every facet of concrete/masonry activity —from complicated formwork to lavish brickwork — are provided in exhaustive detail . . . efficiently presented in the famous Means unit and systems cost format.

The massive unit cost section shows crew, daily output, and man-hour information, with bare costs for materials, labor and equipment for each component. The "most-popular" systems cost section actually pictures the concrete or masonry system you are estimating with drawings and diagrams. Unit costs are shown with each system for quick, tailor-made substitutions.

Cost adjustment factors for geographical areas by major cities enable you to adjust prices for your area of the U.S. or Canada. In most mid-sized construction projects nearly a quarter of all costs are concentrated in concrete and masonry work. You simply can't afford to use incomplete or inadequate cost information in this important area. The Means concrete/masonry cost guide gives you this specialized information quickly and reliably. So why take unnecessary risks with other methods?

Order your copy *now*.

Book No. 63085 . . . . . . . . . $36.95/copy

# Means Site Work Cost Data 1985

**$36.95 per copy** — 4th annual edition, over 375 pages, illustrated

*... the hard-to-find site work cost information you've always wanted in one truly comprehensive resource ...*

*Completely updated for 1985!*

- *complete unit and systems costs*
- *illustrations and explanations*
- *new time-saving features*

Thumb through the new 1985 site work cost guide and you'll be convinced instantly you have the tool that's equal to today's demanding estimating problems.

It's super-inclusive. Perfectly organized. And, of course, fully reliable in the Means tradition.

Its goal is to give you a cost resource that *actually does* cover every aspect of modern site work estimating... exploration, polution controls, soil retention... hidden utilities, security, site enhancement... *you name it*!

You'll marvel at the 200+ pages of unit costs.

You'll appreciate the 50+ site work systems costs which illustrate the work for you.

You'll be delighted with the abundance of descriptions, notes and helpful guidance touching upon just the right information you need.

**Features:**

**comprehensive unit prices**
- crew
- daily output
- man-hours
- units of measure
- material
- labor
- equipment
- total costs

**complete systems prices**
- cost per lineal foot or square foot
- quantity
- description
- illustrations

Book No. 60985 .............. $36.95/copy

**Destined to become the "professional's choice" for uncompromised trade and labor productivity data.**

$89.95 per copy

1st Edition
over 575 pages
hardcover

# Means Man-Hour Standards

Here is the working encyclopedia of labor productivity information construction professionals have been looking forward to... *Means Man-Hour Standards*.

This massive achievement is by far the most comprehensive, easy-to-use reference of its type available anywhere.

And because it's *Means*, you can be sure it's reliable.

You'll find every bit of labor data you may need... all in a superb layout with "quick-find" indexes, and handy visual flip tabs.

The efficient format permits very rapid comparisons of labor requirements for thousands of construction functions in the convenient U.C.I. arrangement.

Specifically, it offers you...

— precise descriptions of over 12,000 construction tasks with wide ranges of related variations and specifications
— detailed crew information with equipment required
— productivity and crew data shown together
— man-hours with units
— introductory information and guidance with each section
— U.C.I. format

With *Means Man-Hour Standards*, you simply turn to the page containing the construction activity you're estimating... and there in one place is the data you need. No shuffling back and forth.

Never before has obtaining and using man-hour information been so easy.

## Table of Contents

Introduction
Using the Data
Divisions
   1. General Conditions
   2. Site Work
   3. Concrete
   4. Masonry
   5. Metals
   6. Wood and Plastics
   7. Moisture and Thermal Protection
   8. Doors, Windows, Glass
   9. Finishes
  10. Specialties
  11. Architectural Equipment
  12. Furnishings
  13. Special Construction
  14. Conveying Systems
  15. Mechanical
  16. Electrical
Abbreviations
Index

Book No. 67140 .......... $89.95/copy

# Interior Cost Data 1985

## Commercial and Industrial Construction

*Partitions — Ceilings — Finishes — Floors — Furnishings*

**$35.95 per copy** — 2nd annual edition, over 400 pages, illustrated

*...the unit and systems cost tool every designer and estimator has been waiting for*

Today, few designers and estimators can keep up with the avalanche of price options for finish work.

All told, thousands of new materials and hardware items for interior construction are introduced each year.

Styles change constantly. Designs take new directions. New construction schemes come into play.

The result is an estimating jungle for designers and builders without the tools to cut through this tangled mass of cost options.

But now, thanks to Means new *Interior Cost Data 1985* you can put these hazards aside for good.

For here, in one monumental resource, are all the prices and help you could ever hope to have for making accurate interior work estimates.

*Interior Cost Data 1985* has drawn together the far reaches of finish work unit costs into **one** easy-to-use source of uncompromised accuracy and thoroughness.

It's all here...materials, equipment, hardware, custom work, furnishings, labor costs...every fact you need to plan for finish costs in new and remodeling commercial and industrial construction.

### Special Features:

- 10,000 unit cost entries
- quick-find illustrations
- extensive labor cost information
- special job requirement adjutments
- custom fabrication
- dust and noise protection
- specialized equipment
- scaffolding, bracing
- security precautions
- preparation work
- guidance, technical support
- unit and systems costs

Book No. 64985 .......... $35.95/copy

# Repair & Remodeling Cost Data 1985

**Accuracy, efficiency, and thoroughness for estimating rebuilding jobs**

$37.95 per copy — 6th annual edition, over 425 pages, illustrated

*For contractors, subcontractors, builders, estimators, engineers, architects, governmental and insurance agencies, corporate facilities managers, firms of all sizes . . .*
*Reliable unit and systems cost guidance for repair and remodeling estimating problems*

## Everything you need to make better rebuilding estimates

- comprehensive unit costs under UCI groupings with crew, man-hours, unit, and bare materials, labor, equipment and totals with adjustable overhead and profit columns
- illustrated systems prices for most widely-used remodeling methods with two additional sections for popular alternative systems and custom installations . . .presented in UNIFORMAT numbering system
- comprehensive labor cost information . . .union rates, crew output, productivity, every needed variation to calculate this vital cost area
- city cost modifiers for 162 U.S. and Canadian metro regions to factor in costs for your locality
- costs for the latest materials, methods, and equipment used in renovation work
- detailed diagrams, illustrations, explanations for ease of understanding and use . . . even a visual table of contents
- adjustable costs for contractor advantages such as owned equipment, low-cost materials purchases, low overhead, minimum daily charges, etc.

The Means *Repair & Remodeling Cost Data* book is the undisputed choice of anyone who must make accurate estimates for renovation projects. It's the most versatile, easy-to-use preliminary cost and bidding reference available from any source.

With over 425 pages containing 10,000 unit and building systems prices, this exciting guidebook handles almost any repair and remodeling challenge. With it you can accurately assemble rapid early-stage estimates as well as comprehensive competitive bids.

Most important of all, the Means repair and remodeling manual is designed to admirably fulfill varied user needs — whether a small-scale builder seeking to protect profits on remodeling work — right up to the most discriminating user needing prices for custom installations.

It's easy to use. Every section provides thorough, easy-to-understand explanations to achieve the desired level of precision. Its superb layout of helpful diagrams, notes, and tips allows you to put the information to work for you quickly.

Book No. 60885 . . . . . . . . . . . . . . . . . . . . . . . . . . . $37.95/copy

# Means Mechanical Cost Data 1985

**NEW 1985 PRICES**
for plumbing, HVAC... all other mechanical construction work.

**$36.95 per copy**  8th annual edition over 425 pages illustrated

## What's new for 1985

Separated from our electrical data, *Means Mechanical Cost Data 1985* offers you the benefits of greatly-expanded unit and systems costs and estimating guidance... devoted *exclusively* to plumbing, HVAC, fire protection, and all related preparation and finishing construction.

This new manual will be even more valuable and important to your work than ever before, providing you with *more* unit cost line items, *more* systems prices, *more* illustrations of parts and materials, and *more* general help covering every aspect of your cost needs.

Whether you need this cost assistance for complete competitive bids, or for help in specific areas of mechanical construction, you'll find all the cost facts you want quickly.

**So many ways to benefit from this storehouse of mechanical cost information**

What's new for 1985...

- up-to-date unit cost entries more illustrations of parts and materials for easy selection
- new "systems" cost groupings for faster, more reliable estimates
- new man-hour labor and productivity data for better comparisons
- more engineering sizing guides, notes, specifications
- updated mechanical square foot costs for 100 typical buildings and installations
- newest products, materials and installation methods included
- extensive energy use and conservation data, heat reclaiming equipment, retrofit and replacement materials cost information
- fixtures for the handicapped
- regional cost adjustment factors for labor and materials
- helpful diagrams, notes, explanations.

**Book No. 61085 ............ $36.95/copy**

# Means Electrical Cost Data 1985

**Completely updated and expanded, with 100% focus on electrical estimating cost solutions...**

$36.95 per copy — 8th annual edition, over 325 pages, illustrated

## Your "all-in-one" electrical estimating guide for 1985

- **latest electrical unit and systems costs**
- **quick-find illustrations and indexing**
- **time-saving design and specifications guidance**

With today's unpredictable costs for electrical construction, you need a fast, convenient way to keep up with it all.

That's what **Means Electrical Cost Data 1985** does for you. It gives you nearly 18,000 unit and systems costs covering every kind of electrical — or related — construction question you're likely to have.

For 1985 there are many additional price entries, new estimating models, and a better man-hour format which helps you in job scheduling as well as estimating.

For every phase of the electrical estimating process you'll have sharp, to-the-point answers. You'll have clear specifications, accurate materials and labor costs, and complete guidance in the dozens of notes, formulas and explanations.

For you, the busy professional, it's the fast, efficient way to estimate electrical work.

In fact, it will more than pay for itself the first time you use it for an electrical estimate!

- provides you with more electrical estimating data for better application to your work
- 10,000 fully current unit costs with additional three dimensional drawings of parts for faster identification
- man-hour format allows you to more easily interface with material deliveries and project work scheduling
- detailed estimating procedural models, many with complete diagrams, give you basic electrical frameworks to adapt to your projects
- expanded design reference tables covering every possible code, requirement, or fact you might need
- includes demolition and remodeling, square foot costs, geographical materials and labor cost factors, crew data, and electrical "systems" costs

**Book No. 62085** .............. **$36.95/copy**

# Building Construction Cost Data 1985

**$34.95 per copy**

43rd anniversary edition
over 425 pages

*America's foremost construction cost information guide with over 20,000 unit prices for labor, materials and installation*

*Building Construction Cost Data,* now in its 43rd year of publication, offers the reliability of over 20,000 thoroughly-researched unit prices, plus the most efficient, easy-to-use UCI format in the industry. Even the most complicated estimates can be accurately prepared in less time than with comparable references.

Astute construction professionals prefer *Building Construction Cost Data* as their primary estimating resource. It's the original, most sought-after book of its kind . . . known and depended upon by its users for unparalleled accuracy and versatility . . . a cost resource prepared with you — the user — in mind.

- accurate, detailed costs for over 20,000 building components help make sure no cost factor is overlooked
- new materials, fixtures, hardware, and equipment items included in each section
- all items updated to reflect 1985 costs and construction techniques
- unit prices divided into material and installation costs with overhead and profit shown separately
- hourly and daily wage rates for installation crews with crew sizes, equipment, and average daily crew output listed
- city cost adjustment factors for material and labor costs in 19 categories for each of 162 major U.S. and Canadian metro areas
- factors to compute historical costs for all cities dating back to 1940
- square foot and cubic foot cost section showing range and median costs for 57 common building types with plumbing, HVAC, and electrical percentages tabulated separately
- easiest-to-use format in contrasting typefaces and shaded columns
- costs are the national average and can be easily adjusted for local wage scales
- special cost advantages such as owned equipment, low-cost materials purchases, and low overhead, can be identified and computed separately
- over 13 index pages for quick item location and cross-reference
- helpful examples, instructions, illustrations, and explanations of how costs were developed for each major division

Book No. 60185 .................... $34.95/copy

## R.S. Means Co., Inc.
156 Construction Plaza    (617) 747-1270
P.O. Box 800    Kingston, MA 02364-0800

### 1985 ORDER FORM

Enclosed is our check for the following publications:

| QTY. | BOOK NO. | TITLE | UNIT PRICE | TOTAL |
|---|---|---|---|---|
| | 60185 | Building Construction Cost Data 1985 | $ 34.95 | $ |
| | 61085 | Means Mechanical Cost Data 1985 | 36.95 | |
| | 62085 | Means Electrical Cost Data 1985 | 36.95 | |
| | 60885 | Repair & Remodeling Cost Data 1985 | 37.95 | |
| | 60585 | Means Square Foot Costs 1985 | 39.95 | |
| | 60385 | Means Systems Costs 1985 | 38.95 | |
| | 60985 | Means Site Work Cost Data 1985 | 36.95 | |
| | 60785 | Residential/Light Commercial Cost Data 1985 | 34.95 | |
| | 64985 | Interior Cost Data 1985 | 35.95 | |
| | 63085 | Concrete & Masonry Cost Data 1985 | 36.95 | |
| | 60285 | Labor Rates for the Construction Industry 1985 | 34.95 | |
| | 61485 | Means Historical Cost Indexes 1985 | 15.95 | |
| | 61185 | Means Construction Cost Indexes *(Yearly Subscription)* | 47.95 | |
| | 67145 | Square Foot Estimating *(Hardcover)* | 36.95 | |
| | 67146 | Square Foot Estimating *(Softcover)* | 31.95 | |
| | 67140 | Means Man-Hour Standards *(Hardcover only)* | 89.95 | |
| | 67152 | Means Scheduling Manual *(Softcover only)* | 32.95 | |
| | 67160 | Estimating for the General Contractor *(Hardcover)* | 35.95 | |
| | 67170 | Estimating for the General Contractor *(Softcover)* | 30.95 | |
| | 50185 | Boating Cost Guide | 29.95 | |
| | | | TOTAL | $ |

Note: Prices subject to change without notice.    Mass. residents add 5% state sales tax:
Postage and Handling extra when billed.

Canadian customers ONLY — Please order books from:
Southam Communications, Ltd.
1450 Don Mills Road
Don Mills, Ontario, Canada, M3B-2X7
*(Canadian Prices are higher.)*

**SEND MY ORDER TO:**
*(Please print)*
Name _____
Company _____
Address _____
City/State/Zip _____    *Please Indicate If Company ☐ Or Home ☐ Address)*

---

## R.S. Means Co., Inc.
156 Construction Plaza    (617) 747-1270
P.O. Box 800    Kingston, MA 02364-0800

### 1986 ORDER FORM

Enclosed is our check for the following publications:

| QTY. | BOOK NO. | TITLE | UNIT PRICE | TOTAL |
|---|---|---|---|---|
| | 60016 | Building Construction Cost Data 1986 | $ 36.95 | $ |
| | 60026 | Means Mechanical Cost Data 1986 | 38.95 | |
| | 60036 | Means Electrical Cost Data 1986 | 38.95 | |
| | 60046 | Repair & Remodeling Cost Data 1986 | 38.95 | |
| | 60056 | Means Square Foot Costs 1986 | 40.95 | |
| | 60066 | Means Systems Costs 1986 | 41.95 | |
| | 60076 | Means Site Work Cost Data 1986 | 38.95 | |
| | 60086 | Residential/Light Commercial Cost Data 1986 | 36.95 | |
| | 60096 | Interior Cost Data 1986 | 36.95 | |
| | 60116 | Concrete & Masonry Cost Data 1986 | 38.95 | |
| | 60126 | Labor Rates for the Construction Industry 1986 | 36.95 | |
| | 60136 | Means Historical Cost Indexes 1986 | 16.95 | |
| | 60146 | Means Construction Cost Indexes *(Yearly Subscription)* | 49.95 | |
| | 60153 | Means Man-Hour Standards *(Hardcover only)* | 89.95 | |
| | 60164 | Means Scheduling Manual *(Softcover only)* | 32.95 | |
| | 60173 | Square Foot Estimating *(Hardcover)* | 36.95 | |
| | 60183 | Square Foot Estimating *(Softcover)* | 31.95 | |
| | 60192 | Estimating for the General Contractor *(Hardcover)* | 35.95 | |
| | 60212 | Estimating for the General Contractor *(Softcover)* | 30.95 | |
| | 60236 | Boating Cost Guide | 29.95 | |
| | | | TOTAL | $ |

Note: Prices subject to change without notice.    Mass. residents add 5% state sales tax.
Postage and Handling extra when billed.

Canadian customers ONLY — Please order books from:
Southam Communications, Ltd.
1450 Don Mills Road
Don Mills, Ontario, Canada, M3B-2X7
*(Canadian Prices are higher.)*

**SEND MY ORDER TO:**
*(Please print)*
Name _____
Company _____
Address _____
City/State/Zip _____    *Please Indicate If Company ☐ Or Home ☐ Address)*